Lecture Notes in Computer Science 4668

Commenced Publication in 1973
Founding and Former Series Editors:
Gerhard Goos, Juris Hartmanis, and Jan van Leeuwen

Joaquim Marques de Sá   Luís A. Alexandre
Włodzisław Duch   Danilo P. Mandic (Eds.)

Artificial Neural Networks – ICANN 2007

17th International Conference
Porto, Portugal, September 9-13, 2007
Proceedings, Part I

Springer

Volume Editors

Joaquim Marques de Sá
University of Porto, Faculty of Engineering
Rua Dr. Roberto Frias, 4200-465 Porto, Portugal
E-mail: jmsa@fe.up.pt

Luís A. Alexandre
University of Beira Interior, Dept. of Informatics
and
IT-Networks and Multimedia Group
Rua Marquês d'Ávila e Bolama, 6201-001 Covilhã, Portugal
E-mail: lfbaa@di.ubi.pt

Włodzisław Duch
Nicolaus Copernicus University, Dept. of Informatics
ul. Grudziadzka 5, 87-100 Torun, Poland
E-mail: wduch@is.umk.pl

Danilo P. Mandic
Imperial College London
Communication and Signal Processing Research Group
Dept. of Electrical and Electronic Engineering
Exhibition Road, London, SW7 2BT, UK
E-mail: d.mandic@imperial.ac.uk

Library of Congress Control Number: 2007934292

CR Subject Classification (1998): F.1, I.2, I.5, I.4, G.3, J.3, C.2.1, C.1.3

LNCS Sublibrary: SL 1 – Theoretical Computer Science and General Issues

ISSN 0302-9743
ISBN-10 3-540-74689-7 Springer Berlin Heidelberg New York
ISBN-13 978-3-540-74689-8 Springer Berlin Heidelberg New York

Springer is a part of Springer Science+Business Media

springer.com

Printed in Germany

Typesetting: Camera-ready by author, data conversion by Scientific Publishing Services, Chennai, India
Printed on acid-free paper SPIN: 12116348 06/3180 5 4 3 2 1 0

Preface

This book includes the proceedings of the International Conference on Artificial Neural Networks (ICANN 2007) held during September 9–13, 2007 in Porto, Portugal, with tutorials being presented on September 9, the main conference taking place during September 10-12 and accompanying workshops held on September 13, 2007. The ICANN conference is organized annually by the European Neural Network Society in co-operation with the International Neural Network Society, the Japanese Neural Network Society, and the IEEE Computational Intelligence Society. It is the premier European event covering all topics related to neural networks and cognitive systems. The ICANN series of conferences was initiated in 1991 and soon became the major European gathering for experts in these fields. In 2007 the ICANN conference was organized by the Biomedical Engineering Institute (INEB - Instituto de Engenharia Biomédica), Porto, Portugal, with the collaboration of the University of Beira Interior (UBI - Universidade da Beira Interior), Covilhã, Portugal and ISEP, Polytechnic Engineering School, Porto, Portugal. From 376 papers submitted to the conference, 197 papers were selected for publication and presentation, following a blind peer-review process involving the Program Chairs and International Program Committee; 27 papers were presented in oral special sessions; 123 papers were presented in oral regular sessions; 47 papers were presented in poster sessions. The quality of the papers received was very high; as a consequence, it was not possible to accept and include in the conference program many papers of good quality. A variety of topics constituted the focus of paper submissions. In regular sessions, papers addressed the following topics: computational neuroscience and neurocognitive studies, applications in biomedicine and bioinformatics, spiking neural networks, data clustering, signal and times series processing, learning theory, advances in neural network learning methods, advances in neural network architectures, data analysis, neural dynamics and complex systems, ensemble learning, self-organization, robotics and control, pattern recognition, text mining and Internet applications, vision and image processing. Special sessions, organized by distinguished researchers, focused on significant aspects of current research, namely: emotion and attention, understanding and creating cognitive systems, temporal synchronization and nonlinear dynamics in neural networks, complex-valued neural networks. Papers presented in poster sessions were organized in the following topics: real-world applications, signal and time series processing, advances in neural network architectures, advances in neural network training, meta learning, independent component analysis, graphs, evolutionary computing, estimation, spatial and spatio-temporal learning. Prominent lecturers gave six keynote speeches at the conference. Moreover, well-known researchers presented seven tutorials on state-of-the-art topics. Four post-conference workshops, entitled “Cognitive Systems”, “Neural Networks in Biomedical Engineering and

Bioinformatics", "What It Means to Communicate" and "Neural Networks of the Future?", concluded the focus of ICANN 2007 on the state-of-the-art research on neural networks and intelligent technologies. An in-depth discussion was held on the prospects and future developments both in theory and practice in those important topics. We would like to thank all the members of the local committee for their contribution to the organization of ICANN 2007. A special thanks to Alexandra Oliveira whose dedication and work quality were a major guarantee of the success of ICANN 2007. We also wish to thank Alfred Hofmann and the LNCS team from Springer for their help and collaboration in the publication of the ICANN 2007 proceedings.

July 2007

Joaquim Marques de Sá
Luís A. Alexandre

Organization

General Chair

Joaquim Marques de Sá
University of Porto, Portugal

Co-chair

Luís A. Alexandre
University of Beira Interior, Portugal

Program Chairs

Włodzisław Duch
Torun, Poland & Singapore, ENNS President

Danilo Mandic
Imperial College London, UK

Honorary Chair

John G. Taylor
Kings College, London, UK

Program Committee

Alessandro Sperduti, University of Padova, Italy
Alessandro Villa, University of Grenoble, France
Amir Hussain, University of Stirling, UK
Andreas Nuernberger, University of Magdeburg, Germany
Andreas Stafylopatis, NTUA, Greece
Andrzej Cichocki, RIKEN Brain Sci. Inst., JP
Bruno Apolloni, University of Milan, Italy
David Miller, University of Pennsylvania, USA
Dragan Obradovic, Siemens Corp. Res., Germany
Erkki Oja, Helsinki University, Finland
Erol Gelenbe, Imperial College London, UK
Hojjat Adeli, Ohio State University, USA
Jacek Mandziuk, Warsaw University, Poland

João Luís Rosa, Catholic University Campinas, Brazil
Jose Dorronsoro, Universidad Aut. de Madrid, Spain
José Príncipe, University of Florida, USA
Jürgen Schmidhuber, TU Munich, DE - IDSIA, Switzerland
Lefteri Tsoukalas, Purdue University, USA
Marios Polycarpou, University of Cyprus, Cyprus
Mark Embrechts, Rensselaer Inst., USA
Michel Verleysen, University of Louvain-la-Neuve, Belgium
Nikola Kasabov, Auckland University, New Zealand
Okyay Kaynak, Bogazici University, Turkey
Olli Simula, Helsinki University, Finland
Peter Andras, University of Newcastle, UK
Péter Érdi, HU & Kalamazoo College, USA
Stan Gielen, University of Nijmegen, The Netherlands
Stefan Wermter, University of Sunderland, UK
Stefanos Kolias, NTUA, Greece
Steve Gunn, University of Southampton, UK
Thomas Martinetz, University of Luebeck, Germany

Local Organizing Committee

Alexandra Oliveira, INEB
Ana Maria Tomé, Aveiro University
Bernardete Ribeiro, Coimbra University
Carlos Soares, Porto University
Daniel Carrilho, Health Techn. -IPP
Fernando Sereno, ESE-IPP
Helena Brás Silva, ISEP-IPP
Hugo Proença, UBI
Jorge Santos, ISEP-IPP
Lígia Carvalho, INEB
Luís Silva, INEB
Paulo Cortez, Minho University
Paulo Fazendeiro, UBI
Petia Georgieva, Aveiro University

Reviewers

Abe	Shigeo	Kobe University
Agell	Núria	Ramon Llull University
Aiolli	Fabio	Pisa University
Alexandre	Frederic	INRIA Lorraine/LORIA-CNRS
Alexandre	Luís	University of Beira Interior
Alhoniemi	Esa	Turku University
Andras	Peter	University of Newcastle

Anguita	Davide	Genoa University
Angulo-Bahon	Cecilio	Technical University of Catalonia
Apolloni	Bruno	University of Milan
Archambeau	Cédric	Université Catholique de Louvain
Arenas	Jerónimo	Universidad Carlos III de Madrid
Atencia	Miguel	Universidad de Málaga
Avrithis	Yannis	National Technical University of Athens
Barbosa	Jorge	University of Porto
Bermejo	Sergi	Universitat Politècnica da Catalunya
Bianchini	Monica	Università di Siena
L Boni	Andrea	University of Trento
Bourlard	Herve	IDIAP Research Institute
Boyer	Domingo	University of Córdoba
Cabestany	Joan	Universitat Politécnica da Catalunya
Colla	Valentina	Scuola Sup. Sant'Anna Pisa
Corchado	Emilio	Universidad de Burgos
Cornford	Dan	Aston University
Corona	Francesco	Helsinki University of Technology
Correia	Miguel	University of Porto
Cortez	Paulo	University of Minho
Crook	Nigel	Oxford Brookes University
Dorronsoro	José	Universidad Autónoma de Madrid
Dounias	Georgios	University of the Aegean
Duch	Wlodzislaw	Nicolaus Copernicus University
Duro	Richard	University of Coruña
Embrechts	Mark	Rensselaer Polytechnic Institute
Érdi	Péter	Henry Luce Foundation
Fazendeiro	Paulo	University of Beira Interior
Franco	Leonardo	Universidad de Málaga
Francois	Damien	Université Catholique de Louvain
Fyfe	Colin	University of Paisley
Garcia-Pedrajas	Nicolas	University of Córdoba
Georgieva	Petia	University of Aveiro
Gielen	Stan	University of Nijmegen
Giudice	Paolo	Istituto Nazionale di Fisica Nucleare
Gonzalez	Ana	Universidad Autónoma de Madrid
Gosselin	Bernard	Faculté Polytechnique de Mons
Grana	Manuel	University Pais Vasco
Gunn	Steve	University of Southampton
Hammer	Barbara	University of Osnabrueck
Heidemann	Gunther	Bielefeld University

Hollmen	Jaakko	Technical University of Helsinki
Honkela	Antti	Helsinki University of Technology
Hoyer	Patrik	Helsinki Institute for Information Technology
Igel	Christian	Ruhr-Universitaet Bochum
Indiveri	Giacomo	UNI-ETH Zurich
Jin	Yaochu	Honda Research Institute Europe
Jutten	Christian	LIS-INPG
Kaban	Ata	University of Birmingham
Kaiser	Marcus	Newcastle University
Karhunen	Juha	Helsinki University of Technology
Karpouzis	Kostas	ICCS-NTUA
Kasderidis	Stathis	Institute of Computer Science - FORTH
Kaynak	Okyay	Bogazici University
Kim	DaeEun	Max Planck Institute for Psychological Research
Kollias	Stefanos	National Technical University of Athens
Koroutchev	Kostadin	Universidad Autónoma de Madrid
Kounoudes	Tasos	SignalGeneriX
Laaksonen	Jorma	Technical University of Helsinki
Lagos	Francisco	Universidad de Málaga
Lang	Elmar	Universität Regensburg
Lansky	Petr	Academy of Sciences of the Czech Republic
Larochelle	Hugo	University of Montréal
Leclercq	Edouard	Université du Havre
Leiviskä	Kauko	University of Oulu
Lendasse	Amaury	Helsinki University of Technology
Likas	Aristidis	University of Ioannina
Loizou	Christos	Intercollege, Limassol Campus
Magoulas	George	Birkbeck College, University of London
Mandic	Danilo	Imperial College London, UK
Mandziuk	Jacek	Warsaw University of Technology
Marques de Sá	Joaquim	University of Porto
Martinetz	Thomas	University of Luebeck
Martinez	Dominique	LORIA
Masulli	Francesco	Polo Universitario di La Spezia G. Marco
Micheli	Alessio	University of Pisa
Moreno	Juan	Universidad Politécnica de Cataluña
Muresan	Raul	SC. NIVIS SRL
Müller	Klaus-Robert	University of Potsdam

Nakayama	Minoru	CRADLE
Navía-Vázquez	Ángel	Universidad Carlos III de Madrid
Neskovic	Predrag	Brown University
Nikolopoulos	Konstantinos	Lancaster University Management School
Nürnberger	Andreas	Otto-von-Guericke Universität Magdeburg
Obradovic	Dragan	Siemens AG
Oja	Erkki	Helsinki University of Technology
Osowski	Stanislaw	Warsaw University of Technology
Parra	Xavier	Technical University of Catalonia
Patan	Krzysztof	University of Zielona Góra
Paugam-Moisy	Helene	Institut des Sciences Cognitives
Peters	Gabriele	Universitaet Dortmund
Peterson	Leif	Dept. of Public Health of the Methodist Hospital
Petrosino	Alfredo	University of Naples "Parthenope"
Polani	Daniel	University of Hertfordshire
Porrmann	Mario	Heinz Nixdorf Institute
Príncipe	José	CNEL - University of Florida
Proença	Hugo	University of Beira Interior
Puzenat	Didier	Université Antilles-Guyane
Reyes	Jose	Universidade da Coruña
Ribeiro	Bernardete	University of Coimbra
Rocha	Miguel	University of Minho
Rosa	João	PUC-Campinas
Rospars	Jean-Pierre	INRA - Laboratoire de Biométrie
Rossi	Fabrice	INRIA Rocquencourt
Ruiz	Francisco	Universitat Politécnica de Catalunya
Sandoval	Francisco	Universidad de Málaga
Santos	Jorge	Instituto Superior de Engenharia do Porto
Scheper	Tjeerd	Oxford Brookes University
Schmidhuber	Jürgen	TU Munich, DE - IDSIA
Schwenker	Friedhelm	University of Ulm
Sereno	Fernando	Escola Superior de Educação do Porto
Serrano	Eduardo	Universidad Autónoma de Madrid
Silva	Luís	INEB - Instituto de Engenharia Biomédica
Simula	Olli	Helsinki University of Technology
Stafylopatis	Andreas	National Technical University of Athens
Steil	Jochen	University of Bielefeld
Suárez	Alberto	Universidad Autónoma de Madrid

Suykens	Johan	Katholieke Universiteit Leuven
Thomaidis	Nikos	University of the Aegean
Tomé	Ana Maria	University of Aveiro
Touzet	Claude	Université de Provence /CNRS
Trentin	Edmondo	Universitá di Siena
Tsakonas	Athanasios	University of the Aegean
Varona	Pablo	Universidad Autónoma de Madrid
Verleysen	Michel	Université Catholique de Louvain
L Vigário	Ricardo	Helsinki University of Technology
Villa	Alessandro	Université de Lausanne
Villmann	Thomas	Clinic for Psychotherapy
Vinciarelli	Alessandro	IDIAP Research Institute
Wennekers	Thomas	University of Plymouth
Wermter	Stefan	University of Sunderland
Wersing	Heiko	Honda Research Institute Europe GmbH
Wyns	Bart	Ghent University
Yearwood	John	University of Ballarat
Zervakis	Michalis	Technical University of Crete

Sponsors

ENNS - European Neural Networks Society
INNS - International Neural Networks Society
JNNS - Japanese Neural Networks Society
IEEE Computational Intelligence Society
EURASIP - European Association for Signal and Image Processing

INEB - Instituto de Engenharia Biomédica, Portugal
UBI - Universidade da Beira Interior, Portugal
ISEP - Instituto Superior de Engenharia do Porto, Portugal
UP - Reitoria da Universidade do Porto, Portugal
DEEC - Departamento de Engenharia Electrotécnica e de Computadores, UP
IPP - Instituto Politécnico do Porto

FCT - Fundação para a Ciência e Tecnologia
FLAD - Fundação Luso-Americana para o Desenvolvimento
Fundação Calouste Gulbenkian

Microsoft Research Cambridge Lab
PT - Portugal Telecom

Table of Contents – Part I

Learning Theory

Advances in Neural Network Learning Methods

Ensemble Learning

Spiking Neural Networks

Advances in Neural Network Architectures

Neural Dynamics and Complex Systems

Data Analysis

Estimation

Spatial and Spatio-Temporal Learning

Evolutionary Computing

Meta Learning, Agents Learning

Complex-Valued Neural Networks (Special Session)

Temporal Synchronization and Nonlinear Dynamics in Neural Networks (Special Session)

Generalization Error of Automatic Relevance Determination

Shinichi Nakajima[1] and Sumio Watanabe[2]

[1] Nikon Corporation, 201-9 Miizugahara, Kumagaya, 360-8559 Japan
nakajima.s@nikon.co.jp, swatanab@pi.titech.ac.jp
http://watanabe-www.pi.titech.ac.jp/~nkj23/index.html
[2] Tokyo Institute of Technology, Mailbox R2-5, 4259 Nagatsuda, Yokohama, 226-8503 Japan

Abstract. The automatic relevance determination (ARD) shows good performance in many applications. Recently, it has been applied to brain current estimation with the variational method. Although people who use the ARD tend to pay attention to one benefit of the ARD, sparsity, we, in this paper, focus on another benefit, generalization. In this paper, we clarify the generalization error of the ARD in the case that a class of prior distributions is used, and show that good generalization is caused by singularities of the ARD. Sparsity is not observed in that case, however, the mechanism that the singularities provide good generalization implies the mechanism that they also provide sparsity.

1 Introduction

The automatic relevance determination (ARD) [1,2], an empirical Bayes (EB) or hierarchical Bayes (HB) approach providing a sparse solution, is known to be useful when we have to estimate a lot of parameters of which the majority are expected to be zero. Recently, with the variational Bayes (VB) approach [3,4], the ARD has been applied to ill-posed brain current estimation from magnetoencephalography (MEG) data, and shown good performance in a real data experiment, as well as in a computer simulation [5,6]. Although there are variations of EB approach, it has been shown that many of them can be described as a unified framework [7]. So, we focus on a similar approach to the one proposed in [5]. Both the ARD and the maximum a posteriori (MAP) estimation avoid ill-posedness by incorporating prior knowledge. However, benefits of the ARD are experimentally observed, one of which is that the ARD provides a sparse solution. Although less people might pay attention to that, there is another benefit of the ARD, generalization ability. In this paper, we focus on generalization, and show that the benefit is caused by singularities. In addition, consideration of the mechanism that the singularities provide good generalization implies the mechanism that they also provide sparsity, which we also discuss in this paper.

It is known that, in general, the Bayesian methods in singular models provide better generalization performance than the MAP estimation [8], which is asymptotically equivalent to the maximum likelihood (ML) estimation, and that, in some cases, the effect of singularities is observed as the James-Stein (JS) type shrinkage [9]. The JS estimator was proposed as an estimator dominating the ML estimator [10], and it has

J. Marques de Sá et al. (Eds.): ICANN 2007, Part I, LNCS 4668, pp. 1–10, 2007.

been shown that the JS estimator can be derived as an EB estimator [11]. Moreover, its asymptotic relation to the ARD with the variational method has been shown [12].

In this paper, focusing on the case that we use a class of prior distributions, we clarify the asymptotic generalization error of the ARD. Then, we clarify the nonasymptotic generalization error in a special case, to show the contribution of prior knowledge when the number of samples is small. We also discuss the mechanism that causes sparsity, the other benefit of the ARD. In Section 2, we describe the model analyzed in this paper, and its VB solution. Then, we analyze its asymptotic generalization error in Section 3, and the nonasymptotic one in Section 4. Discussion including the consideration of sparsity follows in Section 5. Finally, we conclude this paper in Section 6.

2 Automatic Relevance Determination

In brain current estimation, we consider time series data such as

$$Y = VA' + \mathcal{E}, \tag{1}$$

where $Y \in \mathbb{R}^{L \times U}$ is an output matrix observed, $V \in \mathbb{R}^{L \times M}$ is a constant matrix, called the lead field matrix, $A' \in \mathbb{R}^{M \times U}$ is a parameter matrix to be estimated, and $\mathcal{E} \in \mathbb{R}^{L \times U}$ is an observation noise. L, M, and U denote the output dimensionality, the parameter dimensionality, and the length of time series, respectively, and their smaller letters are used as the corresponding indices. In addition, L corresponds to the number of sensors, M to the number of the sites where the current is to be estimated, in brain current estimation. Note that U does not mean the length of the whole time series data but the minimum length of the data used for estimation at one time, as described later. We denote the column and row vectors by smaller letters, for example, $Y = (y_1, \ldots, y_U) = (\tilde{y}_1, \ldots, \tilde{y}_L)^t$, where $y_u \in \mathbb{R}^L$ is the output vector at the time u, and $\tilde{y}_l \in \mathbb{R}^U$ is the time series vector of the l-th output.[1] Here t denotes the transpose of a matrix. By $\mathcal{N}_d(\mu, \Sigma)$ we denote the d-dimensional normal distribution with average μ and covariance matrix Σ, and by $\mathcal{N}_d(\cdot; \mu, \Sigma)$ its probability density. We assume throughout this paper that all the elements of $\mathcal{E}$ are subject to $\mathcal{N}_1(0, \sigma_y^2)$, where $0 < \sigma_y^2 < \infty$ is known. We say that a problem is ill-posed when the rank of V is smaller than the parameter dimensionality, M, so that we cannot determine the unique parameter value without prior knowledge. We assume that $L < M$, so our problem is obviously ill-posed.

In this paper, we consider the following singular model:[2]

$$p(Y|A, b) = \prod_{u=1}^{U} \mathcal{N}_L(y_u; V(b * a_u), \sigma_y^2 I_L), \tag{2}$$

$$\phi(A) = \prod_{u=1}^{U} \mathcal{N}_M(a_u; 0, I_M), \ \phi(b) = \mathcal{N}_M(b; 0, (\mathrm{diag}(c))^2), \tag{3}$$

where $\phi(\cdot)$ denotes a prior distribution, $A \in \mathbb{R}^{M \times U}$ and $b \in \mathbb{R}^M$ are the parameters, and $c \in \mathbb{R}^M$ is a constant vector, of which each element, $c_m > 0$, corresponds to the

[1] A vector with a *tilde* denotes a time series vector throughout this paper.

[2] This model is not singular from the viewpoint of the ML estimation, while it has singularities from the viewpoint of the Bayesian learning [9].

prior deviation of b_m. Here, by I_d we denote the $d \times d$ identity matrix, by $*$ the componentwise product of matrices, for example, $(A * B)_{mu} = A_{mu} B_{mu}$, and by $\mathrm{diag}(\cdot)$ the diagonal matrix consisting of the elements of a vector, for example, $(\mathrm{diag}(c))_{mm} = c_m$. By the transform $B * A \to A'$, where $\mathbb{R}^{M \times U} \ni B = (b, \ldots, b)$, we find that the singular model above is equivalent to the following automatic relevance determination (ARD) [1, 2] of the original linear model, Eq.(1):

$$p(Y|A') = \prod_{u=1}^{U} \mathcal{N}_L(y_u; V a'_u, \sigma_y^2 I_L), \tag{4}$$

$$\phi(A') = \prod_{u=1}^{U} \mathcal{N}_M(a'_u; 0, (\mathrm{diag}(b))^2), \; \phi(b) = \prod_{m=1}^{M} \Gamma(b_m^2/c_m^2; 1/2, 2), \tag{5}$$

where we denote by $\Gamma(\kappa, \nu)$ the Gamma distribution with shape parameter κ and scale parameter ν, and by $\Gamma(\cdot; \kappa, \nu)$ its probability density. Therefore, we call the singular model, Eqs.(2) and (3), the ARD model, where the parameter, A', of the *original* ARD model, Eqs.(4) and (5), corresponds to the componentwise product of the parameters, $B * A$. In addition, note that the deviation of A'_{mu}, which corresponds to b_m, is assumed to be constant for $u = 1, \ldots, U$, as proposed in [5].[3] This assumption essentially affects generalization performance, as shown in the following sections. Hereafter, we assume that we have adopted a coordinate system such that V is general diagonal. This assumption is equivalent to the one that we use a prior distribution such that the components along the right singular vectors of V are independent of each other. However, note that this assumption breaks sparsity, another benefit of the ARD, which will be discussed in Section 5.2.

The variational Bayes (VB) approach provides a tractable posterior distribution, denoted by $r(A, b)$, that approximates the intractable Bayes posterior distribution, by restricting the set of possible posterior distributions [3,4]. We apply it to the ARD, Eqs.(2) and (3), as in [5], restricting the posterior distribution such that A and b are independent of each other, i.e., $r(A, b) = r(A)r(b)$. We suppose that we have n sets of output samples observed, i.e., $Y^n = \{Y^{(i)}; i = 1, \ldots, n\}$, whose average is denoted by $\bar{Y} = n^{-1} \sum_{i=1}^{n} Y^{(i)} = (\bar{y}_1, \ldots, \bar{y}_U) = (\tilde{\bar{y}}_1, \ldots, \tilde{\bar{y}}_L)^t$. In a similar fashion to the case, analyzed in [12], that V is not assumed to be general diagonal and the posterior distribution is restricted such that all the elements of A and b are independent of each other, we obtain the following theorem, whose proof is omitted:

Theorem 1. *When V is general diagonal, the VB posterior distribution is given by*

$$r(A, b|Y^n) = \prod_{m=1}^{M} \left(\mathcal{N}_U(\tilde{a}_m; \hat{\tilde{a}}_m, \hat{\sigma}_{a_m}^2 I_U) \cdot \mathcal{N}_1(b_m; \hat{b}_m, \hat{\sigma}_{b_m}^2) \right), \tag{6}$$

where the m-th element time series vector of the componentwise product of the means, that is, the VB estimator of $\tilde{a}'_m$ in the original linear model, Eq.(1), is

$$\hat{\tilde{a}}'_m = \hat{b}_m \hat{\tilde{a}}_m = \begin{cases} 0 & \text{if } v_m = 0 \\ \mathcal{S}(\tilde{\bar{y}}_m/\|v_m\|; \sigma_y^2 U/\|v_m\|^2) + O_p(n^{-1}) & \text{if } v_m \neq 0 \end{cases}. \tag{7}$$

$$\textit{Here, } \mathcal{S}(\tilde{z}; \chi) = \theta(n\|\tilde{z}\|^2 > \chi) \left(1 - \chi/n\|\tilde{z}\|^2\right) \tilde{z} \tag{8}$$

[3] In [5], the hyperprior of the *inverse* variance, b_m^{-2}, is assumed to be subject to $\Gamma(\kappa, \nu)$, which satisfies the condition for obtaining a stable iterative algorithm in the framework of the VB approach [13]. We conjecture that this difference between the prior distributions less affects generalization, which will be discussed in Section 5.1.

is the positive-part James-Stein type shrinkage (PJS) operator with the degree of shrinkage $\chi > 0$*, where* $\theta(\cdot)$ *denotes the indicator function of an event.*

In the following sections, we compare the generalization error of the ARD with that of the MAP estimator, given by

$$\hat{a}'^{\mathrm{MAP}}_u = (V^t V + \sigma_y^2 I_M / n c_m^2)^{-1} V^t \bar{y}_u, \tag{9}$$

where the prior deviation of the parameter, b in Eq.(5), is assumed to be equal to c, and not estimated from observation.

3 Asymptotic Generalization Error

Although the purpose of an inverse problem is not prediction of a future output but estimation of the parameter itself, it is difficult to compare learning methods fairly in the parameter domain, since performance is strongly affected by prior knowledge. So, some criteria, for example, the goodness-of-fit (GOF) value and χ^2 value [14], are used in the field of brain activity estimation for evaluation of the adequacy of models and learning methods. They evaluate the difference between the observed output and the predictive output, like the training error in the terms of machine learning. Since we would like to count the effect of overfitting, we analyze the generalization error, which evaluates the difference between the true output and the predictive output. The generalization error is defined as the average Kullback-Leibler (KL) divergence between the true distribution, $q(Y)$, and the VB predictive distribution, $p(Y|Y^n) = \langle p(Y|A, b)\rangle_{r(A,b|Y^n)}$:

$$G(n, A'^*) = \left\langle \int q(Y) \log \frac{q(Y)}{p(Y|Y^n)} dY \right\rangle_{q(Y^n)}. \tag{10}$$

Here, $\langle\cdot\rangle_p$ denotes the expectation value over a distribution p, $\langle\cdot\rangle_{q(Y^n)}$ the expectation value over all sets of n samples, $\int dY$ the integral over all the elements of Y, and $A'^* = B^* * A^*$ the true *original* parameter matrix.

Let M^* be the true number of nonzero row vectors of A'^*, namely, $(M - M^*)$ vectors satisfy $\tilde{a}'^*_m = 0$. We, in advance, eliminate the components such that $v_m = 0$, since they never affect learning. Then, we obtain the following theorem:

Theorem 2. *When* V *is general diagonal, the generalization error of the ARD, Eqs.*(2) *and* (3)*, is asymptotically expanded as*

$$2G(n, A'^*) = 2\lambda n^{-1} + O(n^{-3/2}),$$
$$\text{where } 2\lambda = UM^* + U(M - M^*)\frac{\Gamma_{func}((U-2)/2, U/2)}{\Gamma_{func}(U/2)}. \tag{11}$$
$$\text{Here, } \Gamma_{func}(h, x) = \int_x^\infty s^{h-1} e^{-s} ds, \text{ and } \Gamma_{func}(h) = \Gamma_{func}(h, 0) \tag{12}$$

are the upper incomplete Gamma function and the complete one, respectively. We call λ *the generalization coefficient.*

(Outline of the proof) Considering the extent of the VB posterior distribution, Eq.(6), like in [9], we can find that the predictive distribution of the ARD is asymptotically

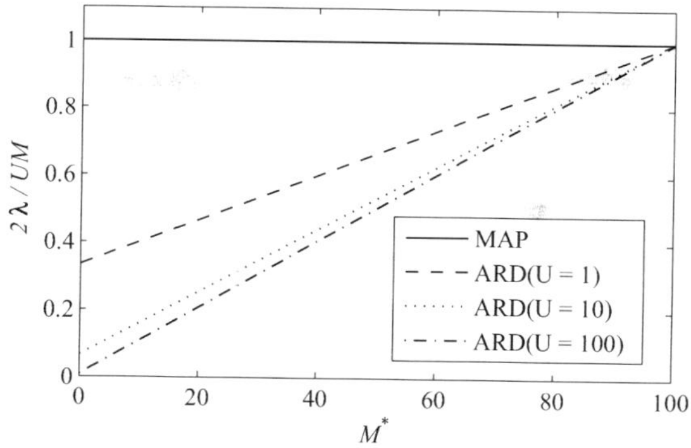

Fig. 1. Asymptotic generalization error when $M = 100, U = 1, 10, 100$

equivalent to the model at the VB estimator. So, we just need to evaluate the squared risk normalized by the noise variance, σ_y^2:

$$2G(n, A'^*) = \sigma_y^{-2}\langle \mathrm{tr}((\hat{A}' - A'^*)^t V^t V(\hat{A}' - A'^*))\rangle_{q(Y^n)} + O(n^{-3/2}), \tag{13}$$

where tr$(\cdot)$ denotes the trace of a matrix. For the M^* nonzero elements, the generalization error is equal to UM^*, since those true values are on regular points. For the other $(M - M^*)$ irrelevant elements, we have

$$2G_{\mathrm{irr}}(n, A'^*) = (M - M^*)\langle(\|\tilde{g}\|^2 - U)^2/\|\tilde{g}\|^2\rangle_{q(\tilde{g})} n^{-1} + O(n^{-3/2}), \tag{14}$$

where $\tilde{g}$ is a U-dimensional random vector subject to $\mathcal{N}_U(0, I_U)$, over which $\langle\cdot\rangle_{q(\tilde{g})}$ denotes the expectation. Using the polar coordinate representation and calculating the expectation in Eq.(14), we have

$$2\lambda = (M - M^*)\frac{4\Gamma_{\mathrm{func}}((U+2)/2, U/2) - 4U\Gamma_{\mathrm{func}}(U/2, U/2) + U^2\Gamma_{\mathrm{func}}((U-2)/2, U/2)}{2\Gamma_{\mathrm{func}}(U/2)}$$
$$= U(M - M^*)\frac{\Gamma_{\mathrm{func}}((U-2)/2, U/2)}{\Gamma_{\mathrm{func}}(U/2)}.$$

Thus, we obtain Theorem 2. (Q.E.D.)

Figure 1 shows the generalization coefficient normalized by the parameter dimensionality, $2\lambda/UM$, in the case that $M = 100$ and $U = 1, 10, 100$. Note that the normalized generalization coefficient is always equal to unity in the MAP estimation, Eq.(9), which is asymptotically equivalent to the ML estimator. The horizontal axis indicates M^*. We see in Fig.1 that the ARD provides much better generalization performance when the model has a lot of irrelevant components. Another important fact is that the larger U is, the better generalization is. From the viewpoint of statistical physics, the increase of U leads to the increase of the state density of singularities, which enhances suppression of overfitting. However, note that setting too large value to U may result in large M^*, since the $(M - M^*)$ irrelevant elements are defined as the ones that never have positive true value during U.

4 Nonasymptotic Generalization Error When $U = 1$

One purpose of brain current estimation is to investigate the brain activity of human or other beings [6]. In that purpose, we can obtain a number of samples. However, in another purpose such as brain computer interface (BCI) [15], where the estimation result is used as an input signal to control some device in realtime, the number of samples is only one. Therefore, nonasymptotic analysis is also important. If we focus on the case that $U = 1$, we can obtain the nonasymptotic solution, again in a similar fashion to the analysis in [12]:

Theorem 3. *When V is general diagonal and $U = 1$, the VB estimator of the m-th element is given by*

$$\hat{a}'_m = \begin{cases} 0 & \text{if } v_m = 0 \\ \mathrm{sign}(y_m) \cdot \max\left(0, \left|\mathcal{S}(\bar{y}_m/\|v_m\|; \sigma_y^2/\|v_m\|^2)\right| - \frac{\sigma_y^2}{nc_m\|v_m\|^2}\right) & \text{if } v_m \neq 0 \end{cases}, \quad (15)$$

where $\mathrm{sign}(\cdot)$ *denotes the sign of a scalar.*[4]

When the number of samples, n, is small, the predictive distribution is not well approximated by the normal distribution, and difficult to be calculated. Since our purpose is not to know the rigorous generalization error of the predictive distribution but to evaluate the accuracy of the estimated parameter, we analyze a *plug-in* generalization error, which we define as the KL divergence between the true distribution, $q(Y)$, and the model at the VB estimator, $p(Y|\hat{A}, \hat{b})$:

$$G'(n, A'^*) = \left\langle \int q(Y) \log \frac{q(Y)}{p(Y|\hat{A},\hat{b})} dY \right\rangle_{q(Y^n)}. \quad (16)$$

We obtain the following theorem:

Theorem 4. *When V is general diagonal and $U = 1$, the plug-in generalization error of the ARD is given by*

$$2G'(n, A'^*) = \textstyle\sum_{m=1}^{M} \langle (\hat{a}'^{(0)}_m(g) - a'^{*(0)}_m)^2 \rangle_{q(g)}, \quad (17)$$

$$\text{where } \hat{a}'^{(0)}_m(g) = \mathrm{sign}(a'^{*(0)}_m + g) \max\left(0, |\mathcal{S}(a'^{*(0)}_m + g; 1)| - \frac{\sigma_y}{\sqrt{n}c_m\|v_m\|}\right),$$

$$a'^{*(0)}_m = \sqrt{n}a'^*_m\|v_m\|/\sigma_y,$$

and g is a random variable subject to $\mathcal{N}_1(0, 1^2)$.

(Proof) Substituting $\hat{a}'^{(0)}_m(g) = \sqrt{n}\hat{a}'_m\|v_m\|/\sigma_y$ in the first term of Eq.(13), which is equal to the *plug-in* generalization error, immediately completes the theorem. (Q.E.D.)

The expectation in Eq.(17) can, relatively easily, be calculated, by creating random variables subject to $\mathcal{N}_1(0, 1^2)$. Figure 2 shows the normalized *plug-in* generalization error, $2G'/UM$, in the case that $M = 5, n = 1, \sigma_y^2 = 1^2$, and $\|v_m\| = 1, c_m = 10$ for $\forall m$. The dashed line ($M^{**} = 2$) corresponds to the case that the true *original* parameter

[4] In this section, the dimensionality of a time series vector is one, so we abbreviate *tilde*.

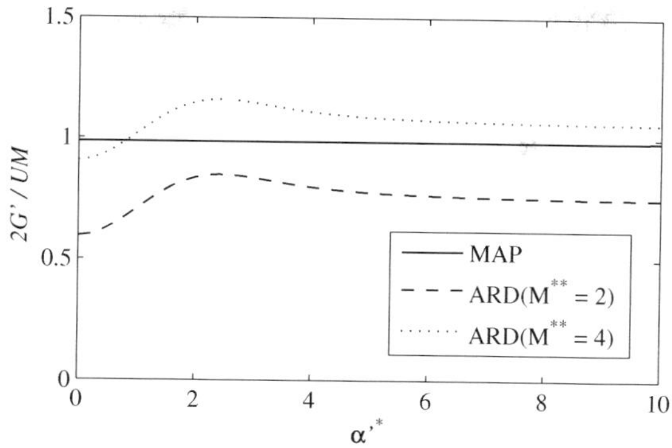

Fig. 2. Nonasymptotic *plug-in* generalization error when $M = 5, U = 1, M^{**} = 2, 4$

is equal to $a'^{*} = (10, 10, \alpha'^{*}, 0, 0)^{t}$, and the dotted line ($M^{**} = 4$) to the case that $a'^{*} = (10, 10, 10, 10, \alpha'^{*})^{t}$, where α'^{*} is indicated by the horizontal axis.[5] We see in Fig. 2 that the generalization performance of the ARD is much better than the MAP estimation when a lot of irrelevant components exist, however, worse when few irrelevant component exists. This is because suppression of overfitting, naturally, accompanies insensitivity to true components with small amplitudes. However, we conjecture in Section 5.1 that the ARD provides no worse generalization error than the MAP estimation when we use a very vague prior and $U \geq 4$. In addition, since we assume a relatively vague prior in Fig. 2, the generalization error when $\alpha'^{*} = 0$ as well as when $\alpha'^{*} = 10$ is well approximated by the asymptotic result given by Theorem 2, for example, the result when $a'^{*} = (10, 10, 0, 0, 0)^{t}$ is approximated by the asymptotic result $2G'/UM = 0.60$ when $M^{*} = 2$, and the result when $a'^{*} = (10, 10, 10, 0, 0)^{t}$ by the asymptotic result $2G'/UM = 0.73$ when $M^{*} = 3$.

5 Discussion

5.1 Generalization

We have shown that the ARD has the benefit of generalization performance. The suppression of overfitting of the MAP estimation disappears in the asymptotic limit, while that of the ARD remains. We can say that this benefit is caused by its singularities, because the asymptotic generalization performance of the Bayesian methods differs from that of the ML estimation only in singular models. By using the algebraic geometrical method [8], the asymptotic generalization error of the rigorous Bayes estimation, to which Bayesian methods are to approximate, has been analyzed in a similar model to the ARD [16]. Considering the result of [16], we can say that, in the ARD, the generalization property of the Bayes estimation is similar to that of the VB approach, clarified in this paper. In general, it has been proved that the asymptotic generalization error of the Bayes estimation is no greater than that of the ML estimation if we use a prior

[5] By M^{**} we denote the number of true components large enough.

distribution having positive values on the singularities, while no smaller if we use the Jeffreys' prior [17], which has zero values on the singularities, and sufficiently small values on their neighborhoods so that the state density of the singularities is comparable to the other points. In fact, if we use a prior distribution such that the *inverse* variance, b_m^{-2}, of $\tilde{a}'_m$ is subject to $\Gamma(\kappa, \nu)$, as in [5, 6], the prior probabilities on a part of the singularities, $\{(\tilde{a}_m, b_m); b_m = 0\}$, is zero, so, it seems possible that the property that we have clarified in this paper differs from that of the method in [5, 6]. However, we conjecture that, when $U \geq 2$, the properties would be similar to each other, since the other part of the singularities, $\{(\tilde{a}_m, b_m); \tilde{a}_m = 0\}$, has a larger state density, and therefore, dominates its performance when the dimensionality of $\tilde{a}_m$ is larger than that of b_m. (See [9].)

We discussed in Section 4 that the ARD does not always provide better generalization performance than the MAP estimation, since suppression of overfitting accompanies insensitivity to true components with small amplitudes. However, we conjecture that, when $U \geq 4$ and we use a very vague prior, the ARD could dominate the MAP estimation. That is for the following reasons: that Theorem 3 says that, even in nonasymptotic cases, the ARD estimator converges to the PJS estimator when $U = 1$ and c_m for $\forall m$ goes to infinity; and that the PJS estimator was proved to dominate the ML estimator when $0 < \chi \leq 2(U-2)$ [18], where χ is the degree of shrinkage, introduced in Eq.(8). Although the asymptotic terms when $U \geq 2$ have, unfortunately, not been derived yet, we expect that they would converge to zero when c_m goes to infinity, as in the case that $U = 1$. Deriving the terms or their bounds and proving the domination of the ARD over the MAP estimation is future work.

5.2 Sparsity

Comparing with the minimum norm maximum likelihood (MNML) estimator,

$$\hat{a}_u'^{\text{MNML}} = (V^t V)^- V^t \bar{y}_u, \tag{18}$$

where $(\cdot)^-$ denotes the Moore-Penrose generalized inverse of a matrix, we can find that the term caused by l^2-norm penalty of the MAP estimator, Eq.(9), is linear to the observed output, $\bar{y}_u$. On the other hand, Theorems 1 and 3 say that both of the asymptotic and the nonasymptotic shrinkage factors of the ARD, i.e., the factor included in the PJS operator, Eq.(8), which depends on χ, and the factor proportional to c_m^{-1} in Eq.(15), are less increasing of $\bar{y}_u$. Assuming for simplicity that $U = 1$, we can say in general that

Proposition 1. *An estimator, $\hat{w} \in \mathbb{R}^M$, is more sparse than $\hat{w}'$, if each element of $\hat{w}$ is written as*

$$\hat{w}_m = (1 - f(|\hat{w}'_m|))\, \hat{w}'_m,$$

where $f(|w_m|)$ is a nonincreasing, nonconstant, and bounded function of $|w_m|$ such that $0 \leq f(|w_m|) \leq 1$.

Let $\hat{w}' = \hat{a}'^{\text{MNML}}$. Then, the MAP estimator, Eq.(9), is one such that $f(|w_m|)$ is constant, and therefore, each component, $\hat{a}_m'^{\text{MAP}}$, uniformly shrinks, while the ARD estimator is sparse.

However, we should remember that we have adopted the coordinate system such that V is general diagonal. Therefore, the sparsity above is observed on that coordinate system, and the sparsity on the original coordinate system, which we prefer, is not necessarily observed, if we use a prior distribution such that the components along the right singular vectors of V are independent of each other. On the other hand, when we apply the ARD, we usually use a prior distribution such that the components on the original coordinate system are independent of each other, which is conjectured to lead to sparsity on the original coordinate system. Further analysis is future work.

According to the discussion above, the following PJS estimator based on the MNML estimator, proposed in [12], is expected to provide a more sparse solution than the ARD:

$$\hat{\tilde{a}}_m'^{\mathrm{PJS}} = \mathcal{S}(\hat{\tilde{a}}_m'^{\mathrm{MNML}}; \sigma_y^2 U / \|v_m\|^2), \tag{19}$$

where we adopt the original coordinate system. Its application to real problems is future work.

5.3 Properties of Variational Bayes Approach

Properties of the VB approach, proposed in [3,4], has recently been clarified. For example, properties on convergence have been clarified [19], bounds of the VB free energy in singular models have been derived [20]. Moreover, solutions in bilinear models have been obtained in closed forms [21,22,12]. The last example means that, in such models, we do not need iterative calculation like expectation-maximization (EM) algorithm. The generalization error has also been clarified in bilinear models, and it has been shown that the VB approach has both of an ML like aspect and a Bayesian aspect [9].

6 Conclusions

In this paper, we have clarified the asymptotic generalization error of the automatic relevance determination (ARD) in the case that we use a class of prior distributions, and the nonasymptotic one in a more special case. Thus, we have concluded that one benefit of the ARD, generalization, is caused by its singularities, and that the larger the time duration that the hyperparameter is assumed to be constant is, the stronger suppression of overfitting is. We have also discussed another benefit of the ARD, sparsity, but further analysis is needed, which is our future work.

Acknowledgments

The authors would like to thank Masanori Osako of ATR for the discussion on [6]. They would also like to thank Yutaka Iwasaki and Hiroshi Ooki of Nikon Corporation for encouragement to research this subject.

References

1. MacKay, D.J.C.: Bayesian Non-linear Modeling for the Energy Prediction Competition. ASHRAE Transactions 100, 1053–1062 (1994)

2. Neal, R.M.: Bayesian Learning for Neural Networks. Springer, Heidelberg (1996)
3. Hinton, G.E., van Camp, D.: Keeping Neural Networks Simple by Minimizing the Description Length of the Weights. In: Proc. of COLT, pp. 5–13 (1993)
4. Attias, H.: Inferring Parameters and Structure of Latent Variable Models by Variational Bayes. In: Proc. of UAI (1999)
5. Sato, M., Yoshioka, T., Kajihara, S., Toyama, K., Goda, N., Doya, K., Kawato, M.: Hierarchical Bayesian Estimation for MEG inverse problem. Neuro Image 23, 806–826 (2004)
6. Osako, M., Yamashita, O., Hiroe, N., Sato, M.: Verification of Hierarchical Bayesian Estimation Combining MEG and fMRI: A Motor Task Analysis (in Japanese). In: Technical Report of IEICE, Tokyo, Japan, vol. NC2006-130, pp. 73–78 (2007)
7. Wipf, D., Ramirez, R., Palmer, J., Makeig, S., Rao, B.: Analysis of Empirical Bayesian Methods for Neuroelectromagnetic Source Localization. In: Advances in NIPS, vol. 19 (2006)
8. Watanabe, S.: Algebraic Analysis for Nonidentifiable Learning Machines. Neural Computation 13, 899–933 (2001)
9. Nakajima, S., Watanabe, S.: Variational Bayes Solution of Linear Neural Networks and its Generalization Performance. Neural Computation 19, 1112–1153 (2007)
10. James, W., Stein, C.: Estimation with Quadratic Loss. In: Proc. of the 4th Berkeley Symp. on Math. Stat. and Prob., pp. 361–379 (1961)
11. Efron, B., Morris, C.: Stein's Estimation Rule and its Competitors—an Empirical Bayes Approach. J. of Am. Stat. Assoc. 68, 117–130 (1973)
12. Nakajima, S., Watanabe, S.: Analytic Solution of Hierarchical Variational Bayes in Linear Inverse Problem. In: Kollias, S., Stafylopatis, A., Duch, W., Oja, E. (eds.) ICANN 2006. LNCS, vol. 4132, pp. 240–249. Springer, Heidelberg (2006)
13. Sato, M.: Online Model Selection Based on the Variational Bayes. Neural Computation 13, 1649–1681 (2001)
14. Hamalainen, M., Hari, R., Ilmoniemi, R.J, Knuutila, J., Lounasmaa, O.V.: Magnetoencephalography — Theory, Instrumentation, and Applications to Noninvasive Studies of the Working Human Brain. Rev. Modern Phys. 65, 413–497 (1993)
15. Blankertz, B., Dornhege, G., Krauledat, M., Curio, G., Muller, K.R.: The Non-invasive Berlin Brain-Computer Interface: Fast Acquisition of Effective Performance in Untrained Subjects (2007) (to appear in Neuro Image)
16. Watanabe, S., Amari, S.: Learning Coefficients of Layered Models When the True Distribution Mismatches the Singularities. Neural Computation 15, 1013–1033 (2003)
17. Watanabe, S.: Algebraic Information Geometry for Learning Machines with Singularities. In: Advances in NIPS, vol. 13, pp. 329–336 (2001)
18. Stein, C.: Estimation of the Mean of a Multivariate Normal Distribution. Annals of Statistics 9, 1135–1151 (1981)
19. Wang, B., Titterington, D.M.: Convergence and Asymptotic Normality of Variational Bayesian Approximations for Exponential Family Models with Missing Values. In: Proc. of UAI, Banff, Canada, pp. 577–584 (2004)
20. Watanabe, K., Watanabe, S.: Stochastic Complexities of Gaussian Mixtures in Variational Bayesian Approximation. Journal of Machine Learning Research 7, 625–644 (2006)
21. Nakajima, S., Watanabe, S.: Generalization Error and Free Energy of Variational Bayes Approach of Linear Neural Networks. In: Proc. of ICONIP, Taipei, Taiwan, pp. 55–60 (2005)
22. Barber, D., Chiappa, S.: Unified Inference for Variational Bayesian Linear Gaussian State-Space Models. In: Advances in NIPS, vol. 19 (2006)

On a Singular Point to Contribute to a Learning Coefficient and Weighted Resolution of Singularities

Takeshi Matsuda and Sumio Watanabe

[1] Department of Computational Intelligence and Systems Science, Tokyo Institute of Technology, Mailbox R2-5, 4259 Nagatsuta-chou, Midori-Ku, Yokohama, Kanagawa, 226-8503, Japan
[2] Precision and Intelligence Laboratory, Tokyo Institute of Technology,Mailbox R2-5, 4259 Nagatsuta-chou, Midori-Ku, Yokohama, Kanagawa, 226-8503, Japan

Abstract. A lot of learning machines which have the hidden variables or the hierarchical structures are the singular statistical models. They have a different learning performance from the regular statistical models. In this paper, we show that the learning coefficient is easily computed by weighted blow up, in contrast, and that there is the case that the learning coefficient cannot be correctly computed by blowing up at the origin O only.

1 Introduction

Learning machines which have its singular Fisher information matrix is called the singular models. For example, layered neural networks, normal mixtures, hidden Markov models, Boltzmann machines and Bayes networks are the singular statistical models. These learning machines are not subject to the conventional statistical theory of the regular statistical models. Recently, the generalization performance of a singular learning machine in Bayes estimation determined by the algebraic geometrical structure of the learning machine was proved [7]. The generalization error

$$G(X^n) = \int q(x)\log\frac{q(x)}{p(x \mid X^n)}dx$$

is equal to

$$G(n) = \frac{\lambda}{n} + o(\frac{1}{n}),$$

where n is the number of training samples and λ is the learning coefficient. We are able to express $G(n)$ with a stochastic complexity $F(n)$ as follows:

$$G(n) = F(n+1) - F(n),$$

where $F(n) = \lambda\log n - (m-1)\log\log n + o(1)$. The constants $(-\lambda)$ and m are the values which depend on the complex function of one variable

J. Marques de Sá et al. (Eds.): ICANN 2007, Part I, LNCS 4668, pp. 11–18, 2007.

$$\zeta(z) = \int H(\omega)^z \varphi(\omega) d\omega,$$

where $H(\omega) = \int q(x) log \frac{q(x)}{p(x|\omega)} dx$ is the Kullback information, $q(x)$ is a true distribution and $p(x \mid \omega)$ is a learning machine. The learning coefficients of some learning machine, for example a three-layer perceptron and a reduced rank regression have been obtained by using blow-up process [2] and [3]. When the origin O is a singular point, we blow up at the origin O. Then it can happen that the singular points appear at the origin O and another point (not origin). Therefore, there is the case where the learning coefficient λ cannot be decided only doing the blow-up at O. In this paper, we introduce the calculation method of the learning coefficients using the weighted blowup and construct the example that the learning coefficient cannot calculate by only doing blow-up at O. The weighted blowup is method to get resolution of singularities using weight of polynomials or analytic functions ,and by this method, we can easily get resolution of singularities.

2 Bayes Learning and Asymptotic Theory

In this section, we summarize the well known statistical framework of Bayes estimation.

Let $X^n = (X_1, X_2, ..., X_n)$ be training samples which are independently taken from the probability distribution $q(x)$. $q(x)$ is called true distribution and defined on the N-dimensional Euclidean space $\mathbf{R}^N$. The integer n is referred to as the number of training samples. A learning machine is represented by a conditional probability density function $p(x|w)$ where w is a d-dimensional parameter. Let $\varphi(w)$ be a priori distribution. Then the Bayes a posteriori distribution is defined by

$$p(w|X^n) = \frac{1}{Z(X^n)} \varphi(w) \prod_{i=1}^{n} p(X_i|w),$$

where $Z(X^n)$ is the normalizing constant. The Bayes predictive distribution is also defined by

$$p(x|X^n) = \int p(x|w)\, p(w|X^n)\, dw,$$

which is the estimated probability density function on $\mathbf{R}^N$ by Bayes learning. The Generalization error $G(n)$ is defined by

$$G(n) = E\Big[\int q(x) \log \frac{q(x)}{p(x|X^n)} dx\Big],$$

where $E[\cdot]$ denotes the expectation value overall sets of X^n. Also we define the stochastic complexity by

$$F(n) = E\Big[-\log Z(X^n)\Big] + n \int q(x) \log q(x) dx.$$

Relation between generalization error and stochastic complexity is as follows;

$$G(n) = F(n+1) - F(n)$$

The stochastic complexity indicates how appropriate the set of $p(x|w)$ and $\varphi(w)$ are for a given training sample a set of X^n. In the learning theory, it is important to clarify the asymptotic behaviors of $G(n)$ and $F(n)$. The following theorem shows the relation between algebraic geometry of the Kullback information and the singular learning machines.

Theorem 1. *When n tends to infinity, the generalization error and the stochastic complexity are respectively given by*

$$G(n) = \frac{\lambda}{n} + o(\frac{1}{n}),$$
$$F(n) = \lambda \log n - (m-1)\log\log n + O(1),$$

where $(-\lambda)$ and m are each equal to the largest pole and its order of the zeta function,

$$\zeta(z) = \int H(w)^z \, \varphi(w) \, dw.$$

Here $H(w)$ is the Kullback information

$$H(w) = \int q(x) \log \frac{q(x)}{p(x|w)} dx.$$

Proof. The proof is given in [7].

3 Preliminary Results and Facts

First, we review the definition of analytic equivalence of Kullback information. Let U and V be open sets in $\mathbf{R}^d$ whose closures are compact. Let $K(w)$ and $H(w)$ be analytic functions on U and V, respectively.

Definition 1. *Two real analytic functions $K(w)$ on U and $H(w)$ on V are analytically equivalent if a bijective analytic map $g : V \to U$ exists*

$$H(w) = K(g(w)) \quad w \in V$$

and the Jacobian $|g'(w)|$ satisfies $\epsilon < |g'(w)| < C$ in V for some $\epsilon, C > 0$.

Theorem1 shows that the learning coefficient is determined by the Kullback information and the a priori distribution. From the definition of analytic equivalence, the following theorem holds.

Theorem 2. *If $K(w)$ and $H(w)$ are analytically equivalent, then two zeta functions*

$$\zeta_1(z) = \int_U K(w)^z dw,$$
$$\zeta_2(z) = \int_V H(w)^z dw$$

have the same largest pole.

Proof. The proof of this theorem is to see [6].

Note that two zeta functions do not have the same second largest pole in general. Now, we will explain the concept of weighted resolution of singularities of Kullback information.

Let $v = (v_1, \ldots, v_n)$ be a set of non-negative integers. For a monomial $x^u = x_1^{u_1} \cdots x_n^{u_n}$,we define the weighted degree $ord_w(x^u)$ with the weight v as

$$ord_w(x^u) =< v, u >= v_1u_1 + \ldots + v_nu_n.$$

A polynomial is said to be quasi-homogeneous if it is a linear combination of the monomials which have the same weighted degree with some weight. An analytic function f is said to have an algebraic isolated singularity at O, if the dimension of a real vector space

$$M(f) = R[[x_1, \cdots, x_n]]/ < \frac{\partial f}{\partial x_1}, \cdots, \frac{\partial f}{\partial x_n} >$$

is finite. The following theorem shows a sufficient condition of the analytic equivalence.

Theorem 3. *Let f be an analytic function*

$$f = f_d + f_{d+1} + f_{d+2} + \cdots, \quad f_d \neq 0$$

where f_d is a quasi-homegeneous polynomial of degree q with weight v. If the weighted degree of x^{u_i} exceeds d and $c_1, \ldots, c_s$ are constants, then f and $f_d + c_1x^{u_1} + \cdots + c_sx^{u_s}$ are analytically equivalent.

Proof. For the proof of this theorem, see [1].

Finaly, we will explain the concept of weighted blow up. The definitiion of a toric variety is to see [5].

We fix the lattice $N = \mathbf{Z}^n$, $\Delta = \{$the first quadrant σ of $\mathbf{R}^n$ and its all faces $\}$. Then we have $T_N(\Delta) = \mathbf{R}^n$, where $T_N(\Delta)$ is a toric variety associated with a fan Δ in $N_R = N \otimes_{\mathbf{Z}} \mathbf{R}$. Let $\Delta(w)$ be a decomposition of Δ for $w = (w_1, \cdots, w_n) \in \sigma \cap N$. We assume $\Delta(w)$ is a fan consisting of all faces of

$$\sigma_1 = \mathbf{R}_{\geq 0}e_1 + \cdots + \mathbf{R}_{\geq 0}e_{i-1} + \mathbf{R}_{\geq 0}w + \mathbf{R}_{\geq 0}e_{i+1} + \cdots + \mathbf{R}_{\geq 0}e_n.$$

Then the corresponding morphism

$$\phi : T_N(\Delta(w)) \to T_N(\Delta)$$

is called a weighted blowup by weight w. Specifically we can calculate as follows. Let $x_1, \cdots, x_n$ be eigen coordinates in $\mathbf{R}^n$, for $\mathbf{Z}_m$(cyclic group). The weighted blow up of $X \subset \mathbf{R}^n$ with weights $a_1, \cdots, a_n$ is a projective birational morphism $\pi : Y \to X$ such that Y is covered by open sets $U_1, \cdots, U_n$, where

$$U_i = \mathbf{R}^n_{y_1, \cdots, y_n} / \mathbf{Z}_{a_i}(-a_1, \cdots, m, \cdots, -a_n).$$

The coordinates in X and in U_i are related by

$$x_i = y_i^{\frac{a_i}{m}}, x_j = y_j \cdot y_i^{\frac{a_j}{m}}, j \neq i.$$

4 Learning Coefficients by Weighted Resolution of Singularities

In this section, we introduce the calculation method of learning coefficients using weighted blowup. First, we prove the following theorem.

4.1 Main Theorem

Theorem 4. *Let H be an analytic function satisfying the following condition:*
(1) $H = H_d + H_{d+1} + \cdots$,
(2) $H_d = x_1^{a_1} \cdots x_n^{a_n}(x_1^{m_1} + \cdots + x_n^{m_n})$,
(3) H_d is a weighted homogeneous polynomial of weighted degree d with weight $v = (p_1, \cdots, p_n)$,
(4) H_d has an algebraic isolated singularity at O,
(5) H_d is non-degenerate.
We consider the following zeta function

$$\zeta(z) = \int H(x)^{\alpha z} \varphi(x) dx,$$

where $x = (x_1, \cdots, x_n)$. Then there exists some a_i, $(1 \leq i \leq n)$ and

$$\begin{cases} \lambda = max\{\frac{p_1 + \cdots + p_n}{(p_1 a_1 + \cdots + p_n a_n + p_i m_i)\alpha}\} \\ m = 1 \end{cases}$$

Proof. Let H'_d be a weighted homogeneous polynomial of weight d. Then we can write H as follow:

$$H = x_1^{a_1} \cdots x_n^{a_n}(g(x) + H'_{d+1} + \cdots),$$

and there exists the weighted homogeneous polynomial $g(x)$ which satisfies

$$g(x) \sim H'_d + H'_{d+1} + \cdots,$$

where $g(x) = x_1^{m_1} + \cdots + x_n^{m_n}$. If two analytic functions are analytically equivalent, then they have the same largest poles. Therefore, we consider the following zeta function $\zeta_1(z)$ to calculate the largest pole and its order of $\zeta(z)$,

$$\zeta_1(z) = \{(x_1^{a_1} \cdots x_n^{a_n}) g(x)\}^{\alpha z} \varphi(x) dx.$$

From the definition of weighted blowup, we get

$$\zeta_1(z) = \sum_{i=1}^{n} \int_{U_i} \{(x_1^{a_1} \cdots x_i^{\frac{p_1 a_1 + \cdots + p_n a_n + p_i m_i}{p_i}} \cdots x_n^{a_n})$$

$$(x_1^{m_1} + \cdots + x_n^{m_n})\} \mid J_i \mid \varphi_i(x) dx,$$

where

$$U_i = \{(x_1, \cdots, x_n) : \mid x_i \mid \leq \mid x_1^{\frac{p_i}{p_1}}, \cdots, \mid x_i \mid \leq \mid x_n^{\frac{p_i}{p_n}} \mid\}$$

$$| J_i |= x_i^{\frac{p_1+\cdots+p_{i-1}+p_{i+1}+\cdots+p_n}{p_i}}.$$

Since

$$\varphi_i(x) = \varphi_i(x_1 x_i^{\frac{p_1}{p_i}}, x_2 x_i^{\frac{p_2}{p_i}}, \cdots, x_n x_i^{\frac{p_n}{p_i}}) = \varphi_i(0,\cdots,0)$$

$$+x_1\varphi_{ix_1}(0,\cdots,0) + x_n\varphi_{ix_n}(0,\cdots,0) + \frac{1}{2!}\{x_1^2\varphi_{ix_1x_1}(0,\cdots,0)$$

$$+\cdots + x_n^2\varphi_{ix_nx_n}(0,\cdots,0) + x_1x_2\varphi_{ix_1x_2}(0,\cdots,0)\cdots +$$

$$+x_nx_1\varphi_{ix_nx_1}(0,\cdots,0)\},$$

we have the following regular function

$$\zeta_1(z) = \frac{\varphi_i(0,\cdots,0)}{z + \frac{p_1+\cdots+p_n}{\alpha(p_1a_1+\cdots+p_na_n+p_im_i)}} + \cdots$$

$$(Re(z) > -\frac{p_1 + \cdots + p_{i-1} + 3p_i + p_{i+1} + \cdots + p_n}{\alpha(p_1a_1 + \cdots + p_na_n + p_im_i)}).$$

Hence we obtain

$$\begin{Bmatrix} \lambda = max\{\frac{p_1+\cdots+p_n}{(p_1a_1+\cdots+p_na_n+p_im_i)\alpha}\} \\ m = 1. \end{Bmatrix}$$

For instance, let $H_1 = x^2 + 2xy^2 = x(x + 2y^2)$, then $p_1 = 2$, $p_2 = 1$, $a_1 = 1$, $a_2 = 0$, $m_1 = 1$ and $m_2 = 2$. Hence we get $\lambda = \frac{3}{4}$. However, we cannot apply this theorem to the following case. Let $H_2 = x^2 + 2xy^2 + y^4$ (degenerate polynomial), then $\lambda = \frac{1}{2}$.

Next, let us study the following analytic function;

$$H(x, y) = x^p + \Sigma_{qk+pl\leq pq} x^k y^l + y^q.$$

Then since $H \not\sim x^p + y^q$, we cannot apply the above theorem. The maps of blow up and weighted blow up are birational maps. Therefore composite mapping is also birational. Using this fact, we get the following theorem.

Theorem 5. *Let H be an analytic function on the 2-dimensional space which satisfying*

$$H = x^p + \sum_{qk+pl<pq} a_{kl}x^k y^l + y^q,$$

and $a = K + L$ be the minimum degree of $a_{kl}x^k y^l = a_{KL}x^K y^L$. We assume K and L are the value of minimun k and l which satisfy $a_{kl} \neq 0$,respectively. Then the pole λ' that is the nearest to the origin calculated by resolution of singularity at O is as follows.

$$\begin{cases} \lambda' = max\{\frac{p-K+L}{pL}, \frac{2}{K+L}, \frac{q-L+K}{Kq}\} \\ m = 1 \end{cases}$$

Proof. It is easy to show that the condition satisfying

$$x^p + y^q \not\sim x^p + \sum_{qk+pl<pq} a_{kl}x^k y^l + y^q$$

is $L < q$. Let $Y \subset \mathbf{R}^2$ be the curve $H = 0$. Y is singular at the origin O. We will blow up Y at the origin. Substitute $x = x, y = xy$ on $Y_{11}(Y = Y_{11} \cup Y_{12})$

$$H_{11} = x^{K+L}(x^{p-(K+L)} + \sum a_{kl}x^{k+l-(K+L)}y^l + x^{q-(K+L)}y^q) \quad \text{on} Y_{11}.$$

Similarly, substitute $x = xy, y = y$ on Y_{12}

$$H_{12} = y^{K+L}(x^p y^{p-(K+L)} + \sum a_{kl}x^k y^{k+l-(K+L)} + y^{q-(K+L)}).$$

Then, clearly we have

$$H_{11} \sim x^{K+L}(x^{p-(K+L)} + a_{KL}y^L),$$
$$H_{12} \sim y^{K+L}(a_{KL}x^K + y^{q-(K+L)}).$$

Therefore, we should consider the following zeta function.

$$\zeta_1(z) = \int_{Y_{11}} x^{K+L}(x^{p-(K+L)} + a_{KL}y^L)x\varphi(x, xy)dxdy+$$
$$\int_{Y_{12}} y^{K+L}(a_{KL}x^K + y^{p-(K+L)})y\varphi(xy, y)dxdy,$$

where

$$Y_{11} = \{(x, y) \in Y; \mid x \mid \leq \mid y \mid\},$$
$$Y_{12} = \{(x, y) \in Y; \mid y \mid \leq \mid x \mid\}.$$

From the previous theorem, we get

$$\begin{cases} \lambda = max\{\frac{p-K+L}{pL}, \frac{2}{K+L}, \frac{q-L+K}{Kq}\} \\ m = 1 \end{cases}$$

It's not always true that λ' is the learning coefficient (i.e. $\lambda \neq \lambda'$).

4.2 Example

(1)Let $H = x^5 + y^2 + \Sigma_{2k+5l<10}a_{kl}x^k y^l$, then $\lambda = \lambda' = 1$.
(2)Let $H = x^5 + y^3 + \Sigma_{3k+5l<15}a_{kl}x^k y^l$. We assume $a_{11} = 0$.

If $\text{multi}_{(A,B)}(H_{11}) \leq \text{multi}_{(0,0)}(H_{11})$, then $\lambda = \lambda'$.

The symbol $\text{multi}_{(A,B)}(H_{11})$ expresses multiplicity of H_{11} at $(A, B) \in \mathbf{R}^2$. For example, we put $a_{12} = -2$ and $a_{21} = 1$. Then we get $\lambda' = \frac{4}{5}$ from the origin singular point O. But $\lambda = \frac{3}{5}$ from the singular point $(0, 1)$ on H_{11}.

In general, if $\text{multi}_{(A,B)}(H_{11}) \leq \text{multi}_{(0,0)}(H_{11})$, then we have $\lambda = \lambda'$.

5 Conclusion

In this paper, we have introduced the calculation method of learning coefficients by weighted blow up and studied the relation between the learning coefficients and resolution of singularities. If the analytic function is analytically equivalent to $x_1^{m_1} + \cdots + x_n^{m_n}$, then we can easily compute the learning coefficient by weighted blow up. Whereas if it is not so, we showed that there is the case that the pole which is obtained by resolution of singularity at the origin O is not the learning coefficient. By the coefficient part of the analytic function, a wide range of singular points appears. It is the problem for the future study to construct mathematical criterion for the higher dimensional above case.

Acknowledgment. This work was supported by the Ministry of Education, Science, Sports, and Culture in Japan, Grant-in-aid for scientific research 18079007.

References

1. Arnol'd, I.V.: Normal forms of functions in neighbourhoods of degenerate critical points. Russian Mathematical Surveys. 29(2), 10–50 (1974)
2. Aoyagi, M., Watanabe, S.: Stochastic complexities of reduced rank regression in Bayesian estimation. Neural Networks 18(7), 924–933 (2005)
3. Aoyagi, M., Watanabe, S.: Resolution of singularities and generalization error with Bayesian estimation for layered neural network. IEICE Trans. J88-D-II(10), 2112–2124 (2005)
4. Atiyah, M.: Resolution of singularities and division of distributions. Communications of Pure and Applied Mathematics 13, 145–150 (1970)
5. Fulton, W.: Introduction to Toric Varieties, vol. 131. Princeton University Press, Princeton (1993)
6. Matsuda, T., Watanabe, S.: Analytic Equivalence of Bayes a Posteriori Distributions. In: Kollias, S., Stafylopatis, A., Duch, W., Oja, E. (eds.) ICANN 2006. LNCS, vol. 4131, pp. 113–121. Springer, Heidelberg (2006)
7. Watanabe, S.: Algebraic Analysis for Non-identifiable Learning Machines. Neural Computation 13(4), 899–933 (2001)

Improving the Prediction Accuracy of Echo State Neural Networks by Anti-Oja's Learning

Štefan Babinec[1] and Jiří Pospíchal[2]

[1] Department of Mathematics, Fac. of Chemical and Food Technology
Slovak University of Technology, 812 37 Bratislava, Slovakia
Phone/FAX: +421 2 52495177
stefan.babinec@stuba.sk

[2] Institute of Applied Informatics, Fac. of Informatics and Information Technologies
Slovak University of Technology, 842 16 Bratislava, Slovakia
Phone: +421 2 60291548, FAX: +421 2 65420587
pospichal@fiit.stuba.sk

Abstract. Echo state neural networks, which are a special case of recurrent neural networks, are studied from the viewpoint of their learning ability, with a goal to achieve their greater prediction ability. A standard training of these neural networks uses pseudoinverse matrix for one-step learning of weights from hidden to output neurons. This regular adaptation of Echo State neural networks was optimized by updating the weights of the dynamic reservoir with Anti-Oja's learning. Echo State neural networks use dynamics of this massive and randomly initialized dynamic reservoir to extract interesting properties of incoming sequences. This approach was tested in laser fluctuations and Mackey-Glass time series prediction. The prediction error achieved by this approach was substantially smaller in comparison with prediction error achieved by a standard algorithm.

1 Introduction

From the point of information transfer during processing, neural networks can be divided into two types: feed-forward neural networks and recurrent neural networks. Unlike the feed forward networks, recurrent neural networks contain at least one cyclical path, where the same input information repeatedly influences the activity of the neurons in a cyclical path. The advantage of such networks is their close correspondence to biological neural networks, but there are many theoretical and practical difficulties connected with their adaptation and implementation. The common problem of all such networks is the lack of an effective supervised training algorithm. This problem was overcome with Echo State neural networks [2]. A very fast algorithm is used in these networks consisting of a calculation of one pseudo-inverse matrix, which is a standard numerical task. But the advantage of one step learning turns into a disadvantage when we try to improve the predictive abilities of the network. The pseudo-inverse matrix approach does not offer any straightforward solution in this case.

J. Marques de Sá et al. (Eds.): ICANN 2007, Part I, LNCS 4668, pp. 19–28, 2007.

One possibility of how to improve the dynamics of such a neural network is an adaptation of the dynamic reservoir. It is not feasible to apply the usual learning approaches because of the large number of neurons in the recurrent part of an Echo State neural network. However, it motivates us to use more robust technique. This led us to the idea of training the reservoir weights by Anti-Oja's learning, which provides a gradual change of the weights and a significantly lower computational cost. In this paper, a one-step learning process, involving the use of a pseudo-inverse matrix, is combined with Anti-Oja's learning, adapting the weights of the dynamic reservoir itself. Connection between "liquid state" computing, related to Echo states was mentioned previously in [4,6]. The predictive abilities of optimized Echo State neural networks were tested on laser-fluctuations and Mackey-Glass time series, which are standard benchmark data.

2 Echo State Neural Network

Echo State neural networks are atypical in architecture and training of recurrent neural networks (RNN). This new approach leads to a fast, simple and constructive supervised learning algorithm for the RNN. The basic idea of Echo State neural networks is an application of a huge reservoir, as a source of dynamic behavior of a neural network, from which neural activities are combined into the required output.

The activity of hidden layer neurons in an RNN is further denoted as $\mathbf{x}_n = (x_1(n), x_2(n), \ldots, x_N(n))$, where $x_i(n)$ is the output of the ith-hidden neuron in time (n), and (N) is the number of hidden neurons. Under certain conditions, each $x_i(n)$ is a function of the networks previous inputs $\mathbf{u}(n), \mathbf{u}(n-1), \ldots$, previously processed by the network. The input vector is denoted as $\mathbf{u}_n = (u_1(n), u_2(n), \ldots, u_K(n))$, where $u_i(n)$ is the input of the ith-input neuron in time (n) and (K) is the number of input neurons. Therefore, there exists such a function, E, so that:

$$\mathbf{x}(n) = E(\mathbf{u}(n), \mathbf{u}(n-1), \ldots). \tag{1}$$

Metaphorically speaking, the state of the neural network $\mathbf{x}(n)$ can be considered as an "echo", or in other words, a reflection of its previous inputs.

2.1 Description of the Neural Network

Neural network consists of (K) input, (N) hidden and (L) output neurons. The state of the neurons in the input layer at time (n) is characterized by the vector $\mathbf{u}_n = (u_1(n), u_2(n), \ldots, u_K(n))$, in the output layer by the vector $\mathbf{y}_n = (y_1(n), y_2(n), \ldots, y_L(n))$ and in the hidden layer by the vector $\mathbf{x}_n = (x_1(n), x_2(n), \ldots, x_N(n))$. The values of all the synaptic weights will be stored in matrices. An input weight matrix will be created: $\mathbf{W}^{in} = (w_{ij}^{in})$ with a size of $N \times K$, a weight matrix between hidden neurons: $\mathbf{W} = (w_{ij})$ with a size of $N \times N$, a matrix of output weights: $\mathbf{W}^{out} = (w_{ij}^{out})$ with a size of $L \times (K+N+L)$, and a matrix of weights from the output back to the reservoir: $\mathbf{W}^{back} = (w_{ij}^{back})$ with a size of $N \times L$. It is notable that, in this type of network, both direct input-output weights, as well as the weights between output neurons are allowed. The

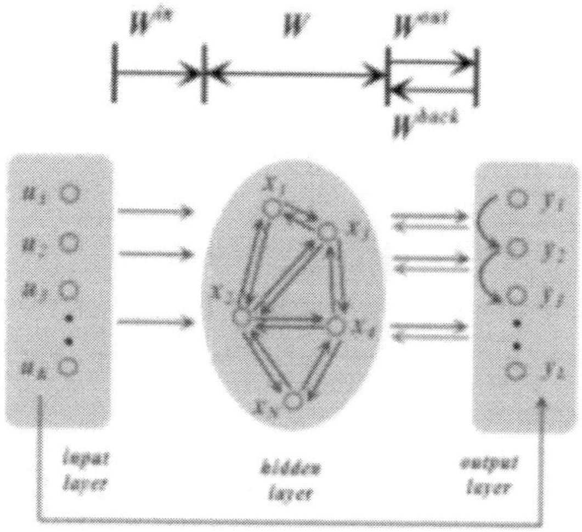

Fig. 1. The typical architecture of Echo State neural networks

structure and topology of Echo State networks can be adjusted according to their current task. It is not necessary, for example, to use sigmoid output neurons, back weights from the output layer to the reservoir may or may not exist, and even the input neurons may not be used. In this application, sigmoid output neurons were not used, however back weights from the output layer to the reservoir were used. No loops were used for output neurons. We can find detailed description of the learning algorithm in [2,3].

3 Motivation

At first we have to realize, how important the values and the structure (interconnection between hidden neurons) of synaptic weights are for the prediction ability of Echo State neural networks. We will do two simple experiments. In the first one, Echo State neural network will be trained independently fifty times and only the values of synaptic weights in its hidden part will be randomly changed. In the second experiment we will do the same thing, but with one difference. Now we will change the network topology of the dynamic reservoir. We can see graphical representation of these experiments in the Fig. 2. These results led us to one important conclusion from the results shown in the Fig. 2. Values of synaptic weights and the structure of the dynamic reservoir have big influence

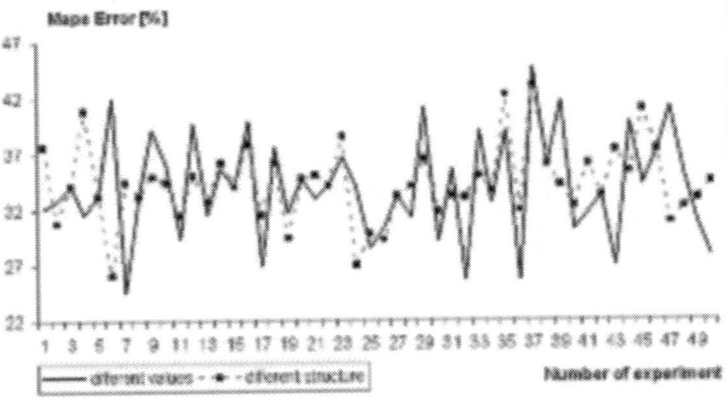

Fig. 2. Influence of the values and the structure of synaptic weights in dynamic reservoir on the prediction quality. The order of experiments is arbitrary.

on the prediction ability of Echo State networks. Therefore, our main goal is to optimize or train the synaptic weights in the dynamic reservoir. However, it is not feasible to apply the usual learning approaches due to the large number of neurons in the recurrent part of an Echo State neural network.

Here we have to realize second important observation about the nature of Echo State neural networks and their dynamics. The neurons in the dynamic reservoir should be slightly interconnected. It means the weight matrix $\mathbf{W}$(synaptic weights matrix of internal neurons) should be sparse. This is a simple method how to encourage a rich variety of dynamics between different internal units.

Now we can explain why are we using Anti-Oja's learning as the supplement of the classic one-step learning algorithm in our experiments. Anti-Oja's learning changes synaptic weights in a way that decreases correlation between hidden neurons. In final consequence, it results in increase of diversity between internal state dynamics of hidden neurons, which will become much richer. We can say that the Anti-Oja's learning quells the effect of some synaptic weights and on the other side, it raises impact of some others. This learning changes values of synaptic weights and the structure of hidden layer, so it optimizes the dynamic reservoir in the required way. Moreover, the computational demands of Anti-Oja's learning are substantially lower than those of standard learning techniques or optimization methods. A more detailed description of Anti-Oja's learning is presented in the next chapter.

3.1 Anti-Oja's Learning Rule

The only difference between Anti-Hebbian and Hebbian learning is in weight update formula, which is in case of Anti-Hebbian learning negative ($w(n+1) = w(n) - \Delta w(n)$). From that reason, we will introduce the description of Hebbian learning. Hebbian learning [1] is the first and the best-known learning rule in neural networks. Hebb first suggested this approach in his book *The Organization of Behavior* [5]. Its essence consists of the following two rules:

- If two neurons, located at the opposite sides of a synapse, are activated concurrently (synchronously), then the value of the synaptic weight will increase.
- If two neurons, located at the opposite sides of a synapse, are activated at different times (asynchronously), then the value of the synaptic weight will decrease.

The synapses of neural networks, which follow the rules named before, are called *Hebbian synapses.* The change in a synaptic weight (w_{kj}) of a Hebbian synapse can be expressed mathematically for a pair of neurons k and j, where j sends data to k, and for a time, noted as n, is as follows:

$$\Delta w_{kj}(n) = F(y_k(n), x_j(n)). \tag{2}$$

Here, $x_j(n)$ represents the activity of a pre-synaptic neuron and $y_k(n)$ the activity of a post-synaptic neuron. F represents the function of post-synaptic and pre-synaptic activities of the already mentioned neurons $y_k(n)$ and $x_j(n)$. These activities are sometimes considered scale-free. A special case of Equation (2) is the following formula:

$$\Delta w_{kj}(n) = \eta y_k(n) x_j(n). \tag{3}$$

In this formula, η is a small positive constant, the so-called learning rate. Equation (3) is the simplest approach to a change of a synaptic weight w_{kj}, expressed as a product of the input and output signals. This rule clearly emphasizes the correlative nature of Hebbian synapses. Sometimes, it is referred to as the activity-product rule. It follows from Equation (3) that repeated application of the input signal (pre-synaptic activity) $x_j(n)$ leads to an exponential increase, which eventually can lead to saturation of w_{kj}. To avoid such a situation, there is a need to limit the growth of the synaptic weight. One of the possibilities is to introduce a nonlinear, forgetting factor into formula (3) that defines the change of w_{kj}. The change of the synaptic weight will be redefined as follows:

$$\Delta w_{kj}(n) = \eta y_k(n) x_j(n) - \alpha y_k(n) w_{kj}(n). \tag{4}$$

Here, α is a new, positive constant and $w_{kj}(n)$ is a synaptic weight in time n. This formula can be further refined into:

$$\Delta w_{kj}(n) = \alpha y_k(n)[c x_j(n) - w_{kj}(n)], \tag{5}$$

where c is equal to η/α. This formula (5) is sometimes called the generalized Hebbian Rule. The meaning of this formula is, that if $x_j(n) < w_{kj}(n)/c$ then the modified synaptic weight $w_{kj}(n+1)$ in time $n+1$ decreases proportionally to a post-synaptic activity $y_k(n)$. On the other hand, when $x_j(n) > w_{kj}(n)/c$, then the synaptic weight $w_{kj}(n+1)$ increases proportionally to $y_k(n)$. We have used in our experiments another modification of Hebbian learning rule called Oja's learning rule:

$$\Delta w_{kj}(n) = \eta y_k(n)[x_j(n) - y_k(n)w_{kj}(n)], \tag{6}$$

where η is small positive constant. There were two reasons for using this last Oja's rule in our experiments. With this rule, the prediction accuracy was on average slightly better, but more importantly, it had less variance.

4 Experiments

Most of the publications about prediction strive to achieve the best prediction, which is then compared with results of other prediction systems on selected data. This paper is different in this aim; its goal was to compare results by original "one-step" learning algorithm, with our new approach. We have used two classic benchmarking data sets.

The first testing set was composed of a time sequence of 1000 samples of laser fluctuations data, and the quality of prediction was measured by an error of prediction in the next 100 steps. A mean absolute percentage error (MAPE) was used to measure the quality of prediction on this testing set, where test values P_i^{real} and predicted values P_i^{calc} are used, and N is the number of couples of values (the length of the predicted time series):

$$MAPE = \frac{\sum_{i=1}^{N} \left| \frac{P_i^{real} - P_i^{calc}}{P_i^{real}} \right|}{N} \times 100. \tag{7}$$

The second testing set was composed of a time sequence of 3000 samples of Mackey-Glass (MG) data, and the quality of prediction was measured by an error of prediction in the independently generated 2000 steps. A normalized root mean square error (NRMSE) was used to measure the quality of prediction on this second testing set, where $d(n)$ and $y(n)$ are real, resp. predicted values. σ^2 is a variance of MG signal and m is a number of independent instances of MG set, on which the trained Echo State neural network was tested. In our case, m equals 50.

$$NRMSE_k = \left(\frac{\sum_{j=1}^{m} (d_j(n+k) - y_j(n+k))^2}{m\sigma^2} \right)^{1/2}. \tag{8}$$

The most used prediction horizon in literature is 84 steps ahead in the series (variable k in equation 8). From that reason, we will use the same value. Different error measures were used here to allow eventual comparison with previous results [2,7].

The learning was divided into two phases. The first phase was aimed to find the best parameters for Echo State neural network for quality prediction results from the training set. No optimization involving Anti-Oja's learning was used in this first phase. The results for laser data are in Table 1 and for Mackey-glass

Table 1. Results of representative experiments in the first phase: quality of the laser prediction for different parameter values

Index	*Size of DR*	*Alpha*	*Average MAPE*	*Best MAPE*
1	200	0.7	38.296%	34.233%
2	200	0.8	33.864%	31.244%
3	250	0.7	36.291%	31.343%
4	250	0.8	34.236%	29.526%
5	300	0.8	35.943%	32.855%
6	300	0.8	35.441%	30.893%

Table 2. Results of representative experiments in the first phase: quality of the Mackey-Glass prediction for different parameter values

Index	*Size of DR*	*Alpha*	*Average* $NRMSE_{84}$	*Best* $NRMSE_{84}$
1	200	0.7	0.00051	0.00045
2	200	0.8	0.00045	0.00039
3	300	0.7	0.00049	0.00042
4	300	0.8	0.00046	0.00038
5	400	0.7	0.00040	0.00036
6	400	0.8	0.00036	0.00034

data in Table 2. In these tables, *DR* represents the dynamic reservoir; *Alpha* is the parameter influencing the ability of the neural network to exhibit echo states. These *DR* and *Alpha* values were chosen in accordance with the proposal used by Jaeger (see [2]). Experiments were carried out in the following way. For each value of *DR* and the parameter *Alpha*, the values of synaptic weights in *DR* were randomly generated 1000 times. This number was estimated to be sufficiently large enough for statistical evaluation of prediction error on a testing set and for each initialization of weights, the error for the testing set was calculated. Further, an average error of all 1000 trials is presented in the columns *Average MAPE* (Table 1) and *Average* $NRMSE_{84}$ (Table 2). Also, the smallest achieved error was recorded in the *Best MAPE* and *Best* $NRMSE_{84}$ columns in the same tables. A clear correlation between Best and Average value columns is apparent from Tables 1 and 2. When a better *Average MAPE* (*Average* $NRMSE_{84}$) was achieved, there is also a better *Best MAPE* (*Best* $NRMSE_{84}$). The best results for laser prediction were achieved with a *DR* consisting of 250 neurons and for the parameter *Alpha* 0.8. For Mackey-Glass prediction, the best results were achieved with a *DR* composed of 400 neurons and for *Alpha* 0.8.

The main experiments were carried out in the second phase with a partial knowledge of behavior of the neural network gained from the first phase. The parameters of the neural network and the initialization of the synaptic weights of the dynamic reservoir were chosen based on the best results from the first phase of the experiments. The adjustment of synaptic weights using Oja's learning rule (6) was carried out for all the samples of the training set. The only exceptions were the last 10 samples in the case of laser prediction. These 10 samples were chosen as the validation set, which was, after the adjustment, used for checking the prediction quality of the samples, which were not available for training the neural network. The validation set for Mackey-Glass data was newly generated and was composed of 500 samples. Using this method, a great number of iterations was carried out, while the use of the validation set monitored the quality of prediction. The representative results are given in Tables 3 and 4. The variable

Table 3. Results of representative experiments for laser prediction in the second phase after optimization using Anti-Oja's learning

Index	*Number of iterations*	η	*MAPE 1*	*MAPE 2*
1	1630	0.001	29.526%	24.331%
2	2650	0.0001		18.098%
3	4830	0.00001		21.567%

Table 4. Results of representative experiments for Mackey-Glass prediction in the second phase after optimization using Anti-Oja's learning

Index	*Number of iterations*	η	$NRMSE_{84}$ *1*	$NRMSE_{84}$ *2*
1	1120	0.001	0.00034	0.00032
2	1786	0.0001		0.00031
3	2700	0.00001		0.00027

Number of iterations. Tells how many iterations were required to produce the smallest error with the validation set. The variable η represent those values of parameter from Equation (6), which were used in the calculations that yielded the best results on the validation set after optimization of the neural network by Anti-Oja's learning. The variables *MAPE 1* / $NRMSE_{84}$ *1* represent the prediction error on the testing set before optimization. This value was the same for all the experiments, since each optimization started with a neural network with the same initialization of weights and other parameters. Variable *MAPE 2* / $NRMSE_{84}$ *2* provides the prediction error on the testing set after optimization using Anti-Oja's learning.

The results indicate that the initialization of weights in the dynamic reservoir has a decisive role in prediction abilities of Echo State neural networks. Another result is that an optimization using Anti-Oja's learning can substantially improve results. These conclusions can be drawn from Figures 3 and 4 as well.

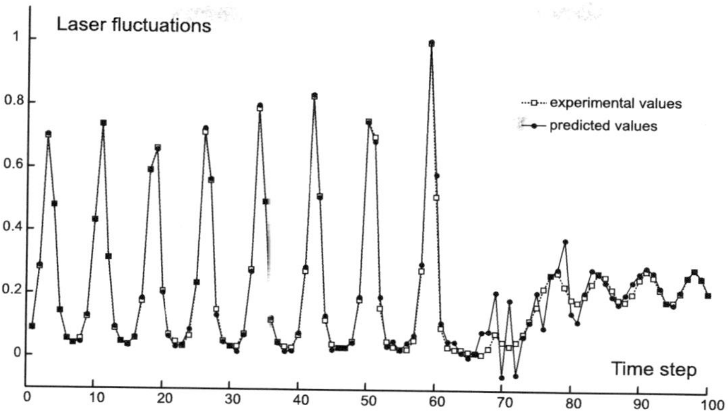

Fig. 3. Testing data: 100 records of laser fluctuations and 100 values predicted by original Echo State neural network with "one-step" learning algorithm (Experiment No.4 from Table 1, DR Size 250, Alpha 0.8, MAPE 29.52 %)

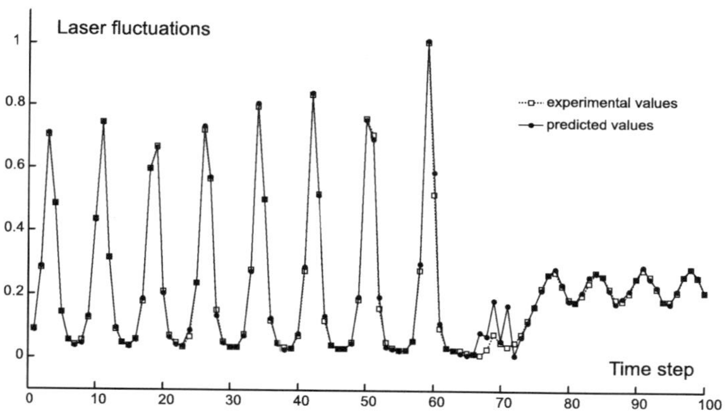

Fig. 4. Testing data: 100 records of laser fluctuations and 100 values predicted by Echo State neural network optimized with Anti-Oja's learning (Experiment No. 2 from Table 3, MAPE 18.09 %)

5 Conclusions

Echo State neural networks are relatively new in the domain of neural networks. Their advantage is a closer connection with biological models inherent to recurrent neural networks and in their usage of the reservoir of dynamic behavior without adjusting the weights within the hidden layer. Echo State neural networks have a substantial advantage over other types of recurrent networks in their one-step learning ability, even though this approach may not be very biologically plausible. As a disadvantage, they can be considered to have a relatively low ability to generalize and in general lack an approach, which would be able to improve, at least partially, a previously learned network.

The problem of improving on a previously learned network does not emerge in the case of common feed-forward or even other recurrent neural networks. If the need arises to improve the network, a simple adjustment of adding more iterations of the back propagation of error can help. However, this does not work for the Echo State networks, when a standard algorithm allows an all or nothing approach, for example: either we shall teach, or we shall not teach the neural network, but nothing between. The already taught neural network cannot be partially amended by this approach.

The work described in this paper tried to solve this problem, where one special approach was used to improve the learning ability of the Echo State neural network. This approach originates in metaphors from nature and it is application of the well-known Anti-Oja's learning. The computational cost of the Anti-Oja's learning is substantial, but hybrid optimization is probably the only feasible way to improve an already trained Echo State network.

We have chosen laser fluctuations and Mackey-Glass time series as a testing data. Our aim was to find out if this approach is able to increase prediction quality of Echo State neural networks. From the results shown in the paper, it is clear, that this aim has been accomplished. Application of Anti-Oja's learning in Echo State neural networks can increase the quality of the network's prediction.

Acknowledgement. This work was supported by Slovak Scientific Grant Agency, Grant VEGA 1/1047/04 and by Slovak Research Development Agency under the contract No. APVT-20-002504.

References

1. Haykin, S.: Neural networks - A comprehensive foundation. Macmillian Publishing, NYC (1994)
2. Jaeger, H.: The Echo State Approach to Analysing and Training Recurrent Neural Networks. German National Research Center for Information Technology, GMD report 148 (2001)
3. Jaeger, H.: Short Term Memory in Echo State Networks. German National Research Center for Information Technology, GMD report 152 (2002)
4. Natschlager, T., Maass, W., Markram, H.: The "liquid computer":A novel strategy for real-time computing on time series. Special Issue on Foundations of Information Processing of TELEMATIK 8(1), 39–43 (2002)
5. Hebb, D.O.: The organization of behavior: A neuropsychological theory. Wiley, New York (1949)
6. Goldenholz, D.: Liquid computing: A real effect. Technical report, Boston University Department of Biomedical Engineering (2002)
7. Babinec, S., Pospichal, J.: Optimization in Echo state neural networks by Metropolis algorithm. In: Matousek, R., Osmera, P. (eds.) Mendel'2004. Proceedings of the 10th International Conference on Soft Computing, pp. 155–160. VUT Brno Publishing (2004)

Theoretical Analysis of Accuracy of Gaussian Belief Propagation

Yu Nishiyama[1] and Sumio Watanabe[2]

[1] Department of Computational Intelligence and Systems Science,
Tokyo Institute of Technology,
4259, Nagatuta, Midori-ku, Yokohama, 226-8503 Japan
nishiyudesu@cs.pi.titech.ac.jp

[2] Precision and Intelligence Laboratory, Tokyo Institute of Technology,
4259, Nagatsuta, Midori-ku, Yokohama, 226-8503 Japan
swatanab@pi.titech.ac.jp

Abstract. Belief propagation (BP) is the calculation method which enables us to obtain the marginal probabilities with a tractable computational cost. BP is known to provide true marginal probabilities when the graph describing the target distribution has a tree structure, while do approximate marginal probabilities when the graph has loops. The accuracy of loopy belief propagation (LBP) has been studied. In this paper, we focus on applying LBP to a multi-dimensional Gaussian distribution and analytically show how accurate LBP is for some cases.

1 Introduction

Belief propagation (BP) is the calculation method which enables us to obtain the marginal probabilities with a tractable computational cost and has been widely studied in the areas such as probabilistic inference for artificial intelligence, turbo codes, code division multiple access (CDMA) systems, low-density parity check (LDPC) codes, and probabilistic image processing. BP is known to provide true marginal probabilities when the graph describing the target distribution has a tree (singly connected) structure, while do approximate marginal probabilities when the graph has loops (cycles).

The accuracy of loopy belief propagation (LBP) has been theoretically studied when the target distributions are discrete distributions [1]-[3] and Gaussian Markov random fields mainly about the means [4]. The accuracy has also been studied from the viewpoint of information geometry [5][6]. In probabilistic image processing, the accuracy has been numerically studied as the accuracy of estimated hyperparameters based on Gaussian graphical models [7][8].

In this paper, we focus on applying loopy belief propagation (LBP) to a multi-dimensional Gaussian distribution whose inverse covariance matrix corresponds to the graph with loops. Then, we mathematically show the differences between true marginal densities and the approximate marginal densities calculated by LBP. To be more specific, we give the exact solutions of messages and approximate marginal densities calculated by LBP and give the Kullback-Leibler (KL)

J. Marques de Sá et al. (Eds.): ICANN 2007, Part I, LNCS 4668, pp. 29–38, 2007.

distances when the inverse covariance matrix corresponds to the graph with a single loop. Next, we develop the results to the more general case that the graph has a single loop and an arbitrary tree structure. Furthermore, we give the expansions of inverse variances of approximate marginal densities up to third-order when the graph is allowed to have an arbitrary structure (i.e., multiloops).

The situation in which LBP is applied to a multi-dimensional Gaussian distribution can be seen for example in probabilistic image processing based on Gaussian graphical models [7][8] and Markov random fields [4]. Besides, clarifying the theoretical properties of LBP in the standard models such as a Gaussian distribution helps to know the properties of more complex LBP and to design LBP algorithms efficiently.

This paper is organized as follows. In section **2**, we review the scheme for belief propagation. In section **3**, we give the main results of this paper. In section **4**, we give the conclusion and future works.

2 Belief Propagation

Here, we review the scheme for Belief propagation. We review the belief propagation for tree structures in section **2.1**, loopy belief propagation in section **2.2**, and the LBP when we apply it to a multi-dimensional Gaussian distribution in section **2.3**.

2.1 Belief Propagation for Tree Structures

Let the target distribution to which we apply LBP be the that of pairwise form

$$p(\mathbf{x}) = \frac{1}{Z} \prod_{\{ij\} \in B} W_{ij}(x_i, x_j), \tag{1}$$

where Z is the normalization constant and B is the set which shows the existences of correlations between x_i and x_j. For example, when $\mathbf{x} = (x_1, x_2, x_3, x_4)^T$ and correlations exist between $\{x_1, x_2\}$, $\{x_2, x_3\}$, and $\{x_1, x_4\}$ respectively, the set B is expressed as $B = \{\{12\}, \{23\}, \{14\}\}$. When the graph given by B is a tree graph, the marginal distributions $\{p_i(x_i)\}$, $\{p_{ij}(x_i, x_j)\}$ are exactly calculated as follows by using normalized messages $\{\mathcal{M}_{i\to j}\}$, $\{\mathcal{M}_{j\to i}\}$, $(\{ij\} \in B)$.

$$p_i(x_i) = \frac{1}{Z_i} \prod_{k \in \mathcal{N}_i} \mathcal{M}_{k\to i}(x_i),$$
$$p_{ij}(x_i, x_j) = \frac{1}{Z_{ij}} \Big(\prod_{k \in \mathcal{N}_i \setminus \{j\}} \mathcal{M}_{k\to i}(x_i) \Big) W_{ij}(x_i, x_j) \Big(\prod_{k \in \mathcal{N}_j \setminus \{i\}} \mathcal{M}_{k\to j}(x_j) \Big). \tag{2}$$

Here, both Z_i and Z_{ij} are the normalization constants and $\mathcal{N}_i$ is the subset of random variables which directly correlate with random variables x_i. $\mathcal{N}_i$ is so-called the set of nearest neighbor variables of x_i.

2.2 Loopy Belief Propagation

Eqs.(2) are exactly correct when the graph given by B is a tree graph. In loopy belief propagation, we also compose marginal distributions by eqs.(2) when the graph given by B has loops. This leap of logic yields the problem of convergence of LBP and gives marginal distributions approximately. Since the marginal distributions $\{p_i(x_i)\}$ and $\{p_{ij}(x_i,x_j)\}$ should satisfy the constraints $\int p_{ij}(x_i,x_j)dx_j = p_i(x_i)$ for consistency, messages $\{\mathcal{M}_{i\to j}\}$ satisfy the equations given by

$$\begin{aligned}
\mathcal{M}_{i\to j}(x_j) &= \frac{1}{\tilde{Z}_{ij}}\int_{-\infty}^{\infty} W_{ij}(x_i,x_j)\prod_{k\in\mathcal{N}_i\backslash\{j\}}\mathcal{M}_{k\to i}(x_i)dx_i,\\
\mathcal{M}_{j\to i}(x_i) &= \frac{1}{\tilde{Z}_{ji}}\int_{-\infty}^{\infty} W_{ij}(x_i,x_j)\prod_{k\in\mathcal{N}_j\backslash\{i\}}\mathcal{M}_{k\to j}(x_j)dx_j.
\end{aligned} \tag{3}$$

Note that there are $2|B|$ equations in total so that eqs.(3) are the decision equations for $2|B|$ messages $\{\mathcal{M}_{i\to j}\}$.

2.3 Gaussian Belief Propagation

We assume that the target probability density in eq.(1) is a multi-dimensional Gaussian probability density whose mean vector is $\mathbf{0}$ and $\mathbf{x}\in R^d$:

$$p(\mathbf{x}) = \sqrt{\frac{\det S}{(2\pi)^d}}\exp\Big\{-\frac{1}{2}\mathbf{x}^T S\mathbf{x}\Big\}. \tag{4}$$

Here, S is an inverse covariance matrix and we denote the components of the matrix S by $(S)_{ij} = s_{i,j}$. Then, $W_{ij}(x_i,x_j)$ in eq.(1) can be expressed as

$$W_{ij}(x_i,x_j) = \exp\Big\{-\frac{1}{2}\Big(\frac{s_{i,i}}{|\mathcal{N}_i|}x_i^2 + s_{i,j}x_ix_j + \frac{s_{j,j}}{|\mathcal{N}_j|}x_j^2\Big)\Big\}, \tag{5}$$

where $|\mathcal{N}_i|$ is the number of elements of subset $\mathcal{N}_i$. For simplicity, we put $\frac{s_{i,i}}{|\mathcal{N}_i|} = \tilde{s}_{i,i}$. We set the condition $|\mathcal{N}_i| > 0$ for $^\forall i \in \{1,\cdots,d\}$.

We assume that the probability densities of messages $\{\mathcal{M}_{i\to j}\}$ are Gaussian densities:

$$\mathcal{M}_{i\to j}(x_j) = \sqrt{\frac{\lambda_{i\to j}}{2\pi}}\exp\Big\{-\frac{\lambda_{i\to j}}{2}x_j^2\Big\}. \tag{6}$$

Then, by substituting eqs.(6) into eqs.(3), we reduce eqs.(3) to the decision equations for the parameters of inverse variances $\{\lambda_{i\to j}\}$ as follows.

$$\tilde{\lambda}_{i\to j} = -\frac{s_{i,j}^2}{s_{i,i} + \sum_{k\in\mathcal{N}_i\backslash\{j\}}\tilde{\lambda}_{k\to i}}. \tag{7}$$

Here, $\{\tilde{\lambda}_{i\to j}\}$ are $\tilde{\lambda}_{i\to j} \equiv \lambda_{i\to j} - \tilde{s}_{j,j}$. After obtaining the values of $\{\tilde{\lambda}_{i\to j}\}$ which satisfy eqs.(7), we compose messages $\{\mathcal{M}_{i\to j}\}$ by eqs.(6). Next, we compose marginal densities $\{\tilde{p}_i(x_i)\}$ and $\{\tilde{p}_{ij}(x_i, x_j)\}$ approximately by substituting eqs.(6) into eqs.(2). That is, we obtain marginal densities by substituting $\{\lambda_{i\to j}\}$ which satisfy eqs.(7) into

$$\tilde{p}_i(x_i) \propto \exp\Big\{-\frac{\sum_{k\in\mathcal{N}_i}\lambda_{k\to i}}{2}x_i^2\Big\}, \quad \tilde{p}_{ij}(x_i,x_j) \propto \exp\Big\{-(x_i,x_j)\frac{\tilde{S}_{i,j}}{2}\begin{pmatrix} x_i \\ x_j \end{pmatrix}\Big\},$$

$$\tilde{S}_{i,j} \equiv \begin{pmatrix} \tilde{s}_{i,i} + \sum_{k\in\mathcal{N}_i\setminus\{j\}} \lambda_{k\to i} & s_{i,j} \\ s_{i,j} & \tilde{s}_{j,j} + \sum_{k\in\mathcal{N}_j\setminus\{i\}} \lambda_{k\to j} \end{pmatrix}. \tag{8}$$

Throughout this paper, we put the inverse variances of approximate marginal densities calculated by LBP as $\Lambda_i (\equiv \sum_{k\in\mathcal{N}_i} \lambda_{k\to i})$.

3 Main Results of This Paper

We consider the differences between true and approximate marginal densities when we apply LBP to a multi-dimensional Gaussian density. We consider the differences in each case (**3.1**, **3.2**, **3.3**). In section **3.1**, we give the exact solutions of messages $\{\mathcal{M}_{i\to j}\}$ (equally the inverse variances $\{\lambda_{i\to j}\}$), approximate marginal densities $\tilde{p}_i(x_i)$ and $\tilde{p}_{ij}(x_i, x_j)$ (equally $\{\Lambda_i\}$ and $\{\tilde{S}_{i,i+1}\}$), and the KL distances when inverse covariance matrix S corresponds to the graph with a single loop. In section **3.2**, we develop the results of **3.1** to the graph with a single loop and tree structures. In section **3.3**, we give the expansions of inverse covariances of marginal densities $\{\Lambda_i\}$ when the covariances are very small. All the proofs in section **3** are shown in [9].

3.1 On the Graphs with a Single Loop

We consider the case in which the inverse covariance matrix S can be described as the graph which have only a single loop as illustrated in Fig.1. Then, without loss of generality, we can consider the case that the set B is $B = \{\{12\}, \{23\}, \cdots, \{d-1d\}, \{d1\}\}$. For the graph with a single loop, we obtain the following theorem.

Theorem 1. *When the inverse covariance matrix S corresponds to the graph given by $B = \{\{12\}, \{23\}, \cdots, \{d-1d\}, \{d1\}\}$, the inverse variances of messages $\{\lambda_{i\to j}\}$ are given by*

$$\lambda_{i\to i+1} = \frac{s_{i,i+1}\Delta_{i,i+1} - s_{i+1,i+2}\Delta_{i+1,i+2} \pm \sqrt{\mathcal{D}}}{2\Delta_{i+1,i+1}},$$

$$\lambda_{i\to i-1} = \frac{s_{i,i-1}\Delta_{i,i-1} - s_{i-1,i-2}\Delta_{i-1,i-2} \pm \sqrt{\mathcal{D}}}{2\Delta_{i-1,i-1}},$$

$$\mathcal{D} \equiv (\det S)^2 + (-1)^d 4 s_{1,2}s_{2,3}\cdots s_{d-1,d}s_{d,1}\det S, \tag{9}$$

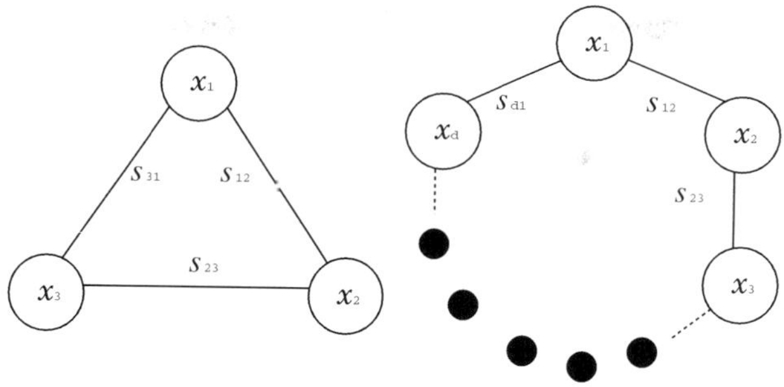

Fig. 1. Examples of a single loop

where $i \in \{1, 2, \cdots, d\}$ *and periodic boundary conditions* $s_{d,d+1} \equiv s_{d,1}$, $s_{1,0} \equiv s_{1,d}$ *hold.* $\Delta_{i,j}$ *are the cofactors of the matrix* S.

Theorem 1 gives immediately the following corollary.

Corollary 1. *When the inverse covariance matrix* S *corresponds to the graph given by* $B = \{\{12\}, \{23\}, \cdots, \{d-1d\}, \{d1\}\}$, *the inverse variances* $\{\Lambda_i\}$ *and the inverse covariance matrices* $\{\tilde{S}_{i,i+1}\}$ *of approximate marginal densities are given by*

$$\Lambda_i = \frac{\det S}{\Delta_{i,i}}\left(1 + \frac{(-1)^d 4 s_{1,2} \cdots s_{d,1}}{\det S}\right)^{\frac{1}{2}}, \quad \tilde{S}_{i,i+1} = \begin{pmatrix} \frac{E_{i,i+1}}{\Delta_{i,i}} & s_{i,i+1} \\ s_{i,i+1} & \frac{E_{i,i+1}}{\Delta_{i+1,i+1}} \end{pmatrix}, \tag{10}$$

where

$$E_{i,i+1} \equiv \frac{\det S + \sqrt{\mathcal{D}}}{2} - s_{i,i+1}\Delta_{i,i+1}. \tag{11}$$

In corollary 1, if $s_{1,2}, s_{2,3}, \cdots, s_{d\,1}$ approach 0, we know that Λ_i and $\tilde{S}_{i,i+1}$ approach

$$\Lambda_i \to \frac{\det S}{\Delta_{i,i}}, \quad \tilde{S}_{i,i+1} \to \begin{pmatrix} \frac{\det S}{\Delta_{i,i}} & 0 \\ 0 & \frac{\det S}{\Delta_{i+1,i+1}} \end{pmatrix} \tag{12}$$

respectively since $E_{i,i+1} \to \det S$. These results are equivalent to the inverse covariance matrices of the true marginal densities. Taking the limit as $s_{1,2}, s_{2,3}, \cdots, s_{d,1} \to 0$ means that the graph with a single loop (e.g., Fig.1) turns to a tree graph and LBP can calculate marginal densities exactly. Hence, corollary 1 agrees with and explains the well-known fact that belief propagation is exactly correct when the graph provided by a target distribution is a tree graph but incorrect when the graph has loops.

When we calculate the KL distances between true and approximate marginal densities in order to know the differences between both densities, we obtain the following theorem. Here, KL and symmetrized KL (SKL) distances are defined as follows.

$$\mathrm{KL}(q(\mathbf{x})||p(\mathbf{x})) \equiv \int q(\mathbf{x}) \log \frac{q(\mathbf{x})}{p(\mathbf{x})} d\mathbf{x}, \tag{13}$$

$$\mathrm{SKL}(q(\mathbf{x}), p(\mathbf{x})) \equiv \mathrm{KL}(q(\mathbf{x})||p(\mathbf{x})) + \mathrm{KL}(p(\mathbf{x})||q(\mathbf{x})). \tag{14}$$

We note that a SKL distance satisfies the definition of distance although a KL distance does not (since $\mathrm{KL}(q||p) \neq \mathrm{KL}(p||q)$).

Theorem 2. *The KL and SKL distances between true marginal densities $p_i(x_i)$ and approximate marginal densities $\tilde{p}_i(x_i)$ calculated by LBP are given by*

$$\mathrm{KL}(p_i||\tilde{p}_i) = -\frac{1}{2} + \frac{1}{2}(1+\epsilon)^{\frac{1}{2}} - \frac{1}{4}\log\Big(1+\epsilon\Big),$$
$$\mathrm{SKL}(p_i, \tilde{p}_i) = -1 + \frac{1}{2}(1+\epsilon)^{-\frac{1}{2}} + \frac{1}{2}(1+\epsilon)^{\frac{1}{2}}. \tag{15}$$

where ϵ is given by

$$\epsilon \equiv \frac{(-1)^d 4 s_{1,2} \cdots s_{d,1}}{\det S}. \tag{16}$$

Similarly, the KL and SKL distances between $p_{ij}(x_i, x_j)$ and $\tilde{p}_{ij}(x_i, x_j)$ are given by

$$\mathrm{KL}(p_{i,i+1}||\tilde{p}_{i,i+1}) = -1 - \frac{1}{2}\log(1 - \bar{\Delta}_{i,i+1}) + \frac{1+\sqrt{1+\epsilon}}{2} + \frac{s_{i,i+1}\Delta_{i+1,i}}{\det S}$$
$$-\frac{1}{2}\log\left\{(\frac{1+\sqrt{1+\epsilon}}{2})^2 - \frac{s_{i,i+1}^2 \Delta_{i,i}\Delta_{i+1,i+1}}{(\det S)^2}\right\},$$

$$\mathrm{SKL}(p_{i,i+1}, \tilde{p}_{i,i+1}) = -2 + \Big[1 + \frac{1}{1-\bar{\Delta}_{i,i+1}}\Big\{(\frac{1+\sqrt{1+\epsilon}}{2} - \frac{s_{i,i+1}\Delta_{i,i+1}}{\det S})^2$$
$$- \frac{s_{i,i+1}^2 \Delta_{i,i}\Delta_{i+1,i+1}}{(\det S)^2}\Big\}^{-1}\Big]\frac{1+\sqrt{1+\epsilon}}{2}, \tag{17}$$

where we put $\bar{\Delta}_{i,i+1} \equiv \frac{\Delta_{i,i+1}^2}{\Delta_{i,i}\Delta_{i+1,i+1}}$.

From the above theorem 2, we can know that both KL and SKL distances go to 0 as ϵ tends to 0 since $\bar{\Delta}_{i,i+1}$ tends to 0. We can say that ϵ is the control parameter which decides the distances between both marginal densities. The parameter ϵ decides the accuracy of LBP in the case that the target distribution is a multi-dimensional Gaussian density whose inverse covariance matrix S forms the graph with a single loop.

3.2 On the Graphs with a Single Loop and Tree Structures

From the results for a single loop in section **3.1**, we can immediately develop the results to the more general case that the graph given by B has a single

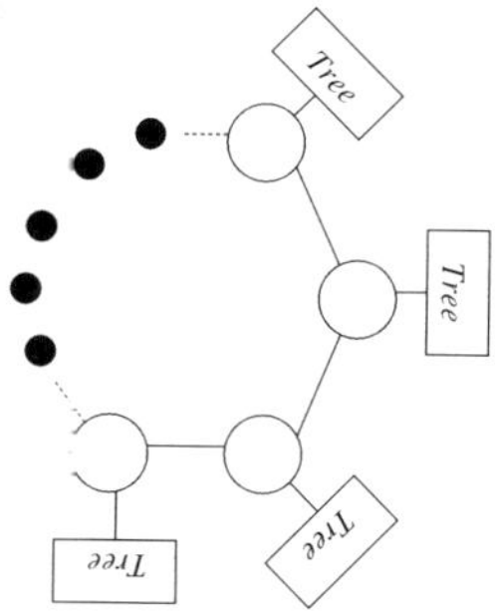

Fig. 2. Graphs with a single loop and tree structures

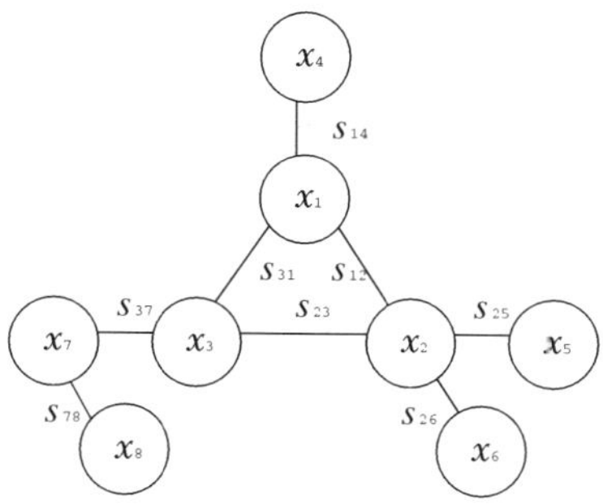

Fig. 3. An example

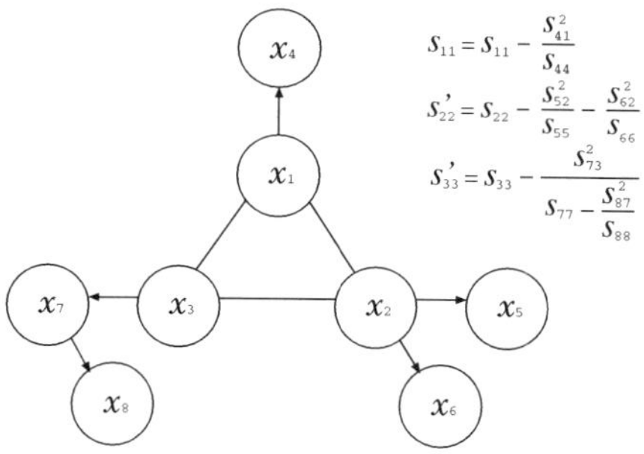

Fig. 4. The same graph as Fig.3

loop and an arbitrary tree structure as illustrated in Fig. 2. For example, we consider the graph shown by Fig.3. Then, in practice, we calculate the inverse variances $\{\Lambda_i\}$. The variances $\{\Delta_i\}$ are obtained by solving eqs.(7) when $d = 8$ and $B = \{\{12\},\{23\},\{31\},\{14\},\{25\},\{26\},\{37\},\{78\}\}$. The graph in Fig.3 can be equivalent to the graph in Fig.4 after replacing inverse variances $\{s_{i,i}\}$ as

$$s_{1,1} \rightarrow s'_{1,1}\left(\equiv s_{1,1} - \frac{s_{4,1}^2}{s_{4,4}}\right), \quad s_{2,2} \rightarrow s'_{2,2}\left(\equiv s_{2,2} - \frac{s_{5,2}^2}{s_{5,5}} - \frac{s_{6,2}^2}{s_{6,6}}\right),$$
$$s_{3,3} \rightarrow s'_{3,3}\left(\equiv s_{3,3} - \frac{s_{7,3}^2}{s_{7,7} - \frac{s_{8,7}^2}{s_{8,8}}}\right) \tag{18}$$

(These replacements are understandable by solving eqs.(7) directly). In Fig.4, the directed edges mean that there exist messages in those directions but not in reverse directions (e.g., message $\mathcal{M}_{1\to 4}(x_4)$ exists but message $\mathcal{M}_{4\to 1}(x_1)$ does not exist between x_1 and x_4). If we focus on variable nodes x_1, x_2, and x_3, these variable nodes form a closed single loop. Then, referring to the result of corollary 1, we obtain

$$\Lambda'_i = \frac{\det S'}{\Delta'_{i,i}}\left(1 - \frac{4s_{1,2}s_{2,3}s_{3,1}}{\det S'}\right)^{\frac{1}{2}} \tag{19}$$

for $i \in \{1, 2, 3\}$. Here, S' and $\Delta'_{i,i}$ are the inverse covariance matrix and the cofactor respectively composed by $\{s'_{i,i}\}$ instead of $\{s_{i,i}\}$. The others $\{\Lambda_i\}$, $i \in \{4, \cdots, 8\}$ are calculated as follows by using the notation of continued fractions.

$$\Lambda'_4 = s_{4,4} - \frac{s_{1,4}^2\, s_{4,1}^2}{\Lambda'_1 + s_{4,4}}, \quad \Lambda'_5 = s_{5,5} - \frac{s_{2,5}^2\, s_{5,2}^2}{\Lambda'_2 + s_{5,5}}, \quad \Lambda'_6 = s_{6,6} - \frac{s_{2,6}^2\, s_{6,2}^2}{\Lambda'_2 + s_{6,6}},$$
$$\Lambda'_8 = s_{8,8} - \frac{s_{7,8}^2\, s_{8,7}^2}{\Lambda'_7 + s_{8,8}}, \quad \Lambda'_7 = s_{7,7} - \frac{s_{8,7}^2}{s_{8,8}} - \frac{s_{3,7}^2\, s_{7,3}^2\, s_{8,7}^2}{\Lambda'_3 + s_{7,7} - s_{8,8}} \tag{20}$$

(These equations are also understandable by solving eqs.(7)).

In a similar way, we obtain $\{\lambda_{ij}\}$, $\{\Lambda_i\}$, $\{\tilde{S}_{i,j}\}$, KL, and SKL distances in various connected graphs which have at most a single loops.

3.3 Expansions of $\{\Lambda_i\}$ at Small Covariances

In sections **3.1** and **3.2**, we address the inverse covariance matrix S whose graph structure is restricted to the graph with a single loop. In this section, we address the inverse covariance matrix S whose graph is allowed to have an arbitrary structure (i.e. the graph has multiloops) but the covariance is very small. To achieve that, we introduce a new parameter s and change the off-diagonal components of the matrix S to $(S)_{ij} = s s_{i,j}$, where s satisfy $0 \leq s \leq 1$. If $s = 0$, the matrix S turns to a diagonal matrix and if $s = 1$, the matrix S turns to the original matrix S in eq.(4). Then, our aim is to obtain the expansions of the inverse variances $\{\Lambda_i\}$ with respect to s:

$$\Lambda_i(s) = a_{i0} - a_{i1}s - a_{i2}s^2 - a_{i3}s^3 + \cdots \tag{21}$$

when s satisfies $s \ll 1$.

In this paper, for simplicity, we derive the terms up to third-order. Before that, we prepare the following theorem for simultaneous equations which $\{\Lambda_i\}$ satisfy.

Theorem 3. *The approximate inverse variances $\{\Lambda_i\}$ calculated by LBP satisfy either of $\sum_{i=1}^{d} 2^{|\mathcal{N}_i|}$ simultaneous equations*

$$|\mathcal{N}_i| - 2 + \frac{2s_{i,i}}{\Lambda_i} = \sum_{j \in \mathcal{N}_i} \pm\sqrt{1 + \frac{4s^2 s_{j,i}^2}{\Lambda_j \Lambda_i}}, \quad i \in \{1, \cdots, d\}, \tag{22}$$

where $\pm$ takes an arbitrary sign for each term in the summation.

There exist $\sum_{i=1}^{d} 2^{|\mathcal{N}_i|}$ simultaneous equations since the sign $\pm$ assigns either of positive or negative for each term. In addition, we note that there exist extra simultaneous equations which have no solution. For example, there is no solution when $|\mathcal{N}_i| \geq 2$ for $^{\exists}i$ and all the signs in the summation are negative.

When $s = 0$ in the matrix S, approximate inverse variance Λ_i should be equal to the true inverse variance $s_{i,i}$ (i.e., $\Lambda_i(0) = s_{i,i}$ for $i \in \{1, \cdots, d\}$) since the matrix S becomes a diagonal matrix and LBP algorithm turns to BP algorithm for tree structures. By imposing these conditions, we obtain the following theorem.

Theorem 4. *For $\sum_{i=1}^{d} 2^{|\mathcal{N}_i|}$ simultaneous equations in Theorem 3, conditions $\Lambda_i(0) = s_{i,i}$, $i \in \{1, \cdots, d\}$ holds if and only if all the signs in the summation are positive. That is, the simultaneous equations are given by*

$$|\mathcal{N}_i| - 2 + \frac{2s_{i,i}}{\Lambda_i} = \sum_{j \in \mathcal{N}_i} \sqrt{1 + \frac{4s^2 s_{j,i}^2}{\Lambda_j \Lambda_i}}, \quad i \in \{1, \cdots, d\}. \tag{23}$$

Then, the solutions of the inverse variances $\{\Lambda_i\}$ which satisfy eqs.(23) are expanded with respect to s as follows.

$$\Lambda_i(s) = s_{i,i} - \sum_{j=1(\neq i)}^{d} \frac{s_{j,i}^2}{s_{j,j}} s^2 + O(s^4). \tag{24}$$

Compared with the result of theorem 4, the inverse variances of true marginal densities are expanded as follows.

$$\frac{\det S}{\Delta_{i,i}} = s_{i,i} - \sum_{j=1(\neq i)}^{d} \frac{s_{j,i}^2}{s_{j,j}} s^2 + \frac{s_{i,i}}{3}[\mathrm{tr}(S_d^{-1} S_o)^3 - \mathrm{tr}\{(S_d)_{i,i}^{-1}(S_o)_{i,i}\}^3] s^3 + O(s^4). \tag{25}$$

Here, matrices S_d and S_o are the diagonal and off-diagonal matrices which satisfy $S = S_d + sS_o$ respectively. Matrices $(S_d)_{i,i}$ and $(S_o)_{i,i}$ are the minor matrices of S_d and S_o with the i-th row and column omitted. From the eqs.(24)(25), we can know that the approximate inverse variances $\{\Lambda_i\}$ calculated by LBP give exact coefficients up to second-order term but yield the differences from third-order term. When we calculate the KL distances, we obtain the following theorem.

Theorem 5. *KL distances from true marginal densities $p_i(x_i)$ to approximate marginal densities $\tilde{p}_i(x_i)$ calculated by LBP are expanded as follows with respect to s for $i \in \{1, 2, \cdots, d\}$.*

$$KL(p_i||\tilde{p}_i) = \left(tr \frac{(S_d^{-1} S_o)^3 - \{(S_d)_{i,i}^{-1}(S_o)_{i,i}\}^3}{6} \right)^2 s^6 + O(s^7). \tag{26}$$

Theorem 5 tells us that approximate marginal densities calculated by LBP are how close to the true marginal densities when covariances in S are very small.

4 Conclusion and Future Works

In this paper, we analytically show how accurate LBP is when the target distribution is a multi-dimensional Gaussian distribution. In particular, we calculate the fixed point, approximate marginal densities, and KL distances exactly when matrix S forms the graph with a single loop. Then, we know that a control parameter ϵ decides the accuracy. Next, we develop the result to the more general case. Furthermore, we show the series expansions of inverse variances $\{\Lambda_i(s)\}$

and KL distances up to third-order with respect to s when the matrix S forms an arbitrary graph. These results contribute to the foundation for understanding theoretical properties underlying more complex LBP algorithms and designing LBP algorithms efficiently. We have several future works, some of which are to derive the convergence rate to a fixed point, to obtain the higher order terms in eqs.(24), and to compare the theoretical results with numerical experiments.

Acknowledgement

This research was partially supported by the Ministry of Education, Science, Sports and Culture, Grant-in-Aid for JSPS Fellows and for Scientific Research 18079007, 2007.

References

1. Weiss, Y.: Belief propagation and revision in networks with loops, Technical Report 1616, MIT AI lab (1997)
2. Aji, S.M., Horn, G.B., McEliece, R.J.: On the convergence of iterative decoding on graphs with a single cycle. In: proc. 1998 ISIT (1998)
3. Weiss, Y.: Correctness of local probability propagation in graphical models with loops. Neural Computation 12(1), 1–41 (2000)
4. Weiss, Y., Freeman, W.: Correctness of belief propagation in graphical models with arbitrary topology. Neural Computation 13(10), 2173–2200 (2001)
5. Ikeda, S., Tanaka, T., Amari, S.: information geometry of turbo and low-density parity-check codes. IEEE Trans. Inf. Theory 50(6), 1097–1114
6. Ikeda, S., Tanaka, T., Amari, S.: Stochastic reasoning, free energy, and information geometry. Neural Computation 16(9), 1779–1810 (2004)
7. Tanaka, K., Shouno, H., Okada, M.: Accuracy of the Bethe approximation for hyperparameter estimation in probabilistic image processing. J. Phys. A, Math. Gen. 37(36), 8675–8696 (2004)
8. Tanaka, K.: Generalized Belief Propagation Formula in Probabilistic Information Processing Based on Gaussian Graphical Model. IEICE D-II J88-D-II(12), 2368–2379 (2005)
9. Nishiyama, Y., Watanabe, S.: Theoretical Analysis of Accuracy of Belief Propagation in Gaussian Models. IEICE Technical Report 107(50), 23–28 (2007)

Relevance Metrics to Reduce Input Dimensions in Artificial Neural Networks

Héctor F. Satizábal M.[1,2,3] and Andres Pérez-Uribe[2]

[1] Université de Lausanne, Hautes Etudes Commerciales (HEC), Institut des Systèmes d'Information (ISI)
hector-fabio.satizabal-mejia@heig-vd.ch
[2] University of Applied Sciences of Western Switzerland (HEIG-VD)(REDS)
andres.perez-uribe@heig-vd.ch
[3] Corporación BIOTEC

Abstract. The reduction of input dimensionality is an important subject in modelling, knowledge discovery and data mining. Indeed, an appropriate combination of inputs is desirable in order to obtain better generalisation capabilities with the models. There are several approaches to perform input selection. In this work we will deal with techniques guided by measures of input relevance or input sensitivity. Six strategies to assess input relevance were tested over four benchmark datasets using a backward selection wrapper. The results show that a group of techniques produces input combinations with better generalisation capabilities even if the implemented wrapper does not compute any measure of generalisation performance.

Keywords: Input Selection, Input Relevance Measures.

1 Introduction

Artificial neural networks make a non linear mapping between an output and an input dataset. These inputs could be high-dimensional depending on the modelling problem. Moreover, it is well known that for a fixed amount of data, the number of input variables cannot be increased indefinitely without a reduction in performance[1] [1]. Thus, the set of variables involved in the modelling process must be reduced by variable selection or preprocessing.

There are several approaches to reduce the input dimensionality [11]. The *filter* approach uses statistics to measure the relevance of each variable [23]. These calculations are independent of the learning machine and are computed as a preprocessing step. The *wrapper* approach [15,19,24] considers the specific learning algorithm as a black box, and some measures of performance are used to evaluate the different combinations of inputs. There are also *embedded* techniques that modify the learning algorithm by the addition of some penalty term to the error function [5,6,20]. These penalty terms are intended to reduce the numeric

[1] This phenomenon has been termed the curse of dimensionality.

J. Marques de Sá et al. (Eds.): ICANN 2007, Part I, LNCS 4668, pp. 39–48, 2007.

value of the weight connections of the network during the training process. As a result, the influence of irrelevant inputs is minimized. In every case, the main idea is to discard irrelevant or noisy information, increasing the generalisation properties of the model.

The principal disadvantage of *embedded* methods is their dependency on the learning algorithm. Instead, *filters* and *wrappers* could be considered as universal feature selectors since they are not based on a specific learning method [11]. Moreover, *wrappers* have more computational cost than *filters*, but they also give better results because of the inclusion of the learning algorithm in the selection process. This high computational cost is due to the performance measure computed at each input evaluation step (cross-validation).

In the *wrapper* approach, testing the whole set of input combinations could be a computationally infeasible approach depending on the quantity of inputs. Instead, a search strategy like *backward* or *forward* selection should guide the searching process. *Backward* selection has been commonly used by several researchers in neural network feature selection [19,21,24]. This selection strategy begins with the complete set of features and ranks each one of the input variables according to some measure of relevance. Then, the least relevant features are pruned one by one [19] in a sequential manner, or could also be pruned in group by the detection of a sufficient large gap in the mentioned relevance ranking [24].

Concerning the metrics of relevance issue, it is well known that neural networks, although are good modelling tools to perform prediction or classification tasks, they do not give any information about the nature of the problem, or the relations between variables and their relevance for the whole process. The resulting model is not a mechanistic representation of the phenomenon to be modelled. Instead, the neural network has a set of parameters that do not directly represent any characteristic of the process generating the data. Many researchers have proposed distinct approaches trying to extract the knowledge that lies behind the parameters of the network [8,17,19,24].

Relevance metrics could be categorized into two broad groups. Metrics based only on calculations over the network parameters [8,17,19], and metrics where the input dataset is fed into the model in order to get a numerical or analytical measure of sensitivity of each input feature [24]. Actually, there are a large number of strategies to measure input relevance in neural networks. Some of these techniques are more effective than others to perform input selection.

In this work, we use a backward elimination approach to test some of the existing methodologies that perform input ranking. Therefore, the goal is to identify the relevance metrics performing the best the task of detecting irrelevant or redundant input features. The backward selection methodology is evaluated over four benchmark datasets from the UCI Machine-Learning Repository [16]. The organization of the paper is as follows: Section 2 describes the relevance metrics under test. Section 3 shows the details of the backward elimination approach implemented. Section 4 presents the results of the set of experiments realized. Finally, in section 5 we give some conclusions of the tests.

2 Relevance Metrics

2.1 Using Network Parameters Only

These methods are based on the idea that the input weights of an artificial neuron represent input gain. The learning algorithm (back-propagation) minimizes the error signal by adjusting the weight connections between neurons such that the output of the net tracks the target signal [12]. Irrelevant or noisy inputs have to be attenuated by small weights in order to minimize the general error signal. Conversely, the weights of relevant inputs must be increased. The most widely used metrics are the *Garson's Algorithm*, the *Overall Connection Weights* metric and the *ANNIGMA* input gain approximation. The equation to calculate the relevance in a single hidden layer and single output neural network is given in each case. Expansion to multiple hidden layers networks should be straightforward.

Garson's Algorithm. This weight based method is derived from the one first proposed by Garson [8]. Some researchers have used this approach to asses input relevance [9,14,17,18]. Input relevance is calculated as shown in equation 1.

$$R_{ik} = \sum_{j=1}^{nh} \left(\frac{|\omega_{ij}|\,|\omega_{jk}|}{\sum_{i'=1}^{ni} \left(|\omega_{i'j}|\,|\omega_{jk}|\right)} \right) = \sum_{j=1}^{nh} \left(\frac{|\omega_{ij}|}{\sum_{i'=1}^{ni} |\omega_{i'j}|} \right) \quad (1)$$

Where ni and nh are the number of inputs and hidden units respectively, ω_{ij} is the weight between input i and hidden unit j, and ω_{jk} is the weight between hidden unit j and the output k. This method uses input-hidden connections.

Overall Connections. This strategy was proposed by Olden and Jackson [17], and contrary to Garson's algorithm, it does not use the absolute value of the connection weights. Instead, the raw input-hidden-output weight product is obtained (see equation 2). The use of the weight sign is founded on the idea that the effect of an input could be cancelled in posterior layer calculations. Moreover, the sign indicates the direction of the contribution of each input. So far, many researchers have used this technique in order to rank inputs according to their relevance [14,18].

$$R_{ik} = \sum_{j=1}^{nh} (\omega_{ij}\omega_{jk}) \quad (2)$$

Where nh is the number of hidden units, ω_{ij} is the weight between input i and hidden unit j, and ω_{jk} is the weight between hidden unit j and the output k.

ANNIGMA. ANNIGMA is an acronym of *Artificial Neural Net Input Gain Measurement Approximation* [19]. This method approximates the activation functions of the artificial neurons by a linear factor. Thus, the analytical output-input ratio is calculated (see equation 3). This ratio (called *Local Gain*) is normalized, and used as a measure of input relevance.

$$LG_{ik} = \sum_{j=1}^{nh} |\omega_{ij}\omega_{jk}| \tag{3}$$

Where nh is the number of hidden units, ω_{ij} is the weight between input i and hidden unit j, and ω_{jk} is the weight between hidden unit j and the output k.

2.2 Using Network Parameters and Input Patterns

These methodologies measure the neural network input sensitivity in a direct or indirect way by the stimulation of the model with a set of input patterns. This approach is the one used by the following three relevance metrics: *Sensitivity Matrix*, *Input Perturbation* and *Input Cancellation*. Equations to calculate the relevance in a single hidden layer and single output neural network are given in each case. Expansion to multiple hidden layers should be straightforward.

Sensitivity Matrix. The neural network sensitivity expresses the amount of change of an output with the variations of a specific input. In other words, the sensitivity is the partial derivative of an output node with respect to an input node. This is also known as the Jacobian matrix [2]. Since the Jacobian matrix is a function of the network inputs, it is possible to find a different value of sensitivity for each input pattern. There are different approaches to integrate these values in a singular value of sensitivity per input [24]. In our case, we use the accumulated absolute value (see equation 4). This direct calculation of input sensitivity has been used in related works to asses input relevance [24], some researchers use the name *Partial Derivatives* for this method [9,14,18].

$$R_{ik} = \sum_{p \in patterns} \left| \frac{\partial O_k^p}{\partial I_i} \right| \quad \text{and} \quad \frac{\partial O_k^p}{\partial I_i} = \frac{\partial f_k}{\partial a_k^p} \sum_{j=0}^{nh} \left(\omega_{jk} \frac{\partial f_j}{\partial a_j^p} \omega_{ij} \right) \tag{4}$$

Where nh is the number of hidden units, ω_{ij} is the weight between input i and hidden unit j, ω_{jk} is the weight between hidden unit j and the output k, $\frac{\partial f_j}{\partial a_j^p}$ is the first derivative of the activation function of the hidden unit j and $\frac{\partial f_k}{\partial a_k^p}$ is the first derivative of the activation function of the output unit k.

Input Perturbation. This strategy is a numerical method to obtain the input sensitivity of the neural network. The output error is calculated after the addition of a small perturbation to one of the inputs of the network. Error variations are accumulated over the whole set of training patterns (see equation 5). In this work, we test two different variations of the technique. The former adds several input perturbations within a range[2] and the input sensitivity is accumulated after each input perturbation. This approach is refered as *noise perturbation metric* in the experiments. The latter adds only two perturbations to the input. These values correspond to the extreme values of the distribution used in the former method. This approach is refered as *fix input perturbation*. The mentioned processes are

[2] Uniformly distributed, 5% of the input range, 100 times.

executed for all the inputs individually to obtain a measure of input relevance. This metric has been used by some researchers in different works [9,14,18].

$$R_{ik} = \sum_{p \in patterns} (R_{ik}^p) \quad \text{and} \quad R_{ik}^p = \left[SE\left(O_k(I_i^p)\right) - SE\left(O_k(I_i^p + \epsilon)\right)\right]^2 \qquad (5)$$

Where $SE()$ is the squared error function, and ϵ is a small perturbation added to the input signal.

Input Cancellation. This method accumulates the change in the output error over the whole set of training patterns after replacing one of the inputs by its average (see equation 6). The relevance of each input is assessed by computing the output error increments.

$$R_{ik} = \sum_{p \in patterns} (R_{ik}^p) \quad \text{and} \quad R_{ik}^p = \left[SE\left(O_k(I_i^p)\right) - SE\left(O_k(\bar{I}_i)\right)\right]^2 \qquad (6)$$

Where $SE()$ is the squared error function, and $\bar{I}_i$ is the average of input i.

3 Description of the Backward Selection Algorithm

The backward elimination process is guided by the aforementioned metrics of input relevance. The back-propagation with momentum algorithm [2] was used to train the networks. Each training phase stops if a fixed number of epochs is reached or, if all the training observations have an output error that is smaller than a fixed threshold (2%-4% of output range). The backward selection algorithm goes as follows:

1. Train a randomly initialized artificial neural network using the complete set of inputs.
2. Compute input relevance for every input.
3. Prune the least relevant input.
4. Continue training for a fixed number of iterations.
5. If all the training observations are correctly mapped:
   - Then, go to step 2.
   - Else, restore last pruned input.
6. Compute input relevance for every input and exit.

Notice that the aforementioned backward selection strategy does not include the calculation of any generalisation performance measure. The process is exclusively guided by the relevance of each variable according to each technique.

4 Experiments

The backward selection strategy described so far was tested on four benchmark datasets. In each case, the performance of the pruned model (the resulting set of inputs after applying backward selection) was obtained using cross-validation. The selection algorithm was executed 50 times with each metric in order to obtain an error distribution. Results are shown using box-plots in order to compare the performance of the different neural networks [4].

4.1 Non Linear Static Function

This dataset was used by Sugeno and Yasukawa [22] to perform input selection in a fuzzy system. The single output y belongs to a static system with two inputs x_1 and x_2:

$$y = \left(1 + x_1^{-2} + x_2^{-1.5}\right)^2 \quad \text{and} \quad 1 \leq x_1, x_2 \leq 5 \tag{7}$$

Sugeno and Yasukawa also added two dummy inputs x_3 and x_4 to check the appropriateness of their identification method. The dataset contains 50 rows and the number of K-folds for the cross validation performance measure was 25.

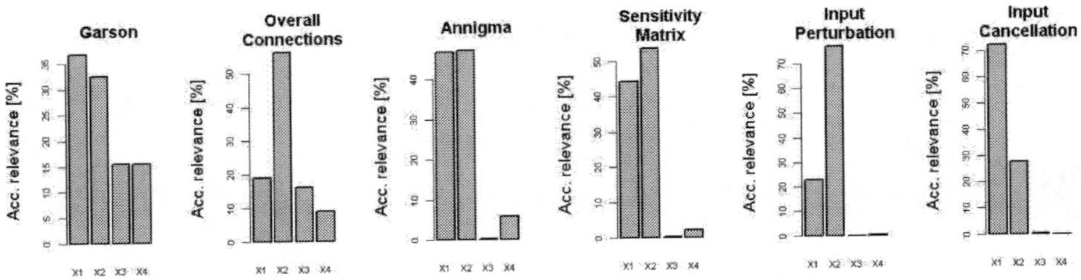

Fig. 1. Accumulated input relevance after 50 iterations of different techniques over the non linear function benchmark. Each bar represents one of the input variables x_1, x_2, x_3 and x_4 (form left to right).

Figure 1 shows the accumulated relevance over 50 iterations of the metrics under study. One can see that the four last metrics (*ANNIGMA*, *Sensitivity Matrix*, *Input Perturbation* and *Input Cancellation*) perform better the identification of irrelevant inputs. Moreover, *Input Perturbation* and *Input Cancellation* reject consistently the two dummy variables.

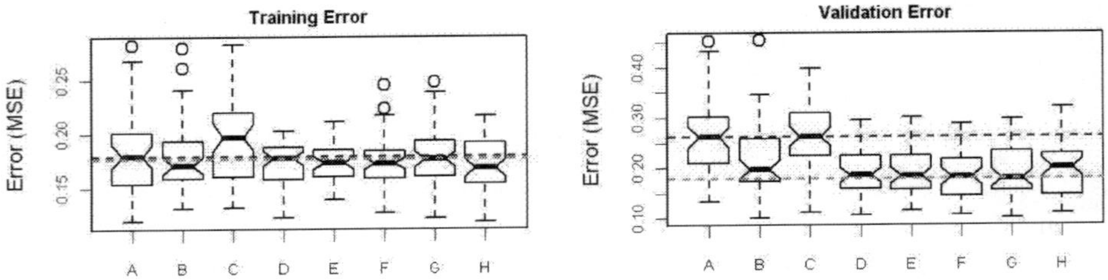

Fig. 2. Different input selection techniques applied to the non linear function dataset. A. All inputs. B. Garson. C. Overall connection weights. D. ANNIGMA. E. Sensitivity matrix. F. Input cancellation. G. Noise perturbation. H. Fix perturbation.

Figure 2 shows that the pruned networks obtained using *ANNIGMA*, *Sensitivity Matrix*, *Input Perturbation* and *Input Cancellation* present a higher generalisation performance. This result validates the higher capability of these methods for identifying irrelevant inputs, as shown in figure 1.

4.2 Wine Recognition Benchmark

These data are the results of a chemical analysis of wines grown in the same region in Italy but derived from three different cultivars [7]. The analysis determined the quantities of 13 constituents found in each of the three types of wines. The dataset contains 178 rows and the number of K-folds for the cross validation performance measure was 16.

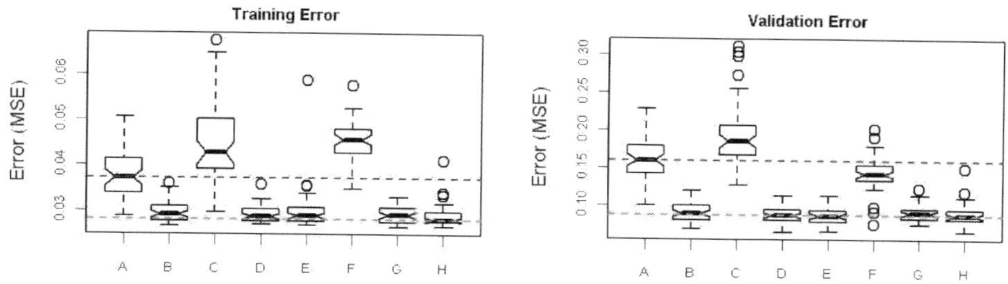

Fig. 3. Different input selection techniques applied to the wine recognition benchmark. A. All inputs. B. Garson. C. Overall connection weights. D. ANNIGMA. E. Sensitivity matrix. F. Input cancellation. G. Noise perturbation. H. Fix perturbation.

Different networks were trained and pruned using backward elimination. Figure 3 shows the results of the cross validation tests made over the resulting pruned networks. In this case, the networks where the inputs were selected using *Garson*, *ANNIGMA*, *Sensitivity Matrix* and *Input Perturbation* present the lowest validation errors. Notice that the *Overall Connections* metric yields networks with a validation error median that is higher than the one obtained with the complete set of inputs. *Overall Connections* and *Input Cancellation* produce networks with high training errors.

4.3 Box-Jenkins Furnace Benchmark

This dataset was presented by Box and Jenkins [3], and it has been used by several researchers in input identification problems [13,22]. The process is a gas furnace with single input $u(t)$ and single output $y(t)$: gas flow rate and CO_2 concentration, respectively. In this dynamic process, the output y_t is expressed as a function of the signals $y_{(t-1)}$, $y_{(t-2)}$, $y_{(t-3)}$, $y_{(t-4)}$, $u_{(t-1)}$, $u_{(t-2)}$, $u_{(t-3)}$, $u_{(t-4)}$, $u_{(t-5)}$ and $u_{(t-6)}$. The dataset has 296 data points and 19 K-folds were used for the cross-validation evaluation.

Figure 4 shows the performance distribution after cross-validation of the final pruned networks using backward elimination. *Sensitivity Matrix*, *Input Perturbation* and *Input Cancellation* present the best validation performance for this specific dataset. Concerning the training performance issue, *Sensitivity matrix* and *Input Perturbation* produce pruned networks with best characteristics than the other metrics.

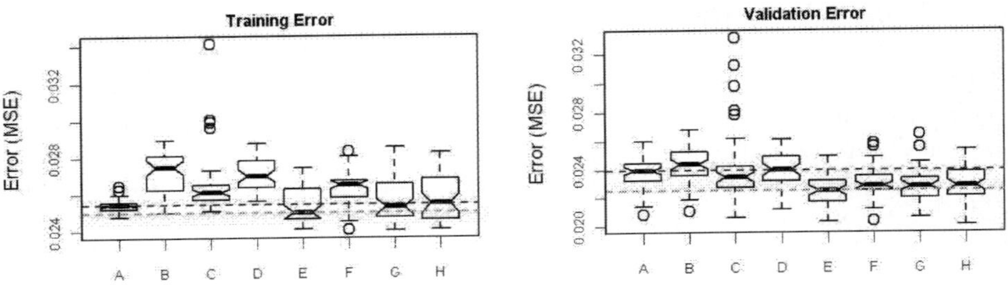

Fig. 4. Different input selection techniques applied to the Box-Jenkins benchmark. A. All inputs. B. Garson. C. Overall connection weights. D. ANNIGMA. E. Sensitivity matrix. F. Input cancellation. G. Noise perturbation. H. Fix perturbation.

4.4 Sonar Data

This is the dataset used by Gorman and Sejnowski in their study on the classification of sonar signals using a neural network [10]. The task was to discriminate between sonar signals bounced off a metal cylinder and those bounced off a roughly cylindrical rock. There are 60 inputs corresponding to the amplitude of the different bounding frequencies, and the output is the object class (metal cylinder or cylindrical rock). This 208-row dataset has been used for several researchers in the feature selection task [19,20]. In our case, we performed a 13 K-folds cross-validation evaluation.

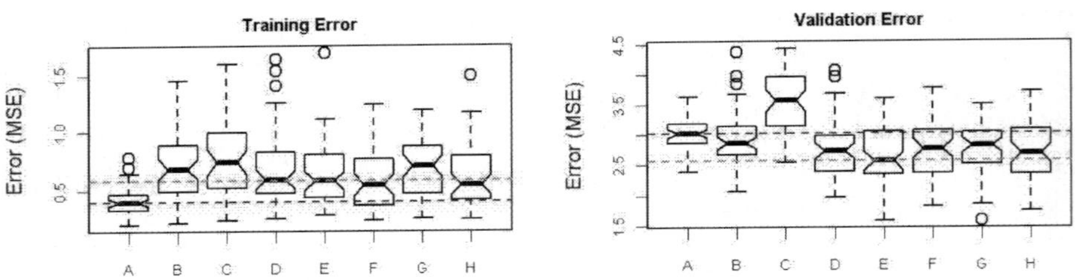

Fig. 5. Different input selection techniques applied to the sonar benchmark. A. All inputs. B. Garson. C. Overall connection weights. D. ANNIGMA. E. Sensitivity matrix. F. Input cancellation. G. Noise perturbation. H. Fix perturbation.

Figure 5 shows that the *Sensitivity Matrix* metric produces pruned networks with the lowest validation error for this dataset. The only metric having a validation error median that is higher than the one obtained using the complete set of inputs is *Overall Connections.* All the set of metrics yields pruned networks having a median training error that is higher than the one that is obtained using the complete set of inputs. The lower training errors were found with the *ANNIGMA*, *Sensitivity Matrix*, *Input Cancellation* and *Fix Input Perturbation* metrics.

5 Conclusions

Wrapper techniques although having a high computational cost are good universal input selectors. Some metrics of input relevance have been developed in order to optimize the searching process of the adequate set of inputs. In this article, six different methodologies for the assessment of input relevance have been tested over four benchmark datasets. Metrics were divided into two groups depending on the use of input information to determine input relevance:

G1. *Garson*, *Overall Connections* and *ANNIGMA* only use network parameters (weight connections) to estimate input relevance.
G2. *Sensitivity Matrix*, *Input Perturbation* and *Input Cancellation* use a set of patterns to stimulate the neuronal model and get the relevance information.

Results converge to show best performances when input relevance is assessed using methodologies belonging to group G2. We could say that within this group the *Sensitivity Matrix* technique presents the best behaviour in most of the cases.

Techniques belonging to group G1 should be used if there are time execution constraints since they have less computational cost. In this case, a set of tests with the specific dataset must be carried out in order to find the most suitable relevance metric. For instance, *Garson* technique behaves well in the wine recognition benchmark (Section 4.2) even if only the input-hidden connections are used to compute relevance. Within the group G1, we found that the *ANNIGMA* metric behaves better than the others in most of the cases (all tests except the Box-Jenkins benchmark. Section 4.3). Notice that *ANNIGMA* was created as a simplification of *Sensitivity Matrix* by replacing the first derivative of the activation functions by a constant term.

An important remark should be made. The relevance metrics were tested in a backward elimination framework Thus, the techniques were evaluated on their ability to detect irrelevant or redundant variables. The ability to rank a set of inputs according to their relevance was not tested.

Acknowledgements

This work is part of a cooperation project between BIOTEC, CIAT, CENICAÑA (Colombia) and HEIG-VD (Switzerland) named *"Precision agriculture and the construction of field-crop models for tropical fruits"*. The economical support is given by several institutions in Colombia (MADR, COLCIENCIAS, ACCI) and the State Secretariat for Education and Research (SER) in Switzerland.

References

1. Bellman, R.: Adaptive control processes. A guided tour, Princeton University Press, New Jersey (1961)
2. Bishop, C.: Neural networks for pattern recognition. Oxford University Press, New York (1995)
3. Box, G., Jenkins, G.: Time Series Analysis, Forecasting and Control. Holden Day, San Francisco (1970)

4. Chambers, J.M., Cleveland, W.S., Kleiner, B., Tukey, P.A.: Graphical Methods for Data Analysis. Wadsworth & Brooks/Cole, 62 (1983)
5. Chapados, N., Bengio, Y.: Input decay: simple and effective soft variable selection. In: Neural Networks. Proc. IJCNN'01. Int. Joint Conference, vol. 2, pp. 1233–1237 (2001)
6. Cibas, T., Fogelman, F., Gallinari, P., Raudys, S.: Variable selection with optimal cell damage. In: ICANN'94. Int. Conf. on Artificial Neural Networks, Amsterdam (1994)
7. Forina, M.: PARVUS - An Extendible Package for Data Exploration, Classification and Correlation. Institute of Pharmaceutical and Food Analysis and Technologies, Via Brigata Salerno, 16147 Genoa, Italy (1991)
8. Garson, G.D.: Interpreting neural network connection weights. Artificial Intelligence Expert 6, 47–51 (1991)
9. Gevrey, M., Dimopoulos, I., Lek, S.: Review and comparison of methods to study the contribution of variables in artificial neural network models. Ecological Modelling 160, 249–264 (2003)
10. Gorman, R., Sejnowski, T.: Analysis of Hidden Units in a Layered Network Trained to Classify Sonar Targets. Neural Networks 1, 75–89 (1988)
11. Guyon, I., Elisseeff, A.: An introduction to variable and feature selection. The Journal of Machine Learning Research 3, 1157–1182 (2003)
12. Haykin, S., Neural Networks, A.: Comprehensive Foundation. Macmillan College Publishing, New York, NY (1994)
13. Jang, J.-S.R.: Input selection for ANFIS learning. In: Proceedings of the Fifth IEEE International Conference on Fuzzy Systems, vol. 2, pp. 1493–1499. IEEE Computer Society Press, Los Alamitos (1996)
14. Kaur, K., Kaur, A., Malhotra, R.: Alternative Methods to Rank the Impact of Object Oriented Metrics in Fault Prediction Modeling using Neural Networks. Transactions on Engineering, Computing and Technology 13, 207–212 (2006)
15. Kohavi, R., John, G.: Wrappers for feature selection. Artificial Intelligence, 232–273 (1997)
16. Merz, C., Murphy, P.: UCI Repository of machine learning databases. Irvine, CA: University of California, Department of Information and Computer Science (1998), `http://www.ics.uci.edu/~mlearn/MLRepository.html`
17. Olden, J., Jackson, D.: Illuminating the "black box": a randomization approach for understanding variable contributions in artificial neural networks. Ecological Modelling 154, 135–150 (2002)
18. Olden, J., Joy, M., Death, R.: An accurate comparison of methods for quantifying variable importance in artificial neural networks using simulated data. Ecological Modelling 178, 389–397 (2004)
19. Schuschel, D., Hsu, C.: A Weight Analysis Based Wrapper Approach to Neural Nets Feature Subset Selection. IEEE Transactions on Systems, Man and Cybernetics 32, 207–221 (2002)
20. Setiono, R., Liu, H.: Neural-network feature selector. IEEE transactions on neural networks 8, 654–662 (1997)
21. Stracuzzi, D., Utgoff, P.: Randomized Variable Elimination. The Journal of Machine Learning Research 5, 1331–1362 (2004)
22. Sugeno, M., Yasukawa, T.: A Fuzzy-Logic-Based Approach to Qualitative Modelling. IEEE Transactions on Fuzzy Systems 1, 7–31 (1993)
23. Witten, I.H., Frank, E.: Data Mining. Practical Machine Learning Tools and Techniques, 2nd edn. Elsevier, Morgan Kaufmann publishers (2005)
24. Zurada, J., Malinowski, A., Usui, S.: Perturbation method for deleting redundant inputs of perceptron networks. Neurocomputing 14, 177–193 (1997)

An Improved Greedy Bayesian Network Learning Algorithm on Limited Data

Feng Liu[1], Fengzhan Tian[2], and Qiliang Zhu[1]

[1] Department of Computer Science, Beijing University of Posts and Telecommunications,
Xitu Cheng Lu 10, 100876 Beijing, China
lliufeng@hotmail.com

[2] Department of Computer Science, Beijing Jiaotong University,
Shangyuan Cun 3, 100044 Beijing, China

Abstract. Although encouraging results have been reported, existing Bayesian network (BN) learning algorithms have some troubles on limited data. A statistical or information theoretical measure or a score function may be unreliable on limited datasets, which affects learning accuracy. To alleviate the above problem, we propose a novel BN learning algorithm MRMRG, Max Relevance and Min Redundancy Greedy algorithm. MRMRG algorithm applies Max Relevance and Min Redundancy feature selection technique and proposes Local Bayesian Increment (LBI) function according to the Bayesian Information Criterion (BIC) formula and the likelihood property of overfitting. Experimental results show that MRMRG algorithm has much better accuracy than most of existing BN learning algorithms when learning BNs from limited datasets.

1 Introduction

During the last two decades, many BN learning algorithms have been proposed. In general, BN learning algorithms take one of the two approaches: the constraint-based method and the search & score method. The constraint-based approach [1],[2] estimates from the data whether certain condition independences hold between variables. Typically, this estimation is performed using statistical or information theoretical measure. The search & score approach [3],[4],[7] attempts to find a graph that maximizes the selected score. Score function is usually defined as a measure of fitness between the graph and the data.

Although encouraging results have been reported, the recent explosion of high dimensional and limited datasets in the biomedical realm and other domains has induced a serious challenge to these BN learning algorithms. The existing algorithms must face higher dimensional and smaller datasets. To further enhance learning efficiency and accuracy, this paper proposes Max-Relevance and Min-Redundancy Greedy BN learning algorithm. MRMRG algorithm applies Max Relevance and Min Redundancy feature selection technology to obtain better efficiency and accuracy on limited datasets, and proposes Local Bayesian Increment function according to BIC approximation formula and the likelihood property of overfitting to learn BNs on limited datasets.

J. Marques de Sá et al. (Eds.): ICANN 2007, Part I, LNCS 4668, pp. 49–57, 2007.

This paper is organized as follows. Section 2 provides a brief review of Bayesian network. Section 3 describes Max Relevance and Min Redundancy feature selection technology. In Section 4, we propose Local Bayesian Increment function. Section 5 represents the details of MRMRG algorithm. At the same time, we also analyze the time complexity of MRMRG. Section 6 shows an experimental comparison among MRMRG, K2 and TPDA. Finally, we conclude our work and present future works.

2 Bayesian Network

A Bayesian network is defined as a pair $B = \{G, \Theta\}$, where G is a directed acyclic graph (DAG) $G = \{V(G), A(G)\}$, with a set of nodes $V(G) = \{X_1, \ldots, X_n\}$, representing a set of random variables and a set of arcs $A(G) \subseteq V(G) \times V(G)$, representing independent relationships that exist between variables. Θ represents the set of parameters that quantify the network. It contains a parameter $\theta_{ijk} = P(x_{ik} \mid \phi_i[j])$ for each possible value x_{ik} of X_i, and $\phi_i[j]$ of Π_i. Here Π_i denotes the parents set of X_i and $\phi_i[j]$ is a particular instantiation of the parents set Π_i.

In the search & score approach, learning BN can be formally expressed as follows: given a complete training dataset D, find a directed acyclic graph G^* such that $G^* = \arg\max_G score(G : D)$, where $score(G : D)$ is the score function measuring the degree of fitness of any candidate DAG G to the dataset D.

In general, the score functions for learning BN have the decomposability property: the value assigned to each structure can be expressed as a sum of local values that depend on each variable and its parents: $score(G : D) = \sum_{X_i \in V(G)} score(X_i, \Pi_i : D)$. The score functions are high-order functions. The higher is the dimension of $\{X_i\} \cup \Pi_i$, the higher are the orders of the score functions. We take K2 score proposed by Cooper and Herskovits in [3] as an example.

We use the following notion: the number of values for the variable X_i is r_i; the number of possible configurations for the parents set Π_i of X_i is q_i; N_{ijk} is the number of instances in the dataset D where X_i takes the value x_{ik} and Π_i takes the value $\phi_i[j]$; $N_{ij} = \sum_{k=1}^{r_i} N_{ijk}$.

K2 score is based on several assumptions (multinomiality, parameter independence, parameter modularity, uniformity of the prior distribution of the parameters given the network structure), and can be expressed as follows:

$$g_{K2}(G : D) = log(p(G)) + \sum_{i=1}^{n} \left(\sum_{j=1}^{q_i} \left(\log \left(\frac{(r_i - 1)!}{(N_{ij} + r_i - 1)!} \right) + \sum_{k=1}^{r_i} \log(N_{ijk}!) \right) \right),$$

where $p(G)$ represents the prior probability of the DAG G.

3 Max-Dependence and MRMR

Definition 1. *In feature selection, **Max-Dependence sc- heme** [5] is to find a feature set S^* with m features, which jointly have the largest dependency on the target class C;* $S^* = \arg\max_S I(S;C)$.

Definition 2. *In feature selection, **Max-Relevance criterion** [5] is to select a feature set S^* with m features satisfying* $S^* = \arg\max_S(\frac{1}{|S|}\sum_{X_i \in S} I(X_i;C))$, *which approximates $I(S^*;C)$ with the mean value of all mutual information values between the features $X_i, i = 1,\ldots,m$ and class C.*

Definition 3. *In feature selection, **Min-Redundancy criterion** [5] is to select a feature set S^* with m features such that they are mutually minimally similar (mutually maximally dissimilar)·* $S^* = \arg\min_S(\frac{1}{|S|^2}\sum_{X_i,X_j \in S} I(X_i;X_j))$.

Definition 4. *In feature selection, **Max-Relevance and Min-Redundancy criterion** [5] is to find a feature set S^* with m features obtained by optimizing the Max-Relevance criterion and the Min-Redundancy criterion simultaneously. Assume that the two conditions are equally important, and use the following formula:* $S^* = \arg\max_S(\sum_{X_i \in S} I(X_i;C) - \frac{1}{|S|}\sum_{X_i,X_j \in S} I(X_i;X_j))$.

We select the feature set $S_m = \{X_1,\ldots,X_m\}$, the class variable C. Using the standard multivariate mutual information $MI(X_1,\ldots,X_m) = \int\int p(x_1,\ldots,x_m) \log\left(\frac{p(x_1,\ldots,x_m)}{p(x_1)\ldots p(x_m)}\right) dx_1 \ldots dx_m$, we get the formula: $I(S_m;C) = MI(S_m,C) - MI(S_m)$. Applying the Max-Dependence scheme, we can get the formula:

$$S^* = \arg\max_{S_m} I(S_m;C) = \arg\max_{S_m}\left(MI(S_m,C) - MI(S_m)\right). \tag{1}$$

Equation (1) is similar to the MRMR feature selection criterion: The second term requires that the features S_m are maximally independent of each other(that is, minimum redundant), while the first term requires every feature to be maximally dependent on C. In practice, the authors Peng & Ding have shown that if one feature is selected at one time, then MRMR criterion is a nearly optimal implementation of Max-Dependence scheme on limited datasets.[6]

4 Local Bayesian Increment Function

Let X and Y be two discrete variables, $\mathbf{Z}$ be a set of discrete variables, and z be an instantiation for $\mathbf{Z}$. $X, Y \notin \mathbf{Z}$.

Definition 5. *According to Moore's recommendation [8] about the chi-squared test, the sub-dataset $D_{\mathbf{Z}=z}$ satisfying the following condition is **locally sufficiently large** for $\{X \cup Y\}$ given $\mathbf{Z} = z$: All cells of $\{X \cup Y\}$ in the contingency table conditioned on $\mathbf{Z} = z$ have expected value greater than 1, and at least 80% of the cells in the contingency table about $\{X \cup Y\}$ on $\mathbf{Z} = z$ have expected value greater than 5.*

Learning on limited datasets, we loose the "locally sufficiently large" condition: If the number of cases in $D_{\mathbf{Z}=z}$ is much larger than the number of values for $\{X \cup Y\}$, for example $\|D_{\mathbf{Z}=z}\| \geq 4 \times (\|X\| \times \|Y\|)$; then we assume that the sub-dataset $D_{\mathbf{Z}=z}$ is "locally sufficiently large" for $\{X \cup Y\}$ given $\mathbf{Z} = z$.

Let D be a dataset of m cases. Let V be a set of n discrete variables, where X_i in V has r_i possible values $(x_{i1}, x_{i2}, \ldots, x_{ir_i})$. B_P and B_S denote BN structures containing just the variables in V. B_S exactly has one edge $Y \rightarrow X_i$ more than B_P. X_i has the parents set Π_i in B_P and the parents set $\Pi_i \cup Y$ in B_S. N_{ijk} is the number of cases in D, which variable X_i is instantiated as x_{ik} and Π_i is instantiated as $\phi_i[j]$. Let $N_{ijk} = \sum_{y} N_{i,\{j \cup y\},k}, N_{ij} = \sum_{k=1}^{r_i} N_{ijk}$. $\hat{\Theta}_i, \hat{\Theta}$ denote the maximum likelihoods of Θ_i, Θ. π_i^l denotes the instantiation of Π_i in the lth case.

Cases occur independently. The prior distribution of possible Bayesian networks is uniform. Given a Bayesian network model, there exist two properties: Parameter Independence and Parameter Modularity. [4]

We apply the BIC formula also used by Steck in [9] : $BIC(B_S) = \log L\left(\hat{\Theta}\right) - \frac{1}{2}\log(m)dim\left(\hat{\Theta}\right) \approx \log(P(D \mid B_S))$ to control the complexity of BN model. BIC adds the penalty of structure complexity to LBI function to avoid overfitting.

Definition 6 (Local Bayesian Increment Function)

$$Lbi(Y, i, \Pi_i) = \log\left(P(B_S, D)/P(B_P, D)\right) \approx BIC(B_S) - BIC(B_P)$$
$$= \log\left(L\left(\hat{\Theta}^{B_S}\right)/L\left(\hat{\Theta}^{B_P}\right)\right) - \frac{1}{2}\log(m)\left[dim\left(\hat{\Theta}^{B_S}\right) - dim\left(\hat{\Theta}^{B_P}\right)\right]$$
$$\log\left(L\left(\hat{\Theta}^{B_S}\right)/L\left(\hat{\Theta}^{B_P}\right)\right) = \log\left(P\left(D \mid \hat{\Theta}^{B_S}\right)\right) - \log\left(P\left(D \mid \hat{\Theta}^{B_P}\right)\right)$$
$$= \sum_{l=1}^{m} \log\left(P\left(x_i^l \mid \hat{\Theta}_i^{B_S}, \pi_i^l \cup y\right)/P\left(x_i^l \mid \hat{\Theta}_i^{B_P}, \pi_i^l\right)\right)$$

According to the likelihood property of overfitting(the marginal likelihood of overfitting for the training dataset is usually no less than non-overfitting), we assume that the log-likelihood does not change on the sub-dataset $D_{\Pi_i=\phi_i[*]}$ which are not "locally sufficiently large" for $\{X \cup Y\}$ (that is, to assume that there is overfitting between X and the parents set $\Pi_i \cup Y$ on the $D_{\Pi_i=\phi_i[*]}$),

$$\sum_{d_l \in D_{\Pi_i=\phi_i[*]}} \log P\left(x^l \mid \hat{\Theta}^{B_S}, \pi_l \cup y\right) = \sum_{d_l \in D_{\Pi_i=\phi_i[*]}} \log P\left(x^l \mid \hat{\Theta}^{B_P}, \pi_l\right). \quad (2)$$

According to (2), we infer the following results:

$$\log\left(L\left(\hat{\Theta}^{B_S}\right)\right) - \log\left(L\left(\hat{\Theta}^{B_P}\right)\right)$$
$$= \sum_{j} N_{ij} \times I_j(X, Y), \text{for } j, D_{\Pi_i=\phi_i[j]} \text{ is "locally sufficiently large"}$$
$$\dim\left(\hat{\Theta}^{B_S}\right) - \dim\left(\hat{\Theta}^{B_P}\right) = (r_y - 1)(r_i - 1)q_i$$

$$Lbi(Y, i, \Pi_i) = \sum_j N_{ij} \times I_j(X, Y) - \frac{1}{2}(r_y - 1)(r_i - 1)q_i \log(m),$$
$$\text{for } j, D_{\Pi_i=\phi_i[j]} \text{ is "locally sufficiently large".}$$

Note: $I_j(X, Y)$ is the mutual information between X and Y on $D_{\Pi_i=\phi_i[j]}$.

5 MRMRG Algorithm

MRMRG algorithm initializes the current parents set Π_i of the variable X_i to NULL, and then adds the variables one by one, which acquire the maximal value for Local Bayesian Increment (LBI) function, into the parents set Π_i from $Pre_i - \Pi_i$, until the result of LBI function is no more than 0. Repeating the above steps for every variable, we can obtain an approximately optimal Bayesian network.

The pseudo-code of MRMRG algorithm sees Fig.1. Pre_i denotes the set of variables that precede X_i. Π_i denotes the current parents set of the variable X_i. $k < 5$.

Input: A set V of n variables, an ordering on the variables, a dataset D containing m cases.
Output: for each variable X_i (i=1, . . . , n), a printout of the parents set Π_i.

Procedure MRMRG()
  For every variable X_i, call the procedure **GSParentsSet** (X_i, Pre_i) and obtain Π_i;
  For every variable X_i, output the parents set Π_i of the variable X_i
Endproc

Procedure GSParentsSet(C, Pre_i)
  Initialize Π_i to NULL and OK to TRUE;
  while OK
    For every variable $X \in Pre_i - \Pi_i$, compute the formula $\left\{ I(X;C) - \frac{1}{|\Pi_i|+1} \sum_{X_j \in \Pi_i} I(X;X_j) \right\}$ (1);
    Sort the variables in $Pre_i - \Pi_i$ by descendant order according to the value of (1);
    Obtain the top k variables $\{Y_1, Y_2, \ldots, Y_k\}$ from the sorted variables set $Pre_i - \Pi_i$;
    $Y_{\max} = \arg\max_{Y_j \in \{Y_1, Y_2, \ldots Y_k\}} \left[Lbi(Y_j, i, \Pi_i) \right]$;
    If $Lbi(Y_{\max}, i, \Pi_i) > 0$ then
      $\Pi_i = \Pi_i \cup \{Y_{\max}\}$;
    Else
      OK=FALSE;
    Endif
  Endwhile
  Return Π_i;
Endproc

Fig. 1. Max-Relevance and Min-Redundancy Greedy BN Learning

Given an ordering on the variables, MRMRG algorithm improves greedy BN learning algorithms (such as K2 algorithm [3]) in the following two ways in order to learn more accurately and efficiently on limited datasets.

Firstly, on limited datasets, the results of traditional scoring functions (such as K2 score [3],BDe score [4], etc) $score(C, \Pi_i \cup X_j)$ have less and less reliability and robustness with the dimension increase of $\Pi_i \cup X_j$, so that the formula $Y = \arg \max_{X_j \in Pre_i - \Pi_i} score(C, \Pi_i \cup X_j)$ cannot obtain the variable Y with the maximal score, even cannot acquire a variable with approximately maximal score sometimes. Since MRMR technology only uses 2-dimensional computation, it has much higher reliability and robustness than traditional scoring functions on limited datasets. Furthermore, according to the discussion in section 2.2, we know that if one feature is selected at one time (that is Greedy search), MRMR technology is nearly optimal implementation scheme of Max-Dependence scheme, which is equivalent to the maximal score method, on limited datasets. We consider that for some variable $X_j \in Pre_i - \Pi_i$, if the value of $\{I(X_j;C) - \frac{1}{|\Pi_i|+1} \sum_{X \in \Pi_i} I(X_j;X)\}$ is the largest, then it is the most probable that the value of the formula $score(C, \Pi_i \cup X_j)$ is the largest. Thus, MRMRG algorithm applies Max-Relevance and Min-Redundancy (MRMR) feature selection technology and replaces $score(C, \Pi_i \cup X_j)$ with the formula $\{I(X_j;C) - \frac{1}{|\Pi_i|+1} \sum_{X \in \Pi_i} I(X_j;X)\}$ to obtain the variable Y which gets the maximal score. Firstly, MRMRG algorithm selects the top k variables from the variables set $Pre_i - \Pi_i$ sorted according to the value of the formula $\{I(X_j;C) - \frac{1}{|\Pi_i|+1} \sum_{X \in \Pi_i} I(X_j;X)\}$ by descendant order. Then, it take the variable Y with the largest value of LBI function among the k variables as the variable with the maximal score.

Secondly, MRMRG algorithm proposes LBI function to replace traditional score increment functions (such as K2 [3],BDe [4]) to control the complexity of Bayesian network and to avoid overfitting. When the dataset D is "sufficiently large" for $\{X \cup Y \cup \Pi_i\}$, LBI function is equivalent to K2 increment function. When the dataset D is not "sufficiently large" for $\{X \cup Y \cup \Pi_i\}$, but there exist sub-datasets $D_{\Pi_i = \phi_i[*]}$ that are "locally sufficiently large" for $\{X \cup Y\}$ given $\Pi_i = \phi_i[*]$, MRMRG algorithm can also apply LBI function to improve accuracy and avoid overfitting (see section 4).

5.1 The Time Complexity of MRMRG

$r = max(r_i), i = 1, \ldots, n$, where r_i is the number of values for the variable X_i. The complexity of the formula $I(X;C) - \frac{1}{|\Pi_i|+1} \sum_{X_j \in \Pi_i} I(X;X_j)$ is $O(n)$. The complexity of computing $Lbi(Y, i, \Pi_i)$ is $O(mnr)$. The while statement loops at most $O(n)$ times, each time it is entered. In the worst case, the complexity of **GSParentsSet**(X_i, Pre_i) is $O(kmn^2r)$. The for statement loops n times. So, in the worst case, the complexity of **MRMRG()** is $O(kmn^3r)$.

6 Experimental Results

We implemented MRMRG algorithm, K2 algorithm[3], TPDA algorithm [2] and presented the comparison of the experimental results for 3 implementations. .

Tests were run on a PC with Pentium4 1.5GHz and 1GB RAM. The operating system was Windows 2000. These programs were developed under Matlab 7.0. 5 Bayesian networks were used. Table 1 shows the characteristics of these networks. The characteristics include the number of nodes, the number of arcs, the maximal number of node parents/children(Max In/Out-Degree), and the minimal/maximal number of node values(Domain Range).

From these networks, we performed these experiments with 200, 500, 1000 training cases each. For each network and sample size, we sampled 20 original datasets and recorded the average results by each algorithm. Let $k = 3$ in Fig.1.

Table 1. Bayesian networks

BN	Nodes Num	Arcs Num	Max In/Out-Degree	Domain Range
Insur	27	52	3/7	2-5
Alarm	37	46	4/5	2-4
Barley	48	84	4/5	2-67
Hailf	56	66	4/16	2-11
Munin	189	282	3/15	1-21

6.1 Comparison of Runtime Among Algorithms

A summary of the time results of the execution of all the 3 algorithms is in Table 2. We normalized the times reported by dividing by the corresponding running time of MRMRG on the same datasets and reported the averages over sample sizes. [10] Thus, a normalized running time of greater than 1 implies a slower algorithm than MRMRG on the same learning task. A normalized running time of lower than 1 implies a faster algorithm than MRMRG.

From the results, we can see that MRMRG has better efficiency than other 3 algorithms K2 and TPDA. In particular, for smaller sample sizes (200, 500, 1000), MRMRG runs several times faster than K2 and TPDA.

6.2 Comparison of Accuracy Among Algorithms

We compared the accuracy of BNs learned by these 3 algorithms according to the BDeu score. The BDeu score corresponds to the posteriori probability of the structure learned.[4] The BDeu scores of which Equivalent Sample Size (ESS) is 10 in our experiments were calculated on a seperate test set sampled from the true BN containing 20000 samples. Tables 3-7 report the results.

From the results, we can see that MRMRG can learn more accurately than TPDA on limited datasets. In particular, MRMRG has better accuracy than K2 on limited datasets.

Table 2. Normalized Runtime

Size	MRMRG	K2	TPDA
200	1.0	2.39	8.82
500	1.0	5.57	7.61
1000	1.0	9.18	4.57

Table 3. Average BDeu(Insur)

Size	MRMRG	K2	TPDA
200	-21.915	-22.993	-31.507
500	-19.032	-19.583	-28.786
1000	-19.386	-19.533	-25.111

Table 4. Average BDeu(Alarm)

Size	MRMRG	K2	TPDA
200	-15.305	-16.069	-24.456
500	-13.858	-13.950	-18.097
1000	-13.319	-13.583	-15.429

Table 5. Average BDeu(Barley)

Size	MRMRG	K2	TPDA
200	-81.794	-83.972	-106.782
500	-76.327	-79.194	-103.783
1000	-76.846	-77.375	-111.069

Table 6. Average BDeu(Hailf)

Size	MRMRG	K2	TPDA
200	-72.138	-73.361	-106.135
500	-71.217	-72.662	-101.382
1000	-70.955	-71.734	-97.374

Table 7. Average BDeu(Munin)

Size	MRMRG	K2	TPDA
200	-65.971	-68.393	-135.103
500	-62.483	-63.250	-125.625
1000	-61.967	-62.837	-140.476

7 Conclusion

Accuracy are the main index in evaluating algorithms for learning BN. MRMRG algorithm greatly reduces the number of high dimensional computations and improves the accuracy of learning. The experimental results indicate that MRMRG has better performance on accuracy than most of existing algorithms on limited datasets.

We are interesting in incorporate ordering-based search method [7] into our MRMRG algorithm to implement MRMRG without the information of the order between nodes.

References

1. Spirtes, P., Glymour, C., Scheines, R.: Causation, Prediction and Search, 2nd edn. MIT Press, Massachusetts (2000)
2. Cheng, J., Greiner, R., Kelly, J., Bell, D., Liu, W.: Learning Belief Networks form Data: An Information Theory Based Approach. Artificial Intelligence 137(1-2), 43–90 (2002)
3. Cooper, G., Herskovits, E.: A Bayesian Method for Constructing Bayesian Belief Networks from Databases. In: Ambrosio, B., Smets, P. (eds.) UAI '91. Proceedings of the Seventh Annual Conference on Uncertainty in Artificial Intelligence, pp. 86–94. Morgan Kaufmann, San Francisco, CA (1991)
4. Heckerman, D., Geiger, D., Chickering, D.M.: Learning Bayesian Networks: the Combination of Knowledge and Statistical Data. Machine Learning. 20(3), 197–243 (1995)

5. Peng, H.C., Ding, C., Long, F.H.: Minimum redundancy maximum relevance feature selection. IEEE Intelligent Systems 20(6), 70–71 (2005)
6. Peng, H.C., Long, F.H., Ding, C.: Feature Selection Based on Mutual Information: Criteria of Max-Dependency, Max-Relevance,and Min-Redundancy. IEEE Transactions on PAMI 27(8), 1226–1238 (2005)
7. Teyssier, M., Koller, D.: Ordering-Based Search: A Simple and Effective Algorithm for Learning Bayesian Networks. In: Chickering, M., Bacchus, F., Jaakkola, T. (eds.) UAI '05. Proceedings of the 21st Conference on Uncertainty in Artificial Intelligence, pp. 584–590. Morgan Kaufmann, San Francisco, CA (2005)
8. Moore, D.S.: Goodness-of-Fit Techniques, 1st edn. Marcel Dekker, New York (1986)
9. Steck, H.: On the Use of Skeletons when Learning in Bayesian Networks. In: Boutilier, C., Goldszmidt, M. (eds.) UAI '00. Proceedings of the 16th Conference on Uncertainty in Artificial Intelligence, pp. 558–565. Morgan Kaufmann, San Francisco,CA (2000)
10. Ioannis, T., Laura, E.B.: The Max-Min Hill-Climbing Bayesian Network Structure Learning Algorithm. Machine Learning 65(1), 31–78 (2006)

Incremental One-Class Learning with Bounded Computational Complexity

Rowland R. Sillito and Robert B. Fisher

School of Informatics, University of Edinburgh, UK

Abstract. An incremental one-class learning algorithm is proposed for the purpose of outlier detection. Outliers are identified by estimating - and thresholding - the probability distribution of the training data. In the early stages of training a non-parametric estimate of the training data distribution is obtained using kernel density estimation. Once the number of training examples reaches the maximum computationally feasible limit for kernel density estimation, we treat the kernel density estimate as a maximally-complex Gaussian mixture model, and keep the model complexity constant by merging a pair of components for each new kernel added. This method is shown to outperform a current state-of-the-art incremental one-class learning algorithm (Incremental SVDD [5]) on a variety of datasets, while requiring only an upper limit on model complexity to be specified.

1 Introduction

The problem of one-class learning (also known interchangeably as "outlier / novelty / anomaly detection") arises in a wide variety of different application domains. The fundamental goal of one-class learning is to generate a rule that distinguishes between examples of a known class of items and examples from previously-unseen novel classes, on the exclusive basis of training examples from the known class.

This problem presents itself in cases where one wishes to distinguish between members of a class for which examples are abundantly available, and members of another rarely observed class. This often arises when attempting to detect abnormal activity, eg. jet engine failure, computer network intrusions, disease symptoms, etc. In each of these domains, anomalous examples may be scarce or entirely absent during training, but their subsequent identification is of crucial importance. A wide variety of different methods have been proposed to address this problem (see [3] for a review). However, almost all existing one-class classification algorithms require all training examples to be available at once, for a single "batch" learning step: if a new example is presented, the classifier must be retrained from scratch.

Since outliers might only be identifiable by their deviation from a normal model, a key problem in one class learning is the choice of model complexity. In some cases training data may be well described by the parameters of a single Gaussian distribution, while in other cases - eg. where the data has multiple

J. Marques de Sá et al. (Eds.): ICANN 2007, Part I, LNCS 4668, pp. 58–67, 2007.

modes or lies on a non-linear manifold - a more complex model is required. It is important to select the correct level of model complexity: if it is too low, the learned normal model may also include the anomalies that we wish to detect; if it is too high, the model may not include the majority of normal examples.

In many cases it would be useful to be able to incrementally train a classifier as data became available, without needing to pre-specify the level of model complexity. In this paper we propose a new technique for performing incremental one-class learning, where only an upper limit on model complexity needs to be specified. While computationally feasible, our algorithm attempts to estimate the underlying p.d.f. (probability density function) of the training data using non-parametric kernel density estimation (with Gaussian kernels), thereby generating a maximally complex one-component-per-example Gaussian mixture model. Once a maximum number of mixture components has been reached, it is kept constant by merging a pair of components for every new component added. We choose pairs of components for merging based on an information theoretic merging-cost function originally proposed by Goldberger and Roweis in [2].

Currently, at least two related techniques exist in the literature. In [9] Yamanishi et al. propose an unsupervised outlier detection procedure "SmartSifter" in which a Gaussian mixture model is trained using an on-line adaptation of the EM (Expectation Maximization) algorithm. A key aspect of their adaptation is the inclusion of "discounting" parameters which ensure that the effect of older training examples on the model parameters is rapidly displaced by new examples. Each new training example is given a score based on the extent to which it changes the model parameters: a comparatively high score, indicating a large change in model parameters, indicates the possibility of an outlier. This algorithm seems inappropriate for comparison with the proposed algorithm as it models a finite window of training data preceding each new example, rather than attempting to incrementally build a complete normal class description.

A more closely related algorithm has been proposed in [5], where Tax and Laskov present an incremental training procedure for the SVDD (Support Vector Data Description) algorithm originally proposed by Tax and Duin in [7]. The SVDD algorithm attempts to find the smallest hypersphere that encloses the training data, and allows more complex hyper-volumes (which may fit the data better) to be obtained by introducing kernel functions which map the training data to a higher dimensional space [7]. In both batch and incremental forms, the SVDD algorithm relies crucially on the correct choice of model complexity parameters. Various methods have been proposed to address this issue in the absence of example outliers, including: procedures for generating synthetic outliers (Tax and Duin [4]) and, more recently, a consistency based approach which takes the simplest possible classifier and increases its complexity parameter until the proportion of correctly recognized training data starts to fall (Tax and Muller [8]). The incremental variant of SVDD does not include any on-line model complexity selection, but it is conceivable that the batch optimization methods could be applied to a pre-existing dataset to optimally parametrize the classifier before using it for on-line training.

We provide a detailed description of the proposed algorithm in Section 2, and then illustrate its performance on a variety of datasets in Section 3, where we also compare its performance with the incremental SVDD algorithm [5] (optimized using the consistency criterion proposed in [8]). Our algorithm is shown to yield equivalent, often better, performance than the incremental SVDD algorithm without requiring any time-consuming parameter optimization.

2 Algorithm

The proposed algorithm is designed to receive a sequence of labeled multivariate training data, and to determine - at any stage in the training process - whether or not new data are outliers. This is achieved by estimating the underlying probability density function that gave rise to the data, and setting a threshold on this density: if a new example has a probability lower than the threshold, it is classified as an outlier.

2.1 Density Estimation

Phase 1: Kernel Density Estimation. Initially the probability density of the data is determined using Kernel Density Estimation. This technique allows us to evaluate the probability of a new example by taking a uniformly weighted combination of a set of Gaussian kernels (with identical covariance matrices Σ) centered on each of the training data. Thus, finding the probability of a new data point z, given a set of N training data $X = \{x_1, \ldots, x_N\}$, simply consists of evaluating the following function:

$$p(z) = \frac{1}{(2\pi)^{\frac{d}{2}} |\Sigma|^{\frac{1}{2}}} \cdot \frac{1}{N} \cdot \sum_{n=1}^{N} e^{-\frac{1}{2}(z-x_n)^T \Sigma^{-1} (z-x_n)} \quad (1)$$

The factor $\frac{1}{(2\pi)^{\frac{d}{2}} |\Sigma|^{\frac{1}{2}}}$ (where d refers to the dimensionality of the data) ensures that the resulting probability distribution integrates to 1. Training the model is straight-forward: when a new training example x_{NEW} is received, it is simply added to the set X:

$$X \rightarrow X \cup \{x_{NEW}\} \quad (2)$$

The only remaining problem is the choice of the covariance matrix Σ. To reduce computational cost, we use a uniform covariance matrix $\Sigma(\sigma) = I^d \cdot \sigma^2$, which leaves only a single parameter σ to be determined. The value of σ is chosen using the "leave-one-out" likelihood criterion proposed by Duin in [1] (recently shown to be a good model selection criterion for multivariate kernel density estimation by Zhang et al. in [10]). This technique allows us to find a parameter that maximizes the likelihood of the dataset, while avoiding the problem of the data likelihood tending towards infinity as $\sigma \rightarrow 0$. For a given value of σ, the "leave-one-out" likelihood function combines the log-likelihoods for every individual example x_n given a model constructed from all others $\forall x \neq x_n$, as follows:

$$LL(\sigma) = \sum_{n=1}^{N} \log \left(\frac{1}{(2\pi\sigma)^{\frac{d}{2}}} \cdot \frac{1}{N-1} \cdot \sum_{\forall x \neq x_n} e^{-\frac{1}{2\sigma^2}(x_n - x)^T (x_n - x)} \right) \quad (3)$$

Every time the training dataset is updated (during the kernel density estimation phase) we evaluate (3) for a range of values surrounding the previous σ, and choose $\sigma = \arg\max_\sigma (LL(\sigma))$.

Phase 2: Mixture Model Merging. Since the computational cost of evaluating (1) scales linearly with the quantity of training data, it eventually becomes infeasible to estimate the p.d.f. of the data in this fashion. Noting that the kernel density estimate is essentially a maximally-complex Gaussian mixture model, we adapt a method proposed by Goldberger and Roweis for reducing the complexity of Gaussian mixture models in [2]: once the maximum feasible model complexity has been reached, we keep it constant by merging a pair of components for each new component added.

Initialization. Once the maximum model complexity has been reached $N = N_{max}$ we initialize a data structure to store a Gaussian mixture model with weights (initially uniform), and covariances (set to the final value estimated in the kernel density phase), and means (the training data) as follows:

$$\begin{aligned} w_{1 \ldots N} &= \tfrac{1}{N} \\ \Sigma_{1 \ldots N} &= I^d \cdot \sigma^2_{final} \\ \mu_{1 \ldots N} &= x_{1 \ldots N} \end{aligned} \quad (4)$$

For each pair of components $G_i = \{w_i, \mu_i, \Sigma_i\}$ and $G_j = \{w_j, \mu_j, \Sigma_j\}$ a merging cost is then calculated using (7) - explained in the next section - forming an $N \times N$ matrix C. This one-off[1] calculation of $\frac{N_{max}(N_{max}-1)}{2}$ different cost values is computationally feasible for values of N_{max} where it is still possible to evaluate (1) in reasonable time.

Merging Strategy. In this stage every new training example still contributes a Gaussian kernel. However the covariance matrix is now fixed to the final estimate obtained in the preceding stage, and we employ a merging strategy to keep the number of mixture components constant. For every new component added, a pair of components (which may include the new one) is merged as follows:

$$\begin{aligned} w_{merge(i,j)} &= w_i + w_j \\ \mu_{merge(i,j)} &= \tfrac{w_i}{w_i+w_j}\mu_i + \tfrac{w_j}{w_i+w_j} \cdot \mu_j \\ \Sigma_{merge(i,j)} &= \tfrac{w_i}{w_i+w_j}\left(\Sigma_i + (\mu_i - \mu_{merge(i,j)})(\mu_i - \mu_{merge(i,j)})^T\right) \\ &\quad + \tfrac{w_j}{w_i+w_j}\left(\Sigma_j + (\mu_j - \mu_{merge(i,j)})(\mu_j - \mu_{merge(i,j)})^T\right) \end{aligned} \quad (5)$$

[1] Subsequently maintaining this cost matrix only requires a fixed number of $N_{max}+1$ cost evaluations for each new training example.

We wish to choose a pair of components to merge in a way that minimizes the resulting change in the p.d.f. encoded by the model. The Kullback-Leibler divergence provides a means of assessing the "damage" caused by replacing a particular pair of components with a single merged component. Essentially, the KL divergence $\mathcal{KL}(P||Q) = \int_{-\infty}^{\infty} p(x) \log \frac{p(x)}{q(x)} dx$ quantifies the expected information loss per sample when an approximating distribution Q is substituted for a true distribution P. For a pair of Gaussian distributions $G_p = \{\mu_p, \Sigma_p\}$ and $G_q = \{\mu_q, \Sigma_q\}$, it can be calculated as follows [2]:

$$\mathcal{KL}(G_p||G_q) = \frac{1}{2}\left(\log \frac{|\Sigma_q|}{|\Sigma_p|} + \mathrm{Tr}(\Sigma_q^{-1}\Sigma_p) + (\mu_p - \mu_q)\Sigma_q^{-1}(\mu_p - \mu_q)^T - d\right) \tag{6}$$

This allows us to quantify the cost of replacing components G_i and G_j (where $i \neq j$) with their merged counterpart $G_{merge(i,j)}$ by calculating a weighted combination (as proposed by Goldberger and Roweis in [2]) of their respective Kullback-Leibler divergences from $G_{merge(i,j)}$ as follows:

$$cost(G_i, G_j) = w_i \mathcal{KL}(G_i||G_{merge(i,j)}) + w_j \mathcal{KL}(G_j||G_{merge(i,j)}) \tag{7}$$

Updating Procedure. When a new training example x_{NEW} arrives, a temporary new component $G_{N_{max}+1} = \{\frac{1}{N_{ex}+1}, x_{NEW}, I^d \cdot \sigma^2_{final}\}$ is created, and the weights of existing components are rescaled by a factor of $\frac{N_{ex}}{N_{ex}+1}$, where N_{ex} is the total number of training examples received before the new one. The cost matrix is augmented with a new row/column for the new component, and a pair of components is chosen such that $\{G_i, G_j\} = \arg\min_{G_i, G_j}(cost(G_i, G_j))$. If $\{G_i, G_j\}$ are both existing components, then G_i is replaced with $G_{merge(i,j)}$ and G_j is replaced with the new component; alternatively if G_j is the new component then G_i is simply replaced with $G_{merge(i,j)}$. The temporary component $G_{N_{max}+1}$ is then removed, and the merging cost matrix C updated accordingly. This procedure requires a fixed total of $N_{max} + 1$ evaluations of (7) for every new training example, as the cost matrix only needs to be updated for entries corresponding to merged/new components.

2.2 Classification Threshold

Given the proposed density estimation method, an important remaining issue is the choice of classification threshold. A naive approach would be to set the threshold at that the level of the least probable (given the current model) training example, thereby correctly classifying all training data as normal. However, it is quite possible that the least probable training example - which will be located in the most sparsely populated region of training data - may have a probability value equivalent to that of the outliers we wish to detect. To avoid this problem, and to make the method robust to potential outliers in the training set, we set the threshold at a value that deliberately misclassifies a certain proportion of the training data as outliers. In the experiments described in the following section we choose a value of 10%, aiming to learn a classifier that filters out 90% of normal data.

3 Experiments

In this section we measure the classification performance of the proposed algorithm on a variety of datasets, showing how classification performance changes as the model is trained on more training examples. To place the performance

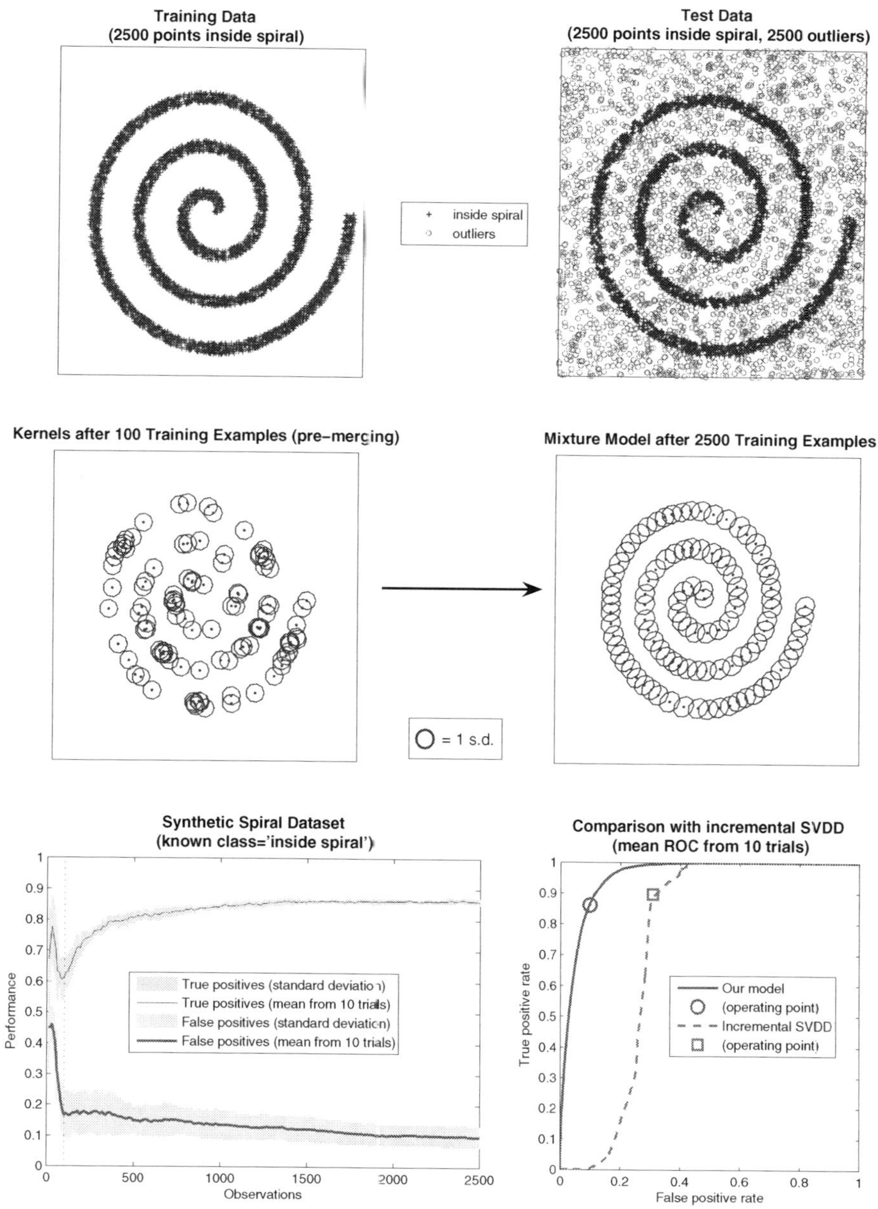

Fig. 1. Results for the synthetic spiral dataset. See text for description.

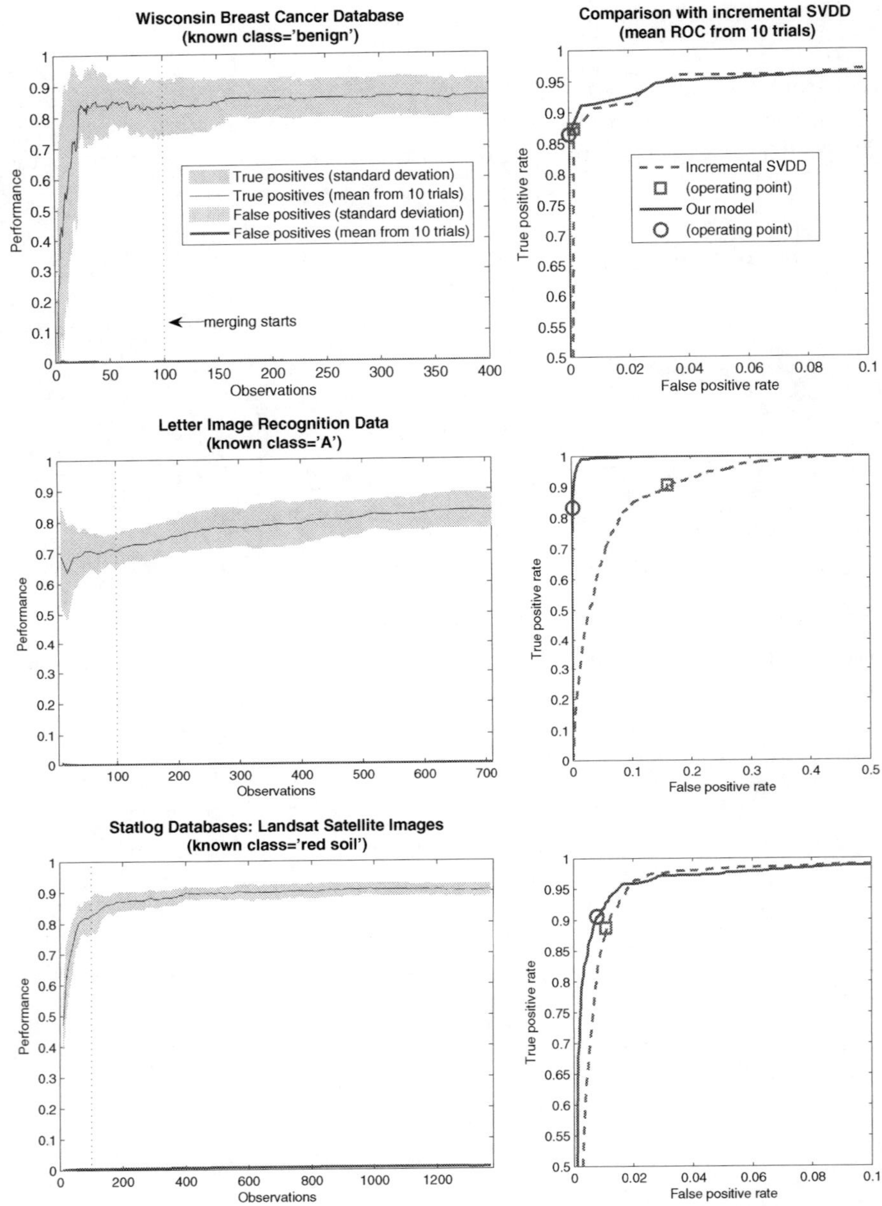

Fig. 2. Results for real datasets. See text for description.

of the proposed algorithm in context we compare its performance to that of the incremental SVDD algorithm [5], making comparisons at the point where both algorithms have been trained on all training examples in a given dataset.

We use a freely available implementation of the incremental SVDD algorithm, *incsvdd*, contained in the *DDtools* MATLAB toolbox [6]. In all tests we use the

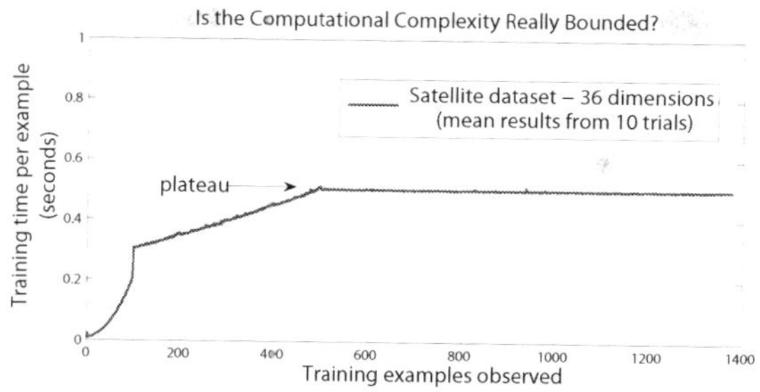

Fig. 3. Measuring computational complexity

radial basis kernel function, and optimize the kernel parameter (for the whole training dataset) using the *consistent_occ* function (also from [6]) which implements the consistency-based model selection criterion proposed in [8]. We initially apply this criterion to a range of 20 linearly spaced values between the shortest and longest Euclidean distances observed within the dataset; to search for potentially better parameter values on a finer scale, we then run a second parameter optimization for a further 20 values surrounding the optimal parameter from the first set. As for the proposed algorithm, we set the SVDD threshold parameter at a level that aims to reject to 10% of the training data.

Synthetic Dataset. An initial experiment was carried out on a synthetic 2 dimensional dataset: we defined a spiral shaped region which we used to divide a set of uniformly distributed random datapoints into a hypothetical normal class of datapoints (points in the spiral region) and outliers (all other points). We used 2500 spiral points for training the algorithm, and a further 2500 points from the spiral along with 2500 outlier points for testing it, as shown in Figure 1.

For this test (as for all subsequent tests) we set the upper limit on the number of mixture components N_{max} to be 100. The middle section of figure 1 shows the configuration of the 100 Gaussian components before the merging phase commences, and at the end of the training process. The resulting model organization appears to accurately reflect the shape of the spiral: indeed, at the end of training the algorithm correctly classifies 88.13% of all test data, with a True Positive rate[2] of $TP = 86.1\%$ and a False Positive rate[3] of $FP = 0.0984\%$. The TP and FP curves shown in the lower left hand section of Figure 1 indicate that the classification performance increased in a stable fashion as more training examples were processed. In this plot, and in subsequent plots of this type, the vertical dotted line indicates the start of the merging phase.

At the end of training, the incremental SVDD algorithm correctly classified 79.44% of the test data (with $TP = 89.69\%$ and $FP = 0.308\%$), misclassifying

[2] Indicating normal examples correctly identified as normal.
[3] Indicating outliers incorrectly classified as normal.

a much larger number of outliers as normal. The ROC[4] curve in the lower right hand section of Figure 1 shows the different TP and FP values obtained as the (training data rejection) threshold is varied for each classifier, indicating that the proposed algorithm outperforms incremental SVDD algorithm across the range of possible thresholds. Both plots in Figure 1 show the mean performance for 10 different random orderings of the training data.

Real Datasets. A series of subsequent experiments were then carried out on three different real-world datasets obtained from the UCI Machine Learning Repository[5]:

1. The *Wisconsin Breast Cancer Database*, which contains 699 (9-dimensional) datapoints, containing 458 normal examples and 241 cases of cancer.
2. The *Letter Recognition Database,* which contains 20,000 (16-dimensional) parametrizations of examples of printed letters, with 26 classes corresponding to the alphabet. We use the 789 examples of the letter 'A' as a hypothetical normal class, and all other classes as outliers.
3. The *STATLOG Landsat Satellite Database*, which contains 6435 (36 dimensional) vectors corresponding multispectral images of 6 different types of ground coverage: we use the 1533 examples of 'red soil' as the normal class.

For each of these datasets we use 90% of examples of the chosen normal class as training data, and the remaining 10% for testing. All subsequent experiments are performed for 10 different testing/training permutations of the normal class. Again, we test our algorithm with a maximum complexity level of 100 components, and compare it to the consistency-optimized incremental SVDD algorithm. The classification results illustrated by the ROC curves in Figure 2, indicate that the proposed algorithm consistently outperforms the incremental SVDD algorithm, although the performance obtained on the Cancer and Satellite datasets is very similar.

Computational Complexity. To confirm the assertion that the proposed algorithm has bounded computational complexity, we recorded the time taken to train our algorithm on each datapoint during the tests on the 36 dimensional Satellite dataset. This is plotted (excluding the point where the merging matrix is first initialized) in Figure 3, indicating that a fixed processing time per example is indeed reached soon after the merging phase commences. Our algorithm takes an average time of 565.56 ± 0.32 seconds to train on the 1379 examples, while the SVDD algorithm takes a significantly shorter time of 6.25 ± 0.34 seconds to train, albeit after a parameter optimization step which takes 482.53 ± 60.21 seconds. Evaluation times for the two algorithms are similar: our algorithm takes 2.88 ± 0.02 seconds to classify 5056 testing examples, while the incremental SVDD algorithm takes 2.01 ± 0.102 seconds to classify the same examples.

[4] Receiver Operating Characteristic.

[5] http://www.ics.uci.edu/~mlearn/MLRepository.html

4 Discussion

We have proposed a simple procedure for incrementally training a one-class classifier to perform outlier detection, without the need for any time-consuming model optimization procedures. Despite its simplicity, the proposed algorithm appears to perform better than the incremental SVDD algorithm, even though the parameters of the latter were being chosen through a lengthy optimization process. The fact that the optimization process proposed in [8] did not find parameters that allowed incremental SVDD to outperform our algorithm does not mean that such parameters could not be found in principle: it does, however, illustrate the the key strength of our algorithm - the fact that it automatically generates models that achieve a useful level of outlier detection performance.

References

1. Duin, R.P.W.: On the choice of smoothing parameters for Parzen estimators of probability density functions. IEEE Trans. Computers C-25, 1175–1179 (1976)
2. Goldberger, J., Roweis, S.: Hierarchical clustering of a mixture model. Advances in Neural Information Processing Systems 17, 505–512 (2005)
3. Hodge, V.J., Austin, J.: A survey of outlier detection methodologies. Artificial Intelligence Review 22, 85–126 (2004)
4. Tax, D.M.J., Duin, R.P.W.: Uniform object generation for optimizing one-class classifiers. Journal of Machine Learning Research 2, 155–173 (2001)
5. Tax, D.M.J., Laskov, P.: Online SVM learning: from classification to data description and back. In: Proc. 13th IEEE NNSP Workshop, IEEE Computer Society Press, Los Alamitos (2003)
6. Tax, D.M.J.: DDtools, the Data description toolbox for Matlab. version 1.5.5
7. Tax, D.M.J., Duin, R.P.W.: Support vector domain description. Pattern Recognition Letters 20, 1191–1199 (1999)
8. Tax, D.M.J., Muller, K.-R.: A consistency-based model selection for one-class classification. In: Proc. ICPR 2004, vol. 3, pp. 363–366 (2004)
9. Yamanishi, K., Takeuchi, J., Williams, G., Milne, P.: On-line unsupervised outlier detection using finite mixtures with discounting learning algorithms. Data Mining and Knowledge Discovery 8, 275–300 (2004)
10. Zhang, X., King, M., Hyndman, R.J.: A Bayesian approach to bandwidth selection for multivariate kernel density estimation. Computational Statistics & Data Analysis 50, 3009–3031 (2006)

Estimating the Size of Neural Networks from the Number of Available Training Data

Georgios Lappas

Technological Educational Institution (TEI) of Western Macedonia
Kastoria Campus, P.O. Box 30, GR 52100 Kastoria, Greece
lappas@kastoria.teikoz.gr

Abstract. Estimating *a priori* the size of neural networks for achieving high classification accuracy is a hard problem. Existing studies provide theoretical upper bounds on the size of neural networks that are unrealistic to implement. This work provides a computational study for estimating the size of neural networks using as an estimation parameter the size of available training data. We will also show that the size of a neural network is problem dependent and that one only needs the number of available training data to determine the size of the required network for achieving high classification rate. We use for our experiments a threshold neural network that combines the perceptron algorithm with simulated annealing and we tested our results on datasets from the UCI Machine Learning Repository. Based on our experimental results, we propose a formula to estimate the number of perceptrons that have to be trained in order to achieve a high classification accuracy.

1 Introduction

On non-seperable problems, finding the best function for minimizing the classification error is an NP-complete problem. Blum and Rivest [11] proved that even for a very small network to find weights and thresholds that learn any given set of training examples is an NP-complete problem. Höffgen and Simon [17] deal with the problem of learning a probably almost optimal weight vector for a neuron, finding that it is an NP-complete problem. Also finding an optimum network configuration for solving combinatorial optimization problems is not an easy task [34]. Thus, establishing an optimal weight configuration of threshold units (in order to minimize the error rate) cannot be executed in efficient time and therefore only approximations of optimum solutions can be achieved.

In order for neural networks to approximate high classification rates, which is equivalent to low error rates, to a specific classification problem it usually requires a certain amount of adjustments, which are apparently unavoidable in the light of the *No Free Lunch Theorems*; cf. [33]. The need for tunning network parameters when the classification accuracy is very important may lead us to lots of experiments for establishing a high classification rate. *A piori* estimation of these parameters may help in reducing the time and effort to find an appropriate neural network for achieving high classification accuracy. An important

J. Marques de Sá et al. (Eds.): ICANN 2007, Part I, LNCS 4668, pp. 68–77, 2007.

adjustment related to neural networks implementations is the number of hidden units (network size) that should be trained in order to approximate a function that best describes the training data.

The problem of finding the smallest network that can realize an arbitrary function given a set of m vectors in n dimensions defines the *circuit complexity* problem [10]. The circuit complexity problem is of great importance for hardware implementations and tries to find tight bounds for the number of units used to realize an arbitrary function. Identifying sequences of Boolean functions $\{f(x_1, ..., x_n)\}_{n=n_0}^{\infty}$ with "superpolynomial" (or exponential) gate number in constant depth circuits of unbounded fan-in gates is a very difficult problem and only slow progress has been made over the past decades. For example, Razborov and Widgerson [27] designed a sequence of functions that requires at least the superpolynomial number $n^{\Omega(logn)}$ of threshold gates in any circuit of depth 3. Furthermore, it has been shown [3,12] that computing the permanent of an $n \times n$ matrix requires a superpolynomial number of gates in any threshold circuit of constant depth; for more information about the computational power of small depth circuits, see [31]. In [10] Beiu surveys a large number of bounds for a number of different network approaches.

Theoretical analysis on lower and upper bounds on VC-dimension [30] has been performed for some type of networks. Baum and Haussler [8,9] show that in a feedforward network with W weights and N computational threshold units the VC-dimension is bounded $VC \leq 2 \cdot W \cdot log_2(e \cdot N)$. Maas [25] has obtained a lower bound $W log_2 W$ for certain types of multilayer feedforward network of threshold units. Other tight bounds for specific type of functions have been obtained in [6,7,14,18,19,29,32]; see Anthony and Bartlett [5] for a review of the field.

Bounding VC-dimensions is a challenging task in mathematical investigations of neural networks which provides a number of sophisticated mathematical tools. The bounds, however, tend to be too large, since they provide such guarantees of generalisation for any probability distribution of training examples and for any training algorithm that minimizes the training error on the training examples. Therefore, the VC-dimension is a very general theoretical measure of pattern classification ability. The proposed bounds are usually unrealistic and with limited practical value for real world problems. Other alternative methods for calculating more realistic bounds should be considered, as in Haussler et al. [16], where a probability distribution was introduced for considering the performance of a classifier that implements a Bayes optimal classification algorithm [13]. Empirical studies investigating the relation of classification ability to parameters of the problem are important for more realistic VC-dimension bounds.

In this work we relate the network learning capacity to the number of existing training samples. We propose a formula of an upper bound of size $S = 8 \cdot \sqrt{2^n/n}$ computational units as sufficient for a small error rate, where $n := \log|\mathcal{S}_L|$ is the number of bits neccessary to enumerate all existing training data. Since in the problems that we used from the UCI Repository $n \leq 12$ and $S \leq 147$, the proposed formula leads to realistic to implement neural networks. With this formula we also show that the size of the neural network depends on the size of existing training samples $\mathcal{S}_L$. This means that if more training data emerge in a problem then a larger network is required. Consequently, this also proves that the number of trained units cannot be constant for all classification problems but is a problem dependent parameter.

2 Methodology

Our experiments are performed with LSA machine [1,2], a neural network that combines the classical perceptron algorithm [28] with a specific type of simulated annealing [15] as the stochastic local search procedure for finding threshold gates of a classification circuit. The simulated annealing procedure employs a logarithmic cooling schedule $c(k) = \Phi / \ln(k+2)$ search strategy where the "temperature" decreases at each step. The simulated annealing-based extension of the perceptron algorithm is activated when the number of misclassified examples increases for the new hypothesis compared to the previous one. In this case, a random decision is made according to the rules of simulated annealing. If the new hypothesis is rejected, a random choice is made among the misclassified examples for the calculation of the next hypothesis. A detailed description on the LSA machine can be found in [1,2].

We will experiment with depth-two circuits where depth-one consists of computational units and depth-two of a voting function that decides the output class. Depth-four networks with the LSA machine lead approximately to the same size of networks for achieving low error rates [21]. The dataset D is divided into two disjoint sets of data: the training dataset $|S_L|$ for training the neurons and the testing dataset $|S_T|$ for evaluating the classification error of the network. Each neuron at depth-one is trained by a sample set of size p randomly drawn from the available training data set S_L.

According to the No-Free-Lunch-Theorems, the performance of learning algorithms is problem-dependent. Four problem-dependent parameters, the network size N, the sample size p for training a unit, the length of inhomogeneous Markov chain k and the constant Φ of the simulated annealing cooling schedule, are required for the LSA machine. The constant Φ determines an escape from local minima in the simulated annnealing process and is expressed in terms of a percentage of the size p of the sample set used to train a neuron, i.e. $\Phi = G \cdot p$, where $G \in (0,1)$. For the estimation of the network size we chose two datasets from the UCI Machine Learning Repository (http://www.ics.uci.edu/~mlearn/MLRepository.html), where one of the datasets induces three classes.

Splice-junction Gene Sequences Database (SJGSD): The database consists of 3190 vectors representing 60 attributes. There are three classes with distribution: 25% for IE (767 instances); 25% for EI (768 instances); and 50% for "Neither" (1655 instances). In order to discriminate between the three classes, we introduce three databases, each related to a single class as positive examples: the "IE database" consists of 767 positive examples (IE class) and 2423 negative examples (union of EI class and "Neither" class); the "EI database" consists of 768 positive examples and 2422 negative examples; the "Neither database" consists of 1655 positive examples and 1535 negative examples.

Wisconsin Breast Cancer Database (WBCD): WBCD is a binary classification problem where each vector represents 9 features. The output indicates either a benign case (positive example) or a malignant case (negative example). The data set consists of 699 samples, where 16 samples have missing values which are discarded in a pre-processing step. The remaining 683 data are divided into 444 benign and 239 malignant cases.

2.1 Estimations of Circuit Size

Classification problems P can be encoded as Boolean functions f_P on n input variables where the sample sets usually provide only a tiny fraction of input-output pairs of f_P. For the splice-junction data we have "DNA windows" of length 60 with the usual DNA information from $\{A, C, G, T\}$. Therefore, in binary notation we have $n = 120$. For the WBCD we have 9 attributes, each taking a value from 1 to 10, i.e. the binary encoding leads us to $n = 36$.

In both cases the sample data provide only a tiny fraction of the theoretically possible number of function values. A priori, we can argue that not all combinations of binary inputs are feasible (or even a small fraction only has indeed a valid interpretation). We take this into account by simply encoding (enumerating) the sample data instead of using the variable number n as the input to the core of the classification circuit: We apply a methodology which is well-known from the synthesis of partially defined Boolean functions [22] in order to obtain some rough estimations of the size of circuits representing the sample data. We denote by $s_L^P := \mid \mathcal{S}_L \mid$ the size of the training data set, i.e. we have $s_L^{\text{splice}} = 2127$ and $s_L^{\text{wbcd}} = 455$. By n_P we denote the number of binary variables calculated from the number of input attributes and the range of values each attribute can take, i.e. $n_{\text{splice}} = 120$ and $n_{\text{wbcd}} = 36$. The s_L^P training data can be enumerated by using $n_L^P := \lceil \log s_L^P \rceil$ binary variables, i.e. we have $n_L^{\text{splice}} = 12$ and $n_L^{\text{wbcd}} = 9$.

For a given problem P, we introduce the classification circuit $\mathcal{C}_P$: The circuit is built from threshold functions of unbounded fan-in (as basic gates; cf. [23,25]) and has minimum size with respect to the gate number $\mathcal{S}(\mathcal{C}_P)$. The number of input nodes is n_P. We now try to approximate $\mathcal{C}_P$ by a composition of two circuits

$$\widehat{\mathcal{C}_P} = \mathcal{C}[n_P \rightarrow n_L^P] + \mathcal{C}[n_L^P], \tag{1}$$

where $\mathcal{C}[n_P \rightarrow n_L^P]$ is a multi-output circuit that calculates the encoding of elements from $\mathcal{S}_L$. The encoding then becomes the binary input to $\mathcal{C}[n_L^P]$. Based on the results about local encoding [22] we assume

$$\mathcal{S}\big(\mathcal{C}[n_P \rightarrow n_L^P]\big) \; < \; \mathcal{S}\big(\mathcal{C}[n_L^P]\big) \;\; \text{and therefore} \;\; \mathcal{S}\big(\widehat{\mathcal{C}_P}\big) \; < \; 2 \cdot \mathcal{S}\big(\mathcal{C}[n_L^P]\big). \tag{2}$$

Actually, $\mathcal{S}(\mathcal{C}[n_P \rightarrow n_L^P])$ depends strongly on the distribution of sample data within the whole domain of feasible samples of P and (2) is valid for sufficiently large n^P and complex P only. Nevertheless, we will focus on $\mathcal{S}(\mathcal{C}[n_L^P])$ alone which seems to be justified by our experimental observation that indeed the chosen problems are associated with complex functions f_P.

To estimate $\mathcal{S}(\mathcal{C}[n_L^P])$, we use the asymptotically optimal design of Boolean functions by liner threshold functions as presented in [23]:

$$\mathcal{S}(f_n) \leq 2 \cdot \sqrt{\frac{2^n}{n}} \cdot \left(1 + \underline{o}(\sqrt{2^n/n})\right), \tag{3}$$

where $f_n = f(x_1, ..., x_n)$ is an arbitrary Boolean function. Moreover, almost all Boolean functions f_n asymptotically require $2 \cdot \sqrt{2^n/n}$ linear threshold gates for their representation [23] (i.e. almost all functions are as complex as the most complex functions).

We computed classification results for depth-two circuits C. The remaining parameter settings are $k = 20,000$, and $\Phi = p/3$. The size p is determined by preliminary experiments on depth-two circuits for $N_{\text{per}} = 50$. The experiments lead us to the following settings: $p_{\text{splice}} := \mid \mathcal{S}_{\text{L}} \mid /4$ for all three classes and $p_{\text{wbcd}} := |\mathcal{S}_{\text{L}}|/6$.

Table 1. Error Rates on Test Sets for Networks of Size N_{per}

N_{per}	Splice-Junction IE	Splice-Junction EI	Splice-Junction Neither	WBCD
30	4.1%	3.4%	6.3%	1.1%
48	4.0%	3.3%	6.0%	0.9%
66	4.0%	3.3%	6.0%	0.9%
80	4.0%	3.3%	6.0%	1.2%
84	4.0%	3.3%	6.0%	1.2%
130	3.9%	3.3%	6.0%	1.1%
154	3.9%	3.3%	6.0%	1.1%
180	3.8%	3.2%	5.9%	1.1%
252	3.8%	3.1%	5.9%	1.0%
300	3.8%	3.1%	5.9%	1.0%

Let R^P denote the maximum of the three best (not necessarily different) classification rates recorded in Table 1 for problem P and circuits $\mathcal{C}$. We set:

$$\mathcal{S}_{\min}^P := \min_{R=R^P} \mathcal{S}(\mathcal{C}^P[e = R]), \tag{4}$$

where $e = R$ means an error rate of R by the given circuit. We now allow a margin of deviation from R^P by Δ. As an estimation of the circuit size that ensures a high classification rate we take

$$\mathcal{S}^P := \max\{\mathcal{S}(\mathcal{C}^P[e = R]) : (R^P < R \leq R^P + \Delta) \& (\mathcal{S}(\mathcal{C}^P[e = R]) \leq \mathcal{S}_{\min}^P)\}. \tag{5}$$

If the max-operation is over an empty set, we take $\mathcal{S}^P := \mathcal{S}^P_{\min}$. Taking the max-operation in (5) gives some confidence that the error rates have already stabilised. The calculation of the corresponding values for the four classification problems is summarised in Table 2. From Table 1 and Table 2 we conclude that

Table 2. Circuit size estimates compared to $\mathcal{S}^P_{\text{pred}}$

Circuit size estimates	Splice-Junction IE	EI	Neither	WBCD
$\mathcal{S}_P$	154	154	154	180
$\mathcal{S}^P_{\text{pred}}$	147	147	147	60

for P = WBCD the differences in the classification rate are very small. For the Splice-junction data, where the size of sample sets is much larger, we observe an improvement and then stabilisation of error rates with increased circuit size.

We compare (see Table 2) the values of $\mathcal{S}_P$ to $\mathcal{S}^P_{\text{pred}} := x \cdot 2 \cdot (2^{n^P_L}/n^P_L)^{1/2}$; cf. (3), where $x = 4$ is chosen on the following grounds: The RHS of (2) doubles the complexity $\mathcal{S}(\widehat{\mathcal{C}_P})$, and we assume that the third factor on the RHS of (3), which summarises the complexity of auxiliary sub-circuits, at most doubles the product of the first two factors for relatively small values of $(2^{n^P_L}/n^P_L)^{1/2}$. Thus, we provide empirical evidence that

$$8 \cdot (2^{n^{cp}_L}/n^{cp}_L)^{1/2} \tag{6}$$

is an appropriate upper bound for the size of a neural network in order to achieve a high generalisation capability of the network.

We note that the simple rules from (4) and (5) generate the same $\mathcal{S}_P$ for $P \in \{\text{IE, EI, Neither}\}$. Furthermore, if P = WBCD and $N_{\text{per}} = 48$ or $N_{\text{per}} = 66$, the error rate in Table 1 is only 0.9%, i.e. $\mathcal{S}^{wbcd}_{\text{pred}} = 60$ from Table 3 seems to be a good estimation.

In the context of machine learning, the RHS of (3) provides an estimation of the size of elements of the hypotheses space (circuits of threshold functions) that represent particular objects (concepts). The method to prove the lower bound for (3) was utilised in [24,25] (cf. also [4]) to obtain a lower bound for a sufficient number of samples such that the error rate on test samples is below ε with probability at least $(1 - \delta)$. The lower bound is expressed in terms of the VC-dimensions of neural nets, which basically equals the number of neurons (threshold gates); see also [26]. The lower bound is of the type

$$\max\{4 \cdot n \cdot (n+k)^2/\varepsilon \cdot \log(13/\varepsilon)\,;\, 4/\varepsilon \cdot \log(2/\delta)\}, \tag{7}$$

where k is the number of gates in neural nets (which represent the hypotheses). We note that $k = \mathcal{S}^P_{\text{pred}}$ (or even $k = 2 \cdot (2^{n^P_L}/n^P_L)^{1/2}$) and ε in the range of values from Table 1 would result in a number of examples much larger than the number of samples available from our datasets, which is relatively independent of δ.

3 Evaluation of the Estimated Network Size on Additional Datasets

We will evaluate in this Section the proposed upper bound on more datasets from the UCI Repository.

Hayes-Roth Datasets (Hayes): The datasets has three classes (named hayes-1, hayes-2 and hayes-3 in the rest of the text), which consist of five numerical attributes. The dataset in UCI Repository consists of a training file of 132 samples and a test file of 28 samples. We merged the two file and used 2/3 of data for training and 1/3 for testing, therefore we reduced the number of training samples and increased the number of test samples, preserving however in our testset the 28 original test samples.

Iris Plant Datasets (Iris): The set consists of 3 classes of 50 instances each, where each class refers to a type of iris plant. One class is linearly separable from the other two; the latter are not linearly separable from each other. There are four attributes containing measurements in cm of sepal length, sepal width, petal length, and petal width, and the task is to identify the three classes: Iris Setosa (linearly seperable), Iris Versicolour, and Iris Virginica. We denote the three classification problems, as iris-1, iris-2 and iris-3.

The Monk's Problems (Monk): The Monk-1 and Monk-2 datasets, used in this research, have a target concept to be identified. Monk-1 consists of 556 samples, monk-2 consists of 601 samples, and both datasets, if we omit the "id" attribute, which is unique for each sample, have 6 attributes.

US Congressional Voting Records Database (Votes): The US Congressional Voting Records (Votes) data set includes votes for each of the U.S. House of Representatives Congressmen on 16 key issues (attributes) identified by the Congressional Quarterly Almanaco (CGA) in 1984. The dataset consists of 435 samples of two classes (267 democrats, 168 republicans). The dataset consists of 16 features, which are key issues for congressman's votes for which Boolean data (as yes, no votes) exist in voting records.

Waveform Datasets (Wave): Waveform is a three class dataset (waveform 1, waveform 2, waveform 3) where each class is a "wave" generated from a combination of 2 of 3 "base" waves. The dataset contains 5,000 samples, with 21 attributes with continuous values from 0 to 6. We used in our research these datasets with their original numerical values.

We will apply the formula to the above described collection of additional datasets. Table 3 shows the estimated circuit size for domains with $|D|$ data, where the available training data are $|S_L| = 2/3 \cdot |D|$. 'LS' indicates a zero error classification for the dataset, and we consider the set as lineary seperable. Therefore such datasets will be excluded from further investigation. From preliminary experiments for $N = 50$ units, $p = |S_L|/2$, $G = |S_L|$, and K=25,000 we have fine-tuned for each dataset the rest of parameters G, p and K. With respect to parameters G, p and K, we compare according to Equation (6) the classification

Table 3. Datasets and Predicted Network Size

Domain	$\|D\|$	$\|S_L\|$	n_L^{cp}	S_{pred}
Hayes-Roth 1	160	106	7	35
Hayes-Roth 2	160	106	7	35
Hayes-Roth 3	160	106	7	LS
Iris 1	150	100	7	LS
Iris 2	150	100	7	35
Iris 3	150	100	7	35
Monks 1	556	370	9	60
Monks 2	601	401	9	60
Votes	435	290	9	60
Waveform 1	5,000	3,333	12	147
Waveform 2	5,000	3,333	12	147
Waveform 3	5,000	3,333	12	147

error from the predicted size of neural network $N = |S|$ from Table 3 to half of circuit sizes $N = |S|/2$ and double circuit sizes $N = 2 \cdot |S|$ from the predicted size. Comparison results are shown in Table 4.

Table 4. Classification Errors for $|S|$, $|S|/2$, $2 \cdot |S|$

Domain	e_T^2 for $\|S\|$	e_T^2 for $\|S\|/2$	e_T^2 for $2 \cdot \|S\|$
Hayes-Roth 1	11.1% ±1.0%	11.9% ±2.1%	11.1 ±1.0%
Hayes-Roth 2	11.5% ±2.0%	11.9% ±2.6%	10.4 ±1.7%
Iris 2	0.7% ±0.1%	0.7% ±0.1%	0.7 ±0.1%
Iris 3	0.7% ±0.1%	0.7% ±0.1%	0.7 ±0.1%
Monks 1	22.7% ±1.3%	23.1% ±2.4%	23.1 ±1.1%
Monks 2	34.2% ±2.6%	34.7% ±3.8%	37.3 ±4.7%
Votes	3.2% ±0.4%	3.4% ±0.1%	3.3 ±0.1%
Waveform 1	14.7% ±0.1%	14.7% ±0.1%	14.8 ±0.1%
Waveform 2	11.7% ±0.2%	11.7% ±0.2%	11.8 ±0.1%
Waveform 3	11.4% ±0.2%	11.6% ±0.2%	11.4 ±0.1%

The comparison shows that our upper bound $8 \cdot (2^{n_L^{cp}}/n_L^{cp})^{1/2})$ for estimating the circuit size does indeed imply the highest classification rates. Except for the 'Hayes-Roth 2' dataset, where a 10.4 classification rate is obtained for double circuit size, the highest classification rate is obtained for N according to (6).

4 Conclusions

Minimising the error of missclassified samples for a network is an NP-hard problem for non-linearly separable problems and only approximations of the optimal weight vector can be found. Relating classifier parameters to problem-dependent attributes allows us to *a priori* estimate values for these parameters and employ methods that reduces significantly the number of required experiments to achieve high classification rates. Our experimental results for parameter setting shows that the VC-dimension bound from equation (7) for a sufficient number of samples seems to be too large, or in other words we obtained, following parameter settings, error rates ε having datasets with considerably much smaller number of available training samples than the sufficient number of samples that equation (7) suggests.

To propose an upper bound for the size of threshold circuits is important for the following reasons: Firstly, it allows us to *a piori* set one of the difficult problem dependent parameters, and hence we can focus on the rest of problem dependent parameters in a complex classification problem. Secondly, it is important for parallel processing theory in terms of resources required to approximate best classification rates. Thirdly, it shows that the number of nodes in neural networks depends on the available number of training examples, and hence it suggests that a constant size neural network is not appropriate for most of the classification problems.

References

1. Albrecht, A., Lappas, G., Vinterbo, S.A., Wong, C.K., Ohno-Machado, L.: Two applications of the LSA machine. In: ICONIP'02. Proc. 9th International Conference on Neural Information Processing, Singapore, pp. 184–189 (2002)
2. Albrecht, A., Wong, C.K.: Combining the perceptron algorithm with logarithmic simulated annealing. Neural Processing Letters 14, 75–83 (2001)
3. Allender, E.: The permanent requires large uniform threshold circuits. In: Cai, J.-Y., Wong, C.K. (eds.) COCOON 1996. LNCS, vol. 1090, pp. 127–135. Springer, Heidelberg (1996)
4. Anthony, M.: Boolean Functions and Artificial Neural Networks. CDAM Research Report LSE-CDAM-2003-01, Department of Mathematics and Centre for Discrete and Applicable Mathematics, The London School of Economics and Political Science, London, UK (2003)
5. Anthony, M., Bartlett, P.L.: Neural Network Learning. Cambridge University Press, UK (1999)
6. Bartlett, P.L.: The Sample Complexity of Pattern Classification with Neural Networks: the Size of the Weights is More Important Than the Size of the Network. IEEE Transactions on Information Theory 44(2), 525–536 (1998)
7. Bartlett, P.L., Maiorov, V., Meir, R.: Almost Linear VC-dimension Bounds for Piecewise Polynomial Networks. Neural Computation 10, 2159–2173 (1998)
8. Baum, E.B.: On the Capabilities of Multilayer Perceptrons. Journal of Complexity 4, 193–215 (1988)
9. Baum, E.B., Haussler, D.: What Size Net Gives Valid Generalization? Neural Computation 1(1), 151–160 (1989)
10. Beiu, V.: Digital Integrated Circuit Implementations. In: Fiesler, E., Beale, R. (eds.) Handbook on Neural Computation, Oxford University Press, Oxford (1997)
11. Blum, A., Rivest, R.L.: Training a 3-Node Neural Network is NP-Complete. Neural Networks 5, 117–127 (1992)
12. Caussinus, H., McKenzie, P., Thérien, D., Vollmer, H.: Nondeterministic NC^1 computation. J. Comput. System Sci. 57, 200–212 (1998)
13. Duda, R.O., Hart, P.E., Stork, D.G.: Pattern Classification. Wiley-Interscience, New York (2001)

14. Goldberg, P.W., Jerrum, M.R.: Bounding the Vapnik-Chervonenkis Dimension of Concept Classes Parametrised by Real Numbers. Machine Learning 18(2/3), 131–148 (1995)
15. Hajek, B.: Cooling schedules for optimal annealing. Mathem. Operat. Res 13, 311–329 (1988)
16. Haussler, D., Kearns, M., Schapire, R.: Bounds on the Sample Complexity of Bayesian Learning Usng Information Theory and the VC Dimension. Machine Learning 14, 83–113 (1994)
17. Höffgen, K.-U., Simon, H.-U.: Robust Trainability of Single Neurons. In: Proc. 5th Annual ACM Workshop on Computational Learning Theory, pp. 428–439. ACM Press, New York (1992)
18. Karpinski, M., Macintyre, A.J.: Polynomial Bounds for VC Dimension of Sigmoidal and General Pfaffian Neural Networks. Journal of Computer and System Sciences 54(1), 169–176 (1997)
19. Koiran, P., Sontag, E.D.: Neural Networks with Quadratic VC Dimension. Journal of Computer and System Sciences 54(1), 190–198 (1997)
20. Kolmogorov, A.N.: On the Representation of Continuous Functions of Several Variables by Superposition of Continuous Functions of one Variable and Addition. Doklady Akademiia Nauk 114(5), 953–956 (1957)
21. Lappas, G., Frank, R.J., Albrecht, A.: A Computational Study on Circuit Size vs. Circuit Depth. International Journal on Artificial Intelligence Tools 15(2), 143–162 (2006)
22. Lupanov, O.B.: On a Method to Design Control Systems - The Local Encoding Approach (in Russian). Problemy Kibernetiki 14, 31–110 (1965)
23. Lupanov, O.B.: On the design of circuits by threshold elements (in Russian). Problemy Kibernetiki 26, 109–140 (1973)
24. Maass, W.: Bounds on the computational power and learning complexity of analog neural nets. In: Proc. 25th Annual ACM Symp. on the Theory of Computing, pp. 335–344. ACM Press, New York (1993)
25. Maass, W.: On the complexity of learning on neural nets. In: Maass, W. (ed.) EuroColt'93. Proc. Computational Learning Theory, pp. 1–17. Oxford University Press, Oxford (1994)
26. Maass, W., Legenstein, R.A., Bertschinger, N.: Methods for estimating the computational power and generalization capability of neural microcircuits. In: Proc. Advances in Neural Information Processing Systems (2005)
27. Razborov, A., Wigderson, A.: $n^{\Omega(logn)}$ Lower Bounds on the Size of Depth 3 Threshold Circuits with AND Gates at the Bottom. In: Inf. Proc. Letters, 45th edn., pp. 303–307 (1993)
28. Rosenblatt, F.: Principles of Neurodynamics. Spartan Books, New York (1962)
29. Sontag, E.D.: VC-dimension of Neural Networks. In: Bishop, C.M. (ed.) Neural Networks and Machine Learning, pp. 69–95. Springer Verlag, Berlin (1998)
30. Vapnik, V., Chervonenkis, A.Y.: On the Uniform Convergence of Relative Frequencies of Events to their Probabilities. Theory of Probability and Its Applications 16(2), 264–280 (1971)
31. Vollmer, H.: Some Aspects of the Computational Power of Boolean Circuits of Small Depth. University Würzburg, Habilitationsschrift (1999)
32. Wenocur, R.S., Dudley, R.M.: Some Special Vapnik-chervonenkis Classes. Discrete Mathematics 33, 313–318 (1981)
33. Wolpert, D.H., Macready, W.G.: No Free Lunch Theorems for Optimization. IEEE Trans. on Evolutionary Computation 1, 67–82 (1997)
34. Yannakakis, M.: Computational Complexity. In: Aarts, E.H.L., Lenstra, J.K. (eds.) Local Search in Combinatorial Optimization, Wiley & Sons, Chichester (1998)

A Maximum Weighted Likelihood Approach to Simultaneous Model Selection and Feature Weighting in Gaussian Mixture

Yiu-ming Cheung and Hong Zeng

Department of Computer Science, Hong Kong Baptist University, Hong Kong
{ymc,hzeng}@comp.hkbu.edu.hk

Abstract. This paper is to identify the clustering structure and the relevant features automatically and simultaneously in the context of Gaussian mixture model. We perform this task by introducing two sets of weight functions under the recently proposed *Maximum Weighted Likelihood* (MWL) learning framework. One set is to reward the significance of each component in the mixture, and the other one is to discriminate the relevance of each feature to the cluster structure. The experiments on both the synthetic and real-world data show the efficacy of the proposed algorithm.

1 Introduction

The finite mixture model has provided a formal approach to address the clustering problems. In this unsupervised domain, there are two key issues. One is the determination of an appropriate number of components (also called *number of clusters* or *model order* interchangeably) in a mixture model. In general, the true clustering structure may not be well described with too few components, whereas the estimated model may "over-fit" the data if it uses too many components. The other issue is how to identify the relevance of observation features with respect to the clustering structure. From the practical viewpoint, some features may not be so important, or even be irrelevant, to the clustering structure. Their participation in the clustering process will prevent a clustering algorithm from finding an appropriate partition. Hence, it is necessary to discriminate the relevance of each feature with respect to the clustering structure in the clustering analysis.

Clearly, the above two issues are closely related. However, most of the existing approaches deal with these two issues separately or sequentially. Some methods typically choose the most influential features prior to a clustering algorithm, e.g., see [1,2]. Although the success of their algorithms has been demonstrated in their application domains, these pre-selected features may not be necessarily suitable to the clustering algorithm that will be ultimately employed. Moreover, some approaches [3,4] wrap the clustering algorithms in an outer layer to evaluate the candidate feature subsets. The performance of such a method is superior to that of the previous approaches, but their search strategies of generating the feature

J. Marques de Sá et al. (Eds.): ICANN 2007, Part I, LNCS 4668, pp. 78–87, 2007.

subset candidates are prone to find a local maxima. Recently, it has been believed that the above two issues should be jointly optimized in a single learning paradigm. A typical example is the work of [5], which introduces the concept of *feature saliency* to measure the relevance of each feature to the clustering structure, as the *feature weight*. It then heuristically integrates the *Minimum Message Length* (MML) criterion into the likelihood, and optimizes this penalized likelihood using a modified Expectation-Maximization (EM) algorithm to obtain the feature weights and the clustering results. Nevertheless, the penalty terms given by the MML criterion are static and fixed for all components at each EM iteration. As a result, they may not be robust enough under a certain environment, in which it is more desirable to implement a dynamic and embedded scheme to control the model complexity.

In this paper, we propose such an approach to Gaussian mixture clustering by formulating the above two issues into a single *Maximum Weighted Likelihood* (MWL) optimization function [6]. We introduce two sets of weight functions to address these two issues, respectively. Consequently, both the model selection and feature weighting are performed automatically and simultaneously. The experiments on both the synthetic and real-world data have shown the efficacy of the proposed algorithm.

2 The MWL Learning Framework

Suppose N i.i.d. observations, denoted as $\mathbf{x_1}, \mathbf{x_2}, \ldots, \mathbf{x_N}$, come from the following mixture model:

$$p(\mathbf{x}_t|\Theta^*) = \sum_{j=1}^{k^*} \alpha_j^* p(\mathbf{x}_t|\theta_j^*) \tag{1}$$

with

$$\sum_{j=1}^{k^*} \alpha_j^* = 1 \quad \text{and} \quad \forall 1 \leq j \leq k^*, \quad \alpha_j^* > 0,$$

where each observation $\mathbf{x_t}$ $(1 \leq t \leq N)$ is a column vector of d-dimensional features, i.e. $\mathbf{x_t} = [x_{1t}, \ldots, x_{dt}]^T$, and $\Theta^* = \{\alpha_j^*, \theta_j^*\}_{j=1}^{k^*}$. Furthermore, θ_j^* denotes the parameter set of the jth probability density function (pdf) $p(\mathbf{x}_t|\theta_j^*)$ in the mixture model, k^* is the true cluster number, and α_j^* is the mixing proportion of the jth component in the mixture. Θ^* is estimated from these N observations by:

$$\hat{\Theta}_{ML} = \arg\max_{\Theta}\{\log p(\mathbf{X_N}|\Theta)\}. \tag{2}$$

where $\mathbf{X_N} = \{\mathbf{x_1}, \mathbf{x_2}, \ldots, \mathbf{x_N}\}$, and $\hat{\Theta}_{ML} = \{\alpha_j, \theta_j\}_{j=1}^{k}$ is a maximum likelihood (ML) estimate of Θ^*.

When the number of components k is known, the ML estimate in (2) could be obtained by the Expectation-Maximization (EM) algorithm. However, from the practical viewpoint, it is difficult or even impossible to know the number of components in advance. Recently, a promising approach called *Rival Penalized Expectation-Maximization* (RPEM for short) [6], where the model order is determined automatically and simultaneously with the parameter estimation, has been developed under the MWL learning framework. The main idea of the MWL framework is to introduce unequal weights in general into the conventional maximum likelihood, which actually provides a new promising way for regularization so that the weighted likelihood does not increase monotonically over the candidate model complexity. Specifically, the weighted likelihood is given bellow:

$$Q(\Theta, \mathbf{X}_N) = \frac{1}{N}\sum_{t=1}^{N} \log p(\mathbf{x}_t|\Theta) = \frac{1}{N\zeta}\sum_{t=1}^{N}\sum_{j=1}^{k} g(j|\mathbf{x}_t, \Theta) \log p(\mathbf{x}_t|\Theta)$$
$$= \frac{1}{N\zeta}\sum_{t=1}^{N} \mathcal{M}(\Theta, \mathbf{x}_t) \tag{3}$$

$$\mathcal{M}(\Theta, \mathbf{x}_t) = \sum_{j=1}^{k} g(j|\mathbf{x}_t, \Theta) \log[\alpha_j p(\mathbf{x}_t|\theta_j)] - \sum_{j=1}^{k} g(j|\mathbf{x}_t, \Theta) \log h(j|\mathbf{x}_t, \Theta) \tag{4}$$

where $h(j|\mathbf{x}_t, \Theta) = \frac{\alpha_j p(\mathbf{x}_t|\theta_j)}{p(\mathbf{x}_t|\Theta)}$ is the posterior probability that $\mathbf{x}_t$ belongs to the jth component in the mixture, k is an estimate of k^* with $k \geq k^*$, and ζ is a constant. The $g(j|\mathbf{x}_t, \Theta)$'s are the weight functions, satisfying the constraints:

$$\forall t, j, \ \sum_{j=1}^{k} g(j|\mathbf{x}_t, \Theta) = \zeta; \ \lim_{h(j|\mathbf{x}_t,\Theta)\to 0} g(j|\mathbf{x}_t, \Theta) \log h(j|\mathbf{x}_t, \Theta) = 0.$$

In the RPEM algorithm of [6], the weight functions are constructed as:

$$g(j|\mathbf{x}_t, \Theta) = (1 + \varepsilon_t) I(j|\mathbf{x}_t, \Theta) - \varepsilon_t h(j|\mathbf{x}_t, \Theta) \tag{5}$$

with

$$I(j|\mathbf{x}, \Theta) = \begin{cases} 1 \ if \ j = c \equiv \arg\max_{1\leq i\leq k} h(i|\mathbf{x}, \Theta); \\ 0 \ otherwise. \end{cases} \tag{6}$$

where ε_t is a small positive quantity. Under this weight construction, given an observation $\mathbf{x}_t$, a positive weight $g(c|\mathbf{x}_t, \Theta)$ is assigned to the log-likelihood of the winning component, i.e., the component with the maximum value of $h(j|\mathbf{x}_t, \Theta)$, so that it is updated to adapt to $\mathbf{x}_t$, meanwhile all rival components are penalized with a negative weight. This intrinsic rival penalization mechanism of the RPEM makes the genuine clusters survive, whereas the "pseudo-clusters" gradually vanish. The updating details of $\hat{\Theta}_{MWL} = \arg\max_\Theta\{Q(\Theta, \mathbf{X}_N)\}$ can be found in [6].

The numerical results have shown its outstanding performance on both of synthetic and real-life data [6], where all features are equally useful in clustering process. Nevertheless, analogous to the most of the existing clustering algorithms, the performance of the RPEM may deteriorate provided that there exist some irrelevant features in feature vectors. In the following, we will therefore perform the feature relevancy analysis and further extend the RPEM accordingly within the MWL framework.

3 Simultaneous Clustering and Feature Weighting

To discriminate the importance of each feature in the cluster structure, we utilize the concept of *feature saliency* defined in [5] as our feature weight ($i.e., w_l, 0 \leq w_l \leq 1, \forall 1 \leq l \leq d$): the lth feature is relevant with a probability w_l that the feature's pdf is dependent of the pdf's of components in the mixture. For those features whose values are distributed among all clusters, we regard its distribution as a common one. We suppose the features are independent of each other, then the pdf of a more general mixture model can be written below as in [5]:

$$\begin{aligned} p(\mathbf{x}_t|\Theta) &= \sum_{j=1}^{k} \alpha_j \prod_{l=1}^{d} p(x_{lt}|\Phi) \\ &= \sum_{j=1}^{k} \alpha_j \prod_{l=1}^{d} [w_l p(x_{lt}|\theta_{lj}) + (1-w_l) q(x_{lt}|\lambda_l)] \end{aligned} \tag{7}$$

where $p(x_{lt}|\theta_{lj}) = \mathcal{N}(x_{l,t}|m_{lj}, s_{lj}^2)$ denotes a Gaussian density function of the relevant feature x_{lt} with the mean m_{lj}, and the variance s_{lj}^2; $q(x_{lt}|\lambda_l)$ is the common distribution of the irrelevant feature. In this paper, we shall limit it to be a Gaussian as well for a general purpose, i.e., $q(x_{lt}|\lambda_l) = \mathcal{N}(x_{lt}|cm_l, cs_l^2)$. Subsequently, the full parameter set of the general Gaussian mixture model is redefined as $\Theta = \{\{\alpha_j\}_{j=1}^k, \Phi\}$ and $\Phi = \{\{\theta_{lj}\}_{l=1,j=1}^{d,k}, \{w_l\}_{l=1}^d, \{\lambda_l\}_{l=1}^d\}$. Note that

$$p(x_{lt}|\Phi) = w_l p(x_{lt}|\theta_{lj}) + (1-w_l) q(x_{lt}|\lambda_l) \tag{8}$$

is a linear mixture of two possible densities for each feature, and the feature weight w_l acts as a regulator to determine which distribution is more appropriate to describe the feature. A new perspective is to regard this form as a *lower level* Gaussian mixture, which resembles the *higher level* Gaussian mixture on which the weights of the genuine clusters are estimated. Hence, the feature weight w_l can be considered as the counterpart of component weight α_j. Subsequently, a similar rewarding and penalizing scheme can be embedded into the likelihood function of (3) for this *lower level* mixture. To this end, we re-write (3) as:

$$\tilde{Q}(\Theta, \mathbf{X}_N) = \frac{1}{N} \sum_{t=1}^{N} \log p(\mathbf{x}_t|\Theta) = \frac{1}{N\zeta} \sum_{t=1}^{N} \tilde{\mathcal{M}}(\Theta, \mathbf{x}_t). \tag{9}$$

To control the complexity of the model to be estimated, we introduce two sets of weight functions, i.e. $\tilde{g}(.|\mathbf{x}_t, \Theta)$ and $\tilde{f}(.|x_{lt}, \Phi)$, into the log-likelihood for the components in the *higher level* and *lower level* mixtures, respectively. Altogether, by inserting the following formulas:

$$p(\mathbf{x}_t|\Theta) = \frac{\alpha_j p(\mathbf{x}_t|\Phi)}{\tilde{h}(j|\mathbf{x}_t, \Theta)};\ p(x_{lt}|\Phi) = \frac{w_l p(x_{lt}|\theta_{lj})}{h'(1|x_{lt}, \Phi)} = \frac{(1-w_l) q(x_{lt}|\lambda_l)}{h'(0|x_{lt}, \Phi)},$$

into (9), we obtain the weighted log-likelihood for the mixture model as follows:

$$\tilde{\mathcal{M}}(\Theta, \mathbf{x}_t) = \sum_{j=1}^{k} \tilde{g}(j|\mathbf{x}_t, \Theta) \log \alpha_j +$$
$$\sum_{j=1}^{k}\sum_{l=1}^{d} \tilde{g}(j|\mathbf{x}_t, \Theta)\Big\{ \tilde{f}(1|x_{lt}, \Phi) \log[w_l p(x_{lt}|\theta_{lj})] + \tilde{f}(0|x_{lt}, \Phi) \log[(1-w_l) q(x_{lt}|\lambda_l)] \Big\} -$$
$$\sum_{j=1}^{k}\sum_{l=1}^{d} \tilde{g}(j|\mathbf{x}_t, \Theta) \tilde{f}(1|x_{lt}, \Phi) \log h^{'}(1|x_{lt}, \Phi) - \sum_{j=1}^{k}\sum_{l=1}^{d} \tilde{g}(j|\mathbf{x}_t, \Theta) \tilde{f}(0|x_{lt}, \Phi) \log h^{'}(0|x_{lt}, \Phi)$$
$$- \sum_{j=1}^{k} \tilde{g}(j|\mathbf{x}_t, \Theta) \log \tilde{h}(j|\mathbf{x}_t, \Theta) \quad (10)$$

where

$$\tilde{h}(j|\mathbf{x}_t, \Theta) = \frac{\alpha_j p(\mathbf{x}_t|\Phi)}{p(\mathbf{x}_t|\Theta)} = \frac{\alpha_j \prod_{l=1}^{d} [w_l p(x_{lt}|\theta_{lj}) + (1-w_l) q(x_{lt}|\lambda_l)]}{\sum_{i=1}^{k} \alpha_i \prod_{l=1}^{d} [w_l p(x_{lt}|\theta_{li}) + (1-w_l) q(x_{lt}|\lambda_l)]},$$

$$h^{'}(1|x_{lt}, \Phi) = \frac{w_l p(x_{lt}|\theta_{lj})}{w_l p(x_{lt}|\theta_{lj}) + (1-w_l) q(x_{lt}|\lambda_l)}, \quad h^{'}(0|x_{lt}, \Phi) = 1 - h^{'}(1|x_{lt}, \Phi).$$

$\tilde{h}(j|\mathbf{x}_t, \Theta)$ indicates the probability that some features in the data points come from the jth density component in the subspace. $h^{'}(1|x_{lt}, \Phi)$ represents the posterior probability that the lth feature conforms to the mixture model. That is, it reflects the prediction for the relevance of the lth feature to the clustering structure.

We design the weight functions for the *higher level* mixture as follows:

$$\tilde{g}(j|\mathbf{x}_t, \Theta^{old}) = I(j|\mathbf{x}_t, \Theta) + \tilde{h}(j|\mathbf{x}_t, \Theta), j = 1, \ldots, k_{max}, \quad (11)$$

where the $I(j|\mathbf{x}_t, \Theta)$ is the indicator function defined in (6). It is clear that this form meets the requirements of weight functions under the MWL framework. The rationale behind this form is that we give an award to the winning component, i.e., the cth one, by assigning a weight whose value is larger than the corresponding $\tilde{h}(c|\mathbf{x}_t, \Theta)$. In contrast, we keep the weights of those rival components exactly equal to their corresponding $\tilde{h}(j|\mathbf{x}_t, \Theta)$'s. That is, we give the winning component an award, but the rival ones not. Hence, this is actually another kind of award-penalization scheme. Such a scheme is able to make the genuine components survive in the learning process, whereas those "pseudo-clusters" will be faded out from the mixture gradually.

In (10), the new weight functions $\{\tilde{f}(1|x_{lt}, \Phi), \tilde{f}(0|x_{lt}, \Phi)\}$ should satisfy the following constraint:

$$\lim_{h^{'}(i|x_{lt}, \Phi) \to 0} \tilde{f}(i|\mathbf{x}_t, \Theta) \log h^{'}(i|x_{lt}, \Phi) = 0, i \in \{0, 1\}.$$

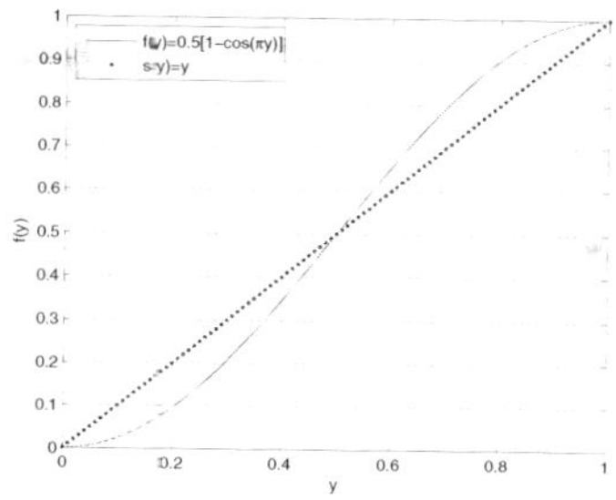

Fig. 1. f(y) vs. s(y). f(y) is plotted with "-"; s(y) is plotted with ".";

Accordingly, an applicable form is presented below:

$$\tilde{f}(1|x_{lt}, \Phi) = f(h^{'}(1|x_{lt}, \Phi)), \tilde{f}(0|x_{lt}, \Phi) = 1 - f(h^{'}(1|x_{lt}, \Phi)),$$

with

$$f(y) = 0.5[1 - \cos(\pi y)], y \in [0, 1]. \tag{12}$$

$f(y)$ is plotted in Fig. 1. An interesting property of $f(y)$ can be observed from Fig. 1:

$$y > f(y) > 0, \ for\ 0 < y < 0.5; \ y < f(y) < 1, \ for\ 0.5 < y < 1.$$

Hence, if $h^{'}(1|x_{lt}, \Phi) > h^{'}(0|x_{lt}, \Phi)$, it implies that the feature seems useful in the clustering. Subsequently, its log-likelihood $\log[w_l p(x_{lt}|\theta_{lj})]$ is amplified by a larger coefficient $\tilde{f}(1|x_{lt}, \Phi)$, meanwhile the log-likelihood of the "common" distribution is suppressed by a smaller coefficient $\tilde{f}(0|x_{lt}, \Phi)$. A reverse assigning scheme is also held for the case when the feature seems less useful. In this sense, the function $f(y)$ rewards the important features that make the greater contribution to the likelihood, and penalizes those of insignificant features. Consequently, the estimation for the whole parameter set is given by:

$$\hat{\Theta}_{MWL} = \arg\max_{\Theta}\{\tilde{Q}(\Theta, \mathbf{X}_N)\}. \tag{13}$$

An EM-like iterative updating by gradient ascent technique is used to estimate the parameter set. Algorithm 1 shows the pseudo-code description of the proposed algorithm. In implementation of this algorithm, $\{\alpha_j\}_{j=1}^k$s must satisfy the constraint: $\sum_{j=1}^k \alpha_j = 1$ and $0 \le \alpha_j < 1$. Hence, we also update $\{\beta_j\}_{j=1}^k$s instead of $\{\alpha_j\}_{j=1}^k$ as shown in [6], and $\{\alpha_j\}_{j=1}^k$ are obtained by: $\alpha_j = e^{\beta_j} / \sum_{i=1}^k e^{\beta_i}, 1 \le j \le k$. As for another parameter w_l with the constraint: $0 \le w_l \le 1$, we update γ_l instead of w_l, and w_l is obtained by the sigmoid function of γ_l: $w_l = \frac{1}{1+e^{-\tau \cdot \gamma_l}}, 1 \le l \le d$. The constant τ ($\tau > 0$) is used to tune the shape of the sigmoid function. To implement a moderate sensitivity for w_l, i.e., an approximately equal increasing or decreasing quantity in both γ_l and w_l, the slope of the sigmoid function around 0 with respect to (w.r.t.) γ_l (namely, around 0.5 w.r.t. w_l) should be roughly equal to 1. By rule of thumb,

Algorithm 1. The complete algorithm.

input : $\{\mathbf{x}_1, \mathbf{x}_2, \ldots, \mathbf{x}_N\}$, k_{max}, η, $epoch_{max}$, *initial* $\mathbf{\Theta}$
output: *The converged* $\hat{\mathbf{\Theta}}$

count $\leftarrow 0$;
while $count \leq epoch_{max}$ **do**
  for $t \leftarrow 1$ **to** N **do**
    step 1: *Calculate* $h^{'}$, $\tilde{h}$ *to obtain* $\tilde{f}$ *and* $\tilde{g}$;
    step 2:

$$\hat{\Theta}^{new} = \hat{\Theta}^{old} + \Delta\Theta = \hat{\Theta}^{old} + \eta \left.\frac{\partial \tilde{\mathcal{M}}(\mathbf{x}_t;\Theta)}{\partial \Theta}\right|_{\hat{\Theta}^{old}}$$

  end
  $count \leftarrow count + 1$;
end

we find that a linear fitting of the two curves: $w_l = \frac{1}{1+e^{-\tau \cdot \gamma_l}}$ and $w_l = \gamma_l + 0.5$ around 0 w.r.t γ_l occurs around $\tau = 4.5$. In the following, we will therefore set τ at 4.5.

For each observation $\mathbf{x}_t$, the changes in the parameters set are calculated as:

$$\Delta\beta_j = \eta_\beta \left.\frac{\partial \tilde{\mathcal{M}}(\mathbf{x}_t;\Theta)}{\partial \beta_j}\right|_{\Theta^{old}} = \eta_\beta \sum_{i=1}^{k_{max}} \frac{\partial \tilde{\mathcal{M}}(\mathbf{x}_t;\Theta)}{\partial \alpha_i} \cdot \left.\frac{\partial \alpha_i}{\partial \beta_j}\right|_{\Theta^{old}}$$
$$= \eta_\beta(\tilde{g}(j|\mathbf{x}_t,\Theta) - \alpha_j^{old}), \quad (14)$$

$$\Delta m_{lj} = \eta \left.\frac{\partial \tilde{\mathcal{M}}(\mathbf{x}_t;\Theta)}{\partial m_{lj}}\right|_{\Theta^{old}}$$
$$= \eta \tilde{g}(j|\mathbf{x}_t,\Theta)\tilde{f}(1|x_{lt},\Phi)\frac{x_{lt} - m_{lj}^{old}}{(s_{lj}^{old})^2}, \quad (15)$$

$$\Delta s_{lj} = \eta \left.\frac{\partial \tilde{\mathcal{M}}(\mathbf{x}_t;\Theta)}{\partial s_{lj}}\right|_{\Theta^{old}}$$
$$= \eta \tilde{g}(j|\mathbf{x}_t,\Theta)\tilde{f}(1|x_{lt},\Phi)\left[\frac{(x_{lt} - m_{lj}^{old})^2}{(s_{lj}^{old})^3} - \frac{1}{s_{lj}^{old}}\right], \quad (16)$$

$$\Delta cm_l = \eta \left.\frac{\partial \tilde{\mathcal{M}}(\mathbf{x}_t;\Theta)}{\partial cm_l}\right|_{\Theta^{old}}$$
$$= \eta \sum_{j=1}^{k_{max}} \tilde{g}(j|\mathbf{x}_t,\Theta)\tilde{f}(0|x_{lt},\Phi)\frac{x_{lt} - cm_l^{old}}{(cs_l^{old})^2}, \quad (17)$$

$$\Delta cs_l = \eta \left.\frac{\partial \tilde{\mathcal{M}}(\mathbf{x}_t;\Theta)}{\partial cs_l}\right|_{\Theta^{old}}$$

$$= \eta \sum_{j=1}^{k_{max}} \tilde{g}(j|\mathbf{x}_t, \Theta)\tilde{f}(0|x_{lt}, \Phi)\left[\frac{(x_{lt} - cm_l^{old})^2}{(cs_l^{old})^3} - \frac{1}{cs_l^{old}}\right], \tag{18}$$

$$\Delta\gamma_l = \eta \left.\frac{\partial\tilde{\mathcal{M}}(\mathbf{x}_t;\Theta)}{\partial\gamma_l}\right|_{\Theta^{old}} = \eta \left.\frac{\partial\tilde{\mathcal{M}}(\mathbf{x}_t;\Theta)}{\partial w_l} \cdot \frac{\partial w_l}{\partial\gamma_l}\right|_{\Theta^{old}}$$

$$= \eta \cdot \tau \cdot \sum_{j=1}^{k_{max}} \tilde{g}(j|\mathbf{x}_t, \Theta)\left[\tilde{f}(1|x_{lt}, \Phi)(1 - w_l^{old}) - \tilde{f}(0|x_{lt}, \Phi)w_l^{old}\right]. \tag{19}$$

Generally speaking, the learning rate of β_js should be chosen as $\eta_\beta > \eta$ to help eliminate the much smaller α_j (we suggest $\eta = 0.1\eta_\beta$).

4 Experimental Results

4.1 Synthetic Data

We generated 1,000 2-dimensional data points from a Gaussian mixture of three components:

$$0.3*\mathcal{N}\left[\begin{pmatrix}1.0\\1.0\end{pmatrix}, \begin{pmatrix}0.15 & 0\\0 & 0.15\end{pmatrix}\right] + 0.4*\mathcal{N}\left[\begin{pmatrix}1.0\\2.5\end{pmatrix}, \begin{pmatrix}0.15 & 0\\0 & 0.15\end{pmatrix}\right] + 0.3*\mathcal{N}\left[\begin{pmatrix}2.5\\2.5\end{pmatrix}, \begin{pmatrix}0.15 & 0\\0 & 0.15\end{pmatrix}\right].$$

In order to illustrate the ability of the proposed algorithm to perform automatic model selection and feature weighting jointly, we appended two additional features to the original set to yield a 4-dimensional one. The last two features are sampled independently from $\mathcal{N}(2; 2.5^2)$ as the Gaussian noise covering the entire data set. Apparently, the last two dimensions do not hold the same mixture structure, thus are not as significant as the first two dimensions in the partitioning process.

We initialized k_{max} to 15, and all β_j's and γ_l's to 0, which is equivalent to setting each α_j to 1/15 and w_l to 0.5, the remaining parameters were randomly initialized. The learning rates are $\eta = 10^{-5}$, $\eta_\beta = 10^{-4}$. After 500 epochs, the mixing coefficients and feature weights are obtained by the proposed algorithm:

$$\hat{\alpha}_6 = 0.4251\ \hat{\alpha}_8 = 0.2893\ \hat{\alpha}_{15} = 0.2856\ \hat{\alpha}_j = 0, j \neq 6, 8, 15$$
$$\hat{w}_1 = 0.9968\ \hat{w}_2 = 0.9964\ \hat{w}_3 = 0.0033\ \hat{w}_4 = 0.0036.$$

The feature weights of the first two dimensions converge close to 1, while those of the last two dimensions are assigned close to 0. It can be seen that the algorithm has accurately detected the underlying cluster structures in the first two dimensions, and meanwhile the appropriate model order and component parameters have been well estimated.

4.2 Real-World Data

We further investigated the performance of our proposed algorithm on serval benchmark databases [7] for data mining. The partitional accuracy of the algorithm without the prior knowledge of the underground class labels and the

Table 1. Results of the 30-fold runs on the test sets for each algorithm, where each data set has N data points with d features from k^* classes

Data Set	Method	Model Order $Mean \pm Std$	Error Rate $Mean \pm Std$
wine	RPEM	2.5 ± 0.7	0.0843 ± 0.0261
$d = 13$	EMFW	3.3 ± 1.4	0.0673 ± 0.0286
$N = 178$	NFW	2.8 ± 0.7	0.0955 ± 0.0186
$k^* = 3$	proposed method	3.3 ± 0.4	$\mathbf{0.0292 \pm 0.0145}$
heart	RPEM	1.7 ± 0.1	0.3167 ± 0.0526
$d = 13$	EMFW	2.5 ± 0.5	0.2958 ± 0.0936
$N = 270$	NFW	2.2 ± 0.4	0.2162 ± 0.0473
$k^* = 2$	proposed method	**fixed at 2**	$\mathbf{0.2042 \pm 0.0379}$
wdbc	RPEM	1.7 ± 0.4	0.2610 ± 0.0781
$d = 30$	EMFW	6.0 ± 0.7	0.0939 ± 0.0349
$N = 569$	NFW	3.0 ± 0.8	0.4871 ± 0.2312
$k^* = 2$	proposed method	2.6 ± 0.6	$\mathbf{0.0834 \pm 0.0386}$
ionosphere	RPEM	1.8 ± 0.5	0.4056 ± 0.0121
$d = 34$	EMFW	3.2 ± 0.6	0.1968 ± 0.0386
$N = 351$	NFW	2.2 ± 0.5	0.3201 ± 0.0375
$k^* = 2$	proposed method	2.9 ± 0.7	0.2029 ± 0.0667

relevancy of each features were measured by the *error rate* index. We randomly split those raw data sets into equal size for the training sets and the testing sets. The process was repeated 30 times, yielding 30 pairs of different training and test sets. For comparison, we conducted the proposed algorithm, the RPEM algorithm, as well as the approach in [5] (denoted as EMFW). To examine the efficacy of the feature weight function $f(.|x_{lt}, \Phi)$, we also conducted the algorithm (denoted as NFW) with the feature weight function in (12) setting to $s(y) = y, y \in [0, 1]$, i.e., no penalization on the *lower level* Gaussian mixture in feature relevancy estimation. The means and standard deviations of the results obtained on the four sets are summarized in Table 1, from which we have the three remarks as follows:

Table 2. The average weighting results of the 30 fold-runs on the real data, where the feature weights for *wdbc* and *ionosphere* are not included as the number of their features is too large to accommodate in this Table

Data set	Feature 1	2	3	4	5	6	7	8	9	10	11	12	13
wine	0.9990	0.8799	0.0256	0.2831	0.2354	0.9990	0.9990	0.0010	0.9900	0.9869	0.7613	0.9990	0.0451
heart	0.0010	0.6776	0.5872	0.0010	0.0010	0.9332	0.5059	0.0010	0.8405	0.3880	0.3437	0.4998	0.7856

Remark 1: In Table 1, it is noted that the proposed method has much lower error rates than the RPEM algorithm that is unable to perform the feature discrimination. Table 2 lists the mean feature weights of the sets obtained by the proposed algorithm in the 30 runs, indicating that only several features have good discriminating power. *Remark 2:* Without the feature weight functions to assign unequal emphases on the likelihood for the *lower level* Gaussian mixture, it is found that the performance of the NFW algorithm is unstable because of no enough penalization to the model complexity. It therefore validates the design of the proposed weight functions for weighting features. *Remark 3:* It is observed

that the method in [5] tends to use more components for the mixture, whereas the proposed algorithm not only gives general lower mismatch degrees, but also produces much more parsimonious models.

5 Conclusion

In this paper, a novel approach to tackle the two challenges for Gaussian mixture clustering, has been proposed under the Maximum Weighted Likelihood learning framework. The model order for the mixture model and the feature weighting are obtained simultaneously and automatically. Experimental results have shown the promising performance of the proposed algorithm on both the synthetic and the real-world data sets.

References

1. Liu, H., Yu, L.: Toward integrating feature selection algorithms for classification and clustering. IEEE Transactions on Knowledge and Data Engineering 17, 491–502 (2005)
2. Dash, M.C., Scheuermann, K., Liu, P.H.: Feature selection for clustering-A filter solution. In: Proceedings of 2nd IEEE International Conference on Data Mining, pp. 115–122. IEEE Computer Society Press, Los Alamitos (2002)
3. Dy, J., Brodley, C.: Feature selection for unsupervised learning. Joural of Machine Learning Research 5, 845–889 (2004)
4. Figueiredo, M.A.T., Jain, A.K., Law, M.H.C.: A feature selection wrapper for mixtures. LNCS, vol. 1642, pp. 229–237. Springer, Heidelberg (2003)
5. Law, M.H.C., Figueiredo, M.A.T., Jain, A.K.: Simultaneous feature selection and clustering using mixture models. IEEE Transactions on Pattern Analysis and Machine Intelligence 26, 1154–1166 (2004)
6. Cheung, Y.M.: Maximum weighted likelihood via rival penalized EM for density mixture clustering with automatic model selection. IEEE Transactions on Knowledge and Data Engineering 17, 750–761 (2005)
7. Blake, C.L., Merz, C.J.: UCI Repository of machine learning databases (1998), http://www.ics.uci.edu/mlearn/MLRepository.html

Estimation of Poles of Zeta Function in Learning Theory Using Padé Approximation

Ryosuke Iriguchi[1] and Sumio Watanabe[2]

[1] Department of Computational Intelligence and Systems Science, Tokyo Institute of Technology, mailbox R2-5, Nagatsuta 4259, Midori-ku, Yokohama-shi, 226–8503 Japan
iri@cs.pi.titech.ac.jp

[2] Precision and intelligence Laboratory, Tokyo Institute of Technology, mailbox R2-5, Nagatsuta 4259, Midori-ku, Yokohama-shi, 226–8503 Japan
swatanab@pi.titech.ac.jp

Abstract. Learning machines such as neural networks, Gaussian mixtures, Bayes networks, hidden Markov models, and Boltzmann machines are called singular learning machines, which have been applied to many real problems such as pattern recognition, time-series prediction, and system control. However, these learning machines have singular points which are attributable to their hierarchical structures or symmetry properties. Hence, the maximum likelihood estimators do not have asymptotic normality, and conventional asymptotic theory for statistical regular models can not be applied. Therefore, theoretical optimal model selections or designs involve algebraic geometrical analysis. The algebraic geometrical analysis requires blowing up, which is to obtain maximum poles of zeta functions in learning theory, however, it is hard for complex learning machines. In this paper, a new method which obtains the maximum poles of zeta functions in learning theory by numerical computations is proposed, and its effectiveness is shown by experimental results.

1 Introduction

Learning machines such as neural networks, Gaussian mixtures, Bayes networks, hidden Markov models, and Boltzmann machines are called singular learning machines, which have been applied to many real problems such as pattern recognition, time-series prediction, and system control. However, these learning machines have hierarchical structures or symmetry properties, which make correspondence between the parameters and probabilistic distributions many-to-one. Because of many-to-one correspondence, the parameters cannot be determined only by the behaviour of the learning machines. In addition, the conventional asymptotic theory of statistical regular models cannot be applied, and hence mathematical basis to obtain predictive accuracy or optimal design of learning models is not enough developed [1].

Furthermore, the parameter set that represents same probabilistic distributions of these learning machines has singular points in parameter space, hence

J. Marques de Sá et al. (Eds.): ICANN 2007, Part I, LNCS 4668, pp. 88–97, 2007.

theoretical optimal model selections or designs require blowing up, an algebraic geometrical method to resolve singular points [2]. Blowing up is used to obtain the poles of zeta functions in learning theory, and then an asymptotic form of stochastic complexity or generalization error of Bayesian learning can be obtained. It is known that the stochastic complexity and the generalization error of Bayesian learning of singular models are smaller than the ones of statistical regular models which have the same number of parameters. In this paper, a new method by which we obtain the maximum poles of zeta functions in learning theory by numerical computations is proposed, and its its effectiveness is shown by experimental results.

2 Bayesian Learning

Assume that n samples $X^n = (X_1, \ldots, X_n)$ are independently obtained from the true distribution $q(x)$. Let $p(x \mid w)$ be a learning machine that has parameter w, and $\varphi(w)$ be a prior distribution of the parameter. In Bayesian learning, a probabilistic distribution in parameter space called the posterior distribution is defined as

$$p(w \mid X^n) \stackrel{\text{def}}{=} \frac{1}{p(X^n)} \varphi(w) \prod_{i=1}^{n} p(X_i \mid w), \tag{1}$$

where $p(X^n)$ is the normalization constant (partition function) (2) which lets integral of (1) by w be 1.

$$p(X^n) \stackrel{\text{def}}{=} \int \varphi(w) \prod_{i=1}^{n} p(X_i \mid w)\, dw. \tag{2}$$

Equation (2) represents the likelihood for the learning machine and prior distribution under given samples, hence also called the marginal likelihood. Averaging out the learning machine by (1), the predictive distribution is defined by

$$p(x \mid X^n) \stackrel{\text{def}}{=} \int p(w \mid X^n) p(x \mid w)\, dw. \tag{3}$$

Also, the Kullback distance between the true distribution $q(x)$ and the predictive distribution (3) is called generalization error $G(X^n)$ (4), and its average is called average generalization error $G(n)$ (5). The generalization error is used to evaluate the precision of learning.

$$G(X^n) \stackrel{\text{def}}{=} \int q(x) \log \frac{q(x)}{p(x \mid X^n)}\, dx, \tag{4}$$

$$G(n) \stackrel{\text{def}}{=} \int q(X^n) G(X^n)\, dX^n. \tag{5}$$

Equation (1) can be rewritten as

$$p(w \mid X^n) = \frac{1}{Z(X^n)} \varphi(w) \exp\left(-nK_n(w)\right),$$

where $K_n(w)$ is the empirical Kullback information

$$K_n(w) \stackrel{\text{def}}{=} \frac{1}{n} \sum_{i=1}^{n} \log \frac{q(X_i)}{p(X_i \mid w)}, \tag{6}$$

and $Z(X^n)$ is the normalization constant,

$$Z(X^n) \stackrel{\text{def}}{=} \int \varphi(w) \exp\left(-nK_n(w)\right) dw. \tag{7}$$

Let the stochastic complexity (free energy) $F(n)$ be

$$F(n) \stackrel{\text{def}}{=} E_{X^n}\left[-\log Z(X^n)\right], \tag{8}$$

then the average generalization error $G(n)$ (5) is equal to the increase of $F(n)$,

$$G(n) = F(n+1) - F(n).$$

That is, to obtain the average generalization error, it is enough to obtain the stochastic complexity. The stochastic complexity is a fundamental quantity used for model selection or hyper-parameter optimization.

Let the Kullback information $K(w)$ be

$$K(w) \stackrel{\text{def}}{=} \int q(x) \log \frac{q(x)}{p(x \mid w)}\, dx \tag{9}$$

and $F^\star(n)$ be

$$F^\star(n) \stackrel{\text{def}}{=} -\log \int \varphi(w) \exp\left(-nK(w)\right) dw,$$

then there exists an n-independent constant C, such that

$$F^\star(n/2) - C \leq F(n) \leq F^\star(n)$$

holds [3].

2.1 Zeta Function in Learning Theory

The zeta function in learning theory $\zeta(z)$ is defined as

$$\zeta(z) \stackrel{\text{def}}{=} \int K(w)^z \varphi(w)\, dw. \tag{10}$$

Then, if $\varphi(w)$ has a compact support, $\zeta(z)$ is holomorphic in $\operatorname{Re} z > 0$, and has poles on the negative-real axis. Let the poles of $\zeta(z)$ be $-\lambda_1, -\lambda_2, \ldots$ $(0 < \lambda_1 < \lambda_2 < \ldots)$ and its order be $m_1, m_2, \ldots$ respectively, then $F^\star(n)$ has an asymptotic expansion,

$$F^\star(n) = \lambda_1 \log n - (m_1 - 1) \log\log n + O(1)$$

as $n \to \infty$ [2]. Therefore, it is important to obtain the maximum pole of $\zeta(z)$.

According to recent works, the following property of the zeta function has been known [4].

Let $-\Lambda(K,\varphi)$ be the maximum pole of the zeta function (10). Then if $|K_1| \leq |K_2|$ and $\mathrm{supp}\,\varphi_1 \supseteq \mathrm{supp}\,\varphi_2$,

$$\Lambda(K_1,\varphi_1) \leq \Lambda(K_2,\varphi_2). \tag{11}$$

Furthermore, for an arbitrary $a \neq 0$,

$$\Lambda(aK,\varphi) = \Lambda(K,a\varphi) = \Lambda(K,\varphi). \tag{12}$$

3 The Proposed Method

In this section, we propose the method to obtain the maximum pole of the zeta function in learning theory by approximating the function. The method is based on Padé approximation with denominator primary expression.

3.1 Padé Approximation

The Padé approximation of degrees (n,m) with center z_0 to given $f(z)$ is the rational function $R_{n,m}(z)$ (13) which satisfies the conditions (14) [5, §5.12]. Note that $b_0 = 1$.

$$R_{n,m}(z) = \frac{\sum_{i=0}^{n} a_i(z-z_0)^i}{\sum_{i=0}^{m} b_i(z-z_0)^i}, \tag{13}$$

$$R_{n,m}^{(k)}(z_0) = f^{(k)}(z_0) \quad (k = 0,1,\ldots,m+n). \tag{14}$$

The coefficients of Padé approximation $a_i\,(1 \leq i \leq n)$ and $b_i\,(1 \leq i \leq m)$ are obtained by resolving a system of linear equations (15). These equations are derived from differentiating (16) $k = 0,1,\ldots,n+m$ times, and then substitute $z = z_0$.

$$a_k - \sum_{j=0}^{k} \frac{f^{(k-j)}(z_0)}{(k-j)!} b_j = 0, \tag{15}$$

$$\sum_{i=0}^{n} a_i(z-z_0)^i = R_{n,m}(z) \sum_{i=0}^{m} b_i(z-z_0)^i, \tag{16}$$

3.2 Proposal Method

We estimate the maximum pole λ_1c of the zeta function $\zeta(z)$. Let $R_{n,1}$ be the Padé approximant of degrees $(n,1)$ with center z_0 to $\zeta(z)$. The estimated pole is referred to as $\hat{\lambda}_n$, and it is the pole of $R_{n,1}$. The pole of $R_{n,1}$ is derived from $1 - b_1(\hat{\lambda} + z_0) = 0$ (17).

$$\hat{\lambda}_n = \frac{1}{b_1} - z_0 = -\frac{\frac{\zeta^{(n)}(z_0)}{n!}}{\frac{\zeta^{(n+1)}(z_0)}{(n+1)!}} - z_0. \tag{17}$$

Thus $\hat{\lambda}_n$ can be estimated using $\zeta^{(n)}(z_0)$ and $\zeta^{(n+1)}(z_0)$.

The Laurent series of $\zeta(z)$ about its maximum pole is

$$\zeta(z) = \sum_{m=1}^{m_1} \frac{c_{1m}}{(z+\lambda_1)^m} + \bar{\zeta}(z),$$

where $\bar{\zeta}(z)$ is a function holomorphic in $\mathrm{Re}(z) > -\lambda_2$ and has poles at $-\lambda_2, -\lambda_3, \ldots$. The Taylor series of $\bar{\zeta}(z)$ about point z_0 is

$$\bar{\zeta}(z) = \sum_{i=0}^{\infty} d_i (z - z_0)^i. \tag{18}$$

Then equation (17) can be rewritten as

$$\begin{aligned} \hat{\lambda}_n + z_0 &= -\frac{(-1)^n \sum_{m=1}^{m_1} \binom{m+n-1}{m-1} \frac{c_{1m}}{(z_0+\lambda_1)^{m+n}} + d_n}{(-1)^{n+1} \sum_{m=1}^{m_1} \binom{m+n}{m-1} \frac{c_{1m}}{(z_0+\lambda_1)^{m+n+1}} + d_{n+1}} \\ &= (z_0+\lambda_1) \frac{\sum_{m=1}^{m_1} \binom{m+n-1}{m-1} \frac{c_{1m}}{(z_0+\lambda_1)^m} + d_n(-\lambda_1 - z_0)^n}{\sum_{m=1}^{m_1} \binom{m+n}{m-1} \frac{c_{1m}}{(z_0+\lambda_1)^m} + d_{n+1}(-\lambda_1 - z_0)^{n+1}}. \end{aligned} \tag{19}$$

The right side of (18) converges in the region $|z - z_0| < |z_0 + \lambda_2|$, hence it converges on $z = -\lambda_1$. That is, $d_n(z_0+\lambda_1)^n \to 0$ as $n \to \infty$. With that, we set aside the term d_n in (19),

$$\hat{\lambda}_n + z_0 = (z_0+\lambda_1) \frac{\frac{c_{11}}{z_0+\lambda_1} + \frac{c_{12}}{(z_0+\lambda_1)^2}(n+1) + \cdots}{\frac{c_{11}}{z_0+\lambda_1} + \frac{c_{12}}{(z_0+\lambda_1)^2}(n+2) + \cdots}. \tag{20}$$

In specific, if the order of the maximum pole $m_1 = 1$, it follows $\hat{\lambda}_n = \lambda_1$.

Let the state density function $v(t)$ be

$$v(t) \stackrel{\text{def}}{=} \int \delta(K(w) - t)\varphi(w)\, dw. \tag{21}$$

Then, the nth-order derivative of $\zeta(z)$(10) can be rewritten as

$$\zeta^{(n)}(z) = \int_0^{\infty} (\log t)^n t^z v(t)\, dt \tag{22}$$

The random samples $\{t_i\}_{i=1}^N$ subject to $v(t)$ are obtained by $t_i = K(w_i)$ if random samples $\{w_i\}_{i=1}^N$ are subject to $\varphi(w)$. With that, the nth-order derivative of $\zeta(z)$ can be estimated by

$$\zeta^{(n)}(z) \approx \frac{1}{N} \sum_{i=1}^{N} (\log t_i)^n {t_i}^z. \tag{23}$$

From (12), the maximum pole of $\zeta(z)$ is invariant for constant multiplication to t, hence $\{t_i\}_{i=0}^N$ is normalized to range $0 < t_i < 1$.

3.3 Determination of z_0 from n

From (20), the degree of the numerator of the Padé approximation n should be set large enough. However, from (23) and denoting $t_1 = \min_{i=1}^{N} t_i$, equation (17) is rewritten as

$$\begin{aligned}\hat{\lambda}_n + z_0 &= -(n+1)\frac{\sum_{i=1}^{N}(\log t_i)^n {t_i}^{z_0}}{\sum_{i=1}^{N}(\log t_i)^{n+1}{t_i}^{z_0}} \\ &\approx -(n+1)\frac{(\log t_1)^n {t_1}^{z_0}}{(\log t_1)^{n+1}{t_1}^{z_0}} = -(n+1)\frac{1}{\log t_1}\end{aligned}$$

as $n \to \infty$, hence correct integration cannot be obtained. This phenomenon can be avoided by settling z_0 from n appropriately. This technique is described in following paragraphs.

Let $s_i = -\log t_i$, equation (23) is rewritten as

$$\zeta^{(n)}(z) \approx \frac{(-1)^n}{N}\sum_{i=1}^{N}\exp(f(s_i)), \tag{24}$$

where

$$f(s) \stackrel{\text{def}}{=} \log\left[s^n e^{-sz}\right]$$

and its saddle point $f'(s^\star) = 0$ is

$$s^\star = \frac{n}{z} \tag{25}$$

and

$$f''(s^\star) = -\frac{n}{{s^\star}^2}.$$

Hence if n is large enough,

$$f(s) \approx f(s^\star) - \frac{1}{2}\frac{n}{{s^\star}^2}(s - s^\star)^2 \tag{26}$$

holds. Given samples $\{t_i\}_{i=1}^{N}$ subject to $v(t)$ (21), equation (26) yields

$$\frac{1}{N}\sum_{i=1}^{N}\exp(f(s_i)) \approx \frac{\exp\left[f(s^\star)\right]}{N}\sum_{i=1}^{N}\exp\left[-\frac{1}{2}\frac{(s_i - s^\star)^2}{\frac{{s^\star}^2}{n}}\right].$$

Therefore, given $\alpha > 0$, denote

$$\sigma \stackrel{\text{def}}{=} \frac{s^\star}{\sqrt{n}}$$

and

$$t^\star \stackrel{\text{def}}{=} \exp\left[-s^\star - \alpha\sigma\right], \tag{27}$$

then it is necessary for (23) to hold that $t^\star$ is greater than or equal to $\min_{i=1}^N t_i$. Hence, substituting (27) and (25) to $t^\star = \min_{i=1}^N t_i$,

$$\min_{i=1}^N t_i = t^\star = \exp\left[-s^\star - \alpha\frac{s^\star}{\sqrt{n}}\right] = \exp\left[-\frac{n+\alpha\sqrt{n}}{z}\right],$$

and solving for z, we obtain

$$z = \frac{n+\alpha\sqrt{n}}{-\log \min_{i=1}^N t_i} \tag{28}$$

which indicates that z_0 is appropriate when it is set as (28).

4 Experiments

For some learning machines whose maximum poles λ_1 have been theoretically obtained, the experiments calculating $\hat{\lambda}_n$ by the proposal method were performed.

4.1 Experimental Condition

We adopted reduced rank regressions, or three-layer neural networks with linear hidden units, which is ones of singular learning machines. Let M be the input dimension, N be the output dimension, and H be the number of hidden units. The parameter w of the reduced rank regression is, letting A be an $H \times M$ matrix and B be an $N \times H$ matrix,

$$w = (A, B). \tag{29}$$

Let the probability distribution of output y be

$$p(y \mid x, w) \stackrel{\text{def}}{=} \frac{1}{(2\pi)^{\frac{N}{2}}} \exp\left(-\frac{\|y - BAx\|^2}{2}\right),$$

and denote the true parameter $N \times M$ matrix R,

$$\bar{K}(w) \stackrel{\text{def}}{=} \|BA - R\|,$$

then it is known that

$$\Lambda(K, \varphi) = \Lambda(\bar{K}, \varphi).$$

Furthermore, let r denote the rank of R, assume $N + r \leq M + H$, $M + r \leq N + H$, $H + r \leq M + N$, then if $M + H + N + r$ is even, $m_1 = 1$ and

$$\lambda_1 = \frac{2(H+r)(M+N) + 2MN - (H+r)^2 - M^2 - N^2}{8}.$$

If $M + H + N + r$ is odd, $m_1 = 2$ and

$$\lambda_1 = \frac{2(H+r)(M+N) + 2MN - (H+r)^2 - M^2 - N^2 + 1}{8}$$

[6].

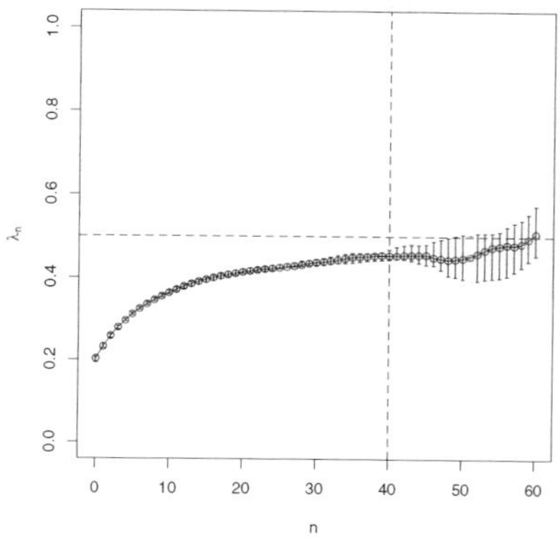

Fig. 1. $N = M = H = 1,\ R = R_1$

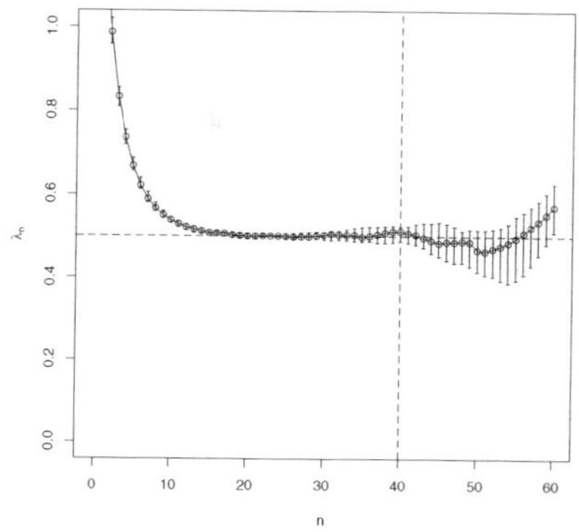

Fig. 2. $N = M = H = 1,\ R = R_2$

Let the prior distribution $\varphi(w)$ be the standard normal distribution, the number of samples N (23) be $N = 100000$, the center of Padé approximation z_0 (17) be

$$z_0 = \frac{40 + 2\sqrt{40}}{-\log t_1} \tag{30}$$

from (28) with $\alpha = 2, n = 40$.

We examined 4 cases.

1. The numbers of input, output, and hidden unit are all 1, the true parameter is $R_1 = 0$, then $w = (w_1, w_2)$ and $\lambda_1 = 1/2,\ m_1 = 2$ (Fig. 1).
2. The numbers of input, output, and hidden unit are all 1, the true parameter is $R_2 = 1$, then $w = (w_1, w_2)$ and $\lambda_1 = 1/2,\ m_1 = 1$ (Fig. 2).
3. The numbers of input, output and hidden units are all 2, the true parameter is
$$R_3 = \begin{pmatrix} 0 & 0 \\ 0 & 0 \end{pmatrix}$$
then $w = (w_1, \ldots, w_8)$ and $\lambda_1 = 3/2,\ m_1 = 1$ (Fig. 3).
4. The numbers of input, output, and hidden units are all 2, the true parameter is
$$R_4 = \begin{pmatrix} 1 & 0 \\ 0 & 0 \end{pmatrix}$$
then $w = (w_1, \ldots, w_8)$ and $\lambda_1 = 2,\ m_1 = 1$ (Fig. 4).

Experimental results of 4 cases are shown in Fig. 1–4.

4.2 Experimental Results

With the conditions described above, the experiments were performed 20 times for each R_i (Fig. 1–4). The vertical and horizontal axes denote $\hat{\lambda}_n$ and the degree of Padé approximation n in (17) respectively. The horizontal dashed lines

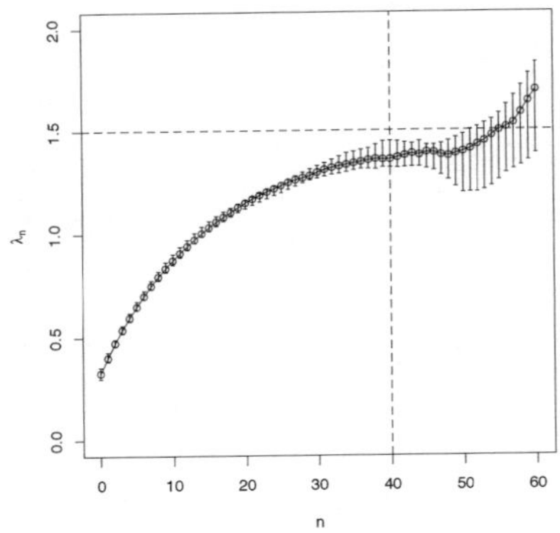

Fig. 3. $N = M = H = 2$, $R = R_3$

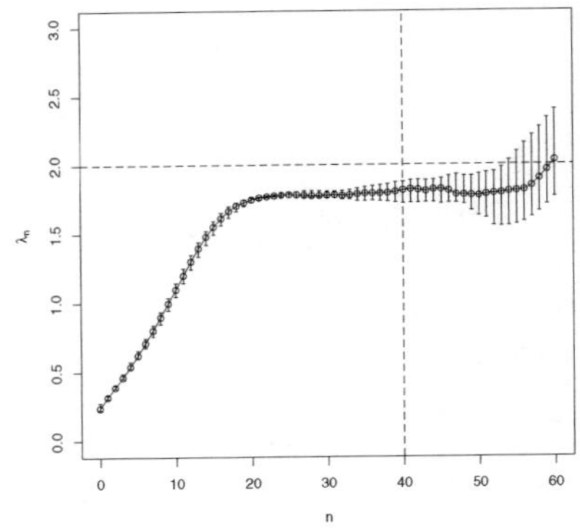

Fig. 4. $N = M = H = 2$, $R = R_4$

denote the theoretical value λ_1, and the vertical dashed lines denote that z_0 is determined by (30) to obtain correct integral with $n = 40$. In addition, the error bar and circle denote lower quartile, median, upper quartile of 20 experimental results respectively.

We determined z_0 to obtain correct integral on $n = 40$, hence for all R_i, the variance is large in $n > 40$.

5 Discussion

With the true distribution $q(x)$ unknown, let n training samples x_j $(1 \le j \le n)$ be obtained and

$$L_n(w) \stackrel{\text{def}}{=} -\frac{1}{n}\sum_{j=1}^{n} \log p(x_j \mid w).$$

If n is large enough, let

$$c = -\frac{1}{n}\sum_{j=1}^{n} \log q(x_j), \tag{31}$$

then $K_n(w)$ (6) and $K(w)$ (9) have relationship

$$0 \le K(w) \approx K_n(w) = L_n(w) - c,$$

therefore

$$L_n(w) \gtrapprox c. \tag{32}$$

Assume that the random samples $t_1 < t_2 < \cdots < t_N$ subject to $v(t)$ (21) with $K(w) = L_n(w)$ be obtained. From equation (32), it can be thought $c \lessapprox t_1$, hence from equation (11),

$$\Lambda(K_n, \varphi) \gtrapprox \Lambda(L_n - t_1, \varphi)$$

holds. Therefore, let $t_i' \stackrel{\text{def}}{=} t_i - t_1$ and apply the proposal method, the lower bound of λ_1 can be obtained.

6 Conclusion

In this paper, the method to obtain the learning coefficient by approximating the zeta function in learning theory with Padé approximation was proposed, and its efficiency was validated. For future works, validation of the estimation of λ_1 with unknown true distribution described §5 can be pointed.

Acknowledgement. This research was partially supported by the Ministry of Education, Science, Sports and Culture, Grant-in-Aid for Scientific Research 18079007, 2007.

References

1. Amari, S., Park, H., Ozeki, T.: Geometrical singularities in the neuromanifold of multilayer perceptrons. In: Dietterich, T.G., Becker, S., Ghahramani, Z. (eds.) NIPS, pp. 343–350. MIT Press, Cambridge (2001)
2. Watanabe, S.: Algebraic analysis for nonidentifiable learning machines. Neural Computation 13(4), 899–933 (2001)
3. Watanabe, S.: Algebraic geometrical methods for hierarchical learning machines. Neural Networks 14(8), 1049–1060 (2001)
4. Watanabe, S.: Algebraic geometry of learning machines with singularities and their prior distributions(technical papers: "ibis 2000"). Journal of Japanese Society of Artificial Intelligence 16(2), 308–315 (2001)
5. Press, W.H., Teukolsky, S.A., Vetterling, W.T., Flannery, B.P.: Numerical Recipes in C: The Art of Scientific Computing, 2nd edn. Cambridge University Press, The Pitt Building Trumpington Street, Cambridge CB2 1RP (1992)
6. Aoyagi, M., Watanabe, S.: Desingularization and the generalization error of reduced rank regression in bayesian estimation. In: Workshop on Information-Based Induction Sciences, Institute of Electronics, Information and Communication Engineers, pp. 237–244 (2004)

Neural Network Ensemble Training by Sequential Interaction

M.A.H. Akhand and Kazuyuki Murase

Graduate School of Engineering, University of Fukui, Japan
{Akhand, Murase}@synapse.his.fukui-u.ac.jp

Abstract. Neural network ensemble (NNE) has been shown to outperform single neural network (NN) in terms of generalization ability. The performance of NNE is therefore depends on well diversity among component NNs. Popular NNE methods, such as bagging and boosting, follow data sampling technique to achieve diversity. In such methods, NN is trained independently with a particular training set that is probabilistically created. Due to independent training strategy there is a lack of interaction among component NNs. To achieve training time interaction, negative correlation learning (NCL) has been proposed for simultaneous training. NCL demands direct communication among component NNs; which is not possible in bagging and boosting. In this study, first we modify the NCL from simultaneous to sequential style and then induce in bagging and boosting for interaction purpose. Empirical studies exhibited that sequential training time interaction increased diversity among component NNs and outperformed conventional methods in generalization ability.

Keywords: Bagging, boosting, negative correlation learning, diversity and generalization.

1 Introduction

Neural network ensemble (NNE) is a collection of several neural networks (NNs) that are trained for the same task to achieve better performance. In order to achieve better performance than single NN, component NNs in an NNE should be diverged to compensate the failure of one NN by others. Because there is no advantage in combining NNs that exhibit identical generalization ability, NNE performance depends largely on the diversity among component NNs [1].

Considerable work has been done in the last decade in designing NNEs [1]-[8]. It is argued that component NNs trained on different training sets are likely to produce more diversity than other approaches [2]-[6]. The most popular methods for designing NNEs based on such data sampling are bagging [4] and boosting [5]. Both the methods create different training sets to train different component NNs in an NNE [3]-[6].

The bagging algorithm [4] creates a new training set for each component NN by forming bootstrap replicates of the original training set. Given a original training set T consisted of m patterns, a new training set T' is constructed by drawing m patterns

J. Marques de Sá et al. (Eds.): ICANN 2007, Part I, LNCS 4668, pp. 98–108, 2007.

uniformly (with replacement) from T. Due to random selection with same original distribution, each training set T' contains only about 63.2% unique patterns from T, with many patterns appearing multiple times while others are left [6].

In boosting [5], the training set T' for each component NN is chosen from the original training set T based on the performance of earlier NN(s). Training patterns that were incorrectly classified by previous component NN(s) are chosen more often than patterns that were correctly classified for creating a new training set. From various boosting algorithms in this study we only concern with most popular one i.e., ada boosting. So in this paper boosting simply means ada boosting [5].

For creating an NNE, both bagging and boosting train a user specified predefined number of NNs. They use independent training strategy where the training of each NN is performed independently with a particular generated training set. To combine component NNs' output for NNE output, bagging uses multiple voting technique, while boosting follows weighted voting where the weight of a NN depends on the accuracy of it [3],[6].

Since bagging and boosting use an independent training strategy, there is no training time interaction among component NNs. In another word, they do not utilize any feedback utilization system from other NNs to train a particular NN. So it is possible that independently trained NNs are not necessarily negatively correlated [7]-[8] and do not make much contribution to the integrated system like NNE. This problem could be removed by introducing training time interaction in bagging and boosting which will provide component NNs to communicate each other during train.

About training time interaction in NNE, negative correlation learning (NCL) [7]-[8] is proposed for simultaneous training. NCL trains NNs simultaneously rather than independently by introducing a correlation penalty term into the error function of a NN, so that an error base training time interaction is established among component NNs. In simultaneous training, NCL demands direct communication among component NNs, that is, in order to train a particular component NN all other NNs must be active to provide their response for this pattern. Also there might be competitions among NNs since all NNs in an NNE are concerned with the whole NNE error [9]. Finally NCL increase computational cost much more.

At present, communication among component NNs in bagging and boosting using NCL scheme is not possible due to three reasons: (i) both the methods use independent training strategy, (ii) training sets are different for different component NNs, and (iii) arrangement of training patterns is different in different sets. In addition, boosting produces training set based on the performance on existing NNs, so that there is no chance to get information of coming NNs.

In this paper we made a modification over standard NCL to a sequential manner instead of the simultaneous manner and then induced it in bagging and boosting for training time interaction among component NNs. The rest of this paper is organized as follows. Section 2 explains the modification of NCL to implement in bagging and boosting as an interactive process to get interactive bagging/boosting from standard bagging/boosting. Section 3 experimentally explains the effect of training time interaction on bagging and boosting on several benchmark problems. Finally, Section 4 concludes the paper with a brief summary and conclusions.

2 Modification of Negative Correlation Learning (NCL) for Interaction and Interactive Bagging/Boosting Methods

This section describes the modification technique of NCL to make it usable for interaction purpose in bagging and boosting and describe interactive bagging and boosting algorithm in detail.

At simultaneous training, NCL introduces a penalty function in error function of individual NN. The error function E_i for the ith component NN in NCL is defined as:

$$E_i = \frac{1}{N}\sum_{n=1}^{N}\left[\frac{1}{2}(F_i(n)-d(n))^2 + \lambda[(F_i(n)-F(n))\sum_{j\neq i}(F_j(n)-F(n))]\right] \quad (1)$$

where N is the size of training set, $F_i(n)$ is the output of the ith component NN and $d(n)$ is the desired output of the NNE for nth training pattern [8]. The first term of Eq. (1) is the empirical risk function of the ith component NN and the second term is known as the correlation penalty function. $F(n)$ in Eq. (1) represents the output of the NNE on the nth training pattern that can be obtained by averaging the outputs of all component NNs in an NNE. If NNE consists with M NNs then $F(n)$ will be:

$$F(n) = \frac{1}{M}\sum_{i=1}^{M} F_i(n) \quad (2)$$

To update the weights of the ith NN, the partial derivative of E_i with respect to F_i is as follows [9].

$$\frac{\partial E_i(n)}{\partial F_i(n)} = F_i(n) - d(n) - 2\lambda(F_i(n) - F(n)) \quad (3)$$

We may rearrange the Eq.(3) by placing the value of $F(n)$ from Eq. (2). The Eq. (3) now becomes

$$\frac{\partial E_i(n)}{\partial F_i(n)} = F_i(n) - d(n) - 2\lambda\left(\tfrac{M-1}{M}F_i(n) - \tfrac{1}{M}\sum_{j\neq i}F_j(n)\right) \quad (4)$$

According to Eq.(4) for nth pattern, $\sum_{j\neq i} F_j(n)$ is the contribution of other NNs (let it F_{IC}) in NCL and is independent of the current NN which is under training. Storing F_{IC} in an intermediate space communication among component NNs may establish when they are trained one after another as like bagging and boosting. After training a component NN, F_{IC} will update for coming component NN(s).

Fig. 1 graphically represents standard NCL and modified NCL with information centre(IC) for M number of NNs. Standard NCL in Fig. 1(a) is a closed loop because at that time training is perform pattern by pattern basis and after training all the component NNs for a particular pattern, again training of all the NNs is perform for next pattern. Due to NNE's output $F(n)$ in Eq. (3) it demands all the exiting NNs' output for a particular training pattern; which is the theme of direct communication. On the other hand in modified NCL in Fig. 1(b), training is performed sequential manner and a particular NN is trained independently based on Eq. (4) collecting F_{IC} from IC (i.e., the intermediate space) only, for the response of previously trained NN(s). At the beginning IC will be empty, so first NN will train independently (for $M=1$) and does not collect F_{IC} from IC and only send its information after training. On the other hand, last NN only collects F_{IC} from IC and does not required to update.

If IC contains $F_{IC}(n)$ at the training of ith NN, then after training the IC will update for original training set by the following way:

$$F_{IC}(n)= F_{IC}(n)+ F_i(n) \tag{5}$$

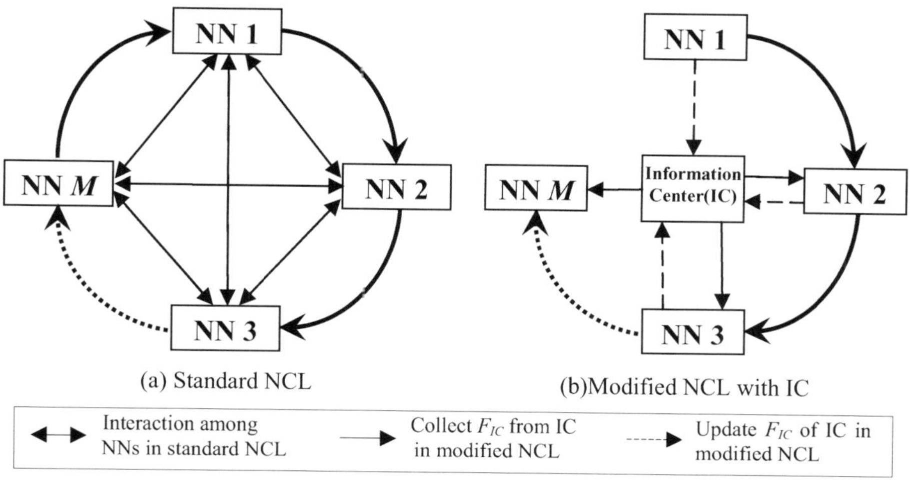

Fig. 1. Graphical representation of component NNs training in standard NCL and in conjunction with information center (IC) to maintain indirect communication in modified NCL

Inducing modified NCL based on IC, with sequential interaction in standard bagging and boosting methods, the following Figs. 2-3 show the pseudo code for interactive bagging and interactive boosting algorithms.

1. Let M as the number of NNs to be trained for NNE
Take original training set $T= \{(x_1, d_1),..., (x_N, d_N)\}$ with class level $d_i \in K = \{1,2,.....,k\}$.
2. *for i=1 to M* {
   a. Construct a NN, NN^i
   b. Make a new training set, T^i by sampling N patterns uniformly at random with replacement from T.
   c. Train NN^i by T^i in conjunction with F_{IC} of IC.
   d. Update IC. }
3. NNE decision for a particular pattern n, $Int\_Bagging(n) = \arg\max_{d \in K} \sum_{i:NN^i(x_n)=d} 1$

Fig. 2. Interactive bagging algorithm

It is seen that a particular NN^i is trained in interactive bagging/boosting by a particular training set T^i in conjunction with F_{IC} of IC. This is different from standard bagging/boosting where NN^i is independently trained by only T^i. Due to the use of F_{IC}, NN^i will maintain interaction indirectly with previously trained NNs. Both interactive bagging and boosting method update F_{IC} in IC after training a NN. So, step 2(c) in both Fig. 2 and Fig. 3 contain the modification over standard cases. Also, step

2(d) in Fig. 2 and step 2(h) in Fig. 3 are the additional steps to update IC that are absent in standard bagging/boosting.

1. Let M as the number of NNs to be trained for NNE.
   Take original training set $T= \{(x_1, d_1),..., (x_N, d_N)\}$ with class level $d_i \in K = \{1,2,.....,k\}$.
   Assign weight for each pattern of T, initially weights are same, i.e., $w_i(n)=1/N$
2. *for i=1 to M* {
   a. Construct a NN, NN^i
   b. Make a new training set, T^i by sampling N patterns at random with replacement from T based on weight distribution $w_i(n)$.
   c. Train NN^i by T^i in conjunction with F_{IC} of IC.
   d. $\varepsilon_i = \sum_{(x_n,d_n)\in T:NN^i(x_n)\neq d_n} w_i(n)$ (weighted error on training set)
   e. $\beta_i = (1-\varepsilon_i)/\varepsilon_i$
   f. *for each* $(x_n,d_n)\in T$,
      if $NN^i(x_n) \neq d_n$ *then* $w_{i+1}(n) = w_i(n).\beta_i$, *otherwise* $w_{i+1}(n) = w_i(n)$
   g. Normalized the weights, $w_{i+1}(n)$.
   h. Update IC. }
3. NNE decision for a particular pattern n, $Int\_Boosting(n) = \arg\max_{d\in K} \sum_{i:NN^i(x_n)=d} \log(\beta_i)$

Fig. 3. Interactive boosting algorithm

3 Experimental Studies

This section experimentally describes the effect of training time interaction through IC among bagging/boosting component NNs on final NNE's performance. For proper observation we implemented two different sets of experiments, one is standard bagging/boosting and another is bagging/boosting with training time interaction, which we defined as interactive bagging/boosting. We have selected several benchmark classification problems from the University of California, Irvine (UCI) machine learning benchmark repository to evaluate performance. These are

Table 1. Characteristics of benchmark data sets

Problem	Total Pattern	Input Attributes Continuous	Input Attributes Discrete	NN architecture Inputs	NN architecture Classes	NN architecture Hidden Node
ACC	690	6	9	51	2	10
CAR	1728	-	6	21	4	10
GCC	1000	7	13	63	2	10
HDC	303	6	7	22	2	5
LMP	148	-	18	18	4	10
PST	90	1	7	19	3	5
ZOO	101	15	1	16	7	10

Australian Credit Card (ACC), Car (CAR), German Credit Card (GCC), Heart Disease Cleveland (HDC), Lymphography (LMP), Postoperative (PST) and Zoo (ZOO) problems. Detailed descriptions of the problems are available at the UCI web site (www.ics.uci.edu/~mlearn/).

We followed benchmark methodology [12] for preprocessing UCI data to use in our experiments. The characteristics of the problems are summarized in Table 1 which shows a diverse test bed. For NNE construction we trained 10 NNs that is reported as standard number [3]. The logistic sigmoid function was used as an activation function. Learning rate of back propagation learning was set 0.1 and NNs was initialized by random weights between -0.5 and 0.5. For interactive bagging/boosting cases λ was fixed as 0.5.

Standard 3-fold cross validations were used in our experiments, where initially available training examples were partitioned into three equal or nearly equal sets and by turn one set was reserved for test set while the remaining two sets were used for training set. Training set is the main acting set on which training and other operation was performed. Testing set was reserved to measure the final performance i.e., generalization ability and diversity, of produced NNE only and all the NNs were blind for this during training. To make fair comparisons, both standard bagging/boosting and interactive bagging/boosting used an equal fixed number of iteration to train all component NNs of NNE and that was 50.

3.1 Experimental Results

Tables 2-3 show the average results of different problems over ten standard 3-fold cross validations (i.e., 3*10 = 30) independent runs. The testing error rate (TER) in the tables refers to the rate of wrong classification produced by the trained NNE on the test set. The diversity in the tables indicates how predictions differ among component NNs on the test set. We used the most commonly used pair wise plain disagreement measure technique [10] to measure the diversity among component NNs. For two NNs i and j, the plain disagreement is equal to the proportion of the patterns on which the NNs make different predictions:

$$div\_plain_{i,j} = \frac{1}{N}\sum_{k=1}^{N} Diff\,(C_i(x_k), C_j(x_k)), \tag{6}$$

Where N is the number of patterns in the testing set, $C_i(x_k)$ is the class assigned by NN i to pattern k, and $Diff(a, b) = 0$, if $a = b$ otherwise $Diff(a, b) = 1$. This measure is

Table 2. Experimental results of bagging

Problem	Standard bagging		Interactive bagging	
	TER	Diversity	TER	Diversity
ACC	0.1481	0.0862	0.1371	0.1957
CAR	0.2688	0.0740	0.2461	0.1867
GCC	0.2500	0.1860	0.2382	0.3225
HDC	0.1693	0.1045	0.1534	0.2131
LMP	0.1700	0.1273	0.1563	0.3125
PST	0.3022	0.1042	0.2889	0.3141
ZOO	0.1487	0.1670	0.1262	0.3478

equal to 0, when both NNs return the same class for each pattern, and it is equal to 1 when the predictions are different for all the cases. The total NNE diversity is the average of all NN pairs in the NNE.

From Tables 2-3, it is observed that standard boosting achieved higher diversity over standard bagging for all the cases because boosting itself enforce diversity by design, whereas bagging has no such mechanism [13]. For ACC problem diversity were 0.0862 and 0.2409 for standard bagging and standard boosting, respectively. Again from Tables 2-3, it is found that in both bagging and boosting, interactive cases achieved higher diversity with respect to standard cases. As an example, for ACC problem, achieved diversity were 0.1957 and 0.4025, for interactive bagging and interactive boosting, respectively, where as, diversity were 0.0862 and 0.2409 for standard bagging and standard boosting, respectively. In addition for ACC problem, due to interaction, with diversity improvement TER was also improved from 0.1481 to 0.1371 for bagging and 0.1677 to 0.1552 for boosting.

Table 3. Experimental results of boosting

Problem	Standard boosting		Interactive boosting	
	TER	Diversity	TER	Diversity
ACC	0.1677	0.2409	0.1552	0.4025
CAR	0.2710	0.1418	0.2510	0.2786
GCC	0.2630	0.3288	0.2597	0.4389
HDC	0.1733	0.2554	0.1607	0.4276
LMP	0.1873	0.2603	0.1624	0.4585
PST	0.3419	0.3506	0.2900	0.4899
ZOO	0.0797	0.3112	0.0927	0.4613

Another important observation from Tables 2-3 is that due to sequential interaction, TER did not improve with diversity improvement manner for both bagging and boosting cases. In addition, for ZOO problem at boosting TER increased slightly, i.e., decreased generalization ability. Although maintaining diversity among component NNs is the important condition for improving the performance of NNEs, it is seen from our results that this not true for all the problems. This indicates that the maintenance of proper diversity is necessary for improving the generalization ability of NNEs. This is something contradictory to the conventional belief, which suggests that if the diversity is more the generalization ability will be better. The reason behind this will clear from coming correct response set analysis in Section 3.2.1.

3.2 Experimental Analysis

This subsection first investigates actual reason to improve diversity and thus TER due to interaction by correct response set analysis; which is introduced in NCL [8]. And then presents variation effect of λ on TER and diversity.

3.2.1 The Effect of Interaction on Correct Response Set

The correct response set S_i of ith individual NN on the testing set consists of all the patterns in the testing set which are classified correctly by the individual NN i. Let Ω_i

denote the size of set S_i, and $\Omega_{i1i2\cdots ik}$ denote the intersection size i.e., the size of set $S_{i1} \cap S_{i2} \cap \cdots \cap S_{ik}$.

For analysis purpose three problems were selected and Table 4 represents average result of 15 (i.e., five 3-fold cross validations) independent runs on test set for that problems for both standard and interactive bagging/boosting methods. Ω_{All} represents $\Omega_{i1i2\cdots i10}$, i.e., the size of test set patterns that's truly classified by all 10 NNs in the NNE and among 10 NNs, no one able to classify Ω_{None} test set patterns.

In bagging only about 28.8% (i.e., 100%-63.2%) patterns differs for different training sets; which is the element to make different function by different NNs. This implies higher values of Ω_{All} and Ω_{None} in standard bagging and achieved lower diversity. For CAR problem, Ω_{All} and Ω_{None} were 374.40 and 110.13, respectively, but for standard boosting these values were 319.80 and 78.20, respectively. Boosting achieved lower values of Ω_{All} and Ω_{None} by creating training set emphasizing on the previously misclassified patterns, and so that achieved higher diversity.

In interactive cases a NN is trained in conjunction with IC to maintain indirect communication with other NNs. Finally, different individual NNs were able to specialize on different parts of the testing set as like NCL, as a result common functional space (Ω_{All}) decreased. At the same time interaction motivated some NNs to unsolved space so that misclassified space by all the NNs (Ω_{None}) also reduced. Finally diversity increased due to reduction of common functional space and accuracy maintained/increased by covering additional space that was uncovered by standard case. In interactive bagging for CAR problem, Ω_{All} and Ω_{None} were 304.07 and 51.40, respectively; both the values are smaller than standard bagging case. For this problem, due to iteration, TER improved 0.269 to 0.247.

Table 4. Correct response set analysis for bagging and boosting

		Standard				Interactive			
		Ω_{All}	Ω_{None}	TER	Diver.	Ω_{All}	Ω_{None}	TER	Diver.
Bagging	ACC	163.47	13.6	0.148	0.087	92.47	5.00	0.138	0.198
	CAR	374.40	110.13	0.269	0.072	304.07	51.40	0.247	0.188
	ZOO	22.93	2.47	0.142	0.159	13.2	0.93	0.134	0.330
Boosting	ACC	76.20	4.67	0.168	0.250	6.20	0.33	0.155	0.418
	CAR	319.80	78.20	0.269	0.142	234.73	24.20	0.252	0.277
	ZOO	13.87	0.67	0.083	0.336	3.33	0.20	0.093	0.475

In the boosting due to interaction, Ω_{All} and Ω_{None} again reduced as like interactive bagging, and some cases Ω_{All} and Ω_{None} closed to zero. For ACC problem, Ω_{All} reduced from 76.20 to 6.20 and Ω_{None} reduced from 4.667 to 0.333. This effect implies on NNE performance and TER improved 0.168 to 0.155 due to interaction. On the other hand, for ZOO problem though Ω_{All} reduced from 13.87 to 3.33 but Ω_{None} was not reduced much more because standard case also achieved lower value. For this reason though diversity improved but TER degraded slightly for ZOO problem.

Finally the actions of interaction are: (i) to reduce the common functional space (Ω_{All}); which promotes diversity and might reduce accuracy at some cases and (ii) to motivate some NNs to cover additional space so that Ω_{None} also reduce; which is the element to improve accuracy of NNE as well as increase diversity. Combining both actions the outcomes are: (i) improve diversity all the cases and (ii) improve TER when Ω_{None} value is high at standard case.

3.2.2 The Variation Effect of λ Value on Interactive Training

Figure 4 presents the average results of 15 (i.e., five 3-fold cross validations) for interactive bagging/boosting when λ varied from 0 to 1 as suggested in standard NCL [8]. At Fig. 4, λ=0 is nothing but standard bagging or standard boosting. The meaning and measuring procedure of TER and diversity are like as before.

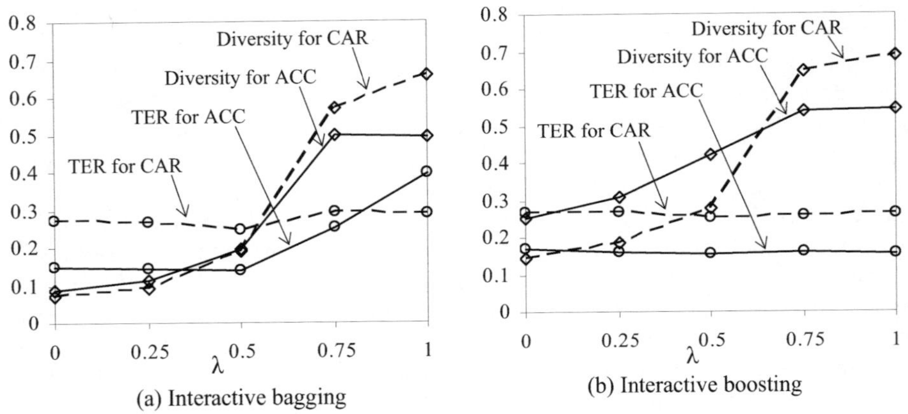

Fig. 4. Variation effect of λ on interactive bagging and boosting

From Fig. 4 it is observed that diversity increased with respect to λ value as like it is reported for standard NCL [8]. Diversity was the lowest level for λ =0, and shown highest value at around λ =1. As an example, for ACC problem, at λ =1 diversity were 0.492 and 0.543 for interactive bagging (Fig. 4(a)) and interactive boosting (Fig. 4(b)), respectively, these values were highest values in both the methods.

For bagging it is also found that for λ >0.5 though diversity improved much more, but TER also increased for some cases, such as ACC problem in Fig. 4(a). For this reason λ was selected as 0.5 for main experiments. On the other hand, boosting shown stable TER with respect to λ variation. Also few cases, better TER was observed at higher values of λ. As an example, for ACC problem at λ =1 TER was 0.156; which was the lowest with respect to other values of λ.

4 Conclusions

Both bagging and boosting followed independent training strategy to train component NNs and the difference in training sets among component NNs is the element to produce diversity. To overcome the deficiency of communication among NNs, this study introduced training time sequential interaction using modified NCL and proposed interactive bagging/boosting. While NCL demands direct communication among component NNs during training throughout, the proposed method utilizes an intermediate space (i.e., IC). The IC gathers the responses of trained NNs and a particular NN is maintained indirect communication with other NNs when it is trained in conjunction with IC only.

Liu [11] proposed bagging with NCL, where NNs are trained by NCL on different training set that was created by bagging technique, i.e., bootstrap samples from original training set. While he has only presented result for NCL and bagging with NCL for a single problem. We embedded NCL in bagging and boosting in this study.

From the analysis of correct response set, we exhibited here the reason why standard boosting shows more diversity with respect to standard bagging and how interaction improves diversity in both methods, and also why TER did not improve with diversity improvement manner. Now it is also noticeable that to improve TER farther more it is required a special technique that covers full space by different NNs, not reducing common functional space.

Another reason why diversity improvement is also appreciable in NNE is that in higher diversity case different pattern will truly classified by different NNs and reduce the chance to misclassify unknown patterns by all the NNs. In this point of view, due to diversity improvement the interactive bagging/boosting is appeared as more justified NNE methods, so that NNE will show stable performance in practical situation when real world unseen patterns will come. In addition, due to simplicity, λ value was fixed at 0.5 for main experiments, and from experimental analysis it was observed that, TER varied with respect to λ. So it might be possible to get better TER after problem-wise tuning of λ value.

Last of all, the main attraction of the proposed method is the training time indirect communication among component NNs when direct communication is not possible, and we successfully implemented it and presented the performance. Similar technique may be applied for other methods where direct communication is not possible.

Acknowledgements. Supported by grants to KM from the Japanese Society for promotion of Sciences, the Yazaki Memorial Foundation for Science and Technology, and the University of Fukui.

References

1. Sharkey, A.J.C., Sharkey, N.E.: Combining Diverse Neural Nets. Knowledge Engineering Review 12, 299–314 (1997)
2. Brown, G., Wyatt, J., Harris, R., Yao, X.: Diversity Creation Methods: A Survey and Categorization. Information Fusion 6, 99–111 (2005)
3. Opitz, D.W., Maclin, R.: Popular Ensemble Methods: An Empirical Study. Journal of Artificial Intelligence Research 11, 169–198 (1999)
4. Breiman, L.: Bagging Predictors. Machine Learning 24, 123–140 (1996)
5. Freund, Y., Schapire, R.E.: Experiments with a New Boosting Algorithm. In: Proc. of the 13th International Conference on Machine Learning, pp. 148–156. Morgan Kaufmann, San Francisco (1996)
6. Bauter, E., Kohavi, R.: An Empirical Comparison of Voting Classification Algorithms: Bagging, Boosting, and Variants. Machine Learning 36, 105–142 (1999)
7. Liu, Y., Yao, X.: Simultaneous training of negatively correlated neural networks in an ensemble. IEEE Trans. on Systems, Man, and Cybernetics 29, 716–725 (1999)

8. Liu, Y., Yao, X.: Ensemble Learning via Negative Correlation. Neural Networks 12, 1399–1404 (1999)
9. Islam, M.M., Yao, X., Murase, K.: A Constructive Algorithm for Training neural Network Ensembles. IEEE Transactions on Neural Networks 14, 820–834 (2003)
10. Tsymbal, A., Pechenizkiy, M., Cunningham, P.: Diversity in Search Strategies for Ensemble Feature Selection. Information Fusion 6, 83–98 (2005)
11. Liu, Y.: Generate Different Neural Networks by Negative Correlation Learning. In: Wang, L., Chen, K., Ong, Y.S. (eds.) ICNC 2005. LNCS, vol. 3610, pp. 149–156. Springer, Heidelberg (2005)
12. Prechelt, L.: Proben1- A Set of Benchmarks and Benching Rules for Neural Network Training Algorithms, Tech. Rep. 21/94, University of Karlsruhe, Germany (1994)
13. Kuncheva, L., Skurichina, M., Duin, R.: An experimental study on diversity for bagging and boosting with linear classifiers. Information Fusion 3, 245–258 (2002)

Improving Optimality of Neural Rewards Regression for Data-Efficient Batch Near-Optimal Policy Identification

Daniel Schneegaß[1,2], Steffen Udluft[1], and Thomas Martinetz[2]

[1] Information & Communications, Learning Systems
Siemens AG, Corporate Technology, D-81739 Munich, Germany
[2] Institute for Neuro- and Bioinformatics
University at Lübeck, D-23538 Lübeck, Germany
daniel.schneegass.ext@siemens.com

Abstract. In this paper we present two substantial extensions of Neural Rewards Regression (NRR) [1]. In order to give a less biased estimator of the Bellman Residual and to facilitate the regression character of NRR, we incorporate an improved, Auxiliared Bellman Residual [2] and provide, to the best of our knowledge, the first Neural Network based implementation of the novel Bellman Residual minimisation technique. Furthermore, we extend NRR to Policy Gradient Neural Rewards Regression (PGNRR), where the strategy is directly encoded by a policy network. PGNRR profits from both the data-efficiency of the Rewards Regression approach and the directness of policy search methods. PGNRR further overcomes a crucial drawback of NRR as it extends the accordant problem class considerably by the applicability of continuous action spaces.

1 Introduction

Neural Rewards Regression [1] has been introduced as a generalisation of Temporal Difference Learning (TD-Learning) and Approximate Q-Iteration with Neural Networks. It further offers the trade between minimising the Bellman Residual as well as approaching the fixed point of the Bellman Iteration. The method works in a full batch learning mode, explicitly finds a near-optimal Q-function without an algorithmic framework except Back Propagation for Neural Networks, and can, in connection with Recurrent Neural Networks, also be used in higher-order Markovian and partially observable environments. The approach is motivated by Kernel Rewards Regression (KRR) [3] for data-efficient Reinforcement Learning (RL) [4]. The RL problem is considered as a regression task fitting a reward function on the observed signals, where the regressor is chosen from a hypothesis space, such that the Q-function can be gained out of the reward function. NRR is formulated similarly. The usage of shared weights with Back Propagation [5,6] allows us to restrict the hypothesis space appropriately.

Policy Gradient (PG) methods [7] convince due to their capability to identify a near-optimal policy without using the loop way through the Value Function. They are known to be robust, convergence guarantees are given under mild conditions, and they behave well in partially observable domains. But PG methods usually base on Monte Carlo and hence high variances estimations of the discounted future rewards. Recent work addressing this problem are e.g. provided by Wang and Dietterich [8] as well as

J. Marques de Sá et al. (Eds.): ICANN 2007, Part I, LNCS 4668, pp. 109–118, 2007.

Ghavamzadeh and Engel [9]. We introduce Policy Gradient Neural Rewards Regression (PGNRR), which, from the PG perspective, reduces the variance of the Value estimator and, from the Rewards Regression perspective, improves the considered problem class and enables generalisation over the policy space.

The remainder of this paper is arranged as follows. After a brief introduction to Reinforcement Learning, we recapitulate NRR and explain the underlying idea in sec. 3. We describe, in which way the RL task is interpreted as a regression problem and how the trade-off between either the two learning procedures as well as the two optimality criteria is realised. Subsequently (sec. 4) we generalise further by applying the new Bellman Residual minimisation technique to NRR. Finally we extend NRR by a policy network for direct policy search (sec. 5) and motivate an advancement in data-efficiency (sec. 6). Preliminary practical results (sec. 7) on a common benchmark problem and a real-world application are shown for the purpose of supporting the theoretical model.

2 Markov Decision Processes and Reinforcement Learning

In RL the main objective is to achieve a policy, that optimally moves an agent within an environment, which is defined by a Markov Decision Process (MDP) [4]. An MDP is generally given by a state space S, a set of actions A selectable in the different states, and the dynamics, defined by a transition probability distribution $P_T : S \times A \times S \rightarrow [0, 1]$ depending on the current state, the chosen action, and the successor state. The agent collects so-called rewards $R(s, a, s')$ while transiting. They are defined by a reward probability distribution P_R with the expected reward $R = \int_{\mathbb{R}} r P_R(s, a, s', r) dr$, $s, s' \in S, a \in A$.

In most RL tasks one is interested in maximising the discounting Value Function

$$V^\pi(s) = \mathbf{E}^\pi_\mathbf{s}\left(\sum_{i=0}^{\infty} \gamma^i R\left(s^{(i)}, \pi(s^{(i)}), s^{(i+1)}\right)\right)$$

for all possible states s where $0 < \gamma < 1$ is the discount factor, s' the successor state of s, $\pi : S \rightarrow A$ the used policy, and $\mathbf{s} = \{s', s'', \ldots, s^{(i)}, \ldots\}$. Since the dynamics of the given state-action space cannot be modified by construction one has to maximise V over the policy space. One typically takes one intermediate step and constructs a so-called Q-function depending on the current state and the chosen action approaching

$$Q^\pi(s, a) = \mathbf{E}_{s'}(R(s, a, s') + \gamma Q^\pi(s', \pi(s'))).$$

We define $V^* = V^{\pi_{\text{opt}}}$ as the Value Function of the optimal policy and Q^* respectively. This is the function we want to estimate. It is defined as

$$Q^*(s, a) = \mathbf{E}_{s'}(R(s, a, s') + \gamma V^*(s')) = \mathbf{E}_{s'}\left(R(s, a, s') + \gamma \max_{a'} Q^*(s', a')\right),$$

which is called the Bellman Optimality Equation. Therefore the best policy is apparently the one using the action $\pi(s) = \arg\max_a Q^*(s, a)$ maximising the (best) Q-function. We define the Bellman Operator T and the variance of the discounted sum of future rewards using T' as $(TQ)(s, a) = \mathbf{E}_{s'}(R(s, a, s') + \gamma \max_{a'} Q(s', a'))$ as well as $(T'Q)(s,a)^2 = \mathbf{Var}_{s'}(R(s,a,s') + \gamma \max_{a'} Q(s',a'))$ for any Q. There are several optimality criteria as approximations to the optimal Q-function. In this work, we concern the Bellman Residual minimisation, provided by the Q-function minimising the distance $\|Q - TQ\|$ (w.r.t. some appropriate norm) and the Temporal-Difference solution $Q = \mathbf{Solve}(TQ)$ given by the fixed point of the Bellman Operator followed by its projection on the Q-function's hypothesis space H_Q. For details we refer to [4,10,11].

3 Neural Rewards Regression

In the standard setting for NRR [1], we describe the Q-function for each possible action as Feed-Forward-Networks $N_a(s) = Q(s,a)$. The reward function to be fitted is hence given as $R(s,a,s') = N_a(s) - \gamma \max_{a'} N_{a'}(s')$, where the max-operator has to be modelled by an appropriate architecture. Alternatively, a monolithic approach with one network $N(s,a) = Q(s,a)$ which takes the state and action as input $R(s,a,s') = N(s,a) - \gamma \max_{a'} N(s',a')$, can also be applied. The second model has the advantage

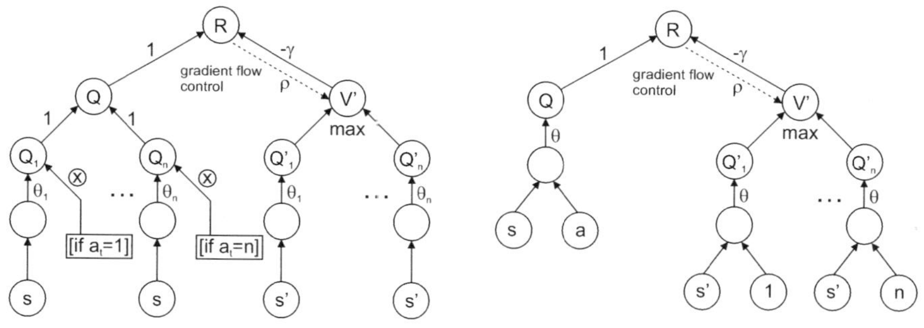

Fig. 1. Both alternatives of the NRR architecture (left: one network for each action, right: the monolithic approach). Connectors with the same parameters have shared weights.

of performing a generalisation over the action space as well. Using the Back Propagation algorithm this problem setting approaches the minimum of the (regularised) sampled Bellman Residual over all l observed transitions

$$L = \sum_{i=1}^{l} F(L_i) + \Omega(\theta) = \sum_{i=1}^{l} F\left(Q(s_i,a_i) - \gamma V(s_{i+1}) - r_i\right) + \Omega(\theta),$$

where θ are the network parameters, F an error function, and Ω an appropriate regulariser. The observed r_i and s_{i+1} have to be unbiased estimates for their expectations. The error function's gradient $\frac{dL}{d\theta} = \sum_{i=1}^{l} F'(L_i) \frac{d}{d\theta}\left(Q(s_i,a_i) - \gamma V(s_{i+1})\right) + \frac{d\Omega(\theta)}{d\theta}$ depends on the current Q-function and the successor Value Function (fig. 1). To obtain the fixed point of the Bellman Iteration instead, one retains $y_i := r_i + \gamma V(s_{i+1})$ and minimises iteratively $L = \sum_{i=1}^{l} F\left(Q(s_i,a_i) - y_i\right) + \Omega(\theta)$, until convergence of Q. Its gradient is then given as $\frac{dL}{d\theta} = \sum_{i=1}^{l} F'(L_i) \frac{d}{d\theta} Q(s_i,a_i) + \frac{d\Omega(\theta)}{d\theta}$. It can be seen, that both gradients differ only in their direction terms, but not in the error term. Blocking the gradient flow through the Value Function part of the network hence reconstructs the latter gradient. Therefore, in the backward path of the Back Propagation algorithm, the error propagation between the R-cluster and the V'-cluster is multiplied by a constant ρ (see fig. 1). The setting $\rho = 1$ leads to the sampled Bellman Residual minimisation, while for $\rho = 0$ one obtains the Bellman Iteration. Any other compromise is a possible trade-off between these two error definitions [12]. Optimality is then defined as the solution of the general equation system

$$\Gamma = \sum_{i=1}^{l} F'\left(Q(s_i,a_i) - \gamma V(s_{i+1}) - r_i\right) \frac{d}{d\theta}\left(Q(s_i,a_i) - \rho\gamma V(s_{i+1})\right) + \frac{d\Omega(\theta)}{d\theta} = 0.$$

A closer look at the approach reveals the similarities to TD-Learning [4,13]. Combining $\rho = 0$, $F(x) = x^2$, and an incremental learning scheme, NRR is identical to that classical approach. But the architecture allows us to apply the whole palette of established learning algorithms for Neural Networks [14].

4 Auxiliared Bellman Residual

It is well-known, that minimising the Bellman Residual on the one hand has the advantage of being a well-controllable learning problem as it is close to a Supervised Learning scheme, but on the other hand, tends to minimise terms of higher-order moments of the discounted sum of future rewards in the stochastic case, if no further, uncorrelated samples of every transition can be given [4,10]. In general the solutions are biased to Q-functions which are smoother for successor states of stochastic transitions. Specifically, if s_{i+1} and r_i are unbiased estimates for the successor states and rewards, respectively, the expression $\left(Q(s_i,a_i) - \gamma V(s_{i+1}) - r_i\right)^2$ is not an unbiased estimate for the true squared Bellman Residual $(Q(s,a) - (TQ)(s,a))^2$, but for $(Q(s,a) - (TQ)(s,a))^2 + (T'Q)(s,a)^2$. As an alternative to the usage of doubled trajectories, which is no option in our setting, Antos et. al. [2] introduced a modified squared Bellman Residual as a better approximation of the true Bellman Residual. They left it unnamed, so that we choose the working name Auxiliared Bellman Residual, defined as

$$L_{\text{aux}} = \sum_{i=1}^{l} \left(Q(s_i,a_i) - \gamma V(s_{i+1}) - r_i\right)^2 - \sum_{i=1}^{l} \left(h(s_i,a_i) - \gamma V(s_{i+1}) - r_i\right)^2$$

The task is then to solve $\hat{Q} = \arg\min_{Q \in H_Q} \max_{h \in H_h} L_{\text{aux}}$. The idea behind is to find an h that fits the Bellman operator over the observations. Its unavoidable error cancels the variance of the original expression down. We obtain

$$\begin{aligned} Z &= \mathbf{E}_{s'}(Q(s,a) - \gamma V(s') - R(s,a,s'))^2 - \mathbf{E}_{s'}(h(s,a) - \gamma V(s') - R(s,a,s'))^2 \\ &= (Q(s,a) - (TQ)(s,a))^2 - \mathbf{Err}(h(s,a),(TQ)(s,a)), \end{aligned}$$

which is the true loss function with an additional error term due to the suboptimal approximation of h, if H_h is unable to fit the Bellman operator arbitrarily precisely. This technique allows us to upper-bound the true Bellman Residual, if the error of

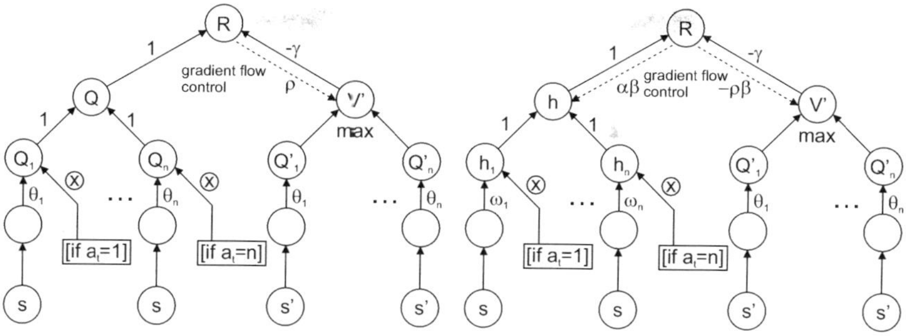

Fig. 2. The Auxiliared Bellman Residual NRR architecture. Both networks are trained simultaneously. The right network is used to obtain an approximation of $T'Q^2$.

h on TQ can be bounded. However, it is easy to see, that $\hat{L} \leq L$ within a saddle point of L_{aux}, if $H_Q = H_h$. Otherwise h would not provide the minimum of $\hat{L}$. Therefore, an optimum of L_{aux} would be provided by any Temporal-Difference fixed point, if it exists, as only in that case Q can fit the Bellman Operator as well as h and $L_{\text{aux}} = 0$. In contrast to the original proposal [2], we hence choose H_h either as a considerable richer function class than H_Q or with prior knowledge of the true Bellman Operator, such that $\hat{L}$ basically provides a better estimate of $T'Q^2$. As any such estimate of the variance is still biased, the method converges to a biased estimator of the true Bellman Residual minimising function $\hat{Q}^* \in H_Q$ within its function space only, but obviously provides a better approximation than the estimator given in sec. 3. In the following we consider the standard setting $F(x) = x^2$ and obtain the gradients

$$\Delta\theta = \Gamma + \beta\rho\gamma \sum_{i=1}^{l} \left(h(s_i, a_i) - \gamma V(s_{i+1}) - r_i\right) \frac{d}{d\theta} V(s_{i+1})$$

$$\Delta\omega = \alpha\beta \sum_{i=1}^{l} \left(h(s_i, a_i) - \gamma V(s_{i+1}) - r_i\right) \frac{d}{d\omega} h(s_i, a_i)$$

where ω are the accordant parameters describing h, $0 \leq \beta \leq 1$ controlling the influence of the auxiliary function h, and $\alpha \geq 1$ the strength of the optimisation of h in comparison with Q. This is incorporated into the NRR architecture as follows. In addition to the original NRR network, responsible for minimising L, we install a second similar, but auxiliary network, where the Q-clusters are replaced by respective h-clusters while the network part for Q' remains the same (see fig. 2). All three Q-function networks are still connected via shared weights. Finally, the auxiliary network has to fulfil two tasks, maximising $\hat{L}$ w.r.t. θ and minimising $\hat{L}$ w.r.t. ω. Because of the negative sign of $\hat{L}$, the gradient through the Q' part of the auxiliary network has to be sent in the opposite direction. The parameter α has to be chosen sufficiently large, such that h can fit the Bellman Operator almost instantaneously.

5 Policy Gradient Neural Rewards Regression

We now consider our second improvement, by introducing PGNRR, which is primarily a further generalisation of NRR to continuous action spaces. In general it is feasible for special function classes only to detect $V(s) = \max_a Q(s, a)$ in polynomial time. One possibility is to apply the idea of wire-fitting [15] or comparable approaches to NRR. Much more general and simpler is an approach combining the regression scheme on the rewards with PG methods. Therefore we replace the ensembles of Q-function

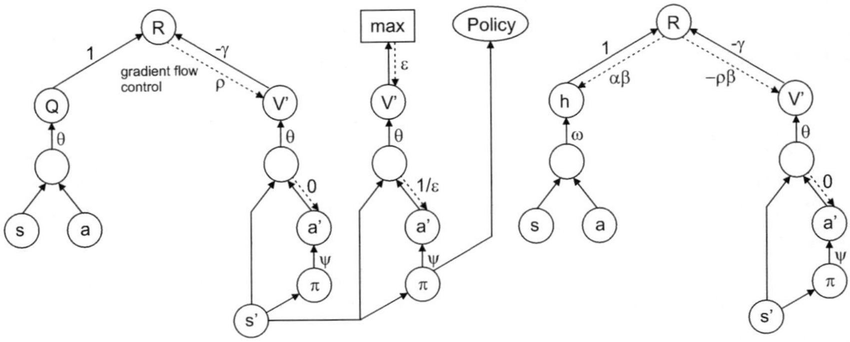

Fig. 3. The Policy Gradient NRR architecture. The additional architectural elements are used to train the policy network, whose task is to perform a near-optimal policy.

networks for the discrete actions by a single network evaluating $Q(s, a)$ for continuous states and actions. For the successor states we evaluate $Q(s', \pi(s'))$ where we assume that $\pi : S \rightarrow A$ with parameters ψ tend to perform the optimal policy (see fig. 3). Hence $Q(s', \pi(s'))$ gets close to $\max_{a'} Q(s', a')$. This is achieved by the maximisation of the Q-function for the successor states, simultaneously with the regression on the rewards. Therefore, a kind of Batch On-Policy-Iteration or Batch Actor-Critic Iteration is performed, by exploiting the intrinsic interrelationship between Q-function and policy. The gradient flow control technique in connection with shared weights is again sufficient to construct the appropriate architecture. In the standard network part for the successor state, the gradient flow through the policy network is simply cut off, such that the policy does not influence the regression on the rewards. In the extended part, a sufficiently small ϵ enables that only the policy contributes to the Q-function's maximisation. The common gradients are given as

$$\Delta\theta = \sum_{i=1}^{l} \Big(\left(Q(s_i, a_i) - \gamma Q(s_{i+1}, \pi(s_{i+1})) - r_i \right)$$
$$\frac{d}{d\theta} \left(Q(s_i, a_i) - \rho\gamma Q(s_{i+1}, \pi(s_{i+1})) \right) + \frac{\epsilon d}{d\theta} Q(s_{i+1}, \pi(s_{i+1}))$$
$$+\beta\rho\gamma \left(h(s_i, a_i) - \gamma Q(s_{i+1}, \pi(s_{i+1})) - r_i \right) \frac{d}{d\theta} Q(s_{i+1}, \pi(s_{i+1})) \Big) + \frac{d\Omega(\theta)}{d\theta}$$

$$\Delta\omega = \alpha\beta \sum_{i=1}^{l} \left(h(s_i, a_i) - \gamma Q(s_{i+1}, \pi(s_{i+1})) - r_i \right) \frac{d}{d\omega} h(s_i, a_i)$$

$$\Delta\psi = \sum_{i=1}^{l} \frac{d}{\epsilon d\psi} \pi(s_{i+1}) \frac{\epsilon d}{d\pi(s_{i+1})} Q(s_{i+1}, \pi(s_{i+1})) = \sum_{i=1}^{l} \frac{d}{d\psi} Q(s_{i+1}, \pi(s_{i+1})).$$

Still, a near-optimal policy is found by using Back-Propagation with shared weights only. After convergence, the policy can be evaluated by the network for π without using the Q-function as interim result.

6 Improving Data-Efficiency

With the problem class extension, we obtain an advancement of data-efficiency as well. This is based on two different arguments. The first one concerns the improvement w.r.t. the approachable optimality criteria. As we are now able to find the solution of a less biased Bellman Residual minimisation, we profit from that optimality criterion, which has several advantages [11,13]. Secondly, it has often been stated in the literature, that PG approaches are better able to profit from prior knowledge as the policies are usually known to be much simpler than the Q-functions and hence appropriate function classes can be given. Beside this argument, the question arises, if the combination of learning a Q-function and a policy, provides an improvement w.r.t. the bias-variance-dilemma [16] without special prior knowledge. This could be achieved by trading between the complexity of both hypothesis spaces $H_Q \subset C_Q$ and $H_\pi \subset C_\pi$, which are subsets of assumed concept spaces. However, there exists a function $\Pi : 2^{C_Q} \rightarrow 2^{C_\pi}$ mapping from all possible Q-function hypothesis spaces to the set of all possible policy hypothesis spaces and for $\Pi(H_Q) = H_\pi$ it holds

$$\forall Q \in H_Q : \exists \pi \in H_\pi : \forall s \in S : \pi(s) = \arg\max_a Q(s, a)$$
$$\forall \pi \in H_\pi : \exists Q \in H_Q : \forall s \in S : \pi(s) = \arg\max_a Q(s, a).$$

That is, $\Pi(H_Q)$ contains a corresponding policy for any $Q \in H_Q$ and vice versa. For illustration, we tested the Cerebellar Model Articulation Controller (CMAC) architecture [4,17] as regressor for the true Q-function together with the Wet-Chicken benchmark problem [18]. The bias-variance-trade-off is controlled by the number of tiles. The policy is chosen out of the respective function class $\Pi(H_Q)$, which can easily be given for this example, and is fitted, such that the average of $Q(s, \pi(s))$ is maximal over the observations. If we decrease the parameters of π, such that $H_\pi \subset \Pi(H_Q)$, then we additionally trade bias against variance over the policy space. In general, it is indeed possible to obtain the maximal values at a point that does not lie on the diagonal, where $H_\pi = \Pi(H_Q)$. Fig. 4 shows the results. However, a detailed theoretical analysis, under which conditions an improvement can be obtained, is part of future work.

7 Benchmark Results

In order to support the theoretical considerations on data-efficiency with empirical evidence, we tested the novel architecture on the Cart-Pole problem as described in [10] and on a simulation of a gas turbine, which has been used for a Reinforcement Learning project to reduce acceleration and pressure intensities in the combustion chamber (RMS) while maintaining high and stable load. A more detailed description can be found in [20]. We have chosen these two benchmarks, because standard NRR already shows competitive results in comparison with TD-Learning [13] on human-designed as well as automatically trained local basis functions, the Adaptive Heuristic Critic [4], Neural Fitted Q-Iteration [19], and the Recurrent Control Neural Network [20] as a model-based approach exploiting the special properties of the considered problem class. Unfortunately a fair comparison of NRR and PGNRR is difficult as each setting redounds to one method's advantage. We decided to rediscretise the continuous action policy. But it is apparent that a better performance in comparison to the natural discrete method is only a sufficient criterion for its prevalence as the continuous method could have found a policy which especially exploits its continuous action character.

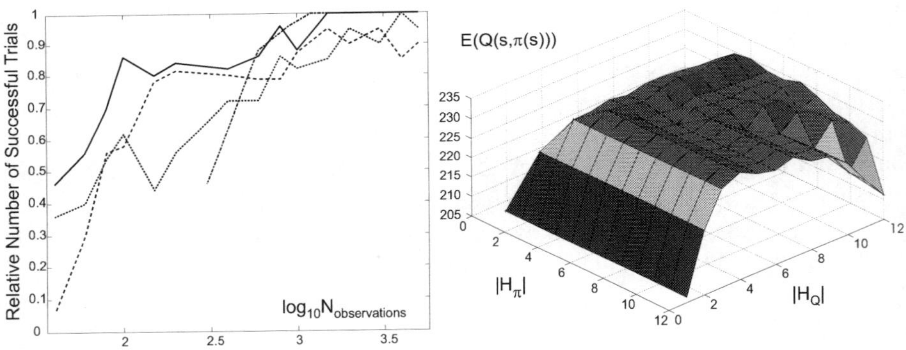

Fig. 4. Left: Performance comparison on the Cart-Pole problem. The figure shows the number of successful learning trials, where the controller balances the pole for at least 100000 steps, dashed: NRR, solid: PGNRR, dotted: PGNRR rediscretised, and dash-dotted: Neural Fitted Q-Iteration [19] for at least 3000 steps. Right: Bias-variance-trade-off for Q-function and policy for a tile-coding example. The maximal values are averaged over 100 test trials.

Nevertheless, for the Cart-Pole problem (see fig. 4), it can be seen that PGNRR performs considerably better than NRR and also its discretised version is quite comparable to the standard NRR. Both methods were trained with 4-layer networks. The most successful setting for NRR was already verified with Q-function networks with 16 and 8 neurons in the first and the second hidden layer, respectively. In accordance with our observations w.r.t. the bias-variance-trade-off, for PGNRR we improved the complexity of the Q-function (30 and 15 hidden neurons) and have chosen a smaller policy network (10 and 5 hidden neurons). As learning algorithm we applied the Vario-Eta method [21], where we observed best results for this problem. We have chosen a batch size of 5 and

Table 1. Performance comparison on a controller for a gas turbine simulation. We opposed our best NRR controller with a PGNRR controller for the same discrete action data set. As the underlying MDP's action space is two-dimensional we applied each a one- and two-action discretisation. Interestingly, the performance of the two discretised controllers is indeed better than on the natural discrete controller.

Gas Turbine	NRR Discrete	PGNRR Simple Discretised	PGNRR Double Discretised	PGNRR Continuous
Average Reward	0.8511 ± 0.0005	0.8531 ± 0.0005	0.8560 ± 0.0005	0.8570 ± 0.0005

$\eta = 0.05$. The input data was rescaled to constrain the inputs between -1 and $+1$. The net training was performed with an exponentially decreasing learning rate. For comparison we plotted the number of successful learning trials with Neural Fitted Q-Iteration [19], where the pole is balanced for at least 3000 steps, apparently the results are an upper bound for the presented setting.

As a second benchmark we applied NRR and PGNRR on a gas turbine simulation [20]. There are two different control variables, the so-called pilot gas and inlet guide vane (IGV). These variables are, beside other control variables and measurements, part of the state space. The actions are to wait and do nothing, to increase and decrease pilot gas, and to increase and decrease IGV each by a certain amount. Only one of these five actions is selectable in each time step. For the continuous version we allow a maximum of the discrete modification simultaneously for both control variables. We used the Recurrent Neural Rewards Regression (RNRR) model as described in [1] and an appropriate Recurrent PGNRR architecture for comparison. The recurrent substructure is necessary, because the state space is not Markovian, specifically the underlying MDP is of higher order.

For both architectures we used the same settings as described for the Cart-Pole problem, but additionally applied weight decay as regularisation with a factor $\lambda = 0.001$. Applying this setting we obtained the best results with the NRR architecture and also reproducible results with PGNRR. For comparison with the standard NRR, we rediscretised the continuous actions independent from each other as well as to the simple discrete case, where the larger relative action was discretised and the appropriate other action not executed. It can be seen in tbl. 1 that with both discretised versions indeed a reward improvement could achieved. Note, that a reward difference of 0.002 and 0.006 corresponds to an additional power output of 0.14 and 0.42 MW, respectively, with stable RMS for a power plant with a maximum power of 200 MW.

8 Conclusion

We introduced two generalisations of Neural Rewards Regression to extend the approachable optimality criteria and the considered problem class substantially. Of course, PGNRR can also be combined with Recurrent Neural Networks [1], so that a very broad problem class is addressed. Future work will mainly concern the broadening of its application for gas turbine control and the theoretical analysis of the improvement of data-efficiency.

References

1. Schneegass, D., Udluft, S., Martinetz, T.: Neural rewards regression for near-optimal policy identification in markovian and partial observable environments. In: Verleysen, M. (ed.) Proc. of the European Symposium on Artificial Neural Networks, pp. 301–306 (2007)
2. Antos, A., Szepesvári, C., Munos, R.: Learning near-optimal policies with bellman-residual minimization based fitted policy iteration and a single sample path. In: Lugosi, G., Simon, H.U. (eds.) COLT 2006. LNCS (LNAI), vol. 4005, pp. 574–588. Springer, Heidelberg (2006)
3. Schneegass, D., Udluft, S., Martinetz, T.: Kernel rewards regression: An information efficient batch policy iteration approach. In: Proc. of the IASTED Conference on Artificial Intelligence and Applications, pp. 428–433 (2006)
4. Sutton, R.S., Barto, A.G.: Reinforcement Learning: An Introduction. MIT Press, Cambridge (1998)
5. Hinton, G.E., McClelland, J.L., Rumelhart, D.E.: Distributed representations. In: Parallel Distributed Processing, pp. 77–109 (1986)
6. Haykin, S.: Neural Networks: A Comprehensive Foundation. Macmillan, New York (1994)
7. Sutton, R.S., McAllester, D., Singh, S., Mansour, Y.: Policy gradient methods for reinforcement learning with function approximation. In: Advances in Neural Information Processing Systems, vol. 12 (2000)
8. Wang, X., Dietterich, T.: Model-based policy gradient reinforcement learning. In: International Conference on Machine Learning (2003)
9. Ghavamzadeh, M., Engel, Y.: Bayesian policy gradient algorithms. In: Advances in Neural Information Processing Systems (2006)
10. Lagoudakis, M.G., Parr, R.: Least-squares policy iteration. Journal of Machine Learning Research, 1107–1149 (2003)
11. Munos, R.: Error bounds for approximate policy iteration. In: Proc. of the International Conference on Machine Learning, pp. 560–567 (2003)
12. Baird III, L.C.: Residual algorithms: Reinforcement learning with function approximation. In: Proc. of the International Conference on Machine Learning, pp. 30–37 (1995)
13. Tsitsikilis, J.N., Van Roy, B.: An analysis of temporal difference learning with function approximation. IEEE Transactions on Automatic Control 42(5), 674–690 (1997)
14. Pearlmutter, B.: Gradient calculations for dynamic recurrent neural networks: A survey. IEEE Transactions on Neural Networks 6(5), 1212–1228 (1995)
15. Baird, L., Klopf, A.: Reinforcement learning with high-dimensional, continuous actions. Technical Report WL-TR-93-1147, Wright Laboratory, Wright-Patterson Air Force Base, OH 45433-7301 (1993)
16. Geman, S., Bienenstock, E., Doursat, R.: Neural networks and the bias/variance dilemma. Neural Computation 4(1), 1–58 (1992)
17. Timmer, S., Riedmiller, M.: Fitted q iteration with cmacs. In: ADPRL. Proceedings of the IEEE International Symposium on Approximate Dynamic Programming and Reinforcement Learning, pp. 1–8. IEEE Computer Society Press, Los Alamitos (2007)
18. Tresp, V.: The wet game of chicken. Siemens AG, CT IC 4, Technical Report (1994)
19. Riedmiller, M.: Neural fitted q-iteration - first experiences with a data efficient neural reinforcement learning method. In: Proceedings of the 16th European Conference on Machine Learning, pp. 317–328 (2005)
20. Schaefer, A.M., Schneegass, D., Sterzing, V., Udluft, S.: A neural reinforcement learning approach to gas turbine control. In: Proc. of the International Joint Conference on Neural Networks (2007) (to appear)
21. Neuneier, R., Zimmermann, H.G.: How to train neural networks. In: Orr, G.B., Müller, K.-R. (eds.) Neural Networks: Tricks of the Trade. LNCS, vol. 1524, pp. 373–423. Springer, Heidelberg (1998)

Structure Learning with Nonparametric Decomposable Models

Anton Schwaighofer[1], Mathäus Dejori[2], Volker Tresp[2], and Martin Stetter[2]

[1] Fraunhofer FIRST, Intelligent Data Analysis Group, Kekulestr. 7, 12489 Berlin, Germany
http://ida.first.fraunhofer.de/~anton
[2] Siemens Corporate Technology, 81730 Munich, Germany

Abstract. We present a novel approach to structure learning for graphical models. By using nonparametric estimates to model clique densities in decomposable models, both discrete and continuous distributions can be handled in a unified framework. Also, consistency of the underlying probabilistic model is guaranteed. Model selection is based on predictive assessment, with efficient algorithms that allow fast greedy forward and backward selection within the class of decomposable models. We show the validity of this structure learning approach on toy data, and on two large sets of gene expression data.

1 Introduction

Recent years have seen a wide range of new developments in the field of graphical models [1,2], in particular in the area of Bayesian networks, and in the use of graphical models for probabilistic inference in complex systems. Graphical models can either be constructed manually, by referring to the (assumed) process that underlies some system, or by learning the graphical model structure from sample data [3]. In this article, we present a novel method for learning the structure of a decomposable graphical model from sample data. This method has been developed particularly for decomposable models on continuous random variables. Yet, through its use of nonparametric density estimates, it can also handle discrete random variables within the same framework.

The major motivation for this work was to develop a structure learning approach that can be used without discretization of the data, but does not make strong distributional assumptions (most previous approaches assume joint Gaussianity [1,4,5], exceptions are [6,7]). In Sec. 4, we will consider the problem of learning the structure of functional modules in genetic networks from DNA microarray measurements. In this important application domain, it has been noted that discretization needs to be conducted very carefully, and may lead to a large loss of information [8].

In the field of molecular biology, it seems particularly reasonable to use decomposable graphical models. Functional modules are considered to be a critical level of biological organization [9]. We see a close relationship between the cliques of a decomposable model (that is, a fully connected subgraph) and the

J. Marques de Sá et al. (Eds.): ICANN 2007, Part I, LNCS 4668, pp. 119–128, 2007.

functional modules of the genetic network. A clique is interpreted as a set of *functionally correlated* genes. The structure of cliques then describes the relationships between individual functional modules. As opposed to simple clustering approaches, the graphical model structure can also represent the fact that some genes contribute to many of these functional modules.

We proceed by first giving a brief introduction to decomposable models and structure learning in Sec. 2. Sec. 3 presents nonparametric decomposable models, which are decomposable models based on nonparametric density estimation, and the according structure learning methodology. In Sec. 4 we demonstrate the performance of the algorithm on toy data, and then apply it to estimate the structure of functional modules in two different microarray data sets.

2 Decomposable Models

A graphical model (or, probabilistic network) describes a family of probabilistic models, based on a directed or undirected graph $G = (V, E)$. Nodes V in the graph represent random variables, whereas the edges E stand for probabilistic dependencies. Common classes of graphical models [1,2] are Bayesian networks (based on graphs with directed edges), Markov networks (based on graphs with undirected edges), decomposable models (undirected chordal graphs) and chain graphs (graphs with both undirected and directed edges).

To define decomposable models, we assume a set of n random variables $\{x_1, \ldots, x_n\}$, represented in an undirected graphical model with nodes $V = \{1, \ldots, n\}$. The *absence* of an edge (i, j) between variables x_i and x_j implies that x_i and x_j are independent, conditioned on all other random variables (conditional independence), denoted by $x_i \perp\!\!\!\perp x_j \,|\, x_{\{1,\ldots,n\}\setminus\{i,j\}}$, the pairwise Markov property. If the graph G describes a decomposable model (see the definition below), the joint density can be written in terms of marginal densities of the random variables contained in cliques of the graph (fully connected subgraphs),

$$p(\boldsymbol{x}) = \frac{\prod_{C \in \mathcal{K}} p(\boldsymbol{x}_C)}{\prod_{S \in \mathcal{S}} p(\boldsymbol{x}_S)}. \tag{1}$$

Here, $\mathcal{K}$ denotes the set of cliques in graph V, and $\mathcal{S}$ the set of separators (that is, the intersections of two neighboring cliques).

This factorization is only possible if and only if G describes a decomposable model, that is, if and only if the graph G is a chordal graph (see Ch. 4 of [2]), or, equivalently, if the graph G can be represented in the form of a join tree[1] The join tree of a graph G is a tree $T = (\mathcal{K}, F)$ with the clique set $\mathcal{K}$ as its node set and edges F, that satisfies the clique intersection property: For any two cliques $C_1, C_2 \in \mathcal{K}$, the set $C_1 \cap C_2$ is contained in every clique on the (unique) path between C_1 and C_2 in T. From the adjacent cliques in the join tree, we can also compute the separators $\mathcal{S}$ required in (1).

[1] Also called the junction tree, or clique tree.

Structure Learning: In general, the problem of finding optimal graphical models from data under a non-trivial scoring function is NP-hard [10]. A large number of (heuristic) approaches has been developed to estimate graphical models from data, such as constraint-based methods [11], frequentist [1] and fully Bayesian approaches [12]. We will subsequently use a scoring function based method, where we combine a function that estimates the model quality (the scoring function) with a suitable search strategy through model space.

3 Nonparametric Decomposable Models

Previous structure learning approaches for continuous variables often assume that all of the involved random variables have a joint multivariate Gaussian distribution. While being computationally attractive, we believe that this can only be observed in very limited domains, and thus wish to develop structure learning for continuous variables with general probability distributions. In our representation we employ kernel density estimates [13] for each clique in the decomposable model.

This choice brings some specific advantages: Provided that a suitable kernel function is chosen, no re-fitting of the density models is necessary after the model structure has been changed. Also, clique density models will remain consistent (that is, corresponding marginal distributions match, see [6] for an approach where this caused numerous problems).

For a set of m samples $\mathcal{D} = \{\boldsymbol{x}^1, \ldots, \boldsymbol{x}^m\}$ (each $\boldsymbol{x}^i \in \mathbb{R}^n$) from a probability distribution over random variables $\{x_1, \ldots, x_n\}$, a kernel density estimate [13] is

$$p(\boldsymbol{x} \mid \mathcal{D}, \boldsymbol{\theta}) = \frac{1}{m} \sum_{i=1}^{m} g(\boldsymbol{x}; \boldsymbol{x}^i, \boldsymbol{\theta}). \tag{2}$$

As the kernel function g, we chose a Gaussian,

$$g(\boldsymbol{x}; \boldsymbol{x}^i, \boldsymbol{\theta}) = (2\pi)^{-n/2} |\mathrm{diag}\, \boldsymbol{\theta}|^{-1/2} \exp\left(-\frac{1}{2} (\boldsymbol{x} - \boldsymbol{x}^i)^\top (\mathrm{diag}\, \boldsymbol{\theta})^{-1} (\boldsymbol{x} - \boldsymbol{x}^i) \right), \tag{3}$$

with the variance along the jth dimension given by θ_j, $j = 1, \ldots, n$.

For the proposed nonparametric decomposable models, we require models for each clique and separator in Eq. 1, and thus model all clique marginal distributions by nonparametric density estimates. Denoting by $\boldsymbol{x}_C$ the vector of random variables appearing in clique C, and $\mathcal{D}_C = \{\boldsymbol{x}_C^1, \ldots, \boldsymbol{x}_C^m\}$, the clique marginal model for clique C is

$$p(\boldsymbol{x}_C \mid \mathcal{D}_C, \boldsymbol{\theta}_C) = \frac{1}{m} \sum_{i=1}^{m} g(\boldsymbol{x}_C; \boldsymbol{x}_C^i, \boldsymbol{\theta}_C), \tag{4}$$

Note that choosing this form of nonparametric density estimates in Eq. 1 automatically ensures an essential property of density models in this context: With constant parameter vector $\boldsymbol{\theta}$, all clique and separator marginal models are consistent, by means of the properties of the Gaussian kernel function. Consistency means that, when considering clique models $p(x_1, x_2)$ ad $p(x_2, x_3)$, the marginals $p(x_2)$ that can be computed from both clique models will match.

Choosing Kernel Parameters: In our experiments, we choose the variance parameters $\boldsymbol{\theta}$ for the nonparametric density models in Eq. 4 by maximizing the leave-one-out log likelihood of the data $\mathcal{D}$,

$$\hat{\boldsymbol{\theta}} = \arg\max_{\boldsymbol{\theta}} \sum_{i=1}^{m} \log p(\boldsymbol{x}^i \mid \mathcal{D} \setminus \boldsymbol{x}^i, \boldsymbol{\theta}), \tag{5}$$

via a conjugate gradient algorithm. Setting the parameters $\boldsymbol{\theta}$ is done once at the beginning of structure learning, $\boldsymbol{\theta}$ remains fixed thereafter.

3.1 Learning: Scoring Functions

For learning the structure of nonparametric decomposable models, we used a scoring-based learning strategy, with model scoring based on predictive assessment. The key idea of predictive assessment model scores is to evaluate the model on data not used for model fitting. The advantage of such scores lies in its computational simplicity (for example, marginal likelihood is most often very difficult and time-consuming to compute) and in its insensitivity to possibly incorrect model assumptions. Commonly used variants of predictive assessment are cross-validation (asymptotically equivalent to maximum likelihood with AIC complexity term, [2]) and prequential validation (ML with BIC complexity term). We chose predictive assessment with 5-fold cross-validation as our scoring function.

The joint density function of a nonparametric decomposable model is given in Eq. 1. Taking logs, Eq. 1 splits into the sum of clique and separator contributions, such that the 5-fold cross-validation can be evaluated separately on cliques and separators. The contribution of a single clique C (resp. separator S) to the total cross-validation log-likelihood is denoted by the clique score $\mathcal{L}(C)$,

$$\mathcal{L}(C) = \mathcal{L}(\mathcal{D}_C) = \sum_{k=1}^{5} \sum_{\boldsymbol{x}_C \in \mathcal{D}_C^k} \log p\left(\boldsymbol{x}_C \mid \mathcal{D}_C \setminus \mathcal{D}_C^k\right) \tag{6}$$

By this we mean that data $\mathcal{D} = \{\boldsymbol{x}^1, \ldots, \boldsymbol{x}^m\}$ are split into 5 disjoint sets $\mathcal{D}^1, \ldots, \mathcal{D}^5$. Parzen density estimates are built from all data apart from $\mathcal{D}^k$, and evaluated on the data in $\mathcal{D}^k$. The subscript C denotes a restriction of all quantities to the random variables contained in clique C.

The overall cross-validation log-likelihood (model score) for the decomposable model, given by its set of cliques $\mathcal{K} = \{C_1, \ldots, C_A\}$ and separators $\mathcal{S} = \{S_1, \ldots, S_B\}$, simply becomes

$$\mathcal{L}(\mathcal{K}, \mathcal{S}) = \sum_{j=1}^{A} \mathcal{L}(C_j) - \sum_{k=1}^{B} \mathcal{L}(S_k) \tag{7}$$

Scores after Model Change: Based on the model score in Eq. 7, it is straightforward to derive the change of model score if an edge is inserted into the model. In particular, the difference of scores can be computed from local changes only, i.e., it is only necessary to consider the cliques involved in the insert operation.

Consider inserting an edge (u, v), thereby connecting cliques C_u and C_v. In the current model G, the contribution of these two cliques and their separator $S_{uv} = C_u \cap C_v$ to the model score is $\mathcal{L}(C_u) + \mathcal{L}(C_v) - \mathcal{L}(S_{uv})$. Inserting edge (u, v) creates a model G' with a new clique $C_w = S_{uv} \cup \{u, v\}$ and separators $S_{uw} = C_u \cap C_w = S_{uv} \cup \{u\}$ and $S_{vw} = C_v \cap C_w = S_{uv} \cup \{v\}$. The change of model score from G to G' thus simply becomes

$$\Delta_{uv} = \mathcal{L}(S_{uv}) + \mathcal{L}(S_{uv} \cup \{u, v\}) - \mathcal{L}(S_{uv} \cup \{u\}) - \mathcal{L}(S_{uv} \cup \{v\}) \qquad (8)$$

One can easily verify that this equation also holds for the case of merging cliques, i.e., the case when C_u and/or C_v are no longer maximal in G' and merge with C_w.

[14] prove that the number of edge scores that need to be re-computed after inserting an edge has a worst case bound of $O(n)$. In practice, most of the edge scores remain unchanged after an edge insertion, and only few edge scores need to be recomputed. For example, in a problem involving 100 variables, 5 edge scores were recomputed on average. We observed in our experiments that the average number of edge scores to recompute seems to grow under linear.

3.2 Learning: Search Strategy

In searching for a model that matches well with data (i.e., has a high scoring function), we traverse the space of decomposable models using a particular search strategy. Commonly used search strategies are greedy forward selection (start from an initially empty graph, iteratively add edges that brings the highest improvement in terms of scoring function) or greedy backward elimination (start from a fully connected graph, iteratively delete edges). In our approach, we use a *combination of forward and backward search*, starting from an empty graph. We either add or delete edges, depending on which operation brings the largest improvement of scoring function.

The search for candidate models still needs to be restricted to the class of decomposable models: In the current decomposable model graph $G = (V, E)$, we can attempt to either insert or delete an edge (u, v) to obtain graph G'. Is edge (u, v) a valid edge, in that the new model G' is still a decomposable model? [14] presented a method that is suitable for use with greedy forward selection. We use an approach inspired by dynamic algorithms for chordal graphs [15]. With this approach, enumerating all valid edges can be performed in $O(n^2 \log n)$ amortized time. As a further advantage over [14], the information about separators and cliques of the graph (required for computing the model score) is always available.

Checking Decomposability of G': To check chordality of G', we define a weight function $w : \mathcal{K} \times \mathcal{K} \rightarrow \mathbb{N}_0$, that assigns each edge $e = (C_1, C_2)$ of the join tree a weight $w(e) = w(C_1, C_2) = |C_1 \cap C_2|$. The following theorem [15] now checks whether we can insert an edge (u, v) while maintaining decomposability:

Theorem 1. *Let G be a chordal graph without edge (u, v). Let T be the join tree of G, and let C_u, C_v be the closest nodes in T such that $u \in C_u, v \in C_v$. Assume $(C_u, C_v) \notin T$. There exists a clique tree T' of G' with $(C_u, C_v) \in T'$ iff the minimum weight edge e on the path between C_u and C_v in T satisfies $w(e) = w(C_u, C_v)$.*

Checking whether deleting edge (u, v) maintains decomposability is a bit easier:

Theorem 2. *Let G be a chordal graph with edge (u, v). $G \backslash (u, v)$ is decomposable if and only if G has exactly one maximal clique containing (u, v).*

Splay Tree Representation for the Join Tree: In the above theorems, the major operation on the join tree is searching for the closest cliques that contain the variables u and v. [16] present a representation for trees that allows a particularly efficient implementation of shortest path searches, with only $O(\log n)$ operations per search. We use this data structure to maintain the join tree T. The representation is based on self-adjusting binary search trees, the so-called splay trees. It can be shown that all standard tree operations (in particular, the link, cut and shortest path search operations required for the decomposability check) have an amortized time bound of $O(\log n)$ for a splay tree with n nodes. A chordal graph on n nodes has a join tree with at most n cliques, thus all join tree operations can be implemented in $O(\log n)$.

4 Experiments

4.1 Toy Data

In a first experiment, we investigate whether the structure learning algorithm presented in the previous sections can indeed recover the true structure of some data. To this aim, we draw 50 samples from a decomposable model over 8 variables, where each clique is modelled by a randomly initialized Gaussian mixture model with 10 components (with consistency between cliques ensured). In Fig. 1 the model score $\mathcal{L}(\mathcal{D} \,|\, \mathcal{K}, \mathcal{S})$, as defined in Eq. 7, is plotted as more and more edges are added to the model. We found that the algorithm recovers the true structure of the generating decomposable model when $\mathcal{L}(\mathcal{D} \,|\, \mathcal{K}, \mathcal{S})$ is at its maximum.

4.2 Learning from DNA Microarray Measurements

We used our method two learn the structure from two gene expression data sets. To also estimate the confidence of each of the learned structures, we used a 20-fold bootstrap scheme: For each of the perturbed bootstrap data sets $\mathcal{D}^{(i)}$, we use the structure obtained when the model score Eq. 7 is at its maximum. In the analysis, we only consider edges that have a confidence of 90% or above

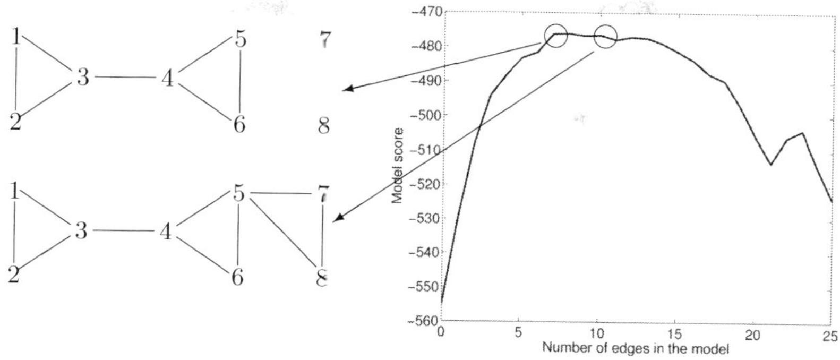

Fig. 1. Test of the structure learning algorithm on a toy model, where 50 samples were drawn from a decomposable model with known structure. The plot shows the model score, as defined in Eq. 7, when edges are added successively. The structure found when the model score is at its maximum is exactly the structure of the generating model (shown top left). Shown bottom left is a structure that also obtains a high model score.

(edges that were found in at least 18 out of the 20 runs). Yet, thresholding by edge confidence may lead to a non-decomposable result model. To again obtain a decomposable model, we implemented the following greedy insertion scheme: Maintain a list L of all edges with a bootstrap confidence estimate of 90% or above. Sort the list by ascending confidence value. Start with an empty re-built model (denoted by G' in the following). Consider now the highest confidence edge in list L. If this edge can be inserted into G' without loosing decomposability of G' (using the test described in Sec. 3.2), then insert the edge into G' and delete it from L. Otherwise, proceed with the next edge in L. The algorithm terminates if L is empty, or if no edge can be inserted into G' without loosing decomposability.

St. Jude Leukemia Data Set. We first applied our approach on data obtained from measuring gene-expression levels in human bone marrow cells of 7 different pedriatric acute lymphoblastic leukemia (ALL) subtypes [17]. Out of the 12.000 measured genes, those are selected that best define the individual subtypes using a Chi-square test. For each of the 7 subtypes the 40 most discriminative genes are chosen yielding to a set of 280 genes. 9 genes show a discriminative behavior for more than one subtype but were included only once resulting in a final dataset of $n = 327$ samples and $p = 271$ genes.

The learned network topology [18] (with restriction to high confidence edges, as described in the previous section) shows a few highly connected genes, with most edges connecting genes that are known to belong to the same ALL subtype. Thus, most genes are conditionally independent from each other, given one of these highly connected genes. Biologically speaking, the expression behavior of many genes only depends on a set of few genes, which therefore are supposed to play a key role in the learned domain. Since the structure is inferred from leukemia data, a high connectivity may indicate a potential importance for leukemogenesis or for tumor development in general. In fact, as shown in Tab. 1, highly connected genes are either known to be genes with an oncogenic characteristic or known to be involved in critical biological processes, like immune response (HLA-DRA), protein degradation (PSMD10), DNA repair (PAPR1) or embryogenesis (SCML2). PSMD10, for example, is known to act as a regulatory subunit of the 26S proteasome. PSMD10 connects most cliques that contain genes which are altered in ALL subtype *hyperdipl>50*. The dominant role of PSMD10 in the *hyperdipl>50* subtype seems reasonable, since the 26S proteasome is involved in general protein degradation and its hyperactivity is likely to be a response to the excessive protein production caused by the hyperdiploidy. HLA-DRA, the most highly connected gene belongs to the major histocompatibility complex class 2 (MCH II) which has been reported to be associated with leukemia and various other cancer types. The other seven members of the MCH complex which are present in the analyzed data set are either part of the same clique as HLA-DRA or part of an adjacent clique. The compact representation of this functional module emphasizes the capability of our approach to learn functional groups within a set of expression data.

Table 1. Genes with highest connectivity in the graphical model structure learned from the ALL data set (expression patterns in bone marrow cells of leukemia)

Gene	Affymetrix ID	# of connections	Annotation
HLA-DRA	37039_at	29	major histocompatibility complex, class II, DR alpha
PSMD10	37350_at	23	proteasome (prosome, macropain) 26S subunit, non-ATPase, 10
PAPR1	1287_at	17	poly (ADP-ribose) polymerase family, member 1
SCML2	38518_at	15	sex comb on midleg-like 2 (Drosophila)

Spira Data Set. We next analyzed gene expression data derived from human epithelial cells of current, former and never smoking subjects (101 in total) taken from [19]. The 96 most discriminative probes were selected, then structure learning was applied as described in the previous section. Tab. 2 lists the three highest degree genes in the learned decomposable model. All of the highest connected genes have detoxifying and antioxidant properties. The NQO1 gene, for example, serves as a quinone reductase in connection with conjugation reactions of hydroquinons involved for example in detoxification pathways. Its high connectivity seems reasonable as it prevents the generation of reactive oxygen radicals and protects cells from oxidative challenges such as the exposure to cigarette smoke. Fig. 2 shows the NQO1 subgraph, plotted with the radius of each gene node proportional to its connectivity. Besides the central role of NQO1, note that multiply appearing genes are grouped into cliques (for example, the three probes that correspond to the UGTA1A6 gene are all put into one clique).

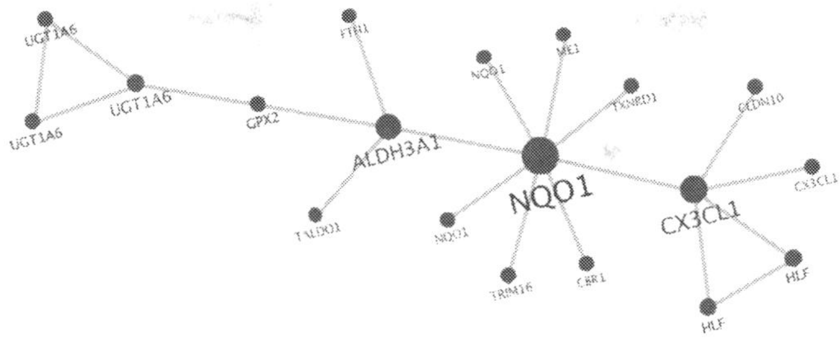

Fig. 2. The NQO1 subgraph obtained from the Spira data set. The area of each gene node is proportional to its degree.

Table 2. The three highest connected genes in the graphical model structure learned from the Spira data set (expression patterns in epithelial cells of smokers and non-smokers)

Gene	Affymetrix ID	# of connections	Annotation
NQO1	210519_s_at	8	NAD(P)H dehydrogenase, quinone 1
CX3CL1	823_at	4	chemokine (C-X3-C motif) ligand 1
MTX1	208581_x_at	4	metaxin 1

5 Conclusions

We presented a novel approach to learning a decomposable graphical model from data with continuous variables. Key issues for this algorithm are non-parametric kernel density estimates for each clique, and an efficient method for restricting search to the class of decomposable models. The method permits working directly with continuous data, without discretization as a pre-processing step. Our experiments on toy data and two gene expression data sets confirmed that the structure learning method does find meaningful structures.

We are currently applying our approach to larger sets of gene expression data. Graphical models are used increasingly in bioinformatics, and we believe that the structure learning approach presented in this article has a wide range of applications there. In particular, we plan to use the results of decomposable model learning as hypothesis structures for more detailed modelling. Currently, detailed modelling of regulatory processes in cells (including dynamical effects) is a very active research topic. These methods are often computationally intensive, and can only be applied to networks with a small number of genes. Thus, it is important to first find hypothesis networks, that are in turn modelled in more detail. We believe that the structure learning method presented in this paper can serve this purpose very well.

References

1. Lauritzen, S.L.: Graphical Models. Oxford Statistical Science Series, vol. 17. Clarendon Press, Oxford (1996)
2. Cowell, R.G., Dawid, A.P., Lauritzen, S.L., Spiegelhalter, D.J.: Probabilistic Networks and Expert Systems. In: Statistics for Engineering and Information Science, Springer, Heidelberg (1999)
3. Heckerman, D.: A tutorial on learning with Bayesian networks. In: Jordan, M.I. (ed.) Learning in Graphical Models, MIT Press, Cambridge (1998)
4. Song, Y., Goncalves, L., Perona, P.: Unsupervised learning of human motion. IEEE Transactions on Pattern Analysis and Machine Intelligence 25(7), 814–827 (2003)
5. Banerjee, O., El Ghaoui, L., d'Aspremont, A., Natsoulis, G.: Convex optimization techniques for fitting sparse gaussian graphical models. In: De Raedt, L., Wrobel, S. (eds.) Proceedings of ICML06, pp. 89–96. ACM Press, New York (2006)
6. Hofmann, R., Tresp, V.: Nonlinear Markov networks for continuous variable. In: Jordan, M.I., Kearns, M.J., Solla, S.A. (eds.) Advances in Neural Information Processing Systems, vol. 10, MIT Press, Cambridge (1998)
7. Friedman, N., Nachman, I.: Gaussian process networks. In: Proceedings of UAI 2000, pp. 211–219. Morgan Kaufmann, San Francisco (2000)
8. Friedman, N., Linial, M., Nachman, I., Pe'er, D.: Using bayesian networks to analyze expression data. Journal of Computational Biology 7, 601–620 (2000)
9. Hartwell, L.H., Hopfield, J.J., Leibler, S., Murray, A.W.: From molecular to modular cell biology. Nature 402, C47 (1999)
10. Bouckaert, R.R.: Properties of Bayesian belief network learning algorithms. In: de Mantaras, R.L., Poole, D.L. (eds.) Proceedings of UAI 94, pp. 102–109. Morgan Kaufmann, San Francisco (1994)
11. de Campos, L.M.: Characterizations of decomposable dependency models. Journal of Artificial Intelligence Research 5, 289–300 (1996)
12. Giudici, P., Green, P.J.: Decomposable graphical Gaussian model determination. Biometrika 86, 785–801 (1999)
13. Silverman, B.W.: Density Estimation for Statistics and Data Analysis. In: Monographs on Statistics and Applied Probability, vol. 26, Chapman & Hall, Sydney, Australia (1986)
14. Deshpande, A., Garofalakis, M., Jordan, M.I.: Efficient stepwise selection in decomposable models. In: Breese, J., Koller, D. (eds.) Proceedings of UAI 2001, Morgan Kaufmann, San Francisco (2001)
15. Ibarra, L.: Fully dynamic algorithms for chordal graphs and split graphs. Technical Report DCS-262-IR. Department of Computer Science, University of Victoria, CA (2000)
16. Sleator, D.D., Tarjan, R.E.: Self-adjusting binary search trees. Journal of the ACM 32(3), 652–686 (1985)
17. Yeoh, E.J., et al.: Classification, subtype discovery, and prediction of outcome in pediatric acute lymphoblastic leukemia by gene expression profiling. Cancer Cell 1(2), 133–143 (2002)
18. Dejori, M., Schwaighofer, A., Tresp, V., Stetter, M.: Mining functional modules in genetic networks with decomposable graphical models. OMICS A Journal of Integrative Biology 8(2), 176–188 (2004)
19. Spira, A., Beane, J., Shah, V., Liu, G., Schembri, F., Yang, X., Palma, J., Brody, J.S.: Effects of cigarette smoke on the human airway epithelial cell transcriptome. Proceedings of the National Academy of Sciences of the United States of America 101(27), 10143–10148 (2004)

Recurrent Bayesian Reasoning in Probabilistic Neural Networks

Jiří Grim and Jan Hora

[1] Institute of Information Theory and Automation
of the Czech Academy of Sciences,
P.O. Box 18, 18208 Prague 8, Czech Republic
[2] Faculty of Nuclear Science and Physical Engineering
Czech Technical University,
Trojanova 13, CZ-120 00 Prague 2, Czech Republic

Abstract. Considering the probabilistic approach to neural networks in the framework of statistical pattern recognition we assume approximation of class-conditional probability distributions by finite mixtures of product components. The mixture components can be interpreted as probabilistic neurons in neurophysiological terms and, in this respect, the fixed probabilistic description becomes conflicting with the well known short-term dynamic properties of biological neurons. We show that some parameters of PNN can be "released" for the sake of dynamic processes without destroying the statistically correct decision making. In particular, we can iteratively adapt the mixture component weights or modify the input pattern in order to facilitate the correct recognition.

1 Introduction

The concept of probabilistic neural networks (PNN) relates to the early work of Specht [18] and others (cf. [13], [19], [15]). In this paper we consider the probabilistic approach to neural networks based on distribution mixtures with product components in the framework of statistical pattern recognition. We refer mainly to our papers on PNN published in the last years (cf. [3] - [12]).

Let us recall that, given the class-conditional probability distributions, the Bayes decision function minimizes the classification error. For this reason, in order to design PNN for pattern recognition, we approximate the unknown class-conditional distributions by finite mixtures of product components. In particular, given a training data set for each class, we compute the estimates of mixture parameters by means of the well known EM algorithm [17,2,16]. The main idea of the considered PNN is to view the components of mixtures as formal neurons. In this way there is a possibility to explain the properties of biological neurons in terms of the component parameters. The mixture-based PNN do not provide a new technique of statistical pattern recognition but, on the other hand, the interpretation of a theoretically well justified statistical method is probably the only way to better understanding of the functional principles of biological neural networks.

J. Marques de Sá et al. (Eds.): ICANN 2007, Part I, LNCS 4668, pp. 129–138, 2007.

The primary motivation of our previous research has been to demonstrate different neuromorphic features of PNN. Simultaneously the underlying statistical method has been modified in order to improve the compatibility with biological neural networks. We have shown that the estimated mixture parameters can be used to define information preserving transform with the aim to design multilayer PNN sequentially [3,4,20]. In case of long training data sequences the EM algorithm can be realized as a sequential procedure which corresponds to one "infinite" iteration of EM algorithm including periodical updating of the estimated parameters [5,8]. In pattern recognition the classification accuracy can be improved by parallel combination of independently trained PNN [9,10]. The probabilistic neuron can be interpreted in neurophysiological terms at the level of functional properties of a biological neuron [10,8] and the resulting synaptical weights can be viewed as a theoretical counterpart of the well known Hebbian principle of learning [14]. The PNN can be trained sequentially while assuming strictly modular properties of the probabilistic neurons [8]. Weighting of training data in PNN is compatible with the technique of boosting which is widely used in pattern recognition [12]. In this sense the importance of training data vectors may be evaluated selectively as it is assumed e.g. in connection with "emotional" learning.

One of the most apparent limitations of the probabilistic approach to neural networks has been the biologically unnatural complete interconnection of neurons with all input variables. The complete interconnection property follows from the basic paradigms of probability theory. For the sake of Bayes formula all class-conditional probability distributions must be defined on the same space and therefore each neuron must be connected with the same (complete) set of input variables. We have proposed a structural mixture model which avoids the biologically unnatural condition of complete interconnection of neurons. The resulting subspace approach to PNN is compatible with the statistically correct decision making and, simultaneously, optimization of the interconnection structure of PNN can be included into the EM algorithm [11,10].

Another serious limitation of PNN arises from the conflicting properties of the estimated mixture parameters and of the well known short-term dynamic processes in biological neural networks. The mixture parameters computed by means of EM algorithm reflect the statistical properties of training data and uniquely determine the performance of PNN. Unfortunately the "static" role of mixture parameters is sharply contrasting with the short-term dynamic properties of biological neurons. Motivated by this contradiction we apply in this paper the recurrent Bayesian reasoning originally proposed in the framework of probabilistic expert systems [7]. The statistical decision-making based on recurrent Bayesian reasoning is invariant with respect to the a priori component weights and makes PNN more compatible with the short-term dynamic properties of biological neural networks. In the following we first summarize basic properties of PNN (Sec. 2) and discuss the biological interpretation of the probabilistic neuron (Sec.3). Sec.4 describes the proposed principles of recurrent Bayesian reasoning and in Sec.5 we apply the proposed PNN to recognition of handwritten numerals.

2 Subspace Approach to Statistical Pattern Recognition

In the following sections we confine ourselves to the problem of statistical recognition of binary data vectors

$$\boldsymbol{x} = (x_1, x_2, \ldots, x_N) \in \mathcal{X}, \quad \mathcal{X} = \{0, 1\}^N$$

to be classified according to a finite set of classes $\Omega = \{\omega_1, \omega_2, \ldots, \omega_K\}$. Assuming probabilistic description of classes we reduce the final decision-making to the Bayes formula

$$p(\omega|\boldsymbol{x}) = \frac{P(\boldsymbol{x}|\omega)p(\omega)}{P(\boldsymbol{x})}, \quad P(\boldsymbol{x}) = \sum_{\omega \in \Omega} P(\boldsymbol{x}|\omega)p(\omega), \quad \boldsymbol{x} \in \mathcal{X}. \tag{1}$$

where $P(\boldsymbol{x}|\omega)$ represent the conditional probability distributions and $p(\omega), \omega \in \Omega$ denote the related a priori probabilities of classes.

Considering PNN we approximate the conditional distributions $P(\boldsymbol{x}|\omega)$ by finite mixtures of product components

$$P(\boldsymbol{x}|\omega) = \sum_{m \in \mathcal{M}_\omega} F(\boldsymbol{x}|m)f(m) = \sum_{m \in \mathcal{M}_\omega} f(m) \prod_{n \in \mathcal{N}} f_n(x_n|m). \tag{2}$$

Here $f(m) \geq 0$ are probabilistic weights, $F(\boldsymbol{x}|m)$ denote the component specific product distributions, $\mathcal{M}_\omega$ are the component index sets of different classes and $\mathcal{N}$ is the index set of variables. In case of binary data vectors we assume the components $F(\boldsymbol{x}|m)$ to be multivariate Bernoulli distributions, i.e. we have

$$f_n(x_n|m) = \theta_{mn}^{x_n}(1 - \theta_{mn})^{1-x_n}, \quad n \in \mathcal{N}, \quad \mathcal{N} = \{1, \ldots, N\}, \tag{3}$$

$$F(\boldsymbol{x}|m) = \prod_{n \in \mathcal{N}} f_n(x_n|m) = \prod_{n \in \mathcal{N}} \theta_{mn}^{x_n}(1 - \theta_{mn})^{1-x_n}, \quad m \in \mathcal{M}_\omega. \tag{4}$$

In order to simplify notation we assume consecutive indexing of components. Hence, for each component index $m \in \mathcal{M}_\omega$ the related class $\omega \in \Omega$ is uniquely determined and therefore the parameter ω can be partly omitted in Eq. (2).

The basic idea of PNN is to view the component distributions in Eq. (2) as formal neurons. If we define the output of the m-th neuron in terms of the mixture component $F(\boldsymbol{x}|m)f(m)$ then the posterior probabilities $p(\omega|\boldsymbol{x})$ are proportional to partial sums of the neural outputs (cf. (1),(2)):

$$p(\omega|\boldsymbol{x}) = \frac{p(\omega)}{P(\boldsymbol{x})} \sum_{m \in \mathcal{M}_\omega} F(\boldsymbol{x}|m)f(m). \tag{5}$$

The well known "beauty defect" of the probabilistic approach to neural networks is the biologically unnatural complete interconnection of neurons with all input variables. In order to avoid the undesirable complete interconnection condition we have introduced the structural mixture model [5]. In particular we make substitution

$$F(\boldsymbol{x}|m) = F(\boldsymbol{x}|0)G(\boldsymbol{x}|m,\phi_m)f(m), \;\; m \in \mathcal{M}_\omega \tag{6}$$

where $F(\boldsymbol{x}|0)$ is a "background" probability distribution usually defined as a fixed product of global marginals

$$F(\boldsymbol{x}|0) = \prod_{n\in\mathcal{N}} f_n(x_n|0) = \prod_{n\in\mathcal{N}} \theta_{0n}^{x_n}(1-\theta_{0n})^{1-x_n}, \quad (\theta_{0n} = \mathcal{P}\{x_n = 1\}) \tag{7}$$

and the component functions $G(\boldsymbol{x}|m,\phi_m)$ include additional binary structural parameters $\phi_{mn} \in \{0,1\}$

$$G(\boldsymbol{x}|m,\phi_m) = \prod_{n\in\mathcal{N}} \left[\frac{f_n(x_n|m)}{f_n(x_n|0)}\right]^{\phi_{mn}} = \prod_{n\in\mathcal{N}} \left[\left(\frac{\theta_{mn}}{\theta_{0n}}\right)^{x_n} \left(\frac{1-\theta_{mn}}{1-\theta_{0n}}\right)^{1-x_n}\right]^{\phi_{mn}}. \tag{8}$$

An important feature of PNN is the possibility to optimize the mixture parameters $f(m), \theta_{mn}$, and the structural parameters ϕ_{mn} simultaneously by means of EM algorithm (cf. [17,5,11,10,16]). Given a training set of independent observations from the class $\omega \in \Omega$

$$\mathcal{S}_\omega = \{\boldsymbol{x}^{(1)}, \ldots, \boldsymbol{x}^{(K_\omega)}\}, \quad \boldsymbol{x}^{(k)} \in \mathcal{X},$$

we obtain the maximum-likelihood estimates of the mixture (6) by maximizing the log-likelihood function

$$L = \frac{1}{|\mathcal{S}_\omega|} \sum_{x\in\mathcal{S}_\omega} \log\left[\sum_{m\in\mathcal{M}_\omega} F(\boldsymbol{x}|0)G(\boldsymbol{x}|m,\phi_m)f(m)\right]. \tag{9}$$

For this purpose the EM algorithm can be modified as follows:

$$f(m|\,\boldsymbol{x}) = \frac{G(\boldsymbol{x}|\,m,\phi_m)f(m)}{\sum_{j\in\mathcal{M}_\omega} G(\boldsymbol{x}|j,\phi_j)f(j)}, \quad m \in \mathcal{M}_\omega, \; \boldsymbol{x} \in \mathcal{S}_\omega, \tag{10}$$

$$f^{'}(m) = \frac{1}{|\mathcal{S}_\omega|}\sum_{x\in\mathcal{S}_\omega} f(m|\,\boldsymbol{x}), \quad \theta^{'}_{mn} = \frac{1}{|\mathcal{S}_\omega|f^{'}(m)} \sum_{x\in\mathcal{S}_\omega} x_n f(m|\boldsymbol{x}), \quad n \in \mathcal{N} \tag{11}$$

$$\gamma^{'}_{mn} = f^{'}(m)\left[\theta^{'}_{mn}\log\frac{\theta^{'}_{mn}}{\theta_{0n}} + (1-\theta^{'}_{mn})\log\frac{(1-\theta^{'}_{mn})}{(1-\theta_{0n})}\right], \tag{12}$$

$$\phi^{'}_{mn} = \begin{cases} 1, \gamma^{'}_{mn} \in \Gamma^{'}, \\ 0, \gamma^{'}_{mn} \notin \Gamma^{'}, \end{cases}, \quad \Gamma^{'} \subset \{\gamma^{'}_{mn}\}_{m\in\mathcal{M}_\omega\; n\in\mathcal{N}}, \quad |\Gamma^{'}| = r. \tag{13}$$

Here $f^{'}(m), \theta^{'}_{mn}$, and $\phi^{'}_{mn}$ are the new iteration values of mixture parameters and $\Gamma^{'}$ is the set of a given number of highest quantities $\gamma^{'}_{mn}$. The iterative equations (10)-(13) generate a nondecreasing sequence $\{L^{(t)}\}_0^\infty$ converging to a possibly local maximum of the log-likelihood function (9).

3 Structural Probabilistic Neural Networks

The main advantage of the structural mixture model is the possibility to cancel the background probability density $F(\boldsymbol{x}|0)$ in the Bayes formula since then the decision making may be confined only to "relevant" variables. In particular, making substitution (6) in (2) and introducing notation $w_m = p(\omega)f(m)$ we can express the unconditional probability distribution $P(\boldsymbol{x})$ in the form

$$P(\boldsymbol{x}) = \sum_{\omega\in\Omega}\sum_{m\in\mathcal{M}_\omega} F(\boldsymbol{x}|m)w_m = \sum_{m\in\mathcal{M}} F(\boldsymbol{x}|0)G(\boldsymbol{x}|m,\phi_m)w_m \tag{14}$$

Further denoting

$$q(m|\boldsymbol{x}) = \frac{F(\boldsymbol{x}|m)w_m}{P(\boldsymbol{x})} = \frac{G(\boldsymbol{x}|m,\phi_m)w_m}{\sum_{j\in\mathcal{M}} G(\boldsymbol{x}|j,\phi_j)w_j}, \quad m\in\mathcal{M}, \quad \boldsymbol{x}\in\mathcal{X} \tag{15}$$

the conditional component weights, we can write

$$p(\omega|\boldsymbol{x}) = \frac{\sum_{m\in\mathcal{M}_\omega} G(\boldsymbol{x}|m,\phi_m)w_m}{\sum_{j\in\mathcal{M}} G(\boldsymbol{x}|j,\phi_j)w_j} = \sum_{m\in\mathcal{M}_\omega} q(m|\boldsymbol{x}). \tag{16}$$

Thus the posterior probability $p(\omega|\boldsymbol{x})$ becomes proportional to a weighted sum of the component functions $G(\boldsymbol{x}|m,\phi_m)$ each of which can be defined on a different subspace. In other words the input connections of a neuron can be confined to an arbitrary subset of input neurons. In this sense the structural mixture model represents a statistically correct subspace approach to Bayesian decision-making which is directly applicable to the input space without any feature selection or dimensionality reduction [10].

In multilayer neural networks each neuron of a hidden layer plays the role of a coordinate function of a vector transform $\boldsymbol{T}$ mapping the input space $\mathcal{X}$ into the space of output variables $\mathcal{Y}$. We denote

$$\boldsymbol{T}:\mathcal{X}\to\mathcal{Y}, \quad \mathcal{Y}\subset R^M, \quad \boldsymbol{y}=\boldsymbol{T}(\boldsymbol{x}) = (T_1(\boldsymbol{x}), T_2(\boldsymbol{x}),\dots,T_M(\boldsymbol{x}))\in\mathcal{Y}. \tag{17}$$

It has been shown (cf. [4], [20]) that the transform defined in terms of the posterior probabilities $f(m|\boldsymbol{x})$:

$$y_m = T_m(\boldsymbol{x}) = \log q(m|\boldsymbol{x}), \quad \boldsymbol{x}\in\mathcal{X}, \quad m\in\mathcal{M} \tag{18}$$

preserves the statistical decision information and minimizes the entropy of the output space $\mathcal{Y}$.

From the neurophysiological point of view the conditional probability $q(m|\boldsymbol{x})$ can be naturally interpreted as a measure of excitation or probability of firing of the m-th neuron given the input pattern $\boldsymbol{x}\in\mathcal{X}$. The output signal y_m (cf. (18)) is defined as logarithm of the excitation $q(m|\boldsymbol{x})$ and therefore logarithm plays the role of activation function of probabilistic neurons. In view of Eqs. (8), (16) we can write

$$y_m = \mathrm{T}_m(\boldsymbol{x}) = \log q(m|\boldsymbol{x}) =$$

$$= \log w_m + \sum_{n\in\mathcal{N}} \phi_{mn} \log \frac{f_n(x_n|m)}{f_n(x_n|0)} - \log[\sum_{j\in\mathcal{M}} G(\boldsymbol{x}|j,\phi_j)f(j)]. \quad (19)$$

Consequently, we may assume the first term on the right-hand side of Eq. (19) to be responsible for the spontaneous activity of the m-th neuron. The second term in Eq. (19) summarizes contributions of input variables x_n (input neurons) chosen by means of binary structural parameters $\phi_{mn} = 1$. In this sense, the term

$$g_{mn}(x_n) = \log f_n(x_n|m) - \log f_n(x_n|0) \quad (20)$$

can be viewed as the current synaptic weight of the n-th neuron at input of the m-th neuron - as a function of the input value x_n. The effectiveness of the synaptic transmission, as expressed by the formula (20), combines the statistical properties of the input variable x_n with the activity of the "postsynaptic" neuron "m". In words, the synaptic weight (20) is high when the input signal x_n frequently takes part in excitation of the m-th neuron and, in turn, it is low when the input signal x_n usually doesn't contribute to the excitation of the m-th neuron. This formulation resembles the classical Hebb's postulate of learning (cf. [14], p.62):

"When an axon of cell A is near enough to excite a cell B and repeatedly or persistently takes part in firing it, some growth process or metabolic changes take place in one or both cells such that A's efficiency as one of the cells firing B, is increased."

Note that the synaptic weight (20) is defined generally as a function of the input variable and separates the weight g_{mn} from the particular input values x_n. The last term in (19) includes the norming coefficient responsible for competitive properties of neurons and therefore it can be interpreted as a cumulative effect of special neural structures performing lateral inhibition. This term is identical for all components of the underlying mixture and, for this reason, the Bayesian decision-making would not be influenced by its accuracy.

Let us recall finally that the estimation of mixture parameters in Eq. (19) is a long-term process reflecting the global statistical properties of training data. Obviously, any modification of mixture parameters may strongly influence the outputs of hidden layer neurons (19) and change the Bayes probabilities (16) with the resulting unavoidable loss of decision-making optimality. Unfortunately the "static" nature of the considered PNN strongly contradicts with the well known short-term dynamic processes in biological neural networks related e.g. to short-term synaptic plasticity or to complex transient states of neural assemblies.

In the next section we show that, in case of recurrent use of Bayes formula, some parameters of PNN can be "released" for the sake of short-term processes without destroying the statistically correct decision making. In particular, we can adapt the component weights to a specific data vector on input or the input pattern can be iteratively adapted in order to facilitate the correct recognition.

4 Recurrent Bayesian Reasoning

Let us recall that the conditional distributions $P(\boldsymbol{x}|\omega)$ in the form of mixtures implicitly introduce an additional low-level "descriptive" decision problem [3] with the mixture components $F(\boldsymbol{x}|m)$ corresponding to features, properties or situations. Given a vector $\boldsymbol{x} \in \mathcal{X}$, the implicit presence of these "elementary" properties can be characterized by the conditional probabilities $q(m|\boldsymbol{x})$ related to the *a posteriori* probabilities of classes by (16).

In Eq. (14) the component weights w_m represent *a priori* knowledge of the decision problem. Given a particular input vector $\boldsymbol{x} \in \mathcal{X}$, they could be replaced by the more specific conditional weights $q(m|\boldsymbol{x})$. In this way we obtain a recurrent Bayes formula

$$q^{(t+1)}(m|\boldsymbol{x}) = \frac{G(\boldsymbol{x}|m,\phi_m)q^{(t)}(m|\boldsymbol{x})}{\sum_{j\in\mathcal{M}} G(\boldsymbol{x}|j,\phi_j)q^{(t)}(j|\boldsymbol{x})}, \quad q^{(0)}(m|\boldsymbol{x}) = w_m, \quad m \in \mathcal{M}. \quad (21)$$

The recurrent computation of the conditional weights $q^{(t)}(m|\boldsymbol{x})$ resembles natural process of cognition as iteratively improving understanding of input information. Formally it is related to the original convergence proof of EM algorithm by Schlesinger [17]. Eq. (21) is a special case of the iterative inference mechanism originally proposed in probabilistic expert systems [7]. In a simple form restricted to two components the iterative weighting has been also considered in pattern recognition [1].

It can be easily verified that the iterative procedure defined by (21) converges [7]. In particular, considering a simple log-likelihood function corresponding to a data vector $\boldsymbol{x} \in \mathcal{X}$

$$\mathcal{L}(\boldsymbol{w},\boldsymbol{x}) = \log[\sum_{m\in\mathcal{M}} F(\boldsymbol{x}|0)G(\boldsymbol{x}|m,\phi_m)w_m] \quad (22)$$

we can view the formula (21) as the EM iteration equations to maximize (22) with respect to the component weights w_m. If we set $w_m^{(t)} = q^{(t)}(m|\boldsymbol{x})$ then the sequence of values $\mathcal{L}(\boldsymbol{w}^{(t)},\boldsymbol{x})$ is known to be nondecreasing and converging to a unique limit $\mathcal{L}(\boldsymbol{w}^*,\boldsymbol{x})$ since (22) is strictly concave as a function of $\boldsymbol{w}$ (for details cf. [7]). Consequently, the limits of the conditional weights $q^*(m|\boldsymbol{x})$ are independent with respect to the initial values $q^{(0)}(m|\boldsymbol{x})$. Simultaneously we remark that log-likelihood function $\mathcal{L}(\boldsymbol{w},\boldsymbol{x})$ achieves an obvious maximum by setting the weight of the highest component function $F(\boldsymbol{x}|m_0)$ to one, i.e. for $w_{m_0} = 1$. We illustrate both aspects of the recurrent Bayes formula (21) in numerical experiments of Sec.5.

Let us note that, in a similar way, the log-likelihood criterion (22) can also be maximized as a function of $\boldsymbol{x}$ by means of EM algorithm. If we denote

$$Q_n^{(t)} = \log\frac{\theta_{0n}}{1-\theta_{0n}} + \sum_{m\in\mathcal{M}} \phi_{mn} q^{(t)}(m|\boldsymbol{x}) log \frac{\theta_{mn}(1-\theta_{0n})}{\theta_{0n}(1-\theta_{mn})} \quad (23)$$

then, in connection with (21), we can compute a sequence of data vectors $\boldsymbol{x}^{(t)}$:

$$x_n^{(t+1)} = \begin{cases} 1, \; Q_n^{(t)} \geq 0, \\ 0, \; Q_n^{(t)} < 0, \end{cases} \tag{24}$$

starting with an initial value $\boldsymbol{x}^{(0)}$ and maximizing the criterion $\mathcal{L}(\boldsymbol{w}, \boldsymbol{x})$. In particular, the sequence $\mathcal{L}(\boldsymbol{w}, \boldsymbol{x}^{(t)})$ is nondecreasing and converges to a local maximum of $P(\boldsymbol{x})$. In other words, given an initial value of $\boldsymbol{x}$, we obtain a sequence of data vectors $\boldsymbol{x}^{(t)}$ which are more probable than the initial data vector and therefore they should be more easily classified. In numerical experiments we have obtained slightly better recognition accuracy than with the exact Bayes decision formula. By printing the iteratively modified input pattern we can see that atypically written numerals quickly adapt to a standard form.

5 Numerical Example

In evaluation experiments we have applied the proposed PNN to the NIST Special Database 19 (SD19) containing about 400000 handwritten digits. The SD19 numerals have been widely used for benchmarking of classification algorithms. In our experiments we have normalized all digit patterns (about 40000 for each numeral) to a 16x16 binary raster. Unfortunately there is no generally accepted partition of SD19 into the training and testing data set. In order to guarantee the same statistical properties of both data sets we have used the odd samples of each class for training and the even samples for testing.

Table 1. Example: Recognition of numerals from the NIST SD19 database. Classification error in % obtained by different methods and for differently complex mixtures.

Experiment No.	1	2	3	4	5	6	7	8
Number of Components	10	100	200	300	400	700	800	900
Number of Parameters	2005	16700	38340	41700	100100	105350	115880	133100
Exact Bayes Formula	16.55	6.87	5.02	4.80	3.84	3.77	3.61	3.50
Modified Weights	22.63	9.22	7.50	7.30	4.93	5.55	5.67	5.24
Iterated Weights	16.50	6.99	5.09	4.88	3.88	3.85	3.70	3.56
Adapted Input Vector	16.62	6.84	4.97	4.75	3.81	3.74	3.58	3.48

By using the training data we have estimated the class-conditional mixtures of different complexity by means of EM algorithm of Sec. 2. In all experiments the number of mixture components was chosen identically in all classes. The corresponding total number of mixture components is given in the second row of Tab.1. The EM algorithm has been stopped by relative increment threshold 10^{-5} implying several tens ($10 \div 40$) of EM iterations. Unlike Eq. (13) the structural parameters ϕ_{mn} have been chosen by using the computationally more efficient thresholding:

$$\phi'_{mn} = \begin{cases} 1, \; \gamma'_{mn} \geq 0.1\gamma'_0, \\ 0, \; \gamma'_{mn} < 0.1\gamma'_0, \end{cases}, \qquad \gamma'_0 = \frac{1}{|\mathcal{M}_\omega| N} \sum_{m \in \mathcal{M}_\omega} \sum_{n \in \mathcal{N}} \gamma'_{mn}. \tag{25}$$

Here γ_0' is the mean value of the individual informativity of variables γ_{mn}'. The coefficient 0.1 is chosen relatively low because all variables are rather informative and the training data set is large enough to get reliable estimates of all parameters. Hence, the threshold value $0.1\gamma_0'$ actually suppresses in each component only the really superfluous variables. For each experiment the resulting total number of component specific variables ($\sum_m \sum_n \phi_{mn}$) is given in the third row of Tab.1.

We have used four different versions of Bayes decision rule. The fourth row correspond to the Bayes rule based on the estimated class-conditional mixtures. The fifth row illustrates the influence of altered component weights. In particular, approximately one half of weights (randomly chosen) has been almost suppressed by setting $w_m = 10^{-8}$ with the remaining weights being equal to $w_m = 10$ without any norming. Expectedly, the resulting classification accuracy is much worse than the standard Bayes rule. The sixth row shows how the "spoiled" weights can be repaired by using the iterative weighting (21). The achieved accuracy illustrates that the classification based on iterated weights is actually independent with respect to the initial weights. The last row illustrates the performance of classification based on iteratively modified input vector. The convergence of the iteration equations (21), (23) is achieved in three or four steps. The classification accuracy of Bayes formula applied to the modified input pattern is even slightly better that in the standard form.

6 Conclusion

Considering PNN in the framework of statistical pattern recognition we obtain formal neurons strictly defined by means of parameters of the estimated class-conditional mixtures. In this respect a serious disadvantage of PNN is the fixed probabilistic description of the underlying decision problem which is not compatible with the well known short-term dynamic processes in biological neural networks. We have shown that if we use the Bayes formula in a recurrent way then some mixture parameters may take part in short term dynamic processes without destroying the statistically correct decision making. In particular, the mixture component weights can be iteratively adapted to a specific input pattern or the input pattern can be iteratively modified in order to facilitate the correct recognition.

Acknowledgement. This research was supported by the project GAČR No. 102/07/1594 of Czech Grant Agency, by the EC project FP6-507752 MUSCLE, and partially by the projects 2C06019 ZIMOLEZ and MŠMT 1M0572 DAR.

References

1. Baram, Y.: Bayesian classification by iterated weighting. Neurocomputing 25, 73–79 (1999)
2. Grim, J.: On numerical evaluation of maximum - likelihood estimates for finite mixtures of distributions. Kybernetika 18, 173–190 (1982)

3. Grim, J.: Maximum-likelihood design of layered neural networks. In: International Conference on Pattern Recognition. Proceedings, pp. 85–89. IEEE Computer Society Press, Los Alamitos (1996)
4. Grim, J.: Design of multilayer neural networks by information preserving transforms. In: Pessa, E., Penna, M.P., Montesanto, A. (eds.) Third European Congress on Systems Science, pp. 977–982. Edizioni Kappa, Roma (1996)
5. Grim, J.: Information approach to structural optimization of probabilistic neural networks. In: Ferrer, L., Caselles, A. (eds.) Fourth European Congress on Systems Science, pp. 527–539. SESGE, Valencia (1999)
6. Grim, J.: A sequential modification of EM algorithm. In: Gaul, W., Locarek-Junge, H. (eds.) Classification in the Information Age. Studies in Classif., Data Anal., and Knowl. Organization, pp. 163–170. Springer, Berlin (1999)
7. Grim, J., Vejvalková, J.: An iterative inference mechanism for the probabilistic expert system PES. International Journal of General Systems 27, 373–396 (1999)
8. Grim, J., Just, P., Pudil, P.: Strictly modular probabilistic neural networks for pattern recognition. Neural Network World 13, 599–615 (2003)
9. Grim, J., Kittler, J., Pudil, P., Somol, P.: Combining multiple classifiers in probabilistic neural networks. In: Kittler, J., Roli, F. (eds.) MCS 2000. LNCS, vol. 1857, pp. 157–166. Springer, Heidelberg (2000)
10. Grim, J., Kittler, J., Pudil, P., Somol, P.: Multiple classifier fusion in probabilistic neural networks. Pattern Analysis & Applications 5, 221–233 (2002)
11. Grim, J., Pudil, P., Somol, P.: Recognition of handwritten numerals by structural probabilistic neural networks. In: Bothe, H., Rojas, R. (eds.) Proceedings of the Second ICSC Symposium on Neural Computation, pp. 528–534. ICSC, Wetaskiwin (2000)
12. Grim, J., Pudil, P., Somol, P.: Boosting in probabilistic neural networks. In: Kasturi, R., Laurendeau, D., Suen, C. (eds.) Proc. 16th International Conference on Pattern Recognition, pp. 136–139. IEEE Comp. Soc, Los Alamitos (2002)
13. Haykin, S.: Neural Networks: a comprehensive foundation. Morgan Kaufman, San Mateo, CA (1993)
14. Hebb, D.O.: The Organization of Behavior: A Neuropsychological Theory. Wiley, New York (1949)
15. Hertz, J., Krogh, A., Palmer, R.G.: Introduction to the Theory of Neural Computation. Addison-Wesley, New York, Menlo Park CA, Amsterdam (1991)
16. McLachlan, G.J., Peel, D.: Finite Mixture Models. John Wiley and Sons, New York, Toronto (2000)
17. Schlesinger, M.I.: Relation between learning and self-learning in pattern recognition (in Russian). Kibernetika (Kiev) 6, 81–88 (1968)
18. Specht, D.F.: Probabilistic neural networks for classification, mapping or associative memory. In: Proc. of the IEEE Intternational Conference on Neural Networks, vol. I, pp. 525–532. IEEE Computer Society Press, Los Alamitos (1988)
19. Streit, L.R., Luginbuhl, T.E.: Maximum-likelihood training of probabilistic neural networks. IEEE Trans. on Neural Networks 5, 764–783 (1994)
20. Vajda, I., Grim, J.: About the maximum information and maximum likelihood principles in neural networks. Kybernetika 34, 485–494 (1998)

Resilient Approximation of Kernel Classifiers

Thorsten Suttorp and Christian Igel

Institut für Neuroinformatik
Ruhr-Universität Bochum, 44780 Bochum, Germany
{thorsten.suttorp, christian.igel}@neuroinformatik.rub.de

Abstract. Trained support vector machines (SVMs) have a slow run-time classification speed if the classification problem is noisy and the sample data set is large. Approximating the SVM by a more sparse function has been proposed to solve to this problem. In this study, different variants of approximation algorithms are empirically compared. It is shown that gradient descent using the improved Rprop algorithm increases the robustness of the method compared to fixed-point iteration. Three different heuristics for selecting the support vectors to be used in the construction of the sparse approximation are proposed. It turns out that none is superior to random selection. The effect of a finishing gradient descent on all parameters of the sparse approximation is studied.

1 Introduction

Support vector machines (SVMs) show excellent classification performance across a wide range of applications. Unfortunately, the good classification results often come along with extremely slow execution times, which scale linearly with the number of support vectors. As shown in [1], with probability tending to 1 with increasing training sample size, the fraction of training patterns that become support vectors is bounded from below by the Bayes optimal classification rate. That is, for noisy data the run-time complexity of an SVM increases linearly with the size of the training set. But in many real-world applications classifiers have to meet strict real-time constraints. For this reason SVMs—although showing superior classification performance—are frequently not considered. One way to address this problem is to first train an SVM and then to approximate the obtained solution by a more sparse kernel classifier [2, 3, 4, 5, 6]. The new solution is a weighted combination of a set of vectors mapped to the feature space plus a bias parameter. The sparse classifier is build incrementally by adding a (support) vector to the set of vectors, adjusting its position, and recomputing all weighting coefficients. In this study, we empirically investigate variants of this algorithm. First, we replace the fixed-point iteration suggested in previous work for placing the new vectors by the improved Rprop algorithm [7], which is an efficient and robust first order gradient method. Second, we propose different heuristics for choosing the new vectors in the incremental procedure and investigate how these techniques influence the final solution. Third, we look at the performance improvements achieved by a computationally expensive finishing

J. Marques de Sá et al. (Eds.): ICANN 2007, Part I, LNCS 4668, pp. 139–148, 2007.

optimization of all parameters of the final sparse solution. In our experiments, we monitor the correlation between the distance measure that serves as the optimization criterion during approximation and the accuracy of the resulting sparse solution on test data. Although a high correlation is assumed in the SVM approximation methods, to the best of our knowledge this has never been validated systematically.

In the next section, we briefly review the basic principles for approximating SVMs. In Sect. 3 different variants are proposed. The experiments for evaluating these approximation algorithms are presented in Sect. 4. Finally, the results are discussed.

2 Background

We first concisely describe soft margin SVMs and then the approximation technique proposed in [3,4,5,6].

2.1 Support Vector Machines

We consider L_1-norm soft margin SVMs for binary classification [8]. Let (x_i, y_i), $1 \leq i \leq \ell$, be consistent training examples, where $y_i \in \{-1, 1\}$ is the label associated with input pattern $x_i \in \mathcal{X}$. Support vector machines map input patterns to a feature space $\mathcal{F}$, in which the transformed data is linearly separated. The transformation $\phi : \mathcal{X} \to \mathcal{F}$ is implicitly done by a positive semi-definite kernel function $k : \mathcal{X} \times \mathcal{X} \to \mathbb{R}$ computing a scalar product in the feature space $k(x_i, x_j) = \langle \phi(x_i), \phi(x_j) \rangle$. Patterns are classified by the sign of a function f of the form

$$f(x) = \langle w, \phi(x) \rangle + b = \sum_{i=1}^{\ell} \alpha_i k(x_i, x) + b \ . \tag{1}$$

The coefficients $\alpha_i \in \mathbb{R}$ defining the weight vector $w = \sum_{i=1}^{\ell} \alpha_i \phi(x_i)$ and b are determined by solving the quadratic optimization problem

$$\min_{w,b} \quad H[f] = \sum_{i=1}^{\ell} [1 - y_i f(x_i)]_+ + \frac{1}{2C} \|w\|^2 \ , \tag{2}$$

where $[z]_+ = 0$ if $z < 0$ and $[z]_+ = z$ otherwise. The first part penalizes patterns that are not classified correctly with a particular margin (i.e., distance from the separating hyperplane in $\mathcal{F}$). The second part regularizes the solution, in the sense that minimizing the norm of the weight vector corresponds to minimizing the norm of the function $\Psi(x) = \langle w, \phi(x) \rangle$ in $\mathcal{F}$. The regularization parameter $C \in \mathbb{R}^+$ controls the trade-off between the two objectives. The x_i with $\alpha_i \neq 0$ are called support vectors (SVs), their number is denoted with #SV. For solving the dual quadratic optimization problem we use a sequential minimal optimization approach based on second order information as proposed in [9].

2.2 Approximation of SVMs

For many practical applications the calculation of (1) obtained by SVM learning is computationally too expensive. This problem is tackled in [3] by approximating $\Psi(\cdot) = \sum_{i=1}^{\ell} \alpha_i k(x_i, \cdot)$ by a function $\Psi'(\cdot) = \sum_{i=1}^{L} \beta_i k(z_i, \cdot)$ with $L \ll \#\text{SV}$ and $\beta_i \in \mathbb{R}$. The approximation technique proposed there is based on minimizing the distance function

$$\rho^2 := \|\Psi - \Psi'\|_{\mathcal{F}}^2 \; . \tag{3}$$

The reduced set $\{z_1, \dots, z_L\}$ of approximating vectors (AVs) can be constructed by iteratively adding single vectors to a given solution. The following algorithm implements this idea. The different steps are discussed below.

Algorithm 1. Approximation of SVMs

initialize $\Psi_1 := \Psi$
for $i = 1, \dots, L$ **do**
  determine z_i by minimizing $\rho^2(z_i) = \|\Psi_i - \beta_i \phi(z_i)\|_{\mathcal{F}}^2$
  calculate optimal $\beta_1, \dots, \beta_i$
  $\Psi_{i+1} \leftarrow \Psi - \sum_{j=1}^{i} \beta_j \phi(z_j)$
adjust $z_1, \dots, z_L$ and $\beta_1, \dots, \beta_L$ by global gradient descent
determine optimal offset b'

In [5] the authors note that instead of minimizing $\rho^2(z)$ it is possible to directly minimize the distance between Ψ and its orthogonal projection onto span$(\phi(z))$, which is given by $\left\| \frac{(\Psi \cdot \phi(z))}{\phi(z) \cdot \phi(z)} \phi(z) - \Psi \right\|_{\mathcal{F}}^2 = \|\Psi\|_{\mathcal{F}}^2 - \frac{(\Psi \cdot \phi(z))^2}{\phi(z) \cdot \phi(z)}$, and it therefore suffices to minimize $-\frac{(\Psi \cdot \phi(z))^2}{\phi(z) \cdot \phi(z)}$. This expression reduces for kernels with $k(z, z) = 1$ for all z (e.g. Gaussian kernels) to

$$E(z) := -(\Psi \cdot \phi(z))^2 = -\left(\sum_{i=1}^{\ell} \alpha_i k(x_i, z) \right)^2 \; . \tag{4}$$

Minimization of this distance measure can be realized using gradient methods or fixed-point iteration [5,3,4]. For all of these techniques starting points for the optimization have to be selected.

Fixed-Point Iteration. Determining a single approximation vector z can be performed by means of fixed-point iteration as proposed in [4]. This method is based on the observation that for an extremum of $(\Psi \cdot \phi(z))^2$ the condition $\nabla_z (\Psi \cdot \phi(z))^2 = 0$ has to be fulfilled, which implies $\sum_{i=1}^{\ell} \alpha_i \nabla_z k(x_i, z) = 0$. For kernels with $k(x_i, z) = k(\|x_i - z\|^2)$, such as Gaussian kernels, this reduces to $\sum_{i=1}^{\ell} \alpha_i k'(\|x_i - z\|^2)(x_i - z) = 0$, which is equivalent to

$$z = \frac{\sum_{i=1}^{\ell} \alpha_i k'(\|x_i - z\|^2) x_i}{\sum_{i=1}^{\ell} \alpha_i k'(\|x_i - z\|^2)} \; . \tag{5}$$

Based on this equation, the vector z is computed iteratively. For Gaussian kernels $k(x_i, z) = \exp(-\gamma\|x_i - z\|^2)$ with $\gamma \in \mathbb{R}^+$ the iteration step $t+1$ reads

$$z^{(t+1)} = \frac{\sum_{i=1}^{\ell} \alpha_i \exp(-\gamma\|x_i - z^{(t)}\|^2) x_i}{\sum_{i=1}^{\ell} \alpha_i \exp(-\gamma\|x_i - z^{(t)}\|^2)} \quad . \tag{6}$$

Determining Optimal Coefficients. In each iteration of the approximation algorithm the coefficients β_i have to be determined anew. The optimal coefficients $\beta = (\beta_1, \ldots, \beta_L)$ for approximating $\Psi = \sum_{i=1}^{\ell} \alpha_i k(x_i, \cdot)$ by $\Psi' = \sum_{i=1}^{L} \beta_i k(z_i, \cdot)$ for linear independent $\phi(z_1), \ldots, \phi(z_L)$ can be computed as $\beta = (K^z)^{-1} K^{zx} \alpha$, where $K^z_{ij} := (\phi(z_i) \cdot \phi(z_j))$ and $K^{zx}_{ij} := (\phi(z_i) \cdot \phi(x_j))$, see [4].

Global Gradient Descent. For further minimizing the distance function ρ^2 gradient descent can be applied to all parameters of the sparse solution [5]. The derivative of ρ^2 with respect to a component of an AV is given by

$$\frac{\partial \rho^2}{\partial (z_i)_k} = 4\gamma \left[\sum_{j=1}^{L} \beta_j \beta_i K^{zz}_{ij} ((z_j)_k - (z_i)_k) - \sum_{j=1}^{\ell} \alpha_j \beta_i K^{zx}_{ij} ((x_j)_k - (z_i)_k) \right] \tag{7}$$

and the derivative with respect to a coefficient β_i by

$$\frac{\partial \rho^2}{\partial \beta_i} = 2 \sum_{j=1}^{L} \beta_l K^{zz}_{ij} - 2 \sum_{j=1}^{\ell} \alpha_j K^{zx}_{ij} \quad . \tag{8}$$

Determining Optimal Offset. The offset b used in the original SVM $f(x) = \Psi(x) + b$ is not necessarily optimal for the approximation $f'(x) = \Psi'(x) + b'$. We consider the differences of $f(x)$ and $\Psi'(x)$: $b'(x) = \sum_{i=1}^{\ell} \alpha_i k(x_i, x) + b - \sum_{i=1}^{L} \beta_i k(z_i, x)$ for all SVs of the original SVM and compute their mean

$$b' = \frac{1}{\#\mathrm{SV}} \sum_{x \in \{x_i | 1 \le i \le \ell \wedge \alpha_i \neq 0\}} b'(x) \quad . \tag{9}$$

3 Resilient Approximation of SVMs

In order to increase the performance of the SVM approximation algorithm, we consider new heuristics for choosing the starting points for determining the vectors in the reduced set. Further, we replace the fixed-point iteration by a more robust gradient descent technique.

3.1 Techniques for Choosing Initial Vectors

Constructing a reduced set of AVs relies on the successive addition of single vectors. The choice of initial vectors for gradient descent or fixed-point iteration is crucial for the quality of the final solution. Here, three different techniques for choosing initial vectors from the set of SVs of the original SVM are proposed. The techniques *random selection* and *kernel-clustering* ensure that the fraction of positive and negative SVs of the original SVM (i.e., x_i with $y_i = 1$ and $y_i = -1$, respectively) equals the fraction of positive and negative SVs chosen for the incremental update of the approximation. Let n_{pos} denote the number of positive SVs and n_{neg} the number of negative SVs of the original SVM. Then, the number of initial vectors to be chosen from the positive SVs is given by $L_{\text{pos}} := \max\{1, \lfloor \frac{n_{\text{pos}}}{\#\text{SV}} \cdot L \rfloor\}$ and the number from the negative ones by $L_{\text{neg}} := L - L_{\text{pos}}$.

Random Selection. The starting points are selected uniformly at random from the original SVs. No particular assumptions about the order of drawing are made.

α-proportional Selection. The α-proportional selection has its origin in stochastic universal sampling, which is known from evolutionary algorithms [10]. A weighted roulette wheel containing all original SVs is simulated. The slots are sized according to the corresponding $|\alpha_i|$, and L equally spaced markers are placed along the outside of the wheel. The wheel is spun once and the slots that are hit by the markers define the L initial vectors. This heuristic is based on the idea that vectors with big $|\alpha_i|$ are more important than those with small ones.

Kernel-Clustering. Approximated SVMs realize qualitatively different solutions compared to their original SVMs (see below). The AVs tend to lie rather in the center of the training examples than on the boundaries. This inspires us to choose initial vectors for incrementally updating the approximation by the means of clustering.

The well-known K-means-algorithm generates K initial centers at the beginning and then iterates the following procedure until convergence: All vectors are assigned to their closest center, and the new center of each cluster is set to the mean of the vectors assigned to it. In this process it can happen that one cluster remains empty during the iterations. There are some techniques for dealing with this situation. We choose the vector that has the greatest distance from its cluster center to start a new cluster.

The K-means-algorithm can be transferred from clustering data in input space to clustering data in kernel-induced feature space [11]. The kernel function determines the distance of the vectors to the cluster centers. The preimage of the centers cannot be calculated in general. Therefore, the algorithm finally provides pseudo centers, which are given by the input vectors that are closest to the real centers in the feature space. For approximating an SVM, positive and negative SVs are clustered independently into the predetermined number of clusters (L_{pos} and L_{neg}). In [12] the efficient clustering of a large sample set is considered.

3.2 Resilient Minimization of the Distance Function

Although fixed-point iteration has been favored, the optimization of $E(z)$ can also be done using a general gradient descent algorithm [2]. The partial derivatives of $E(z)$ are given by

$$\frac{\partial E(z)}{\partial z_k} = -2 \sum_{i=1}^{\ell} \alpha_i k(x_i, z) \cdot \sum_{i=1}^{\ell} \alpha_i \frac{\partial k(x_i, z)}{\partial z_k} \ . \tag{10}$$

We employ the efficient Rprop (resilient backpropagation) algorithm for gradient-based optimization [13, 7]. It considers only the signs of the partial derivatives of the error function E to be optimized and not their amount. In each iteration t of Rprop, each objective parameter $z_k^{(t)}$ is increased or decreased depending on whether the sign of the partial derivative $\partial E(z^{(t)})/\partial z_k^{(t)}$ of the objective function with respect to the parameter is positive or negative. The amount of the update is equal to the adaptive individual step size $\Delta_k^{(t)}$, that is, we have

$$z_k^{(t+1)} = z_k^{(t)} - \operatorname{sign}\left(\partial E(z^{(t)})/\partial z_k^{(t)}\right) \cdot \Delta_k^{(t)} \ . \tag{11}$$

Prior to this update, the step size is adjusted based on changes of sign of the partial derivative in consecutive iterations. If the partial derivative changes its sign, indicating that a local minimum has been overstepped, then the step size is multiplicatively decreased; otherwise, it is increased: if $\partial E(z^{(t-1)})/\partial z_k^{(t-1)} \cdot \partial E(z^{(t)})/\partial z_k^{(t)}$ is positive then $\Delta_k^{(t)} = \eta^+ \Delta_k^{(t-1)}$, if the expression is negative then $\Delta_k^{(t)} = \eta^- \Delta_k^{(t-1)}$, where $\eta^+ > 1$ and $\eta^- \in]0, 1[$. The iRprop$^+$ algorithm used in this study implements weight backtracking. It partially retracts "unfavorable" previous steps. Whether a parameter change was "unfavorable" is decided based on the evolution of the partial derivatives and the overall error [7]. The standard parameters $\eta^+ = 1.2$ and $\eta^- = 0.5$ reliably give good results and the initial step sizes Δ_k can be chosen small, so that no parameter-tuning is required when applying resilient backpropagation.

4 Experiments

Experimental Setup. We applied the algorithms to the banana data set [14], which is ideal for visualizing the sparse SVM solutions. Optimal parameters (C, γ) for SVM learning were determined by grid search and 5-fold cross-validation on the training data. Second, we used the spam-database data set available from UCI repository [15] having 1,813 positive examples (spam) and 2,788 negative ones. We transformed every feature to zero mean and unit variance. Third, we considered the MNIST handwritten digit database [16], which was split into two classes containing the digits {0,1,2,3,4} and {5,6,7,8,9}, respectively. Forth, we used the connect-4 opening database containing 67,557 game states [15]. For binary classification the "draw" examples were removed resulting in 61,108 data points. The data were split roughly into two halves making up training and test data. Parameters (C, γ) for the problems spam-database, MNIST and connect-4 were taken from [9].

For each benchmark problem, an SVM was trained and then a predefined number of approximation trials was conducted for different numbers of AVs. All three techniques for selecting initial vectors in combination with iRprop$^+$ and fixed-point iteration were considered. A final gradient descent was performed for the two "small" problems banana and spam-database.

Results. We compared different SVM approximation techniques, which differ in the method used for minimizing the distance function and the method for choosing the initial vectors. Surprisingly, all techniques for selecting initial vectors in combination with iRprop$^+$ showed almost the same behavior. Because no significant difference could be observed we confine ourselves to discussing the results with random selection only. Fixed-point iteration performed badly on spam-database, MNIST, and connect-4 in that way that the algorithm was not able to reach the target number of approximation vectors without getting stuck. The repeated choice of different initial vectors did not help.

The 2-dimensional banana data set provided insight into the characteristics of the resulting SVM approximation. Compared to the original SVM qualitatively different solutions (see Fig. 1) were realized. The vectors of the sparse solution lay rather in the center of the training examples than on the boundaries.

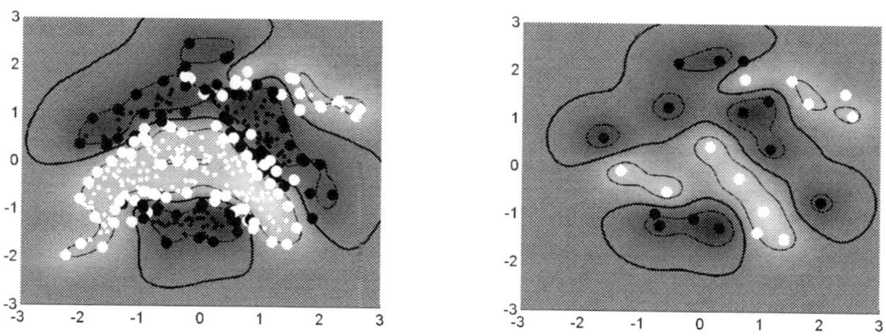

Fig. 1. Approximation of an SVM on the banana data set. The shading codes for the answer of the classifier. The colors white and black correspond to positive and negative examples, respectively. The big points mark the SVs and AVs, respectively. Left image: all training data and the 160 support vectors of the original SVM. Right image: typical approximated SVM with 25 AVs.

The results of the SVM approximation are depicted in Figs. 2-5. In the left and middle pictures the number of AVs is plotted against ρ^2 and the accuracy on the test data, respectively. In addition to the median selected quantiles are given. The fraction of the number of AVs to the number of SVs of the original SVM is reported in brackets. The horizontal line in the middle plots gives the accuracy of the original SVM on the test data. The plots on the right show ρ^2 against the accuracy on the training data. The clusters that can be recognized correspond to solutions with the same number of AVs.

On the banana data set sparse solutions with 20 AVs (e.g., 1/8 of the original amount of SVs) were obtained that performed better on the test data than the original SVM. On spam-database, MNIST and connect-4 data the performance of the original SVM was not fully achieved, but only a small fraction of the number of original SVs led to competitive results (Figs. 3-5).

The final gradient descent on all parameters, which was only tested on banana and spam-database, improved the quality of the approximation. Further, the gradient descent decreased the variance of the solutions in both objectives, distance and accuracy on the test data. As expected, the distance measure ρ^2 was clearly correlated to the accuracy of the approximated SVM on the test data.

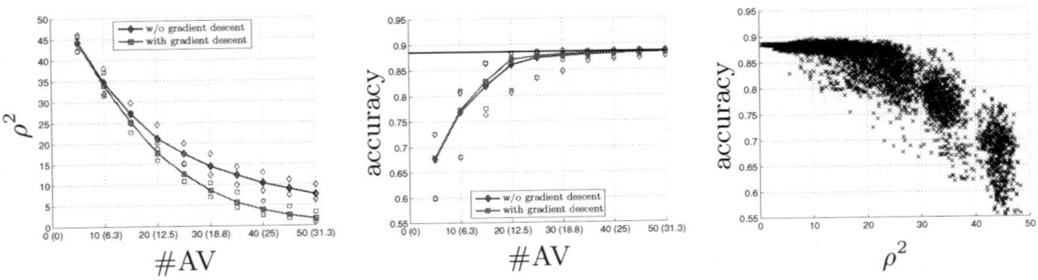

Fig. 2. Approximation results on banana (100 trials); median, 10% and 90% quantiles are given. The fraction of AVs to SVs is reported in brackets.

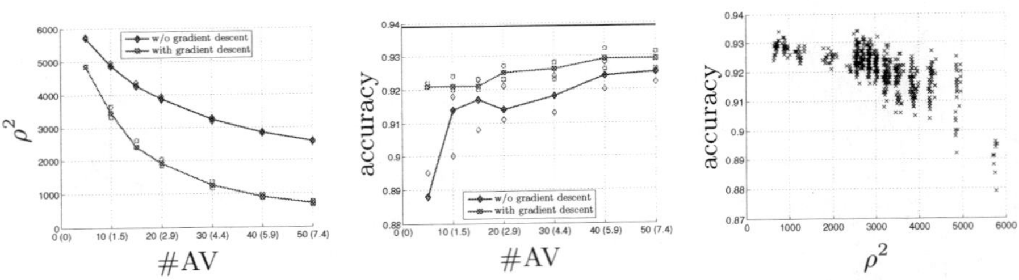

Fig. 3. Approximation results on spam-database (25 trials); median, 10% and 90% quantiles are given

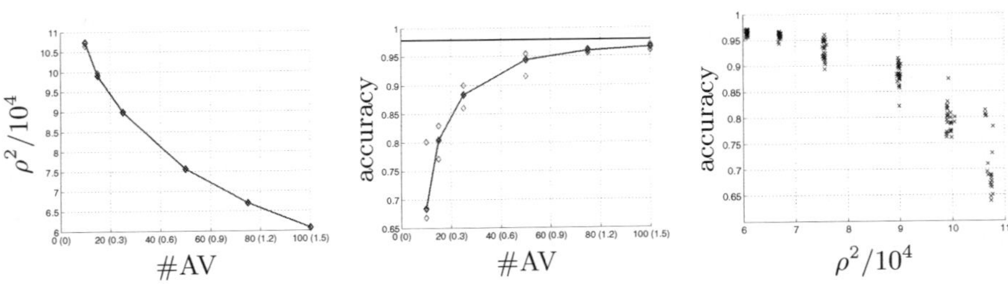

Fig. 4. Approximation results on MNIST (10 trials); median, 20% and 80% quantiles are given

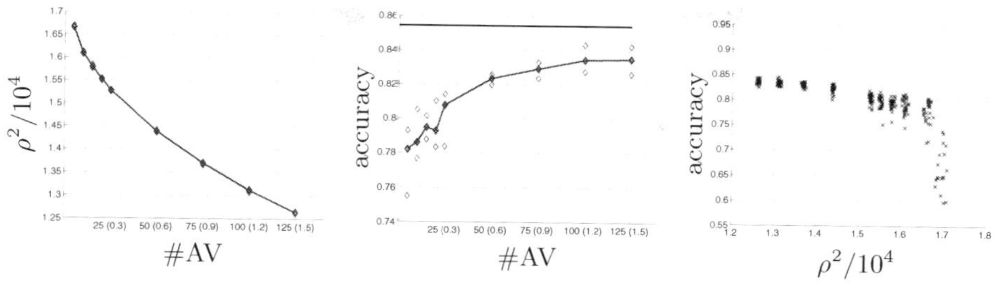

Fig. 5. Approximation results on connect-4 (10 trials); median, 20% and 80% quantiles are given

5 Discussion

The approximation of kernel classifiers allows for the use of SVM training to build classifiers for real-world applications requiring fast decisions. The performance of the sparse solution may stay behind the original classifier, but the computational complexity is drastically reduced. We found that the SVM approximation provides qualitative different solutions compared to the original SVM. The sparse classifiers are not built on vectors at the decision boundary. The SVM strategy to consider the training patterns at the boundary fails to produce sparse classifiers if the problem is noisy. The newly proposed variant that utilizes the improved Rprop algorithm extends the applicability of SVM approximation compared to fixed-point iteration. The latter technique totally failed on some benchmark problems.

We considered different techniques for choosing initial vectors for the iterations. It turned out that the choice of the selection method for initial vectors did not affect the quality of the final solution. We further analyzed the correlation of the distance function and the classification rates on the test data sets. These two variables were correlated justifying the distance function used for minimization. In accordance to the results in [5], in our experiments a final optimization of all parameters of the sparse models increased the performance. Additionally, we found that the variance of the results produced by the approximation algorithms is reduced.

Thus, we recommend the improved Rprop algorithm for resilient minimization of the distance function in combination with random selection of initial vectors. This algorithm combines simplicity and reliability on all problems considered in this paper. Wherever applicable, a finishing gradient descent on the final solution is recommended to improve the results of the final classifier.

Acknowledgment

Part of the work was funded by L-1 Identity Solutions AG, Bochum, the European Commission, and the State of Northrhine-Westfalia (Germany) as part of the project "Zukunftswettbewerb Ruhrgebiet" (contract no. 0403z010).

References

1. Steinwart, I.: Sparseness of support vector machines. Journal of Machine Learning Research 4, 1071–1105 (2003)
2. Burges, C.J.C.: Simplified support vector decision rules. In: ICML 1996. Proceedings of the 13th International Conference on Machine Learning, pp. 71–77 (1996)
3. Burges, C.J.C., Schölkopf, B.: Improving the accuracy and speed of support vector machines. In: Mozer, M., Jordan, M., Petsche, T. (eds.) Advances in Neural Information Processing Systems, Cambridge, MA, vol. 9, pp. 375–381 (1997)
4. Schölkopf, B., Knirsch, P., Smola, A.J., Burges, C.J.C.: Fast approximation of support vector kernel expansions, and an interpretation of clustering as approximation in feature space. In: Levi, P., Ahlers, R.J., May, F., Schanz, M. (eds.) DAGM-Symposium, pp. 124–132. Springer, Heidelberg (1998)
5. Schölkopf, B., Mika, S., Burges, C.J.C., Knirsch, P., Müller, K.R., Rätsch, G., Smola, A.J.: Input space versus feature space in kernel-based methods. IEEE Transactions on Neural Networks 10(5), 1000–1017 (1999)
6. Romdhani, S., Torr, P., Schölkopf, B., Blake, A.: Efficient face detection by a cascaded support-vector machine expansion. Proceedings of the Royal Society A: Mathematical, Physical and Engineering Sciences 460, 3283–3297 (2004)
7. Igel, C., Hüsken, M.: Empirical evaluation of the improved Rprop learning algorithm. Neurocomputing 50(C), 105–123 (2003)
8. Cortes, C., Vapnik, V.: Support-vector networks. Machine Learning 20(3), 273–297 (1995)
9. Glasmachers, T., Igel, C.: Maximum-gain working set selection for support vector machines. Journal of Machine Learning Research 7, 1437–1466 (2006)
10. Baker, J.E.: Reducing bias and inefficiency in the selection algorithm. In: Grefenstette, J.J. (ed.) Proceedings of the Second International Conference on Genetic Algorithms, pp. 14–21 (1987)
11. Girolami, M.: Mercer kernel-based clustering in feature space. IEEE Transactions on Neural Networks 13(3), 780–784 (2002)
12. Zhang, R., Rudnicky, A.I.: A large scale clustering scheme for kernel k-means. In: Proceedings of the International Conference on Pattern Recognition, pp. 289–292 (2002)
13. Riedmiller, M.: Advanced supervised learning in multi-layer perceptrons – From backpropagation to adaptive learning algorithms. Computer Standards and Interfaces 16(5), 265–278 (1994)
14. Rätsch, G., Onoda, T., Müller, K.-R.: Soft margins for AdaBoost. Machine Learning 42(3), 287–320 (2001)
15. Newman, D.J., Hettich, S., Blake, C.L., Merz, C.J.: UCI repository of machine learning databases (1998)
16. LeCun, Y., Bottou, L., Bengio, Y., Haffner, P.: Gradient-based learning applied to document recognition. Proceedings of the IEEE 86(11), 2278–2324 (1998)

Incremental Learning of Spatio-temporal Patterns with Model Selection

Koichiro Yamauchi and Masayoshi Sato

Graduate School of Information Science and Technology, Hokkaido University, Kita-ku, Kita, 14 Jyou Nishi 9 Chyou, Hokkaido Japan

Abstract. This paper proposes a biologically inspired incremental learning method for spatio-temporal patterns based on our recently reported "Incremental learning through sleep (ILS)" method. This method alternately repeats two learning phases: awake and sleep. During the awake phase, the system learns new spatio-temporal patterns by rote, whereas in the sleep phase, it rehearses the recorded new memories interleaved with old memories. The rehearsal process is essential for reconstructing the internal representation of the neural network so as not only to memorize the new patterns while keeping old memories but also to reduce redundant hidden units. By using this strategy, the neural network achieves high generalization ability.

The most attractive property of the method is the incremental learning ability of non-independent distributed samples without catastrophic forgetting despite using a small amount of resources. We applied our method to an experiment on robot control signals, which vary depending on the context of the current situation.

Keywords: incremental learning, spatio-temporal patterns, model selection, RBF

1 Introduction

Generally, a compact learning machine, which has no redundant parameters, confers a high level of generalization [1] [2]. Moreover, a compact learning machine also saves resources.

However, because the appropriate number of parameters is initially unknown, it has to be found through trial and error learning using whole samples. This trial and error learning task is usually called "model selection."

To deal with spatio-temporal patterns practically, learning machines need incremental learning in which they learn the temporal patterns in an on-line manner. In this learning scheme, the machines are asked to memorize new instances immediately after their presentation without forgetting old memories. Moreover, the spatio-tempral patterns, such as sensory inputs, presented in an on-line manner are never independently and identically distributed instances (iid), which is an essential condition to approximate off-line learning effects in an on-line manner. In such environments, learning with "model-selection" to get the simplest data model is even harder.

J. Marques de Sá et al. (Eds.): ICANN 2007, Part I, LNCS 4668, pp. 149–158, 2007.

We have reported a hybrid learning system, called Incremental Learning through Sleep1 (ILS1), that is capable of both incremental learning and model selection [3]. ILS1 employees a normal radial basis function network (RBF)[4] as the core learning machine. The RBF allows nested internal representations using several radial functions, so that the number of parameters is usually less than in other learning architectures such as the generalized regression neural network (GRNN) [5] [6], receptive field weighted regression (RFWR) [7] and normalized gaussian networks (NGNet)[8]. Instead, ILS1 needs a rehearsal period, where the RBF learns samples while pruning redundant hidden units.

ILS1 is, however, for learning static patterns, not spatio-temporal patterns. In this paper, we extend ILS1 to the learning of spatio-temporal patterns and show an example of such incremental learning.

Section 2 outlines extended ILS1 for learning of spatio-temporal patterns, and Sections 3 and 4 explain the two phases in the learning procedure. Section 5 shows comparison our system with other systems and an example of system behavior to an incremental learning task of spatio-temporal patterns.

2 Outline

The system consists of ILS1 with recurrent connections (see 1). ILS1 is designed to build a compact data model through incremental learning. Note that the local internal representation of RBF is suitable for the incremental learning[9]. Moreover, RBF enables a nested internal representation using several radial functions so that it is also suitable for building a compact data model. Although other models such as GRNN[5], EFuNN[10] and RFWR[7] are also suitable for incremental learning, their internal representation makes it hard to build a nested one so that the amount of resources they require is usually larger than that of RBF.

Actually, the model selection of RBF is a time-consuming trial and error process. To avoid this inconvenience, the system has two learning phases: awake and sleep. During the awake phase, the system quickly learns new instances by rote, while it achieves model-selection to get a compact RBF during the sleep phase. We believe that the sleep phase can be made to coincide with the out-of-service period so this time-consuming process should not cause inconvenience to the user.

To perform the above tasks, the system consists of three networks: a main network (M-Net), a fast-learning network (F-Net), and a slow-learning network (S-Net) (Fig. 1). The M-Net and S-Net consist of radial basis function network. The M-Net is a fixed network that keeps the network structure acquired during the latest sleep phase. The M-Net is also used to get precise output values for the recognition during the preceding awake phase and for getting pseudo instances for the next sleep phase.

Spatio-temporal patterns are presented to the system during the awake phase, whereas its outputs are feed back as part of the inputs of the network through a "context-layer," which consists of delay lines with taps. Therefore, past system

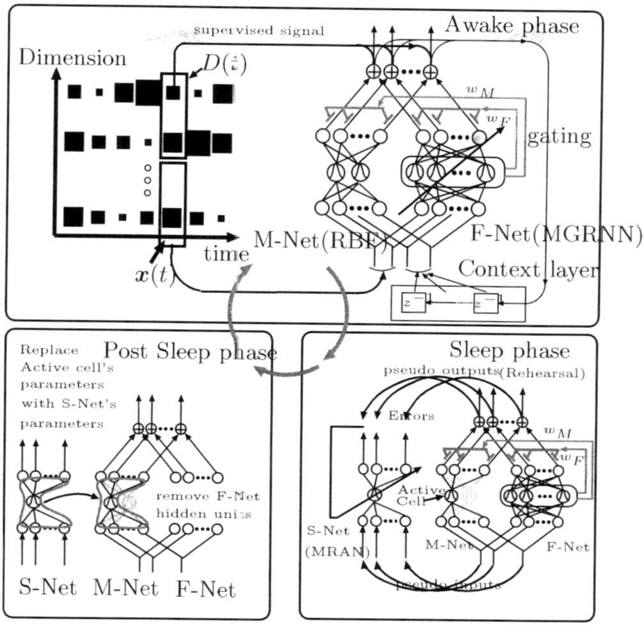

Fig. 1. Structure and behavior of system: The system achieves incremental learning of spatio-temporal pattenrs

outputs are converted into spatial patterns representing the current context, which are to be a part of inputs for the F- and M-Nets.

If the M-Net error for the current input exceeds a threshold, the F-Net learns the current input-output pair. The F-Net consists of an MGRNN [6]. This procedure achieves one-pass learning of new inputs like that of nearest neighbors and smoothly interpolates the outputs between instances.

The system output is the weighted sum of the F- and M-Net outputs. The weight is determined according to the outputs of the F-Net's hidden units. If the outputs are large, the F-Net's weight is large while M-Net's weight is small. As a result, the system output becomes close to the desired one immediately after encountering the novel instance.

During the sleep phase, the S-Net should learn the instances stored in the F- and M-Net with reduction its number of hidden units. To achieve this, the S-Net learns the pseudo instances generated by the F- and M-Nets in online manner. The desired output for the S-Net is the weighted sum of the M- and F-Net outputs. The S-Net learns the pseudo instances so as to minimize not only its error but also its parameters. Reducing the parameters in this way should improve the S-Net's generalization ability. The S-Net's learning method is a modified version of MRAN [11], which uses a growing strategy that prunes redundant cells, as will be described later. The modified MRAN performs stable learning, even if the distribution of inputs is unstable. The S-Net continues learning until its outputs are sufficiently close to those of the pseudo outputs. Note that the pseudo instances are generated as i.i.d samples.

After the S-Net learning(post sleep phase in Fig 1), the M-Net's parameters are basically replaced by the S-Net's parameters. As a result, the M-Net becomes a more suitable network for the recognition. After that, all hidden F- and S-Net units are removed.

The F-Net then begins to learn new instances again. In the following, $\boldsymbol{f}_M^*[\boldsymbol{X}(t), \boldsymbol{\theta}_M]$, $\boldsymbol{f}_f^*[\boldsymbol{X}(t), \boldsymbol{\theta}_f]$ and $\boldsymbol{f}_s^*[\boldsymbol{X}(t), \boldsymbol{\theta}_s]$ denote the respective output vectors of the M-Net, F-Net and S-Net. Here, $\boldsymbol{\theta}_*$ and $\boldsymbol{X}(t)$ denote the parameter vector of the networks and the input vector assigned to the network at time t, respectively. $\boldsymbol{X}(t)$ consists of the current input vector $\boldsymbol{x}(t)$ and past output vectors $\boldsymbol{x}_d(t)$. Therefore, $\boldsymbol{X}(t) = (\boldsymbol{x}^T(t), \boldsymbol{x}_d^T(t))^T$ and $\boldsymbol{x}_d(t) = (\boldsymbol{f}^*[\boldsymbol{x}(t-\tau)]^T, \boldsymbol{f}^*[\boldsymbol{x}(t-\tau-1)]^T, \cdots, \boldsymbol{f}^*[\boldsymbol{x}(t-1)]^T)^T$, where $\boldsymbol{f}^*[\boldsymbol{x}(t)]$ denotes the output vector of the system explained in section 3.

The M-Net and S-Net outputs are the same as that of RBF. For example, the i-th output of the M-Net is

$$f_{Mi}^*[\boldsymbol{X}(t), \boldsymbol{\theta}_M] = \sum_{M\alpha} c_{i\alpha}^M \phi_{M_\alpha}[\boldsymbol{X}(t)], \quad \phi_{M_\alpha}[\boldsymbol{X}(t)] \equiv \exp\left(-\frac{\|\boldsymbol{X}(t) - \boldsymbol{u}_{M\alpha}\|^2}{2\sigma_{M\alpha}^2}\right). \tag{1}$$

Here, $c_{i\alpha}^M$ is the connection strength between the $M\alpha$-th hidden unit and the i-th output cell and $\phi_{M_\alpha}[\boldsymbol{x}]$ denotes the output value from the $M\alpha$-th hidden unit, $\boldsymbol{t}_{M\alpha}$ and $\sigma_{M\alpha}$ are the reference vector and kernel width.

On the other hand, the F-Net output is the MGRNN output:

$$f_{Fi}^*[\boldsymbol{X}(t), \boldsymbol{\theta}_F] = \frac{\sum_{F\alpha} c_{i\alpha}^F \phi_{F_\alpha}[\boldsymbol{X}(t)]/\sigma_{F_\alpha}}{\sum_{F_j} \phi_{F_j}[\boldsymbol{X}(t)]/\sigma_{F_j}} \tag{2}$$

where $\phi_{F_j}(\boldsymbol{x})$ is the output of each hidden unit, which has the same form as Eq(1).

3 Awake Phase

During the awake phase, the system recognizes familiar inputs by calculating the sum of the outputs from the F-Net and M-Net while the F-Net learns novel instances quickly. Let $\boldsymbol{f}^*[\boldsymbol{x}(t)]$ and $\boldsymbol{D}[\boldsymbol{x}(t)]$ be the final output vector of the system and the desired output vector to $\boldsymbol{x}(t)$, respectively, where $\boldsymbol{f}^*[\boldsymbol{x}(t)]$ is the weighted sum of the F-Net and M-Net outputs:

$$\boldsymbol{f}^*[\boldsymbol{x}(t)] = w_F[\boldsymbol{X}(t)]\boldsymbol{f}_F^*[\boldsymbol{X}(t), \boldsymbol{\theta}_F] + w_M[\boldsymbol{X}(t)]\boldsymbol{f}_M^*[\boldsymbol{X}(t), \boldsymbol{\theta}_M] \tag{3}$$

The weight $w_F[\boldsymbol{x}]$ and $w_M[\boldsymbol{x}]$ are the weights for the F- and M-Net outputs and they vary depending on the activity of hidden F-Net units. Therefore,

$$w_F[\boldsymbol{X}] = \frac{1}{1 + \exp(-A\{\sum_{F_j} \phi_{F_j}[\boldsymbol{X}] - \theta_d\})}, \quad w_M[\boldsymbol{X}] = 1 - w_F[\boldsymbol{X}] \tag{4}$$

where A denotes a gain that determines the smoothness of the weighting function and θ_d is the threshold for determining the trusted input region for the F-Net's output. We set A and θ_d to 100 and 0.95, which are determined through numerical tests.

The F-Net learns current new data $(\boldsymbol{X}(t), \boldsymbol{D}(t))$ by adding new $N+1$-th hidden unit if $\|\boldsymbol{D}(t) - \boldsymbol{f}^*[\boldsymbol{X}(t)]\| > \epsilon_E$, where ϵ_E is a threshold. The initial parameters of the new hidden unit are

$$c^F_{iN+1} = D_i(t), \qquad \boldsymbol{u}_{F_{N-1}} = \boldsymbol{X}(t),$$
$$\sigma_{F_{N+1}} = \lambda_F \min\left\{\min_{\alpha \neq N-1} \|\boldsymbol{X}(t) - \boldsymbol{u}_{F_\alpha}\|, \min_\beta \|\boldsymbol{X}(t) - \boldsymbol{u}_{M_\beta}\|\right\}, \qquad (5)$$

where λ_F denotes overlap ratio for the F-Net. Note that the overlap ratio is for determining the kernel width: σ^F_{N+1}. A larger λ_F makes σ^F_{N+1} wider. According to the MGRNN heuristic, λ_F is set to 0.5. The kernel width of the old F-Net unit, which is the nearest to the new allocated hidden unit, needs to be the same width as that of the new kernel. Therefore, if the F_j-th F-Net unit is the nearest unit, $\sigma_{F_j} := \sigma_{F_{N+1}}$ else if the nearest unit is a M-Net unit, we do not touch the M-Net kernel width to keep its memory.

The above equation is repeated whenever a novel instance appears.

4 Sleep Phase

The learning during the sleep phase aims to modify the M-Net's parameters using the S-Net so as to fit the new instances stored in the F-Net.

Actually, however, the M-Net itself plays the role of the buffer and the S-Net learns the pseudo instances instead of the M-Net. This is for avoiding the lost of old memory in the M-Net due to the learning. Therefore, the F-Net and M-Net generate pseudo instance $(\hat{\boldsymbol{X}}, \boldsymbol{D}^*(\hat{\boldsymbol{X}}))$, where $\boldsymbol{D}^*(\hat{\boldsymbol{X}})$ denotes the pseudo desired output vector. Then the S-Net learn the pseudo instances in online learning manner. This is for achieving buffer less learning. Therefore, if the S-Net learning is offline learning, the system needs a buffer to store all the pseudo instances. This is not suitable for our system in terms of saving storage space. So, S-Net learns the pseudo instances in online learning manner.

Prior to the S-Net learning, this system copies all or some of the M-Net's parameters to the S-Net as the initial parameters. To avoid wasting computational power, the M-Net copies the parameters $\boldsymbol{c}^M_j, \boldsymbol{u}^M_j, \sigma^M_j$ associated with only the hidden units which are activated during the adjacent awake phase to the S-Net. During the awake phase, each hidden unit in the M-Net checks its "activated-flag" when the unit is activated. If $\phi_{M_\alpha}(\boldsymbol{X})$ exceeds the threshold ϵ_o, then $ActivatedFlag_i = true$. If $ActivatedFlag_i = true$, then the i-th M-Net hidden unit is regarded as "active unit." Below, we call the hidden unit an 'active unit.'

Note that the active units in the M-Net represent the memory that will be interfered with by the learning of the new instances in the F-Net. Therefore, the system realizes effective learning using the pseudo instances generated from the active units.

4.1 Pseudo Patterns

During the sleep phase, S-Net learns pseudo patterns. Note that the system does not need to reconstruct the spatio-temporal patterns presented in previous awake phases as the pseudo patterns. Instead, the system recalls $(\hat{\boldsymbol{X}}, \boldsymbol{D}^*(\hat{\boldsymbol{X}}))$ as follows.

At first, the system randomly chooses one hidden unit from all the hidden units of the F-Net and active units in the M-Net every time one pseudo input is generated. Using this strategy, the system generates independent and identically distributed (iid) pseudo inputs for the S-Net online learning. As a result, the S-Net learning approximates offline learning which prevent catastrophic forgetting.

The pseudo input vector $\hat{\boldsymbol{X}}$ generation algorithm is varied depending on which network, F- or M-Net the hidden unit belongs to:

$$\hat{\boldsymbol{X}} = \begin{cases} \boldsymbol{u}_{F_j} & \text{if F-Net's hidden unit} \\ \boldsymbol{u}_{M_j} + l\ \sigma_{M\max_j} \boldsymbol{\kappa} & \text{if M-Net's active unit} \end{cases} \quad (6)$$

where $\boldsymbol{\kappa}$ is a random vector with unit length, and l is a random value in the interval [0,1]. Here, $\sigma_{M\max\ j}$ denotes maximum Euclid distance between the j-th hidden unit center and instances. If the j-th M-Net hidden unit center is the closest to the center of the k-th F-Net's hidden unit and $\sigma_{M\max\ j} < \|\boldsymbol{u}_{Mj} - \boldsymbol{u}_{Fk}\|$, then $\sigma_{M\max\ j}$ is updated to be the current distance: $\sigma_{M\max\ j} := \|\boldsymbol{u}_{Mj} - \boldsymbol{u}_{Fk}\|$. This parameter setting is achieved just before a removing process for all F-Net hidden units at the end of sleep phase.

The output vector $\boldsymbol{D}^*(\hat{\boldsymbol{X}})$ is basically the weighted sum of the F- and M-Net outputs without non-active hidden units. Precisely, $\boldsymbol{D}^*(\hat{\boldsymbol{X}})$ should be the outputs minus the M-Net output without active hidden units:

$$\boldsymbol{D}^*(\hat{\boldsymbol{X}}) \equiv w_F[\hat{\boldsymbol{X}}]\boldsymbol{f}_F^*[\hat{\boldsymbol{X}}] + w_M[\hat{\boldsymbol{X}}]\boldsymbol{f}_M^*[\hat{\boldsymbol{X}}, \boldsymbol{\theta}_{Mactive}] - \boldsymbol{f}_M^*[\hat{\boldsymbol{X}}, \boldsymbol{\theta}_{Mnon-active}]. \quad (7)$$

where $\boldsymbol{\theta}_{Mactive}$ and $\boldsymbol{\theta}_{Mnon-active}$ are the parameters of the M-Net associated with the active and non-active hidden units, respectively. This is because, the parameters of the active M-Net's hidden units are to be replaced by the parameters of the S-Net (see 4.3), so the S-Net output must fit to the subtracted value.

4.2 S-Net Learning (Sleep Phase)

The S-Net learns the pseudo instances $(\hat{\boldsymbol{X}}, \boldsymbol{D}^*(\hat{\boldsymbol{X}}))$. Actually, we can employ various learning methods for the S-Net. The conditions for the S-Net learning method are

- It is an online learning method, and
- The method reduces redundant parameters by a pruning or merging strategy.

The first condition is needed for bufferless learning; and the second is the essential condition for our system.

In this paper, we use a modified version of the MRAN learning algorithm [11], which is an online learning with a pruning strategy. In the pruning strategy, the

S-Net prunes any hidden units whose contribution ratio is smaller than a threshold. The contribution ratio of the S_j-th hidden unit is the relative magnitude of outputs, as per the equation below.

$$r_j(\hat{\boldsymbol{X}}) = \|\boldsymbol{c}_j \phi_{S_j}(\hat{\boldsymbol{X}})\| / \max_i \|\boldsymbol{c}_i \phi_{S_i}(\hat{\boldsymbol{X}})\|, \quad (8)$$

where $\boldsymbol{c}_j$ is the connection strength vector between the S_j-th hidden unit and the output units. Note that this equation estimates the contribution ratio for only one input vector $\hat{\boldsymbol{X}}$. To estimate the contribution ratio for all input space roughly, the Minimum RAN algorithm estimates Eq.(8) using several different input vectors. Therefore, if $r_j(\hat{\boldsymbol{X}})$ is less than a threshold for M consecutive pseudo input $\hat{\boldsymbol{X}}$, the hidden unit is pruned. This estimation method is the same as that of the original MRAN [11].

The summarized pruning algorithm is given in Fig 2. In the box, δ denotes a performance-threshold and this is also determined through preliminary experiments to minimize Akaike's Information Criterion (AIC) [12], which evaluates the statistical suitability of linear models for a dataset. According to AIC, the model of minimal number of parameters which are essential for the learning of the dataset is good. Although the AIC is for evaluating linear models, we can use it for rough evaluation of the parameter δ.

```
for each pseudo instance (X̂, D*(X̂))
  for each S_j-th hidden unit
   if ||X̂ − u_Sα|| < σ_Sα then
     Calculate the contributing ratio r_Sj(X̂) with Eq. 8
     if r_Sj(X̂) < δ for M consecutive X̂ then
        prune the S_j-th hidden unit;
```

Fig. 2. Pruning algorithm

In the experiment, we set $\delta = 0.4$, to which the averaged ILS1 performance is almost the best over several datasets.

The S-Net's learning method is that of M-RAN proposed by Yingwei et al. [11] with the modified pruning algorithm, but the extended Kalman filter (EKF) used in the original M-RAN is replaced by the standard gradient descent method.

The leaning process is repeated $N_{optimize}$ times, where $N_{optimize}$ is set N_o times the number of hidden F-Net units and M-Net's active units. In the experiment described later, N_o was set to 200.

4.3 Post Sleep Phase

After learning with the S-Net, the system removes both active M-Net hidden units and all F-Net hidden units and their associated parameters, after which all parameters of the S-Net are moved to the M-Net. Then, the system restarts the awake phase.

5 Experiments

5.1 Comparison ILS1 with Other Systems Using Static Patterns

First, we compared the original ILS1, which has no recurrent connection, with other incremental learning systems, GRNN-editing, RFWR[7], EFuNN[10] and AOSVR[13] on the servo dataset in the UCI machine learning repository [1]. Note that GRNN-editing corresponds to a variation of nearset-neighbor editing for regression.

ILS1 learned 20 instances in each awake phase, and its performance was evaluated with the mean error immediately after both awake- and sleep- phases. GRNN-editing and EFuNN learned instances in the same manner, whereas AOSVR and RFWR learned them one by one.

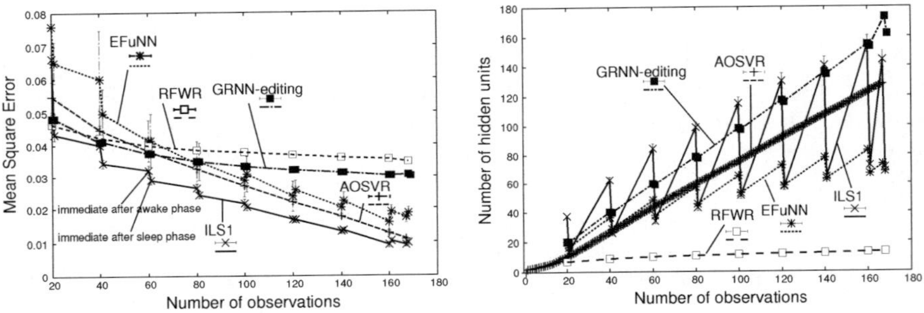

Fig. 3. Mean square error (left) and number of hidden units (right) vs number of observations in servo dataset: The error bars indicate 95% confidence intervals

Each instance was presented to the systems only once (one-pass learning). We evaluated the performance of ILS1 immediately after every awake and sleep phase with $E_{all}(t) = \sum_k^{\text{all samples}} SqureErr(\boldsymbol{x}_k, \boldsymbol{\theta}(t))/$ Sample-Size. Note that $E_{all}(t)$ reflects both the ratio of forgetting and the generalization ability. $E_{all}(t)$ was also used to evaluate other systems. Fig.3 shows the error and cell-number vs number of observations. We can see from figure that the number of ILS1 hidden units became small immediately after each sleep phase. Although the number was not the smallest, the error of ILS1 was the smallest of all.

5.2 Learning Spatio-temporal Patterns

The system was applied to a virtual robot control task. The experiment, used a robot, with one front tire and two rear tires (see Fig.4). The rear tires were always rotated at a fixed speed. The angle of the front tire was controlled by the output value from the system. Therefore, the network controls the direction of movement.

[1] http://www.ics.uci.edu/~mlearn/MLRepository.html

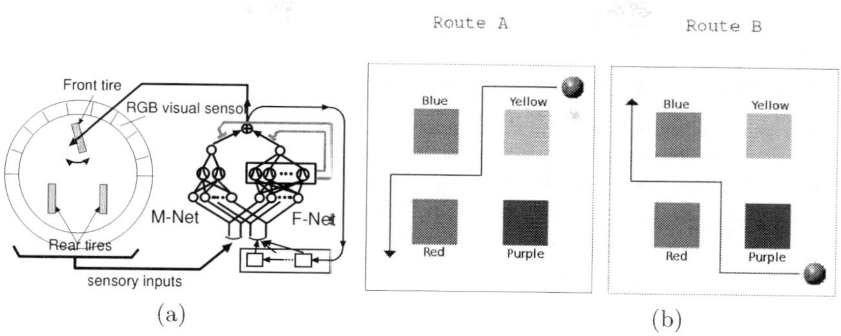

Fig. 4. Structure of the robot (a) and routes for learning (b)

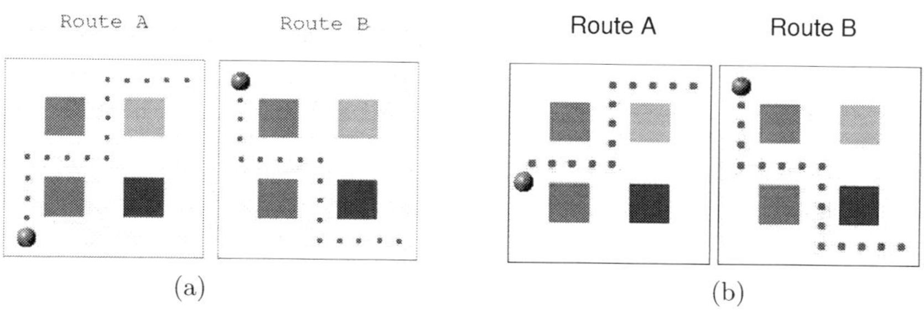

Fig. 5. Resultant routes after learning in the case of proposed system (a) and MRAN with the context layer (b)

The robot has 17 sets of RBG visual sensors. Each visual sensor yields an output value if an object is located in the corresponding direction. Our system learned to drive the robot along the two routes shown in Fig. 4(b). Note that the direction of the last corner depends on the starting point. Therefore, the system has to recognize the differences in the contexts at the last corner.

The learning was executed in the order of routes A and B. For simplicity, the system performances were examined after learning only by using the M-Net. The new system executed one-pass learning of each route. During the 1st awake phase, the system executed one-pass learning of route A, and it rehearsed the learned one during the 1st sleep phase. After the learning of route A, the system executed the learning and rehearsal of route B through the 2nd awake and sleep phases.

We also compared our method with a learning system using MRAN with a context layer. In the case of MRAN, there was no rehearsal mechanism such as in ILS1. The MRAN with the context layer repeated the learning of route A until the mean square error was less than 0.015, and then, it repeated the learning route B incrementally until the mean squared error was less than 0.015. Figure 5 shows the MRAN behaviors after learning. We can see that the MRAN with the context layer cannot recall the first learned route A correctly but ours can recall it correctly.

6 Conclusion

We presented a learning system for incremental learning of spatio-temporal patterns. The system is based on our previous system ILS1, [3], that simultaneously achieves quick adaptation and model selection by using two learning phases: awake and sleep. The experimental results revealed that the system outperformed other learning systems in solving function approximation problems. Using this method, the system can reduce its resources for learning.

Incremental learning of spatio-temporal patterns usually interferes with past learning results. Our method overcomes this problem despite using a small amount of resources.

References

1. Murata, N., Yoshizawa, S., Amari, S.-i.: Network information criterion - determining the number of hidden units for an artificial neural network model. IEEE Transctions on Neural Networks 5(6), 865–872 (1994)
2. van de Laar, P., Heskes, T.: Pruning using parameter and neuronal metrics. Neural Computation 11, 977–993 (1999)
3. Yamauchi, K., Hayami, J.: Incremental learning and model selection for radial basis function network through sleep. IEICE Transactions on Information and Systems E90-D(4), 722–735 (2007)
4. Poggio, T., Girosi, F.: Networks for approximation and learning. In: Proceeding of the IEEE International Conference on Neural Networks, vol. 78(9), pp. 1481–1497. IEEE Computer Society Press, Los Alamitos (1990)
5. Specht, D.F.: A general regression neural network. IEEE Transactions on Neural Networks 2(6), 568–576 (1991)
6. Tomandl, D., Schober, A.: A modified generalized regression neural network (mgrnn) with a new efficient training algorithm as a robust "black-box"-tool for data analysis. Neural Networks 14, 1023–1034 (2001)
7. Schaal, S., Atkeson, C.G.: Constructive incremental learning from only local information. Neural Computation 10(8), 2047–2084 (1998)
8. Moody, J., Darken, C.J.: Fast learning in neural networks of locally-tuned processing units. Neural Computation 1, 281–294 (1989)
9. Yamauchi, K., Yamaguchi, N., Ishii, N.: Incremental learning methods with retrieving interfered patterns. IEEE Transactions on Neural Networks 10(6), 1351–1365 (1999)
10. Kasabov, N.: Evolving fuzzy neural networks for supervised/unsupervised online knowledge-based learning. IEEE Transactions on Systems, Man, and Cybernetics 31(6), 902–918 (2001)
11. Yingwei, L., Sundararajan, N., Saratchandran, P.: A sequential learning scheme for function approximation using minimal radial basis function neural networks. Neural Computation 9, 461–478 (1997)
12. Akaike, H.: A new look at the statistical model identification. IEEE Transactions on Automatic Control AC-19(6), 716–723 (1974)
13. Ma, J., Theiler, J., Perkins, S.: Accurate on-line support vector regression. Neural Computation 15, 2683–2703 (2003)

Accelerating Kernel Perceptron Learning

Daniel García, Ana González*, and José R. Dorronsoro*

Dpto. de Ingeniería Informática and Instituto de Ingeniería del Conocimiento
Universidad Autónoma de Madrid, 28049 Madrid, Spain

Abstract. Recently it has been shown that appropriate perceptron training methods, such as the Schlesinger–Kozinec (SK) algorithm, can provide maximal margin hyperplanes with training costs $O(N \times T)$, with N denoting sample size and T the number of training iterations. In this work we shall relate SK training with the classical Rosenblatt rule and show that, when the hyperplane vector is written in dual form, the support vector (SV) coefficients determine their training appearance frequency; in particular, large coefficient SVs penalize training costs. Under this light we shall explore a training acceleration procedure in which large coefficient and, hence, large cost SVs are removed from training and that allows for a further stable large sample shrinking. As we shall see, this results in a much faster training while not penalizing test classification.

1 Introduction

The standard formulation of SVM training seeks to maximize the margin of a separating hyperplane by solving the problem

$$\min \frac{1}{2}\|W\|^2 \text{ subject to } \; y_i(W \cdot X_i + b) \geq 1, i = 1, \ldots, N, \tag{1}$$

where we assume a linearly separable training sample $\mathcal{S} = \{(X_i, y_i) : i = 1, \ldots, N\}$ with $y_i = \pm 1$. Any pair (W, b) verifying the restrictions in (1) is said to be in canonical form. However, in practice the problem actually solved is the simpler dual problem of maximizing

$$L_D(\alpha) = \sum_i \alpha_i - \frac{1}{2} \sum_{i,j} \alpha_i \alpha_j y_i y_j X_i \cdot X_j, \tag{2}$$

subject to the restrictions $\alpha_i \geq 0$, $\sum_i \alpha_i y_i = 0$. The optimal weight W^o can be then written in the so–called dual form $W^o = \sum \alpha_i^o y_i X_i$ and patterns for which $\alpha_i^o > 0$ are called support vectors (SV). When the problem is not linearly separable, the "hard" margins in (1) are replaced by "soft" margins where slack variables ξ_i are introduced such that $y_i(W \cdot X_i + b) \geq 1 - \xi_i, i = 1, \ldots, N$, allowing for erroneously classified vectors when $\xi_i > 1$. A penalty term of the form $C \sum_i \xi_i^k$, where C is an appropriately chosen parameter, is added to the weight norm. Usual choices for k are 1 or 2.

* All authors have been partially supported by Spain's TIN 2004–07676.

J. Marques de Sá et al. (Eds.): ICANN 2007, Part I, LNCS 4668, pp. 159–168, 2007.

Problem (2) is in fact a quadratic programming problem, that may be solved using quadratic programming algorithms, either of a general type or specialized for SVMS such as Platt's SMO algorithm [9], its refinement LIBSVM [3] or Joachim's SVMlight [6] or, simply, by gradient ascent [13] in (2). Recently it has been shown that (2) can also be solved by perceptron methods, such as the Relaxed On–line Maximum Margin Algorithm (ROMMA, [7]) the Schlesinger–Kozinec (SK) algorithm [4], the Huller, a related on–line algorithm [2], or even "convex" versions of Rosenblatt's standard perceptron [5]. All these methods work iteratively selecting a single pattern that is used to update the current weight. Although it seems that no exhaustive comparisons of training times among these methods have been made, it may be said [6] that SVMlight outperforms LIBSVM and LIBSVM does so with SMO. The situation has been even less studied for perceptron methods, but partial results in [4] seem to indicate that the SK algorithm is faster than SMO for moderate precision computations of the optimal hyperplane, while SMO may be slightly faster for greater precision. Both clearly outperform ROMMA but SVMlight may be faster by about an order of magnitude. In any case, this larger speed comes at the cost of a bigger complexity of the methods involved: SVMlight and LIBSVM are quite costly to implement, SMO is less so and ROMMA and, particularly, SK are very simple. Moreover, perceptron methods may be better suited for on–line learning, an appealing possibility for very large training samples.

In this work we shall offer a proposal for a faster training of the SK algorithm. As discussed in the next section, for a size N sample and a number T of iterations, its training time measured in dot products or kernel operations is $T \times N$, as any sample pattern may eventually become a support vector (SV). To speed things up, both T and N should be reduced and, as in other methods, a simple way of doing so is by shrinking the training sample, as the number of final SVs is usually a fraction of the full sample size. This can be done, for instance, by removing at a certain training iteration all patterns X_i for which $\alpha_i = 0$ at that time. However, this will not reduce the number of iterations needed and, moreover, for this to work, perceptron training may have to be at a rather advanced convergence stage so that the number of SVs has stabilized: if done too early, it may result in a bad test set accuracy; in turn, this implies that a potentially large number of full sample iterations have to be performed.

A way to achieve an early stabilization of the number of SVs is to remove from the training sample patterns whose coefficients become "too large". In fact, a reason for the appearance of "late" small coefficient SVs is the need to balance large coefficient SVs that, as we shall see, cannot be correctly classified. Thus, removing large coefficient patterns makes the initial problem linearly separable. Recent results [1,8] on the so–called reduced convex hull show that the generalization capability of the resulting hyperplane is essentially the same of the one obtained solving the original full sample non–linearly separable problem. Moreover, and as we shall also see, an SV coefficient measures how often it is selected as an updating pattern in perceptron training and their removal results then in a smaller number of iterations.

The paper is organized as follows. In section 2 we shall briefly review both the SK algorithm and the convex Rosenblatt rule for linear and non–linear problems and consider kernel versions for them, while in section 3 we shall analyze the coefficient structure of the final SVs and propose our acceleration procedure. Section 4 will experimentally relate the coefficients of the SVs obtained by standard SK training with the number of their training appearances and will numerically compare the convergence speed of SK training with that of the proposed acceleration procedure. The paper ends with some conclusions and pointers to further work.

2 The SK Algorithm and Rosenblatt's Rule

We consider first the linearly separable case. While in (1) the bias term b is handled explicitly, we will work for simplicity in an alternative homogeneous setting defining $\tilde{W} = (W, b)$, $\tilde{X}_i = (X_i, 1)$. Then $\tilde{W} \cdot \tilde{X}_i = W \cdot X_i + b$, and we can consider the following minimization problem

$$\min \frac{1}{2}\|\tilde{W}\|^2 \text{ subject to } y_i \tilde{W}\tilde{X}_i \geq 1. \tag{3}$$

Although formally similar to (1), the solution of (3) is slightly different, as we try to minimize $\|\tilde{W}\|^2 = \|W\|^2 + b^2$ instead of just $\|W\|^2$. It can be shown, however, that if $\tilde{W}^o$ and W^o, b^o are the optimal values for (3) and (1) respectively, we have $\|\tilde{W}^o\|^2 = \|W^o\|^2 + (b^o)^2$. In particular, for large dimension vectors, $\|\tilde{W}^o\|^2$ should be a reasonably good approximation of $\|W^o\|^2$. From now on we shall work on this homogeneous setting, dropping for the time being the upper $\tilde{}$ and writing W, X_i instead of $\tilde{W}$, $\tilde{X}_i$.

Considering the dual formulation of (3) and writing the optimal W^o as $W^o = \sum \alpha_i^o y_i X_i$, it follows from the Karush–Kuhn–Tucker conditions that $\sum \alpha_i^o = \|W^o\|^2$. Setting now $W^* = W^o/\|W^o\|^2 = \sum \alpha_i^* y_i X_i$, we have $\sum \alpha_i^* = 1$; that is, $W^* \in C(\tilde{\mathcal{S}})$, where $C(A)$ denotes the convex hull of a set A and $\tilde{\mathcal{S}} = \{y_i X_i\}$. Moreover, since W^o is in canonical form, if $W' = \sum_i \alpha_i y_i X_i$ is another vector in $C(\tilde{\mathcal{S}})$, we have

$$W^o \cdot W' = \sum_i \alpha_i y_i W^o \cdot X_i \geq \sum_i \alpha_i = 1,$$

and, as a consequence, $1 \leq W^o \cdot W' \leq \|W^o\| \cdot \|W'\| = \|W'\|/\|W^*\|$. Thus, $\|W^*\| \leq \|W'\|$ for any $W' \in C(\tilde{\mathcal{S}})$. In other words, if m^o is the optimum margin, W^o the optimum margin vector in canonical form and we set $W^* = W^o/\|W^o\|^2$, W^* is the solution of

$$\|W^*\| = \min\{\|W\| : W \in C(\tilde{\mathcal{S}})\}, \tag{4}$$

and we have $m^o = 1/\|W^o\| = \|W^*\|$. A consequence of the above is that for any vector $W \in C(\tilde{\mathcal{S}})$ we have $m(W) \leq m^o = m(W^*) = \|W^*\| \leq \|W\|$ and, therefore, defining $g(W) = \|W\| - m(W)$, we have $0 = g(W^*) \leq g(W)$ for all W. Thus, a procedure that minimizes $g(W)$ leads to a maximum margin hyperplane.

Algorithm 1. Standard kernel perceptron training

```
W_0 = X_l; α_l^0 = 1; α_j^0 = 0; D_j^0 = y_j W_0 · X_j;
for t = 1,2,... do
  l(t) = select (S, D^{t-1}) ;
  λ^t = lambda (X_{l(t)}, ||W_{t-1}||^2, D^{t-1}) ;
  (α^t, ||W_t||^2, D^t) = update(λ^t, l(t), α^{t-1}, ||W_{t-1}||^2, D^{t-1}) ;
end for
```

We turn our attention to non–linearly separable problems. As mentioned in the introduction, we will work with slack margin variables ξ_i for which we have $y_i W \cdot X_i \geq 1-\xi_i, i = 1,\ldots,N$, and with a quadratic penalty function $J(W,\xi) = \|W\|^2 + C\sum_i \xi_i^2$. We will consider extended weight vectors $\tilde{W}$, defined by $\tilde{W} = \left(W, \sqrt{C}\xi_1,\ldots,\sqrt{C}\xi_N\right)$, and extended patterns $\tilde{X}_i = \left(X_i, 0,\ldots,\frac{y_i}{\sqrt{C}},\ldots,0\right)$. It is then easy to check that $J(W,\xi) = \|\tilde{W}\|^2$ and $y_i\tilde{W}\cdot\tilde{X}_i = y_i W\cdot X_i + \xi_i$. Therefore, minimizing $J(W)$ under the restrictions $y_i W \cdot X_i \geq 1-\xi_i$ is equivalent to solving

$$\tilde{W}^o = \text{argmin}\{\|\tilde{W}\|^2 \ : \ y_i\tilde{W}\cdot\tilde{X}_i \geq 1, \ i = 1,\ldots,N\}, \tag{5}$$

while the equivalent convex hull formulation would now be

$$\tilde{W}^* = \arg\min\{\|\tilde{W}\|^2 \ : \tilde{W} \in C(\tilde{\mathcal{S}})\}. \tag{6}$$

Let α_i^* denote the optimal coefficients obtained solving (6). Then $(W^*, \sqrt{C}\xi^*) = \tilde{W}^* = \sum \alpha_i^* y_i \tilde{X}_i = (W^*, \alpha_1^*/\sqrt{C},\ldots,\alpha_N^*/\sqrt{C})$ and, therefore, $\sqrt{C}\xi_i^* = \alpha_i^*/\sqrt{C}$; that is, $\alpha_i^* = C\xi_i^*$. Moreover, the equivalence observed in the linear setting between the solutions $\tilde{W}^o$ and $\tilde{W}^*$ of (5) and (6) becomes now $\tilde{W}^o = \tilde{W}^*/\|\tilde{W}^*\|^2$ and, hence, $\xi_i^o = \xi_i^*/\|\tilde{W}^*\|^2 = \alpha_i^*/C\|\tilde{W}^*\|^2$. As a consequence, the restrictions $y_i W^o \cdot X_i \geq 1-\xi_i^o$ can be also written as

$$y_i W^o \cdot X_i = y_i \frac{W^*}{\|\tilde{W}^*\|^2} \cdot X_i \geq 1 - \frac{\alpha_i^*}{C\|\tilde{W}^*\|^2} = 1 - \frac{\alpha_i^*}{C\|W^*\|^2 + \sum(\alpha_i^*)^2}.$$

The SK and convex perceptron algorithms work in the above setting trying to minimize either (4) or (6). Both start at a random $W_0 = X_l$ and the updating pattern $X_l = X_{l(t)}$ at step t is chosen so that

$$l = \arg\min_i\{y_i W_{t-1} \cdot X_i\} = \arg\min_i\{m(W_{t-1}; X_i)\}, \tag{7}$$

where $m(W;X) = yW \cdot X/\|W\|$ denotes the margin of X with respect to W; the weight update has then the form

$$W_t = (1-\lambda^t)W_{t-1} + \lambda^t y_{l(t)} X_{l(t)}, \tag{8}$$

that guarantees $W_t \in C(\tilde{\mathcal{S}})$. For convex perceptrons [5] we just take $\lambda_t = 1/t$; then $t \times W_t = \sum_1^t y_{l(i)} X_{l(i)}$ is simply the weight vector given by the standard Rosenblatt's rule and (8) can be seen as a convex version of that rule. The

Algorithm 2. Accelerated kernel perceptron training

```
W_0 = X_l ; α_l^0 = 1; α_j^0 = 0; D_j^0 = y_j W_0 · X_j ; fl0 = fl1 = 0 ;
for t = 1,2,... do
  l(t) = select (S, D^{t-1}) ;
  λ^t = lambda (X_{l(t)}, ||W_{t-1}||^2, D^{t-1}) ;
  (α^t, ||W_t||^2, D^t) = update(λ^t, l(t), α^{t-1}, ||W_{t-1}||^2, D^{t-1}) ;
  if  (α_{l(t)}^t > C||W_t||^2) and (fl0 == 0)  then
    η = α_{l(t)}/(α_{l(t)} - 1) ;
    (α^t, ||W_t||^2, D^t) = update(η, l(t), α^{t-1}, ||W_{t-1}||^2, D^{t-1}) ;
    remove X_{l(t)} ; fl1 = 1 ;
  end if
  if  (number of SVs is stable) and (fl1 == 1)  then
    remove all X_i for which α_i^t == 0 ; fl0 = 1 ;
  end if
end for
```

SK algorithm uses $\lambda^t = \arg\min_\lambda\{\|(1-\lambda)W_{t-1} + \lambda X_{l(t)}\|\}$, that guarantees $\|W_t\| \le \|W_{t-1}\|$. It can be easily seen that

$$\lambda^t = \min\left(1, \frac{\| W_{t-1}\|^2 - y_{l(t)} W_{t-1} \cdot X_{l(t)}}{\|W_{t-1}\|^2 - 2y_{l(t)} W_{t-1} \cdot X_{l(t)} + \|X_{l(t)}\|^2}\right) \tag{9}$$

It is clear that the above formulation lends itself naturally to a kernel treatment, as the only operations other than linear vector combinations are dot products. Thus, we will assume first that $X = (\phi(x), 1)$ is a nonlinear transformation associated to a positive definite kernel $k(x, x')$; for the linearly separable case we can write the dot products as $X \cdot X' = 1 + \phi(x) \cdot \phi(x') = 1 + k(x, x') = K(x, x')$, which we extend to $K'(x, x') = K(x, x') + \delta_{x,x'}/C$ for the non–linearly separable case. Working for instance in the non–linearly separable setting and writing now the extended weight vector $\tilde{W}_t$ as $\tilde{W}_t = \sum_j \alpha_j^t y_j \tilde{X}_j = \sum_l \alpha_j^t y_j \tilde{\Phi}(x_j)$, we have

$$\alpha_j^t = \left(1 - \lambda^t\right) \alpha_j^{t-1} + \lambda^t \delta_{j\ l(t)}. \tag{10}$$

Moreover, using the notation $D_j^t = y_j \tilde{W}_t \cdot \tilde{X}_j$, the updating index l is then chosen as $l = l(t) = \arg\min_i \left\{D_i^{t-1}\right\}$ and it can be seen that $D_j^0 = y_j K'(x_j, x_{l_0})$. Furthermore, $\|\tilde{W}_0\|^2 = K'(x_{l_0}, x_{l_0})$ and it follows that

$$\begin{aligned} D_j^t &= (1 - \lambda^t) D_j^{t-1} + \lambda^t y_l y_j K'(x_l, x_j), \\ \|\tilde{W}_t\|^2 &= (1 - \lambda^t)^2 \|\tilde{W}_{t-1}\|^2 + 2(1 - \lambda^t)\lambda^t D_l^t + (\lambda^t)^2 K'(x_l, x_l). \end{aligned} \tag{11}$$

The memory requirements of the SK and convex perceptron algorithms are just O(N) for a size N sample, even if we assume for simplicity that we cache the self–dot products $K'(x_l, x_l)$, $l = 1, \ldots, N$. It thus follows that at each iteration the only kernel operations are those needed to update the D_j^t. If there is variable

Table 1. Starting sample size, # of different SVs and coefficient correlation for the SK and convex perceptron algorithms and correlation for the SK method between pattern coefficients and # of updating appearances

Dataset	sample s.	# diff. SVs	coeff. corr.	coeff.–freq. corr.
Austral. Cr.	690	13	0.9961	0.9949
Breast C. W.	629	14	0.9966	0.9961
German Cr.	1000	19	0.9947	0.9883
Heart Dis.	240	9	0.9975	0.9903
Pima Diab.	622	6	0.9940	0.9793
Primary Tumor	678	2	0.9945	0.9915
Ripley	1125	19	0.9966	0.9868
Satimage	6436	101	0.9856	0.9772
Segment	2310	0	0.9994	0.9989
Sick	3773	29	0.9972	0.9947
Thyroid	7200	35	0.9925	0.9878
Vehicle	846	11	0.9968	0.9946
Vowel	990	2	0.9991	0.9986

sample size N_t at each iteration, the total number of kernel operations for a T iteration training run is $\sum_1^T N_t$. In standard SK training, $N_t = N$ for all t, which results in $T \times N$ kernel operations.

3 Perceptron Training Acceleration

We have just seen that writing $\Lambda^* = C\|\tilde{W}^*\|^2 = C\|W^*\|^2 + \sum(\alpha_i^*)^2$, we then have $y_i W^o \cdot X_i = 1 - \frac{\alpha_i^*}{\Lambda^*}$ for any SV X_i. Therefore, large α_i^* patterns also require large slacks and their classification will be wrong if $\alpha_i^* > \Lambda^*$. We can look to the α_i^* from another point of view. Notice that, for convex perceptrons the coefficient of a given SV represents the number of times it has been selected as the updating vector. In fact, if training requires T iterations, and each final SV X_i has appeared T_i times as the updating pattern, the final weight obtained through the convex Rosenblatt's rule has the form

$$W^* = \frac{1}{T}\sum_1^T y_t X_t = \sum_1^N \frac{T_i}{T} y_i X_i = \sum \alpha_i^* y_i X_i;$$

thus, for convex perceptrons, the coefficient of a given SV represents the number of times it has been selected as the updating vector. Therefore, large α^* SVs take up a large part of convex perceptron training. This interpretation cannot be made in a straight way for SK perceptrons. However, and as we shall numerically illustrate in section 4, both the SK and the convex Rosenblatt's rule updates lead to essentially the same final SVs and the same coefficients for them. Moreover, there is a large correlation for SK perceptrons between the final α^* and the appearance frequency of their corresponding SVs. Hence, we can conclude that,

Table 2. Final test accuracies for the standard and accelerated SK algorithm with a 0.05 and 0.01 precisions

Dataset	SK std. (0.05)	SK acc. (0.05)	SK std. (0.01)	SK acc. (0.01)
Austral. Cr.	86.15 ± 4.26	85.31 ± 3.76	86.06 ± 4.10	85.28 ± 3.77
Breast C. W.	96.93 ± 2.01	96.50 ± 2.24	96.93 ± 2.01	96.53 ± 2.24
German Cr.	75.20 ± 4.09	75.70 ± 3.47	75.32 ± 4.06	75.82 ± 3.71
Heart Dis.	81.40 ± 6.58	82.42 ± 7.24	81.07 ± 6.42	82.82 ± 7.41
Pima Diab.	77.21 ± 4.41	76.02 ± 5.19	77.31 ± 4.30	76.07 ± 4.99
Primary Tumor	85.71 ± 4.08	84.44 ± 4.14	85.86 ± 4.14	84.50 ± 4.13
Ripley	88.88 ± 2.49	88.21 ± 2.36	88.69 ± 2.44	88.13 ± 2.36
Satimage	93.75 ± 0.79	93.16 ± 0.79	93.75 ± 0.82	93.15 ± 0.78
Segment	100.00 ± 0.00	100.00 ± 0.00	100.00 ± 0.00	100.00 ± 0.00
Sick	97.12 ± 0.67	97.29 ± 0.68	97.12 ± 0.65	97.22 ± 0.74
Thyroid	97.39 ± 0.54	97.53 ± 0.60	97.31 ± 0.59	97.47 ± 0.61
Vehicle	97.69 ± 1.40	97.52 ± 1.41	97.69 ± 1.40	97.54 ± 1.40
Vowel	100.00 ± 0.00	100.00 ± 0.00	100.00 ± 0.00	100.00 ± 0.00

for them, large α^* SVs also take up a large part of their training. This suggests that SK training could be accelerated by removing from the training sample those SVs X_i which attain a large α_i coefficient. We will do so when at an iteration t the coefficient α_l^t of the X_l just used as the updating pattern verifies $\alpha_l^t > C\|\tilde{W}_t\|^2$; notice that since $C\|\tilde{W}_t\|^2 \geq C\|\tilde{W}^*\|^2 = \Lambda^*$, the previous condition makes it quite likely that $\alpha_l^* > \Lambda^*$ when standard SK training ends.

Besides removing those costly to train SVs, a reduction in sample size also makes successive weight updates faster, even if we take into account that any SV removal requires to update the current weight vector so that it remains in the sample's convex hull. To do so, we observe that the α_i updates required by such a removal have the same form of the standard convex updates (10). In fact, the removal of a certain X_l requires to change its weight α_l to 0 and to reassign the other weights as $\alpha_i' = \alpha_i/(1-\alpha_l)$ so that we still have a convex combination. If we choose $\eta = \alpha_l/(\alpha_l - 1)$, we can write this as an update $\alpha_j' = (1-\eta)\alpha_j + \eta\delta_{j,l}$ as in (10). As a consequence, the other required updates of $\|\tilde{W}_t\|^2$ and D_j^t can be done too according to (11) Moreover, after the removal of large coefficient patterns, the sample becomes linearly separable and the number of SVs stabilizes, allowing to shrink the training sample by the removal of the cero coefficient patterns. The pseudocode of the accelerated kernel perceptron training can be seen in algorithm 2. The extra cost of a large coefficient SV removal is of $|\mathcal{S}_t|$ kernel operations, with $\mathcal{S}_t$ the sample at iteration t, which simply doubles the cost of a standard iteration. On the other hand, the removal of a cero coefficient pattern requires no kernel operations. The gain derived from a smaller number of overall iterations over a smaller sample clearly compensates the previous extra cost, even if we take into account that, as the training sample is changed, so is the maximum margin and some of the work done in the previous iterations to maximize it may be lost.

Table 3. Number of kernels operations (in thousands) for the standard and accelerated SK algorithm and $\gamma = 0.05$, and reduction achieved by the second method

Dataset	# KOs std. SK	# KOs acc. SK	reduct. %
Austral. Cr.	8,141 ± 772	3,245 ± 437	39.86
Breast C. W.	2,761 ± 311	1,048 ± 169	37.98
German Cr.	20,324 ± 867	9,926 ± 804	48.84
Heart Dis.	1,662 ± 221	2,093 ± 912	125.93
Pima Diab.	18,317 ± 914	4,717 ± 402	25.76
Primary Tumor	9,427 ± 583	3,870 ± 518	41.05
Ripley	38,504 ± 1,423	9,039 ± 944	23.48
Satimage	404,812 ± 16,181	135,064 ± 10,344	33.36
Segment	639 ± 106	639 ± 106	100.00
Sick	63,507 ± 2,966	39,100 ± 5,695	61.57
Thyroid	253,926 ± 18,998	141,175 ± 36,585	55.60
Vehicle	3,946 ± 328	3,512 ± 465	89.01
Vowel	2,060 ± 314	2,057 ± 313	99.84

4 Numerical Experiments

In this section we shall explore the effect of the previous acceleration procedure over 13 datasets (see table 1) taken from the UCI problem database [11] or from the LIBSVM repository [3]. Some of these data sets, namely, those of the Heart Disease, Wisconsin breast cancer or Thyroid disease, originally contain data from more than two classes. We have reduced them to 2–class problems considering their patterns as coming from either healthy or sick patients. We have also chosen as the minority one class 1 in the Vehicle dataset, class 0 for the Vowel dataset, class damp soil for the Satimage dataset, class 1 for the Segment database and classes 1, 2, 3, 4 and 6 for the Primary Tumor database. We shall work with the gaussian kernel $k(x, x') = \exp\left(-\|x - x'\|^2/\sigma^2\right)$, and normalize the original patterns to componentwise 0 mean and 1 variance. We shall also use common values $\sigma^2 = 50$ and $C = 10$ for all datasets; the test accuracies reported here are comparable with those achieved by other methods. We have stopped training either at the first iteration t at which $g(W_t) \leq \gamma \| W_t \|$ or after a maximum number N_{max} of iterations. In our experiments we have used 0.05 and 0.01 for the γ precision values and $N_{max} = 100,000$.

For each dataset we have performed a 5×10 cross validation procedure, dividing 5 times the full datasets randomly into 10 subsets with approximately equal sizes and using 9 of these subsets for training and the remaining one for test. We check first that large coefficient SVs for the SK algorithm do correspond to patterns appearing more often as updating vectors. Table 1, that also gives the original training sample sizes, shows for each data set the difference between the number of SVs obtained by the SK and convex perceptron algorithms, the correlation between their final coefficients and, only for the SK method, the correlation of the SV coefficients and their updating appearance frequencies.

Table 4. Number of kernels operations (in thousands) for the standard and accelerated SK algorithm and $\gamma = 0.01$, and reduction achieved by the second method

Dataset	# KOs std. SK	# KOs acc.	reduct. %
Austral. Cr.	42,240 ± 5,554	7,359 ± 994	17.42
Breast C. W.	14,318 ± 2,241	1,819 ± 254	12.71
German Cr.	89,371 ± 1,619	43,244 ± 3,655	48.39
Heart Dis.	8,531 ± 2,045	9,806 ± 896	114.95
Pima Diab.	69,093 ± 193	13,601 ± 1,837	19.69
Primary Tumor	43,105 ± 3,889	8,208 ± 1,297	19.04
Ripley	112,500 ± 96	17,502 ± 2,982	15.56
Satimage	579,150 ± 151	218,845 ± 9,034	37.79
Segment	3,969 ± 710	3,874 ± 713	97.62
Sick	318,105 ± 23,115	63,369 ± 5,283	19.92
Thyroid	648,000 ± 108	178,508 ± 25,882	27.55
Vehicle	21,937 ± 3,369	7,447 ± 1,019	33.95
Vowel	12,419 ± 2,755	4,137 ± 647	33.32

The table values have been computed over a single full sample training and for $\gamma = 0.01$. As it can be seen, both methods give essentially the same final SVs for all problems except, may be, Satimage, and there is a very high correlation between the SV coefficients. It can be concluded that both methods give very similar hyperplanes. Moreover, there is a high correlation between SK weight coefficients and the number of appearances of the corresponding SVs.

The standard and accelerated SK methods' behaviour is described in the next three tables. For each dataset we have started removing large coefficient SVs as soon as the margin of the separating hyperplane is > 0. Once at least one such SV has been removed, we have considered the number of SVs as stable if it does not change in 1,000 subsequent operations. Once this happen, we remove all cero coefficient patterns and do not allow any further elimination of large coefficient SVs. The first table, 2, gives a comparison of test accuracies and their standard deviations for $\gamma = 0.05$ and 0.01. Both values are very similar and a t test fails to reject an equal mean value hypothesis at levels above 20% for all datasets except for the Primary Tumor, Satimage (both slightly in favour of standard SK) and Thyroid (in favour of accelerated SK) problems. Therefore, it can be said that the proposed acceleration method does not affect the generalization capability of the resulting hyperplanes. Table 3 gives the number of kernel operations of each method for the precision factor $\gamma = 0.05$, as well as the reduction percentage of kernel operation (KOs). The number of KOs in the accelerated method is 60% below the number of KOs in standard training for 8 data sets and below 40% for 5 of them; there is no gain for the Segment and Vowel problems and the accelerated method is actually slower for the relatively small Heart Disease dataset. Table 4 gives similar results for $\gamma = 0.01$; now the number of KOs in the accelerated method is below 40% for 10 of the 13 data sets and below 20% for 6 of them. The situation does not change for the Heart Disease and Segment problems but clearly improves for Vowel.

5 Conclusions

Although their overall speed is not yet comparable to the most advanced SVM training methods (such as SVMlight), kernel perceptron methods, particularly the SK algorithm, are much simpler to implement and may be better suited to on–line applications. Moreover, there should be room to improve their speed performance. We have proposed here a procedure to accelerate the SK algorithm by removing from the training sample those SVs with large coefficients (that will not be correctly classified anyway) and, later on, those training patterns which are likely not to become SVs and, thus, that will not be part of the final hyperplane. While very simple, the proposed method achieves a clear acceleration on most of the datasets studied, particularly when a large precision for the separating hyperplane is wanted, and points out that it may be worthwhile to further pursue this research, particularly for on–line learning, exploring other ideas, such as an updating vector selection that offers a larger reduction of the weight norm or a working set more tightly controlled.

References

1. Bennett, K., Bredensteiner, E.: Geometry in learning. In: Gorini, C., Hart, E., Meyer, W., Phillips, T. (eds.) Geometry at Work. Mathematical Association of America (1997)
2. Bordes, A., Bottou, L.: The Huller: a simple and efficient online SVM. In: Gama, J., Camacho, R., Brazdil, P.B., Jorge, A.M., Torgo, L. (eds.) ECML 2005. LNCS (LNAI), vol. 3720, pp. 505–512. Springer, Heidelberg (2005)
3. Chang, Ch., Lin, Ch.: LIBSVM: a library for support vector machines (2001), `http://www.csie.ntu.edu.tw/~cjlin/LIBSVM`
4. Franc, V., Hlavac, V.: An iterative algorithm learning the maximal margin classier. Pattern Recognition 36, 1985–1996 (2003)
5. García, D., González, A., Dorronsoro, J.R.: Convex Perceptrons. In: Corchado, E., Yin, H., Botti, V., Fyfe, C. (eds.) IDEAL 2006. LNCS, vol. 4224, pp. 578–585. Springer, Heidelberg (2006)
6. Joachims, T.: Making Large-Scale Support Vector Machine Learning Practical. In: Schölkopf, B., Burges, C., Smola, A. (eds.) Advances in Kernel methods, pp. 169–184. MIT Press, Cambridge (1999)
7. Li, Y., Long, P.M.: The Relaxed Online Maximum Margin Algorithm. Machine Learning 46, 361–387 (2002)
8. Mavroforakis, M., Theodoridis, S.: A Geometric Approach to Support Vector Classification. IEEE Trans. on Neural Networks 17, 671–682 (2006)
9. Platt, J.C.: Fast training of support vector machines using sequential minimal optimization. In: Schölkopf, B., Burges, C., Smola, A. (eds.) Advances in Kernel methods, pp. 185–208. MIT Press, Cambridge (1999)
10. Schölkopf, B., Smola, A.J.: Learning with Kernels. MIT Press, Cambridge (2001)
11. UCI-benchmark repository of machine learning data sets: University of California Irvine, `http://www.ics.uci.edu`
12. Vapnik, V.N.: The Nature of Statistical Learning Theory. Springer, Berlin (1995)
13. Vijayakumar, S., Wu, S.: Sequential support vector classifiers and regression. Proc. Int. Conf. Soft Computing, 610–619 (1999)

Analysis and Comparative Study of Source Separation Performances in Feed-Forward and Feed-Back BSSs Based on Propagation Delays in Convolutive Mixture

Akihide Horita, Kenji Nakayama, and Akihiro Hirano

Graduate School of Natural Science and Technology, Kanazawa University
Kakuma-machi, Kanazawa, 920-1192 Japan
horita@leo.ec.t.kanazawa-u.ac.jp, nakayama@t.kanazawa-u.ac.jp

Abstract. Feed-Forward (FF-) and Feed-Back (FB-) structures have been proposed for Blind Source Separation (BSS). The FF-BSS systems have some degrees of freedom in the solution space, and signal distortion is likely to occur in convolutive mixtures. On the other hand, the FB-BSS structure does not cause signal distortion. However, it requires a condition on the propagation delays in the mixing process. In this paper, source separation performance in the FB-BSS is theoretically analyzed taking the propagation delays into account. Simulation is carried out by using white signals and speech signals as the signal sources. The FF-BSS system and the FB-BSS system are compared. Even though the FB-BSS can provide good separation performance, there exits some limitation on location of the signal sources and the sensors.

1 Introduction

Two kinds of approaches have been proposed for Blind Source Separation (BSS). One of them is a Feed-Forward (FF-) BSS, and the other is a Feed-Back (FB-) BSS [1]-[3]. In the FF-BSS, there exits some degrees of freedom in determining a separation block, and signal distortion is likely caused. Several methods have been proposed to suppress the signal distortion [4],[6],[7],[8]. Mainly, additional conditions are imposed on the learning process for source separation. On the other hand, the FB-BSS has no degree of freedom, and distortion free outputs can be obtained as a result of source separation. Therefore, the FB-BSS can provide good performances in both source separation and signal distortion. However, the FB-BSS requires some condition on propagation delays in a mixing process [5]. The source separation performance is degraded if the propagation delays do not satisfy this condition.

In this paper, source separation performance of the FB-BSS is theoretically analyzed based on the propagation delays. Simulation results obtained by using white signals and speech signals will be shown to confirm the theoretical analysis and to compare the FF-BSS and the FB-BSS. Finally, usefulness of the FB-BSS is discussed based on relative locations of the sources and the sensors.

J. Marques de Sá et al. (Eds.): ICANN 2007, Part I, LNCS 4668, pp. 169–179, 2007.

2 Feed-Forward BSS

2.1 Network and Input-Output Equations

A block diagram of the FF-BSS is shown in Fig.1(Left). In the separation block, an FIR filter, shown in Fig.1(Right), is used.

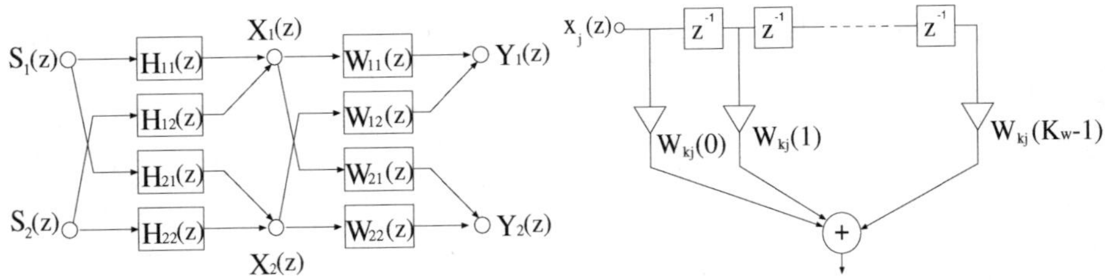

Fig. 1. (Left) A block diagram of FF-BSS. (Right) FIR filter used for $W_{kj}(z)$.

The sources $s_i(n), i = 1, 2, \cdots, N$ are convoluved with impulse responses of the mixing block $h_{ji}(n)$, and are observed as $x_j(n), j = 1, 2, \cdots, N$. The separation block outputs $y_k(n)$ are convolution sums of $w_{kj}(n)$ and $x_j(n)$.

$$x_j(n) = \sum_{i=1}^{N} \sum_{m=0}^{K_h-1} h_{ji}(m) s_i(n-m) \tag{1}$$

$$y_k(n) = \sum_{j=1}^{N} \sum_{l=0}^{K_w-1} w_{kj}(l) x_j(n-l) \tag{2}$$

2.2 Learning Algorithm

The conventional learning algorithm [2],[3] is applied. $w_{kj}(n, l)$ are updated by

$$w_{kj}(n+1, l) = w_{kj}(n, l) + \Delta w_{kj}(n, l) \tag{3}$$

$$\Delta w_{kj}(n, l) = \eta\{w_{kj}(n, l) - \sum_{q=0}^{K_w-1} \varphi(y_k(n)) y_p(n-l+q) w_{pj}(n, q)\}, p \neq j \tag{4}$$

$\varphi(y_k(n))$ is a probability density function of $y_k(n)$. In the above algorithm, the signal distortion is likely caused. A technique to suppress the signal distortion has been proposed [8]. First, $w_{kj}(n+1, l)$ are updated following Eqs.(3) and (4). Second, $w_{jj}(n+1, l)$ are modified as follows:

$$w_{jj}(n+1, l) = (1-\alpha)\tilde{w}_{jj}(n+1, l) + \alpha \bar{w}_{jj}(n+1, l), 0 < \alpha \leq 1 \tag{5}$$

$\tilde{w}_{jj}(n+1, l)$ are updated by Eqs.(3) and (4). $\bar{w}_{jj}(n+1, l)$ are determined by the conditions of complete source separation and signal distortion free [8].

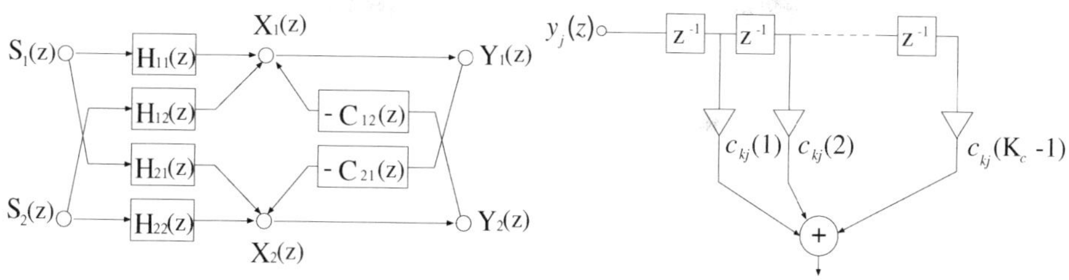

Fig. 2. (Left) Block diagram of FB-BSS. (Right) FIR filter used for $C_{21}(z)$ and $C_{12}(z)$.

3 Feed-Back BSS

3.1 Block Diagram and Input-Output Relations

A block diagram of the FB-BSS is shown in Fig.2(Left) [1]. The separation block employs an FIR filter shown in Fig.2(Right).

Since a feed-back loop in a discrete time system needs at least one sample delay, a direct path from the input to the output is not used in the FIR filter, that is $c_{kj}(0) = 0$. The outputs of the separation block are expressed by

$$y_k(n) = x_k(n) - \sum_{\substack{j=1 \\ \neq k}}^{N} \sum_{l=1}^{K_c-1} c_{kj}(l) y_j(n-l) \tag{6}$$

3.2 Learning Algorithm

The update equation of $c_{kj}(n,l)$ was proposed by Jutten et al by using the Kullback-Leibler mutual information, and is shown here [1]. $f(y_k(n))$ and $g(y_j(n-l))$ are different odd functions. One of them is at least a nonlinear function.

$$c_{kj}(n+1,l) = c_{kj}(n,l) + \Delta c_{kj}(n,l) \tag{7}$$

$$\Delta c_{kj}(n,l) = \mu f(y_k(n)) g(y_j(n-l)) \tag{8}$$

4 Derivation of Learning Algorithm for FB-BSS Based on Propagation Delay

A learning algorithm is derived based on the propagation delays in the mixing process [5]. For simplicity, the FB-BSS having 2 sources, 2 sensors and 2 outputs, as shown in Fig.2(Left), is taken into account. It is assumed that the propagation delays of $H_{11}(z)$ and $H_{22}(z)$ are less than those of $H_{21}(z)$ and $H_{12}(z)$. This means that the sensor of $X_1(z)$ locates close to $S_1(z)$, and that of $X_2(z)$ locates close to $S_2(z)$. Thus, this assumption is actually acceptable.

The outputs of the separation block can be expressed by [5]

$$\begin{bmatrix} Y_1(z) \\ Y_2(z) \end{bmatrix} = \frac{1}{1 - C_{12}(z)C_{21}(z)} \begin{bmatrix} 1 & -C_{12}(z) \\ -C_{21}(z) & 1 \end{bmatrix} \begin{bmatrix} H_{11}(z) & H_{12}(z) \\ H_{21}(z) & H_{22}(z) \end{bmatrix} \begin{bmatrix} S_1(z) \\ S_2(z) \end{bmatrix}$$

$$= \frac{1}{1 - C_{12}(z)C_{21}(z)} \begin{bmatrix} H_{11}(z) - C_{12}(z)H_{21}(z) & H_{12}(z) - C_{12}(z)H_{22}(z) \\ H_{21}(z) - C_{21}(z)H_{11}(z) & H_{22}(z) - C_{21}(z)H_{12}(z) \end{bmatrix} \begin{bmatrix} S_1(z) \\ S_2(z) \end{bmatrix} \quad (9)$$

When source separation is complete, $C_{kj}(z)$ and $y_k(n)$ have the following two kinds of solutions (a) and (b).

$$\boldsymbol{h}_{ji} = [h_{ji}(0), h_{ji}(1), \cdots, h_{ji}(K_h - 1)]^T \quad (10)$$

$$\boldsymbol{s}_i(n) = [s_i(n), s_i(n-1), \cdots, s_i(n - K_h + 1)]^T \quad (11)$$

(a) Non-diagonal elements are zero.

$$C_{12}(z) = \frac{H_{12}(z)}{H_{22}(z)} \qquad C_{21}(z) = \frac{H_{21}(z)}{H_{11}(z)} \quad (12)$$

$$y_1(n) = \boldsymbol{h}_{11}^T \boldsymbol{s}_1(n) \qquad y_2(n) = \boldsymbol{h}_{22}^T \boldsymbol{s}_2(n) \quad (13)$$

(b) Diagonal elements are zero.

$$C_{12}(z) = \frac{H_{11}(z)}{H_{21}(z)} \qquad C_{21}(z) = \frac{H_{22}(z)}{H_{12}(z)} \quad (14)$$

$$y_1(n) = \boldsymbol{h}_{12}^T \boldsymbol{s}_2(n) \qquad y_2(n) = \boldsymbol{h}_{21}^T \boldsymbol{s}_1(n) \quad (15)$$

From the assumption regarding the delay of $H_{ji}(z)$, $C_{21}(z)$ and $C_{12}(z)$ in (a) have a positive delay, that is, they are causal systems. On the other hand, $C_{21}(z)$ and $C_{12}(z)$ in (b) have a negative delay, resulting in non-causal systems, which cannot be realizable. For this reason, $S_1(z)$ cannot be cancelled at $X_1(z)$. In the same manner, $S_2(z)$ cannot be cancelled at $X_2(z)$. This means that the diagonal elements of Eq.(9) cannot be cancelled by adjusting $C_{jk}(z)$. On the other hand, $S_2(z)$ and $S_1(z)$ can be cancelled at $X_1(z)$ and $X_2(z)$, respectively. In other words, the non-diagonal elements of Eq.(9) can be cancelled. Combining these properties, the power of the separation block outputs can be set as a cost function. Applying the gradient method, the same learning algorithm as shown in Eqs.(7) and (8) can be derived [5].

5 Analysis of Source Separation Based on Propagation Delay in Mixing Process

5.1 Effects of Propagation Delay on Learning FB-BSS

It is also assumed that, in the FB-BSS shown in Fig.2, $S_1(z)$ and $S_2(z)$ are separated at $Y_1(z)$ and $Y_2(z)$, respectively. This does not lose generality. Source separation performance of the FB-BSS is determined by the following conditions.

1. $S_2(z)$ and $S_1(z)$ should be cancelled at $X_1(z)$ and $X_2(z)$, respectively.
2. $S_1(z)$ and $S_2(z)$ should be preserved at $X_1(z)$ and $X_2(z)$, respectively.

As discussed in Sec.4, the learning of the FB-BSS is equivalent to minimize the output powers. If the delay of $C_{12}(z)H_{21}(z)$ is large enough compared to that of $H_{11}(z)$, then the $S_1(z)$ component cannot be cancelled at $X_1(z)$, and can be separated at $Y_1(z)$. This means that the power of $Y_1(z)$ can be minimized by cancelling the $S_2(z)$ components at $X_1(z)$. Situation is the same as in minimizing the power of $Y_2(z)$. In these cases, $C_{kj}(z)$ converge to the optimal solutions given by Eq.(12). When difference between the delays of $C_{12}(z)H_{21}(z)$ and $H_{11}(z)$ is not large enough, correlation between $C_{12}(z)H_{21}(z)S_1(z)$ and $H_{11}(z)S_1(z)$ will be increased. As a result, some cancellation between them can be possible. In other words, the power of $Y_1(z)$ can be minimized by reducing not only the $S_2(z)$ component but also the $S_1(z)$ component. In this situation, $C_{kj}(z)$ are shifted from the optimal solutions given by Eq.(12) toward the undesirable solutions given by Eq.(14), resulting in poor source separation performances.

First, the condition (1) is taken into account. As shown in Fig.2, the transfer functions $C_{kj}(z)$ require at least one sample delay, corresponding to z^{-1}. Therefore, in order to cancel $S_2(z)$ at $X_1(z)$, the difference between the delays of H_{21} and $H_{11}(z)$ should be equal to or larger than one sample delay. If this condition is not sufficiently satisfied, $S_2(z)$ cannot be well cancelled. The same situation is held in $X_2(z)$.

Next, the condition (2) is taken into account. Preserving the $S_1(z)$ component in $X_1(z)$, at the same time, in $Y_1(z)$, is highly dependent on the difference between the delays of $C_{12}(z)H_{12}(z)$ and $H_{11}(z)$. Here, the signal is simply denoted $u(n)$ for convenience. The sampling frequency of $u(n)$ is denoted f_s. In order to consider the delay difference less than the sampling period $T = 1/f_s$, the sampling frequency f_s is converted to $Kf_s, K > 1$. This up-sampling is carried out by inserting zero samples and band limitation. Let $u_z(n)$ be the up-sampled signal by inserting $K-1$ zero samples between the $u(n)$ samples. The frequency component of $u(n)$ is assumed to be distributed in $0 \sim f_s/m$. $u_z(n)$ is band limited by using the ideal filter, whose pass band is $0 \sim f_s/m$ and the sampling frequency is Kf_s. An impulse response of this ideal filter is given by

$$\phi(n) = \frac{2}{Km} \frac{\sin\left(\frac{2\pi}{Km}n\right)}{\frac{2\pi}{Km}n} \tag{16}$$

$\phi(n)$ is regarded as an interpolation function. Let the band limited signal be $u_K(n)$, which is a convolution sum of $u_z(n)$ and $\phi(n)$ as shown below.

$$u_K(n) = \sum_{k=0}^{n} \phi(n-k)u_z(k) \tag{17}$$

5.2 Canceling $u_K(n)$ by Using $u_K(n-l)$

In this case, the delay difference is l samples under the sampling frequency of Kf_s. We will consider cancelation of $S_1(z)$ at $X_1(z)$ depending on delay difference. In this case, $u_K(n)$ and $u_K(n-l)$ can be regarded as $h_{11}(n) * s_1(n)$ and $c_{12}(n) * h_{21}(n) * s_1(n)$, respectively. The operation $*$ is a convolution sum.

Since amplitude and phase responses of the signals can be adjusted by $C_{jk}(z)$, an effect of the delay difference is only taken into account in this section.

Cancelation of $u_K(n)$ by using $u_K(n-l)$ can be expressed by

$$\begin{aligned} u_K(n) - u_K(n-l) &= \sum_{k=0}^{n} \phi(n-k)u_z(k) - \sum_{k=0}^{n} \phi(n-k-l)u_z(k) \\ &= \sum_{k=0}^{n} [\phi(n-k) - \phi(n-k-l)]u_z(k) \end{aligned} \quad (18)$$

From the above equation, the cancelation of $u_K(n)$ can be evaluated by $[\phi(n-k)-\phi(n-k-l)]$, which is equivalent to $\phi(l)$. $\phi(l)$ becomes zero at $l = Km/2$, and takes small value after that. Therefore, as a rule of thumb, the delay difference, with which $u_K(n)$ can be cancelled by using $u_K(n-l)$, is less than a $Km/2$ sample delay under the sampling frequency of Kf_s, which is equivalent to an $m/2$ sample delay with the f_s sampling frequency.

An example of $\phi(n)$ is shown in Fig.3, in which $m = 4$ and $K = 4$. In this

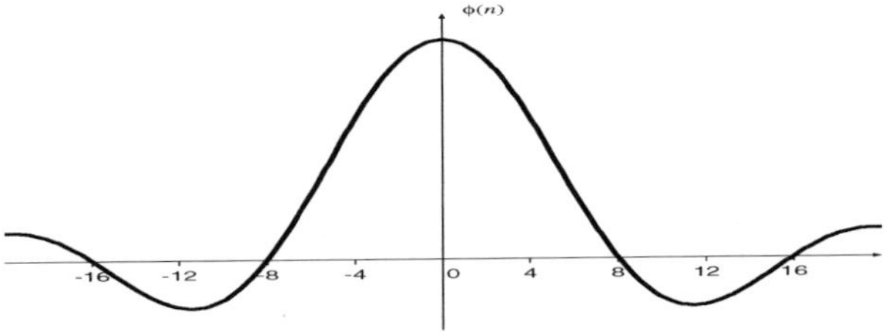

Fig. 3. Example of interpolation function $\phi(n)$ with $m = 4$ and $K = 4$

figure, one scale on the horizontal axis indicates T/K sec and $4(=K)$ scales corresponds to T sec. $\phi(n)$ takes zero at 8 samples($\times T/K$) and 2 samples($\times T$), and takes small values after these points.

(Example 1)
Let $f_s = 8$kHz and the frequency components of the signal are distributed under 1kHz. Since $f_s/m = 1kHz$ and $m = 8$, if the delay difference between $C_{12}(z)H_{12}(z)$ and $H_{11}(z)$ is larger than $m/2 = 4$ samples, then cancellation of $S_1(z)$ at $X_1(z)$ is small. Since $C_{kj}(z)$ have 1 sample delay, by assigning 3 samples delay between $H_{21}(z)$ and $H_{11}(z)$, the above conditions (1) and (2) can be simultaneously satisfied. In this case, the source separation performance is mainly determined by the condition(2).

(Example 2)
Let $f_s = 8$kHz and the frequency component of the signal is distributed under 4kHz. Since $f_s/m = 4$kHz and $m = 2$, if the delay difference between $C_{12}(z)H_{12}(z)$ and $H_{11}(z)$ is more than $m/2 = 1$ samples, then the cancelation

of $S_1(z)$ at $X_1(z)$ is small. Since $C_{kj}(z)$ has 1 sample delay, the delay difference between $H_{21}(z)$ and $H_{11}(z)$ is not necessary. However, in order to satisfy the condition(1), at least 1 sample delay is required. In this case, the source separation performance is mainly determined by the condition(1).

5.3 Cancelation of $u_K(n)$ Evaluated by Correlation

The cancelation of $u_K(n)$ by using $u_K(n-l)$ can be also evaluated by the mean square of $\phi(n-k) - \phi(n-k-l)$ as follows:

$$E[(\phi(n-k) - \phi(n-k-l))^2]$$
$$= E[\phi^2(n-k)] - 2E[\phi(n-k)\phi(n-k-l)] + E[\phi^2(n-k-l)] \quad (19)$$

The delay difference of interest is a few samples delay, that is l is a small integer. Therefore, $E[\phi^2(n-k)]$ and $E[\phi^2(n-k-l)]$ are almost the same. They are denoted $E[\phi^2(n-k)]$ here. The above equation is further rewritten as

$$2E[\phi^2(n-k)] - 2E[\phi(n-k)\phi(n-k-l)]$$
$$= 2E[\phi^2(n-k)]\left(1 - \frac{E[\phi(n-k)\phi(n-k-l)]}{E[\phi^2(n-k)]}\right) \quad (20)$$

Since $E[\phi^2(n-k)]$ is the signal power, it is not related to the cancelation of $u_K(n)$. From the above equation, if a correlation of $u_K(n-k)$ and $u_K(n-k-l)$, that is $E[\phi(n-k)\phi(n-k-l)]/E[\phi^2(n-k)]$ is close to 1, then $S_1(z)$ is well cancelled at $X_1(z)$. Simulations in Sec.6.2, the correlation of $h_{11}(n) * s_1(n)$ and $h_{21}(n) * s_1(n)$ is used to evaluate the cancelation of $S_1(z)$ at $X_1(z)$.

On the contrary, in the FF-BSS, the propagation delays in the mixing process do not affect the source separation performance.

5.4 Effects of Fractional Propagation Delays

The propagation delay in the mixing process is not always an integer times of the sampling period $T = 1/f_s$. It may be sometime a fractional period. This means the transfer functions $H_{ji}(z)$ in the mixing process cannot be expressed by a rational function of z^{-1}. On the other hand, the FIR filters used in the separation block are implemented with the sampling frequency of $f_s = 1/T$, and their transfer function are expressed by a power series of z^{-1}. Therefore, if the propagation delay is not an integer times of T, the FIR filter cannot realize the inverse function of the mixing process, as a result, the source separation performance is degraded. This point is also investigated through simulation.

6 Simulations and Discussions

6.1 Simulation Setup

2-channel and 3-channel models are used. Simulation results only for the 2-channel model are demonstrated due to page limitation. The sampling frequency

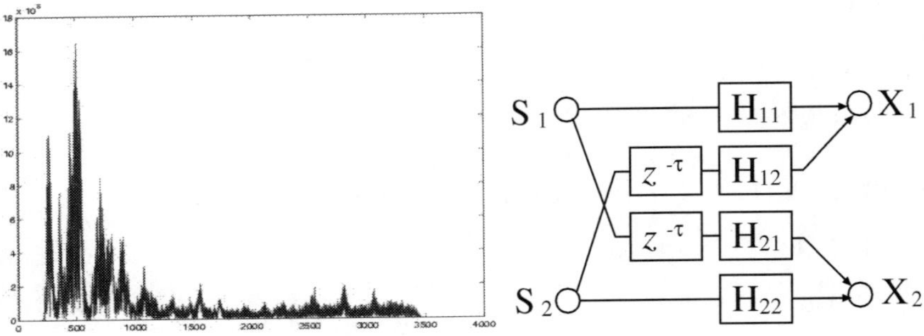

Fig. 4. (Left) Spectrum of speech signal. (Right) Mixing process with delay difference.

is f_s = 8kHz. White signals and speech signals, whose spectrum is shown in Fig.4(Left), are used as the sources. Figure 4(Right) shows a mixing process.

The delay difference between $H_{jj}(z)$ and $H_{ji}(z), j \neq i$ is realized by τ. Two kinds models for $H_{ji}(z)$ are used. One of them is an instantaneous mixing process shown in Eq.(21) and the other is an approximately actual room environment. The source separation is evaluated by the Signal-to-Interference Ratio (SIR [dB]) given by Eq.(24). $A_{ki}(z)$ is a transfer function from the ith signal source to the kth output of the separation block.

$$\boldsymbol{H}(z) = \begin{bmatrix} 1 & 0.9 \\ 0.9 & 1 \end{bmatrix} \tag{21}$$

$$\sigma_s^2 = \frac{1}{2\pi} \sum_{i=1}^{N} \int_{-\pi}^{\pi} |A_{ii}(e^{j\omega})S_i(e^{j\omega})|^2 d\omega \tag{22}$$

$$\sigma_i^2 = \frac{1}{2\pi} \sum_{k=1}^{N} \sum_{\substack{i=1 \\ \neq k}}^{N} \int_{-\pi}^{\pi} |A_{ki}(e^{j\omega})S_i(e^{j\omega})|^2 d\omega \tag{23}$$

$$SIR = 10 \log_{10} \frac{\sigma_s^2}{\sigma_i^2} \quad [dB] \tag{24}$$

6.2 Instantaneous and Propagation Delay Model

Figure 5 shows the SIR and the correlation by using Eq.(21) and the white signal sources. 'Delay difference' means τ in Fig.4. $C_{12}(z)$ has one sample delay as shown in Fig.2(Right), the delay difference at $X_1(z)$ becomes $\tau + 125\mu s$. The cancelation of $S_1(z)$ at $X_1(z)$ should be evaluated by $\tau + 125\mu s$. The band width of the white signal is 4kHz, and $m = 2$ as discussed in Sec.5.1. The interpolation function is given by

$$\phi(n) = \frac{1}{K} \frac{\sin\left(\frac{2\pi}{2K}n\right)}{\frac{2\pi}{2K}n} \tag{25}$$

$\phi(n)$ takes zero at every K samples, which correspond to one sample with the 8kHz sampling frequency. Since the white signals are sampled by 8kHz, there is

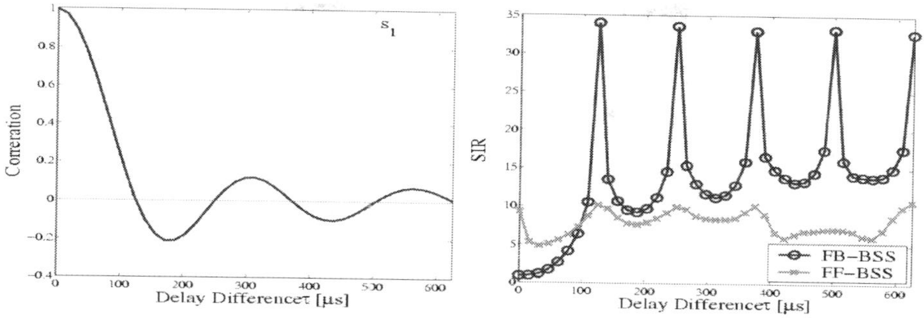

Fig. 5. White signal sources. (Left) Correlation between $h_{11}(n) * s_1(n)$ and $h_{21}(n) * s_1(n)$. (Right) SIR with respect to τ.

no correlation between the samples. The correlation shown in Fig.5(Left) is also similar to $\phi(n)$ given by Eq.(25). As described in Sec.5.1, the rule of thumb for the delay difference, with which the condition(2) is satisfied, is the $m/2$ sample delay, which is one sample delay under the 8kHz sampling frequency. Since $C_{12}(z)$ and $C_{21}(z)$ include one sample delay, the condition(2) is satisfied by using τ=0. However, the condition(1) requires $\tau \geq$ one sample delay. Therefore, the SIR converges to high level at τ=125μs. Like this, in the case of the white signals, the condition(1) is dominant to achieve a high SIR. This situation is exactly the same as Example 2 in Sec.5.1.

The SIR has peaks at every $\tau = 125\mu s$. This is due to no correlation at these points. Furthermore, the mixing process can be expressed as a rational function of z^{-1}, whose inverse function can be approximated by the separation block, which is implemented with the 8kHz sampling frequency.

6.3 Convolutive and Propagation Delay Model

The mixing process similar to actual audio environment is used. $H_{ji}(z)$ have no delay difference. The delay difference is also realized by τ in Fig.4.

Figure 6 shows simulation results for the speech signal sources. The correlation becomes zero around 400μs. From the analysis in Sec.5.1, $125\mu s \times (m/2) = 400\mu s$, and $m \simeq 6.4$. This is due to the frequency component of the speech signals, which is mainly distributed under 8kHz/6.4$\simeq$ 1.25kHz as shown in Fig.4(Left). The SIR is gradually increased until τ=350μs. The delay difference at $X_1(z)$ becomes $(350 + 125)\mu s = 475\mu s$, which approximately corresponds to $m/2$ sample delay difference. In this case, the condition(2) is dominant. This case corresponds to Example 1 in Sec.5.1.

The SIR of the FF-BSS is not affected by the delay difference τ, and it is almost constant with respect to τ.

The simulation results for the 3-channel model are almost similar to those of the 2-channel model.

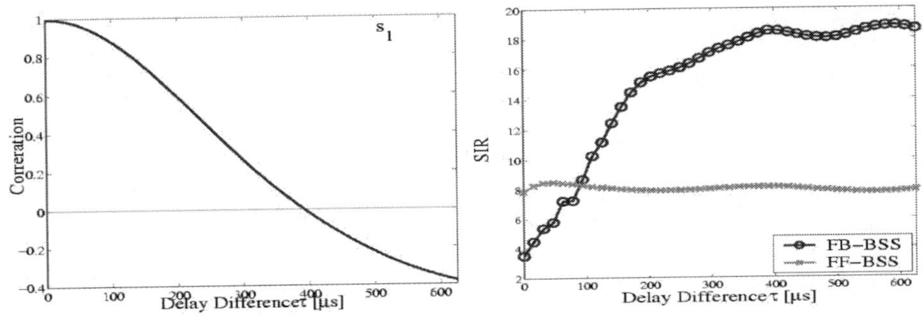

Fig. 6. Speech signal sources. (Left) Correlation between $h_{11}(n) * s_1(n)$ and $h_{21}(n) * s_1(n)$. (Right) SIR with respect to τ.

6.4 Delay Difference Based on Location of Signal Sources and Sensors

A relation between the delay difference and location of the sources and the sensors is investigated taking sound propagation time into account. A layout of them shown in Fig.7(Left) is taken into account. Relations of the delay differences and the layouts are shown in Fig.7(Right). The FB-BSS can provide better performances with τ=125μs, which is generated by $L_2 = 5, 9, 25$cm for $(L_1, D) = (300, 100)$cm, $(100, 100)$cm and $(100, 300)$cm, respectively.

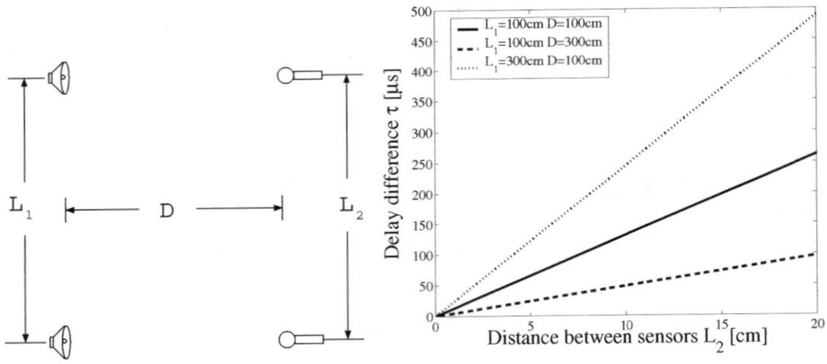

Fig. 7. (Left) Layout of signal sources and sensors. (Right) Delay difference τ depending on layout of signal sources and sensors.

7 Conclusions

The source separation performance of the FB-BSS is theoretically analyzed based on the propagation delay difference in the mixing process. It is determined by two conditions, canceling the interferences and preserving the desired source, which are highly dependent on the delay difference in the mixing process. Simulation results using white signal sources and speech signal sources support the theoretical analysis. Furthermore, a relation between the delay difference and the layout of the sources and sensors is discussed. The layout conditions, which are useful for the FB-BSS or the FF-BSS, are derived.

References

1. Nguyen Thi, H.L., Jutten, C.: Blind source separation for convolutive mixtures. Signal Processing 45(2), 209–229 (1995)
2. Cichocki, A., Amari, S., Adachi, M., Kasprzak, W.: Self-adaptive neural networks for blind separation of sources. In: Proc. ISCAS'96, Atlanta, pp. 157–161 (1996)
3. Amari, S., Chen, T., Cichocki, A.: Stability analysis of learning algorithms for blind source separation. Neural Networks 10(8), 1345–1351 (1997)
4. Murata, N., Ikeda, S., Ziehe, A.: An approach to blind source separation based on temporal structure of speech signals. Neurocomputing 41, 1–24 (2001)
5. Nakayama, K., Hirano, A., Horita, A.: A learning algorithm for convolutive blind source separation with transmission delay constraint. In: Proc. IJCNN'2002, pp. 1287–1292 (May 2002)
6. Saruwatari, H., Takatani, T., Yamajo, H., Sishikawa, T., Shikano, K.: Blind separation and deconvolution for real convolutive mixture of temporally correlated acoustic signals using SIMO-model-based ICA. In: ICA'03, pp. 549–554 (April 2003)
7. Horita, A., Nakayama, K., Hirano, A., Dejima, Y.: Analysis of signal separation and signal distortion in feedforward and feedback blind source separation based on source spectra. In: IEEE&INNS, Proc., IJCNN2005, Montreal, July-August, pp. 1257–1262 (2005)
8. Horita, A., Nakayama, K., Hirano, A., Dejima, Y.: A learning algorithm with distortion free constraint and comparative study for feedforward and feedback BSS. In: Proc. EUSIPCO2006, Florence, Italy (September 2006)

Learning Highly Non-separable Boolean Functions Using Constructive Feedforward Neural Network

Marek Grochowski and Włodzisław Duch

Department of Informatics, Nicolaus Copernicus University,
Grudziądzka 5, Toruń, Poland
grochu@is.umk.pl

Abstract. Learning problems with inherent non-separable Boolean logic is still a challenge that has not been addressed by neural or kernel classifiers. The k-separability concept introduced recently allows for characterization of complexity of non-separable learning problems. A simple constructive feedforward network that uses a modified form of the error function and a window-like functions to localize outputs after projections on a line has been tested on such problems with quite good results. The computational cost of training is low because most nodes and connections are fixed and only weights of one node are modified at each training step. Several examples of learning Boolean functions and results of classification tests on real-world multiclass datasets are presented.

1 Introduction

Many algorithms and neural network architectures for learning parity problem and other difficult Boolean functions were presented in [1,2,3,4,5]. These methods are suitable only for the parity problem and will not work for other complex problems. Biological neural networks are able to solve quite complex learning problems inherent in optimization of behavior or understanding linguistic patterns. Finding general algorithm capable of solving a larger set of problems of similar complexity as the parity problem is still a challenge. Sequential constructive methods using computational geometry algorithms are the most promising in this respect. Several interesting algorithms of this type, including the irregular partitioning, Carve, target switch, oil spot, and sequential window learning algorithm, have been reviewed and compared in [6]. Most of them work only for binary problems and therefore pre-processing using Gray coding is used.

As shown recently [7] complexity of classification problems may be characterized by the concept of k-separability: the dataset $\boldsymbol{x}$ of points belonging to two classes is called k-separable if a direction $\boldsymbol{w}$ exist such that k is the minimum number of intervals containing data from a single class after projection on a line $y_i = \boldsymbol{x}_i \cdot \boldsymbol{w}_i$. For example, linearly separable two-class data points form just two intervals (open-ended on left and right side), while the data that cannot be linearly separated may require 3 or more intervals containing vectors

J. Marques de Sá et al. (Eds.): ICANN 2007, Part I, LNCS 4668, pp. 180–189, 2007.

from a single class only. The n-bit parity problems require at least $k = n + 1$ intervals. This concept allows for precise characterization of the complexity of learning. Although usually the smallest k is of interest sometimes higher k's may be preferred if the margins between projected clusters are larger, or if among k clusters some have very small number of elements. Problems that may be solved by linear projection on at least k-clusters belonging to alternating classes are called k-separability problems [7]. For the parity $k = n + 1$ is sufficient and it is quite likely (although it has not been proven) that all Boolean functions may be learned using linear projection on no more than $n + 1$ intervals.

The difficulty of learning Boolean functions grows quickly with the minimum k required to solve a given problem. Linear (2-separable) problems are quite simple and may be solved with linear SVM or any other variant of LDA model. Kernel transformations may convert some data distributions into linearly separable distributions in higher dimensional space. This strategy does not work for parity-like and other difficult Boolean functions, with virtually no generalization from learned examples. Gaussian-based kernels set each data point as a separate support vector, and in case of parity all points have closest neighbors from the opposite class. Nearest neighbor methods and decision trees have the same problem, linear methods fail completely while multilayer perceptrons, although in principle may be used [1,2,3,4,5], in practice require special network architectures and convergence of the learning procedure is almost impossible to achieve. As already shown in [8] (see also [9,10]) some problems require at least $O(n^2)$ parameters using networks with localized functions and only $O(n)$ parameters when non-local functions are used. The n-parity problem may be trivially solved using a periodic function with a single parameter [7] while the multilayer perceptron (MLP) networks need $O(n^2)$ parameters and learn it only with great difficulty.

For many complicated problems often a simple linear mapping exists that leaves only trivial non-linearities that may be separated using window-like neurons. For example XOR, the simplest non-linearly separable problem, is solved by a network with one node implementing window-like transfer function:

$$\tilde{M}_i(\boldsymbol{x}; \boldsymbol{w}, a, b) = \begin{cases} 1 \text{ if } \boldsymbol{w}\boldsymbol{x} \in [a, b] \\ 0 \text{ if } \boldsymbol{w}\boldsymbol{x} \notin [a, b] \end{cases} \tag{1}$$

This function is suitable for learning all 3-separable data, and the number of such Boolean functions for 3 or more bits is much greater than of the linearly separable functions [7]. There are many advantages of using window-type functions in neural networks, especially in difficult, highly non-separable classification problems [8].

Problems requiring $k = 3$ are already slightly more difficult for non-local transformations (for example MLPs) and problems with high k quickly become intractable for general classification algorithms. Although some k-separable problems may also be solved using complex models it is rather obvious that simplest linear solutions, or solutions involving smooth minimal-complexity non-linear mappings combined with interval non-linearities, should show better generalization and such solutions should be easier to comprehend.

An initial data transformation by the hidden layer may create an image of the data that projected on some direction $\boldsymbol{w}$ will produce pure clusters of samples that belong to a single class and can be easily separated from vectors from the opposite class. Unfortunately the standard learning algorithm cannot be used because it is not known a priori to which interval a given vector should be assigned, so labels cannot be used directly as targets for learning. In this paper a simple and fast network with constructive algorithm and hidden nodes with transfer functions of type (1) is proposed to solve this problem. This network is described in the next section, and the Section 3 contains some experimental results on learning the parity problem and other difficult Boolean functions. Some cross-validation test results on benchmark multiclass problems are also presented.

2 The Network Architecture and Training

Different neural network approaches may be used to search for k-separable solutions [7,11]. For difficult Boolean functions what is needed is a combination of projection and clustering, with localized functions capturing interesting clusters in projections. For 3-separable two-class problems the transformation provided by the network should lead to 3 intervals: $[-\infty, a], [a, b], [b, +\infty]$ to $-1, +1, -1$ values. This may be implemented using a single transfer function of type (1) that separates instances from the interval $[a, b]$ from the rest, grouping vectors from a single class into a cluster. For backpropagation algorithm (and other gradient descent based methods) soft windowed-type functions should be used. Many types of transfer functions can be used for realization of (1) (for taxonomy of a neural transfer functions see [12], [8]). For example, the bicentral function may be used in a product form:

$$M_i(\boldsymbol{x}; \boldsymbol{w}, t, a, \beta) = \sigma(\beta(\boldsymbol{wx} - t - a))(1 - \sigma(\beta(\boldsymbol{wx} - t + a))$$

or taken as a difference of two sigmoidal functions:

$$M_i(\boldsymbol{x}; \boldsymbol{w}, a, b, \beta) = \sigma(\beta(\boldsymbol{wx} - a)) - \sigma(\beta(\boldsymbol{wx} - b))$$

.

These two functions differ for $b < a$, with the second one producing negative output $M(b < a) < 0$. This property may be useful for "unlearning" instances misclassified by previous hidden nodes. Hard-window type function (1) is obtained by setting large value of the slope β of sigmoidal functions at end of the learning process.

$$M_i(\boldsymbol{x}; \Gamma_i) \overset{\beta \to \infty}{\longrightarrow} \tilde{M}_i(\boldsymbol{x}; \Gamma_i)$$

or by introduction of an additional threshold parameter:
$\tilde{M}_i(\boldsymbol{x}; \Gamma_i) = \operatorname{sgn}(M_i(\boldsymbol{x}; \Gamma_i) - t_i)$. An interesting new window-type function that may be used as a transfer function is:

$$M_i(\boldsymbol{x}; \boldsymbol{w}, a, b, \beta) = \frac{1}{2}\left(1 - \tanh(\beta(\boldsymbol{wx} - a)) \tanh(\beta(\boldsymbol{wx} - b))\right) . \quad (2)$$

It is easy to show that for all possible values of β there is $M(wx = b) = M(wx = a) = 0.5$. Hence the interval boundaries during the learning phase, when the slope β is small, are exactly the same as boundaries taken when the value of β reaches large values – in the bicentral function case interval boundaries for $\beta \to \infty$ may be different than $[t - a, t + a]$.

For more complex problems ($k > 3$) the network with weight sharing based on several such nodes may be used. A standard network with window functions should minimize an error measure:

$$E(\boldsymbol{x}; \Gamma) = E_{\boldsymbol{x}} ||y(\boldsymbol{x}; \Gamma) - c(\boldsymbol{x}))|| \tag{3}$$

where $y(\boldsymbol{x}; \Gamma)$ is network output, $c(x) \in \{0, 1\}$ denote class of given vector x and Γ is a set of all parameters that are changed during training (weights, biases, etc.). This expectation is calculated over all vectors, and any norm may be used, including the most popular mean square error or cross entropy measures. The minimum is reached for parameters Γ and if we could find good solution for the whole network the problem would be solved. For complex problems (larger k) this would require global minimization. This form of error does not give any control over purity of the clusters, but adding for each hidden node with window-like transfer function an extra term:

$$E(\boldsymbol{x}; \Gamma; a, b, \lambda) = E_{\boldsymbol{x}} ||y(\boldsymbol{x}; \Gamma) - c(\boldsymbol{x}))|| + \lambda E_{y \in [a,b]} ||y(\boldsymbol{x}; \Gamma) - c(\boldsymbol{x}))|| \tag{4}$$

will allow to get minimum error for a pure cluster, with λ controlling the tradeoff between the covering and the purity. This approach requires constructive network, with nodes trained one at a time. All experiments described below were done with the following error function:

$$\begin{aligned} E(\boldsymbol{x}; \Gamma, \lambda_1, \lambda_2) = &\frac{1}{2} \sum_{\boldsymbol{x}} (y(\boldsymbol{x}; \Gamma) - c(\boldsymbol{x}))^2 + \\ &+ \lambda_1 \sum_{\boldsymbol{x}} (1 - c(\boldsymbol{x})) y(\boldsymbol{x}; \Gamma) - \lambda_2 \sum_{\boldsymbol{x}} c(\boldsymbol{x}) y(\boldsymbol{x}; \Gamma) \end{aligned} \tag{5}$$

First term in (5) is the standard mean square error (MSE) measure, second term with λ_1 (penalty factor) increases the total error for vectors x_i from class $c(x_i) = 0$ that falls into group of vectors from class 1 (it is a penalty for "unclean" clusters), third term with λ_2 (reward factor) decreases the value of total error for every vector x_i from class 1 that was correctly placed inside created clusters (it is a reward for large clusters).

The network used here has one hidden layer and one sigmoidal or linear node in the output layer; it's architecture is shown in Fig. 1.

All connection weights between the output and the hidden layer have fixed strength 1, simply summing the outputs filtered through the window functions. Every hidden node should separate a large group of vectors from class $c = 1$. The weight sharing requirement to obtain linear projections may relaxed, allowing different projection directions for different neurons, although initialization from

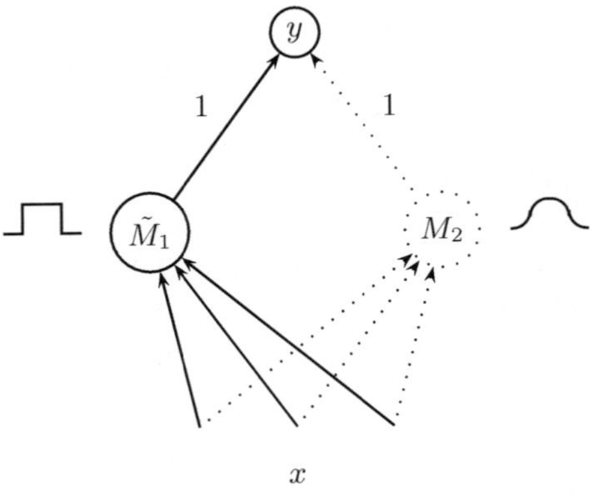

Fig. 1. Example of a network with two hidden neurons. Only parameters of the second node are adapted (dotted line), the first node M_1 has been frozen with large value of β.

the same direction may frequently need to linear k-separable solution. Learning starts with an empty hidden layer and in every phase of the training one new unit is added, initialized and trained using the backpropagation algorithm until convergence is reached. Weight sharing is not used to allow for different projections that may lead to partially overlapping clusters. Node weights are initialized with small random values, while biases a and b (for sigmoidal type of neurons) are estimated using $a = (\boldsymbol{wx})_{min} + \frac{1}{3}|(\boldsymbol{wx})_{max} - (\boldsymbol{wx})_{min}|$ and $b = (\boldsymbol{wx})_{min} + \frac{2}{3}|(\boldsymbol{wx})_{max} - (\boldsymbol{wx})_{min}|$. The input vectors correctly handled by the first neuron do not contribute to the error, therefore the weights of this neuron are kept frozen during further learning. Next node is added and learning procedure is repeated on the remaining data. If number of cases correctly classified by a given new node drops below certain minimum the learning procedure stops and this node is removed from the network.

3 Results

Reward and Penalty. There are 65536 possible 4 dimensional Boolean functions and 192 of them are 6-separable functions [7]. The remaining functions are k-separable with $k < 6$. Hence for learning this functions at least 2 hidden nodes of type (1) are needed. The network have been applied to 192 functions that are 6-separable, using various values of λ_1 and λ_2. Figure 2 shows the effect of reward and penalty on average error, the number of learning cycles and the numbers of nodes needed in the final network to find a perfect solution. While reward λ_2 grows the error becomes larger. Penalty factor has less drastic influence on error, only for values close to zero the error slightly grows but for almost all λ_1 there was no training error. By increasing the penalty factor the learning procedure becomes slower, more cycles are needed for convergence. Reward and penalty factors effect also the complexity of evolved network architecture. With

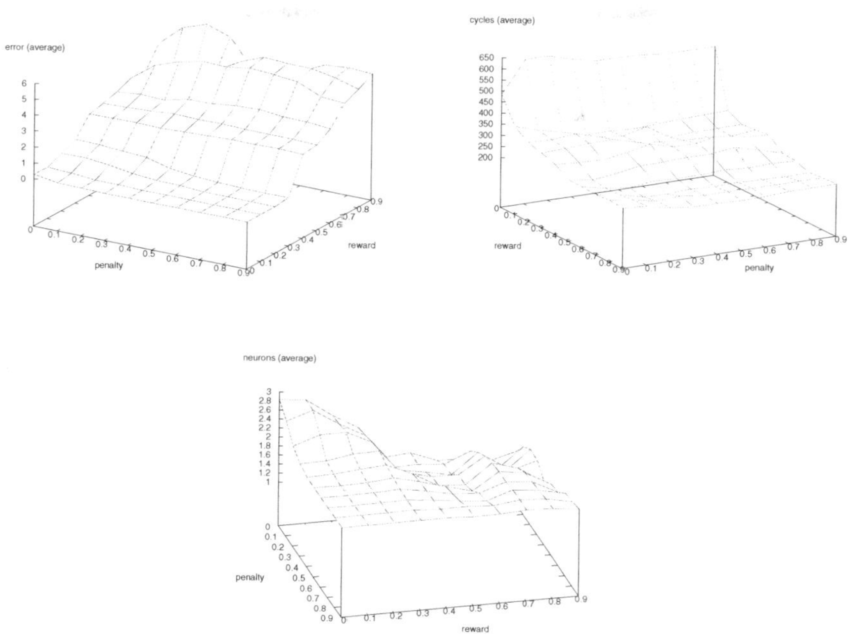

Fig. 2. Influence of penalty and reward factor on error (left), number of cycles (right) and number of neurons (bottom)

small values of λ_1 and λ_2 the network ends with more neurons to achieve perfect solution. For values of penalty and reward $\lambda_1 \in [0.4, 0.8]$ and $\lambda_2 \in [0.1, 0.3]$ the network makes no mistakes, the convergence is fast and the network needs less neurons than the network without additional penalty and reward terms. Hence for further experiments $\lambda_1 = 0.5$ and $\lambda_2 = 0.2$ was taken.

Parity. Many models were presented to handle parity problem [1,2,3] but these models cannot be applied to other problems of similar complexity. Our network has easily learned all parity problems with dimensions less then 8, producing a very simple model in a small number of trials (initializations). For functions with higher dimension greater number of network initializations are needed to find the best solution (lowest k, with smallest number of clusters). In most cases weights of a single hidden unit converges to direction parallel to diagonal of n-dimension hypercube. Input vectors projected on this direction for parity problem gives well separated clusters of ones and zeros, thus this direction seems to be the desired solution. First hidden unit usually finds the largest cluster of ones. For example for the 6-dimensional parity problem projection of data on diagonal direction gives the following order of labels:

$$0111111000000000000000\overline{11111111111111111111}0000000000000001111110$$

The line placed above the middle sequence of 1 shows the range of interval $[a, b]$ created by the first hidden node. Next node tries to find another direction and similar interval that contain remaining vectors from class 1. Projection on weights of the second and the third hidden neuron give

$$1000001111111111000000000000\overline{1111111111}0000000000001111111111000001$$
$$01111100000000000011111111110000\overline{11}00000111111111100000000000111110.$$

This is not k-separability solution as the projection directions were not aligned by weight sharing. With the increased dimension of the problem complexity of created network grows and the correct solution are harder to find. This is presented in Fig 3 where results of the training on parity problems are compared with results obtained for parity-like functions created from the parity function by perturbation of a few labels, and for the randomly generated labels with a probability $p(C = 1|\boldsymbol{x}) = 0.5$ for a given vector $\boldsymbol{x}$. Results were averaged over 100 randomly generated Boolean functions. While optimal solutions for some parity functions have not been found increasing the number of initializations should eventually find it.

In [6] sequential constructive methods based on computational geometry algorithms have been applied to the parity problems up to 9 bits. Computational time for all of them grows exponentially with the size of the problem. Most algorithms need a large number of neurons and may not generalize well. The sequential window learning algorithm was able to find models of minimal complexity. This algorithm also uses window-like neurons, but works only for binary data and employs learning algorithm based on computational learning techniques. While a detailed comparison of our approach with sequential constructive methods is certainly worthwhile the lack of software implementations for these methods makes such comparison a longer-term project.

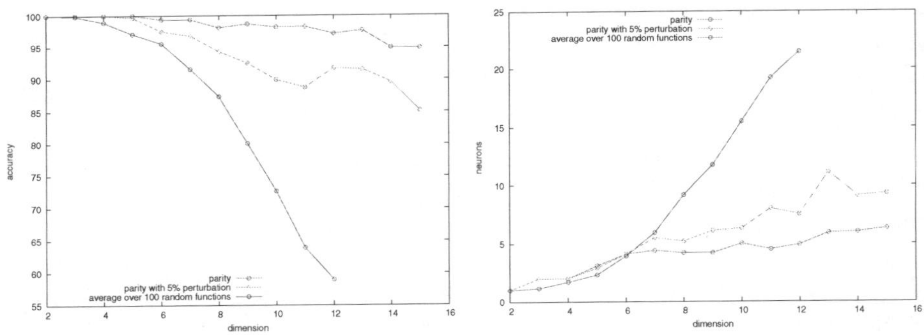

Fig. 3. Influence of dimensionality on the average training accuracy (left) and the average number of neurons (right) for some Boolean problems

Monk 1 Problem. The three monk problems are artificial, small problems designed to test machine learning algorithms [13]. The task is to determine whether an object described by six symbolic features is "a monk" or not. The first problem defines "being a monk" as having the head shape = body shape, or jacket color = *red*, independent of the value of other features. There are 432 combinations of the 6 symbolic attributes, and for the first problem 124 cases were

Table 1. Accuracy of the 10x10CV tests. Network parameters are $\lambda_1 = 0.5$, $\lambda_2 = 0.2$.

dataset	1-NN	Naive Bayes	SVM	c3sep
Appendicitis	81.3±1.5	85.3±1.0	86.5±0.3	85.3±1.0
Australian	78.0±0.2	80.0±0.3	85.0±0.1	86.3±0.6
Flag	50.1±1.1	41.1±1.1	51.1±1.1	53.6±1.8
Glass	68.2±1.7	47.4±1.7	66.7±0.9	61.1±1.3
Ionosphere	85.2±1.2	82.2±0.2	85.2±0.2	85.1±1.5
Iris	95.9±0.5	94.9±0.2	95.5±0.3	95.7±1.0
Pima-diabetes	70.5±0.5	75.2±0.5	70.3±1.0	76.3±0.4
Promoters	78.5±1.8	85.81±1.3	93.1±1.5	74.7±5.6
Sonar	86.8±1.8	67.8±1.2	84.2±1.1	77.9±2.4
Wine	95.1±0.8	98.1±0.3	95.1±0.2	97.1±0.8

Table 2. Average number of hidden neurons created in the 10x10CV classification test, and the number of support vectors used by SVM

	support vectors	neurons (total)	average number of neurons per class							
Appendicitis	32.1	4.2	2.2	2.0						
Australian	207.2	8.8	4.5	4.3						
Flag	315.2	26.7	5.1	6.1	4.0	2.1	2.7	3.6	1.1	1.6
Glass	295.8	14.0	1.4	3.9	1.0	2.8	1.9	2.8		
Ionosphere	63.9	7.9	3.0	4.9						
Iris	43.4	5.0	1.0	2.0	2.0					
Pima-diabetes	365.3	9.1	4.2	4.8						
Promotores	77.2	3.69	1.91	1.78						
Sonar	109.7	8.5	4.5	4.0						
Wine	63.3	4.0	1.0	2.0	1.0					

randomly selected for the training set. In [14] an MLP network solution of this problem requiring 4 neurons has been presented. Our network required only two window-like neurons to learn this problem and achieved 100% accuracy on the test. Among sequential constructive methods only the sequential window learning algorithm with Hamming clustering has correct solution, but this required conversion of 6 symbolic features to 15 binary ones.

Real World Problems. Creation of a separate network classifier for each class of data allows to handle real-world multi-class datasets. The classification test was performed on a few datasets obtained from UCI repository [15]. Three other well known classifiers were used for comparison of results. Table 3 presents averaged results over 10 runs of 10-fold cross-validation tests. Accuracy of our networks is listed in the last column of the table called "c3sep". Every node in the hidden layer was initialized 10 times and the data was normalized before training.

Table 2 shows the number of neurons that were created for a given datasets for each of the classes. For most datasets the network was able to find very simple solutions with only a few neurons, achieving quite good accuracy, comparable to much more complex systems. It should be stressed that the goal here is to find the simplest model, not the highest accuracy solution. The number of support vectors for SVM models with comparable accuracy is much higher.

4 Discussion and Further Work

The approach presented here has been able to learn quite difficult Boolean functions using a very simple model. Other learning systems, such as SVMs, decision trees or similarity-based methods (including RBF networks) cannot deal with such problems. There is a widespread belief that neural networks and kernel classifiers, being universal approximators, should be able to learn any difficult problem. They may learn it but will not be able to generalize, turning themselves into look-up tables, because they do not use the correct underlying model to represent data.

The error function Eq. 5 with additional penalty and reward terms used to train the constructive network shows more advantages when dealing with complex logical problems. For many benchmark classification problems linear solutions are almost optimal and there is not much to be gained. Small computational costs of the constructive network training are a great advantage and allow for exploration of many models created from different initializations, enabling search for the simplest models (in the k-separability sense) that can solve the problem. Results obtained for the parity problem and the real-world data are encouraging, although the real power of solutions based on k-separability targets for learning should be seen only for data with inherent complex logics, such as those arising from the natural language processing and problems in bioinformatics (in preparation). The ability to solve such problems should open the door to many new applications.

Efficient learning of Boolean functions of high complexity (large k) is still a great challenge, although first steps towards this goal have been made here. Substitution of back-propagation learning by global minimization algorithms may lead to better results for complex functions with large number of bits. One disadvantage of the present approach based on constructive network architecture with modified error function is that the network is forced to create pure clusters only for vectors from class $c(x) = 1$, while for many Boolean problems a better solution may be obtained by looking for clusters of both classes simultaneously. Another approach to improve learning is by additional transformation of input data before training, for example extracting new features using the kernel functions approach and adding them to the original dataset. The network trained on such data should be able to choose original or transformed features of the data. Many more ideas and applications are being currently explored.

Acknowledgments. We are grateful for the support by the Polish Committee for Scientific Research, research grant 2005-2007.

References

1. Iyoda, E., Nobuhara, H., Hirota, K.: A solution for the n-bit parity problem using a single translated multiplicative neuron. Neural Processing Letters 18(3), 233–238 (2003)
2. Stork, D.G., Allen, J.: How to solve the n-bit parity problem with two hidden units. Neural Networks 5, 923–926 (1992)
3. Wilamowski, B., Hunter, D.: Solving parity-n problems with feedforward neural network. In: Proc. of the Int. Joint Conf. on Neural Networks (IJCNN'03), Portland, Oregon, vol. I, pp. 2546–2551. IEEE Computer Society Press, Los Alamitos (2003)
4. Sahami, M.: Learning non-linearly separable boolean functions with linear threshold unit trees and madaline-style networks. In: National Conference on Artificial Intelligence, pp. 335–341 (1993)
5. Liu, D., Hohil, M.E., Smith, S.H.: N-bit parity neural networks: new solutions based on linear programming. Neurocomputing 48, 477–488 (2002)
6. Muselli, M.: Sequential constructive techniques. In: Leondes, C. (ed.) Optimization Techniques of Neural Network Systems, Techniques and Applications, vol. 2, pp. 81–144. Academic Press, San Diego, CA (1998)
7. Duch, W.: k-separability. In: Kollias, S., Stafylopatis, A., Duch, W., Oja, E. (eds.) ICANN 2006. LNCS, vol. 4131, pp. 188–197. Springer, Heidelberg (2006)
8. Duch, W., Jankowski, N.: Survey of neural transfer functions. Neural Computing Surveys (1999)
9. Bengio, Y., Delalleau, O., Roux, N.L.: The curse of highly variable functions for local kernel machines. Advances in Neural Information Processing Systems 18, 107–114 (2006)
10. Bengio, Y., Monperrus, M., Larochelle, H.: Non-local estimation of manifold structure. Neural Computation 18, 2509–2528 (2006)
11. Duch, W.: Towards comprehensive foundations of computational intelligence. In: Duch, W., Mandziuk, J. (eds.) Challenges for Computational Intelligence, Springer, Heidelberg (2007) (in print)
12. Duch, W., Jankowski, N.: Taxonomy of neural transfer functions. In: International Joint Conference on Neural Networks, Como, Italy, vol. III, pp. 477–484. IEEE Computer Society Press, Los Alamitos (2000)
13. S.T., et al.: The monk's problems: a performance comparison of different learning algorithms. Technical Report CMU-CS-91-197, Carnegie Mellon University (1991)
14. Duch, W., Adamczak, R., Grąbczewski, K.: A new methodology of extraction, optimization and application of crisp and fuzzy logical rules. IEEE Transactions on Neural Networks 12, 277–306 (2001)
15. Merz, C., Murphy, P.: UCI repository of machine learning databases (1998-2004), `http://www.ics.uci.edu/~{}mlearn/MLRepository.html`

A Fast Semi-linear Backpropagation Learning Algorithm

Bertha Guijarro-Berdiñas, Oscar Fontenla-Romero, Beatriz Pérez-Sánchez, and Paula Fraguela

Facultad de Informática, Universidad de A Coruña, Campus de Elviña 5, 15071 A Coruña, Spain

Abstract. Ever since the first gradient-based algorithm, the brilliant *backpropagation* proposed by Rumelhart, a variety of new training algorithms have emerged to improve different aspects of the learning process for feed-forward neural networks. One of these aspects is the *learning speed.* In this paper, we present a learning algorithm that combines linear-least-squares with gradient descent. The theoretical basis for the method is given and its performance is illustrated by its application to several examples in which it is compared with other learning algorithms and well known data sets. Results show the proposed algorithm improves the learning speed of the basic backpropagation algorithm in several orders of magnitude, while maintaining good optimization accuracy. Its performance and low computational cost makes it an interesting alternative even for second order methods, specially when dealing large networks and training sets.

1 Motivation

Among the many variants of neural network architectures that exist, feed-forward neural networks, and particular those based on the MultiLayer Perceptron (MLP), are one of the most popular models with successful applications in many fields. It is well-known that the power of these networks comes from having several layers of adaptive weights and nonlinear activation functions such as the sigmoid or hyperbolic tangent. Moreover, usually the performance of a network is evaluated with the sum-of-squares error function that compares the network's output with a desired signal. Under these circumstances there is not a closed-form solution to find the weight values that minimizes the sum-of-squares error function [9]. Therefore the usual approach is to use the derivatives of the error function with respect to the weight parameters in gradient-based optimization algorithms for finding the minimum of the error function.

Ever since the first gradient-based algorithm, the brilliant *backpropagation* proposed by Rumelhart [1], a variety of new training algorithms have emerged to improve different aspects of the learning process for feed-forward neural networks. One of these aspects is the *learning speed* (i.e, the velocity during learning towards the convergence to the minimum error), which is generally slow. Some

J. Marques de Sá et al. (Eds.): ICANN 2007, Part I, LNCS 4668, pp. 190–198, 2007.

of the newly proposed algorithms that try to improve this aspect are modifications of the classic backpropagation such as adding a momentum term [1,2], an adaptive step size [3] or using stochastic learning [4].

Others are second order methods that use the second derivatives of the error function. Some of the most relevant examples of these types of methods are the quasi-Newton, Levenberg-Marquardt [5] and the conjugate gradient algorithms [6]. Although second order methods are among the fastest learning algorithms, they are not practical for large neural networks trained in batch mode due to their computational cost.

Finally, it is also possible to find methods based on linear least-squares [7,8,9,10,11]. These methods are mostly based on measuring the error of an output neuron *before* the nonlinear transfer function instead of *after* it, as is the usual approach. Usually, they are proposed as initialization methods for multi-layer networks or as training methods for one-layer networks.

Specifically, in [10] a method is described to solve a one-layer non-linear neural network using linear least squares. In this paper, we extend this approach to networks consisting of multiple layers of weights (MLPs) by using the algorithm in [10] for the output layer and optimizing the preceding layers using gradient descent. As a consequence, we improve the learning speed of the basic backpropagation algorithm in several orders of magnitude, while maintaining its optimization accuracy.

In section 2, the method for learning the weights of the network is presented. Section 3 compares the performance of the proposed method with some other well-known training methods. In Section 4 these results are discussed. Finally, section 5 offers some conclusions and recommendations.

2 The Proposed Algorithm

Without loss of generality, in this article we will consider a two-layer MLP like the one shown in Fig. 1 with the variable names described below.

Constants I, K, J and S denote respectively, the number of inputs, hidden units, outputs and training samples. Each layer of the network consists of a linear matrix $\mathbf{W}^{(n)}$ of weights $w_{ji}^{(n)}$ connecting neuron j in layer n with neuron i in layer $n-1$, thus the superscript $n = 1, \ldots, N$ is used to refer to each layer. These weight matrices are followed by nonlinear mappings $f_j^{(n)}$, usually chosen to be sigmoid-type functions.

The input vectors to each layer n of the MLP are represented by $\mathbf{x}^{(n)}$. Notice that the bias of each layer has been incorporated into the weight matrix by adding constant inputs $x_{0s}^{(n)} = 1, \forall n$.

In addition, for all $j = 1, ..., J; s = 1, ..., S$, we will denote by z_{js} the inputs to the non-linearities of the output layer; y_{js} the real output produced by the network, and by d_{js} the desired response provided in the training set. Finally, in the following we will consider the MSE between the real $\mathbf{y}$ and the desired output $\mathbf{d}$ of the network as the training optimization criterion.

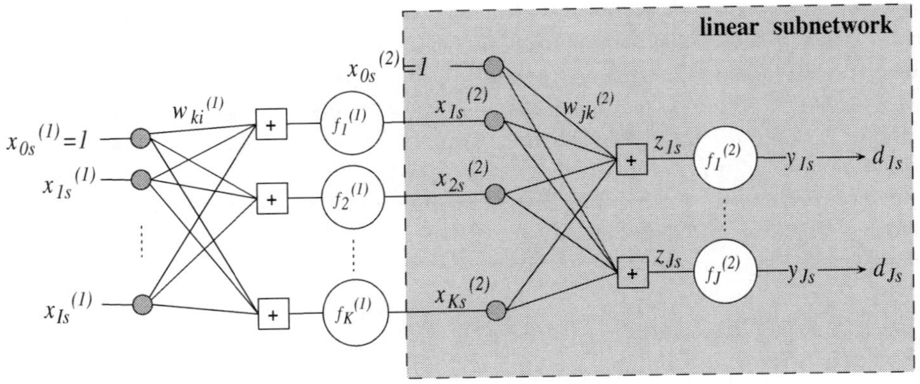

Fig. 1. Architecture of a two-Layer MLP

2.1 One-Layer Linear Learning

Take, for instance, the one-layer neural network corresponding to the shadowy part of Fig. 1. In [10,11,12], the authors considered the approximate least squares optimization of a one-layer nonlinear network assuming the MSE before the nonlinearity as the criterion to be optimized by means of the following theorem:

Theorem 1. *Minimization of the MSE between* $\mathbf{d}$ *and* $\mathbf{y}$ *at the output of the nonlinearity* f *is equivalent (up to first order) to minimizing a MSE between* $\mathbf{z}$ *and* $\bar{\mathbf{d}} = f^{-1}(\mathbf{d})$*, where the inverse function is evaluated at each entry separately. Mathematically, this is given by*

$$\min_{\mathbf{W}} \sum_{s=1}^{S} \sum_{j=1}^{J} \left(f_j \left(\sum_{k=0}^{K} w_{jk} x_{ks} \right) - d_{js} \right)^2 \approx \min_{\mathbf{W}} \sum_{s=1}^{S} \sum_{j=1}^{J} \left(\sum_{k=0}^{K} w_{jk} x_{ks} - f_j^{(-1)}(d_{js}) \right)^2 . \tag{1}$$

According to this new error criterion the weights can be optimized by solving a system of $J \times S$ linear equations defined by:

$$\frac{\partial MSE}{\partial w_{jp}} = 2 \sum_{s=1}^{S} \left(\sum_{k=0}^{K} w_{jk} x_{ks} - f_j^{(-1)}(d_{js}) \right) x_{ps} = 0; \quad p = 0, 1, \ldots, K; \quad \forall j . \tag{2}$$

The use of this system presents two main advantages: 1) the global optimum of the training set is obtained, and 2) there is a considerable savings in training time with respect to other gradient-based optimization techniques.

2.2 Combining Linear and Non-linear Learning

Theorem 1 can be used as a basis to provide the desired signal before the non-linearity for every layer of the network and, therefore, linearly find the optimal weights for each layer. However, such an algorithm can only be used as an initialization method and not as a learning method. The reason is that it would calculate the weight matrices independently for each layer, and therefore it would jump from one region to another of the weight space instead of smoothly converging to the solution [12]. The problem arises when both nonlinear layers must be optimized at the same time.

As an alternative, in this research, only the last layer of the network is optimized using the linear method previously described whereas the other layers are optimized using a standard backpropagation gradient descent with momentum and an adaptive learning rate. The proposed algorithm for a multilayer feedforward neural network is as follows:

Step 1: Set the initial weights $\mathbf{W}^{(n)} \forall n$.
Step 2: Using the current weights, propagate the signal forward to calculate the outputs of each layer.
Step 3: Evaluate the value of the MSE between $\mathbf{y}$ and $\mathbf{d}$ and calculate the gradient of the network's performance with respect to its weights, that is, $\delta_{jp}^{(n)} = \frac{\partial MSE}{\partial w_{jp}^{(n)}}, \forall j, p, n$.
Step 4: Convergence checking. If $|MSE_{previous} - MSE| < \epsilon$ or $||\boldsymbol{\delta}|| < \epsilon'$ stop and return the weights[1] The constants ϵ and ϵ' are thresholds to control convergence.
Step 5: Obtain the adjustment of each weight according to the gradient descent with momentum, i.e.:

$$\Delta w_{jp}^{(n)}(t) = \mu \times \Delta w_{jp}^{(n)}(t-1) + \rho \times (1-\mu) \times \delta_{jp}^{(n)}; \quad n = 1, \ldots, N-1; \quad \forall j, p \quad (3)$$

where μ is the momentum constant, and ρ is the learning rate.
Step 6: Evaluate the value of MSE' (MSE between $\mathbf{z}$ and $\bar{\mathbf{d}}$) and update $\mathbf{W}^{(N)}$ using the linear system of equations in 2.
Step 7: Update the weights of the hidden layers $\mathbf{W}^{(n)}, n = 1, \ldots, N-1$ according to the increments calculated in Step 5:

$$W^{(n)} = W^{(n)} + \Delta W^{(n)}, n = 1, \ldots N-1 \quad (4)$$

Step 8: Adapt the learning rate ρ and continue from Step 2.

3 Experimental Results

In this section the proposed method (newGDX) is illustrated by its application to two system identification problems of different size, and its performance is compared with four popular learning methods. Two of these methods are of complexity $O(n)$: the gradient descent (GD) and the gradient descent with adaptive momentum and step sizes (GDX), in which the proposed method is based. The other two methods are the scaled conjugated gradient (SCG) (complexity of $O(n^2)$), and the Levenberg-Marquardt (LM) (complexity of $O(n^3)$).

[1] Some other stopping criteria are allowed such as a maximum time or number of training epochs.

For each experiment all the learning methods shared the following conditions:

– They were carried out in MATLAB®.
– The logistic function was used for hidden neurons, while for output neurons the linear function was used.
– The input data set was normalized.
– Each experiment was repeated five times, using a different set of initial weights for each one. This initial set was the same for all the algorithms, and was obtained by the Nguyen-Widrow [13] initialization method.

3.1 K.U. Leuven Competition Data

In this case, we compare the algorithms over a small size problem. The K.U. Leuven time series [14] prediction competition data were generated from a computer simulated 5-scroll attractor, resulting from a generalized Chua's circuit which is a paradigm for chaos. The aim of the neural network is to predict the current sample using only 4 previous data points. For this problem a 4-8-1 topology was used. The number of available training patterns was 2000 and the learning process was allowed to run for a maximum of 3000 epochs.

In this case, the experiments were run on a 2.60GHz Pentium 4 processor with 512MB of RAM memory.

Fig. 2a shows the mean MSE training error curves obtained by each of the tested methods for the 5 simulations. As can observed, already in the second epoch the proposed method obtains an error very close to its minimum. Also, in Fig. 2b are shown the box-whisker plots corresponding to the minimum MSE values obtained by each method during training. In this graphic the box corresponds to the interquartile range, the bar inside the box represents the median, the whiskers extend to the farthest points that are not outliers, and outliers are represented by the plus sign.

Also, in table 1 some performance measures are shown that allows for the comparison of the algorithms. The first column ($M1$) measures corresponds to the MSE of each method at the 2nd epoch of training. The second column ($M2$) is the number of epochs it takes the other methods to obtain the MSE that the *newGDX* algorithm exhibits at the 2nd epoch. As these measures are calculated over the 5 simulations they all are provided in terms of mean and corresponding standard deviation.

Finally, table 2 shows the mean time $Tepoch_{mean}$ (in seconds) per epoch of every algorithm. The *Ratio* column contains the relation between the $Tepoch_{mean}$ of each algorithm and the one proposed.

3.2 Lorenz Time Series

In this second experiment, the Lorenz Time Series [15] corresponding to a system with chaotic dynamics was employed to test the algorithm over a large size problem. In this case, the goal of the network is to predict the current sample based on the eight previous samples. For this data set a 8-100-1 topology was

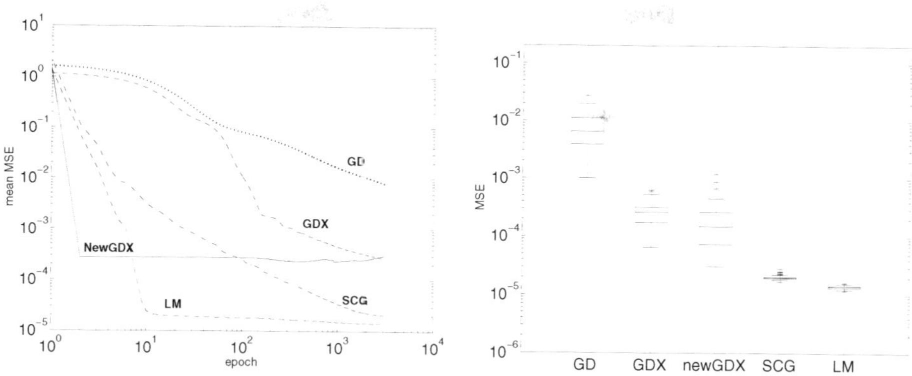

(a) Mean MSE training error curves. (b) Boxplot of the minimum MSE values obtained during training.

Fig. 2. Results of the learning process for the K.U. Leuven Competition Data

Table 1. Performance measures for the K.U. Leuven Competition Data

Algorithm	M1	M2
GD	1.53 ± 2.07	NaN^1
GDX	1.09 ± 1.42	504 ± 489
newGDX	$2.72 \times 10^{-4} \pm 2.24 \times 10^{-4}$	2 ± 0
SCG	$1.10 \times 10^{-1} \pm 1.03 \times 10^{-1}$	31.1 ± 21.5
LM	$6.88 \times 10^{-2} \pm 1.03 \times 10^{-1}$	4.70 ± 1.30

[1] Not a Number: the algorithm never reaches this minimum along the 1000 iterations

used. It is important to remark here that the objective was not to obtain the network's topology that best solves the problem, but rather, to deal with large networks. Also, 150000 samples were employed for the learning process. In this case, and due to the large size of both the data set and the neural networks, the number of iterations was reduced to 1000. Neither the GD nor the LM methods were used. The results of the GD will not be presented because the method performed poorly, and the LM is impractical in these cases as it is computationally highly demanding.

In this case, the experiments were run on a Compaq HPC 320 with an Alpha EV68 1 GHz processor and 4GB of memory.

Fig. 3.2 shows the mean MSE training error curves obtained by each of the tested methods for the 5 simulations. As can observed, in this case the proposed method obtains an error very near to its minimum around the 20th epoch. Fig. 3.2 shows the boxplot corresponding to the minimum MSE values obtained by each method during training.

Table 2. CPU time comparison for the K.U. Leuven Competition Data

Algorithm	$Tepoch_{mean}$	$Tepoch_{std}$	Ratio
GD	0.0140	1.05×10^{-2}	0.96
GDX	0.0147	1.50×10^{-2}	1.01
newGDX	0.0146	1.12×10^{-2}	—
SCG	0.0276	1.18×10^{-2}	1.89
LM	0.0402	1.22×10^{-2}	2.75

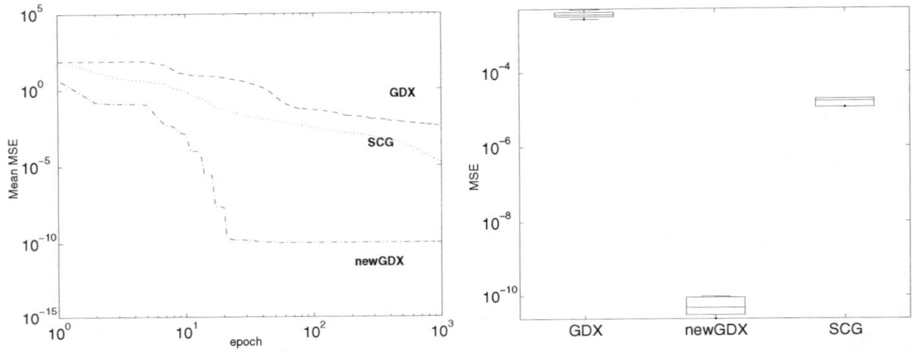

(a) Mean MSE training error curves.

(b) Boxplot of the minimum MSE values obtained during training.

Fig. 3. Results of the learning process for the Lorenz Time Series

Table 3. Performance measures for the Lorenz Time Series

Algorithm	M1	M2
GDX	5.61 ± 1.70	NaN
newGDX	$1.83 \times 10^{-10} \pm 2.24 \times 10^{-10}$	18.8 ± 1.64
SCG	$4.26 \times 10^{-2} \pm 1.98 \times 10^{-2}$	NaN

Also, in table 3 the $M1$ and $M2$ performance measures are shown that allows the comparison of the algorithms. In this case, they refer to the $20th$ epoch of training.

Finally, table 4 shows the mean time (in seconds) per epoch of every algorithm.

4 Discussion

From Figs. 2 and 3 we can see that the proposed method obtains its minimum error (or an approximate one) at a very early epoch. Regarding its accuracy, it improves the one exhibited by the first order methods in which it is based.

Table 4. CPU time comparison for the Lorenz Time Series

Algorithm	$Tepoch_{mean}$	$Tepoch_{std}$	*Ratio*
GDX	9.75	0.04	—
newGDX	10.03	0.06	1.03
SCG	21.01	0.63	2.15

Although in the Leuven experiments the second order methods overcome the proposed one, it is important to notice what happens when the size of the problem increases, like in the Lorenz experiment. In this case, the Scale Conjugate Gradient would probably need much more training time, and the Levenberg-Marquardt algorithm is not directly applicable due to its high computational request.

Also from tables 1 and 3, it can be concluded that when our method reaches an error value near its minimum the other algorithms are in a minimum several orders of magnitude higher. In the Leuven experiment, as shown by measure $M2$, only the Levenberg-Marquardt can quickly reach (and even improve) the proposed algorithm. This is not the case in the Lorenz experiment where none reach our minimum along the 1000 iterations used. This conclusion, together with the short time per epoch needed by the proposed method, as shown in tables 2 and 4, definitely makes it a fast learning algorithm.

5 Conclusions and Future Work

The analyzed results allow us to confirm that the proposed method offers an interesting combination of speed, reliability and simplicity. Although on occasions its accuracy is improvable, it generally obtains good approximations that even overcome those provided by classic or second algorithms. These features makes the proposed algorithm suitable for those situations when the speed of the method is important in reaching a good solution, although it might not be the best one.

Acknowledgements

We would like to acknowledge support for this project from the Xunta de Galicia (project PGIDT05TIC10502PR). Also, we thank the Supercomputing Center of Galicia (CESGA) for allowing us the use of the high performance computing servers.

References

1. Rumelhart, D.E., Hinton, G.E., William, R.J.: Learning representations of back-propagation errors. Nature 323, 533–536 (1986)

2. Vogl, T.P., Mangis, J.K., Rigler, A.K., Zink, W.T., Alkon, D.L.: Accelerating the convergence of back-propagation method. Biological Cybernetics 59, 257–263 (1988)
3. Jacobs, R.A.: Increased rates of convergence through learning rate adaptation. Neural Networks 1(4), 295–308 (1988)
4. LeCun, Y., Bottou, L., Orr, G.B., Müller, K.-R.: Efficient BackProp. In: Orr, G.B., Müller, K.-R. (eds.) Neural Networks: Tricks of the Trade. LNCS, vol. 1524, Springer, Heidelberg (1998)
5. Hagan, M.T., Menhaj, M.: Training feedforward networks with the Marquardt algorithm. IEEE Transactions on Neural Networks 5(6), 989–993 (1994)
6. Beale, E.M.L.: A derivation of conjugate gradients. In: Lootsma, F.A. (ed.) Numerical methods for nonlinear optimization, pp. 39–43. Academic Press, London (1972)
7. Biegler-König, F., Bärmann, F.: A Learning Algorithm for Multilayered Neural Networks Based on Linear Least-Squares Problems. Neural Networks 6, 127–131 (1993)
8. Yam, J.Y.F., Chow, T.W.S, Leung, C.T.: A New method in determining the initial weights of feedforward neural networks. Neurocomputing 16(1), 23–32 (1997)
9. Bishop, C.M.: Neural Networks for Pattern Recognition. Oxford University Press, New York (1995)
10. Castillo, E., Fontenla-Romero, O., Alonso Betanzos, A., Guijarro-Berdiñas, B.: A global optimum approach for one-layer neural networks. Neural Computation 14(6), 1429–1449 (2002)
11. Fontenla-Romero, O., Erdogmus, D., Principe, J.C., Alonso-Betanzos, A., Castillo, E.: Linear least-squares based methods for neural networks learning. In: Kaynak, O., Alpaydın, E., Oja, E., Xu, L. (eds.) ICANN 2003 and ICONIP 2003. LNCS, vol. 2714, pp. 84–91. Springer, Heidelberg (2003)
12. Erdogmus, D., Fontenla-Romero, O., Principe, J.C., Alonso-Betanzos, A., Castillo, E.: Linear-Least-Squares Initialization of Multilayer Perceptrons Through Backpropagation of the Desired Response. IEEE Transactions on Neural Networks 16(2), 325–337 (2005)
13. Nguyen, D., Widrow, B.: Improving the learning speed of 2-layer neural networks by choosing initial values of the adaptive weights. In: Proceedings of the International Joint Conference on Neural Networks. vol. 3, pp. 21–26 (1990)
14. Suykens, J.A.K., Vandewalle, J. (eds.): Nonlinear Modeling: advanced black-box techniques. Kluwer Academic Publishers, Boston (1998)
15. Lorenz, E.N.: Deterministic nonperiodic flow. Journal of the Atmospheric Sciences 20, 130–141 (1963)

Improving the GRLVQ Algorithm by the Cross Entropy Method

Abderrahmane Boubezoul, Sébastien Paris, and Mustapha Ouladsine

Laboratory of Sciences of Information's and of System, LSIS UMR 6168, University Paul Cézanne, Aix-Marseille III Av Escadrille de Normandie Niemen 13397 Marseille Cedex 20, France
{abderrahmane.boubezoul,sebastien.paris,mustapha.ouladsine}@lsis.org

Abstract. This paper discusses an alternative approach to parameter optimization of prototype-based learning algorithms that aim to minimize an objective function based on gradient search. The proposed approach is a stochastic optimization method called the Cross Entropy (CE) method. The CE method is used to tackle the initialization sensitiveness problem associated with the original generalized Learning Vector Quantization (GLVQ) algorithm and its variants and to locate the globally optimal solutions. We will focus our study on a variant which deals with a weighted norm instead of the Euclidean norm in order to select the most relevant features. The results in this paper indicate that the CE method can successfully be applied to this kind of problems and efficiently generate high quality solutions. Also, highly competitive numerical results on real world data sets are reported.

1 Introduction

Prototype based learning has been an ongoing research problem for last decades and it has been approached in various ways. It has been gaining more interest lately due to its ability to generate fast and intuitive classification models with good generalization capabilities. One prominent algorithm of Prototype based learning algorithms is a Learning Vector Quantization (LVQ) introduced by Kohonen ([1]), this algorithm and its variants have been intensively studied because of their robustness, adaptivity and efficiency. The idea of LVQ is to define class boundaries based on prototypes, a nearest neighbor rule and a winner-takes-it-all paradigm. The standard LVQ has some drawbacks: i) basically LVQ adjusts the prototypes using heuristic error correction rules, ii) it does not directly minimize an objective function, thus it cannot guarantee the convergence of the algorithm which leads to instability behavior especially in the case of overlapped data, iii) the results are strongly dependent on the initial positions of the prototypes. In order to improve the standard LVQ algorithm several modifications were proposed by Kohonen and by other researchers (see[2]). Standard LVQ does not distinguish between more or less informative features due to the usage of the Euclidean distance, to improve that, extensions from various authors are suggested (see [2] [3] and [4]). The previous approaches obey to heuristic update for

J. Marques de Sá et al. (Eds.): ICANN 2007, Part I, LNCS 4668, pp. 199–208, 2007.

relevance and prototypes vectors are adapted using simple perceptron learning which may cause problems for non linear separable data. For these reasons another variant based on minimisation of cost function using stochastic gradient descent method was proposed by Sato and Yamada (see [5]). Hammer and *al* suggested to modify the GLVQ cost function by using weighted norm instead of Euclidean distance. This algorithm, called Generalized Relevances LVQ, showed in several tasks competitive results compare to SVM and it had been proved that GLRVQ can be considered as a large margin classifier (see [6]).

Although the GRLVQ algorithm guarantees convergence and shows better classification performances than other LVQ algorithms, it suffers from initialization sensitiveness due to the presence of numerous local minima incurred as a result of the use of gradient descent method especially for multi-modal problems. The same authors proposed another algorithm (Supervised Relevance Neural Gas SRNG) to tackle initialization sensitiveness (see[7]). They propose to combine the GRLVQ with the neighborhood oriented learning in the neural gas (NG). The algorithm mnetionned above required choosing several parameters such as learning rate, size of an update neighborhood. A suitable choice of these parameters values may not always be evident and also changed from one data set to another. In this paper, we present an initialization insensitive GRLVQ which is based on the well-known and efficient cross entropy (CE) method [8]. We refer to this new algorithm as the Cross Entropy Method GLVQ (CEMGRLVQ).

The rest of the paper is structured as follows. In section 2, we introduce the basics of classification and prototype learning; we review both the formulation of GLVQ and GRLVQ. In section 3, we explain how the CE method can be considered as a stochastic global optimization procedure for GRLVQ algorithm, In section 4, we present the results of numerical experiments using our proposed algorithm on some benchmark data sets and compare with the results obtained using the GRLVQ and GLVQ stochastic gradient descent method. We conclude in section 5.

2 Problem Statement

Machine learning can be formulated as searching for the most adequate model describing a given data. A learning machine may be seen as a function $f\left(\boldsymbol{\mathcal{X}};\boldsymbol{\theta}\right)$ which transforms objects from the the input space $\boldsymbol{\mathcal{X}}$ to the output space $\boldsymbol{\mathcal{Y}}$: $f(\boldsymbol{\mathcal{X}};\boldsymbol{\theta}) : \boldsymbol{\mathcal{X}} \longrightarrow \boldsymbol{\mathcal{Y}}$.

The data domain $\boldsymbol{\mathcal{X}}$ and the set of target values $\boldsymbol{\mathcal{Y}}$ are determined by the definition of the problem for which $f(\boldsymbol{\mathcal{X}};\boldsymbol{\theta})$ is constructed. The output of the function can be a continuous value (for regression application) or can predict a class label of the input object (for classification application). The learning model $f(\boldsymbol{\mathcal{X}};\boldsymbol{\theta})$ usually depends on some adaptive parameters $\boldsymbol{\theta}$ sometimes also called free parameters. In this context, learning can be seen as a process in which a learning algorithm searches for these parameters $\boldsymbol{\theta}$, which solve a given task.

In supervised learning, the algorithm learns form the data set $\boldsymbol{\mathcal{S}}$ often called the "training set" and consists of N samples, $(\boldsymbol{x}, y)$ pairs, drawn independently identically from a probability distribution $p(\boldsymbol{x}, y)$. We define the collected data by $\boldsymbol{S} \triangleq \{(\boldsymbol{x}_1, y_1), (\boldsymbol{x}_2, y_2), \ldots, (\boldsymbol{x}_N, y_N)\}$ where $\boldsymbol{x}_i \triangleq [x_{1i}, x_{2i}, \ldots, x_{di}]^T \in \boldsymbol{\mathcal{X}}$ and $\boldsymbol{y} \triangleq \{y_i\}_{i=1,\ldots,N}$, $y_i \in \{1, \ldots, L\}$. N, L denote the number of records in the data set and the number of classes respectively. In real applications $\boldsymbol{x}_i \in \boldsymbol{\mathcal{X}} \subset \mathbb{R}^d$ is a multidimensional real vector.

In the Bayes decision theory the sanity behavior of the classifier is usually measured by classification error, the so-called overall loss (see [?]):

$$\Re(\boldsymbol{\theta}) \triangleq \mathbb{E}_X \left[L(f(\boldsymbol{x}; \boldsymbol{\theta})|\boldsymbol{x})\right] = \int_{\boldsymbol{x} \in \boldsymbol{\mathcal{X}}} L(f(\boldsymbol{x}; \boldsymbol{\theta})|\boldsymbol{x}) p(\boldsymbol{x}) d\boldsymbol{x}, \tag{1}$$

where $L(f(\boldsymbol{x}; \boldsymbol{\theta})|\boldsymbol{x})$ is the expected loss defined by:

$$L(f(\boldsymbol{x}; \boldsymbol{\theta})|\boldsymbol{x}) \triangleq \int_{y \in \mathcal{Y}} \ell(f(\boldsymbol{x}; \boldsymbol{\theta})|y) p(y|\boldsymbol{x}) dy, \tag{2}$$

where $\ell(f(\boldsymbol{x}; \boldsymbol{\theta})|y)$ denotes the individual loss.We assume training data drawn by some underlying joint distribution $p(\boldsymbol{x}, y)$ which is in practice unknown. In real applications, only a finite number of samples are available. The loss functional $\Re(\boldsymbol{\theta})$ is usually replaced by the so-called empirical loss functional computed as follows:

$$\Re_{emp}(\boldsymbol{\theta}) = \frac{1}{N} \sum_{i=1}^{N} \sum_{k=1}^{L} \ell(f(\boldsymbol{x}_i; \boldsymbol{\theta})|y_i = k) \mathbf{1}(y_i = k), \tag{3}$$

$\mathbf{1}(.)$ is an indicator function such that $\mathbf{1}(true) = 1$ if the condition between the parentheses is satisfied or $\mathbf{1}(false) = 0$ if not.

2.1 GLVQ Algorithm

It has been shown by Diamantini in [9] that traditional LVQ (LVQ1) proposed by Kohonen does not minimize neither explicit risk function, nor the Bayes risk. In [10] Juang proposed a learning scheme called minimum classification error (MCE) approach that minimizes the expected loss in Bayes decision theory by a gradient-descent procedure (Generalized Probabilistic Descent). The MCE criterion is defined using specific discriminant functions.

Let us consider $\boldsymbol{w}_{kj} \in \boldsymbol{\mathcal{X}}$ is the jth prototype among P_k vectors associated to class $k = 1, \ldots, L$ and $\boldsymbol{\theta}_k \triangleq \{\boldsymbol{w}_{kj} | j = 1, \ldots, P_k\} = \boldsymbol{W}_k$ is the collection of all prototypes of the class k. The collection of all prototypes representing all classes are defined as $\boldsymbol{\theta} \triangleq \bigcup_{k=1}^{L} \boldsymbol{\theta}_k = \boldsymbol{W}$ and $P = \sum_{l=1}^{L} P_l$ denotes the total number of prototypes.

As proposed by Sato and Yamada, the generalized LVQ learning algorithm use the following discriminant function $\mu_k(\boldsymbol{x}; \boldsymbol{\theta})$ defined by:

$$\mu_k(\boldsymbol{x}; \boldsymbol{\theta}) \triangleq \frac{d_k(\boldsymbol{x}; \boldsymbol{\theta}) - d_l(\boldsymbol{x}; \boldsymbol{\theta})}{d_k(\boldsymbol{x}; \boldsymbol{\theta}) + d_l(\boldsymbol{x}; \boldsymbol{\theta})}, \tag{4}$$

where $d_k(\boldsymbol{x}; \boldsymbol{\theta})$ is the square euclidean distance between $\boldsymbol{x}$ and the closest prototype belonging to the same class of $\boldsymbol{x}$ and $d_l(\boldsymbol{x}; \boldsymbol{\theta})$ is the squared Euclidean distance between the input vector $\boldsymbol{x}$ and its best matching prototype vector $\boldsymbol{w}_{lr}$ with a different class label. This discriminant function $\mu_k(\boldsymbol{x};\boldsymbol{\theta}) \in [-1, 1]$ ensure that if $\boldsymbol{x}$ is correctly classified to class k then $\mu_k(\boldsymbol{x};\boldsymbol{\theta}) \leq 0$.

Introducing (4) in (3), the empirical loss minimized by the MCE criterion is rewritten by:

$$\Re_{emp}(\boldsymbol{\theta}) = \frac{1}{N}\sum_{i=1}^{N}\sum_{k=1}^{L}\ell(\mu_k(\boldsymbol{x}_i;\boldsymbol{\theta}))\mathbf{1}(y_i = k), \tag{5}$$

where $\ell(\mu_k(\boldsymbol{x}_i;\boldsymbol{\theta})) = \ell(f(\boldsymbol{x}_i;\boldsymbol{\theta})|y = k)$.

$\ell(\mu_k(\boldsymbol{x};\boldsymbol{\theta}))$ is a smoothed loss function instead of the $0-1$ loss function of the Bayes decision theory, for example the sigmoid function defined by:

$$\ell(z;\xi) \triangleq \frac{1}{1+e^{-\xi z}}, \tag{6}$$

where ξ is a scalar (which usually increases with time). From (5), the general GLVQ cost function is expressed as:

$$F_{GLVQ}(\boldsymbol{X};\boldsymbol{\theta}) = \sum_{i=1}^{N}\ell(\mu_k(\boldsymbol{x}_i;\boldsymbol{\theta})) = \sum_{i=1}^{N}\frac{1}{1+e^{-\xi(t)\mu_k(\boldsymbol{x}_i;\boldsymbol{\theta})}}, \tag{7}$$

where $\boldsymbol{X} \triangleq \{\boldsymbol{x}_i\}_{i=1,\ldots,N}$. Although the GLVQ algorithm ensures convergence and exhibits better classification performance than other LVQ algorithms.

2.2 Attribute Weighing and GRLVQ Algorithm

In general, prototype based classifiers are dependant on the metric which is used to measure proximity, the performance of this type of algorithms depends essentially on that choice. The usual choice is the euclidean metric which is not always appropriate because it supposes that all the attributes contribute equally in the classification. We will use the same idea as Hammer and *al* in [9], they suggested to replace the Euclidean metric by Weighted Euclidean metric by introducing input weights $\boldsymbol{\lambda} \triangleq [\lambda_1, \lambda_2, \ldots, \lambda_d]^T \in \mathbb{R}^d$, $\lambda_i \geq 0$ to allow a different scaling of the inputs. They substitute the Euclidean metric by its scaled variant in the GLVQ formulation

$$d_\lambda(\boldsymbol{a}, \boldsymbol{b}) \triangleq (\boldsymbol{a}-\boldsymbol{b})^T\boldsymbol{\lambda}\mathbf{I}(\boldsymbol{a}-\boldsymbol{b}), \tag{8}$$

where $(\boldsymbol{a}, \boldsymbol{b}) \in \boldsymbol{\mathcal{X}}$ are two column vectors, $\mathbf{I}$ is identity matrix.

From (7) the new formulation of the misclassification measure $\mu_k^\lambda(\boldsymbol{x};\boldsymbol{\theta})$ can be expressed as

$$\mu_k^\lambda(\boldsymbol{x};\boldsymbol{\theta}) = \frac{d_k^\lambda(\boldsymbol{x};\boldsymbol{\theta}) - d_l^\lambda(\boldsymbol{x};\boldsymbol{\theta})}{d_k^\lambda(\boldsymbol{x};\boldsymbol{\theta}) + d_l^\lambda(\boldsymbol{x};\boldsymbol{\theta})}, \tag{9}$$

where $d_k^\lambda(\boldsymbol{x};\boldsymbol{\theta})$ and $d_l^\lambda(\boldsymbol{x};\boldsymbol{\theta})$ are now the corresponding weighted distances. Now the GRLVQ cost function is computed as follows

$$F_{GRLVQ}(\boldsymbol{X};\boldsymbol{\theta}) = \sum_{i=1}^{N}\frac{1}{1+e^{-\xi(t)\mu_k^\lambda(\boldsymbol{x}_i;\boldsymbol{\theta})}}. \tag{10}$$

In the next section we will present the Cross Entropy method to tackle the initialization sensitiveness problem and to estimate the attribute's weights.

3 Cross Entropy Method

The general optimization problem introduced in the GLVQ and GRLVQ formulation can be viewed as a task of finding the best set of parameters $\boldsymbol{W}$ and $\boldsymbol{\lambda}$ that minimize the corresponding objective function. Optimization problems are typically quite difficult to solve; they are often $\mathcal{NP}$-hard, *i.e.* $\min_{\boldsymbol{\theta}\in\boldsymbol{\Theta}}\{F(\boldsymbol{\theta})\}$, where $\boldsymbol{\theta} \in \boldsymbol{\Theta}$ represents the vector of the input variables, $F(\boldsymbol{\theta})$ is the scalar objective function and $\boldsymbol{\Theta}$ is the constraint set. Many approaches exist to solve this kind of problems (Gradient based procedures, random search, Meta heuristics, model-based methods,...) (see [11]). As an alternative to the stochastic gradient descent algorithm we consider the CE method approach because it offers good results for multi-extremal functional optimization (see [12]). This method requires neither special form of the objective function and its derivatives nor heuristic assumptions.

We view the problem of minimization of the cost function (7) as a continuous multi-extremal optimization problem with constraints. The Cross Entropy method was introduced by Rubinstein (see [8]) as an efficient method for the estimation of rare-event probabilities and has been successfully applied to a great variety of research areas (see for example [13]. The main ideas of the CE method are described in [13] and [14]. For this reason, in this paper we only present features of CE that are relevant to the problem hereby studied.

The central idea of CE method is related to the association of a stochastic problem to the original optimization problem, called parameterized associated stochastic problem (ASP) characterized by a density function $p(.;\boldsymbol{v}), \boldsymbol{v} \in \boldsymbol{\mathcal{V}}$. The stochastic problem is solved by identifying the optimal importance sampling density p^* which minimizes the Kullback-Leibler distance (also called the cross-entropy) between p and p^*. The CE method can be viewed as an iterative method that involves two major steps until convergence is reached:

1. Generating samples according to $p(.;\boldsymbol{v})$ and choosing the elite of these samples.
2. Updating the parameter $\boldsymbol{v}$ of the distribution family on the basis of the elite samples, in order to produce a better solution in the next iteration.

In this paper we will give a brief introduction of the CE method, the reader can refer to [14] for a more detailed description of the CE method.

Consider the following general cost function minimization problem. Let $\boldsymbol{\Theta}$ be a constraint set and $\boldsymbol{\theta} \in \boldsymbol{\Theta}$ is a given vector. Let us denote the desired minimum of the function $F(\boldsymbol{\theta})$ by $\gamma^\star$. The cost function minimization problem can be formulated by:

$$\gamma^\star = \min_{\boldsymbol{\theta}\in\boldsymbol{\Theta}}\{F(\boldsymbol{\theta})\}. \tag{11}$$

In other words, we seek an optimal solution $\gamma^\star$ satisfying $F(\boldsymbol{\theta}^\star) \leq F(\boldsymbol{\theta})$, $\forall \boldsymbol{\theta} \in \boldsymbol{\Theta}$. The CE method transforms the deterministic optimisation problem (11) to stochastic problem using a family of probability density functions (pdfs) $p\left(.;\boldsymbol{v}\right)$ which depends on a reference parameter $\boldsymbol{v} \in \mathcal{V}$. Consequently, the associated stochastic estimation problem is

$$l(\gamma) \triangleq \mathbb{P}_{\boldsymbol{v}}\left(F(\boldsymbol{\Theta}) \leq \gamma\right) = \mathbb{E}_{\boldsymbol{v}}\left[\boldsymbol{I}_{F(\boldsymbol{\Theta})\leq\gamma}\right] = \int \boldsymbol{I}_{F(\boldsymbol{\theta})\leq\gamma} p\left(.;\boldsymbol{v}\right) d\boldsymbol{\theta}. \tag{12}$$

At each iteration t of the CE method algorithm V random samples are drawn on the basis of $p\left(.;\boldsymbol{v}_{t-1}\right)$, then the new value of $\boldsymbol{v}_t$ is updated according to $\boldsymbol{v}_{t-1}$ and the best elite samples. The update formula is especially simple if $p\left(.;\boldsymbol{v}\right)$ belongs to Natural Exponential Function (NEF)(Gaussian, Truncated Gaussian, Binomial,. . .).

Instead of updating the parameter vector $\widehat{\boldsymbol{v}}_{t-1}$ to $\widehat{\boldsymbol{v}}_t$ directly the smoothed updating procedure is often used:

$$\widehat{\boldsymbol{v}}_t = \alpha\widetilde{\boldsymbol{v}}_t + (1-\alpha)\widehat{\boldsymbol{v}}_{t-1}, \tag{13}$$

with $0 \leq \alpha \leq 1$ and $\widetilde{\boldsymbol{v}}_t$ is the solution of (12). This smoothed adaptation (13) is used to reduce the probability of the algorithm to get stuck in a local minima.

3.1 Cross Entropy Method for GRLVQ Optimisation

In order to link the GRLVQ cost function minimization (10) with the CE theory, we define the parameter $\boldsymbol{\theta}$ as $\boldsymbol{\theta} \triangleq \{\boldsymbol{W}, \boldsymbol{\lambda}\}$. Both $\boldsymbol{W}$ and $\boldsymbol{\lambda}$ are sampled from two sampling distributions which belong to the NEF family in order to simplify parameters update.

In this paper we consider a Truncated Gaussian and Dirichlet distributions to estimate the prototypes and attribute's weights, respectively. That is, for each $\boldsymbol{W}^l \triangleq \{w^l_{pq}\}_{\substack{p=1,\ldots,d\\ q=1,\ldots,P}}$ (a $d \times P$ matrix) with $l = 1,\ldots,V$, each components w^l_{pq} are drawn from a Truncated Gaussian such $w^l_{pq} \sim \mathcal{N}t\left(m^t_{pq}, (\sigma^t_{pq})^2, \boldsymbol{a}_p, \boldsymbol{b}_p\right)$ where m^t_{pq} denotes the average of pq^{th} components at iteration t, $(\sigma^t_{pq})^2$ the variance and $\boldsymbol{a}_p, \boldsymbol{b}_p$ denote a lower and upper bounding box for each dimension containing all data respectively. In practice, we choose $\boldsymbol{a}_p = C \min_{j=1,\ldots,N}\{x_{jp}\}$ and $\boldsymbol{b}_p = C \max_{j=1,\ldots,N}\{x_{jp}\}$ with $C \geq 1$. To sample attribute's weights we use the Dirichlet distribution. Let $\boldsymbol{\lambda} \triangleq [\lambda_1, \lambda_2, \ldots, \lambda_d]^T$ a random vector whose elements sum to 1. The density function of the Dirichlet distribution with the parameter vector $\boldsymbol{\delta}$ is:

$$\boldsymbol{\lambda}(\boldsymbol{\delta}) \sim \mathcal{D}ir(\delta_1, \delta_2, \ldots, \delta_d) = \frac{\Gamma(\sum_d \delta_d)}{\prod_d \Gamma(\delta_d)} \prod_d \lambda_d^{\delta_d - 1}, \tag{14}$$

where $\lambda_d > 0, \sum_d \lambda_d = 1$. The parameters $\boldsymbol{\delta}$ of the Dirichlet distribution can be estimated by maximizing the log-likelihood function of given data. The log-likelihood is convex in $\boldsymbol{\delta}$ which guarantee a unique optimum. To estimate $\boldsymbol{\delta}$, we consider the simple and efficient iterative gradient descent method described in minka's paper (see [15]). The parameters $\left(\boldsymbol{M}^t \triangleq \{m^t_{pq}\}, \boldsymbol{\delta}^t \triangleq \{\delta^t_p\}, \boldsymbol{\Sigma}^t \triangleq \{(\sigma^t_{pq})^2\}\right)$ at iteration t are updated *via* the sample mean and sample standard deviation of elite samples. The algorithm is summarized as follows:

1. Choose some initial $\{\boldsymbol{M}^0, \boldsymbol{\Sigma}^0, \boldsymbol{\delta}^0\}$ for $p = 1, \ldots, d, q = 1, \ldots, P$. Set $t = 1$ (level counter).
2. Draw samples $\boldsymbol{W}^l \sim \mathcal{N}t\left(\boldsymbol{M}^{(t-1)}, \boldsymbol{\Sigma}^{(t-1)}, \boldsymbol{a}_p, \boldsymbol{b}_p\right)$, $\boldsymbol{\lambda}^l \sim \mathcal{D}ir(\boldsymbol{\delta}^{(t-1)})$, $l = 1, \ldots, V$.
3. Compute $S^l = F_{GRLVQ}(\boldsymbol{X}; \boldsymbol{\theta}^l)$ scores by applying eq.(10) $\forall l$.
4. Sort S^l in ascending order and denote by $\mathcal{I}$ the set of corresponding indices. Let us denote $\left(\widetilde{m}^{(t-1)}_{pq}, (\widetilde{\sigma}^{(t-1)}_{pq})^2\right)$ the mean and the variance of the best $\lceil \rho V \rceil$ prototypes elite samples of $\{\boldsymbol{W}^{\mathcal{I}(l)}\}, l = 1, \ldots, \lceil \rho V \rceil$ respectively. The $\widetilde{\boldsymbol{\delta}}^t$ is the corresponding the Dirichlet parameter fitted on $\{\boldsymbol{\delta}^{\mathcal{I}(l)}\}$.
5. $\widehat{\boldsymbol{M}}^t = \alpha \widetilde{\boldsymbol{M}}^t + (1-\alpha)\widehat{\boldsymbol{M}}^{t-1}$, $\widehat{\boldsymbol{\Sigma}}^t = \beta_t \widetilde{\boldsymbol{\Sigma}}^t + (1-\beta_t)\widehat{\boldsymbol{\Sigma}}^{(t-1)}$ and $\widehat{\boldsymbol{\delta}}^t = \alpha \widetilde{\boldsymbol{\delta}}^t + (1-\alpha)\widehat{\boldsymbol{\delta}}^{(t-1)}$
6. If convergence is reached or $t = T$ (T denote the final iteration), then **stop**; otherwise set $t = t + 1$ and reiterate from step 2.

Fig. 1. Cross Entropy Method for GRLVQ optimisation: CEMGRLVQ

According to Rubinstein et *al* in [13] it is found empirically that the algorithm is quite robust under the choice of the initial parameters N,ρ and β, as long as ρ is not too small, provided that the initial variances are chosen large enough, ensuring at the beginning to cover all solutions space.

To prevent the algorithm from being trapped in local optima Kroese in [12] proposed to use dynamic smoothing (15) where at each iteration the variance $(\sigma^t_{pq})^2$ is updated using a smoothing parameter:

$$\beta_t = \beta_0 - \beta_0 \left(1 - \frac{1}{t}\right)^c, \tag{15}$$

where c is a small integer (typically between 5 and 15) and β_0 is a large smoothing constant (typically between 0.8 and 0.99). By using β instead of α the convergence to the degenerate case has polynomial speed instead of exponential.

4 Experimental Results

In this section, we applied both the CEMGLVQ($\boldsymbol{\lambda}$ are not updated and keept constant)/CEMGRLVQ algorithm with some machine learning algorithms (GLVQ, GRLVQ) to show the effectiveness of the proposed method. Following standard procedure in experiments with other published works, each data set is initially normalized and algorithm parameters are selected as described in the papers related to the algorithms described above. To obtain meaningful comparisons against other published algorithms and to assess the effectiveness of the proposed algorithm, we used a stratified 10 fold cross-validation. We tested these algorithms in real world data sets taken from the public UCI repository [16] (see Table 1). For comparison we used different data sets which cover a broad spectrum of properties occurring frequently in real world applications. In particulary, we used waveform data set because it contains features with only noise and intrinsic within-class multi-modal structure. The parameters in the algorithms were set as follows:

- GLVQ: $\xi = 0.5$, $\varepsilon_k = 0.05$, $\varepsilon_l = 0.01$ where ε_k and ε_l are learning rates for nearest correct and wrong prototype, respectively.
- GRLVQ: $\xi = 0.1$, $\varepsilon_k = 0.05$, $\varepsilon_l = 0.01$, $\varepsilon_\lambda = 1e - 5$. ε_λ is the weights learning rate. For these two algorithms, the prototypes for different classes were randomly initialized around the center of the corresponding classes. Number of prototypes per class is indicated in Table.1.
- For the CEMGLVQ: $V = 5N$ where N denote the number of train samples, $\xi = 0.1$, $\rho = 3.10e - 3$, $\beta_0 = 0.95$, $h = 5.10e - 6$, $q = 10$, $\epsilon = 5.10e - 9$ et $c = 10$, $C = 1.2$. It is found empirically that when α is between 0.6 and 0.9 it gives best results in our case we choose $\alpha = 0.7$.

For all algorithms the number of iterations is set to $K = 600$. We based our comparison on two criteria: the error rate and optimal value of the cost function. Results are presented in Tables (2 and 3) respectively. We see in the Table 2 that our algorithm outperforms the GLVQ algorithm and shows slightly better performances than GRLVQ. In Table 3 we see that the CEMGRLVQ gives the lowest value of the cost function which leads to high quality solutions of the optimization problem. Compared to other algorithms based on gradient descent, the CEMGRLVQ is more demanding on computational cost of running. The theoretical complexity of CE is an open problem still under investigation, this complexity is partially depending on the studied problem (see [17]). The computational complexity of our algorithm is $O(KVN)$ in each iteration, However, the CE method can be accelerated by parallelizing all cost function evaluations.

Table 1. List real world data sets used in comparaison between algorithms. Data sets indicated by † contain features with only noise, while indicating by $*$ contain intrinsic within-class multi-modal structure.

Name	# Features	# Patterns	# Classes	# Prototypes per class
Liver	6	345	2	4
Pima	8	768	2	4
Glass	9	214	6	4
WBC	10	699	2	4
$Waveform^{*\dagger}$	21	500	3	4

Table 2. Recognition rates on real data sets. The format of the numbers in this table is $M \pm S$, where M is the mean recognition rate, S is the standard deviation. For each data set, the best method is described by the boldface.

Name	GLVQ (%)	CEMGLVQ (%)	GRLVQ (%)	CEMGRLVQ (%)
Liver	57.35±6.82	65.28±4.42	56.17±7.26	**68.38±4.04**
Pima	70.65±4.56	72.5±5.16	**75.65±3.57**	74.60±4.20
Glass	60.55±9.60	61.11±9.77	51.66±12.29	**64.11±4.19**
WBC	92.94±4.08	96.08±3.98	94.65±4.21	**96.23±2.64**
$Waveform^{*\dagger}$	79.00±6.09	79.33±6.62	79.66±8.43	**83.33±3.54**

Table 3. Optimal Cost Function Values. The format of the numbers in this table is $M \pm S$, where M is the mean Cost Function Value, S is standard deviation.

Name	GRLVQ	CEMGRLVQ
Liver	154.7883±0.1351	**153.8191±0.0495**
Pima	341.2171±0.2107	**338.9825±0.1580**
Glass	97.4918±0.1792	**95.9039±0.0770**
WBC	302.2815±0.0933	**301.9468±0.0722**
$Waveform^{*\dagger}$	133.9954±0.0826	**133.3087±0.0502**

5 Attribute Weighing Estimation

Feature ranking is hard to compare because of the differences in data interpretation. In this section, we will be concerned with the Iris data set. This is one of the best known databases in the pattern recognition literature. The data set contains three classes (Iris Setosa, Iris Versicolour, and Iris Virginica) of 50 instances each. There are four input attributes (sepal length, sepal width, petal length, and petal width). The relevance vector that resulted after the experiment is [.2294 .1516 .2483 .3708] this result points out that the most important features are the latest two features. Our ranking is similar to GRLVQ results [.2353 .2193 .2704 .2750].

6 Conclusion

In this paper we considered solving the GRLVQ initialization sensitiveness problem. We formulated it as a stochastic optimization problem and applied the Cross Entropy method to solve it. The suggested method was shown to be apt to deal well with such problems. It produced recognition rates that were fairly superior to other proposed methods. The main benefit of the proposed method is its robustness to the exact choice of its hyper parameters which is not the case of the other methods (learning rate and size of an update neighborhood (SRNG)). In general, the CE method is efficient and easy to implement. The CE method can be seen as a stochastic method, which gives CE method the ability to find more global optima than deterministic methods. In future work we will concentrate on the investigation of more appropriate cost function which do not only take into account the two best matching prototypes (form the correct and the wrong class) but prototypes cooperation (for example H2MGLVQ [18]).

References

1. Kohonen, T.: Learning vector quantization for pattern recognition. Technical Report TKK-F-A601 (1986)
2. Kohones, T.: Bibliography on the self-organizing map (som) and learning vector quantization(lvq) (2002)
3. Bojer, T., Hammer, B., Schunk, D., Toschanowitz, K.V.: Releavance determination in learning vector quantization. In: the European Symposium on Artificial Neural networks, pp. 271–276 (2001)
4. Hammer, B., Villman, T.: Estimating relevant input dimensions for self-organizing algorithms. Advances in Self-Organizing Maps, 173–180 (2001)
5. Sato, A.S., Yamada, K.: A formulation of learning vector quantization using a new misclassification measure. In: The 14th International Conference on Pattern Recognition (1998)
6. Hammer, B., Strickert, M., Villmann, T.: Relevance lvq versus svm. In: Rutkowski, L., Siekmann, J.H., Tadeusiewicz, R., Zadeh, L.A. (eds.) ICAISC 2004. LNCS (LNAI), vol. 3070, pp. 592–597. Springer, Heidelberg (2004)
7. Hammer, B., Strickert, M., Villmann, T.: Supervised neural gas with general similarity measure. Neural Processing Letters (2004)
8. Rubinstein, R.Y.: Optimization of computer simulation models with rare events. European Journal of operational Research 99, 89–112 (1997)
9. Diamantini, C., Spalvieri, A.: Certain facts about kohonen's lvq1 algorithm. IEEE Transactions on Circuits and Systems I(47), 425–427 (1996)
10. Juang, B.H., Katagiri, S.: Discriminative learning for minimum error classification. IEEE Transactions on Signal Processing 40(2), 3043–3054 (1992)
11. Fu, M.C., Glover, F.W., April, J.: Simulation optimization: A review, new developements, and applications. In: WSC2005. The 37th Winter Simulation Conference, pp. 83–95 (2005)
12. Kroese, D., Porotsky, S., Rubinstein, R.: The cross-entropy method for continuous multi-extremal optimization. Methodology and Computing in Applied Probability 8, 383–407 (2006)
13. Rubinstein, R.Y., Kroese, D.P.: The Cross-Entropy Method: A Unified Approach to Combinatorial Method, Monte-Carlo Simulation, Randomized Optimization and Machine Learning. Springer, Heidelberg (2004)
14. Boer, P.D., Kroese, D., Mannor, S., Rubinstein, R.Y.: A tutorial on the cross-entropy method. Annals of Operations Research, 19–67 (2005)
15. Minka, T.: Estimating a dirichlet distribution (2003)
16. Newman, D.J., Hettich, S., Blake, C.L., Merz, C.J.: Uci repository of machine learning databases (1998)
17. Wu, J., Chung, A.: Cross entropy: A new solver for markov random field modeling and applications to medical image segmentation. In: Duncan, J.S., Gerig, G. (eds.) MICCAI 2005. LNCS, vol. 3749, pp. 229–237. Springer, Heidelberg (2005)
18. Qin, A., Suganthan, P.: Initialization insensitive lvq algorithm based on cost-function adaptation. Pattern Recognition 38(5), 773–776 (2005)

Incremental and Decremental Learning for Linear Support Vector Machines

Enrique Romero, Ignacio Barrio, and Lluís Belanche

Departament de Llenguatges i Sistemes Informàtics
Universitat Politècnica de Catalunya, Barcelona, Spain

Abstract. We present a method to find the exact maximal margin hyperplane for linear Support Vector Machines when a new (existing) component is added (removed) to (from) the inner product. The maximal margin hyperplane with the new inner product is obtained in terms of that for the old inner product, without re-computing it from scratch and the procedure is reversible. An algorithm to implement the proposed method is presented, which avoids matrix inversions from scratch. Among the possible applications, we find feature selection and the design of kernels out of similarity measures.

1 Introduction

Support Vector Machines (SVMs) are widely used tools for classification and regression problems based on margin maximization [1]. The standard formulation of SVMs uses a fixed data set to construct a linear combination of simple functions depending on the data. In many application domains, however, the number of examples or the number of variables may vary with time. In these cases, adaptive solutions that efficiently work in these changing environments are of great interest. Procedures for exact incremental (on-line) learning for Support Vector Machines (SVMs) have been introduced in [2,3] for classification and regression problems respectively. In [4,5], the dynamic adaptation of the kernel parameter was studied. In [6], the entire SVM solution path is obtained for every value of the regularization parameter.

In this work the situation is that of adapting the inner product for linear SVMs in an incremental manner. Specifically, suppose that we have obtained the maximal margin hyperplane of a linear SVM, and we would like to add/remove a new/existing component to/from the inner product. Can the maximal margin hyperplane with the new inner product be obtained in terms of that for the old inner product? A positive answer to this question would allow to add new variables to the data and obtain the linear SVM solution with these new variables without computing it from scratch. Analogously, existing variables could be removed from the SVM solution.

Similar to previous works [2,5], the main idea is the preservation of the Karush-Kuhn-Tucker (KKT) conditions to find the (exact) solution and guide the process. The solution will be updated in response to the perturbation

J. Marques de Sá et al. (Eds.): ICANN 2007, Part I, LNCS 4668, pp. 209–218, 2007.

induced by the addition/removal of the new/existing component to/from the inner product. The new/existing component will not be added/removed at once, but it will be added/removed in suitable increments that allow to control the migrations among the subsets of support vectors (vectors in the margin, violating the margin or exceeding the margin). For small increments (no migrations among the subsets of support vectors), the analysis leads to the solution of a linear equations system. When a migration occurs, the dimensions of the matrices involved in this linear equations system must be modified. The addition and removal of the new/existing component can be treated in the same way, so that the procedure is reversible. Direct applications include time-varying learning environments, feature selection and kernel design from arbitrary similarity measures.

2 Background

To fix notation, let $\mathcal{X} = \{(x_1, y_1), \ldots, (x_L, y_L)\}$ be a data set, and consider the classification task given by $\mathcal{X}$, where each instance $x_i = (x_i^1, \ldots, x_i^N) \in \mathbb{R}^N$ and $y_i \in \{-1, +1\}$. SVMs can be described as follows [1]: the input vectors are mapped into a (usually high-dimensional) inner product space through some non-linear mapping ϕ, chosen *a priori*. In this space (the *feature space*), an optimal separating hyperplane is constructed. By using a (positive definite) kernel function $K(u, v)$ the mapping can be implicit, since the inner product defining the hyperplane can be evaluated as $\langle \phi(u), \phi(v) \rangle = K(u, v)$ for every two vectors $u, v \in \mathbb{R}^N$. In the SVM framework, an optimal hyperplane means a hyperplane with maximal normalized margin for the examples of every class. The normalized margin is the minimum distance to the hyperplane.

When the data set is not linearly separable (neither in the input space nor in the feature space), some tolerance to noise is introduced in the model. Using Lagrangian theory, the maximal margin hyperplane for a binary classification problem given by a data set $\mathcal{X}$ is a linear combination of simple functions depending on the data: $f(x) = b + \sum_{i=1}^{L} y_i \alpha_i K(x_i, x)$, where $K(u, v)$ is a kernel function and the coefficients vector $(\alpha_i)_{i=1}^{L}$ is the (1-norm soft margin) solution of a constrained optimization problem in the dual space [1,2]:

$$\begin{array}{lll} \text{Minimize}(\boldsymbol{\alpha}, b) & W = \frac{1}{2} \sum_{i,j=1}^{L} y_i \alpha_i y_j \alpha_j K(x_i, x_j) + b \sum_{i=1}^{L} y_i \alpha_i - \sum_{i=1}^{L} \alpha_i & \\ \text{subject to} & 0 \leqslant \alpha_i \leqslant C \qquad i = 1, \ldots, L & \end{array} \quad (1)$$

for a certain constant C. Note this is not the standard formulation, since dual and primal variables are present at the same time. This allows to deal with the bias term in a similar way than with the rest of the variables. The regularization parameter C allows to control the trade-off between the margin and the training errors. The *hard margin* hyperplane is obtained with $C = \infty$.

An example is well classified if and only if its functional margin $y_i f(x_i)$ with respect to f is positive. The KKT conditions uniquely define the solution $\{\boldsymbol{\alpha}, b\}$ of (1):

$$g_i = \frac{\partial W}{\partial \alpha_i} = y_i f(x_i) - 1 \begin{cases} > 0 & \alpha_i = 0 \\ = 0 & 0 \leqslant \alpha_i \leqslant C \\ < 0 & \alpha_i = C \end{cases} \tag{2}$$

$$h = \frac{\partial W}{\partial b} = \sum_{i=1}^{L} y_i \alpha_i = 0 \tag{3}$$

Following [2,5], the training examples can be partitioned into three different categories: the set $\mathcal{S}$ of *margin support vectors* on the margin ($g_i = 0$), the set $\mathcal{E}$ of *error support vectors* violating the margin ($g_i < 0$) and the set $\mathcal{R}$ of *reserve vectors* exceeding the margin ($g_i > 0$).

An incremental algorithm for on-line training of SVMs for classification problems (1) was presented in [2]. When new examples are considered, the KKT conditions are preserved on all previously seen data and satisfied for the new data. This is done in a sequence of analytically computable steps. The whole procedure is reversible, thus allowing "decremental unlearning" and leave-one-out error estimation. In [5], the model is extended to adapt the SVM solution to changes in regularization and kernel parameters. A similar method for constructing ϵ-insensitive SVMs for regression problems is described in [3]. In [6], the entire path of the solutions of SVMs is obtained for every value of the regularization parameter in a similar way, exploting the fact that the coefficients α_i in the solution of (1) are piece-wise linear in C.

3 Incremental and Decremental Inner Product Learning for Linear SVMs

3.1 Problem Setting

Suppose that we are working with linear SVMs (wherein $K(x_i, x_j)$ is the *inner product* $\langle x_i, x_j \rangle = \sum_{k=1}^{m} x_i^k x_j^k$) and we would like to solve the following situations:

1. We have $m < N$ components in our inner product and we want to add a new component: $\langle x_i, x_j \rangle^{\text{new}} = \sum_{k=1}^{m+1} x_i^k x_j^k = \langle x_i, x_j \rangle + x_i^{m+1} x_j^{m+1}$.
2. We have $m+1 \leq N$ components in our inner product and we want to remove an existing component: $\langle x_i, x_j \rangle^{\text{new}} = \sum_{k=1}^{m} x_i^k x_j^k = \langle x_i, x_j \rangle - x_i^{m+1} x_j^{m+1}$.

The problem in both cases is to obtain the solution of (1) with the new inner product $\langle x_i, x_j \rangle^{\text{new}}$ in terms of that for the old one $\langle x_i, x_j \rangle$ (i.e., without recomputing it from scratch). A solution to these problems is described in the next sections.

3.2 Sketch of the Solution

We start from the solution $\{\boldsymbol{\alpha}, b\}$ of (1) with $\langle x_i, x_j\rangle$ (which satisfy the KKT conditions). In order to obtain the solution of (1) with the new inner product, we follow the strategy of computing the increments $\{\Delta\boldsymbol{\alpha}, \Delta b\}$ such that $\{\boldsymbol{\alpha}+\Delta\boldsymbol{\alpha}, b+\Delta b\}$ satisfy the KKT conditions of (1) with $\langle x_i, x_j\rangle^{\text{new}}$. The KKT conditions will be preserved by changing the parameters in response to the perturbation induced by the addition/removal of the new/existing component to/from the inner product. In this process, examples of the different categories ($\mathcal{S}$, $\mathcal{E}$ and $\mathcal{R}$) may change their state. In order to control these changes, the new/existing component will not be added/removed at once, but added/removed in steps of a certain amplitude Δp such that it leads to the minimum number of category changes, which can be controlled. With the previous notation, the addition and removal of the new/existing component can be treated in the same way. During the process, the inner product will be

$$\langle x_i, x_j\rangle' = \langle x_i, x_j\rangle + p \cdot x_i^{m+1} x_j^{m+1}. \tag{4}$$

Initially, $p = 0$. When adding a new component, p must reach $p = 1$ with increments $\Delta p > 0$. When removing an existing component, p must reach $p = -1$ with increments $\Delta p < 0$. For every Δp (different at every step), the solution will be recalculated so that the KKT conditions (2,3) are preserved on all data (thus controlling the migrations among $\mathcal{S}$, $\mathcal{E}$ and $\mathcal{R}$).

Therefore, for a given perturbation Δp, our objective is to determine the necessary changes in the solution of (1) and the migrations among $\mathcal{S}$, $\mathcal{E}$ and $\mathcal{R}$ so that the KKT conditions are preserved on all data. Let us define

$$W' = \frac{1}{2}\sum_{i,j=1}^{L} y_i\alpha_i y_j\alpha_j \langle x_i, x_j\rangle' + b\sum_{i=1}^{L} y_i\alpha_i - \sum_{i=1}^{L}\alpha_i$$

with $\{\boldsymbol{\alpha}, b\}$ satisfying (2,3), and

$$W^+ = \frac{1}{2}\sum_{i,j=1}^{L} y_i\alpha_i^+ y_j\alpha_j^+ \langle x_i, x_j\rangle^+ + b^+\sum_{i=1}^{L} y_i\alpha_i^+ - \sum_{i=1}^{L}\alpha_i^+$$

with $\alpha_i^+ = \alpha_i + \Delta\alpha_i$, $b^+ = b + \Delta b$ and $\langle x_i, x_j\rangle^+ = \langle x_i, x_j\rangle' + \Delta p\; x_i^{m+1} x_j^{m+1}$.

3.3 Calculation of the Increments

The increments $\{\Delta\boldsymbol{\alpha}, \Delta b\}$ can be computed as a function of Δp, as explained next. Two situations may arise, tackled in the rest of the section:

1. The solution changes, but no example changes its state among the different categories ($\mathcal{S}$, $\mathcal{E}$ and $\mathcal{R}$). This is the typical situation when $|\Delta p|$ is small. In this case, only the coefficients of the support vectors and the bias term must be updated.

2. There exists one or more examples that migrate among $\mathcal{S}$, $\mathcal{E}$ and $\mathcal{R}$. These changes can be controlled if Δp is chosen so that the minimum number of migrations occurs. In this case, $\mathcal{S}$, $\mathcal{E}$ and $\mathcal{R}$ must be updated, but recalculating the solution is not necessary.

1. Small $|\Delta p|$ (no migrations among $\mathcal{S}$, $\mathcal{E}$ and $\mathcal{R}$). Suppose that $|\Delta p|$ is small enough so that no example changes its state: $\Delta\alpha_i = 0$ for every $i \notin \mathcal{S}$, $\alpha_i = 0$ for every $i \in \mathcal{R}$ and $\alpha_i = C$ for every $i \in \mathcal{E}$.

Since the new solution must satisfy the KKT conditions, we have:

$$g_i^+ = \frac{\partial W^+}{\partial \alpha_i^+} = y_i f^+(x_i) - 1 = y_i \left(\sum_{j=1}^{L} y_j \alpha_j^+ \langle x_i, x_j \rangle^+ + b^+ \right) - 1 = 0 \qquad \forall i \in \mathcal{S}$$

$$h^+ = \frac{\partial W^+}{\partial b^+} = \sum_{i=1}^{L} y_i \alpha_i^+ = 0.$$

Therefore, for every $i \in \mathcal{S}$ (recall that also $g_i' = \frac{\partial W'}{\partial \alpha_i} = 0$)

$$g_i^+ - g_i' = y_i \left(\sum_{j=1}^{L} y_j \alpha_j \Delta p \; x_i^{m+1} x_j^{m+1} \right) + y_i \left(\sum_{j=1}^{L} y_j \Delta\alpha_j \langle x_i, x_j \rangle^+ \right) + y_i \Delta b = 0.$$

Since no example changes its state, we have

$$0 = y_i \left(\sum_{j \in \mathcal{S}} y_j \alpha_j \Delta p \; x_i^{m+1} x_j^{m+1} \right) + y_i \left(\sum_{j \in \mathcal{E}} y_j C \Delta p \; x_i^{m+1} x_j^{m+1} \right) +$$
$$y_i \left(\sum_{j \in \mathcal{S}} y_j \Delta\alpha_j \langle x_i, x_j \rangle^+ \right) + y_i \Delta b.$$

In addition (recall that also $h' = \frac{\partial W'}{\partial b} = 0$),

$$h^+ - h' = \sum_{i=1}^{L} y_i \Delta\alpha_i = \sum_{i \in \mathcal{S}} y_i \Delta\alpha_i = 0.$$

The above analysis can be summarized as ($x_{\mathcal{S}_1}, \ldots, x_{\mathcal{S}_{L_\mathcal{S}}}$ are the examples in $\mathcal{S}$)

$$\left(\frac{\mathcal{Q}'}{\Delta p} + \mathcal{U} \right) \cdot \begin{pmatrix} \Delta b \\ \Delta\alpha_{\mathcal{S}_1} \\ \vdots \\ \Delta\alpha_{\mathcal{S}_{L_\mathcal{S}}} \end{pmatrix} = -\,\mathcal{U} \cdot \begin{pmatrix} 0 \\ \alpha_{\mathcal{S}_1} \\ \vdots \\ \alpha_{\mathcal{S}_{L_\mathcal{S}}} \end{pmatrix} - \mathcal{V} \qquad (5)$$

where $\mathcal{Q}'$, $\mathcal{U}$ are $(L_\mathcal{S} + 1) \times (L_\mathcal{S} + 1)$ symmetric matrices

$$\mathcal{Q}' = \begin{pmatrix} 0 & y_{\mathcal{S}_1} & \cdots & y_{\mathcal{S}_{L_\mathcal{S}}} \\ y_{\mathcal{S}_1} & \mathcal{Q}'_{\mathcal{S}_1\mathcal{S}_1} & \cdots & \mathcal{Q}'_{\mathcal{S}_1\mathcal{S}_{L_\mathcal{S}}} \\ \vdots & \vdots & \ddots & \vdots \\ y_{\mathcal{S}_{L_\mathcal{S}}} & \mathcal{Q}'_{\mathcal{S}_{L_\mathcal{S}}\mathcal{S}_1} & \cdots & \mathcal{Q}'_{\mathcal{S}_{L_\mathcal{S}}\mathcal{S}_{L_\mathcal{S}}} \end{pmatrix} \qquad \mathcal{U} = \begin{pmatrix} 0 & 0 & \cdots & 0 \\ 0 & \mathcal{U}_{\mathcal{S}_1\mathcal{S}_1} & \cdots & \mathcal{U}_{\mathcal{S}_1\mathcal{S}_{L_\mathcal{S}}} \\ \vdots & \vdots & \ddots & \vdots \\ 0 & \mathcal{U}_{\mathcal{S}_{L_\mathcal{S}}\mathcal{S}_1} & \cdots & \mathcal{U}_{\mathcal{S}_{L_\mathcal{S}}\mathcal{S}_{L_\mathcal{S}}} \end{pmatrix} \tag{6}$$

$$\mathcal{Q}'_{\mathcal{S}_i\mathcal{S}_j} = y_{\mathcal{S}_i} y_{\mathcal{S}_j} \left\langle x_{\mathcal{S}_i}, x_{\mathcal{S}_j} \right\rangle' \qquad \mathcal{U}_{\mathcal{S}_i\mathcal{S}_j} = y_{\mathcal{S}_i} y_{\mathcal{S}_j} x_{\mathcal{S}_i}^{m+1} x_{\mathcal{S}_j}^{m+1} \tag{7}$$

and $\mathcal{V}$ is the vector

$$\mathcal{V} = \left(0, \mathcal{V}_{\mathcal{S}_1}, \cdots, \mathcal{V}_{\mathcal{S}_{L_\mathcal{S}}}\right)^t \qquad \mathcal{V}_{\mathcal{S}_i} = C\, y_{\mathcal{S}_i} x_{\mathcal{S}_i}^{m+1} \sum_{j \in \mathcal{E}} y_j x_j^{m+1}. \tag{8}$$

Therefore, given Δp, the increments of the parameters can be computed by solving the linear equations system (5). Note that, contrary to the situation in [2,5], the solution of (5) is not a linear function of Δp.

2. Large $|\Delta p|$ (controlling the migrations among $\mathcal{S}$, $\mathcal{E}$ and $\mathcal{R}$). When $|\Delta p|$ is large, migrations among $\mathcal{S}$, $\mathcal{E}$ and $\mathcal{R}$ may occur after solving (5). However, we can choose Δp such that the minimum number of migrations occurs. In this case several migrations may occur simultaneously, but an example can only migrate from its current set to a neighbor set (that is, between $\mathcal{S}$ and $\mathcal{E}$ or between $\mathcal{S}$ and $\mathcal{R}$). These migrations are determined by the KKT conditions:

1. From $\mathcal{E}$ to $\mathcal{S}$: One error support vector becomes a margin support vector. This happens when g_i (that was negative) becomes 0.
2. From $\mathcal{S}$ to $\mathcal{E}$: One margin support vector becomes an error support vector. This happens when its coefficient becomes C.
3. From $\mathcal{R}$ to $\mathcal{S}$: One reserve vector becomes a margin support vector. This happens when g_i (that was positive) becomes 0.
4. From $\mathcal{S}$ to $\mathcal{R}$: One margin support vector becomes a reserve vector. This happens when its coefficient becomes 0.

When a migration occurs, $\mathcal{Q}'$, $\mathcal{U}$ and $\mathcal{V}$ must be updated, since not only their components but also their dimension change. However, recomputing the solution is not necessary as a consequence of the migration: if a new support vector is inserted in $\mathcal{S}$, its coefficient remains the same (0 if migrated from $\mathcal{R}$ or C if migrated from $\mathcal{E}$); if an example is deleted from $\mathcal{S}$, its coefficient (which also equals 0 or C) will not vary until it is inserted in $\mathcal{S}$ again.

4 Implementation

4.1 Efficient Computation of the Linear Equations System

An efficient way to solve (5) would be desirable. This section describes an efficient way to update $(\mathcal{Q}'/\Delta p + \mathcal{U})^{-1}$ when:

1. $|\Delta p|$ is small enough so that no example changes its state.
2. Migrations among $\mathcal{S}$, $\mathcal{E}$ and $\mathcal{R}$ occur.

1. Small $|\Delta p|$ (no changes in the dimension). Note that $\mathcal{U} = \boldsymbol{u} \cdot \boldsymbol{u}^t$, with $\boldsymbol{u} = \left(0, y_{\mathcal{S}_1} x_{\mathcal{S}_1}^{m+1}, \cdots, y_{\mathcal{S}_{L_S}} x_{\mathcal{S}_{L_S}}^{m-1}\right)^t$. Therefore, the inverse of $\mathcal{Q}'/\Delta p + \mathcal{U}$ can be efficiently computed by applying the Sherman-Morrison-Woodbury matrix inversion formula:

$$(\mathcal{Q}'/\Delta p + \mathcal{U})^{-1} = (\mathcal{Q}'/\Delta p + \boldsymbol{u} \cdot \boldsymbol{u}^t)^{-1} = \Delta p \left(\mathcal{Q}'^{-1} - \frac{\Delta p \; \mathcal{Q}'^{-1} \boldsymbol{u}\boldsymbol{u}^t \mathcal{Q}'^{-1}}{1 + \Delta p \; \boldsymbol{u}^t \mathcal{Q}'^{-1} \boldsymbol{u}} \right) \quad (9)$$

As a consequence, for every Δp such that $\mathcal{S}$ does not change, only $\mathcal{Q}'^{-1}$ is needed to compute $(\mathcal{Q}'/\Delta p + \mathcal{U})^{-1}$. Note that $\mathcal{Q}'^{-1}$ can also be updated in the same way when $p_{new} = p + \Delta p$, since $\mathcal{Q}'_{new} = \mathcal{Q}' + \Delta p \, \mathcal{U}$. Thus,

$$(\mathcal{Q}'_{new})^{-1} = \mathcal{Q}'^{-1} - \frac{\Delta p \; \mathcal{Q}'^{-1} \boldsymbol{u}\boldsymbol{u}^t \mathcal{Q}'^{-1}}{1 + \Delta p \; \boldsymbol{u}^t \mathcal{Q}'^{-1} \boldsymbol{u}} \quad (10)$$

2. Large $|\Delta p|$ (inserting/deleting examples in/from $\mathcal{S}$). When a new example is inserted/deleted in/from $\mathcal{S}$, matrix $\mathcal{Q}'^{-1}$ can be updated incrementally using block matrices techniques, thus avoiding the computation of $\mathcal{Q}'^{-1}$ from scratch, as explained next.

When a new margin support vector $x_{\mathcal{S}_{L_S+1}}$ is inserted in $\mathcal{S}$, matrix $\mathcal{Q}'^{-1}$ is expanded as

$$\mathcal{Q}'^{-1} \leftarrow \begin{pmatrix} & & 0 \\ & \mathcal{Q}'^{-1} & \vdots \\ 0 & \cdots & 0 \end{pmatrix} + \frac{1}{\gamma} \begin{pmatrix} \mathcal{B} \\ 1 \end{pmatrix} \cdot \begin{pmatrix} \mathcal{B}^t & 1 \end{pmatrix} \quad (11)$$

where

$$\mathcal{B} = -\mathcal{Q}'^{-1} \cdot \mathcal{Q}'_{*L_S+1}$$

$$\gamma = \mathcal{Q}'_{\mathcal{S}_{L_S+1}\mathcal{S}_{L_S+1}} - (\mathcal{Q}'_{*L_S+1})^t \cdot \mathcal{Q}'^{-1} \cdot \mathcal{Q}'_{*L_S+1}$$

with

$$\mathcal{Q}'_{*L_S+1} = \left(y_{L_S+1}, \mathcal{Q}'_{\mathcal{S}_1 \mathcal{S}_{L_S+1}}, \ldots, \mathcal{Q}'_{\mathcal{S}_{L_S} \mathcal{S}_{L_S+1}} \right)^t.$$

When an example $x_{\mathcal{S}_k}$ is deleted from $\mathcal{S}$, matrix $\mathcal{Q}'^{-1}$ is contracted as

$$\mathcal{Q}'^{-1}_{\mathcal{S}_i\mathcal{S}_j} \leftarrow \mathcal{Q}'^{-1}_{\mathcal{S}_i\mathcal{S}_j} - \frac{\mathcal{Q}'^{-1}_{\mathcal{S}_i\mathcal{S}_k} \mathcal{Q}'^{-1}_{\mathcal{S}_k\mathcal{S}_j}}{\mathcal{Q}'^{-1}_{\mathcal{S}_k\mathcal{S}_k}} \quad (12)$$

for every $\mathcal{S}_i, \mathcal{S}_j \in \{0, \mathcal{S}_1, \ldots, \mathcal{S}_{L_S}\}$ such that $\mathcal{S}_i, \mathcal{S}_j \neq \mathcal{S}_k$. The index 0 refers to the first row/column.

AddRemoveComponent (data set $\mathcal{X}$, partitions $\{\mathcal{S},\mathcal{E},\mathcal{R}\}$, $\{\boldsymbol{\alpha}, b\}$ satisfying (2,3))
  $p \leftarrow 0$
  Compute $\mathcal{Q}'^{-1}$
  repeat
    Find the maximum $|\Delta p|$ (*addition*: $\Delta p \in (0, 1-p]$; *removal*: $\Delta p \in [-1-p, 0)$) such that, after solving (5) with (9), the number M of migrations among $\mathcal{S}$, $\mathcal{E}$ and $\mathcal{R}$ is minimum
    Update $\mathcal{Q}'^{-1}$ with (10)
    $\{\boldsymbol{\alpha}, b\} \leftarrow \{\boldsymbol{\alpha}, b\} + \{\Delta\boldsymbol{\alpha}, \Delta b\}$, where $\{\Delta\boldsymbol{\alpha}, \Delta b\}$ is the solution of (5)
    if $M \neq 0$ **then**
      Update $\mathcal{Q}'^{-1}$ with (11) or (12) depending on whether it must be expanded or contracted (i.e., an example must be inserted/deleted in/from $\mathcal{S}$)
    end if
    $p \leftarrow p + \Delta p$
  until $|p| = 1$
end AddRemoveComponent

Fig. 1. An algorithm to add/remove a component to/from the inner product

4.2 A Complete Algorithm

The previous analysis leads to algorithm in figure 1. Several remarks are in order:

1. If succesive additions/removals have to be done, the first computation of $\mathcal{Q}'^{-1}$ is only necessary in the first step (for the rest, it can be a parameter).
2. The minimum number M of migrations among $\mathcal{S}$, $\mathcal{E}$ and $\mathcal{R}$ may be 0. In this case, $|p + \Delta p| = 1$ and the algorithm stops.
3. After updating $\mathcal{Q}'^{-1}$, recomputing the solution is not necessary (section 3.3).
4. After adding/removing one component, the values of $\{\mathcal{S},\mathcal{E},\mathcal{R}\}$, $\{\boldsymbol{\alpha}, b\}$ and $\mathcal{Q}'^{-1}$ can be used as parameters to a new call to **AddRemoveComponent**. Therefore, in the whole process of adding/removing iteratively several components to the inner product, only one matrix inversion has to be made from scratch (the one previous to the first call to the function).
5. To find the maximum $|\Delta p|$ we perform a binary (dichotomous) search in the corresponding real interval. The first cut-point is $\frac{1-p}{2}$ (addition) or $\frac{-1-p}{2}$ (removal). If no migrations occur, the search continues in the left subinterval, otherwise in the right subinterval. This process iterates until the value of $|\Delta p|$ converges to machine precision.

5 Applications

5.1 Feature Selection

A first and direct application of the algorithm described in section 4.2 is to perform feature selection with linear SVMs. Classical search algorithms for feature selection, like forward selection, backward elimination, or plus-l take-away-r

work by adding and removing features one at a time. Therefore, the algorithm described in section 4.2 can be easily used to add and remove input features in a prescribed way. Explicit feature selection search methods for linear SVMs (see [7], for example), can also benefit from this algorithm.

5.2 Basis Selection Guided by the Margin

Suppose that an explicit transformation of the input space is performed with a set of predefined basis functions $\Phi = \{\phi_j \mid \phi_j : \mathbb{R}^N \to \mathbb{R}\}_{j=1}^{M}$: for every $(x_i, y_i) \in \mathcal{X}$, consider the vector $(z_i, y_i) \in \mathbb{R}^M \times \mathbb{R}$, with $z_i = (\phi_1(x_i), \ldots, \phi_M(x_i))$. The solution of a linear SVM in this new space would give a non-linear model in the original input space based on margin maximization. The problem, however, would be to select an appropiate subset of basis functions from Φ, since it seems clear that some of them may be useless for the problem at hand. In order to select this subset, a search process can be performed, which is equivalent to perform feature selection in the new space. The algorithm described in section 4.2 can be used to that end, adding and removing basis functions to/from the partially obtained solutions.

5.3 Kernel Out of a Similarity

A kernel function $K(x, y)$ can be seen as a form of *similarity measure* between objects x and y. There is a vast literature in the design of similarity measures in data analysis [8]. One of the main advantages is to be able to cope with data heterogeneity and special values (like missing values) with a clear semantics. However, the need to fulfill the positivity condition prevents their use with SVMs. Though there are some works on proving positivity for certain similarity measures [9], they are limited to certain simple measures and even then the property may be spoiled in presence of missing data.

A kernel can be defined from a similarity as follows. Given a similariry $s : X \times X \to \mathbb{R}$ and Z a finite subset of X, the function

$$K(x, y) = \sum_{z_i \in Z} s(x, z_i)\, s(y, z_i) \tag{13}$$

is a positive kernel. The kernel defined in (13) is the sum-product aggregation of the similarities between x, y by using their respective similarities to a third object $z_i \in Z \subseteq X$. The semantics is then that x, y are similar if both are consistently similar to a set of reference objects Z. We call the elements of this set Z the *kernel vectors*. This process can be expressed in terms of the data, if $X = \{x_1, \ldots, x_L\}$ is the set of training vectors. In practice, the elements in Z (the kernel vectors) can be chosen to minimize (1) while keeping their number at a minimum, allowing more compact and computationally cheaper models. This can be done explicitly with a search algorithm that progressively adds/removes vectors to/from $Z \subseteq X$. In fact, this search process is similar to a feature selection search process.

6 Conclusions and Future Work

A method to find the exact maximal margin hyperplane for linear Support Vector Machines when a new (existing) component is added (removed) to (from) the inner product has been presented. The maximal margin hyperplane with the new inner product is obtained from the solution using the old inner product and the procedure is reversible. We have presented a full algorithm that implements the proposed method. Applications of the algorithm include time-varying learning environments wherein descriptive variables arrive one at a time, basis function selection with linear SVMs and kernel design from similarity measures, avoiding the need for positivity.

As future work, the method can be extended to non-linear SVMs where the objective would be to find solutions combining several kernels. In this new scenario, starting from the SVM solution with a certain kernel K_1, and given a function K_2 such that $K_1 + K_2$ is a kernel, we would like to obtain the solution with the new kernel $K^{\text{new}} = K_1 + K_2$. Similarly to the addition of components for linear kernels presented in this work, the new "component" K_2 can be added in suitable increments controlling the migrations among the support vectors.

Acknowledgments

This work was supported by the Consejo Interministerial de Ciencia y Tecnología (CICYT), under projects CGL2004-04702-C02-02 and TIN2006-08114.

References

1. Vapnik, V.N.: The Nature of Statistical Learning Theory. Springer, New York (1995)
2. Cauwenberghs, G., Poggio, T.: Incremental and Decremental Support Vector Machine Learning. In: Advances in Neural Information Processing Systems, vol. 12, pp. 409–415. MIT Press, Cambridge (2000)
3. Martín, M.: On-Line Support Vector Machine Regression. In: Elomaa, T., Mannila, H., Toivonen, H. (eds.) ECML 2002. LNCS (LNAI), vol. 2430, pp. 282–294. Springer, Heidelberg (2002)
4. Cristianini, N., Campbell, C., Shawe-Taylor, J.: Dynamically Adapting Kernels in Support Vector Machines. In: Advances in Neural Information Processing Systems, vol. 11, pp. 204–210. MIT Press, Cambridge (1999)
5. Diel, C., Cauwenberghs, G.: SVM Incremental Learning, Adaptation and Optimization. In: International Joint Conference on Neural Networks, vol. 4, pp. 2685–2690 (2003)
6. Hastie, T., Rosset, S., Tibshirani, R., Zhun, J.: The Entire Regularization Path for the Support Vector Machine. Journal of Machine Learning Research 5, 1391–1415 (2006)
7. Guyon, I., Weston, J., Barnhill, S., Vapnik, V.N.: Gene Selection for Cancer Classification using Support Vector Machines. Machine Learning 46(1-3), 389–422 (2002)
8. Chandon, J.L., Pinson, S.: Analyse Typologique. Théorie et Applications. Masson (1981)
9. Gower, J.C., Legendre, P.: Metric and Euclidean Properties of Dissimilarity Coefficients. Journal of Classification 3, 5–48 (1986)

An Efficient Method for Pruning the Multilayer Perceptron Based on the Correlation of Errors

Cláudio M.S. Medeiros and Guilherme A. Barreto

[1]Department of Mechatronics, Federal Center of Technological Education (CEFET), Av. 13 de Maio, 2081, Fortaleza, Ceará, Brazil
[2]Department of Teleinformatics Engineering, Federal University of Ceará (UFC), Av. Mister Hull, S/N, Center of Technology, Campus of Pici, Fortaleza, Ceará, Brazil

Abstract. In this paper we present a novel method for pruning redundant weights of a trained multilayer Perceptron (MLP). The proposed method is based on the correlation analysis of the errors produced by the output neurons and the backpropagated errors associated with the hidden neurons. Repeated applications of it leads eventually to the complete elimination of all connections of a neuron. Simulations using real-world data indicate that, in terms of performance, the proposed method compares favorably with standard pruning techniques, such as the Optimal Brain Surgeon (OBS) and Weight Decay and Elimination (WDE), but with much lower computational costs.

1 Introduction

Even nowadays, two decades after the rediscovery of the back-propagation algorithm, and despite all the available literature on the MLP network, a beginner soon becomes aware of the difficulties in finding an optimal architecture for real-world applications. In fact, this is a hard task also for an experienced practitioner. An architecture that is too small will not be able to learn from data properly, no matter what training algorithm is used for this purpose. Otherwise, an architecture having too many hidden neurons (and, hence, weight connections) are prone to fit too much of the noise on the training data.

Hence, a crucial step in the design of a MLP is related with the *network model selection* problem [1]. This problem, which still is a research topic of interest [2,3,4,5], can be roughly defined as the task of finding the smallest architecture that generalizes well, making good predictions for new data. The generalization of a neural network architecture can be assessed by changing the number of adaptive parameters (weights and biases) in the network.

Among the several ways to implement this in practice, we list the following three as possibly the commonest approaches. (*i*) *Exhaustive search plus early stopping*: the performances of several networks having different number of hidden neurons are evaluated during training on an independent validation set. Training of each network is stopped as soon as its generalization error begins to increase. The optimal architecture is the one providing the smallest generalization error. (*ii*) *Growing algorithms*: one can start training a network with a

J. Marques de Sá et al. (Eds.): ICANN 2007, Part I, LNCS 4668, pp. 219–228, 2007.

small number of hidden neurons and add neurons during the training process, with the goal of arriving at an optimal network structure. This is the approach used by the Cascade-Correlation architecture [6]. (*iii*) Pruning algorithms, we can train a network with a relatively large number of hidden neurons and then prune out the least significant connections, either by removing individual weights or by removing complete units. This is the approach used by the *Optimal Brain Surgeon* (OBS) and the *Weight Decay and Elimination* (WDE).

The OBS method [7] performs weight pruning based on the so-called *weight saliencies*, defined as $S_i = \frac{1}{2}\frac{\omega_i^2}{[\mathbf{H}^{-1}]_{ii}}$, where ω_i denotes the i-th weight (or bias) and $[\mathbf{H}^{-1}]_{ii}$ is the i-th diagonal entry of the inverse of the Hessian matrix. The WDE algorithm originates from a regularization method that modifies the error function by the introduction of a term that penalize large weights [8].

These methods require significant computational efforts. For example, even if we restrict the exhaustive search to a specific class of architectures (e.g. one-hidden-layered MLP) it is still a burdensome task. The OBS algorithm demands the computationally-intensive inversion of the Hessian matrix of the error function. The WDE algorithm, by its turn, requires the specification of a user-defined regularization parameter (λ) [1]. In this paper, we introduce an efficient method for pruning unnecessary weights of a trained multilayer Perceptron (MLP) without the need of matrix inversions and any additional regularization parameter. The proposed method is based on the correlation analysis between the errors of the output neurons and the errors backpropagated to the hidden neurons. Repeated application of the method leads eventually to the complete elimination of all connections of a neuron. Computer simulations using real-world data are carried out to compare the proposed method with OBS and WDE algorithms.

The remainder of the paper is organized as follows. In Section 2 we briefly review the back-propagation algorithm. In the section 3 the proposed method is introduced and its main properties are discussed. Simulations are then presented in Section 4. We conclude the paper in Section 5 with a summary of the achievements and suggestions for further developments.

2 The Back-Propagation Algorithm in a Nutshell

We describe next the backpropagation algorithm used for training a fully connected, one-hidden-layered MLP. At time step t, the activation of a hidden neuron is computed as

$$u_i^{(h)}(t) = \sum_{j=1}^{P} w_{ij}(t)x_j(t) - \theta_i(t) = \sum_{j=0}^{P} w_{ij}(t)x_j(t), \quad i = 1, \ldots, Q \qquad (1)$$

where w_{ij} is the synaptic weight connecting the input j to the hidden neuron i, $\theta_i(t)$ is the threshold of the hidden neuron i, Q $(2 \leq Q < \infty)$ is the number of hidden neurons and P is the dimension of the input vector (excluding the threshold). For simplicity, we set $x_0(t) = -1$ and $w_{i0} = \theta_i^{(h)}(t)$.

The output of neuron i is then defined by

$$y_i^{(h)}(t) = \varphi_i\left[u_i^{(h)}(t)\right] = \varphi_i\left[\sum_{j=0}^{P} w_{ij}(t)x_j(t)\right], \qquad (2)$$

where $\varphi_i(\cdot)$ is a sigmoidal function. Similarly, the output values of the output neurons are given by

$$y_k^{(o)}(t) = \varphi_k\left[u_k^{(o)}(t)\right] = \varphi_k\left[\sum_{i=0}^{Q} m_{ki}(t)y_i^{(h)}(t)\right], \qquad (3)$$

where m_{ki} is the synaptic weight connecting the hidden neuron i to the output neuron k ($k = 1, \ldots, M$), and $M \geq 1$ is the number of output neurons. We set $y_0(t) = -1$ and $m_{k0} = \theta_k^{(o)}(t)$, where $\theta_k^{(o)}(t)$ is the threshold of neuron k.

The backward pass starts at the output layer by propagating the error signals, $e_k^{(o)}(t) = d_k(t) - y_k^{(o)}(t)$, where $d_k(t)$ is the target output value for neuron k, toward the hidden layer. The so called *local gradient* of neuron k is given by

$$\delta_k^{(o)}(t) = \varphi_k'\left[u_k^{(o)}(t)\right] e_k^{(o)}(t). \qquad (4)$$

where $\varphi_k'\left[u_k^{(o)}(t)\right] = \partial\varphi_k/\partial u_k^{(o)}$. Similarly, the local gradient $\delta_i^{(h)}(t)$ of the hidden neuron i is computed as

$$\delta_i^{(h)}(t) = \varphi_i'\left[u_i^{(h)}(t)\right] \sum_{k=1}^{M} m_{ki}(t)\delta_k^{(o)}(t) = \varphi_i'\left[u_i^{(h)}(t)\right] e_i^{(h)}(t), \quad i = 0, \ldots, Q, \qquad (5)$$

where the term $e_i^{(h)}(t)$ plays the role of a *backpropagated* or *projected* error signal for the hidden neuron i, since such "hidden" error signals are linear combinations of the "true" error signals computed for the output neurons.

Finally, the synaptic weights of the output neurons are updated according to the following rule

$$m_{ki}(t+1) = m_{ki}(t) + \eta\delta_k^{(o)}(t)y_i^{(h)}(t), \quad i = 0, \ldots, Q, \qquad (6)$$

where $0 < \eta \ll 1$ is the learning rate. The weights of the hidden neurons are, by their turn, adjusted through a similar learning rule

$$w_{ij}(t+1) = w_{ij}(t) + \eta\delta_i^{(h)}(t)x_j(t), \quad j = 0, \ldots, P. \qquad (7)$$

One complete presentation of the entire training set during the learning process is called an *epoch*. Many epochs may be required until the convergence of the back-propagation algorithm is verified. Thus, it is good practice to randomize the order of presentation of training examples from one epoch to the next, in order to make the search in the weight space stochastic over the learning cycles.

A simple (and naive) way of evaluating convergence is through the average squared error

$$\varepsilon_{train} = \frac{1}{2N}\sum_{t=1}^{N}\sum_{k=1}^{M}\left[e_k^{(o)}(t)\right]^2 = \frac{1}{2N}\sum_{t=1}^{N}\sum_{k=1}^{M}\left[d_k(t) - y_k^{(o)}(t)\right]^2, \tag{8}$$

computed at the end of a training run using the training data vectors. If it falls below a prespecified value then convergence is achieved. The generalization performance of the MLP should be evaluated on a testing set, which contains examples not seen before by the network.

3 The Proposed Methodology

Initially, the user should train a MLP network with a relative large number of hidden neurons. The network is trained to give the lowest possible value for ε_{train}. Recall that large MLPs are prone to overfitting due to local optimization of their cost function, which can lead to a poor generalization performance. In principle, the application of a pruning procedure would discard redundant connections of an overparameterized network, thus providing a smaller model with equivalent or better generalization performance than the original one.

The main idea behind the pruning procedure to be described is to maintain those connections that produce higher correlations between the errors in a given layer and the errors that are backpropagated to the preceding layer. Connections associated to lower error correlations are candidates to be discarded.

3.1 Pruning Hidden-to-Output Layer Weights

The method starts with the training data set being submitted once again to the trained network. No weight updating is allowed from this stage on. Once all the N training examples are presented, construct the following error matrices

$$\mathbf{E}_o = \begin{bmatrix} e_1^{(o)}(1) & e_2^{(o)}(1) & \cdots & e_M^{(o)}(1) \\ e_1^{(o)}(2) & e_2^{(o)}(2) & \cdots & e_M^{(o)}(2) \\ \vdots & \vdots & \vdots & \vdots \\ e_1^{(o)}(N) & e_2^{(o)}(N) & \cdots & e_M^{(o)}(N) \end{bmatrix} \text{ and } \mathbf{E}_h = \begin{bmatrix} e_0^{(h)}(1) & e_1^{(h)}(1) & \cdots & e_Q^{(h)}(1) \\ e_0^{(h)}(2) & e_1^{(h)}(2) & \cdots & e_Q^{(h)}(2) \\ \vdots & \vdots & \vdots & \vdots \\ e_0^{(h)}(N) & e_1^{(h)}(N) & \cdots & e_Q^{(h)}(N) \end{bmatrix} \tag{9}$$

The rows of the matrix $\mathbf{E}_o$ correspond to the errors generated by the output neurons for a given training example. Hence, this matrix is denoted *the matrix of output errors*, in contrast to the matrix $\mathbf{E}_h$, whose rows correspond to the backpropagated errors associated to the hidden neurons. In particular, the first column of $\mathbf{E}_h$ corresponds to backpropagated errors associated with the thresholds $m_{k0} = \theta_k^{(o)}$, $k = 1, \ldots, M$.

Table 1. Procedure for Pruning Hidden-to-Output Layer Weights

```
1. Set l = 1;                      // set the ranking index to 1
2. WHILE l ≤ L DO                  // begin the pruning loop
   2.1. Set a = m_{r_l};           // save current value
   2.2. Set m_{r_l} = 0;           // set the weight to zero
   2.3. Compute J_train;           // Performance Index on training set
   2.4. IF J_train < J_tol,        // We set J_train = CR_train
        THEN Set m_{r_l} = a;      // recover former value
             Set l = L + 1;        // stop pruning
        ENDIF
   2.5. Set l = l + 1;             // continue pruning
 ENDWHILE
```

The second step consists in the computation of the following matrix product

$$\mathbf{C}_{oh} = \mathbf{E}_o^T \mathbf{E}_h, \quad (10)$$

where the superscript T denotes the transpose of a matrix. Note that the (k, i)-th entry of $\mathbf{C}_{oh}$, denoted by $\mathbf{C}_{oh}[k, i]$, corresponds to the scalar product (correlation) of the k-th column of $\mathbf{E}_o$ with the i-th column of $\mathbf{E}_h$,

$$\mathbf{C}_{oh}[k, i] = \sum_{t=1}^{N} e_k^{(o)}(t) e_i^{(h)}(t), \quad (11)$$

for $k = 1, \ldots, M$, and $i = 0, \ldots, Q$.

The third step requires the sorting of the entries $\mathbf{C}_{oh}[k, i]$ in ascending order

$$\mathbf{C}_{oh}[\mathbf{r}_1] < \mathbf{C}_{oh}[\mathbf{r}_2] < \cdots < \mathbf{C}_{oh}[\mathbf{r}_L], \quad (12)$$

where the vector $\mathbf{r}_l = (k_l, i_l)$ contains the coordinates of the entry occupying position l in the ranking, and $L = \dim(\mathbf{C}_{oh}) = M \times (Q + 1)$ is the number of entries in $\mathbf{C}_{oh}$.

The fourth step involves the execution of the pruning procedure shown in Table 1. In this table, J_{train} denotes a given index used to evaluate the network performance on the training data, such as the average squared error ε_{train} or the classification rate CR_{train}. The constant J_{tol} is a user-defined value.

If $J_{train} = \varepsilon_{train}$, then J_{tol} is the maximum error value permitted for the training data. So, we eliminate a given connection $m_{\mathbf{r}_l}$ only if the value of J_{train}, computed after the elimination of that connection, remains lower than J_{tol}.

If $J_{train} = CR_{train}$ instead, then J_{tol} is the minimum recognition rate allowed for the training data. In this case, we eliminate a given connection $m_{\mathbf{r}_l}$ only if the value of J_{train}, computed after the elimination of that connection, still remains higher than the prespecified J_{tol}. This is our choice for Table 1.

3.2 Pruning Input-to-Hidden Layer Weights

For pruning the weights that connect the input units to the hidden neurons, w_{ij}, we need to propagate backwards the errors of the hidden neurons, $e^{(h)}(t)$, in order to obtain the errors projected onto the input units

Table 2. Procedure for Pruning Input-to-Hidden Layer Weights

1. Set $l = 1$;	`// set the ranking index to 1`
2. **WHILE** $l \leq L$ **DO**	`// begin the pruning loop`
2.1. Set $b = w_{\mathbf{r}_l}$;	`// save current value`
2.2. Set $w_{\mathbf{r}_l} = 0$;	`// set the weight to zero`
2.3. Compute J_{train};	`// Performance Index on training set`
2.4. **IF** $J_{train} < J_{tol}$,	`// We set` $J_{train} = CR_{train}$
THEN Set $w_{\mathbf{r}_l} = b$;	`// recover former value`
Set $l = L + 1$;	`// stop pruning`
ENDIF	
2.5. Set $l = l + 1$;	`// continue pruning`
ENDWHILE	

$$e_j^{(i)}(t) = \sum_{i=1}^{Q} w_{ij}(t)\delta_i^{(h)}(t), \quad j = 0, \ldots, P. \tag{13}$$

After the presentation of N training vectors, the resulting errors can be organized into an error matrix

$$\mathbf{E}_i = \begin{bmatrix} e_0^{(i)}(1) & e_1^{(i)}(1) & \cdots & e_P^{(i)}(1) \\ e_0^{(i)}(2) & e_1^{(i)}(2) & \cdots & e_P^{(i)}(2) \\ \vdots & \vdots & \vdots & \vdots \\ e_0^{(i)}(N) & e_1^{(i)}(N) & \cdots & e_P^{(i)}(N) \end{bmatrix}_{N\times(P+1)} . \tag{14}$$

Similarly to the procedure described in the previous section, we need to compute the following matrix product

$$\mathbf{C}_{hi} = \mathbf{E}_h^T \mathbf{E}_i, \tag{15}$$

where the (i, j)-th entry of $\mathbf{C}_{hi}$, denoted by $\mathbf{C}_{hi}[i, j]$, is given by

$$\mathbf{C}_{hi}[i, j] = \sum_{t=1}^{N} e_i^{(h)}(t) e_j^{(i)}(t), \tag{16}$$

for $i = 1, \ldots, Q$, and $j = 0, \ldots, P$. Then, we sort the entries $\mathbf{C}_{hi}[i, j]$ in ascending order

$$\mathbf{C}_{hi}[\mathbf{s}_1] < \mathbf{C}_{hi}[\mathbf{s}_2] < \cdots < \mathbf{C}_{hi}[\mathbf{s}_L], \tag{17}$$

where the vector $\mathbf{s}_l = (i_l, j_l)$ contains the coordinates of the entry occupying position l in the ranking, and $L = \dim(\mathbf{C}_{hi}) = M \times (P + 1)$ is the number of entries in $\mathbf{C}_{hi}$.

The final step involves the execution of the pruning procedure shown in Table 2. For this table, we also use $J_{train} = CR_{train}$.

For obvious reasons, from now on we refer to the proposed method as the CAPE method (**C**orrelation **A**nalysis of back-**P**ropagated **E**rrors).

Table 3. Numerical results of the application of the pruning methods

Data Set	Architecture	Q	N_c	CR_{train}	CR_{test}	ε_{train}	ε_{test}
Iris	Original	9	75	99.05	93.33	0.0127	0.0858
	CAPE	4	16	99.05	93.33	0.1765	0.2422
	OBS	4	18	99.05	93.33	0.1969	0.2105
	WDE	9	75	99.05	93.33	0.3277	0.3274
Wine	Original	9	156	100	97.47	0.0007	0.0403
	CAPE	4	30	100	94.94	0.2573	0.3569
	OBS	4	39	100	89.87	0.1836	0.2696
	WDE	3	53	100	94.94	0.0025	0.0594

4 Simulations and Discussion

For evaluating the proposed pruning method, initially we have made some preliminary tests using two benchmarking pattern classification data sets (Iris and Wine). The Iris data set is composed of 150 4-dimensional pattern vectors distributed into 3 classes. The Wine data set is composed of 178 13-dimensional vectors also distributed into 3 classes. For the Iris (Wine) data set, 35 (33) patterns per class are randomly selected for training purposes and the remaining ones are used for testing.

Weights and biases are randomly initialized within the range -0.5 to $+0.5$. All neurons use the hyperbolic tangent activation function and the inputs are normalized to the range of network activations. The output target vectors use the 1-out-of-M binary encoding. The learning rate was set to $\eta = 0.001$ for all simulations.

Network pruning is carried out through application of the CAPE method, as described in Section 3. Numerical results are shown in Table 3. In this table, N_c is the number of connections, CR_{train} and CR_{test} stand for the classification rate for the training and testing data, respectively.

For the Iris problem, CR_{train} and CR_{test} were preserved with the application of CAPE. For the Wine data set, only CR_{train} was preserved, but the achieved CR_{test} is still acceptable. It is worth noting that the CAPE method effectively reduced the number of connections (N_c) in both cases, achieving equivalent or better generalization performances than the standard OBS and WDE methods.

As a more demanding classification task, we have used a biomedical data set kindly made available for this research by the *Group of Applied Research in Orthopaedics* (GARO) of the *Centre Médico-Chirurgical de Réadaptation des Massues*, Lyon, France. The task consists in classifying patients as belonging to one out of three categories: Normal (100 patients), Disk Hernia (60 patients) or Spondylolisthesis (150 patients).

Each patient is represented in the database by six biomechanical attributes derived from the shape and orientation of the pelvis and cervical, thoracic, and lumbar spine: pelvic incidence, pelvic tilt, sacral slope, pelvic radius, lumbar lordosis angle and grade of spondylolisthesis. Readers interested in more

Table 4. Numerical results of the successive application of the pruning methods

	Q	N_c	CR_{train}	CR_{test}	ε_{train}	ε_{test}	AIC
Architecture 1	24	243	89.68	87.50	0.1132	0.1274	490.36
CAPE	19	126	92.06	86.96	0.1662	0.1273	255.59
OBS	22	134	92.06	86.96	0.1777	0.1598	271.46
PWM	21	121	89.68	84.24	0.1733	0.1544	245.51
WDE	24	219	78.57	80.98	0.4237	0.4176	439.72
Architecture 2	18	183	96.03	86.41	0.0616	0.1891	371.57
CAPE	14	98	95.24	88.59	0.1089	0.1632	200.43
OBS	14	91	95.24	87.50	0.1529	0.1827	185.76
PWM	16	96	96.03	88.59	0.1453	0.1918	195.86
WDE	16	158	85.71	90.76	0.1747	0.1132	319.49
Architecture 3	13	133	93.65	81.52	0.0765	0.2311	271.14
CAPE	13	103	93.65	83.70	0.1211	0.2113	210.22
OBS	13	102	93.65	80.43	0.1267	0.2410	208.13
PWM	13	96	93.65	78.80	0.1356	0.2348	196.00
WDE	12	108	65.08	69.02	0.4323	0.4292	217.68

details about these and other related attributes as well as their relationships to pathologies of the vertebral column are referred to [9].

Numerical features were measured by an orthopedic specialist from X-ray images of the vertebral column. From a total of 310 resulting pattern vectors, the training set is formed by randomly selecting 42 pattern vectors per class and the remaining 184 examples are used for testing purposes. The learning rate is set to 0.001. For this simulation, we set $PI_{tol} = CR_{tol} = 85\%$. This value serves as a reference for the minimum acceptable recognition rate for training and testing purposes. That is, even if a pruned network has produced a CR_{train} value higher than CR_{tol}, its CR_{test} value should also be higher than CR_{tol}.

Network pruning is carried out progressively through successive applications of the CAPE method. Numerical results are shown in Table 4. In this table, AIC stands for Akaike's Information criterion[1] and the acronym PWM stands for *Pruning by Weight Magnitude*, which is a pruning method based on the elimination of small magnitude weights [1]. Weights are sort in increasing order of magnitude. Starting from the smallest weight, a given weight is pruned as long as its elimination does not decrease CR_{train} to a value below CR_{tol}.

The CAPE algorithm has achieved a "pruned Architecture 1" with $Q = 19$ hidden neurons and $N_c = 126$ connections, but with less connections than a fully-connected MLP architecture with the same number of hidden neurons (i.e. $N_c = 193$). The OBS method has ended with a pruned network with $Q = 22$ hidden neurons and $N_c = 134$. The final classification rates (CR_{train} and CR_{test}) of both methods were coincidentally the same.

Important facts are worth mentioning with respect to the PWM method. Firstly, despite the fact that it has ended with the network with the smallest

[1] The AIC has the follow structure $AIC = -2\ln(\varepsilon_{train}) + 2N_c$ [8].

number of connections ($N_c = 121$) for $Q = 21$ hidden neurons, its CR_{train} and CR_{test} values were the worst ones. Secondly, due the small number of connections it has achieved the lowest AIC value, wrongly indicating this architecture as the best one for this problem. These results indicate that the AIC index is possibly not the best model selection method for classification problems, since it is based on the average squared error only, not on classification rates.

Since the performance of the Architecture 1 pruned by the CAPE method remained at acceptable levels, we start to wonder what would have happened if we had started the pruning process with a fully-connected, one-hidden-layered MLP, denoted Architecture 2, with a number of hidden neurons close to the final value achieved for Architecture 1 pruned by the CAPE method. Our rationale in doing that is to verify if there were still room for further pruning of connections. The training parameters for the Architecture 2 are the same as before.

The application of the CAPE and the OBS method to Architecture 2 led to the same number of hidden neurons ($Q = 14$), with the latter achieving less connections than the former. However, the classification rate CR_{test} for the CAPE method was higher. The AIC model, in this case, indicated the OBS method as the one providing the best pruned architecture, despite the fact that the CAPE method has produced the highest classification rate. This was mainly due to the smaller number of connections achieved by the OBS method.

The PWM method did quite well in terms of classification rates, ending with a pruned architecture with $Q = 16$ number of neurons and $N_c = 96$ connections. It is worth mentioning, however, that the PWM is highly sensitive to weight initialization, and the numerical results shown in Table 4 corresponds to the best one obtained after 10 training trials. The CAPE and the OBS method are much less sensitive to weight initialization.

As the recognition rate CR_{test} for the pruned Architecture 2, irrespective of the pruning method, also remained above the permitted value $CR_{tol} = 85\%$, we decided to start pruning from another fully-connected, one-hidden-layered MLP, with $Q = 13$. However, no method was able to reduce the number of hidden neurons, only the number of connections. Anyway, all the pruned networks achieved recognition rates CR_{train} and CR_{test} below the permitted value. Thus, one can infer that the architectures best suited to the problem are the pruned ones obtained from Architecture 2.

The best results obtained for the application of the WDE algorithm to the Architectures 1 and 2 are also shown in Table 4. We set $\lambda = 0.001$ after some experimentation. The performance of the WDE algorithm was very poor for all the architectures. For the sake of completeness, we also implemented the *Exhaustive search plus early stopping* method for model selection purposes of a fully-connected one-hidden-layered MLP. After a long period of trials we ended with a network with $Q = 12$ ($N_c = 123$) and $CR_{test} = 83.57$. Despite the fact that the number of hidden neurons found by this method is lower than the one provided by the CAPE method ($Q = 14$ and $N_c = 98$), the latter achieved an equivalent recognition rate with less connection weights.

5 Conclusions

In this paper we introduced the CAPE method, an easy-to-apply and efficient procedure for pruning unnecessary weights of a trained multilayer Perceptron (MLP). The CAPE method is based on the correlation analysis of the errors produced by the output neurons and the backpropagated errors associated with the hidden neurons.

Simulations using real-world data indicate that, in terms of performance, the CAPE method compares favorably with standard pruning techniques, such as the OBS, WDE and Early Stopping, but demands much lower computational efforts. A pruning method based on the magnitude of the weights was also evaluated. It can give good results sometimes, but it is highly sensitive to weight initialization in comparison with the OBS and the CAPE methods.

Currently, we are evaluating the CAPE method in a number of benchmarking classification problems, in order to gather more evidence of its superior performance for model selection.

Acknowledgements. The authors thank CAPES for its financial support via a PRODOC grant.

References

1. Bishop, C.M.: Neural Networks for Pattern Recognition. Oxford University Press, Oxford (1995)
2. Seghouane, A.-K., Amari, S.-I.: The AIC criterion and symmetrizing the Kullback-Leibler divergence. IEEE Transactions on Neural Networks 18(1), 97–106 (2007)
3. Curry, B., Morgan, P.H.: Model selection in neural networks: Some dificulties. European Journal of Operational Research 170(2), 567–577 (2006)
4. Nakamura, T., Judd, K., Mees, A.I., Small, M.: A comparative study of information criteria for model selection. International Journal of Bifurcation and Chaos 16(8), 2153–2175 (2006)
5. Xiang, C., Ding, S.Q., Lee, T.H.: Geometric interpretation and architecture selection of the MLP. IEEE Transactions on Neural Networks 16(1), 84–96 (2005)
6. Fahlman, S.E., Lebiere, C.: The cascade-correlation learning architecture. In: Touretzky, D.S. (ed.) Advances in Neural Information Processing Systems, vol. 2, pp. 524–532. Morgan Kaufmann, San Mateo (1990)
7. Hassibi, B., Stork, D.G.: Second order derivatives for network pruning: Optimal brain surgeon. In: Hanson, S.J., Cowan, J.D., Giles, C.L. (eds.) Advances in Neural Information Processing Systems, vol. 5, pp. 164–171. Morgan Kaufmann, San Mateo, CA (1993)
8. Principe, J.C., Euliano, N.R., Lefebvre, W.C.: Neural and Adaptive Systems. John Wiley & Sons, West Sussex, England (2000)
9. Berthonnaud, E., Dimnet, J., Roussouly, P., Labelle, H.: Analysis of the sagittal balance of the spine and pelvis using shape and orientation parameters. Journal of Spinal Disorders & Techniques 18(1), 40–47 (2005)

Reinforcement Learning for Cooperative Actions in a Partially Observable Multi-agent System

Yuki Taniguchi, Takeshi Mori, and Shin Ishii

Graduate School of Information Science,
Nara Institute of Science and Technology (NAIST)
Takayama 8916-5, Ikoma, 630-0192, Japan
{yuki-t, tak-mori, ishii}@is.naist.jp

Abstract. In this article, we apply a policy gradient-based reinforcement learning to allowing multiple agents to perform cooperative actions in a partially observable environment. We introduce an auxiliary state variable, an internal state, whose stochastic process is Markov, for extracting important features of multi-agent's dynamics. Computer simulations show that every agent can identify an appropriate internal state model and acquire a good policy; this approach is shown to be more effective than a traditional memory-based method.

1 Introduction

The achievement of cooperative actions in multi-agent systems by machine learning techniques is an interesting problem both in technological and biological viewpoints [7]. Reinforcement learning (RL) [8] is one possible approach to this problem, in which the agents are modeled to acquire cooperative policies autonomously through the interactions with other agents, and has been widely studied recently [5] [11]. Many RL techniques applied to multi-agent systems assume completely observable environments, formulated as Markov decision processes (MDPs), where every agent can perceive the complete state of other agents and select actions depending on the state [5] [11] [6]. Many real-world problems such as autonomous control of multiple robots cannot be formulated as MDPs, however, because the robot sensors are often attached only to each robot so that the robot cannot perceive the complete state of the other robots.

Such problems can be formulated as partially observable Markov decision processes (POMDPs) [4], and have often been solved by RL using belief state techniques, in which the value function that represents the goodness of the state is estimated over the belief space [9] [12]. However, those techniques suffer from several difficulties even with an effective approximation [3]; the dimensionality of the belief space is usually high. Moreover, the state transition model is necessary for estimation of the belief state, but the state transition (dynamics) of other agents are often unknown in many real-world multi-agent problems. These difficulties make the RL process unstable and inefficient.

Recently, a policy gradient algorithm with finite state controllers (FSCs) has been proposed for solving POMDPs, called IState-GPOMDP [1]. The FSC is a

J. Marques de Sá et al. (Eds.): ICANN 2007, Part I, LNCS 4668, pp. 229–238, 2007.

probabilistic policy possessing an internal state; it is a probabilistic automaton which represents the probability of an action selection (controller) and a next internal-state (own dynamics). In the IState-GPOMDP, the transition probability of the internal state embedded in the FSC is identified by the policy gradient algorithm together with the optimization of the policy. The important dynamic characters of the target state space are extracted by learning of the transition probability of the internal-state, which is performed in an irrelevant manner to the underlying dimensionality of the target state space. Such a feature extraction is favorable especially in a multi-agent setting, in which the true state space has often high dimensionality. In the robotic soccer problem [6], for example, the state space is composed of the relative location (distance and angle) between the soccer players. Then, the dimensionality of the state space can be huge as the number of players increases. Because the effective dimensionality of such a high dimensional system is often much smaller than the dimensionality of the whole state space, as to reflect the dependence between the state valiables, feature extraction by the IState-GPOMDP can be more effective than directly reconstructing the true state space as done by the belief state-based methods.

In this article, in order to allow multiple agents to perform cooperative actions in a partially observable multi-agent environment, we apply the IState-GPOMDP and show that its learning process is faster and stabler than that by a memory-based policy gradient method. Moreover, we show the important characters of the agent's reward-maximization can be extracted from interactions with other agents as the transition models of the internal state. The IState-GPOMDP has been applied only to single-agent partially observable systems or multi-agent problems where the internal state is unnecessary to be resolved [1]. This is the first study to apply the policy gradient algorithm with FSCs to a partially observable multi-agent system.

2 Partially Observable Multi-agent Systems and FSCs

Each agent is assumed to select its actions according to its own parameterized policy, which is optimized so as to maximize the long term average reward. As a natural model for multi-agent cooperative problems, we consider a finite POMDP, consisting of a set of real states $\mathcal{S}$, a set of actions $\mathcal{A}$ which agent can take, a set of observations $\mathcal{O}$ which each agent can perceive, and a numerical reward $r \in \mathcal{R}$. For simplicity, real states, actions, observations and rewards are represented individually for each agent. In the partially observable multi-agent setting, at time t, the real state $s_t \in \mathcal{S}$ cannot be perceived directly by the agent, but instead, the observation o_t is drawn from the probability $P(o_t|s_t)$ conditioned on the state $s_t \in \mathcal{S}$. Because the uncertainty is introduced by this observation process, the reactive policy, which depends only on the observation, does not exhibit the best performance. Then, we employ the model in which an internal state $y_t \in \mathcal{Y}$ is added to the POMDP definition above as an additional input to the policy; such a policy (controller) is called a finite state controller (FSC). The stochastic process over the internal state is represented by the transition probability $P(y_{t+1}|y_t, o_t)$, where y_t and y_{t+1} are the internal states at time t and $t+1$, respectively. In this case the policy is the mapping from observation o_t and internal state y_t to action a_t. Figure 1(a) represents the graphical model of a single-agent POMDP employing an internal state. In this model, the colored nodes: the observation, the action and the internal state, can be observed at each time by the agent. On the other hand, Figure 1(b) shows the graphical model of a

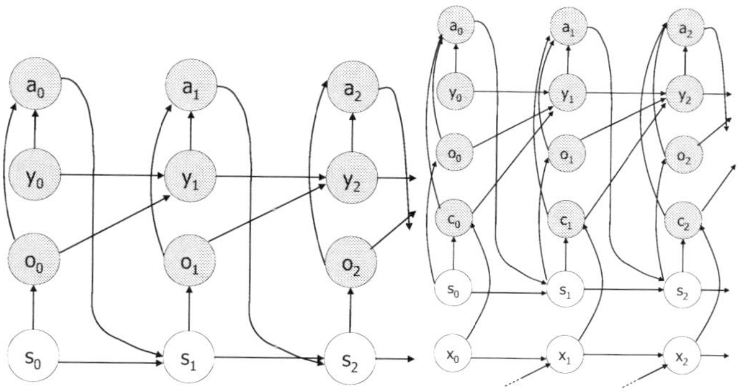

(a) (b)

Fig. 1. (a) Graphical model of a POMDP with an internal state for a single agent system. The current action a_t and the following internal state y_{t+1} depend on the current internal state y_t and the observation o_t. (b) Graphical model modified to deal with a multi-agent system. The current action a_t and the following internal state y_{t+1} depend not only on the current internal state y_t and the observation o_t but also on the relative observation c_t.

two-agents POMDP, which is the simplest control model of partially observable multi-agent systems. In this model, a 'relative' observation of the other agent $c \in \mathcal{C}$ and a real state of the other agent $x \in \mathcal{X}$ are added to the model in Figure 1(a). Here, the relative observation is assumed to be dependent on the real states of the two agents, so that at time t c_t is drawn from the relative observation probability $P(c_t|s_t, x_t)$. According to the graphical model in Figure 1(b), the FSC is represented by the internal-state transition probability $P(y_{t+1}|y_t, o_t, c_t)$ and the action selection probability $P(a_t|y_t, o_t, c_t)$.

3 IState-GPOMDP for Cooperative Actions

In this section, we explain the IState-GPOMDP proposed by Aberdeen and Baxter [1], and apply it to our partially observable multi-agent system. The IState-GPOMDP is a policy gradient-based RL method which does not seek to estimate the value function, but it adjusts the policy parameters θ and the internal-state transition parameters ϕ directly to maximize the average reward. In our particular setting, two agents learn the parameters individually. For each agent, the internal-state transition probability and the action selection probability are represented as

$$\begin{aligned} &P(y_{t+1}|y_t, o_t, c_t; \phi) \\ &P(a_t|y_t, o_t, c_t; \theta). \end{aligned} \quad (1)$$

The objective of each agent is to seek the parameters ϕ and θ that maximize the long term average reward:

$$\eta(\phi, \theta) := \lim_{T\to\infty} \frac{1}{T} E_{\phi,\theta}\left[\sum_{t=1}^{T} r_t\right], \quad (2)$$

where $E_{\phi,\theta}$ denotes the expectation with respect to the trajectory $(s_0, y_0, o_0, c_0, a_0), (s_1, y_1, o_1, c_1, a_1), \ldots$ prescribed by the parameters ϕ and θ. The internal state and the action are assumed to be drawn from the parametrized Bolzmann probabilities as

$$\begin{aligned} P(y_{t+1} = l|y_t = i, o_t = j, c_t = k) &:= \frac{\exp(\phi_{i,j,k,l})}{\sum_{y\in|\mathcal{Y}|}\exp(\phi_{i,j,k,y})} \\ P(a_t = m|y_t = i, o_t = j, c_t = k) &:= \frac{\exp(\theta_{i,j,k,m})}{\sum_{a\in|\mathcal{A}|}\exp(\theta_{i,j,k,a})}. \end{aligned} \quad (3)$$

In this method, the policy gradient $\nabla\eta$ is estimated by gathering information through the interactions with the other agent along the trajectory. The gradient estimate Δ_t at time t is given by the product of the immediate reward r_t and the eligibility trace at that time. The following table shows the pseudocode of the IState-GPOMDP applied to the multi-agent control problem to achieve cooperative actions by an agent_A and an agent_B.

0: while
1: until the terminal condition, i.e., while $t < T$
2: For agent_A, observe o_t from $P(o_t|s_t)$ and c_t from $P(c_t|s_t, x_t)$.
3: Draw y_{t+1} from $P(y_{t+1}|y_t, o_t, c_t, \phi)$ and a_t from $P(a_t|y_t, o_t, c_t, \theta)$.
4: Update the estimation of the policy gradient Δ_t with respect to ϕ and θ.
5: Repeat $2 \sim 4$ for agent_B.
6: Perform action a_t for agent_A and one for agent_B.
7: $t++$.
8: end
9: $\epsilon\Delta_t$ is added to the parameters ϕ and θ, where ϵ is the learning rate.
10: end

In this algorithm, the convergence of ϕ and θ to a global or local optimum is guaranteed under some moderate conditions [10] [1]. The number of possible values of the multinomial internal-state is the hyperparameter of this RL scheme and is set for each agent in advance.

4 Computer Simulation

In this section, we propose an instance of partially observable multi-agent control problems, called the synchronized-cooperative goal (SCG) problem. In this problem, the agent needs to synchronize with the other agent by taking cooperative actions, in order to maximize the long term reward. Then, we apply the IState-GPOMDP with the multi-agent setting to this problem.

4.1 Synchronized-Cooperative Goal Problem

In the SCG problem, depicted in Figure 2, each agent is required to synchronize with the other agent to receive the positive reward. Two agents move from a start column to a goal column. Agent_A moves on the upper passage, and agent_B on the lower passage. There are three passage settings whose start and

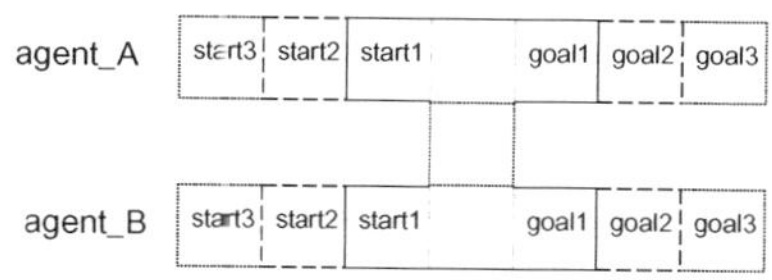

Fig. 2. The synchronized-cooperative goal problem. A agent_A moves on the upper passage and a agent_B moves on the lower passage.

goal columns are (start1, goal1), (start2, goal2) and (start3, goal3). The upper and lower passages are independently and randomly selected from these passage settings before each learning episode starts. $\mathcal{A}=$ {GO, WAIT} is the set of actions of the two agents. When the agent takes the GO action, it moves to the right column, and when it takes the WAIT action, it stays at the same column. When the start and the goal are (start2, goal2) for example, the agent must take the GO action four times to reach the goal. Each agent cannot observe the passage setting and the current state of either agent, but can observe their own observation and the 'relative' observation at each time step as follows. $\mathcal{O}=$ {NOT_CENTER, CENTER} is the set of observations about their own state. The agent can observe CENTER when being at the central column, or NOT_CENTER at another column. $\mathcal{C}=$ {NOT_SYNCH, SYNCH} is the set of relative observations; the two agents can observe SYNCH when they are simultaneously at the central column on their passage, or NOT_SYNCH when either agent is at another column. When either agent reaches the goal, a single episode ends. In this case, if both agents reach the goal at the same time after they have observed SYNCH, the reward $r_t = +1$ is given to both agents. If one agent has reached the goal before the other agent reaches or has reached without having observed SYNCH, the reward $r_t = -0.1$ is given to both agents. After that, a new setting is selected randomly for both agents, and the agents are initialized to be placed at their start columns to start the next episode. Since the observations of the agents are restricted, this problem is an imperfect perception problem, which is difficult to be solved especially in multi-agent settings [7].

Figure 3 shows an example of optimal control in a single episode. In this optimal control, each agent continues to select the GO action until reaching the central column, that is, whenever the pair of observations (o_t, c_t) is (NOT_CENTER,

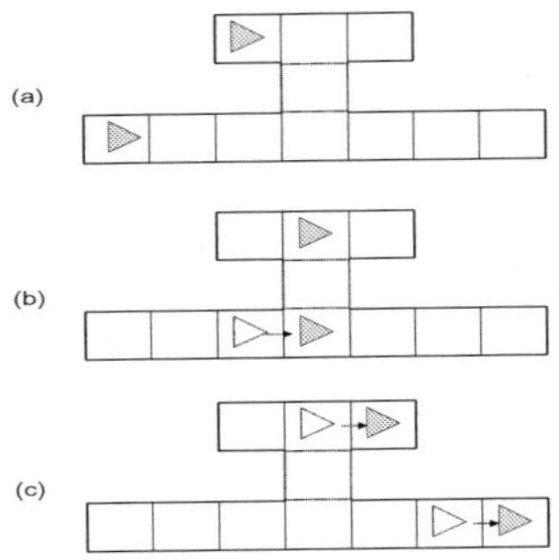

Fig. 3. Example cooperative actions based on an optimal policy for the synchronized-cooperative goal problem. (a) The upper and lower passages are selected randomly. Then, both agents are initialized to be at their start columns. (b) Both agents move to their central columns. The agent_A waits at the central column and collects information to synchronize the following actions. (c) The terminal state of this episode. Both agents reach the goal at the same time, then both agents receive the positive reward.

NOT_SYNCH). Once the agent observes (CENTER, NOT_SYNCH) or (CENTER, SYNCH), it should select the GO action or the WAIT action appropriately. When agent_A arrives at the central column earlier than agent_B, agent_A should wait and collect information of the number of time steps until agent_B comes. While agent_B takes the GO action after observing SYNCH at the central column, agent_A should take the WAIT action to reach the goal simultaneously with agent_B based on the counting of time steps before observing SYNCH. To successfully complete the task, each agent has to extract the characters of the task, namely, what kind of passage settings the two agents are performing, and where they are.

4.2 Experimental Results

We introduce the internal state model (section 2), and apply the IState-GPOMDP (section 3), to our SCG problem above. We examine the performance of the IState-GPODP of various numbers of internal-state values and that of a traditional POMDP-RL method. In the following simulation experiments, the number of training episodes in a single RL run was set to $T = 10^5$. The initial state of the internal state was not a learning parameter, but set randomly before each learning episode. The multinomial internal state has full connectivity between the multinomial elements (see Figures 5 and 6), and the parameters representing the connectivity were estimated. To achieve the SCG problem optimally, the number of internal-state values should be four or more.

Figure 4 shows the learning processes of the IState-GPOMDP and a memory-based policy gradient method. According to the latter approach, the stochastic policy with memories: $P(a_t|o_t, c_t, o_{t-1}, c_{t-1}, \dots, o_{t-|\mathcal{G}|}, c_{t-|\mathcal{G}|}; \theta)$ is optimized by GPOMDP [2], where $|\mathcal{G}|$ is the number of memories, that is, the policy is set to be dependent on observations and relative observations in the previous time steps. In this case, the policy was also parametrized as the Bolzmann probability, similar to equation (3). In Figure 4, the learning processes are averaged over 50 runs and smoothed over every 1000 time steps, showing the IState-GPOMDPs was superior to the memory-based policy gradient method. In the IState-GPOMDP, the larger the number of internal-state values became, the larger the average reward became whereas slightly heavier the computation cost was. In the memory-based method, on the other hand, the increase in the number of memories made the computation effort large but did not increase the average reward. Table 1 shows the comparison of the success rates among 50 learning runs; the success rate is the ratio of episodes in all of which the agents received positive rewards in 1000 time steps with the learned parameters. In the IState-GPOMDP, as the number of internal-state values increased, the success rate increased. In contrast, the memory-based policy gradient method could not achieve the optimal policy with any number of memories.

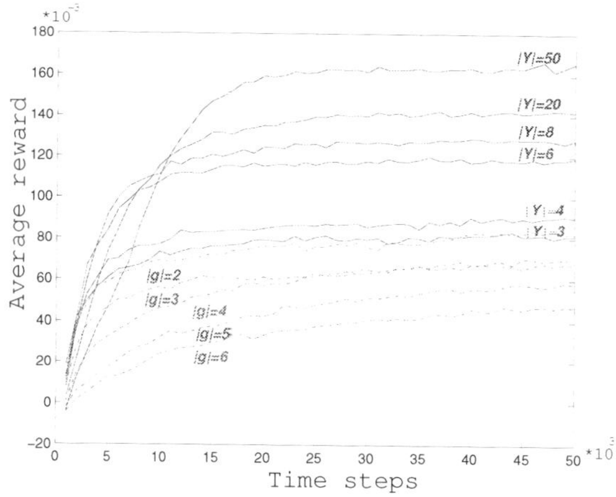

Fig. 4. The learning processes of the IState-GPOMDP and the memory-based reactive policy method with various numbers of internal-state values and memories, respectively

Figure 5 shows example internal-state transition models of agent_A and agent_B acquired by the IState-GPOMDP with four internal-state values (nodes), i.e., $|\mathcal{Y}| = 4$. Each arrow expresses the transition to the internal state with a maximum probability: $y_{t+1} = argmax_{y_{t+1}} P(y_{t+1}|y_t, o_t, c_t)$. The label $(\cdot, \cdot)$ attached to the arrow such as (α, GO) means the pair of observation and action, where the first argument, α, β, or γ, simply denotes the pair of observations (o_t, c_t): $\alpha \equiv$ (NOT_CENTER, NOT_SYNCH), $\beta \equiv$(CENTER, NOT_SYNCH), or $\gamma \equiv$(CENTER, SYNCH), and the second argument, GO or WAIT, means the action with maximum probability: $a_t = argmax_{a_t} P(a_t|y_t, o_t, c_t)$. For example, the label (γ, GO) attached to the arrow from the internal state 1 to 2 in Figure 5(a) means the quintuplet of the internal-state: $y_t = 1$, the observations (o_t, c_t) =(CENTER, SYNCH), the action a_t =GO, and the next internal-state:

Table 1. The performance comparison in terms of success rate by the IState-GPOMDP and the memory-based method. $|\mathcal{G}|$ denotes the number of memories, and 'average' and 'maximum' denote respectively the average and the maximum of the success rate after 10^5 learning episodes over the 50 runs.

Method	$\lvert\mathcal{Y}\rvert$ or $\lvert\mathcal{G}\rvert$	average	maximum	standard deviation
IState-GPOMDP	3	0.4772	0.9263	0.2715
	4	0.5379	1.0	0.2413
	6	0.7064	1.0	0.2040
	8	0.7791	1.0	0.1576
	20	0.8683	1.0	0.1366
	50	0.9406	1.0	0.1040
memory-based method	2	0.4641	0.8517	0.3261
	3	0.4993	0.7308	0.1912
	4	0.4918	0.8187	0.2161
	5	0.4456	0.8145	0.2037
	6	0.3517	0.6989	0.2035

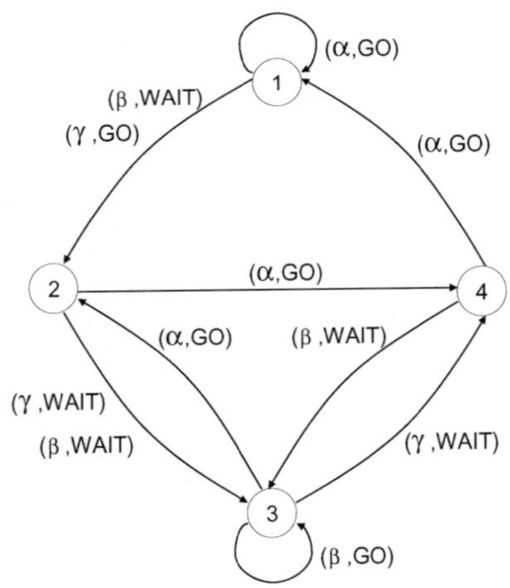

(a)agent_A

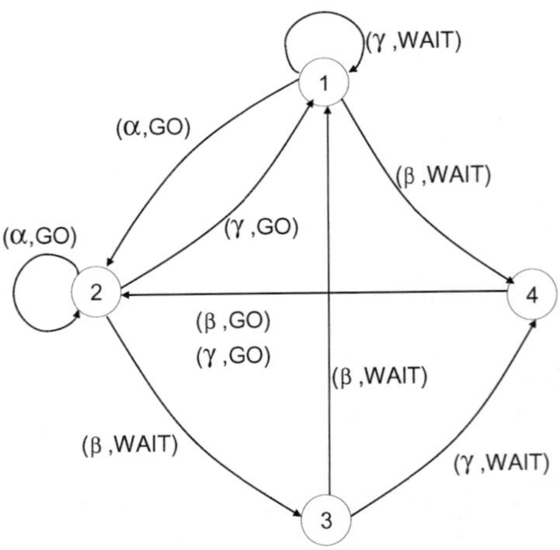

(b)agent_B

Fig. 5. Example internal state transition models when each agent observed (o_t, c_t), y_{t+1} =argmax$_{y_{t+1}} P(y_{t+1}|y_t, o_t, c_t)$ and a_t =argmax$_{a_t} P(a_t|y_t, o_t, c_t)$, in the synchronized-cooperative goal problem. $\{\alpha, \beta, \gamma\}$ represents the pair of observations, i.e., $\alpha\equiv$ (o_t, c_t) =(NOT_CENTER, NOT_SYNCH), $\beta\equiv$ (o_t, c_t) =(CENTER, NOT_SYNCH), and $\gamma\equiv$ (o_t, c_t) =(CENTER, SYNCH).

$y_{t+1} = 2$. In this figure, internal-state transitions whose probability was less than 0.01 are omitted. These models realize one of the optimal policy pairs.

Figure 6 shows example internal-state models with 6 internal-state values (nodes), i.e., $|\mathcal{Y}| = 6$. The model of agent_A uses all 6 nodes, but that of agent_B uses effectively 5 nodes, by isolating the 6-th node.

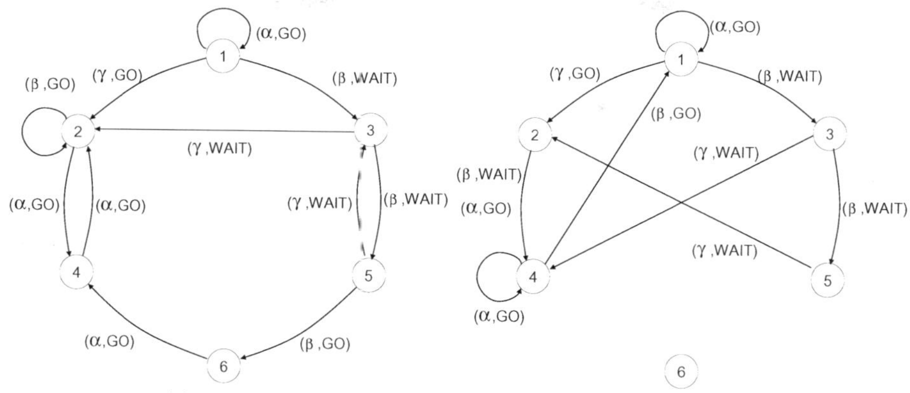

(a)agent_A (b)agent_B

Fig. 6. Example internal state transition models with 6 internal-state values (nodes)

5 Discussion

We applied the IState-GPOMDP to the SCG problem, which is a partially observable multi-agent control problem. We compared the performance of the IState-GPOMDPs with various numbers of internal-state values and that of the memory-based policy gradient method with various numbers of memories, and showed that learning processes of the IState-GPOMDP were faster and stabler than those by the memory-based policy gradient method. According to our experiments, as the number of internal-state values increased, the policy obtained by the IState-GPOMDP was improved, while the learning speed became slightly slow. In the memory-based approach, on the other hand, the performance was not improved by the increase in the number of memories.

When using the Boltzmann policy, the numbers of learning parameters are of $O(|\mathcal{Y}|^2)$ in the IState-GPOMDP and of $O((|\mathcal{O}+\mathcal{C}|)^{|\mathcal{G}|})$ in the memory-based method. In the latter method, the large number of parameters makes the learning slow as depicted in Figure 4. Actually, in the SCG problem, setting of $|\mathcal{G}| = 10$ makes the number of parameters be 118,098, which would lead to the computational difficulty. Accordingly, the IState-GPOMDP has much scalability to deal with high-dimensional multi-agent systems such as the robotic soccer problem [6].

In our present SCG problem, to receive this, the agents need to have observed SYNCH, which is a type of subgoal setting. When we relaxed the positive reward condition into just reaching the goal at the same time, the IState-GPOMDPs could hardly achieve the optimal policies, because the lack of the subgoal setting would make the search space of the optimal policies much larger. We will study an algorithm to achieve optimal policies even without such a subgoal setting.

6 Concluding Remarks

In this article, we applied an RL technique to solving multi-agent control problems based on the formulation of POMDPs. We showed that important features for the agent's reward-maximization can be extracted from interactions with other agents as the transition models of the internal state. We applied this method to a newly-proposed SCG problem and showed the learning process is faster and stabler than that by a memory-based policy gradient method.

Although the objective of the two agents was symmetric in the current SCG problem, our method can be applied to asymmetric problems in which two agents have asymmetric reward settings. Moreover, our method can be applied to control problems constituted by three or more agents. Such extension of the applicability will be shown in our future study.

References

1. Aberdeen, D., Baxter, J.: Scaling Internal State Policy-Gradient Methods for POMDPs. In: Proceedings of the 19th International Conference on Machine Learning, pp. 3–10 (2002)
2. Baxter, J., Bartlett, P.L.: Infinite-Horizon Policy-Gradient Estimation. Journal of Artificial Intelligence Research 15, 229–256 (2001)
3. Hauskrecht, M.: Value-function approximations for partially observable Markov decision processes. Journal of Artificial Intelligence Research 13, 33–99 (2000)
4. Kaelbling, L.P., Littman, M.L, Cassandra, A.R.: Planning and acting in partially observable stochastic domains. Artificial Intelligence 101, 99–134 (1998)
5. Littman, M.L.: Markov games as a framework for multi-agent reinforcement learning. In: Proceedings of the 11th International Conference on Machine Learning, pp. 157–163 (1994)
6. Stone, P., Sutton, R., Singh, S.: Reinforcement Learning for 3 vs. 2 Keepaway. RoboCup-2000: Robot soccer world cup IV 249–258 (2000)
7. Stone, P., Veloso, M.: Multiagent Systems: A Survey from a Machine Learning Perspective. Autonomous Robotics 8(3) (2000)
8. Sutton, R., Barto, A.: An introduction to reinforcement learning. MIT Press, Cambridge (1998)
9. Thrun, S.: Monte Carlo POMDPs. Advances in Neural Information Processing Systems 12, 1064–1070 (2000)
10. Williams, R.J.: Simple statistical gradient-following algorithms for connectionist reinforcement learning. Machine Learning 8, 229–256 (1992)
11. Whitehead, S.D.: A complexity analysis of cooperative mechanisms in reinforcement leaning. In: Proc. of the 9th National Conf. on Artificial Intelligence, vol. 2, pp. 607–613 (1991)
12. Yoshimoto, J., Ishii, S., Sato, M.: System identification based on on-line variational Bayes method and its application to reinforcement learning. In: Kaynak, O., Alpaydın, E., Oja, E., Xu, L. (eds.) ICANN 2003 and ICONIP 2003. LNCS, vol. 2714, pp. 123–131. Springer, Heidelberg (2003)

Input Selection for Radial Basis Function Networks by Constrained Optimization

Jarkko Tikka

Helsinki University of Technology, Laboratory of Computer and Information Science, P.O. Box 5400, FI-02015 HUT, Finland
tikka@mail.cis.hut.fi
http://www.cis.hut.fi/tikka

Abstract. Input selection in the nonlinear function approximation is important and difficult problem. Neural networks provide good generalization in many cases, but their interpretability is usually limited. However, the contributions of input variables in the prediction of output would be valuable information in many real world applications. In this work, an input selection algorithm for Radial basis function networks is proposed. The selection of input variables is achieved using a constrained cost function, in which each input dimension is weighted. The constraints are imposed on the values of weights. The proposed algorithm solves a log-barrier reformulation of the original optimization problem. The input selection algorithm was applied to both simulated and benchmark data and obtained results were compelling.

1 Introduction

Radial basis function (RBF) networks have been widely utilized in regression problems. The advantages of RBF networks are that the training of networks is relatively fast and they are capable on universal approximation with nonrestrictive assumptions [1]. Fastness of the training is a consequence of simple structure of the RBF networks. They have only one hidden layer, in which each node corresponds to a basis function and a mapping from the hidden layer to the output layer is linear. The activation of hidden node is evaluated by the distance between an input vector and a center of the basis function. The usual choice for the basis function is the radially symmetric Gaussian function.

Many approaches are proposed to optimize the number of basis functions and widths of the Gaussian functions. In [2], the widths are fixed to be same and centers of the basis functions are selected from the input vectors using a regularized forward selection. Unsupervised clustering techniques, such as k-means and Gaussian mixture models trained with the EM algorithm, are another alternative to determine the centers of basis functions [3,4]. After the centers are selected, the widths of the Gaussian functions are found, for example, using the p-nearest neighbor rule [3] or weighting the standard deviation of the data in each cluster [5]. Nevertheless, these studies do not consider the input selection at all.

J. Marques de Sá et al. (Eds.): ICANN 2007, Part I, LNCS 4668, pp. 239–248, 2007.

The disadvantage of RBF networks is their black-box characteristics. Basically, the network includes all the input variables and, in addition, importances of the inputs are not clear at all. However, interpretation or understanding of the underlying process can be increased by selecting the input variables. In addition to interpretability, the rejection of non-informative inputs can improve the generalization capability of the network [6].

Several approaches exist to perform input selection [6]. In the filter approach, the input selection procedure is independent from the final non-linear model. The inputs can be selected, for example, based on mutual information [7] or linear models [8,9], and the final non-linear model is trained using the selected subset of inputs. The wrapper methodology takes into account the nonlinear model in the input selection [10]. Time consuming strategy is to fix the number of input variables and search through all the possible combinations. Computationally more efficient search strategies are presented in [10]. In the embedded methods, the input selection and training of the model are carried out simultaneously [6]. For instance, in the case of classification and support vector machines, a radius margin bound is used as a cost function and weights of the inputs are optimized by gradient descent algorithm [11]. Another alternative is to built the model using all the available inputs and use the backward elimination of the least significant inputs. In [12], several different criteria for the ranking of inputs are suggested.

In this work, an input selection algorithm for RBF networks is proposed. The algorithm is based on the weighted Euclidean distance, thus each input dimension has its own weight. The sum of weights are constrained such that some of the weights tend to be zero and corresponding inputs are rejected from the final model. The optimal weights are calculated by solving a constrained optimization problem, which takes into account the non-linear dependency between the inputs and the output. The proposed algorithm belongs to the class of embedded input selection methods.

The article is organized as follows. The ordinary RBF networks are briefly presented in Sect. 2 followed by the input selection algorithm for RBF networks in Sect. 3. The results of experiments on a simulated data set and Boston Housing benchmark data set are shown in Sect. 4. Finally, conclusions are given in Sect. 5.

2 Radial Basis Function Networks

Let us assume that there are N measurements available from an output variable y_j and input variables $\boldsymbol{x}_j = [x_{j,1}, \ldots, x_{j,d}]$, $j = 1, \ldots, N$. In a regression problem, the task is to estimate the values of output y_j as accurately as possible using the inputs $\boldsymbol{x}_j$. If the dependency is presumed to be non-linear, an unknown function can be estimated using artificial neural networks. In the case of RBF networks with Gaussian basis functions the model can be written as

$$\hat{y}_j = \sum_{n=1}^{N} \alpha_n K(\boldsymbol{c}_n, \boldsymbol{x}_j) + \alpha_0, \text{ where } K(\boldsymbol{c}_n, \boldsymbol{x}_j) = \exp\left(-\frac{\|\boldsymbol{c}_n - \boldsymbol{x}_j\|^2}{\sigma_n^2}\right) . \quad (1)$$

Usually, the training of RBF networks consists of three stages. First, the centers of Gaussian basis functions $\boldsymbol{c}_n$ are placed using some unsupervised learning algorithm. Second, the widths of basis functions σ_n are computed. Third, the parameters α_0 and $\alpha_n, n = 1, \ldots, N$ are estimated by minimizing the mean squared error (MSE). However, in this work the basis function is placed on each training data point $\boldsymbol{x}_n, n = 1, \ldots, N$ and the widths of Gaussian basis functions have common value $\sigma_n = \sigma$. The parameters α_0 and $\alpha_n, n = 1, \ldots, N$ are estimated by minimizing the regularized cost function

$$J = \frac{1}{N}\sum_{j=1}^{N}(y_j - \hat{y}_j)^2 + \gamma \sum_{n=1}^{N} \alpha_n^2 \ , \tag{2}$$

where the second term controls the smoothness of nonlinear mapping. For the given values of σ and γ the parameters α_0 and α_n are found by solving the system of linear equations.

3 Input Selection for RBF Networks

The disadvantage of model (1) is that it includes all the available input variables. In addition, it is nearly impossible to distinguish irrelevant and relevant inputs from each other. In [13], the problem is circumvented by using a Mahalanobis-like distance in place of the Euclidean distance. The distance is evaluated using a genetic algorithm in order to minimize the error criterion of the network. Nevertheless, the approach does not necessarily select inputs.

In this work, a weighted Euclidean distance is used

$$d_w(\boldsymbol{c}_n, \boldsymbol{x}_j) = \sqrt{\sum_{i=1}^{d} w_i (c_{n,i} - x_{j,i})^2}, \quad w_i \geq 0, \quad i = 1, \ldots, d \ . \tag{3}$$

The constraints $w_i \geq 0$ guarantee that the distance $d_w(\boldsymbol{c}_n, \boldsymbol{x}_j)$ is real-valued and nonnegative. The output of RBF network with distance (3) is

$$\hat{y}_j(\boldsymbol{w}) = \sum_{n=1}^{N} \alpha_n K_w(\boldsymbol{c}_n, \boldsymbol{x}_j) + \alpha_0 \text{ and } K_w(\boldsymbol{c}_n, \boldsymbol{x}_j) = \exp\left(-d_w(\boldsymbol{c}_n, \boldsymbol{x}_j)^2\right) \ . \tag{4}$$

Same weights w_i are used in all the basis functions. The basis functions are ellipsoidal, whose principal axes are parallel to the coordinate axes. The basis function is located to each training data point as in (1).

The goal is to estimate the weights w_i by minimizing the MSE between the observations y_j and the outputs of network $\hat{y}_j(\boldsymbol{w})$. However, if $\boldsymbol{w} \to \boldsymbol{0}$ the output of network is constant α_0, which equals to the mean of the observations y_j. On the other hand, if $\boldsymbol{w} \to \boldsymbol{\infty}$ the output of network interpolates exactly the observations y_j. In order to achieve a smooth mapping there has to be a constraint on the values of weights w_i. This leads to consider the following optimization problem

$$\begin{aligned} \underset{\boldsymbol{w}}{\text{minimize}} \quad & E(\boldsymbol{w}) = \frac{1}{N}\sum_{j=1}^{N}(y_j - \hat{y}_j(\boldsymbol{w}))^2 + \gamma\sum_{n=1}^{N}\alpha_n^2 \\ \text{such that} \quad & \sum_{i=1}^{d} w_i \leq t \quad \text{and} \quad w_i \geq 0, \quad i = 1, \ldots, d \ . \end{aligned} \tag{5}$$

The regularization term for the parameters α_n is also needed in this case, since the basis function is located to each data point. The constraint $\sum_{i=1}^{d} w_i < t$ shrinks the values of weights toward zero. It is known that the constraint of this type tends to set some of the coefficients w_i exactly to zero with appropriate choice of the parameter t [14,15]. This means that the corresponding input variables x_i are dropped from the model.

Problem (5) can be transformed into an unconstrained problem using the barrier function method [15]. With a logarithmic barrier function, it can be written as

$$\begin{aligned} \underset{\boldsymbol{w}}{\text{minimize}} \quad & J(\boldsymbol{w}) = E(\boldsymbol{w}) + \mu B(\boldsymbol{w}), \quad \text{where} \quad \mu > 0 \quad \text{and} \\ & B(\boldsymbol{w}) = -\log\left(t - \sum_{i=1}^{d} w_i\right) - \frac{1}{d}\sum_{i=1}^{d}\log(w_i) \ , \end{aligned} \tag{6}$$

where μ is a predefined small constant. The objective function $J(\boldsymbol{w})$ is differentiable with respect to all the parameters α_0, $\alpha_n, n = 1, \ldots, N$, and $w_i, i = 1, \ldots, d$.

In this work, the unconstrained problem (6) is solved in two phases. First, the values of weights w_i are fixed and the parameters α_0 and α_n are optimized by solving the system of linear equations. Second, the obtained values of α_0 and α_n are fixed and the weights w_i are optimized. These two steps are repeated until the convergence is achieved.

The objective function $J(\boldsymbol{w})$ cannot be solved in the closed from with respect to the weights w_i. The solution is determined using Levenberg-Marquardt optimization algorithm [4,16]. The derivative of $J(\boldsymbol{w})$ with respect to w_k is

$$\nabla J(\boldsymbol{w})_k = -\frac{2}{N}\sum_{j=1}^{N} e_j \frac{\partial \hat{y}_j(\boldsymbol{w})}{\partial w_k} + \mu\frac{\partial B(\boldsymbol{w})}{\partial w_k} \ , \tag{7}$$

where $e_j = y_j - \hat{y}_j(\boldsymbol{w})$. The second partial derivative of the MSE part $E(\boldsymbol{w})$ with respect w_k and w_l is

$$\frac{\partial^2 E(\boldsymbol{w})}{\partial w_l \partial w_k} = -\frac{2}{N}\sum_{j=1}^{N} \frac{\partial e_j}{\partial w_l}\frac{\partial \hat{y}_j(\boldsymbol{w})}{\partial w_k} + e_j\frac{\partial^2 \hat{y}_j(\boldsymbol{w})}{\partial w_l \partial w_k} \ . \tag{8}$$

Algorithm 1

1: Set $k = 0$, $\lambda^k = 1$, $\mu = 10^{-6}$, and initialize $\boldsymbol{w}^k$

2: Evaluate the search direction $\boldsymbol{p}^k$ by solving
$\left[\tilde{\boldsymbol{H}}(\boldsymbol{w}^k) + \lambda^k \boldsymbol{I}\right] \boldsymbol{p}^k = -\nabla J(\boldsymbol{w}^k)$

3: Determine the step length
$\delta = \max\left\{0 \le \delta \le 1 : \sum_{i=1}^{d} w_i^k + \delta p_i^k < t, \;\; w_i^k + \delta p_i^k > 0, \;\; i = 1, \ldots, d\right\}$

4: Calculate the ratio
$r^k = \dfrac{J(\boldsymbol{w}^k) - J(\boldsymbol{w}^k + \delta \boldsymbol{p}^k)}{J(\boldsymbol{w}^k) - \tilde{J}(\boldsymbol{w}^k + \delta \boldsymbol{p}^k)}$

5: If $r^k > 0.75$, set $\lambda^k = \lambda^k/2$

6: If $r^k < 0.25$, set $\lambda^k = 2\lambda^k$

7: If $J(\boldsymbol{w}^k + \delta \boldsymbol{p}^k) < J(\boldsymbol{w}^k)$, set $\boldsymbol{w}^{k+1} = \boldsymbol{w}^k + \delta \boldsymbol{p}^k$, $\lambda^{k+1} = \lambda^k$, and $k = k + 1$

8: Go to step 2 or terminate if the stopping criterion is fulfilled

Following [4,16], the second term is neglected and the element (l, k) of the approximated Hessian matrix is

$$\tilde{H}(\boldsymbol{w})_{l,k} = \frac{2}{N} \sum_{j=1}^{N} \frac{\partial \hat{y}_j(\boldsymbol{w})}{\partial w_l} \frac{\partial \hat{y}_j(\boldsymbol{w})}{\partial w_k} + \mu \frac{\partial^2 B(\boldsymbol{w})}{\partial w_l \partial w_k} \; . \tag{9}$$

Only the first partial derivatives of model (4) are required in the evaluation of (7) and (9), which reduces the computational complexity.

The algorithm to optimize weights $\boldsymbol{w}$ for the given values of γ and t and the fixed values of parameters α_0 and α_n is summarized in Algorithm 1. It starts by initializing the trust region parameter λ^0 and weights w_i^0 such that the starting point is strictly feasible, i.e. $\sum w_i^0 < t$, and $w_i^0 > 0$. The algorithm continues by evaluating the search direction $\boldsymbol{p}^k$ and determining the step length δ such that the next iterate $\boldsymbol{w}^k + \delta \boldsymbol{p}^k$ stays strictly in the feasible region (steps 2 and 3). The trust region parameter λ is adjusted according to the ratio r^k, which is the ratio between the actual and the predicted decrease in the value of objective function $J(\boldsymbol{w})$. In step 7, the new iterate is accepted if it leads to reduction in the value of objective function. Steps 2-7 are repeated as long as a stopping criterion is fulfilled, which can be the maximum number of iterations or a relative change in the value of objective function.

3.1 Input Selection Algorithm for RBF Network

The two phase algorithm for solving the optimization problem is summarized in Algorithm 2. Let us assume that the regularization parameter γ and the shrinking parameter t are given. The algorithm starts by initializing the values of weights $w_i^m, m = 0$ and evaluating the kernel matrix K_{w^m}. The initialization $w_i^m = 0.9t/d$ corresponds to model (1) with the width $\sigma^2 = d/0.9t$. The next

Algorithm 2. Input selection algorithm for RBF network

1: Set $m = 0$, initialize the weights $w_i^m = 0.9t/d, \quad i = 1, \ldots, d$, and evaluate the values of kernel matrix $K_{w^m}(\boldsymbol{x}_n, \boldsymbol{x}_j), \; n = 1, \ldots, N$ and $j = 1, \ldots, N$

2: Use the weights w_i^m to determine the values of α_0^{m+1} and α_n^{m+1}, $n = 1, \ldots, N$ by minimizing

$$E(\boldsymbol{w}) = \frac{1}{N} \sum_{j=1}^{N} (y_j - \hat{y}_j(\boldsymbol{w}))^2 + \gamma \sum_{n=1}^{N} \alpha_n^2$$

3: Use the obtained values α_0^{m+1} and α_n^{m+1}, $n = 1, \ldots, N$ to determine the weights w_i^{m+1} by **Algorithm 1**

4: Update the values of kernel matrix $K_{w^{m+1}}(\boldsymbol{x}_n, \boldsymbol{x}_j)$ and evaluate the value of cost function (6) using the the parameters α_0^{m+1}, α_n^{m+1}, and w_i^{m+1}, set $m = m + 1$

5: Go to step 2 or terminate if the stopping criterion is fulfilled

step is to estimate the parameters α_0^{m+1} and α_n^{m+1} by minimizing the regularized error function $E(\boldsymbol{w})$ for the fixed values of weights w_i^m. New values for the weights w_i^{m+1} are computed using Algorithm 1 using in the previous step obtained values of α_0^{m+1} and α_n^{m+1} (step 3). Algorithm 1 is terminated if the relative decrease in the value of objective function during the last five iterations is less than 10^{-4}. The maximum number of iterations is 10 in Algorithm 1. Algorithm 2 continues by updating the kernel matrix $K_{w^{m+1}}$ and evaluating the value of objective function (6) using the new values of parameters α_0^{m+1}, α_n^{m+1} and the weights w_i^{m+1} (step 4). Steps 2-4 are repeated until the stopping criterion is achieved. The iteration is terminated if the relative decrement is less than 10^{-4} during the last five iterations or 200 iterations are performed. Algorithm 2 produces the sequence of decreasing values for the objective function.

4 Experiments

4.1 Simulated Data

In this experiment, the performance of Algorithm 2 is illustrated using a simulated data set. In the case of simulated data, the assessment of quality of results is straightforward, since the underlying phenomenon is completely known. The values of each of five input variables $\boldsymbol{x} = [x_1, \ldots, x_5]$ were independently drawn from the uniform distribution in the range $x_i \in [-3, 3]$. The target was one dimensional sinc function and noisy samples from that model were generated as

$$y_j = \mathrm{sinc}(x_{j,1}) + \varepsilon_j, \quad j = 1, \ldots, N \; , \tag{10}$$

where ε_j were independent samples from the normal distribution $\varepsilon_j \sim \mathrm{N}(0, 0.15^2)$. Thus, only first input was relevant. The sizes of the training and the validation sets were $N_t = 200$ and $N_v = 2000$, respectively.

Algorithm 2 is evaluated in twenty points of the regularization parameter γ and the shrinking parameter t, which were logarithmically equally spaced in the ranges $\gamma \in [10^{-4}, 1]$ and $t \in [0.3, 4]$, respectively. In all, 400 RBF networks were trained using Algorithm 2. The validation error was the MSE for the validation set and the minimum was achieved using the parameter values $\gamma = 0.013$ and $t = 0.779$. The standard deviation of the validation errors $e_j = y_j - \hat{y}_j(\boldsymbol{w})$ was 0.133, which corresponds very well to the standard deviation of the noise.

On the left panel of Fig. 1, the evolution of the weights w_i, $i = 1, \ldots, 5$ during the training are presented as a function of the iterations m in Algorithm 2. The stopping criterion was fulfilled after 62 iterations. In the end, only the weight w_1 was nonzero, which corresponds to the input variable x_1. This means that Algorithm 2 detected the relevant input and discarded irrelevant ones from the model. The right panel of Fig. 1 presents the samples of the first dimension of input variables (gray dots), the target sinc function (solid line), and estimated function (dashed line). The peak of the sinc function is slightly underestimated and there is also visible overestimation with large values of x_1. Otherwise the approximation is nearly perfect.

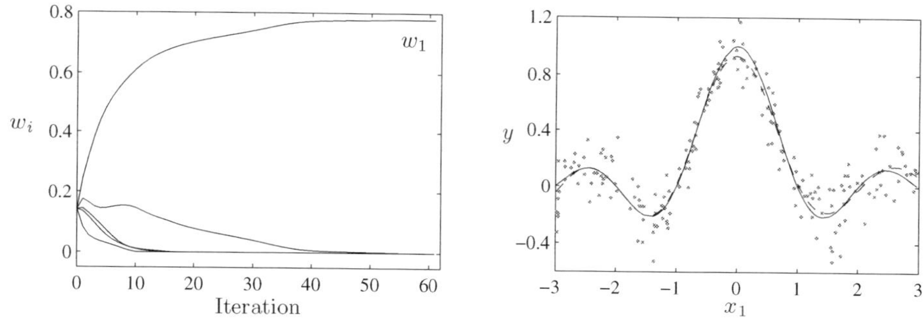

Fig. 1. Development of weights w_i as a function of iterations in Algorithm 2 for the simulated data (*left panel*) using the values of parameters γ and t, which produce the minimum validation error. On the *right panel*, the training samples $\boldsymbol{x}_1$ (*gray dots*), the unnoisy sinc function (*solid line*), and the estimated sinc function (*dashed line*).

4.2 Boston Housing Data

The purpose of this experiment is to illustrate the performance of Algorithm 2 and compare it to the ordinary RBF networks defined by (1) using a real data set. The used data are called Boston Housing data set and it can be downloaded from the UCI Machine Learning Repository[1]. The data contain 506 samples from $d = 13$ input variables x_i, $i = 1, \ldots, 13$ and from a single output y. The data set was randomly divided to the training set ($N_t = 400$) and to the test set ($N_{test} = 106$). All the inputs and the output were scaled to have zero mean and unit variance to make the weights comparable.

The training of RBF networks was carried out using a 10-fold cross validation (CV). The ordinary RBF was evaluated using 20 values of the regularization

[1] http://www.ics.uci.edu/~mlearn/MLRepository.html

Table 1. MSEs for Boston Housing data. Minimum CV error (2. column), the test error of the minimum CV error model (3.column), and the optimal test error (4. colum).

Model	CV error	Test error	Opt. test error
Ordinary RBF	0.127	0.143	0.126
Sparse RBF	0.130	0.149	0.112

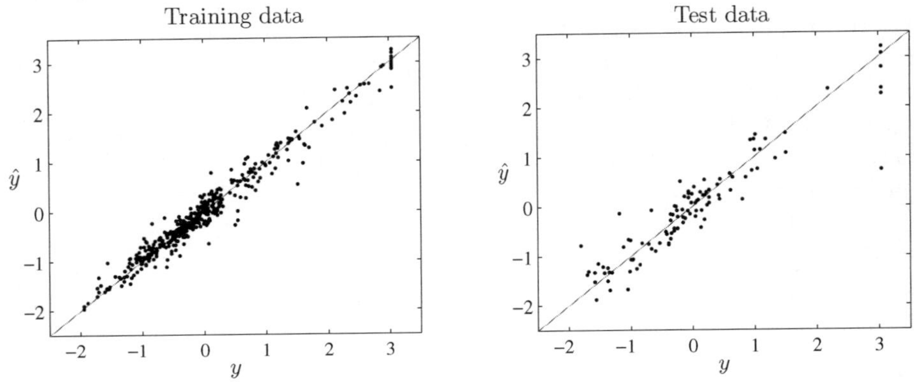

Fig. 2. The values of estimated output $\hat{y}_j(\boldsymbol{w})$ plotted versus the actual values y_j for the training data (*left*) and the test data (*right*)

parameter γ_o and the widths of Gaussian basis function σ^2, which were logarithmically equally spaced in the ranges $\gamma_o \in [10^{-9}, 5 \cdot 10^{-5}]$ and $\sigma^2 \in [4, 500]$. Algorithm 2 was calculated using 20 logarithmically equally spaced points of the regularization parameter γ_a and the shrinking parameter t ($\gamma_a \in [2 \cdot 10^{-6}, 0.03]$ and $t \in [0.2, 7]$).

The minimum CV errors and the corresponding test errors for the ordinary RBF network and the sparse RBF network, i.e. Algorithm 2, are shown in the second and in the third column of Table 1. The test errors were also calculated for all the combinations of parameters and the results are shown in the last column of Table 1. In practice, the both methods are equally accurate in the prediction. However, it is notable that the most accurate CV model did not produce the best possible test error in either model.

In Fig. 2, the estimated outputs versus the actual outputs are shown for the minimum CV error model for the training data (*left*) and the test data (*right*). The predictions are overall very good in the training data. In the test set, some of the highest values of the output are evidently underestimated otherwise the approximation is excellent.

On the left panel of Fig. 3, the weights w_i of the minimum CV error model are shown. All the weights are non-zero, but the values are not equal. Thus, the model highlights differences in importances of the inputs in the prediction. In [9] proposed method selects also all the input variables in the case of Boston

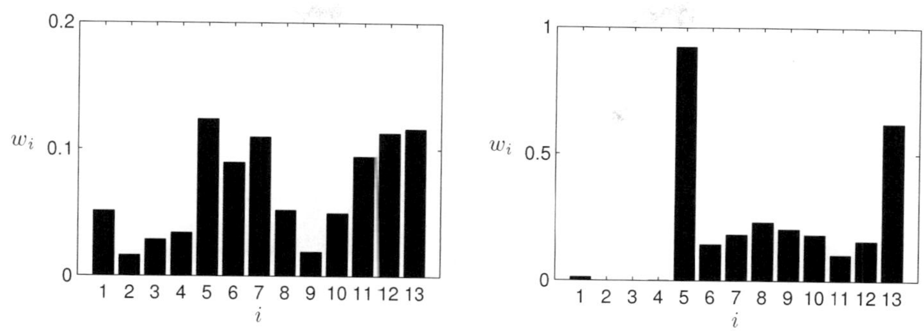

Fig. 3. The values of weights w_i in the minimum CV error model (*left*) and in the model producing the optimal test error (*right*)

Housing data. On the right panel of Fig. 3, the weights w_i of the optimal test error model are illustrated. The inputs x_2 and x_3 are rejected from this model. Also, the weights of the first and fourth inputs w_1 and w_4 are nearly zero. The validation error of the optimal test error model (0.146) was just slightly worse than the minimum CV error (0.130). Thus, it would probably be reasonable to make compromise between the CV error and number of selected inputs in the model selection as it is done in [8].

5 Summary and Conclusions

In this work, an input selection algorithm for the RBF networks was proposed. The algorithm solves a constrained optimization problem. The weighted Euclidean distance is used in the basis functions and the constraint structure on weights favors sparsity in terms of the input variables. The proposed algorithm solves the weights and the parameters of the output layer in two phases. Sparsity in terms of the inputs makes the models more interpretable compared to the ordinary RBF networks. The method was applied to the simulated and real world data set and the results were convincing in both cases.

In the case of large training data, it is not feasible to place the basis function in each training data point. The centers of basis functions can then be selected using some unsupervised technique. After that, the centers are kept fixed and the proposed algorithm can be applied without any modifications. However, the regularization parameter γ is not necessarily needed anymore. That would decrease computational complexity in the model selection phase, since only the value of shrinking parameter t would have to be validated. The disadvantage of unsupervised approach is that the selection of centers are based only on the input data. The resulting centers may be suboptimal solution with respect to the prediction accuracy. Probably better results would be achieved if the centers of the basis functions were optimized using the embedded approach. That is, the selection of centers of basis functions would be incorporated into the training process. Further work is also required that the weights and the parameters of output layer would be optimized simultaneously.

References

1. Park, J., Sandberg, I.W.: Approximation and Radial-Basis-Function Networks. Neural Computation 5, 305–316 (1993)
2. Orr, M.J.L.: Regularization in the Selection of Radial Basis Function Centers. Neural Computation 7, 606–623 (1995)
3. Moody, J., Darken, C.J.: Fast Learning in Networks of Locally-Tuned Processing Units. Neural Computation 1, 218–294 (1989)
4. Bishop, C.M.: Neural Networks for Pattern Recognition. Oxford University Press, Oxford (1995)
5. Benoudjit, N., Verleysen, M.: On the Kernel Widths in Radial-Basis Function Networks. Neural Processing Letters 18, 139–154 (2003)
6. Guyon, I., Elisseeff, A.: An Introduction to Variable and Feature Selection. Journal of Machine Learning Reasearh 3, 1157–1182 (2003)
7. Herrera, L.J., Pomares, H., Rojas, I., Verleysen, M., Guilén, A.: Effective Input Variable Selection for Function Approximation. In: Kollias, S., Stafylopatis, A., Duch, W., Oja, E. (eds.) ICANN 2006. LNCS, vol. 4131, pp. 41–50. Springer, Heidelberg (2006)
8. Tikka, J., Lendasse, A., Hollmén, J.: Analysis of Fast Input Selection: Application in Time Series Prediction. In: Kollias, S., Stafylopatis, A., Duch, W., Oja, E. (eds.) ICANN 2006. LNCS, vol. 4132, pp. 161–170. Springer, Heidelberg (2006)
9. Bi, J., Bennett, K.P., Embrechts, M., Breneman, C.M., Song, M.: Dimensionality Reduction via Sparse Support Vector Machines. Journal of Machine Learning Research 3, 1229–1243 (2003)
10. Kohavi, R., John, G.H.: Wrappers for Feature Selection. Artificial Intelligence 97, 273–324 (1997)
11. Weston, J., Mukherjee, S., Chapelle, O., Pontil, M., Poggio, T., Vapnik, V.: Feature Selection for SVMs. Advances in Neural Information Processing Systems 13 (2000)
12. Rakotomamonjy, A.: Variable Selection Using SVM-based Criteria. Journal of Machine Learning Research 3, 1357–1370 (2003)
13. Valls, J.M., Aler, R., Fernández, O.: Using a Mahalanobis-Like Distance to Train Radial Basis Neural Networks. In: Cabestany, J., Prieto, A.G., Sandoval, F. (eds.) IWANN 2005. LNCS, vol. 3512, pp. 257–263. Springer, Heidelberg (2005)
14. Tibshirani, R.: Regression Shrinkage and Selection via the LASSO. Journal of the Royal Statistical Society, Series B (Methodological) 58, 267–288 (1996)
15. Bazaraa, M.S., Sherali, H.D., Shetty, C.M.: Nonlinear Programming Theory and Algorithms. John Wiley & Sons, Inc., West Sussex, England (1979)
16. Nørgaard, M., Ravn, O., Poulsen, N.K., Hansen, L.K.: Neural Networks for Modelling and Control of Dynamic Systems. Springer, Heidelberg (2003)

An Online Backpropagation Algorithm with Validation Error-Based Adaptive Learning Rate

Stefan Duffner and Christophe Garcia

Orange Labs, 4, Rue du Clos Courtel, 35512 Cesson-Sévigné, France
{stefan.duffner,christophe.garcia}@orange-ftgroup.com

Abstract. We present a new learning algorithm for feed-forward neural networks based on the standard Backpropagation method using an adaptive global learning rate. The adaption is based on the evolution of the error criteria but in contrast to most other approaches, our method uses the error measured on the *validation* set instead of the *training* set to dynamically adjust the global learning rate. At no time the examples of the validation set are directly used for training the network in order to maintain its original purpose of validating the training and to perform "early stopping". The proposed algorithm is a heuristic method consisting of two phases. In the first phase the learning rate is adjusted after each iteration such that a minimum of the error criteria on the validation set is quickly attained. In the second phase, this search is refined by repeatedly reverting to previous weight configurations and decreasing the global learning rate. We experimentally show that the proposed method rapidly converges and that it outperforms standard Backpropagation in terms of generalization when the size of the training set is reduced.

1 Introduction

The Backpropagation (BP) algorithm [1] is probably the most popular learning algorithm for multilayer perceptron (MLP)-type neural architectures due to its simplicity and effectiveness. However, the choice of the learning rate used when updating the weights is crucial for the successful convergence and the generalization capacity of the network. A too small learning rate leads to slow convergence and a too high learning rate to divergence. Moreover, in the latter case the network is likely to overfit to the training data when using an *online* Backpropagation algorithm as it might specialize to the examples presented at the beginning of the training. Numerous solutions for the dynamic adaptation of the learning rate have been proposed in the literature. Most of them focus on the acceleration of the training process rather than their generalization performance. They can roughly be divided into two groups: global and local adaption techniques. The former is referring to methods adjusting an overall learning rate for the whole network and the latter to the adaptation of independent learning rates for each weight.

A method for global adaptation has been proposed by Chan et al. [2] where the angle between the last weight update and the current gradient is calculated. If it

J. Marques de Sá et al. (Eds.): ICANN 2007, Part I, LNCS 4668, pp. 249–258, 2007.

is less than 90° the learning rate is increased otherwise it is decreased. Salomon et al. [3] proposed an evolutionary based adaption of the learning rate. At each iteration, two weight updates, one with increased and with decreased learning rate, are performed separately. The resulting network that performs better is retained and used as a starting point for the next iteration. A heuristic method, the so-called "bold driver" method, has been employed by Battiti et al. [4] and Vogl et al. [5]. Here the learning rate is adjusted according to the evolution of the error criteria E. If E decreases the learning rate is slightly increased, otherwise it is drastically decreased. Hsin et al. [6] propose to use a weighted average of the cosines between successive weight updates, and Plagianakos et al. [7] calculate a two point approximation to the secant equation underlying quasi-Newton methods in order to obtain a dynamic learning rate and additionally make use of an acceptability condition to ensure convergence. LeCun et al. [8] calculate a global learning rate by an online estimation of the largest eigenvalue of the Hessian matrix. They show that the optimal learning rate is approximately the inverse of this largest eigenvalue. Finally, the approach of Magoulas et al. [9] estimate the local shape of the error surface by the Lipschitz constant and set the learning rate accordingly. They also applied this technique to calculate a separate dynamic learning rate for each weight [10].

Local learning rate adjustment methods have been very popular due to their efficiency and generally higher convergence speed. A very well-know technique is the Delta-Bar-Delta method introduces by Jacobs et al. [11]. Here, the learning rates are adjusted according to sign changes of the exponential averaged gradient. Similarly, Silva and Almeida [12] proposed a method where the learning rates are increased if the respective gradients of the last two iterations have the same size and decreased otherwise. The RPROP method introduced by Riedmiller et al. [13] uses a step size which doesn't depend on the gradient magnitude but which is increased or decreased according to gradient sign changes.

Finally, many methods do not use an explicit learning rate but first calculate a descent gradient direction and then perform a line search such that the error criteria is minimized in the direction of the gradient [14, 15, 16].

Note that most of the existing adaptive learning algorithms are *batch* learning algorithms, i.e. the weights are updated after all examples have been presented to the network. On the other hand, *online* algorithms update the weights after the presentation of each example. They generally converge faster when the input space is large compared to the number of examples (e.g. in image processing tasks) or in more complex architectures like convolutional neural networks (CNN) that use shared weights. Thus, for many real world applications the online Backpropagation algorithm or its variants are still the best choice. There are also some adaptive online algorithms in the literature. For example Schraudolph [17], Harmon et al. [18] and Almeida et al. [19] proposed methods similar to the Incremental Delta-Bar-Delta approach introduced by Sutton et al. [20], an extension of the Delta-Bar-Delta technique for stochastic training. However, these algorithms mainly aim at a faster convergence rather than an increased generalization capacity.

We present a heuristic learning algorithm that improves generalization capacity and shows good convergence speed. It is an online Backpropagation method with adaptive global learning rate. The learning rate adaption is based on the idea of the so-called "bold driver" method [5, 4], i.e. the learning rate is initialised with a very small value and increased or decreased according to the evolution of the error criteria E of the past iterations. The main difference with respect to previous works is that the error is not measured on the *training* but on the *validation* set. Some further optimizations have been made in order to ensure fast convergence. The aim of this heuristic approach is not only to accelerate convergence compared to standard online Backpropagation but also to improve generalization.

The remainder of this paper is organized as follows: Section 2 describes the training algorithm with its two phases. In section 3, experimental results are presented and finally in section 4 we draw our conclusions.

2 The Training Algorithm

The proposed training method is an extension of the standard Backpropagation algorithm [1]. Backpropagation is a gradient descent technique that minimizes an error criteria E which is usually the mean squared error (MSE) of the N output values o_{ij} with respect to its desired values d_{ij} of the neural network:

$$E = \frac{1}{NP} \sum_{j=1}^{P} \sum_{i=1}^{N} (o_{ij} - d_{ij})^2 \quad , \tag{1}$$

where P is the number of examples in the training set.

Training is performed *online*, i.e. the weights of the neural network are updated after presentation of each training example. Classically, at iteration t a given weight $w_{ij}(t)$ is updated by adding a $\Delta w_{ij}(t)$ to it:

$$\Delta w_{ij}(t) = -\epsilon(t) \cdot \frac{\partial E(t)}{\partial w_{ij}(t)} \quad , \tag{2}$$

where ϵ is the learning rate.

The training database is usually divided into a *training set* and a *validation set*. The training, i.e. the Backpropagation, is only performed on the training set. The purpose of the validation set is to determine when to stop training in order to obtain a neural network that generalizes sufficiently well. This technique is commonly known as "early stopping". In the proposed algorithm the validation set has a second role. It is used to control the adjustment of the learning rate $\epsilon(t)$ after each training iteration t. Note, that the learning rate adjustment is performed only once per iteration. Our approach basically consists of two phases:

1. the main learning phase and
2. the refinement phase.

In the following sections we will detail these steps.

2.1 The Main Learning Phase

The adaption of the learning rate in our approach is similar to the "bold driver" method [5, 4] where the learning rate is initialized with a very small value (e.g. 10^{-10}) and adjusted after each training iteration according the difference of the error criteria E between the current and the preceding iteration.

The proposed method applies this idea to the validation set instead of the training set in order to reduce overfitting. Moreover, the procedure is slightly modified to be more tolerant to error increases as the validation error is more likely to oscillate than the training error. Let us consider the typical runs of the error curves of a training and validation set when using standard Backpropagation. Fig. 1 illustrates this in a simplified manner. When applying the technique

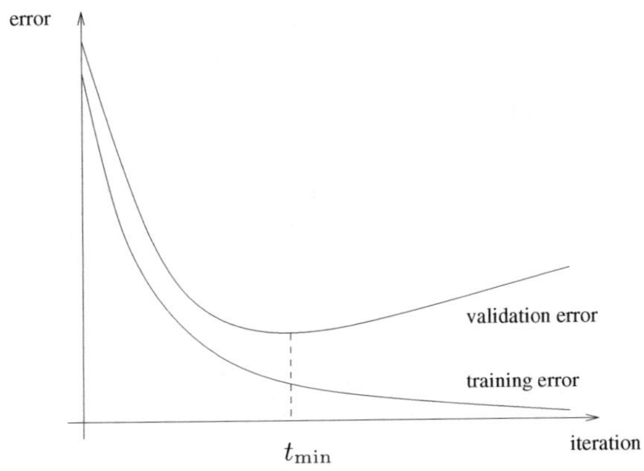

Fig. 1. The typical evolution of the error criteria evaluated on the training and validation set

of "early stopping" the weight configuration at iteration $t_{\min}$, i.e. where the validation error is minimal, is retained as the network is supposed to show the highest generalization performance at this point. Further training likely leads to overfitting. The purpose of the first phase of the proposed algorithm is thus to reach the point $t_{\min}$ more quickly.

To this end, the normalized difference between the error criteria of the current and the preceding iteration is calculated:

$$\delta(t) = \frac{E_v(t) - E_v(t-1)}{E_v(t)} \quad . \tag{3}$$

$E_v(t)$ is the error criteria at iteration t calculated on the whole validation set (cf. Eqn. 1).

The algorithm further requires a running average $\bar{\delta}(t)$ of the preceding values of δ:

$$\bar{\delta}(t) = \alpha \cdot \delta(t) + (1-\alpha) \cdot \bar{\delta}(t-1) \quad , \tag{4}$$

where $0 < \alpha \leq 1$ (e.g. $\alpha = 0.1$).

The principal learning rate updating rule is the following:

$$\epsilon(t) = \begin{cases} d \cdot \epsilon(t-1) & \text{if } \delta(t) \cdot \bar{\delta}(t-1) < 0 \text{ and } |\bar{\delta}(t-1)| > \theta \,, \\ u \cdot \epsilon(t-1) & \text{otherwise} \,. \end{cases} \tag{5}$$

where u and d are positive constants, $0 < d < 1 < u$, and θ is a threshold to allow for small error oscillations. In our experiments we used $u = 1.1$, $d = 0.5$ and $\theta = 0.01$. Thus, the learning rate adaption is based on the signs of the error differences of the current and the preceding iterations. If the sign changes the learning rate is decreased otherwise it is increased. The principle of this procedure is similar to the Delta-Bar-Delta method [11] but the calculation is not based on gradients.

2.2 Refinement Phase

If the training has passed iteration $t_{\min}$ where E_v is minimal the network is likely to overtrain. In fact, as the gradient descent is performed in discrete steps the actual minimum $E_{\min}$ of the error surface of the validation set is likely to be missed and lies between the weight configurations of two successive training iterations. Fig. 2 illustrates this. Clearly, there are two cases to differentiate:

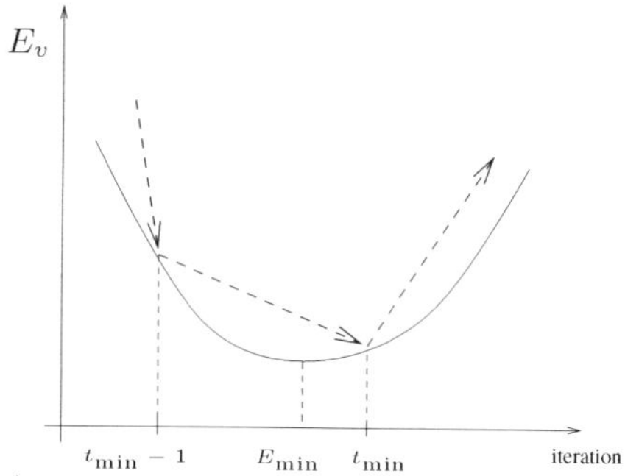

(a) The actual minimum has been missed *before* iteration $t_{\min}$

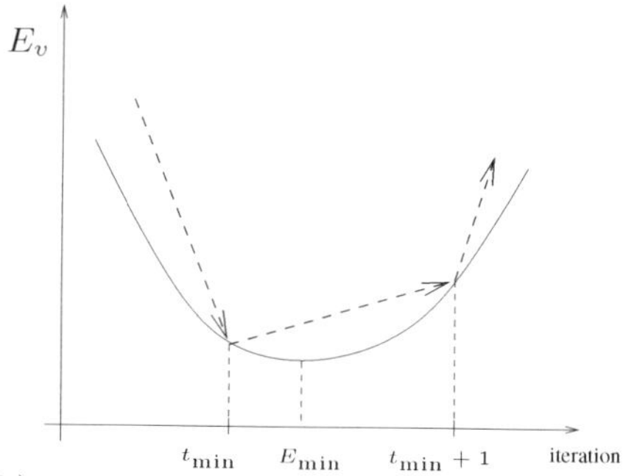

(b) The actual minimum has been missed *after* iteration $t_{\min}$

Fig. 2. The two possible cases that can occur when the minimum on the validation set is reached

either the minimum $E_{\min}$ has been missed before or after iteration $t_{\min}$. Now we assume that the validation error surface is relatively smooth and that no other local minimum lies between iterations $t_{\min} - 1$ and $t_{\min}$ or between iterations $t_{\min}$ and $t_{\min} + 1$ respectively. In order to try to attain a smaller error the network reverts to the weight configuration at iteration $t_{\min} - 1$, decreases the learning rate and training is continued. Note that for training only the examples of the training set are used. Thus, it is uncertain if the actual minimum can be attained at all. If no smaller error has been found for a certain number of iterations T the "real" minimum is more likely to have occurred "after" iteration

$t_{\min}$, (see Fig. 2(b)). In this case, the network reverts to iteration $t_{\min}$, the learning rate is again decreased and training continues. If a smaller error is reached during this process the temporary minimum is retained and the training continues normally. Otherwise the reverting procedure is repeated while always retaining the absolute minimum and the respective weight configuration found so far. Algorithm 1 summarizes the overall training procedure. Note that the computational overhead of the algorithm compared to standard backpropagation with fixed learning rate is negligible as the error on the validation set needs to be calculated anyway to do the "early stopping".

Fig.3 illustrates a typical evolution of the error criteria $E_v(t)$ during the training process using the proposed learning algorithm. Because of the initialization with a very small value the error stays nearly constant at the beginning but drops very quickly at some point due to the exponential increase of the learning rate, and finally it converges to a minimum. In general, the main part of the minimization is done in the first phase and the error decrease in the refinement phase is relatively small.

Algorithm 1. The basic learning algorithm

Initialize weights and individual learning rates
Set $\epsilon(0) := 10^{-10}$, $\delta(0) := 0$ and $\bar{\delta}(0) := 0$
Calculate $E_v(0)$
$t := 0$
repeat
    Do one training iteration
    $t := t + 1$
    Calculate $\delta(t) = \frac{E_v(t) - E_v(t-1)}{E_v(t)}$
    if $\delta(t) \cdot \bar{\delta}(t-1) < 0$ and $|\bar{\delta}(t-1)| > \theta$ **then**
        $\epsilon(t) = d \cdot \epsilon(t-1)$
    else
        $\epsilon(t) = u \cdot \epsilon(t-1)$
    end if
    $\bar{\delta}(t) = \alpha \cdot \delta(t) + (1 - \alpha) \cdot \bar{\delta}(t-1)$
    if $E_v(t) < E_{\min}(t)$ **then**
        save the current weight configuration
        $t_{\min} := t$
    end if
    if $t - t_{\min} > T$ **then**
        Revert to weight configuration at $t_{\min} - 1$ (or $t_{\min}$)
    end if
until $t = t_{\max}$

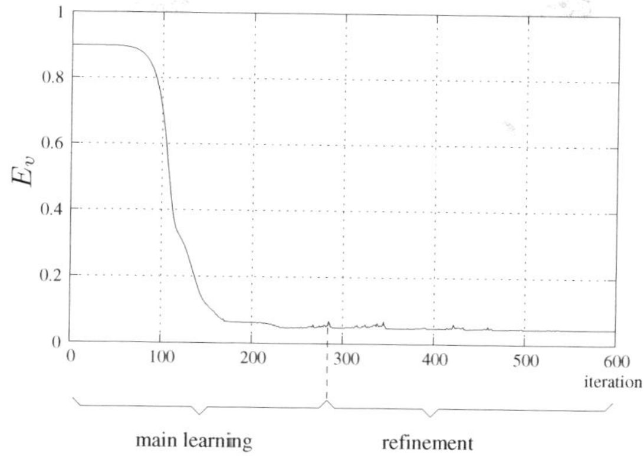

Fig. 3. A typical evolution of the error criteria on the validation set using the proposed learning algorithm

3 Experimental Results

We evaluated the proposed learning algorithm on a MLP trained to classify the examples of the well-known NIST database of handwritten digits. The database contains 3823 training and 1797 test examples of 8x8 matrices. From the training set 500 examples were selected randomly and used for validation. The MLP we used for the experiments had 64 input, 10 hidden and 10 output neurons, fully inter-connected. The neurons all had sigmoid activation functions.

To ensure that the neural network is well-conditioned we additionally use *fixed local* learning rates that are distributed stepwise from the last layer to the first layer according to the incoming connections of each neuron. Thus, the output neuron(s) have the highest and the input the lowest local learning rate. The overall learning rate is just the product of the fixed local and the dynamic global learning rate.

In the first experiment, we compare the convergence properties of the proposed algorithm to the ones of standard Backpropagation. Fig. 4 shows the resulting error curves evaluated on the validation set. The different curves for the Backpropagation algorithm have been obtained by using different global learning rates ($10^{-3}, 10^{-4}$ and 10^{-5}). The global learning rate of the proposed method was initialized with the value 10^{-7}. Note that our approach converges more slowly at the beginning but catches up quickly and finishes stable on the same level or even lower than Backpropagation.

Fig. 5 illustrates 0that our method is not sensitive to different initializations of the global learning rate. The curves show the validation error curves for three different runs with initial learning rates of 10^{-6}, 10^{-8} and 10^{-10} respectively. Note that the point of time where the minimum is reach increases only *linearly* when the initial learning rate is decreased *exponentially*. This is another side effect of the exponential learning rate update rule. All the runs converge to approximately the same solution, and the recognition rates are about the same for all networks.

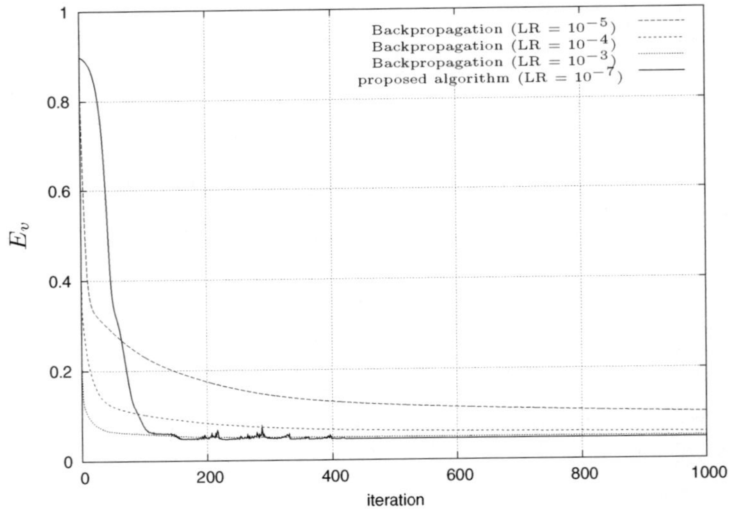

Fig. 4. The evolution of the validation error on the NIST database using Backpropagation and the proposed algorithm

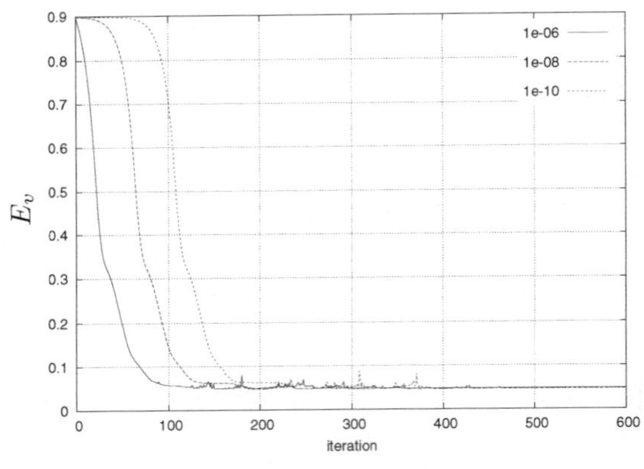

Fig. 5. The validation error curves with different initial global learning rates

The final experiment demonstrates that the algorithm not only converges faster but also improves the generalization performance of the resulting neural networks. To this end the training set was gradually reduced and the respective recognition rates on the test set were calculated and compared to the standard Backpropagation as well as to the bold driver method [5, 4]. Table 1 shows the overall results. One can see that the proposed method performs slightly better with training set sizes 3323 and 1000 and clearly outperforms the other algorithms when only 600 and 100 training examples are used.

Table 2 shows the respective results with a neural network with 40 hidden neurons. The recognition rates of the proposed method are slightly better then

Table 1. Recognition rate (in %) with varying training set size (10 hidden neurons)

training set size algorithm	3323	1000	600	100
Backpropagation	94.30	93.42	78.31	73.88
bold driver [5, 4]	93.41	91.32	83.74	72.75
proposed algorithm	94.50	93.71	**85.29**	**78.10**

Table 2. Recognition rate (in %) with varying training set size (40 hidden neurons)

training set size algorithm	3323	1000	600	100
Backpropagation	95.71	93.89	86.58	80.31
bold driver [5, 4]	94.97	93.20	86.45	79.96
proposed algorithm	95.77	93.81	87.06	80.47

for the other algorithms albeit the difference is less significant. However, convergence speed is still superior as illustrated in Fig. 4.

4 Conclusion

We presented a learning algorithm with adaptive global learning rate based on the online Backpropagation method. The adaption is performed in two successive phases, the main learning and a refinement phase. In the main learning phase the learning rate is increased or decreased according to the evolution of the validation error. This leads to a fast and robust convergence to the minimum where "early stopping" is performed. In the second phase the network reverts to preceding weight configurations in order to find a minimum close to the one found in the first step. We experimentally show that this method converges faster than Backpropagation while exhibiting a superior generalization capacity.

References

[1] Rumelhart, D.E., McClelland, J.L.: Parallel Distributed Processing: Explorations in the Microstructure of Cognition. Foundations, vol. 1. MIT Press, Cambridge, MA, USA (1986)

[2] Chan, L.W., Fallside, F.: An adaptive learning algorithm for backpropagation networks. Computer Speech and Language 2, 205–218 (1987)

[3] Salomon, R.: Improved convergence rate of back-propagation with dynamic adaptation of the learning rate. In: Schwefel, H.-P., Männer, R. (eds.) Parallel Problem Solving from Nature. LNCS, vol. 496, pp. 269–273. Springer, Heidelberg (1991)

[4] Battiti, R.: Accelerated backpropagation learning: Two optimization methods. Complex Systems 3, 331–342 (1989)

[5] Vogl, T., Mangis, J., Rigler, J., Zink, W., Alkon, D.: Accelerating the convergence of the back-propagation method. Biological Cybernetics 59, 257–263 (1988)
[6] Hsin, H.C., Li, C.C., Sun, M., Sclabassi, R.: An adaptive training algorithm for back-propagation neural networks. In: International Conference on Systems, Man and Cybernetics, vol. 2, pp. 1049–1052 (1992)
[7] Palgianakos, V., Vrahatis, M., Magoulas, G.: Nonmonotone methods for backpropagation training with adaptive learning rate. In: International Joint Conference on Neural Networks, vol. 3, pp. 1762–1767 (1999)
[8] LeCun, Y., Simard, P., Pearlmutter, B.: Automatic learning rate maximization by on-line estimation of the hessian's eigenvectors. In: Hanson, S., Cowan, J., Giles, L. (eds.) Advances in Neural Information Processing Systems, vol. 5, Morgan Kaufmann Publishers, San Mateo, CA (1993)
[9] Magoulas, G.D., Vrahatis, M.N., Androulakis, G.S.: Effective back-propagation with variable stepsize. Neural Networks 10, 69–82 (1997)
[10] Magoulas, G.D., Vrahatis, M.N., Androulakis, G.S.: Improving the convergence of the backpropagation algorithm using learning rate adaptation methods. Neural Computation 11(7), 1769–1796 (1999)
[11] Jacobs, R.A.: Increased rates of convergence through learning rate adaption. Neural Networks 1, 295–307 (1988)
[12] Silva, F.M., Almeida, L.B.: Acceleration techniques for the backpropagation algorithm. In: Proceedings of the EURASIP Workshop 1990 on Neural Networks, London, UK, pp. 110–119. Springer, Heidelberg (1990)
[13] Riedmiller, M., Braun, H.: A direct adaptive method for faster backpropagation learning: The RPROP algorithm. In: Proc. of the IEEE Intl. Conf. on Neural Networks, San Francisco, CA, pp. 586–591. IEEE Computer Society Press, Los Alamitos (1993)
[14] Luenberger, D.G.: Introduction to linear and nonlinear programming. Addison-Wesley, Reading (1973)
[15] Fahlman, S.E.: Faster-learning variations on back-propagation: An empirical study. In: Touretzky, D.S., Hinton, G.E., Sejnowski, T.J. (eds.) Connectionist Models Summer School, pp. 38–51. Morgan Kaufmann Publishers, San Mateo, CA (1988)
[16] Leonard, J., Kramer, M.: Improvement of the backpropagation algorithm for training neural networks. Computers & Chemical Engineering 14(3), 337–341 (1990)
[17] Schraudolph, N.: Local gain adaptation in stochastic gradient descent. In: ICANN. Ninth International Conference on Artificial Neural Networks, vol. 2, pp. 569–574 (1999)
[18] Harmon, M., Baird III, L.: Multi-player residual advantage learning with general function approximation. Technical report, Wright Laboratory, WL/AACF, Wright-Patterson Air Force Base, OH (1996)
[19] Almeida, L.B., Langlois, T., Amaral, J.D., Plakhov, A.: Parameter adaptation in stochastic optimization. In: Saad, D. (ed.) On-Line Learning in Neural Networks, Cambridge University Press, Cambridge (1999)
[20] Sutton, R.S.: Adapting bias by gradient descent: an incremental version of beltabar-delta. In: Proceedings of the Tenth International Conference on Machine Learning, Cambridge, MA, pp. 171–176 (1992)

Adaptive Self-scaling Non-monotone BFGS Training Algorithm for Recurrent Neural Networks

Chun-Cheng Peng and George D. Magoulas

School of Computer Science and Information Systems
Birkbeck College, University of London
Malet Street, London WC1E 7HX, UK
{ccpeng, gmagoulas}@dcs.bbk.ac.uk

Abstract. In this paper, we propose an adaptive BFGS, which uses a self-adaptive scaling factor for the Hessian matrix and is equipped with nonmonotone strategy. Our experimental evaluation using different recurrent networks architectures provides evidence that the proposed approach trains successfully recurrent networks of various architectures, inheriting the benefits of the BFGS and, at the same time, alleviating some of its limitations.

Keywords: BFGS, self-scaling factor, nonmonotone strategy, line-search, adaptive learning, recurrent networks.

1 Introduction

There are a relatively large number of applications, where Quasi-Newton (QN) methods have been used for training static Artificial Neural Networks (ANNs), e.g. [1][3][4][9][12]-[15][21][22][25]-[27][30][31][33][35]-[37][40][41][44]-[47][48][50]. QN methods exploit the idea of building up curvature information as the iterations of the training method are progressing. This is achieved by using the objective function values and its gradient to estimate the Hessian matrix. Thus, a new approximated Hessian matrix B_{k+1} is required at each iteration to satisfy the quasi-Newton condition $B_{k+1}s_k = y_k$, where s_k and y_k are the changes in function variable and in gradient, respectively.

There are several techniques in neural networks to update B_{k+1} such as the Powell-Symmetric-Broyden update formula:

$$B_{k+1} = B_{k+1} + \frac{1}{s_k^T s_k}\left(\left(y_k - B_k s_k\right)s_k^T + s_k\left(y_k - B_k s_k\right)^T\right) - \frac{\left(y_k - B_k s_k\right)^T s_k}{\left(s_k^T s_k\right)^2} s_k s_k^T, \quad (1)$$

the Davidon-Fletcher-Powell formula:

$$\begin{aligned} B_{k+1} &= B_{k+1} + \frac{1}{s_k^T B_k s_k} B_k s_k s_k^T B_k + \frac{1}{y_k^T s_k} y_k y_k^T + \left(s_k^T B_k s_k\right) w_k w_k^T, \\ w_k &= \frac{1}{y_k^T s_k} y_k - \frac{1}{s_k^T B_k s_k} B_k s_k \end{aligned} \quad (2)$$

and the Broyden-Fletcher-Goldfarb-Shanno (BFGS) update [21]:

J. Marques de Sá et al. (Eds.): ICANN 2007, Part I, LNCS 4668, pp. 259–268, 2007.

$$B_{k+1} = B_{k+1} - \frac{1}{s_k^T B_k s_k} B_k s_k s_k^T B_k + \frac{1}{y_k^T s_k} y_k y_k^T, \tag{3}$$

which is the most commonly used update rule for training neural networks [19].

There are several scaling approaches in the optimisation literature, such as those presented in [2][38], where the scaling factor ρ_k can be defined as [50]

$$\rho_k = \frac{y_k^T s_k}{s_k^T B_k s_k} \tag{4}$$

The aim of scaled BFGS is to tune the Hessian approximations at each iteration when eigenvalues become large. Numerical evidence shows that methods that apply a scaling factor for B_{k+1} are superior to the original QN methods. In other words, when ρ_k is sufficiently large, the eigenvalues of B_{k+1} are relative small, with strong self-correcting property [50].

As indicated in [24][44][45] the QN method for training ANNs is very fast to converge but in many cases it converges to a local minimum; the same holds for the BFGS method. Although several efforts have been made to reduce the memory requirement of updating the Hessian approximation [5][26][31][35][43][48], the limitations caused by monotone line search still exists and the problem of local minima still limits the applicability of these methods. Furthermore, despite the theoretical advantages of the self-scaling approaches for the Hessian approximation in numerical optimisation, these are rarely introduced to training ANNs [37].

Lastly, the literature of Recurrent Neural Networks (RNNs) includes very few attempts to train RNNs using QN methods with limited results [5][7][8][17][24][28]. In [49], the drawbacks of first-order gradient learning algorithms for RNNs were discussed and second-order approaches to speed up the convergence and to tackle the issue of weight vanishing were proposed. In [10], RNNs for sequence processing were discussed. Local methods combined with Back-Propagation-Through-Time (BPTT) proved efficient only for some local feedback RNNs and for short-term memorisation tasks [10][11]. When the task requires learning long-term dependencies, i.e. a late output depends on a very early input which has to be remembered, RNNs trained with gradient-descent methods treat past information as noise and gradually ignore it. Time-weighted pseudo-Newton has been proposed as an alternative but it remains to be verified whether it would work on other more general situation [10]. Learning algorithms for *fixed point* and *nonfixed point* RNNs were discussed in [39] and an extended Real-time Recurrent Learning (RTRL) was proposed. Ref [29] summarises ten different kinds of weight updating based on gradient descent that could be applied to RNNs, while [18] reviews twelve methods from optimisation and global minimisation but not all of them appear to be suitable for RNNs. In the area of time-series modelling, BPTT and RTRL for learning RNNs have been popular choices [6][16], while simulated annealing has provided promising results but training time is relatively higher [16].

In this paper we propose an adaptive BFGS, which uses a self-adaptive scaling factor for the Hessian matrix and is equipped with nonmonotone strategy. The rest of this paper is organised as follows. Section 2 presents the new algorithm, while Section 3 presents the datasets used in our experiments. Section 4 discusses our experimental results and, finally, Section 5 presents concluding remarks and future works.

2 The Adaptive Self-scaling Nonmonotone BFGS

Training a neural network architecture can be expressed as an unconstrained optimisation problem

$$\min f(x),\ x \in R^n, \tag{5}$$

where $f : R^n \rightarrow R$ represents the cost function and its gradient $g = \nabla f(x)$ is available. Let us consider the current approximation to the solution of the above problem is x_k and if $g \neq 0$, then, in some way, a line-search method finds a step-length α_k along a search direction d_k, and computes the next approximation x_{k+1} as follows:

$$x_{k+1} = x_k + \alpha_k d_k. \tag{6}$$

Traditional optimisation strategies for RNNs are monotone, i.e. they compute a step length that reduces the function value at each iteration. Unfortunately, even when an algorithm is globally convergent, there is no guarantee that the method would be efficient in the sense that it may be trapped in a local minimum point and never jump out to a better one under ill conditions [20], such as poorly initialised weights in the case of RNNs. To overcome the disadvantages and accelerate convergence, we explore the use of a nonmonotone approach, which was first proposed in [22]. Taking the M previous f values into consideration, the following condition can be set:

$$f(x_{k+1}) \leq \max_{0 \leq j \leq M_k} \left[f(x_{k-j}) \right] + \gamma \alpha_k g_k^T d_k, \tag{7}$$

where M_k is named nonmonotone learning horizon [41] and constant $\gamma \in (0,1)$. Furthermore, [41] proposed a dynamic and efficient way to determine the size of M, by a local estimation of the Lipschitz constant, which could provide helpful information on the morphology of a function. Thus, a poorly user-defined value for the nonmonotone learning horizon could be avoided by applying this approach.

A high level description of the proposed algorithm follows:

Algorithm: Adaptive Self-scaling Non-monotone BFGS (ASCNMBFGS)

STEP 0. Initialize a point x_0, counter k=0, a symmetric positive definite matrix B_0, and boundary of nonmonotone learning horizon M

STEP 1. If $g_k = 0$, stop;

STEP 2. Determine the line search direction by $d_k = -B_k^{-1} g_k$;

STEP 3. Adapt M^k by the following conditions:

$$M^k = \begin{cases} M^{k-1}+1, & \text{if } \Lambda^k < \Lambda^{k-1} < \Lambda^{k-2} \\ M^{k-1}-1, & \text{if } \Lambda^k > \Lambda^{k-1} > \Lambda^{k-2} \\ M^{k-1} & \text{, otherwise,} \end{cases} \text{ where } \Lambda^k = \frac{\|g_k - g_{k-1}\|}{x_k - x_{k-1}};$$

STEP 4. Find a step length α_k by the line search approach below:

For $0 < \lambda_1 < \lambda_2$ and $\sigma, \delta \in (0,1)$, at each iteration, one chooses a parameter l_k such that the step length $\alpha_k = \overline{\alpha}_k \cdot \sigma^{l_k}$, where $\overline{\alpha}_k \in (\lambda_1, \lambda_2)$, satisfies

$$f(x_k + \alpha_k d_k) \le \max_{0 \le j \le M^k} \left[f(x_{k-j}) \right] + \delta \cdot \alpha_k \cdot g_k^T \cdot d_k$$

STEP 5. Generate a new iteration point by $x_k = x_k + \alpha_k d_k$;

STEP 6. Update the Hessian approximation B_k by the following self-scaling BFGS

$$B_{k+1} = \rho_k \left[B_k - \frac{B_k s_k s_k B_k^T}{s_k^T B_k s_k} \right] + \frac{y_k y_k^T}{y_k^T s_k},$$

where $s_k = x_{k+1} - x_k$, $y_k = g_{k+1} - g_k$ and ρ_k is the self-scaling factor which satisfies Eq. (4)

STEP 7. Let $k = k+1$, go to STEP 1.

There are three heuristic parameters in the above algorithm, i.e. M^k in Step 3 and σ, δ in Step 4. In practice, the nonmonotone learning horizon of Step 3 requires accumulating information from a few iterations in order to be initialised; in the meantime, our approach uses the Charalambous line-search method [11]. We also set an upper boundary of M^k. In Step 4, σ plays the role of tuning parameter for the step length, while δ controls the amount of change which is defined by the current gradient and search direction.

The proposed algorithm is a modified version of BFGS, which combines the usage of adaptive nonmonotone learning [41] with the self-scaling utility proposed in [50]; where the former takes the benefits of Lipschitz constant taking into account the local shape of a given function, while the latter has been proved to have the ability of solving unconstrained nonconvex optimisation problems; these characteristics look eminently suitable for training RNNs.

In order to illustrate the differences between the classic BFGS and our approach, we visualise the learning behaviour of the two methods in the parity-5 problem using two well-known RNN architectures, a Feedforward Time Delay (FFTD) network and Elman's Layered Recurrent Network (LRN). Figure 1 shows convergence behaviours of our algorithm, ASCNMBFGS, and of the original BFGS, by solid-red and dashed-blue lines, respectively. In this example, we see the BFGS is trapped in neighbourhoods of local minima but the ASCNMBFGS manages to reach lower function values.

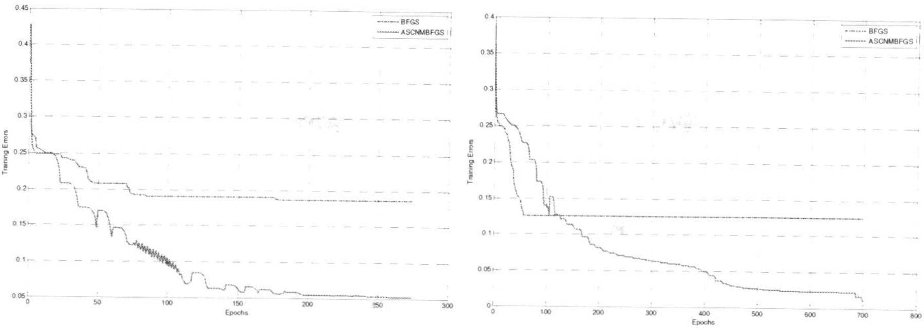

Fig. 1. Convergence behaviour of BFGS and ASCNMBFGS for training a FFTD network (left) and a LRN (right) on the parity-5 problem

3 Description of Applications

The first application is the well-known class of n-bit parity problems, which are typically nonlinear separable and have been widely used to verify performances of novel training algorithms [9][12]-[15][23][25][27][30][33][37][42][46]. We have also focused on temporal sequence processing tasks, and we have chosen two representative applications [32][42] in order to compare the algorithms generalisation performance, i.e. the ability of the various RNNs trained with these algorithms to perform well when tested on unknown input patterns.

(1) *Parity-N Problem*: given one two-valued, such as{0,1} (or {-1,1}), N-bit long input string I and an appropriate parity function, a one-bit output string O should be generated containing either 0 (-1) or 1 (+1). For example, assuming the input range is {0,1}, if there is an odd number of bits of 1s in I, then $O = 1$; otherwise, $O = 0$. We have tested the parity-5 and the parity-10 with input range {0,1}.

(2) *Sequence Learning*: given a grammar consisting of a set of letters, we could derive different length of sequences. In the particular application, the letter sequences consist of combinations of the strings {*ba*}, {*dii*} and {*guuu*} and all possible permutations of these 3 strings are legal, with each letter represented by 4-bits [32]. The task is to predict successive letters in a random sequence of 1000 words and each word consists of one of the above 3 consonant-vowel combinations. A sequence of 2993 letters is used for training and a sequence of 9 letters for testing as in [32].

(3) *Reading Aloud*: this task concerns learning the mapping of a set of orthographic representation to their phonological forms [42]. Both subsets of orthography and phonology have 3 different parts, i.e., onset, vowel and coda, with 30, 27 and 48, and 23, 14 and 24 possible characters, respectively. There are 105-dimensional input patterns and 61-dimensional output patterns, while the training dataset has 2998 patterns. Ideally, a specially designed RNN architecture with 100 hidden nodes, which is described in detail in [42], is needed to solve this problem.

4 Presentation of Experimental Results

All simulations in this paper were coded in Matlab 7.2 running on Windows XP platform. In this section, *Conv.* stands for the number of runs that reached the termination criteria and *Epochs* denotes the average number of epochs.

(1) *Parity-N Problem*: The stopping criterion for both parity-5 and parity-10 was set to a mean-squared error (MSE) of 0.01 within 2000 epochs and 4000 epochs respectively. In all cases, we made 100 runs from different initial weights using three different RNN architectures, i.e. FFTD, LRN and NARX. All RNNs used 7 hidden nodes for the parity-5 and 10 hidden nodes for the parity-10, and the heuristic parameters of the ASCNMBFGS were set to $3 \le M^k \le 15$, $\sigma = 0.9$ and $\delta = 0.01$.

Table 1. Average performance of the two algorithms on the parity-5

Algorithm	FFTD		LRN		NARX	
	Conv.	*Epoch*	*Conv.*	*Epoch*	*Conv.*	*Epoch*
BFGS	20	220	16	419	98	37
ASCNMBFGS	73	561	73	407	99	79

Table 2. Average performance of the two algorithms on the parity-10

Algorithm	FFTD		LRN		NARX	
	Conv.	*Epoch*	*Conv.*	*Epoch*	*Conv.*	*Epoch*
BFGS	6	1230	5	1483	67	244
ASCNMBFGS	58	1945	52	1736	90	283

As shown in Tables 1 and 2, the new method exhibit good average performance. It requires on average more epochs than the BFGS to train the RNNs but it always outperforms with regards to the number of converged runs out of 100.

(2) *Sequence Learning* (SL): in this case, we adopted the approach followed in [32] and trained the networks within 23 epochs keeping the same settings for the BFGS and ASCNMBFGS heuristic parameters as in the parity-N problem (in order to test the robustness of the method). We made 100 different runs for each of the three RNN architectures to estimate the average generalisation performance of the two algorithms. It is worth mentioning that the results in Table 3 were achieved using RNNs with 10 hidden nodes, while [32] requires 16 hidden nodes to produce an average MSE of 25% using a first-order training method.

(3) *Reading Aloud* (RA): this task requires a special RNN architecture and is computationally expensive in training. As our purpose was to test the new method on standard RNN architectures, we didn't deploy the special architecture proposed in [42], neither 1900 training epochs. Instead, we used a NARX network with only 5 hidden nodes that produced good results in previous tests, and trained it for 500 epochs. Fig 2 presents examples of convergence behaviour for BFGS and the ASCNMBFGS. The computational cost was still very high but we produced good solutions (see Table 4) using the ASCNMBFGS and BFGS algorithms. The heuristic

parameters were set to the same values as in the parity problems. It is worth mentioning the error reported in [42] is about 0.83 %, using 100 hidden nodes and 3 times more training epochs.

Table 3. Simulation results for the SL problem using 3 different RNN architectures

Nets	Algorithms	Average MSE generalisation (%) (testing on unknown data)
FFTD	ASCNMBFGS	15.457
	BFGS	17.437
LRN	ASCNMBFGS	15.070
	BFGS	16.492
NARX	ASCNMBFGS	14.902
	BFGS	18.673

Table 4. Results (%) for the RA problem

Generalisation MSE (testing on unknown data)	
ASCNMBFGS	BFGS
0.1039	1.4355
0.4255	1.8626

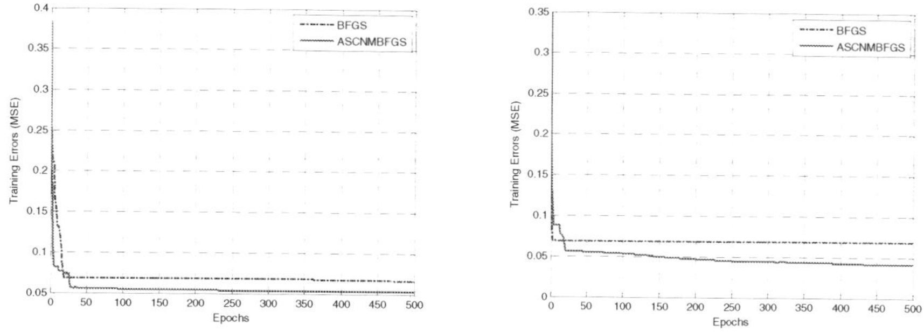

Fig. 2. Behaviours of BFGS and our method for training NARX networks on the reading aloud problem

5 Conclusion and Future Works

In this paper, we proposed a modified BFGS algorithm, called Adaptive Self-sCaling NonMonotone BFGS (ASCNMBFGS) that employs a self-scaling factor and adaptive nonmonotone learning. Simulation results show that our algorithm improves the convergence of the classic BFGS on three different recurrent neural network architectures, i.e. FFTD, LRN and NARX. We are currently working on fine tuning the heuristic parameters of our method and on performing a large scale study to fully explore the behaviour of the new algorithm and identify possible limitations.

References

1. Abraham, A.: Meta learning evolutionary artificial neural networks. Neurocomputing 56, 1–38 (2004)
2. Al-Baali, M.: Numerical experience with a class of self-scaling quasi-Newton algorithms. J. Optim. Theory Appl. 96, 533–553 (1998)
3. Alpsan, D., Towsey, M., Ozdamar, O., Tsoi, A.C., Ghista, D.N.: Efficiency of modified backpropagation and optimisation methods on a real-world medical problem. Neural Networks 8(6), 945–962 (1995)
4. Asirvadam, V.S., McLoone, S.F., Irwin, G.W.: Memory efficient BFGS neural-network learning algorithms using MLP-network: a survey. In: Proc. IEEE Int'l Conf. Control Application, pp. 586–591. IEEE Computer Society Press, Los Alamitos (2004)
5. Bajramovic, F., Gruber, C., Sick, B.: A comparison of first- and second- order training algorithms for dynamic neural networks. In: Proc. IEEE Int'l Joint Conf. Neural Networks, pp. 837–842. IEEE Computer Society Press, Los Alamitos (2004)
6. Baldi, P.: Gradient descent learning algorithm overview: a general dynamical systems perspective. IEEE Trans. Neural Networks 6(1), 182–195 (1993)
7. Becerikli, Y., Konar, A.F., Samad, T.: Intelligent optimal control with dynamic neural networks. Neural Networks 16, 251–259 (2003)
8. Becerikli, Y., Oysal, Y., Konar, A.F.: Trajectory priming with dynamic fuzzy networks in nonlinear optimal control. IEEE Tr. Neu. Net. 15(2), 383–394 (2004)
9. Beigi, H.S.M.: Neural networks learning through optimally conditioned quadratically convergent methods requiring no line search. In: Proc. 36th Midwest Symposium on Circuits and Systems, pp. 109–112 (1993)
10. Bengio, Y., Simard, P., Frasconi, P.: Learning long-term dependencies with gradient descent is difficult. IEEE Trans. Neural Networks 5(2), 157–166 (1994)
11. Charalambous, C.: A conjugate gradient algorithm for the efficient training of artificial neural networks. IEE Proceedings Part G 139, 301–310 (1992)
12. Chella, A., Gentile, A., Sorbello, F., Tarantino, A.: Supervised learning for feed-forward neural networks: a new minimax approach for fast convergence. In: Proc. IEEE Int'l Conf. Neural Networks, pp. 605–609. IEEE Computer Society Press, Los Alamitos (1993)
13. Chen, O.T.-C., Sheu, B.J.: Optimization schemes for neural network training. In: Proc. IEEE Int'l Conf. Neural Networks, pp. 817–822. IEEE Computer Society Press, Los Alamitos (1994)
14. Chen, K., Xu, L., Chi, H.: Improved learning algorithms for mixture of experts in multiclass classification. Neural Networks 12, 1229–1252 (1999)
15. Denton, J.W., Hung, M.S.: A comparison of nonlinear methods for supervised learning in multilayer feedforward neural nets. Eu. J. Ope. Res. 93, 358–368 (1996)
16. Dietterich, T.G.: Machine learning for sequential data: a review. In: Caelli, T.M., Amin, A., Duin, R.P.W., Kamel, M.S., de Ridder, D. (eds.) SPR 2002 and SSPR 2002. LNCS, vol. 2396, pp. 15–30. Springer, Heidelberg (2002)
17. dos Santos, E.P., von Zuben, F.J.: Efficient second-order learning algorithms for discrete-time recurrent neural networks. In: Medsker, L.R., Jain, L.C. (eds.) Rec. neural nets: design and applications, pp. 47–75. CRC Press, USA (2000)
18. Duch, W., Korczak, J.: Optimisation and global minimisation methods suitable for neural networks. Neural Computing Surveys 2, 163–212 (1999)
19. Fischer, M.M., Staufer, P.: Optimisation in an error backpropagation neural network environment with a performance test on a spectral pattern classification problem. Geographical Analysis 31(2), 89–108 (1999)

20. Gill, P.E., Murray, W., Wright, M.H.: Practical Optimization. Academic Press, London (1981)
21. Gordienko, P.: Construction of efficient neural networks: algorithms and tests. In: Proc. IEEE Int'l Joint Conf. Neural Networks, pp. 313–316. IEEE Computer Society Press, Los Alamitos (1993)
22. Grippo, L., Lampariello, F., Lucidi, S.: A nonmonotone line search technique for Newton's method. SIAM J. Numerical Analysis 23, 707–716 (1986)
23. Gruber, C., Sick, B.: Fast and efficient second-order training of the dynamic neural network paradigm. In: Proc IEEE Int'l J. Conf. Neu. Nets., vol. 4, pp. 2482–2487. IEEE Computer Society Press, Los Alamitos (2003)
24. Hui, L.C.K., Lam, K.-Y., Chea, C.W.: Global optimisation in neural network training. Neural Computing and Applications 5, 58–64 (1997)
25. Irwin, G., Lightbody, G., McLoone, S.: Comparison of gradient based training algorithms for multilayer perceptrons. In: Proc. IEE Colloquium Advances in Neural Networks for Control and Systems, 11/1-11/6 (1994)
26. Ishikawa, T., Tsukui, Y., Matsunami, M.: Optimization of electromagnetic devices using artificial neural network with quasi-Newton algorithm. IEEE Trans. Magnetics 32(3), 1226–1229 (1996)
27. Karjala, T.W., Himmelblau, D.M., Miikkulainen, R.: Data rectification using recurrent (Elman) neu. nets. In: Proc. IEEE Int'l J. Conf. Neu. Nets., pp. 901–906. IEEE Computer Society Press, Los Alamitos (1992)
28. Kremer, S.C.: Spatiotemporal connectionist networks: a taxonomy and review. Neural Computation 13, 249–306 (2001)
29. Lightbody, G., Irwin, G.W.: A novel neural internal model control structure. In: Proc. American Control Conf., pp. 350–354 (1995)
30. Lightbody, G., Irwin, G.W.: Multi-layer perceptron based modelling of nonlinear systems. Fuzzy Sets and Systems 79, 93–112 (1996)
31. Likas, A., Stafylopatis, A.: Training the random neural network using quasi-Newton methods. European J. Operational Research 126, 331–339 (2000)
32. McLeod, P., Plunkett, K., Rolls, E.T.: Introduction to connectionist modelling of cognitive processes, pp. 148–151. Oxford University Press, Oxford (1998)
33. McLoone, S., Asirvadam, V.S., Irwin, G.W.: A memory optimal BFGS neural network training algorithm. In: Proc. IEEE Int'l J. Conf. Neu. Nets., pp. 513–518. IEEE Computer Society Press, Los Alamitos (2002)
34. McLoone, S., Irwin, G.W.: Fast parallel off-line training of multilayer perceptrons. IEEE Tr. Neural Networks 8(3), 646–653 (1997)
35. McLoone, S., Irwin, G.: A variable memory quasi-Newton training algorithm. Neural Processing Letters 9, 77–89 (1999)
36. Mizutani, E.: Powell's dogleg trust-region steps with the quasi-Newton augmented Hessian for neural nonlinear least-squares learning. In: Proc. IEEE Int'l Joint Conf. Neural Networks, pp. 1239–1244. IEEE Computer Society Press, Los Alamitos (1999)
37. Nawi, N.M., Ransing, M.R., Ransing, R.S.: An improved learning algorithm based on the Broyden-Fletcher-Goldfarb-Shanno (BFGS) method for back propagation neural networks. In: Proc. 6th Int'l Conf. Intel. Sys. Design and Applications, pp. 152–157 (2006)
38. Nocedal, J., Yuan, Y.: Analysis of a self-scaling quasi-Newton method. Math. Program 61, 19–37 (1993)
39. Pearlmutter, B.A.: Gradient calculations for dynamic recurrent neural networks: a survey. IEEE Trans. Neural Networks 6(5), 1212–1228 (1995)

40. Phua, P.K.H., Ming, D.: Parallel nonlinear optimization techniques for training neural networks. IEEE Trans. Neural Networks 14(6), 1460–1468 (2003)
41. Plagianakos, V.P., Magoulas, G.D., Vrahatis, M.N.: Deterministic nonmonotone strategies for effective training of multi-layer perceptrons. IEEE Trans. Neural Networks 13(6), 1268–1284 (2002)
42. Plaut, D., McClelland, J., Seidenberg, M., Patterson, K.: Understanding normal and impaired reading: computational principles in quasi-regular domains. Psychological Review 103, 56–115 (1993)
43. Saini, L.M., Soni, M.K.: Artificial neural network based peak load forecasting using Levenberg-Marquardt and quasi-Newton methods. In: Proc. Generation, Transmission and Distribution, pp. 578–584 (2002)
44. Setiono, R., Hui, L.C.K.: Some n-bit parity problems are solvable by feed-forward networks with less than n hidden units. In: Proc. IEEE Int'l Joint Conf. Neural Networks, pp. 305–308. IEEE Computer Society Press, Los Alamitos (1993)
45. Setiono, R., Hui, L.C.K.: Use of a quasi-Newton in a feedforward neural network construction algorithm. IEEE Trans. Neural Networks 6(1), 273–277 (1995)
46. Sorensen, P.H., Norgaard, M., Ravn, O., Poulsen, N.K.: Implementation of neural network based non-linear predictive control. Neurocomputing 28, 37–51 (1999)
47. Sperduti, A., Starita, A.: Supervised neural networks for the classification of structures. IEEE Trans. Neural Networks 8(3), 714–735 (1997)
48. Tsoi, A.C.: Recurrent neural network architectures: an overview. In: Giles, C.L., Gori, M. (eds.) Adaptive Processing of Sequences and Data Structures. LNCS (LNAI), vol. 1387, pp. 1–26. Springer, Heidelberg (1998)
49. Tsoi, A.C.: Gradient based learning algorithms. In: Giles, C.L., Gori, M. (eds.) Adaptive Processing of Sequences and Data Structures. LNCS (LNAI), vol. 1387, pp. 27–62. Springer, Heidelberg (1998)
50. Yin, H.X., Du, D.L.: The global convergence of self-scaling BFGS algorithm with nonmonotone line search for unconstrained nonconvex optimization problems. Acta Mathematica Sinica, Springer, Heidelberg (2006), (Online) http://www.springerlink.com/content/l7h30w8t42l7mh64/fulltext.pdf

Some Properties of the Gaussian Kernel for One Class Learning

Paul F. Evangelista[1], Mark J. Embrechts[2], and Boleslaw K. Szymanski[2]

[1] United States Military Academy, West Point, NY 10996
[2] Rensselaer Polytechnic Institute, Troy, NY 12180

Abstract. This paper proposes a novel approach for directly tuning the gaussian kernel matrix for one class learning. The popular gaussian kernel includes a free parameter, σ, that requires tuning typically performed through validation. The value of this parameter impacts model performance significantly. This paper explores an automated method for tuning this kernel based upon a hill climbing optimization of statistics obtained from the kernel matrix.

1 Introduction

Kernel based pattern recognition has gained much popularity in the machine learning and data mining communities, largely based upon proven performance and broad applicability. Clustering, anomaly detection, classification, regression, and kernel based principal component analysis are just a few of the techniques that use kernels for some type of pattern recognition. The kernel is a critical component of these algorithms - arguably the most important component.

The gaussian kernel is a popular and powerful kernel used in pattern recognition. Theoretical statistical properties of this kernel provide potential approaches for the tuning of this kernel and potential directions for future research. Several heuristics which have been employed with this kernel will be introduced and discussed.

Assume a given data set $\mathbf{X} \in \mathbb{R}^{N \times m}$. $\mathbf{X}$ contains N instances or observations, $\mathbf{x}_1, \mathbf{x}_2, ..., \mathbf{x}_N$, where $\mathbf{x}_i \in \mathbb{R}^{1 \times m}$. There are m variables to represent each instance i. For every instance there is a label or class, $y_i \in \{-1, +1\}$. Equation 1 illustrates the formula to calculate a gaussian kernel.

$$\kappa(i,j) = e^{\frac{-\|\mathbf{x}_i - \mathbf{x}_j\|^2}{2\sigma^2}} \tag{1}$$

This kernel requires tuning for the proper value of σ. Manual tuning or brute force search are alternative approaches. An brute force technique could involve stepping through a range of values for σ, perhaps in a gradient ascent optimization, seeking optimal performance of a model with training data. Regardless of the method utilized to find a proper value for σ, this type of model validation is common and necessary when using the gaussian kernel. Although this approach is feasible with supervised learning, it is much more difficult to tune σ for unsupervised learning methods. The one-class SVM, originally proposed by Tax and

J. Marques de Sá et al. (Eds.): ICANN 2007, Part I, LNCS 4668, pp. 269–278, 2007.

Duin [16] and also detailed by Scholkopf et. al. [12], is a good example of an unsupervised learning algorithm where training validation is difficult due to the lack of positive instances. The one-class SVM trains with all negative instances or observations, and based upon the estimated support of the negative instances, new observations are classified as either inside the support (predicted negative instance) or outside of the support (predicted positive instance). It is quite possible, however, that there are very few or no positive instances available. This poses a validation problem. Both Tax and Duin [16] and Scholkopf et. al. [12] state that tuning the gaussian kernel for the one-class SVM is an open problem.

2 Recent Work

Tax and Duin [16] and Scholkopf et. al. [12] performed the groundbreaking work with the one-class SVM. Stolfo and Wang [15] successfully apply the one-class SVM to the intrusion data set that we use in this paper. Chen et. al. [3] uses the one-class SVM for image retrieval. Shawe-Taylor and Cristianini [14] provide the theoretical background for this method.

Tax and Duin [17] discuss selection of the σ parameter for the one-class SVM, selecting σ based upon a predefined error rate and desired fraction of support vectors. This requires solving the one-class SVM for various values of σ and the parameter C, referred to in this paper as $1/\nu N = C$. This method relies on the fraction of support vectors as an indicator of future generalization. Tuning of the parameter C does not significantly impact the ordering of the decision values created by the one-class SVM; tuning of σ influences the shape of the decision function and profoundly impacts the ordering of the decision values. When seeking to maximize the area under the ROC curve (AUC), the ordering of the decision values is all that matters. The technique in this paper is very different from the one which is mentioned in [17]. We use the kernel matrix directly, therefore the one-class SVM does not need to be solved for each change in value of σ. Furthermore, tuning the kernel matrix directly requires the tuning of only one parameter.

3 The One-Class SVM

The one-class SVM is an anomaly detection model solved by the following optimization problem:

$$\min_{R\in\mathbb{R},\zeta\in\mathbb{R}^N,c\in F} \quad R^2 + \frac{1}{vN}\sum_i \zeta_i \tag{2}$$

$$\text{subject to} \quad \| \Phi(\mathbf{x}_i) - c \|^2 \leq R^2 + \zeta_i \text{ and } \zeta_i \geq 0 \text{ for } i = 1, ..., N$$

The lagrangian dual is shown below in equation 3.

$$\max_{\alpha} \sum_i \alpha_i \kappa(\mathbf{x}_i, \mathbf{x}_i) - \sum_{i,j} \alpha_i \alpha_j \kappa(\mathbf{x}_i, \mathbf{x}_j) \tag{3}$$

$$\text{subject to} \quad 0 \leq \alpha_i \leq \frac{1}{vN} \text{ and } \sum_i \alpha_i = 1$$

Scholkopf et. al. point out the following reduction of the dual formulation when modeling with gaussian kernels:

$$\min_{\alpha} \sum_{i,j} \alpha_i \alpha_j \kappa(\mathbf{x}_i, \mathbf{x}_j) \tag{4}$$

$$\text{subject to} \quad 0 \leq \alpha_i \leq \frac{1}{vN} \text{ and } \sum_i \alpha_i = 1$$

This reduction occurs since we know that $\kappa(\mathbf{x}_i, \mathbf{x}_i) = 1$ and $\sum_i \alpha_i = 1$. Equation 4 can also be written as $\min \boldsymbol{\alpha}'\mathbf{K}\boldsymbol{\alpha}$. Shawe-Taylor and Cristianini [14] explain that $\boldsymbol{\alpha}'\mathbf{K}\boldsymbol{\alpha}$ is the weight vector norm, and controlling the size of this value improves the statistical stability or regularization of the model.

All training examples with $\alpha_i > 0$ are support vectors, and the examples which also have a strict inequality of $\alpha_i < \frac{1}{vN}$ are considered non-bounded support vectors.

In order to classify a new test instance, $\mathbf{v}$, we would evaluate the following decision function:

$$f(v) = \kappa(\mathbf{v}, \mathbf{v}) - 2\sum_j \alpha_j \kappa(\mathbf{v}, \mathbf{x}_j) + \sum_{j,k} \alpha_k \alpha_j \kappa(\mathbf{x}_k, \mathbf{x}_j) - R^2$$

Before evaluating for a new point, R^2 must be found. This is done by finding a non-bounded support vector training example and setting the decision function equal to 0 as detailed by Bennett and Campbell [1]. If the decision function is negative for a new test instance, this indicates a negative or healthy prediction. A positive evaluation is an unhealthy or positive prediction, and the magnitude of the decision function in either direction is an indication of the model's confidence. It is useful to visualize the decision function of the one class SVM with a small two dimensional dataset, shown in figure 1. The plots corresponding to a small σ value clearly illustrate that overtraining is occurring, with the decision function wrapped tightly around the data points. Large values of sigma simply draw an oval around the points without defining the shape or pattern.

4 Method

The behavior of the gaussian kernel is apparent when examined in detail. The values lie within the (0,1) interval. A gaussian kernel matrix will have ones along the diagonal (because $\| \mathbf{x}_i - \mathbf{x}_i \| = 0$). Additionally, a value too small for σ will force the matrix entries towards 0, and a value too large for σ will force matrix entries towards 1. There is also a property of all kernels, which we will refer to as the fundamental premise of pattern recognition, which simply indicates that for good models, the following relationship consistently holds true:

$$(\kappa(i,j)|(y_i = y_j)) > (\kappa(i,j)|(y_i \neq y_j)) \tag{5}$$

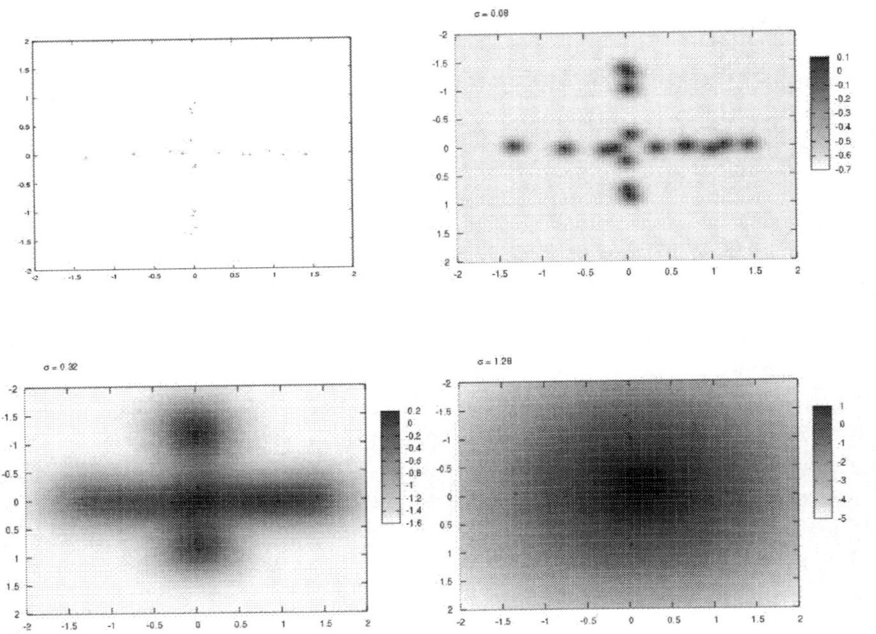

Fig. 1. Visualization of one class SVM for three different values for σ. The source data, named the cross data due to its pattern, is shown in the top left.

Consistent performance and generalization of the fundamental premise of pattern recognition is the goal of all kernel based learning. Given a supervised dataset, a training and validation split of the data is often used to tune a model which seems to consistently observe the fundamental premise. However, in an unsupervised learning scenario positive labeled data is limited or non-existent, and furthermore, models such as the one-class SVM have no use for positive labeled data in the training data.

A first approach towards tuning a kernel matrix for the one-class SVM might lead one to believe that the matrix should take on very high values, indicating that all of the kernel entries for the training data is of one class and therefore should take on high values. Although this approach would first seem to be consistent with the fundamental premise in equation 5, this approach would be misguided. The magnitude of the values within the kernel matrix is not an important attribute. The important attribute is actually the spread or the variance of the entries in the kernel matrix. At first this may seem to be anomalous with equation 5, however a closer examination of the statistics of a kernel matrix illustrates why the variance of the kernel matrix is such a critical element in model performance.

Shawe-Taylor and Cristianini point out that small values of σ allow classifiers to fit any set of labels, and therefore overfitting occurs [14]. They also state that large values for σ impede a classifiers ability to detect non-trivial patterns because the kernel gradually reduces to a constant function. The following mathematical discussion supports these comments for the one-class SVM. Considering again the one-class SVM optimization problem, posed as $\min \boldsymbol{\alpha}'\mathbf{K}\boldsymbol{\alpha}$ assuming the a gaussian kernel, if the sigma parameter is too small and $\kappa(i,j) \to 0$, the optimal solution is $\alpha_i = 1/N$. Equation 4, the objective function, will equate to $1/N$ (since $\sum_i (1/N)^2 = 1/N$). If the sigma parameter is too large and $\kappa(i,j) \to 1$, the optimal solution is the entire feasible set for $\boldsymbol{\alpha}$. Given these values for the variables and parameters, the objective function will now equate to 1. The brief derivation for the case when $\kappa(i,j) \to 1$ follows:

$$\alpha'\mathbf{K}\alpha = \sum_{i=1}^{N} \alpha_i^2 + \sum_{i=1}^{N} \sum_{j=i+1}^{N} 2\alpha_i\alpha_j\kappa(i,j)$$

$$= \sum_{i=1}^{N} \alpha_i^2 + \sum_{i=1}^{N} \alpha_i \sum_{j=1\dots i-1,i+1\dots N} \alpha_j$$

$$= \sum_{i=1}^{N} \alpha_i^2 + \sum_{i=1}^{N} \alpha_i(1-\alpha_i) = \sum_{i=1}^{N} \alpha_i^2 + \sum_{i=1}^{N} \alpha_i - \sum_{i=1}^{N} \alpha_i^2 = \sum_{i=1}^{N} \alpha_i = 1$$

The objective function bounds are $(1/N, 1)$, and the choice of σ greatly influences where in this bound the solution lies.

4.1 The Coefficient of Variance

In order to find the best value for σ, a heuristic is employed. This heuristic takes advantage of the behavior of the one-class SVM when using the gaussian kernel. The mean and the variance of the non-diagonal kernel entries, $\kappa(i,j)|i \neq j$, play a crucial role in this heuristic. We will refer to the mean as $\bar{\kappa}$ and the variance as s^2. For any kernel matrix where $i, j \in \{1, ..., N\}$, there are $N^2 - N$ off diagonal kernel entries. Furthermore, since all kernel matrices are symmetrical, either the upper or lower diagonal entries only need to be stored in memory, of which there are $l = (N^2 - N)/2$. From here forward, the number of unique off diagonal kernel entries will be referred to as l.

It is first necessary to understand the statistic used in this heuristic, the coefficient of variance. The coefficient of variance is commonly referred to as 100 times the sample standard deviation divided by its mean, or $\frac{100s}{\bar{x}}$. This statistic describes the relative spread of a distribution, regardless of the unit of scale. Due to the scale and behavior of $\bar{\kappa}$ and s, this coefficient of variance monotonically increases for gaussian kernels as σ ranges from 0 to ∞. Using the sample variance rather than the standard deviation, different behavior occurs. The monotonic increase of the coefficient of variance occurs because when σ is small, $s > \bar{\kappa}$; however, as σ increases, there is a cross-over point and then $s < \bar{\kappa}$. However, the variance of $\kappa(i,j)|i \neq j$ is always smaller than the mean of $\kappa(i,j)|i \neq j$. This property is what makes the variance of a kernel matrix such a critical component for the direct tuning method. The proof follows. For the sake of notation simplicity, $x_k \in (0,1)$, $k \in [l]$, will represent off diagonal kernel entries, that is entries $\kappa(i,j)|i \neq j$.

$$VAR(x_k) \leq \bar{x} \implies \frac{\sum_k x_k^2 - 2\bar{x}\sum_k x_k + l\bar{x}^2}{l-1} \leq \frac{(l-1)\bar{x}}{l-1}$$

$$\implies \frac{l\bar{x}^2 - 2l\bar{x}^2 - (l-1)\bar{x} + \sum_k {x_k}^2}{l-1} \leq 0$$

$$\implies \sum_k x_k^2 - l\bar{x}^2 - (l-1)\bar{x} \leq 0 \implies \sum_k x_k^2 - \sum_k x_k + \frac{\sum_k x_k}{l}(1 - \sum_k x_k) \leq 0$$

$$\implies l\sum_k x_k^2 - \sum_k x_k(l-1) - (\sum_k x_k)^2 \leq 0$$

$$\implies \sum_k x_k^2 - (\sum_k x_k)^2 + (l-1)\sum_k x_k^2 - (l-1)\sum_k x_k \leq 0$$

$$\implies \underbrace{\sum_k x_k^2 - (\sum_k x_k)^2}_{\text{always } \leq 0} + (l-1)\underbrace{\left(\sum_k x_k^2 - \sum_k x_k\right)}_{\text{always } \leq 0 \text{ for } 0 \leq x_k \leq 1} \leq 0$$

The fact that the variance of $\kappa(i,j)|i \neq j$ is always smaller than the mean of $\kappa(i,j)|i \neq j$ indicates that $s^2/\bar{\kappa}$ is a fraction. Furthermore, as σ ranges from 0 to ∞, this fraction ascends to a global max. In order to protect against division by zero and roundoff error, a small value, ϵ, can be added to the denominator. The results in the following objective function for optimization which can be solved quickly with a gradient ascent algorithm. Solving the optimization problem in equation 6 leads to the best choice for σ.

$$\max_{\sigma} \frac{s^2}{\bar{\kappa} + \epsilon} \; \forall (i \neq j) \tag{6}$$

such that

$$\kappa(i,j) = e^{\frac{-\|\mathbf{x}_i - \mathbf{x}_j\|^2}{2\sigma^2}}, \; \bar{\kappa} = \frac{\sum_{i=1}^{N}\sum_{j=i+1}^{N} \kappa(i,j)}{l}, \; s^2 = \frac{\sum_{i=1}^{N}\sum_{j=i+1}^{N} (\kappa(i,j) - \bar{\kappa})^2}{l-1}$$

Figure 2 illustrates the impact of sigma on the kernel matrix. The dataset used to create figure 2 was the ionosphere data which is available from the UCI repository. These data are all of the negative class, and it is evident that both the kernel element values and the dispersion of these values changes as σ changes. The optimal value for σ for this dataset is 1, and the color visualization of the kernel matrix when $\sigma = 1$ clearly indicates that there is dispersion in the kernel matrix. However, it is also noticeable that the central tendency of the kernel entries for this optimal σ is a small value, closer to zero than one. This is consistent for all of the datasets. This is why it is important to use a metric that detects relative dispersion and is not biased by the magnitude of the kernel entries.

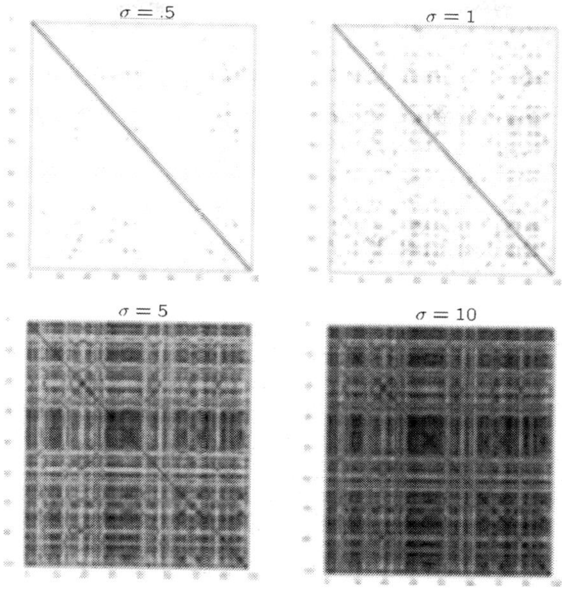

Fig. 2. Color visualization of a kernel matrix with various values for σ. This visualization involves 100 observations where kernel entries close to 0 are light; darker entries represent values closer to unity.

4.2 Why Direct Tuning of the Kernel Matrix Works

As mentioned previously, it is desirable to minimize $\boldsymbol{\alpha}'\mathbf{K}\boldsymbol{\alpha}$. $\boldsymbol{\alpha}$ is sparse when there are few support vectors, and this sparseness is typically desirable to minimize complexity and improve regularization. However, meaningless sparse solutions for α can occur as all $\kappa(i,j) \to 0$. Meaningful sparse solutions for $\boldsymbol{\alpha}$ occur when the kernel matrix entries are not concentrated either towards 0 or 1, but are showing good dispersion and there is payoff in the optimization to select the few instances which clearly define the margin of the density approximated by the one-class SVM. Although the variance, s^2, is a good indicator of the dispersion in the kernel matrix, it is biased towards a larger σ since variance is affected by unit of scale or magnitude. $s^2/(\bar{\kappa}+\epsilon)$ is robust against unit of scale. This statistic illustrates the dispersion of a distribution regardless of scale. For the one class SVM, the optimal value for σ and maximum value for $s^2/(\bar{\kappa}+\epsilon)$ typically occurs as σ increases from 0 and the kernel entries first begin to illustrate dispersion. When there is maximal dispersion in the kernel matrix, the values assigned to $\boldsymbol{\alpha}$ will reflect the behavior and relationships of observations, which is the purpose of the statistical learning. Maximal dispersion of the kernel matrix also supports the fundamental premise of pattern recognition. When σ is not properly tuned, the values assigned to $\boldsymbol{\alpha}$ will be erroneously based upon the behavior of the optimization problem or the optimization algorithm used to solve the problem.

5 Experimental Results

In order to evaluate the performance of the direct tuning heuristic, several experiments were conducted. The data included three benchmark sets: banana, chessboard, and ionosphere (from UCI repository). Additionally, two computer intrusion datasets named Schonlau and Sick, after their respective creators, are also examined. The Schonlau data involves determining authenticity of a user based on UNIX commands. This data was originally discussed in ([4,5,13]) and the actual data used in this paper was also discussed in ([6,7,8]). The Sick data was originally examined in [11]. The parameter ν is set to .5 for every experiment.

Dataset	dimensions	positive	negative	comment
Banana	2	50	50	see ([9])
Chessboard	2	50	50	www.cs.wisc.edu/ ~olvi/data/check1.txt
Ionosphere	34	126	225	UCI Repository
Schonlau	26	231	4769	www.schonlau.net
Sick	137	250	5143	see ([11])

For each dataset, one half of the negative class was used for training the one-class SVM and the test data comprised of the other half of the negative class and all members of the positive class. The same split remained consistent for all experiments for the purposes of control and consistency. The performance measure utilized is the area under the receiver operating characteristic curve (AUC).

The three benchmark datasets clearly illustrated optimal performance when $s^2/(\bar{\kappa}+\epsilon)$ is optimal. For each of the benchmark solutions, a gradient ascent algorithm was tested and converged on the optimal σ within 4-6 iterations depending on the initial values and dataset. An initial value of 1 for σ seemed to work well since most of the optimal values for σ ranged between .1 and 10.

The two computer intrusion datasets did not clearly indicate this same type of ideal performance. The Schonlau dataset performed the worst by far, and the Sick dataset did indicate a region of good performance. The value of σ seemed indifferent for the Sick data.

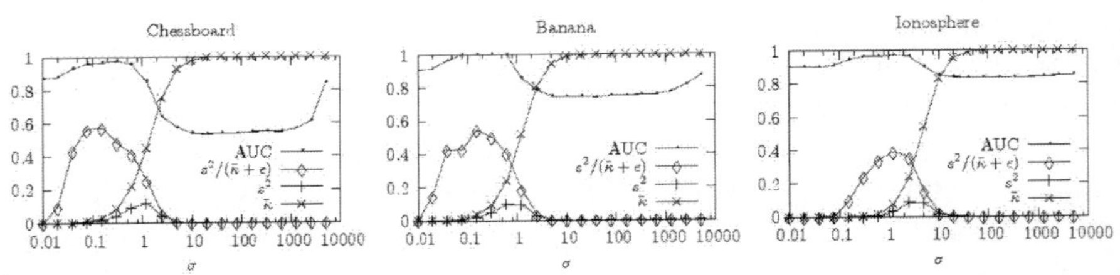

Fig. 3. Experimental results for three benchmark datasets

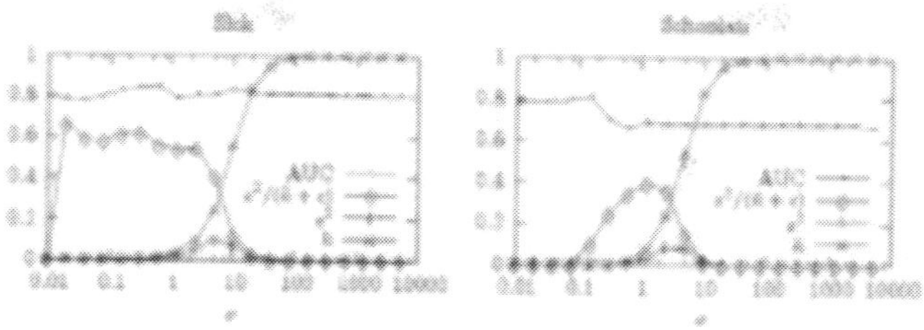

Fig. 4. Experimental results for two computer intrusion datasets

6 Conclusions

Direct tuning of the gaussian kernel matrix is a novel and promising approach for the necessary tuning of the powerful gaussian kernel. The heuristic proposed is grounded and supported with underlying theory. The significance of tuning the gaussian kernel for unsupervised learning applies directly to ensemble methods. Techniques such as bagging [2], random subspace selection [10], and fuzzy aggregation techniques [6,7] can be employed for unsupervised learning with the gaussian kernel.

Future research could involve automated tuning approaches for supervised learning, however this approach will clearly have to outperform the traditional technique of validation. Direct tuning for supervised learning would be faster, but first it needs to be shown that it is also just as accurate as validation.

References

1. Bennett, K.P., Campbell, C.: Support Vector Machines: Hype or Hallelujah. SIGKDD Explorations 2(2) (2001)
2. Breiman, L.: Bagging predictors. Machine Learning 24(2), 123–140 (1996)
3. Chen, Y., Zhou, X., Huang, T.S.: One-Class SVM for Learning in Image Retrieval. In: Proceedings of IEEE International Conference on Image Processing, Thessaloniki, Greece, IEEE Computer Society Press, Los Alamitos (2001)
4. DuMouchel, W., Ju, W.H., Karr, A.F., Schonlau, M., Theus, M., Vardi, Y.: Computer Intrusion: Detecting Masquerades. Statistical Science 16(1), 1–17 (2001)
5. DuMouchel, W., Schonlau, M.: A Fast Computer Intrusion Detection Algorithm Based on Hypothesis Testing of Command Transition Probabilities. In: The Fourth International Conference of Knowledge Discovery and Data Mining, August 1998, pp. 189–193 (1998)
6. Evangelista, P.F., Bonissone, P., Embrechts, M.J., Szymanski, B.K.: Fuzzy ROC Curves for the One Class SVM: Application to Intrusion Detection. In: Proceedings of the International Joint Conference on Neural Networks, Montreal, Canada, August 2005 (2005)
7. Evangelista, P.F., Bonissone, P., Embrechts, M.J., Szymanski, B.K.: Unsupervised Fuzzy Ensembles and Their Use in Intrusion Detection. In: Proceedings of the European Symposium on Artificial Neural Networks, Bruges, Belgium, April 2005 (2005)

8. Evangelista, P.F., Embrechts, M.J., Szymanski, B.K.: Computer Intrusion Detection Through Predictive Models. Smart Engineering System Design: Neural Networks, Fuzzy Logic, Evolutionary Programming, Data Mining and Complex Systems, St. Louis, Missouri, November 2004, pp. 489–494 (2004)
9. Frossyniotis, D., Likas, A., Stafylopatis, A.: A Clustering Method Based on Boosting. Pattern Recognition Letters 25, 641–654 (2004)
10. Ho, T.K.: The Random Subspace Method for Constructing Decision Forests. IEEE Transactions on Pattern Analysis and Machine Intelligence 20(8), 832–844 (1998)
11. Hofmann, A., Horeis, T., Sick, B.: Feature Selection for Intrusion Detection: An Evolutionary Wrapper Approach. In: International Joint Conference on Neural Networks, Budapest, Hungary (2004)
12. Schölkopf, B., Platt, J.C., Taylor, J.S., Smola, A.J., Williamson, R.C.: Estimating the Support of a High Dimensional Distribution. Neural Computation 13, 1443–1471 (2001)
13. Schonlau, M., Theus, M.: Detecting Masquerades in Intrusion Detection Based on Unpopular Commands. Information Processing Letters 76(1-2), 33–38 (2000)
14. Shawe-Taylor, J., Cristianini, N.: Kernel Methods for Pattern Analysis. Cambridge University Press, Cambridge (2004)
15. Stolfo, S., Wang, K.: One Class Training for Masquerade Detection. In: 3rd IEEE Conference Data Mining Workshop on Data Mining for Computer Security, Florida, 19 November 2003, IEEE Computer Society Press, Los Alamitos (2003)
16. Tax, D.M.J., Duin, R.P.W.: Support Vector Domain Description. Pattern Recognition Letters 20, 1191–1199 (1999)
17. Tax, D.M.J., Duin, R.P.W.: Support Vector Data Description. Machine Learning 54, 45–66 (2004)

Improved SOM Learning Using Simulated Annealing

Antonino Fiannaca[1], Giuseppe Di Fatta[2], Salvatore Gaglio[1,3], Riccardo Rizzo[1], and Alfonso M. Urso[1]

[1]ICAR-CNR, Consiglio Nazionale delle Ricerche, Palermo, Italy
[2]School of Systems Engineering, Univeristy of Reading, UK
[3]Dipartimento di Ingegneria Informatica, Universitá di Palermo, Italy

Abstract. Self-Organizing Map (SOM) algorithm has been extensively used for analysis and classification problems. For this kind of problems, datasets become more and more large and it is necessary to speed up the SOM learning. In this paper we present an application of the Simulated Annealing (SA) procedure to the SOM learning algorithm. The goal of the algorithm is to obtain fast learning and better performance in terms of matching of input data and regularity of the obtained map. An advantage of the proposed technique is that it preserves the simplicity of the basic algorithm. Several tests, carried out on different large datasets, demonstrate the effectiveness of the proposed algorithm in comparison with the original SOM and with some of its modification introduced to speed-up the learning.

1 Introduction

The Self-Organizing Map (SOM) algorithm is used to build a mapping from an high dimensional data space to a low-dimensional representation space. A specific characteristic, distinguishing SOM from other data mining techniques, is the neighborhood preservation of lattice map. Nowadays, datasets used in data mining become larger time after time and a fast analysis is required. Typically, techniques used for speeding up SOM algorithm, work in two different ways: modifying the characteristic of neurons to adjust influence of input in the lattice map [1] [2] or modifying the standard algorithm to address the variation of learning parameters [3] [4]. The first approach acts on lower abstraction layer than second ones, because they redefine the SOM neurones; otherwise the second one changes only the learning rule depending on network evolution. This work deals with an application for unsupervised clustering of high dimensional datasets using a technique preserving the simplicity and the functionality of classic SOM algorithm: in this point of view, the proposed approach works at the higher abstraction level. Our solution, during training phase, introduces an optimization technique, the simulated annealing (SA), to drive the system toward the global minimum. Then we have carried out studies both on the shapes and on the ideal values that learning rate factor α could assume during the training phase. However the variation of learning rate factor can not satisfy the previous

J. Marques de Sá et al. (Eds.): ICANN 2007, Part I, LNCS 4668, pp. 279–288, 2007.

goal, thus an optimization technique, driving the system to final optimal solution, is been used. The simulated annealing [5] [6] could be an attempt to find global optimum with respect to above-mentioned constraints. SA has been also adopted in SOM to improve the detection of winning neurons [7] or to select a representative pattern for each update during training phase [8]. In [7] the SA, replaced by a deterministic implementation (Deterministic Annealing), does not imply an improvement in term of computational complexity but just with regard to quality of learning; that happens because the selection of winning neurons, solved by Kohonen in simplest way, here is steered by a minimization of a cost function where the SA is strongly used. Instead, in [8], learning process requires nearly the same execution time than standard SOM and, moreover, the evolution of training should avoid local minima even though the learning could advantage a few patterns. In this work we introduce a SOM learning algorithm that adds SA as evaluation and directional criterion for the map evolution. The proposed algorithm, called Fast Learning SOM ($FLSOM$) is faster than the regular SOM and it is compared with SOM, PLSOM, HabSOM and ConsSOM(see section 2).

The paper has the following structure: the next section describes some related works aiming to speed up learning of SOM; the section 3 reports the basic SOM algorithm and describes the proposed algorithm; the section 4 reports the experimental results. Finally some conclusions are reported in section 5.

2 Related Work

Many different mechanisms have been proposed in the literature for improving SOM performances. The key issue is to determine which parameters need to be considered, in order to obtain a SOM that both achieves a clustering in a short time and creates a data projection strongly related to the distribution of data in the input space. One such attempt was done in auto-SOM [4], introducing a complex algorithm, based on Kalman filters coupled with a recursive parameter estimation method depending on [9] [10], to guide the weight vectors toward the center of their respective Voronoi cells in input spaces. Using this algorithm, it is possible to automatically estimate the learning parameters during the training of SOMs. Indeed the introduction of Kalman filters leads the network to a good training, but it is more computationally expensive than the classic SOM. Notice that in [4] authors confirm that the incremental learning SOM algorithm is faster than the auto-SOM, because the former needs fewer learning steps than the latter during training process. Moreover, they also assert that standard techniques often get stuck in local minima. A recent improvement to the SOM original algorithm is the parameterless SOM (PLSOM) [3] [11]: this technique is based on the standard SOM algorithm and removes both classic learning rate function and neighborhood size function. Usually these functions are decreased over time and do not take into account the evolution of the network during learning process. PLSOM evaluates the adaptation of input stimuli and calculates the learning factor and neighborhood size depending on the local quadratic fitting error of the map to the input space. Unfortunately only the local error is used during the evolution of the map; this means that both the first data inputs and the initialization of the map, play a key role in map evolution and, moreover, the selection of the learning rate is modified without evaluating weights in most of neurons. An adaptation of habituation mechanism in SOMs was proposed in [1] and results was compared to conscience algorithm [2]. These algorithms are based mainly on the identification of neurons that win frequently and thus on the introduction of an handicap for these neurons. This way, each neuron is selected with almost the same probability of the other ones. In this context, the habituation mechanism is more flexible than conscience-learning because it can be used to manage the learning processing a fine grain way. Otherwise, the conscience mechanism speed up learning process using an *a priori* knowledge to estimate a value of probability in order to catch winning neuron. The algorithm presented in [1], compared to standard SOM implementation, is not more computationally expensive, although the algorithm adds an habituation parameter. This parameter is a function of a local error so that the habituation is increased when the network is not ordered. The introduction of the habituation causes the deceleration of the training during the refinement phase of the learning.

3 SOM Adaptive Learning Rate Algorithm

In this section the learning algorithm of the FLSOM is introduced. Although the SOM learning procedure is well known it is necessary to briefly highlight some points of the original algorithm to better understand the proposed one.

3.1 Self-Organizing Map Basic Algorithm

Self-Organizing Maps [12] are neural structures capable of building maps of the input data, which preserve neighborhood relationships. Although the SOM algorithm is well known, it is necessary to report some formulas in order to better understand the modified algorithms. The incremental learning algorithm [13] trained for epochs is used; this version of the learning algorithm is reported in table 1.

The main formulas of the standard SOM are reported below. In (step 3.(a).ii) of Table 1 the winner unit, called best matching unit (bmu) is selected according to:

$$bmu = \arg\left(\min_{i \in N} \|x - w_i\|\right) . \tag{1}$$

In (step 3.(a).iii) neural weights are updated using the following rule:

$$w_i(t+1) = w_i(t) + \alpha(t) h_{ci}(t) \left[x - w_i(t)\right] , \tag{2}$$

where h_{ci} is the neighborhood kernel around the best matching unit. One of the most common shapes of the kernel is the gaussian shape:

$$h_{ci}(t) = \exp\left(\frac{d\left(r_{bmu}, r_i\right)}{2\sigma^2(t)}\right) , \tag{3}$$

where the term $d(r_{bmu}, r_i)$ stands for the distance between the bmu unit and the generic unit i on the SOM lattice.

Table 1. Incremental learning algorithm trained for epochs

1. Initialize the SOM neurons with random weights.
2. Set epoch counter $p = 1$ and step counter $t = 1$.
3. If stop condition is not verified
   (a) If learning dataset is not empty
      i. Get randomly a pattern $x(t)$ from learning dataset.
      ii. Find $bmu(x(t))$.
      iii. Train / Update weights of bmu and its neighborhood.
      iv. Remove $x(t)$ from input learning dataset.
      v. $t = t + 1$.
   (b) $p = p + 1$.
4. End of learning after p epochs.

The learning parameter $\alpha(t)$ and the neighborhood radius $\sigma(t)$ are decreasing functions of time of the same kind and follow the same law.

$$\alpha(t) = \alpha_{MAX} \left(\frac{\alpha_{MIN}}{\alpha_{MAX}} \right)^{\left(\frac{t}{t_{max}}\right)}, \tag{4}$$

$$\sigma(t) = \sigma_{MAX} \left(\frac{\sigma_{MIN}}{\sigma_{MAX}} \right)^{\left(\frac{t}{t_{max}}\right)}. \tag{5}$$

3.2 The Fast Learning SOM Algorithm (FLSOM)

Simulated annealing (SA) is an optimization method typically used for large scale problems, where a global minimum is hidden by many local minima. In the present work, we use this heuristic to improve the quality of learning process of the SOM, preserving its unsupervised characteristic. The adopted approach provides an adaptive learning rate factor $\alpha(t, QE)$, steered by the simulated annealing heuristic over the current resolution of the map. Using a SOM trained for epochs, at the end of each learning epoch, the Quantization Error (QE) can be identified with the parameter "temperature" T and the evolution of the network can be identified with a perturbation of the system. Notice that the algorithm parameter that control temperature schedule is automatically adjusted according to algorithm progress; an analogous criterion, the adaptive simulated annealing, was widely analysed and developed in [14] and [15]. The QE of a SOM is defined as the euclidean distance between a data vector and its best matching unit according to:

$$QE = \|x - m_c(x)\| \,, \tag{6}$$

where x is the data vector input and $m_c(x)$ is the *bmu*. The system evolution can be delineated by the QE progression because, if the QE at the end of each learning epoch is smaller than the QE computed in previous epoch, then the projection of the samples on the SOM map is closer to the original positions in the input space. Thus we obtain a linear cooling schedule as $QE_{new} = QE_{old} - \Delta QE$, where ΔQE is the variation of the total energy of the system.

The pseudocode of the algorithm is given in table 2. In this algorithm the term *Training(SOM)* refers to a learning epoch shown in table 1; the result of the training is a candidate SOM that is tested using QE. At the beginning all parameters, including the range of the learning rate, are initialized and the first epoch of the algorithm is executed (steps 1, 2). At the end of each learning epoch the QE is calculated and if the difference between this QE and that one calculated at the end of previous epoch is under a threshold δ then the learning process stops (steps 5, 5.(d)). Each learning epoch generates a perturbation of the status of neurons in the map. If this perturbation satisfies low-energy criteria according to the simulated annealing (step 5.(f)), then the current configuration is accepted and a new perturbation is calculated. Otherwise, if the perturbation does not satisfy low-energy criteria, then the first perturbation will be used for the next epoch (step 5.(g).i). If the previous configuration is better than the current one, then the previous one is restored (step 5.(g).ii). The size of perturbations depends on the evolution of network resolution or, in other words, by the ratio of the current QE and the maximum QE (step 5.(f).2). The values of the learning rate factor are adapted according to these equations (steps 5.(f).iii, 5.(f).iv).

4 Experimental Results

The proposed FLSOM is evaluated against the standard SOM and most of its variations as *PLSOM*, *HabSOM* and *ConsSOM*. In order to compare the five different algorithms, five data set have been tested and the quality of the approximation, the "smoothness" of the lattice and the entropy are used for comparison.

4.1 Quality Criteria

Three evaluation criteria are used to measure the quality of the map and to compare the results of the algorithms: quantization error (QE) already defined, regularity degree [16] (RD) and maximization of entropy (EN). The first one measures the resolution of the map, the second one the local distortion and the last one the frequency of input distribution over the map. The RD can be calculated at the end of each learning epoch and it is useful for evaluating the topological organization of the lattice during the training. This parameter is easily calculated as the position of neurons with respect to their neighbours.

Table 2. FLSOM algorithm

1. Initialize the $SOM_{current}$ with random weights, the SOM parameters: $\alpha_{MAX}, \alpha_{MIN}$, and the epoch counter $p = 1$
2. Start with first learning epoch: $SOM(p) = Training(SOM_{current})$
3. Set $QE_{MAX} = QE(p)$
4. Initialize $\Delta QE(p) = QE_{MAX}$
5. While $\Delta QE(p) \geq \delta$
   (a) $p = p + 1$
   (b) Run a new learning epoch:
   $SOM(p) = Training(SOM_{current})$
   (c) Calculate $QE(p)$
   (d) Calculate $\Delta QE(p) = QE(p) - QE(p-1)$
   (e) Get a random value $0 < rand < 1$
   (f) if the configuration satisfies low-energy criteria (i.e. $e^{-\frac{\Delta QE}{QE}} < rand$), or the configuration is better than the one of previous epoch ($\Delta QE(p-1) > \Delta QE(p)$)
      i. use current state i.e. set $SOM_{current} = SOM(p)$
      ii. Calculate $\alpha_{inc}(QE) = \Delta\alpha * \left|1 - \frac{QE(p)}{QE_{MAX}}\right|$
      iii. Set $\alpha_{MAX} = \alpha_{MAX} + \alpha_{inc}(QE)$
      iv. Set $\alpha_{MIN} = \alpha_{MIN} + \alpha_{inc}(QE)$
   (g) Else
      i. if ($e^{-\frac{\Delta QE}{QE}} > rand$)) *i.e. configuration does not satisfy low-energy criteria*
      Use the initial vales of α_{MAX} and α_{MIN} in eq. 4
      ii. if ($\Delta QE(p-1) < \Delta QE(p)$) *i.e. previous configuration is better than the current configuration*
      Set $SOM_{current} = SOM(p-1)$
6. End of learning after p epochs

According to this criterion, an ideal value of the regularity degree should be close to zero (true for a quite flat map). In real applications configurations with low values of degree of regularity are better. The EN ensures us that the quantization intervals are used with the same frequency during the quantization of the input signal. If we have N neural units, we obtain a input manifold divided into V_i intervals where $i = 0, 1, ..., N$. After the training, the input pattern v should fall in interval V_i with a probability:

$$p(V_i) = \frac{1}{N}. \tag{7}$$

So that information-theoretic entropy will be maximized:

$$H = -\sum_{i=1}^{N} p(V_i) * log(p(V_i)) = logN. \tag{8}$$

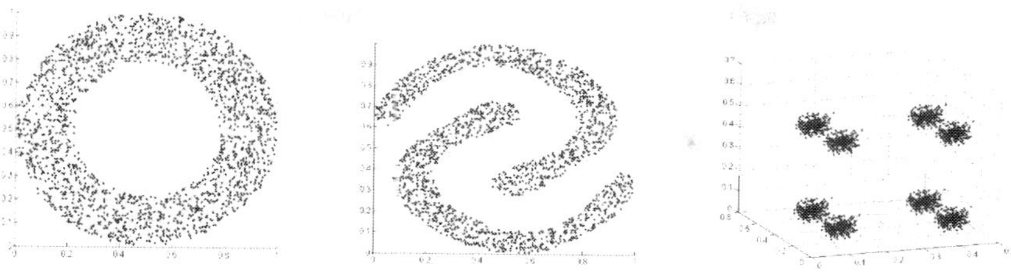

(a) Ring (b) Two-C (c) Blobs-3D

Fig. 1. Artificial datasets

4.2 Evaluation of the Proposed Algorithm

The approach is evaluated using the five SOMs algorithms cited in previous subsection. To allow comparison among SOM implementations, the algorithm shown in table 2 has been modified in step 5, where the stop condition depending on δ has been replaced by a stop condition depending on maximum number of learning epochs. To choose this number, several tries have been run over all used datasets. The number selected, 18 *epochs*, was sufficient to approximate a nearly complete evolution, for all SOMs implementations and for all datasets, according to algorithm shown in table 2 with a very little $\delta = 0.05$. The weights of all SOM networks are initialized with random values over the input space and all maps have a 80×80 square lattice. The training phase for each epoch is done with $\alpha_{MAX} = 0.75$, $\alpha_{MIN} = 0.15$, $\sigma_{MAX} = 7$, $\sigma_{MIN} = 2$. In the FLSOM algorithm the values of the learning parameters are dynamically increased up to $\alpha_{MAX} = 1$ and $\alpha_{MIN} = 0.75$; the HabSOM has $\beta = 0.99$, $ab_{MAX} = 1.0$ and tr =3.0E-5; the ConsSOM has B =1.0E-4 and $C = 5.0$. These parameter values are those that give the best results for the datasets used.

4.3 Validation of the Proposed Learning Process

The validation of the FLSOM has been carried out using three artificial datasets, shown in figure 1, and two datasets from real world. In detail the first one, here called *Ring* (fig. 1(a)), and the second one, here called $Two - C$ (fig. 1(b)), are artificial datasets in two-dimensional space and provide two thousand input signals; the third one $Blobs - 3D$ (fig. 1(c)) is an artificial dataset that draws eight blobs in three-dimensional space and provide two thousand and four hundred input signals; the fourth one, the well know *Iris*, is a real dataset in four-dimensional space and provide one hundred fifty instances and there are three clusters, each has 50 instances; the last dataset [17] is a real dataset reduced according to [18] in twenty-dimensional space and provide three hundred twenty-five input signals. All the reported figures are averaged over 100 runs of the algorithms. For each epoch of each SOM implementation, the means of QE and EN have been calculated.

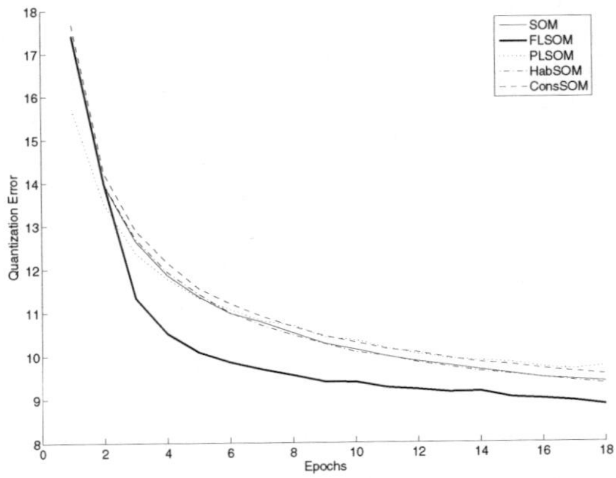

Fig. 2. Quantization error versus number of epochs in *Ring* dataset. The FLSOM algorithm reaches the stop condition with the smallest value of QE in 117.65 s, while the standard algorithm stops on epoch 18 in 128.60 s.

Figure 2 shows the average evolution of quantization error during the training process of *Ring* dataset for all SOMs. The chart clearly shows the effectiveness of the FLSOM algorithm: it reaches a lower QE value in a smaller number of epochs. Moreover execution time is not strongly influenced by the proposed algorithm: in fact the FLSOM learning process stops in 117.65 s, while the standard SOM stops in 128.60 s. Results of QE and RD for all datasets are given in table 3. In this table are shown, for all SOMs, the number of epochs (≤ 18) necessary to reach the required values scored by the worst of all SOMs implementations. The best values for each dataset are emphasized with bold type. These results

Table 3. For all SOMs, the number of epochs necessary to reach required QE value and required RD value

	Ring	**Two-C**	**Blobs-3D**	**Iris**	**NCI-325**
Required **QE**	9.50	8.60	15.80	6.10	8.10
SOM	16	17	17	17	17
FLSOM	**9**	**10**	**3**	**5**	**5**
PLSOM	18+	13	18+	8	18+
HabSOM	16	18	15	17	16
ConsSOM	18	18+	18+	17	18+
Required **RD**	0.00045	0.00043	0.00046	0.120	0.075
SOM	10	**8**	8	18	18
FLSOM	16	14	13	**10**	**9**
PLSOM	**7**	**8**	9	18+	18+
HabSOM	10	9	**7**	18	17
ConsSOM	10	9	18	18	18

Table 4. For all SOMs, the value of Entropy reached after 18 epochs

Ideal Entropy for all SOMs: 8.764					
	Ring	**Two-C**	**Blobs-3D**	**Iris**	**NCI-325**
SOM	5.489	5.485	5.805	**4.808**	5.166
FLSOM	5.489	**5.490**	5.809	4.807	**5.547**
PLSOM	5.474	5.477	5.773	4.794	5.012
HabSOM	**5.493**	5.480	**5.810**	4.801	5.172
ConsSOM	5.483	5.481	5.772	4.803	5.169

state that the FLSOM is better than the other implementations with regard to resolution of the map (i.e. QE value). In other words, the FLSOM is faster than the other algorithms.

Regarding RD measure, FLSOM appears to performing worse than the others SOMs, because it needs the highest number of epochs to reach the same value of RD. Actually, this result is not as bad as it seems: analyzing the values of RD reached after 18 epochs by all the SOM implementation.

The average entropies calculated after 18 epochs, for all SOMs, are shown in table 4. The best values for each dataset are emphasized with bold type. Even though values of entropy are closer for all implementations, the FLSOM scores always one of highest results.

5 Conclusions

In this paper FLSOM, a technique that uses the Simulated Annealing as a method to select a candidate SOM during the training process, has been proposed. The SOM training process modifies the learning rate factor in an adaptive way. Results of experimental tests, carried out on both artificial and real datasets, comparing the proposed method with standard and modified SOMs demonstrate the good performances obtained by FLSOM in terms of convergence time, and resolution of the maps. On the other hand, the performances obtained by FLSOM in terms of local distortion and frequency of input distribution over the map are comparable with those obtained by standard and modified SOMs, especially when real world datasets are considered.

References

1. Rizzo, R., Chella, A.: A Comparison between Habituation and Conscience Mechanism in Self-Organizing Maps. IEEE Transactions on neural networks 17(3), 807–810 (2006)
2. DeSieno, D.: Adding a conscience to copetitive learning. In: Proc. ICNN'88, International conference on Neuroal Networks, pp. 117–124. IEEE Computer Society Press, Piscataway, NJ (1988)
3. Berglund, E., Sitte, J.: The parameterless self-organizing map algorithm. IEEE Transactions on neural networks 17(2), 305–316 (2006)

4. Haese, K.: Auto-SOM: recursive parameter estimation for guidance of self-organizing feature maps. Neural Comput. 13(3), 595–619 (2001)
5. Kirkpatrick, S., Gelatt, C., Vecchi, M.: Optimization by Simulated Annealing. Science 220(4598), 671–680 (1983)
6. Metropolis, N., Rosenbluth, A.W., Rosenbluth, M.N., Teller, A.H., Teller, E.: Equation of state calculations by fast computing machines. J. Chem. Phys. 21(6), 1087–1092 (1953)
7. Graepel, T., Burger, M., Obermayer, K.: Self-organizing maps: generalizations and new optimization techniques. Neurocomputing 21, 173–190 (1998)
8. Douzono, H., Hara, S., Noguchi, Y.: A Clustering Method of Chromosome Fluorescence Profiles Using Modified Self Organizing Map Controlled by Simulated Annealing. In: IJCNN'00. IEEE-INNS-ENNS International Joint Conference on Neural Networks, vol. 4 (2000)
9. Haese, K.: Kalman filter implementation of self-organizing feature maps. Neural Comput. 11(5), 1211–1233 (1999)
10. Haese, K.: Self-organizing feature map with self-adjusting learning parameters. IEEE Transactions on Neural Network 9(6), 1270–1278 (1998)
11. Berglund, E., Sitte, J.: The parameter-less SOM algorithm. In: Proc. ANZIIS, pp. 159–164 (2003)
12. Kohonen, T.: Self-Organizing Maps. Springer Verlag, Berlin (1995)
13. Van Hulle, M. (ed.): Faithful Representations and Topographic Maps: From Distortion- to Information-Based Self-Organization. John Wiley, New York (2000)
14. Ingber, L.: Very fast simulated re-annealing. Journal of Mathl. Comput. Modelling 12(8), 967–973 (1989)
15. Ingber, L.: Adaptive simulated annealing (ASA): lessons learned. J. Control and Cybernetics 25(1), 33–54 (1996)
16. Goppert, J., Rosenstiel, W.: Regularized SOM-Training: A Solution to the Topology-Approximation dilemma, University of Tbingen (1996)
17. National Cancer Institute, DTP AIDS antiviral screen dataset [online], `http://dtp.nci.nih.gov/docs/aids/aids/data.html`
18. Di Fatta, G., Fiannaca, A., Rizzo, R., Urso, A., Berthold, M.R., Gaglio, S.: Context-Aware Visual Exploration of Molecular Databases. In: Perner, P. (ed.) ICDM 2006. LNCS (LNAI), vol. 4065, pp. 136–141. Springer, Heidelberg (2006)

The Usage of Golden Section in Calculating the Efficient Solution in Artificial Neural Networks Training by Multi-objective Optimization

Roselito A. Teixeira[1], Antônio P. Braga[2], Rodney R. Saldanha[3], Ricardo H.C. Takahashi[4], and Talles H. Medeiros[2]

[1] Centro Universitário do Leste de Minas Gerais, Coronel Fabriciano, MG, Brazil
roselito@unilestemg.br

[2] Federal University of Minas Gerais - Department of Electronic Engineering Belo Horizonte, MG, Brazil
talles, apbraga@cpdee.ufmg.br

[3] Federal University of Minas Gerais - Department of Electrical Engineering Belo Horizonte, MG, Brazil
rodney@cpdee.ufmg.br

[4] Federal University of Minas Gerais - Department of Mathematics Belo Horizonte - MG, Brazil
taka@mat.ufmg.br

Abstract. In this work a modification was made on the algorithm of Artificial Neural Networks (NN) Training of the Multilayer Perceptron type (MLP) based on multi-objective optimization (MOBJ), to increase its computational efficiency. Usually, the number of efficient solutions to be generated is a parameter that must be provided by the user. In this work, this number is automatically determined by an algorithm, through the usage of golden section, being generally less when specified, showing a sensible reduction in the processing time and keeping the high generalization capability of the obtained solution from the original method.

1 Introduction

The process of obtaining a well trained NN, is not always a simple task, and generally requires a great effort from the designer in determining the parameters that will make the Neural Network present high generalization capability. Typically, NN can suffer from the effect of *over-fitting* or *under-fitting*. This effect appears to under or oversize the NN, which minimizes its generalization capability. To maximize this characteristic, it is necessary to establish a trade-off between bias and variance of the models, *bias* × *variance dilemma* [1].

The reduction of a neural model variance can even be minimized by the number of free parameters or controlling the weight vector variance norm. Some algorithms such as *pruning* [2] adjust the neural architecture, others, penalize the solution of high norms, like regularization techniques [3]. In the latter case, a cost function given in addition to Eq. (1) must be minimized:

$$J(\mathbf{x}; w) = E_D(w, \mathbf{x}) + \lambda E_S(w), \quad (1)$$

J. Marques de Sá et al. (Eds.): ICANN 2007, Part I, LNCS 4668, pp. 289–298, 2007.

denoted w, the weight vector, $\mathbf{x}$ is a input pattern, $E_D(w, \mathbf{x})$ is training error, $E_S(w)$ is a regulated term and λ is a regulated parameter. This must be adjusted in accord to the data quality, being less for noiseless data or bigger when the data is contaminated by noise. However, to attain this value is not a simple task and an incorrect value can make it be *to under* or *over-fitting*. A common regulated term is the weight vector norm, as Eq. (2),

$$E_S(w) = ||w||^2 \tag{2}$$

it results in the function of cost given for the Eq. 3

$$J(\mathbf{x}; w) = \frac{1}{N_T} \sum_{j=1}^{N_T} [d_j - f(w, \mathbf{x}_j)]^2 + \lambda \frac{1}{2} ||w||^2 \tag{3}$$

being N_T the size of training set, $\mathbf{x}_j$ and d_j are, respectively, j-the input and output pattern of this set.

Differently from the methods above, the multi-objective method consists originally in [4] controlling the weight vector norm through the simultaneous minimization of two cost functions, described in the Equations (4) and (5), solving a problem of multiple objectives. It is noticed the absence of regulated parameter.

$$J_1(\mathbf{x}; w) = \frac{1}{N_T} \sum_{i=1}^{N_T} [d_i - f(w, \mathbf{x}_i)]^2 \tag{4}$$

$$J_2(w) = ||w|| \tag{5}$$

The generated solution shows *under-fitting* solution in one extreme and *over-fitting* solution on the other extreme. The best solution is located between the two extremities [4]. The chosen solution combined with the best trade-off between the norm $\|w\|$ and the error e^2 is performed on the next stage of the multi-objective method, after determining some Pareto-optimal solution. The chosen solution is expected to present a suitable adjustment to the underlying function $f_g(\mathbf{x})$, resulting in high generalization capability. The formal procedure employed in choosing the best solution is named the *decider*.

Whenever the method is capable to find a solution with high generalization capability, the computational cost is high because it has to generate a great number of efficient solution, from the multi-objective optimization point. From this solutions, only one is chosen as final one, and the others are discarded. Therefore, the implementation of a strategy capable of generating a reduced number of efficient solutions can significantly reduce the training time of the method originally considered. In this work, a technique called golden section is used to reduce the number of solutions obtained in MOBJ algorithms, reducing the training time and keeping the quality of the final solution. The original (MOBJ) and golden section (MOBJgs) algorithms are described as follows.

2 The MOBJ Algorithm

The MOBJ algorithm employs an approach for neural model complexity control which is independent from dimension, trading-off the bias and variance. The first phase in multi-objective optimization, is to get the Pareto-optimal set [5], which is constituted of efficient solution w^*. The following phase is to choose the best solution (high generalization capability). The formal procedure employed in choosing the best solution is named the *decider*. In literature, many algorithms can be found to deal with multi-objective problems [5].

Following these phases, a description is made of the implemented method to solve the multi-objective optimization problem for training the Neural Networks.

2.1 Description of MOBJ Algorithm

The multi-objective algorithm is implemented using the method ε-constraint [5] and through this, the optimization is rewritten in the form of mono-objective being solved by the elipsoidal algorithm [6]. Soon, a constrained mono-objective optimization problem is solved for different values for ε. The parametric variation of ε allows the generation of the Pareto-optimal set.

The MOBJ algorithm obtains a number ζ of solutions, were ζ is provided by the user. Each one of these solutions is a non-dominated or efficient solution, constituting the Pareto optimal set. The method ε-constraint to problem of training of Neural Networks formula is developed as follows:

$$\begin{aligned} &\min_{w \in \mathcal{W}} \; J_1(\mathbf{x}; w) \\ &\text{sunbectto a}: \; J_2(w) \leq \varepsilon \end{aligned} \qquad (6)$$

where $\mathcal{W}$ is an espace of weight vector. Through the problem (Pε) described in the Eq. (6) it is possible to generate w^* completely, even for non-convex Pareto set. As

$$J_1(\mathbf{x}; w) = \frac{1}{N_T} \sum_{j=1}^{N_T} (d_j - f(\mathbf{x}_j; w))^2$$

and

$$J_2(w) = \|w\|,$$

Eq. (6) it's

$$\begin{aligned} &\min_{w \in \mathcal{W}} \; \frac{1}{N_T} \sum_{j=1}^{N_T} (d_j - f(\mathbf{x}_j; w))^2 \\ &\text{subjectto a}: \; \|w\| \leq \varepsilon \end{aligned} \qquad (7)$$

For MOBJ algorithm (Pε) to calculate ζ solution, it is necessary that the user informs this number and this will have to be made into the parametric variation ε. This is made originally from linearly form spaced one δ_ε to be informed. Also, the number of neurons in the hidden layer must be informed. After that, a decider must choose between the candidates, the final solution, using a validation set. The MOBJ algorithm originally proposed uses an efficient decider, being briefly described in the following subsection.

2.2 The Decision Process

In the previous sub-section a method to attain the Pareto-optimal set w^* was presented. The next phase is to choose the best solution $w^* \in \mathcal{W}^*$ as final solution. This solution, therefore, allows an adequate fit to the underlying function $f_g(\mathbf{x})$, resulting in high generalization capability.

The decider is based on the validation set error for decision making. A validation set $\sqcup_v = \{\mathbf{x}_{V_i}, d_{V_i} + \xi_i\}_{i=1}^{N_V}$ [1] with N_V patterns is presented for efficient solution set obtained in the first phase, and the decider picks up the solution with smallest validation error from the available Pareto set samples. The efficiency of the decider is proven in [7], through a theorem which shows that the Pareto-optimal set has a solution that makes the adequate fit and that this solution can be selected using a validation set.

The decision rule is given by the Eq. (8),

$$w^* = \arg \min_{w \in \mathcal{W}^*} e_V \tag{8}$$

where e_V is given by the Eq. (9).

$$e_V = \frac{1}{N_V} \sum_{i=1}^{N_V} [(d_{V_i} + \xi_i) - f(\mathbf{x}_{V_i}; w)]^2 \tag{9}$$

being ξ the present noise in the data and $f(\mathbf{x}_V; w)$ the output of the Neural Network.

The vectorw^* to minimize the function given for the Eq. (9) is the best solution of the Pareto-optimal set and its high generalization capability is guaranteed.

2.3 Computational Aspects for the MOBJ Algorithm

In the described approach in the Sub-section 2.1, the user must inform the number of desired efficient solution ζ and the difference of norm δ_ε between them. For example, for $\zeta = 30$ and $\delta_\varepsilon = 0.5$, one gets 30 solution, of norm 0 until norm 15, spaced of 0.5. Thus, the computational cost of generating 30 solutions, (30 neural networks). One knows that it is between that where high generalization capability can be found, which must be chosen by the decider.

However, alternatives for reducing the number of solution generated can significantly reduce the computational cost of the algorithm. With this objective, an alteration of the original method is considered in this work using golden section, as it details in the following section.

3 The MOBJ Algorithm with Golden Section

This proposal should act according to the generation of norm values ε to optimize Eq. (7). In the method originally considered,

$$\varepsilon = k\delta\varepsilon, \quad k = 1, 2, \ldots, \zeta$$

which it always brings to a number ζ of solution to be generated.

[1] $\mathbf{x}_{V_i}, d_{V_i}$ is, respectively, input an ouput validation pattern and ξ_i is a gaussian noise.

According to the new proposal, the user must inform only the maximum limit of norm and the algorithm will have to look for the optimal value norm using validation data set, generating the smallest possible number of efficient solution. The best solution must have to be of an inferior norm to a specified maximum limit. The values are generated according to the described algorithm in the following sub-section.

3.1 Description of Algorithm

Considering a validation data set with N_V patterns, the error relating these patterns for a weight vector w is given by the Eq. (9):

$$e_V = \frac{1}{N_V}\sum_{i=1}^{N_V}[(d_{V_i} + \xi_i) - f(\mathbf{x}_{V_i}; w)]^2.$$

Considering α and β the minimum limit and the maximum for a norm. Here, α it is always equal to zero and β is specified by the user.

With these limits, one more calculates two values of norm according to the Equations (10) and (11).

$$\varepsilon_1 = \alpha + (\beta - \alpha) * 0,382 \tag{10}$$

$$\varepsilon_2 = \alpha + (\beta - \alpha) * 0,618 \tag{11}$$

The optimization process is carried through according to Eq. (7) for $\varepsilon = \varepsilon_1$ and $\varepsilon = \varepsilon_2$, obtaining two NN, denoted here for w_1 and w_2.

It is evaluated Eq. (9) in the points w_1 and w_2 considering the validation set, obtaining e_{V1} and e_{V2}. The following process is repeated until $\varepsilon_1 \approx \varepsilon_2$:

---

if $e_{V1} > e_{V2}$ **then**
  $\alpha = \varepsilon_1, \varepsilon_1 = \varepsilon_2, e_{V1} = e_{V2}$
  $\varepsilon_2 = \alpha + (\beta - \alpha) * 0,618$
  It is effected according to optimization Eq. (7) for $\varepsilon = \varepsilon_2$, obtaining a eficient solution w;
  Evaluate Eq. (9) to w, obtaining a new e_{V2};
else
  $\beta = \varepsilon_2, \varepsilon_2 = \varepsilon_1, e_{V2} = e_{V1}$
  $\varepsilon_1 = \alpha + (\beta - \alpha) * 0,382$
  It is effected according to optimization Eq. (7) for $\varepsilon = \varepsilon_1$, obtaining a eficient solution w;
  Evaluate Eq. (9) to w, obtaining a new e_{V1};
end if

With this proposal, a reduced number of solution is generated and the distance between the efficient solution is not more uniform. After the generation of the efficient solution, comes the decision process already described in subsection 2.2.

The next step, the method with golden section (MOBJgs) is compared with the original method (MOBJ) in different learning tasks.

4 Classification and Regression Problems

In this section, the original and gold section methods were applied to the classification and regression problems. The objective of this section is to verify the computational cost of each method and to compare the generalization capability of these solution.

4.1 Classification Problem

In this simulation, a classification problem of two classes is boarded, whose patterns had been random sampled in two normal distributions with two variables being that, the first class is centered in the point $(2, 2)$ and second in the point $(4, 4)$. Both classes present variances equal to 0.8^2. Each class is constituted by 100 patterns being 60 for training and 40 for the validation. Neural Network with architecture $2 \times 50 \times 1$ and activation functions $f_a(\cdot) = tanh(\cdot)$ for hidden and output neurons.

For the training by MOBJ algorithm, the following parameters had been specified: $\delta_\varepsilon = 0.5$ and $\zeta = 30$. For the MOBJgs algorithm, only the upper bound of norm was specified, being $\beta = 15$. The Figure 1 illustrate the obtained solutions for the methods.

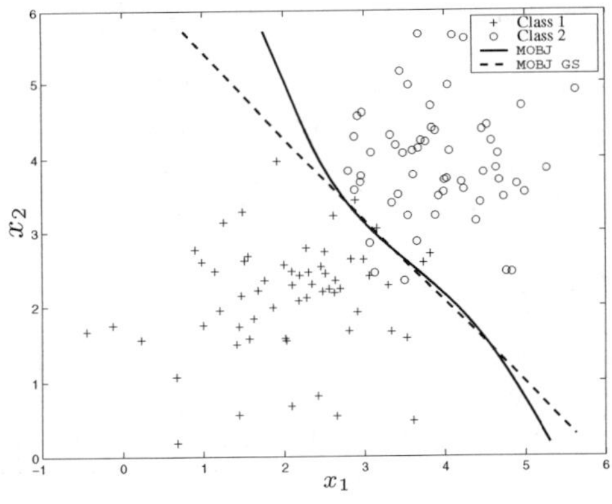

Fig. 1. Decision Frontiers obtained by MOBJ method with and without Golden Section

The Figure 2 shows the efficient solution generated by each method. One notices that MOBJ solution is linearly spaced and the MOBJgs solution (with golden section) is inferior and has concentrated number next to the final solution. The Figure 3 shows a zoom of the region of concentration of the MOBJgs solution which are next to final solution MOBJ.

4.2 Regression Problem

In this section, a regression problem is solved with NN of architecture 1-50-1 and activation functions $f_a(\cdot) = tanh(\cdot)$ and linear for, respectively, hidden and output layers. The training and validation sets used in each case are constituted of 80 samples made

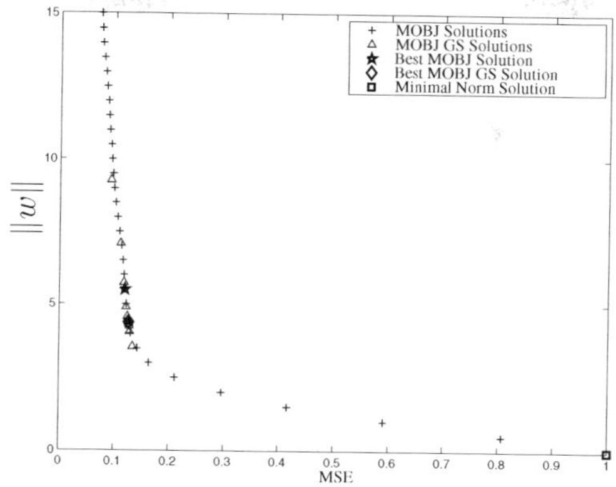

Fig. 2. Pareto set obtained by MOBJ method with and without Golden Section

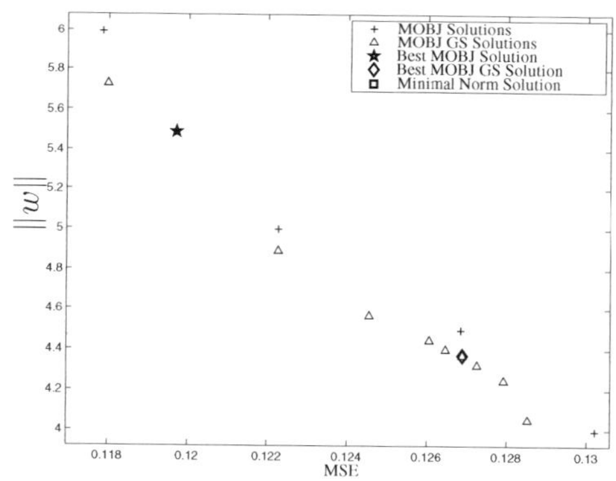

Fig. 3. Zoom in the region of concentration of the MOBJgs solutions

on the underlying function given by the Eq. (12) and to these samples were added a noise term normally distributed with mean zero and variance $\sigma^2 = 0.15^2$.

$$f(x) = 4.26(e^{-x} - 4e^{-2x} + 3e^{-3x}) + \xi; \qquad (12)$$

The Figure 4 illustrates the solutions for the methods MOBJ and MOBJgs. The Figure 5 illustrates the efficient solution generated by the methods. The Figure 6 brings a magnifying part of the concentration region of the MOBJgs solution. It is noticed the proximity of the final solution generated by the methods.

5 Discussions

Considers the training and validation sets previously used in Section 4.1, referring to the classification problem. Comparing the algorithms, 05 tests for each user, parameters set and the length of time taken to obtain the solution, calculated the processing time,

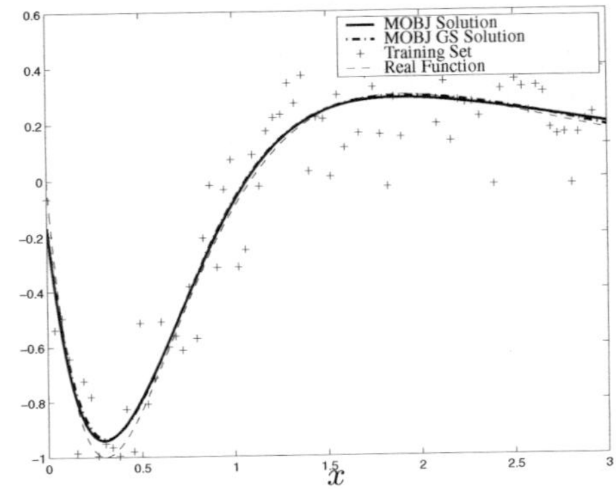

Fig. 4. Solution obtained by MOBJ method with and without Golden Section

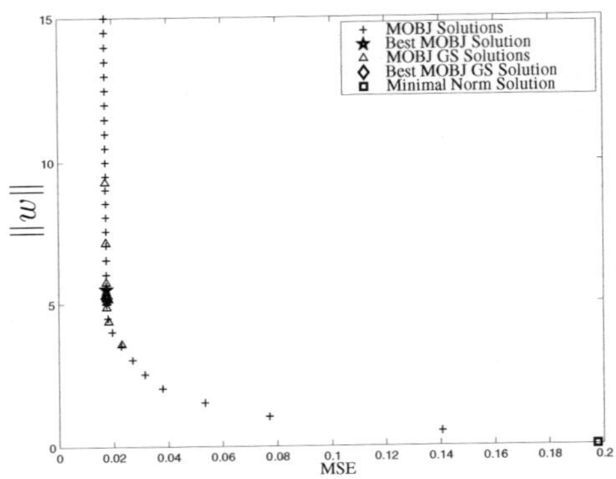

Fig. 5. Pareto set obtained by MOBJ method with and without Golden Section

the shunting line standard of this time σ_t, the Error in relation to a test set with 500 patterns and the shunting line standard of the error σ_e. The Table 1 show results for MOBJ algorithm. One notices that in each in case, the solution of the greater norm presents this equal value $\delta_\varepsilon \times \zeta$.

With the shown results above it can be verified, as he expected it was , that the training time increases with the number of solution ζ gotten and also with the distance between the norms δ_ε. This justifies the fact that the generation of 20 solution with increment of equal norm the 1 (maximum norm=20) to be faster than the generation of 30 solution by increasing the equal norm to 0.5 (maximum norm=15).

The Table 2 show results for MOBJgs algorithm, whose parameter has limited norm β must be informed. One notices that in each in case, this value was equal to the fact $\delta_\varepsilon \times \zeta$ to facilitate the comparison with the original method. It can be noticed in the Table 2 that the number of solution N generated for the method was always around 12, which decreases the training time for the method with golden section. However, for higher values of β, the time increases in relation to the smallest values of β for the same

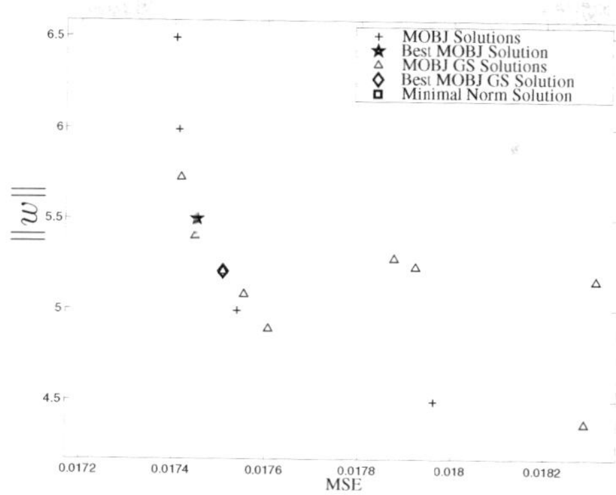

Fig. 6. Zoom in the region of concentration of the MOBJgs solution

Table 1. Simulation for MOBJ algorithm without golden section

$\delta\varepsilon$	ζ	T_m	σ_T	Error	σ_E
0,5	10	7.5842	0.1931	4.70	0.1871
1,0	10	18.8624	0.1266	5.10	0.2000
0,5	30	59.3906	1.0710	4.66	0.1517
1,0	15	42.0690	0.8642	4.92	0.0837
0,5	40	107.6530	4.6660	4.82	0.3701
1,0	20	68.4126	6.4909	5.12	0.2168

Table 2. Simulation for MOBJ algorithm with golden section

β	N	T_m	σ_T	Error	σ_E
5	10	9.0816	1.0529	4.48	0.2049
10	11	26.7938	2.9598	4.70	0.1225
15	12	35.6750	1.3244	4.58	0.2387
15	12	35.7096	4.2527	4.64	0.1673
20	13	46.0844	4.6654	4.56	0.2074
20	13	50.5968	6.8982	4.62	0.1924

method, due to the interval being bigger. One also notices in the two tables that the test errors are close, which shows the efficiency in attaining solution of high generalization capability of both methods.

The MOBJ algorithm with golden section was capable to generate final solution with norms close to the generated ones for the original algorithm. This comparison can be made observing the Figure 2 for the classification problem and the Figure 5 for the regression problem.

6 Conclusions

- Reduction of the training time of NN with high generalization capability:
  In the original method, the user must inform the number ζ of solution to be generated and in the modified method, this number is automatically determined generally, is less than the one adjusted by the user. The training time increases on increase in the number of solution to be generated. Therefore, few solution imply reduced time to get them.
- Reduction for a parameter user only, in relation to MOBJ method, originally considered with two parameters:
  In the original method, the user must inform the number ζ of solution to be generated and the difference δ_ε of norm between them. For the MOBJgs method, only the limit β of norm for the solution must be informed, which eliminates the necessity of having to adjust the two parameters. As can be seen in the Table 2, the quality of the final solution regarding its generalization capability is a little sensible to the value of β, which demonstrates the robustness of the algorithm in relation to this parameter.

It stands out in the method MOBJgs, the solution quality generated in relation to the statistics representation of the validation set. One notices that the attainment of ε in the optimization process makes use of this set and these values are guided by the point of the minimum function given for the Eq. (9). This does not occur in the original MOBJ algorithm, where the validation set is used only in the decision phase, not intervening in the solution generation process, which only depends on the training set.

References

1. Geman, S., Bienenstock, E., Doursat, R.: Neural networks and bias/variance dilemma. Neural Computing 4(1), 1–58 (1992)
2. Hinton, G.E.: Deterministic boltzmann machine learning performs steepest descent in weight-space. Neural Computation 1, 143–150 (1989)
3. Girosi, F., Jones, M., Poggio, T.: Regularization theory and neural networks architectures. Neural Computation 7(2), 219–269 (1995)
4. Teixeira, R.A., Braga, A.P., Saldanha, R.R., Takahashi, R.H.C.: Improving generalization of mlps with multi-objective optimization. Neurocomputing 35(1-4), 189–194 (2000)
5. Chankong, V., Haimes, Y.Y.: Multiobjective Decision Making: Theory and Methodoly, vol. 8. Elsevier, Horth-Holland (1983)
6. Shor, N.Z.: Cut-off method with space extension in convex programming problems. Cybernetics 12, 94–96 (1977)
7. Teixeira, R A.: Treinamento de Redes Neurais Artificiais Atravs de Otimizao Multi-Objetivo: Uma Nova Abordagem para o Equilbrio entre a Polarizao e a Varincia. PhD thesis, Escola de Engenharia da UFMG, Agosto (2001)

Designing Modular Artificial Neural Network Through Evolution

Eva Volna

Faculty of Science, University of Ostrava, 30th dubna st. 22, 70103 Ostrava 1, Czech Republic
eva.volna@osu.cz

Abstract. The purpose of this article is to make a contribution to the study of modular structure of neural nets, in particular to describe a method of automatic neural net modularization. The problem specific modularizations of the representation emerge through the iterations of the evolutionary algorithm directly with the problem. We used the probability vector to construct n – bit vectors, which represented individuals in the population (in our approach they describe an architecture of a neural network). All individuals in every generation are pseudorandomly generated from the probability vector that is associated with this generation. The probability vector is updated on the basis of best individuals in a population, so that next generations are getting progressively closer to best solutions. The process is repeated until the probability vector entries are close to zero or to one. The resulting probability vector then determines an optimal solution of the given optimization task.

Keywords: Adaptation, modular neural network architecture, probability vector, evolutionary algorithms.

1 Introduction

The presented method is a method of automatic neural net modularization. The method is based on principles of evolutionary algorithms and builds a modular architecture of the neural network. The problem specific modularizations of the representation emerge through the iteration of the evolutionary algorithm directly with the problem. The method is based on the article [5] and uses a population of chromosomes and there is an evolution of the probability vector on the basis of the best evaluated individuals in this algorithm, which are selected on the ground of the speed and quality of learning of the given tasks. New individuals in the next generation are generated from the updating probability vector. The best individual is included to the next generation automatically. The termination condition is depended on the saturation parameter. The modular network for a given task emerges as a winning architecture, since it has only a small interference between modules, which otherwise inhibits the convergence to a good solution.

J. Marques de Sá et al. (Eds.): ICANN 2007, Part I, LNCS 4668, pp. 299–308, 2007.

2 Evolutionary Algorithm with Hill – Climbing

Population P consists of p individuals: $P = \{\alpha_1, \alpha_2,..., \alpha_p\}$, where p is a cardinality of a population, e.g. p is equal to a number of chromosomes in P. Every individual in a population represents one (multilayer) neural network with one hidden layer of units and it is described by its chromosome, which is a binary vector with a constant length:

$$\alpha_i = (e_{11}, ..., e_{1n}, ..., e_{m1}, ..., e_{mn})_i \in \{0,1\}^{mn}, \quad (i=1,...p), \tag{1}$$

where	m	is a number of hidden units;
	n	is a number of output units;
	$e_{ij} = 0,$	if a connection between the *i-th* hidden unit and *j-th* output unit does not exist;
	$e_{ij} = 1,$	if a connection between the *i-th* hidden unit and *j-th* output unit exists.

There is every chromosome randomly generated with probability 0.5 in the initial population. Chromosome of the individual (e.g the neural network architecture with m hidden units a n output units) is shown in Fig. 1, where $e_{ij} = 0$, if the connection between the *i-th* hidden unit and the *j-th* output unit doesn't exist and $e_{ij} = 1$, if one exists ($i = 1,...,m$; $j = 1,...n$). Connections between input and hidden units are not included in the chromosomes, because they are not necessary for a creation of modular network architecture.

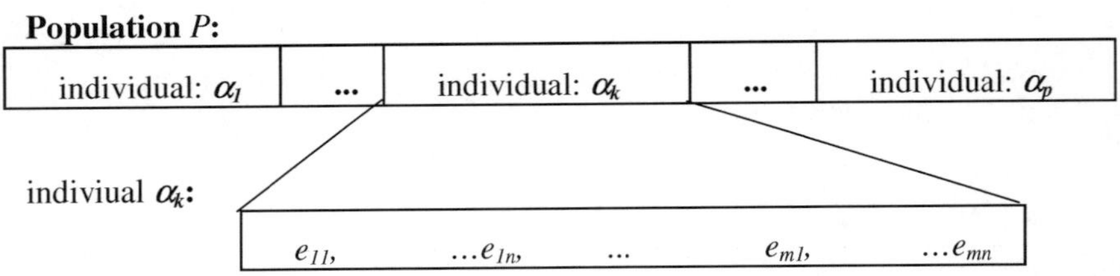

Fig. 1. A population of individuals

Every individual (e.g. its neural architecture) is then partially adapted by backpropagation [2] and evaluated on basis of its adaptation. There is number of epochs a very important criterion in the described method, because modular architectures start to learn faster than single connectionist networks with the same architecture of units [3]. Our goal is a composing such neural network architecture that is able to learn a given problem in the best quality and fast.

Our task is to look for such mn – bit vector $\alpha_{opt} \in \{0,1\}^{mn}$ over the domain (e.g. a search space $S = \{0,1\}^{mn}$) that corresponds to a global minimum of the function f ($f : \{0,1\}^{mn} \to R$). An objective function f assigns to mn – bit vector $\alpha = (e_{11}, ..., e_{mn})$ a real number $E \in R$ (E is error after a partial adaptation of backpropagation):

$$\alpha_{opt} = \arg \min_{\alpha \in \{0,1\}^{mn}} f(\alpha). \tag{2}$$

Function f „represents" a background, where chromosomes (= individuals) exist. A successfulness rate of each individual is its function value. We look for a minimum of the object function, because chromosome is more successful, when this function value is smaller. Because of it we define a fitness evaluation (F) of population $P \subseteq \{0,1\}^{mn}$ with p chromosomes (R_+ represents positive real numbers) $F : P \rightarrow R_+$ that qualifies the following:

$$\forall \alpha_1, \alpha_2 \in P : f(\alpha_1) \leq f(\alpha_2) \Rightarrow F(\alpha_1) \geq F(\alpha_2) \geq 0. \tag{3}$$

A backpropagation error is a fitness function parameter. A fitness function value of the *i-th* individual F_i is calculated as follows:

$$F_i = \frac{\sum_{k=1}^{con} f_{ik}}{con} \tag{4}$$

where $i = 1, \ldots, p,$ p is number of individuals in a population;

$f_{ik} = \frac{1}{E_{ik}}$ is a fitness value of the *i-th* individual in the *k-th* adaptation;

$k = 1, \ldots, con,$ con is a define natural constant, $con>1$;

E_{ik} is the backpropagation error of the *i-th* individual in the *k-th* adaptation.

Constanta *con* guarantees that a fitness value F_i is calculated from the statistical significant number of partial network adaptations, therefore one should be adequately great value [4].

There are not crossover and mutation operators used in the described method, this algorithm looks like hillclimbing algorithms [5]. This algorithm is based on the probability vector emergency. The probability vector is updated on the basis of well-evaluated individuals in the population. Entries $0 \leq w_{ij} \leq 1$ of the probability vector

$$w = (w_{11}, \ldots, w_{1n}, \ldots, w_{m1}, \ldots, w_{mn}) \in [0,1]^{mn}, \tag{5}$$

(m is a number of hidden units; n is a number of output units) determine probabilities of appearance of '*1*' entries in given positions. A solution R of the optimization problem can be expressed by respect to the evolution of a probabilistic vector $w \in [0,1]^{mn}$ as follows: R: $[0,1]^{mn} \rightarrow \{0,1\}^{mn}$. The solution has a limit as $t \rightarrow \infty$:

$$w_{opt} = \alpha_{opt} = \lim_{t \rightarrow \infty} \left(\arg \min_{\alpha \in P_t} f(\alpha) \right), \tag{6}$$

where P_t is a population of individuals α_i $(i = 1, \ldots, p)$ in the time t.

If the probability vector entries are close either to zero or to one, then this resulting probability vector $w \in [0,1]^{mn}$ determines an binary vector $\alpha \in \{0,1\}^{mn}$ and we can write: $\mathbf{w}' \leftarrow \mathbf{w} + \lambda(\boldsymbol{\alpha} - \mathbf{w})$, e.g. the new probability vector $\mathbf{w}'$ is near upon the best solution $\boldsymbol{\alpha}$.

Entries of the probability vector are calculated in every next generation as follows:

1. We calculate F_{avg} , e.g the average fitness value of the population in the given generation:

$$F_{avg} = \frac{\sum_{i=1}^{p} F_i}{p}, \tag{7}$$

where p is a number of individual in the population;
F_i is a fitness function value of the i-th individual, see (4).

2. We choose a set of q individuals with $F_i \geq F_{avg}$, e.g. $\alpha_1, \alpha_2, \ldots, \alpha_q$ $(1 \leq q \leq p)$, where p is a number of individual in the population.
3. Entries of the probability vector of the population in the given $w'_k \in [0,1]$ are calculated as follows:

$$w'_k = (1 - \lambda)w_k + \lambda w''_k \tag{8}$$

where $k = 1, \ldots, mn$ — mn is a number of the probability vector $\mathbf{w}$ entries;
λ is a constant $(0 < \lambda < 1)$;
w_k is a value of the k-th entry of the probability vector in the last generation.
w''_k is a value of the k-th bit of the probability vector $\mathbf{w}$ that is calculated as follows:

$$w''_k = \frac{\sum_{i=1}^{q} (e_k)_i}{q} \tag{9}$$

where $(e_k)_i$ is a value of the k-th bit of the chromosome of the individual α_i $(i = 1, \ldots, q)$ and it is true $F_i \geq F_{avg}$ for these individuals.

If the best solution $\boldsymbol{\alpha}$ is in time $t \rightarrow \infty$, e.g. $\lim_{t\rightarrow\infty} \mathbf{w} = \boldsymbol{\alpha}$, then we can write $\mathbf{w}' \leftarrow (1 - \lambda)\,\mathbf{w} + \lambda\,\boldsymbol{\alpha}$, which is after edits: $\mathbf{w}' \leftarrow \mathbf{w} + \lambda(\boldsymbol{\alpha} - \mathbf{w})$. The best individual in the population is included to the next population automatically. Values of the chromosomes of the rest of individuals α_i $(i = 2, \ldots, p)$ are calculated for the next generation as follows: for instance if $w_k = 0(1)$, then $(e_k)_i = 0(1)$; for $0 < w_k < 1$ the corresponding $(e_k)_i$ is determined randomly by

$$\left(e_k\right)_i = \begin{cases} 1 & if \quad random < w_k \\ 0 & \text{otherwise} \end{cases} \tag{10}$$

where *random* is a random number with uniform distribution from the interval [0, 1).

The process of the evolutionary algorithm is ended, if the saturation parameter $\tau(\boldsymbol{w})$[1] is greater then defined value.

3 An Outline of the Convergence Manner of the Method

The differential equation (8) determines an evolution time of the probability vector. Its structure is relatively simple and therefore it is possible to derive from its solutions general properties that are independent on the type of optimized objective functions. First, we introduce some simplifications of (8). Let us postulate that only one best solution is recorded, e.g. the saturation parameter $\tau(\boldsymbol{w})$ is greater than define value and the population $P = \{\alpha_1, ..., \alpha_p\}$ consists from almost identical individuals. Then the components of the probability vector $\boldsymbol{w} = (w_1, ..., w_k, ..., w_{mn})$ can be expressed as following:

$$w_k \leftarrow (1-\lambda)\, w_k + \lambda \frac{\sum_{i=1}^{q} \left(e_k\right)_i}{q} \quad \wedge \quad \frac{\sum_{i=1}^{q} \left(e_k\right)_i}{q} \sim \left(e_k\right)_i \tag{11}$$

where $\alpha_i = (e_1, ..., e_k, ..., e_{mn})_i \in P$; and after modification: $\boldsymbol{w} \leftarrow \boldsymbol{w} + \lambda(\boldsymbol{\alpha} - \boldsymbol{w}) \wedge \alpha_1 \sim \alpha_2 \sim ... \sim \alpha_q = \alpha$.. We get a difference scheme of updating of the probability vector initiated at $t = 0$ by $\boldsymbol{w}^{(0)} = (1/2, ..., 1/2)$. The time derivation of the probability vector $\boldsymbol{w}(t)$ can be expressed this way [5]

$$w' = \frac{\partial w}{\partial t}. \tag{12}$$

The initial condition is $\boldsymbol{w}(0) = (1/2,...,1/2)$. Let more at time t is $\boldsymbol{w} = (w_1, ..., w_k, ..., w_{mn})$, and $w_k = \frac{\sum_{i=1}^{q} \left(e_k\right)_i}{q}$ is the average value of the best q individuals $\alpha_i = (e_1, ..., e_k, ...,e_{mn})_i$ in the population $P = \{\alpha_1, ..., \alpha_p\}$ and it is true $F_i \geq F_{avg}$ $(i = 1, ...,q \leq p)$, where F_{avg} is calculated by the formula (7) a F_i by the formula (4).

The differential equation (12) determines the time of evolution probability vector $\boldsymbol{w}$ starting from $t = 0$. Since entries of $\boldsymbol{w}$ are not mutually coupled, it can be solved independently for each entry w_i. Therefore, without loss of generality, we shall postulate $mn = 1$ (mn is a number of the probability vector $\boldsymbol{w}$ entries), that is

[1] $\tau(\boldsymbol{w})$ = a number of entries (w_i) of the probability vector $\boldsymbol{w}$ that are less than or equals w_{eff} ; greater than or equals $(1 - w_{eff})$, where w_{eff} is a small positive number.

differential equation (12) corresponds to single unknown variable $w = w(t)$ specified by the initial condition $w(0) = 1/2$. Let us assume that α is a constant, then we can solve the differential equation (12) as follows:

$$
\begin{aligned}
\frac{dw}{dt} &= \lambda\,(\alpha - w(t)) \\
\frac{\frac{dw}{dt}}{\alpha - w(t)} &= \lambda \\
\ln(\alpha - w(t)) &= -\lambda\, t \\
\alpha - w(t) &= e^{-\lambda t} \\
\ln(\alpha - w(t)) - \ln(\alpha - w(0)) &= -\lambda\, t \\
\ln(\alpha - w(t)) &= -\lambda\, t + \ln(\alpha - w(0)) \\
\alpha - w(t) &= e^{-\lambda t + \ln(\alpha - w(0))} = e^{-\lambda t}(\alpha - w(0)) \\
w(t) &= e^{-\lambda t}(w(0) - \alpha) + \alpha \quad \wedge \quad w(0) = \frac{1}{2} \\
w(t) &= \alpha + \left(\frac{1}{2} - \alpha\right) e^{-\lambda t}
\end{aligned}
\tag{13}
$$

For $\alpha = 0(1)$ we get two different solutions [5]:

$$
\begin{aligned}
w_0(t) &= \frac{1}{2} e^{-\lambda t} \\
w_1(t) &= 1 - \frac{1}{2} e^{-\lambda t}
\end{aligned}
\tag{14}
$$

Both of solutions satisfy the same initial condition $w(0) = 1/2$, but asymptotically they are different, while the solution $w_0(t)$ converges to zero (e.g. $\lim_{t\to\infty} w_0(t) = 0$), the second solution $w_1(t)$ converges to one (e.g. $\lim_{t\to\infty} w_1(t) = 1$), so $\lim_{t\to\infty} w_\alpha(t) = \alpha$. The assumption that vector α is kept constant in the course of the whole evolution of system is very restrictive, it does not correspond to real situations, at best it may be satisfied only for some of its entries starting from some time t_0.

Therefore, we turn our attention to the forthcoming more complex situation when the assumption of the constant character is not use. Let us solve the differential equation (12) as follows:

$$\frac{dw}{dt} = \lambda\left(\alpha(t) - w(t)\right)$$
$$\frac{dw}{dt} + \lambda\, w(t) = \lambda\alpha(t)$$
$$\frac{dw}{dt} e^{\lambda t} + e^{\lambda t} \cdot \lambda\, w(t) = \lambda\, \alpha(t) e^{\lambda t}$$
$$\frac{d\left(w(t)\cdot e^{\lambda t}\right)}{dt} = \lambda\alpha(t) e^{\lambda t}$$
$$w(t)\cdot e^{\lambda t} - w(0) = \int_0^t \lambda\alpha(\tau) e^{\lambda\tau}\, d\tau \tag{15}$$
$$w(t) = e^{-\lambda t}\left(w(0) + \int_0^t \lambda\alpha(\tau) e^{\lambda\tau}\, d\tau\right) \quad \wedge \quad w(0) = \frac{1}{2}$$
$$w(t) = \frac{1}{2} e^{-\lambda t} + \lambda e^{-\lambda t} \int_0^t \lambda\alpha(\tau) e^{\lambda\tau}\, d\tau$$

We used a mean-value theorem of integral calculus [7] during solution (15):

$$\int_a^a f(x)\, g(x)\, dx = \langle f \rangle_a^b \int_a^b g(x)\, dx \tag{16}$$

where $\langle f \rangle_a^a$ is the mean value of function f on the interval [a, b]: $\langle f \rangle_a^a = \frac{1}{b-a}\int_a^b f(x)\, dx$.

Two different types of transformation are used. Therefore, we get two different solutions of $w(t) = \frac{1}{2} e^{-\lambda t} + \lambda e^{-\lambda t} \int_0^t \lambda\alpha(\tau) e^{\lambda\tau}\, d\tau$ [5]:

- The first type (denoted by A) of transformation is achieved straightforwardly by applying the mean-value theorem, we get $w_A(t) = \frac{1}{2} e^{-\lambda t} + \langle \alpha \rangle_0^t \left(1 - e^{-\lambda t}\right)$, and $\langle \alpha \rangle_0^t$ denotes mean value α on the time interval [0, *t*], where $0 \le t \le t_0$. The solution randomly changes its values on the time interval $[0, t_0]$ by the formula $w = \frac{\sum_{i=1}^{q} e_i}{q}$, where e_i is a value of positions in the *i-th* chromosome for given (independent) entry of the probability vector w $(1 \le i \le q \le p)$, q is number of best individuals in the population P, and it is true $F_i \geq F_{avg}$ $i = 1, ..., q \leq p$), where F_{avg} is calculated by the formula (7) a F_i by the formula (4).

- The second type (denoted by B) of transformation can be derived for the time t greater or equal to the time t_0, $t \geq t_0$, and we get the following solution: $w_B(t) = \frac{1}{2} e^{-\lambda t} + \langle \alpha \rangle_0^{t_0} e^{-\lambda t}\left(e^{-\lambda t_0} - 1\right) + \langle \alpha \rangle_{t_0}^{t} e^{-\lambda t}\left(1 + e^{-\lambda(t-t_0)}\right)$, where $\langle \alpha \rangle_{t_0}^{t}$ is the mean value of α on the time interval $[t_0, t]$, and the solution is determined for $\forall t \geq t_0$.

These two different forms of the general solution have the same functional value for $t = t_0$, $w_A(t_0) = w_B(t_0)$. Time evolution of the probability vector $w(t)$ is fully determined by the mean value $\langle \alpha \rangle_0^{t_0}$ and $\langle \alpha \rangle_{t_0}^{t}$, and moreover its asymptotic properties are determined by the existence or nonexistence of $\langle \alpha \rangle_{t_0}^{\infty}$, e.g. $\lim_{t \to \infty} w_B(t) = \lim_{t \to \infty} \langle \alpha \rangle_{t_0}^{t} \sim \langle \alpha \rangle_{t_0}^{\infty}$. If the entity does not exist, then the probability $w(t)$ asymptotically manifests chaotic oscillations.

4 Experiments

The illustrative calculations are presented for two tasks: (1) the shift detection task, and (2) the rotation problem. The method produced a good separation of models in neural network but not only for the presented tasks.

In the shift detection task, a system is shown a binary pattern and the same pattern shifted one bit to the left or right. The system must detect (a) the pattern, and (b) the direction of the shift. The model is a feed-forward network with nine input nodes, nine hidden nodes, and seven output nodes. Nine input nodes are organized into 3x3 matrix, and patterns are represented by selectively activating specific input nodes. Four shapes are formed by activating different combinations of input nodes. These shapes are defined as different combinations of cells within grid that is for each shape shifted one bit to left or right. If the input layer is considered to be a one dimensional retina, then the shapes are nine bits long and the first and last bits concur. Thus, there were twelve different combinations of shape and its shift directions, and each combination is represented by a binary pattern. The seven output nodes are divided into two subsets of four and three nodes. Nodes in the "shape" subset are responsible for indicating the identity of the input. Each input is associated with one of the four "shape" nodes, and one of the three directions: (a) to left, (b) without shift, (c) to right. The system is considered to correctly recognize and locate an input.

The second model is also a feed-forward network with nine input nodes that are organized into 3x3 matrix, nine hidden nodes, and eight output nodes. Four shapes are also formed by activating different combinations of cells within grid that is for each shape rotated 0°, 90°, 180°, or 270°. Thus, there were sixteen different combinations of shape and their rotations, where each combination is represented by a binary pattern. The eight output nodes are divided into two subsets of fours nodes. Nodes in the "shape" subset are responsible for indicating the identity of the input. Each input is associated with one of the four "shape" nodes, and one of the four rotations: (a) 0°, (b) 90°, (c) 180°, (d) 270°. The system is considered to correctly recognize and locate an input.

Parameters of the experimental part are the following: number of individuals is 100; partial training if backpropagation are 150 epochs; learning rate is 1; momentum is 0; *con* from the formula (4) is 100, λ from the formula (8) is 0.2; the saturation parameter $\tau(\mathbf{w}) = 0.99$.

The evolution of the best individuals in the population is shown in Tab. I, from which is evidently seen, the emergence of modularity of the best individual from the population during a calculation. These results support also the fact that the system was created dynamically during a learning process. Calculation was terminated in the moment, when a saturation parameter was greater than the defined value, e.g. for this particular task was calculation terminated in the 498-th generation (the shift detection task) and in the 353-th generation (the rotation problem), see Table 1. Other numerical simulations give similar results.

Table 1. Number of modules during calculation

	the shift detection task			**the rotation problem**		
generation	*number of modules solving the problem „shape"*	*number of modules solving the problem „shift"*	*number of interferences between modules*	*number of modules solving the problem „shape"*	*number of modules solving the problem „rotation"*	*number of interferences between modules*
1	1	1	12	0	0	13
100	3	4	7	1	3	9
200	4	2	8	4	3	6
300	4	3	7	6	3	4
400	5	4	5		*generation*: 353	
		generation: 498		7	4	2
	6	4	4			

5 Conclusion

Both the theoretical discussion and numerical simulation reflect the modular structure significance as a tool of a negative influence interference rejection on neural network adaptation. As the hidden units in the unsplit network are perceived as some input information processing (e.g. from a retina) for output units, where a multiple pattern classification is realized on the basis of diametrically distinct criteria (e.g. neural network has to classify patterns according to their form, location, colors, ...), so in the beginning of an adaptation process the interference can be the reason that output units also get further information about an object classification than the one which is desired from them. This negative interference influence on running the adaptive process is removed just at the modular neural network architecture, which is proved also by results of the performed experiment. The winning modular network architecture was the product of emergence using evolutional algorithms. The neural network serves here as a special way of solving the evolutional algorithm, because of its structure and properties it can be slightly transformed into an individual in evolutionary algorithm.

The method is like a mix between a *genetic algorithm* [6] and a *hill-climbing with learning* [5]. The method uses a population of chromosomes (*genetic algorithm*), but there is an evolution of this population (*genetic algorithm*) and the probability vector too (*hill-climbing with learning*) on the basis of the best evaluated individuals (*genetic algorithm*) in this algorithm. New individuals in the next generation are generated from the updating probability vector. The best individual is included to the next generation automatically (elitism in *genetic algorithm*). The ending condition is depends on the saturation parameter (*hill-climbing with learning*).

Acknowledgements

This work was supported by internal grant of the Faculty of Science of the University of Ostrava 31/1079.

References

1. Volna, E.: Neural structure as a modular developmental system. In: Sinčák, P., Vaščák, J., Kvasnička, V., Pospíchal, J. (eds.) Intelligent technologies – theory and applications, pp. 55–60. IOS Press, Amsterdam (2002)
2. Fausett, L.V.: Fundamentals of neural networks. Prentice-Hall, Inc., Englewood Cliffs, New Jersey (1994)
3. Rueckl, J.G.: Why are "What" and "Where" processed by separate cortical visual systems? A computational investigation. Journal of Cognitive Neuroscience 2, 171–186 (1989)
4. Anděl, J.: Matematická statistika. I. vydání. SNTL Praha (in Czech) (1978)
5. Kvasnička, V., Pelikán, M., Pospíchal, J.: Hill climbing with learning (an abstraction of genetic algorithm). Neural network world 5, 773–796 (1996)
6. Goldberg, D.E.: Genetic algorithm in search optimalization and machine learning. Addison-Wesley, Reading, Massachusets (1989)
7. Rektorys, K.: Přehled užité matematiky. SNTL Praha (in Czech) (1981)

Averaged Conservative Boosting: Introducing a New Method to Build Ensembles of Neural Networks

Joaquín Torres-Sospedra, Carlos Hernández-Espinosa, and Mercedes Fernández-Redondo

Departamento de Ingenieria y Ciencia de los Computadores, Universitat Jaume I, Avda. Sos Baynat s/n, C.P. 12071, Castellon, Spain
{jtorres, espinosa, redondo}@icc.uji.es

Abstract. In this paper, a new algorithm called *Averaged Conservative Boosting (ACB)* is presented to build ensembles of neural networks. In *ACB* we mix the improvements that *Averaged Boosting* (*Aveboost*) and *Conservative Boosting* (*Conserboost*) made to *Adaptive Boosting* (*Adaboost*). In the algorithm we propose we have applied the conservative equation used in *Conserboost* along with the averaged procedure used in *Aveboost* in order to update the sampling distribution used in the training of *Adaboost*. We have tested the methods with seven databases from the *UCI repository*. The results show that the best results are provided by our method, *Averaged Conservative Boosting*.

1 Introduction

One technique commonly used to create a classification system based on artificial neural networks (ANN) consists on training different networks and combine their output vectors in a suitable manner. This procedure, known as ensemble of neural networks, increases the generalization capability [1] and [2]. Figure 1 shows a basic ensemble of neural networks diagram.

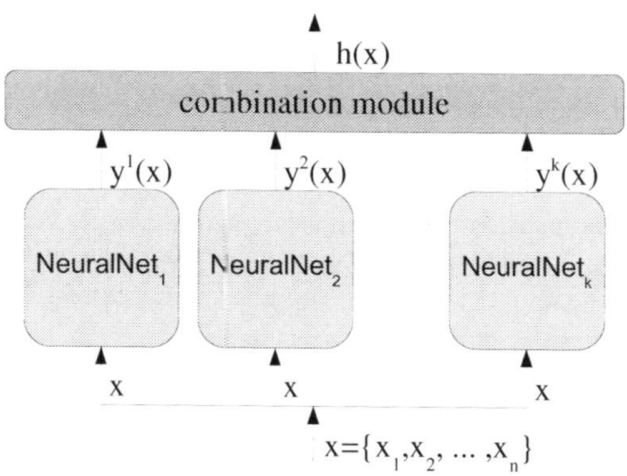

Fig. 1. Basic Ensemble Diagram

J. Marques de Sá et al. (Eds.): ICANN 2007, Part I, LNCS 4668, pp. 309–318, 2007.

Adaptive Boosting (*Adaboost*) is one of the best performing methods to create an ensemble of neural networks [3]. *Adaboost* is a method that constructs a sequence of networks which overfits the training set used to train a neural network with hard to learn patterns. A sampling distribution is used to select the patterns we use to train the network. After training a network the values of the sampling distribution are updated. The probability of selecting a pattern increases if the network that has been trained does not classify correctly the pattern whereas the probability is decreased if the pattern is correctly classified. We can see a *Adaboost* basic training diagram in figure 2.

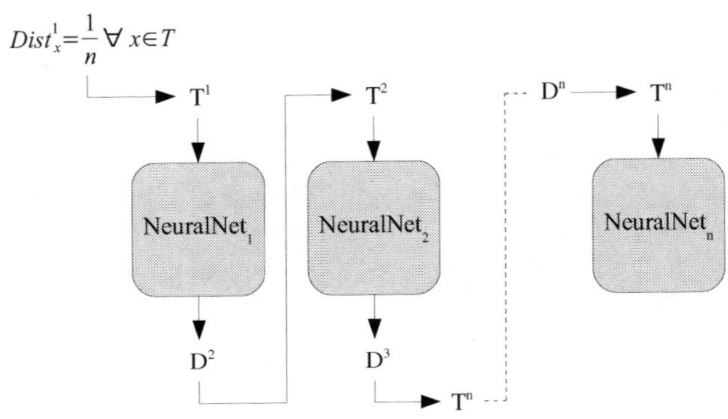

Fig. 2. Adaboost training basic diagram

Adaboost has been deeply studied and successfully improved by authors like Breiman [4], Kuncheva [5] or Oza [6]. Better ensembles have been built by modifying the equation used to update the sampling distribution or by adding new constraints to the original algorithm. Oza proposed an averaged version of *Adaboost* called *Averaged Boosting* (*Aveboost*) in [6]. This version applied an averaged equation which depends on the equation used in *Adaboost*, on the distribution values of the previous network and on the number of networks that have been trained. Kuncheva proposed a relaxed version of *Adaboost* called *Conservative Boosting* (*Conserboost*) [5]. In addition, this method allows the reinitialitation of the *Sampling Distribution* when a constraint is not reached. Although these variants of boosting performed better than *Adaboost* in general, the results depended on the database and on the ensemble size.

We propose a method called *Averaged Conservative Boosting* (*ACB*) which mixes *Aveboost* [6] and *Conserboost* [5] in order to get a more robust boosting method. We introduce a version of *Adaboost* in which the sampling distribution is updated mixing the averaged equation of *Aveboost* and the relaxed equation of *Conserboost*. For this reason, we have built ensembles of 3, 9, 20 and 40 multilayer feedforward (MF) networks on seven databases from the *UCI repository* to test the performance of (*ACB*). The methods are described in 2 whereas the experimental setup and results are in subsection 3.

2 Theory

In this section we briefly review *Adaptive Boosting*, *Averaged Boosting* and *Conservative Boosting*. Finally we describe the *Averaged Conservative Boosting* algorithm.

2.1 Adaboost Review

In *Adaboost*, the successive networks are trained with a training data set T' selected at random from the original training data set T, the probability of selecting a pattern from T is given by the *sampling distribution* associated to the network $dist_{net}$. The sampling distribution associated to a network is calculated when the previous network learning process has finished. *Adaboost* is described in algorithm 1.

Algorithm 1. AdaBoost $\{T, V, k\}$

Initialize Sampling Distribution $Dist$:
  $Dist_x^1 = 1/m \; \forall x \in T$
for $net = 1$ to k **do**
  Create T' sampling from T using $Dist^{net}$
  MF Network Training T', V
  Calculate missclassified vector:

$$miss_x^{net} = \begin{cases} 1 \text{ if } h^{net}(x_x) \neq d(x_x) & x \text{ is incorrecly classified} \\ 0 \text{ otherwise} & x \text{ is correcly classified} \end{cases}$$

  Calculate error:
  $\epsilon_{net} = \sum_{x=1}^{m} Dist_x^{net} \cdot miss_x^{net}$
  Update sampling distribution:

$$Dist_x^{net+1} = Dist_x^{net} \cdot \begin{cases} \frac{1}{(2\epsilon_{net})} & \text{if } miss_x^{net} \\ \frac{1}{2(1-\epsilon_{net})} & \text{otherwise} \end{cases}$$

end for

Adaboost uses a specific method to combine the output provided by the networks of the ensemble. This combination method is described in equation 1 (*boosting combiner*).

$$h(x) = \underset{c=1,\ldots,classes}{\arg\max} \sum_{net:h^{net}(x)=c}^{k} \log \frac{1-\epsilon_{net}}{\epsilon_{net}} \tag{1}$$

2.2 Averaged Boosting and Conservative Boosting

Oza proposed in [6] *Averaged Boosting* (*Aveboost*). *Aveboost* is a method based on *Adaboost* in which the sampling distribution related to a neural network is based on the sampling distribution of the previous network, on the equation used to update the distribution on *Adaboost* and on the number of networks previously trained.

Conservative Boosting (*Conserboost*), the method proposed by Kuncheva [5], uses a softer equation to update the sampling distribution in which the value of the sampling distribution is only updated for the missclassified patterns. In addition, this method allows the reinitialization of the sampling distribution.

The step related to update the sampling distribution is described in algorithm 2 for the case of *Aveboost* whereas 3 shows it for the case of *Conserboost*.

Algorithm 2. Aveboost $\{T, V, k\}$

...

Update sampling distribution:

$$C_x^{net} = Dist_x^{net} \cdot \begin{cases} \frac{1}{(2\epsilon_{net})} & \text{if } miss_x^{net} \\ \frac{1}{2(1-\epsilon_{net})} & \text{otherwise} \end{cases}$$

$$Dist_x^{net+1} = \frac{t \cdot Dist_x^{net} + C_x^{net}}{t+1}$$

...

Algorithm 3. Conserboost $\{T, V, k\}$

...

Update sampling distribution:

if $\epsilon_{net} \geq 0.5$ **then**

Reinitialize the sampling distribution:

$$Dist_x^{net} = 1/m \, \forall x \in T$$

end if

Calculate the β coefficient:

$$\beta_{net} = \sqrt{\frac{1-\epsilon_{net}}{\epsilon_{net}}}$$

Update the sampling distribution:

$$Dist_x^{net+1} = Dist_x^{net} \cdot (\beta_{net})^{miss_x^{net}}$$

Normalize the sampling distribution

$$\sum_{x=1}^{Npat} Dist_x^{net+1} = 1$$

...

2.3 The ACB Algorithm

The new algorithm proposed, *Averaged Conservative Boosting* (*ACB*), is based on *Adaptive Boosting*, on *Averaged Boosting* and on *Conservative Boosting*. Since we have got good results with *Aveboost* and *Conserboost* separately, we have mixed them in order to get a robuster method. We have build an *averaged* equation to update the sampling distribution based on the *relaxed* equation of *Conserboost*. Moreover we have allowed the reinitialitation of the sampling distribution.

To construct an ensemble with the *ACB* algorithm we have used the algorithm described in algorithm 4.

As we can see in algorithm 4, we have mixed the equation to update the sampling distribution of *Aveboost* with the equation of *Conserboost*. In each iteration, the sampling distribution is increased for the missclassified patterns with the averaged conservative equation whereas the probability value is kept for the correctly classified patterns. Moreover, we can also see that the sampling distribution is reinitialized when $\epsilon_{net} \geq 0.5$ so

Algorithm 4. ACB $\{T, V, k\}$

Initialize Sampling Distribution $Dist$:
  $Dist_x^1 = 1/m \; \forall x \in T$
for $net = 1$ to k **do**
  Create T' sampling from T using $Dist^{net}$
  MF Network Training T' , V
  Calculate missclassified vector:
  $miss_x^{net} = \begin{cases} 1 \text{ if } h^{net}(x_x) \neq d(x_x) \\ 0 \text{ otherwise} \end{cases}$
  Calculate error:
  $\epsilon_{net} = \sum_{x=1}^{m} Dist_x^{net} \cdot miss_x^{net}$
  if $\epsilon_{net} \geq 0.5$ **then**
    Reinitialize the sampling distribution:
    $Dist_x^{net} = 1/m \; \forall x \in T$
  end if
  Calculate the β coefficient:
  $\beta_{net} = \sqrt{\frac{1-\epsilon_{net}}{\epsilon_{net}}}$
  Update the sampling distribution:
  $C_x^{net} = Dist_x^{net} \cdot (\beta_{net})^{miss_x^{net}}$
  $Dist_x^{net+1} = \frac{t \cdot Dist_x^{net} + C_x^{net}}{t+1}$
  Normalize the sampling distribution
  $\sum_{x=1}^{Npat} Dist_x^{net+1} = 1$
end for

the distribution is not overfitted with hard to learn patterns. Then, the sampling distribution is updated with the averaged conservative equation we propose in this paper. Finally, the sampling distribution is normalized to get a $[0..1]$-ranged probability distribution.

In our experiments we have usec the *Boosting combiner* (1) and the *Average Combiner* (2) to combine the networks of the ensembles generated by the *Averaged Conservative Boosting* algorithm.

$$h = \arg\max_{class=1 \ldots Nclasses} \{\bar{y}_{class}\} \tag{2}$$

where,

$$\bar{y} = \frac{1}{k} \sum_{net=1}^{k} y_{class}^{net} \tag{3}$$

3 Experimental Testing

In this section, we firtsly describe the experimental setup and the datasets we have used in our experiments. Then, we show the results we have obtained with all the boosting methods previously described on the different datasets. Finally, we compare these results in order to get the best performing method.

In our experiments we have trained ensembles of 3, 9, 20 and 40 MF networks with *Adaptive Boosting*, *Averaged Boosting*, *Conservative Boosting* and *Averaged Conservative Boosting* on the seven problems described in subsection 3.1 using the training parameters described in table 1.

Moreover, we have generated 10 different partitions of data at random in training, validation and test sets and repeat the whole learning process 10 times. With this procedure we can obtain a mean performance of the ensemble for each database and an error in the performance calculated by standard error theory.

3.1 Datasets

We have used seven different classification problems from the *UCI repository of machine learning databases* [7] to test the performance of the methods. The seven databases we have used are: Balance Scale Database (bala), Australian Credit Approval (cred), Ecoli Database (ecoli), Solar Flare Databases (flare), Heart Disease Databases (hear), Haberman's Survival Data (survi) and Wisconsin Breast Cancer Database (wdbc).

The seven databases we have selected for our experiments are commonly used in the bibliography.

Table 1 shows the *MF* training parameters (number of hidden units, maximum number of iterations, the adaptation step and the momentum rate) used in our experiments along with the results we obtained with a single *Multilayer Feedforward* network. These parameters has been set after carrying out a trial and error procedure in which we have tested some configurations of the *MF* network on the validation set.

Table 1. MF training parameters

Database	*Hidden Nodes*	*Iterations*	*Step*	*Momentum*	*Performance*
bala	20	5000	0.1	0.05	87.6 ± 0.6
cred	15	8500	0.1	0.05	85.6 ± 0.5
ecoli	5	10000	0.1	0.05	84.4 ± 0.7
flare	11	10000	0.6	0.05	82.1 ± 0.3
hear	2	5000	0.1	0.05	82.0 ± 0.9
survi	9	20000	0.1	0.2	74.2 ± 0.8
wdbc	6	4000	0.1	0.05	97.4 ± 0.3

3.2 Results

In this subsection we present the results we have obtained with the ensembles of MF networks trained with the boosting methods described in Section 2. In the case of *ACB* we have applied two different combiners: the *Average combiner* and the *Boosting combiner*. Table 2-6 show the results we have obtained with ensembles of 3, 9, 20 and 40 networks trained with *Adaboost*, *Aveboost*, *Conserboost* and *ACB* respectively.

Table 2. Adaptive Boosting results

Database	*3 Nets*	*9 Nets*	*20 Nets*	*40 Nets*
bala	94.5 ± 0.8	95.3 ± 0.5	96.1 ± 0.4	95.7 ± 0.5
cred	84.9 ± 1.4	84.2 ± 0.9	84.5 ± 0.8	85.1 ± 0.9
ecoli	85.9 ± 1.2	84.7 ± 1.4	86 ± 1.3	85.7 ± 1.4
flare	81.7 ± 0.6	81.1 ± 0.7	81.1 ± 0.8	81.1 ± 0.7
hear	80.5 ± 1.8	81.2 ± 1.4	82 ± 1.9	82.2 ± 1.8
survi	75.4 ± 1.6	74.3 ± 1.4	74.3 ± 1.5	73 ± 2
wdbc	95.7 ± 0.6	95.7 ± 0.7	96.3 ± 0.5	96.7 ± 0.9

Table 3. Averaged Boosting results

Database	*3 Nets*	*9 Nets*	*20 Nets*	*40 Nets*
bala	95.8 ± 0.4	96.1 ± 0.6	96.2 ± 0.4	96.5 ± 0.5
cred	85.9 ± 0.7	86.4 ± 0.4	86.6 ± 0.8	85.5 ± 0.9
ecoli	85.3 ± 1	86.5 ± 1.2	86.2 ± 1.2	86 ± 1.1
flare	81.8 ± 0.8	82 ± 0.7	82.4 ± 0.7	80.7 ± 1.1
hear	83.2 ± 1.6	84.9 ± 1.3	83.9 ± 1.4	83.6 ± 1.5
survi	75.1 ± 1.2	74.4 ± 1.2	74.8 ± 1.2	74.6 ± 1.1
wdbc	95.6 ± 0.5	96.6 ± 0.4	95.8 ± 0.6	96 ± 0.5

Table 4. Conservative Boosting results

Database	*3 Nets*	*9 Nets*	*20 Nets*	*40 Nets*
bala	94.7 ± 0.8	95.4 ± 0.6	95.7 ± 0.6	96.2 ± 0.7
cred	86.5 ± 0.6	87.1 ± 0.7	85.9 ± 0.7	86 ± 0.7
ecoli	85.4 ± 1.3	86.2 ± 1.2	86.9 ± 1.2	87.8 ± 1.1
flare	82.1 ± 1	82.2 ± 0.9	82.8 ± 0.6	82.4 ± 0.6
hear	83.2 ± 1.4	83.2 ± 1.3	83.1 ± 1.6	83.9 ± 0.9
survi	75.6 ± 1.1	74.4 ± 1.5	72.8 ± 1.3	73.3 ± 1.5
wdbc	97 ± 0.6	96.6 ± 0.7	96.4 ± 0.6	96.3 ± 0.5

Table 5. The new Averaged Conservative Boosting with the average combiner results

Database	*3 Nets*	*9 Nets*	*20 Nets*	*40 Nets*
bala	95.8 ± 0.4	96.2 ± 0.5	96.2 ± 0.4	96.5 ± 0.6
cred	86.9 ± 0.9	87.2 ± 0.6	86.6 ± 0.8	86.4 ± 0.9
ecoli	86 ± 1	87.5 ± 0.6	88.4 ± 0.7	86.9 ± 1.2
flare	82 ± 0.8	82.4 ± 0.7	82.4 ± 0.7	82.4 ± 0.7
hear	82.7 ± 1.5	84.6 ± 1.3	84.2 ± 1.4	83.9 ± 1.3
survi	75.6 ± 1.2	74.6 ± 1.3	74.8 ± 1.4	74.8 ± 1.6
wdbc	96.9 ± 0.4	96.9 ± 0.6	96.7 ± 0.6	96.6 ± 0.4

Table 6. The new Averaged Conservative Boosting with the boosting combiner results

Database	*3 Nets*	*9 Nets*	*20 Nets*	*40 Nets*
bala	95.8 ± 0.4	96.6 ± 0.4	96 ± 0.5	96.1 ± 0.5
cred	87.1 ± 0.8	87 ± 0.6	86.7 ± 0.7	86.5 ± 0.7
ecoli	86.3 ± 0.9	87.4 ± 0.7	87.4 ± 0.5	87.8 ± 0.7
flare	82.1 ± 0.7	82.1 ± 0.6	82.7 ± 0.6	82.9 ± 0.6
hear	82.9 ± 1.5	84.1 ± 1.3	84.1 ± 1.4	84.1 ± 1.4
survi	74.6 ± 1.5	73.9 ± 1.5	74.9 ± 1.4	75.1 ± 1.5
wdbc	96.2 ± 0.6	96.5 ± 0.6	96.5 ± 0.6	96.4 ± 0.5

3.3 General Measurements

We have also calculated the *Increase of Performance* (IoP) and the *Percentage of Error Reduction* (PER) of the ensembles with respect to a single network. The IoP is an absolute measurement whereas the PER is a relative one. We have used equation 4 to calculate the ensemble IoP value and equation 5 to calculate the ensemble PER value.

$$IoP = Error_{singlenet} - Error_{ensemble} \tag{4}$$

$$PER = 100 \cdot \frac{Error_{singlenet} - Error_{ensemble}}{Error_{singlenet}} \tag{5}$$

A positive value of the IoP means that the performance of the ensembles is better than the performance of a single network on the dataset. The PER value ranges from 0%, where there is no improvement by the use of a particular ensemble method with respect to a single network, to 100%. There can also be negative values on the IoP and on the PER, which means that the performance of the ensemble is worse than the performance of the single network.

Furthermore, we have calculated the mean IoP and the mean PER with respect to the *Single Network* across all databases for each method to get some global measurements to compare the methods. The results are shown on table 7.

Table 7. Mean IoP and mean PER among all databases

	mean IoP				mean PER			
method	*3 Nets*	*9 Nets*	*20 Nets*	*40 Nets*	*3 Nets*	*9 Nets*	*20 Nets*	*40 Nets*
adaboost	0.75	0.45	0.99	0.73	−1.56	−2.97	3.34	−0.67
conserboost	1.61	1.68	1.44	1.78	9.64	8.62	6.74	8.36
aveboost	1.34	1.92	1.78	1.36	2.09	10.08	5.79	4.22
acb-averaging	1.82	2.29	2.28	2.01	10.96	14.11	13.17	10.7
acb-boosting	1.68	2.02	2.12	2.21	7.01	10.67	10.76	10.84

3.4 Discussion

The main results show that *Averaged Conservative Boosting* performs better than *Adaboost*, *Aveboost* and *Conserboost*. Moreover, we can also see that the error in performance among the ten experiments tends to be lower in most of the cases in *ACB*. In figure 3 we can graphically see an example.

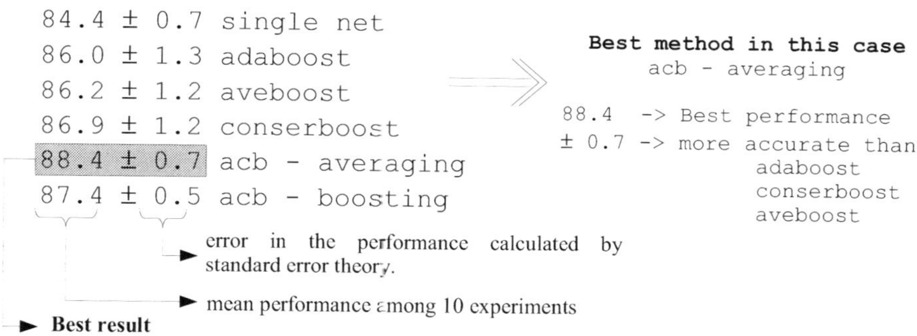

Fig. 3. Comparing the results - 20-network ensemble for database ecoli

According to the global measurements, *ACB* is the best performing method being the *Average Combiner* the most appropriate combiner when we have an ensemble of 3, 9 or 20 networks whereas the *Boosting combiner* should be used to combine 40-network ensembles.

4 Conclusions

In this paper we have presented *Averaged Conservative Boosting*, a boosting algorithm based on *Adaboost*, *Aveboost* and *Conserboost*. We have trained ensembles of 3, 9, 20 and 40 networks with *Adaboost*, *Aveboost*, *Conserboost* and *ACB* to cover a wide spectrum of the number of networks in the ensemble. Moreover, we have applied to *ACB* two different combiners to get the final hypothesis of the ensemble.

The main results and the global measurements show that *ACB* is the best performing method. Although there are some cases in which *Aveboost* and *Conserboost* performs worse than *Adaboost*, our results show that *ACB* with the *Output Average* performs better than *Adaboost* in all the cases. The most appropriate method to combine the outputs in *ACB* is the *Average combiner* for the ensembles of 3, 9 and 20 networks and the *Boosting combiner* is the best choice for the 40-network ensembles.

Moreover, we can see that *Adaboost*, *Aveboost* and *Conserboost* clearly depend on the number of networks in the ensemble whereas in *ACB* the number of networks has a lower impact on the ensemble performance.

Acknowledgments

This research was supported by the project number *P1·1B2004-03* entitled *'Desarrollo de métodos de diseño de conjuntos de redes neuronales'* of Universitat Jaume I - Bancaja in Castellón de la Plana, Spain.

References

1. Tumer, K., Ghosh, J.: Error correlation and error reduction in ensemble classifiers. Connection Science 8(3-4), 385–403 (1996)
2. Raviv, Y., Intratorr, N.: Bootstrapping with noise: An effective regularization technique. Connection Science, Special issue on Combining Estimators 8, 356–372 (1996)
3. Freund, Y., Schapire, R.E.: Experiments with a new boosting algorithm. In: International Conference on Machine Learning, pp. 148–156 (1996)
4. Breiman, L.: Arcing classifiers. The Annals of Statistics 26(3), 801–849 (1998)
5. Kuncheva, L., Whitaker, C.J.: Using diversity with three variants of boosting: Aggressive. In: Roli, F., Kittler, J. (eds.) MCS 2002. LNCS, vol. 2364, Springer, Heidelberg (2002)
6. Oza, N.C.: Boosting with averaged weight vectors. In: Windeatt, T., Roli, F. (eds.) MCS 2003. LNCS, vol. 2709, pp. 15–24. Springer, Heidelberg (2003)
7. Newman, D.J., Hettich, S., Blake, C.L., Merz, C.J.: UCI repository of machine learning databases (1998), `http://www.ics.uci.edu/~mlearn/MLRepository.html`

Selection of Decision Stumps in Bagging Ensembles*

Gonzalo Martínez-Muñoz, Daniel Hernández-Lobato, and Alberto Suárez

Computer Science Department, Universidad Autónoma de Madrid
C/ Francisco Tomás y Valiente, 11
Madrid 28049, Spain
gonzalo.martinez@uam.es,daniel.hernandez@uam.es,alberto.suarez@uam.es

Abstract. This article presents a comprehensive study of different ensemble pruning techniques applied to a bagging ensemble composed of decision stumps. Six different ensemble pruning methods are tested. Four of these are greedy strategies based on first reordering the elements of the ensemble according to some rule that takes into account the complementarity of the predictors with respect to the classification task. Subensembles of increasing size are then constructed by incorporating the ordered classifiers one by one. A halting criterion stops the aggregation process before the complete original ensemble is recovered. The other two approaches are selection techniques that attempt to identify optimal subensembles using either genetic algorithms or semidefinite programming. Experiments performed on 24 benchmark classification tasks show that the selection of a small subset ($\approx 10-15\%$) of the original pool of stumps generated with bagging can significantly increase the accuracy and reduce the complexity of the ensemble.

1 Introduction

Numerous experimental studies show that pooling the decisions of classifiers in an ensemble can lead to improvements in the generalization performance of weak learners. Ensemble methods take advantage of instabilities in the algorithm used to induce a base learner. These instabilities are exploited to generate a collection of diverse classifiers that are expected to be complementary with respect to the classification task considered. Ensemble classification is then achieved by a majority voting scheme. In this context, complementarity means that the errors of the different hypothesis are not correlated, so that, when the classifiers are pooled, correct decisions are amplified. In this manner, the ensemble can achieve a better classification accuracy than a single learner.

In bagging ensembles [1] diverse classifiers are built by inducing each hypothesis from of a different bootstrap sample from the training data [2]. Typically, the

* This work has been supported by *Consejería de Educación de la Comunidad Autónoma de Madrid*, *European Social Fund*, and the *Dirección General de Investigación*, grant TIN2004-07676-C02-02.

J. Marques de Sá et al. (Eds.): ICANN 2007, Part I, LNCS 4668, pp. 319–328, 2007.

generalization error of a bagging classification ensemble decreases monotonically with the size of the ensemble. Asymptotically, this error tends to a constant level, which is considered the best result bagging can achieve. Nevertheless, this asymptotic level is reached only after a large number of hypotheses are included in the ensemble. To reduce the memory and processing requirements, the selection of a subset of classifiers from the original ensemble was first proposed in [3]. Several later investigations show that ensemble pruning can produce very good results [3,4,5,6,7,8,9]. In particular, by pruning the original ensemble, the speed and storage requirements can be significantly reduced without a significant deterioration, and sometimes with an improvement, of the generalization performance.

This article presents an empirical study of several ensemble pruning techniques applied to bagged decision stumps. Decision stumps are binary classification trees consisting in a single decision (i.e., the tree has a single internal node, the root node and two leaves). Stumps can be seen as classification rules based on one question with only two possible answers. Classification is achieved by making a class assignment for each different answer.

The paper is organized as follows: In Section 2 a short review of the different pruning techniques used in the experiments is given. In Section 3 we describe the experimental procedure carried out to compare the pruning procedures in the different classification problems considered and present the results of these experiments. Finally, Section 4 summarizes the conclusions of this work.

2 Methods

In this section we describe several heuristics designed to extract a subensemble with good generalization properties from an initial bagging ensemble. The ensemble pruning methods considered are of two types. A first family of methods is based on altering the order in which the classifiers in the original ensemble are aggregated. The original bagging ensemble is used as a pool of hypothesis from which a sequence of subensembles of increasing size is constructed. Starting from a subensemble of size $u - 1$, a subensemble of size u is built by incorporating the classifier from the original bagging ensemble that is expected to maximize the generalization performance of the augmented subensemble. The final subensemble is obtained by selecting the first τ classifiers from the ordered ensemble. The value of τ is either fixed beforehand to achieve the desired amount of pruning or is estimated using the training data. The various pruning techniques considered differ by the heuristic rule that is used to guide the ordered aggregation. Heuristics based on individual properties of the classifiers have been proved not useful. Successful ordering heuristics need to take into account the complementarity between classifiers with respect to the classification task. The following is a description of the four different ordering heuristics that have been investigated in this work.

Reduce Error (RE): This heuristic was first proposed in [3]. It first selects the classifier with the lowest classification error on the training set. Then, classifiers are sequentially incorporated one by one, in such a way that the classification error of the partially aggregated subensemble estimated on the training set is as low as possible. The algorithm designed in [3] includes the possibility of *backfitting* to correct the mistakes made by this greedy strategy. Nonetheless, *backfitting* has been shown not to reduce the generalization error significantly in bagging ensembles [7]. Furthermore, it dramatically increases the running time of the algorithm. For these reasons no backfitting was implemented in this study.

Complementariness Measure (CC): The heuristic introduced in [5] preferentially incorporates classifiers whose predictions are complementary to those of the existing subensemble. At each step in the aggregation this criterion selects the classifier that correctly classifies more examples in which the partially aggregated subensemble errs. This measure of disagreement is in fact the amount by which that classifier shifts the ensemble decision toward the correct classification.

Margin Distance Minimization (MD): This heuristic was first proposed in [5]. It is based on defining a signature vector $\mathbf{c}^t$ for each classifier t in the original ensemble. This vector has as many components as there are instances in the training set. Component c_i^t is 1 if classifier t correctly classifies the ith instance, and -1 otherwise. The average signature vector of the ensemble is defined as $\langle \mathbf{c} \rangle = T^{-1} \sum_{t=1}^{T} \mathbf{c}^t$. Note that the ith training example is correctly classified if the ith component of $\langle \mathbf{c} \rangle$ is strictly positive. As a result, if the average signature vector of a subensemble is in the first quadrant (all its components are strictly positive), it will correctly classify all examples in the training set. The greedy strategy progressively incorporates into the partially aggregated subensemble those classifiers that reduce the most the distance of the subensemble signature vector $\langle \mathbf{c} \rangle$ to a fixed point $\mathbf{o}$. This point is set to be in the first quadrant with all components equal to a small value p (e.g. $p \approx 0.1$) so that vector components corresponding to examples that are simple to classify quickly reach a value close to p and hence have less influence in future aggregation decisions. The value of the parameter p does not significantly affect the performance of the algorithm provided that it is small $p \in [0.05, 0.2]$ [5].

Boosting Based Ordering (BB): This approach was introduced in [8] and is based on the instance reweighting scheme proposed in the Adaboost algorithm [10]. The method is similar to boosting, except that, instead of generating new hypotheses from the weighted data at each iteration, the classifier with the lowest weighted error over the training set is chosen and aggregated to the partial subensemble. If there is no classifier whose weighted training error is below 50% the weights are set to be uniform. Unlike Adaboost, if the selected classifier has zero error over the training set the algorithm still continues to aggregate classifiers until all elements have been incorporated.

A second family of methods attempts to directly identify optimal subensembles using either global optimization heuristics, such as genetic algorithms, or semidefinite programming on a relaxed version of the problem:

Genetic Algorithm (GA): Following the approach described in [4], a population of individuals is evolved to identify quasi-optimal subensembles. The individuals in the population represent candidate solutions to the optimization problem. A subensemble is represented by a binary string chromosome whose length is equal to the number of classifiers in the original ensemble, T: $\mathbf{b} \in \{0,1\}^T$. The allele value b_t indicates whether classifier t is included in the subensemble ($b_t = 1$) or not ($b_t = 0$). The configuration parameters of the GA were adjusted following the recommendations of [11], using the values proposed in [4] and information from exploratory experiments. A population of 100 individuals with diagonal initialization is considered. Uniform crossover (probability 0.65) and single bit-flip mutation (probability $5 \cdot 10^{-3}$) are used to generate variability in the population. The fitness of an individual is measured as the accuracy of the corresponding subensemble on the training set. The probability that an individual is selected for reproduction is proportional to its fitness. Elitism is used. The GA is halted after a fixed number of epochs (200) have elapsed.

Semi-definite Programming (SDP): This pruning technique has been recently introduced in [6]. It is based on building a matrix $\mathbf{G}$ whose element G_{ij} is the number of common errors between classifiers with labels i and j. The diagonal term G_{ii} is the error of the ith classifier. The elements on this matrix are then normalized $\tilde{G}_{ii} = N^{-1}G_{ii}$ and $\tilde{G}_{ij,\, i\neq j} = \frac{1}{2}\left(G_{ij}G_{ii}^{-1} + G_{ji}G_{jj}^{-1}\right)$, where N is the size of the training set. Intuitively, $\sum_i \tilde{G}_{ii}$ measures the ensemble strength and $\sum_{i\neq j} \tilde{G}_{ij}$ measures its diversity. The subensemble selection problem of size k is formulated as an integer programming problem. The goal is to minimize $arg_{\mathbf{x}}\, min\, \mathbf{x}^T\tilde{\mathbf{G}}\mathbf{x}$ s.t. $\sum_i x_i = k$ and $x_i \in \{0,1\}$, where the binary variable x_i indicates whether classifier ith should be included or not in the subensemble. The solution to this problem can be approximated very accurately by carrying out a convex semi-definite programming (SDP) relaxation.

3 Experiments

Experiments on 24 datasets from the UCI repository [12] have been performed to compare the performance of the different ensemble pruning methods when applied to bagged decision stumps. These datasets include synthetic and real-world problems from different fields of application, with different numbers of classes and attributes. Since stumps produce a binary decision, they are at a disadvantage when applied to multiclass problems, even when used in combination with bagging. In any case, it is interesting to investigate whether the pruned ensembles can achieve significant error reduction on the multiclass problems. The characteristics of the classification tasks are summarized in Table 1.

The results reported are averages over 100 executions. For each real-world dataset, these 100 executions correspond to 10×10-fold cross-validation. For the synthetic datasets (*Led24*, *Ringnorm*, *Twonorm* and *Waveform*) random sampling was applied to generate each of the training and testing sets. Specifically, for each dataset, the following steps were carried out: (i) Generate the training and testing sets by 10-fold-cv or random sampling (see Table 1) and generate 100 bagged decision stumps. The stumps were created using the CART growing tree algorithm [13] and bootstrap samples. The full ensemble generalization error is estimated in the unseen test set. (ii) Order the decision stumps using: reduce-error (RE), complementariness measure (CC), Margin Distance Minimization using $p = 0.075$ (MD75) and boosting based pruning (BB). The values of the heuristics are calculaded using information only from the training set. (iii) The error of ordered ensembles is computed on the training and the test set for subensembles of sizes 1 to 100. (iv) Apply to the same original ensemble the selection approaches: GA and SDP. The SDP algorithm is set to select 21 decision stumps. Pruned subensembles with this number of stumps generally have a good performance in the classification problems investigated. The number of decision stumps selected by GA is not fixed.

Fig. 1 displays the curves that trace the dependence of the ensemble generalization error with respect to the number of stumps for randomly ordered bagging ensembles and for ensembles ordered with the different heuristics: RE, CC, MD75 and BB. The plots on the left (right) correspond to training (test) error curves. Note that in (randomly-ordered) bagging the error exhibits a monotonic decrease and eventual saturation at a constant level as the number of classifiers included in the ensemble increases. The effect of ordered aggregation is that the error initially decreases faster than in the original curve, it attains a minimum at intermediate subensemble sizes and then increases to the asymptotic error level of the complete bagging ensemble. If the aggregation process is stopped at the minimum of the error curve, the corresponding subensemble is smaller and has a better generalization accuracy than the original complete bagging ensemble. Plots are shown only for a selection of datasets, but they are representative of the qualitative behavior of error curves in all the problems investigated. The average error of the optimal subensembles with 21 classifiers selected by SDP approach is displayed in the same figure for comparison. Finally, the average error and the average size of the subensemble selected by GA are reported. Fig. 1 shows that the learning curves exhibit a qualitatively similar evolution of the error as a function of the number of stumps in both the training and test sets. In particular, the position of the minimum error for the ordering heuristics in the

Table 1. Characteristics of the datasets and testing method

Dataset	Train	Test	Attr.	Cls.	Dataset	Train	Test	Attr.	Cls.
Audio	226	10-fold-cv	69	24	Liver	345	10-fold-cv	6	2
Australian	690	10-fold-cv	14	2	New-thyroid	215	10-fold-cv	5	3
Breast W.	699	10-fold-cv	9	2	Ringnorm	300	5000 cases	20	2
Diabetes	768	10-fold-cv	8	2	Segment	2310	10-fold-cv	19	7
Ecoli	336	10-fold-cv	7	8	Sonar	208	10-fold-cv	60	2
German	1000	10-fold-cv	20	2	Tic-tac-toe	958	10-fold-cv	9	2
Glass	214	10-fold-cv	9	6	Twonorm	300	5000 cases	20	2
Heart	270	10-fold-cv	13	2	Vehicle	846	10-fold-cv	18	4
Horse-Colic	368	10-fold-cv	21	2	Votes	435	10-fold-cv	16	2
Ionosphere	351	10-fold-cv	34	2	Vowel	990	10-fold-cv	10	11
Labor	57	10-fold-cv	16	2	Waveform	300	5000 cases	21	3
Led24	200	5000 cases	24	10	Wine	178	10-fold-cv	13	3

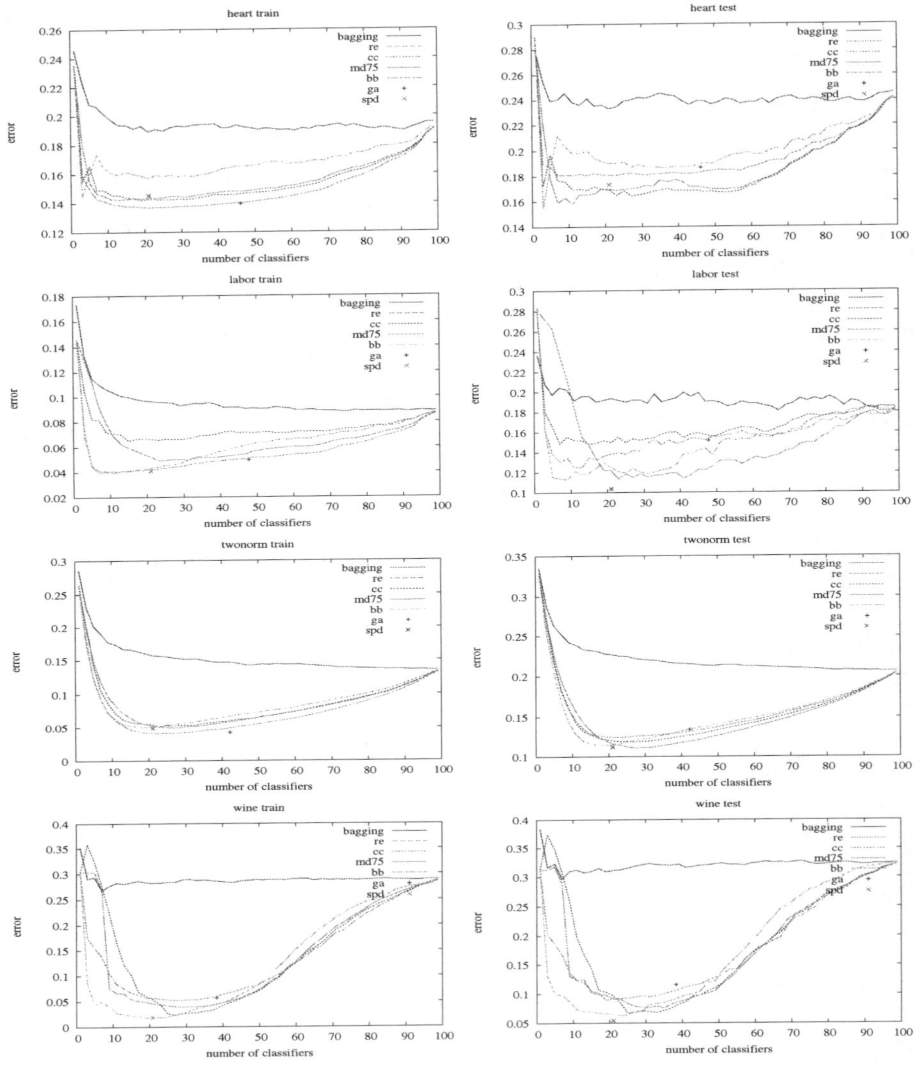

Fig. 1. Average train (*left column*) and test (*right column*) errors for *Heart* (*top row*), *Labor Negotiations* (*second row*), *Twonorm* (*third row*) and *Wine* (*last row*) datasets

training error curves tends to coincide with the position of the minimum in the test error curves. This fact makes it possible to estimate the optimum number of stumps to use in the final subensemble using information only from the training set. Note that the curves shown in Fig. 1 correspond to average values and not to single executions. Nonetheless, the proximity of the positions of the minima in the training and in the test sets is also observed in single runs. The coincidence of the minima is apparent in problems where data is abundant, which implies that

Table 2. Errors for full bagging (100 stumps) and the different ordering heuristics and selection procedures

Dataset	bagging	RE		CC		MD75		BB		GA	SDP
		Min	21	Min	21	Min	21	Min	21		
audio	53.5±2.8	53.5±2.8	53.5±2.8	53.5±2.8	53.5±2.8	53.5±2.8	53.5±2.8	53.5±2.8	53.5±2.8	53.5±2.8	53.5±2.8
australian	14.5±3.8	14.5±3.8	14.5±3.8	14.5±3.8	14.5±3.8	14.5±3.8	14.5±3.8	14.5±3.8	14.5±3.8	14.5±3.8	14.5±3.8
breast	7.0±3.7	**5.5±3.0**	**5.6±3.0**	**4.9±2.9**	**5.4±2.9**	**5.5±2.8**	**5.4±2.9**	**5.4±3.0**	**6.0±2.9**	**5.9±3.2**	**5.4±2.8**
diabetes	27.8±4.1	**26.3±4.3**	**26.2±4.3**	**26.4±4.1**	28.5±5.5	**26.3±4.2**	**26.0±4.4**	**26.5±4.1**	**26.9±4.7**	**27.0±4.3**	**26.9±4.7**
ecoli	35.4±1.8	35.5±2.0	35.4±1.8	35.3±2.3	37.6±5.1	35.4±1.9	35.5±1.9	35.5±2.0	35.4±1.8	35.4±1.8	36.6±4.1
german	30.0±0.0	31.0±2.8	30.0±0.0	31.0±2.9	30.0±0.0	31.0±2.9	30.0±0.0	31.0±2.9	30.0±0.0	30.0±0.0	30.0±0.0
glass	51.4±6.7	**38.0±7.8**	**37.8±8.0**	**36.4±7.4**	52.1±5.2	**36.0±7.5**	**36.0±8.1**	**38.9±8.4**	**48.1±9.8**	**38.3±8.2**	**39.2±8.2**
heart	24.2±10.1	**18.0±8.2**	**18.1±8.1**	**16.9±6.6**	**16.9±6.4**	**16.7±6.8**	**17.0±6.8**	**17.1±7.2**	**19.0±7.4**	**18.7±9.2**	**17.3±6.9**
horse-colic	18.6±6.3	18.7±6.2	18.6±6.3	18.7±6.2	18.6±6.3	18.6±6.3	18.6±6.3	18.7±6.2	18.6±6.3	18.6±6.3	18.6±6.3
ionosphere	17.2±5.2	**10.3±4.8**	17.0±4.7	**10.4±4.8**	16.4±5.1	**13.5±5.1**	17.3±5.1	**10.3±4.8**	17.2±5.2	**10.4±5.5**	17.5±5.3
labor	18.5±15	**11.4±13**	**13.8±13**	**13.5±13**	15.0±13	**12.3±14**	**12.2±13**	**9.9±12**	**12.5±14**	**15.2±14**	**10.4±12**
led24	77.6±4.8	**55.2±8.4**	**58.0±8.4**	**60.7±7.2**	**71.0±8.1**	**67.4±6.4**	**74.5±5.8**	**72.2±4.8**	78.0±4.6	**63.8±8.7**	**60.6±8.0**
liver	37.3±7.6	**33.3±8.2**	**33.6±8.1**	**32.5±7.5**	**30.9±7.1**	**33.1±7.5**	**32.1±7.4**	**32.2±8.0**	**32.7±7.8**	**34.3±7.7**	**32.4±8.3**
new-thyroid	22.9±4.4	**18.7±5.0**	**19.1±5.0**	**18.2±4.1**	**18.3±4.1**	**19.2±4.6**	**20.7±4.5**	**19.3±4.3**	21.7±4.5	**19.3±4.7**	**19.6±4.2**
ringnorm	43.9±2.6	**34.5±1.5**	**40.3±1.2**	**35.2±1.5**	**43.0±1.5**	**35.5±1.5**	**42.7±1.7**	**35.0±1.6**	**42.9±1.4**	**39.8±1.7**	43.3±1.6
segment	55.0±5.2	**44.0±4.0**	**45.2±5.3**	**43.3±2.5**	**50.1±7.4**	**43.3±2.5**	**52.6±6.6**	**50.8±7.0**	57.4±4.7	**43.1±2.1**	**44.9±5.3**
sonar	25.7±9.6	27.4±11.0	27.1±9.9	27.6±9.6	28.3±9.7	26.8±10.3	27.0±10.0	27.1±10.1	27.7±9.4	26.8±9.2	27.5±9.7
tic-tac-toe	30.1±4.8	30.1±4.8	30.1±4.8	30.2±4.7	30.5±4.6	30.1±4.8	30.1±4.8	30.2±4.7	30.4±4.7	30.1±4.8	30.4±4.7
twonorm	20.6±5.5	**12.2±2.0**	**12.4±2.0**	**11.6±1.9**	**12.0±1.8**	**11.3±2.2**	**11.6±1.8**	**11.3±1.9**	**11.5±2.2**	**13.4±2.9**	**11.2±1.8**
vehicle	59.6±2.7	**44.8±4.8**	**45.3±5.3**	**45.3±5.1**	**54.3±6.1**	**44.8±4.7**	**49.1±4.9**	**53.1±7.1**	**58.2±3.8**	**48.3±7.8**	**45.7±5.3**
votes	4.4±3.0	4.4±3.0	4.4±3.0	4.4±3.0	5.5±3.4	4.4±3.0	4.4±3.0	4.4±3.0	4.6±3.1	4.4±3.0	4.8±3.3
vowel	74.6±2.1	**67.2±3.2**	**68.3±3.1**	**70.8±3.7**	75.8±4.5	**69.5±3.1**	**72.9±3.4**	**72.9±2.9**	77.6±4.3	**67.5±4.2**	**70.5±3.8**
waveform	39.2±5.0	**27.5±4.8**	**29.1±6.2**	**27.8±5.0**	**29.2±5.9**	**28.9±5.0**	**30.3±5.2**	**28.0±4.9**	**30.4±6.7**	**32.4±7.1**	**29.3±6.3**
wine	32.4±7.6	**8.6±9.5**	**9.2±10.2**	**5.2±5.2**	**9.7±11.9**	**7.0±6.1**	**9.2±6.7**	**6.2±5.5**	**6.5±5.6**	**11.5±11.7**	**5.4±5.2**
Average	34.2±5.2	27.9±5.4	28.9±5.4	28.1±4.9	31.1±5.5	28.5±5.0	30.1±5.0	29.3±5.1	31.7±5.1	29.5±5.6	29.0±5.2

Table 3. Number of selected stumps for the different datasets and algorithms

Dataset	**RE**	**CC**	**MD75**	**BB**	**GA**
audio	1.0±0.0	1.0±0.0	1.0±0.0	1.0±0.0	50.0±5.0
australian	1.0±0.0	1.0±0.0	1.0±0.0	1.0±0.0	49.9±4.7
breast	12.2±19.3	8.8±9.3	11.7±10.0	6.6±5.9	49.0±5.0
diabetes	1.8±1.6	4.9±8.3	3.9±7.8	2.6±5.0	46.7±6.3
ecoli	1.0±0.0	1.1±0.4	1.0±0.0	1.0±0.0	50.8±4.7
german	1.0±0.1	1.1±0.6	1.1±0.5	1.1±0.4	50.6±4.5
glass	12.6±16.1	14.4±17.5	16.9±15.0	42.5±24.1	47.7±5.4
heart	14.4±12.0	20.0±14.0	19.4±13.9	14.0±18.2	46.1±5.7
horse-colic	1.0±0.1	1.0±0.1	1.0±0.0	1.0±0.1	50.0±4.8
ionosphere	2.0±0.2	2.2±2.0	8.8±8.6	2.1±0.6	44.4±6.3
labor	5.8±5.6	8.4±7.7	17.4±9.9	5.9±3.6	47.8±5.6
led24	22.7±11.7	34.0±18.8	37.6±20.8	23.6±23.1	45.3±6.3
liver	26.6±13.6	28.3±12.9	24.6±7.3	19.5±10.9	42.9±6.5
new-thyroid	3.3±2.4	14.7±26.7	17.2±28.0	11.7±23.9	46.9±6.2
ringnorm	3.7±5.6	3.6±1.6	3.8±2.9	3.4±1.3	41.1±7.5
segment	32.5±25.1	26.5±21.3	33.0±19.8	35.2±28.6	49.2±5.0
sonar	11.3±6.8	12.9±13.0	14.7±5.8	8.9±4.6	45.2±5.7
tic-tac-toe	1.0±0.0	1.0±0.2	1.0±0.0	1.0±0.2	50.0±5.2
twonorm	24.8±8.9	24.8±9.5	26.4±6.7	20.1±6.7	42.3±7.1
vehicle	14.4±11.5	24.9±17.2	18.8±16.0	19.0±19.8	41.0±7.1
votes	1.0±0.0	1.0±0.0	1.0±0.0	1.0±0.0	49.6±5.3
vowel	32.0±19.4	38.8±26.7	26.7±20.7	50.0±26.8	46.5±5.2
waveform	18.2±11.2	22.1±14.5	23.8±13.9	16.6±11.8	41.3±7.2
wine	16.9±10.4	22.3±9.1	33.4±10.2	20.0±7.7	38.3±7.7
Average	10.9±7.6	13.3±9.6	14.4±9.1	12.9±9.3	46.4±5.8

the sampling fluctuations are small (e.g. *Twonorm*). It is less clear in problems where few examples are available for either training or testing (e.g. *Labor*). This is in contrast with ensembles composed of more complex classifiers (e.g. CART trees). In such ensembles, the minimum in the training error curves is generally achieved for smaller subensembles than in the test curves [7].

Table 2 displays the test error in the different classification tasks, averaged over the 100 executions. The results obtained with the ordering heuristics, which produce a collection of subensembles of increasing size (i.e. RE, CC, MD75 and BB), are presented in two columns: the leftmost column indicates the test error for subensembles of a fixed size (21 stumps). The column immediately to the right displays the average test error of subensembles whose size is determined from the position of the minimum of the training error (column labeled with *Min*). Note that the values displayed in columns *Min* of the Table and the values of the test curves shown in Fig. 1 do not generally coincide: the former is an average for different subensemble sizes and the latter is an average for a fixed number of classifiers. Results for the selection approaches GA and SDP are also shown in Table 2. Values of the test error that are significantly better than bagging (using a paired t-test with $p - value < 0.01$) are highlighted in boldface. Errors for cases where bagging performs significantly better (at a $p - value < 0.01$) are underlined. In general, pruned bagging ensembles composed of decision stumps perform better than the complete original ensemble. The largest improvements are obtained using the reduce error and complementariness ordering heuristics to aggregate decision stumps until a minimum in the training error is reached. The improvements in accuracy are fairly large in some domains: In *Wine* the error drops from 32.4 of full bagging to 5.2 of complementariness. In *Twonorm* the generalization error nearly halves. In *Heart* the error achieved by MD75 (Min) is 16.7 and bagging obtains 24.2, etc. As anticipated, the generalization error in classification tasks with more than 2 classes is rather large. Notwithstanding, the pruning procedures manage to significantly reduce the generalization error in multiclass classification tasks: *Glass, Led24, New-thyroid, Segment, Vehicle, Vowel, Waveform* and *Wine*. In any case, the performance in multiclass problems generally remains rather poor.

Table 3 presents the average number of stumps selected for each problem by the different ordering heuristics, using the minimum error in the training dataset and by the GA stumps selection approach. SDP subensembles have a fixed pre-specified size of 21 decision stumps. The tabulated figures show that the amount of pruning varies accross methods and datasets. However, on average, the number of selected stumps is fairly small (10–15%) for all ordering heuristics. The smallest subensembles are generally obtained using the reduce error ordering heuristic. This method also obtains low generalization errors. Note that for several datasets (*Audio, Australian, E-coli, ...*) the number of selected stumps is on average 1 and that, on these problems, the average generalization error of full bagging and just one stump is very similar. For these classification tasks bagging in not able to generate enough diversity and most of the decision stumps in the ensemble are actually equal. For most of the problems analyzed the genetic approach selects subensembles with approximately 50% of the original bagged stumps. Thus, GA tends to select suboptimal subsensembles that are too large.

4 Conclusions

This paper presents an empirical evaluation of different techniques used to prune ensembles of decision stumps generated with bagging. Six ensemble pruning heuristics are tested: reduce error (RE), complementariness (CC), margin distance minimization (MD75), boosting based pruning (BB), genetic algorithms (GA) and semidefinite programing (SDP). The first four heuristics are greedy approaches that determine the order in which the classifiers generated in bagging are incorporated into the ensemble according to some rule that exploits the complementarity of the base learners. Pruning is performed by selecting the first τ classifiers in the ordered ensemble either by specifying a fixed subemsemble size (21 in this investigation) or by estimating the optimum number of classifiers in some manner. For decision stumps the minimum of the curve that displays the dependence of the test error with the size of the subensemble is fairly close to the minimum in the corresponding curve for the training error. This is probably related to the fact that decision stumps are simple classifiers and do not tend to overfit the data. Hence, the optimal stopping point for the ordered aggregation is close to the minimum of the error curve in the training dataset. The GA and SDP approaches attempt to identify a single subensemble that is optimal according to some estimate of the generalization performance. The GA approach solves the selection problem by a genetic algorithm without prescribing a target size for the optimal subensemble. In SDP a convex semi-definite programming relaxation is used to approximately maximize a goal function dependent on both accuracy and diversity for a prespecified ensemble size. For most of the datasets the pruning techniques investigated significantly reduce the generalization error of bagged decision stumps by selecting a fairly small subset of classifiers. The best performances are for ordered bagging, where the order of aggregation is determined by either the reduce error or complementariness heuristic, and aggregation stops at the minimum of the error curve for the training data. SDP obtains good overall results despite the fact that the number of classifiers needs to be fixed *a priori*. The performance of GA in terms of generalization error and percentage of pruning achieved is significantly worse than the other methods.

References

1. Breiman, L.: Bagging predictors. Machine Learning 24(2), 123–140 (1996)
2. Efron, B., Tibshirani, R.J.: An Introduction to the Bootstrap. Chapman & Hall/CRC (1994)
3. Margineantu, D.D., Dietterich, T.G.: Pruning adaptive boosting. In: Proc. 14th International Conference on Machine Learning, pp. 211–218. Morgan Kaufmann, San Francisco (1997)
4. Zhou, Z.H., Tang, W.: Selective ensemble of decision trees. In: Wang, G., Liu, Q., Yao, Y., Skowron, A. (eds.) RSFDGrC 2003. LNCS (LNAI), vol. 2639, pp. 476–483. Springer, Heidelberg (2003)
5. Martínez-Muñoz, G., Suárez, A.: Aggregation ordering in bagging. In: Proc. of the IASTED International Conference on Artificial Intelligence and Applications, pp. 258–263. Acta Press (2004)
6. Zhang, Y., Burer, S., Street, W.N.: Ensemble pruning via semi-definite programming. Journal of Machine Learning Research 7, 1315–1338 (2006)
7. Martínez-Muñoz, G., Suárez, A.: Pruning in ordered bagging ensembles. In: Proceedings of the 23rd International Conference on Machine Learning, pp. 609–616 (2006)
8. Martínez-Muñoz, G., Suárez, A.: Using boosting to prune bagging ensembles. Pattern Recognition Letters 28(1), 156–165 (2007)
9. Zhou, Z.H., Wu, J., Tang, W.: Ensembling neural networks: Many could be better than all. Artificial Intelligence 137(1-2), 239–263 (2002)
10. Freund, Y., Schapire, R.E.: A decision-theoretic generalization of on-line learning and an application to boosting. In: Proc. 2nd European Conference on Computational Learning Theory, pp. 23–37 (1995)
11. Eiben, A.E., Smith, J.E.: Introduction to evolutionary computing. Springer, Berlin (2003)
12. Blake, C.L., Merz, C.J.: UCI repository of machine learning databases (1998)
13. Breiman, L., Friedman, J.H., Olshen, R.A., Stone, C.J.: Classification and Regression Trees. Chapman & Hall, New York (1984)

An Ensemble Dependence Measure

Matthew Prior and Terry Windeatt

Centre for Vision Speech and Signal Processing
University of Surrey, Surrey, GU2 7XH, UK
{m.prior,t.windeatt}@surrey.ac.uk

Abstract. Ensemble methods in supervised classification problems have been shown to be superior to single base classifiers of comparable performance, particularly when used in conjunction with multi-layer perceptron base classifiers. An ensemble's performance is related to the accuracy and diversity of its component classifiers. Intuitively, diversity seems to be a desirable quality for a collection of non-optimal classifiers. Despite much interest being shown in diversity, little progress has been made in linking generalisation performance to any specific diversity metric.

With the agglomeration of even modestly accurate statistically independent classifiers it can be shown theoretically that ensemble accuracy can be forced close to optimality. Despite this theoretical benchmark, real world ensembles fall far short of this performance. The root of this problem is the lack of statistical independence amongst the base classifiers. We investigate a measure of statistical dependence in ensembles, D, and its relationship to the Q diversity metric and pairwise correlation and also examine voting patterns in real world ensembles. We show that, whilst Q is relatively insensitive to changes in the ensemble configuration D measures correlations between the base classifiers effectively. The experiments are based on several two class problems from the UCI data sets and use bootstrapped ensembles of relatively weak, multi-layer perceptron, base classifiers.

1 Introduction

Most theoretical studies of multiple classifier ensemble performance base their analysis on ensembles of homogeneous accuracy and assume statistical independence for the base classifiers. Though the output of each individual classifier in the ensemble is unaffected by the output of any other, the classifiers share a common derivation through their training sets. This commonality undermines their independence. The result is that, for real world ensembles, performance falls far short of that which theory predicts.

We let the random variables, $\chi = \{X_1, X_2, ...X_I\}$, represent the probability of individual base classifiers correctly classifying an input pattern n. For a majority vote ensemble E, containing I statistically independent base classifiers, all with the probability of correct classification equal to p, then the we can model the classification behaviour of the ensemble as a sequence of I Bernoulli trials through the use of the property of ergodicity. The probability that the ensemble

J. Marques de Sá et al. (Eds.): ICANN 2007, Part I, LNCS 4668, pp. 329–338, 2007.

correctly classifies an input pattern is the value for the binomial cumulative distribution function at the point where more than half of of the ensemble correctly classifies an input pattern,

$$\Pr_{E_{homogenous}}(correct|E) = \sum_{j=\lceil \frac{I}{2} \rceil}^{I} \binom{I}{j} p^j (1-p)^{I-j}, \tag{1}$$

where $\Pr(\bullet)$ is the probability of an event, $\lceil \bullet \rceil$ represents the first integer value larger than $\bullet$ and $\binom{b}{c} = \frac{b!}{c!(b-c)!}$.

Our work is motivated from the perspective of investigating the diversity and independence of classifier ensembles under those conditions where base classifiers of only relatively weak performance are available.

To describe the situation where an ensemble performs optimally, Kuncheva coins the phrase: pattern of success[7]. Under the pattern of success, performance exceeds that predicted by the assumption of independence of the base classifiers. With the characteristic that none of the voting patterns of the classifiers needlessly re-enforce an already correct decision. The maximum level of performance gain attainable given the pattern of success is considerable. In the case of a 3 member homogeneous majority vote ensemble with base classifier accuracy of 0.6, the ensemble performance is 0.7 for independent classifiers, but can rise to 0.9 under the pattern of success. We can also see that the ensemble can outperform the independence level of accuracy without achieving the ultimate pattern of success, we call ensembles with this quality supra-independent.

There is no known method for generating classifiers that match the pattern of success. In the absence of such a method, we observe that given a sufficient number of independent classifiers it is theoretically possible, though in practice difficult, to construct ensembles of arbitrary accuracy. If the weaker condition of independence in ensemble construction can be achieved, then good results should be obtainable. To this end we have propose a measure, D, of dependence that can be applied to ensembles to quantify their proximity to independence. We further investigate the behaviour of this and other metrics when base classifier complexity is modulated within varying sized ensembles of bootstrapped, randomly initialised, multi-layer perceptrons (MLP).

2 The Ensemble Dependence Measure, *D*

If we extend the model of theoretical ensemble performance to allow for ensembles with heterogeneous accuracies, we can calculate the probability of a correct ensemble prediction as follows:

$$\Pr_{E_{heterogeneous}}(correct|E) = \sum_{m \in M} \prod_{i} p_{m_i}. \tag{2}$$

Where, $M = \{m : \sum_{i=1}^{I} V(e_i) \geq \lceil \frac{I}{2} \rceil\}$ and is the set of all voting patterns of the ensemble for which the majority vote correctly labels the input pattern. $V(e_i)$

is the truth vote output for the ith classifier in the ensemble. The truth vote is 1 if the classifier votes for the correct label for the input pattern and 0 if it votes incorrectly. p_{m_i} is the probability that the ith classifier in voting pattern m correctly labels the input.

If we relate this theoretical model to the more common real world situation of having dependence between the classifiers we get

$$\Pr(correct|E) = \Pr_{E_{heterogeneous}}(correct|E) - D(E), \tag{3}$$

where $D(\bullet)$ represents a term whose value is the difference between $\Pr_{E_{heterogeneous}}$ and the empirical accuracy of the ensemble, $\Pr(correct|E)$. D is thus a measure of the statistical dependence present in the ensemble.

Where

$$D(\bullet)\ is \left\{ \begin{array}{c} < 0\ if\ the\ ensemble\ is\ supra-independent \\ 0\ if\ the\ ensemble\ is\ independent \\ > 0\ if\ the\ ensemble\ is\ dependent \end{array} \right\}.$$

D is related to the Kullback-Leibler divergence in that it is a measure of the difference between the maximum entropy approximation of the ensemble accuracy and the estimated value of the actual ensemble accuracy.

3 Out-of-Bootstrap Estimation

In order to inject diversity into the ensembles there must be some mechanisms to generate base classifier boundaries that are dissimilar[2]. This was achieved by using bagging[1] to manipulate the distribution of the training set and by random initialisation of network weights in the MLP base classifiers.

Bagging brings with it the inclusion of bootstrapping into the training process. By training each base classifier on a sample with replacement, B, of the same size as the training set, T_r, each classifier, on average, sees a training set which contains examples of around two thirds of all the available training patterns. Such classifiers can be treated as being derived from a set of i.i.d. random variables. This leads to an improvement in error rate of the ensemble over the base classifiers, primarily through the action of variance reduction [4].

Bootstrapping offers a second advantage, namely the availability of an out-of-bootstrap estimate[10] of the generalisation error of the base classifier. This is derived from those samples not in its training set and is found by applying the learned base classifier, e_i, to $T_r \notin B_i$. Though this itself is not directly useful for assessing ensemble performance, an ensemble out-of-bootstrap estimate based on a set of classifiers of about $\frac{1}{3}$ of the ensemble size. By selectively applying patterns from the training set only to those classifiers which did not see that pattern during the training phase it is possible to form a generalisation error predictor, G.

$$G(E) = \frac{1}{I} \sum_{i=1}^{I} \left(1 - \frac{\sum_{T_{rn} \notin B_i} \Omega(e_i(T_{r_n}), L_n)}{|\{T_r \setminus B_i\}|} \right), \quad (4)$$

where I is the number of members in the ensemble, $|\bullet|$ represents the cardinality of a set, Ω is a membership indicator function such that $\Omega(w, z) = \begin{cases} 1 : w \in z \\ 0 : w \notin z \end{cases}$, $\setminus$ indicates set subtraction and L_n is the true label for pattern n. This out-of-bootstrap estimator, G, has been shown to be useful in predicting ensemble generalisation performance[3,9,8].

4 Diversity Measures

In the absence of a known Bayesian optimal classifier, ensembles succeed in improving performance only if the classifiers within the ensemble exhibit diversity [5]. That is to say that the individuals should offer differing opinions on the classification applicable to a given set of input patterns. Ensembles exhibiting values at the extremes of diversity, those in which all members exhibit so little diversity that the same classification is given by each base classifier and those for which the members voting patterns approximate randomness across the input set, are clearly suboptimal. There is however no agreement on how to set an optimal level of diversity or on which measure of diversity is most appropriate.

Q is highlighted in [6] as being the diversity measure that exhibits the greatest number of desirable properties so we selected this to compare with D. ρ is the pairwise correlation coefficient and was used to provide a comparative measure of dependence. The pairwise diversity measures are shown in Table 2, they are based on the probabilities of the output predictions of a pair of classifiers.

Consider two classifiers $\{e_1, e_2\}$, the likelihoods for the four possible outcomes from the result of classification of an input pattern are shown in Table 1. We

Table 1. Probabilities of paired classifier outputs

	e_1 correct	e_1 incorrect
e_2 correct	a	b
e_2 incorrect	c	d

can then define the following measures between pairs of classifiers $\{e_j, e_k\}$ as shown in Table 2. Values of these measures for ensembles are generated from the average value across all pairs of classifiers in the ensemble, for instance in the case of Q for an ensemble with size I:

$$Q_{ensemble} = \frac{\sum_{\{(j,k):j \neq k\}} Q_{j,k}}{\binom{I}{2}},$$

Table 2. Pairwise Diversity Measures

Name	Symbol	Definition
Q measure	Q	$Q_{j,k} = \frac{ad-bc}{ad+bc}$
Correlation coefficient	ρ	$\rho_{j,k} = \frac{ad-bc}{\sqrt{(a+b)(c+d)(a+c)(b+d)}}$

5 Experimental Method

We took six two class data sets from the UCI repository , {Cylinder Bands [BAN], Ionosphere [ION], Promoters [PRO], Sonar [SON], Wisconsin [WIS] and Vote [VOT] } and used these to investigate the behaviour of the various measures. Because the difference between D and the empirical training accuracy, A_{Train}, can converge rapidly to $1 - A_{Train}$ when base classifier error rates are low and the ensemble size increases, we added label noise to the following data sets to make the behaviour of D more observable, { ION,VOT,WIS }.

To see how the measures responded to classifiers of differing complexity, we constructed bootstrapped ensembles of varying sizes. These were comprised of base classifiers formed from relatively simple neural networks. The individual classifiers were combined by majority voting. The architecture and number of iterations of training for these networks were also varied to further adjust the range of complexity of the final classifiers. The networks were trained with the Levenberg-Marquardt algorithm and random initialisation was applied. The procedure was as follows.

1. If necessary, 20% label noise was added to the data set. The labelled data set $S = \{s_1, s_2, ...s_N\}$ was randomly divided into two equal parts, these formed the training and test sets, $\{T_r, T_e\}$. This was repeated 10 times, to form paired training and test sets.
2. Ensembles of classifiers, $E = \{e_1, e_2...e_I\}$, were constructed from neural networks, e, with each network being randomly initialised and trained on a bootstrap sample T_r^*. This was repeated to form an ensemble of size I trained from bootstrap samples $\{T_{r_1}^*, ..., T_{r_I}^*\}$ for all training/test set pairs.
3. An out of bootstrap generalisation estimate, $G(E)$, was formed using the ensemble E with each of the members e_i only voting on those samples, approximately $\frac{1}{3}$ of the total, which were not used to train them, i.e. the training samples omitted from the ith bootstrap, $T_r/T_{r_i}^*$ as defined in eqn. 3. Ties were broken randomly.
4. A measure of the dependence present in the ensemble, $D(E)$, was constructed from the empirical probabilities of the e_i^{th} base classifiers correctly classifying the members of T_r, as indicated in eqn. 2.
5. The diversity measures indicated in Table 2 were calculated from T_r.

This procedure was then repeated 50 times on the 10 train/test sets pairs, giving 500 results for each combination of ensemble size, network architecture and number of iterations. Ensembles with the parameters given in Table 3 were tested.

Table 3. Multi-layer Perceptron Ensemble Specifications

Hidden Neurons	Iterations	Ensemble Size	Designation
1	1	{3,5,7,11,15}	$\{E_1, E_2, E_3, E_4, E_5\}$
1	3	{3,5,7,11,15}	$\{E_6, E_7, E_8, E_9, E_{10}\}$
2	3	{3,5,7,11,15}	$\{E_{11}, E_{12}, E_{13}, E_{14}, E_{15}\}$

6 Results

The theoretical accuracy of the ensemble converges to a value of 1 as the base classifiers become more accurate and as the ensemble size grows. Beyond this point of convergence D reduces to the training error and has no more useful information to give. It is should be noted that this would be true for any other derived measure, or theoretical assumption, which relies on the independence of the base classifiers.

The graphs in Figure 1 show D, G, Q and the test and training errors. For small ensembles, $\{E_1, E_6, E_{11}\}$, D indicates suitably low values of dependence. As the ensemble size is increased D clearly records the growth in dependence. For the highest complexity ensembles, such as, $\{E_{13}, E_{14}, E_{15}\}$, D begins to suffer from some observable saturation. This is particularly noticeable in the case of the SON data set where the training accuracy approaches 0.90 on a data set with 20% label noise and the theoretical ensemble accuracy approaches 1, with the result that the plot of D exhibits a pronounced knee at the 0.1 level.

We also measured the correlation, on a per trial basis, between the diversity measures D, G, Q,ρ and the test set accuracy at each of the complexity levels. For the correlation measurements we stratified the correlations with the training sets. The average mean stratified correlation figures, Table 4, suggest that the training error, followed by D and then G are the most accurate predictors of generalisation performance, with Q and ρ having almost no correlation at all with test error. This said, none of the predictors showed a strong overall correlation. The explanation for these may lie in the correlation graphs.

For all the data sets we observe that in general the mean stratified correlation coefficient for all the measures tends to drop significantly as the complexity of the classifier ensembles increase, Figure 2 shows the results for the ION and SON data sets. Ultimately these seem to converge towards zero, resulting in the situation where there is little correlation at all with the test error.

A case in point is G, in Figure 1 it can be seen that this measure tracks the test error very well indeed, but the resulting correlation figures are not indicative of this. It is our conjecture that, as the base classifiers become more powerful there is a reduction in the bias component of the error for the ensemble. This leaves the dominant error mechanism to be the variance of the classifiers, which begins to approximate a constant summed with a noise process and hence the correlation tends towards zero, with the result that the generated errors correlate poorly between test and training sets.

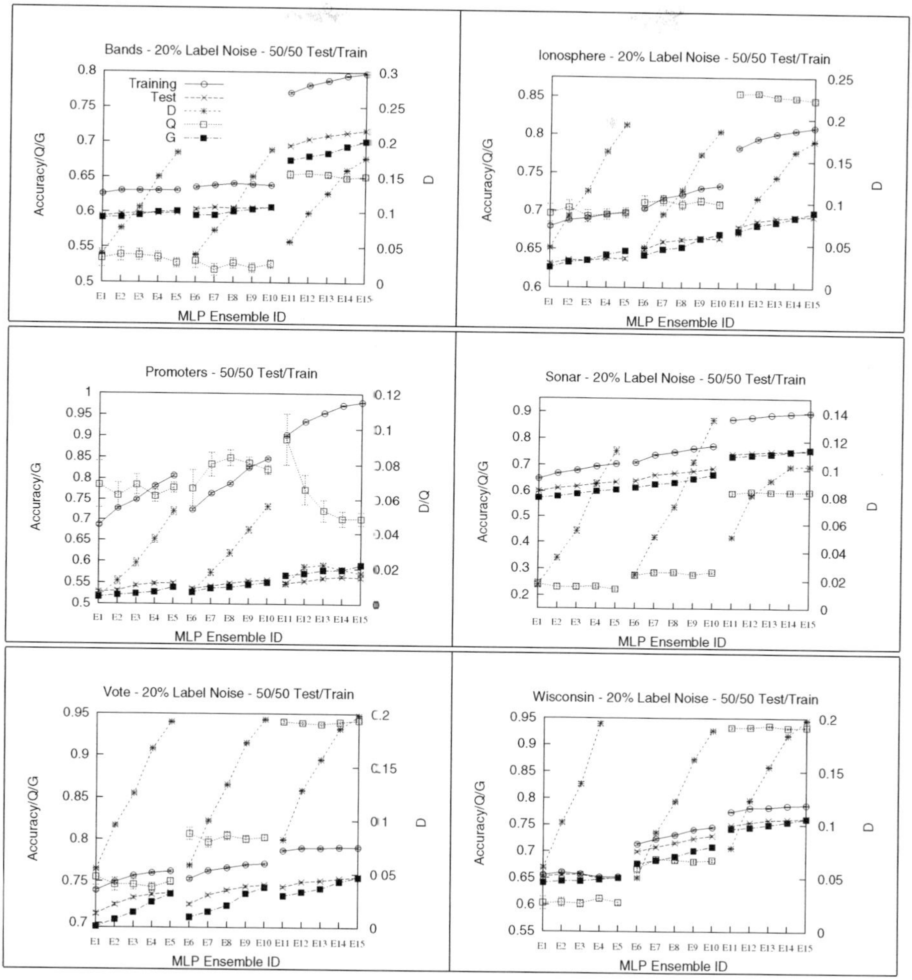

Fig. 1. The mean value of D, G, Q, training and test accuracies for the series of ensemble configurations, $\{E_1 ... E_5\}$,$\{E_6 ... E_{10}\}$, $\{E_{11} ... E_{15}\}$. Values for G, Q and training and test accuracies are shown on the left hand Y-axes, except in the case of the promoters data set where Q is shown on the right hand axis. Values for D are shown on the right hand Y-axes. Each data-point is calculated from the results for 500 ensembles. Bars show the standard error of the mean and are generally close enough to the mean as to be not visible.

We also examined the ensemble voting performance. In [7] Kuncheva presents a theoretical analysis of a 3 member majority vote ensemble with homogeneous accuracy. In that example all voting patterns of the ensemble are examined, in about 43% of the cases the ensemble outperforms the accuracy of the majority vote independent classifiers and so can be said to show supra-independent performance. As D is the difference between the independence based theoretical

Table 4. Mean stratified Pearson correlation coefficients for test error with measures D, G, Q, ρ and training error, TR, on the three series of classifiers $\{E_{1..5}, E_{6..10}, E_{11..15}\}$. The three series means are on the first row of each cell, the overall mean is beneath the series mean.

	TR	D	G	Q	ρ
BAN	0.66,0.68,0.37 0.57	0.45,0.53,0.37 0.45	0.25,0.31,0.24 0.27	-0.42,-0.44,0.11 -0.25	0.47,0.46,-0.21 -0.24
ION	0.67,0.52,0.22 0.47	0.55,0.45,0.22 0.41	0.41,0.36,0.10 0.29	-0.02,0.05,0.02 0.02	-0.01,-0.11,-0.03 -0.05
PRO	0.18,0.10,0.01 0.09	0.15,0.09,0.02 0.09	0.07,0.05,0.02 0.05	-0.07,-0.05,0.00 -0.04	0.02,-0.00,0.00 -0.01
SON	0.64,0.51,0.16 0.43	0.44,0.39,0.15 0.33	0.34,0.29,0.08 0.24	-0.18,-0.11,-0.05 -0.12	0.14,-0.01,-0.04 0.03
VOT	0.56,0.39,-0.05 0.30	0.39,0.36,0.12 0.29	0.36,0.31,0.13 0.27	0.15,0.20,0.18 0.18	-0.19,-0.23,-0.211 -0.21
WIS	0.95,0.81,0.29 0.68	0.81,0.71,0.32 0.62	0.77,0.64,0.25 0.55	-0.29,0.18,0.19 0.02	0.51,-0.04,-0.25 0.07
All	0.42	0.37	0.28	-0.03	-0.07

accuracy and the measured accuracy it can be used to detect those ensembles in which better than independent voting patterns have occurred.

Table 5. Probability of bootstrapped, randomly initialised MLP ensembles of sizes {3,5,7,11,15} and {3}, exhibiting measured values of D which indicate independence or supra-independence. Based on 7,500 ensembles for each data set.

	Size	BAN	ION	PRO	SON	VOT	WIS	Mean
independent	{3,5,7,11,15}	0.003	0.002	0.051	0.009	0.001	0.001	0.011
supra-independent	{3,5,7,11,15}	0.007	0.010	0.240	0.073	0.003	0.005	0.056
combined	{3,5,7,11,15}	0.010	0.012	0.291	0.082	0.004	0.006	0.067
independent	{3}	0.010	0.005	0.037	0.021	0.003	0.005	0.014
supra-independent	{3}	0.031	0.039	0.383	0.182	0.015	0.023	0.112
combined	{3}	0.041	0.044	0.420	0.203	0.018	0.028	0.126

Our empirical results in Table 5 show that the real world likelihood for these patterns of voting is highly dependent on the data set and classifier characteristics. Generally these supra-independent classifiers are rare, averaging out at around 7% on a wider range of ensemble sizes. But the results for the PRO data set strongly affect this value, which would otherwise be much lower. PRO manages to reach values in line with the theoretical predictions for 3 member ensembles. Looking at the graph of test error, Figure 1, suggests that the ensemble classification performance is generally poor on this sparse data set even though the classifiers have nearly perfectly learned the training set.

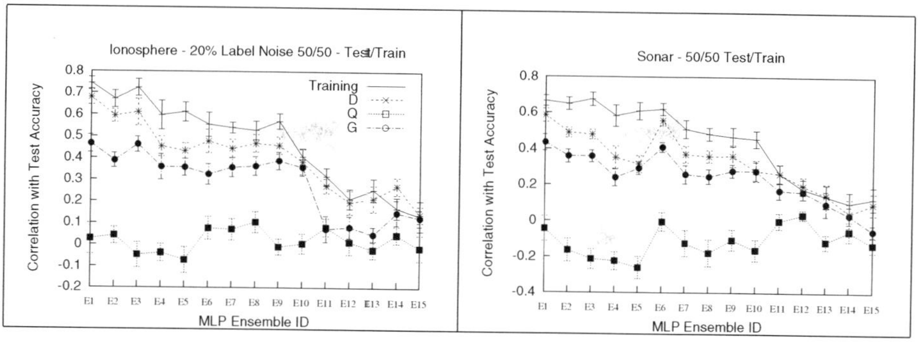

Fig. 2. The mean stratified Pearson correlation coefficient of D, G, Q and Training Accuracy with the Test Error for all ensemble configurations, $\{E_1, E_2, ...E_{15}\}$. Each data-point is calculated from the results for 500 ensembles. Bars show the standard error of the mean.

7 Conclusions

Theoretically both Q and D have a value of zero when an ensemble exhibits independence, but away from this common point their behaviour is very different. Figure 1 shows that Q shows little, if any, response to changes in ensemble size for a given base classifier complexity. Yet D climbs rapidly as the ensemble enlarges and fails to achieve the exponential increase in accuracy predicted by theory. This highlights that diversity and independence are two different quantities. We might expect that given a sample from a population of base classifiers, Q would estimate the diversity present in the population. As the sample, or ensemble, size is increased we would expect that the estimate would become more accurate, but not that the underlying value of diversity would change. Thus Q is useful for estimating the diversity present in a classifier generation mechanism, but less so for characterising the performance of an ensemble as we vary its size. On the other hand, D measures the departure from theoretical independence of an ensemble and can directly show the degradation caused by dependence as the ensemble is increased. When measuring correlations, D has the advantage over ρ that it is able to capture all higher order interactions between the base classifiers and is not limited to characterising pairwise effects. From this perspective we feel that D represents a useful measure to understand the relationship between size and accuracy for ensembles.

During our experimental trials on some 45,000 bagged MLP ensembles covering six data sets, we recorded the percentages of voting patterns which exhibited the property of supra-independence, or independence. For 3 member ensembles the average is around 10%, significantly lower than the 43% projected from a theoretical analysis based on a uniform distribution of voting patterns. Further work is required to see if these beneficial patterns of voting can be used to increase ensemble accuracy.

References

1. Breiman, L.: Bagging predictors. Machine Learning 24(2), 123–140 (1996)
2. Brown, G., Wyatt, J., Harris, R., Yao, X.: Diversity creation methods: a survey and categorisation. Information Fusion 6(1), 5–20 (2005)
3. Bylander, T.: Estimating generalization error on two-class datasets using out-of-bag estimates. Machine Learning 48(1-3), 287–297 (2002)
4. Fumera, G., Roli, F., Serrau, A.: Dynamics of variance reduction in bagging and other techniques based on randomisation. In: Oza, N.C., Polikar, R., Kittler, J., Roli, F. (eds.) MCS 2005. LNCS, vol. 3541, pp. 316–325. Springer, Heidelberg (2005)
5. Kuncheva, L.: That elusive diversity in classifier ensembles. In: Perales, F.J., Campilho, A., Pérez, N., Sanfeliu, A. (eds.) IbPRIA 2003. LNCS, vol. 2652, Springer, Heidelberg (2003)
6. Kuncheva, L., Whitaker, C.: Ten measures of diversity in classifier ensembles: limits for two classifiers. In: Proc. IEE Workshop on Intelligent Sensor Processing vol. 6, pp. 1–6 (2001)
7. Kuncheva, L., Whitaker, C., Shipp, C., Duin, R.: Limits on the majority vote accuracy in classier fusion. IEEE Transactions on Pattern Analysis and Machine Intelligence (2000)
8. Prior, M., Windeatt, T.: Over-fitting in ensembles of neural network classifiers within ecoc frameworks. In: Oza, N.C., Polikar, R., Kittler, J., Roli, F. (eds.) MCS 2005. LNCS, vol. 3541, pp. 286–295. Springer, Heidelberg (2005)
9. Prior, M., Windeatt, T.: Parameter tuning using the out-of-bootstrap generalisation error estimate for stochastic discrimination and random forests. In: International Conference on Pattern Recognition, pp. 498–501 (2006)
10. Rao, J., Tibshirani, R.: The out-of-bootstrap method for model averaging and selection. Prepint, University of Toronto Statistics Department (1996)

Boosting Unsupervised Competitive Learning Ensembles

Emilio Corchado[1], Bruno Baruque[1], and Hujun Yin[2]

[1] Department of Civil Engineering. University of Burgos, Spain
escorchado@ubu.es, bbaruque@ubu.es
[2] School of Electrical and Electronic Engineering. University of Manchester, UK
h.yin@manchester.ac.uk

Abstract. Topology preserving mappings are great tools for data visualization and inspection in large datasets. This research presents a combination of several topology preserving mapping models with some basic classifier ensemble and boosting techniques in order to increase the stability conditions and, as an extension, the classification capabilities of the former. A study and comparison of the performance of some novel and classical ensemble techniques are presented in this paper to test their suitability, both in the fields of data visualization and classification when combined with topology preserving models such as the SOM, ViSOM or ML-SIM.

Keywords: topology preserving mappings, boosting, bagging, unsupervised learning.

1 Introduction

Several tools can be used to treat the high amounts of data that industrial and business operations processes. A very useful one is the unsupervised leaning, in the field of artificial neural networks (ANNs). For unsupervised learning only the input and the network's internal dynamics are the two elements required. No external mechanism is used to check on the weight setting mechanism.

This paper is based in one of the two major methods of unsupervised learning: competitive learning, where the output neurons of a neural network compete among themselves for being the one to be active. This too mirrors the reality of what happens in the brain in that there are finite resources for learning and so one neuron's gain means another's loss.

The principal problem of the models based on competitive learning is, as happens with all ANNs, their instability. This research is focused on the comparison and study of some novel and classical ensemble extension versions of some competitive learning models based on the topology preserving concept. Therefore the extended models are the Self-organising Map (SOM) [1], the Visualization-induced Self-Organizing Maps (ViSOM) [2] and the Maximum Likelihood Scale Invariant Map (ML-SIM) [3] in combination with some simple ensemble techniques such as the

J. Marques de Sá et al. (Eds.): ICANN 2007, Part I, LNCS 4668, pp. 339–348, 2007.

Bagging [4] or the AdaBoost [5]. The aim is to verify if the performance of these unsupervised connectionist models can be improved by means of these ensemble meta-algorithms.

2 Self-Organizing Maps

The Self-Organizing Map (SOM) algorithm [1] is based on a type of unsupervised learning called competitive learning; an adaptive process in which the neurons in a neural network gradually become sensitive to different input categories, sets of samples in a specific domain of the input space [6]. Its aim is to produce a low dimensional representation of the training samples while preserving the topological properties of the input space.

This study focuses on an interesting extension of this algorithm. It is called Visualization Induced SOM (ViSOM) [2] and it is proposed to directly preserve the local distance information on the map, along with the topology. The ViSOM constrains the lateral contraction forces between neurons and hence regularises the interneuron distances so that distances between neurons in the data space are in proportion to those in the input space [7].

The difference between the SOM and the ViSOM hence lies in the update of the weights of the neighbours of the winner neuron as can be seen from Eqs (1) and (2).

Update of neighbourhood neurons in SOM:

$$w_k(t+1) = w_k(t) + \alpha(t)\eta(v,k,t)\big(x(t) - w_v(t)\big) \tag{1}$$

Update of neighbourhood neurons in ViSOM:

$$w_k(t+1) = w_k(t) + \alpha(t)\eta(v,k,t)\left(\left[x(t) - w_v(t)\right] + \left[w_v(t) - w_k(t)\right]\left(\frac{d_{vk}}{\Delta_{vk}\lambda} - 1\right)\right) \tag{2}$$

where w_v is the winning neuron, α the learning rate of the algorithm, $\eta(v,k,t)$ is the neighbourhood function where v represents the position of the winning neuron in the lattice and k the positions of the neurons in the neighbourhood of this one, x is the input to the network and λ is a "resolution" parameter, d_{vk} and Δ_{vk} are the distances between the neurons in the data space and in the map space respectively.

Another example of a topographic mapping algorithm is the Maximum Likelihood Scale Invariant Map (ML-SIM) [3]. It is similar to the SOM but in this case training is based on the use of a particular Exploratory Projection Pursuit (EPP) model [9] called Maximum Likelihood Hebbian Learning (MLHL) Network [9] [10]. The competitive learning and a neighbourhood function are then used in a similar way as in the SOM. The distinctiveness is that in this case the winner's activation is then fed back through its weights and this is subtracted from the inputs to calculate the error or residual. Then the MLHL algorithm is used to update the weights of all nodes in the neighbourhood of the winner, which can be expressed as,

$$\mathbf{e}(t) = x(t) - w_v(t)\cdot y_v, (y_v = 1) \tag{3}$$

$$w_k(t+1) = \alpha(t) \cdot \eta(v,k,t) \cdot sign(\mathbf{e}(t) - w_v(t)) \,|\, \mathbf{e}(t) - w_v(t) \,|^{p-1}, \forall i \in N_c \quad (4)$$

These three models can be adapted for classification of new samples using a semi-supervised procedure [11]. This added feature can also serve as a measure of the stability of the trained network. A high accuracy in the classification rate implies that the neurons of the network are reacting in a more consistent way to the classes of the samples that are presented. As a consequence, the map should represent the data distribution more accurately.

3 Unsupervised Competitive Learning Ensembles

The ultimate goal of constructing an ensemble is to improve the performance obtained of a single working unit. When talking about classification it is generally accepted that the sets of patterns misclassified by the different classifiers would not necessarily overlap. This suggests that different classifier designs potentially offer complementary information about the patterns to be classified and could be harnessed to improve the performance of the selected classifier [12]. Many ensemble models and theories have been developed in the previous years that range from a quite simple technique like Bagging [4] or AdaBoost [5] to the more sophisticated ones such as LPBoost [13] or and many other variants. These techniques have been mainly applied to models designed specifically for classification, specially supervised classifiers [14]. In the present study the central idea is to verify the improvements that an ensemble technique can provide in the multi-dimensional data visualization field over an unsupervised learning process such as the Competitive Learning.

3.1 Bagging and AdaBoosting

Boosting meta-algorithms consists on training a simple classifier in several stages by incrementally adding new capacities to the current learned function. In the case of the present work the decision taken was to begin by implementing simpler boosting algorithm to initially study its effect on some topology preserving algorithms. Bagging and AdaBoost are the two boosting algorithms for ensemble training applied in this work.

Bagging (or bootstrap aggregating) [4] is one of the simplest techniques for ensemble construction. Its aim is to improve classification and regression models in terms of stability and classification accuracy, also reducing variance and helping to avoid overfitting. It consists on training each of the classifiers composing the ensemble separately using a different subset of the main training dataset. This is accomplished by using re-sampling with replacement over the training set. The classification results are obtained by (weighted or majority) voting among its composing classifiers. The technique provides the ensemble with a balance between variability and similarity.

The idea of AdaBoost [5] is to train several different classifiers, one each stage, in slightly different datasets by re-sampling with replacement. The difference is that it is taken into accounts which of the training samples are not correctly classified by the current classifier. When a sample is not well classified its probability is increased, so there are more chances that it will be presented to the next trained classifier as input. That way, the ensemble concentrates in the samples that are harder to classify, improving its learning capabilities. There have been proposed two slightly different versions of the algorithm [15]. AdaBoost.M1 is recommended for datasets with samples that only belong to two different classes while AdaBoost.M2 is recommended for dataset with more than two different classes. Although this meta-algorithm has been previously used for supervised classification [14], a short number of studies [16], [17], [19] involve unsupervised one such as in the present work.

3.2 Summarizing Some Applied Ensembles Techniques

The models used in this work are mainly designed as visualization tools. Constructing classical ensembles can be considered as a good option when trying to boost their classification capabilities, stabilizing its learning algorithm and avoiding overfitting; but when dealing with its visualization feature an ensemble is not directly displayable [18]. Representing all the networks in a simple image can only be useful when dealing with only 1-dimensional maps [17] but gets messy when visualizing 2-D maps. To overcome this problem some "ensemble combination" algorithms have been devised during this work in order to obtain a unique network that somewhat represents the information contained in the different networks composing the ensemble. Our objective is that this "combination" unites good classification accuracy with truthful representation of data for visual inspection. This part of the work has two approaches, which were inspired by SOM bagging [19] in one hand and by SOM fusion [20] on the other.

For all the tests involving this combination of networks the procedure is the same. A simple n-fold cross-validation is used in order to employ all data available for training and testing the model and having several executions to calculate an average of its performance. In each step of the cross-validation first, an ensemble of networks must be obtained. The way the ensemble is trained does not affect the way the combination is computed. In the case of this study this has been done using the bagging or the adaboost meta-algorithm. Then the computation of the combination is performed. Finally, both the ensemble and the combination generated from it are tested employing the test fold.

The different options studied for combining the network of the ensemble into a single network, summarizing its main characteristics are described in the following paragraphs. The first version is an existing one while the second and third, 'Superposition' and 'Superposition + Re-labelling', are novel techniques presented for the first time in this study.

Fusion: This method involves comparing the networks neuron by neuron in the input space [20]. This implies that all the networks in the ensemble must have the same size. First, it searches for the neurons that are closer in the input space (selecting only one neuron in each network of the ensemble), then it "fuses" them to obtain the final neuron in the "fused" map. This process is repeated until all the neurons have been fused. A more detailed description of this procedure can be found in [20]. Here the labelling of the neurons of the fused network, employing again the training dataset, is done in order to get a clear visualization of the map.

Superposition: In order to obtain a visual illustration (in the form of a 2-dimensional map) of the information the whole ensemble contains, this procedure has been designed during the development of this work. It consists of "superposing" the maps formed by the networks composing the ensemble into a final map, on a neuron by neuron comparison (as is done in fusion). Note that the weights of each network in the ensemble are initialized in a way that makes the neurons in the same position of two (or more) networks comparables. A description of the process could be:

1. Selection of the neuron in the same position in all maps (1 neuron for each map).
2. Creation of a neuron in the same position in the final map. Its inter-neuronal weights are the average of the inter-neuronal weights of the neurons selected in 1. Its frequency in recognizing a class is the addition of the frequency of the neurons selected in 1 for each class recognized (This is used in the classification stage).
3. Labelling the neuron in that position according to its most frequently recognized class (This is used in the representation of the map).
4. Repeating 1-3 until all the neurons have been processed.

This way ensures that the resultant "summarized" or "superposed" map represents visually what the majority of the maps composing the ensemble represent in a neuron-by-neuron basis. When using the resultant "superposed" map for classification purposes it returns the class represented by the neuron that is activated when the new sample is presented to the network.

Superposition + Re-labelling: This method has two main phases. The first one is the superposition explained before. The second one consists of testing which class actually recognizes better each neuron after the superposition, instead of relying on the recognition of the neurons in the same position done previously in the individual testing of each of the ensemble networks. So, after the superposition phase, the same dataset used for training is presented to the resultant network of the superposition to check which class is more consistently recognized by each neuron. Usually less number of neurons responds to this re-labelling, giving as a result a more compact map.

4 Experiment Details

Several experiments have been performed to check the suitability of using the previously described boosting and combining techniques under the frame of the mentioned topology preserving models. The datasets selected, the very well known Iris dataset and Wisconsin Breast Cancer dataset, were obtained from the UCI repository [20]. Visualization results are displayed in Fig.1 and Fig.2.

Fig. 1 displays maps representing the iris dataset. The circles represents "iris-setosa", the squares the "iris-virginica" and the triangles the "iris-versicolor". As it can be seen, the dataset's inner structure remains the same in the single map and in the three combinations having a clearly separable group (circles) and two non-linearly separable groups (triangles and squares).

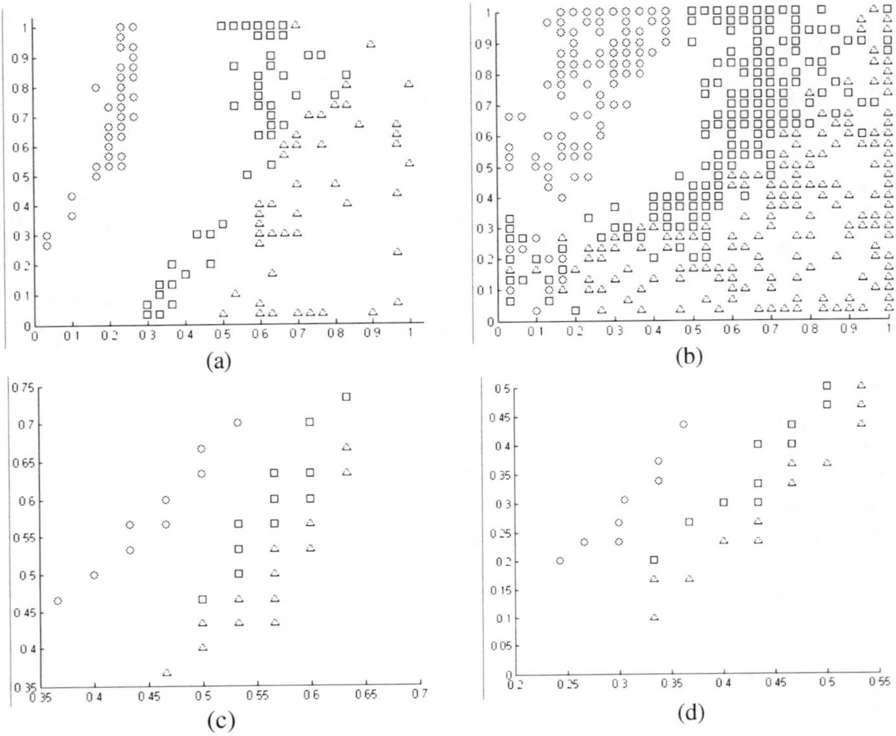

Fig. 1. Iris dataset represented by a **ViSOM** of 20x20 neurons. Fig 1(a) is obtained from a **single ViSOM**. Fig 1(b) is obtained from the '**superposition**' of an ensemble of **10 maps** with the same characteristics as Fig 1(a). Fig (c) and (d) are respectively the **'superposition+re-labelling'** and the **'fusion'** of the same ensemble.

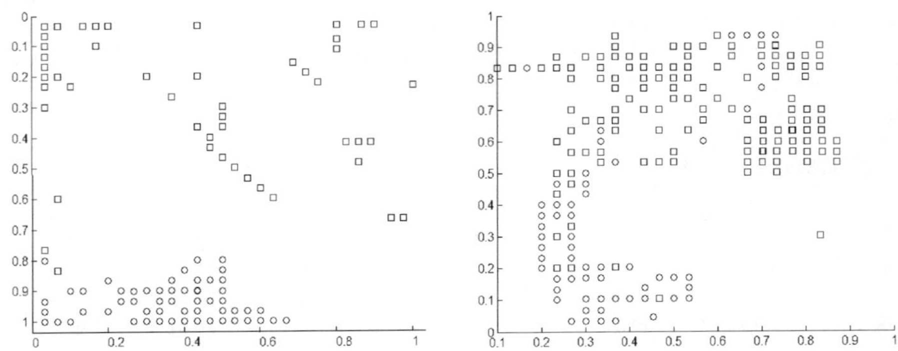

Fig. 2. Wisconsin Breast Cancer dataset represented by a **ViSOM** of 30x30 neurons. Fig 2(a) is obtained from a **single ViSOM**. Fig 2(b) is obtained from the '**superposition**' of an ensemble of **10 maps** with the same characteristics as Fig 2(a). Fig 2(c) and Fig 2(d) are respectively the **'superposition+re-labelling'** and the **'fusion'** of the same ensemble as Fig 2(b).

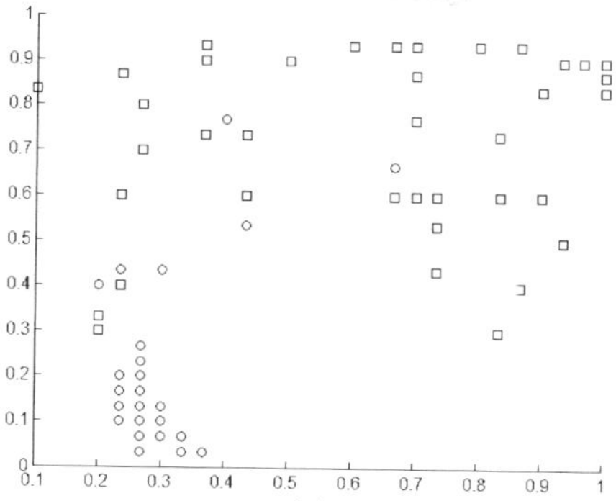
(c)

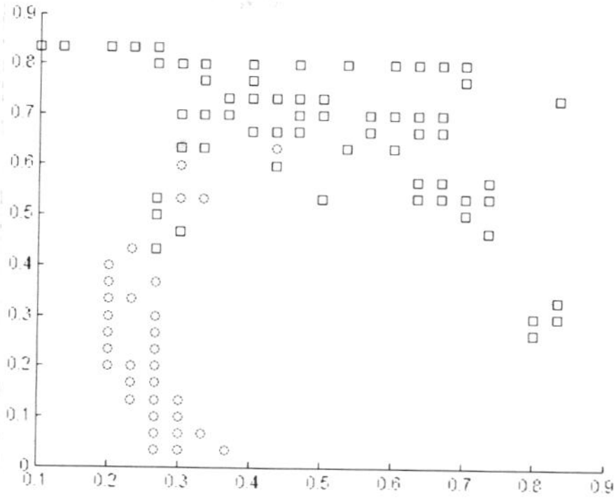
(d)

Fig. 2. *(continued)*

Fig. 2 displays maps representing the Wisconsin breast cancer dataset. The circles represent the "benign" cases, while the squares represent the "malignant" cases. As it can be seen a very condensed cloud of points corresponding to the "benign" cases appears at the bottom of all images, while a much more sparse cloud representing the "malignant" cases occupies the rest of the map.

As it can be seen, the visualization obtained by the combination methods (Fig. 1 b,c,d, Fig.2 b,c,d) keeps the internal structure of the dataset, being able to reproduce that structure with minor variations. The advantage of using an ensemble (by bagging or boosting) is the added stability to the analysis. This means that when dealing with few and scattered data points disposed over the input space the ensemble of maps should be able to reconstruct the structure of the dataset by combining the same "areas" of the different maps that compose the ensemble. If there is an area where no neurons (or very few) were activated in one map, but the neurons in the same position were activated in another map (because those maps were trained in slightly different dataset), there will not be a blank space (or very scattered disposition of neurons) in the resultant combination. This can be easily seen in the "superposition" combination where each neuron represents the data that activated that same neuron in all the maps of the ensemble. The "superposition + re-labelling" makes possible to obtain a map with the same main disposition but much reduced in complexity, as fewer neurons are now represented. This can serve to eliminate some distortions that can appear when are simply superposing all the maps in the ensemble, at cost of lower definition in the representation. This is more obvious in the cancer dataset in Fig 2. The "fusion" combination obtains very similar results to this second technique, as it follows the same combination and then re-labelling idea. The only difference comes by the criteria used to combine the neurons of each map. The classification accuracy obtained in the experiments is shown in Tables 1 – 4.

Table 1. Percentage of correct recognition of samples of the **Iris** dataset using ensembles of 10 networks employing a 10-fold cross validation for the tests. All ensembles were trained using the **bagging** meta-algorithm.

Model	**Best Single Netwk.**	**Ensemble (weighted voting)**	**Superp.**	**Superp. + ReLabelling**	**Fusion**
SOM (5x5)	94%	97.33%	84%	94%	94.6%
SOM (10x10)	73.3%	97.3%	80.6%	82%	83.3%
SOM (20x20)	48.6%	90.6%	66.6%	69.3%	58%
ViSOM (10x10)	93.3%	92.6%	61.3%	87.3%	94%
ViSOM (20x20)	87.3%	97.3%	80.6%	89.3%	94%
ViSOM (30x30)	75.3%	94.6%	80.6%	82.6%	85.3%
ML-SIM (20x20)	75.3%	76%	45%	74%	76.6%

Table 2. Percentage of correct recognition of samples of the **Iris** dataset using ensembles of 10 networks employing a 10-fold cross validation for the tests. All ensembles were trained using the **AdaBoost.M2** meta-algorithm.

Model	**Best Single Netwk.**	**Ensemble (weighted voting)**	**Superp.**	**Superp. + ReLabelling**	**Fusion**
SOM (5x5)	95.3%	97.3%	77.3%	90.6%	96.0%
SOM (10x10)	75.3%	95.3%	80.6%	83.3%	83.3%
SOM (20x20)	58%	90%	71.3%	71.3%	61.3%
ViSOM (10x10)	92%	94.6%	79.3%	97.3%	90.6
ViSOM (20x20)	86.6%	94.6%	86.6%	88.6%	88.6%
ML-SIM (20x20)	75.3%	77.3%	44%	79.3%	78.6%

Table 3. Percentage of correct recognition of samples of the **Wisconsin Breast Cancer** dataset using ensembles of 10 networks employing a 10-fold cross validation for the tests. All ensembles were trained using the **bagging** meta-algorithm.

Model	**Best Single Netwk.**	**Ensemble (weighted voting)**	**Superp.**	**Superp. + ReLabelling**	**Fusion**
SOM (5x5)	97%	97%	89.4%	96.1%	96.2%
SOM (10x10)	93%	96%	93.4%	93.3%	94.7%
SOM (20x20)	77.6%	96%	83.4%	90%	82.4%
ViSOM (10x10)	94.1%	97%	96.1%	93.6%	94.4%
ViSOM (20x20)	80.9%	95.3%	92.3%	83%	84%
ViSOM (30x30)	77.3%	93.3%	91.7%	78.3%	79.6%
ML-SIM (20x20)	73.9%	94.5%	70.1%	85.7%	89.2%

Table 4. Percentage of correct recognition of samples of the **Wisconsin Breast Cancer** dataset using ensembles of 10 networks employing a 10-fold cross validation for the tests. All ensembles were trained using the **AdaBoost.M1** meta-algorithm.

Model	**Best Single Netwk.**	**Ensemble (weighted voting)**	**Superp.**	**Superp. + ReLabelling**	**Fusion**
SOM (5x5)	96%	95.8%	93.8%	96%	96.7%
SOM (10x10)	91.6%	96.4%	94.3%	92.8%	94.5%
SOM (20x20)	79.6%	96.6%	94.4%	86.1%	89.3%
ViSOM (10x10)	85.4%	96.9%	93.8%	94%	93.6%
ViSOM (20x20)	84.9%	96.3%	95.1%	86.8%	87.7%
ViSOM (30x30)	77.9%	95.1%	93.3%	83.8%	84.5%
ML-SIM (20x20)	73%	95.4%	66.1%	87.2%	87.5%

There are several aspects worth of noting in relation with these experiments and the results presented from Table 1 to Table 4. The most obvious one is that the best model, both for visualization and for classification is the ViSOM. The second best model is the SOM and the last one the ML-SIM. This was expected, taking into account that the ViSOM was specially developed for data visualization while ML-SIM was specially designed for working with radial-based datasets. The second conclusion is that the amount of neurons composing the map is directly proportional to the clarity and definition of the map, but inversely proportional to the classification accuracy of the network.

Talking about the ensemble algorithms, it can be concluded that the visualization capabilities of the models are slightly enhanced when using the simple superposition technique, as more neurons are represented in only one map, although this can sometimes make the data structures in the map a little less defined. On the contrary, the classification side of the models is really benefited from the ensemble inclusion. Even when the ensemble is constructed using networks of low classification accuracy the ensemble is able to obtain a greater accuracy. This was expected from all the previous work done in ensemble and classification. The combinations of the ensemble networks into a unique one are far more dependant of the accuracy of the networks employed to create them but still they manage to get a classification accuracy that is most of the times equal or superior to the accuracy of the better network of the ensemble, specially the superposition + re-labeling. This can be taken into account when a unique, but more stable network is needed; both for classification and visualization.

5 Conclusions and Future work

This study presents an interesting mixture of techniques for representation of multi-dimensional data in 2-D maps. They are based in the combination of several maps trained over slightly different datasets to form an ensemble of networks with self-organizing capabilities. The training of the ensemble of networks has been tested by using the bagging and boosting techniques for their comparison. The purpose of this combination of maps is to avoid the overfitting and to improve the stability of previously developed models based in unsupervised and competitive learning. They can be especially useful when a reduced dataset is available. As an ensemble of maps is impossible to represent, several different combination algorithms are proposed and tested. They have been applied for the first time in the case of the ViSOM and compared with the SOM and the ML-SIM performance. A novel ensemble combination technique has also been presented, and tested in a very successfully way. As an added effect, the classification capabilities of the models are also increased. These techniques have been tested in two widely known real datasets. Future work will include far exhaustive testing of the presented combination of techniques using several more complex datasets, as well as adapting the present model to other novel boosting meta-algorithms to check if more improvements can be obtained.

Acknowledgments

This research has been supported by the MCyT project TIN2004-07033.

References

1. Kohonen, T.: The Self-Organizing Map. Neurocomputing 21, 1–6 (1998)
2. Yin, H.: Data Visualisation and Manifold Mapping Using the Visom. Neural Networks 15, 1005–1016 (2002)
3. Corchado, E., Fyfe, C.: The Scale Invariant Map and Maximum Likelihood Hebbian Learning. In: International Conference on Knowledge-Based & Intelligent Information & Engineering System (KES) (2002)
4. Breiman, L.: Bagging Predictors. Machine Learning 24, 123–140 (1996)
5. Freund, Y., Schapire, R.E.: A Decision-Theoretic Generalization of on-Line Learning and an Application to Boosting. Journal of Computer and System Sciences 55, 119–139 (1997)
6. Kohonen, T., Lehtio, P., Rovamo, J., Hyvarinen, J., Bry, K., Vainio, L.: A Principle of Neural Associative Memory. Neuroscience 2, 1065–1076 (1977)
7. Yin, H.: Visom - a Novel Method for Multivariate Data Projection and Structure Visualization. Neural Networks, IEEE Transactions on 13, 237–243 (2002)
8. Corchado, E., MacDonald, D., Fyfe, C.: Maximum and Minimum Likelihood Hebbian Learning for Exploratory Projection Pursuit. Data Mining and Knowledge Discovery 8, 203–225 (2004)
9. Fyfe, C., Corchado, E.: Maximum Likelihood Hebbian Rules. In: European Symposium on Artificial Neural Networks (ESANN) (2002)
10. Kraaijveld, M.A., Mao, J., Jain, A.K.: A Nonlinear Projection Method Based on Kohonen's Topology Preserving Maps. Neural Networks, IEEE Transactions on 6, 548–559 (1995)
11. Heskes, T.: Balancing between Bagging and Bumping. In: Advances in Neural Information Processing Systems 9 - Proceedings of the 1996 Conference vol. 9, pp. 466–472 (1997)
12. Demiriz, A., Bennett, K.P., Shawe-Taylor, J.: Linear Programming Boosting via Column Generation. Machine Learning 46(1-3), 225–254 (2002)
13. Schwenk, H., Bengio, Y.: Boosting Neural Networks. Neural Computation 12, 1869–1887 (2000)
14. Freund, Y., Schapire, R.E.: Experiments with a New Boosting Algorithm. In: International Conference on Machine Learning, pp.148–156 (1996)
15. Gabrys, B., Baruque, B., Corchado, E.: Outlier Resistant PCA Ensembles. In: Gabrys, B., Howlett, R.J., Jain, L.C. (eds.) KES 2006. LNCS (LNAI), vol. 4253, pp. 432–440. Springer, Heidelberg (2006)
16. Corchado, E., Baruque, B., Gabrys, B.: Maximum Likelihood Topology Preserving Ensembles. In: Corchado, E., Yin, H., Botti, V., Fyfe, C. (eds.) IDEAL 2006. LNCS, vol. 4224, pp. 1434–1442. Springer, Heidelberg (2006)
17. Kaski, S.: Data Exploration Using Self-Organizing Maps. Department of Computer Science and Engineering. Helsinki University of Technology. Espoo, Finland (1997)
18. Petrakieva, L., Fyfe, C.: Bagging and Bumping Self Organising Maps. Computing and Information Systems (2003)
19. Georgakis, A., Li, H., Gordan, M.: An Ensemble of SOM Networks for Document Organization and Retrieval. In: Int. Conf. on Adaptive Knowledge Representation and Reasoning (AKRR'05), p. 6 (2005)
20. Newman, D.J., Hettich, S., Blake, C.L., Merz, C.J.: UCI Repository of Machine Learning Databases. University of California, Irvine, Dept. of Information and Computer Sciences (1998), http://www.ics.uci.edu/~mlearn/MLRepository.html

Using Fuzzy, Neural and Fuzzy-Neural Combination Methods in Ensembles with Different Levels of Diversity

Anne M.P. Canuto and Marjory C.C. Abreu

Informatics and Applied Mathematics Department, Federal University of RN Natal, RN - Brazil, 59072-970 Telephone:+55-84-3215-3813
anne@dimap.ufrn.br, marjory.abreu@gmail.com

Abstract. Classifier Combination has been investigated as an alternative to obtain improvements in design and/or accuracy for difficult pattern recognition problems. In the literature, many combination methods and algorithms have been developed, including methods based on computational Intelligence, such as: fuzzy sets, neural networks and fuzzy neural networks. This paper presents an evaluation of how different levels of diversity reached by the choice of the components can affect the accuracy of some combination methods. The aim of this analysis is to investigate whether or not fuzzy, neural and fuzzy-neural combination methods are affected by the choice of the ensemble members.

Keywords: Ensembles, Diversity in Ensembles, Fuzzy Sets, Neural Networks and Fuzzy Neural Networks.

1 Introduction

In the literature, the use of multi-classifier system (MCS), also known as committees or ensembles, has been widely used for several pattern recognition tasks. In the last decade, for instance, a large number of papers [2,12] have proposed the combination of multiple classifiers for designing high performance classification systems. Several combination methods and algorithms have been developed in the literature, including methods based on computational intelligence [5,9,15,16,18].

One of the most important steps in the design of a multi-classifier system (MCS) is the choice of the components (classifiers). This step is very important to the overall performance of a MCS since the combination of a set of identical classifiers will not outperform the individual members. The ideal situation would be an ensemble that should have diversity among its base classifiers. Diversity has been recognized as a very important feature in classifier combination and has been addressed by several authors, as in [11,13,17]. According to [4], there is a tendency to decrease diversity when using ensembles with the same type of classifiers, when compared with heterogeneous ensembles. In addition, in [4], it has been detected that some combination methods are more sensible to variations of the choice of the ensemble members than others.

In this paper, it is aimed to investigate the importance of the choice of the components of an ensemble (ensemble members) in the accuracy of ensembles

J. Marques de Sá et al. (Eds.): ICANN 2007, Part I, LNCS 4668, pp. 349–359, 2007.

combined by computational intelligence (CI-based) methods (fuzzy connectives, neural networks and fuzzy neural networks). In order to do this, these combination methods will be analyzed, when using ensembles with different levels of diversity (variation in the ensemble members). As a result of this analysis, it is aimed to define which combination methods are strongly affected by the choice of the ensemble members.

2 Related Works

The use of Computational Intelligence (CI) as an aggregation (combination) tool provides several advantages due to these methods can inherit several important features of the CI methods. In the literature, some papers have proposed the combination of multiple classifiers using CI-based combination methods, such as in [2,4,5,8,9,15,16,18].

The use of neural networks as a combination method in an ensemble has been widely used in the literature [2,4,18]. The most common neural network model used as a combination method is Multi-layer Perceptron (MlP). In [18], for instance, an investigation of MlP-based combiner applied to a neural-based ensemble was presented. On the other hand, in [4], a MlP-based combiner was applied in a non-neural ensemble.

The use of fuzzy connectives as a tool to combine the outputs of classifiers was first proposed in [7]. Lately, some combination methods and algorithms based on fuzzy sets have been developed in the literature [5,8,9,15,16]. In [9,15,16], some fuzzy combination methods were compared with some non-fuzzy methods in ensembles of classifiers. In [9], the authors studied the potential of fuzzy combination methods for ensembles of classifiers designed by AdaBoost. The results involving sums of ranks and statistical comparisons showed that, in general, fuzzy methods fared better than non-fuzzy methods. In the literature, fuzzy neural network is rarely used as a combination method in ensembles.

In [2], a MLP-based fuzzy neural network has been used as a combination method and it has shown to overcome the accuracy of a MlP neural network and some fuzzy combiners.

Unlike the aforementioned works, the main focus of this paper is not to analysis the benefits, in terms of accuracy, of using CI-based combination methods in ensembles. But it is aimed to analyze the influence of the choice of the ensemble members in the accuracy of CI-based combination methods. As already mentioned, one of the most important steps in the design of an ensemble is the choice of the components (classifiers). In doing this analysis, it is aimed to help designers in the choice of the individual classifiers and combination methods of an ensemble.

3 Ensembles

The goal of using ensembles is to improve the performance of a pattern recognition system in terms of better generalization and/or in terms of increased efficiency and clearer design [2,10]. There are two main issues in the design of an ensemble: the ensemble components and the combination methods that will be used.

With respect to the first issue, the correct choice of the set of base classifiers is fundamental to the overall performance of an ensemble. The ideal situation would be a set of base classifiers with uncorrelated errors - they would be combined in such a way to minimize the effect of these failures. In other words, the base classifiers should be diverse among themselves. Once a set of classifiers has been created, the next step is to choose an effective way of combining their outputs. There are a great number of combination methods reported in the literature [2,3,10]. According to their functioning, there are in the literature, three main strategies of combination methods: fusion-based, selection-based, and hybrid methods [10]. CI-based methods are considered as fusion-based combination methods.

3.1 Diversity in Ensembles

As already mentioned, there is no gain in the combination (MCS) that is composed of a set of identical classifiers. The ideal situation, in terms of combining classifiers, would be a set of classifiers that present uncorrelated errors. In other words, the ensemble must show diversity among the members in order to improve the performance of the individual classifiers. Diversity can be reached in three different ways:

- Variations of the parameters of the classifiers (e.g., varying the initial parameters, such as the weights and topology, of a neural network model).
- Variations of the training dataset of the classifiers (e.g., the use of learning strategies such as Bagging and Boosting or the use of feature selection methods).
- Variations of the types of classifier (e.g., the use of different types of classifiers, such neural networks and decision trees, as members of an ensemble - hybrid ensembles).

In this paper, variations of the diversity are captured using different types of classifiers and different parameters.

4 CI-Based Combination Methods

4.1 Neural Networks

This combination method employs neural networks to combine the output of the classifiers. The neural network combiner (NNC) uses the output generated by each classifier as its input. The input of the neural network combiner can contain either only the top output or all the outputs of the classifiers. When used as a non-confidence based method, only the outputs of the classifiers are used as input while the confidence measures are included as input when used as a confidence based method. Usually, the choice of the configuration for the NNC (number of layers, number of neurons per layer, initial parameters) is performed in the same way as a neural network model, through checking its performance for individual test cases.

In this paper, a MlP-based combination method will be used. In this method, only the outputs of the classifiers will be used and a validation set is used to train the neural network combiner.

4.2 Fuzzy Neural Networks

As already mentioned, very little effort has been done to combine the output of classifiers using fuzzy neural networks. In this paper, the fuzzy Multi-layer Perceptron (F-MlP) proposed in [2] will be used as a combination method: As the neural network combiner, all the outputs of the classifiers will be used as input for the fuzzy neural combiner. Also, this combiner will be trained using a validation set.

This fuzzy multilayer perceptron clamps desired membership values to the output nodes during the training phase instead of choosing binary values as in a winner-take-all method. Then, the errors may be back-propagated with respect to the desired membership values as the fuzzy desired output. In addition, this fuzzy Multi-layer Perceptron added a parameter in the weight update equation. This parameter describes the degree of ambiguity of a pattern during training and hence allows avoidance of a situation where ambiguous patterns exert too great a degree of influence in the weight updating process.

4.3 Fuzzy Connectives

The use of fuzzy connectives as a tool to combine the outputs of classifiers started to be investigated in [7]. Since then, some fuzzy set connectives have been used for the purpose of aggregation, which can vary from very simple connectives, such as: fuzzy min or max, to more complex ones, such as: Zimmermann operators [19], fuzzy integral [8,7], among others. The next subsections will describe some fuzzy connectives that will be used in this paper, which are: fuzzy integral (Sugeno and Choquet), Zimmermann operators as well as BADD defuzzyfication.

4.3.1 Sugeno Fuzzy Integral

The Sugeno integral combines objective evidence scores for a hypothesis with the priori expectation (importance) of that evidence to the hypothesis. Assuming that: h: X → [0,1] is a Ω-measurable function, the Sugeno fuzzy integral, ε, with respect to a fuzzy measure μ over an arbitrary set X of the function h can be computed by:

$$\varepsilon = \max_{i=1,\dots,n}[\min_{x \notin A} h(X_i), \mu(A_i)]$$

where: $A_i = \{x_1, x_2, \dots, x_i\}$; $\mu(A_1) = \mu^1$; $\mu(A_i) = \mu^i + \mu(A_{i-1}) + \lambda\mu^i\mu(A_{i-1})$ and $h(x_1) \leq h(x_2) \leq \dots \leq h(x_i)$

In an intuitive way, fuzzy integral elements can be seen as: h(x) is the degree to which the concept of h is satisfied by x; $\min_{x \in A} h(x)$ is the degree to which the concept h is satisfied by all the elements in A; $\mu(A_i)$ is the degree to which the subset of objects A satisfies the concept measured by μ (densities of the elements of the subset A); $\min_{x \in A} h(x), \mu(A_i)$ indicates the degree to which A satisfies both the criteria of the measure μ and $\min_{x \in A} h(x)$ and max takes the largest value of $\min_{x \in A} h(x), \mu(A_i)$.

4.3.2 Choquet Integral

Unlike the Sugeno integral, the Choquet integral uses a measure which is additive. So it is defined as:

$$\rho = \sum_{i=1}^{n} [h(x_i) - h(x_{i-1})] \mu_i^n$$

where:

$$h(x_0) = 0 \text{ and } \quad \mu_i^j = \begin{cases} \mu(\{x_i, x_{i+1}, \ldots, x_j\}) & \text{if } i \le j \\ 0 & \text{otherwise} \end{cases}$$

Incorporating Choquet and Sugeno fuzzy integral as a combination method, X corresponds to the set of classifiers, where each classifier x_i produces a partial evaluation (classifier's outputs) for each class k in accordance to an input pattern A, which are represented by $h_k(x_i)$; μ is the importance (confidence) of each classifier for each class. The class with the largest integral value is considered as the final decision.

4.3.3 Zimmermann and Zysno Operator

Zimmermann and Zysno have introduced the so-called γ-operators in [19]. It is a compensatory operator and it can be defined as follows.

$$y(x) = \left(\prod_{i=1}^{C} y(x_i) \right)^{1-\gamma} \left(1 - \prod_{i=1}^{C} (1 - y(x_i) \right)^{\gamma}$$

where $y_i(x_i)$ is the output of the ith classifier. The parameter γ controls the degree of compensation between the union and intersection parts of the operator. If γ = 0, no compensation is present, then it equals to the product and provides the truth values for the connective and. If γ = 1, full compensation is present, then this formula equals the generalized algebraic sum and provides the truth values for the connective or.

4.3.4 BADD

The BADD defuzzification strategy [18] assumes that data space is partitioned into L overlapping regions and each classifier is associated with one of these regions. Decisions from the classifiers are combined to the following rule:

$$y(x) = \frac{\left(\sum_{i=1}^{L} (m(x_i))^z \, y(x_i) \right)}{\sum_{i=1}^{L} (m(x_i))^z}$$

where z is a parameter, and m is the membership degree of given x in the ith region of the data space. The membership degrees are calculated based on the Euclidian distance between the weight vector and the input pattern.

5 Experimental Setting Up

In order to analyze the influence of the ensemble members in the accuracy of the CI-based combination methods, an empirical comparison of ensembles using several structures and sizes is performed. As it can be seen, all combination methods are trainable methods. In order to define the training set of these combination methods, approximately 10% of the instances of a database were taken out to create a validation set. These combination methods were then trained using the validation set. All methods were trained using a 10-fold cross validation method. Also, two different databases are used in this investigation, which are described as follows.

- Database A: It is a protein database which represents a hierarchical classification, manually detailed, and represents known structures of proteins. The main protein classes are all-α, all-β, α/β, α+β and small. It is an unbalanced database, which has a total of 582 patterns, in which 111 patterns belong to class all-α, 177 patterns to class all-β, 203 patterns to α/β, 46 patterns to class α+β and 45 patterns to class small.
- Database B: It is a primate splice-junction gene sequences (DNA) with associated imperfect domain theory, obtained from [1]. A total of 3190 Instances using 60 attributes were used. These attributes describe sequences of DNA used in the process of creation of proteins.

Among other aspect, in this experimental work, the impact of the size of the ensembles and the types of its base classifiers are investigated. With respect to this first issue, experiments were conducted using three different ensemble sizes, using 3, 5 and 7 base classifiers. The choice of a small number of classifiers is due to the fact that diversity strongly affects the accuracy of ensembles when using less than ten classifiers [10]. In this sense, it is believed that the accuracy of the combination methods used in ensembles with few members is more sensitive to variations in the ensemble members than using larger numbers of members. Regarding the second aspect, seven different types of learning algorithms are used, which are: MLP (Multi-layer Perceptron), Fuzzy MLP, K-NN (nearest neighbor), RBF (radial basis function), SVM (Support Vector Machine), J48 decision tree and JRIP (Optimized IREP). The choice of the aforementioned classifiers was due to the different learning bias that they use during their functioning.

For each ensemble size, different configurations were used, some of them using base classifiers built with the same type of learning algorithm (homogeneous structures or non-hybrid systems - NH) and the remaining ones employed base classifier generated with different learning methods (heterogeneous structures or hybrid systems - HYB). In order to make the choice of the heterogeneous structures of the ensembles more systematic, ensembles 3 (HYB 3), 5 (HYB 5) and 7 (HYB 7) different types of classifiers were taken into consideration (for ensembles sizes 3, 5 and 7). As there are several possibilities for each hybrid structure, this paper presents the average of the accuracy delivered by all possibilities of the corresponding hybrid structure. In this way, it becomes easier to make a better analysis of the results.

All the classification methods used in this study, apart from Fuzzy MLP, were obtained from the WEKA machine learning visual package (http:\\www.cs.waikato.ac.nz/~ml/weka/). All combination methods were implemented in Java language.

6 Results

Before starting the investigation of the accuracy of the ensembles, it is important to analyze the accuracy of the individual classifiers. As already mentioned, variations of the same classifiers were obtained using different setting parameters. As seven variations of each classifier were used in this investigation, for simplicity reasons, only the average accuracies of classifiers are shown in Table 1.

Table 1. Classifier accuracy (CA) and standard deviation (SD) of the Individual Classifiers

	K-NN	SVM	MLP	F-MLP	RBF	DT	JRIP
A	69.45±6.19	**83.90**±5.55	78.71±3.45	83.67±3.46	81.86±3.63	78.17±4.52	74.51±5.21
B	75.85±2.81	81.89±3.84	84.69±4.11	**85.13**±2.68	81.73±2.65	79.99±3.64	81.88±3.67

According to the average accuracy provided by the classifiers, it can be seen that all classifiers have delivered a similar pattern of accuracy for both databases. The SVM classifier has provided the largest accuracy for database A, while Fuzzy MLP has delivered the largest accuracy for database B.

6.1 Ensembles with Three Base Classifiers

Table 2 shows accuracy and standard deviation of ensembles with three base classifiers applied to both databases. In this Table, six different combination methods were analyzed, in which four of them are fuzzy (Sugeno and Choquet integrals, Zimmermann operators and BADD defuzzyfication), one Multi-layer Perceptron and one Fuzzy Multi-layer Perceptron.

For each combination method, their performance using one type of classifier (non-hybrid structure - NH) and three types of classifiers (hybrid structure – HYB 3) are investigated. As already mentioned, values presented in Table 2 represents the average accuracy and standard deviation of all possibilities. The third column of Table 2 shows the difference in accuracy (*Dif*) delivered by the combination method when varying the ensemble members. It is calculated by the difference between the largest and the smallest accuracies. This value aims to define the variation in accuracy when using different ensemble members. In this sense, the methods which have high values of *Dif* have a strong variation when using different ensemble members. As a consequence, it can be stated that they are strongly affected by the choice of the ensemble members.

Table 2. Accuracy (Acc) and standard deviation (SD) of Hybrid and Non-hybrid Ensembles with three base classifiers

	Acc±SD	**Acc±SD**		**Acc±SD**	**Acc±SD**	
	NH	HYB 3	*Dif*	NH	HYB 3	*Dif*
MIP	88.33±7.31	90.25±2.82	1.92	90.92±2.12	92.73±2.77	1.81
F-MIP	92.69±6.14	93.15±3.53	0.47	95.02±3.01	95.34±2.28	0.32
Choquet	94.74±2.98	93.17±2.34	1.57	95.36±2.48	95.47±2.56	0.11
Sugeno	95.28±3.24	95.65±3.62	0.38	91.23±2.33	92.33±2.13	1.10
BADD	95.36±3.26	95.43±3.14	0.12	94.89±2.66	94.67±2.33	0.22
Zimmermann	91.23±2.13	91.53±2.36	0.35	93.58±2.61	93.47±2.81	0.11

As it can be observed from Table 2, the accuracies of the ensembles were, on average, larger than the corresponding individual classifiers. In analyzing the accuracy of the combination methods, the largest accuracies delivered by all combination methods were reached when using a hybrid structure (HYB 3), in most of the cases.

Of the combination methods, the largest average accuracy (both databases) was provided by ensembles combined by BADD, followed by Choquet, Fuzzy MlP, Zimmermann, Sugeno and MlP.

When analyzing the difference in accuracy delivered by the combination methods, the smallest difference was always reached by a fuzzy combination method. Finally, the smallest average difference in accuracy (both databases) delivered by the combination methods is reached by BADD (0.17), followed by Zimmermann (0.23), Fuzzy MlP (0.4), Sugeno (0.74), Choquet (0.84) and MlP (1.87). Based on this result, it is possible to state that ensembles with fuzzy combiners are less affected by variations in the ensemble members than the neural combiner. Ensembles with fuzzy-neural combiner are less affected by variations in the ensemble members than neural ensembles and it is in a similar situation than fuzzy combiners.

6.2 Ensembles with Five Base Classifiers

Table 3 shows accuracy and standard deviation of ensembles with five base classifiers applied to both databases. For each combination method, their accuracy using one type of classifier (NH), three types of classifiers (HYB 3) and five types of classifiers (HYB 5) will be investigated.

The accuracies of the ensembles were, on average, larger than the corresponding ensembles with three base classifiers. As in the previous section, the largest accuracies delivered by all combination methods were always reached when using the totally hybrid structure (HYB 5), followed by the partial hybrid structure (HYB 3) and by the non-hybrid structure (NH), for both databases.

Table 3. Accuracy (Acu) and standard deviation (SD) of Hybrid and Non-hybrid Ensembles with five base classifiers

Ensembles with Five Base Classifiers						
	MlP	F-MlP	Choquet	Fuzzy	BADD	Zimmermann
NH	88.47±4.88	92.27±4.48	93.96±3.14	96.12±3.57	95.28±3.24	91.84±2.24
HYB3	89.19±2.56	93.61±3.9	95.74±2.67	96.33±3.49	95.66±3.62	91.37±2.53
HYB5	92.28±2.75	94.38±3.37	93.74±3.75	91.36±4.56	96.12±3.57	91.76±2.18
Dif	3.81	2.11	2.00	4.97	0.84	0.47
Database B						
NH	92.99±2.11	94.97±2.23	95.64±2.77	92.47±2.26	94.81±2.51	93.48±2.67
HYB3	93.45±2.35	95.33±2.19	95.74±2.99	92.56±2.28	95.12±2.39	93.86±2.69
HYB5	94.15±2.44	95.64±2.2	95.65±2.81	92.87±2.74	95.22±2.51	93.18±2.15
Dif	1.06	0.67	0.10	0.40	0.41	0.68

In relation to the fuzzy combiners, a similar scenario to the previous section happened here, in which the largest accuracy was reached by the same fuzzy combiner (BADD). Also, the smallest difference in accuracy was reached by a fuzzy combiner (Zimmermann), although it is not the same combiner of the previous section. In relation to the neural and fuzzy-neural combiners, the scenario was similar to the previous section, in which the neural combiner provided the worst accuracy and the second largest difference in accuracy (it was the highest difference in the previous section). In the same way, the fuzzy-neural combiner provided the third largest accuracy and the fourth smallest difference in accuracy (it was the third smallest difference in the previous section).

6.3 Ensembles with Seven Base Classifiers

Table 4 shows the accuracy and standard deviation of ensembles with seven base classifiers applied to both databases. For each combination method, their accuracy using one type of classifier (NH), three types of classifiers (HYB 3), five types of classifiers (HYB 5) and seven types of classifiers (HYB 7) will be investigated.

The accuracies of the ensembles were, on average, larger than the corresponding ensembles with five base classifiers. As in the prev99ious section, the largest accuracies delivered by all combination methods were reached when using a totally hybrid structure (HYB 7), followed by the two partially hybrid structures (HYB 5 and HYB 3) and by the non-hybrid structure (NH).

Table 4. Accuracy (Acu) and standard deviation (SD) of Hybrid and Non-hybrid Ensembles with seven base classifiers

	Ensembles with Seven Base Classifiers					
	MlP	F-MlP	Choquet	Fuzzy	BADD	Zimmermann
NH	85.64±5.98	92.25±4.16	95.84±2.36	92.58±2.56	95.24±2.74	93.96±2.61
HYB3	86.55±2.59	92.48±2.33	95.86±2.69	92.33±2.74	96.49±2.19	93.15±3.61
HYB5	88.94±2.66	92.34±3.17	95.1±1.96	93.49±2.12	96.77±2.14	97.05±3.12
HYB 7	92.64±4.3	94.36±3.59	93.49±2.11	92.99±2.61	93.49±2.11	95.34±1.54
Dif	7.0	2.11	1.61	1.16	3.28	3.90
			Database B			
NH	92.64±2.85	95.56±2.95	97.17±2.74	91.28±4.3	96.33±3.49	91.56±2.2
HYB3	94.04±2.53	95.8±2.59	91.74±2.94	91.89±4.91	91.36±4.56	91.26±2.66
HYB5	94.98±2.5	95.94±1.97	96.39±2.31	91.64±4.85	91.28±4.3	91.39±2.94
HYB 7	95.19±1.97	96.66±2.68	91.68±2.19	92.33±2.12	91.89±4.91	91.79±2.91
Dif	2.55	1.1	5.49	1.05	5.05	0.53

In relation to the fuzzy combiners, the largest accuracy was again reached by a fuzzy combiner (Choquet), which is not the same of the previous sections. In this context, the smallest difference in accuracy was again reached by a fuzzy combiner (Sugeno), although it is not the same combiner of the previous sections.

There is an increase in the accuracy of the fuzzy-neural combiner, when compared with ensembles with fewer members, reaching the second highest accuracy. In addition, the fuzzy-neural combiner reached the third largest accuracy (similar to ensembles with three base classifiers). In relation to the neural combiner, it had a similar behavior than ensembles using fewer members.

7 Conclusion

In this paper, an investigation of different ensemble structures and sizes was performed. The main aim of this paper was to investigate which combination methods are strongly affected by the choice of the ensemble members. This investigation was done using two different datasets. Through this analysis, it could be observed that the largest accuracies were almost always reached by the hybrid structures, for all three ensemble sizes analyzed. Of the combination methods, the largest average accuracy was reached by ensembles combined by Fuzzy methods. In addition to this, ensembles combined by neural network are more sensitive to variations in the ensemble members (higher Dif values) than the fuzzy and fuzzy neural methods, for all ensemble sizes. However, the use of fuzzy theory on a neural network model (fuzzy MlP) was important to smooth this problem, making this type of ensemble equally affected by variation in the ensemble members than the fuzzy combiners.

References

1. Blake, C.L., Merz, C.J.: UCI Repository of machine learning databases. University of California, Department of Information and Computer Science, Irvine, CA (1998), http://www.ics.uci.edu/~mlearn/MLRepository.html
2. Canuto, A.: Combining neural networks and fuzzy logic for applications in character recognition. PhD thesis, University of Kent (2001)
3. Canuto, A.M.P., Oliveira, L., Xavier Jr., J., Santos, A., Abreu, M.: Performance and Diversity Evaluation in Hybrid and non-hybrid Structures of Ensembles. In: 5th Int. Conf. on Hybrid Intelligent Systems, pp. 285–290 (2005)
4. Canuto, A.M.P., Abreu, M., Oliveira, L.M., Xavier Jr., J.C., Santos, A.M.: Investigating the influence of the choice of the ensemble members in accuracy and diversity of selection-based and fusion-based methods for ensembles. Pattern Recognition Letters 28, 472–486 (2007)
5. Canuto, A.M.P., Fairhurst, M., Howells, G.: Fuzzy Connectives as a Combination Tool in a Hybrid Multi-Neural System. Int. Journal of Neural Systems 13(2), 67–76 (2003)
6. Choquet, G.: Cooperation of modularized neural networks by fuzzy integral with owa operators. Ann. Inst. Fourier 5, 131–295 (1953)
7. Cho, S.-B., Kim, J.H.: Multiple network fusion using fuzzy logic. IEEE Transactions on Neural Networks 6(2), 497–501 (1995)
8. Czyz, J., Sadeghi, M., Kittler, J., Vandendorpe, L.: Decision fusion for face authentication. In: Proc. First Int. Conf. on Biometric Authentication, pp. 686–693 (2004)
9. Grabisch, M.: Fuzzy integral in multicriteria decision making. Fuzzy Sets and Systems 69, 279–298 (1995)
10. Kuncheva, L.I.: 'Fuzzy' vs 'Non-fuzzy' in combining classifiers designed by boosting. IEEE Transactions on Fuzzy Systems 11(6), 729–741 (2003)
11. Kuncheva, L.I.: Combining Pattern Classifiers. Methods and Algorithms. Wiley, Chichester (2004)
12. Kuncheva, L., Whitaker, C.: Measures of diversity in classifier ensembles. M. Learning 51, 181–207 (2003)
13. Lemieux, A., Parizeau, M.: Flexible multi-classifier architecture for face recognition systems. In: The 16th International Conference on Vision Interface (2003)
14. Shipp, C.A., Kuncheva, L.I.: Relationships between combination methods and measures of diversity in combining classifiers. Inf. Fusion 3(2), 135–148 (2002)
15. Sugeno, M.: Theory of Fuzzy Integrals and its Applications. PhD thesis, Tokyo Institute of Technology (1974)

16. Torres-Sospedra, J., Hernandes-Espinosa, C., Fernandes-Redondo, M.: A Comparison of Combination Methods for Ensembles of RBF Networks. In: IJCNN 2005, pp. 1137–1141 (2005)
17. Torres-Sospedra, J., Hernandes-Espinosa, C., Fernandes-Redondo, M.: A Research on Combination Methods for Ensembles of Multilayer Feedforward. In: IJCNN 2005, pp. 1125–1130 (2005)
18. Tsymbal, A., Pechenizkiy, M., Cunningham, P.: Diversity in search strategies for ensemble feature selection. Inf. Fusion, Special issue on Diversity in MCSs 6(1), 83–98 (2005)
19. Verikas, A., Lipnickas, A., Malmqvist, K., Bacauskiene, M., Gelzinis, A.: Soft Combination of neural classifiers: A comparative study. Pattern Recognition Letters 20, 429–444 (1999)
20. Zimmermann, H., Zysno, P.: Latent connectives in human decision making. Fuzzy Sets and Systems 4, 37–51 (1980)

SpikeStream: A Fast and Flexible Simulator of Spiking Neural Networks

David Gamez

Department of Computer Science, University of Essex, Colchester, C04 3SQ, UK
daogam@essex.ac.uk

Abstract. SpikeStream is a new simulator of biologically structured spiking neural networks that can be used to edit, display and simulate up to 100,000 neurons. This simulator uses a combination of event-based and synchronous simulation and stores most of its information in databases, which makes it easy to run simulations across an arbitrary number of machines. A comprehensive graphical interface is included and SpikeStream can send and receive spikes to and from real and virtual robots across a network. The architecture is highly modular, and so other researchers can use its graphical editing facilities to set up their own simulation networks or apply genetic algorithms to the SpikeStream databases. SpikeStream is available for free download under the terms of the GPL.

Keywords: Spiking neural networks, SpikeStream, event-based simulation, synchronous simulation, biologically inspired robotics.

1 Introduction

SpikeStream is a new fast and flexible neural simulator with the following key features:

- Written in C++ using Qt for the graphical user interface.
- Database storage.
- Parallel distributed operation.
- Sophisticated visualisation, editing and monitoring tools.
- Modular architecture.
- Variable delays.
- Dynamic synapses.
- Dynamic class loading.
- Live operation.
- Spike exchange with external devices over a network.

The first part of this paper outlines the different components of the SpikeStream architecture and sets out the features of the GUI in more detail. Next, the performance of SpikeStream is documented along with its communication with external devices. The last part of this paper suggests some applications for SpikeStream, compares it with other simulators and describes its limitations. Documentation and source code for SpikeStream are available for free download at http://spikestream.sourceforge.net.

J. Marques de Sá et al. (Eds.): ICANN 2007, Part I, LNCS 4668, pp. 360–369, 2007.

2 Architecture

SpikeStream is built with a modular architecture that enables it to operate across an arbitrary number of machines and allows third party applications to make use of its editing, archiving and simulation functions. The main components of this architecture are a number of databases, the graphical SpikeStream Application, programs to carry out simulation and archiving functions, and dynamically loaded neuron and synapse classes.

2.1 Databases

SpikeStream is based around a number of databases that hold information about the network model, patterns and devices. This makes it very easy to launch simulations across a variable number of machines and provides a great deal of flexibility in the creation of connection patterns. The SpikeStream databases are as follows:

- *Neural Network*. Each neuron has a unique ID and connections between neurons are recorded as a combination of the presynaptic and postsynaptic neuron IDs. The available neuron and synapse types along with their parameters and class libraries are also held in this database.
- *Patterns*. Holds patterns that can be applied to the network for training or testing.
- *Neural Archive*. Stores archived neuron firing patterns. Each archive contains an XML representation of the network and data in XML format.
- *Devices*. The devices that SpikeStream can exchange spikes with over the network.

These databases are edited by SpikeStream Application and used to set up the simulation run. They could also be edited by third party applications - to create custom connection patterns or neuron arrangements, for example - without affecting SpikeStream's ability to visualise and simulate the network.

2.2 SpikeStream Application

An intuitive graphical user interface has been written for SpikeStream (see Fig. 1) with the following features:

- *Editing*. Neuron and connection groups can be created and deleted.
- *3D Visualisation*. Neuron and connection groups are rendered in 3D using OpenGL and they can be rotated, selectively hidden or shown, and their individual details displayed. The user can drill down to information about a single synapse or view all of the connections simultaneously.
- *Simulation*. The simulation tab has controls to start and stop simulations and vary the speed at which they run. Neuron and synapse parameters can be set, patterns and external devices connected and noise injected into the system.
- *Monitoring*. Firing and spiking patterns can be monitored and variables, such as the membrane potential, graphically displayed.
- *Archiving*. Archived simulation runs can be loaded and played back.

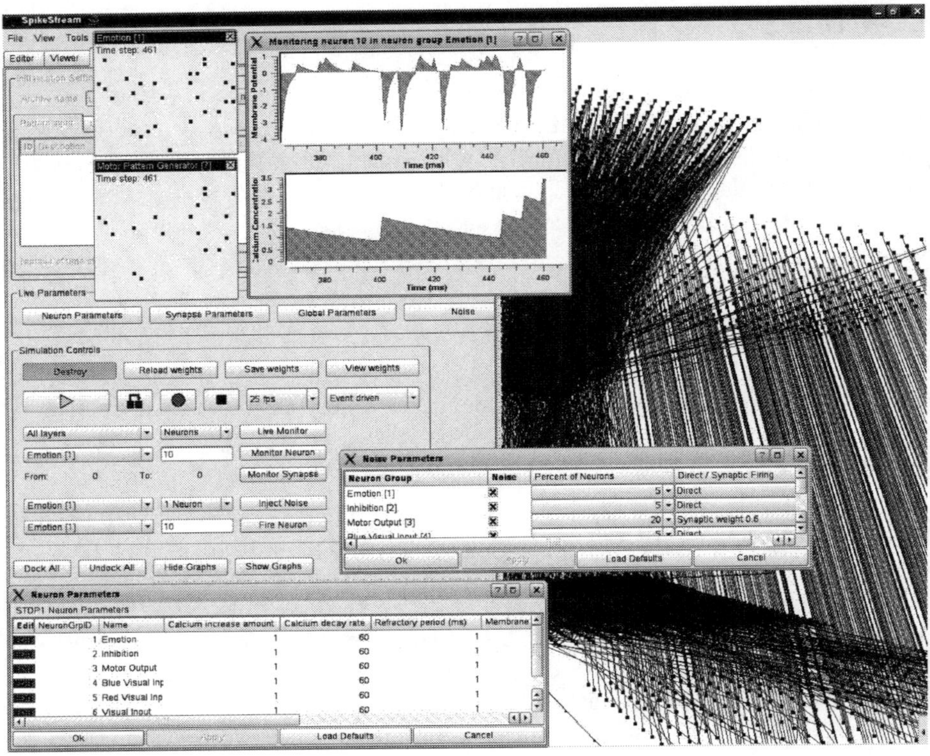

Fig. 1. SpikeStream graphical user interface

2.3 SpikeStream Simulation

The SpikeStream simulator consists of a number of processes that are launched and coordinated using PVM, with each process simulating an individual neuron group using a combination of event-based and synchronous simulation based on SpikeNET [7]. In common with synchronous simulations, the simulation period is divided into steps with an arbitrarily small time resolution and each neuron group receives lists of spikes from other layers at each time step. However, only the neuron and synapse classes that receive a spike are updated, which substantially cuts down on the amount of processing required. Since the main overhead is calculating the neurons' state and sending the spikes, the simulator's update speed depends heavily on the level of network activity and at high levels the performance becomes the same as a synchronous simulator. In theory, SpikeStream's run speed should be relatively independent of the time step resolution, since the calculation of each time step is efficient and the network should emit the same number of spikes per second independently of the time step resolution. In practice, the setting of this value can affect the number of spikes emitted by the network because higher values reduce the number of spikes arriving during a neuron's refractory period and alter the network dynamics (see Table 2).

One difference between SpikeStream and SpikeNET is that messages are sent rather than requested at each time step, which cuts down the message passing by 50% in a fully recurrent architecture. The spikes themselves are a compressed version of the presynaptic and postsynaptic neuron IDs, which enables each spike to be uniquely routed to an individual synapse class. Variable delays are created by copying emitted spikes into one of 250 buffers and at each time step only the spikes in the current buffer are sent. Unlike the majority of neural simulation tools, SpikeStream is designed to operate in live as well as simulation time so that it can control real and virtual robots in close to real time and process input from live data sources (see section 4). Although SpikeStream is primarily an event-driven simulator, it can be run synchronously to accommodate neuron models that generate spontaneous activity.

2.4 SpikeStream Archiver

During a simulation run, the firing patterns of the network can be recorded by SpikeStream Archiver, which stores spike messages or lists of firing neurons in XML format along with a simple version of the network model.

2.5 Neuron and Synapse Classes

Neuron and synapse classes are implemented as dynamically loaded libraries, which makes it easy to experiment with different neuron and synapse models without recompiling the whole application. Each dynamically loadable class is associated with a parameter table in the database, which makes it easy to change parameters during a simulation run. The current distribution of SpikeStream includes neuron and synapse classes implementing Brader et al.'s STDP learning rule [3].

3 Performance

3.1 Tests

The performance of SpikeStream was measured using three networks based on the benchmarks put forward by Brette et. al. [4]. The size and connectivity of these networks is given in Table 1 and they were divided into four layers containing 80% excitatory and 20% inhibitory neurons. At the beginning of each simulation run the networks were driven by a random external current until their activity became self sustaining and then their performance was measured over repeated runs of 300 seconds. A certain amount of fine tuning was required to make each network enter a self-sustaining state that was not highly synchronized.

The neuron model for these tests was based on the Spike Response Model [11], with the voltage V_i at time t for a neuron i that last fired at $\hat{t}$ being given by:

$$V_i(t) = \sum_j \sum_f w_{ij} e^{-\frac{(t - t_j^{(f)})}{\tau_m}} - e^{n-(t-\hat{t}_i)^m} \mathrm{H}'(t - \hat{t}_i) , \qquad (1)$$

where ω_{ij} is the synaptic weight between i and j, f is the last firing time of neuron j and H' is given by:

$$H'(t-\hat{t}_i) = \begin{cases} \infty, & if\ 0 \le (t-\hat{t}_i) < \rho \\ 1, & otherwise \end{cases} . \quad (2)$$

For all networks the membrane time constant, τ_m, was set to 3, the refractory parameters m and n were set to 0.8 and 3, the connection delay was set to 1 ms and the absolute refractory period, ρ, was set to 3. V_i had a resting vale of 0 and when it exceeded the threshold the neuron was fired and the contributions from previous spikes were reset. The remaining neuron and synapse parameters are given in Table 1.

Table 1. Size of test networks and their parameters

Parameter	Small network	Medium network	Large network
Neurons	4000	10,000	19,880
Connections	321985	1,999,360	19,760,878
ω_{ij} (excitatory)	0.11	0.11	0.11
ω_{ij} (inhibitory)	-1.0	-0.6	-0.6
Threshold	0.1	0.15	0.25

The first two networks were tested on one and two Pentium IV 3.2 GHz machines connected using a megabit switch with time step values of 0.1 and 1.0 ms. The third network could only be tested on two machines because its memory requirements exceeded that available on a single machine. All of the tests were run without any learning, monitoring or archiving.

3.2 Results

Fig. 2 plots the amount of time taken to simulate one second of biological time for each of the test networks. In this graph the performance difference between 0.1 and 1.0 ms time step resolution is partly due to the fact that ten times more time steps were processed at 0.1 ms resolution, but since SpikeStream is an event-based simulator, the processing of a time step is not a particularly expensive operation. The main cause of this speed up were changes in the networks' dynamics brought about by the lower time step resolution, which reduced the average firing frequency of the networks by the amounts given in Table 2.

Table 2. Average firing frequencies in biological time at different time step resolutions

Time step resolution	Small network	Medium network	Large network
0.1 ms	109 Hz	72 Hz	40 Hz
1.0 ms	79 Hz	58 Hz	30 Hz

The differences in average firing frequency shown in Table 2 suggest that the relationship between real and biological time needs to be combined with other performance measurements for event-based simulators.

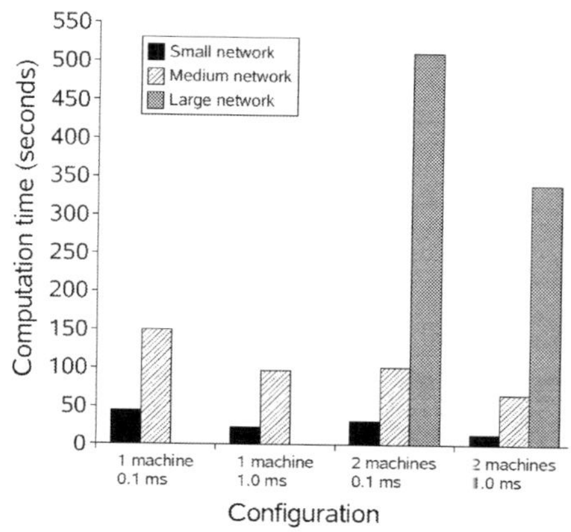

Fig. 2. Time taken to compute one second of biological time for one and two machines using time step resolutions of 0.1 and 1 ms

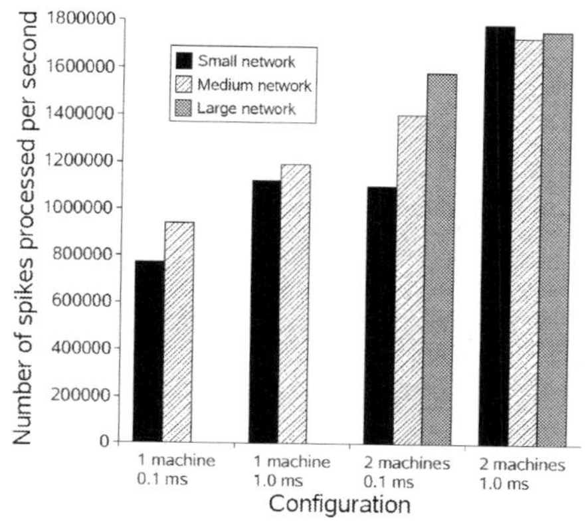

Fig. 3. Number of spikes processed per second of real time for one and two machines using time step resolutions of 0.1 and 1 ms

To address this issue, the number of spikes processed in each second of real time was also measured and plotted in Fig. 3. This graph shows that SpikeStream can handle between 800,000 and 1.2 million spike events per second on a single machine and between 1.2 million and 1.8 million spike events per second on two machines - an observation that should stay reasonably constant with different networks and activity levels. Fig. 2 and Fig. 3 both show that the performance increased when the processing load was distributed over multiple machines, but with network speed as a key limiting factor, multiple cores are likely to work better than multiple networked machines.[1]

4 External Devices

SpikeStream can pass spikes over a network to and from external devices, such as cameras and real and virtual robots, in a number of different ways:

1. *Synchronized TCP*. Spikes are exchanged with the device at each time step and both only move forward when they have received their spikes.

[1] Brette et. al. [4] give the performance of other neural simulators on their benchmark networks, which can be compared with the SpikeStream results in Fig. 2 and Fig. 3. It is worth noting that the run time of a SpikeStream simulation is not affected by the speed of its databases.

2. *Loosely synchronized UDP*. Spikes are sent and received continuously to and from the external device with the rate of the simulation controlled by the rate of arrival of the spike messages.
3. *Unsynchronized UDP*. Spikes are sent and received continuously from the external device. This option is designed for live work with robots.[2]

The main external device that has been used and tested with SpikeStream is the SIMNOS virtual robot created by Newcombe [9]. SIMNOS is a humanoid robot with an internal structure inspired by the human musculoskeletal system and it is simulated in soft real time using Ageia PhysX [1]. Within this physics simulation environment SIMNOS' skeletal body components are modelled as jointed rigid bodies and its muscles are modelled using spring-damper systems.

Visual data (available with a wide variety of pre-processing methods) and muscle lengths and joint angles are encoded by SIMNOS into spikes using a selection of methods developed by Newcombe [9] and passed across the network to SpikeStream using synchronized TCP. SIMNOS also receives muscle length data from SpikeStream in the form of spiking neural events, which are used to control the virtual robot. Together SIMNOS and SpikeStream provide an extremely powerful way of exploring sensory and motor processing and integration. More information about SIMNOS is available at www.cronosproject.net.

5 Applications

5.1 General Application Areas

Some potential applications of SpikeStream are as follows:

- *Biologically inspired robotics*. Spiking neural networks developed in SpikeStream can be used to process sensory data from real or virtual robots and generate motor patterns. A good example of this type of work is that carried out by Krichmar et. al. on the Darwin series of robots [12].
- *Genetic algorithms*. The openness of SpikeStream's architecture makes it easy to write genetic algorithms that edit the database and run simulations using PVM.
- *Models of consciousness and cognition*. Dehaene and Changeux [6] and Shanahan [17] have built models of consciousness and cognition based on the brain that could be implemented in SpikeStream.
- *Neuromorphic engineering*. SpikeStream's dynamic class loading architecture makes it easy to test neuron and synapse models prior to their implementation in silicon. Initial work has also been done on enabling SpikeStream to read and write AER events, which would enable it to be integrated into AER chains, such as those developed by the CAVIAR project [5].
- *Teaching*. Once installed SpikeStream is well documented and easy to use, which makes it a good tool for teaching students about biologically structured neural networks and robotics.

[2] This method of spike streaming has not been fully implemented in SpikeStream 0.1.

5.2 Recent Experiments

SpikeStream is currently being used to develop a neural network that uses analogues of imagination, emotion and sensory motor integration to control the eye movements of the SIMNOS virtual robot. When this network is online it spontaneously generates eye movements to different parts of its visual field and learns the association between an eye movement and a visual stimulus using Brader et. al.'s [3] spike time dependent learning rule. When it has learnt these associations, its emotion system is hardwired so that blue objects inhibit motor output and visual input. This switches the network into 'imagination' mode, in which it explores sensory motor patterns until it finds one that positively stimulates its emotional system. This removes the inhibition, and SIMNOS' eye is then moved to look at the red stimulus. This work is inspired by other simulations of the neural correlates of consciousness, such as Dehaene and Changeux [6] and Shanahan [17].[3]

6 Other Spiking Simulators

SpikeStream simulates biologically structured spiking networks of up to 100,000 neurons with each neuron treated as a point without the complex dendritic structure of a biological neuron. This sets it apart from simulators, such as NEURON [16], GENESIS [10] and NCS [15], that work with complex dendritic trees. SpikeStream also differs from rate-based simulators, such as Topographica [19], and synchronous simulators, such as NEST [8], which update all the neurons at each time step and often run for a fixed period of simulation time. The spiking biological aspect of SpikeStream also differentiates it from the many simulators written for conventional neural networks, which are often trained by back-propagation and have input, output and hidden layers.

The closest simulator to SpikeStream is Delorme and Thorpe's SpikeNET [7]. This simulator runs substantially faster than SpikeStream,[4] but its extra performance comes at the cost of a number of limitations These include a lack of synaptic delay, the fact that each neuron can only fire once, a lack of recurrent connections, no graphical interface, a simple and inflexible neural model and synaptic weights that are shared between neurons in an array. All of these limitations were the motivation for creating a new simulator based on the SpikeNET architecture that was more flexible and easier to use.

SpikeStream also has similarities with SpikeSNNS [13], which is a spiking version of the Stuttgart Neural Network Simulator. This was also influenced by SpikeNET, but has a much slower performance and the SNNS interface is somewhat outdated and difficult to use.

---

[3] Videos of this network in operation are available at www.davidgamez.eu/videos/index.html.

[4] SpikeStream can simulate 100,000 neurons firing at 1Hz in real time with 10 connections per neuron and SpikeNET can simulate approximately 400,000 neurons firing at 1Hz in real time with 50 connections per neuron. These measurements for SpikeNET were obtained using a substantially slower machine and so the performance of SpikeNET would probably be at least 800,000 neurons per second firing at 1 Hz today.

Other spiking simulators include the Amygdala library [2] and Mvaspike [14], which lack graphical interfaces and are not designed for robotic use, and the Spiking Neural Simulator developed by Smith [18], which can simulate a spiking network for a fixed period of time, but lacks many of the features included with SpikeStream. With many of these simulators it would be possible to use parts of SpikeStream, for example its graphical interface and database, to set up and run simulations on a different simulation engine.

7 Limitations

The flexibility and speed of SpikeStream come at the price of a number of limitations, some of which will be resolved in future releases of the software:

1. Neurons are treated as points. Each connection can have a unique delay, but there is none of the complexity of a full dendritic tree.
2. The connection delay is a function of the time step, not an absolute value, and the maximum number of time steps is limited to 250. This design was motivated by a desire to keep the Connections table in the database as small as possible.
3. SpikeStream currently only works with rectangular layers of neurons. Although the editing and visualisation have been partly extended to deal with three-dimensional neuron groups, more work is required to enable three dimensional live monitoring.
4. Any two neurons can only have a single connection between them. This restriction exists because the ID of each connection in the database is formed from a combination of the presynaptic and postsynaptic neuron IDs.

8 Conclusion

This paper has outlined the architecture and performance of SpikeStream, which can simulate medium sized networks of up to 100,000 neurons. This simulator is modular, flexible and easy to use and can interface with real and virtual robots over a network. SpikeStream is available for free download under the terms of the GPL licence.

Acknowledgements

Many thanks to Owen Holland for feedback, support and advice about SpikeStream and to the blind reviewers for their helpful comments and suggestions. The interface between SIMNOS and SpikeStream was developed in collaboration with Richard Newcombe, who designed the spike conversion methods. Thanks also to Renzo De Nardi and Hugo Gravato Marques in the Machine Consciousness lab at Essex and to everyone at the 2006 Telluride workshop. This work was funded by the Engineering and Physical Science Research Council Adventure Fund (GR/S47946/01).

References

1. Ageia PhysX, http://www.ageia.com/developers/api.html
2. Amygdala simulator, http://amygdala.sourceforge.net/
3. Brader, J.M., Senn, W., Fusi, S.: Learning real world stimuli in a neural network with spike-driven synaptic dynamics. Neural Computation (submitted, 2006)
4. Brette, R., Rudolph, M., Carnevale, T., Hines, M., Beeman, D., Bower, J.M., Diesmann, M., Morrison, A., Goodman, P.H , Harris Jr., F.C., Zirpe, M., Natschlaeger, T., Pecevski, D., Ermentrout, B., Djurfeldt, M., Lansner, A., Rochel, O., Vieville, T., Muller, E., Davison, A.P., El Boustani, S., Destexhe, A.: Simulation of networks of spiking neurons: A review of tools and strategies. J Comp. Neurosci (in press)
5. CAVIAR project, http://www.imse.cnm.es/caviar/
6. Dehaene, S., Changeux, J.P.: Ongoing Spontaneous Activity Controls Access to Consciousness: A Neuronal Model for Inattentional Blindness. Public Library of Science Biology 3(5), e141 (2005)
7. Delorme, A., Thorpe, S.J.: SpikeNET: An Event-driven Simulation Package for Modeling Large Networks of Spiking Neurons. Network: Computational in Neural Systems 14, 613–627 (2003)
8. Diesmann, M., Gewaltig, M.-O.: NEST: An Environment for Neural Systems Simulations. In: Macho, V. (ed.) Forschung und wissenschaftliches Rechnen, Heinz-Billing-Preis, GWDG-Bericht (2001)
9. Gamez, D., Newcombe, R., Holland, O., Knight, R.: Two Simulation Tools for Biologically Inspired Virtual Robotics. In: Proceedings of the IEEE 5th Chapter Conference on Advances in Cybernetic Systems, Sheffield, pp. 85–90. IEEE Computer Society Press, Los Alamitos (2006)
10. GENESIS simulator, http://www.genesis-sim.org/GENESIS/
11. Gerstner, W., Kistler, W.: Spiking Neuron Models. Cambridge University Press, Cambridge (2002)
12. Krichmar, J.L., Seth, A.K., Nitz, D.A., Fleischer, J.G., Edelman, G.M.: Spatial navigation and causal analysis in a brain-based device modeling cortical-hippocampal interactions. Neuroinformatics 3, 197–221 (2005)
13. Marian, I.: A biologically inspired computational model of motor control development. MSc Thesis, Department of Computer Science, University College Dublin, Ireland (2003)
14. Mvaspike simulator, http://www-sop.inria.fr/odyssee/softwares/mvaspike/
15. NCS simulator, http://brain.cse.unr.edu/ncsDocs/
16. NEURON simulator, http://www.neuron.yale.edu/neuron/
17. Shanahan, M.P.: A Cognitive Architecture that Combines Internal Simulation with a Global Workspace. Consciousness and Cognition 15, 433–449 (2006)
18. Spiking Neural Simulator, http://www.cs.stir.ac.uk/~lss/spikes/snn/index.html
19. Topographica Neural Simulator, http://topographica.org/Home/index.html

Evolutionary Multi-objective Optimization of Spiking Neural Networks

Yaochu Jin[1], Ruojing Wen[2], and Bernhard Sendhoff[1]

[1] Honda Research Institute Europe
Carl-Legien-Str. 30, 63073 Offenbach, Germany
[2] Department of Computer Science, University of Karlsruhe
Zirkel 2, 76131 Karlsruhe, Germany

Abstract. Evolutionary multi-objective optimization of spiking neural networks for solving classification problems is studied in this paper. By means of a Pareto-based multi-objective genetic algorithm, we are able to optimize both classification performance and connectivity of spiking neural networks with the latency coding. During optimization, the connectivity between two neurons, i.e., whether two neurons are connected, and if connected, both weight and delay between the two neurons, are evolved. We minimize the the classification error in percentage or the root mean square error for optimizing performance, and minimize the number of connections or the sum of delays for connectivity to investigate the influence of the objectives on the performance and connectivity of spiking neural networks. Simulation results on two benchmarks show that Pareto-based evolutionary optimization of spiking neural networks is able to offer a deeper insight into the properties of the spiking neural networks and the problem at hand.

1 Introduction

Spiking neural networks (SNNs) are believed to be biologically more plausible [1,2] and computationally more powerful than analog neural networks [3]. However, the computational power of SNNs has yet to be demonstrated, mainly due to the fact that an efficient supervised learning algorithm still lacks. In contrast to analog neural networks, for which various sophisticated supervised learning algorithms have been developed [4], only a very limited number of supervised learning algorithms are available for training SNNs, which can be attributed to the discontinuous nature of spiking neurons.

SpikePop [5] is the first supervised learning algorithm that has been developed based on the error backpropagation principle widely used in training analog neural networks. However, SpikeProp has several weaknesses. First, the performance of the learning algorithm is quite sensitive to parameter initialization. If a neuron is silent after initialization, no training is possible for the weights of its incoming connections. As a result, the neuron will never be able to produce any spikes. Second, SpikeProp is only applicable to latency-based coding. Third, SpikeProp works only for SNNs where neurons spike only once in the simulation time. Fourth, SpikeProp has been developed for training the weights only. To address these weaknesses, a few algorithms extending the SpikeProp have been suggested [6,7,8].

J. Marques de Sá et al. (Eds.): ICANN 2007, Part I, LNCS 4668, pp. 370–379, 2007.

Both SpikeProp and its variants are gradient-based learning algorithms, where simplifications have to be made to apply the gradient method. To resolve this problem inherent to the SpikeProp learning algorithms, evolutionary algorithms [9,10], which have shown to be very successful in training analog neural networks [11], have also been employed for training spiking neural networks. For example, the connectivity and the sign of the connectivity (the neuron is excitatory if the sign is positive and inhibitory if negative) of a spike response model (SRM) [12] are evolved for vision-based robot control using a genetic algorithm (GA). Similar work has been reported in [13] and [14]. In [13], an adaptive GA is adopted to evolve the weights of the SRM model for robot navigation, while in [14], a parallel deferential evolution has been employed to evolve the weights of the SRM. Both weights and delays of an integrate-and-fire (IAF) model with dynamic synapses [15] and a SRM are evolved using an evolution strategy [16]. It is found that the performance of the SNNs are comparable to that of the analog feedforward neural network on two benchmark problems.

Encouraged by the success of the Pareto-based approach to machine learning [17], this work employs a Pareto-based multi-objective genetic algorithm to evolve the connectivity, weights and delays of the SRM. Different to single objective optimization, Pareto-based multi-objective learning using multi-objective evolutionary algorithms [18] is able to achieve a number of Pareto-optimal solutions rather than one single optimal solution. By analyzing the trade-off between different learning objectives, such as the performance and complexity, we are able to gain a deeper insight into the neural network model as well as the problem to learn [19], by, e.g., identifying interpretable models and models that can generalize on unseen data from the Pareto-optimal solutions.

Several objectives concerning learning performance and connectivity can be taken into account in optimizing spiking neural networks for supervised learning such as classification. In addition to minimizing the classification error, we can also minimize the number of connections, or the total length of synaptic connections, which can be replaced by the sum of the delays in the SNN, assuming that the delay between two neurons is proportional to the connection length between them.

2 Spiking Neural Network Model

The spiking network model adopted in this work is the same as the one in [5], which has a feedforward architecture consisting of neurons described by the SRM [12]. The network architecture is shown in Fig. 1(a), where multiple synaptic terminals can exist between two neurons, refer to Fig. 1(b). For clarity, we assume that the network has only one output neuron.

Assume neuron k receives spikes from N pre-synaptic neurons that fire at time instant t_{ik}. For the sake of simplicity, we assume each neuron emit at most one spike during the simulation interval, though this is not a restriction from our evolutionary learning method. The internal state of neuron k at time t, $x_k(t)$, which represents the membrane potential, can be described by

$$x_k(t) = \sum_{i=1}^{N} \sum_{l=1}^{m^i} w_{ik}^l y_i^l(t), \quad (1)$$

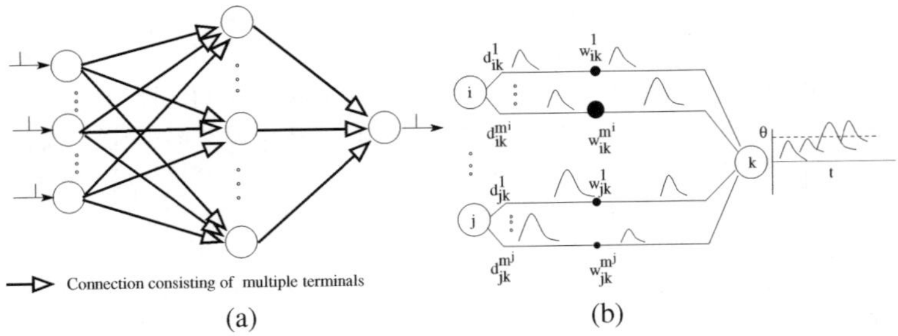

Fig. 1. Feedforward spiking neural networks with multiple synaptic terminals between two neurons (a) The feedforward architecture. (b) Two synaptic connections with multiple terminals.

where m^i is the number of synaptic terminals between neurons i and k, w_{ik}^l is the synaptic strength, and $y_i^l(t)$ is the unweighted contribution of neuron i to neuron k through the l-th synaptic terminal, which can be expressed by

$$y_i^l(t) = \epsilon(t - t_i - d_{ik}^l), \tag{2}$$

where ϵ is a spike response function modeling the post-synaptic potential, t_i is the firing time of neuron i, and d_{ik}^l is the synaptic delay of the l-th synaptic terminal. The spike response function can be described as follows:

$$\epsilon(t) = \begin{cases} \frac{t}{\tau} e^{1-\frac{t}{\tau}}, & \text{if } t \geq 0 \\ 0, & \text{if } t < 0 \end{cases}, \tag{3}$$

where τ is the membrane potential decay time constant.

When the membrane potential of neuron k crosses a predefined threshold θ, it will emit a spike. After firing, there is a refractory time during which no spikes can be generated again. As we assumed that at most one spike will be generated during the simulation interval, the refractory time is set to be larger than the simulation interval.

3 A Multi-objective Genetic Algorithm for Pareto Learning

Evolutionary algorithms (EAs) are well suited for Pareto-based multi-objective learning of spiking neural networks for several reasons. First, EAs do not require explicit gradient information for updating weights, unlike most supervised learning methods. Second, not only weights, but synaptic delays and connectivity of the network such as the number of terminals between two neurons can also be evolved. Third, evolutionary algorithms have shown to be very powerful for multi-objective optimization [18], thus for multi-objective learning.

3.1 Genetic Representation of SNNs

In this work, a genetic algorithm using gray coding has been adopted. Each individual consists of two chromosomes encoding a connection matrix and a weight matrix, respectively. An element in the connection matrix (c_{ik}) is an integer equal to or greater than zero, representing the connectivity from neuron i to neuron k. We investigate both single-terminal and multi-terminal synaptic connections. In case of single-terminal connections, there is at most one connection between two neurons. There is no connection from neuron i to neuron k if $c_{ik} = 0$. If $c_{ik} > 0$, then it represents the synaptic delay between two neurons. In the multi-terminal case, $c_{ik} > 0$ represents the number of terminals between two neurons, and the synaptic delays of the terminals range from $1 to c_{ik}$. The weight matrix (w_{ik}) encodes the strength of the connections between neuron i and neuron j.

3.2 Objectives

Existing supervised learning algorithms for training spiking neural networks minimize the error on training data only. In contrast, a multi-objective learning algorithm can take into account more than one objective with respect to training performance as well as the connectivity of the SNN. In this work, we adopt one of the two error functions, i.e., either the root mean square error (RMSE) or the classification error in percentage, to optimize the performance of the SNN. Besides, either the number of connections or the sum of delays is minimized for optimizing the connectivity of the SNN.

3.3 Genetic Variations and Pareto-based Selection

The uniform crossover and bit-flip mutation are applied at a probability of p_c and p_m, respectively, to evolve the connectivity and weights of the SNN. To select parent individuals for the next generation, we employ the crowded tournament selection method proposed in NSGA-II [20]. First, the offspring and the parent populations are combined. Then, a non-dominated rank and a local crowding distance are assigned to each individual in the combined population. In non-dominated ranking, the non-dominated solutions are found out and assigned a rank of 1. These solutions consist of the first non-dominated front. After that, the non-dominated solutions with rank 1 are removed temporarily from the combined population. Then, non-dominated solutions of the rest individuals in the population are identified, which consist of the second non-dominated front. A rank of 2 is assigned to these solutions. This procedure repeats until all individuals are assigned to a non-dominated front. In the next step, a crowding distance is calculated for each individual with regard to the non-dominated front it belongs to. The crowding distance of a solution is the distance between its two neighboring solutions on the same non-dominated front in the objective space. A large distance is assigned to the two boundary solutions on each non-dominated front. Here, the larger the crowding distance is, the less crowded around the solution.

After non-dominated sorting and calculation of the crowding distance, selection begins. During selection, two solutions are chosen randomly. The solution with the better (lower) rank wins the tournament. If the two solutions have the same rank, the one with the larger crowding distance wins. If two solutions have the same rank and the same crowding distance, choose a winner randomly. This tournament procedure continues until the parent population for the next generation is filled up.

A diagram of the multi-objective genetic algorithm for optimization of the connectivity and weight of SNNs is shown in Fig. 2. The code of our algorithm is developed within the SHARK environment [21].

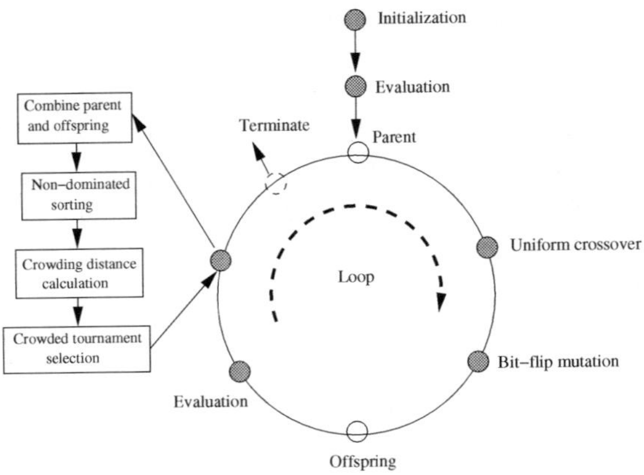

Fig. 2. A multi-objective GA for optimizing the connectivity and weights of SNNs

4 Empirical Studies

4.1 Experimental Setup

The multi-objective learning algorithm has been tested on two benchmark problems from the UCI Machine Learning Repository, namely, the Wisconsin breast cancer diagnosis data set and the Prima Indian diabetes data set. The cancer data consists of 699 instances with 9 input attributes, and the diabetes data contains 768 examples with 8 inputs, all of which are normalized between 0 and 1. The output of both data sets is either 0 or 1, meaning benign (negative) or malignant (positive). The data are divided into training and test sets. Of the cancer data, 525 data pairs are used for training and 174 pairs for test. Of the diabetes data, 576 examples for training and 192 for test.

Temporal coding is used for both inputs and output, where the inputs are scaled between 0 ms and 6 ms, and the outputs are normalized between 10 ms and 16 ms for the cancer data, while for the diabetes data, the input is scaled between 0 ms and 16 ms, and the output is set to 20 ms for negative cases and 26 for positive ones. For example, given an input value of 0.1, a spike is generated at time 0.6 ms after the reference time. For setting the reference time, an reference input is included in the network, which serves as the clock, as in [5]. Thus, the number of input neurons equals the number of attributes plus one. For an output value of 0, the output neuron should emit a spike at time 16 ms, while for an output value of 1, the neuron should emit a spike at time 10 ms. The simulation period is set to 50 ms with a step-size of 0.1 ms, the membrane potential decay time constant (τ) is set to 11, and the threshold θ is set to 1000.

Both parent and offspring populations consist of 100 individuals, each of which is composed of two chromosomes, one representing the connectivity and the other the weights. In the single-terminal case, each element of the connection matrix is represented by a 4-bit gray-coded string, which means that the value of the connectivity elements is between 0 and 15. As mentioned in the previous section, a connection value of 0 means that the two neurons are not connected, while a non-zero integer encodes the delay in millisecond between the two neurons. In other words, the maximum delay the genetic algorithm can evolve is 15 ms. In the multi-terminal case, the 4-bit string means that a minimum of zero to a maximum of 15 terminals exist between two neurons. For example, if this value is zero, no terminals exist between the corresponding two neurons. If this value is three, there are three terminals between the neurons, and the synaptic delay of the three terminals is 1 ms, 2 ms, and 3 ms, respectively. Each synaptic weight is encoded by a 10-bit gray-coded string that is initialized between -10 and 40. The maximum number of hidden neurons is set to 10.

The crossover probability is set to 0.6, and the mutation rate is set to be inversely proportional to the total length of the chromosome. Each evolutionary optimization has been run for 300 generations and the results are averaged over ten independent runs.

4.2 Results from Networks with Single-Terminal Connections

We first perform simulation studies for the cancer data when the network has single-terminal connections. We consider four different objective setups, where the two objectives are: Number of connections and classification error (Case 1), number of connections and root mean square error (RMSE) (Case 2), sum of delays and classification error (Case 3), and sum of delays and RMSE (Case 4). The learning results for cases 1) and 2) are presented in Fig. 3(a), where the asterisks denote the results for case 1) and

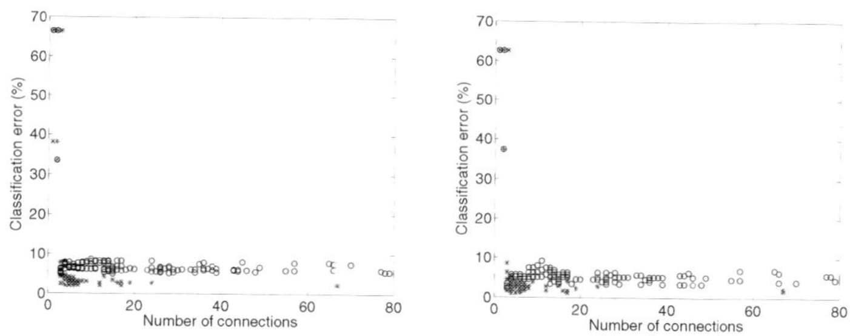

(a) (b)

Fig. 3. Results from 10 independent runs for Case 1 (denoted by asterisks), and Case 2 (denoted by circles) from the cancer data. (a) Training, and (b) test.

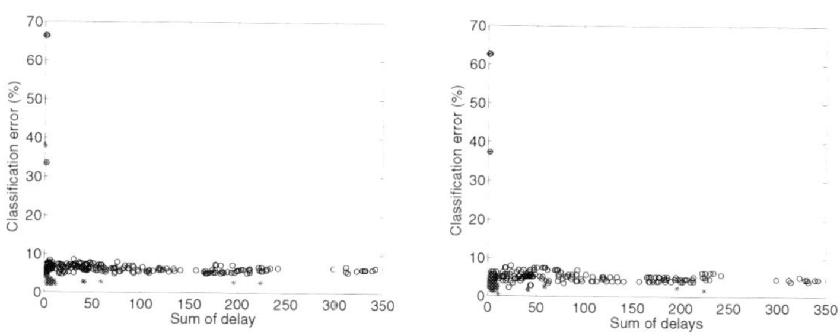

(a) (b)

Fig. 4. Results from 10 independent runs for Case 3 (denoted by asterisks), and Case 4 (denoted by circles) from the cancer data. (a) Training, and (b) test.

the circles the results for case 2). We can see from the figure that, different to single objective learning, we obtain a number of non-dominated solutions in each run, which trade the complexity against the performance. Besides, we note that smaller classification errors have been obtained in case 1) than in case 2). This is also true on the test data, refer to Fig. 3(b). In case the RMSE is used as the second objective, the algorithm tends to obtain more complex networks. We also notice that there is a clear knee point on the Pareto front, i.e., the classification error decreases rapidly when the number of connections increases to about 5. After that, improvement in performance is minor as the complexity further increases. Thus, the achievement of the non-dominated front helps us to choose the most economic network for a given problem. Note that in the figures, results from 10 independent runs are merged, thus some of the solutions become dominated by others, though the results from different runs do not vary much. The minimal classification error we obtained on the test data is 1.2%, which is a little better than those reported in [5] and [16], where the test error is 1.8% and 2.4%, respectively. Our results are also comparable to those of analog neural networks with two hidden layers [22], where the mean classification error on the test data is 1.15% and 2.87%, respectively, when direct connections between input and output nodes are present or absent. In [22], the analog neural networks are trained by RPROP, which is a very efficient gradient-based learning algorithm [23].

The results for case 3) and case 4) are given in Fig. 4. The performance of the achieved non-dominated NNs is quite similar to cases 1) and 2). To get an impression on the relationship between the two different measures for performance and connectivity, we re-plot the results of Fig. 3 and Fig. 4 in terms of RMSE in Fig. 5 and Fig. 6, respectively. From the figures, the following observations can be made. First, it does not make much difference whether RMSE or classification error is used for optimizing the performance, neither does it, whether the number of connections or the sum of delays is adopted for optimizing the connectivity. Second, in all cases, no overfitting has occurred, which is of great interest compared to serious overfittings in analog networks with high complexity [19].

Simulations have also been conducted for the diabetes data. For this data set, we show only the results for Cases 1 and 2, refer to Fig. 7(a) for the training and Fig. 7(b) for the test data. We note that for the diabetes data, the discrepancy between the results

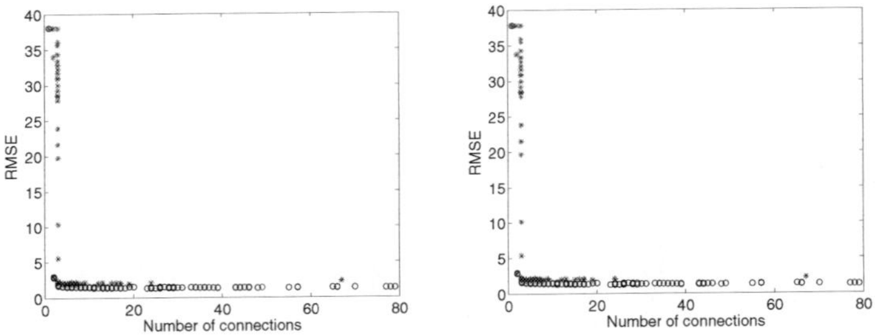

(a) (b)

Fig. 5. Re-plots of the results from the cancer data for Case 1 (denoted by asterisks), and Case 2 (denoted by circles). (a) Training, and (b) test.

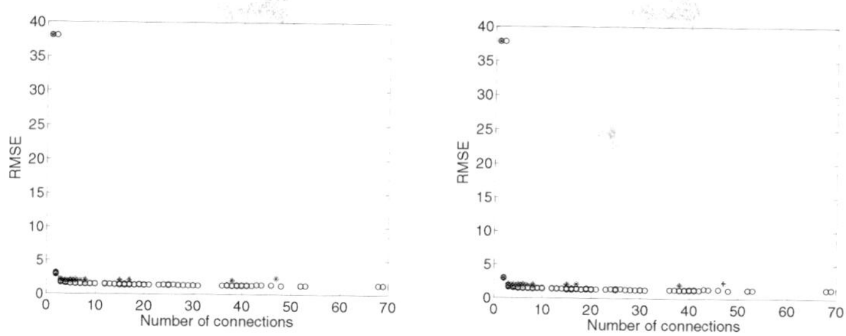

(a) (b)

Fig. 6. Re-plots of the results from the cancer data for Case 1 (denoted by asterisks), and Case 2 (denoted by circles). (a) Training, and (b) test.

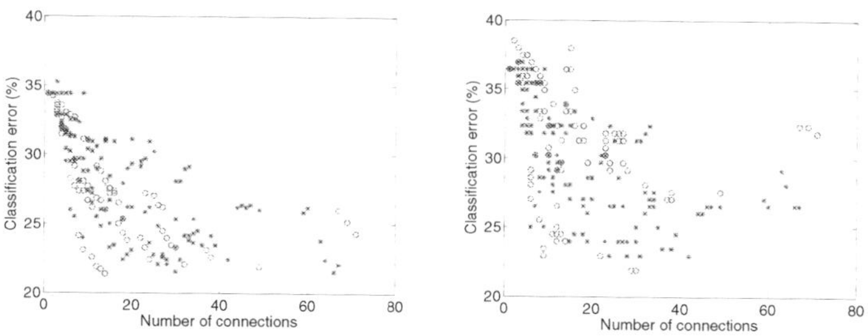

(a) (b)

Fig. 7. Results from 10 independent runs from the diabetes data for Case 1 (denoted by asterisks), and Case 2 (denoted by circles). (a) Training, and (b) test.

from different runs is much larger than that of the cancer data, which suggests that the diabetes data is more difficult to classify than the cancer data. Since results on the diabetes data using SNNs have not been reported elsewhere, we compare our results with those in [22]. In that work, the mean classification error over 60 runs on the test data is 25.83% when an analog network with two hidden layers are used. Thus, the performance of the SNN is better than that of the analog neural network in some of the runs are. Finally, similar to the cancer data, no serious overfitting has been observed for the diabetes data as the complexity of the network increases.

4.3 Results from Networks with Multi-terminal Connections

We now discuss briefly the results on SNNs with multiple terminals between two connections. The results for the cancer data are presented in Fig. 8, where the number of connections and the classification error are minimized. Since there can be multiple terminals between two neurons, the number of connections becomes much smaller than the single terminal cases, though the performance on the training and test data is similar.

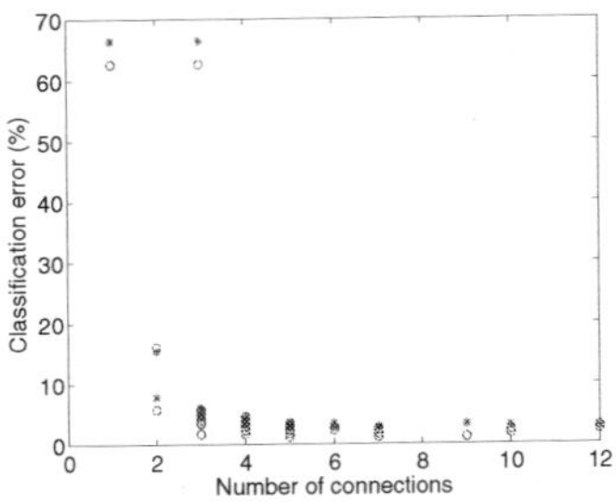

Fig. 8. Results on the cancer data from SNNs with multiple terminals. Asterisks denote the training data and circles the test data.

5 Conclusion

In this paper, we suggest a method for optimizing the performance and connectivity of SNNs using a Pareto-based genetic algorithm. Compared to existing supervised learning algorithms for SNNs, the Pareto-based approach considers both learning performance and connectivity of the SNNs without aggregating the two objectives. Either the RMSE or classification error has been minimized for optimizing the performance, and either the number of connections or the sum of delays has been minimized for optimizing connectivity.

One main merit to use the Pareto-based learning is that we can obtain a number of Pareto-optimal solutions that trade off performance against complexity. By exploiting the tradeoff solutions, we are able to gain an insight into the learning properties of SNNs, e.g., the minimal complexity needed for a given problem. Besides, we are able to investigate whether overfitting occurs when the complexity of the network increases. In our study, no overfitting has been observed, no matter how high the complexity of the network is. This is a very interesting finding, however, further experiments must be performed to confirm this observation.

No conclusion can be made on whether the RMSE or classification error should be used for classification. We have also shown that complexity of the SNNs are comparable, when the number of connections or the sum of delays is used to optimize connectivity.

One of the future work is to compare SNNs using other coding schemes than temporal coding and to consider SNNs that can emit multiple spikes within the simulation time. In addition, studying the relationship between functionality and connectivity of large scale spiking networks using sophisticated network simulation software like NEST [24] is our further research target.

References

1. Thorpe, S.J., Delorme, A., Van Rullen, R.: Spike-based strategies for rapid processing. Neural Networks 14(6-7), 715–726 (2001)
2. Bohte, S.M.: The evidence for neural information processing with precise spike-timing: A survey. Natural Computing 3(2), 195–206 (2005)

3. Maass, W.: Networks of spiking neurons: The throd generation of neural network models. Neural Networks 10(9), 1659–1671 (1997)
4. Reed, R.D., Marks II, R.J.: Neural Smithing - Supervised Leanring in Feedforward Artificial Neural Networks. MIT Press, Cambridge (1999)
5. Bohte, S.M.: Error-backpropagation in temporally encoded networks of spiking neurons. Neurocomputing 48, 17–37 (2002)
6. Booij, O., Nguyen, H.: A gradient decent rule for spiking neurons emiiting multiple spikes. Information Processing Letters 95(6), 552–558 (2005)
7. Schrauwen, B., Van Campenhout, J.: Extending SpikeProp. In: IJCNN, pp. 471–476 (2004)
8. Schrauwen, B., Van Campenhout, J.: Backpropagation for population-temporal coded spiking neural networks. In: IJCNN, pp. 3463–3470 (2006)
9. Goldberg, D.E.: Genetic Algorithms in Search, Optimization, and Machine Learning. Addison-Wesley, Reading (1989)
10. Schwefel, H.-P.: Evolution and Optimum Search. John Wiley, Chichester (1994)
11. Yao, X.: Evolving artificial neural networks. Proceedings of the IEEE 87(9), 1423–1447 (1999)
12. Gerstner, W.: Time structure of the activity in neural networks models. Physical Review 51(1), 738–758 (1995)
13. Hagras, H., et al.: Evolving spiking neural network controllers for autonoumous robots. In: IEEE Intternational Conference on Robotics and Automation, vol. 5, pp. 4620–4626. IEEE Computer Society Press, Los Alamitos (2004)
14. Pavlidis, N.G., et al.: Spiking neural network training using evolutionary algorithms. In: IJCNN, pp. 2190–2194 (2005)
15. Tsodyks, M., Pawelzik, K., Markram, H.: Neural networks with dynamic synapses. Neural Computation 10, 821–835 (1998)
16. Belatreche, A., Maguire, L.P., McGinnity, M.: Advances in design and application of spiking neural networks. Soft Computing 11, 239–248 (2007)
17. Jin, Y. (ed.): Multi-Objective Machine Learning. Springer, Heidelberg (2006)
18. Deb, K.: Multi-objective Optimization Using Evolutionary Algorithms. Wiley, Chichester (2001)
19. Jin, Y., Sendhoff, B.: Pareto-based multi-objective machine learning: An overview and case studies. IEEE Transactions on Systems, Man, and Cybernetics, Part C: Applications and Reviews (accepted, 2007)
20. Deb, K., Agrawal, S., Pratap, A., Meyarivan, T.: A fast elitist non-dominated sorting genetic algorithm for multi-objective optimization: NSGA-II. In: Parallel Problem Solving from Nature, vol. VI, pp. 849–858 (2000)
21. `http://shark-project.sourceforge.net/`
22. Prechelt, L.: Proben1 - A set of neural network benchmark problems and benchmark rules. Technical report, Falkutät für Informatik, Universität Karlsruhe (1994)
23. Riedmiller, M., Braun, H.: A direct adaptive method for faster backpropgation learning: The RPROP algorithm. In: IEEE International Conference on Neural Networks, vol. 1, pp. 586–591. IEEE Computer Society Press, Los Alamitos (1993)
24. `http://www.nest-initiative.org/`

Building a Bridge Between Spiking and Artificial Neural Networks

Florian Kaiser and Fridtjof Feldbusch

University of Karlsruhe
CES – Chair for Embedded Systems
Karlsruhe, Germany
{fkaiser,feldbusch}@ira.uka.de

Abstract. Spiking neural networks (SNN) are a promising approach for the detection of patterns with a temporal component. However they provide more parameters than conventional artificial neural networks (ANN) which make them hard to handle. Many error-gradient-based approaches work with a time-to-first-spike code because the explicit calculation of a gradient in SNN is - due to the nature of spikes - very difficult. In this paper, we present the estimation of such an error-gradient based on the gain function of the neurons. This is done by interpreting spike trains as rate codes in a given time interval. This way a bridge is built between SNN and ANN. This bridge allows us to train the SNN with the well-known error back-propagation algorithm for ANN.

1 Introduction

Neural networks are a mainstream tool within machine learning for solving classification and regression problems. They provide very good results for pattern detection on static input like objects in pictures. Spiking neurons are biophysical models of real neurons that exchange information via spikes. Coming from biology, spiking neural networks (SNN) give rise to high expectations for pattern detection on input with a temporal component like speech. However learning in SNN is a difficult task. First, there is the number of different parameters that have to be adjusted like weight and delay of synapses, threshold, membrane time constant, etc. Second, there is the difficulty to compute an error-gradient due to spike trains. A good overview of different approaches is given in [1]. Often [2,3,4] learning algorithms use the 'time-to-first-spike' code because here the spike time of a single spike is a continuous function of time and thus an error-gradient can be used for learning.

In this paper we approach learning in SNN from a rate code view and we present an estimation of the error-gradient based on a gain function. Furthermore we use the well known error gradient back-propagation algorithm [5] to train a multilayer feed-forward SNN. To derive the algorithm, we need to know - speaking in terms of ANN - what the base and transfer function of the SNN is. If we use the term activity instead of spike rate, both functions together are known as the gain function of a neuron. The gain function depends on the model describing the spiking neuron. In our work, we use the Spike Response Model (SRM) [6] with an abstract epsilon kernel.

Our aim for the future is to learn the number of spikes in a sequence of time intervals and thus to reintroduce time into the code.

J. Marques de Sá et al. (Eds.): ICANN 2007, Part I, LNCS 4668, pp. 380–389, 2007.

2 The Neuron Model

As neuron model we use a variant of the simplified Spike Response Model SRM_0 [6] without the linear response to externally induced currents known as κ_0 kernel. The ϵ_0 kernel describes the post-synaptic potentials (PSP), the η_0 kernel the re-polarization (reset) and hyper-polarization (relative refractory period).

$$u_i(t) = \sum_{t_i^f} \eta_0 \left(t - t_i^f\right) + \sum_j w_{ij} \sum_{t_j^f} \epsilon_0 \left(t - t_j^f - d_{ij}\right) \tag{1}$$

The spike times of the post-synaptic neuron i are denoted by t_i^f, the pre-synaptic ones by t_j^f. The axonal and synaptic delay between neurons j and i is described by d_{ij}, the synaptic strength by w_{ij}. In contrast to the SRM_0 described in [6], we use the sum of the η kernel over all past fire times t_i^f and not just the last fire time.

The neuron i fires a spike if its current potential u_i, the sum over all kernels, exceeds the threshold ϑ_i.

$$t = t_i^f \quad \Leftrightarrow \quad u_i(t) \geq \vartheta_i(t)$$

After firing the neuron is reset (see below) and enters an absolute refractory state, where it can not fire again for some time Δ^{arp}. This is modeled by setting the threshold to infinity for this period.

$$\vartheta_i(t) = \begin{cases} \infty & \text{if } 0 < t < \Delta^{\text{arp}} \\ \vartheta_0 & \text{otherwise} \end{cases}$$

Mind that it is possible, if $\Delta^{\text{arp}} = 0$ and the membrane potential is high enough, that the neuron fires two or more spikes at exactly the same time. Gerstner and Kistler [6] prevents this by the additional requirement for firing $\frac{\partial u_i(t)}{\partial t} \geq 0$. We however prevent it by setting $\Delta^{\text{arp}} > 0$.

As PSP we use the exponential function

$$\epsilon_0(t) = \exp\left(\frac{-t}{\tau_{\text{m}}}\right) \theta(t)$$

where τ_{m} is the membrane time constant and θ the Heaviside function. An example for a similar shaped ϵ kernel can be found in [7]. This simple definition of the PSP makes the derivation of the gain function in section 3 easier.

For the η_0 kernel, we use the fixed threshold interpretation according to [6] resulting in the subtraction of a reset potential u_{reset} from the membrane potential u_i.

$$\eta_0(t) = -u_{\text{reset}} \exp\left(\frac{-t}{\tau_{\text{m}}}\right) \theta(t)$$

When $u_{\text{reset}} > \vartheta_i$ the neuron has a relative refractory period, because in this case more incoming spikes (or PSP) are required to trigger a spike again after firing. The reset "decays" over time like the PSP. Usually this is parameterized by the time constant τ_{recov}. In our model we choose $\tau_{\text{recov}} = \tau_{\text{m}}$. This is not necessary for the derivation of the gain function, but as before it makes the derivation easier. In the following we refer to the time constant as τ instead of τ_{m}.

3 The Gain Function

The gain function of a neuron describes its output spike rate according to some input activity. We define the input activity as the spike count n in a certain time interval T.

Before starting to develop the gain function we have to discuss the computation mode of the neurons. According to König [8] a neuron can operate in two ways: As integrator or as coincidence detector. This behavior is defined by the membrane time constant τ. For instance if τ is much smaller than the inter-spike interval then the neuron only fires a spike if enough spikes arrive at the more or less same time. On the opposite side if τ is larger than the inter-spike interval the exact timing of the spikes does not matter. The neuron is then sensitive to activity but looses its role as coincidence detector. Since we want the spiking neurons to be sensitive to the activity, we have to do the following fuzzy simplifications leading to a loss of timing information. This information has to be reintroduced later.

Simplification. *The computation mode of the neuron is integration only.*

Or in other words, the membrane time constant τ has to be high enough.

With this in mind the first step in developing the gain function is to rewrite the function describing the membrane potential u (see eq. (1)) to use the spike count instead of spike trains. Because the spike count n says nothing about the spike distribution, we make the following simplification.

Simplification. *All spikes have the same distance $\Delta^{isi} = T/n$ to their neighbors. A spike train s is then given by $s = \{\frac{1}{2}\Delta^{isi} + k \cdot \Delta^{isi} | k = 0, \ldots, n-1\}$.*

Additionally and without loss of generality we set the synaptic delay to zero, $d_{ij} = 0$. With the simplifications above and using geometric series, the temporal summation of all PSP from a synapse can be written as follows:

$$\begin{aligned}\textstyle\sum_{t_j^f} \epsilon_0\left(t - t_j^f\right) &= \textstyle\sum_{t_j^f} \exp\left(-\frac{t-t_j^f}{\tau_i}\right)\\ &= \exp\left(-\frac{t-\Delta_j^{\mathrm{isi}}(n_j-\frac{1}{2})}{\tau_i}\right)\textstyle\sum_{f=0}^{n_j-1} \exp\left(-\frac{f\Delta_j^{\mathrm{isi}}}{\tau_i}\right)\\ &\overset{\mathrm{geo}}{=} \exp\left(-\frac{t-\Delta_j^{\mathrm{isi}}(n_j-\frac{1}{2})}{\tau_i}\right)\frac{1-\exp\left(-\frac{n_j\Delta_j^{\mathrm{isi}}}{\tau_i}\right)}{1-\exp\left(-\frac{\Delta_j^{\mathrm{isi}}}{\tau_i}\right)}\\ &= \exp\left(-\frac{t-(T-\frac{1}{2}\Delta_j^{\mathrm{isi}})}{\tau_i}\right)\frac{1-\exp\left(-\frac{T}{\tau_i}\right)}{1-\exp\left(-\frac{T}{n_j\tau_i}\right)}\ .\end{aligned} \tag{2}$$

The indices i and j refer to the post- and pre-synaptic neurons. The same transformation can be applied to the sum of η kernels. The geometric series can clearly be seen in Figure (1).

With this transformation and for $t = T$, equation (1) can then be rewritten as

$$u_i(T) = -u_{\mathrm{reset}} \exp(-\frac{1}{2}\frac{\Delta_i^{\mathrm{isi}}}{\tau_i})\frac{1-\exp(-\frac{T}{\tau_i})}{1-\exp(-\frac{T}{n_i\tau_i})} + \sum_j w_{ij}\exp(-\frac{1}{2}\frac{\Delta_j^{\mathrm{isi}}}{\tau_i})\frac{1-\exp(-\frac{T}{\tau_i})}{1-\exp(-\frac{T}{n_j\tau_i})}$$

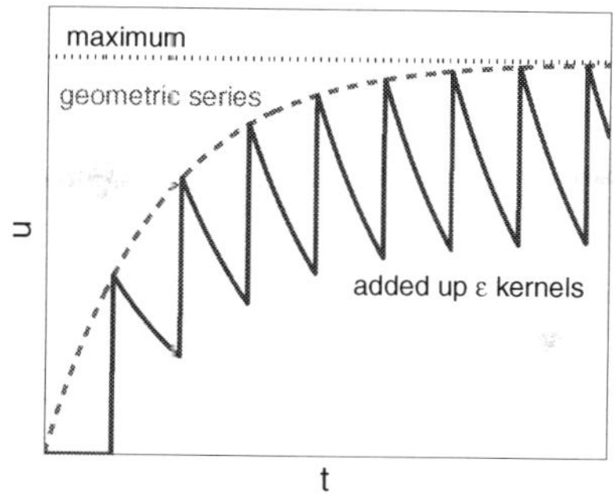

Fig. 1. Sum of equidistant ϵ kernels. The maximum line (··) denotes the limit of the local maxima for infinite many spikes. The right part on the right side of equation (2) describes the maxima of the added up ϵ kernels (–), while the left part the decrease between such two maxima. The maxima lie on the geometric series function (--).

and with the approximations gotten from limit analysis

$$\frac{a^{\frac{1}{2n}}}{1-a^{\frac{1}{n}}} \approx -\frac{1}{\ln(a)} n \qquad , \quad 0 < a < 1 \tag{3}$$

$$\frac{1}{1-a^{\frac{1}{n}}} \approx -\frac{1}{\ln(a)} n + \frac{1}{2} \qquad , \quad 0 < a < 1 \tag{4}$$

and $\Delta^{\text{isi}} = T/n$ further simplified to

$$\begin{aligned} u_i(T) &\overset{(3)}{\approx} -u_{\text{reset}} \tfrac{1-\exp(-T/\tau_i)}{T/\tau_i} n_i + \textstyle\sum_j w_{ij} \tfrac{1-\exp(-T/\tau_i)}{T/\tau_i} n_j \\ &\overset{(4)}{\approx} -u_{\text{reset}} \tfrac{1}{1+\frac{1}{2}\frac{T}{\tau_i}} n_i + \textstyle\sum_j w_{ij} \tfrac{1}{1+\frac{1}{2}\frac{T}{\tau_i}} n_j \quad . \end{aligned} \tag{5}$$

Equation (5) can now easily be solved for the output spike count n_i. However, n_i depends not only on the input spike counts n_j but on the membrane potential u_i of the neuron as well. To get rid of the membrane potential we have to do several estimations.

Let us temporarily assume, that all PSP are excitatory, the reset potential is equal to the threshold and that the absolute refractory period is small enough to ensure that the neuron is not refractory at time T. Then the following conditions hold:

$$u_i(T) \geq 0 \tag{6}$$

$$u_i(T) < \vartheta_i \tag{7}$$

Based on these conditions and given the input activities n_j, we can compute an upper and a lower bound for the output activity n_i

$$u_i(T) \geq 0 \Leftrightarrow n_i \leq \sum_j \frac{w_{ij}}{u_{\text{reset}}} n_j$$

$$u_i(T) < \vartheta \Leftrightarrow n_i > -\frac{\vartheta_i}{u_{\text{reset}}} \left(1 + \frac{1}{2}\frac{T}{\tau_i}\right) + \sum_j \frac{w_{ij}}{u_{\text{reset}}} n_j$$

However, to give a good estimation based on these boundaries and to minimize the average failure, we choose as gain function the function

$$n_i = -\frac{1}{2}\frac{\vartheta_i}{u_i^{\text{reset}}}\left(1+\frac{1}{2}\frac{T}{\tau_i}\right)+\sum_j \frac{w_{ij}}{u_i^{\text{reset}}}n_j \qquad (8)$$

that lies exactly in the middle between those bounds.

In practice there will be inhibitory PSP, arbitrary reset potentials and non-equidistant distributed spikes. Hence we have to discuss their effects. Reset potentials greater than the threshold and inhibitory PSP behave nearly in the same way as excitatory PSP according to equation (2), but with a reversed algebraic sign. With inhibitory PSP and larger reset potentials, condition (6) and its associated upper bound can not strictly be uphold, because the membrane potential u_i can of course be smaller than zero. This would mean, that n_i is negative. However, since n_i can not be smaller than 0, this situation is easy to deal with. A further problem is the timing of inhibitory PSP, because they only have an effect on succeeding PSP, not preceding ones. In the case that all excitatory PSP take place at the beginning of an interval and the inhibitory ones at the end, this would lead to false approximation of n_i. The effect of non-equidistant spikes is a timing problem as well and analytical hard to grasp too. However, the simulated output n_i of a neuron where inhibitory PSP, arbitrary reset potentials and non-equidistant distributed spikes were used, shown in Figure 2, indicate that the approximation (8) of the gain function is quite robust. Mind that the assumption about the computation mode is still uphold.

Speaking in terms of artificial neural networks, equation (8) describes a linear base function (LBF) quite similar to the one used in sigmoidal neurons. Artificial neurons

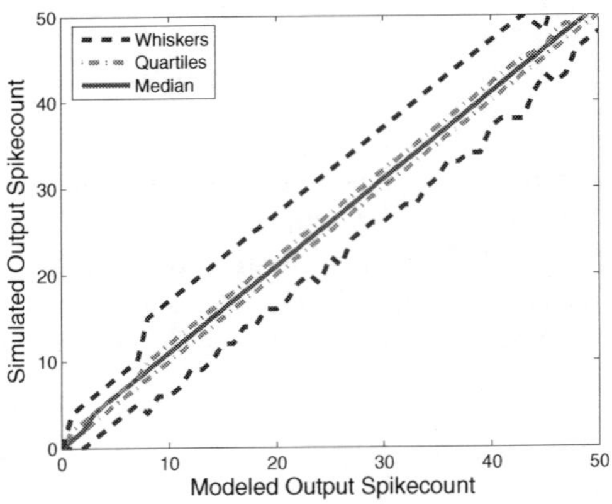

Fig. 2. Simulated spike count of the output neuron vs. modeled spike count according to equation (8). Five input neurons were connected to one output neuron. The parameters were randomly chosen to match equation (8) for a given n_i. The simulation results are given as "continuous" box plot. The whiskers were set for $2.75 \cdot$ inter-quartile range so that the number of outliers is less then 5%. The input spikes were randomly distributed in a $T = 200$ time interval. Further parameters: Input spike rate $n_j \in [1, 50], j = 1 \ldots 5$, membrane time constant $\tau \in [20, 200]$, threshold $\vartheta \in (0.5, 1.5)$, reset potential $u^{\text{reset}} \in (0.5, 2.0)$ and synaptic weight $w_{ij} \in [-2, 2]$.

also have a squashing function often called transfer function. Here, this is implemented by the absolute refractory period Δ^{arp}. For a given time interval T and a Δ^{arp}, the maximum spike rate can not exceed $\left\lfloor \frac{T}{\Delta^{\text{arp}}} \right\rfloor + 1$. The result is a saturated limited function as squashing function.

4 The Back-Propagation Algorithm

The back-propagation algorithm developed in the following is an online version of the gradient descent error back-propagation algorithm proposed by [5].

Given a training example $\left(t^{\text{in}}, t^{\text{out}}\right)$, an output spike rate n^{out} from the simulation and the error function $E\left(t^{\text{out}}, n^{\text{out}}\right)$, a parameter is adapted in the opposite direction of the error gradient $\frac{\partial E}{\partial param}$

$$param^{\text{new}} = param^{\text{old}} - \eta_{\text{param}} \frac{\partial E}{\partial param} \tag{9}$$

where η_{param} is the learning rate. The error gradient for neuron i is computed as

$$\frac{\partial E}{\partial param_i} = \delta_i \frac{\partial \, \text{net}_i}{\partial param_i} \tag{10}$$

with

$$\delta_i = \frac{\partial s}{\partial \, \text{net}_i} \begin{cases} 2\frac{\left(n_i^{\text{out}} - t_i^{\text{out}}\right)}{n_{\max}} & \text{, if } i \text{ output neuron} \\ \sum_{k \in \text{succ}(i)} \delta_k \frac{w_{ki}}{u_k^{\text{reset}}} & \text{, otherwise} \end{cases} \tag{11}$$

net_i is the net (or base) function of neuron i according to equation (8), s the squashing function, n_i^{out} and n_i the output after the transfer function and t_i^{out} the desired target output. The division by the maximal possible output rate $n_{\max}$ normalizes the back-propagated error.

Before computing $\frac{\partial \, \text{net}_i}{\partial param_i}$ and the resulting changes, we use the transformations

$$\hat{u}_i^{\text{reset}} = \frac{1}{u_i^{\text{reset}}} \quad \text{and} \quad \hat{\tau}_i = \frac{T}{\tau_i}$$

to avoid problems with nonlinear changes and complicated derivatives. Now, we get the formulas

$$\frac{\partial \, \text{net}_i}{\partial w_{ij}} = \hat{u}_i^{\text{reset}} n_j \tag{12}$$

$$\frac{\partial \, \text{net}_i}{\partial \hat{u}_i^{\text{reset}}} = -\frac{1}{2}\vartheta_i \left(1 + \frac{1}{2}\hat{\tau}_i\right) + \sum_j w_{ij} n_j \tag{13}$$

$$\frac{\partial \, \text{net}_i}{\partial \hat{\tau}_i} = -\frac{1}{4}\vartheta_i \hat{u}_i^{\text{reset}} \tag{14}$$

$$\frac{\partial \, \text{net}_i}{\partial \vartheta_i} = -\frac{1}{2}\hat{u}_i^{\text{reset}} \left(1 + \frac{1}{2}\hat{\tau}_i\right) \tag{15}$$

that allow us to compute error gradient for a specific parameter according to equation (10).

As the net function, eq. (8), indicates we have to normalize the resulting error gradients with n_{max} to make the training independent of the chosen coding range. This normalization of the net function is additional to the normalization of the error in equation (11). After computing the changes u_i^{reset} and τ_i have to be transformed back again.

In contrast to the back-propagation algorithm in ANN the presented algorithm adapts not only synaptic weights w_{ij} but other neural parameters ($\hat{u}_i^{\text{reset}}$, $\hat{\tau}_i$ and ϑ_i) of the SRM as well.

5 Experimental Setup

The benchmark suite PROBEN1 [9] was used for the test of the back-propagation algorithm. PROBEN1 is a collection of revised benchmarks from the UCI Machine Learning Repository consisting of classification and regression problems.

Input attributes of a benchmark problem can be real-valued, integer-valued, ordinal, nominal or even missing. Luckily in PROBEN1 each input data is already overhauled to fit a coding range of [0, 1]. For the rate coding SNN the input and output data had to be rescaled to fit a coding range [0, n_{max}] based on spike counts. Since spike counts are integer-valued the scaled data were rounded. We set n_{max} to 25. The spikes in the spike trains were equidistant distributed.

To test the classification performance we chose three different problems: The "Wisconsin breast cancer" (cancer), "Pima Indian diabetes" (diabetes) and "splice junction" (gene) problem.

In the cancer problem a diagnosis of breast cancer has to be made based on 9 real-valued inputs. The benchmark consists of 699 examples where 65.5% of the examples are benign. The diabetes problem deals with the diagnosis of diabetes of Pima Indians based on 8 real-valued inputs. It consists of 768 examples where 65.1% are diabetes positive. In the last classification problem gene, the job is to detect intron/exon boundaries in nucleotide sequences of length 60. There are three classification outcomes: Intron/exon (donor), exon/intron (acceptor) and none. Each nucleotide is a 4-valued nominal attribute, therefore each example has 120 binary inputs. The benchmark consists of 3175 examples where 25% are donors and 25% are acceptors.

For the purpose of cross-validation to prevent over-fitting, the data for each benchmark problem was split in three disjoint sets for training, validation and test. The training set got 50% of the data and the validation and test set each 25%. Where the data size was not divisible by 4 the training and validation set were prioritized.

In his paper about PROBEN1 L. Prechelt used the RPROP back-propagation algorithm [10] to train the networks. However in this case it makes no sense to use RPROP, because all neural and synaptic parameters except the weights can not be negative. The adaptation would only depend on the back-propagation term δ_i which is the same for all parameters of a given neuron. Therefore we used the simple back-propagation algorithm described in the previous section in equation (9). The learning rates were set to $\eta_\tau = 1$, $\eta_w = 0.01$, $\eta_\vartheta = 1$ and $\eta_{u_{\text{reset}}} = 0.01$. Further the domains were restricted to $(0, \infty)$ for τ and $[0.5, 1.5]$ for u_{reset} and ϑ to keep these parameters in reasonable ranges.

Table 1. Classification results for networks with shortcuts. Rows with white background show the results for the SNN, rows with gray background the results reported in [9]. *Architecture*: network topology input-hidden1-...-hiddenN-output, *error*: error function of the best run, *class.*: related classification error, *epochs*: range of epochs over 10% best runs.

	Architecture	Validation		Test		Epochs
		error	class.	error	class.	
cancer1	9-4-2-2	3.90	2.29	2.82	1.15	92 - 111
[9]		1.53	1.71	1.05	1.15	75 - 205
diabetes1	8-2-2-2	20.24	33.33	22.59	37.50	3 - 36
[9]		15.07	19.79	16.47	25.00	65 - 525
gene1	120-2-3	8.03	11.08	8.63	13.24	1 - 11
[9]		9.71	12.72	10.11	13.37	30 - 1245

Table 2. Classification results for networks w/o shortcuts. Explanation from Table 1 applies.

	Architecture	Validation		Test		Epochs
		error	class.	error	class.	
cancer1	9-4-2-2	4.64	4.00	2.83	2.87	13 - 28
[9]		1.89	-	1.32	1.38	116 ± 123
diabetes1	8-2-2-2	23.10	36.98	22.93	36.46	3 - 44
[9]		15.93	-	16.99	24.10	201 ± 119
gene1	120-2-3	14.91	47.23	14.87	46.78	2 - 13
[9]		8.19	-	8.66	16.67	124 ± 53

Table 3. Classification results for networks w/o hidden layers. Explanation from Table 1 applies.

	Architecture	Validation		Test		Epochs
		error	class.	error	class.	
cancer1	9-2	12.41	4.57	10.16	6.32	7 - 17
[9]		2.91	-	3.52	2.93	129 ± 13
diabetes1	8-2	20.18	32.81	22.35	36.46	4 - 36
[9]		16.30	-	17.22	25.83	209 ± 50
gene1	120-3	8.10	11.34	8.39	12.48	1 - 11
[9]		9.58	-	9.92	13.64	47 ± 6

We used the mean squared error (MSE) as error function. The learning stopped after maximal 300 epochs or if for ten epochs no new best validation error was achieved. In [9] a "generalization loss" criterion was applied instead. However in this case it led to premature stops, so the simpler stopping criterion above was chosen.

The net topologies were taken from [9]. We tested three different types of multilayer feed-forward topologies for each benchmark. One topology with all possible shortcuts, one without any shortcut and the last without any hidden layers at all (linear network).

To get reliable results we performed 60 runs for each experiment.

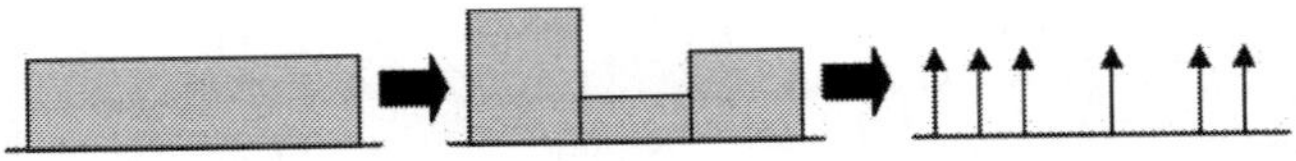

Fig. 3. Break-up of a spike count on a time interval in several smaller spike counts. For infinitesimal small time intervals one get the single spikes.

6 Results and Discussion

The classification results are shown in Tables 1 to 3. The column "Architecture" describes the layer topology including input and output layer. For instance the neural network with shortcuts for the cancer classification problem (Table 1) has two hidden layers. For the validation set there are two columns in the tables. One for the validation error computed by the MSE function depending on the number of spikes and one for the classification error in percent on the validation set. Only the results for the best run are shown where the best run is the one with the best validation error. The same description holds for the test set. The last column shows the range of the training epochs of the 10% best results over the 60 runs. The shown epochs are short by 10 because the stopping criterion is considered.

The tables show a mixture of successful and failed classification results compared to the benchmark results from PROBEN1. For instance, the linear network and the network with shortcuts perform on the gene benchmark even better than the ANN in PROBEN1. However, the neural network without shortcuts utterly fails on this benchmark. Then, why did some classifications fail? The approximation of the gain function can not be the only reason. First because there are successful classifications, e.g. the cancer problem on the NN with shortcuts. And secondly, because there are successful classifications on the linear neural network *and* the neural network without shortcuts. So the reason can not be the error back-propagation over several layers based on a skewed error provoked by the approximation. Looking at the number of training epochs, poorly chosen training parameters η_{param} and initialization of network parameters seems more plausible. Whether this guess is correct or not is part of our current research.

7 Conclusion

In this paper we showed an approximation of the gain function of a simplified version of the Spike Response Model as used by an other group [7]. Based on this gain function a spiking neural network was trained under the assumption that information is encoded by rate or frequency of spikes. This approach allowed the usage of an error back-propagation algorithm for the training.

The results showed that this approach for learning works in principle. However there are some open questions. For instance, how does a different discretization, a consequence of a different chosen absolute refractory period, affect the learning and classification performance? How much contributes the approximation error to the training error? What are good training parameters and initialization values? Can this approach applied to other neuron models like the Leaky-Integrate-&-Fire Model as well? To find the answer to these questions is the goal of our future work.

The project does not end here however. Even the possible improvements of the back-propagation algorithm will result in quite rough estimations of the position of single spikes due to used rate coding. To achieve better estimations of exact positions of spikes we have to bring time back in game. Our idea is to split up the single interval where spikes are counted in a sequence of smaller intervals. The result is a sequence of spike rates and, if the intervals get small enough, even single spikes (see Figure 3). If we consider the synaptic delays we will get a neural network similar to time-delayed neural networks (TDNN) [11] but based on spiking neurons. For TDNN there already exists a back-propagation algorithm [12] that does not unfold the network like Waibel et al. [11] did. Combining our findings with this algorithm, the result will be a powerful back-propagation algorithm for the estimation of neural and synaptic parameters.

References

1. Kasinski, A., Ponulak, F.: Comparison of supervised learning methods for spike time coding in spiking neural networks. Int. J. of Applied Mathematics and Computer Science, University of Z.Gora 16, 101–113 (2006)
2. Bohte, S.M., Poutre, H.L., Kok, J.N.: Error-backpropagation in temporally encoded networks of spiking neurons. Neurocomputing 48, 17–37 (2002)
3. Belatreche, A., Maguire, M.M.L.P., Wu, Q.X.: A method for supervised learning of spiking neural networks. In: Proc. of IEEE Cybernatics Intelligence - Challenges and Advances, pp. 39–44. IEEE Computer Society Press, Los Alamitos (2003)
4. Ruf, B., Schmitt, M.: Learning temporally encoded patterns in networks of spiking neurons. Neural Processing Letters 5, 9–18 (1997)
5. Rumelhart, D.E., Hinton, G.E., Williams, R.J.: Learning representations by back-propagating errors. Nature 323, 533–536 (1986)
6. Gerstner, W., Kistler, W.: Spiking Neuron Models. Cambridge University Press, Cambridge (2002)
7. Marian, I.: A biologically inspired model of motor control of direction. Master's thesis, Department of Computer Science, University College Dublin (2002)
8. König, P., Engel, A.K., Singer, W.: Integrator or coincidence detector? the role of the cortical neuron revisited. Trends in Neuroscience 19, 130–137 (1996)
9. Prechelt, L.: Proben1 - a set of neural network benchmark problems and benchmarking rules. Technical report, Fakultät für Informatik, Universität Karlsruhe (1994)
10. Riedmiller, M., Braun, H.: Rprop - a fast adaptive learning algorithm. Technical report, Fakultät für Informatik, Universität Karlsruhe (1992)
11. Waibel, A., Hanazawa, T., Hilton, G., Shikano, K., Lang, K.J.: Phoneme recognition using time-delay neural networks. IEEE Transactions on Acoustics, Speach, and Signal Processing 37, 328–339 (1989)
12. Wan, E.A.: Finite impulse response neural networks for autoregressive time series prediction. In: Weigend, A., Gershenfeld, N. (eds.) Proceedings of the NATO Advanced Workshop on Time Series Prediction and Analysis, Santa Fe, NM, May 14-17, 1993, Addison-Wesley, Reading (1993)

Clustering of Nonlinearly Separable Data Using Spiking Neural Networks

Lakshmi Narayana Panuku and C. Chandra Sekhar

Department of Computer Science and Engineering,
Indian Institute of Technology Madras, Chennai 600 036, India
{panuku,chandra}@cs.iitm.ernet.in

Abstract. In this paper, we study the clustering capabilities of spiking neural networks. We first study the working of spiking neural networks for clustering linearly separable data. Also, a biological interpretation has been given to the delay selection in spiking neural networks. We show that by varying the firing threshold of spiking neurons during the training, nonlinearly separable data like the ring data can be clustered. When a multi-layer spiking neural network is trained for clustering, subclusters are formed in the hidden layer and these subclusters are combined in the output layer, resulting in hierarchical clustering of the data. A spiking neural network with a hidden layer is generally trained by modifying the weights of the connections to the nodes in the hidden layer and the output layer simultaneously. We propose a two-stage learning method for training a spiking neural network model for clustering. In the proposed method, the weights for the connections to the nodes in the hidden layer are learnt first, and then the weights for the connections to the nodes in the output layer are learnt. We show that the proposed two-stage learning method can cluster complex data such as the interlocking cluster data, without using lateral connections.

1 Introduction

Spiking neural networks (SNNs) are a class of ANNs that is increasingly receiving the attention as both a computationally powerful and biologically plausible mode of computation [1] [2]. SNNs model the precise time of the spikes fired by a neuron, as opposed to the conventional neural networks which model only the average firing rate of the neurons. It is proved that the neurons that convey information by individual spike times are computationally more powerful than the neurons with sigmoidal activation functions [3]. In particular, a computational model in which the values of the input variables are represented by the explicit times at which action potentials occur, rather than by the firing rates of neurons, has been proposed [4]. This model has successfully been used to give simple interpretation to mammalian olfactory systems.

In [5], a Hebbian based learning mechanism has been proposed for SNNs with multi-delay paths between neurons in adjacent layers. This learning mechanism has been observed to select the connections with matching delays. This approach was extended in [6] by considering the firing times, and a learning algorithm that performs unsupervised clustering was proposed. This model encodes the input patterns in the delays of its synapses and is shown to reliably find the centers of clusters in high dimensional spaces.

J. Marques de Sá et al. (Eds.): ICANN 2007, Part I, LNCS 4668, pp. 390–399, 2007.

However, this method is limited in both cluster capacity as well as precision. Bohte *et al.* [7], presented a method for encoding the input data to enhance the precision, capacity and clustering capability of a network of spiking neurons similar to [6] in a flexible and scalable manner. With such encoding, it has been shown that SNNs are able to form clusters at low cost in terms of neurons while enhancing capacity and precision. By extending the network to multiple layers and adding lateral excitatory connections with a self-organizing map like learning rule, it has been shown that correct clustering of nonlinearly separable data such as the interlocking cluster data can be achieved.

In this paper, we first give a biological interpretation to the working of SNNs, while presenting our studies on clustering linearly separable data. When an SNN with multiple delay paths between neurons in adjacent layers is trained to cluster linearly separable data, the Hebbian based winner-takes-all learning rule is observed to select the appropriate delays by driving the weights of those delay terminals to w_{max} and the weights of the remaining delay terminals to 0. This observation is in agreement with the fact that, in biological neural networks different axonal connections will have different signal transmission delays [8]. This observation also re-emphasizes the studies made by Gerstner *et al.* [5], on the auditory system of barn owl, where the unsupervised Hebbian learning rule selects connections with matching delays from a broad distribution of axons with random delays. When an SNN with fixed threshold units (as in [7]) is trained to cluster the nonlinearly separable data like the ring data and the interlocking cluster data, all the examples have been clustered into one group. We propose a model in which the threshold of a spiking neuron is initialized to a small, positive value and is increased gradually as the learning proceeds until it reaches a maximum value. It is demonstrated that the proposed method properly clusters the ring data. When a three-layer SNN is trained to cluster the ring data, it is observed that a multi-layer SNN with a suitable choice of neurons in the layers could perform hierarchical clustering. By decreasing the neuronal population for subsequent layers, hierarchical clustering with decreasing granularity is achieved. A three-layer SNN with lateral connections in the hidden layer is trained in [7] for clustering the interlocking cluster data. We propose, a two-stage learning method to cluster the interlocking cluster data without using lateral connections , there by gaining computational advantage over [7]. In general, in a multi-layer SNN all the layers are trained simultaneously. In the proposed method the layers are trained sequentially. The performance of these two methods for clustering the interlocking cluster data is studied.

This paper is organized as follows: The architecture of a spiking neural network with multiple delay connections is described in Section 2. The learning rule for clustering and the encoding mechanism for converting the real valued inputs into time vectors are also presented in this section. Clustering of linearly separable data using the population encoding and the SNN model is described in Section 3. In Section 4, an adaptive threshold based method is proposed for clustering the nonlinearly separable data. The results of the proposed method on the ring data and the interlocking cluster data are presented. Section 5 discusses the hierarchical clustering capabilities of the multi-layer SNN and a two-stage learning method, and presents the results for clustering the interlocking cluster data.

2 Networks of Spiking Neurons with Multiple Delay Connections

The spiking neural network architecture consists of a fully connected feedforward network of spiking neurons with connections implemented as multiple delayed synaptic terminals, as shown in Fig. 1(a). A connection from pre-synaptic neuron i to post-synaptic neuron j consists of a fixed number of synaptic terminals. Each terminal serves as a subconnection that is associated with a different delay and weight. The delay d^k of a synaptic terminal k is the difference between the firing time of the pre-synaptic neuron, and the time when the post-synaptic potential (PSP) resulting from terminal k starts raising.

The time varying impact of the pre-synaptic spike on a post-synaptic neuron is described by a spike response function, $\varepsilon(.)$, also referred to as the PSP. A neuron j in the network generates a spike when the value of its internal state variable x_j, the "membrane potential", crosses a threshold ϑ. The internal state variable $x_j(t)$ is defined as follows:

$$x_j(t) = \sum_{i \in \Gamma_j} \sum_{k=1}^{m} w_{ij}^k \, \varepsilon\left(t - t_i - d^k\right) \tag{1}$$

where Γ_j is the set of pre-synaptic neurons for the neuron j, m is the number of synaptic terminals between the neurons i and j, w_{ij}^k is the weight of the kth synaptic terminal between the neurons i and j, and t_i is the firing time of the pre-synaptic neuron i. The time at which $x_j(t)$ crosses the threshold ϑ with a positive slope is the firing time of the neuron j, denoted by t_j.

2.1 Learning

For clustering, the weights of the terminals of the connections between the input neurons and the winning neuron, i.e., the output neuron that fires first, are modified using a time-variant of Hebbian learning. The learning rule for the weight w_{ij}^k of the synaptic terminal with delay d^k is as follows:

$$\Delta w_{ij}^k = \eta L(\Delta t) = \eta\left[(1-b)\exp\left(-(\Delta t - c)^2/\beta^2\right) + b\right], \tag{2}$$

where Δt denotes the time difference between the onset of PSP at the kth synaptic terminal, t_{PSP}, and the firing time of the winning neuron, t_j, i.e, $\Delta t = t_{PSP} - t_j = (t_i + d^k) - t_j$. The parameter c determines the position of the peak, b determines the negative update given to a neuron for which Δt is significantly different from c and β sets the width of the positive part of the learning function (Fig. 1(b)). The weights are limited to the range 0 to w_{max}, the maximum value that a weight can take. For a connection, the number of consecutive delayed synaptic weights that are positively reinforced is determined by the parameter β.

2.2 Encoding

When the input variables are continuous valued attributes, it is necessary to encode the values of an input variable into firing times of neurons in the input layer of SNN.

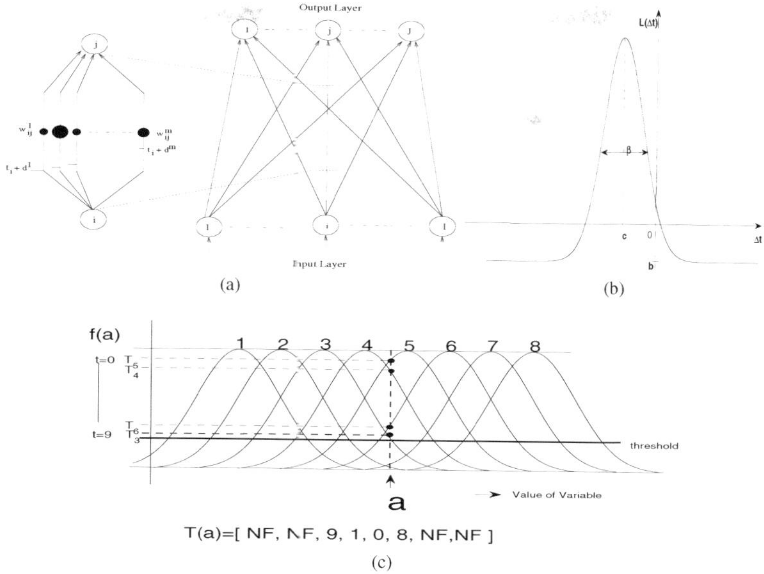

Fig. 1. (a) Architecture of a spiking neural network. (b) Learning function $L(\Delta t)$. (c) Encoding an input variable with value a into firing times, $T(a)$, using overlapping GRFs.

Bohte *et al.* [7], proposed the population coding method that encodes an input variable using multiple overlapping Gaussian receptive fields (GRFs). Each GRF has a center and a width. Let the value of an input variable be a. The activation value of the GRF with center μ and width σ is given by

$$f(a) = \exp\left(-(a-\mu)^2/(2*\sigma^2)\right) \tag{3}$$

The firing time of the neuron associated with this GRF is inversely proportional to $f(a)$. In our experiments the firing time of an input neuron is in the range 0 to 9. For a highly stimulated GRF, with value of $f(a)$ close to 1.0, the firing time $t = 0$ is assigned. When the activation value of the GRF is small, the firing time t is high indicating that the neuron fires later. The firing times of the neurons associated with different GRFs for an input variable with the value a are shown in Fig. 1(c). The population coding is helpful in increasing the distance between the encoded time vectors associated with the input data, and hence the separability of the clusters in the data. For multi-dimensional data, the value of each input variable is encoded separately.

The range of values for each input variable is determined. Each of the input variables is encoded using GRFs covering the range of values for that variable. For a range $[I_{min}...I_{max}]$ of a variable, $n(>2)$ GRFs are used. For the ith GRF, the center of the Gaussian is set to $\mu = I_{min} + ((2i-3)/2)((I_{max} - I_{min})/(n-2))$. One GRF is placed outside the range at each of the two ends. The width of the GRF is set to $\sigma = (1/\gamma)((I_{max} - I_{min})/(n-2))$, where γ controls the extent of overlap between the GRFs. By decreasing the width of the receptive field and increasing the number of GRFs, narrow clusters can also be separated.

While converting the activation values of GRFs into firing times, a threshold has been imposed on the activation value. A receptive field that gives an activation value less than this threshold will be marked as not-firing (NF) and the corresponding input neuron will not contribute to the post-synaptic potential.

For clustering the data with different scales (i.e., data containing wide and narrow clusters), GRFs with different widths have been used for encoding. This technique is called multi-scale encoding. The width σ_t of the narrow Gaussians is set to $(1/\gamma)(I_{max} - I_{min})/(n - 2)$, with $\gamma = 1.5$. The width σ_b of the broad Gaussians is set to $(1/\gamma)(I_{max} - I_{min})/(n + 1)$, with $\gamma = 0.5$. The narrow Gaussians are distributed as outlined earlier. The broad Gaussians are all evenly placed with their centers inside the respective data ranges, with the center of the ith Gaussian placed at $I_{min} + i.((I_{max} - I_{min})/(n + 1))$.

3 Clustering of Linearly Separable Data Using SNN

To understand how the Hebbian based winner-takes-all learning selects the matching delays, an SNN with multiple delay connections has been trained to cluster two linearly separable clusters. The data is encoded with three Gaussians along the x-axis and three Gaussians along the y-axis, resulting in a six dimensional encoded time vector, as shown in Fig. 2(a). The encoded time vectors for some randomly selected examples are shown in Table 1. An SNN having six input neurons, two output neurons and nine delay connections (with 1 milli-second resolution) has been trained to cluster the data. The resulting weights of the multi-delay connections from the first three input neurons (i_1, i_2, i_3) to the two output neurons (j_1, j_2) are shown in Fig. 2(b).

Let us consider the input neurons i_1, i_2, and i_3. The observations can be extended to i_4, i_5, and i_6. Input neuron i_2 fires first for the points belonging to $cluster1$ and also for the points belonging to $cluster2$, providing no discriminatory information. Hence the final weight distributions of the connections from i_2 to both the output neurons are similar (see Fig. 2(b)). For the points of $cluster1$, i_1 fires earlier than i_3 and for the points of $cluster2$, i_3 fires earlier than i_1. After training, this pattern is seen in the final weights of the connections, resulting in proper clustering of the data.

In biological neural networks, the signal transmission delays occur because considerable portions of the axons are very fine myelinated or unmyelinated. Therefore, the signal transmission delays are different for different axonal connections [8]. From

Table 1. Encoded firing time vectors for some examples

Cluster	Example No.	i_1	i_2	i_3	i_4	i_5	i_6
1	1	3	1	8	5	0	8
	2	4	1	8	5	0	7
	3	3	1	8	3	1	NF
2	4	8	1	3	8	1	4
	5	8	1	3	8	1	3
	6	NF	1	3	8	0	4

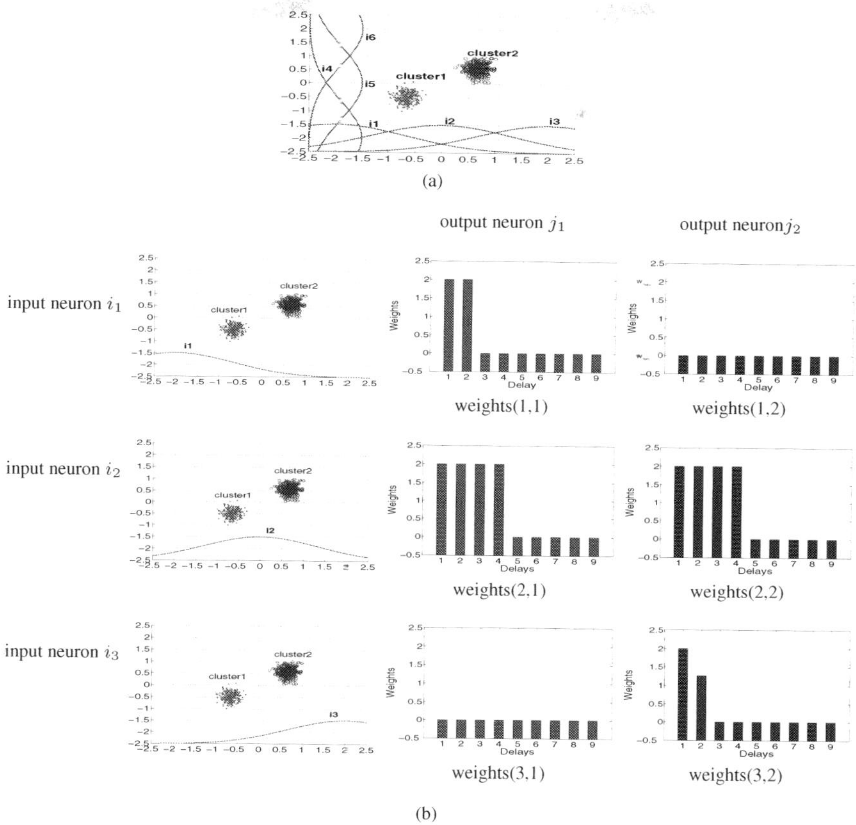

Fig. 2. (a) GRFs for encoding the attributes of 2-D linearly separable clusters. (b) Final weights of the delay connections from input neurons to output neurons.

Fig. 2(b), it is observed that the Hebbian based winner-takes-all learning selects the appropriate delays for every connection, effectively storing the patterns present in the data. It is also seen that the learning rule is driving the individual weights of the synaptic terminals to one of the extremes, 0 or w_{max}. It has also been observed that the formation of final clusters depends on the encoding mechanism, i.e., how well the Gaussians represent the input data.

4 Clustering of Nonlinearly Separable Data Using SNN

Clustering is an unsupervised method for grouping similar data. Clustering algorithms attempt to organize unlabeled feature vectors into clusters or natural groups such that samples within a cluster are more similar to each other than to samples belonging to different clusters. Many clustering algorithms have been proposed in the literature [9]. Most of them work well when the underlying clusters exhibit linearly separable shapes [10]. We consider two examples of highly nonlinearly separable data to demonstrate the

Table 2. Performance of a two-layer SNN in clustering (a) the ring data and (b) the interlocking cluster data

(a)

c	$\Delta\vartheta$	ϑ_{max}	Performance (in %)
-2.0	0.01	5.0	80.47
-2.0	0.001	5.0	82.30
-2.0	0.01	10.0	100.0
-2.0	0.001	10.0	77.73
-2.5	0.01	5.0	100.0
-2.5	0.001	5.0	78.27

(b)

c	$\Delta\vartheta$	ϑ_{max}	Performance (in %)
-2.0	0.001	30.0	66.9
-2.0	0.001	40.0	67.0
-2.5	0.01	30.0	70.6
-2.5	0.01	40.0	67.0
-2.5	0.001	30.0	82.9
-2.5	0.001	40.0	83.0

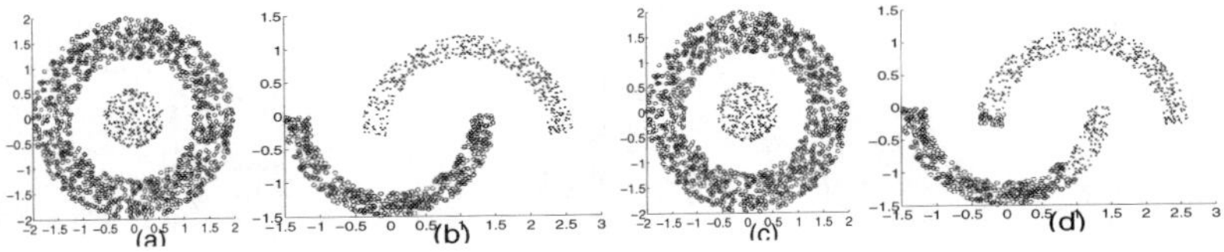

Fig. 3. Scatter plot of (a) the ring data and (b) the interlocking clusters data. Clusters formed by a two-layer SNN for (c) the ring data and (d) the interlocking cluster data.

capabilities of the SNNs in clustering nonlinearly separable data. In the first example, we consider the ring data consisting of data points in 2-dimensional space as shown in Fig. 3(a). The data points in the inner cluster are centered at origin and the points in the outer cluster form an annular ring. The variance of the data in the two clusters is different and the mean of the data in each of the clusters is almost the same. It is seen that the inner and outer clusters of the ring data can be separated by a circular shaped curve. We consider another example of nonlinearly separable data corresponding to two interlocking clusters, shown in Fig. 3(b). The data of two clusters is separable by an 'S' shaped curve.

In [7], an SNN model with fixed threshold units is considered for the task of clustering. This model could cluster linearly separable data. However for clustering the interlocking cluster data, Bohte *et al.* used lateral connections in the hidden layer. When an SNN without lateral connections and having fixed threshold units is trained to cluster nonlinearly separable data like the ring data and the interlocking cluster data, all the examples have got clustered into one group. To overcome this limitation, we propose a varying threshold method. Using this method, we are able to cluster the ring data and the interlocking cluster data without using lateral connections, there by reducing the computational cost. In the proposed method, the threshold of a spiking neuron is initialized to a small, positive value. During learning, the threshold of the wining neuron is increased by $\Delta\vartheta$. The threshold of all the neurons in a layer are limited to a maximum value, ϑ_{max}. Here $\Delta\vartheta$ is a tunable parameter. High values of $\Delta\vartheta$ make the spiking neurons equivalent to fixed threshold spiking neurons. On the other hand, very low values of $\Delta\vartheta$ require large number of iterations to train the network. In our studies, three values of $\Delta\vartheta$ $(0.01, 0.001$ and $0.0001)$ have been used.

An SNN with two layers (input layer and output layer) has been considered for clustering the ring data and the interlocking cluster data. The ring data is encoded using 5 Gaussians along each dimension and an SNN with 10 neurons in the input layer and 2 neurons in the output layer is trained to cluster this data. The performance of the proposed method for different parameter values is given as the accuracy of assigning the data points to their clusters, in Table 2(a). By empirically selecting the suitable parameters, the ring data is properly clustered as shown in Fig. 3(c). The interlocking cluster data is encoded using 9 narrow and 3 broad Gaussians along each dimension. An SNN with 24 neurons in the input layer and 2 neurons in the output layer is trained on this data. It has been observed that the interlocking cluster data is not properly clustered (see Fig. 3(d)) and the best performance is given by the parameters $c = -2.5$, $\vartheta = 0.001$ and $\vartheta_{max} = 40$ (Table 2(b)). This suggests the need for considering a multi-layer SNN. In the following section, we consider a multi-layer SNN and propose a two-stage learning method to cluster the interlocking cluster data, and also explore the hierarchical clustering in SNNs.

5 Hierarchical Clustering Using SNN

We study how a three-layer SNN can perform hierarchical clustering. In a three-layer SNN, it is shown that the neurons in the hidden layer represent the subclusters of the clusters. These subclusters are combined in the output layer to form final clusters, yielding a decreasing level of granularity. A three-layer SNN with 10 neurons in the hidden layer and 2 neurons in the output layer is trained to cluster the ring data. Each neuron in the hidden layer represents a subcluster, as shown with a different colour in Fig. 4(a). The average firing times of the neighbouring neurons for the boundary points are also given in Fig. 4(a). It is observed that the neighbouring neurons are firing in synchrony, in terms of firing times, for the boundary points between the subclusters represented by these neurons. This synchronization information is used in the next layer to bind the subclusters together to form the final clusters (Fig. 4(b)). The parameters used in training the network and the performance are given in Table 3(a). In [7], lateral connections with a self-organizing map like learning rule have been used in the hidden layer to bind the neurons together. Here we demonstrate that using the proposed varying threshold method, the binding of neurons in the hidden layer can be done without using the lateral connections.

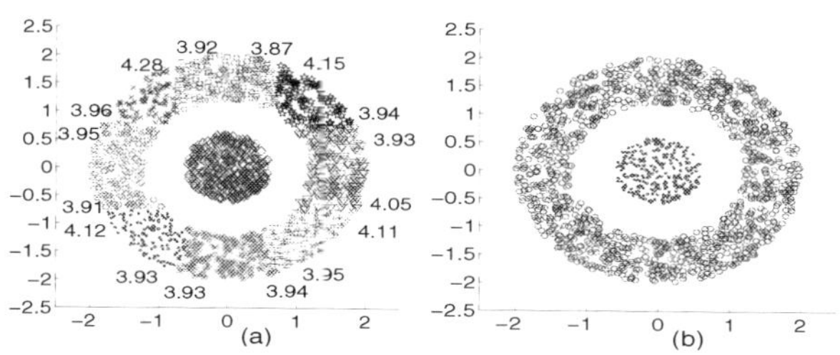

Fig. 4. Hierarchical clustering for the ring data in a three-layer SNN: (a) formation of subclusters in the hidden layer and (b) formation of clusters in the output layer

Table 3. Parameters used in training SNN and the performance for clustering (a) the ring data and (b) the interlocking clusters data

(a)

Data	Parame -ters	Hidden Layer	Output Layer	Performa -nce (%)
Ring data	neurons	10	2	100.00
	c	-2.85	-2.5	
	β	1.60	1.67	
	ϑ_{max}	5.0	15.0	
	$\Delta\vartheta$	0.005	0.0001	

(b)

Data	Parame -ters	Hidden Layer	Output Layer	Performa -nce (%)
Inter-locking cluster data	neurons	10	2	87.10
	c	-2.0	-1.0	
	β	1.50	1.50	
	ϑ_{max}	30.0	3.0	
	$\Delta\vartheta$	0.005	0.0001	

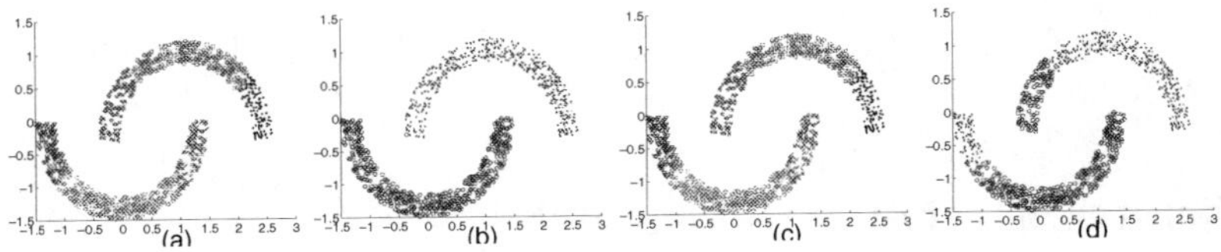

Fig. 5. Clustering the interlocking cluster data using a three-layer SNN with parameters as in Table 4. (a) subclusters formed for two-stage learning, (b) clusters formed for two-stage learning, (c) subclusters formed for one-stage learning, (b) clusters formed for one-stage learning.

Table 4. Parameters used and the performance obtained by two-stage learning and one-stage learning for clustering the interlocking cluster data

Data	Parameters	Hidden Layer	Output Layer	Performance (%)	
				one-stage	two-stage
Inter-locking cluster data	neurons	8	2	77.33	99.30
	c	-2.0	-1.0		
	β	1.67	1.50		
	ϑ_{max}	56.0	3.0		
	$\Delta\vartheta$	0.005	0.0001		

A three-layer SNN is generally trained by modifying the weights of the connections to the nodes in the hidden layer and the connections to the nodes in the output layer simultaneously. Following this method, when a three-layer SNN, with 8 neurons in the hidden layer and 2 neurons in the output layer, is trained to cluster the interlocking cluster data, proper clusters are not formed (Table 3(b)). To overcome this limitation, we propose a two-stage learning method for training a spiking neural network model for clustering. In the proposed method, the weights of the connections to the nodes in the hidden layer are learnt first. Once the subclusters are formed, the weights of the connections to the nodes in the output layer are learnt to form the final clusters. Using this method, the interlocking cluster data is properly clustered (see Figs. 5(a)&(b)). The parameters that have given the best performance for one-stage learning may not give the best performance for the two-stage learning. The parameters used in the two-stage learning are given in Table 4. To compare the performance, layers of the three-layer SNN, with the same parameters, are trained simultaneously and the resulting clusters are shown in Figs. 5(c)&(d) and the performance is given in Table 4. Though the subclusters are formed in the hidden layer, proper clusters are not formed in the output layer.

6 Conclusions

In this work, we have proposed a method in which the firing threshold of a spiking neuron is gradually increased, starting with a small positive value, until they reach a maximum value. Using this method, it is possible to cluster nonlinearly separable data like the ring data. A biological interpretation to the delay selection in SNNs has been given. Hierarchical clustering in spiking neural networks has been demonstrated by training a multi-layer spiking neural network. A two-stage learning method has been proposed to cluster complex data like the interlocking cluster data without using lateral connections. In the two-stage learning, the formation of subclusters for 2-dimensional data is manually verified. However, for higher dimensional data, it is necessary to ensure the formation of subclusters automatically.

References

1. Maass, W., Bishop, C.M.: Pulsed Neural Networks. MIT-Press, London (1999)
2. Gerstner, W., Kistler, W.M.: Spiking Neuron Models. Cambridge University Press, Cambridge (2002)
3. Maass, W.: Fast Sigmoidal Networks via Spiking Neurons. Neural Computation 9, 279–304 (1997)
4. Hopfield, J.J.: Pattern Recognition Computation using Action Potential Timing for Stimulus Representations. Nature 376, 33–36 (1995)
5. Gerstner, W., Kempter, R., Van Hemmen, J.L., Wagner, H.: A Neuronal Learning Rule for Sub-millisecond Temporal Coding. Nature 383, 76–78 (1996)
6. Natschlager, T., Ruf, B.: Spatial and Temporal Pattern Analysis via Spiking Neurons. Network: Comp. Neural Systems 9, 319–332 (1998)
7. Bohte, S.M., Poutre, H.L., Kok, J.N.: Unsupervised Clustering with Spiking Neurons by Sparse Temporal Coding and Multilayer RBF Networks. IEEE Transactions on Neural Networks 13, 426–435 (2002)
8. Wu, J.: Introduction to Neural Dynamics and Signal Transmission Delay. Walther de Gruyter, Berlin (2001)
9. Jain, A.K., Dubes, R.C.: Algorithms for Clustering Data. Prentice-Hall, Englewood Cliffs (1988)
10. Yang, X., Song, Q., Cao, A.: Clustering Nonlinearly Separable and Unbalanced Data Set. In: International Conference on Intelligent Systems, pp. 491–496 (2004)

Implementing Classical Conditioning with Spiking Neurons

Chong Liu and Jonathan Shapiro

School of Computer Science
The University of Manchester
Oxford Road, Manchester M13 9PL, United Kingdom
chong.liu@cs.manchester.ac.uk, jonathan.shapiro@manchester.ac.uk

Abstract. In this paper, we attempt to implement classical conditioning with spiking neurons instead of connectionist neural networks. The neuron model used is a leaky linear integrate-and-fire model with a learning algorithm combining spike-time dependent Hebbian learning and spike-time dependent anti-Hebbian learning. Experimental results show that the major phenomena of classical conditioning, including Pavlovian conditioning, extinction, partial conditioning, blocking, inhibitory conditioning, overshadow and secondary conditioning, can be implemented by the spiking neuron model proposed here and further indicate that spiking neuron models are well suited to implementing classical conditioning.

1 Introduction

Traditionally, classical conditioning [1] is mainly modelled by connectionist neural networks, e.g. Rescorla-Wagner model [2] and Sutton-Barto model [3]. These models are high level abstraction of neuron models which ignore the temporal dynamics of real neurons. However, the dynamics of biological neurons may offer more powerful computational abilities [4]. In this paper, we attempt to see if "more biological" neuron models can be used to implement classical conditioning.

Experiments of Rao and Sejnowski [5] show that the temporally asymmetric window of Hebbian plasticity can be reproduced by a Temporal Difference (TD) learning rule in conjunction with dendritic backpropagating action potentials.

In this paper, we explore the possibility of the inverse problem, viz. is it possible to use spike-time dependent Hebbian learning and spiking neuron models to realize TD learning and further to model classical conditioning?

The neuron model and learning algorithm used to implement classical conditioning in our experiment are introduced in Sec. 2. Then, the results of the simulation are shown in Sec. 3. Finally, Sec. 4 concludes the paper by discussing the robustness of the model and pointing out possible future work.

1.1 Classical Conditioning

Classical conditioning [1] involves different training and testing procedures and various behavioral phenomena. Its study begins with Pavlov's experiments on

J. Marques de Sá et al. (Eds.): ICANN 2007, Part I, LNCS 4668, pp. 400–410, 2007.

dogs in 1890s. Normally, dogs will salivate (Unconditional Response, UR) when they receive food (Unconditional Stimulus, US). After repeatedly fed just after a bell (Conditional Stimulus, CS) is rung, the dog salivates whenever the bell sounds (Pavlovian classical conditioning or Conditional Response, CR). Following Pavlov's experiments, a great number of scientists did similar experiments and found a variety of other phenomena, e.g. extinction, partial conditioning, blocking, inhibitory conditioning, overshadow and secondary conditioning.

Since the discovery of classical conditioning, a lot of researchers attempted to model it. In 1972, Rescorla and Wagner [2] proposed the Rescorla-Wagner rule, which has successfully explained Pavlovian conditioning, extinction and blocking, but failed in the secondary conditioning. In 1988, Sutton and Barto [3] used TD learning to provide a successful explanation of various phenomena of classical conditioning including secondary conditioning.

Both of these two models used connectionist neural networks. The input and output of the neurons in connectionist paradigm are represented as numbers, which change instantaneously in discrete time and are usually thought to represent the firing rate of neurons.

1.2 Spiking Neuron Models

The goal of this paper is to represent classical conditioning using spiking neuron models. The spiking neuron model [4, 6] is a kind of neuron model which emphasizes that the timing of spikes may carry information that is lost in connectionist firing rate models. Data gathered in recent years from neurobiological experiments [7, 8, 9] also support that information is transmitted through spikes and the timing of these spikes is used to transmit information and perform computation. This realization has stimulated a significant growth of research activity in spiking neuron models.

2 Model

In this section, we use a spiking neuron model, which will be described in Sec. 2.1, and a learning algorithm combining both spike-time dependent Hebbian learning and spike-time dependent anti-Hebbian learning, which will be described in Sec. 2.2, to simulate classical conditioning.

2.1 Neuron Structure

In this model, we use only one neuron and regard both US and CS as inputs of the neuron.

1. *Neuron Architecture*
   There is a single neuron whose output represents the response to stimuli as shown in Fig. 1. Both CS and US are considered as inputs of the neuron. US is regarded as a special stimulus with a big fixed weight (set to 1 in the

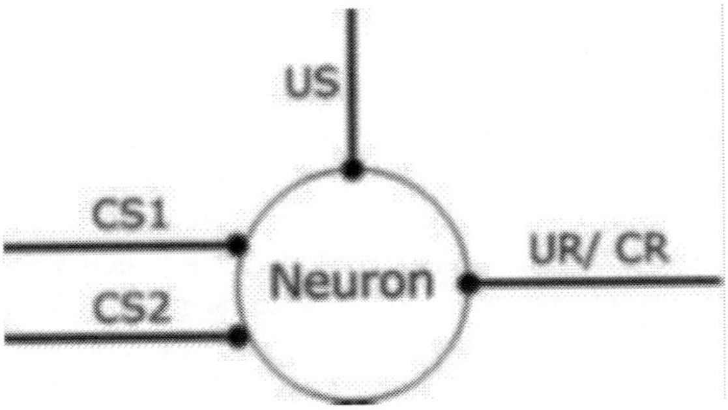

Fig. 1. Neuron architecture. Only one neuron is used. The inputs of the neuron are comprised by both CS and US; the output of the neuron is UR/ CR.

simulation). The output of the neuron is UR/ CR depending on whether it is produced by US or CS. All of these inputs and outputs are expressed in the form of spikes.

2. *Neuron Model*
Leaky linear integrate-and-fire model is used to represent the dynamics of the neuron in the simulation for simplicity. In leaky linear integrate-and-fire model, the membrane potential of the postsynaptic neuron $u(t)$ before firing can be expressed as

$$\tau_{\mathrm{m}} \frac{du(t)}{dt} = -u(t) + RI(t) \tag{1}$$

where τ_{m} refers to the membrane time constant of the postsynaptic neuron, $I(t)$ is the current triggered by spikes from other neurons or external sources at time t and R measures the resistance of the neuron.

When the the membrane potential $u(t)$ of the postsynaptic neuron reaches the firing threshold υ at time $\hat{t}$, a spike will be fired by the postsynaptic neuron. Immediately after the firing moment, its membrane potential is reset to $u_{\mathrm{r}} < \upsilon$. Then, the dynamics of the postsynaptic neuron develop along the trajectory described by (1) until its next spike.

In the numerical simulation, these parameters of the neuron model are set as follows: $\tau_{\mathrm{m}} = 10\mathrm{ms}$, $R = 10\mathrm{k}\Omega$, $\upsilon = 1\mathrm{mV}$, $u_{\mathrm{r}} = 0\mathrm{mV}$.

3. *Synapse Model*
An α-function [6] is used in the simulation to model neuron synapse. Suppose $I(t)$ is the current triggered by spikes from other neurons or external sources at time t, also the current that charges the membrane potential of the postsynaptic neuron, then it can be expressed as

$$I(t) = \frac{1}{\tau_{\mathrm{s}}} \sum_{j} \sum_{k} w_j e^{-\frac{t - t_j^k - d_j}{\tau_{\mathrm{s}}}} \theta(t - t_j^k - d_j) \tag{2}$$

where, τ_{s} is time constant, w_j and d_j are respectively synaptic weight and transmission delay of the jth connection, and t_j^k is the kth input spike time from the jth synapse. $\theta(s)$ is the Heaviside step function and satisfies,

$$\theta(s) = \begin{cases} 0 \text{ if } s \leq 0 \\ 1 \text{ if } s > 0 \end{cases}$$

In the numerical experiment, set $\tau_s = 10$msec, $d_j = 0$msec.

2.2 Learning Algorithm

To achieve results of classical conditioning, spike-time dependent Hebbian learning is used between inputs, while spike-time dependent anti-Hebbian learning is used between input and output.

It has been found that long-term potentiation (LTP) [10] of the cortico-striatal synapse is induced given the coincident occurrence of cortical input and postsynaptic depolarization in a striatal neuron with a phasic dopamine release, and long-term depression (LTD) [11] given the coincident occurrence of cortical input and postsynaptic depolarization in a striatal neuron without a phasic dopamine release. Since it is generally believed that dopamine is involved in reward learning, this suggests that Hebbian learning happens when rewards are present and anti-Hebbian learning happens when rewards are not present.

Although spike-time dependent Hebbian learning is well-known in the biological neural system, the curve of spike-time dependent anti-Hebbian learning is also observed in the cerebellum of electric fish [12,13], viz. if a presynaptic spike arrives at the synapse before the postsynaptic action potential, the synapse is depressed; if the timing is reversed the synapse is potentiated.

For simplicity, we use a square learning window as shown in Fig. 2 to simulate the curves of spike-time dependent Hebbian learning and spike-time dependent anti-Hebbian learning. US has a big fixed weight affected neither by CS nor by UR/ CR. CS has a small weight at the beginning but it can be either potentiated or depressed by US, UR/ CR and other CS.

Suppose there are N stimuli including both CS and US. W^{in} and W^{out} denote the learning window of input Hebbian learning and that of output anti-Hebbian learning respectively as Fig. 2 shows. The weight increment of the ith stimulus between time t and time $t+T$ (T is a period of time) is denoted as $\Delta w_i(t, t+T)$. Then, the mathematic formula of our learning algorithm can be expressed as (3) for CS and (4) for US. If the ith stimulus is a CS, then

$$\begin{aligned}\Delta w_i(t, t+T) = &\sum_{j=1 \cap j \neq i}^{N} \alpha \int_t^{t+T} dt' \int_t^{t+T} dt'' w_j(t) W^{\text{in}}(t''-t') S_i^{\text{in}}(t'') S_j^{\text{in}}(t') \\ &+ \alpha \int_t^{t+T} dt' \int_t^{t+T} dt'' W^{\text{out}}(t''-t') S_i^{\text{in}}(t'') S^{\text{out}}(t')\end{aligned} \tag{3}$$

The first term is the weight update due to the Hebbian learning effect of other inputs as shown by the solid line in Fig. 2, and the second term is the weight update due to the anti-Hebbian learning effect of the output as shown by the dashed line in Fig. 2. Where α is the learning rate (set to 0.01 in the simulation), and

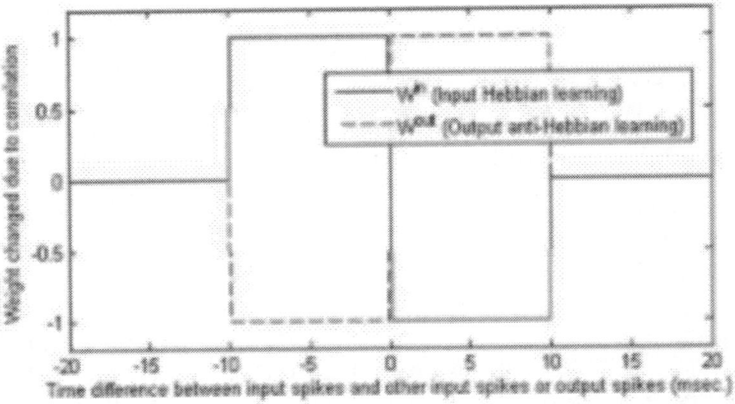

Fig. 2. Square learning window. As the solid line shows, the weight associated with an input (excluding US) will be reinforced if a spike of the input is followed by spikes of other inputs; it will be weakened if a spike of the input is preceded by spikes of other inputs. As the dashed line shows, the weight associated with an input (excluding US) will be weakened if a spike of the input is followed by spikes of the output; it will be reinforced if a spike of the input is preceded by spikes of the output.

$$S^{\text{in}}(t) = \begin{cases} 1 \text{ if } t = \hat{t} \ (\hat{t} \text{ is the arrival time of spikes of inputs}) \\ 0 \text{ otherwise} \end{cases}$$

$$S^{\text{out}}(t) = \begin{cases} 1 \text{ if } u(t) \geq \upsilon \\ 0 \text{ otherwise} \end{cases}$$

where $u(t)$ is the membrane potential and υ is the firing threshold of the neuron.

If the ith stimulus is a US, then

$$\Delta w_i(t, t+T) = 0 \tag{4}$$

3 Simulation Results

Although we have successfully simulated a variety of classical conditioning phenomena, we only discuss the four most important of them here, viz. Pavlovian conditioning, extinction, blocking and secondary conditioning, due to the limitation of the length of the paper.

3.1 Pavlovian Conditioning

In the setting of Pavlovian conditioning experiment, CS1 has 15 spikes from the 16th millisecond to the 30th millisecond, followed by US which also has 15 spikes from the 31st millisecond to the 45th millisecond, and no CS2 is presented as Fig. 3a & b depict. Before learning, UR, induced by US, is shown in Fig. 3c; the initial weights respectively associated with CS1 and associated with CS2 are both

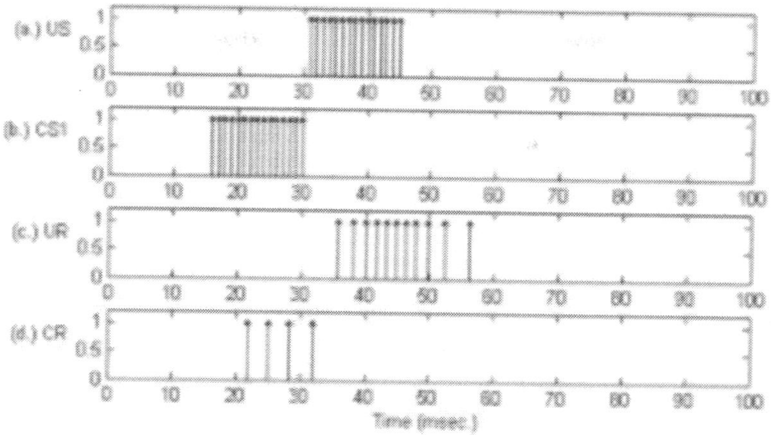

Fig. 3. Inputs and output (Pavlovian conditioning). a;b) CS1 is followed by US; CS2 is not present. c) The output spikes of the neuron when only US is present before learning. d) The output spikes of the neuron without the presence of US after learning.

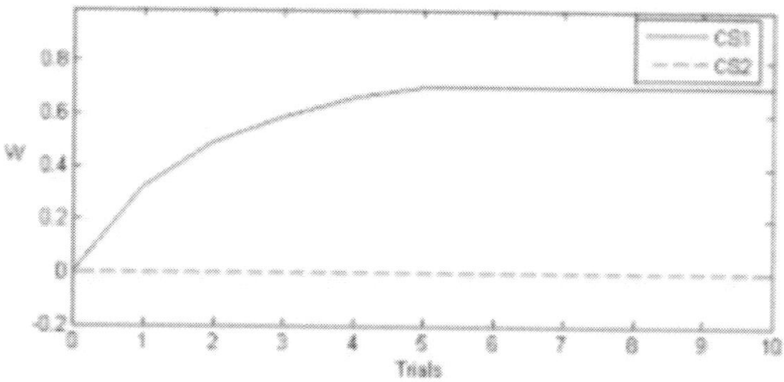

Fig. 4. Weight updates during learning (Pavlovian conditioning). The weight associated with CS1 keeps increasing until it reaches a saturation value near 0.7 and the weight associated with CS2 keeps unchanged. CS1 has been successfully associated with US after learning.

set to 0. During learning, the weight associated with CS1 keeps increasing until it reaches a saturation value near 0.7 and the weight associated with CS2 keeps unchanged as Figure 4 presents. After learning, CS1 alone can produce output spikes (CR) which precedes UR and predicts the approach of US as shown in Fig. 3d: Pavlovian conditioning has been formed.

3.2 Extinction

Before the experiment of extinction, Pavlovian learning was first conducted for 10 trials to associate CS1 with US. Then US is removed. During learning, the

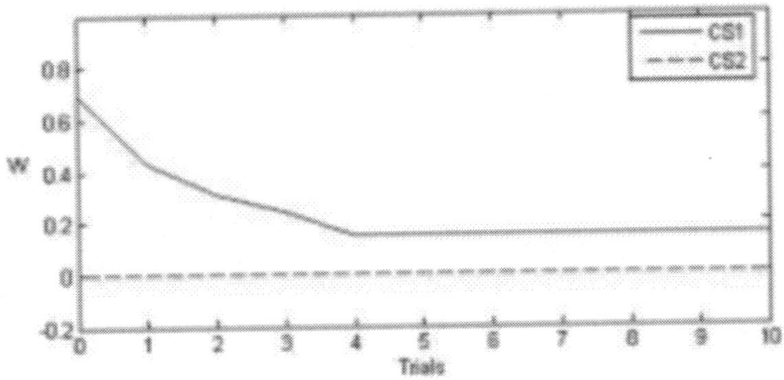

Fig. 5. Weight updates during learning (extinction). The weight associated with CS1 gradually decreases until it arrives at about 0.15. The association of CS1 with US has been extincted after learning.

weight associated with CS1 gradually decreases until it arrives at about 0.15 as Fig. 5 presents. Although the final weight associated with CS1 after learning has not gone down to 0, the extinction of conditioning has been realized because no output spikes are produced at this point, viz. the association of CS1 with US has been extincted.

3.3 Blocking

Like the experiment of extinction, CS1 was first associated with US by using Pavlovian learning for 10 trials before the simulation of blocking. Then CS2, which has the exact same time trajectory with CS1, is presented as Fig. 6a, b & c depict. During learning, the weight associated with CS1 slightly goes down and the weight associated with CS2 slightly goes up, but blocking has been realized because CS2 cannot produce any spikes with a maximum weight about 0.2 as shown in Fig. 7. Comparing Fig. 6d with Fig. 3d, we can see CR produced by both CS1 and CS2 in blocking is almost the same with that produced by CS1 alone in Pavlovian conditioning after learning, which further demonstrates that CS2 has been blocked.

3.4 Secondary Conditioning

Like the experiment of blocking and extinction, Pavlovian learning was first used for 10 trials to associate CS1 with US. Then CS2, which precedes CS1, is presented and at the same time US is removed as shown in Fig. 8. From the learning curve depicted in Fig. 9, it can be seen that the weight associated with CS1 gradually reduces until to zero, and the weight associated with CS2 goes up at the beginning but goes down to a very small amount (about 0.2) eventually. After learning, there are no output spikes: both CS1 and CS2 get extincted.

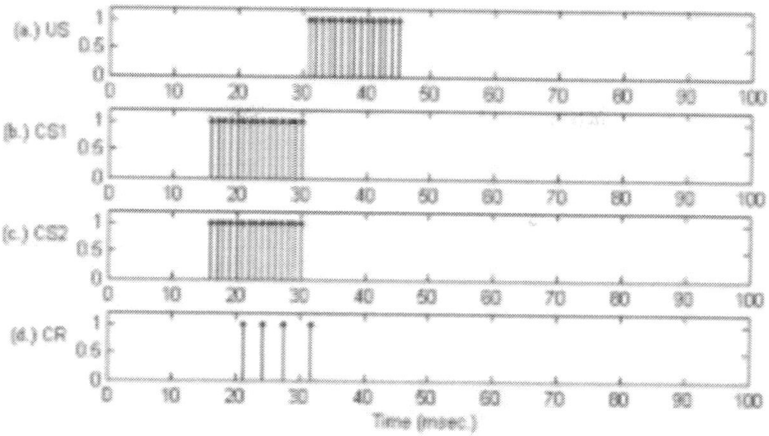

Fig. 6. Inputs and output (blocking). a-c) The two CS, which have the exact same time trajectory, are followed by US. d) The output spikes of the neuron without the presence of US after learning.

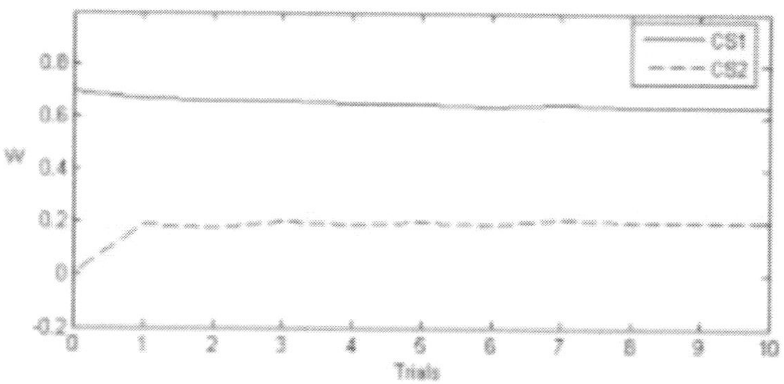

Fig. 7. Weight updates during learning (blocking). The weights associated with CS1 and CS2 slightly change, but blocking has been realized because CS2 cannot produce any spikes with a maximum weight about 0.2.

4 Discussion and Future Work

In this paper, we have used a very simple spiking neuron model to successfully implement the four most important classical conditioning. Because of the simplicity and effectiveness of the spiking neuron model, it seems that spiking neuron models are well suited to implement classical conditioning.

4.1 Robustness

Although only one set of experimental scenario is presented in the paper, the results still hold with different settings of inputs and different initial weights. In

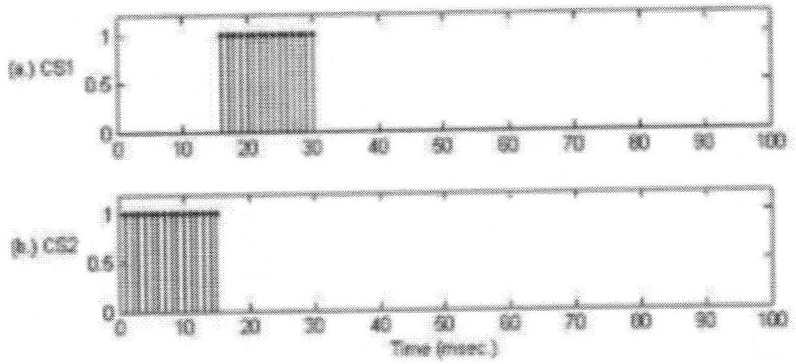

Fig. 8. Inputs and output (secondary conditioning). CS1 is followed by CS2; US is not present. There are no output spikes after learning.

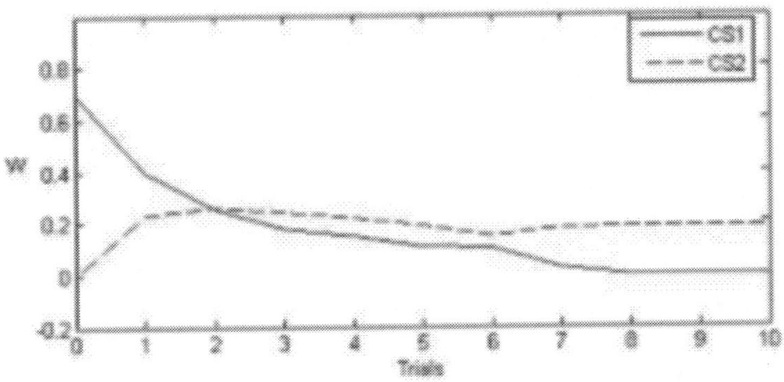

Fig. 9. Weight updates during learning (secondary conditioning). The weight associated with CS1 gradually reduces until to zero, and the weight associated with CS2 goes up at the beginning but goes down to a very small amount (about 0.2) eventually: both CS1 and CS2 get extincted after learning.

the simulation of Pavlovian conditioning, for instance, all spikes of CS1 precede all spikes of US and the first spike of US is just after the last spike of CS1. In fact, spikes of CS1 and US may be overlapped or uncontigeous. As long as some spikes of CS1 precede spikes of US within the learning window, the result does not change, though the weight associated with CS1 may be more or less. As far as the initial weights associated with conditional stimuli are concerned, they are set to 0 in the simulation. However, they could be very small positive values or very big positive values (even more than 1). The value of initial weights only affects the speed of learning but has no influence on the result.

4.2 Future Work

The classical conditioning is a simple and typical example of general reinforcement learning scenarios. Since this work has proved the effectiveness of spiking neuron models to implement classical conditioning, a straightforward future work is to expand it to model more complex reinforcement learning scenarios with wider range of action, richer input sets and more complicated input-output relations.

On the other hand, not only does the amount of rewards matter, the timing of rewards is also important, especially in a dynamic environment. For example, only when a predator can predict the timing of the appearance of its preys in one place will it be able to make an optimal decision, whether continuing waiting in the place or giving up and trying other places. In contrast, it may wait too long time in a hopeless place or leave just before the arrival of its preys if it cannot predict the timing of rewards. However, the timing information of rewards is lost in classical reinforcement learning algorithms because they use a value function as target instead of a reward function. While discounted functions used in present reinforcement learning algorithms can discount future rewards, it fails to predict even the rough timing of rewards. It may be possible to utilize the temporal feature of the learning dynamics of spiking neuron models to predict the time to rewards.

References

1. Dayan, P., Abbott, L.F.: Theoretical Neuroscience: Computational and Mathematical Modeling of Neural Systems. MIT Press, Cambridge (2001)
2. Rescorla, R.A., Wagner, A.R.: A theory of pavlovian conditioning: The effectiveness of reinforcement and non-reinforcement. In: Black, A.H., Prokasy, W.F. (eds.) Classical Conditioning II: Current Research and Theory, Aleton-Century-Crofts, pp. 64–69 (1972)
3. Sutton, R.S., Barto, A.G.: Time-derivative models of pavlovian conditioning. In: Gabriel, M., Moore, J.W. (eds.) Learning and Computational Neuroscience, pp. 497–537. MIT Press, Cambridge (1990)
4. Maass, W., Bishop, C.M.: Pulsed Neural Networks. MIT Press, Cambridge (1998)
5. Rao, R.P.N., Sejnowski, T.J.: Spike-timing-dependent hebbian plasticity as temporal difference learning. Neural Computation 13, 2221–2237 (2001)
6. Gerstner, W., Kistler, W.M.: Spiking Neuron Models. Cambridge University Press, Cambridge (2002)
7. Thorpe, S.J., Imbert, M.: Biological constraints on connectionist models. In: Pfeifer, R., Schreter, Z., Fogelman-Soulié, F., Steels, L. (eds.) Connectionism in Perspective, pp. 63–92. Elsevier, Amsterdam (1989)
8. Jen, P.H., Sun, X.D., Lin, P.J.: Frequency and space representation in the primary auditory cortex of the frequency modulating bat Eptesicus fuscus. Journal of Comparative Physiology 165(1), 1–14 (1989)
9. Rieke, F., Warland, D., de Ruyter van Steveninck, R., Bialek, W.: Spikes — Exploring the Neural Code. MIT Press, Cambridge (1996)
10. Wickens, J.R., Begg, A.J., Arbuthnott, G.W.: Dopamine reverses the depression of rat corticostriatal synapses which normally follows high-frequency stimulation of cortex in vitro. Neuroscience 70(1), 1–5 (1996)

11. Calabresi, P., Pisani, A., Mercuri, N.B., Bernardi, G.: The corticostriatal projection: from synaptic plasticity to dysfunctions of the basal ganglia. Trends in Neurosciences 19(1), 19–24 (1996)
12. Bell, C.C., Han, V.Z., Sugawara, Y., Grankt, K.: Synaptic plasticity in a cerebellum-like structure depends on temporal order. Nature 387, 278–281 (1997)
13. Han, V.Z., Grant, K., Bell, C.C.: Reversible associative depression and nonassociative potentiation at a parallel fiber synapse. Neuron 27, 611–622 (2000)

Deformable Radial Basis Functions

Wolfgang Hübner and Hanspeter A. Mallot

Eberhard–Karls University, Tübingen, Germany
wolfgang.huebner@uni-tuebingen.de

Abstract. Radial basis function networks (RBF) are efficient general function approximators. They show good generalization performance and they are easy to train. Due to theoretical considerations RBFs commonly use Gaussian activation functions. It has been shown that these tight restrictions on the choice of possible activation functions can be relaxed in practical applications. As an alternative difference of sigmoidal functions (SRBF) have been proposed. SRBFs have an additional parameter which increases the ability of a network node to adapt its shape to input patterns, even in cases where Gaussian functions fail.

In this paper we follow the idea of incorporating greater flexibility into radial basis functions. We propose to use splines as localized deformable radial basis functions (DRBF). We present initial results which show that DRBFs can be evaluated more effectively then SRBFs. We show that even with enhanced flexibility the network is easy to train and convergences robustly towards smooth solutions.

1 Introduction

Radial basis functions (RBF) have been used as general function approximators with great success. The learning (approximation) problem is to derive an estimate of an unknown function f from a set of noisy observations $y_i = f(x_i) + \eta(\sigma)$[1]. The radial basis solution is given by a function expansion of the form

$$f(x) \approx \sum_j^N w_j \phi(s_j^2(x - x_j)^2, \boldsymbol{p}_j) = \sum_j^N w_j \phi_j(x) \quad . \tag{1}$$

Unknowns in equation (1) are the (synaptic) weights (w_j), the number of nodes (N), the parameters of the affine transform (s_j, x_j), and additional node parameters ($\boldsymbol{p}_j$). It is also necessary to make a choice on the activation function ϕ and the estimator used to determine the network parameters. Commonly a least–square estimator is used which leads to the objective function

$$E(X,P) = \sum_i [y_i - \sum_j w_j \phi(x_i, \boldsymbol{p}_j)]^2, \quad X = \{(x_i, y_i)\}, P = \{\boldsymbol{p}_j\} \quad . \tag{2}$$

[1] $\eta(\sigma)$ is a normally distributed noise term with standard deviation σ.

J. Marques de Sá et al. (Eds.): ICANN 2007, Part I, LNCS 4668, pp. 411–420, 2007.

Least–square estimators are known to be strongly affected by outliers. In order to overcome this limitation robust estimators, e.g. M–estimators, have been used in combination with radial basis networks [3]. M-estimators reject samples that strongly disagree with the current net state provided the estimated net state becomes increasingly certain.

Determination of the net parameters requires non–linear optimization[2]. Usually this step is divided into two parts. First, the number of nodes and an initial guess about node parameters is determined by some heuristic or cluster algorithm. In a second step equation (2) is minimized in order to adapt the nodes locally to the input samples. Examples are selforganized growing networks [8] for batch learning or the GAP ("Growing and Pruning") algorithm for iterative learning [4,5]. In [10] a third step has been proposed for RBF–learning. In the third step the net parameters are determined by minimizing the net equation with respect to all parameters simultaneously.

The choice of ϕ is mainly guided by the ability of the activation function to act as a local predictor, i.e. the function must converge towards a smooth solution. Smoothness means that the solution gives a good approximation of the generating function f and does not interpolate noisy samples. For practical applications mostly Gaussians are used. The use of Gaussians is justified by theoretical considerations originating from approximation theory [2]. Although Gaussians work well in many situations they are ineffective when discontinuities or partially constant functions have to approximated. In these cases networks grow unnecessarily large and the solution is corrupted by ripples.

In order to overcome these limitations it has been proposed to use a difference of two sigmoid functions (SRBF) [6,3],

$$\phi(x,\nu) = \frac{1}{1+e^{-(x+\nu)}} - \frac{1}{1+e^{-(x-\nu)}} = f^{+}(x,\nu) - f^{-}(x,\nu) \quad . \tag{3}$$

Equation (3) includes an additional shape parameter ν. Depending on ν, ϕ can take the form of a Gauss–function as well as the shape of a box–function. Examples are shown in figure 1c.

Here we extend the idea of shape adaption to add more flexibility to radial basis functions. We use localized splines as activation functions, referred to as deformable radial basis functions (DRBF). Compared to SRBFs DRBFs have less computational load and a larger repertoire of shapes (see figure 1b). The aim of the paper is to investigate the possibility of adding more flexibility to radial basis functions without loosing generalization performance. The paper is structured as follows. In section 2 we describe the construction of localized splines and discuss the necessity to restrict flexibility in order to generate smooth solutions. Section 3 shows initial results from the DRBF algorithm. The paper closes with a discussion about potential pitfalls and future steps.

[2] The term generalized radial basis functions (GRBF) has been introduced in [2] especially because of this non–linear optimization step.

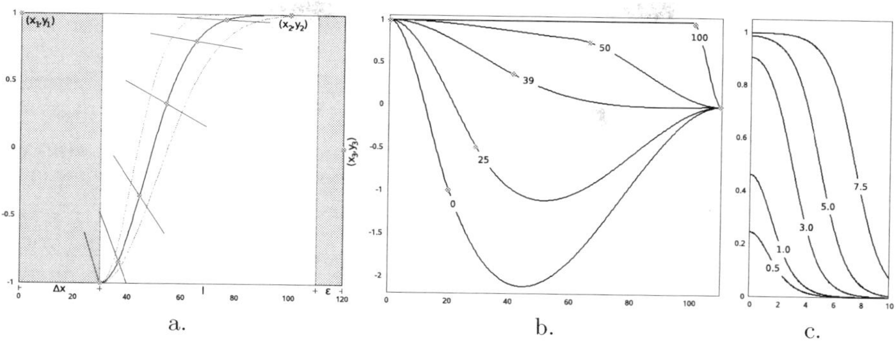

Fig. 1. Construction of deformable radial basis functions using a cubic spline with three control points (dots in figure a and b). a) The free control point is forced on the path shown in the middle ($c = 0.7$, see equation 20). Gray paths result from changing c about ± 0.2. As can be seen the path is quite insensitive to c when y_2 is close to -1. The slope at the control point is illustrated by line segments. b) The function shape transforms smoothly from wave forms to a box function when the control point moves along the trajectory shown in a. c) Shapes generated by a difference of sigmoid function when varying the shape factor ν in equation (3).

2 Construction of Deformable Radial Basis Functions

2.1 C^1 Splines

In this section we describe the construction of deformable radial basis functions based on localized splines. Splines are piecewise concatenated functions (usually polynoms). In general a spline consists of n polynomials of degree m

$$\mathrm{Sp}_k(x) = \sum_{i=0}^{m} a_{k,i}(x - x_k)^i \qquad x_k \leq x < x_{k+1},\ k = 1, \ldots, n \quad . \tag{4}$$

For n polynomials n+1 control (support) points are required to determine the shape of the function. At the control point locations (x_k) the interpolation conditions

$$\mathrm{Sp}_k(x_k) = a_{k,0} = y_k \quad \text{and} \tag{5}$$

$$\mathrm{Sp}_k(x_{k+1}) = \sum_{i=0}^{m} a_{k,i}(x_{k+1} - x_k)^i = y_{k+1} \tag{6}$$

must be fulfilled. Additional conditions are defined for the differentiability of the spline function,

$$\mathrm{Sp}'_k(x_k) = a_{k,1} = y'_k \quad \text{and} \tag{7}$$

$$\mathrm{Sp}'_k(x_{k+1}) = \sum_{i=1}^{m} i a_{k,i}(x_{k+1} - x_k)^{i-1} = y'_{k+1} \quad . \tag{8}$$

If necessary higher order derivatives can be defined analog to equations (7) and (8). The spline has $n(m+1)$ degrees of freedom which are determined by a system of linear equations setup through equations (5-8). It is common practice to determine some of the differentiation conditions implicitly, e.g. the inclusion of second derivatives in cubic splines [7]. As a drawback the increased differentiability can produce solutions with high curvature ("oscillations"). Therefore, implicit conditions are not helpful for our application.

For building radial basis functions we first define the control point locations (x_k) and the control point intervals $h_k = x_{k+1} - x_k$. Since the control points will be determined by least–square–optimization, it is necessary to preserve the point order during optimization, i.e. $x_1 < x_2 < \dots x_{n+1}$, i.e. the location intervals need to be positive[3].

Next we define the boundary conditions for the spline. At the left boundary the control points are fixed at $x_1 = 0$, $y_1 = 1$ and $y'_1 = 0$. This ensures that x_1 can be used as the center of symmetry, which allows the combination of the function with affine transformations without generating discontinuities. The right boundary is used to generate a local support function. Therefore control values are set to $y_{n+1} = 0$ and $y'_{n+1} = 0$. Altogether, the free parameters for a C^1–spline are given by $(x_2, \dots, x_n, y_2, \dots, y_n, y'_2, \dots, y'_n)$.

In order to adapt the shape of the spline it is necessary to express the polynomial coefficients as a function of the control points. This can be done independently for each polynomial by solving a system of linear equations. Here we give the solution for a cubic spline.

$$\boldsymbol{\nu_k} = H_k \boldsymbol{a_k} \rightarrow \boldsymbol{a_k} = H_k^{-1} \boldsymbol{\nu_k} \tag{9}$$

$$\boldsymbol{\nu_k}^\top = (y_{k+1}, y'_{k+1}, y'_k, y_k)^\top, \quad \boldsymbol{a_k} = (a_{k,3}, a_{k,2}, a_{k,1}, a_{k,0})^\top \tag{10}$$

$$H_k = \begin{pmatrix} h_k^3 & h_k^2 & h_k & 1 \\ 3h_k^2 & 2h_k & 1 & 0 \\ 0 & 0 & 1 & 0 \\ 0 & 0 & 0 & 1 \end{pmatrix} \qquad H_k^{-1} = \begin{pmatrix} -2h_k^{-3} & h_k^{-2} & h_k^{-2} & 2h_k^{-3} \\ 3h_k^{-2} & -h_k^{-1} & -2h_k^{-1} & -3h_k^{-2} \\ 0 & 0 & 1 & 0 \\ 0 & 0 & 0 & 1 \end{pmatrix} \tag{11}$$

Equation (9) describes the coefficients of polynom k as a function of the control points and the desired derivative at the control point. A C^2 spline can be calculated in the same way using polynomials of degree 5.

Now it is possible to calculate the partial derivatives used to minimize objective function (2) defined in the previous section,

$$\frac{\partial \mathrm{Sp}_k}{\partial .} = \frac{\partial a_{k,0}}{\partial .} + \sum_{i=1}^{3} \frac{\partial a_{k,i}}{\partial .}(x - x_k)^i + i a_{k,i}(x - x_k)^{i-1} \frac{\partial x_k}{\partial .} \quad . \tag{12}$$

$\partial .$ is a placeholder for the parameters used to define the spline shape, i.e. the elements of $\boldsymbol{\nu}_k$. Partial derivatives with respect to the control point locations

[3] In practice this is not a big problem. Otherwise this condition can be enforced by directly optimizing the interval lengths h_k and ensuring them to be strictly greater 0. The control point locations have then to be calculated iteratively from the intervals.

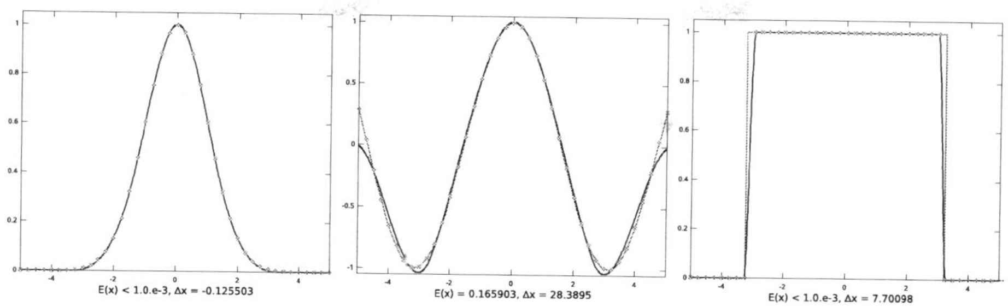

Fig. 2. Approximation of basic shapes with one node. The results show that the solutions greatly differ in the Δx value. This suggests that there exists no simple path in parameter space where all shapes are equally well represented. Samples are illustrated as gray dots. The sample distance is 0.25. Solutions are illustrated by black lines.

(x_k) are required because the control point locations are a function of the control point intervals. Partial derivatives for the coefficients of polynom k are calculated from equation (9),

$$\frac{\partial a_{k,i}}{\partial .} = \frac{\partial H_k^{-1}}{\partial .}\boldsymbol{\nu}_k + H^{-1}\frac{\partial \boldsymbol{\nu}_k}{\partial .} \quad . \tag{13}$$

The matrix formulation is used here only for simplifying the notation. With respect to computational complexity it is worth to expand the derivatives to their full length, which can save a lot of unnecessary multiplications. Depending on the choice of $\boldsymbol{\nu}_k$ most derivatives are 0.

From the construction of the spline it follows directly that the derivative of the spline is continuous in $x \in [0, \infty)$, but not necessarily smooth. The differentiability at the control point location with respect to y_k and y'_k is guaranteed by

$$\frac{\partial \mathrm{Sp}_k}{\partial y_k}(x_k) = \frac{\partial a_{k,0}}{\partial y_k} = 1\,, \tag{14}$$

$$\frac{\partial \mathrm{Sp}_{k-1}}{\partial y_k}(x_k) = \frac{\partial a_{k,1}}{\partial y_k}h_k^2 + \frac{\partial a_{k,2}}{\partial y_k}h_k^3 = 3h_k^{-2}h_k^2 - 2h_k^{-3}h_k^3 = 1\,, \tag{15}$$

$$\frac{\partial \mathrm{Sp}_k}{\partial y'_k}(x_k) = \frac{\partial a_{k,0}}{\partial y'_k} = 0\,, \tag{16}$$

$$\frac{\partial \mathrm{Sp}_{k-1}}{\partial y'_k}(x_k) = \frac{\partial a_{k,1}}{\partial y'_k}h_k^2 + \frac{\partial a_{k,2}}{\partial y'_k}h_k^3 = -h_k^{-1}h_k^2 + 2h_k^{-2}h_k^3 = 0\,. \tag{17}$$

2.2 Generating Smooth Shapes

In the following we only consider the case of three control points, i.e. only one control point determines the shape of the function ($\boldsymbol{p} = (x_2, y_2, y'_2)^\top$). The spline construction is illustrated in figure 1. A direct estimation of the free parameters is not a practical way, especially because the parameter space includes many shapes which are not smooth. These shapes can lead to a local oscillating behavior where samples are correctly interpolated but the solution shows high curvature between sample locations[4]. A second problem is that the high degree of flexibility leads to a large number of local minima. As a consequence local adaptation becomes sensitive to the start solution which can lead to an unpredictable behavior when combined with iterative learning methods.

Usually the problem of high curvature is addressed by adding a regularizer. In terms of Bayesian–inference the regularizer comes in in form of a prior penalizing non–smooth solutions. In radial basis function networks the prior is needed to control the effective number of nodes [9]. With respect to the spline functions, it would become necessary to combine two different prior terms, one controlling the network growth and a second one controlling the variability of single nodes.

Instead of using a regularizer we directly restrict the variability of the spline function. There are several possibilities for the spline to generate high curvatures (see figure 1a). If x_2 is moved close to x_1, i.e. $h_1 \to 0$, a strong peak is generated close to the center. If x_2 is moved close to x_3, i.e. $h_2 \to 0$, the shape depends on y_2. In case y_2 is close to 1 a box function is produced, which is a shape we want the spline to show. On the other hand, if y_2 is smaller than y_1 again a peaked shape is generated. Similar problems occur if y_2 takes large values. In order to exclude these extreme cases we restrict the parameter space in the following way. The first derivative at the control point is calculated from the position of the control point and the control point to the left, i.e. the slope is similar to a linear spline,

$$y_2' = \frac{y_2 - y_1}{x_2 - x_1} = \frac{y_2 - 1}{x_2} \quad . \tag{18}$$

The control point location (x_2) is restricted to the interval $[\Delta x, l + \Delta x]$,

$$x_2(\lambda) = \Delta x + t(\lambda), \quad t(\lambda) = l(\frac{1}{2} + \frac{1}{2}\sin(\frac{\pi\lambda}{l} - \frac{\pi}{2})) \quad , \tag{19}$$

with $l = x_3 - \Delta x - \epsilon$ (see figure 1a). The control point value (y_2) also becomes dependent on t,

$$y_2(\lambda) = 1 - 2e^{-(\frac{t}{\sigma})^2}, \quad \sigma = \frac{cl\sqrt{2}}{3}, \quad c = \text{const.} \tag{20}$$

Equations (19) and (20) force the control point on a curve shown in figure 1a. According to equation (19) x_2 moves periodically through the interval when λ is increased or decreased. We choose the periodic function instead of an asymptotic function in order to make the numerical optimization more stable[5]. If an asymptotic function is used the shape of the spline does not change with respect to λ if the constant section of the function is reached. If λ becomes large once during optimization it is very likely that the conjugate gradient method is trapped in this solution. Another critical variable is the choice of the constant x_3. If x_3

[4] It is important to note that also networks with Gaussian or sigmoidal activation functions can show oscillating behavior, in case the scale factor becomes too small. Usually, nodes which converge to such a solution are removed in a pruning step, e.g. using the significance measure described in [4] or by using prior knowledge [9].

[5] For optimization we use the conjugate gradient algorithm [7].

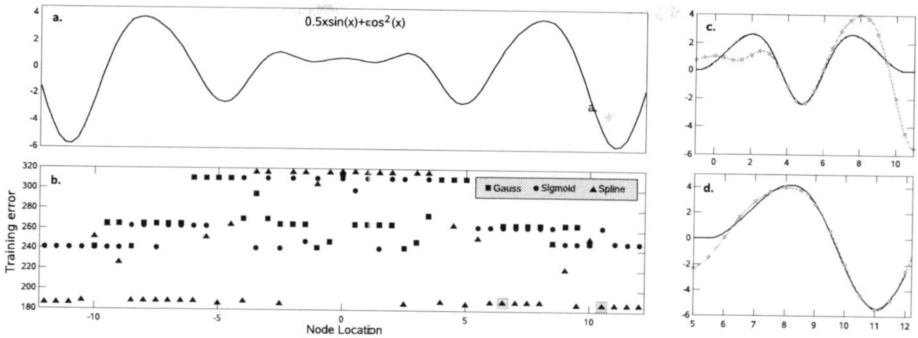

Fig. 3. Convergence of a single node without sample noise. a) Test function. b) Remaining training error after a single node has been initialized with a Gaussian shape and than locally adapted to the samples. c,d) Two examples of DRBF nodes with good error reduction. The examples are taken from the highlighted positions in the bottom row in panel b.

is too small it can happen that the optimization misses several local minima. Here we set x_3 to 100 which resulted in a stable convergence, also the number of iterations is increased.

In general it is not guaranteed (as shown in the result section) that a good solution has to be located on one path in parameter space. Therefore, the above description is too restrictive. Allowing some tolerance on c does not help, as illustrated in figure 1a, because if y_2 is close to -1 the shape is not very sensitive on c. A better way to increase the variety of shapes is to also adjust Δx[6]. Therefore the spline function is controlled by two parameters, i.e. $\boldsymbol{p} = (\Delta x, \lambda)^\top$ which are determined by optimization. As shown in figure 1b, possible shapes are smooth. The only exception is in the upper right corner, where the spline converges towards a box function.

3 Results

Before turning to the full network equation we first examine the convergence of a single node. As has been shown in the previous section it is not guaranteed that every possible shape is equally well represented in the restricted parameter space. Therefore we fitted the shape of a single node to three different basic shapes, a Gaussian function, a cosine function, and a box function. These three shapes are considered as primitive shapes since the spline smoothly changes between these forms when passing through the control path. The results are shown in figure 2. As has been expected the very different solutions for Δx illustrate that there exists no path on which all shapes are represented equally well.

[6] With respect to figure 1a it seems necessary to ensure that Δx stays in a certain interval. This can be achieved in the same way as e.g. x_2 is bounded. Here we let Δx move freely, which doesn't caused problems during the experiments shown in the following section.

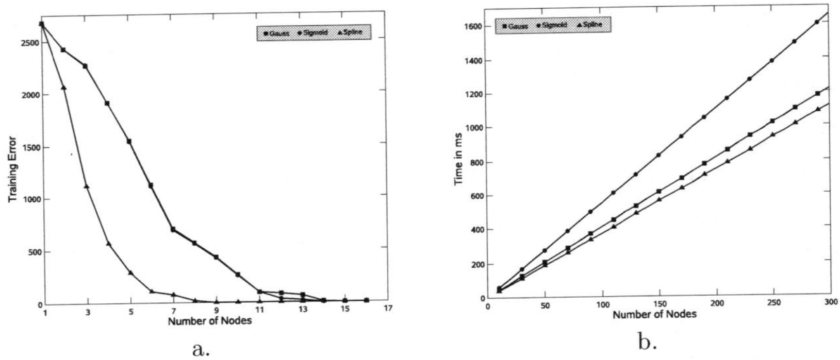

Fig. 4. a) Learning rate for different network types using the test function shown in figure 3. b) Evaluation time of networks with different sizes. As can be seen the computational complexity to evaluate a DRBF is comparable to the Gaussian. The speedup between DRBF and SRBF is approximately 1.5.

Next we tested if a single node converges to a stable and smooth solution under varying start conditions. This point is crucial when the method is combined with growing network algorithms or iterative learning methods. In order to make the test comparable between the three network types, we fitted the sigmoidal function and the spline function to a Gaussian with standard deviation of 1. The Gaussian shape and a scale factor of 0.5 is used as a start condition in all following tests. The training set consists of equally distributed samples taken from $f(x) = 0.5x\sin(x) + \cos^2(x)$ without noise (see figure 3a)[7]. In order to test the local convergence we shifted one node for each network type through the interval $[-12, 12]$ and evaluated the remaining training error (residuum) after minimizing equation (2). Results for the three network types are shown in figure 3b. The spline function is comparable in the learning progress to the Gaussian and the sigmoidal. In some cases, the spline outperforms the other networks. As can be seen in figures 3c and 3d, the reason for the improved learning performance is the increased flexibility of the spline function.

Next we tested the learning progress when using a network composed of multiple nodes. In order to test the learning progress we used a quite simple growing criterion. We placed a node at the location of the sample with the highest remaining training error. After adding a new node, equation (2) is minimized until the full network has converged. This cycle has been repeated 16 times. As shown in figure 4a the spline functions adapt faster to the samples. Figure 4b shows the computational load necessary to evaluate networks of different sizes. The spline net offers a speedup of approximately 1.5 compared to the sigmoid functions, i.e. the runtime complexity of the spline net is comparable to the Gaussians.

Finally we tested the behavior of the spline function for approximating the test function from noisy samples. As a training set we generated samples at 150 randomly chosen locations and added normally distributed noise with a

[7] The function has been used in [3] as a test function.

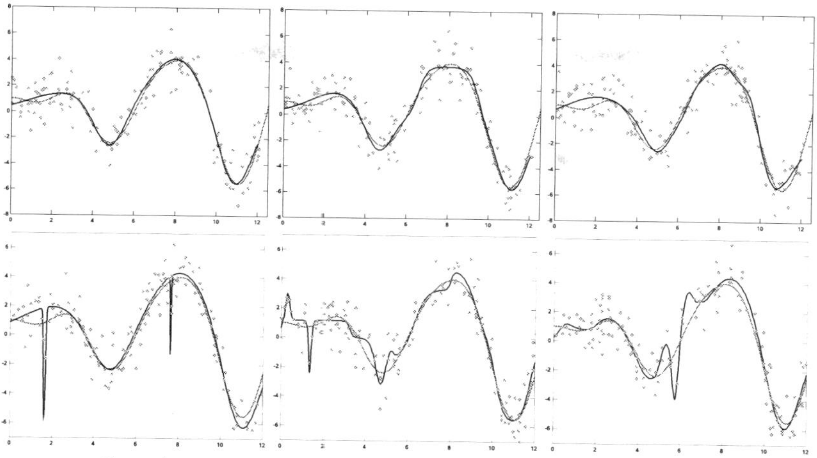

a. Gaussians b. Sigmoidals c. Splines

Fig. 5. Approximation of the test function using noisy samples. The networks consisted of 4 nodes (top row) or 12 nodes (bottom row). As can be seen all network types generate comparable smooth solutions or fail under the same conditions.

standard deviation of 1. The maximum network size has been fixed to 4 and 12 neurons using the same growing criterion as above. Figure 5 shows one example per network type. It can be seen that the spline net, also it has greater flexibility, produces smooth solutions. Overfitting occurs in all networks. The negative example is added here to illustrate that this is not an effect of the increased flexibility of the spline functions.

4 Summary and Discussion

We presented a method to use splines as a basis to define deformable radial basis functions (DRBF). By restricting the variability of the spline function we could show that the DRBF converges towards smooth solutions comparable to Gaussian and sigmoid activation functions. The presented data also suggests that DRBFs can improve the learning performance of radial basis function networks without loosing generalization performance.

We presented initial data and case examples to illustrate the behavior of DRBFs. Next steps include the expansion to multiple dimensions and a more extensive evaluation with real world problems and benchmark tests. In principal there are two alternatives to apply the method to multidimensional input spaces. In [3] a multidimensional function is generated by taking a product of one sigmoid function per dimension. The drawback of this method is that the number of parameters required grows very fast with the number of dimensions. An alternative is to stick to affine transforms. This method offers another option to increase the flexibility of radial basis functions by modifying the distance calculation in the affine transform (distance modifier functions). A discussion and initial results about this idea have been presented in [1].

Acknowledgement. This work is part of the GNOSYS project, funded by the Comission of the European Union (FP6–003835–GNOSYS).

References

1. Falcao, A., Langlois, T., Wichert, A.: Flexible kernels for RBF networks - Brain inspired cognitive systems - selected papers from the 1st international conference on brain inspired cognitive systems. Neurocomputing 69, 2356–2359 (2006)
2. Poggio, T., Girosi, F.: A theory of networks for approximation and learning. A.I. Memo No.1140 (1989)
3. Lee, C., Chung, P., Tsai, J., Chang, C.: Robust radial basis function neural networks. Systems, Man and Cybernetics, Part B, IEEE Transactions on 29, 674–685 (1999)
4. Huang, G.B., Saratchandran, P., Sundararajan, N.: An efficient sequential learning algorithm for growing and pruning RBF (GAP-RBF) networks. IEEE Transactions on Systems, Man and Cybernetics, Part B 34(6), 2284–2292 (2004)
5. Huang, G., Saratchandran, P., Sundararajan, N.: A generalized growing and pruning RBF (GGAP-RBF) neural network for function approximation. IEEE Trans. on Neural Networks 16(1), 57–67 (2005)
6. Geva, S., Sitte, J.: A constructive method for multivariate function approximation by multilayer perceptrons. IEEE Trans. on Neural Networks 3(4), 621–624 (1991)
7. Press, W.H., Teukolsky, S.A., Vetterling, W.T., Flannery, B.P.: Numerical Recipes in C++, 2nd edn. Cambridge Unviversity Press, Cambridge (2002)
8. Marsland, S., Shapiro, J., Nehmzow, U.: A self-organising network that grows when required. Neural Networks 15, 1041–1058 (2002)
9. MacKay, D.: Bayesian interpolation. Neural Computation 4(3), 415–447 (1992)
10. Schwenker, F., Kestler, H., Palm, G.: Three learning phases for radial–basis–function networks. Neural Networks 14(4-5), 439–458 (2001)

Selection of Basis Functions Guided by the L2 Soft Margin

Ignacio Barrio, Enrique Romero, and Lluís Belanche

Departament de Llenguatges i Sistemes Informàtics
Universitat Politècnica de Catalunya, Barcelona, Spain

Abstract. Support Vector Machines (SVMs) for classification tasks produce sparse models by maximizing the margin. Two limitations of this technique are considered in this work: firstly, the number of support vectors can be large and, secondly, the model requires the use of (Mercer) kernel functions. Recently, some works have proposed to maximize the margin while controlling the sparsity. These works also require the use of kernels. We propose a search process to select a subset of basis functions that maximize the margin without the requirement of being kernel functions. The sparsity of the model can be explicitly controlled. Experimental results show that accuracy close to SVMs can be achieved with much higher sparsity. Further, given the same level of sparsity, more powerful search strategies tend to obtain better generalization rates than simpler ones.

1 Introduction

Margin maximization has proven a good approach for classification tasks, and the Support Vector Machine (SVM) [1] is a state-of-the-art technique that shows very good performance. Given a training set $\{x_i, t_i\}_{i=1}^N$ where $x_i \in \mathbb{R}^D$ and $t_i \in \{+1, -1\}$, linear SVMs find a hyperplane that maximizes the margin in the input space. Nonlinearities can be added to these linear SVMs by mapping the input data with a set of basis functions into a feature space, with as many dimensions as basis functions, where the plane is found. The most usual way to add nonlinearities is by means of kernel functions.

An advantage of using (Mercer) kernels in SVMs is that an induced set of basis functions can be used without the need to explicitly consider them. The kernel is usually designed to serve as a good similarity function for a given problem [2]. This advantage sometimes involves a drawback, since the most natural similarity function for a given problem may not satisfy Mercer's condition, therefore designing a specific kernel, besides requiring some expertise, may reduce the quality of the function [3].

Another drawback of SVMs is that, although the solution is sparse in the number of support vectors, this number tends to grow with the number of data [4]. This is problematic for applications requiring high classification speed, since the cost of each prediction is proportional to the number of support vectors.

A number of methods have been proposed to overcome these drawbacks while maintaining identical functional form to SVMs [5,6,7]. Basically, these methods consider a finite set of basis functions (features) and they select a subset of those basis functions so that the solution is sparse in the feature space. These works have shown that very good

J. Marques de Sá et al. (Eds.): ICANN 2007, Part I, LNCS 4668, pp. 421–430, 2007.

performance can be achieved without the requirement of kernel functions. However these methods do not maximize the Euclidean margin.

In order to maximize the margin and find a sparser representation than SVMs, a new constraint can be added to the SVM training problem [8], so that the number of *expansion vectors*[1] is preset by the user. This approach is similar to finding a "constrained" feature space where the margin of the training data is maximized. Another method [9] has been proposed to select the expansion vectors with forward selection while maximizing the L_2 soft margin (penalizing the square of slack variables). Both works require the use of kernels.

The objective of this work is to find sparse models that maximize the margin without the requirement of using kernels. To overcome the requirement of kernels, unlike classical SVMs, we will consider an explicit finite set of basis functions (features). Then we could compute the parameters of a linear SVM where the inputs to the SVM are the inputs of the problem non-linearly mapped into the feature space, but the solution of this linear SVM would not be sparse.

In order to find sparse models, similar to [9], we propose the use of *Search Strategies Guided by the L_2 soft margin* (SSGL2) to select a subset from a set of candidate basis functions without the requirement of using kernels. Although the optimization problem posed is minimized when the whole set of basis functions is included, we seek a tradeoff between margin maximization and subset size. The solution is sparse in the feature space. In fact, we are trying to solve a feature selection problem where each feature is the output of a basis function. The search strategies used require the addition or removal of one basis function at a time. Based on an algorithm recently proposed to train linear L_2 soft margin SVMs [10], we can make use of the current solution to efficiently select the next basis function to add or remove.

Experiments using kernel functions show that, with a much reduced subset of basis functions, the model can be competitive and perform very similar to a linear SVM with the whole set of basis functions as features and to classical nonlinear SVMs. This way, the cost of predictions is much lower. Regarding the search strategies, more complex ones require fewer basis functions to achieve good performance than simpler ones.

This work is organized as follows. In section 2 we mention some alternatives to SVMs that tackle the aforementioned drawbacks. In section 3, we briefly review a method to solve linear L_2 soft margin SVMs. In section 4 we enumerate two popular search strategies. In section 5 we propose the SSGL2 for the selection of basis functions. An experimental study is carried out in section 6, comparing the SSGL2 to other methods. Finally, we draw some conclusions in section 7.

2 Related Methods

A number of methods have been proposed to select a subset of basis functions from a set of candidates without the need of being kernels. These candidates are usually, but not necessarily, centered at the input data. Some of these methods are the Relevance Vector Machine (RVM) [5], 1-norm SVMs (1-NSVM) [7] or Kernel Matching Pursuits

[1] *Expansion vectors* is the term used in [8] to refer to the vectors associated with each kernel evaluation in the model.

(KMP) [6]. On the one hand, RVM and 1-NSVM consider the whole set of basis functions and minimize a cost function that makes most of the weights become zero. On the other hand, KMP follows a search process to select a subset of basis functions that minimize some cost function (usually the sum-of-squares error). None of these methods maximizes the Euclidean margin.

Two methods have recently been proposed to maximize the margin using a reduced number of expansion vectors. Both methods require the use of kernels. The first method [8] adds a new constraint to the SVM problem so that the number of expansion vectors is preset by the user. The new problem is non-convex and a local minimum is found with a gradient based algorithm.

The second method [9] uses forward selection to select a subset of vectors that maximize the L_2 soft margin or, equivalently, optimize

$$\min_{\beta} f(\beta) = \frac{\lambda}{2}\beta^T K\beta + \frac{1}{2}\sum_{i\in I(\beta)} (y_i(\beta) - t_i)^2, \tag{1}$$

where K is the $N \times N$ kernel matrix and $y_i(\beta) = \sum_{j=1}^{N} \beta_j K(x_i, x_j)$ and $I(\beta) = \{i : t_i y_i(\beta) < 1\}$ is the active set. The bias term is omitted for simplicity. The selection technique is based on previous work on optimizing linear L_2 soft margin SVMs [10].

3 Optimizing Linear L_2 Soft Margin SVMs

Linear SVMs find a hyperplane that maximizes the margin in the input space $\mathbb{R}^D$. The objective is to find the parameters $\mathrm{w} = (\omega_1, \omega_2, .., \omega_D)^T$ that solve the problem:[2]

$$\begin{aligned} &\min_{\mathrm{w}} \frac{1}{2}\|\mathrm{w}\|^2 + \frac{C}{2}\sum_{i=1}^{N} \xi_i^2 \\ &\text{subject to} \quad t_i(\mathrm{w}^T x_i) \geq 1 - \xi_i \quad \forall i \in \{1, .., N\}, \end{aligned} \tag{2}$$

where C is a regularization parameter and $x_i = (x_{i1}, x_{i2}, .., x_{iD})^T$ are the input vectors. The output function is $y(x) = \mathrm{w}^T x$.

This problem has an equivalent formulation as

$$\min_{\mathrm{w}} f(\mathrm{w}) = \frac{\lambda}{2}\mathrm{w}^T\mathrm{w} + \frac{1}{2}\sum_{i\in I(\mathrm{w})} (\mathrm{w}^T x_i - t_i)^2, \tag{3}$$

where $\lambda = 1/C$ and $I(\mathrm{w}) = \{i : t_i \mathrm{w}^T x_i < 1\}$. The active set in the final solution, $I(\mathrm{w})$, corresponds to the support vectors. Note that f is a piecewise quadratic function strictly convex, thus it has a unique minimizer. Furthermore, f is continuously differentiable [10]. An iterative algorithm has been proposed [10] to optimize (3). Basically, each iteration k consists of two steps. In the first one, the optimal parameters $\overline{\mathrm{w}}$ for the current active set $I(\mathrm{w}_k)$ can be found as a regularized least squares solution. In the second step a line search is done, following $\mathrm{w}_{k+1} = \mathrm{w}_k + \delta^*(\overline{\mathrm{w}} - \mathrm{w}_k)$, to decrease the "full" objective function f. The algorithm has been theoretically shown to converge to the solution of (2) in a finite number of steps.

[2] For the sake of simplicity we will treat the bias term as part of the parameters, w.

4 Search Strategies

A search process has four main elements: an objective function which directs the search, a search strategy that decides how to continue exploring new states, an initial state where the search starts from and a stopping criterion. Two popular search strategies are:

- **PTA(l, r):** Plus l and Take Away r [11]. At every step, l elements are added one at a time (always the one that maximizes the objective function) and then r elements are removed one at a time (always the one that, after removing it, the objective function is maximized). When $l > r$, it is an increasing method, and when $l < r$, it is a decreasing one. Note that **Forward Selection (FS)** is PTA(1,0).
- **SFFS:** Sequential Forward Floating Selection [12] was originally designed for feature selection. At every step, an element is added and then zero or more elements are removed one at a time while the value of the objective function is better than the best value achieved until this moment with the same number of elements.

5 Search Strategies Guided by the L_2 Soft Margin

If we transform the input data into a new feature space $\phi(x) = (\phi_1(x), \phi_2(x), ..\phi_m(x))^T$, the hyperplane that maximizes the margin in this new space can be found by extending (3):

$$\min_{\mathrm{w}} f(\mathrm{w}) = \frac{\lambda}{2}\mathrm{w}^T\mathrm{w} + \frac{1}{2}\sum_{i \in I(\mathrm{w})} (y_i(\mathrm{w}) - t_i)^2, \qquad (4)$$

where $\mathrm{w} = (\omega_1, \omega_2, .., \omega_m)^T$, $y_i(\mathrm{w}) = \mathrm{w}^T\phi(x_i)$ and $I(\mathrm{w}) = \{i : t_i y_i(\mathrm{w}) < 1\}$ is the active set. Given a set of M candidate basis functions, we could tackle the classification problem by including in the model all of them (*i.e.* $m = M$) and solving (4). Since the model with the whole set of candidate basis functions has more flexibility than all the other subsets, the loss function (4) is minimized further. Although that solution can obtain good generalization rates, the cost of the predictions is very high when M is large. In order to reduce the cost, we could select a random subset of $m < M$ basis functions, obtanining a Reduced SVM [13,14], but that subset could be very suboptimal.

In order to obtain good generalization at a low prediction cost, we intend to fulfill a tradeoff between the maximization of the L_2 soft margin and the number of basis functions. As a practical approach to achieve this goal we propose to use a search process, following some search strategy (section 4) guided by the maximization of the L_2 soft margin or, equivalently, the minimization of (4). We will refer to these methods as *Search Strategies Guided by the L_2 soft margin* (SSGL2). More powerful search strategies like SFFS should reduce (4) more than simpler ones like FS. Therefore, a better generalization is expected. However, they will also require more additions and removals, so the learning stage will be slower.

The L_1 soft margin is more commonly used with SVMs than the L_2 soft margin. One of the reasons for the common use of the L_1 soft margin is that it produces sparser solutions in the number of support vectors. This reason does not affect this work, because we are working in the feature space and the sparsity in the number of features is explicitly controlled during the search process. We propose to maximize the L_2 soft margin for computational convenience.

Note that the SSGL2 are guided by (4), and not by (1) as proposed in [9]. The main difference is that the regularization term is simpler in (4) and only depends on the values of the coefficients. This allows the use of non-kernel basis functions.

5.1 Implementation

The search strategies described in section 4 only require the addition and removal of elements (the "best" element at every step). Following is the pseudocode for addition and removal of basis functions with SSGL2 based on the algorithm proposed in [10] and described in section 3.

AddBestBasisFunction (a subset of m basis functions ϕ, the parameters w, a set of candidate basis functions $\{\varphi_i\}$)

1. **for each** candidate basis function φ_i not in ϕ **do**
2. set ϕ_+ the subset obtained by adding φ_i to ϕ and compute the parameters $\overline{\mathrm{w}}$ for the current active set $I(\mathrm{w})$ as a regularized least squares solution
3. w_+ := LineSearch($\phi_+, \mathrm{w}, \overline{\mathrm{w}}$)
4. compute (4) for ϕ_+, w_+
5. **end for**
6. set ϕ, w the subset and parameters obtained by adding to ϕ the φ_i that minimizes (4) in the previous loop
7. w:=TrainSVMKeerthiDeCoste(ϕ, w)
8. **return** (ϕ, w)

end AddBestBasisFunction

RemoveWorstBasisFunction (a subset of m basis functions ϕ, the parameters w)

9. **for each** basis_function ϕ_i in ϕ
10. set ϕ_- the subset obtained by removing ϕ_i from ϕ and compute the parameters $\overline{\mathrm{w}}$ that minimize (4) for the current $I(\mathrm{w})$ as a regularized least squares solution
11. set w$'$ the parameters obtained by removing ω_i from w
12. w_- := LineSearch($\phi_-, \mathrm{w}', \overline{\mathrm{w}}$)
13. compute (4) for ϕ_-, w_-
14. **end for**
15. set ϕ, w the subset and parameters obtained by removing from ϕ the ϕ_i that minimizes (4) in the previous loop
16. w:=TrainSVMKeerthiDeCoste(ϕ, w)
17. **return** (ϕ, w)

end RemoveWorstBasisFunction

TrainSVMKeerthiDeCoste (a subset of basis functions ϕ, the parameters w)

18. **while** not convergence **do**
19. compute the parameters $\overline{\mathrm{w}}$ that minimize (4) for the current $I(\mathrm{w})$
20. w:=LineSearch(ϕ, w, $\overline{\mathrm{w}}$)
21. **end while**
22. **return**(w)

end TrainSVMKeerthiDeCoste

For each candidate basis function, for efficiency reasons, the optimal parameters are approximated (steps 2-3 or 10-12 in the pseudocode) by performing only one iteration of the algorithm in section 3: the regularized least squares stage is computed incrementally (steps 2 or 10) and the cost of the line search (steps 3 or 12) is $O(N \log N)$ [10]. After a basis function has been included, the whole algorithm in section 3 is run until convergence (steps 7 or 16).

Next we explain how to perform steps 2 and 3 incrementally. Suppose the current model has m basis functions and the active set $I(\mathrm{w})$ has N_{SV} elements and denote $d : \{1, .., N_{SV}\} \rightarrow \{1, .., N\}$ the function that given an index k in the active set returns the corresponding index in the training set. Denote Φ the $N_{SV} \times m$ design matrix with elements $\Phi_{kj} = \phi_j(x_{d(k)})$ and denote ϕ_* the column vector corresponding to a new basis function with elements $\phi_{*k} = \phi_{m+1}(x_{d(k)})$. Let $\Sigma = (\lambda I + \Phi^T \Phi)^{-1}$. Under a fixed active set, Σ^{-1} is a partitioned matrix. In that case, we can compute the parameters after adding ϕ_* (step 2) as $\overline{\mathrm{w}} = ((\mathrm{w} - \omega_* \Sigma \Phi^T \phi_*)^T, \omega_*)^T$ where $\omega_* = (\lambda + \phi_*^T \phi_* - \phi_*^T \Phi \Sigma \Phi^T \phi_*)^{-1}$. Similarly, we can compute the parameters decrementally after removing ϕ_i (step 10) as $\overline{\mathrm{w}} = \mathrm{w} - \omega_i \Sigma_{ii}^{-1} \Sigma_i$. The parameter $\overline{\omega}_i$, which is now zero, should be removed.

The cost of an addition is $O(N_{SV} \times m \times M + N \times \log N \times M)$, where M is the number of candidate basis functions. The cost of a removal is $O(N_{SV} \times m^2 + N \times \log N \times m)$ —note that we have to compute (4) for each candidate removal (step 13 in the pseudocode).

5.2 Stopping Criteria

The addition of more basis functions, when using the appropiate regularization parameter, is usually beneficial for the accuracy of the model, but some heuristic stopping criterion should be used to decide when to stop the search.

A first stopping criterion comes when the requirements of memory are strict and precise: the size of the subset is fixed beforehand, and when the model first reaches that number of basis functions, the process is stopped.

However, in many cases, the user will prefer the method to automatically select the number of basis functions. For those cases, we propose this second stopping criterion: let F_n be the error (4) of the best model found with n basis function; given some k, ϵ, if the current model has $m + k$ basis functions and $F_m - F_{m+k} < \epsilon N$ the process is stopped. By modifying k and ϵ one can roughly control the size of the subset.

6 Experimental Study

The goal of the following experiments is threefold: to observe the size of the subsets required to achieve good results, to confirm that more powerful SSGL2 produce better solutions than simpler ones and to compare the performance of SSGL2 to other related methods.

Although the use of kernels is not necessary in the proposed model, in order to compare the SSGL2 to the SVM, $M = N$ candidate Radial Basis Functions (RBF) centered at the training set input vectors were considered. We used $\varphi_i(x) = \exp(-\gamma \|x - x_i\|^2)$. Furthermore, a bias candidate, $\varphi_0(x) = 1$, was used.

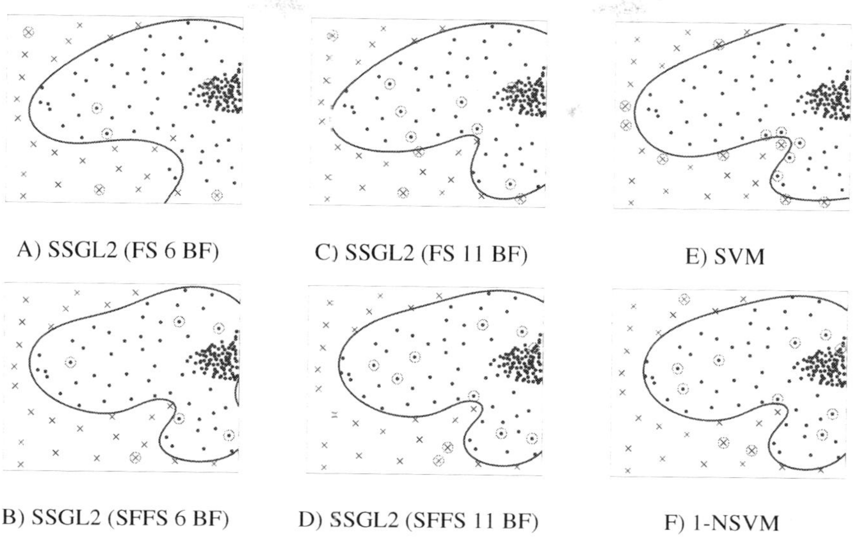

Fig. 1. Basis functions selected by different methods on the toy problem

6.1 Toy Example

In a first experiment, we created a toy data set with two input dimensions and some nonlinearities. C was set to a high value: 10000.

Figure 1 shows the centers of the basis functions selected in the solutions of different methods. The decision boundary generated is also shown. Figures 1A to 1D correspond to SSGL2, concretely FS and SFFS. FS required 11 basis functions to classify all the data correctly, while SFFS required only 6 (both solutions are shown in the figure). For illustrative purposes, Figure 1E shows the solution of a L_1 soft margin SVM using Gaussian kernels (the selected basis functions are the support vectors), and Figure 1F shows the solution of a 1-NSVM. The classical SVM selects points that are close to the decission surface, but that is not the case for the SSGL2 and 1-NSVM.

6.2 Benchmark Comparisons

We used 12 classification benchmark problems, provided by G. Rätsch and available at http://ida.first.fraunhofer.de/projects/bench. These data sets are split in 100 training/test partitions. The results reported in this work show averages over the first 10 partitions.

Settings. The input vectors were linearly scaled to $[-1, +1]$. The RBF width was the same for all the methods: we set $\gamma = (0.3D)^{-1}$, following [15], where D is the dimension of the input vectors.

We used the second stopping criterion described in section 5.2. We set $k = 10$ and $\epsilon = 0.005$. These values were chosen after some preliminary experiments. Furthermore, if the model reached 100 basis functions, the process was also stopped.

In order to choose the regularizer parameter, C, a 5-fold cross-validation process was performed on each training set partition. Values for C were tried ranging from 2^{-4} to 2^{14} multiplying by 2^2 (that is, $\{2^{-4}, 2^{-2}, 2^0, .., 2^{12}, 2^{14}\}$).

Table 1. For each data set: average accuracy obtained in the test sets (upper row) and number of basis functions of the model (lower row)

Data set	N	D	FS	PTA(2,1)	SFFS	ABF	SVM	1-NSVM	SpSVM-2
Banana	400	2	89.0	89.0	89.0	89.1	89.0	88.7	88.9
			29.3	27.2	22.0	400	99.3	18.8	19.8
Breast Cancer	200	9	72.9	72.5	72.4	71.8	70.7	69.7	71.9
			43.6	44.3	42.6	200	130	43.6	7.6
Diabetis	468	8	76.7	76.9	76.8	76.5	75.2	75.8	75.8
			32.3	28.5	25.0	468	267	31.7	18.3
Flare-Solar	666	9	66.5	66.3	66.2	65.4	66.5	66.3	66.6
			24.6	24.2	21.6	666	494	13.9	11.6
German	700	20	76.3	76.8	76.7	76.8	76.3	76.5	76.2
			52.2	46.8	44.1	700	459	108	43.6
Heart	170	13	79.1	79.6	79.0	79.0	79.8	78.7	80.0
			64.4	62.8	56.6	170	116	39.9	22.6
Image	1300	18	96.5	96.6	96.3	96.7	97.2	96.6	96.6
			65.9	55.9	35.6	1300	147	71.7	96
Ringnorm	400	20	97.8	97.7	97.6	98.0	97.5	97.8	97.7
			16.2	15.9	15.1	400	86.7	18.7	15.1
Titanic	150	3	77.8	77.8	77.8	77.8	77.8	77.9	77.3
			23.4	22.7	21.4	150	74.6	8.8	5.3
Waveform	400	21	89.8	89.6	89.5	89.6	89.6	89.1	89.4
			38.6	41.3	37.3	400	134	39.1	15.1
Splice	1000	60	86.8	86.6	86.4	88.6	88.9	85.1	85.1
			99.6	95.8	90.6	1000	841	540	86.1
Twonorm	400	20	97.2	97.1	97.1	97.2	97.3	97.0	96.9
			40.8	43.9	38.9	400	155	15.9	11.6

Models. We run three SSGL2 (FS, PTA(2,1) and SFFS) and compared them to a linear (in the explicit feature space) L_2 soft margin SVM, where the features ϕ_i are all the candidate basis functions (ABF). In other words, ABF is a model minimizing (4) with $m = N$ (plus the bias term). We also compared them to classical L_1 soft margin SVM with Gaussian kernels (labelled SVM) and to 1-NSVM.

Finally, we compared them to SVMs with reduced complexity [9] (labelled SpSVM-2). We used the code provided at http://www.kyb.tuebingen.mpg.de/bs/people/chapelle/primal/ by O. Chapelle. The whole set of candidates was considered for addition at each iteration. The number of basis functions (up to 100) for this method was selected with 5-fold cross-validation.

Results. For each task, the upper row in Table 1 shows the percentages of correctly classified test data (the average over the 10 partitions) and the lower row shows the number of basis functions used by each method. The number of training data N and the number of input variables D for each task are also shown. The SSGL2 obtained, in most cases, very similar accuracy than ABF while using only a subset of basis functions. Their performance was also very similar to SVM, 1-NSVM and SpSVM-2. The SSGL2 found more compact models than SVM and similar to 1-NSVM and SpSVM-2.

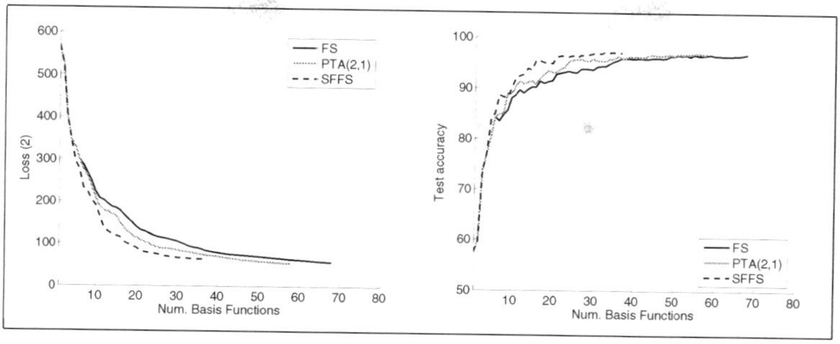

Fig. 2. The vertical axes show L_2 soft margin loss (4) (left) and test accuracy (right) on a partition of the *Image* data set. The horizontal axes show the number of basis functions of the model.

Table 2. Average number of additions plus removals required by the SSGL2

	Ban.	Bre.	Dia.	Fla.	Ger.	Hea.	Ima.	Rin.	Tit.	Wav.	Spl.	Two.
FS	29	44	32	25	52	64	66	16	23	39	100	41
PTA(2,1)	82	133	86	73	140	188	168	48	68	124	287	132
SFFS	269	272	180	118	412	456	540	83	115	264	1066	168

Comparing the search strategies of SSGL2, SFFS usually required a lower number of basis functions to obtain similar accuracy than FS and PTA(2,1).

Figure 2 shows the performance of the models found by the SSGL2 on the first partition of the *Image* data set. The plot on the left shows the L_2 soft margin loss (4) in the training set, while the plot on the right shows the accuracy of the models on the test set. Similar results were obtained for the rest of the data sets. Given the same number of basis functions, the models found by SFFS performed better than those found by FS. Again, we can see that the number of basis functions of the final solution is lower for SFFS than for FS. In contrast, the number of additions and removals required by SFFS was much higher than by FS (see Table 2). PTA(2,1) was in an intermediate position.

7 Conclusions and Future Work

A method has been described to select a subset of basis functions from a set of candidates by means of a search process guided by the L_2 soft margin. Being an explicit search, we can explicitly control the sparsity of the solution.

In the experiments, the SSGL2 found compact and competitive models. SFFS found very good subsets but it required a high number of operations. PTA(2,1) and FS required much less operations but their subsets were not so good. Choosing the search strategy implies a tradeoff between accuracy and computational cost. In orther to satisfy this tradeoff, other search strategies may be considered.

The SSGL2 can be extended in two ways that kernel methods cannot (or at least not so easily). First, any similarity function can be used without the restriction to be a kernel function. Second, a set of candidate basis functions (and a model) can contain different sorts of basis functions.

Acknowledgments

This work was supported by the Consejo Interministerial de Ciencia y Tecnología (CICYT), under projects CGL2004-04702-C02-02 and TIN2006-08114.

References

1. Vapnik, V.: The Nature of Statistical Learning Theory. Springer, Heidelberg (1995)
2. Schölkopf, B., Tsuda, J.P.V.K.: Kernel methods in computational biology. MIT Press, Cambridge (2004)
3. Balcan, M.F., Blum, A.: On a theory of learning with similarity functions. In: Proceedings of the 23rd International Conference on Machine Learning, pp. 73–80. ACM Press, New York (2006)
4. Steinwart, I.: Sparseness of support vector machines. Journal of Machine Learning Research 4, 1071–1105 (2003)
5. Tipping, M.: Sparse Bayesian learning and the relevance vector machine. Journal of Machine Learning Research 1, 211–244 (2001)
6. Vincent, P., Bengio, Y.: Kernel matching pursuit. Machine Learning 48(1-3), 165–187 (2002)
7. Bradley, P.S., Mangasarian, O.L.: Feature selection via concave minimization and support vector machines. In: 15th International Conf. on Machine Learning, pp. 82–90. Morgan Kaufmann, San Francisco (1998)
8. Wu, M., Schölkopf, B., Bakir, G.: Building sparse large margin classifiers. In: 22nd International Conf. on Machine learning, pp. 996–1003. ACM Press, New York (2005)
9. Keerthi, S., Chapelle, O., DeCoste, D.: Building Support Vector Machines with Reduced Classifier Complexity. Journal of Machine Learning Research 8, 1–22 (2006)
10. Keerthi, S., DeCoste, D.: A modified finite Newton method for fast solution of large scale linear SVMs. Journal of Machine Learning Research 6, 341–361 (2005)
11. Kittler, J.: Feature selection and extraction. In: Young, F. (ed.) Handbook of Pattern Recognition and Image Processing, Academic Press, London (1986)
12. Pudil, P., Novovičová, J., Kittler, J.: Floating Search Methods in Feature Selection. Pattern Recognition Letters 15(11), 1119–1125 (1994)
13. Lee, Y.J., Mangasarian, O.L.: Rsvm: Reduced support vector machines. In: SIAM International Conference on Data Mining (2001)
14. Lin, K.M., Lin, C.J.: A study on reduced support vector machines. IEEE Transactions on Neural Networks 14(6), 1449–1559 (2003)
15. Schölkopf, B., Sung, K., Burges, C., Girosi, F., Niyogi, P., Poggio, T., Vapnik, V.: Comparing support vector machines with Gaussian kernels to radial basis function classifiers. IEEE Transactions on Signal Processing 45(11), 2758–2765 (1997)

Extended Linear Models with Gaussian Prior on the Parameters and Adaptive Expansion Vectors

Ignacio Barrio, Enrique Romero, and Lluís Belanche

Departament de Llenguatges i Sistemes Informàtics
Universitat Politècnica de Catalunya, Barcelona, Spain

Abstract. We present an approximate Bayesian method for regression and classification with models linear in the parameters. Similar to the Relevance Vector Machine (RVM), each parameter is associated with an expansion vector. Unlike the RVM, the number of expansion vectors is specified beforehand. We assume an overall Gaussian prior on the parameters and find, with a gradient based process, the expansion vectors that (locally) maximize the evidence. This approach has lower computational demands than the RVM, and has the advantage that the vectors do not necessarily belong to the training set. Therefore, in principle, better vectors can be found. Furthermore, other hyperparameters can be learned in the same smooth joint optimization. Experimental results show that the freedom of the expansion vectors to be located away from the training data causes overfitting problems. These problems are alleviated by including a hyperprior that penalizes expansion vectors located far away from the input data.

1 Introduction

In supervised learning, we are given a set of training data $\{x_i, t_i\}_{i=1}^N$ where $x_i \in \mathbf{R}^D$. We will consider regression tasks, where $t_i \in \mathbf{R}$, and binary classification tasks where $t_i \in \{+1, -1\}$. The objective is to infer a function $y(x)$ that underlies the training data and makes good predictions on unseen input vectors. It is common to express this function as an extended linear model with M fixed basis functions:

$$y(x; w) = w^T \phi(x), \tag{1}$$

where $\phi(x) = (\phi_1(x), \phi_2(x), .., \phi_M(x))^T$ are the outputs of the basis functions and $w = (\omega_1, \omega_2, .., \omega_M)^T$ are the model parameters. When each basis function is associated with a vector, we can write

$$\phi(x) = (k(x, \hat{x}_1), k(x, \hat{x}_2), .., k(x, \hat{x}_M))^T, \tag{2}$$

where $\hat{x}_i$ are the expansion vectors.

Within Bayesian learning, the Relevance Vector Machine (RVM) [1] considers a model where the expansion vectors correspond with all the training input vectors. An individual Gaussian prior distribution is assumed on each parameter, so that following approximate Bayesian inference leads to sparse solutions. The expansion vectors associated to non-zero parameters (known as relevance vectors) belong to the training set,

J. Marques de Sá et al. (Eds.): ICANN 2007, Part I, LNCS 4668, pp. 431–440, 2007.

and are found by marginal likelihood maximization. The η-RVM is an extension of the RVM that, besides selecting the relevance vectors, adapts the hyperparameters of the basis functions— *e.g.* the widths of Radial Basis Functions (RBF)— by marginal likelihood maximization. As Tipping stated, there is an important limitation in the η-RVM, since the learning stage requires an interleaved two-stage training procedure that leaves open the question of how to combine the optimization.

We present a method that finds a model with the form (1, 2) where the expansion vectors (the number of which is fixed beforehand) do not belong to the training set. We place an overall Gaussian prior on the parameters, and find the expansion vectors by marginal likelihood maximization with a gradient based process. We will refer to this method as *Linear Model with Adaptive expansion Vectors maximizing the Evidence* (LMAVE). One might argue that, since the expansion vectors are adaptive, the model is no longer linear. However, we use the term *linear model* because the expansion vectors are not parameters but hyperparameters. This work is inspired in previous works on sparsification of Support Vector Machines [2] and Gaussian processes [3]. These works do not require the expansion vectors to be part of the training data.

This manner of finding the expansion vectors has been suggested in [4,5], but it has not been studied in depth. However, it should include a number of enhancements with respect to the RVM, for example, the expansion vectors do not necessarily belong to the training data, so better estimates can be found. Furthermore, the expansion vectors and other hyperparameters are learned in one joint optimization. The computational demands are lower than for the RVM, which makes this method suitable for large data sets. However, experimental results show that, at least for RBFs, the freedom of the expansion vectors to be away from the training data can cause overfitting problems.

We propose a hyperprior that penalizes expansion vectors located far away from the input data and show that a substantial improvement is achieved. We refer to this method as *Linear Model with Adaptive expansion Vectors maximizing the Posterior* (LMAVP).

2 Linear Models with Overall Gaussian Prior

This section introduces inference with extended linear models where an overall Gaussian prior distribution is assumed on the parameters. For regression we follow previous work on Bayesian interpolation [6]. For classification we use a generalization of the linear model.

2.1 Regression

Consider the targets to be deviated from the real underlying function by *i.i.d.* additive zero-mean Gaussian noise with variance σ_ν^2. The likelihood of the parameters follows a Gaussian distribution: (to save notation, we omit the conditioning on the input vectors, x)

$$p(t|w, \sigma_\nu^2) \sim \mathcal{N}(\Phi w, \sigma_\nu^2), \tag{3}$$

where Φ is the $N \times M$ design matrix with elements $\Phi_{ij} = k(x_i, \hat{x}_j)$.

In order to control the smoothness of $y(x;w)$, we assume a zero-mean overall Gaussian prior distribution over w with variance σ_w^2:

$$p(w|\sigma_w^2) \sim \mathcal{N}(0, \sigma_w^2), \tag{4}$$

Following Bayes' rule, the posterior parameter distribution is written $p(w|t, \sigma_\nu^2, \sigma_w^2) = p(t|w, \sigma_\nu^2)p(w|\sigma_w^2)/p(t|\sigma_w^2, \sigma_\nu^2)$. This is a product of two Gaussians (the likelihood and the prior) divided by a normalizing constant (the marginal likelihood) and it can be rewritten as

$$p(w|t, \sigma_\nu^2, \sigma_w^2) \sim \mathcal{N}(\mu, \Sigma), \tag{5}$$

where

$$\Sigma = (\sigma_\nu^{-2}\Phi^T\Phi + \sigma_w^{-2}I)^{-1} \quad \text{and} \quad \mu = \sigma_\nu^{-2}\Sigma\Phi^T t. \tag{6}$$

The marginal likelihood (also known as the evidence) is given by $p(t|\sigma_w^2, \sigma_\nu^2) = \int p(t|w, \sigma_\nu^2)p(w|\sigma_w^2)dw$. This is a convolution of Gaussians, which is also a Gaussian:

$$p(t|\sigma_w^2, \sigma_\nu^2) = \mathcal{N}(0, C) = (2\pi)^{-N/2}|C|^{-1/2}\exp\Big(-\frac{1}{2}t^T C^{-1} t\Big), \tag{7}$$

where $C = \sigma_\nu^2 I + \sigma_w^2\Phi\Phi^T$. Since matrix C is sized $N \times N$, it is convenient to use established matrix identities and rewrite the logarithm of the marginal likelihood as

$$\log p(t|\sigma_w^2, \sigma_\nu^2) = -\frac{1}{2}\Big[N\log 2\pi - \log|\Sigma| + N\log\sigma_\nu^2 + \\ + M\log\sigma_w^2 + \sigma_\nu^{-2}\|t - \Phi\mu\|^2 + \sigma_w^{-2}\|\mu\|^2\Big].$$

It is common to include extra hyperparameters controlling some aspect of the basis functions (*e.g.* RBF widths). We will refer to all the hyperparameters (including also σ_w^2 and σ_ν^2) as Θ.

In order to estimate the most probable hyperparameters, Θ_{MP}, we can use Bayes' rule: $p(\Theta|t) \propto p(t|\Theta)p(\Theta)$. If a flat (improper) hyperprior $p(\Theta)$ is assumed for all the hyperparameters, then the marginal likelihood $p(t|\Theta)$ (7) should be maximized.

2.2 Binary Classification

In binary classification we model the probability of success $p(t_i = +1|w)$. We map $y(x;w)$ to the unit interval by applying the cumulative distribution function for a Gaussian (probit), $\Psi(z) = \int_{-\infty}^{z}\mathcal{N}(x|0,1)dx$. The likelihood is

$$p(t|w) = \prod_{i=1}^{N} p(t_i|w^T\phi(x_i)) = \prod_{i=1}^{N}\Psi(t_i w^T\phi(x_i)). \tag{8}$$

We use an overall Gaussian prior for the parameters (4). Since the likelihood is not Gaussian, unlike the regression case, we cannot arrive at analytical solutions for the posterior. Instead, we approximate the posterior distribution as[1]

$$p(w|t, \sigma_w^2) \propto p(w|\sigma_u^2)\prod_i p(t_i|w) \approx p(w|\sigma_w^2)\prod_i q(t_i|w). \tag{9}$$

[1] We use p to denote exact quantities and q to denote approximations.

We follow Expectation Propagation (EP) [7] and choose the approximation terms to be Gaussian, parameterized by (m_i, v_i, s_i): $q(t_i|w) = s_i \exp(-\frac{1}{2v_i}(t_i w^T \phi(x_i) - m_i)^2)$. Then the approximate posterior is also Gaussian: $q(w|t, \sigma_w^2) = \mathcal{N}(m_w, V_w)$.

EP chooses the approximation terms such that the posterior using the exact terms and the posterior using the approximation terms are close in Kullback-Leibler (KL) divergence. Qi *et al.* [8] presented an iterative algorithm to choose the approximation terms following EP in $O(INM^2)$ complexity (I is the number of iterations). The algorithm was designed for the RVM, but it is also applicable to linear models with an overall Gaussian prior. It is omitted here for brevity.

Once the approximation terms are found, the marginal likelihood can be approximated by:

$$p(t|\Theta) \approx q(t|\Theta) = \int \prod_i p(w|\sigma_w^2) q(t_i|w) dw = |V_w|^{1/2} \sigma_w^{-M} \exp(B/2) \prod_i s_i \quad (10)$$

where $B = (m_w)^T V_w^{-1} m_w - \sum_i (m_i^2/v_i)$.

3 Linear Models with Adaptive Expansion Vectors

The Linear Model with Adaptive expansion Vectors (LMAV) can be seen as a linear model with an overall Gaussian prior (section 2) where the expansion vectors are considered as hyperparameters. The number of expansion vectors is fixed beforehand. In order to optimize the expansion vectors and the rest of hyperparameters, the LMAV maximizing the evidence (LMAVE) uses a gradient based algorithm that maximizes the evidence (7, 10).

3.1 Derivatives with Respect to the Expansion Vectors

Regression Tasks. For regression tasks, we can decompose the computation of the derivatives of the marginal likelihood (7) with respect to the expansion vectors into two steps [1]. For convenience we compute the derivative of the logarithm $\mathcal{L} = \log p(t|\Theta)$:

$$\frac{\partial \mathcal{L}}{\partial \hat{x}_{ij}} = \sum_{n=1}^{N} \sum_{m=1}^{M} \frac{\partial \mathcal{L}}{\partial \phi_{nm}} \frac{\partial \phi_{nm}}{\partial \hat{x}_{ij}} = \sum_{n=1}^{N} \sum_{m=1}^{M} A_{nm} \frac{\partial \phi_{nm}}{\partial \hat{x}_{ij}},$$

where $\hat{x}_i = (\hat{x}_{i1}, \hat{x}_{i2}, .., \hat{x}_{iD})$, $A = \sigma_\nu^{-2}[(t - \Phi\mu)\mu^T - \Phi\Sigma]$ and $\phi_{nm} = k(x_n, \hat{x}_m)$.

Classification Tasks. For classification tasks, the approximation terms $q(t_i|w)$ have the same form as a likelihood term in a regression problem (3), that is, we convert the classification problem into a regression problem with targets m_i and input dependent (heteroscedastic) noise with variances v_i. Following this interpretation, the covariance matrix and the mean for the posterior $q(w|t, \sigma_w)$ can be rewritten as

$$V_w = (\sigma_w^{-2} I + H^T \Lambda^{-1} H)^{-1} \text{ and } m_w = V_w H^T \Lambda^{-1} m_o, \quad (11)$$

where $m_o = (m_1, .., m_N)^T$, $\Lambda = \text{diag}(v_1, .., v_N)^T$ and H is an $N \times M$ matrix with elements $H_{ij} = t_i k(x_i, \hat{x}_j)$.

We can use (11) to compute the derivatives of the marginal likelihood (10) with respect to the hyperparameters. For the gradient computation, the parameters m_i, v_i, s_i can be considered fixed [9] and we can write: (notice $\mathcal{L} = \log q(t|\Theta)$)

$$\begin{aligned} \frac{\partial \mathcal{L}}{\partial \hat{x}_{ij}} = & -\frac{1}{2} tr(V_u E \frac{\partial H}{\partial \hat{x}_{ij}}) + \frac{1}{2} m_w^T E \frac{\partial H}{\partial \hat{x}_{ij}} m_w - \\ & -\frac{1}{2} m_w^T E \frac{\partial H}{\partial \hat{x}_{ij}} V_w E m_o + m_w^T \frac{\partial H^T}{\partial \hat{x}_{ij}} \Lambda^{-1} m_o, \end{aligned} \quad (12)$$

where $E = 2H^T \Lambda^{-1}$.

3.2 Computational Cost

In regression tasks, computing the gradient requires the matrix multiplication $\Phi\Sigma$ in the calculation of matrix A, which has complexity $O(M^2 \times N)$. In addition, the optimization of M expansion vectors has $M \times D$ dimensions. However, the matrix $\partial\Phi/\partial\hat{x}_{ij}$ only has one non-zero column, which lowers the cost of some computations. The cost of LMAVE for regression is $O(n \times (N \times M \times D + M^2 \times N))$, where n is the number of epochs of the gradient based algorithm.

In classification tasks, each epoch requires the EP approximation and the computation of the gradient. Computing the gradient requires the matrix multiplication $V_w E$ in (12), which has complexity $O(M^2 \times N)$. This complexity is lower than the EP approximation $O(I \times M^2 \times N)$, where I is the number of iterations of the EP algorithm (usually less than 10, so it can be considered as a constant). The optimization of M expansion vectors has $M \times D$ dimensions, but again the matrix $\partial H/\partial\hat{x}_{ij}$ only has one non-zero column. The cost of LMAVE for binary classification is $O(n \times (N \times M \times D + I \times M^2 \times N))$. We have not included the cost of optimizing additional hyperparameters.

3.3 Introducing a Non-flat Hyperprior

The plot on the left in Figure 1 shows an illustrative example of LMAVE with ten expansion vectors. The data are shown with dots. The initial and the final locations of the expansion vectors are marked with crosses (lower and upper row respectively). The function produced by the LMAVE is also shown. We trained the LMAVE with RBFs $k(x, \hat{x}_i) = \exp(-\gamma \sum_{j=1}^{D} (x_j - \hat{x}_{ij})^2)$ with adaptive inverse squared width, γ. There is an input range where no data are provided. We find that an expansion vector is located precisely in that range and the function values become large near that expansion vector. A similar effect happens with the expansion vector located on the right of the data.

Where the data are scarce, the prior assumptions should take high importance. The Gaussian prior assumption on the parameters favours small function values, but the freedom of the expansion vectors to be located anywhere can produce large function values. That is, the prior assumption on the parameters is not enough to control the smoothness.

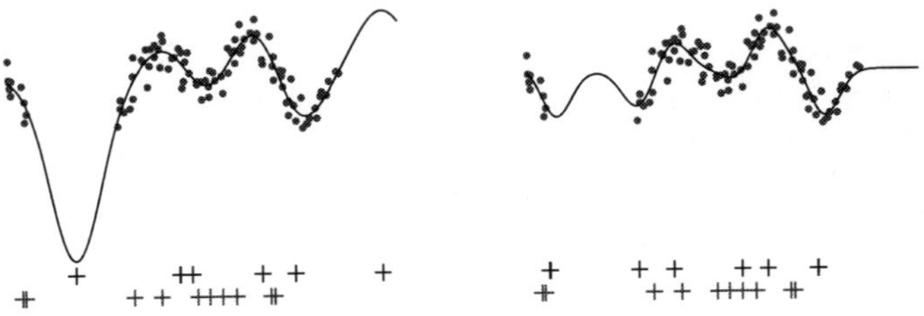

Fig. 1. Comparison between LMAVE (left) and LMAVP (right). See text.

The LMAVE assumes a flat hyperprior on the expansion vectors. To tackle situations like the one in Figure 1 we can include a hyperprior that penalizes expansion vectors located far away from the input data. Taking advantage of $k(x, \hat{x}_i) \in [0, 1]$ being bounded and localized at the expansion vectors, we propose the following hyperprior (note that $k(x, \hat{x}_i)$ depends on γ):

$$p(\hat{x}_i, \gamma) \propto \exp\Big(-\frac{1}{2}\frac{(1 - \max_{j=1}^{N}(k(x_j, \hat{x}_i)))^2}{s^2}\Big). \tag{13}$$

It is exclusively designed for the RBFs described above. Note that we have saved notation and this hyperprior should be written $p(\hat{x}_i, \gamma | \{x_j\}_{j=1}^N, k(\cdot, \cdot))$. This hyperprior satisfies a tradeoff[2] between flexibility on the expansion vectors to model the data and constraint to be located near the input data (the concept 'near' depends on the RBF width). Throughout this work we set $s = 0.1$. This value was chosen after some preliminary experiments.

In order to find maximum a posteriori estimates of the hyperparameters we maximize $\log p(t|\Theta) + \sum_{i=1}^{M} \log p(\hat{x}_i, \gamma)$. The complexity of computing the gradient of the hyperpriors is negligible when compared to the gradient of the evidence, therefore the cost is the same as LMAVE. We call this method *LMAV maximizing the posterior* (LMAVP). The plot on the right in Figure 1 shows the performance of LMAVP on the same data set. The expansion vectors are located near the training data and large function values are avoided.

4 Comparison with Related Methods

4.1 Relevance Vector Machines

The RVM [1] considers a linear model with one expansion vector coincident with each input vector. Unlike the LMAV, an individual Gaussian prior is assumed for each parameter, and following approximate Bayesian inference, most of the parameters tend to zero, so in practice the model can be expressed as an expansion of a subset of input vectors. Unlike the LMAV, the size of the model adapts to the problem at hand. This can be an advantage, but it can become a drawback for applications very restrictive in prediction time and memory, where the level of sparsity may be not enough.

[2] The term s^2 in the denominator controls this tradeoff.

Unlike the RVM, the expansion vectors in LMAV do not necessarily belong to the training data, so better estimates can be found —but if they are located far away, problems can appear (see Figure 1). Furthermore, the optimization of expansion vectors and other hyperparameters can be performed in a smooth joint optimization.

RVM implementations [10,11,12] either store the entire $N \times N$ design matrix or recompute all its columns at each iteration, while the LMAV recomputes an $N \times M$ design matrix at each iteration. Furthermore, the cost of an iteration of LMAV depends only linearly on the number of training data (it depends on other variables, see section 3.2), while this dependence is at least quadratic in RVM implementations. This makes the LMAV especially interesting for large data sets (and low number of expansion vectors).

Unlike the RVM, the LMAV requires the basis functions to be differentiable with respect to the expansion vectors. Furthermore, a bad initialization of the expansion vectors may lead to a bad local maximum.

4.2 Other Related Methods

Sparse large margin classifiers (SLMC) [2] are a sparse alternative to Support Vector Machines (SVMs) [13] for classification. A gradient based algorithm learns the locations of the expansion vectors (the number of which is set beforehand) so that the margin in some feature space is maximized. SLMC makes use of kernels, which allows to work in infinite feature spaces. Unlike SLMC, the LMAV does not use kernels, and it is limited to finite feature spaces. This limitation sometimes turns into an advantage, since the LMAV can use basis functions that do not satisfy Mercer's condition (required to be kernel). Furthermore, the Bayesian framework allows to tune other hyperparameters, such as RBF widths, that are not easily tuned for SLMC.

A Sparse Pseudo-input Gaussian process (SPGP) [3] is a Gaussian process (GP) with a particular covariance function parameterized by the pseudo-inputs (the counterpart to the expansion vectors). This covariance defines a prior over functions. Following approximate Bayesian inference, the pseudo-inputs are optimized by marginal likelihood maximization. The pseudo-inputs and the hyperparameters are found in a gradient based smooth joint optimization. The LMAV is closely related to SPGP, but it is based on a different prior over functions (a different covariance matrix C in (7)), with the advantage that the basis functions $k(x, \hat{x}_i)$ do not need to be covariance functions.

Much work in the literature has been devoted to the location of RBF centers. *Generalized radial basis functions* (GRBF) [14] find the centers with a gradient based algorithm minimizing some cost function based on regularization techniques. In [15] not only the centers are found with this method but also the RBF widths. The authors report overfitting problems. Similarly, the hidden parameters in a feed-forward neural network are commonly trained with gradient based algorithms [16]. LMAV can be considered as a sort of neural network. However, the main difference between both methods is that in neural networks all the parameters (output and hidden ones) are usually trained to minimize the same cost function, whereas in LMAV the expansion vectors (the counterpart to hidden layer parameters) are found by integrating out the (output) parameters. This should bring an improvement over maximum likelihood approaches.

Table 1. Results on regression benchmarks. The test MSE is shown for each method.

Data set		CPU	bank-32nh	pumadyn-32nh	kin-40k	kin-40k	
N		1024	1024	1024	2000	30000	
D		21	32	32	8	8	
η-RVM	$M/N(\%)$	1.8	40.7	2.5	12.6	-	
	Error (MSE)	0.090	1.707	0.602	0.0043	-	
$M/N = 2\%$	SPGP	0.080	1.262	0.586	0.0071	$M = 50$	0.0061
	LMAVE	0.110	2.570	0.704	0.0054		0.0043
	LMAVP	0.093	1.458	0.627	0.0054		0.0045
$M/N = 5\%$	SPGP	0.083	1.259	0.597	0.0054	$M = 100$	0.0039
	LMAVE	0.086	3.944	0.846	0.0043		0.0030
	LMAVP	0.077	1.861	0.642	0.0041		0.0029
$M/N = 10\%$	SPGP	0.079	1.264	0.616	0.0045	$M = 200$	0.0033
	LMAVE	0.088	3.032	0.987	0.0040		0.0015
	LMAVP	0.072	2.058	0.636	0.0031		0.0016

5 Experimental Study

In this section we compared the LMAVE and LMAVP to some related methods. Conjugate gradients was used to optimize all the hyperparameters. We used $n = 200$ epochs.

5.1 Regression Benchmarks

We used four benchmark data sets.[3] To check performance, for the *kin-40k* task we used 20 disjoint sets with 2000 cases each. Ten sets were used for training and ten for test. For each of the other tasks we used eight disjoint sets with 1024 cases each (four sets for training and four for test). To show the utility of LMAV on large data sets, we also split the *kin-40k* data set into 30000 training cases and 10000 test cases.

We used RBFs with multiple adaptive widths $k(x, \hat{x}_i) = \exp(-\sum_{j=1}^{D} \gamma_j (x_j - \hat{x}_{ij})^2)$. We compared the LMAVE and LMAVP to SPGP with covariance function $K(x_1, x_2) = \theta_0 k(x_1, x_2)$ and to η-RVM (RVM with adaptive RBF widths). The number of expansion vectors for LMAV and SPGP was 2%, 5% and 10% of the number of training data (except for the large data set). For SPGP we used the code provided by E. Snelson at www.gatsby.ucl.ac.uk/~snelson. For the η-RVM we followed the implementation proposed by Tipping [1].

Table 1 shows the results. For each task, the number of training data, N, and the input dimension, D, are shown. The mean squared error (MSE) is reported for each method. LMAVP was competitive with SPGP and the RVM, and it outperformed the LMAVE. The hyperprior alleviated overfitting problems in most cases but it was not strong enough for the *bank-32nh* problem. In the large data set, the RVM is hopeless, since it requires too much storage and computation, while the LMAV required from 16 minutes (with $M = 50$) to 110 minutes (with $M = 200$) on a Pentium IV 3.0 GHz.

[3] The *kin-40k* data set is available at http://ida.first.fraunhofer.de/~anton/data.html. The rest of data sets are available at http://www.cs.toronto.edu/~delve

Table 2. Results on classification tasks. The test error rates are shown for each method.

Data set		Banana	Breast	Titanic	Waveform	Image
N		400	200	150	400	1300
D		2	9	3	21	18
SVM	N_{SV}	86.7	112.8	70.6	158.9	172.1
	Error(%)	11.8	28.6	22.1	9.9	2.8
$M/N_{SV} = 5\%$	SLMC	16.5	27.9	26.4	9.9	5.2
	LMAVE	26.3	30.8	22.6	12.8	2.9
	LMAVP	26.1	29.9	22.7	12.5	3.5
$M/N_{SV} = 10\%$	SLMC	11.0	27.9	22.4	9.9	3.6
	LMAVE	11.5	29.5	22.7	12.5	2.6
	LMAVP	11.0	28.7	22.8	12.1	3.0
RVM	$M/N_{SV}(\%)$	13.2	5.6	92.5	9.2	20.1
	Error(%)	10.8	29.9	23.0	10.9	3.9

5.2 Classification Benchmarks

Five classification tasks available at http://ida.first.fraunhofer.de/projects/bench were used. These data sets were split in 100 training/test partitions. Similar to [1,2] the results reported in this work show averages over the first 10 partitions of each data set.

In this experiment we used RBFs with a single width $k(x, \hat{x}_i) = \exp(-\gamma \sum_{j=1}^{D}(x_j - \hat{x}_{ij})^2)$. We compared the LMAVE and LMAVP to the RVM, SLMC and SVM. The number of expansion vectors for SLMC, LMAVE and LMAVP is 5% and 10% of the number of support vectors found by the SVM.

Results are shown in Table 2. The results for RVM are taken from [1], where cross-validation was used to find the RBF width. The results for SVM and SLMC are taken from [2], where the regularization parameter and the RBF width were the ones provided by G. Rätsch at the aforementioned website. For LMAVE and LMAVP, the RBF width was jointly optimized with the expansion vectors (see section 3).

The LMAVE and LMAVP were competitive both with the RVM and SLMC. The difference in accuracy between LMAVP and LMAVE was not as notable as in regression tasks.

6 Conclusions

Experimental results show that the freedom of the expansion vectors to be located away from the training data can cause overfitting problems to the LMAVE. Since the prior on the parameters is not enough to tackle overfitting, we have included a particular hyperprior (designed for radial basis functions) penalizing expansion vectors located far away from the input data. This approach (LMAVP) results in a significant improvement over the LMAVE, especially in regression tasks, and is competitive with the RVM. The number of expansion vectors required to achieve good performance is usually very low. This fact makes this method very interesting for large data sets, since the memory and computational demands are lower than for the RVM. A study on different hyperpriors may be potentially interesting (also for non-radial basis functions).

In this work, the expansion vectors are initialized to random training input vectors, but other scenarios are possible, for example we could use a clustering algorithm such as K-means.

The (Gaussian) noise and (probit) 'slack' assumptions taken can be rather stringent, and the presence of outliers can decrease performance. Furthermore, the optimization of hyperparameters based on the evidence (LMAVE) and on maximum a posteriori estimates (LMAVP) are just approximations to full Bayesian inference, which consists on integrating out all the hyperparameters. As such, they are not inmune to overfitting.

Acknowledgments

The authors thank Joaquín Quiñonero Candela for providing the code for the η-RVM. This work was supported by the Consejo Interministerial de Ciencia y Tecnología (CICYT), under projects CGL2004-04702-C02-02 and TIN2006-08114.

References

1. Tipping, M.: Sparse Bayesian learning and the relevance vector machine. Journal of Machine Learning Research 1, 211–244 (2001)
2. Wu, M., Schölkopf, B., Bakir, G.: Building sparse large margin classifiers. In: 22nd International Conf. on Machine learning, pp. 996–1003. ACM Press, New York (2005)
3. Snelson, E., Ghahramani, Z.: Sparse Gaussian Processes using Pseudo-inputs. Advances in Neural Information Processing Systems 18, 1257–1264 (2006)
4. Minka, T.: Bayesian linear regression (1998), note obtainable from `http://research.microsoft.com/~minka/papers/`
5. Quiñonero Candela, J., Rasmussen, C.E.: A Unifying View of Sparse Approximate Gaussian Process Regression. Journal of Machine Learning Research 6, 1935–1959 (2005)
6. MacKay, D.J.C.: Bayesian Interpolation. Neural Computation 4, 415–447 (1992)
7. Minka, T.P.: Expectation Propagation for approximate Bayesian inference. In: 17th Conference in Uncertainty in Artificial Intelligence, pp. 362–369 (2001)
8. Qi, Y.A., Minka, T.P., Picard, R.W., Ghahramani, Z.: Predictive automatic relevance determination by expectation propagation. In: 21st International Conference on Machine Learning (2004)
9. Seeger, M.: Expectation propagation for exponential families (2005), note obtainable from `www.kyb.tuebingen.mpg.de/~seeger/`
10. Quiñonero Candela, J., Winther, O.: Incremental Gaussian processes. Advances in Neural Information Processing Systems 15, 1001–1008 (2003)
11. Tipping, M., Faul, A.: Fast Marginal Likelihood Maximisation for Sparse Bayesian Models. In: Ninth International Workshop on Artificial Intelligence and Statistics (2003)
12. D'Souza, A., Vijayakumar, S., Schaal, S.: The Bayesian backfitting relevance vector machine. In: 21st International Conference on Machine Learning (2004)
13. Vapnik, V.: The Nature of Statistical Learning Theory. Springer, Heidelberg (1995)
14. Poggio, T., Girosi, F.: A theory of networks for approximation and learning. Technical Report AI-1140, MIT Artificial Intelligence Laboratory, Cambridge, MA (1989)
15. Wettschereck, D., Dietterich, T.G.: Improving the performance of radial basis function networks by learning center locations. Advances in Neural Information Processing Systems 4, 1133–1140 (1992)
16. Bishop, C.M.: Neural Networks for Pattern Recognition. Oxford University Press Inc., New York (1995)

Functional Modelling of Large Scattered Data Sets Using Neural Networks

Q. Meng[1], B. Li[2], N. Costen[2], and H. Holstein[3]

[1] Dept. of Computer Science,Loughborough Univ.
q.meng@lboro.ac.uk
[2] Dept. of Computing and Mathematics, Manchester Metropolitan Univ.
b.li@mmu.ac.uk
[3] Dept. of Computer Science, Univ. of Wales, Aberystwyth

Abstract. We propose a self-organising hierarchical Radial Basis Function (RBF) network for functional modelling of large amounts of scattered unstructured point data. The network employs an error-driven active learning algorithm and a multi-layer architecture, allowing progressive bottom-up reinforcement of local features in subdivisions of error clusters. For each RBF subnet, neurons can be inserted, removed or updated iteratively with full dimensionality adapting to the complexity and distribution of the underlying data. This flexibility is particularly desirable for highly variable spatial frequencies. Experimental results demonstrate that the network representation is conducive to geometric data formulation and simplification, and therefore to manageable computation and compact storage.

1 Introduction

Digitalisation techniques such as range scanning devices are widely used for 3D model acquisition. Laser scanners can acquire a cloud of millions of irregularly distributed 3D points from the surface of a digitised object. However, sampling complex geometry almost always yields surface imperfections, in particular "holes" such as shown in Fig. 2. Faithful reproduction of incomplete meshes in the presence of holes and data noise is an ubiquitous problem for scientific visualisation and engineering modelling.

Inspired by advances of using Radial Basis Functions (RBFs) for solving function approximation and pattern classification problems, implicit modelling with RBFs has recently emerged as an effective technique in computer graphics for 3D modelling [1,5,6,9]. However, because RBFs are often globally supported in nature, functional approximation has to be calculated by solving a dense nonlinear system with a high order of computation complexity. This limited earlier RBF approaches as they failed to interpret large data sets of several thousands of points typical of real-world applications. Ferrari et al. [2] introduced an alternative solution to the problem by using an RBF network. However, the network structure and neuron location were fixed *a priori*, based on pre-knowledge of the training data. The lack of adaptivity to the underlying data could necessitate

J. Marques de Sá et al. (Eds.): ICANN 2007, Part I, LNCS 4668, pp. 441–449, 2007.

large numbers of fine-scale neurons to meet a desired accuracy. A dynamic Kohonen network has also been proposed [10]. It allows some structural flexibility, but with a high cost on neuron number which is unfeasible for modelling millions of point data.

To solve the problem of surface reconstruction from a large collection of irregularly sampled noisy range data, we propose a new error-driven self-organising hierarchical RBF network. The network has a multi-layer structure allowing a bottom-up coarse-to-fine approximation. However, this hierarchy is not fixed *a priori*. Multi-layer subnets are constructed by error-driven active learning. They aim to achieve global optimality as well as conquer large mapping errors in subdivisions. For each subnet, neurons can be located, removed or adjusted iteratively in full dimension according to the distribution and complexity of the underlying data. Such adaptivity is preferable to a pre-fixed grid structure [2], particularly for highly varying spatial frequencies. Adaptive sequential learning has benefitted from concepts found in the self-organising *resource allocating network* [7]. Additionally, a greedy *neighbourhood extended Kalman filter* (NEKF) method introduced for network updating leads to a significant computation reduction. The algorithm is straightforward to implement.

2 The Self-organising Hierarchical RBF Network

A typical RBF network has one hidden layer composed of K Gaussian kernels $\phi_k(\mathbf{x})$. The network output $f(\mathbf{x})$ takes the form

$$f(\mathbf{x}) = a_0 + \sum_{k=1}^{K} a_k \phi_k(\mathbf{x}) \ , \tag{1}$$

where $\phi_k(\mathbf{x}) = \exp\left(-\frac{1}{\sigma_k^2}\|\mathbf{x} - \boldsymbol{\mu}_k\|^2\right)$, $\|.\|$ denotes the Euclidean norm, $\boldsymbol{\mu}_k$ locates the *centre* of the kth neuron, standard deviation σ_k indicates its *width* (coverage), a_k is its *weight* connecting to the output and a_0 is the bias term.

2.1 Hierarchical Architecture

Approximation by a single RBF network leaves nonuniform mapping errors across the data space, requiring a large dynamic range and number of neurons to model the smooth bases as well as the finest local features. For modelling efficiency, we propose a multi-layer RBF network using an error-driven active learning strategy, as shown in Fig. 1.

The network can be described as a hierarchical pool of multi-layer RBFs. If there are M_l subnets at the lth layer, an L-layer hierarchical RBF network operates as the sum of RBF subnets $f_{l,m}(x)$,

$$\mathbf{H}_L(\mathbf{x}) = \sum_{l=0}^{L} \sum_{m=1}^{M_l} f_{l,m}(\mathbf{x}) \ . \tag{2}$$

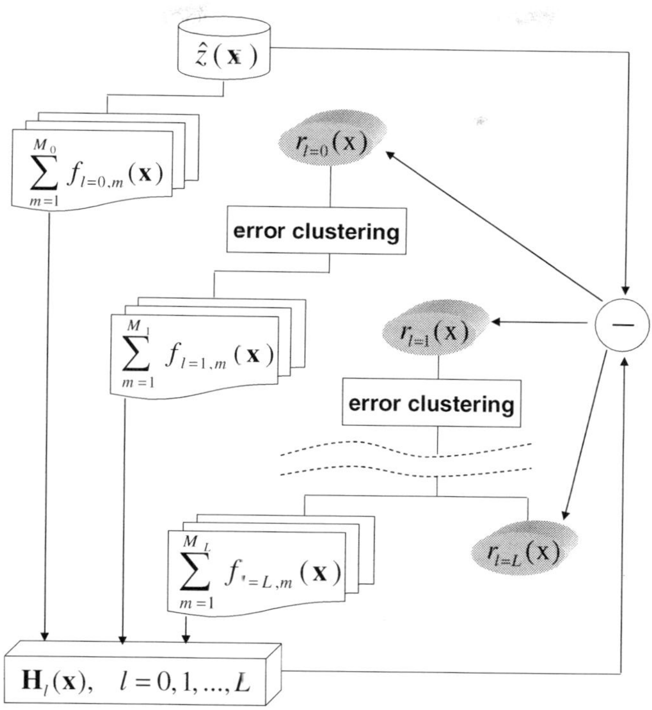

Fig. 1. Architecture of the error-driven hierarchical RBF network

The network facilitates a bottom-up coarse-to-fine approximation in the hierarchy. Neuron resolution scale increases towards high layers. The base layer contains one RBF network which is constructed using the entire training data and covers the whole data space. Higher layer RBFs are constructed by error-driven active learning. First, we calculate mapping errors remaining at the lth layer as

$$r_l(\mathbf{x}) = \hat{z}(\mathbf{x}) - \mathbf{H}_l(\mathbf{x}) \ , \tag{3}$$

where $\hat{z}(\mathbf{x})$ stands for actual measurement values. Then, significant error residuals are clustered and merged into M_{l+1} subdivisions [3]. Subsequently, these clusters are used to train M_{l+1} subnets in the higher layer $l+1$. At this stage, the targeted data space is largely reduced into a set of subregions, and the error residuals are used for the high layer training which reduces the dynamic range of neurons.

Each RBF subnet in the hierarchy is constructed by self-organising adaptive learning as described in Section 2.2. When the mapping accuracy of a subnet tends to a steady value, the subnet training stops. After construction of all RBF subnets in a current layer, mapping error residuals will be recalculated from all obtained layers, and new error clusters will be generated for training the next layer. Such hierarchical RBF construction terminates when a desired mapping accuracy is achieved.

2.2 Adaptivity of the RBFs

The adaptivity of the network derives from the flexibility in the process of network construction. To achieve network adaption to the underlying data, for each subnet, at each learning step: 1) the network grows one neuron, where necessary, based on the "novelty" of the input observation; 2) if the "novelty" criteria are not satisfied, resulting in no newly added neuron at this instance, a subset of neurons are adjusted in full parameter dimensions in accordance with a neighbourhood extended Kalman filter (NEKF) algorithm; 3) "pseudo" neurons, that consistently make little contribution to the network output, are pruned for network compactness.

Network growth. A RBF subnet starts by adopting the first training point as the first neuron. It grows by inserting a new neuron if the current observation satisfies the following three novelty conditions:

Condition 1: the input of the observation is far away from all existing neurons.
Condition 2: the network prediction error for the current observation is significant.
Condition 3: the RMS prediction error within a sliding window is significant.

When the three conditions are met, a new neuron is inserted into the current network of K units, with the following parameters at the position coincident at input $\mathbf{x}_n$:

$$\begin{aligned} \boldsymbol{\mu}_{K+1} &= \mathbf{x}_n \\ a_{K+1} &= e_n \\ \sigma_{K+1} &= \psi \left\| \mathbf{x}_n - \boldsymbol{\mu}_{nr} \right\| \ , \end{aligned} \tag{4}$$

where ψ is an *overlap factor* between the newly inserted neuron and its nearest neighbour. If an observation does not satisfy one of the novelty criteria, no new neuron is added. Instead, a neighbourhood EKF algorithm, described below, is utilised to adjust the network parameters to best fit the current observation.

Parameter updating. For updating network parameters, the Extended Kalman filters (EKF) [8] are preferable to the classical LMS or gradient descent (GD) methods [4]. The EKF usually updates all network parameters for all neurons at each learning step; we therefore refer to it as *global EKF* (GEKF). The computational complexity of the GEKF is $O(A^2)$ per learning step, where A denotes the number of parameters [8].

To minimise computational cost of the global EKF, we utilise a local approach named *neighbourhood EKF* (NEKF). At each learning step, only a subset of K_n neighbouring neuron(s) around the nth observation have their parameters updated. A neighbouring neuron is selected if it is 1) the nearest neighbour to the current observation, or it is 2) within a distance threshold, scaled by neuron resolution η_n, from the current nth observation. The computational cost of NEKF is reduced from $O((4K)^2)$ to a subset $O\left((4K_n)^2\right)$.

Neuron pruning. Network size can become too great under this growth strategy alone, possibly leading to overfiting. To avoid this, we promote network compactness by applying a pruning strategy. Pruning removes those neurons that make an insignificant contribution to the network output over a number of consecutive training observations. The pruning process is defined as follows:

For the current observation $(\mathbf{x}_n, z_n)$, compute the outputs of each hidden unit $o_k^n,\ k = 1, 2, ..., K$:

$$o_k^n = a_k \exp\left(-\frac{\|\mathbf{x}_n - \boldsymbol{\mu}_k\|^2}{\sigma_k^2}\right) . \tag{5}$$

Calculate the normalised output values r_k^n over the largest absolute hidden unit output:

$$r_k^n = \frac{o_k^n}{\max(o_1^n, o_2^n, ..., o_K^n)} . \tag{6}$$

If $r_k^n < \rho$ for W consecutive observations, then the kth node is removed, where ρ is the *pruning threshold.*

3 Application to 3D Point-Cloud Surface Reconstruction

The proposed hierarchical RBF network was implemented in C++. Testing was conducted on 3D surface reconstruction from range images [11]. The images were acquired at pixel resolution of 200×200. The scattered data are typically irregularly sampled with arbitrary topology and varying spatial densities. They contain measurement noise and holes. In order to apply the hierarchical RBF network, valid surface data stored in a 200×200 matrix associated with their locations were normalised and re-sampled into a random order. Such a normalised random sequence was used as a training set of N 2D observation inputs $\mathbf{x}_n = [x_n, y_n]$ and outputs z_n.

3.1 Surface Reconstruction and Mesh Repair

Fig. 2 shows examples of surface reconstruction. The Gaussian RBF networks provide extraordinary interpolation and extrapolation capabilities to: 1) smoothly reconstruct surfaces from non-uniformly sampled data with highly variable spatial frequencies; 2) smoothly blend and repair raw noisy data to fill irregular holes that are large compared to the geometric variation in surfaces; 3) smoothly extend surfaces where there is no data, by extrapolation.

3.2 Progressive Reconstruction by Error-Driven Active Learning

The top row of Fig. 3 shows the angel reconstruction with three progressive layers of error-driven active learning. Corresponding error clusters, presented as error bars in different colours, are shown underneath. Un-clustered small mapping

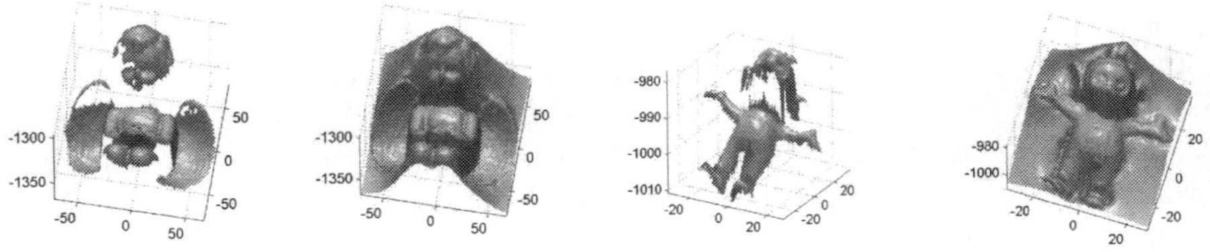

(a) angel

Fig. 2. Incomplete mesh and reconstructed RBF mesh

errors are displayed in black. The quality of reconstruction increases with the number of layers. Incremental surface optimization is obvious in most difficult regions with significant local errors in relation to residual clusters. Remarkable non-uniformly distributed mapping errors exist at layer 1, especially at the edges, due to data discontinuity. These are largely mended in layer 2, using 13 subnets and further reduced by layer 3, with 8 subnets.

3.3 Neuron Adaptivity

Neurons can be added or removed according to the novelty criteria. They are adjustable on all parameters: location, width and weight. Fig. 4 illustrates neuron distribution with three layers of the angel image reconstruction. Neurons are represented by circles in normalised space with a display ratio of 3:1. For intuitive visualisation, the neurons are displayed in 3D space with their corresponding vertical z-values evaluated from the network. The distribution and density of neurons in layer 1 are highly adaptive to the spatial complexity of the underlying angel data. For layer 2 and 3, neurons in subnets are located in subregions associated with error clusters in Fig. 3. Layer 1 contains more of

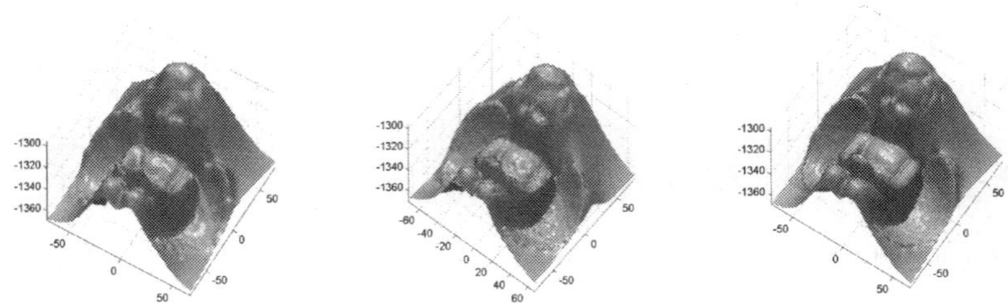

(a) RBF meshes on layer 1, 2 and 3

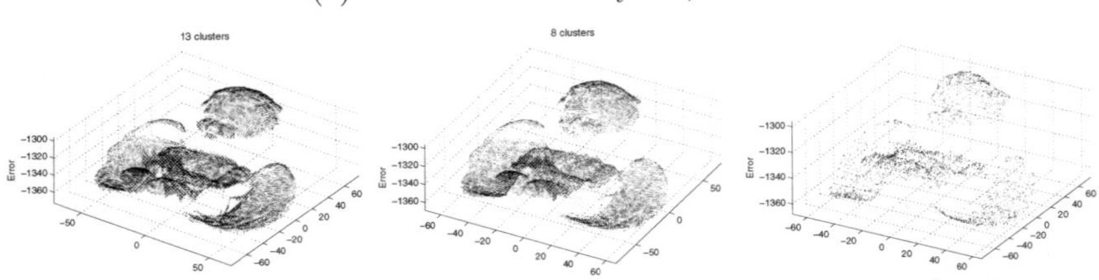

(b) error clusters on layer 1, 2 and 3

Fig. 3. Progressive surface reconstruction

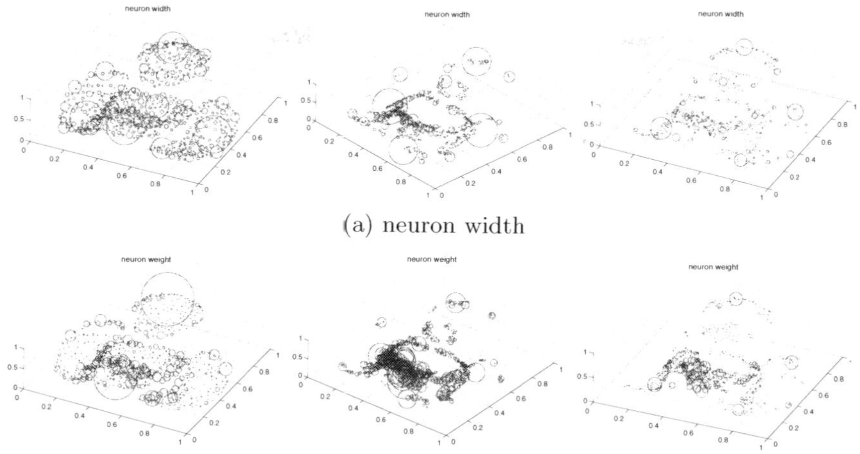

(a) neuron width

(b) neuron weight

Fig. 4. Adaptive neuron distribution of angel reconstruction in normalised space: 711, 405 and 176 Gaussians in layer 1, 2 and 3 respectively

the larger neurons than layers 2 and 3 to produce a coarse global base. Within each subnet, neurons with large widths or high weights can often be generated initially for smooth bases. Although there is an inherent tendency by the greedy algorithm to favour absorption of lower frequencies before higher ones, smaller neurons consistently refine the smoothness towards fidelity to local data so as to guarantee a coarse-to-fine optimality for high frequency details.

3.4 Pruning Effectiveness

Fig. 5 illustrates the pruning effectiveness using the example of angel reconstruction at layer 1. As shown in Fig.5(a), at the start of training, neurons are consistently added to the network, and the training error is reduced rapidly, as shown in Fig.5(b). As the RMS error tends to a steady value after 5000 training steps with about 520 neurons, the pruning strategy works effectively to remove "pseudo" neurons, so overcoming the risk of overfiting.

3.5 Data Compression

Table 1 presents data compression from the reconstruction examples in Fig. 2, in which N stands for the number of 3D points in each range scan, and K denotes the number of Gaussian neurons in the resulting hierarchical RBF network. Storage compression ratio is calculated from $3N : 4K$, since each Gaussian neuron has 4 parameters. Reconstruction accuracy achieved by the resulting RBF network is indicated through the average mapping error $\bar{e} = \frac{1}{N}\sum_{n=1}^{N} |z_n - H([x_n, y_n])|$ of all scan data in normalised space.

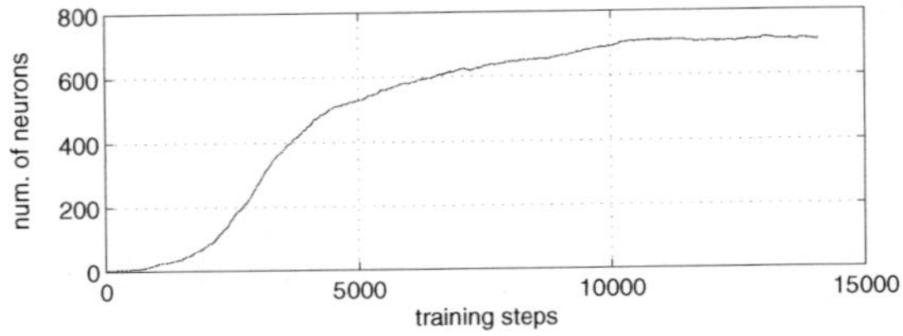

(a) neuron adding and pruning

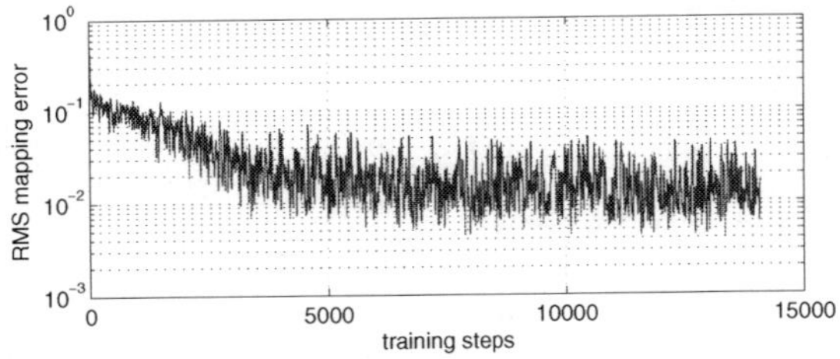

(b) RMS mapping error

Fig. 5. Sequential training of the angel data

Table 1. Data compression and accuracy

range image	num. of points N in range image	num. of Gaussians K in RBF network	storage compression ratio	normalised mapping error $\bar{e}$
angel	14,089	1292	8.2	.0052
teletubby	6841	685	7.5	.0044

4 Conclusion

We have presented an adaptive hierarchical RBF network for functional modelling of large amounts of unstructured data. The network employs error-driven multi-layer active learning to achieve global optimality as well as conquer large mapping errors in subdivisions. RBFs are used as the computational substrate. Each RBF subnet is constructed by heuristic adaptive learning. Neurons are dynamically located, adapting to the fidelity of underlying data. The full range of network parameters, corresponding to location, weight and width of each neuron, are refined iteratively. A pruning strategy is used to promote network compactness. Compared to approaches using pre-defined fixed network structures, adaptive learning provides significant flexibility, particularly suited to data with highly variable spatial frequencies. In addition, we developed a neighbourhood EKF method for network updating which led to a remarkable computation reduction.

The hierarchical RBF network has been tested in a real-world application of water-tight point-cloud surface reconstruction. Experimental results demonstrate the proposed network offers a new approach to the problems of mesh reconstruction, geometric formulation, data simplification and compression.

References

1. Carr, J., Beatson, R., Cherrie, J., Mitchell, T., Fright, W., McCallum, B., Evans, T.: Reconstruction and representation of 3D objects with radial basis functions. In: ACM SIGGRAPH, pp. 67–76. ACM Press, New York (2001)
2. Ferrari, S., Maggioni, M., Borghese, N.A.: Multiscale approximation with hierarchical radial basis functions networks, IEEE Trans. on Neural Networks 15(1), 178–188 (2004)
3. Johnson, S.C.: Hierarchical clustering schemes. Psychometrika 32, 241–254 (1967)
4. Karayiannis, N.: Reformulated radial basis neural networks trained by gradient descent. IEEE Trans. on Neural Networks 10(3), 657–671 (1999)
5. Kojekine, N., Hagiwara, I., Savchenko, V.: Software tools using CSRBF for processing scattered data. Computer & Graphics 27(2), 463–470 (2003)
6. Morse, B., Yoo, T., Rheingans, P., Chen, D.T., Subramanian, K.R.: Interpolating implicit surfaces from scattered surface data using compactly supported radial basis functions. In: Proceedings of Shape Modeling International (2001)
7. Platt, J.: A resource-allocating network for function interpolation. Neural Comput. 3(2), 213–225 (1991)
8. Simon, D.: Training radial basis neural networks with the extended Kalman filter. Neurocomputing 48, 455–475 (2002)
9. Turk, G., O'Brien, J.: Modelling with implicit surfaces that interpolate. ACM Trans. on Graphics 21(4), 855–873 (2002)
10. Varady, L., Hoffmann, M., Kovacs, E.: Improved free-form modelling of scattered data by dynamic neural networks. Journal for Geometry and Graphics 3(2), 177–181 (1999)
11. `http://sampl.ece.ohio-state.edu/data/3ddb/rid/minolta`, range image database at Ohio SAMPL

Stacking MF Networks to Combine the Outputs Provided by RBF Networks

Joaquín Torres-Sospedra, Carlos Hernández-Espinosa, and Mercedes Fernández-Redondo

Departamento de Ingenieria y Ciencia de los Computadores, Universitat Jaume I, Avda. Sos Baynat s/n, C.P. 12071, Castellon, Spain
{jtorres, espinosa, redondo}@icc.uji.es

Abstract. The performance of a Radial Basis Functions network (RBF) can be increased with the use of an ensemble of RBF networks because the RBF networks are successfully applied to solve classification problems and they can be trained by gradient descent algorithms. Reviewing the bibliography we can see that the performance of ensembles of Multilayer Feedforward (MF) networks can be improved by the use of the two combination methods based on *Stacked Generalization* described in [1]. We think that we could get a better classification system if we applied these combiners to an *RBF* ensemble. In this paper we satisfactory apply these two new methods, *Stacked* and *Stacked+*, on ensembles of *RBF* networks. Increasing the number of networks used in the combination module is also successfully proposed in this paper. The results show that training 3 *MF* networks to combine an RBF ensemble is the best alternative.

1 Introduction

A *Radial Basis Functions* (*RBF*) network is a commonly applied architecture which is used to solve classification problems. This network can also be trained by gradient descent [2,3]. So with a fully supervised training, it can be an element of an ensemble of neural networks. Previouses comparisons showed that the performance of *RBF* networks was better than *Multilayer Feedforward* (*MF*) networks. Being the *Simple Ensemble* the best method to train an ensemble of *RBF* networks.

Among the methods of combining the outputs of an ensemble of neural networks, the two most popular are the *Majority voting* and the *Output average* [4]. Last year, two new combination methods based on the idea of *Stacked Generalization* called *Stacked* and *Stacked+* were successfully proposed in [1]. These new methods consist of training a single *MF* to combine the networks.

In this paper, we want to apply these combiners to a ensemble of *Radial Basis Functions* trained with the *Simple Ensemble*. Moreover, we want to increase the number of networks used to combine the ensemble in our experiments in order get a better combiner. Finally, we compare the results we have got with these two new combiners with the results we previously got with other classical combiners.

To test *Stacked* and *Stacked+* with *RBF* ensembles, we have selected nine databases from the UCI repository. This paper is organized as follows. The concepts related to the ensembles training and combination are briefly commented in section 2 whereas the experimental setup, results and the discussion are in section 3.

J. Marques de Sá et al. (Eds.): ICANN 2007, Part I, LNCS 4668, pp. 450–459, 2007.

2 Theory

2.1 Training a Neural Network

There are conceptual similarities in the training process of an *RBF* network and an *MF* network since a gradient descent method can be applied to train both networks [5,6].

In our experiments we have trained the networks for few iterations. In each iteration, the network trainable parameters have been adapted with BackPropagation over the training set. These parameters are the weights for the case of the *MF* networks whereas the weigths and the centers of the gaussian units are the trainable parameters for the case of the *RBF* network. At the end of the iteration the Mean Square Error (MSE) has been calculated over the Validation set. When the learning process has finished, we assign the network configuration of the iteration with minimum MSE to the final network.

For this reason the original learning set L used in the learning process is divided into two subsets: The first set is the training set T which is used to train the networks and the second set is the validation set V which is used to finish the training process.

Algorithm 1. Neural Network Training $\{T\ ,V\}$

for $i = 1$ to $iterations$ **do**
  Train the network with the patterns from the training set T
  Calculate MSE over validation set V
  Save epoch weights and calculated MSE
end for
Select epoch with minimum MSE
Assign best epoch configuration to the network
Save network configuration

2.2 The Multilayer Feedforward Network

The *Multilayer Feedforward* architecture is the most known neural network architecture. This kind of networks consists of three layers of computational units. The neurons of the first layer apply the identity function whereas the neurons of the second and third layer apply the sigmoid function. It has been proved that *MF* networks with one hidden layer and threshold nodes can approximate any function with a specified precision [7,8]. Figure 1 shows the diagram of the *Multilayer Feedforward* network.

2.3 The Radial Basis Function Network

An RBF network has two layer of neurons. The first one, in its usual form, is composed of neurons with Gaussian transfer functions (GF). The second layer is composed of neurons with linear transfer functions. This network can be the base classifier of an ensemble of neural networks since gradient descent can be applied to train it [2,3]. A basic RBF network can be seen in figure 2.

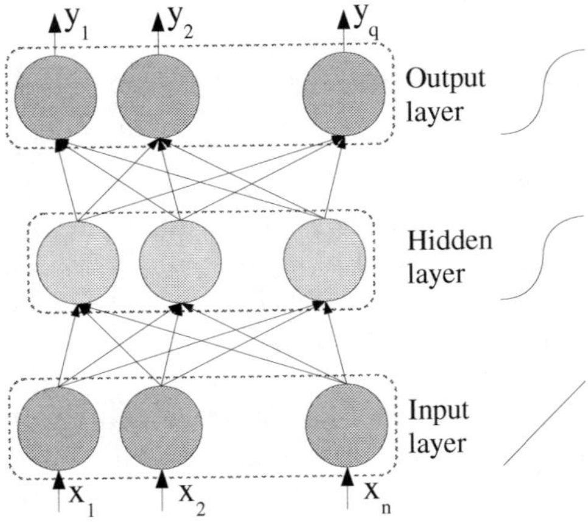

Fig. 1. A MF network

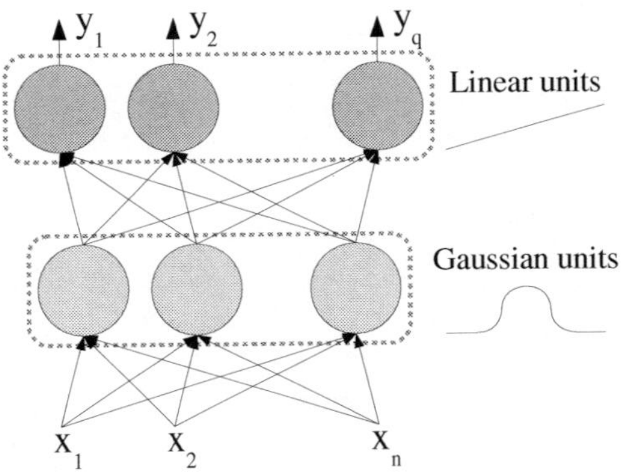

Fig. 2. An RBF network

2.4 Combining Networks with Stacked Generalization

Stacked Generalization was introduced by Wolpert [9]. Some authors have adapted the Wolpert's method to use with neural networks [10,11,12]. In [1] two combination methods based on *Stacked Generalization* were proposed, *Stacked* and *Stacked+*. In these methods a single combination network was trained to combine the expert networks of the ensemble. In both kinds of networks, expert and combination, the neural network architecture choosen was the *MF* network.

The use of the original pattern input vector is the difference between *Stacked* and *Stacked+*. *Stacked* uses the output provided by the expert networks on patterns from the training set whereas *Stacked+* uses the output provided by the experts along with the original pattern input vector to train the combination network.

In this paper we want to combine ensembles of RBF networks with *Stacked* and *Stacked+*. Moreover we want to increase the number of combination networks from one to three and nine as an ensemble of combiners to test if the system could be improved by adding more combination networks. With this procedure we want to combine the expert networks of an ensemble of *RBF* networks with an ensemble of *MF* networks. Finally, we have applied the output average in order to combine the combination networks. Figure 3 show the diagram of *Stacked* and *Stacked+* we have used in our experiments.

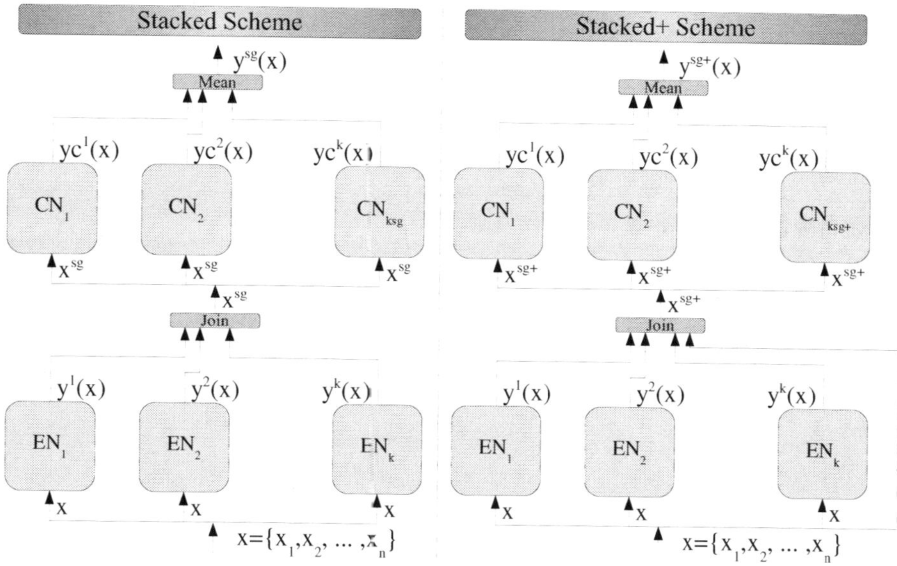

Fig. 3. Stacked and Stacked+ diagrams

3 Experimental Setup

In this section we describe the experimental setup and the datasets we have used in our experiments. Then, we show the main results we have obtained with the combination methods on the different datasets. Moreover, we calculate two general measurements in order to compare the methods. Finally, we discus about the results we have got.

In our experiments we have used ensembles of 3 and 9 *RBF* networks previously trained with *Simple ensemble* on nine different classification problems. Moreover, we have trained 1, 3 and 9 *MF* networks with *Stacked* and *Stacked+* in order to combine the networks of the ensemble. In addition, we have generated 10 different partitions of data at random in training, validation and test sets and repeat the whole learning process 10 times in order to get a mean performance of the ensemble an error in the performance by standard error theory.

3.1 Databases

The datasets we have used in our experiments and their characteristics are described in this subsection. We have applied the new combination methods, *Stacked* and *Stacked+*, to nine classification problems from the UCI repository [13]. The datasets we have used are the following ones:

Balance Scale Database (bala)
The aim is to determine if a balance is scaled tip to the right, tip to the left, or balanced. This dataset contains 625 instances, 4 attributes and 3 classes.

Cylinder Bands Database (band)
Used in decision tree induction for mitigating process delays know as "cylinder bands" in rotogravure printing. This dataset contains 512 instances, 19 attributes and 2 classes.

BUPA liver disorders (bupa)
The aim of this dataset is to try to detect liver disorders. This dataset contains 345 instances, 6 attributes and 2 classes.

Australian Credit Approval (cred)
This dataset concerns credit card applications. This dataset contains 653 instances, 15 attributes and 2 classes.

Glass Identification Database (glas)
The aim of the dataset is to determinate if the glass analysed was a type of 'float' glass or not for Forensic Science. This dataset contains 2311 instances, 34 attributes and 2 classes.

Heart Disease Databases (hear) The aim of the dataset is to determinate the presence of heart disease in the patient. This dataset contains 297 instances, 13 attributes and 2 classes.

The Monk's Problem 1 (mok1)
Artificial problem with binary inputs. This dataset contains 432 instances, 6 attributes and 2 classes.

The Monk's Problem 2 (mok2)
Artificial problem with binary inputs. This dataset contains 432 instances, 6 attributes and 2 classes.

Congressional Voting Records Database (Vote)
Classification between Republican or Democrat. All attributes are boolean. This dataset contains 432 instances, 6 attributes and 2 classes.

Table 1 shows the training parameters (*number of clusters*, *iterations*, *adaptation step* and the *width of the gaussian units*) of the expert networks and the performance of a single network on each database. Moreover we have added to this table the performance of the ensembles of 3 and 9 RBF networks previously trained with *Simple Ensemble* in order to see if the new combination methods proposed increase the performance of the classification systems.

Table 2 shows the training parameters we have used to train the combination networks (*hidden units*, *adaptation step*, *momentum rate* and *number of iterations*) with the new two combiners, *Stacked* and with *Stacked+*.

Table 1. Expert networks training parameters

	Experts - RBF parameters				Experts performance		
	clusters	*iterations*	*step*	*width*	*1 net*	*3 nets*	*9 nets*
bala	60	6000	0.005	0.6	90.2	89.68	89.68
band	40	10000	0.01	1	74	73.82	73.27
bupa	40	8000	0.01	0.4	70.1	71.86	72.43
cred	30	10000	0.005	2	86	87.15	87.23
glas	110	20000	0.01	0.4	93	93.2	93
hear	20	15000	0.005	2	82	83.9	83.9
mok1	30	20000	0.005	0.8	98.5	99.63	99.63
mok2	45	60000	0.005	0.6	91.3	91.5	91.38
vote	5	5000	0.01	1.8	95.4	96.25	96.25

Table 2. Combination network parameters - Stacked and Stacked+

		Stacked - MF parameters				*Stacked+ - MF parameters*			
dataset	*experts*	*hidden*	*step*	*mom*	*ite*	*hidden*	*step*	*mom*	*ite*
bala	3	22	0.4	0.2	1500	10	0.4	0.05	3500
	9	3	0.1	0.01	3000	11	0.1	0.01	6500
band	3	7	0.4	0.1	500	26	0.05	0.01	2500
	9	30	0.1	0.1	500	30	0.05	0.05	500
bupa	3	3	0.4	0.1	750	4	0.2	0.1	750
	9	6	0.2	0.05	750	19	0.4	0.05	750
cred	3	21	0.4	0.2	500	27	0.05	0.01	750
	9	21	0.4	0.2	500	25	0.4	0.1	3500
glas	3	4	0.1	0.001	7500	4	0.4	0.1	7500
	9	4	0.4	0.2	7500	4	0.4	0.2	7500
hear	3	4	0.4	0.2	5000	2	0.2	0.05	3500
	9	17	0.1	0.2	1500	3	0.4	0.1	7500
mok1	3	2	0.4	0.2	7500	4	0.4	0.2	7500
	9	2	0.4	0.2	7500	3	0.4	0.2	7500
mok2	3	2	0.1	0.1	1000	2	0.4	0.1	7500
	9	9	0.4	0.1	7500	1	0.4	0.1	7500
vote	3	28	0.4	0.2	500	30	0.05	0.01	750
	9	26	0.4	0.1	500	12	0.4	0.05	500

3.2 Results

The main results we have obtained with the application of stacking methods are presented in this subsection. Tables 3 and 4 shows the results we have obtained combining ensembles of 3 and 9 networks with *Stacked* and *Stacked+*.

In [14] the complete results of the combination of RBF ensembles with 14 different combination methods are published. Although we have omitted these results to keep the length of the paper short, the general measurements related to these combination methods appear in subsection 3.3. These methods are: *Majority Vote* (vote), *Winner*

Table 3. Results of the 3 RBF Network Ensemble

	stacked	*stacked3*	*stacked9*	*stacked+*	*stacked3+*	*stacked9+*
bala	92.88±0.72	92.96±0.70	92.96±0.70	93.44±0.70	93.36±0.70	93.44±0.68
band	74.36±1.03	74.55±0.94	74.36±0.88	74.73±1.03	75.27±0.95	75.45±0.99
bupa	72.00±1.15	71.43±1.09	71.00±1.11	72.00±1.21	72.14±1.19	72.29±1.19
cred	86.85±0.58	86.85±0.58	86.85±0.58	87.23±0.74	87.00±0.75	87.15±0.73
glas	93.80±1.25	93.20±1.04	93.40±1.08	93.00±1.09	93.00±1.09	93.00±1.09
hear	83.22±1.63	82.88±1.68	83.05±1.64	83.39±1.36	83.73±1.34	83.22±1.44
mok1	99.63±0.38	99.63±0.38	99.63±0.38	99.63±0.38	99.63±0.38	99.63±0.38
mok2	91.25±1.26	91.25±1.26	91.25±1.26	91.50±1.15	91.38±1.13	91.38±1.13
vote	95.38±0.88	95.50±0.84	95.63±0.77	96.13±0.44	96.13±0.44	96.13±0.44

Table 4. Results of the 9 RBF Network Ensemble

	stacked	*stacked3*	*stacked9*	*stacked+*	*stacked3+*	*stacked9+*
bala	92.88±0.70	93.04±0.66	92.88±0.69	94.08±0.64	93.84±0.66	93.76±0.66
band	73.82±1.02	74.00±1.27	74.00±1.27	74.73±1.20	74.73±1.20	74.55±1.21
bupa	72.29±1.21	71.86±1.31	72.14±1.21	71.57±1.12	71.71±1.16	71.29±1.19
cred	86.46±0.63	86.46±0.63	86.46±0.63	86.46±0.63	86.38±0.64	86.38±0.62
glas	92.80±0.85	93.40±0.90	93.20±1.04	93.60±0.93	93.60±0.93	93.40±0.90
bala	82.88±1.78	82.88±1.78	83.05±1.60	83.22±1.39	83.22±1.37	82.88±1.42
band	99.63±0.38	99.63±0.38	99.63±0.38	99.63±0.38	99.63±0.38	99.63±0.38
bupa	91.25±1.24	91.25±1.24	91.25±1.24	91.63±1.24	91.63±1.24	91.75±1.18
cred	96.13±0.68	96.13±0.68	96.13±0.68	95.88±0.46	96.00±0.49	96.00±0.49

Takes All (wta), *Borda Count* (borda), *Bayesian Combination* (bayesian), *Weighted Average* (w.ave), *Choquet Integral* (choquet), *Choquet Integral with Data-Depend Densities* (choquet.dd), *Weighted Average with Data-Depend Densities* (w.ave.dd), *BADD Defuzzification Startegy* (badd), *Zimmermann's Compensatory Operator* (zimm), *Dinamically Averaged Networks versions 1 and 2* (dan and dan2), *Nash vote* (nash).

3.3 General Measurements

We have also calculated the percentage of error reduction (PER) of the results with respect to a single network to get a general value for the comparison among all the methods we have studied. We have used equation 1 to calculate the PER value.

$$PER = 100 \cdot \frac{Error_{singlenetwork} - Error_{ensemble}}{Error_{singlenetwork}} \tag{1}$$

$$IoP = Performance_{ensemble} - Performance_{singlenet} \tag{2}$$

The PER value ranges from 0%, where there is no improvement by the use of a particular ensemble method with respect to a single network, to 100%. There can also be negative values, which means that the performance of the ensemble is worse.

Furthermore, we have calculated the mean increase of performance (IoP) with respect to *Single Network* and the mean percentatge of error reduction (PER) across all databases for the methods proposed in this paper, *Stacked* and *Stacked+*, and for the combination methods that appear in [14]. The PER is calculated by equation 1 whereas the IoP with respect a single network is calculated by equation 2. Table 5 shows the results of the mean PER and the mean IoP.

Table 5. General Measurements

	mean PER		mean IoP	
combiner	*3 experts*	*9 experts*	*3 experts*	*9 experts*
average	13.06	12.63	0.72	0.69
stacked	14.85	14.43	0.98	0.84
stacked3	13.85	15.48	0.85	0.9
stacked9	14.34	15.19	0.83	0.91
stacked+	16.27	17.27	1.11	1.14
stacked3+	16.43	17.28	1.18	1.13
stacked9+	16.37	16.58	1.18	1.01
vote	13.51	13.92	0.83	0.85
wta	12.71	12.63	0.57	0.62
borda	13.42	13.83	0.82	0.84
bayesian	11.65	12.91	0.8	0.88
w.ave	14.16	13.61	0.71	0.57
choquet	12.47	11.7	0.55	0.62
choquet.dd	11.77	12.09	0.51	0.65
w.ave.dd	13.86	12.42	0.81	0.66
badd	13.06	12.63	0.72	0.69
zimm	−146.23	−168.3	−10.53	−11.72
dan	11.02	6.46	0.58	0.24
dan2	12.27	7.44	0.55	0.28
nash	13.57	12.67	0.82	0.69

3.4 Discussion

The main results (tables 3-4) show that *Stacked* and *Stacked+* get an improvement in a wide majority of problems: *bupa*, *cred*, *glas*, *hear*, *mok1* and *vote*.

The results show that the improvement in performance of training an ensemble of nine combination networks *Stacked9/Stacked9+* (instead of three *Stacked3/Stacked3+*) is low. Taking into account the computational cost the best alternative might be an ensemble of three combination networks *Stacked3/Stacked3+*.

Comparing the results of the different traditional combination methods with *Stacked* and *Stacked+*, we can see that there is an improvement by the use of these new methods. For example, in databases *band* and *bala* the results with the methods based on *Stacked Generalization* are quite good. The largest difference between simple average and other method is around 4.0% in the problem *bala* and around 1.5% in the problem *band*.

Comparing the general measurements, the mean *PER* and the mean *IoP*, we can see that *Stacked* and *Stacked+* are the best alternative to combine an ensemble of *RBF* networks. *Stacked+* with 3 combination networks is the best way to combine ensembles of 3 and 9 RBF networks according to the values of the general measurements.

4 Conclusions

In this paper, we have presented experimental results by the use of *Stacked* and *Stacked+*, two new methods based on *Stacked Generalization*, in order to combine the outputs of an ensemble of RBF networks, using nine different databases.

We have trained ensembles of 3 and 9 combination networks (MF) to combine a previously trained ensemble of expert networks (RBF). The results show that, in general, there is a reasonable improvement by the use of *Stacked* and *Stacked+* in a wide majority of databases.

In addition, we have calculated the mean percentage of error reduction over all databases. According to the values of the mean performance of error reduction, the new combination methods, *Stacked* and *Stacked+* are the best methods to combine ensembles of *RBF* networks.

Finally, taking into account the computational cost and the values of the general measurements we can conclude that training 3 combination networks, as an ensemble of *MF* networks, should be considered to be the best alternative when we combine ensembles of *RBF* networks.

Acknowledgments

This research was supported by the project number *P1·1B2004-03* entitled *'Desarrollo de métodos de diseño de conjuntos de redes neuronales'* of Universitat Jaume I - Bancaja in Castellón de la Plana, Spain.

References

1. Torres-Sospedra, J., Hernndez-Espinosa, C., Fernndez-Redondo, M.: Combining MF networks: A comparison among statistical methods and stacked generalization. In: Schwenker, F., Marinai, S. (eds.) ANNPR 2006. LNCS (LNAI), vol. 4087, pp. 302–9743. Springer, Heidelberg (2006)
2. Hernndez-Espinosa, C., Fernndez-Redondo, M., Torres-Sospedra, J.: First experiments on ensembles of radial basis functions. In: Roli, F., Kittler, J., Windeatt, T. (eds.) MCS 2004. LNCS, vol. 3077, pp. 253–262. Springer, Heidelberg (2004)
3. Torres-Sospedra, J., Hernndez-Espinosa, C., Fernndez-Redondo, M.: An experimental study on training radial basis functions by gradient descent. In: Schwenker, F., Marinai, S. (eds.) ANNPR 2006. LNCS (LNAI), vol. 4087, pp. 302–9743. Springer, Heidelberg (2006)
4. Drucker, H., Cortes, C., Jackel, L.D., LeCun, Y., Vapnik, V.: Boosting and other ensemble methods. Neural Computation 6, 1289–1301 (1994)
5. Karayiannis, N.B.: Reformulated radial basis neural networks trained by gradient descent. IEEE Transactions on Neural Networks 10, 657–671 (1999)
6. Karayiannis, N.B., Randoph-Gips, M.M.: On the construction and training of reformulated radial basis function neural networks. IEEE Transactions on Neural Networks 14, 835–846 (2003)

7. Bishop, C.M.: Neural Networks for Pattern Recognition. Oxford University Press, Inc., New York, NY, USA (1995)
8. Kuncheva, L.I.: Combining Pattern Classifiers: Methods and Algorithms. Wiley-Interscience, Chichester (2004)
9. Wolpert, D.H.: Stacked generalization. Neural Networks 5, 1289–1301 (1994)
10. Ghorbani, A.A., Owrangh, K.: Stacked generalization in neural networks: Generalization on statistically neutral problems. In: IJCNN 2001. Proceedings of the International Joint conference on Neural Networks, Washington DC, USA, pp. 1715–1720. IEEE Computer Society Press, Los Alamitos (2001)
11. Ting, K.M., Witten, I.H.: Stacked generalizations: When does it work? In: International Joint Conference on Artificial Intelligence proceedings, vol. 2, pp. 866–873 (1997)
12. Ting, K.M., Witten, I.H.: Issues in stacked generalization. Journal of Artificial Intelligence Research 10, 271–289 (1999)
13. Newman, D.J., Hettich, S., Blake, C.L., Merz, C.J.: UCI repository of machine learning databases (1998), `http://www.ics.uci.edu/~mlearn/MLRepository.html`
14. Torres-Sospedra, J., Hernandez-Espinosa, C., Fernandez-Redondo, M.: A comparison of combination methods for ensembles of RBF networks. In: IJCNN 2005. Proceedings of International Conference on Neural Networks, Montreal, Canada, vol. 2, pp. 1137–1141 (2005)

Neural Network Processing for Multiset Data

Simon McGregor

Centre for Computational Neuroscience and Robotics, UK
sm66@sussex.ac.uk

Abstract. This paper introduces the notion of the *variadic neural network* (VNN). The inputs to a variadic network are an arbitrary-length list of n-tuples of real numbers, where n is fixed. In contrast to a recurrent network which processes a list sequentially, typically being affected more by more recent list elements, a variadic network processes the list simultaneously and is affected equally by all list elements. Formally speaking, the network can be seen as instantiating a function on a *multiset* along with a member of that multiset. I describe a simple implementation of a variadic network architecture, the *multi-layer variadic perceptron* (MLVP), and present experimental results showing that such a network can learn various variadic functions by back-propagation.

1 Introduction

Consider the following imaginary scenario. You are considering an individual for a loan; this person has been rated by a number of credit rating agencies (who do not all agree on a single credit rating) and you want to decide whether or not they are likely to repay the loan, based on those credit ratings. Fortunately, you have some preexisting data on loan repayment, linked to credit ratings; unfortunately, individuals are rated by different numbers of agencies and your system doesn't record which credit agency gave which rating - all you have is a list of different ratings for each individual.

In problems like this one, the data points are *multisets*: variable-size, unordered lists with possible repetition (also known as *bags*). Computer science uses the term *variadic* to describe a function v is on a variable-length list of inputs. I will define a *symmetric* variadic function $v : S* \rightarrow T*$ as one where each element of its input is drawn from the same set S, where its output is a list of the same length as its input, and where permutation of its input results only in like permutation of its output.

Existing machine learning methods tend to treat multisets as vectors in a finite-dimensional space [2,5] (for instance, the multisets of credit ratings described above would be treated as vectors in a 5-dimensional space, if credit ratings can take one of 5 separate values; a "bag of words" approach to document processing treats a document as a high-dimensional vector with each dimension representing a word stem). However, this approach suffers from a problem when the multisets are drawn from the real numbers; the dimensionality of the implied vector space becomes infinite and the natural metric on real numbers is ignored

J. Marques de Sá et al. (Eds.): ICANN 2007, Part I, LNCS 4668, pp. 460–470, 2007.

by the vectorisation. This paper describes an approach which operates directly on multisets of real numbers (or tuples of real numbers) allowing standard neural network learning methods to be applied. The approach is a prelimary exploration both of new methods for machine learning on multisets and of new data domains for neural network methods.

I discuss the limitations of conventional networks, propose a modified architecture specifically for multiset data, describe a training algorithm for this architecture based on backpropagation, present experimental results indicating the new architecture can learn, and in the conclusion outline future work which should be done on this topic. An appendix contains the derivation of the backprop algorithm.

2 Limitations of Conventional Neural Networks

In a conventional feed-forward neural network the signals between nodes are real numbers and the architecture of the network dictates how many real inputs and outputs the network has. To provide additional inputs to the network, new links have to be added and the network must be retrained. How could one go about training a conventional network on multiset data? Below are some possible approaches.

1. Choose a maximum number n_{max} of ratings per individual and train a discriminator network with n_{max} separate inputs. When an individual has fewer ratings than n_{max}, feed the unused inputs with a -1 signal to denote that they are null; alternatively, have an additional n_{max} inputs which denote whether the corresponding rating exists or not.
2. Put the ratings for an individual in an arbitrary sequential order and train a recurrent discriminator network using these sequences.
3. Perform some sort of preprocessing which collapses an individual's ratings into a fixed number n_k of indicator variables, e.g. the mean, standard deviation and count of ratings for each individual. Then train a discriminator network with n_k inputs.

Unfortunately, each of the above described solutions comes with significant drawbacks:

1. If a new individual is encountered with $n' > n_{max}$ ratings, a new network with n' inputs must be trained or some of the ratings simply ignored, because there is no way for the existing network to process the additional inputs. Furthermore, the network will probably not generalise between different numbers of inputs and will be influenced by the order in which the inputs are presented.
2. Recurrent networks tend to be more strongly influenced by recent inputs than older ones and to be influenced by the order of presentation of the inputs.

3. There is no guarantee that the preprocessing will capture all the relevant information about the data: for instance, two sets of credit ratings can have the same mean, standard distribution and count, but still be significantly different from one another.

Kernel-based learning methods are often more flexible than conventional ones in the structure of the data they can process (e.g. string [3] and tree [1] kernels exist for learning with structured data) but this author is not aware of any known kernels developed specifically to respect the properties of multisets.

3 The Variadic Network

This paper proposes an alternative solution for applying neural network learning methods to multiset data. A variation of the standard multi-layer perceptron architecture is presented which allows a network to process an arbitrary-size multiset of real input tuples using a fixed network topology and a fixed number of network parameters. Unlike recurrent networks, which are affected by the order of presentation of input elements, the proposed architecture is constrained to treat all input elements symmetrically.

Most network training methods take advantage of the continuous nature of their signal domain, but there is nothing in principle to limit the signals carried by a network's links to being real scalars. This paper will consider the case where the network's signal domain is $\mathbb{R}*$, i.e. arbitrary-length lists of real numbers.

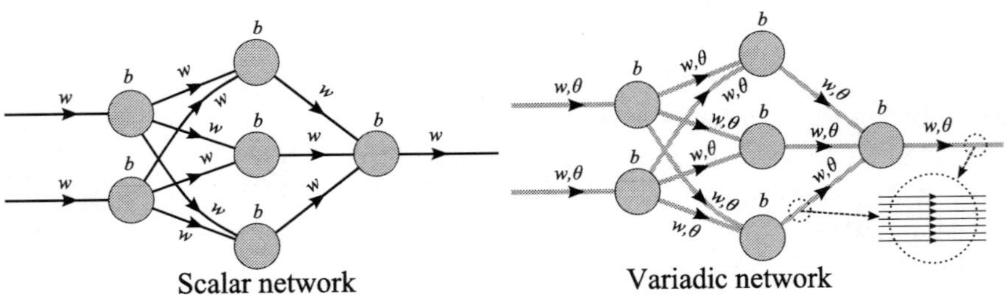

Fig. 1. Above: a scalar network with two real scalar inputs and one real scalar output. Below: a variadic network with two real vector inputs, both of arbitrary but identical dimension, and one real vector output of the same dimension. In the variadic network, each link has an extra scalar parameter θ.

I will use the term *scalar network* to describe a conventional feedforward sigmoidal network with scalar value signals. When a variadic network as described below is provided only with scalar input (i.e. vectors of length 1) it behaves like a scalar network, except that the interaction-weight parameters are redundant equivalents of the weight parameters. In other words, scalar networks are a sort of special case of variadic networks, where the data are always of dimensionality 1.

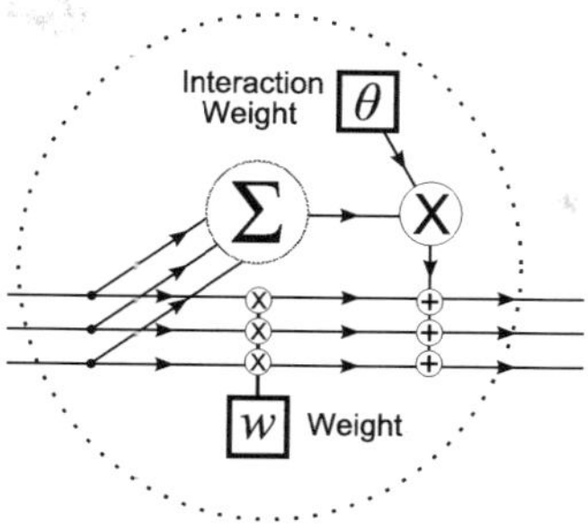

Fig. 2. The interior of a variadic link. The multiplicative weight parameter w is the same as in a conventional network but there is also an interaction-weight parameter θ which allows interaction between different elements of the signal vector.

3.1 The Multiplicative Link, Variadicised

In the MVLP, an interaction term between vector elements is introduced into the network links. In order to allow for interaction between the different elements of a vector, the link's vector output is set equal to a scalar multiple of its input, plus for each element a scalar multiple of the L^1-norm (i.e. the simple sum of elements) of its input. This new sort of link allows for the construction of networks which perform non-trivial variadic computations.

3.2 The Sigmoidal Neuron, Variadicised

A conventional sigmoidal neuron takes a number of real inputs (typically including a constant bias term), sums them to give an activation and outputs some sigmoidal function of the activation. Suppose instead that the inputs to the neuron were vectors of real numbers. In this paper we will consider neurons which simply sum the vectors to give a vector activation, add a bias, and output a vector of "squashed" (sigmoided) activations. Note that the vector inputs of the neuron can be of arbitrary dimension, providing all are of the same dimension.

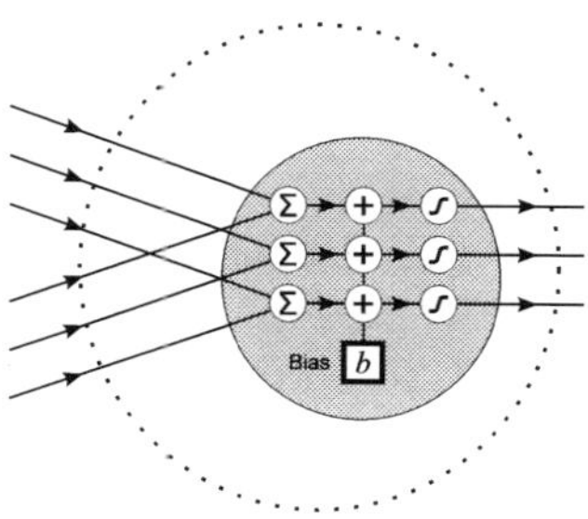

Fig. 3. The interior of a variadic neuron. The neuron performs a pointwise biased sigmoidal function such as tanh(·) like an array of conventional neurons.

Consequently, for a variadic neuron with n vector inputs $\{\mathbf{x_1}, \mathbf{x_2}, ...\}$ each of dimension d, corresponding scalar weights $\{w_1, w_2, ...\}$ and interaction-weights $\{\theta_1, \theta_2, ...\}$, and neuron bias b, the activation $\mathbf{A}$ and output $\mathbf{y}$ are given by: -

$$\mathbf{A} = \sum_{i=1}^{n} \left(w_n \mathbf{x_n} + \theta_n (\mathbf{1}^d \cdot \mathbf{x}_n) \mathbf{1}^d + b\mathbf{1}^d \right)$$

$$\mathbf{y} = (y_1, y_2, ...y_d) : y_i = f(A_i)$$

where $\mathbf{1}^d$ is the d-dimensional vector consisting only of ones, and f is some sigmoidal function such as $\tanh(\cdot)$.

3.3 Scalar Outputs from Variadic Networks

Functions such as mean and standard deviation, or the credit rating loan repayment example, have a scalar range, whereas variadic networks as described above produce vector outputs. Scalar training data is provided by setting every element of a target vector to the desired scalar target value, and scalar output is considered to be the mean of the vector output.

4 Backprop Training a MLVP

Backpropagation training requires only that the partial derivative of network error with respect to each parameter of the network can be calculated. This is easy for MLVPs; if network error is defined as sum-square error over all individual vector elements, the partial derivative of network error with respect to node bias and link weight is effectively the same as for a scalar network, as per [7]. We need only do a little additional work to determine the partial derivative of network error with respect to each interaction-weight parameter.

The basic backprop algorithm is given below; the mathematical calculation of the partial derivative terms is given in an appendix. For the algorithm below, a simple 3-layer architecture is assumed with only input nodes, sigmoidal hidden nodes and linear output nodes. The network's cost function is assumed to be the sum-squared error over all elements of all output nodes.

5 Algorithm

As in a scalar MLP, the network's output is calculated using a forward pass through the network which is also used as part of the training process.

```
k = dimensionality of inputs
foreach node in hidden_nodes and output_nodes
    for i = 0 to k-1
        node.activation[i] = node.bias
foreach node in input_nodes and hidden_nodes
    // calculate the sum of the output for this node
```

```
        sum = 0
        for i = 0 to k-1
            if node is an input
                node.output[i] = respective network input element
            else
                // f is node transfer function
                node.output[i] = f(node.activation[i])
            sum += node.output[i]
        foreach link in node.output_links
            target = link's target node
            for i = 0 to k-1
                target.activation[i] += sum * link.θ
                target.activation[i] += node.output[i] * link.w
foreach node in output_nodes
    for i = 0 to k-1
        node.output[i] = f(node.activation[i])
```

Again, as for the scalar MLP, in order to calculate partial derivatives of network cost with respect to network parameters, a backward pass is made through the network based on comparing the actual network output to its target output.

```
k = dimensionality of inputs
foreach node in output_nodes
    // calculate partial derivative of error wrt activation vector
    // EBG = error / bias gradient, i.e. ∂E/∂B
    node.EBG = 0
    for i = 0 to k-1
        // EAG = error / activation gradient, i.e. ∂E/∂A
        node.EAG[i] = node.output[i] - node.target[i]
        node.EBG += node.EAG[i]
foreach node in hidden_nodes
    for i = 0 to k-1
        // EOG = error / output gradient, i.e. ∂E/∂y
        node.EOG[i] = 0
    foreach link in node.output_links
        target = link's target node
        for i = 0 to k-1
            node.EOG[i] += target.EBG * link.θ
            node.EOG[i] += target.EAG[i] * link.w
    // EBG = error / bias gradient, i.e. ∂E/∂B
    node.EBG = 0
    for i = 0 to k-1
        // EAG = error / activation gradient, i.e. ∂E/∂A
        // f' is derivative of transfer function
        node.EAG[i] = node.EOG[i] * f'(node.activation)
        node.EBG += node.EAG[i]
foreach link in network_links
    // EWG = error / weight gradient, i.e. ∂E/∂w
    link.EWG = 0
    // EθG = error / interaction-weight gradient, i.e. ∂E/∂θ
    link.EθG = 0
```

```
for i = 0 to k-1
    link.EWG += link.to_node.EAG[i] * link.from_node.output[i]
    link.EθG += link.to_node.EBG * link.from_node.output[i]
```

Once the derivative terms have been calculated, the network's weights, interaction-weights and biases can be updated using any gradient-based optimisation method, such as simple gradient descent with momentum, RPROP [6], or scaled conjugate gradients (SCG) [4]. Training can be done in online or batch mode as preferred.

6 Training on Sample Problems

The variadic network architecture as described above was tested on several sample function approximation problems as a proof of concept. The results presented here are based on a single hidden layer with 10 nodes, in batch mode training using the RPROP algorithm [6], which is a simple second-order gradient method. The training algorithm parameters were as follows: $\eta_0 = 0.05, \eta_- = 0.5, \eta_+ = 1.2, L_{min} = 10^{-8}, L_{max} = 50$. Bias, weight and interaction-weight parameters were initialised with a normal distribution whose mean was zero and whose variance was inversely proportional to the node fan-in. The sample problems were chosen as being well-known interesting statistical or geometric variadic functions.

6.1 Sample Problem Descriptions

Each sample problem had 500 randomly generated points as inputs. Each point consisted of an n-dimensional vector (or in the case of the angle problem, 2 n-dimensional vectors) with n uniformly chosen between 2 and 10, and the vectors uniformly chosen from the n-dimensional 2-unit hypercube centred on the origin (i.e. their elements were uniformly chosen between -1 and 1).

Standard deviation. The target output was the overall standard deviation of the network input X, treated as a statistical sample, i.e.

$$f(x_i) = \sqrt{E(X^2) - E(X)^2}$$

Softmax. The target output was the softmax (multiple logistic) function applied to the network input, i.e.

$$f(x_i) = \frac{e^{x_i}}{\sum_{j=1}^{n} e^{x_j}}$$

Angle. The target output was the angle in radians between the two network inputs $\mathbf{x}$ and $\mathbf{y}$, treated as n-dimensional vectors, i.e.

$$f(x_i) = \arccos\left(\frac{\mathbf{x} \cdot \mathbf{y}}{|\mathbf{x}||\mathbf{y}|}\right)$$

6.2 Sample Problem Results

30 runs of 200 training epochs each were performed for each sample training problem. During each run, network generalisation was tested on another 500-point randomly generated dataset. For all three problems, training mean square training error (MSE) fell significantly (by 2-3 orders of magnitude) over training runs, and test MSE was very close to training MSE (the dataset seems to have been too large for significant overfitting).

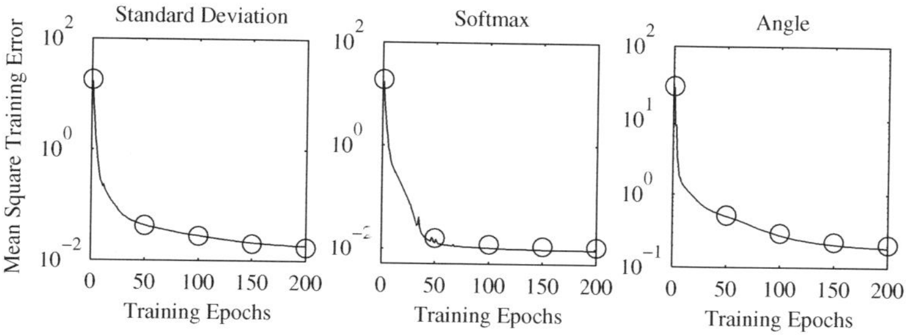

Fig. 4. Training error over time for the three target functions, averaged over 30 runs. Note logarithmic y axis. Circles represents generalisation error of the trained network. Left: standard deviation target function. Centre: softmax target function. Right: angle target function.

7 Conclusion

I have defined a type of network, variadic networks, in which the signals are vectors of arbitrary dimension, the network has a fixed number of parameters independent of the dimension of the vector, the vector elements are treated symmetrically, and the signals can interact across dimensions. A simple MLP extension based on this idea - the MVLP - has been shown capable of learning sample problems.

This approach is a novel and promising method for learning on data comprising variable-length but unordered "bags" of tuples of real numbers.

8 Future Work

Space limitations preclude detailed discussion, but some preliminary work is underway on the following topics: universality and convergence behaviour of the proposed architecture; improved training performance using more sophisticated training methods and network topologies; construction of multiset kernels for kernel learning methods; generalisation behaviour for different input dimension.

References

1. Collins, M., Duffy, N.: Convolution kernels for natural language. In: Dietterich, T.G., Becker, S., Ghahramani, Z. (eds.) Advances in Neural Information Processing Systems 14, MIT Press, Cambridge,MA (2002)
2. Joachims, T.: Text categorization with support vector machines: learning with many relevant features. In: Nédellec, C., Rouveirol, C. (eds.) Machine Learning: ECML-98. LNCS, vol. 1398, pp. 137–142. Springer, Heidelberg (1998)
3. Lodhi, H., Saunders, C., Shawe-Taylor, J., Cristianini, N., Watkins, C.: Text classification using string kernels. J. Mach. Learn. Res. 2, 419–444 (2002)
4. Møller, M.F.: A scaled conjugate gradient algorithm for fast supervised learning. Neural Netw. 6(4), 525–533 (1993)
5. Petrovsky, A.: Multi-attribute classification of credit cardholders: multiset approach. International Journal of Management and Decision Making (IJMDM) 7(2-3), 166–179 (2006)
6. Riedmiller, M., Braun, H.: Rprop- a fast adaptive learning algorithm (1992)
7. Rumelhart, D.E., Hinton, G.E., Williams, R.J.: Learning representations by back–propagating errors. Nature 323, 533–536 (1986)

A Derivation of Backprop Algorithm

Unfortunately the use of vector signals makes the standard matrix notation for network algebra somewhat unwieldy.

Instead, consider every network parameter and signal as a function on nodes or links. Each node n has a vector activation $\mathbf{A}(n) \in \mathbb{R}^k$, where k is the dimensionality of the current network input, a vector output $\mathbf{y}(n) \in \mathbb{R}^k$, and a scalar bias $B(n) \in \mathbb{R}$. Each link l has $\mathbf{y}(a)$ as its input, a signal $\mathbf{S}(l) \in \mathbb{R}^k$ as its output, a weight $w(l)$ and interaction-weight $\theta(l)$.

The activation function $\mathbf{A}$ is defined differently for input and non-input nodes. For each input node n_{in}, $\mathbf{A}(n_{\text{in}})$ is set to some arbitrary input vector with common dimensionality k. (These inputs determine the dimensionality of every other signal in the network.) For a non-input node n,

$$\mathbf{A}(n) = B(n)\mathbf{1^k} + \sum_{l \in \text{inp}(n)} \mathbf{S}(l)$$

where $\mathbf{1^k}$ is the k-dimensional vector whose every element is 1, and $\text{inp}(n)$ is the set of links coming into node n.

We'll assume that the input and output nodes are linear, i.e. that $\mathbf{y}(n_{\text{in/out}}) = \mathbf{A}(n_{\text{in/out}})$. For hidden nodes n_{hid} the output is defined by a node transfer function $\mathbf{f} : \mathbb{R}^k \to \mathbb{R}^k$ based on a pointwise application of a sigmoidal scalar transfer function $f : \mathbb{R} \to \mathbb{R}$ such as the logistic or hyperbolic tangent function:

$$\mathbf{f}(\mathbf{x})_i = f(x_i), i \in \{1 \ldots k\}$$

$$\mathbf{y}(n_{\text{hid}}) = \mathbf{f}(\mathbf{A}(n_{\text{hid}}))$$

The output signal $\mathbf{S}(l)$ for a link l from node a to b is defined as

$$\mathbf{S}(l) = w(l)\mathbf{y}(a) + \theta(l)\mathbf{1}^{\mathbf{k}}(\mathbf{y}(a) \cdot \mathbf{1}^{\mathbf{k}})$$

Assigning vector targets $\mathbf{t}(n_{\text{out}})$ for every output neuron, we can define a sum-squared network error function E

$$E = \frac{1}{2} \sum_{n_{\text{out}}} (\mathbf{t}(n_{\text{out}}) - \mathbf{y}(n_{\text{out}})) \cdot (\mathbf{t}(n_{\text{out}}) - \mathbf{y}(n_{\text{out}}))$$

and calculate partial derivatives of error with respect to all biases, weights and interaction-weights as follows.

As in classical backprop, calculating the derivative of network error with respect to each node's activation and applying the chain rule allow us to determine the derivative of error with respect to the network's modifiable parameters.

For the output $\mathbf{y}(n_{\text{out}})$ of an output node n_{out}, we have

$$\frac{\partial E}{\partial \mathbf{y}(n_{\text{out}})} = \mathbf{y}(n_{\text{out}}) - \mathbf{t}(n_{\text{out}})$$

and since $\mathbf{y}(n_{\text{out}}) = \mathbf{A}(n_{\text{out}})$ for linear output nodes,

$$\frac{\partial E}{\partial \mathbf{A}(n_{\text{out}})} = \mathbf{y}(n_{\text{out}}) - \mathbf{t}(n_{\text{out}})$$

For elements of the signal $\mathbf{S}(l)$ on an arbitrary link l between node a and node b, we have

$$\begin{aligned}\left(\frac{\partial E}{\partial \mathbf{S}(l)}\right)_i &= \left(\frac{\partial E}{\partial \mathbf{A}(b)}\right)_i \frac{\partial \mathbf{A}(b)_i}{\partial \mathbf{S}(l)_i} \\ &= \left(\frac{\partial E}{\partial \mathbf{A}(b)}\right)_i\end{aligned}$$

So far this is essentially the same as classical backprop. Where the difference arises is in the chaining back to hidden nodes n_{hid}.

$$\left(\frac{\partial E}{\partial \mathbf{y}(n_{\text{hid}})}\right)_i = \sum_{l \in \text{cutp}(n_{\text{hid}})} \sum_{j=1}^{k} \left(\frac{\partial E}{\partial \mathbf{S}(l)_j} \frac{\partial \mathbf{S}(l)_j}{\partial \mathbf{y}(n_{\text{hid}})_i}\right)$$

where

$$\frac{\partial \mathbf{S}(l)_j}{\partial \mathbf{y}(n_{\text{hid}})_i} = w(l)\delta_{ij} + \theta(l)$$

with δ_{ij} here being the Kronecker delta function, so that

$$\sum_{j=1}^{k} \left(\frac{\partial E}{\partial \mathbf{S}(l)_j} \frac{\partial \mathbf{S}(l)_j}{\partial \mathbf{y}(n_{\text{hid}})_i}\right) = \theta(l) \sum_{j=1}^{k} \frac{\partial E}{\partial \mathbf{S}(l)_j} + w(l) \frac{\partial E}{\partial \mathbf{S}(l)_i}$$

The activation function f in all nodes is pointwise, which gives us

$$\left(\frac{\partial E}{\partial \mathbf{A}(n_{\text{hid}})}\right)_i = \left(\frac{\partial E}{\partial \mathbf{y}(n_{\text{hid}})}\right)_i f'(\mathbf{A}(n_{\text{hid}})_i)$$

And finally, the $\frac{\partial E}{\partial \mathbf{A}}$ and $\frac{\partial E}{\partial \mathbf{S}}$ terms can be used to calculate the partial derivatives of error with respect to the network bias, weight and interaction-weight parameters.

$$\frac{\partial E}{\partial B(n)} = \sum_{i=1}^{k} \left(\frac{\partial E}{\partial \mathbf{A}(n)}\right)_i \frac{\partial \mathbf{A}(n)_i}{\partial B(n)} = \sum_{i=1}^{k} \frac{\partial E}{\partial \mathbf{A}(n)_i}$$

$$\frac{\partial E}{\partial w(l)} = \sum_{i=1}^{k} \left(\frac{\partial E}{\partial \mathbf{S}(l)}\right)_i \frac{\partial \mathbf{S}(l)_i}{\partial w(l)} = \sum_{i=1}^{k} \left(\frac{\partial E}{\partial \mathbf{S}(l)}\right)_i \mathbf{y}(a)_i$$

$$\frac{\partial E}{\partial \theta(l)} = \sum_{i=1}^{k} \left(\frac{\partial E}{\partial \mathbf{S}(l)}\right)_i \frac{\partial \mathbf{S}(l)_i}{\partial \theta(l)} = \sum_{i=1}^{k} \frac{\partial E}{\partial \mathbf{S}(l)_i} \sum_{j=1}^{k} \mathbf{y}(a)_j$$

The Introduction of Time-Scales in Reservoir Computing, Applied to Isolated Digits Recognition*

Benjamin Schrauwen**, Jeroen Defour, David Verstraeten, and Jan Van Campenhout

Electronics and Information Systems Department, Ghent University, Belgium
Benjamin.Schrauwen@UGent.be

Abstract. Reservoir Computing (RC) is a recent research area, in which a untrained recurrent network of nodes is used for the recognition of temporal patterns. Contrary to Recurrent Neural Networks (RNN), where the weights of the connections between the nodes are trained, only a linear output layer is trained. We will introduce three different time-scales and show that the performance and computational complexity are highly dependent on these time-scales. This is demonstrated on an isolated spoken digits task.

1 Introduction

Many real-world difficult tasks that one would like to solve using machine learning are temporal in nature. In the field of neural networks, several approaches have been proposed that introduce the notion of time into the basic concept of stateless, feed-forward networks, where the main objective is of course to give the network access to information from the past as well as the present. One well known method is to feed the signals of interest through a delay line, which is then used as input to the network (Time Delay Neural Networks). These however introduce additional parameters into the model and impose an artificial constraint on the time window. A more natural way of dealing with temporal input signals is the introduction of recurrent, delayed connections in the network, which allow the network to store information internally. These Recurrent Neural Networks (RNN) are a theoretically very powerful framework, capable of approximating arbitrary finite state automata [1] or even Turing machines [2]. Still, wide-scale deployment of these networks is hindered by the difficult and computationally costly training, caused in part by the fact that the temporal gradient information gets washed out as it is back-propagated into the past [3].

Reservoir Computing (RC) offers an intuitive methodology for using the temporal processing power of RNN without the hassle of training them. Originally introduced independently as the Liquid State Machine [4] or Echo State Networks [5], the basic concept is to randomly construct an RNN and to leave the

* This research is partially funded by FWO Flanders project G.0317.05.

** Corresponding author.

J. Marques de Sá et al. (Eds.): ICANN 2007, Part I, LNCS 4668, pp. 471–479, 2007.

weights unchanged. A separate linear regression function is trained on the response of the reservoir to the input signals using pseudo-inverse. The underlying idea is that a randomly constructed reservoir offers a complex nonlinear dynamic transformation of the input signals which allows the readout to extract the desired output using a simple linear mapping.

Evidently, the temporal nonlinear mapping done by the reservoir is of key importance for its performance. One of the appealing properties of this type of networks is the fact that they are governed by only a few global parameters. Generally, when solving a task using RC, the search for optimal reservoir dynamics is done by adjusting global scaling parameters such as the input scaling or spectral radius[1]. However, as we will show, optimizing the temporal properties of the entire system can also be very influential to the performance and the computational complexity. Since a reservoir is a dynamic system that operates at a certain time-scale, the precise adjustment of the internal temporal behavior of the reservoir to both the input signal and the desired output signal is important. In this contribution, we present an overview of the different ways in which the dynamic behavior of reservoirs has been described in literature, and we investigate the interplay between different temporal parameters for each of these models when applied to signal classification tasks.

2 Reservoir Computing and Time-Scales

Although there exist some variations on the global description of an RC system, we use this setup:

$$\mathbf{x}[t+1] = f\left(W_{\text{res}}^{\text{res}}\mathbf{x}[t] + W_{\text{inp}}^{\text{res}}\mathbf{u}[t]\right)$$
$$\widehat{\mathbf{y}}[t+1] = W_{\text{res}}^{\text{out}}\mathbf{x}[t+1] + W_{\text{inp}}^{\text{out}}\mathbf{u}[t] + W_{\text{bias}}^{\text{out}},$$

with $\mathbf{u}[t]$ denoting the input, $\mathbf{x}[t+1]$ the reservoir state, $\mathbf{y}[t+1]$ the expected output, and $\widehat{\mathbf{y}}[t+1]$ the actual output[2]. All weights matrices to the reservoir ($W_{\star}^{\text{res}}$) are initialized at random, while all connections to the output ($W_{\star}^{\text{out}}$) are trained. The non-linearity f is a hyperbolic tangent.

In this system we can define three different time-scales: the time-scale of the input, the internal state, and the output. Traditionally these are all the same, but in this paper we will show that performance can be improved and that computational demands can be decreased by setting these time-scales correctly.

2.1 Input Time-Scale and Integrator Nodes

In [6], the notion of input time-scales and the link to node integration was introduced. In this contribution we will look at three ways to add an integrator to a node: after the non-linearity (as used in [6]), before the non-linearity (as

[1] The largest absolute eigenvalue of the connection matrix.

[2] We denote discrete time with $[t]$ and continuous time with (t).

used in continuous time RNN), and over the non-linearity (which we introduce). We will first discuss the integrator after the non-linearity.

The input time-scale and integrator can be introduced by starting from a continuous time differential equation:

$$\dot{\mathbf{x}} = \frac{1}{c}\left(-a\mathbf{x} + f\left(W_{\text{inp}}^{\text{res}}\mathbf{u} + W_{\text{res}}^{\text{res}}\mathbf{x}\right)\right)$$

where a denotes the retainment rate (which in this work we set to 1, meaning that no information is leaked), and c is a scaling factor for the temporal dynamics. If this is discretely approximated by using Euler discretization we get:

$$\mathbf{x}((t+1)\delta_2) = (1-\lambda)\mathbf{x}(t\delta_2) + \lambda f\left(W_{\text{inp}}^{\text{res}}\mathbf{u}(t\delta_2) + W_{\text{res}}^{\text{res}}\mathbf{x}(t\delta_2)\right)$$

where δ_2 is the Euler step and $\lambda = \delta_2/c$. Note that the Euler time step δ_2 determines the sample rate of the input, while c can be used to scale the speed of the dynamics.

Although the retainment rate was the major research topic of [6] and λ was ignored, this work focuses on λ and sets the retainment rate to 1. This is because that parameter was previously not thoroughly investigated, and when changing λ the effective spectral radius of the reservoir does not change, while, when changing a, it does. When changing the spectral radius, the dynamic regime of the reservoir, which is very important for the performance of the reservoir, changes. This coupling between time-scale settings and dynamic regime settings is not desired.

When the integrator is placed before the non-linearity, which is common practice in continuous time RNN, we end up with these equations:

$$\begin{aligned}\mathbf{z}((t+1)\delta_2) &= (1-\lambda)\mathbf{z}(t\delta_2) + \lambda\left(W_{\text{inp}}^{\text{res}}\mathbf{u}(t\delta_2) + W_{\text{res}}^{\text{res}}\mathbf{x}(t\delta_2)\right)\\ \mathbf{x}((t+1)\delta_2) &= f(\mathbf{z}((t+1)\delta_2)).\end{aligned}$$

This has the advantage of being stable: if an integrator starts to blow up, it is constrained by the non-linearity, which is not the case in the previous model.

Yet another, empirical model, is possible by placing the integrator *over* the non-linearity:

$$\mathbf{x}((t+1)\delta_2) = f\left((1-\lambda)\mathbf{x}(t\delta_2) + \lambda\left(W_{\text{inp}}^{\text{res}}\mathbf{u}(t\delta_2) + W_{\text{res}}^{\text{res}}\mathbf{x}(t\delta_2)\right)\right).$$

The advantage of this is that an integrator can never blow up and that it has no computational overhead since the integrators can be incorporated in the W matrix by increasing the diagonal elements. But a drawback is that the integrator does leak away even with $a = 1$. The leak is due to the contracting property of the non-linear mapping of the hyperbolic tangent upon itself. This has as a consequence that the overall amplitude of the reservoir dynamics scales down when λ goes to 0.

Theoretically there are some strong and weak points concerning these three possible models, but we will have to experimentally investigate which of these is somehow 'optimal'.

The input of the reservoir is in practice always a sampled signal. We denote the sampling time-scale of the input δ_1. The resampling factor from input to internal time-scale is thus equal to δ_2/δ_1. This resampling should approximate the ideal Nyquist, non-aliasing resampling operation. We use the Matlab `resample` operation for this.

2.2 Output Time-Scale

In this work we also introduce an output time-scale

$$\widehat{\mathbf{y}}((t+1)\delta_3) = W_{\text{res}}^{\text{out}}\mathbf{x}((t+1)\delta_3) + W_{\text{inp}}^{\text{out}}\mathbf{u}(t\delta_3) + W_{\text{bias}}^{\text{out}}.$$

The reservoir states and inputs are thus resampled to this new time-scale before training and applying the linear readout. The resampling factor from internal to output is δ_3/δ_2. We will call this reservoir resampling.

At a first glance this time-scale does not seem to be very important, but as we will show in the experiments, changing this time-scale can have a drastic effect on performance.

3 Experimental Setup and Results

For all of the following experiments, we used the Reservoir Computing Toolbox[3] [7], which allows us to do all the simulations in the Matlab environment. With this toolbox, we will study the task of speech recognition of isolated digits. The Lyon passive ear model [8] is used to frequency-convert the spoken digits into 77 frequency channels. The task of recognizing isolated spoken digits by RC has already been studied [7], and the results have been very positive. The dataset we will use (a subset of the TI48 dataset) contains 500 spoken isolated digits, the digits 0 to 9, where every digit is spoken 10 times by 5 female speakers. We can evaluate the performance of the reservoir by calculating the Word Error Rate (WER), which is the fraction of incorrectly classified words as a percentage of the total number of presented words. Because of the randomness of the reservoirs, we will do every simulation 20 times with different stochastically generated reservoirs, and use 10-fold cross validation to obtain a decent statistical result.

Ridge regression (least squares optimization where the norm of the weights is added as a penalty) was used to train the readout, where the regularization parameter was set by doing a grid search. To classify the ten different digits, ten outputs are trained which should be 1 when the digit is uttered, -1 if not. The temporal mean of the ten classifiers is taken over the complete sample, and a Winner-Take-All is applied to this. The winner output represents the classified digit.

The settings of the reservoir, like reservoir size which is 200 and spectral radius which is 0.7, are in this paper intentionally chosen to be non-optimal so

[3] Which is an open-source Matlab toolbox for Reservoir Computing, freely available at http://www.elis.ugent.be/rct .

Table 1. Minimal classification errors for all cases

	input resampling	reservoir resampling
integrator after non-linearity	2.15%	0.48%
integrator before non-linearity	2.32%	0.48%
integrator over non-linearity	**1.36%**	**0.40%**

we still have a margin of error left to evaluate the resampling techniques on. If everything is set optimally we can very easily get zero training error!

In [6] the intricate interplay between optimal values of a and the spectral radius is investigated. When changing λ, we notices that there is no such intricate dependency to the spectral radius. The optimal spectral radius is not very dependent on the value of λ.

3.1 Input Resampling vs. Integration

First we will study the influence of input resampling versus integration for the different integrator options on the performance of the reservoir. To be able to handle a broad range of parameters, we vary both these parameters in a logarithmic way, with base 10. Figure 1 shows the results of this experiment.

For these experiments, the internal time-scale δ_2 is set to 1 and the input time-scale δ_1 is changed. This is similar to keeping δ_1 constant and changing δ_2. Note that when keeping λ constant and changing δ_2, c changes. We can observe a diagonal area on Figure 1 which corresponds to an optimal performance. This optimal performance is actually achieved with a fixed value of c, but we perceive it as a λ value which changes linearly with δ_1, but this is due to the fact that $\log_{10}(\lambda) = \log_{10}(\delta_2)/\log_{10}(c)$.

For all three integrator cases we see that the optimal performance is attained with the least resampling (bottom of the plots). However, if we resample more, the error only slightly increases. This creates a trade-off between computational complexity and performance.

The optimal performance for the three different integrator settings are given in Table 1. We see that the integrator over the non-linearity performs optimally, which is nice, because this introduces no extra computational requirements.

3.2 Reservoir Resampling vs. Integration

The second experiment studies the output time-scale compared to the internal integrator. The results of this experiment are shown in Figure 2. For these experiments, the input time-scale is set to $\log_{10}(\delta_1) = -2$ and the internal time-scale $\delta_1 = 1$. The setting of the input time-scale is not critical since the conclusions of the results also apply to other input time-scale settings.

These figures are a bit more complex to interpret. The upper part, with $\log_{10}(\delta_3) = 0$, has no reservoir resampling and is thus equal to a slice of Figure 1 where $\log_{10}(\delta_1) = -2$. When increasing the resampling of the reservoir states, we see that for the region of low integration (close to 0) there is a significant

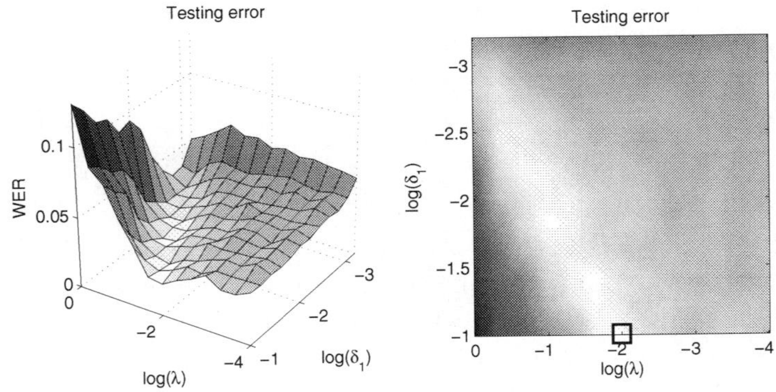

(a) integrator after the non-linearity

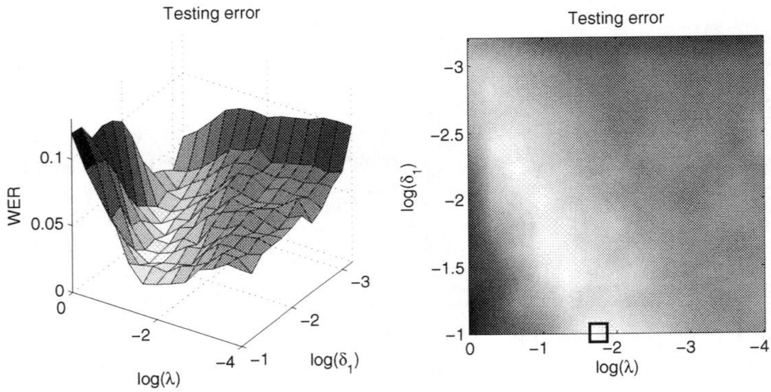

(b) integrator before the non-linearity

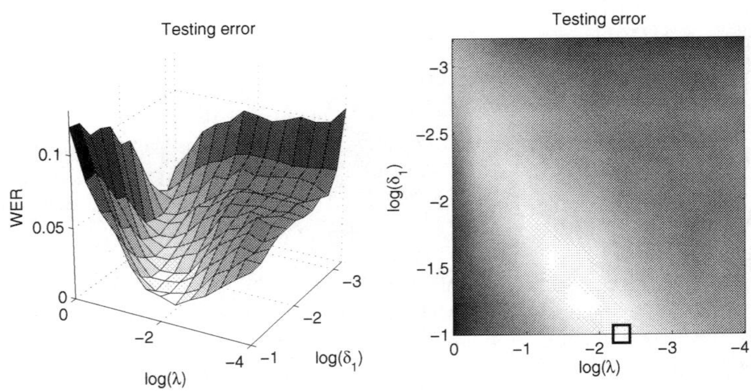

(c) integrator over the non-linearity

Fig. 1. WER results for input resampling versus integration for the three integrator options. The small squares denote minimal error.

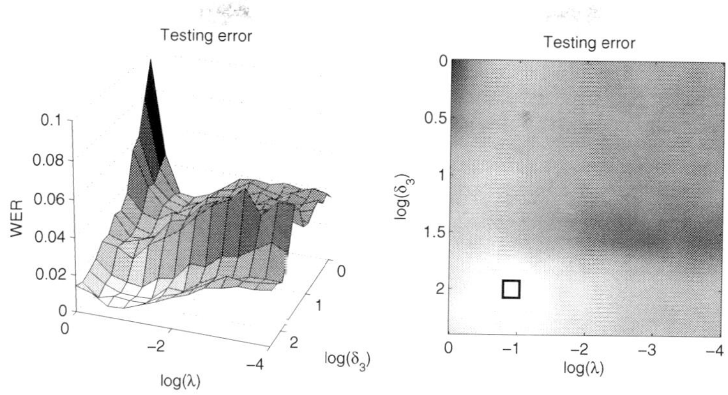

(a) integrator after the non-linearity

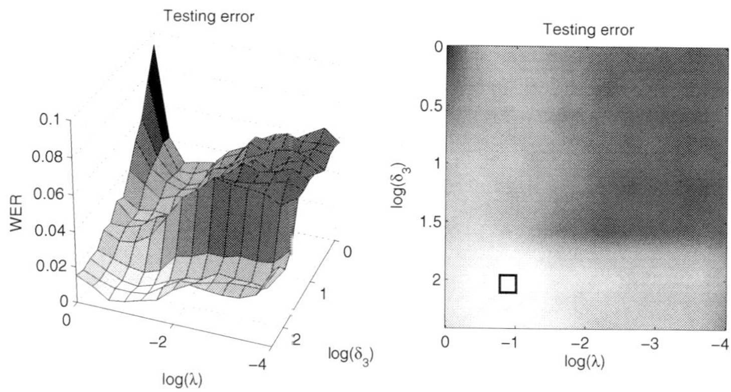

(b) integrator before the non-linearity

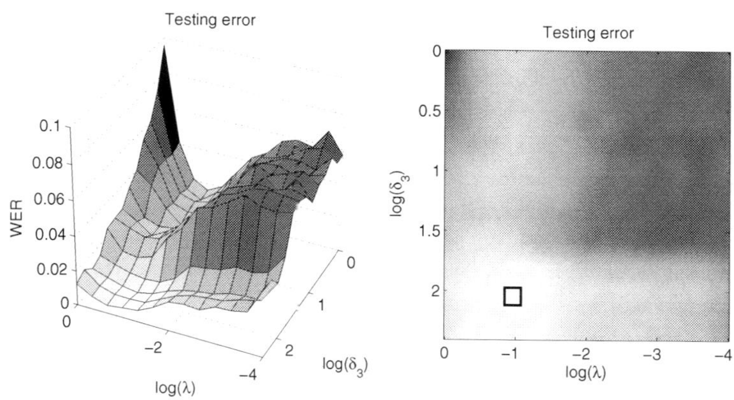

(c) integrator over the non-linearity

Fig. 2. WER results for reservoir resampling versus integration for the three integrator options. The small squares denote minimal error.

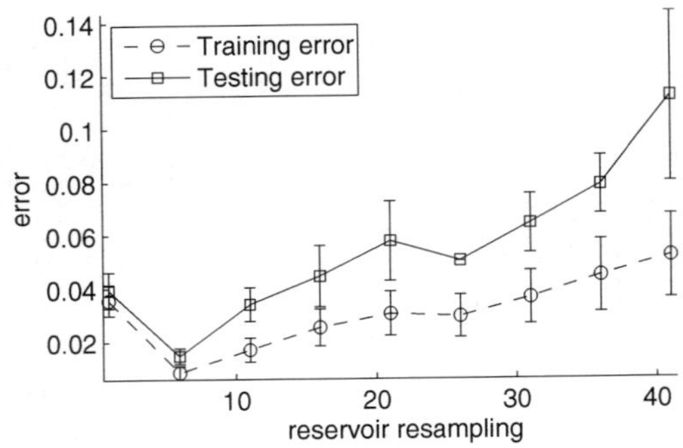

Fig. 3. Classification error for a signal classification task with respect to the output time-scale. Note that the time-scale is not logarithmic in this figure.

decrease of the error. But in the region where the integration is optimal there is initially no improvement.

The bottom part of the figures show a drastic drop in performance when $\log_{10}(\delta_3)$ is larger than 1.5. With such a high input and reservoir resampling, there is actually just one time step left! For this task we thus have optimal performance when reducing all the reservoir's dynamics to a single point in state space: the centroid of the dynamics in state space. This is not completely awkward since the post-processing of the linear classifier's output is anyway by taking its temporal mean before applying Winner-Take-All. The major drawback of this drastic reservoir resampling is that all temporal information is lost. With less reservoir resampling, the reservoir is able to already give a prediction of the uttered word even if it is for example only partially uttered. We thus trade-off performance to the ability of on-line computation.

One might think that when averaging out all the reservoir's dynamics, it has no real purpose. But when training a linear classifier to operate on the temporal average of the frequency-transformed input, so without using a reservoir, we end up with an error of 3%. This is quite good, but still an order of a magnitude worse than when using a reservoir of only 200 neurons.

These conclusions are, however, partly due to the fact that here the desired class output remains constant during the whole input signal. Figure 3 shows that this is not generally the case. Here, the task is to do signal classification, whereby the input signal is constructed by concatenating short signal pieces of 50 time steps, whereby every piece is either a noisy sawtooth or a noisy square wave with the same period. The task is then to classify the signal type at every time step. As the results show, in this case there is a trade-off between the amount of resampling of the reservoir responses and the amount of information available to the linear readout to do the classification. In this case only a small amount of reservoir resampling is needed to attain optimal performance. In state space this can be seen as taking the centroid of a small temporal region of the trajectory. This temporal averaging out of the dynamics seems to significantly increase the classification performance of the RC system.

4 Conclusion and Future Work

It was previously already mentioned that setting input time-scales is possible [6], however, in this work we show that setting this time-scale is critical for getting optimal performance. We showed that there is a clear link between node integration and resampling. When using input resampling the computational complexity decreases linearly with resampling, while only slightly influencing the performance. We introduced three types of node integration and showed how they perform on an isolated spoken digits classification task. The integrator scheme introduced in this work performs optimally and can be implemented without overhead.

Next we showed that an output time-scale can also be introduced. Although that initially this might not seem to be influential on performance, we experimentally show that a large performance gain can be achieved by optimally setting this resampling. Here we get a speed-up and an improvement in performance. This result suggests that for classification tasks it is not the actual precise dynamics that are important, but more the temporally filtered dynamics which is the local centroid of the dynamics in state space.

There are many other options for future research based on this work. We will investigate how this resampling scheme is applicable to more complex tasks like continuous speech, where classification needs to be performed on-line, and where there are multiple time-scales presents. To solve this we might need to introduce hierarchical or heterogeneous reservoirs with different parts working at different time-scales. Ultimately these temporal modules should be created autonomously.

References

1. Omlin, C.W., Giles, C.L.: Constructing deterministic finite-state automata in sparse recurrent neural networks. In: IEEE International Conference on Neural Networks (ICNN'94), Piscataway, NJ, pp. 1732–1737. IEEE Computer Society Press, Los Alamitos (1994)
2. Kilian, J., Siegelmann, H.T.: The dynamic universality of sigmoidal neural networks. Information and Computation 128, 48–56 (1996)
3. Hammer, B., Steil, J.J.: Perspectives on learning with recurrent neural networks. In: Proceedings of ESANN (2002)
4. Maass, W., Natschläger, T., Markram, H.: Real-time computing without stable states: A new framework for neural computation based on perturbations. Neural computation 14(11), 2531–2560 (2002)
5. Jaeger, H.: Short term memory in echo state networks. Technical Report GMD Report 152, German National Research Center for Information Technology (2001)
6. Jaeger, H., Lukosevicius, M., Popovici, D.: Optimization and applications of echo state networks with leaky integrator neurons. Neural Networks (to appear, 2007)
7. Verstraeten, D., Schrauwen, B., D'Haene, M., Stroobandt, D.: A unifying comparison of reservoir computing methods. Neural Networks (to appear, 2007)
8. Lyon, R.: A computational model of filtering, detection and compression in the cochlea. In: Proceedings of the IEEE ICASSP, pp. 1282–1285. IEEE Computer Society Press, Los Alamitos (1982)

Partially Activated Neural Networks by Controlling Information

Ryotaro Kamimura

Information Science Laboratory, Information Technology Center, Tokai University,
1117 Kitakaname Hiratsuka Kanagawa 259-1292, Japan
ryo@cc.u-tokai.ac.jp

Abstract. In this paper, we propose partial activation to simplify complex neural networks. For choosing important elements in a network, we develop a fully supervised competitive learning that can deal with any targets. This approach is an extension of competitive learning to a more general one, including supervised learning. Because competitive learning focuses on an important competitive unit, all the other competitive units are of no use. Thus, the number of connection weights to be updated can be reduced to a minimum point when we use competitive learning. We apply the method to the XOR problem to show that learning is possible with good interpretability of internal representations. Then, we apply the method to a student survey. In the problem, we try to show that the new method can produce connection weights that are more stable than those produced by BP. In addition, we show that, though connection weights are quite similar to those produced by linear regression analysis, generalization performance can be improved by changing the number of competitive units.

1 Introduction

A hierarchical neural network is a very powerful architecture in learning complex patterns. However, as the complexity of a network is larger, the complexity in learning becomes intractable. Thus, we need to focus upon some important elements or parts of networks to make learning easier. We introduce here a partially activated network in which a smaller number of units as well as connection weights are actually used or activated in learning. With this partially activated network, the complexity of neural networks can be significantly reduced.

Partial activation is realized by condensing information content into a smaller number of elements. Information condensation or maximization can be realized by competitive learning; that is, competitive learning can be introduced to simplify networks, because it is a method that focuses upon important units. By having them focus on a single competitive unit, complex neural networks can be significantly simplified. Competitive learning has been one of the most important techniques in neural networks. Many problems have been pointed out, and a number of studies have proposed methods to overcome those problems: [1], [2], conscience learning [3], frequency-sensitive learning [4], rival penalized competitive learning [5], lotto-type competitive learning [6] and entropy maximization [7]. We have so far applied competitive learning to actual problems with good interpretation of connection weights [8], [9]. However, the

J. Marques de Sá et al. (Eds.): ICANN 2007, Part I, LNCS 4668, pp. 480–489, 2007.

method is limited to unsupervised learning, and it cannot be used to estimate continuous targets. In this context, we extend competitive learning to supervised hierarchical network architecture and propose a new supervised competitive learning method called *controlled competitive learning* (CCL). The new method includes the property of hierarchical networks to train continuous targets as well as the property of good interpretation of connection weights originating from competitive learning. In this CCL, conventional competitive learning is guided by error signals transferred from output layers. Thus, we can incorporate teacher information into a framework of competitive learning.

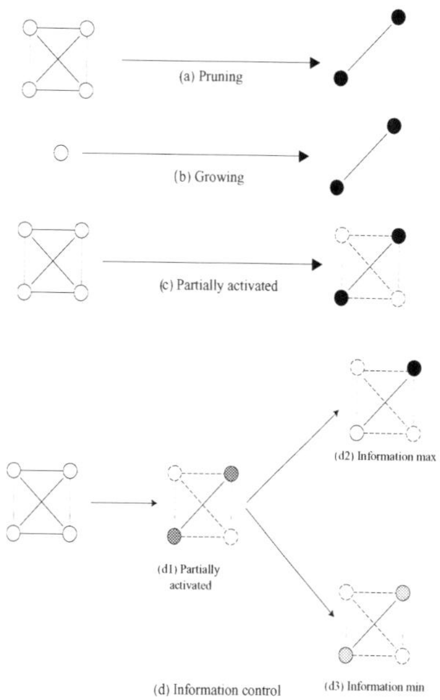

Fig. 1. Three types of network simplification and information control

2 Partially Activated Networks

2.1 Concept of Partially Activated Networks

In neural networks, the simplification of network complexity has been one of the most important issues, because simplification makes networks generalize better and thus learning is accelerated. There have been three types of simplification procedures. As shown in Figure 1(a), network pruning is one of the major approaches to simplification [10]. Figure 1(b) shows a concept of network growing in which the number of neurons gradually increases [11]. In Figure 1(c), a simplification procedure is shown that is not very popular for use in neural networks. A network keeps its architecture throughout the learning and testing mode. However, only elements necessary for given situations are actually used or activated; that is, a network activates a small number of units as well as connection weights absolutely necessary in learning. We call this type of network a

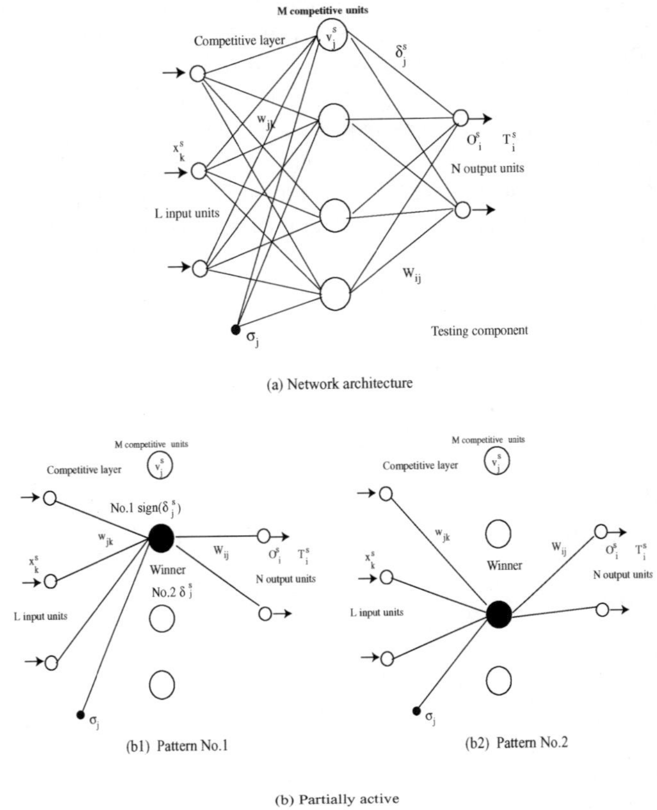

Fig. 2. A concept of partial activation and an implementation of controlled competitive learning by a hierarchical network

partially activated network. In this paper, we try to build this type of network. One of the major merits of this method is that, because elements are not completely deleted in a network, they are immediately activated when necessary. In Figure 2(b), we try to show that a procedure of information control is applied to partially activated elements in a network. In Figure 1(b), information is maximized for the partially activated elements.

2.2 Information Acquisition Procedure

We consider information content stored in competitive unit activation patterns. For this purpose, let us define information to be stored in a neural system. As shown in Figure 2(a), a network is composed of a competitive and an output layer. In the competitive layer, information is maximized to realize competitive processes. In the output layer, errors between targets and outputs are minimized. Information stored in a system is represented by decrease in uncertainty [12]. Uncertainty decrease, that is, information I, is defined by

$$I = -\sum_{\forall j} p(j) \log p(j) + \sum_{\forall s} \sum_{\forall j} p(s) p(j \mid s) \log p(j \mid s), \tag{1}$$

where $p(j)$, $p(s)$ and $p(j|s)$ denote the probability of firing of the jth unit, the probability of firing of the sth input pattern and the conditional probability of firing of the jth unit, given the sth input pattern, respectively.

Let us define mutual information in a neural network. As shown in Figure 2(a), a network is composed of input units x_k^s and competitive units v_j^s. The jth competitive unit receives a net input from input units, and an output from the jth competitive unit can be computed by

$$v_j^s = \exp\left(-\frac{\sum_{k=1}^{L} (x_k^s - w_{jk})^2}{\sigma_j^2} \right), \tag{2}$$

where σ represents the Gaussian width, L is the number of input units, and w_{jk} denote connection weights from the kth input unit to the jth competitive unit. The output is increased as connection weights come closer to input patterns. To compute mutual information, we suppose that the normalized activity v_j^s represents a probability with which a neuron fires. The conditional probability $p(j \mid s)$ is approximated by

$$p(j \mid s) \approx \frac{v_j^s}{\sum_{m=1}^{M} v_m^s}, \tag{3}$$

where M denotes the number of competitive units. Since input patterns are supposed to be given uniformly to networks, the probability of the jth competitive unit is computed by

$$p(j) = \frac{1}{S} \sum_{s=1}^{S} p(j \mid s), \tag{4}$$

where S is the number of input patterns. Information I is computed by

$$I = -\sum_{j=1}^{M} p(j) \log p(j) + \frac{1}{S} \sum_{s=1}^{S} \sum_{j=1}^{M} p(j \mid s) \log p(j \mid s). \tag{5}$$

As information becomes larger, specific pairs of input patterns and competitive units become strongly correlated.

2.3 Information Maximization Procedure

In actual learning, we must focus upon a winning neuron. Figure 2(b1) and (b2) show the second and the third neuron win the competition. For this purpose, we must completely maximize information content by update rules obtained by directly differentiating mutual information. However, the computation of the direct method is heavy, and we try to use a simplified version of information maximization with the winner-takes-all algorithm. We now realize competitive processes, that is, information maximization processes in competitive layers. As shown in Figure 2(a), a network is composed of a competitive and an output layer. In the competitive layer, a competitive unit becomes a winner by the winner-take-all algorithm. For this, at least, we need a mechanism to increase competitive unit activations. The jth competitive unit receives a net input from input units, and an output from the jth competitive unit can be computed by

$$v_j = \exp\left(-\frac{\sum_{k=1}^{L}(x_k - w_{jk})^2}{\sigma_j^2} \right), \tag{6}$$

where σ represents the Gaussian width, L is the number of input units, and w_{jk} denotes connection weights from the kth input unit to the jth competitive unit. In the following formulation, we omit the superscript s to simplify the equations. We must increase this unit activity v_j as much as possible. For that, we must differentiate this unit function. We have

$$\Delta w_{jk} = \phi_j (x_k - w_{jk}), \tag{7}$$

where

$$\phi_j = \frac{2v_j}{\sigma_j^2}. \tag{8}$$

With this equation, all competitive units gradually reach a maximum point. For the selection of a winner, we must introduce the winner-take-all algorithm. For the winner j^*, only this update rule is applied.

$$\Delta w_{jk} = \begin{cases} \phi_j (x_k - w_{jk}), & \text{if} \quad j = j^* \\ 0, & \text{otherwise.} \end{cases}$$

We think that competitive learning is a special computational method in which information is supposed to be maximized (winner-takes-all).

2.4 Information Control Procedure

We extend competitive learning to supervised learning. For this, we need to control information content in competitive units; that is, information should be maximized or minimized, depending upon input-target relations. The output from the output unit is computed by

$$O_i^s = \sum_{i=1}^{N} W_{ij} v_j, \tag{9}$$

where W_{ij} denotes connection weights from the jth competitive unit to the ith output unit, O_i^s represents the actual outputs from the ith output unit, given the sth input pattern, and T_i^s denotes corresponding targets. We now compute the error δ

$$\delta_i = T_i - O_i. \tag{10}$$

The error is computed by

$$E = \sum_{i=1}^{N} \delta_i^2, \tag{11}$$

where T_i denotes targets for the ith output unit. For hidden-output connections, connections only to a winner, j^*, are updated by

$$\Delta W_{ij^*} = \delta_i v_j^*. \quad (12)$$

For input-hidden connections, only connections to the winner j^* are also to be updated. Thus, we have update rules for the winner:

$$\Delta w_{jk} = \Phi_j(x_k - w_{jk}). \quad (13)$$

In the equation, ϕ_j is set to zero for losers, and for the winner we have

$$\Phi_j = \text{sign}(\delta_j))\phi_j \quad (14)$$

, where

$$\delta_j = \sum_{i=1}^{N} \delta_i W_{ij} \quad (15)$$

. For the bias, we update them to decrease errors:

$$\Delta\sigma_j = \Phi_j \sum_{k=1}^{L} (x_k - w_{jk})^2/\sigma^3 \quad (16)$$

Thus, connection weighs to the winner are updated according to the error signals from the output layer: if the signal is positive, information can be maximized; if the signal is negative, information is minimized.

3 Results and Discussion

3.1 XOR Problem

We first applied our method to the well-known XOR problem to make certain that our method could really acquire input-target relations. When the number of competitive

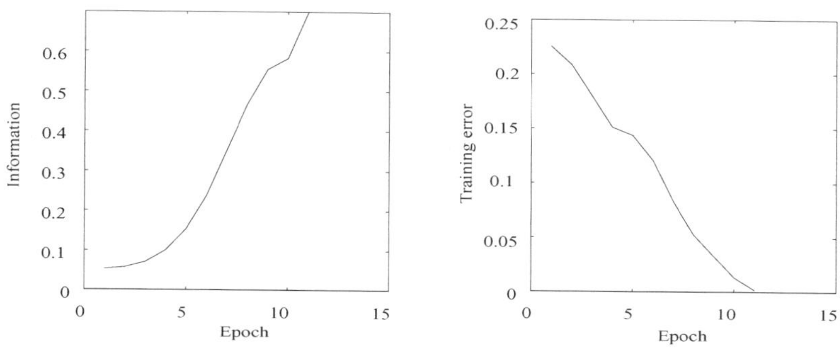

(a) Information (b) Training error

Fig. 3. Information content (a) and training error in MSE (b) as a function of the number of epochs.

units was larger than two, the network could immediately learn the relations; however, when the number of competitive units was set to two, the network could not learn relations without difficulty. Figure 3(a) shows information as a function of the number of epochs. Information rapidly increases and is close to 0.7. Figure 3(b) shows training errors in error rates. As can be seen in the figure, error rates become zero only with 11 epochs. Figure 4(b) shows an example of rates obtained by our method. The result shows that a network can, with difficulty, learn the XOR problem and information is increased.

3.2 Application to a Student Survey

Next, we conducted a survey on the content of an elemental information processing class. Because information technology changes rapidly, universities must carefully take into account students' preferences regarding such technology and related subjects; and, especially since the students' preferences seem to vary widely, there is an urgent need to determine what the students' preferences or needs are for information science education. In the survey, we presented several examples of new courses, and the students were asked to make decisions on taking the given courses. Because of the the limited number of pages of this paper, the detailed procedures of the experiments are omitted.

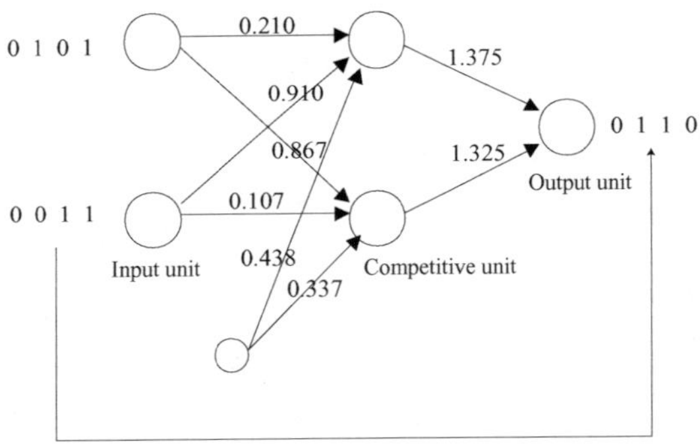

Fig. 4. A network architecture with final connection weights for the XOR problem

Figure 5(a) and (b) show information and training errors in MSE. We can see that information fluctuates in the first place and comes close to zero in the later stage of learning. Because information is controlled to acquire input-target relations, information is not necessarily increased. As shown in Figure 5(b), training errors in MSE rapidly decrease and are close to zero at the end. Figure 5(c) shows connection weights obtained by the new method and correlation coefficients obtained by regression analysis. As can be seen in the figure, two coefficients are quite similar to each other, and connection weights are significantly stable. On the other hand, as shown in Figure 5(d), connection weights by the back-propagation method are greatly different from regression coefficients. The correlation coefficient is 0.008. In addition, learning is quite unstable, and final connection weights are completely different for each trial.

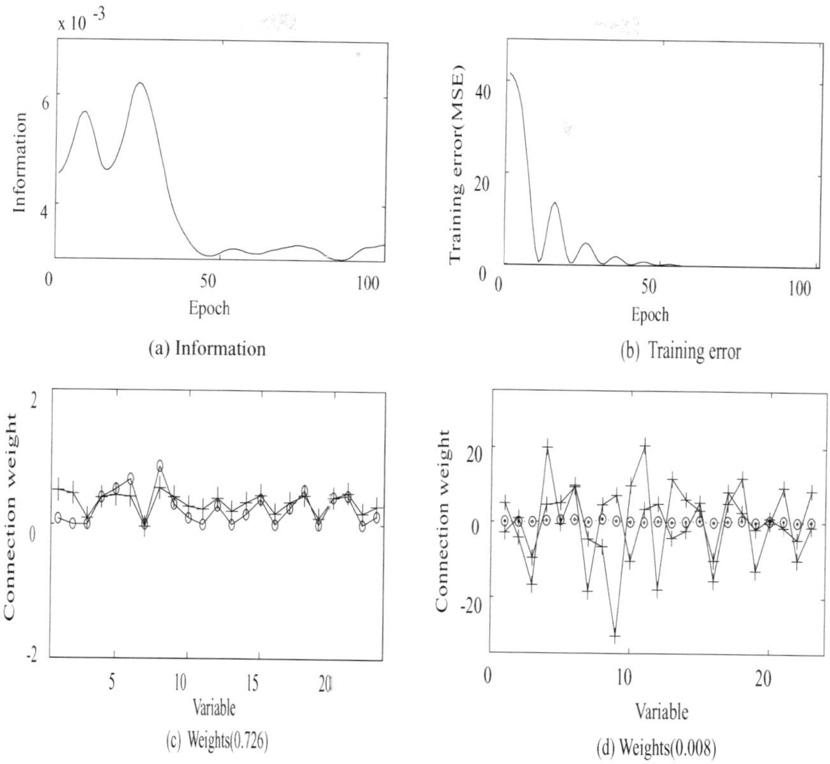

Fig. 5. Information (a) and typical training errors in MSE (b) as a function of the number of epochs. Connection weights and regression coefficients by the new method (c) and by back-propagation (d).

Figure 6 shows generalization errors in MSE. Because the number of input patterns is small, we use one-leave-out cross validation, and generalization errors are averaged over 18 patterns. As can be seen in the figure, back-propagation provides the worst case, and the partially activated network shows the best performance. On the other hand, the radial basis function network ranks in the middle.

We applied the method to another student survey for a new sub-major on information processing. As discussed for the previous results, information is not increased, but decreased in this case. Training errors are close to zero with about 50 epochs. Connection weights are quite similar to regression coefficients by regression analysis. On the other hand, connection weights by the radial-basis network with stochastic gradient and the regression analysis are quit different from the regression coefficients in terms of the magnitude of connection weights. Because only little improvement can be seen in generalization, we tried to increase the number of competitive units. Figure 7 shows generalization errors as a function of the number of epochs. Generalization errors are decreased as the number of competitive units is increased and reach their lowest value when the number of competitive units is eight. Figure 7(b) shows regression coefficients and connection weights for the largest competitive unit. As can be seen in the figure,

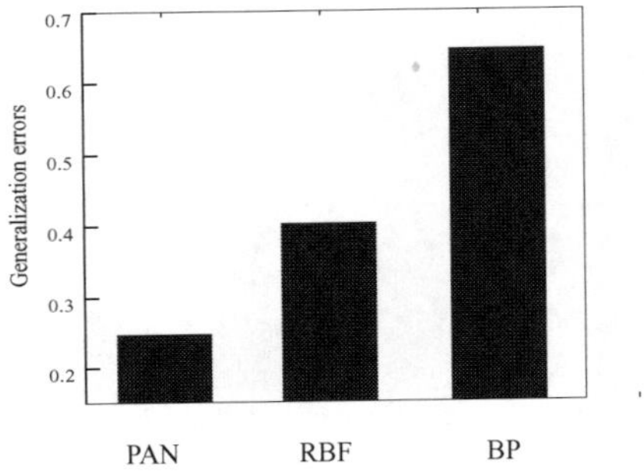

Fig. 6. Generalization errors by three methods

connection weights are significantly similar to regression coefficients. This means that, even if the number of competitive units is increased, similar connection weights are kept in a network.

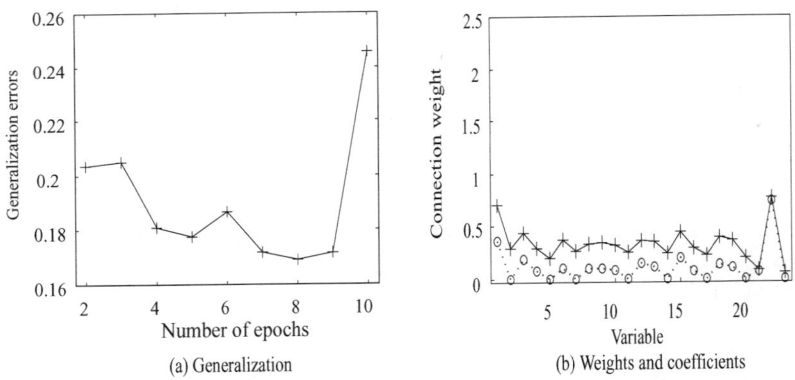

Fig. 7. Generalization errors as a function of the number of competitive units (a) and connection weights (b)

4 Conclusion

In this paper, we have introduced a partially activated network in which only a fraction of elements in a neural network is used in learning. For choosing important elements, we have proposed a new competitive learning called *controlled competitive learning*. This is an extension of conventional competitive learning to supervised hierarchical networks. In addition, we extend competitive learning to continuous targets. By this extension, we can broaden the scope of competitive learning and apply it to many practical problems. These competitive learning methods have been used to control information content in competitive units. Information-theoretic methods have been exclusively related to information maximization or information minimization. Now it is possible to control information content, depending upon given problems. By applying the method to a student survey, we have found that representations obtained by our methods are quite similar to those obtained by regression analysis. In addition, our method can produce very stable learning processes compared with back-propagation approaches to this problem. For generalization, we have shown the possibility that performance can be improved by using partially activated networks. We have borrowed a concept of competitive learning to activate important elements. However, if we can broaden the concept of competitive learning to include more realistic situations, the new approach presented in this paper can certainly be more practically useful.

References

1. Rumelhart, D.E., Zipser, D.: Feature discovery by competitive learning. In: Rumelhart, D.E., G.E.H., et al. (eds.) Parallel Distributed Processing vol. 1, pp. 151–193, MIT Press, Cambridge (1986)
2. Grossberg, S.: Competitive learning: from interactive activation to adaptive resonance. Cognitive Science 11, 23–63 (1987)
3. DeSieno, D.: Adding a conscience to competitive learning. In: Proceedings of IEEE International Conference on Neural Networks, San Diego, pp. 117–124. IEEE Computer Society Press, Los Alamitos (1988)
4. Ahalt, S.C., Krishnamurthy, A.K., Chen, P., Melton, D.E.: Competitive learning algorithms for vector quantization. Neural Networks 3, 277–290 (1990)
5. Xu, L.: Rival penalized competitive learning for clustering analysis, RBF net, and curve detection. IEEE Transaction on Neural Networks 4(4), 636–649 (1993)
6. Luk, A., Lien, S.: Properties of the generalized lotto-type competitive learning. In: Proceedings of International conference on neural information processing, pp. 1180–1185. Morgan Kaufmann Publishers, San Mateo, CA (2000)
7. Hulle, M.M.V.: The formation of topographic maps that maximize the average mutual information of the output responses to noiseless input signals. Neural Computation 9(3), 595–606 (1997)
8. Kamimura, R.: Information theoretic competitive learning in self-adaptive multi-layered networks. Connection Science 13(4), 323–347 (2003)
9. Kamimura, R.: Information-theoretic competitive learning with inverse euclidean distance. Neural Processing Letters 18, 163–184 (2003)
10. Reed, R.: Pruning algorithms-a survey. IEEE Transactions on Neural Networks 5, 740–747 (1993)
11. Fritzke, B.: Growing cell structures – a self-organizing network for unsupervised and supervised learning. Neural Networks 7(9), 1441–1460 (1994)
12. Gatlin, L.L.: Information Theory and Living Systems. Columbia University Press (1972)

CNN Based Hole Filler Template Design Using Numerical Integration Techniques

K. Murugesan[1] and P. Elango[2,*]

[1] Department of Mathematics, National Institute of Technology
Tiruchirappalli – 620 015, Tamil Nadu, India
murugu@nitt.edu

[2] Research Scholar, Department of Computer Applications
National Institute of Technology
Tiruchirapppalli - 620 015, Tamil Nadu, India
elanalin_74@yahoo.com

Abstract. This paper presents, a design method for the template of a hole-filler used to improve the pe rformance of the handwritten character recognition using numerical integration algorithms, based on the dynamic analysis of a cellular neural network (CNN). This is done by analyzing the features of the hole-filler template and the dynamic process of CNN using popular numerical integration algorithms to obtain a set of inequalities satisfying its output characteristics as well as the parameter range of the hole-filler template. Simulation results are presented for Euler, Modified Euler and RK methods and compared. It was found that RK Method performs well in terms of settling time and computation time for all step sizes.

Keywords: Cellular neural network, Hole-filler template, Numerical Integration algorithms, Euler method, Modified – Euler method, RK-method.

1 Introduction

Cellular Neural Networks (CNN) is analog, continuous time, nonlinear dynamic systems and formally belongs to the class of recurrent neural networks. Since their introduction in 1988 by Chua and Yang[10], they have been the subjects of intense research. The Cellular Neural Network (CNN) is an artificial neural network of the nearest neighbour interaction type. It has been widely used [11-19] for image processing, pattern recognition, moving object detection, target classification signal processing, augmented reality and solving partial differential equations etc. The cellular neural network CMOS array was implemented by Anguita et al [1-5] and Dalla Betta et al [6]. The design of a cellular neural network template is an important problem, and has received wide attention [7-9] in the recent past. This paper reports an efficient algorithm exploiting the latency properties of Cellular Neural Networks along with popular numerical integration algorithms. Here, simulation results are also presented and a compared with three different algorithms.

[*] Corresponding author.

J. Marques de Sá et al. (Eds.): ICANN 2007, Part I, LNCS 4668, pp. 490–500, 2007.

In this article, we describe the dynamic behavior of CNN in section II, Hole-filler template design ideas in Section III, Numerical Integration algorithms and its description is given in Section IV, and simulation results in Section V.

2 Dynamic Analysis of CNN

The dynamic equation of cell $C(i, j)$ in an M x N cellular neural network is given by Chua and Yang [10] and is shown in Figure-1.(a)-(b).

$$C\frac{dx_{ij}(t)}{dt} = -\frac{1}{R_x}x_{ij}(t) + \sum_{C(k,l)\in N_r(i,j)} A(i,j;k,l)\,y_{kl}(t) + \sum_{C(k,l)\in N_r(i,j)} B(i,j;k,l)\,u_{kl} + I \quad (1)$$

$$y_{ij}(t) = \tfrac{1}{2}\left[\,|x_{ij}(t)+1| - |x_{ij}(t)| - 1\right],\ 1 \le i \le M,\ 1 \le j \le n \quad (2)$$

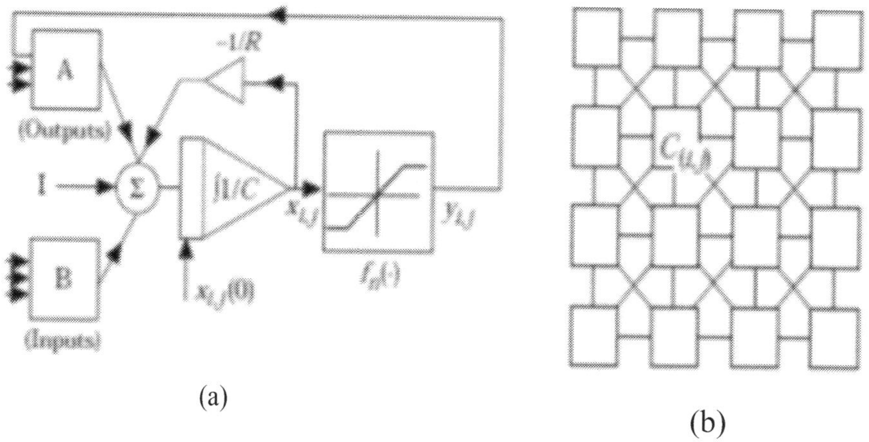

Fig. 1. CNN cell

where x_{ij}, y_{ij} and u_{ij} are the state voltage, output voltage and input voltage respectively and they are functions of time *t*. R_x is a linear resistance, C is a linear capacitor, and $A(i, j;k,l)$ and $B(i, j;k,l)$ are the transconductances of the output and input voltages of $C(k,l)$ with respect to $C(i, j)$ called the cloning templates of CNN. $N_r(i, j)$ denotes the rth-neighbour of $C(i, j)$ and I is an independent current source. From equation (2) one can discern that the output voltage is nonlinear. Now let us rewrite the cell equation (1) as follows:

$$C\frac{dx_{ij}(t)}{dt} = -f\left[x_{ij}(t)\right] + g(t) \tag{3}$$

where

$$f\left[x_{ij}(t)\right] = \frac{1}{R_x} x_{ij}(t), \tag{4}$$

$$g(t) = \sum_{\substack{C(k,l)\in N_r(i,j)\\ C(k,l)\neq C(i,j)}} A(i,j;k,l)\, y_{kl}(t) + \sum_{C(k,l)} B(i,j;k,l)\, u_{kl} + I \tag{5}$$

3 Hole-Filler Template Design

The Hole-Filler is a cellular neural network, which fills up all the holes and remains unaltered outside the holes in a bipolar image. Let $R_x = 1$, $C = 1$ and let +1 stand for the black pixel and –1 for the white one. We shall discuss the images having holes enclosed by the black pixels, when the bipolar image is input with $U = \{u_{ij}\}$ into CNN. The initial state values are set as $x_{ij}(0) = 1$. From the equation (2) the output values are $y_{ij}(0) = 1, 1 \le i \le M, 1 \le j \le N$

Now let us consider the templates *A, B* and the independent current source *I* are given as

$$A = \begin{bmatrix} 0 & a & 0 \\ a & b & a \\ 0 & a & 0 \end{bmatrix}, \quad a > 0, \quad b > 0, \quad B = \begin{bmatrix} 0 & 0 & 0 \\ 0 & 4 & 0 \\ 0 & 0 & 0 \end{bmatrix}, \quad I = -1 \tag{6}$$

Where the template parameters *a* and *b* are to be determined. In order to make the outer edge cells become the inner ones, normally auxiliary cells are added along the outer boundary of the image, and their state values are set to zeros by circuit realization, resulting in the zero output values. The state equation (1) can be rewritten as

$$\frac{dx_{ij}(t)}{dt} = -x_{ij}(t) + \sum_{C(k,l)\in N_r(i,j)} A(i,j;k,l)\, y_{ij}(t) + 4u_{ij}(t) - I \tag{7}$$

For the cell $C(i,j)$, we call the cells $C(i+1,j), C(i-1,j), C(i,j+1)$ and $C(i,j-1)$ to be the non-diagonal cells. Here, several possible cases are to be considered

Case 1: The input value $u_{ij} = +1$ for cell $C(i, j)$, signaling the black pixel. Because the initial state value of the cell $C(i, j)$ has been set to 1, $x_{ij}(0) = 1$, and from equation (2) its initial output value is also $y_{ij}(0) = 1$. According to the hole-filler demands, its eventual output should be $y_{ij}(\infty) = 1$. To obtain this result we set

$$\frac{dx_{ij}(t)}{dt} \geq 0 \tag{8}$$

Substituting this input $u_{ij} = 1$ and equation (6) into equation (7), we obtain

$$\frac{dx_{ij}(t)}{dt} = -x_{ij}(t) + a\left[y_{(i-1)j}(t) + y_{(i+1)j}(t) + y_{i(j-1)}(t) + y_{i(j+1)}(t)\right] + by_{ij}(t) + 3 \tag{9}$$

Combining equations (8) and (9) and considering the minimum value of $x_{ij}(t) = 1$ this case yields

$$a\left[y_{(i-1)j}(t) + y_{(i+1)j}(t) + y_{i(j-1)}(t) + y_{i(j+1)}(t)\right] + by_{ij}(t) + 2 \geq 0 \tag{10}$$

To facilitate our discussion, two sub cases namely cell $C(i, j)$ inside as well as outside the hole (boundary) are distinguished.

Sub case 1.1: The cell $C(i, j)$ is inside the holes. Since $x_{ij}(0) = 1$, from equation (2) its initial output value $y_{ij}(0) = 1$. Considering equations (8) and (2), $y_{ij}(t) \geq 1$. According to the hole-filler demands, its initial output of non-diagonal black pixels (grey level) should not be changed inside the holes. The weighting coefficients of ***a*** and ***b*** are equal to +4 and +1, respectively. Since $A(i, j; i, j) > 1/R_x$, the parameter ***b*** is found to be $b > 1$, that is

$$\begin{aligned} 4a + b + 2 &\geq 0 \\ b &> 1 \end{aligned} \tag{11a}$$

Sub case 1.2: The cell $C(i, j)$ is outside the holes. To satisfy equation (10), we need to check only the minimum value on the left-hand side of equation (10). This is true when there are four non-diagonal white pixels around the cell $C(i, j)$, where the weighting coefficient of ***a*** *in* equation (10) is –4. Since $y_{ij}(t) \geq 1$, the weight of ***b*** is equal to 1. Combining this with $b > 1$ gives

$$\begin{aligned} -4a+b+2 &\geq 0 \\ b &> 1 \end{aligned} \tag{11b}$$

Case 2: The input value of cell $C(i, j)$ is $u_{ij} = 1$, signaling the white pixel. Substituting this input value in equation (7) gives

$$\frac{dx_{ij}(t)}{dt} = -x_{ij}(t) + a\left[y_{(i-1)j}(t) + y_{(i+1)j}(t) + y_{i(j-1)}(t) + y_{i(j+1)}(t)\right] + by_{ij}(t) - 5 \tag{12}$$

Sub case 2.1: The cell $C(i, j)$ is inside the holes. Since $x_{ij}(0) = 1$, from equation (2) its initial output value is $y_{ij}(0) = 1$. According to the hole-filler demands, the holes should be filled by the black pixels, whereas its initial black pixels remain unaltered:

$$\frac{dx_{ij}(t)}{dt} \geq 0 \tag{13}$$

Combining equations (12) and (13) and considering $x_{ij}(t) \geq 1$ yields

$$a\left[y_{(i-1)j}(t) + y_{(i+1)j}(t) + y_{i(j-1)}(t) + y_{i(j+1)}(t)\right] + by_{ij}(t) - 6 \geq 0 \tag{14}$$

where we use the minimum value of $x_{ij}(t)$ in equation (12). Since the cell is inside the holes, its initial output of non-diagonal black pixels remain unchanged. The coefficients of ***a*** *and* ***b*** are equal to +4 and +1, respectively. Combining this with $b > 1$ gives

$$\begin{aligned} 4a+b-6 &\geq 0 \\ b &> 1 \end{aligned} \tag{15}$$

Sub case 2.2: The cell $C(i, j)$ is outside the holes. Since $x_{ij}(0) = 1$, from equation (2) its initial output value is $y_{ij}(0) = 1$. According to the hole-filler demands, the final output of this cell should be white, and in this case $y_{ij}(\infty) \leq -1$.

$$\frac{dx_{ij}(t)}{dt} < 0 \tag{16}$$

Combining equations (12) and (16) and considering $x_{ij}(t) \leq 1$ we get

$$a\left[y_{(i-1)j}(t) + y_{(i+1)j}(t) + y_{i(j-1)}(t) + y_{i(j+1)}(t)\right] + by_{ij}(t) - 6 < 0 \tag{17}$$

where we use the maximum value of $x_{ij}(t)$ in equation (12)

The initial condition is $y_{ij}(0)=1$. Now the problem arises how the output of cell $C(i, j)$ be changed to -1 and where does this change to begin. First we consider the situation where the change begins from the inside of the bipolar image. If the maximum value on the left-hand side in equation (17) is less than zero, equation (17) holds. Inside the image and outside the holes, the maximum weights of ***a*** and ***b*** *are* +4 and +1, respectively. This case was described by equation (15). In fact, the change of the output of the cell $C(i, j)$ is like a wave propagating from the edges to the inside of the image and it is verified from the simulated result. Therefore, we should first consider the edge cell $C(i, j)$, $i=1$ or M, $j=1$ or *N*. For this the maximum weight of ***a*** in equation (17) is +3, which is also the maximum weight of ***a*** outside the holes. The maximum weight of ***b*** is +1, occurring at the initial time:

$$\begin{aligned} &3a+b-6 < 0 \\ &b>1 \end{aligned} \qquad (18)$$

Combining Cases 1 and 2, we obtain

$$\begin{aligned} &3a+b-6<0, \\ &4a+b-6\geq 0, \\ &4a+b+2\geq 0, \end{aligned} \qquad (19)$$

4 Numerical Integration Algorithms

The CNN is amenable to a system of nonlinear differential equations. Therefore, it is necessary to discretize the differential equation for performing simulations. For computational purpose, a normalized time differential equations describing CNN is used by Nossek et al [17].

$$\begin{aligned} f'(x(\pi\tau)) := \frac{dx_{ij}(\pi\tau)}{dt} &= -x_{ij}(\pi\tau) + \sum_{C(k,l)\in N_r(i,j)} A(i,j;k,l)\,y_{kl}(\pi\tau) \\ &+ \sum_{C(k,l)\in N_r(i,j)} B(i,j;k,l)\,u_{kl} + I \end{aligned} \qquad (20)$$

$$y_{ij}(\pi\tau) = \frac{1}{2}\left(\left|x_{ij}(\pi\tau)+1\right| - \left|x_{ij}(\pi\tau)-1\right|\right)$$

Where τ is the normalised time. Well-established numerical integration techniques are used for implementing hole-filler template design. These methods can be derived using the definition of the definite integral

$$x_{ij}((n+1)\tau) - x_{ij}(\pi\tau) = \int_{\tau_n}^{\tau_{n+1}} f'(x(\pi\tau))\,d(\pi\tau) \qquad (21)$$

Euler Method

Euler's method is the simplest of all algorithms for solving ODEs. It is an explicit formula which uses the Taylor-Series expansion to calculate the approximation. We divide the time interval [a, b] into N – steps with $t_i = a + h_i$ where i = 0, 1....N and with initial condition $y(0) = y_0$.

$$w_{i+1} = w_i + hf(t_i, w_i) \tag{22}$$

Where steps size $h = \dfrac{b-a}{N}$

Modified - Euler Method

The modified Euler method starts with an Euler step, giving a provisional value for w_{i+1} at the next time t_{i+1}

$$w'_{i+1} = w_i + hf(t_i, w_i) \tag{23}$$

The step actually taken looks like an Euler step, but with ***f*** replaced by the average of ***f*** at the starting point of the step and ***f*** at the provisional point

$$w_{i+1} = w_i + \frac{h}{2}[f(t_i, w_i) + f(t_{i+1}, w'_{i+1})] \tag{24}$$

Runge – Kutta Method

The Runge-Kutta methods have become very popular, both as computational techniques as well as subject for research. This method was derived by Runge about the year 1894 and extended by Kutta a few years later. Runge-Kutta (RK) algorithms have always been considered superb tools for numerical integration of Ordinary Differential Equations (ODE's). The fact that RK methods are self-starting, easy to program, and show extreme accuracy and versatility in ODE problems has led to their continuous analysis and use in computational research.

$$\begin{aligned}
k_1 &= hf\,(t_i, w_i) \\
k_2 &= hf\,(t_i + \frac{h}{2}, w_i + \frac{1}{2}k_1) \\
k_3 &= hf\,(t_i + \frac{h}{2}, w_i + \frac{1}{2}k_2) \\
k_4 &= hf\,(t_{i+1}, w_i + k_3) \\
w_{i+1} &= w_i + (\frac{1}{6}(k_1 + 2k_2 + 2k_3 + k_4))
\end{aligned} \tag{25}$$

Notice that the first two steps look like the modified Euler method for reaching the mid point $t = t_i + \dfrac{h}{2}$.

5 Simulated Results

This Hole-filler template has been simulated using Pentium IV Machine with 3.0 GHz. speed using different Numerical integration algorithms of hand written character / images given in Figures-3 and Figure-4. The Settling time T_s and computation time T_c for different step sizes are presented in the Table-1 for comparison. The settling time T_s is the time from start of computation until the last cell leaves the interval **[-1.0, 1.0]** which is based on a specified limit (e.g., |dx / dt| < 0.01). The computation time T_c is the time taken for settling the network and adjusting the cell for proper position once the network is settled. The simulation shows the desired output for every cell. We use +1 and -1 to indicate the black and white pixels, respectively. We have marked the selected template parameters ***a*** *and* ***b***, which are restricted to the shaded area as shown in Figure-2 for the simulation.

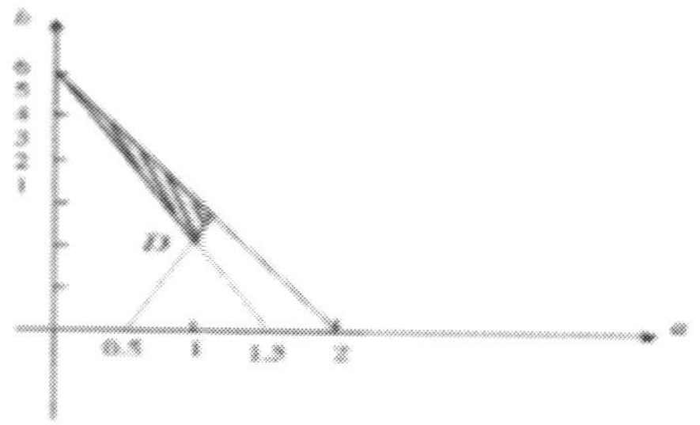

Fig. 2. The parameter range of template

(a) (b)

Fig. 3. Image before and after hole-filling

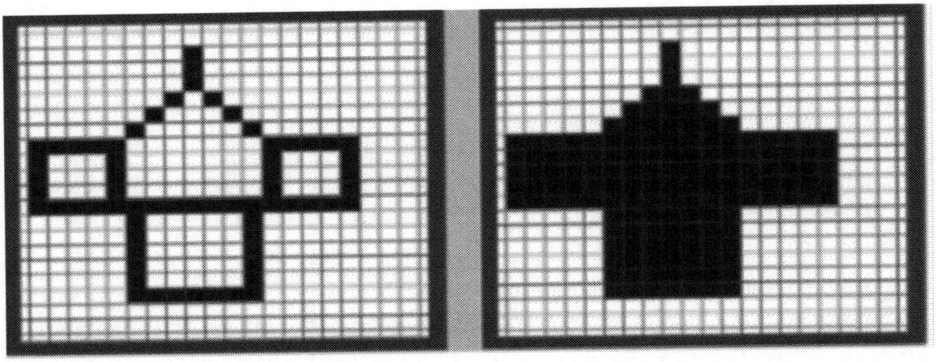

(a) (b)

Fig. 4. Image before and after hole filling (with Cell)

Table 1. Simulated results of hole-filler template design

Step size (h)	Euler Method		Modified Euler		RK-Method	
	Settling Time (T_s)	Computation Time (T_c)	Settling Time (T_s)	Computation Time (T_c)	Settling Time (T_s)	Computation Time (T_c)
0.5	2.5	6.5	2.4	6.0	2.4	5.8
0.6	14.7	15.5	13.7	15.0	12.5	11.4
0.7	28.3	32.5	27.4	32.0	27.2	30.0
0.8	30.7	35.0	30.0	34.6	29.6	32.4
0.9	32.6	36.8	32.0	36.6	31.6	34.2
1.0	33.6	37.9	33.0	37.6	32.8	36.0
1.5	36.8	44.8	36.2	44.7	36.0	43.1
2.0	43.2	47.4	43.0	46.4	42.8	46.2
2.5	45.6	50.6	44.5	50.4	44.3	49.3
3.0	49.3	53.5	49.0	52.1	48.2	52.0

Example

The template A, whose parameters are within the shaded area of Figure-2, and template B and I are given as follows:

$$A = \begin{bmatrix} 0 & 1.0 & 0 \\ 1.0 & 3 & 1.0 \\ 0 & 1.0 & 0 \end{bmatrix}, \quad B = \begin{bmatrix} 0 & 0 & 0 \\ 0 & 4 & 0 \\ 0 & 0 & 0 \end{bmatrix}, \quad I = -1.0$$

For the simulation, the input images are given in Figure-3(a), Figure-4(a) with cell and the output images for the CNN based image edge template are shown in Figure-3(b), Figure-4(b). The simulated results are represented in the form of a graph in Figure- 5 for comparison and conclusion. From the simulation results, the following conclusions are drawn. RK algorithm always takes least settling time and computation time for all step sizes compared to the other two algorithms. Modified-Euler method is comparatively better than Euler method for settling time as well as computation time. Hence where ever we need a simple algorithm, Modified Euler method serves as a good candidate. RK method exhibits superiority over the other the rest of the algorithms with little more complexity in the algorithm.

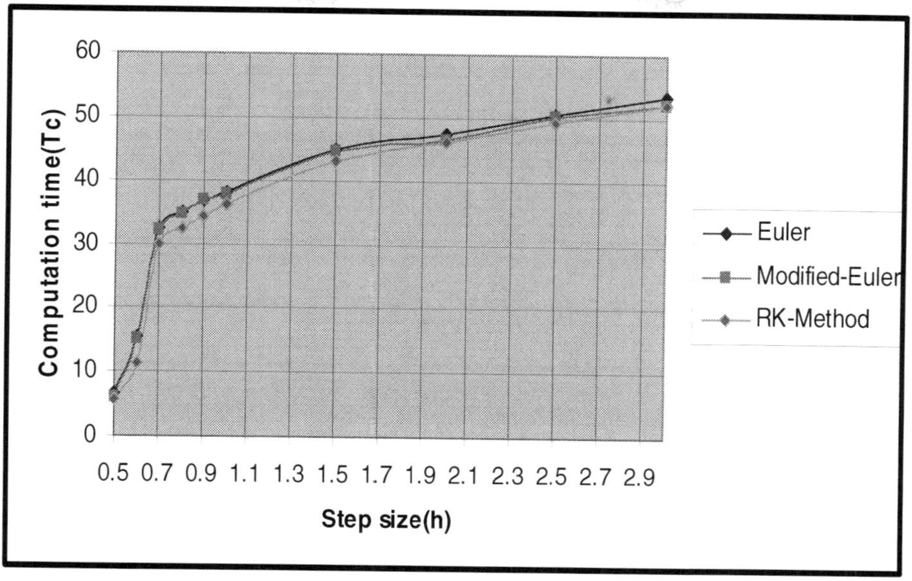

Fig. 5. Comparison of algorithms

6 Conclusion

It is shown that the cellular neural network based hole-filler template could be designed from its dynamic behavior using different numerical algorithms, and also the template for other cellular neural network can similarly be designed. The hole is filled and the outside image remains unaffected that is, the edge of the images are preserved in tact. The templates of the cellular neural network are not unique and this is important in its implementation. From the graph in Figure-5 one can conclude that RK-method always yields less settling time and computation time for all step sizes compared to Euler and Modified-Euler Method. These algorithms has wide application for image/ character recognition.

References

1. Anguita, M., Pelayo, F.J., Ros, E., Palomar, D., Prieto, A.: VLSI implementations of CNNs for image processing and vision tasks: single and multiple chip approaches. In: IEEE International Workshop on Cellular Neural Networks and their Applications, pp. 479–484. IEEE Computer Society Press, Los Alamitos (1996)
2. Anguita, M., Pelayo, F.J., Fernandez, F.J., Prieto, A.: A low-power CMOS implementation of programmable CNN's with embedded photosensors. IEEE Transactions on Circuits Systems I: Fundamental Theory and Applications 44(2), 149–153 (1997)
3. Anguita, M., Pelayo, F.J., Ros, E., Palomar, D., Prieto, A.: Focal-plane and multiple chip VLSI approaches to CNNs. Analog Integrated Circuits and Signal Processing 15(3), 263–275 (1998)

4. Anguita, M., Pelayo, F.J., Rojas, I., Prieto, A.: Area efficient implementations of fixed template CNN's. IEEE Transactions on Circuits and Systems I: Fundamental Theory and Applications 45(9), 968–973 (1997)
5. Anguita, M., Fernandez, F.J., Diaz, A.F., Canas, A., Pelayo, F.J.: Parameter configurations for hole extraction in cellular neural networks. Analog Integrated Circuits and Signal Processing 32(2), 149–155 (2002)
6. Dalla Betta, G.F., Graffi, S., Kovacs, M., Masetti, G.: CMOS implementation of an analogy programmed cellular neural network. IEEE Transactions on Circuits and Sysems - part–II 40(3), 206–214 (1993)
7. Boroushaki, M., Ghofrani, M.B., Lucas, C.: Simulation of nuclear reactor core kinetics using multilayer 3-D cellular neural networks. IEEE Transactions on Nuclear Science 52(3), 719–728 (2005)
8. Chua, L.O., Thiran, P.: An analytic method for designing simple cellular neural networks. IEEE Transactions on Circuits and Systems 38(11), 1332–1341 (1991)
9. Fajfar, F., Bratkovic, T., Tuma, T., Puhan, J.: A rigorous design method for binary cellular neural networks. International Journal of Circuit Theory and Applications 26(4), 365–373 (1998)
10. Chua, L.O., Yang, L.: Cellular neural networks: theory. IEEE Transactions on Circuits and Systems 35(10), 1257–1272 (1988)
11. Crounse, K.R., Chua, L.O.: Methods for image processing and pattern formation in cellular neural networks. IEEE Transactions on Circuits and Systems I: Fundamental Theory and Applications 51(5), 939–947 (2004)
12. Dogaru, R., Julian, P., Chua, L.O., Glesner, M.: The simplicial neural cell and its mixed-signal circuit implementation: an efficient neural-network architecture for intelligent signal processing in portable multimedia applications. IEEE Transactions on Neural Networks 13(4), 995–1008 (2002)
13. Galan, R.C., Jimenez-Garrido, F., Dominguez-Castro, R., Espejo, S., Roska, T., Rekeczky, C., Petras, I., Rodriguez-Vazquez, A.: A bio-inspired two-layer mixed-signal flexible programmable chip for early vision. IEEE Transactions on Neural Networks 14(5), 1313–1336 (2003)
14. Kozek, T., Roska, T., Chua, L.O.: Genetic algorithm for CNN template learning. IEEE Transactions on Circuits and Systems I: Fundamental Theory and Applications 40(6), 392–402 (1993)
15. Lee, C.C., de Pineda, J.: Time-multiplexing CNN simulator. In: IEEE International Symposium on Circuits and Systems, pp. 407–410. IEEE Computer Society Press, Los Alamitos (1994)
16. Lee, C.C., de Pineda, J.: Single-layer CNN simulator. In: International Symposium on Circuits and Systems, vol. 6, pp. 217–220 (1994)
17. Nossek, J.A., Seiler, G., Roska, T., Chua, L.O.: Cellular neural networks: theory and circuit design. International Journal of Circuit Theory and Applications 20, 533–553 (1992)
18. Murugesh, V., Murugesan, K.: Comparison of Numerical Integration in Raster CNN Simulation. In: Manandhar, S., Austin, J., Desai, U., Oyanagi, Y., Talukder, A.K. (eds.) AACC 2004. LNCS, vol. 3285, pp. 115–122. Springer, Heidelberg (2004)
19. Murugesan, K., Gopalan, N.P., Gopal, D.: Error free Butcher algorithm for linear electrical circuits. ETRI journal 27(2), 195–205 (2005)

Impact of Shrinking Technologies on the Activation Function of Neurons*

Ralf Eickhoff[1], Tim Kaulmann[2], and Ulrich Rückert[2]

[1] Chair of Circuit Design and Network Theory, Dresden University of Technology, Germany
ralf.eickhoff@tu-dresden.de
[2] Heinz Nixdorf Institute, System and Circuit Technology, University of Paderborn, Germany
kaulmann,rueckert@hni.upb.de

Abstract. Artificial neural networks are able to solve a great variety of different applications, e.g. classification or approximation tasks. To utilize their advantages in technical systems various hardware realizations do exist. In this work, the impact of shrinking device sizes on the activation function of neurons is investigated with respect to area demands, power consumption and the maximum resolution in their information processing. Furthermore, analog and digital implementations are compared in emerging silicon technologies beyond 100 nm feature size.

1 Introduction

As device dimensions are continuously shrinking more complex systems can be integrated on a single chip in emerging technologies than today and these systems can benefit from neural inspired processing techniques in several aspects. For shrinking device dimensions the requirements during fabrication extremely increase because more and more quantum mechanical effects impact the device characteristics [1]. However, classical designed systems strongly depend on reliable devices with small parameter deviations in order to operate according to their specifications. Here, neural inspired methodologies can relax these technology requirements, e.g. reducing the high demands for lithography [1], and can significantly reduce production costs if these novel design techniques allow changes and deviations in the device characteristics. Therefore, neural processing techniques can help to overcome the challenges of emerging technologies such as fault tolerant and robust operation due to more unreliable devices than today [2]. Moreover, these techniques also allow the implementation of novel classes of circuits with reduced complexity, i.e. less transistors per function compared to classical circuits [3].

Many popular models of biological neurons apply a nonlinear function on the neuron activation potential [4], where the activation is achieved by a weighted

* This work was supported by the Graduate College 776 - Automatic Configuration in Open Systems - funded by the German Research Foundation (DFG).

J. Marques de Sá et al. (Eds.): ICANN 2007, Part I, LNCS 4668, pp. 501–510, 2007.

sum of the input signals. By using this computing technique various problems, e.g. approximation and classification tasks or synthesis of Boolean functions, can be accomplished and due to its simplicity and its direct representation by differential amplifiers or threshold logic, this model is highly suitable for hardware integration. In this work, the effects of shrinking device dimensions and technology scaling on the activation function of neurons are investigated where, especially, the processing resolution within different Complementary Metal Oxide Semiconductor (CMOS) technologies is analyzed. To analyze the scaling impact on neural processing, analog and digital representations of the activation function are investigated in CMOS technologies beyond 100 nm. These digital and analog realizations are compared to each other with respect to their size, the power consumption and the maximal processing resolution.

The remainder is organized as follows. Section 2 briefly introduces scaling trends of actual and future CMOS technologies whereby in Section 3 the scaling impact on analog and digital implementations of activation functions is investigated. Section 4 concludes the paper.

2 CMOS Scaling Theory

CMOS technology is the most popular today's technology in which many integrated circuits are fabricated. This results from its high integration density allowing low cost fabrication. Especially, as device dimension is continuously shrinking more complex system, e.g. large neural networks, multi-processor systems or several systems, can be integrated on a single chip. The basic idea of scaling a Metal Oxide Semiconductor Field Effect Transistor (MOSFET) was developed by [5] in 1974 where all geometrical dimensions and voltages of a transistor are scaled down by a factor $\alpha_L > 1$ whereas the doping and charge densities are increased by the same factor. Applying this scaling behavior to the device, all electrical fields in the transistor remain constant[1] why this theory is called *ideal scaling*.

However, the observed scaling behavior of CMOS processes deviates from the ideal scaling theory because several effects are influencing the MOSFET device characteristics. Because in early processes the carrier velocities and, thus, the operation speed could be increased with larger electrical fields, the voltages have been scaled down much slower than the geometrical dimensions in order to keep the electrical field as high as possible. This improved the operation speed of MOSFETs at the beginning. Moreover, in actual CMOS processes the gate oxide thickness cannot be scaled down with the same factor anymore because only a few atomic layers are used and, therefore, scaling reaches its limit for the oxide thickness. Also other effects such as leakage currents or quantum-mechanical effects have let to a deviation from ideal scaling behavior [6]. Thus, besides a geometrical scaling factor α_L a scaling factor $\alpha_U > 1$ for the voltages and the gate oxide thickness can be introduced in order to accommodate this trend

[1] Therefore, the device characteristic does not change although one obtains smaller transistors.

which leads to a more generalized scaling theory [6]. For further information about scaling theory, its limits and impacts the reader is referred to [6,7,1].

3 Shrinking Threshold Elements

If artificial models of neurons are integrated in hardware, several noise sources in analog implementations or the limited resolution in digital systems affect their nonlinear transfer functions and their behavior will deviate from the original modeling. Often these effects can be included in parameter variations. E.g., Fig. 1 shows these effects of limited resolution and additional noise sources on a classification problem where as a neural network a multilayer perceptron with sigmoidal transfer functions [8] is used. The limited resolution of the activation function integrated in hardware causes an uncertainty region, in which an element cannot be reliable assign to one class and which may lead to misclassification. This effect additionally contributes to the error criteria and impacts system performance where ideal system performs would operate exactly according to the initial design (in Fig. 1 the solid line in the middle of uncertainty). Therefore, besides area and power requirements each neuron obtains a maximal processing resolution, which is affected by the used technology and the architecture of the threshold function.

Fig. 2 shows the sigmoidal transfer function of a neuron realized by MOSFETs. For this purpose, a simple differential amplifier is used, whose transfer function provides the sigmoidal shape and its transfer function $V_{out} = f(V_{in})$ can be described by the tanh function [9]. This threshold element can be used in a multilayer perceptron to perform the threshold operation of a neuron. Besides the popular sigmoidal function, the architecture in Fig. 2 is also able to approximate the Heaviside function if a sufficient large gain A of the amplifier is achieved. Because the activation functions typically use small slopes A ($A \approx 1 - 10$), only small gains are needed for the differential amplifier avoiding cascode stages [9] that might be necessary for the Heaviside function.

If the circuit of Fig. 2 obtains sufficiently high gain and, therefore, acts as comparator, it can also be used in pulsed neural networks [4,10], where information is processed by pulses. The pulse shape is usually generated by using a comparator if the membrane potential of the neuron exceeds a threshold [10]. Furthermore, the circuit of Fig. 2 also allows to act as a two quadrant multiplier if V_{bi} is not hold constant but driven by an input as well. In this case, the weight adaption of the synapse functionality can be performed by the element additionally. Therefore, the circuit of Fig. 2 is one of the most essential elements for integrated hardware of neural networks because it can be used for time-dependent and time-independent models of their biological counterparts.

The overall area, the power consumption and the speed depend of the threshold element in Fig. 2 is primary determined by the operating conditions and its requirements. Because MOSFETs introduce thermal noise sources and flicker noise, also the processing resolution is affected and will deteriorate by scaling. For this analysis, flicker and thermal noise are considered which can be modeled

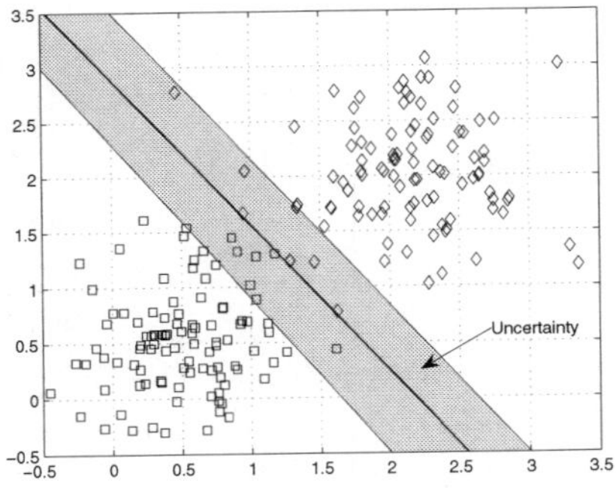

Fig. 1. Affine transformation by a neuron and its uncertainty

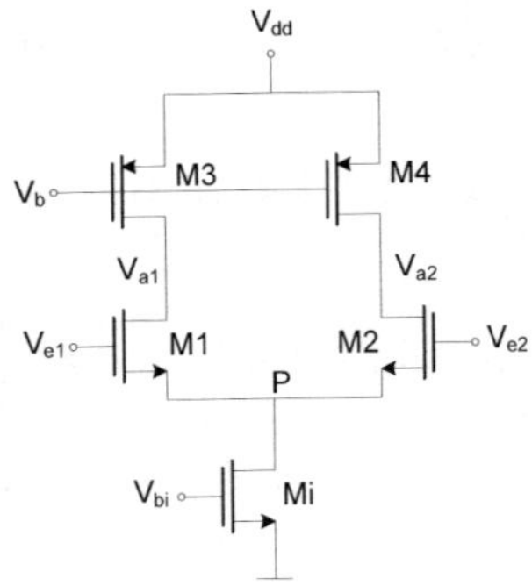

Fig. 2. Performing an activation function using MOS transistors

as additional voltage and current sources between gate and source or drain and source of each transistor Mj, $j \in \{1, 2, 3, 4, i\}$ in Fig. 2 [9].

In order to determine the resolution of the threshold element the noise power at the input has to be calculated. For this analysis it is assumed that the differential amplifier is ideally symmetrical and this leads to input referred noise voltage $\overline{V^2}_n$ of

$$\overline{V^2}_n = 2\overline{V^2}_{n1} + \frac{2g_{m3}^2(r_{o1}\|r_{o3})^2}{A^2}\overline{V^2}_{n3} = 2\overline{V^2}_{n1} + \frac{2g_{m3}^2}{g_{m1}^2}\overline{V^2}_{n3}. \tag{1}$$

In (1) $\overline{V^2}_n$ describes the power density spectrum of the input noise. $\overline{V^2}_{nj}$ denotes the noise voltage originating from the corresponding transistor $j \in \{1, 2, 3, 4, i\}$ including thermal and flicker noise[2]. Further, g_{mj} denotes its transconductance, r_{oj} is the drain-source resistance of the equivalent transistor and A is the small-signal gain of the threshold element where for this configuration holds

$$A = \frac{V_{\text{out}}}{V_{\text{in}}} = \frac{V_{a1} - V_{a2}}{V_{e1} - V_{e2}} = -g_{m1}(r_{o1}\|r_{o3}) \tag{2}$$

Usually, the small signal gain A can be set to the slope of the activation function of the threshold element for $x = 0$.

Additionally, the signal power at the input has to be determined and leads to a signal-to-noise-ratio (SNR) that can be transferred in a corresponding processing resolution using information theory [11]. Assuming a sinusoidal input signal with amplitude v_m the SNR at the input is proportional to

$$\text{SNR} \propto g_{m1}C_L \quad \text{resp.} \quad \text{SNR} \propto 1/g_{m3} \tag{3}$$

In (3) the load capacitance of the threshold element is also influencing the SNR. As devices dimensions scale down also the load capacitance scales with technology and impacts on the SNR. But this load can be evaluated if a layered

[2] Additional noise sources such as shot noise or avalanche noise can be considered by this term too.

structure of identical threshold elements is assumed. This is fulfilled if the architecture in Fig. 2 is used for the threshold operation in a multilayer perceptron. Thus, the load C_L is determined by the output capacitance of the threshold element and the input capacitance of the following stage. The output capacitance of the threshold element is primary determined by the drain-bulk capacitance C_{DB} of the transistors, whereas the input capacitance of the following stage depends on the used architecture. In a multilayer perceptron the output of the threshold element is usually multiplied by a weight in order to determine the output of the whole network as a linear weighed superposition or to weight the output and to serve as an input for the next hidden layer. Therefore, a multiplication stage has to be provided and it can be assumed that for this multiplication a Gilbert cell is used [9]. The input capacitance of the Gilbert cell is determined by the gate-source capacitance C_{GSn} of the n-channel transistors and their gate-drain capacitance C_{GDn}. If the gain of the Gilbert cell remains low, input capacitance is dominated by the gate-source capacitance and, at all, the load capacitance C_L can be approximated by

$$C_L \approx 2C_{DB} + C_{GSn}. \tag{4}$$

Consequently, the total SNR at the output can be calculated and can be transferred in a corresponding resolution using information theory. To keep track of the scaling impact, only thermal noise is considered during the analysis presented here, whereas the later shown results do include flicker noise too. Further it is assumed that the threshold element of Fig. 2 is operating in a layered network, where its output signal is multiplied by a weight in order to apply (4). Taken all these conditions into account, the SNR at the output can be calculated and the impact of scaling technology can be investigated. A generalized scaling with two different scaling factors (cf. Section 2), α_L for geometrical dimensions and α_U for the voltage and the gate oxide length respectively, is considered where these scaling factors are derived from data in [12]. This leads to the $\widehat{\text{SNR}}$ at the output that includes the impact of downscaling technologies

$$\widehat{\text{SNR}} = \frac{V_{dd}^2 g_{m1}(r_{o1} \parallel r_{o3}) C_L \alpha_u}{16\alpha_u^2 \alpha_L^2 \gamma k_B T} \tag{5}$$

$$= \frac{1}{\alpha_u \alpha_L^2} \cdot \text{SNR} \tag{6}$$

$$\approx k - 0,5542 \tag{7}$$

where k_B denotes the Boltzmann constant, T the absolute temperature, γ models the noise influence and k represents the equivalent resolution in bit. SNR represent the initially noise performance of the circuit without any down-scaling, whereas $\widehat{\text{SNR}}$ depicts the noise properties of the same circuit if it is scaled down or implemented in a new technology. The corresponding resolution in (7) can be obtained from information theory [11] where the SNR is transferred in a resolution by

$$\text{SNR}\big|_{\text{db}} = 10 \log_2 \left(\frac{3}{2}\right) + 10 \log_2 \left(2^{2k}\right) \approx 1.76 + 6 \cdot k \tag{8}$$

with k the corresponding resolution. The results of (7) are obtained if the mean of the scaling factors derived from the data in [12] is applied to (6) and (8) is used to determine the corresponding resolution.

As can be concluded from (7), the maximal resolution is decreased by 0.5 bit in every technology generation. This degradation increases only slightly and the resolution decreases by 0.6 bit per technology process if flicker noise is also considered where the flicker noise model from [9] has been used for this analysis.

Table 1. Simulation results of an analog neuron using transistor model of [12]

CMOS [nm]	130	90	65	45	32
Vdd [V]	1,3	1,2	1,1	1,0	0,9
Area [μm^2]	5,89	1,84	0,74	0,18	0,05
Power [mW]	1,2038	0,4416	0,1837	0,0480	0,0153
Gain (A)	6,76	4,82	4,35	3,17	2,08
SNR (output)	$4,99 \cdot 10^6$	$1,70 \cdot 10^6$	$7,95 \cdot 10^5$	$1,82 \cdot 10^5$	$4,46 \cdot 10^4$
SNR (input)	$2,50 \cdot 10^6$	$8,48 \cdot 10^5$	$3,98 \cdot 10^5$	$9,10 \cdot 10^4$	$2,23 \cdot 10^4$
k (theory) [bit]	10,54	9,72	9,13	8,27	7,37
k [bit]	11,13	10,35	9,80	8,74	7,72
Δk (theory) [bit]	-	0,58	0,52	0,63	0,60
Δk [bit]	-	0,78	0,55	1,06	1,01
FOM $= k/(P \cdot A)$	1,6	12,7	71,6	1013	9226

Moreover, to verify the theoretical results simulations are performed under HSpice using the predictive technology models from [12]. The threshold element of Fig. 2 is designed to drive an expected load capacitance within a specified time interval and to achieve an output swing of the full supply voltage. Therefore, the area consumption of the neuron can be calculated from the geometrical dimension of all transistors. Table 1 shows the simulation results for several technology generations and the theoretically derived resolution where these results do include the impact of flicker noise. The gain A denotes the slope of the activation function for $x = 0$ and two different SNR are mentioned in Table 1. The SNR at the output is transferred in its corresponding resolution k and its decrease Δk for every technology process. Because the neurons obtain different gains, the SNR at the output is also transferred to a SNR at the input where the different designs can be compared to each other with respect to their noise properties by this measurement [9].

As can be concluded from Table 1, the power and area consumption significantly decreases over the technology generations due to shrinking device dimensions. Nevertheless, the SNR at the input is reduced, which degrades the resolution of the neural processing. The theoretically derived resolution was too conservative whereas its reduction over the technology processes was assumed too optimistically. This discrepancy results from the fact that during the analysis of (7) it is assumed that the transconductance of transistor M1 and M3 are equal. However, for the simulations the transconductance of M1 was designed to be larger than the one of M3, which leads to an improved SNR and, thus, higher resolutions. Moreover, ESpice utilizes more complex models for flicker noise [13] and the technology processes do not follow the scaling rules ideally, e.g. the threshold voltage remains nearly constant over technologies. Therefore, a larger degradation of Δk between technologies arises compared to the theoretical derived results. Nonetheless, the neuron benefits from shrinking device dimensions because the area and power consumptions are significantly decreased whereas only slight degradation of the resolution occurs.

Furthermore, the impact of the technology scaling on digital implementations of the activation function has been analyzed in comparison to the implementation in Fig. 2. For this investigation, the threshold element has been implemented by the transfer function in [14] on an actual 0.12 μm CMOS technology by ST. The sigmoidal shape of the activation function is achieved by a piece-wise linear approximation in two different sections. Table 2 compares the impact of technology scaling on the digital and analog implementation of a sigmoidal activation function in terms of their area consumption, their electrical power and the resolution k. As figure-of-merit (FOM) also the same quantity FOM $= K/(A \cdot P)$ as in Table 1 is been used.

Table 2. Properties of analog and digital neurons

	Area [μm^2]		**Power** [mW]		**max. k** [bit]		**FOM**	
CMOS [nm]	analog	digital	analog	digital	analog	digital	analog	digital
130	5,89	7140,89	1,2038	3,2946	11	5	1.56	$1.70 \cdot 10^{-4}$
90	1,84	3422,56	0,4416	1,4576	10	5	12.3	$8.02 \cdot 10^{-4}$
65	0,74	1785,22	0,1837	0,6969	9	5	65.8	$3.21 \cdot 10^{-3}$
45	0,18	855,64	0,0480	0,3037	8	5	928.3	$1.54 \cdot 10^{-2}$
32	0,05	432,68	0,0153	0,1382	7	5	8363.2	$6.69 \cdot 10^{-2}$

As can be concluded from Table 2, the analog version significantly outperforms the digital realization with respect to the consumed area and power. As the FOM shows, the analog system performs much better than the digital system. Especially, less area is consumed by the analog version, which is based on the very simple design (see Fig. 2) and the circuit needs only a few transistors. In comparison, the digital part needs at least one multiplier, which increases the area demands significantly although the needed resolution and, thus, the number of transistors can be very low compared to other implementations [14]. At all, much more transistors are utilized in the digital design and, therefore, more power is been consumed. However, the difference in power consumption is essentially smaller than for the area gap. In the analog realization of the threshold element, its power is primary determined by the static current flowing from supply voltage to ground. This current can be neglected in a digital design where the power consumption is essentially determined by recharging the load capacitances. Therefore, the difference in power consumption is much smaller compared to the area demands. However, this domination of dynamic power consumption will change with the technologies where the leakage currents become more important in CMOS technology [15]. Here, the digital design will make up if the leakage power dominates the whole amount of power consumption.

Nevertheless, for both implementations, the overall system performance increases with emerging technologies that can be concluded from the figure-of-merit. The total size and the power of the threshold element mainly benefit from device scaling which completely cancels the effect of decreasing resolution for analog systems with respect to this quality measurement. For digital systems, the resolution remains constant that further improves the FOM. However, if scaling comes to its limits [15] and leakage power significantly increases, the FOM can become also saturated.

The resolution of the digital activation function remains constant over the technology processes, which results from the approximation type of the threshold element from [14]. Because of the piece-wise linear approximation of the sigmoidal function, there exists a lower bound of the absolute error between the implemented function and the logistic function, which cannot be decreased even if high resolutions are used inside the architecture for calculating the activation values digitally. Thus, this minimum absolute error results into a maximum resolution of 5 bit. For a comparison to additional types of digital realized activation functions, the digital implementation of the logistic function is also realized by a memory representation and by Boolean functions. For these realizations, the analog version achieves less area and consumes less power as well possessing an equal resolution. But in contrast to [14], these digital implementations can achieve an arbitrary high resolution, which is only limited by the area and power constraints because both implementations have no theoretical minimum of the absolute error. Therefore, an analog neuron outperforms its digital counterparts in emerging technology processes and represents the most efficient realization form of an artificial neuron in terms of area, power and resolution even if device dimensions are shrinking further and systems become more susceptible to noise and parameter variations.

4 Conclusion

In this work, the scaling effects of emerging CMOS technologies on the activation functions of neurons are investigated. Shrinking device size allows integrating more transistors on the same area than in today's technology, which enables the realization of highly complex systems. Because neural principles operate massively parallel and consist of elementary building blocks, these concepts are intended to provide ideas for novel architectures in those technologies that are able to handle this complexity and also to provide low power and reliable operation.

One drawback of the scaling process is the increased susceptibility of systems to noise because signal amplitudes are shrinking too. As was shown in this work, the scaling trend has only limited impact on the analog representation of activation functions with respect to their achievable processing resolution. The signal-to-noise-ratio decreases much slower than the scaling process improves area and power consumption. Consequently, the results identify the limits of neural systems and determine the maximum resolution, which analog implemented threshold functions can process. Therefore, the weight storage can be adapted to these limited processing capabilities of the activation function and inherent training techniques have to be used to guarantee certain processing qualities.

Moreover, analog implementations outperform their digital realizations, which cannot provide such low power operation and area requirements. However, digital neurons show their benefits in their processing resolution. In contrast to their analog counterparts the resolution is not significantly affected by the scaling process and it can be hold constant over technology generations. As was also shown, the performance of the digital systems mainly depends on the used architecture of the activation function.

References

1. Isaac, R.D.: The future of cmos technology. IBM Journal of Research and Development 44(3), 369–378 (2000)
2. Compañó, R.: Technology roadmap for nanoelectronics. Technical Report 2nd edn. European Commission (2000)
3. Beiu, V., Taylor, J.G.: On the circuit complexity of sigmoid feedforward neural networks. Neural Netw. 9(7), 1155–1171 (1996)
4. Arbib, M.A.: The Handbook of Brain Theory and Neural Networks, 2nd edn. The MIT Press, Cambridge, MA, USA (2002)
5. Dennard, R.H., Gaensslen, F.H., Rideout, V.L., Bassous, E., LeBlanc, A.R.: Design of ion-implanted MOSFET's with very small physical dimensions. IEEE Journal of Solid-State Circuits 9(5), 256–268 (1974)
6. Frank, D., Dennard, R., Nowak, E., Solomon, P., Taur, Y., Wong, H.S.P.: Device scaling limits of Si MOSFETs and their application dependencies. Proceedings of the IEEE 89, 259–288 (2001)
7. Taur, Y.: CMOS design near the limit of scaling. IBM Journal of Research and Development 46(2/3), 213–222 (2002)
8. Haykin, S.: Neural Networks. A Comprehensive Foundation, 2nd edn. Prentice Hall, New Jersey, USA (1999)
9. Razavi, B.: Design of Analog CMOS Integrated Circuits. McGraw-Hill, New York (2000)
10. Mead, C.: Analog VLSI and neural systems. Addison-Wesley Longman Publishing Co., Inc., Boston, MA, USA (1989)
11. Shannon, C.E.: A mathematical theory of communication. Bell System Technical Journal 27, 379–423, 623–656 (1948)

12. Zhao, W., Cao, Y.: New generation of predictive technology model for sub-45nm design exploration. In: ISQED '06: Proceedings of the 7th International Symposium on Quality Electronic Design, Washington, DC, USA, pp. 585–590. IEEE Computer Society Press, Los Alamitos (2006)
13. Simoen, E., Claeys, C.: On the flicker noise in submicron silicon MOSFETs. Solid-State Electronics 43(5), 865–882 (1999)
14. Beiu, V., Peperstraete, J.A., Vandewalle, J., Lauwereins, R.: VLSI Complexity Reduction by Piece-Wise Approximation of the Sigmoid Function. In: Verleysen, M. (ed.) Proc. of the ESANN, Bruges, Belgium, pp. 181–186 (April 1994)
15. Semiconductor Industry Association: International Technology Roadmap for Semiconductors - Edition (2005), `http://public.itrs.net/`

Rectangular Basis Functions Applied to Imbalanced Datasets

Vicenç Soler and Marta Prim

Dept. MiSE, Universitat Autònoma de Barcelona, Campus UAB
08193 Bellaterra, Spain
{vicenc.soler, marta.prim}@uab.cat

Abstract. Rectangular Basis Functions Networks (RecBFN) come from RBF Networks, and are composed by a set of Fuzzy Points which describe the network. In this paper, a set of characteristics of the RecBF are proposed to be used in imbalanced datasets, especially the order of the training patterns. We will demonstrate that it is an important factor to improve the generalization of the solution, which is the main problem in imbalanced datasets. Finally, this solution is compared with other important methods to work with imbalanced datasets, showing our method works well with this type of datasets and that an understandable set of rules can be extracted.

Keywords: RecBF, Imbalanced Datasets, Fuzzy Logic.

1 Introduction

Working with imbalanced datasets [1] is always difficult due to the problem of generalization. Except for problems with a very clear difference between the classes, it is not easy to define a boundary to separate the different classes involved in.

Rectangular Basis Function Networks (RecBFN) come from RBF Networks and were presented by M.Berthold and K.P.Huber in several papers, e.g. [3][4]. They refer to a system that is able to obtain a fuzzy model from the training data, by means of the execution of an algorithm called DDA/RecBF. Just defining the fuzzy output variable and the training data, the system can extract a set of fuzzy variables and rules. Every neuron is represented by a Fuzzy Point (FP).

The RecBFN has been used to build function approximators and precise classifiers, and good results are obtained under a set of balanced data, but no so good under conditions of low or very low balancing of the data. As happen with the common learning methods, learning an imbalanced dataset often produces a system that is not able to generalize the minor-class patterns in the test phase, just memorize them. In the case of the RecBF, due to the existence of conflicting areas, the RecBFN classify much better the major-class patterns (patterns belonging to the class with bigger quantity of patterns in the training dataset) than the minor-class ones (the less representative class in the dataset).

The focus of this paper will be in finding characteristics of the RecBF which can be applied to work with imbalanced datasets.

J. Marques de Sá et al. (Eds.): ICANN 2007, Part I, LNCS 4668, pp. 511–519, 2007.

This paper is structured in the following sections: in section 2 RecBF is explained, in section 3 we can see how to apply RecBF to the imbalanced datasets, in section 4 the results are compared with other methods as in section 5 (with a real dataset), and finally the conclusions are exposed.

2 The Rectangular Basis Function

The DDA/RecBF[3] constructs a set of Membership Functions (MF) from datasets that provide a set of linguistic labels that can be used to concisely express relationships and dependencies in the dataset. The algorithm creates hyper-rectangles (called Fuzzy Points –FP-) belonging to every class, from the input dataset and a defined fuzzy output variable. From these FPs, a set of MFs will be extracted for every variable.

Each FP created by the DDA/RecBF[3] algorithm is characterized by being composed by a support-region and a core-region. In terms of MF, a trapezoidal MF is composed of four points (a,b,c,d): the interval [a,d] defines the support-region and the [b,c] one, the core-region.

```
∀ FP^k_i { //reset the CR
  (a,b,c,d)^k_i = (a,-,-,d)^k_i
}
∀ pattern x {
  k = determine the class of x
  IF ∃ FP^k that covers x THEN
    Call covered (FP^k, x)
  ELSE Call commit(x,k)
  ∀ FP^i_j of another class that covers x
      Call shrink (FP^i_j, x)
  }
}
```

(a)

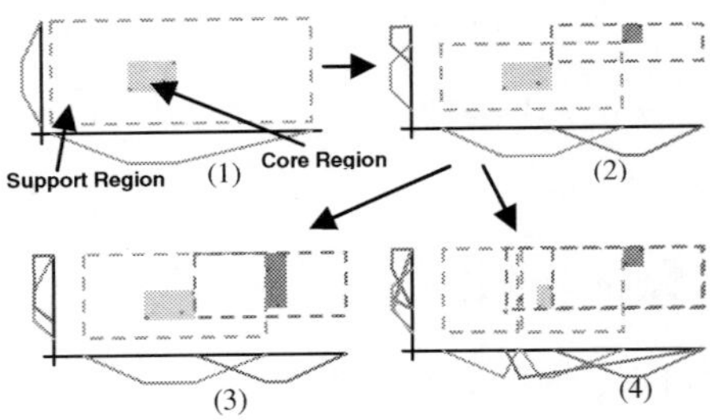

(b)

Fig. 1. (a) One Epoch in the DDA/RecBF algorithm. The shrink operation reduces the support-region of that FP^i_j to not include the pattern *x*. (b) An example of the execution of the DDA/RecBF algorithm. (1) shows 3 patterns from one class: 1 RecBF, (2) shows 2 patterns from another class: creation of a new RecBF and shrink the existing one, (3) and (4) show the different RecBFs created when the inclusion of new pattern is done, just varying the x coordinate: outside and inside the core-region of the other class. The x and y axis show the different MF created.

DDA/RecBF algorithm (shown in Fig. 1(a)) is based on three procedures, which are executed for every training pattern: *covered* is executed when the support-region of a RecBF covers a pattern, *commit* creates a new pattern if the previous condition is false and, finally, the procedure *shrink* solve possible conflicts of the pattern with respect to RecBFs of a different class. Fig. 1(b) shows how the patterns are passed to the DDA/RecBF algorithm and how the different FPs are created. However, in our case we work with imbalanced datasets, and to avoid granulation of the membership functions of the minor-class, it is absolutely necessary to generalize this class, because the main problem is when the method has to classify/test patterns belonging to the minor-class, not shown during the training process.

3 RecBF Applied to Imbalanced Problems

In this section a set of characteristics to use RecBF with imbalanced datasets are exposed.

From now on, the imbalanced problems will be restricted to a 2-class problem. The major-class will be the class with higher number of patterns and the minor-class the other one (reflecting the quantity of examples in the imbalanced dataset included in every class).

3.1 Patterns Ordered by Class

As a result of out research, we can affirm that an important factor is the order of the training dataset. If the dataset is ordered by class, the number of FPs of the second one decreases. This produces more general membership functions and, thus, the fuzzy system will try to generalize the solution rather than to specialize some FPs in the patterns of the minor-class.

To see the difference between ordered or not ordered patterns, an example of the execution of the DDA/RecBF algorithm with unordered and ordered patterns is shown in the lines of Fig.2.

In the case of unordered patterns (Fig.2 (a)), as the FPs of the minor-class are being shrunk, new FPs are being created if needed. Therefore, a high number of FPs of the minor-class is created.

In Fig.2 (b) (sorted patterns by class), the core-region restricts the support-region of new FPs, and the support-region determines if the action *covered* or *commit* is applied. Therefore, after one epoch (the core-regions are reset at every epoch), when the patterns of the major-class are trained again, there are not core-regions defined (reset) anywhere. Thus, with ordered patterns by class, the DDA/RecBF minimizes the quantity of FPs of the second class and let the first class organize itself from the restrictions imposed by the FPs of the second class.

3.2 The Shrink Operation

As mentioned above, the order of the training patterns has influence on the output. If we have a 2-class problem, when the patterns of the 2nd class are showed, only the FPs of the 1st class are shrunk. This fact produces a reduced number of FPs of the 2nd class and a big overlapping of the core-regions.

This problem can be solved with a last shrink over every input pattern. This action reduces the overlapping area between the core-regions of the FPs. That is, for every epoch, at the final, a new shrink operation for every pattern and every FP will be done. This operation is called *reshrink* and consists on applying the *shrink* operation once more at the final of every epoch in order to reduce one degree the overlapping between classes.

(a)	(b)
1. One pattern of the major-class is shown. 1.1. New FP of the major-class. 2. One pattern of the minor-class is shown: 2.1. New FP of the minor-class and 2.2. The FP of the major class is shrunk. 3. More patterns of the major-class: 3.1. New FP if the pattern is outside the support-region of the existing FP of the major-class. 3.2. The FP of the minor-class is shrunk.	1. Patterns of the major-class are shown. 1.1. A new FP of the first class is created. 2. Patterns of the minor-class are shown: 2.1. If exists overlapping: 2.1.1. A new FP is created of the minor-class inside the FP of the major-class. 2.2. If does not exist overlapping: 2.2.1. A new FP is created outside the limits of the major-class. 2.3. The FP of the major-class is shrunk if necessary. 2.4. Thus, just a FP is created inside the overlapping area. 3. Complete the epoch. 4. More patterns of the major-class: 4.1. New FPs are created of the major-class depending on the new support-region. 4.2. The FPs of the minor-class are shrunk

Fig. 2. (a) Execution of the DDA/RecBF algorithm with unordered patterns, and (b) execution of the DDA/RecBF algorithm with patterns sorted by class

The proof will be done in the following lines, first without including the *reshrink* operation, and later including it. In this proof, we suppose that we start the 2nd epoch, because the 1st epoch always produces just 1 FP for every class.

1) Without reshrink. The only way to shrink the core-regions of the FPs of the 2nd class is when the patterns of the 1st class are shown. If the limits are commanded by patterns of the 1st class, the shrinking of the FP of the 2nd class will set its support-region outside the limits of the patterns of the 2nd (as show Fig.3 (a)), and this fact will produce the execution of the *commit* operation rather than the *covered* one. The old FP will be deleted in the next epoch, because it will not contain any pattern (A=0). If in any limit of any variable exist patterns of the 2nd class, this FP will have its $A \neq 0$ and will not be deleted (Fig.3 (b)). Depending on the factor of loosing less volume, the FPs will be shrunk to one of both limits of the variable.

2) With reshrink. As Fig.4 shows, the only difference respect to Fig. 3 is in both (a). The main difference consists on a new FP of the 2nd class created thanks to the *reshrink* operation.

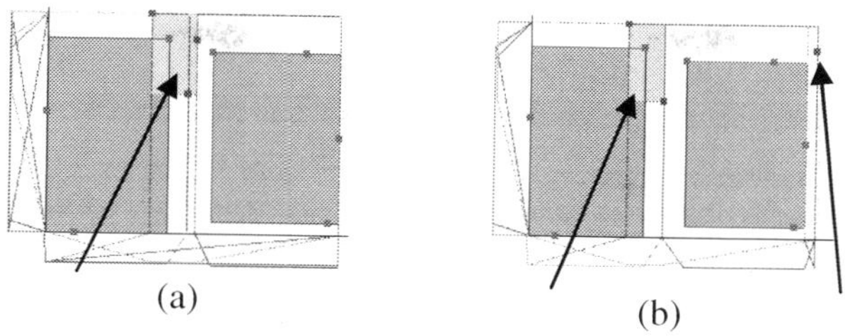

Fig. 3. DDA 2-class after 2 epochs with ordered patterns by class. (a) 1 FP is created and pointed by the arrow (b) The x value of one pattern of the 2nd class is changed and set to the right limit (i.e., just with x value > patterns of the 1st class). Then, 2 FP are created and pointed by the arrows.

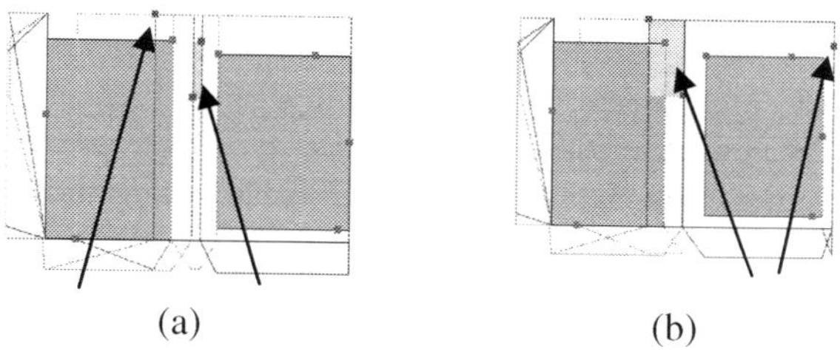

Fig. 4. DDA 2-class after 2 epochs with ordered patterns by class and applying a new shrink at the end of each epoch. (a) and (b) contain the same datasets than the respective graphics in figure 4. The only difference is that a new FP is created in (a) due to (6) at the end of each epoch.

Finally, the number of FP created corresponding to every graphic, has been: Fig. 3 (a): (3 of 1st class, 1 of 2nd class), and (b): (3,2), Fig. 4 (a): (3,2) and (b): (3,3) . So, applying *reshrink* creates more FP. but just to decrease the overlapping between the existing FPs.

3.3 Results in Number of Fuzzy Points

In this section, the characteristics explained above are summarized in the Table 1, in order to show the number of FPs created at the different scenarios. The scenarios planned are 3: unordered patterns, sorted by class and sorted by class plus adding the *reshrink* operation. The *reshrink* operation only has sense with ordered patterns, because it is used to reduce the overlapping produced in this case.

We have chosen 19 datasets, shown in Table 1. These datasets were chosen by other authors and we will use them to compare our results with other methods. The datasets are taken from the UCI repository and some of them have more than 2 classes. To convert them to a 2-class dataset, one class is chosen as the minor one and the rest of them are joined. Between parentheses is reflected the chosen class, as was chosen by other authors.

Table 1. Datasets properties: the number of patterns belonging to each class, the imbalanced rate and the number of attributes. Between parentheses is shown the class chosen to be the minor one.

dataset	#major	#minor	rate	#attrib.	dataset	#major	#minor	Rate	#attrib.
Abalone(19)	4145	32	1:130	8	Hepatitis(1)	123	32	1:4	19
Abalone'(9/18)	689	42	1:16	8	Ionosphere	225	126	1:2	34
Anneal(5)	831	67	1:12	31	Segmentation(1)	180	30	1:6	19
BC	201	85	1:2	9	Segment(1)	1980	330	1:6	19
BCW	444	239	1:2	9	Soybean(12)	639	44	1:15	35
Car(3)	1659	69	1:24	6	Vehicle(1)	634	212	1:3	18
Diabetes	500	268	1:2	8	Vehicle'(4)	647	199	1:3	18
Glass(7)	185	29	1:6	9	Yeast(ME2)	1433	51	1:28	8
Haberman	225	81	1:3	3	Yeast'(CYT/POX)	463	20	1:23	8
Heart dis.	150	120	1:1	13					

Table 2 shows the average and standard deviation of the number of FPs created for every scenario and class, under different selection of the patterns for training/testing. As the goal is to reduce the number of FP of the minor-class to improve the final generalization in imbalanced datasets, the first class is the major-class and the second one is the minor-class. In Table 2 it is shown that ordering the patterns by class reduces in a high degree the quantity of FPs created for every class, and that the introduction of the *reshrink* operation increases the quantity of FPs of just the minor-class in order to reduce the existing overlapping.

Table 2. Number of FPs created for every scenario and class. μ is the average and σ is the standard deviation.

scenario:	unordered				ordered				ordered+reshrink			
class:	major-class		minor-class		major-class		minor-class		major-class		minor-class	
dataset	μ	σ	μ	σ	μ	σ	μ	σ	μ	σ	μ	σ
abalone	22.3	2.4	21.5	2.6	4.8	2.6	1	0	4.7	1.7	1.6	1.0
bc	6.7	0.4	6.2	0.6	8.5	1.0	1.6	0.4	8.9	0.7	7.0	2.4
bcw	10.0	1.2	11.8	1.6	4.0	1.3	1	0	3.8	1.3	3.5	0.5
car	2.4	0.4	2.4	0.9	2	0	2	0	2	0	2.5	0.8
diabetes	47.2	2.7	48.3	3.5	5.0	1.9	2	0	4.2	2.1	5.8	1.9
glass	14.0	1.5	13.1	2.0	5.2	2.3	1	0	5.8	2.4	1.8	0.6
hepatitis	7.2	1.3	6.4	1.4	6.2	1.8	1	0	5.6	1.4	3.5	2.1
hypothyroid	17.7	1.9	2.5	0.8	3.8	2.6	1	0	3.7	2.9	1.5	0.7
ionosphere	27.5	5.0	18.9	3.0	2.4	1.4	1	0	2.3	1.0	3.1	1.0
segmentation	19.6	2.0	17.1	2.2	5.1	2.8	1	0	5.3	2.4	3.1	1.9
sick	54.3	8.9	19.1	2.0	6.5	3.3	1.8	0.3	6.4	2.7	1.8	0.3
soybean	5.4	0.6	4.9	0.6	5.6	1.1	1.8	0.3	6.1	0.9	4	2.2
vehicle	72.3	4.4	70.6	4.5	4.9	2.5	1	0	4.3	1.7	3.6	1.8
yeast	23.7	1.5	23.2	1.7	4.3	1.1	1.1	0.3	4.3	1.2	3.7	2.1
average:	23.6	2.5	19.0	2.0	4.9	1.8	1.3	0.1	4.8	1.6	3.3	1.4

So, the conclusions are that it is better to sort the patterns by class, first the major-class and then the minor-class and the *reshrink* operation will be chosen depending on the overlapping existing in a dataset.

4 Comparison with Other Methods for Imbalanced Datasets

In this section, the conclusions achieved in section 3 will be applied and compared with other methods for imbalanced datasets.

The conclusions from the previous section address the problem of generalization in imbalanced datasets to train the DDA/RecBF algorithm using patterns sorted by class (first the major-class and second the minor-class) and the *reshrink* operation if the overlapping degree is too high.

These conclusions will be applied to the method explained in [5] and in [6], which was developed by the authors of this paper to work with imbalanced datasets. This method creates a fuzzy system able to work with imbalanced datasets by means of a recombination of the membership functions of every class and finding the fuzzy rules by means of a genetic algorithm.

The metric used to compare the results with other specialized methods is the *g-means* metric (1), suggested by [2], which is the most used measure to evaluate results in imbalanced datasets, where acc^+ is the accuracy classification on the positive instances, and acc^- the accuracy on the negative ones.

$$g = \sqrt{acc^+ \cdot acc^-} \tag{1}$$

Table 3. (a) Results of our method for different datasets splitting the dataset either randomly or with cross-validation. (b) The quantity of datasets which results were improved, for every compared method.

(a)

	random		CV	
Dataset (trainnig:test)	μ(g)	σ(g)	μ(g)	σ(g)
Abalone(7:3)	72.23	5.72		
Abalone'			69.68	1.56
Anneal(7:3)	99.95	0.05		
BC(7:3)	62.32	2.10	60.59	0.67
BCW			94.46	0.46
Car(7:3)	96.19	0.37		
Diabetes			70.85	0.98
Glass(7:3)	91.90	2.13	90.34	2.32
Glass(6:1)	89.89	10.39		
Haberman(7:3)	67.07	1.02		
Heart dis. (9:1)	80.64	8.97	79.84	1.24
Hepatitis(7:3)	74.55	6.11		
Hepatitis_avg(7:3)	78.57	2.63		
Ionosphere(9:1)	79.33	7.16	83.67	0.53
Segmentation(6:1)	99.95	0.15		
Segment(7:3)	94.85	1.16	95.83	0.25
Soybean(7:3)	100	0		
Soybean_avg(7:3)	100	0		
Vehicle(7:3)	68.13	2.56	67.09	0.82
Vehicle'			86.71	1.28
Yeast(6:1)	82.34	3.44		
Yeast'			63.72	4.25

(b)

Method	improved
SDC(7:3)	1 of 7
KBA(6:1)	3 of 5
PP(CV10)	2 of 6
aPP(CV10)	1 of 5
EAg	1 of 5
Databoost-IM(CV10)	2 of 7
kNN et al.(CV5)	2 of 2
RD(CV5)	2 of 2

Table 3 shows the final results for the g-means metric. Table 3(a) shows the results of the fuzzy system of the method exposed in [5] and [6]. There are different results depending on the type of splitting data (training/test): by a random selection or cross-validation (CV). The comparison of our method with the results of each other method was done under the same conditions that were provided by the authors of the other methods. Table 3(b) compares those results with the 8 specialized methods: SDC [7], KBA [8], Parallel Perceptrons [9], alternative PP [10], Expert Agents [11], Databoost-IM [12], kNN (et al.) [13] and Restricted Decontamination [14]. The column *improved* shows the quantity of datasets which results have been improved by our method using the characteristics explained in section 3.

5 Application on a Real Dataset

In this section we will apply our method to a well-known imbalanced problem in medicine: the Down syndrome detection in fetus. It is known by everybody that the number of fetus born with Down syndrome is significantly smaller than the ones born without that syndrome. The ratio is about 1:300 and the current method used in medicine detects up to 60-70'% of true positives with an 8-10% of false positives. In our case, it was used a dataset of 8995 cases with just 30 positive ones, 5 input variables (maternal age, maternal weight, age of the fetus (FA) expressed in days and 2 hormonal markers: AFP and hCG). The results of applying our method produced a fuzzy system of just 9 rules (7 negatives and 2 positives) and a rate of 66% of true positives with a 4% of false positives.

This result is better than the obtained by the current methods for this dataset, which was similar to the usual success rates. In addition, thanks to the reduced set of rules obtained, clear information about the system found could be given to the medical staff. We transformed the fuzzy sets involved in every rule in an understandable way. The trapezoidal fuzzy sets, with their points a,b,c,d, have been transformed to “approximately among b & c” and the triangular ones (a,b,c) to “approximately around b”. In Figure 5 an example of rule is shown.

R1:	if	Age	is approximately among	43 & 44	&	
		Weight	is approximately among	41 & 42.5	&	
		FA	is approximately among	134 & 135	&	
		AFP	is approximately among	41.91 & 41.99	&	
		hCG	is approximately among	2.32 & 2.59	then	Down=yes

Fig. 5. Rule 1 of the system found. It is one of the two positive rules.

Finally, the medical staff agreed with the meaning of the rules and it was comprehensible for them.

6 Conclusions

In this paper have been exposed some characteristics of the Rectangular Basis Function that could be used to work with imbalanced datasets in a fuzzy system environment. The study done reflects that the RecBF can be used as a first approximator of the fuzzy variables of the final fuzzy system, and that the order of the patterns is basic to achieve a general solution, that is, having a low quantity of membership functions of the minor class avoids specialization and improves generalization in the testing phase.

These characteristics have been applied to a developed method, implemented by us, which is able to find a fuzzy system that works well with imbalanced datasets. The results show that the solution approximates the results of other methods designed to work with imbalanced datasets. In addition, it have been tested to a real problem (Down syndrome detection in fetus), with excellent results both in improving the results of the method used by hospitals and in extracting a set of understandable rules.

References

1. Japkowicz, N., Stephen, S.: The Class Imbalance Problem: A Systematic Study. Intelligent Data Analysis 6(5), 429–450 (2002)
2. Kubat, M., Matwin, S.: Addressing the Curse of Imbalanced Training Sets: One-Sided Selection". In: Proceedings of the 14th International Conference on Machine Learning (1997)
3. Berthold, M., Huber, K.P.: Constructing Fuzzy Graphs from Examples. Intelligent Data Analysis 3, 37–53 (1999)
4. Berthold, M., Huber, K.P.: Introduction of Mixed Fuzzy Rules. International Journal of Fuzzy Systems 3, 289–382 (2001)
5. Soler, V., Roig, J., Prim, M.: Fuzzy Rule Extraction using Recombined RecBF for Very-Imbalanced Datasets. In: ICANN-2005, Warsaw, Poland, vol. 2, pp. 685–690 (2005)
6. Soler, V., Cerquides, J., Sabrià, J., Roig, J., Prim, M.: Imbalanced Datasets Classification by Fuzzy Rule Extraction and Genetic Algorithms
7. Akbani, R., Kwek, S., Japkowicz, N.: Applying Support Vector Machines to Imbalanced Datasets. In: Boulicaut, J.-F., Esposito, F., Giannotti, F., Pedreschi, D. (eds.) ECML 2004. LNCS (LNAI), vol. 3201, pp. 39–50. Springer, Heidelberg (2004)
8. Wu, G., Chang, E.Y.: KBA: Kernel Boundary Alignment Considering Imbalanced Data Distribution. IEEE Trans. on knowledge and data eng. 17(6), 786–795 (2005)
9. González, A., Cantadory, I., Dorronsoro, J.R.: Discriminant Parallel Perceptrons. ICANN 2, 13–18 (2005)
10. Cantador, I., Dorronsoro, J.R.: Balanced Boosting with Parallel Perceptrons. In: Cabestany, J., Prieto, A.G., Sandoval, F. (eds.) IWANN 2005. LNCS, vol. 3512, pp. 208–216. Springer, Heidelberg (2005)
11. Kotsiantis, S., Pintelas, P.: Mixture of expert agents for handling imbalanced data sets. Annals of Mathematics, Computing & TeleInformatics 1(1), 46–55 (2003)
12. Guo, H., Viktor, H.L.: Learning from imbalanced data sets with boosting and data generation: the DataBoost-IM approach. SIGKDD Explorations 6(1), 30–39 (2004)
13. Barandela, R., Sánchez, J.S., García, V., Rangel, E.: Strategies for learning in class imbalance problems. Pattern Recognition 36(3), 849–851 (2003)
14. Barandela, R., Rangel, E., Sánchez, J.S., Ferri, F.J.: Restricted decontamination for the imbalanced training sample problem. In: Sanfeliu, A., Ruiz-Shulcloper, J. (eds.) CIARP 2003. LNCS, vol. 2905, pp. 424–431. Springer, Heidelberg (2003)

Qualitative Radial Basis Function Networks Based on Distance Discretization for Classification Problems

Xavier Parra and Andreu Català

Knowledge Engineering Research Group - GREC
Technical University of Catalonia - UPC
Av. Víctor Balaguer, s/n - 08800 Vilanova i la Geltrú - Spain
{xavier.parra,andreu.catala}@upc.edu

Abstract. This paper presents a radial basis function neural network which leads to classifiers of lower complexity by using a qualitative radial function based on distance discretization. The proposed neural network model generates smaller solutions for a similar generalization performance, rising to classifiers with reduced complexity in the sense of fewer radial basis functions. Classification experiments on real world data sets show that the number of radial basis functions can be reduced in some cases significantly without affecting the classification accuracy.

1 Introduction

Qualitative Reasoning is a scientific discipline that has developed techniques that allow, until a certain degree, the formalization of several human capacities that are useful to perceive, analyze, understand and model real problems. In addition, measurements are often crucial during the process of understanding the world around us. Moreover, the use of numbers, orders of magnitude and categories are needed to measure and represent the reality. It is precisely in this sense that one of the goals of qualitative reasoning is to obtain models that allow the treatment of problems where they appear variables described in different scales. The work presented is related with the establishment of a relationship between Qualitative Reasoning and the Radial Basis Function Networks (RBF). In this paper it is shown that the classifier's complexity can be reduced when using qualitative information.

Indeed, it is quite difficult to get this aim because neural networks use to treat data in a purely numerical way; weights and bias are represented by real numbers, activation functions (usually Gaussian or linear for RBFs) give a real value as response, patterns are normally represented by real values and, in general, all the variables and elements involved in the learning and test processes have a real value [1]. Besides, in some cases the learning algorithms work with a degree of accuracy that it is not needed by the problem. This can be the case of some classification problems, in which inputs and outputs are qualitative.

J. Marques de Sá et al. (Eds.): ICANN 2007, Part I, LNCS 4668, pp. 520–528, 2007.

The main interest of this work is related with the study of the effect that certain qualitative transformations have on the performance of the neural networks. A qualitative version of the RBF networks is proposed through the use of a qualitative radial function. This version allows obtaining good results when working with classification problems, in which the relationships among the different elements can be established from qualitative comparisons. Obviously, it can be expected that more difficulties turn up if we attempt to approximate a continuous function with this qualitative version of RBF networks because of the accuracy that normally it is required in this type of problem.

This paper is organized as follows: section 2 gives the basic concepts of RBF networks, highlights the importance of stepwise activation functions for these kinds of learning algorithms, and describes the forward selection algorithm to determine the centers for the treatment of qualitative information. An example of the proposed method is given in Section 3, and results obtained from quantitative and qualitative methods are compared. The qualitative model of the RBF networks for classification tasks is tested with different benchmark data sets from [2] and [3]. Finally, the paper ends with several conclusions that synthesize the present work and briefly outlines some proposals for future research.

2 Qualitative Radial Functions

In this section a methodology, allowing Radial Basis Function networks (RBF) to be used with qualitative radial functions, is proposed. Before building appropriate radial functions for this kind of discrete spaces, let us remind ourselves of the basic concepts of Radial Basic Function networks, introduced in [4].

2.1 Radial Basis Function Networks

RBF are a type of artificial neural network for application to problems of supervised learning. RBF can be applied to problems of regression, classification and times series prediction. In this paper only the case of classification is considered.

Several reasons make RBF especially interesting. On the one hand, they are universal classifiers, and on the other, the training process associated to these kind of neural networks is usually much faster than for other neural architectures, such as MLP or SVM, and, in addition, it is possible to extract rules from RBF architecture. RBF are associated with a simple architecture of three layers [4] as shown in Fig. 1:

Each layer is fully connected to the following one and the hidden layer is composed of a number of nodes with radial activation functions called radial basis functions. Each of the input components fits forward to the radial functions, whereas the outputs of these functions are linearly combined with a set of weights into the output of the network.

The characteristic feature of radial functions is that their response decreases, or increases, monotonically with distance from a central point named center of the radial function. These functions involve two parameters, the center and a

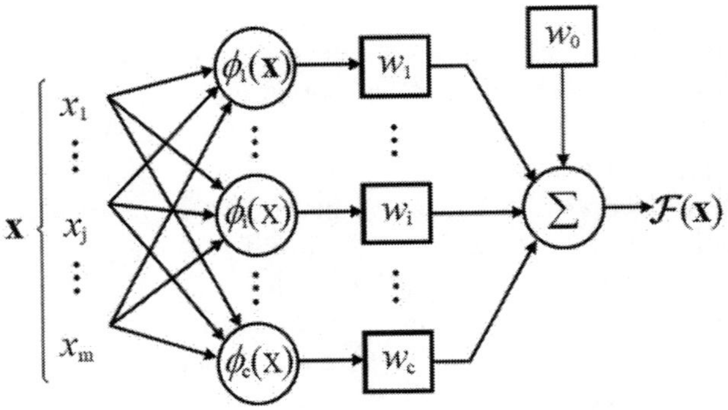

Fig. 1. Radial Basis Function Network Architecture

distance scale. Radial functions in the hidden layer have a structure represented as follows:

$$\Phi_i(\mathbf{x}) = \varphi((\mathbf{x} - \mathbf{c}_i)^T R(\mathbf{x} - \mathbf{c}_i)) \qquad (1)$$

where φ is the radial function used, $\{\mathbf{c}_i | i = 1, \ldots, c\}$ is the set of radial function centers and R is a metric. So, the term $(\mathbf{x} - \mathbf{c})^T R(\mathbf{x} - \mathbf{c})$ denotes the square of the distance between the input $\mathbf{x}$ and the center $\mathbf{c}$ according to the metric defined by R.

Usual radial functions are Gaussian, Cauchy's, multiquadric, inverse multiquadric, spline and logarithm. The most widely used radial function is the Gaussian one, in the particular case when the metric is $R = \mathbf{I}/r^2$, with $\mathbf{I}$ being the identity matrix and r the radius of the radial function:

$$\Phi(\mathbf{x}) = \exp\left(-\frac{(\mathbf{x} - \mathbf{c})^T(\mathbf{x} - \mathbf{c})}{r^2}\right) = \exp\left(-\frac{d^2(\mathbf{x}, \mathbf{c})}{r^2}\right) \qquad (2)$$

where d is the Euclidean distance in $\mathbb{R}^n$. Gaussian function monotonically decreases with distance from the centre $\mathbf{c}$, and in this sense, it is said that its response is local. The output of the RBF network is:

$$F(\mathbf{x}) = w_0 + \sum_{i=1}^{c} \exp\left(-\frac{d^2(\mathbf{x}, \mathbf{c}_i)}{r^2}\right) \qquad (3)$$

2.2 Distance Discretization

In order to simplify the complexity of the classical RBF networks, it is established a qualitative alternative to the classical Gaussian function used during the learning process. This new type of radial function will be called Qualitative Gaussian Radial Function (Fig. 2.) and depends on two parameters α and β. These two parameters allow converting the range of inputs into a finite set of intervals, where variables will be defined in a common absolute orders of magnitude space with granularity 3, OM(3) [5]. The absolute orders of magnitude models works with a finite set of symbols or qualitative labels obtained via a partition of the real line, where any element of the partition is a basic label (e.g. negative large, positive small or positive medium). These models provide a mathematical structure which unifies sign algebra and interval algebra through a continuum of qualitative structures built from the rougher to the finest partition of the real line.

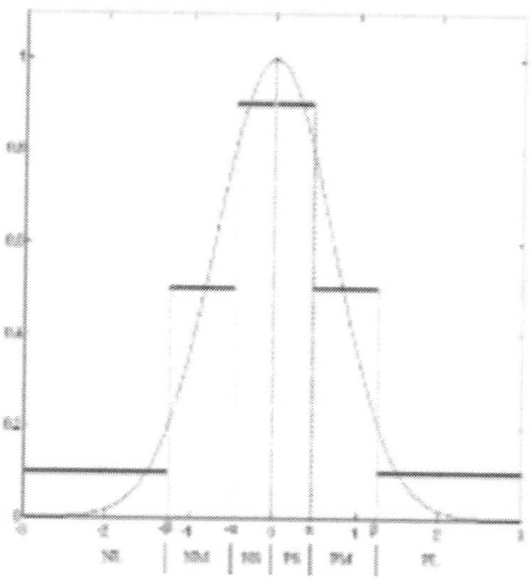

Fig. 2. Qualitative Gaussian Radial Function

A Gaussian function can be expressed as:

$$y = \exp\left(\frac{-x^2}{r^2}\right) \tag{4}$$

where r is a radial width (in Fig. 2., r takes a value equal to 1). The values of α and β depend on the radial width value, and for the study presented in this paper, they have been determined as those values for which the function declines a 20% and a 80%, respectively.

Furthermore, in these points, the first derivative of the Gaussian function is equal to 0.5. Thus, with the positive and negative value of α and β, the space of distances between the pattern selected as centre and the rest of patterns is split up into six qualitative classes (large, medium and small, negative or positive). So that, the following expressions for α and β from (4) can be obtained:

$$x = \pm r\sqrt{-\ln y} \Rightarrow \begin{cases} \text{if} \quad y = 0.8 \rightarrow \alpha = 0.47r \\ \text{if} \quad y = 0.2 \rightarrow \beta = 1.26r \end{cases} \tag{5}$$

2.3 Forward Selection Algorithm

There are several methods to determine the parameters for a RBF [4][6]; it is the set of centers for the basis functions within the entire set of patterns, their radii and the weights. Standard ways of optimizing networks involve the optimization by gradient descent, which needs differentiability and a fixed architecture in advance (number of layers and nodes). Nevertheless RBF has a simpler alternative based in subset selection. Subset selection compares the errors produced when different subsets of centers are chosen. Usually the entire set of patterns has too many subsets, so a heuristic method must be implemented. One of these is the forward selection method. It consists in starting with an empty subset, to which is added one center at a time, the one that most reduces the approximation error. Forward selection ends when some chosen criterion, such as Generalized Cross Validation, stops decreasing. In forward selection method, weights are not selected, because they are directly determined by the chosen centers and radii.

The learning algorithm used during the RBF network training is based on the optimization of the variance increase explained by the desired output [6]. This algorithm sets up the pattern that will be included as new center of radial basis function after quantifying the group of outputs depending on the dispersion of the inputs; in this way determines the patterns goodness and then selects the largest one.

The forward selection advantages in front of standard ways of optimizing networks are that a fixed number of centers in advance is not required, and, moreover, it is a model easily tractable in the case of qualitative input data, and its computational requirements are lower. In addition, forward selection does not require differentiability, and this is crucial when considering a qualitative patterns description by means of orders of magnitude. In such a discrete structure the radial functions are neither continuous nor differentiable.

The aim of the learning algorithm is to fix an RBF network as smaller as possible, and if that learning algorithm converges then the resulting RBF network is unique and perfectly determined. However, it is not possible to ensure that this solution is smaller than any other, so it is possible that a different set of training patterns rises to a smaller RBF network.

3 Experiments and Results

In order to evaluate the performance of the neural network model proposed, we perform simulations over four classification problems and compare RBF networks with the original Gaussian function and RBF networks with the Qualitative Gaussian function. First, the four classification problems and the experiments setup will be shortly introduced. Further on, the results obtained from the experiments will be presented and discussed.

3.1 Classification Problems

For the experiments we use four benchmark data sets namely `iris` from [2], and `breast cancer`[1], `thyroid` and `heart` from [3]. Each data set corresponds to a

[1] This breast cancer domain was obtained from the University Medical Centre, Institute of Oncology, Ljubljana, Yugoslavia. Thanks go to M. Zwitter and M. Soklic for providing the data.

Table 1. Pattern distribution over data subsets

Data set	Training	Test	Total
Iris	95	52	147
Breast Cancer	189	77	266
Thyroid	140	75	215
Heart	170	100	270

real world classification problem. The `iris` data set contains 147 instances with four inputs and one output. All four input attribute are real-valued while the output variable, i.e. the type of iris plant, is a nominal variable with three different classes: *Setosa*, *Versicolour* and *Virginica*. These classes have been represented using 0, 1 and 2, respectively. One class is linearly separable from the other two; the later are not linearly separable from each other. The `breast cancer` data set contains 266 instances with nine inputs and one output. Some of the nine input attributes are linear and some are nominal, while the output variable, i.e. the presence of recurrence events, is a binary attribute: *no-recurrence-events* and *recurrence-events*. These two classes have been represented using +1 and −1, respectively. Each class is not linearly separable from each other. The `thyroid` data set contains 215 instances with five inputs and one output. All five input attributes are real-valued while the output variable, i.e. the type of thyroidism, is a binary attribute: *euthyroidism* and *hyperthyroidism or hypothyroidism*. These two classes have been represented using +1 and −1, respectively. Each class is not linearly separable from each other. The `heart` data set contains 270 instances with thirteen inputs and one output. All thirteen input attributes are continuously valued, while the output variable, i.e. the presence of heart disease, is a binary attribute: *healthy* and *sick*. These two classes have been represented using +1 and −1, respectively. Each class is not linearly separable from each other.

Several simulations have been carried out for each classification problem. The data set available has been partitioned into two subsets: training set and test set, with a relation of mostly 60% and 40% respectively. Table 1 shows the pattern distribution in each data subset. We use thirty different realizations[2] for each data set. On each realization we train 100 classifiers with different radial widths (100 logarithmically equally spaced values between 1 and 100) and then compute its test classification accuracy. The performance function used to stop the training process is the mean squared error (MSE). For all the simulations the goal value for the average squarred error has been fixed to 0.001.

To study and analyze the effect that the use of Qualitative Gaussian functions has over RBF performance, two different kinds of trainings have been done. The first one corresponding to the use of a standard Gaussian function (GRBF), and the second related with the use of the qualitative version of the Gaussian function proposed in this work (QGRBF). In summary, for each classification problem

[2] For the `breast cancer`, `thyroid` and `heart` data sets we use the first thirty realizations from [3]; for the `iris` data set we randomly generate thirty realizations.

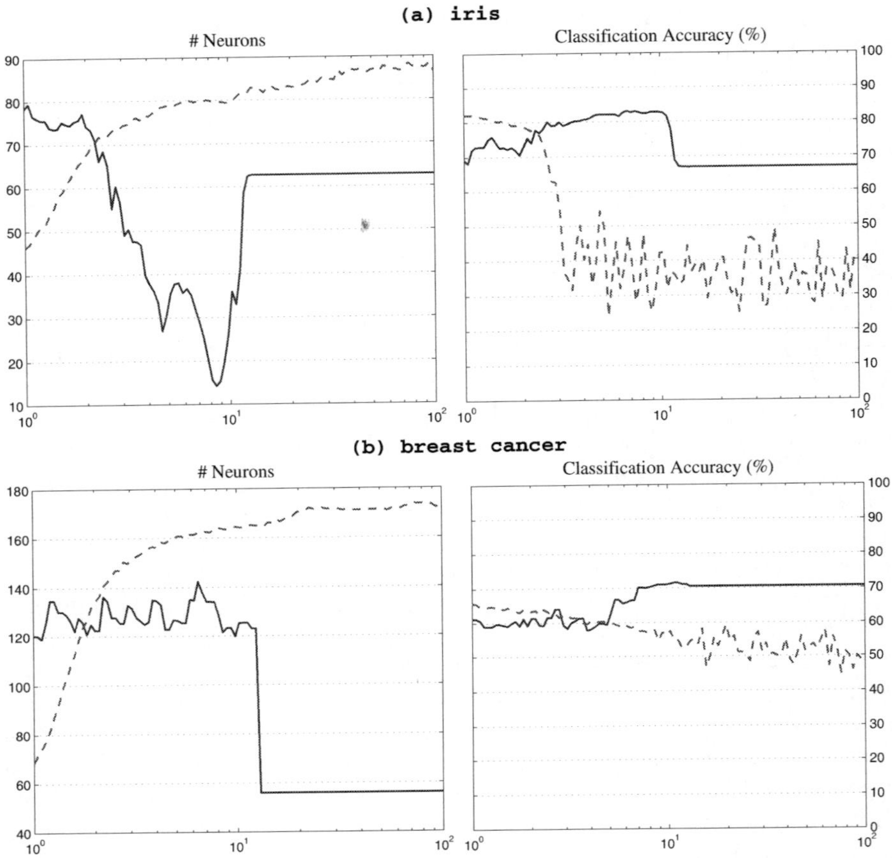

Fig. 3. Number of neurons (left) and classification accuracy (right) using RBF networks with **Qualitative Gaussian** functions (solid line) and **Gaussian** functions (dashed line) for (a) `iris` and (b) `breast cancer` classification problems

we simulate 6000 different classifiers, corresponding to 2 radial functions, 30 realizations and 100 radial widths.

3.2 Results and Discussion

The criterion used to compare the results obtained with the GRBF and the QGRBF networks is the classifier's complexity. The complexity of a classifier is directly related with the size, i.e. the number of neurons or processing elements, of the classifier. Figures 3 and 4 show the relationship between the radial width used during training and the number of processing elements determined by the training process for the GRBF and QGRBF networks[3]. Our experimental results suggest that the Qualitative Gaussian function is in some cases a promising

[3] Mean values over the thirty different realizations are used to generate the figures.

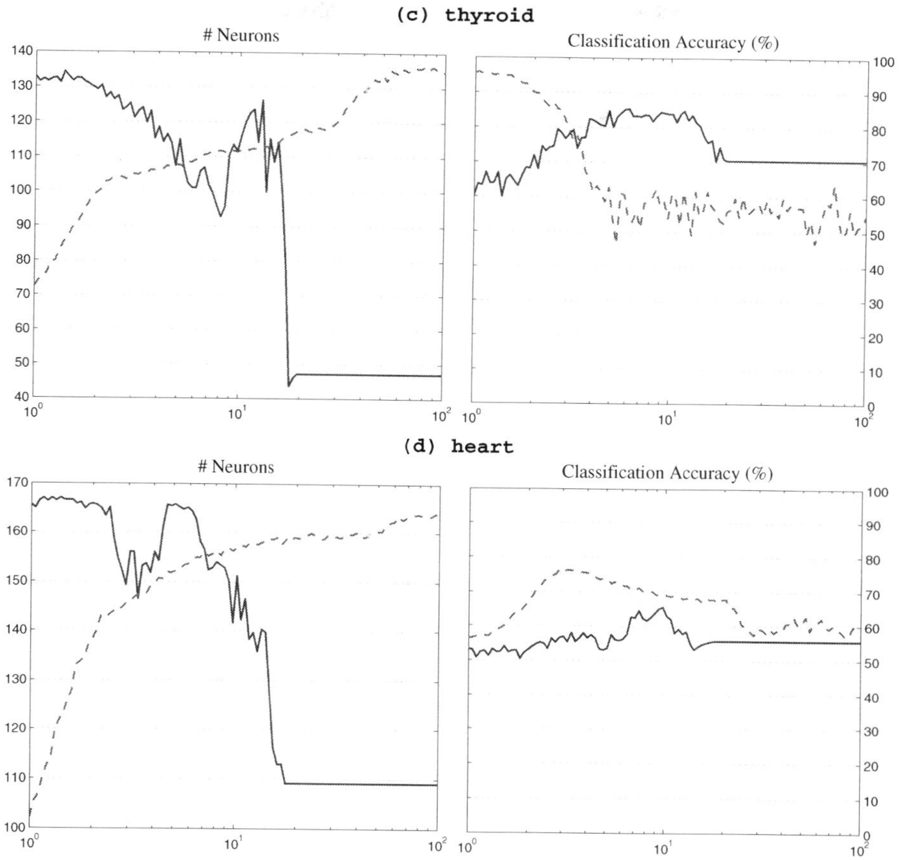

Fig. 4. Number of neurons (left) and classification accuracy (right) using RBF networks with **Qualitative Gaussian** functions (solid line) and **Gaussian** functions (dashed line) for (c) `thyroid` and (d) `heart` classification problems

approach for classifier complexity reduction in RBF networks. Moreover, given that the Qualitative Gaussian function is less computational complex[4], the RBF networks obtained by using such a qualitative function are less computational complex too. Finally, a lower computational complexity means an easier implementation, given that hardware requirements can be significantly lower too.

In general, the number of neurons needed when the width is approximately less than 1.1 is smaller if it is used a GRBF network. However, for a GRBF network and when the width value is larger than this value, the number of neurons is in many cases larger than using the QGRBF network. In general, the classification accuracy over the test data subset is lower for larger networks, probably due to

[4] Note that the activation of a Qualitative Gaussian function can be completely determined with at most three comparisons in front of a Gaussian function which needs an exponential function assessment.

overfitting during training. At this point it is important to note that the only difference between the GRBF nets and the QGRBF nets is in the radial function used, so that, if it is used the Qualitative Gaussian function it is possible in some cases to get a better RBF network in the sense of better classifier's complexity (smaller size and easier implementation).

4 Summary

In this work we have proposed a qualitative version of the Gaussian radial function used in RBF networks. We have found that the use of such a qualitative radial function works quite well without adversely affecting generalization performance in a significant way, but reduces in some cases the classifier's complexity. This has been shown through experiments on four benchmark data sets. The use of qualitative reasoning techniques as an alternative methodology in front of the common quantitative techniques can improve the general system performance in classification tasks as shown by the preceding experiments. The qualitative techniques applied to the learning algorithm of the RBF networks with regard to the qualitative selection of new centers, improve in many cases the general performance of the classical algorithm, e.g. getting better networks in the sense of a reduced classifier's complexity.

Our future work will concentrate on applying the same method to other classification problems and real world applications. We will use different model selection criteria (i.e. estimates of the prediction error) to avoid or to limit the overtraining. Moreover, it is interesting to see how the qualitative radial function can be applied in different scenarios (e.g. regression problems o time series). Continuing in the area of the qualitative reasoning also we will study and analyze the application of input discretization methods and the use of absolute orders of magnitude and qualitative distance measures.

References

1. Català, A., Morcego, B.: Qualitative Aspects of Connectionist Systems. Current Trends in Qualitative Reasoning and Applications. CIMNE Monograph 33 (1995)
2. Newman, D.J., Hettich, S., Blake, C.L., Merz, C.J.: UCI Repository of machine learning databases. University of California, Dept. of Information and Computer Science, Irvine, CA (1998), http://www.ics.uci.edu/~mlearn/MLRepository.html
3. Rätsch, G.: Benchmark Repository. Technical report, Intelligent Data Analysis Group, Fraunhofer-FIRST (2005)
4. Broomhead, D.S., Lowe, D.: Multivariable Functional Interpolation and Adaptive. Network. Complex Systems 2, 321–355 (1988)
5. Piera, N.: Current Trends in Qualitative Reasoning and Applications. CIMNE Monograph 33 (1995)
6. Chen, S., Cowan, C.F.N., Grant, P.M.: Orthogonal least square learning algorithm for radial basis function networks. IEEE Transactions on Neural Networks 2(2), 302–309 (1991)

A Control Approach to a Biophysical Neuron Model

Tim Kaulmann[1], Axel Löffler[2], and Ulrich Rückert[1]

[1] Heinz Nixdorf Institute, Dept. System and Circuit Technology, University of Paderborn, Fuerstenallee 11, 33102 Paderborn, Germany
{kaulmann, rueckert}@hni.upb.de
[2] Robert Bosch GmbH, Robert-Bosch-Strasse 2, 71701 Schwieberdingen, Germany
axel.loeffler@de.bosch.com

Abstract. In this paper we present a neuron model based on the description of biophysical mechanisms combined with a regulatory mechanism from control theory. The aim of this work is to provide a neuron model that is capable of describing the main features of biological neurons such as maintaining an equilibrium potential using the NaK-ATPase and the generation of action potentials as well as to provide an estimation of the energy consumption of a single cell in a) quiescent mode (or equilibrium state) and b) firing state, when excited by other neurons. The same mechanism has also been used to model the synaptic excitation used in the simulated system.

1 Introduction

Neuron models describe the behaviour of the biological nervous cell at various levels of abstraction. The models available today have different accuracy – from simple integrate and fire neurons to highly complex compartment models – depending on the task that has to be investigated and on computing power. Electrical models as described by Koch and Segev [1] or MacGregor [3] can be used to simulate compartments of a neural network and provide good results compared to biophysical measurements.

These models do not directly predict the energy consumption for the information processing in neural networks. Thus, other approaches have been published. Laughlin et. al. [2] published a model of an energy consuming cell, depending on the data rate of neural information transmission. The energy can be estimated by the rate of ATP transduction derived from ion-currents. Destexhe et. al. [5] used a kinetic model for receptor-binding and provided an analytical solution for the energy consumption in case of a rect-impulse as transmitter profile. A publication from Daut [4] uses known models from neural computation with physiological measurements and describes the energy consumption of a cardiac cell based on the most important processes. This work already uses coupled control loops which are also utilised in this paper.

This publication takes some ideas from the publication cited above but instead uses a bottom-up approach to describe the energy consumption of a neuron.

J. Marques de Sá et al. (Eds.): ICANN 2007, Part I, LNCS 4668, pp. 529–538, 2007.

Therefore, a system model is developed from the basic biophysical equations in Sec. 2. Afterwards, we extend the basic model to a model for a weak and a strong synapse, later used in our simulations as well as to the capability of the action potential generation. The simulation results of a small neural system are shown in Sec. 3.

Although there are many different sodium- and potassium-channels involved in the generation of the steady-state membrane voltage as well as other types of ions, we will mainly focus on two channels. One reason for omitting Ca^{2+}, Mg^{2+} and anions in this first approach is their small contribution to the membrane potential (compared to the influence of sodium and potassium). The other reason is that the action potential generation can be described by the two channels by extending their model to voltage-gated channels sufficiently (see Sec. 2.3). Of course, this model can also be extended to incorporate the effect of calcium channels.

2 Neuron Model

Starting from the Nernst equation in (1) where parameter P denotes a relative permeability of sodium channels compared to potassium channels, we can also give the Nernst-potentials of the single ion types (2), where n can either be Na or K.

$$V = -\frac{\mathrm{RT}}{\mathrm{F}} \ln \frac{P c_{\mathrm{Na,int}} + c_{\mathrm{K,int}}}{P c_{\mathrm{Na,ext}} + c_{\mathrm{K,ext}}} \tag{1}$$

$$V_n = -\frac{\mathrm{RT}}{\mathrm{F}} \ln \frac{c_{n,\mathrm{int}}}{c_{n,\mathrm{ext}}} \tag{2}$$

Parameter $\mathrm{R} = \mathrm{k_B N_A}$ describes the universal gas constant, derived from the Boltzmann constant $\mathrm{k_B}$ and the Avogadro constant $\mathrm{N_A}$. The absolute temperature is denoted with T, and $\mathrm{F} = \mathrm{qN_A}$ describes the Faraday constant, where q is the charge of a single electron. $c_{n,\mathrm{ext}}$ and $c_{n,\mathrm{int}}$ describe the extracellular and the intracellular concentration of the different ion types. For further calculations we assume the extracellular ion concentration as constant.

The value for the relative permeability P will be determined after the derivation of the system equations.

Let I_n be the inward current of potassium and sodium ions passing through a cell membrane (by passive transport), we can write the currents according to Ohm's law with their specific conductances g_{Na} and g_{K} (3)

$$I_n = -g_n \cdot (V - V_n) \tag{3}$$

where n can either be Na or K.

From a biophysical point of view, the change in intracellular ion concentrations c_n can be expressed by (4), which links to the electrical view from (3), where V_i is the intracellular volume of a single cell,

$$I_n = q\mathrm{N_A} V_i \frac{d}{dt} c_n \tag{4}$$

and leads to the expression of change in concentration depending on the equilibrium potentials (5).

$$\frac{d}{dt}c_n = -\frac{g_n}{q\mathrm{N_A}V_i}\left(V - V_n\right) \tag{5}$$

The Nernst-potential can be linearised by Taylor expansion at the operating point $c_{n,0}$ of the neuron (6), (7). The concentration $c_{n,0}$ represents the intracellular ion concentration at the steady-state.

$$V_{n,0} \approx -\frac{RT}{F}\ln\left(\frac{c_{n,0}}{c_{n,\mathrm{ext}}}\right) - \frac{RT}{F}\frac{1}{c_{n,0}}\left(c_n - c_{n,0}\right) \tag{6}$$

$$\begin{aligned} V \approx & -\frac{RT}{F}\ln\left(\frac{c_{\mathrm{K},0} + Pc_{\mathrm{Na},0}}{c_{\mathrm{K,ext}} + Pc_{\mathrm{Na,ext}}}\right) \\ & -\frac{RT}{F}\frac{1}{c_{\mathrm{K},0} + Pc_{\mathrm{Na},0}} \cdot \left[\left(c_\mathrm{K} - c_{\mathrm{K},0}\right) + P\left(c_\mathrm{Na} - c_{\mathrm{Na},0}\right)\right] \end{aligned} \tag{7}$$

We can reduce these equations further with the Taylor-coefficients given in (9) to the following compact form

$$\begin{aligned} V_\mathrm{K} &= V_{\mathrm{K},0} + V_{\mathrm{K},1} \cdot \left(c_\mathrm{K} - c_{\mathrm{K},0}\right) \\ V_\mathrm{Na} &= V_{\mathrm{Na},0} + V_{\mathrm{Na},1} \cdot \left(c_\mathrm{Na} - c_{\mathrm{Na},0}\right) \\ V &= V_0 + V_1 \cdot \left[\left(c_\mathrm{K} - c_{\mathrm{K},0}\right) + P\left(c_\mathrm{Na} - c_{\mathrm{Na},0}\right)\right] \end{aligned} \tag{8}$$

$$\begin{aligned} V_{\mathrm{K},0} &= -\tfrac{RT}{F}\ln\left(\tfrac{c_{\mathrm{K},0}}{c_{\mathrm{K,ext}}}\right) & V_{\mathrm{K},1} &= -\tfrac{RT}{F}\tfrac{1}{c_{\mathrm{K},0}} \\ V_{\mathrm{Na},0} &= -\tfrac{RT}{F}\ln\left(\tfrac{c_{\mathrm{Na},0}}{c_{\mathrm{Na,ext}}}\right) & V_{\mathrm{Na},1} &= -\tfrac{RT}{F}\tfrac{1}{c_{\mathrm{Na},0}} \\ V_0 &= -\tfrac{RT}{F}\ln\left(\tfrac{c_{\mathrm{K},0}+Pc_{\mathrm{Na},0}}{c_{\mathrm{K,ext}}+Pc_{\mathrm{Na,ext}}}\right) & V_1 &= -\tfrac{RT}{F}\tfrac{1}{c_{\mathrm{K},0}+Pc_{\mathrm{Na},0}} \end{aligned} \tag{9}$$

and give a simple equation of the passive ion transport mechanisms in vector notation (10)

$$\frac{d}{dt}\underline{c} = \underline{M} \cdot \left(\underline{c} - \underline{c}_0\right) + \underline{I_0} \tag{10}$$

$$\underline{M} = \begin{bmatrix} -\frac{g_\mathrm{K}}{qN_AV_i}\left(V_1 - V_{\mathrm{K},1}\right) & -\frac{g_\mathrm{K}}{qN_AV_i}\left(PV_1\right) \\ -\frac{g_\mathrm{Na}}{qN_AV_i}\left(V_1\right) & -\frac{g_\mathrm{Na}}{qN_AV_i}\left(PV_1 - V_{\mathrm{Na},1}\right) \end{bmatrix} \tag{11}$$

$$\underline{I_0} = \frac{1}{qN_A V_i} \begin{bmatrix} -g_{\mathrm{K}} \left(V_0 - V_{\mathrm{K},0}\right) \\ -g_{\mathrm{Na}} \left(V_0 - V_{\mathrm{Na},0}\right) \end{bmatrix} \tag{12}$$

where $\underline{c} = \left[c_{\mathrm{K}}\, c_{\mathrm{Na}}\right]^{\mathrm{T}}$. The potential V_0 describes the equilibrium potential. As (10) describes the passive transport mechanisms at equilibrium, we receive the vector $\underline{I_0}$ that describes the quiescent current of sodium and potassium ions diffusing through the cell membrane. This current has to be compensated by the NaK-ATPase to keep the membrane voltage stable. This will be discussed in Sec. 2.1.

The relative permeability P can be derived from the requirement for the ion currents in equilibrium state. The passive potassium current flowing from the intracellular space to the extracellular space has to be approx. 2/3 of the sodium current, flowing from extracellular space to intracellular space, when the NaK-ATPase shall be able to balance the concentration gradient by active transport of 3 Na^+ ions to the extracellular space and of 2 K^+ ions to the intracellular space (13).

$$3I_{\mathrm{K},0} + 2I_{\mathrm{Na},0} = 0 \tag{13}$$

With (13) and (3) we receive an additional formula for determining the equilibrium potential V_0 from an electrical point of view (14).

$$V_0 = \frac{g_{\mathrm{K}} V_{\mathrm{K},0} + \frac{2}{3} g_{\mathrm{Na}} V_{\mathrm{Na},0}}{g_{\mathrm{K}} + \frac{2}{3} g_{\mathrm{Na}}} \tag{14}$$

Both solutions for the equilibrium potential – (9) and (14) – can be combined and solved for the relative permeability P which is mainly dependent on the ion conductances g_n. This enables us to model the action potential generation by modulation of the ion channel permeability and to implement voltage-gated channels in our model.

$$P\left(\frac{g_{\mathrm{Na}}}{g_{\mathrm{K}}}\right) = -\frac{c_{\mathrm{K},0} - c_{\mathrm{K,ext}} \cdot \exp(-\frac{V_0 F}{RT})}{c_{\mathrm{Na},0} - c_{\mathrm{Na,ext}} \cdot \exp(-\frac{V_0 F}{RT})} \quad \text{with} \quad V_0 = \frac{V_{\mathrm{K},0} + \frac{2}{3}\frac{g_{\mathrm{Na}}}{g_{\mathrm{K}}} V_{\mathrm{Na},0}}{1 + \frac{2}{3}\frac{g_{\mathrm{Na}}}{g_{\mathrm{K}}}} \tag{15}$$

2.1 NaK-ATPase as a Controller

In this section, we will look at the NaK-ATPase as a controller of a closed loop controlled system, given by the neuron. The nomenclature of the control theory in this section is taken from [6] with the difference that we use matrix $\underline{M}$ as the system matrix and not as preliminary filter (see Fig. 1). As the neuron has two control variables (Na and K ion flow), the system is stabilised by means of a state control using the pole placement method.

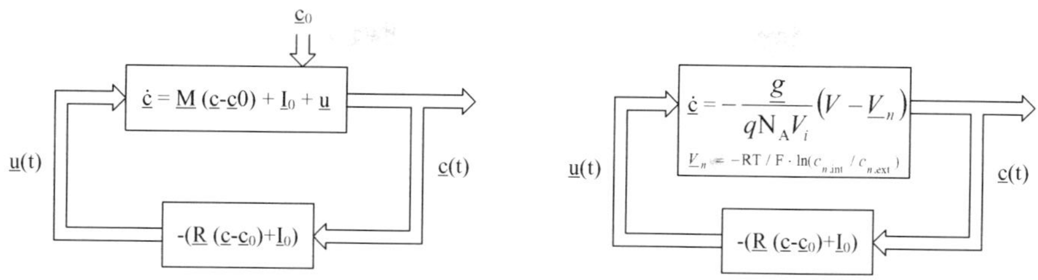

(a) Linearised system. (b) Nonlinear system.

Fig. 1. Nomenclature for control unit design of the closed loop control system

We can directly derive the control vector $\underline{R}$ from the eigenvalues and eigenvectors of system matrix $\underline{M}$

$$\underline{R} = \underline{V} \cdot (\mathrm{diag}(\lambda_1, \lambda_2) - \mathrm{diag}(\lambda_{R1}, \lambda_{R2})) \cdot \underline{V}^{-1} \tag{16}$$

where $\underline{V}$ is the matrix of eigenvectors of system matrix $\underline{M}$ and λ_n denotes the eigenvalues of $\underline{M}$. To stabilise the system, the dominant poles are moved to the left half-plane by setting λ_{Rn} to appropriate values. These values can be chosen by numerical simulation (e. g. $\lambda = -5 \cdot 10^4$ in the following simulations).

In Fig. 2a, the trajectories of the ion concentration of the neuron model with closed control loop are shown. Starting from different initial ion concentrations, the controller derived from the linear system stabilises the nonlinear system at the operating point $c_{\mathrm{K},0}$, $c_{\mathrm{Na},0}$. Figure 2b shows the trajectory of the ion concentration with the open control loop.

2.2 Modeling the Synapse

The synapse model discussed in this section is based upon the basic model already discussed in Sec. 2. The excitation of a synapse leads to an increased

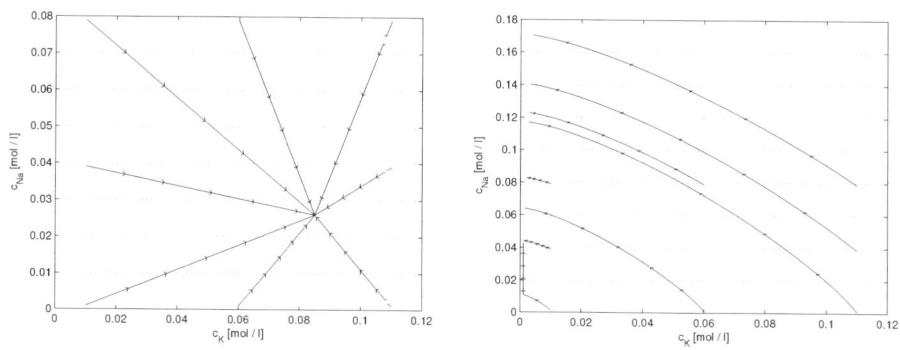

(a) (b)

Fig. 2. Trajectories of the ion concentration in the basic nonlinear neuron model with a) closed control loop and b) open control loop, starting from different initial values amount of neurotransmitter available at the receptors with the effect of a conductance change in the Na and K channels at the neuron. In this model, the additionally opened channels are assumed to be not ion specific, which can be expressed by (17)

$$\begin{aligned} \hat{g}_{\mathrm{K}}(t) &= g_{\mathrm{K}} + G(t) \\ \hat{g}_{\mathrm{Na}}(t) &= g_{\mathrm{Na}} + G(t) \\ G(t) &= g_c A(t) \end{aligned} \tag{17}$$

where g_c denotes the conductivity of a single channel and $A(t)$ denotes the time-dependent amount of additionally opened channels. The additionally opened (or closed) channels depend on the amount of ligated or free neurotransmitter. The transmitter(T)-receptor(R) binding can be described by the reaction kinetics given in (18).

$$\text{(free transmitter)} \quad R + T = B \underset{\beta}{\overset{\alpha}{\rightleftharpoons}} A = RT^{\star} \quad \text{(ligated transmitter)} \tag{18}$$

We further receive differential equations for A and B:

$$\frac{d}{dt} A = \alpha B - \beta A \tag{19}$$

$$\frac{d}{dt} B = \beta A - \alpha B \tag{20}$$

Now we change the probability α of the transition by introducing a time-dependent neurotransmitter profile. The term α is replaced by $\alpha \cdot q(t)$, where $q(t)$ describes the time response of the transmitter profile.

$$\frac{d}{dt} A = \alpha q(t)\,(A + B - A) - \beta A \tag{21}$$

Multiplication of (21) with g_c and application of $G(t) = g_c A(t)$ results in a differential equation with time-variant coefficients for the additional conductance $G(t)$ mainly depending on the transmitter profile

$$\frac{d}{dt} G(t) = \alpha q(t) \left(\underbrace{g_c\,(A + B)}_{G_S} - \underbrace{g_c A(t)}_{G(t)} \right) - \beta \underbrace{g_c A(t)}_{G(t)} \tag{22}$$

where G_S is a saturation conductance of the channels, which describes the conductance when all channels are in open state. The solution to this differential equation is given in (23).

$$\begin{aligned} G(t) = &\left[G_0 + \alpha G_S \int_0^t q(t'') \exp\left(\beta t'' + \alpha \int_0^{t''} q(t')dt' \right) dt'' \right] \\ &\cdot \exp\left(- \left(\beta t + \alpha \int_0^t q(t')dt' \right) \right) \end{aligned} \tag{23}$$

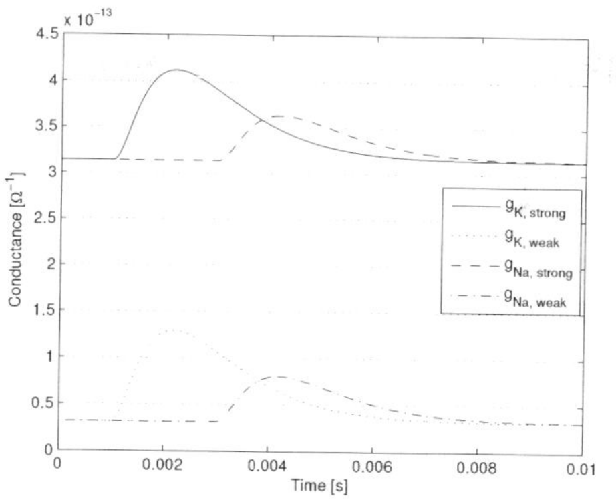

Fig. 3. The time dependent conductances g_K and g_{Na} at the synapse. At $t = 1\,\text{ms}$, the strong synapse is excited, at $t = 3\,\text{ms}$, the weak synapse is excited.

In our simulations we are using the α-function for the shape of the neurotransmitter pulse. For a fast synapse [9], we use the parameters $\alpha = 2\,\text{ms}^{-1}$ and $\beta = 1\,\text{ms}^{-1}$.

2.3 Action Potential Generation

The action potential is generated by a change in ion conductance, which is dependent on the membrane potential (voltage-gated channels). Thus, time dependent ion specific conductances are introduced (24) which can be derived from a Hodgkin-Huxley model [7]. This is in fact an approximation of the Hodgkin-Huxley model, but it preserves the basic functionality.

$$\begin{aligned} \hat{g}_{\text{K}}(t) &= g_{\text{K}} + G_{\text{K}}(t) \quad \text{with} \quad G_{\text{K}}(t) = \bar{g}_{\text{K}} n^4 \\ \hat{g}_{\text{Na}}(t) &= g_{\text{Na}} + G_{\text{Na}}(t) \quad \text{with} \quad G_{\text{K}}(t) = \bar{g}_{\text{Na}} m^3 h \\ &\text{where} \quad m, n, h = f(t) \end{aligned} \tag{24}$$

The time dynamics of the coefficients $m(t)$, $n(t)$ and $h(t)$ can be represented as inhomogeneous differential equations which are similar in their structure but different in their parameters. Thus, only the solution for coefficient $m(t)$ will be shown here as an example. Similar to the discussion in the previous section, we can start from a biochemical reaction with the reaction kinetics shown in (25) where α_m and β_m describe the probability for the transition from the open state of an ion channel to the closed state and V_{Th} describes the threshold voltage.

$$1 - m \underset{\beta_m(V - V_{\text{Th}})}{\overset{\alpha_m(V - V_{\text{Th}})}{\rightleftharpoons}} m \tag{25}$$

This leads to the inhomogeneous differential equation (with piecewise constant coefficients α_m and β_m) given in (26)

$$\frac{d}{dt}m = \alpha_m (V(t) - V_{\mathrm{Th}}) - m [\alpha_m (V(t) - V_{\mathrm{Th}}) + \beta_m (V(t) - V_{\mathrm{Th}})] \tag{26}$$

with its solution given in (27).

$$m(t) = m_0 \mathrm{e}^{-(\alpha_m + \beta_m)t} + \frac{\alpha_m}{\alpha_m + \beta_m} \tag{27}$$

As parameters α_m and β_m are also voltage dependent, these equations have to be solved partially. The parameters for determining the functions m, n and h are taken from [8]. A typical time-course for $m(t)$, $n(t)$, and $h(t)$ is given in Fig. 4a. The resulting conductance of the K channels and the Na channels is shown in Fig. 4b.

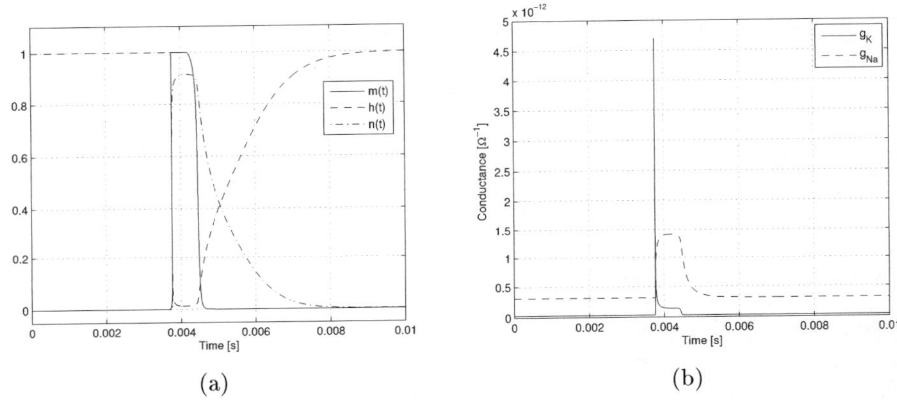

Fig. 4. Time course of a) $m(t)$, $n(t)$ and $h(t)$ and b) the resulting K channel and Na channel conductance for an action potential generated at $t = 3.7\,\mathrm{ms}$

3 Simulation Results

The biophysical neuron model has been set up as a Matlab/Simulink model, where we integrated the basic neuron model as presented in Sec. 2 with the NaK-ATPase as active transport mechanism, implemented as a closed loop controller. Additionally, two synapses and their response to a pre-synaptic action potential have been modeled to show the effect of the voltage-gated ion channels and the action potential generation of our model. Both synapses have been connected to a small dendritic tree. The strength of each synapse can be adjusted by the conductance parameter g_c for each individual block. All of the implemented blocks (neuron, synapses and dendrite) are based upon the basic closed control loop model. Only small adaptations have to be made for the synapse and the dendrite (see discussion in previous sections).

Figure 5a depicts the trajectory of the sodium and potassium concentration for a neuron without action potential generation. The neuron is excited by external stimulus with an α-function at a weak and a strong synapse. In Fig. 5b the trajectory for a system with activated action potential generation is shown.

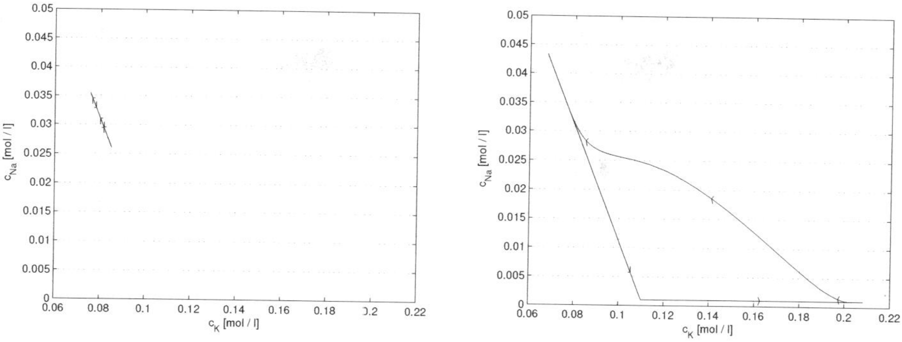

(a) System without AP generation.

(b) System with AP generation.

Fig. 5. Trajectories of the sodium and potassium concentration in a neuron model a) without action potential generation and b) with action potential generation

Figure 6a shows the membrane potential of the full neuron model with axonal element and small dendritic tree is shown on a timescale of 10 ms. The neuron is fed by two synapses, one weak and one strong connection, which are excited with an α-function at 1 ms and 3 ms. The crossing of the threshold voltage leads to action potential generation by voltage-gated ion channels as described in Sec. 2.3. In Fig. 6b, the corresponding power dissipation is depicted. As it can be expected, the power dissipation from the NaK-ATPase reaches its maximum at the action potential generation, when the pump has to counterbalance the voltage-induced ion influx and efflux. We received a dissipation power of 2.4 fW for the steady state and a peak of 87 nW for an action potential. This, of course, is linear adjustable by changing the geometry and conductance parameters.

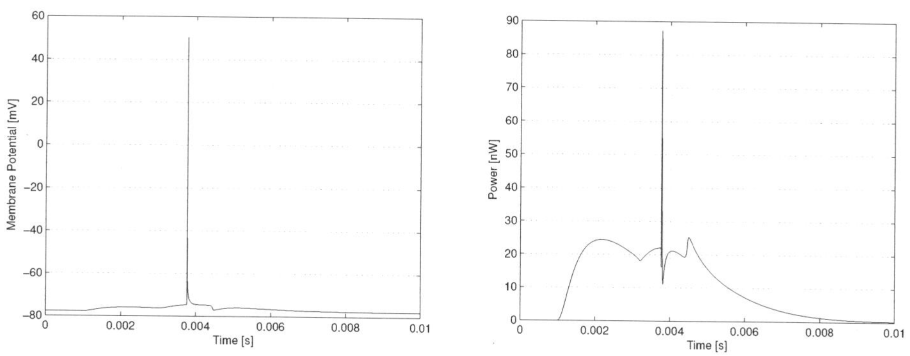

(a) Membrane potential.

(b) Dissipation Power.

Fig. 6. Simulation result of a neuron with two synapses (one weak and one strong synapse) on a small dendritic tree. Excitation of the neuron occurs at $t = 1$ ms and $t = 3$ ms with an α-function at the synaptic inputs.

4 Conclusion

We have presented a neuron model that incorporates a control approach to the mechanism provided by the NaK-ATPase in biological neurons. This model can also be used to estimate the power that a single neuron dissipates for maintaining the steady-state as well as for the generation of action potentials. We could show by simulation that the chosen control approach for the NaK-ATPase is capable of not only stabilising the linearised system, but also the nonlinear system described by diffusion equations. The same mechanism as in the basic model has also been used to model the synaptic excitation by introducing additionally opened channels to the basic model. We have integrated the basic model as well as its extensions for synaptic excitation and the action potential generation into Matlab/Simulink to provide an easy interface for further research on biophysical models. Future work is going to extend this model to incorporate the Ca^{2+} channel and to use this model for building networks within the Matlab/Simulink environment.

References

1. Koch, C., Segev, I.: Methods in Neuronal Modeling (1998)
2. Laughlin, S.B., de Ruyter van Steveninck, R.R., Anderson, J.C.: The metabolic cost of neural information. Nature Neuroscience 1, 36–41 (1998)
3. MacGregor, R., Lewis, E.: Neural Modeling (1977)
4. Daut, J.: The living cell as an energy-transducing machine. A minimal model of myocardial metabolism. Biochimica et Biophysica Acta (BBA) - Reviews on Bioenergetics 895, 41–62 (1987)
5. Destexhe, A., Mainen, Z.F., Sejnowski, T.J.: An efficient Method for Computing Synaptic Conductances Based on a Kinetic Model of Receptor Binding. Neural Computation 6, 14–18 (1994)
6. Föllinger, O.: Regelungstechnik (1994)
7. Hodgkin, A., Huxley, A.: A quantitative description of membrane current and its application to conduction and excitation in nerve. J. Phys. 117, 500–544 (1952)
8. Destexhe, A.: Conductance-based integrate-and-fire models. Neural Computation 9, 503–514 (1997)
9. Chapeau-Blondeau, F., Chambet, N.: Synapse Models for Neural Networks: From Ion Channel Kinetics to Multiplicative Coefficient w_{ij}. Neural Computation 7, 713–734 (1995)

Integrate-and-Fire Neural Networks with Monosynaptic-Like Correlated Activity

Héctor Mesa and Francisco J. Veredas

Dpto. Lenguajes y Ciencias de la Computación,
Universidad de Málaga, Málaga 29071, Spain
fvn@lcc.uma.es

Abstract. To study the physiology of the central nervous system it is necessary to understand the properties of the neural networks that integrate it and conform its functional substratum. Modeling and simulation of neural networks allow us to face this problem and consider it from the point of view of the analysis of activity correlation between pairs of neurons. In this paper, we define an optimized integrate-and-fire model of the simplest network possible, the monosynaptic circuit, and we raise the problem of searching for alternative non-monosynaptic circuits that generate monosynaptic-like correlated activity. For this purpose, we design an evolutionary algorithm with a crossover-with-speciation operator that works on populations of neural networks. The optimization of the neuronal model and the concurrent execution of the simulations allow us to efficiently cover the search space to finally obtain networks with monosynaptic-like correlated activity.

1 Searching for Monosynaptic-Like Circuits

To understand the functionality of the central nervous system (CNS) it is precise to comprehend the characteristics of the neural circuits that support its properties. As it has been demonstrated by neurophysiological studies, across the visual pathway of mammals the properties of the neural responses undergo several important transformations: neuronal receptive fields become more complex [8], averaged firing rates decrease, neural activity becomes more variable [10] and the correlated activity among neurons becomes less precise [11,12,15]. To identify the factors responsible of these transformations, it is necessary to analyze the connectivity into the different neuronal networks disposed from the retina to the visual cortex.

To attack the problem in a progressive way, we started from the simplest neuronal net: the excitatory monosynaptic circuit, i.e. a direct excitatory connection beetween two neurons, without any other neuron intermediating them. The starting point for the afterward studies of more complex networks is the understanding of the different physiological and topological factors that determine the correlated activity between these two monosynaptically connected neurons. Physiological studies on correlated activity in the visual pathway of mammals

J. Marques de Sá et al. (Eds.): ICANN 2007, Part I, LNCS 4668, pp. 539–548, 2007.

conclude that strong monosynaptic connections present a precise correlated activity. This correlation can be shown by cross-correlation analysis. The cross-correlogram of spikes of two neurons with a strong monosynaptic connection is well characterized by the presence of a significant peak displaced to the right of $0\,ms$ [6] as a direct consequence of the synaptic transmission delay (see figure 1). Other way, in physiological experiments of simultaneous intracellular recording of two neurons, the presence of a significant peak in their cross-correlogram is usually interpreted as revealing the existence of a monosynaptic connection [2,1]. Different neurophysiological features imply variations in that correlogram [11,12,15,17] across the visual pathway.

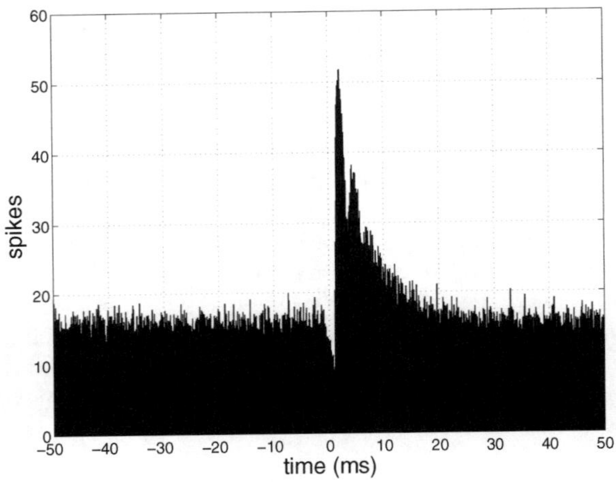

Fig. 1. Cross-correlogram from a simulation of a strong monosynaptic connection between two IAF neurons. This is the histogram of the time difference between pre- and post-synaptic spikes.

The question that arises now is the following: is it likely to find a non-monosynaptic connection between two neurons that gives a correlogram similar to the monosynaptic correlogram? The answer to this question would have important implications for the concern of which are the circuits that give support to some properties found in the CNS, as the shape of the receptive fields of neurons in talamo-cortical or cortico-cortical pathways of the visual system [2,1]. Could a monosynaptic-like correlogram gotten in the analysis of the neural activity in those pathways be generated by some sort of polysynaptic circuit? If so, maybe some conclusions about the circuitry holding some properties in the visual system should be revisited. To discard or consider this possibility, in this article we present the results of searching for circuits of integrate-and-fire (IAF) neurons with monosynaptic-like correlated activity. For this purpose, we used an evolutionary approach and designed specific genetic operators adapted to the peculiarities of this problem. As a main result, giving an initial set of biological and computational restrictions, it was possible to reach the purposed objective of finding a polysynaptic network with monosynaptic-like correlation.

1.1 Problem Description

To deal with the search of circuits with monosynaptic correlated activity, the physiological and topologycal factors that contribute to this correlated activity should be established. In [17] it was concluded that, in agreement with previous physiological studies [5,14], it is the time-course of the EPSPs (Excitatory Post-Synaptic Potentials) the main factor that determines the shape of the monosynaptic peak: rise and decay times of the EPSP, and jitter of transmission delay. To these factors we added other two topological issues: the synaptic efficacy (connection weight) which influences the peak amplitude, and the synaptic latency that affects to the position of the peak.

On the other way, in [18] some evolutionary and genetic strategies were used to search for circuits with monosynaptic-like correlated activity but, giving a set of realistic biological and computational restrictions, it was not possible to find a circuit alternative to the monosynaptic one. In that study the problem of crossover disruption in the genetic algorithm was raised as a severe impediment for the effective evolution of the populations of IAF networks.

In this paper, we use specific methods to 1) increase the computational efficiency in searching for circuits by means of the optimization of the neuron model, the calculation of the cross-correlogram by Fast Fourier Transformation (FFT) and the use of techniques of parallel programming; and 2) avoid crossover disruption in the genetic algorithm. For the latter, we have designed a new crossover operator with speciation that allows, for each generation, to cluster the population into species of individuals —i.e. IAF networks— with the enough genotypic similarity to avoid crossover disruption. Besides, we have designed an evolutionary strategy in batch mode that applies each genetic operator in independent stages, starting from the best results of each previous stage, to ensure the convergence of the population towards good solutions. Finally, heuristic techniques have allowed us to refine the populations of IAF networks and ensure their convergence to solutions nearer to the objective.

2 The Integrate-and-Fire Neuron Model

IAF models are a particular case of simplified neuronal models. In their earlier designs they derived from the pioneering work of Louis Lapicque. Alternative versions of this first original model have been proposed in the literature [9,13]. The traditional form of an IAF neuron consists of a first order differential equation (eq. 1) with a subthreshold integration domain [where the neuron integrates its inputs $I(t)$] and a threshold potential (not explicitly represented in the equations) for action potential generation.

$$C_m \frac{dV_m(t)}{dt} = I(t) - \frac{[V_m(t) - V_{rest}]}{R_m}, \tag{1}$$

where C_m is the neuronal membrane capacitance, V_m the membrane potential, R_m the membrane resistance, V_{rest} the resting potential and $I(t)$ the synaptic input current. For the purposes of this paper, $I(t)$ has been modeled as in eq. 2.

$$I(t) = \sum_j \omega_j g(\hat{t}_j)[E_{syn} - V_m(t)] + noise, \tag{2}$$

where ω_j represents the connection weight (or synaptic efficacy) between neuron j (presynaptic) and the current neuron (postsynaptic), $g(s)$ is the synaptic conductance, $\hat{t}_j$ is the time relative to the last spike of the presynaptic neuron j (also taking into account the existence of a synaptic latency and jitter of transmission delay), E_{syn} is the reversal potential or synaptic equilibrium and, finally, $noise$ represents the background synaptic noise [following a Gaussian distribution with parameters (μ_n, σ_n)].

Synaptic conductance has been modeled (eq. 3) by a two-fold equation consisting of two alpha functions with different time-constants, τ_r and τ_d, for independently determining its rise and decay time, respectively.

$$g(t) = \begin{cases} \frac{t}{\tau_r} e^{1-\frac{t}{\tau_r}} & if\ t \leq \tau_r \\ \frac{\tau_d - \tau_r + t}{\tau_d} e^{1-\frac{\tau_d - \tau_r + t}{\tau_d}} & if\ t > \tau_r. \end{cases} \tag{3}$$

To complete the neuron model, it was included an absolute refractory period (of 2.5 ms) and an after-hyperpolarization potential where the membrane potential decays to after an action potential occurs. This after-hyperpolarization potential is calculated as a factor of 0.2 times the membrane potential before the spike initiation. These two physiological parameters are not shown in the equations for the sake of legibility.

In computer simulations of the neuron model above, synaptic conductances have to be updated at each simulation step, which constitutes a crucial task from the point of view of computational efficiency. The basic problem consists of computing the overall conductance g_{total} in a more efficient manner than the simple convolution of spike trains s_i with conductance functions g_i of all synapses at time t_x:

$$g_{total}(t_x) = \sum_{i=0}^{N} \omega_i \int_0^{t_x} s_i(\tau) g_i(t_x - \tau) d\tau \tag{4}$$

$$s(t) = \sum_{i=0}^{N} \omega_i s_i(t), \tag{5}$$

where N is the total number of synapses.

Starting from conductance equation 3, in [16] the $\mathcal{Z}$-transformation of the discretization of $g(t)$ was used to get a recursive expression of $g[n]$ as a function of only a few previous terms. This recursive function allows us to state each value of the overall conductance at step n as a function of a reduced number of the previous values of the conductance and the presynaptic spike trains, which significantly increases the computational efficiency of the model [16].

3 An Evolutionary Algorithm to Search for IAF Circuits

Starting from the biological and parametric restrictions supported in [17,18], for this article we applied four genetic operators —selection, mutation, crossover-with-speciation and heuristics— for the evolution of populations of IAF neural networks modelled as in section 2. Each network consisted of up to six IAF neurons, of which one neuron was fixed as the reference neuron and another one was established as the target neuron for cross-correlation analysis between both. Networks were fully connected, with no self-connectivity, and the monosynaptic connection from the reference neuron to the target neuron was strictly removed. Only one neuron of each network, distinct of the reference and target neurons, was assigned an inhibitory nature; the rest had an excitatory nature. This excitatory/inhibitory nature of the neurons was not changed during the evolution of the searching algorithm.

In each generation of the evolutionary algorithm (EA), each IAF network —an individual of the population— was simulated to get $2,000\,spikes$ from the target neuron and the cross-correlogram (from reference to target) was computed. To increase the efficiency of these computations, the correlogram was calculated by the FFT of the spike trains of the neurons, which reduced the computational costs in a 20% of time spent with classical methods. The fitness value of each network was computed as a function of the Euclidean distance to a reference monosinaptic correlogram —obtained by an average of ten simulations of a strong monosynaptic connection between two IAF neurons. In different stages of this evolutionary algorithm we used diverse fitness functions —with simple or weighted (at the region of the monosynaptic peak) Euclidean distances— to facilitate the convergence of the population.

The genome of each individual (IAF network) of the population was encoded as a vector of real numbers that stores the weights of the network, the rise and decay times of the synaptic conductance curves, the synaptic latencies and transmission jitters (see table 1). These parameters constitute the principal factors that contribute to the shape of the monosynaptic correlogram [17]. The initial population of the EA was generated in a random manner, with the only restriction of filtering those random individuals producing less than $2,000\,spikes$ in the target neuron for a simulation of $300,000\,steps$ ($30,000\,ms$).

As the selection operator for crossover we used a type of the roulette wheel method [7]. The mutation operator was applied to the genome by establishing a mutation rate for each factor (weight, rise and decay time of the conductance, delay and jitter).

One of the main genetic operators designed in this study is the crossover operator with a speciation method. The high complexity of the individuals of the populations makes classical crossover operators prevent convergence [18]. To

Table 1. Scheme of the genome of one individual (IAF network) of the population, for n neurons, with $n \geq 3$. ω_{ij} is the synaptic weight from neuron i to neuron j; τ_r^i and τ_d^i are the rise and decay times of synaptic conductance of neuron i, respectively; δ_i and ξ_i represent the mean and standard deviation of a Gaussian distribution of transmission delays for neuron i, i.e. its latency and jitter.

$\longleftarrow$					$\mathbf{n(n-1)-1}$			$\longrightarrow$	
w_{13}	$\ldots$	w_{1n}	w_{21}	w_{23}	$\ldots$	w_{2n}	$\ldots$	$w_{n(n-1)}$	$\rightsquigarrow 1$

	$\longleftarrow$	$\mathbf{n}$	$\longrightarrow$	$\longleftarrow$	$\mathbf{n}$	$\longrightarrow$	
$\rightsquigarrow 1$	τ_r^1	$\ldots$	τ_r^n	τ_d^1	$\ldots$	τ_d^n	$\rightsquigarrow 2$

	$\longleftarrow$	$\mathbf{n}$	$\longrightarrow$	$\longleftarrow$	$\mathbf{n}$	$\longrightarrow$
$\rightsquigarrow 2$	δ_1	$\ldots$	δ_n	ξ_1	$\ldots$	ξ_n

avoid this problem, we used a set of techniques known as speciation [4] consisting in forming groups or species of individuals with similar genotypes or phenotypes and restricting the crossover operator to be held between individuals in the same group. To implement this crossover-with-speciation operator, a genotypic distance was chosen to conform the groups and three essential tools were specifically designed for this problem: 1) an evaluation function to estimate the overall fitness of a group; 2) a function to compute the size of a group as being proportional to the fitness of the individuals constituting it; and 3) a cluster function to classify the individuals of the population into species. As the cluster function for this problem it was selected a modality of the ART2 network [3].

Finally, a set of specific heuristic operators was designed with the main objective of affecting the synaptic weights of the IAF networks in a supervised manner. In this way, we tried to control the overall activity of the network —i.e. its excitation/inhibition ratio— to affect the levels of basal background activity, as well as the correlogram peak amplitude, and approach them to the shape of the reference monosynaptic correlogram.

4 Executions and Results

To search for circuits with monosynaptic-like correlated activity, we launched four batches of executions of the EA applying each different genetic operator in an independent manner: for batch #1 we started from a random population and applied mutation and heuristics; on the best 150 individuals resulting from batch #1, in batch #2 we used mutation, crossover-with-speciation and heuristics; on the best 150 individuals obtained from batch #2, in batch #3 we applied again mutation, crossover-with-speciation and heuristics; finally, starting from the best 150 individuals of batch #3 we used mutation and crossover-with-speciation in batch #4 to get the resultant final individuals. In this way, we pretended the population converged towards better solutions over generations —batch by batch— in a controlled manner. The programming source of the EA was written in C++

(with GSL 1.0 library) and it was paralleled to allow the concurrent evaluation of each individual (IAF network) as an independent process. The executions of the different batches were launched in a cluster of computers with OpenMosix consisting of three AMD®Athlon XP 2400+ computers, giving an averaged computation time of 3 hours for 100 generations of the EA with populations of 150 individuals.

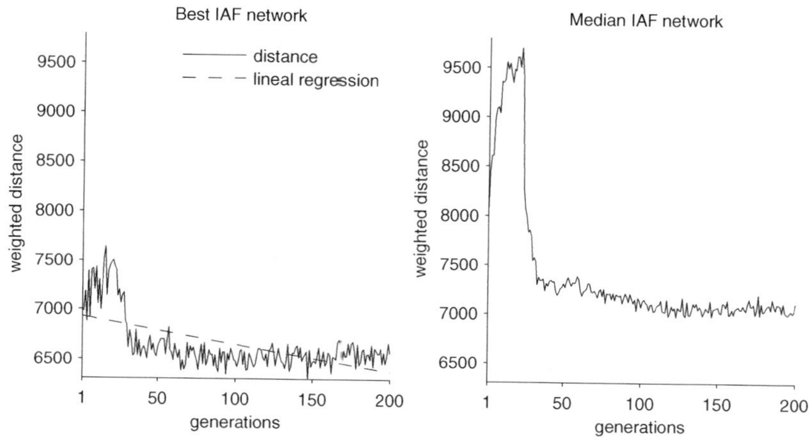

Fig. 2. Execution of the EA with the crossover-with-speciation operator (see text)

Best results with respect to population convergence were obtained with the crossover-with-speciation operator. In figure 2 we show the results of an execution of this operator at batch #2, i.e. over a population of individuals previously refined by mutation and heuristics (batch #1). Figure 2 (left) shows the evolution of the weighted distance from the correlogram of the best IAF network at each generation to the reference monosynaptic circuit, as well as a linear regression to illustrate the convergence to better solutions. On the right of the figure the evolution of the median IAF network of the population is shown. The initial rise at both distance curves is a consequence of the preliminary genetic diversity of the population, which tends to be homogenized along the generations.

After executing the different genetic operators designed as in batches above, we have got a polysynaptic circuit with a correlogram (figure 3, left) very similar to the reference monosynaptic correlogram (figure 3, right). Although subtle differences between both correlograms can be noticed visually, the computed absolute distance between them falls into the range of mutual distance among correlograms from different executions of exactly the same reference monosynaptic circuit (so that the effect of background synaptic noise introduces variability). For both correlograms in figure 3, the peak position, amplitude and width are almost identical, and also their basal line amplitudes coincide.

In figure 4 a scheme of the resulting circuit (best individual at the final population) of the EA is shown. This IAF network produces the correlogram of

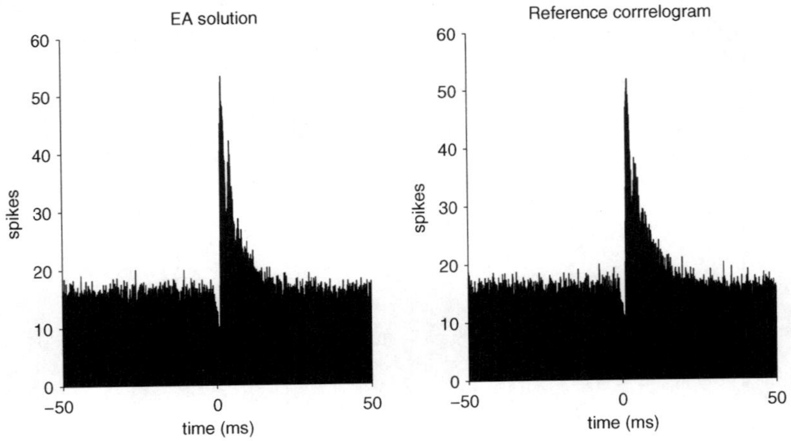

Fig. 3. Correlogram of the best IAF network of the final population (left) and the monosynaptic reference correlogram (right)

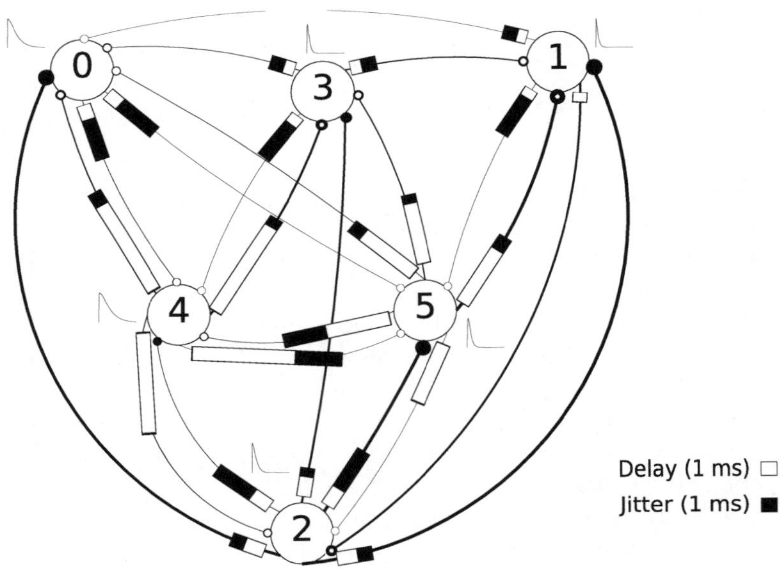

Fig. 4. Scheme of the resultant polysynaptic IAF network with monosynaptic-like correlated activity between neurons #0 (reference) and #1 (target). Little empty circles on a postsynaptic neuron represent excitatory synapses arriving it; solid circles stand for inhibitory synapses —being #2 the only inhibitory neuron. Line thickness means synaptic weight. Length of white and black rectangles near a presynaptic neuron is proportional to synaptic latency and jitter, respectively. Time-course (125 ms) of the synaptic conductance curve is also shown next to each postsynaptic neuron.

Table 2. Parameters of an IAF polysynaptic network with monosynaptic-like correlated activity (see figure 4 for a scheme of the circuit). τ_m represents the membrane time-constant; V_{thres} is the threshold potential; ϑ is an after-hyperpolarization factor; ρ is an absolute refractory period (see text for details and explanations for the rest of the parameters).

Neuronal parameters

$\Delta t\,(ms)$	$\tau_m\,(ms)$	$C_m\,(nF)$	$V_{rest}\,(mV)$	$V_{thres}\,(mV)$	ϑ	$\rho\,(ms)$
0.1	1	0.1	-70	-40	0.2	2.5

Synaptic parameters

#neuron	$\delta\,(ms)$	$\xi\,(ms)$	$\tau_r\,(ms)$	$\tau_d\,(ms)$	$\mu_n\,(nA)$	$\sigma_n\,(nA)$
0	1.01172	0.922166	0.91984	16.1245	2.8	0.5
1	0.601871	0.950428	0.358214	3.77895	2.6	0.5
2	1.28104	0.060946	0.33425	7.25367	2.6	0.5
3	0.693697	0.588336	2.04599	3.94252	2.6	0.5
4	6.49785	2.73206	2.11559	18.583	2.6	0.5
5	4.04503	2.86628	3.50525	4.98314	2.6	0.5

Synaptic weights ($\mu Siemens$)

pre \ post	0	1	2	3	4	5
0	-	-	0	0	0.744677	0.055858
1	0.100444	-	3.39608	0	0	0.047406
2	−4.51582	−5	-	−3.2723	−0.967395	−4.56097
3	1.29592	1.98352	0	-	0.126307	0
4	2.03464	0	1.32564	3.36806	-	0.491359
5	1.01654	4.91448	0.367558	2.03333	1.16246	-

figure 3 (left) in the absence of a monosynaptic connection between the reference neuron (#0, in the figure) and the target neuron (#1, in the figure). In table 2 the physiological and topological parameters for this circuit are illustrated.

5 Conclusions

In this paper we have shown the results of using evolutionary algorithms to search for polysynaptic circuits that generate correlated activity with a monosynaptic-like cross-correlogram shape. We used a set of optimization techniques to design an efficient IAF neuron model. Moreover, the use of parallel programming methods and the increase of the efficiency in computing the cross-correlogram, allowed us to extend the efficacy in inspection of the search space for this problem.

The design of some specific genetic operators —specially that of crossover-with-speciation— and the use of heuristics as well as the execution of the evolutionary algorithms in a batch supervised mode, concluded with the finding of a polysynaptic network configuration that fulfills ours main proposed objective of giving a monosynaptic-like correlated activity. Although severe biological restrictions were initially considered to set a realistic range for each physiological and topological parameter in the search space, a pair of open questions remains for future work: the biological plausibility of the resultant IAF network and its implications in the study of the circuitry of the central nervous system. These questions could have important implications for the concern of which are the circuits that give support to some properties found in the CNS: up to this moment, in physiological experiments of intracellular recordings the presence of a significant peak in a cross-correlogram has been usually interpreted as revealing the existence of a monosynaptic connection. From the findings of this article we can conclude that at least an alternative polysinaptic circuitry could be responsible of getting that sort of correlated activity.

References

1. Alonso, J., Martínez, L.: Functional connectivity between simple cells and complex cells in cat striate cortex. Nat. Neurosci. 1, 95–403 (1998)
2. Alonso, J.M., Usrey, W.M., Reid, R.C.: Precisely correlated firing in cells of the lateral geniculate nucleus. Nature 383, 815–819 (1996)
3. Carpenter, G., Grossberg, S.: ART 2: Self-organization of stable category recognition codes for analog input patterns. Applied Optics 26, 4919–4930 (1987)
4. Deb, K., Spears, W.M.: Speciation methods. In: Bäck, T., Fogel, D.B., Michalewicz, Z. (eds.) Evolutionary Computation 2. Advanced algorithms and operators, ch. 4, pp. 93–100. Institute of Physics Publishing, Bristol, United Kingdom (2000)
5. Fetz, E., Gustafsson, B.: Relation between shapes of post-synaptic potentials and changes in firing probability of cat motoneurones. J. Physiol. 341, 387–410 (1983)
6. Gerstein, G., Perkel, D.: Mutual temporal relationships among neuronal spike trains. Statistic techniques for display and analysis. Biophys. J. 12, 453–473 (1972)
7. Goldberg, D.: Genetic algorithms in search, optimization and machine learning. Addison-Wesley, Reading (1989)
8. Hubel, D.H., Wiesel, T.N.: Receptive fields, binocular interaction and functional architecture in the cat's visual cortex. J. Physiol (Lond.) 160, 106–154 (1962)
9. Jack, J., Nobel, D., Tsien, R.: Electric current flow in excitable cells. Oxford University Press, Oxford, United Kingdom, revised paperback (1983) edition (1975)
10. Kara, P., Reinagel, P., Reid, R.C.: Low response variability in simultaneously recorded retinal, thalamic, and cortical neurons. neuron 27, 635–646 (2000)
11. Levick, W., Cleland, B., Dubin, M.: Lateral geniculate neurons of cat: retinal inputs and physiology. Invest. Ophthalmol. 11, 302–311 (1972)
12. Mastronarde, D.N.: Two clases of single-input X-cells in cat lateral geniculate nucleus. II. Retinal inputs and the generation of receptive-field properties. J. Neurophysiol. 57, 7757–7767 (1987)
13. Stein, R.: The frequency of nerve action potentials generated by applied currents. Proc. Roy. Soc. Lond. B 167, 64–86 (1967)
14. Surmeier, D., Weinberg, R.: The relationship between cross-correlation measures and underlying synaptic events. Brain Research 331(1), 180–184 (1985)
15. Usrey, W., Reppas, J., Reid, R.: Specificity and strength of retinogeniculate connections. J. Neurophysiol. 82, 3527–3540 (1999)
16. Veredas, F., Mesa, H.: Optimized synaptic conductance model for integrate-and-fire neurons. In: 10th IASTED International Conference Artificial Intelligence and Soft Computing, Palma de Mallorca, Spain, August 2006, pp. 97–102 (2006)
17. Veredas, F., Vico, F., Alonso, J.: Factors determining the precision of the correlated firing generated by a monosynaptic connection in the cat visual pathway. J. Physiol. 567(3), 1057–1078 (2005)
18. Veredas, F., Vico, F., Alonso, J.: Evolving networks of integrate-and-fire neurons. Neurocomputing 69(13–15), 1561–1569 (2006)

Multi-dimensional Recurrent Neural Networks

Alex Graves[1], Santiago Fernández[1], and Jürgen Schmidhuber[1,2]

[1] IDSIA , Galleria 2, 6928 Manno-Lugano, Switzerland
[2] TU Munich, Boltzmannstr. 3, 85748 Garching, Munich, Germany
{alex,santiago,juergen}@idsia.ch

Abstract. Recurrent neural networks (RNNs) have proved effective at one dimensional sequence learning tasks, such as speech and online handwriting recognition. Some of the properties that make RNNs suitable for such tasks, for example robustness to input warping, and the ability to access contextual information, are also desirable in multi-dimensional domains. However, there has so far been no direct way of applying RNNs to data with more than one spatio-temporal dimension. This paper introduces multi-dimensional recurrent neural networks, thereby extending the potential applicability of RNNs to vision, video processing, medical imaging and many other areas, while avoiding the scaling problems that have plagued other multi-dimensional models. Experimental results are provided for two image segmentation tasks.

1 Introduction

Recurrent neural networks (RNNs) were originally developed as a way of extending neural networks to sequential data. The addition of recurrent connections allows RNNs to make use of previous context, as well as making them more more robust to warping along the time axis than non-recursive models [15]. Access to contextual information and robustness to warping are also important when dealing with multi-dimensional data. For example, a face recognition algorithm should use the entire face as context, and should be robust to changes in perspective, distance etc. It therefore seems desirable to apply RNNs to such tasks.

However, the standard RNN architectures are explicitly one dimensional, meaning that in order to use them for multi-dimensional tasks, the data must be pre-processed to one dimension, for example by presenting one vertical line of an image at a time to the network. One successful use of neural networks for multi-dimensional data has been the application of convolution networks [10] to image processing tasks such as digit recognition [14]. One disadvantage of convolution nets is that because they are not recurrent, they rely on hand specified kernel sizes to introduce context. Another disadvantage is that they do not scale well to large images. For example, sequences of handwritten digits must be pre-segmented into individual characters before they can be recognised by convolution networks [10].

Various statistical models have been proposed for multi-dimensional data, notably multi-dimensional HMMs. However, multi-dimensional HMMs suffer from two serious drawbacks: (1) the time required to run the Viterbi algorithm, and thereby calculate the optimal state sequences, grows exponentially with the size of the data exemplars; (2) the

J. Marques de Sá et al. (Eds.): ICANN 2007, Part I, LNCS 4668, pp. 549–558, 2007.

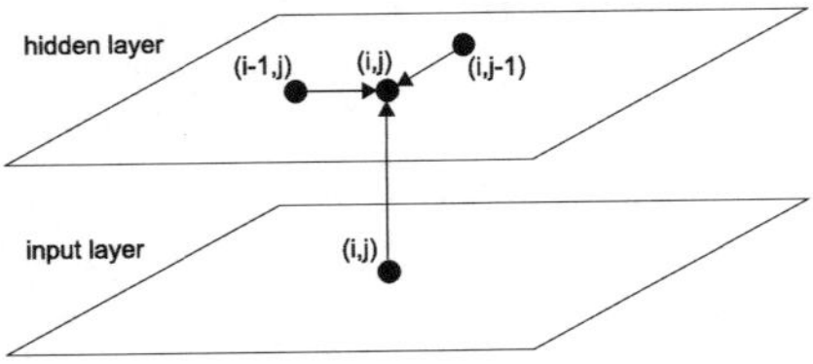

Fig. 1. 2D RNN Forward pass

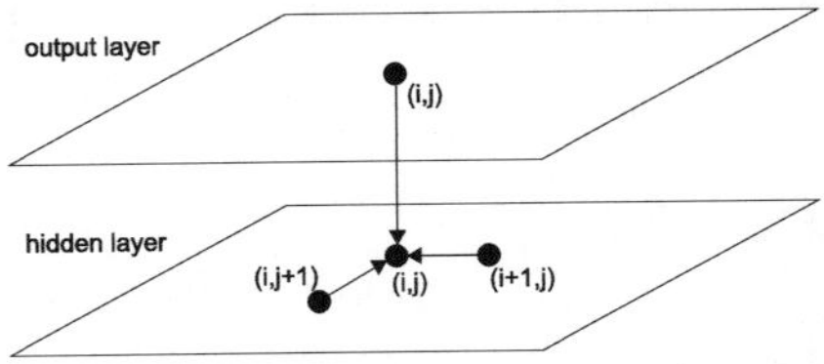

Fig. 2. 2D RNN Backward pass

number of transition probabilities, and hence the required memory, grows exponentially with the data dimensionality. Numerous approximate methods have been proposed to alleviate one or both of these problems, including pseudo 2D and 3D HMMs [7], isolating elements [11], approximate Viterbi algorithms [9], and dependency tree HMMs [8]. However, none of these methods exploit the full multi-dimensional structure of the data.

As we will see, multi dimensional recurrent neural networks (MDRNNs) bring the benefits of RNNs to multi-dimensional data, without suffering from the scaling problems described above.

Section 2 describes the MDRNN architecture, Section 3 presents two experiments on image segmentation, and concluding remarks are given in Section 4.

2 Multi-dimensional Recurrent Neural Networks

The basic idea of MDRNNs is to replace the single recurrent connection found in standard RNNs with as many recurrent connections as there are dimensions in the data. During the forward pass, at each point in the data sequence, the hidden layer of the network receives both an external input and its own activations from one step back along all dimensions. Figure 1 illustrates the two dimensional case.

Note that, although the word *sequence* usually denotes one dimensional data, we will use it to refer to data examplars of any dimensionality. For example, an image is a two dimensional sequence, a video is a three dimensional sequence, and a series of fMRI brain scans is a four dimensional sequence.

Clearly, the data must be processed in such a way that when the network reaches a point in an n-dimensional sequence, it has already passed through all the points from which it will receive its previous activations. This can be ensured by following a suitable ordering on the set of points $\{(x_1, x_2, ..., x_n)\}$. One example of a suitable ordering is $(x_1, \ldots, x_n) < (x'_1, \ldots, x'_n)$ if $\exists\, m \in (1, \ldots, n)$ such that $x_m < x'_m$ and $x_i = x'_i \;\forall\, i \in (1, \ldots, m-1)$. Note that this is not the only possible ordering, and that its realisation for a particular sequence depends on an arbitrary choice of axes. We will return to this point in Section 2.1. Figure 3 illustrates the ordering for a 2 dimensional sequence.

The forward pass of an MDRNN can then be carried out by feeding forward the input and the n previous hidden layer activations at each point in the ordered input sequence, and storing the resulting hidden layer activations. Care must be taken at the sequence boundaries not to feed forward activations from points outside the sequence.

Note that the 'points' in the input sequence will in general be multivalued vectors. For example, in a two dimensional colour image, the inputs could be single

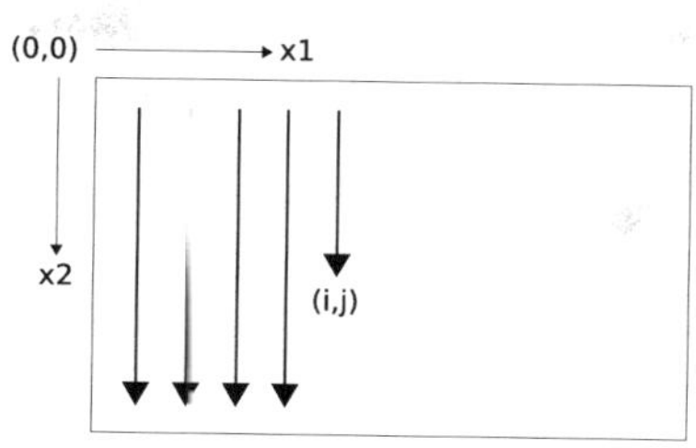

Fig. 3. 2D sequence ordering. The MDRNN forward pass starts at the origin and follows the direction of the arrows. The point (i,j) is never reached before both (i-1,j) and (i,j-1).

pixels represented by RGB triples, or blocks of pixels, or the outputs of a preprocessing method such as a discrete cosine transform.

The error gradient of an MDRNN (that is, the derivative of some objective function with respect to the network weights) can be calculated with an n-dimensional extension of backpropagation through time (BPTT) [15]. As with one dimensional BPTT, the sequence is processed in the reverse order of the forward pass. At each timestep, the hidden layer receives both the output error derivatives and its own n 'future' derivatives. Figure 2 illustrates the BPTT backward pass for two dimensions. Again, care must be taken at the sequence boundaries.

At a point $\mathbf{x} = (x_1, \ldots, x_n)$ in an n-dimensional sequence, define $i_j^{\mathbf{x}}$ and $h_k^{\mathbf{x}}$ respectively as the activations of the jth input unit and the kth hidden unit. Define w_{kj} as the weight of the connection going from unit j to unit k. Then for an n-dimensional MDRNN whose hidden layer consists of H summation units, where θ_k is the activation function of hidden unit k, the forward pass for a sequence with dimensions $(X_1, X_2, \ldots, X_n)$ can be summarised as follows:

for $x_1 = 0$ to $X_1 - 1$ **do**
for $x_2 = 0$ to $X_2 - 1$ **do**
...
for $x_n = 0$ to $X_n - 1$ **do**
for $k = 1$ to H **do**
$a_k = \sum_j in_j^{(x_1,\ldots,x_n)} w_{kj}$
for $i = 1$ to n **do**
if $x_i > 0$ **then**
$a_k \mathrel{+}= \sum_j h_j^{(x_1,\ldots,x_i-1,\ldots,x_n)} w_{kj}$
$h_k^{\mathbf{x}} = \theta_k(a_k)$

Algorithm 1. MDRNN Forward Pass

Defining $\hat{o}_j^{\mathbf{x}}$ and $\hat{h}_k^{\mathbf{x}}$ respectively as the derivatives of the objective function with respect to the activations of the jth output unit and the kth hidden unit at point $\mathbf{x}$, the backward pass is:

for $x_1 = X_1 - 1$ to 0 **do**
for $x_2 = X_2 - 1$ to 0 **do**
...
for $x_n = X_n - 1$ to 0 **do**
for $k = 1$ to H **do**
$e_k \leftarrow \sum_j \hat{o}_j^{(x_1,\ldots,x_n)} w_{jk}$
for $i = 1$ to n **do**
if $x_i < X_i - 1$ **then**
$e_k \mathrel{+}= \sum_j \hat{h}_j^{(x_1,\ldots,x_i+1,\ldots,x_n)} w_{jk}$
$\hat{h}_k^{\mathbf{x}} \leftarrow \theta'_k(e_k)$

Algorithm 2. MDRNN Backward Pass

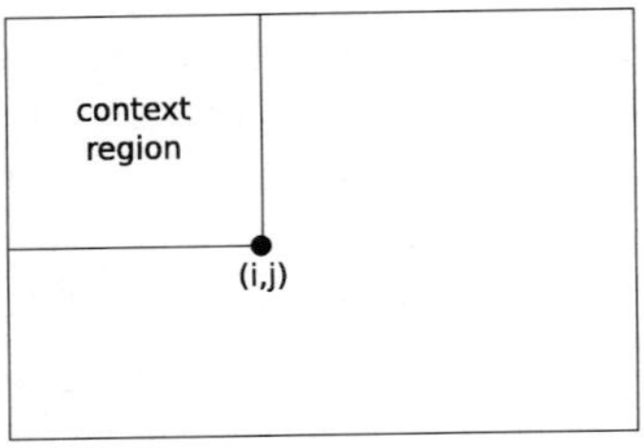

Fig. 4. Context available at (i,j) to a 2D RNN with a single hidden layer

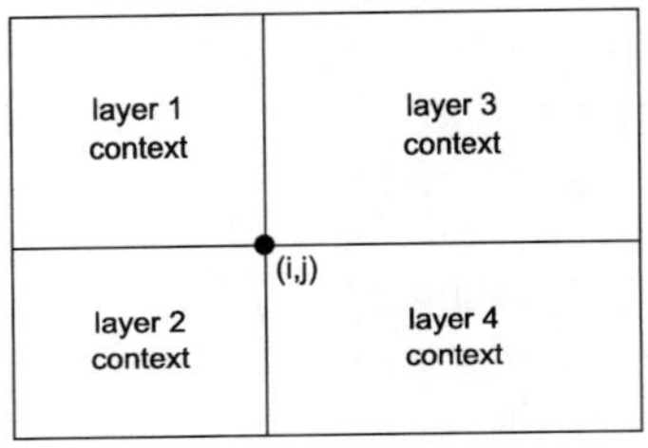

Fig. 5. Context available at (i,j) to a multi-directional 2D RNN

Since the forward and backward pass require one pass each through the data sequence, the overall complexity of MDRNN training is linear in the number of data points and the number of network weights.

2.1 Multi-directional MDRNNs

At a point $(x_1, ..., x_n)$ in the input sequence, the network described above has access to all points $(x'_1, ..., x'_n)$ such that $x'_i \leq x_i \forall\ i \in (1, ..., n)$. This defines an n-dimensional 'context region' of the full sequence, as illustrated in Figure 4. For some tasks, such as object recognition, this would in principle be sufficient. The network could process the image as usual, and output the object label at a point when the object to be recognized is entirely contained in the context region.

Intuitively however, we would prefer the network to have access to the surrounding context in all directions. This is particularly true for tasks where precise localization is required, such as image segmentation. For one dimensional RNNs, the problem of multi-directional context was solved by the introduction of bidirectional recurrent neural networks (BRNNs) [13]. BRNNs contain two separate hidden layers that process the input sequence in the forward and reverse directions. The two hidden layers are

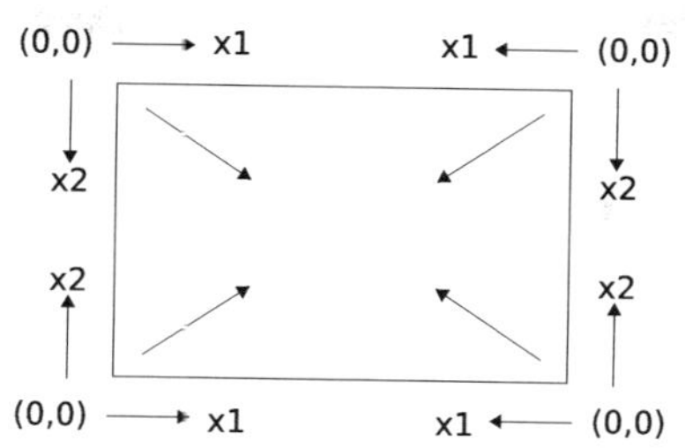

Fig. 6. Axes used by the 4 hidden layers in a multi-directional 2D network. The arrows inside the rectangle indicate the direction of propagation during the forward pass.

connected to a single output layer, thereby providing the network with access to both past and future context.

BRNNs can be extended to n-dimensional data by using 2^n separate hidden layers, each of which processes the sequence using the ordering defined above, but with a different choice of axes. The axes are chosen so that each one has its origin on a distinct vertex of the sequence. The 2 dimensional case is illustrated in Figure 6. As before, the hidden layers are connected to a single output layer, which now has access to all surrounding context (see Figure 5).

If the size of the hidden layers is held constant, the multi-directional MDRNN architecture scales as $O(2^n)$ for n-dimensional data. In practice however, the computing power of the network tends to be governed by the overall number of weights, rather than the size of the hidden layers. This is presumably because the data processing is shared between the layers. Since the complexity of the algorithm is linear in the number of parameters, the $O(2^n)$ scaling factor can be removed by simply using smaller hidden layers for higher dimensions. Furthermore, the complexity of a task, and therefore the number of weights likely to be needed for it, does not necessarily increase with the dimensionality of the data. For example, both the networks described in this paper have less than half the weights than the one dimensional networks we have previously applied to speech recognition [4,3]. For a given task, we have also found that using a multi-directional MDRNN gives better results than a uni-directional MDRNN with the same overall number of weights, as previously demonstrated in one dimension [4].

For a multi-directional MDRNN, the forward and backward passes through an n-dimensional sequence can be summarised as follows:

1: For each of the 2^n hidden layers choose a distinct vertex of the sequence, then define a set of axes such that the vertex is the origin and all sequence co-ordinates are ≥ 0
2: Repeat Algorithm 1 for each hidden layer
3: At each point in the sequence, feed forward all hidden layers to the output layer

Algorithm 3. Multi-Directional MDRNN Forward Pass

1: At each point in the sequence, calculate the derivative of the objective function with respect to the activations of output layer
2: With the same axes as above, repeat Algorithm 2 for each hidden layer

Algorithm 4. Multi-Directional MDRNN Backward Pass

2.2 Multi-dimensional Long Short-Term Memory

So far we have implicitly assumed that the network can make use of all context to which it has access. For standard RNN architectures however, the range of context that can practically be used is limited. The problem is that the influence of a given input on the hidden layer, and therefore on the network output, either decays or blows up exponentially as it cycles around the network's recurrent connections. This is usually referred to as the *vanishing gradient problem* [5].

Long Short-Term Memory (LSTM) [6,2] is an RNN architecture specifically designed to address the vanishing gradient problem. An LSTM hidden layer consists of multiple recurrently connected subnets, known as memory blocks. Each block contains a set of internal units, known as cells, whose activation is controlled by three multiplicative units: the input gate, forget gate and output gate. The effect of the gates is to allow the cells to store and access information over long periods of time, thereby avoiding the vanishing gradient problem.

The standard formulation of LSTM is explicitly one-dimensional, since the cell contains a single self connection, whose activation is controlled by a single forget gate. However we can easily extend this to n dimensions by using instead n self connections (one for each of the cell's previous states along every dimension) with n forget gates.

3 Experiments

3.1 Air Freight Data

The Air Freight database [12] is a ray-traced colour image sequence that comes with a ground truth segmentation based on textural characteristics (Figure 7). The sequence is 455 frames long and contains 155 distinct textures. Each frame is 120 pixels high and 160 pixels wide.

The advantage of ray-traced data is the true segmentation can be defined directly from the 3D models. Although the images are not real, they are realistic in the sense that they have significant lighting, specular effects etc.

We used the sequence to define a 2D image segmentation task, where the aim was to assign each pixel in the input data to the correct texture class. We divided the data at random into a 250 frame train set, a 150 frame test set and a 55 frame validation set. Note that we could have instead defined a 3D task where the network processed the entire video as one sequence. However, this would have left us with only one exemplar.

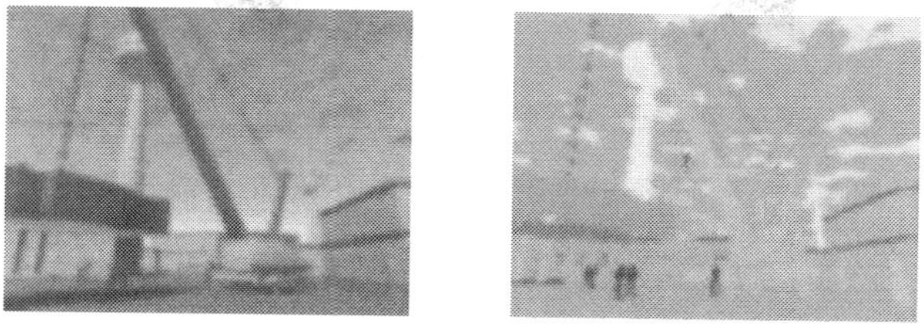

Fig. 7. Frame from the Air Freight database, showing the original image (left) and the colour-coded texture segmentation (right)

For this task we used a multi-directional MDRNN with 4 LSTM hidden layers. Each layer consisted of 25 memory blocks, each containing 1 cell, 2 forget gates, 1 input gate, 1 output gate and 5 peephole weights. This gave a total 600 hidden units. The input and output activation functions of the cells were both tanh, and the activation function for the gates was the logistic sigmoid in the range $[0, 1]$. The input layer was size 3 (one each for the red, green and blue components of the pixels) and the output layer was size 155 (one unit for each textural class). The network contained 43,257 trainable weights in total. As is standard for classification tasks, the softmax activation function was used at the output layer, with the cross-entropy objective function [1]. The network was trained using online gradient descent (weight updates after every training sequence) with a learning rate of 10^{-6} and a momentum of 0.9.

The final pixel classification error rate was $7.1\,\%$ on the test set.

3.2 MNIST Data

The MNIST database [10] of isolated handwritten digits is a subset of a larger database available from NIST. It consists of size-normalized, centered images, each of which is 28 pixels high and 28 pixels wide and contains a single handwritten digit. The data comes divided into a training set with 60,000 images and a test set with 10,000 images. We used 10,000 of the training images for validation, leaving 50,000 for training.

The usual task on MNIST is to label the images with the corresponding digits. This is benchmark task for which many algorithms have been evaluated.

We carried out a slightly modified task where each pixel was classified according to the digit it belonged to, with an additional class for background pixels. However, the original task can be recovered by simply choosing the digit whose corresponding output unit had the highest cumulative activation for the entire sequence.

To test the network's robustness to input warping, we also evaluated it on an altered version of the MNIST test set, where elastic deformations had been applied to every image (see Figure 8).

We compared our results with the convolution neural network that has achieved the best results so far on MNIST [14]. Note that we re-implemented the convolution network ourselves, and we did not augment the training set with elastic distortions, which gives a substantial improvement in performance.

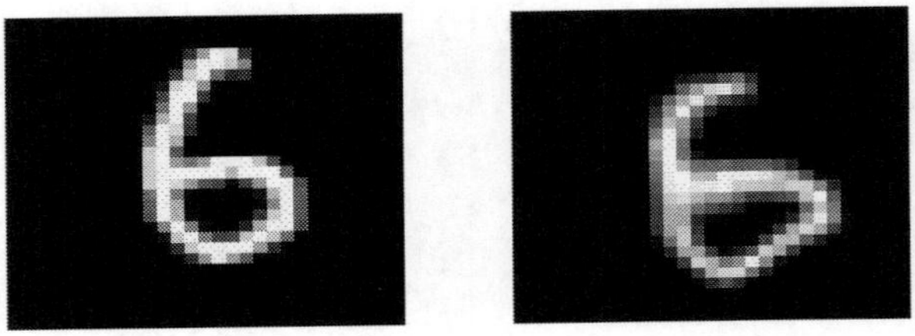

Fig. 8. MNIST image before and after deformation

The MDRNN for this task was identical to that for the Air Freight task with the following exceptions: the sizes of the input and output layers were now 1 (for grayscale pixels) and 11 (one for each digit, plus background) respectively, giving 27,511 weights in total, and the gradient descent learning rate was 10^{-5}.

For the distorted test set, we used the same degree of elastic deformation used by Simard [14] to augment the training set ($\sigma = 4.0$, $\alpha = 34.0$), with a different initial random field for every sample image.

Table 1 shows that the MDRNN matched the convolution net on the clean test set, and was considerably better on the warped test set. This suggests that MDRNNs are more robust to input warping than convolution networks.

Table 1. Image error rates on MNIST (pixel error rates in brackets)

Algorithm	Clean Test Set	Warped Test Set
MDRNN	0.9 % (0.4 %)	7.1 % (3.8 %)
Convolution	0.9 %	11.3 %

3.3 Analysis

One benefit of two dimensional tasks is that the operation of the network can be easily visualised. Figure 9 shows the network activations during a frames from the Air Freight database. As can be seen, the network segments this image almost perfectly, in spite of difficult, reflective surfaces such as the glass and metal tube running from left to right. Clearly, classifying individual pixels in such a surface requires the use of context.

An indication of the network's sensitivity to context can be found by analysing the derivatives of the network outputs at a particular point in the sequence with respect to the inputs at all other points. Collectively, we refer to these derivatives as the *sequential Jacobian*. Figure 10 shows the absolute value of the sequential Jacobian of an output during classification of an image from the MNIST database. It can be seen that the network responds to context from across the entire image, and seems particularly attuned to the outline of the digit.

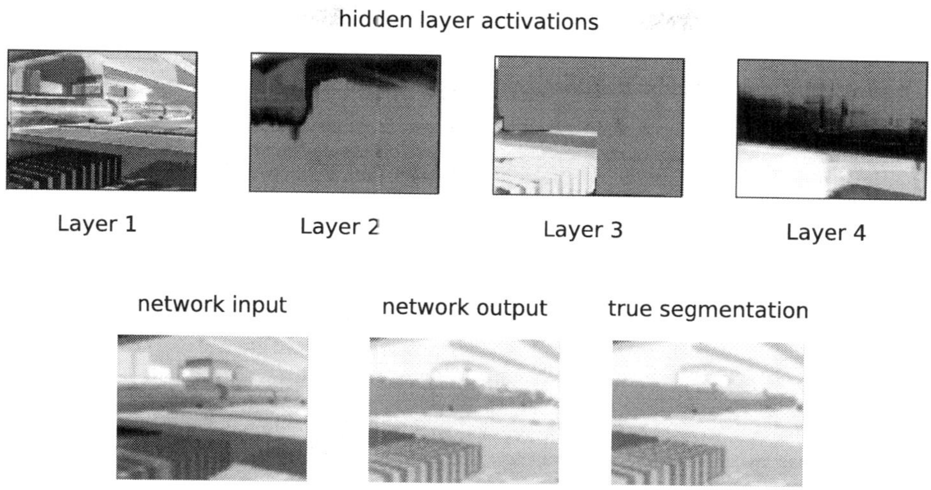

Fig. 9. 2D MDRNN applied to an image from the Air Freight database. The hidden layer activations display one unit from each of the layers. A common behaviour is to 'mask off' parts of the image, exhibited here by layers 2 and 3.

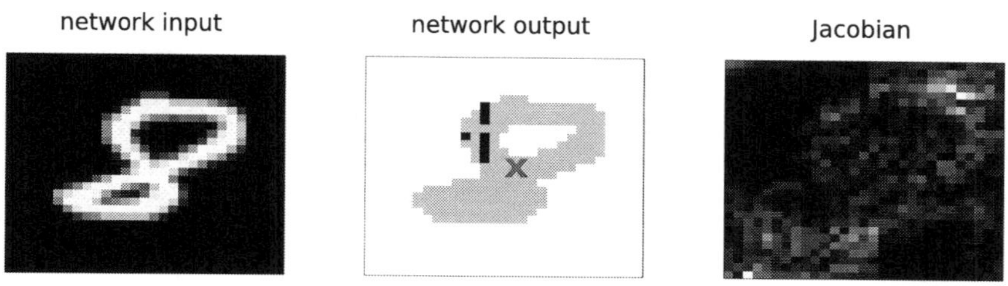

Fig. 10. Sequential Jacobian of a 2D RNN for an image from the MNIST database. The white outputs correspond to the class 'background' and the light grey ones to '8'. The black outputs represent misclassifications. The output pixel for which the Jacobian is calculated is marked with a cross. Absolute values are plotted for the Jacobian, and lighter colours are used for higher values.

4 Conclusion

We have introduced multi-dimensional recurrent neural networks (MDRNNs), thereby extending the applicabilty of RNNs to n-dimensional data. We have added multi-directional hidden layers that provide the network with access to all contextual information, and we have developed a multi-dimensional variant of the Long Short-Term Memory RNN architecture. We have tested MDRNNs on two image segmentation tasks, and found that it was more robust to input warping than a state-of-the-art digit recognition algorithm.

Acknowledgments

This research was funded by SNF grant 200021-111968/1.

References

1. Bridle, J.S.: Probabilistic interpretation of feedforward classification network outputs, with relationships to statistical pattern recognition. In: Fogleman-Soulie, F., Herault, J. (eds.) Neurocomputing: Algorithms, Architectures and Applications, pp. 227–236. Springer, Berlin (1990)
2. Gers, F., Schraudolph, N., Schmidhuber, J.: Learning precise timing with LSTM recurrent networks. Journal of Machine Learning Research 3, 115–143 (2002)
3. Graves, A., Fernández, S., Gomez, F., Schmidhuber, J.: Connectionist temporal classification: Labelling unsegmented sequence data with recurrent neural networks. In: Proceedings of the International Conference on Machine Learning, ICML 2006, Pittsburgh, USA (2006)
4. Graves, A., Schmidhuber, J.: Framewise phoneme classification with bidirectional LSTM and other neural network architectures. Neural Networks 18(5-6), 602–610 (2005)
5. Hochreiter, S., Bengio, Y., Frasconi, P., Schmidhuber, J.: Gradient flow in recurrent nets: the difficulty of learning long-term dependencies. In: Kremer, S.C., Kolen, J.F. (eds.) A Field Guide to Dynamical Recurrent Neural Networks, IEEE Computer Society Press, Los Alamitos (2001)
6. Hochreiter, S., Schmidhuber, J.: Long Short-Term Memory. Neural Computation 9(8), 1735–1780 (1997)
7. Hülsken, F., Wallhoff, F., Rigoll, G.: Facial expression recognition with pseudo-3d hidden markov models. In: Proceedings of the 23rd DAGM-Symposium on Pattern Recognition, London, UK, pp. 291–297. Springer, Heidelberg (2001)
8. Jiten, J., Mérialdo, B., Huet, B.: Multi-dimensional dependency-tree hidden Markov models. In: ICASSP 2006, 31st IEEE International Conference on Acoustics, Speech, and Signal Processing, Toulouse, France, May 14-19, 2006, IEEE Computer Society Press, Los Alamitos (2006)
9. Joshi, D., Li, J., Wang, J.Z.: Parameter estimation of multi-dimensional hidden markov models: A scalable approach. p. III, pp. 149–152 (2005)
10. LeCun, Y., Bottou, L., Bengio, Y., Haffner, P.: Gradient-based learning applied to document recognition. Proceedings of the IEEE 86(11), 2278–2324 (1998)
11. Li, J., Najmi, A., Gray, R.M.: Image classification by a two-dimensional hidden markov model. IEEE Transactions on Signal Processing 48(2), 517–533 (2000)
12. McCarter, G., Storkey, A.: Air Freight Image Segmentation Database, `http://homepages.inf.ed.ac.uk/amos/afreightdata.html`
13. Schuster, M., Paliwal, K.K.: Bidirectional recurrent neural networks. IEEE Transactions on Signal Processing 45, 2673–2681 (1997)
14. Simard, P.Y., Steinkraus, D., Platt, J.C.: Best practices for convolutional neural networks applied to visual document analysis. In: ICDAR '03: Proceedings of the Seventh International Conference on Document Analysis and Recognition, Washington, DC, USA, p. 958. IEEE Computer Society Press, Los Alamitos (2003)
15. Williams, R.J., Zipser, D.: Gradient-based learning algorithms for recurrent networks and their computational complexity. In: Chauvin, Y., Rumelhart, D.E. (eds.) Back-propagation: Theory, Architectures and Applications, pp. 433–486. Lawrence Erlbaum Publishers, Hillsdale, NJ (1995)

FPGA Implementation of an Adaptive Stochastic Neural Model

Giuliano Grossi and Federico Pedersini

Dipartimento di Scienze dell'Informazione
Università degli Studi di Milano
Via Comelico 39, I-20135 Milano, Italy
{grossi,pedersini}@dsi.unimi.it

Abstract. In this paper a FPGA implementation of a novel neural stochastic model for solving constrained NP-hard problems is proposed and developed. The hardware implementation allows to obtain high computation speed by exploiting parallelism, as the neuron update and the constraint violation check phases can be performed simultaneously.

The neural system has been tested on random and benchmark graphs, showing good performance with respect to the same heuristic for the same problems. Furthermore, the computational speed of the FPGA implementation has been measured and compared to software implementation. The developed architecture features dramatically faster computations with respect to the software implementation, even adopting a low-cost FPGA chip.

1 Introduction

The recent advances in design and manufacturing of integrated circuits, such as the programmable chips for fast prototyping (FPGA), allow to design hardware architectures that synthesize large neural networks consisting of hundreds of neural units. A direct consequence of these technological advances is the growing interest in computational models that take advantage of the massive parallelism achievable in hardware.

Here we derive a circuit of a neural model [1] based on a stochastic state transition mechanism and we apply it to the optimization of the cost function of particular NP-hard constrained combinatorial optimization problems ([2,3]). The model considers quadratic pseudo-boolean cost functions, i.e., mappings from the family of subsets of a finite ground set to the set of integers or reals. It focuses on such functions because there are a large number of combinatorial optimization problems which arise naturally or can be formulated easily as pseudo-boolean optimization problems [4].

The proposed model, described in Section 2, is based on a penalization strategy which adaptively modifies the weights, in such a way that the probability to find non-admissible solutions continuously decreases up to negligible values. The network performs a constraint-satisfaction search process that begins with "weak" constraints and then proceeds by gradually strengthening them with a

J. Marques de Sá et al. (Eds.): ICANN 2007, Part I, LNCS 4668, pp. 559–568, 2007.

penalty mechanism. We show that the network evolves, according to this strategy, until a solution is reached, that satisfies the problem constraints.

A simplified version of the algorithm has been developed in order to synthesize a hardware architecture implementing the algorithm. The designed hardware architecture is customizable to the network size and to the kind of problem. An implementation of this architecture, for the solution of the Maximum Independent Set on networks of up to 256 neurons, has been developed on a low-cost FPGA chip (Xilinx *Spartan 3*). The hardware design and some performance evaluation are presented in Section 3.

2 The Model

In this section we recall the adaptive stochastic recurrent network model presented in [1] starting from some preliminaries and definition on the contrained optimization.

2.1 Constrained Pseudo-Boolean Optimization

A family of functions that often plays an important role in optimization models are the so-called *pseudo-Boolean functions* [4]. Their polynomial representation, with particular regard to the quadratic and symmetric one, corresponds in a natural way to the objective (or cost) function of many optimization problems. Our aim is twofold: to use these functions as cost, but also to discriminate admissible solutions from non-admissible ones, by introducing suitable pseudo-Boolean penalty functions. To this purpose, we will use equality quadratic constraints based on two independent variables and expressed as Boolean functions.

Let $\mathbf{B}^n$ be a vector space of n-tuples of elements from $\mathbf{B} = \{0, 1\}$. Mappings $f : \mathbf{B}^n \to \mathbf{R}$ are called pseudo-Boolean functions. Let $V = \{1, \ldots, n\}$ be a set containing n elements: since there is a one-to-one correspondence between the subsets of V and the binary vectors of $\mathbf{B}^n$, these functions are *set functions*, that is, mappings that associate a real value to every subset S of V.

In the following, we will represent such pseudo-Boolean functions by means of multi-linear polynomials having the form:

$$f(x_1, \ldots, x_n) = \sum_{S \subseteq V} c_S \prod_{k \in S} x_k,$$

where c_S are integer coefficients and, by convention, $\prod_{k \in \varnothing} x_k = 1$. The size of the largest subset $S \subseteq V$ for which $c_S \neq 0$ is called the *degree* of f, and is denoted by deg (f). Naturally, a pseudo-Boolean function f is *quadratic* if deg $(f) \leq 2$.

For many optimization problems P the set of admissible solutions Sol_P is a proper subset of the search space $\mathbf{B}^n$ on which the pseudo-Boolean functions are defined, i.e., $\mathrm{Sol}_P \subset \mathbf{B}^n$. In order to use our space-independent general technique to find optima, we need to introduce constraints as base ingredients to succeed in finding an admissible solution.

Frequently, for an admissible solution $S \in \mathrm{Sol}_P$, the constraints are given in terms of binary Boolean functions $\phi : \mathbf{B}^2 \to \mathbf{B}$ evaluated on a fixed set of pairs $T_S = \{(x_i, x_j) | (x_i, x_j) \in \mathbf{B}^2 \ \wedge \ i \neq j \in [1..n]\}$ of size m. Thus, a given constraint for S is satisfied when the following conjunction:

$$\bigwedge_{\mathbf{b}_k \in T_S} \phi(\mathbf{b}_k) = true.$$

By denoting with g the multi-linear polynomial associated to ϕ, such that $g : \mathbf{B}^2 \to \mathbf{B}$ and $g(\mathbf{b}_k) = 1 \Leftrightarrow \phi(\mathbf{b}_k) = true$, the previous satisfiability condition is transformed into the following maximization condition:

$$\sum_{(x_i, x_j) \in T_S} g(x_i, x_j) = m, \tag{1}$$

In order to illustrate the feasibility of mapping a broad class of combinatorial problems into a discrete neural model, we formalize a *maximization* version of a problem P with the following linear objective function subject to quadratic constraints:

$$\begin{aligned} \text{maximize} \quad & \sum_{i=1}^{n} x_i \\ \text{subject to} \quad & \sum_{(x_i, x_j) \in T_S} g(x_i, x_j) = m. \end{aligned}$$

The same can be easily derived for *minimization* problems.

To derive an energy function whose maximum value corresponds to the "best" solution for P, we combine both the cost and the penalty function in the same objective function, obtaining the general form

$$\begin{aligned} f(x_1, \ldots, x_n) &= \alpha \sum_{i=1}^{n} x_i + \sum_{(x_i, x_j) \in T_S} \gamma_k g_k(x_i, x_j) \\ &= \alpha \sum_{i=1}^{n} x_i + \sum_{(x_i, x_j) \in T_S} \gamma_k (a_k x_i + b_k x_j + c_k x_i x_j + d_k) \\ &= \sum_{i=1}^{n} \lambda_i x_i + \sum_{i<j} w_{ij} x_i x_j + C, \end{aligned} \tag{2}$$

where $\alpha, \gamma_1, \ldots, \gamma_m$ are positive integer parameters, useful in finding better solutions (we will explain their role later) while preserving optima; $(w_{ij})_{n \times n}$ is an integer symmetric matrix with null diagonal, $\boldsymbol{\lambda}$ is an integer vector and C an integer constant.

In the next section, we will interpret $(w_{ij})_{n \times n}$ and $\boldsymbol{\lambda}$, respectively, as weights and thresholds of a neural network used to find optimal solutions.

2.2 Examples

In this section we recall some combinatorial optimization problems, which arise naturally in the area of algorithmic graph theory, and can be easily formulated as pseudo-Boolean optimization problems.

All graphs $G = \langle V, E\rangle$ considered are arbitrary undirected graphs, where $V = \{1, \ldots, n\}$ is the set of vertices and $E \subseteq V \oplus V$ (not ordered pairs) is the set of edges. Two distinct vertices i and j are called *adjacent* if they are connected by an edge; the set $\mathcal{N}(i)$ denotes the *neighbours* of vertex i, i.e., the vertices that are adjacent to i.

Max Clique (Max Independent Set). It consists in finding the largest cardinality subset of vertices which are pairwise connected, i.e., adjacent. This problem can be formulated as

$$\begin{aligned} \text{maximize} \quad & \alpha \sum_{i \in V} x_i \\ \text{subject to} \quad & \sum_{\{i,j\} \in E^c} (\bar{x}_i \vee \bar{x}_j) = |E^c|, \end{aligned}$$

where E^c denotes the complement of E.

Min Vertex Cover. A vertex cover for a graph G is a subset of vertices such that each edge has at least one end-point in the subset. The objective function of this problem is:

$$\begin{aligned} \text{minimize} \quad & \alpha \sum_{i \in V} x_i \\ \text{subject to} \quad & \sum_{\{i,j\} \in E} (x_i \vee x_j) = |E|. \end{aligned}$$

Max k-colorable Induced Subgraph. This problem can be translated into the search of a maximum independent set (problem (3)) on the expanded graph $\tilde{G} = \langle \tilde{V}, \tilde{E}\rangle$ built as follows: for each $i \in V$ the set $\{i_1, \ldots, i_k\} \subseteq \tilde{V}$ and for each $\{i, j\} \in E$ the sets $\{\{i_p, j_p\} \mid 1 \leq p \leq k\}$, and $\{\{i_p, i_q\} \mid 1 \leq p, q \leq k \ \wedge \ p \neq q\}$ belong to $\tilde{E}$ (see Fig. 1 for an example).

2.3 Architecture and Dynamics

The network architecture is derived from the problem (for instance, the graph topology), with the set of neurons (or *units*) isomorphic to the set of binary variables representing the solutions, and the connections between neurons derived according to the logical dependencies between variable pairs. The absence of such a connection between a variable pair implies neither strength in the synaptic connection between neurons, nor constraints between the variables themselves.

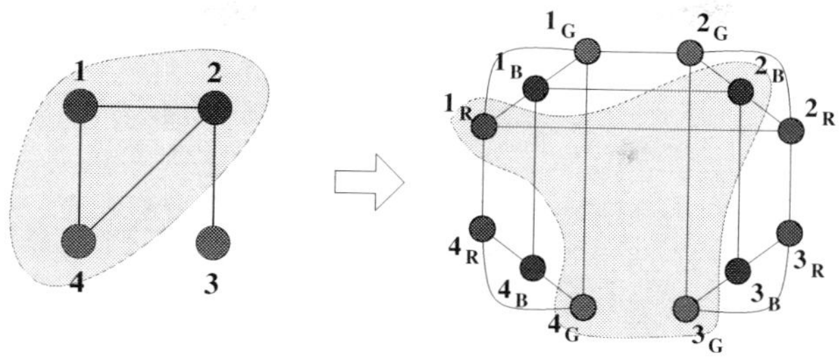

Fig. 1. Graph 3-colorable problem mapped into the maximum independent set problem

Each unit i $(1 \leq i \leq n)$ is stochastic, assuming the state $x_i = 1$ with probability $\phi_\beta(h_i)$ and $x_i = 0$ with probability $1 - \phi_\beta(h_i)$, where

$$h_i = \lambda_i + \sum_{j \in \mathcal{N}(i)} w_{ij} x_j \quad \text{and} \quad \phi_\beta(h) = \frac{1}{1 + e^{-\beta h}}, \tag{3}$$

which is the usual *Glauber* [5] or *logistic* sigmoid-shaped function with parameter $\beta > 0$.

We adopt as energy the quadratic function expressed in (2) and we use the stochastic activation function described above, in order to derive stochastic dynamics able to lead the system status in the saturation region of the sigmoid function.

We can therefore interpret the pseudo-Boolean quadratic function

$$E(x_1, \ldots, x_n) = \sum_{i \in V} \lambda_i x_i + \sum_{i<j} w_{ij} x_i x_j \tag{4}$$

as energy of the network that must be maximized by means of "thermal fluctuations".

By adopting the saturated-nonlinear activation function (3) and letting the network evolve with asynchronous dynamics, the resulting dynamic system locally minimizes the network energy with high probability. However, we are interested in a process that, while minimizing the energy, also guarantees that the final stable state corresponds to a valid solution. For this reason, the proposed process alternates the following two phases called *units updating* and *weights strengthening*:

Units updating *(Phase 1.)*. Asynchronously update the units $i \in [1..n]$ with the following stochastic dynamics:

$$x_i = \begin{cases} 1 & \text{with probability } \phi_\beta(h_i) \\ 0 & \text{with probability } 1 - \phi_\beta(h_i) \end{cases}.$$

Weights strengthening *(Phase 2.)*. Increase the weights w_{ij} and the thresholds λ_i, λ_j of the connections between the units i and j that violate the problem constraints $g(x_i, x_j)$. Many different penalization strategies can be defined: here we update only those connections that cause violations in the current state. This strategy can be formalized as follows ($k \in [1..m]$):

$$\text{if} \quad g(x_i, x_j) = 0 \quad \Rightarrow \quad \gamma_k = \gamma_k + 1. \tag{5}$$

Since for each i, the relationship between the (3) and the (4) is expressed by

$$\begin{aligned} h_i(x_1, \ldots, x_n) &= \Delta E_i = \frac{\partial}{\partial x_i} E(x_1, \ldots, x_n) \\ &= E(x_1, \ldots, x_{i-1}, 1, x_{i+1}, \ldots, x_n) - E(x_1, \ldots, x_{i-1}, 0, x_{i+1}, \ldots, x_n). \end{aligned}$$

thus, flipping the state of the unit i produces a variation of energy $\Delta E_i = h_i(x_1, \ldots, x_n)$, which does not depend on x_i. This energy dependency from h_i is just used to force the convergence toward feasible solution, because h_i is a function of those constraints containing the variable x_i. Thus, changing the absolute value of γ_k (as in (5)) results in a changing of the sign of h_i.

The algorithm ASNM (Adaptive Stochastic Neural Model), that iterates the simulation of the neural network behaviour and the penalization, is sketched in the Algorithm 1.

Algorithm 1. ASNM

```
Require: An initial matrix W, a vector λ and an integer α > 0
  while (there are violations) do
    for all i = 1 to n do
      h_i = λ_i − Σ_{k∈N(i)} w_ik x_k;
      x_i = { 1, with probability φ(h_i)
            { 0, with probability 1 − φ(h_i)
    end for
    for all k = 1 to m do
      if (g_k not satisfied) then
        γ_k = γ_k + 1
      end if
    end for
  end while
Ensure: admissible solution (characteristic vector) x
```

It has to be proved that this algorithm stops when a stable state of a suitable network is reached, where the related subset of vertices is a maximal solution of the problem.

This statement is summarized in the following

Theorem 1. *Let* $\mathbf{x}(t)$ *be the state found by the algorithm at time* t *and let* $Z(t)$ *be the random variable denoting the number of violations at the same time. Then,*

$$\lim_{t \to +\infty} \mathbf{E}[Z(t)] = 0,$$

where $\mathbf{E}[\cdot]$ *denotes the expectation of a random variable.*

Moreover, regarding the optimality of the feasible solution found it can be stated that:

Corollary 1. *Let $\boldsymbol{x}$ be the stable state of the network found by the* ASNM *algorithm. Then $\boldsymbol{x}$ represents, with high probability, an optimal solution, i.e., a suboptimal solution but not subset of another one.*

3 The Hardware Architecture

3.1 Description of the Architecture

Figure 2 shows the block diagram of the proposed architecture. Depending on the kind of problem to be solved, a different *Violation Detection Logic* has to be used, according to the problem's constraints. Figure 3 shows the circuit used to detect the constraint violations for Maximum Independent Set problems. The instance of the particular problem to be solved is represented by the *Adjacency Matrix* $A = \{a_{ij}\}$. In this architecture, A is stored in a memory bank organized as a N-by-N matrix. The coefficients a_{ij} have to be stored in the matrix before the algorithm starts. The dynamically-evolving solution vector is stored in the Status Register (*statusreg*), a N-bit register containing the value of each node,

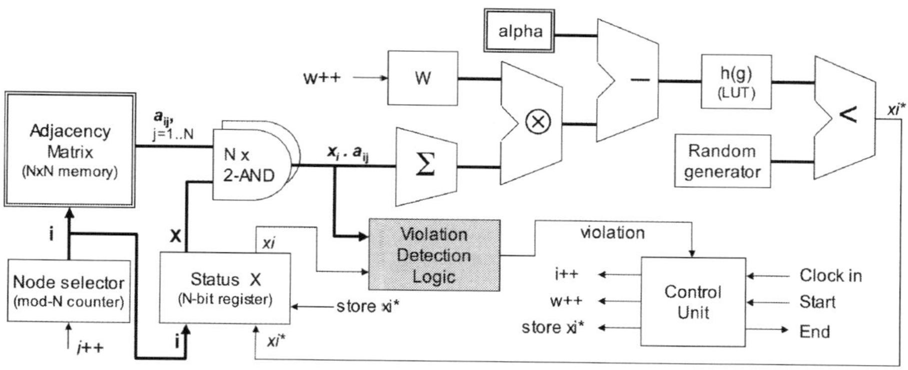

Fig. 2. Block diagram of the proposed architecture. Depending on the kind of problem to solve, the corresponding Violation Detection Logic has to be inserted. The Adjacency Matrix and α are provided as input, prior to algorithm execution.

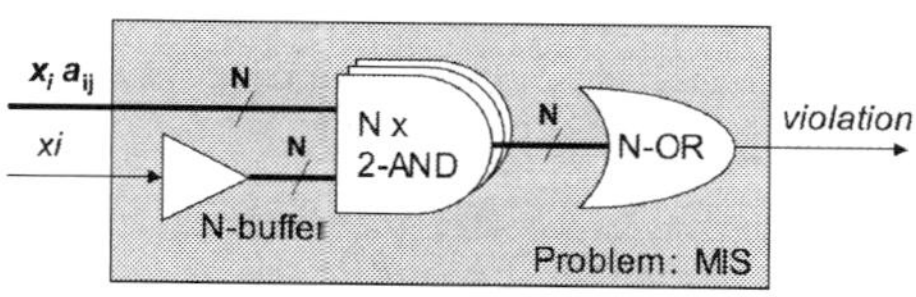

Fig. 3. The Violation Detection Logic corresponding to the Maximum Independent Set problem

x_i. The solution vector initalization depends on the problem to solve: for the MIS, it is initialized to $x_i = 1, \forall i$.

Each iteration of the main computation cycle (the "while" cycle in the ASNM algorithm) is carried out by scanning twice the graph nodes. The first scanning corresponds to the *Units Updating* phase: the current solution is updated, computing a new value for $x_i, \forall i$. In the second one, called *Violation Test* phase, the whole graph is visited, in order to detect any constraint violation. The scannings are performed by letting the Node Selector range from 0 to $N-1$, whereby representing i.

During the *Units Updating* phase, a bank of N 2-input AND gates computes the N products $a_{ij}x_j, \forall j$. These signals are fed to the Addition Unit, that computes $\sum_j a_{ij}x_j$ by counting the "1"s contained in the input. Since the result of the sum can range from 0 to $N = 2^m$, the output of this unit needs $m+1$ bit for its representation. The multiplier then computes $\sum_j Wa_{ij}x_j$, where the value of W is stored in an m-bit register-counter. At each iteration, the value of W is incremented by 1, thus increasing the weight for all nodes. As the computed product is represented by $2m$ bits, the constant α is $2m$ bits wide as well. α is provided as a constant in input, like the matrix A. The subtractor computes then $h_i = \alpha - \sum_j Wa_{ij}x_j$, for the $i-th$ node. From the $2m$-bit representation of h_i, an 8-bit representation is extracted, that best represents the range of the transition zone in the sigmoid function $g(h_i)$; by using 8 bits, the significant range of h_i is therefore quantized into 256 intervals, largely enough for an accurate evaluation of $g(\cdot)$. h_i is the input of the Look-up Table (implemented with a 256x8 ROM) containing a sampling of the function $g(\cdot)$. Its output represents the value of $g(h_i)$, also quantized into 256 steps. Each value of $g(h_i)$ is then compared to an 8-bit random value generated by a pseudo-random sequence generator. The period of the generator has to be large enough to avoid repetition of the sequences before the end of the computation. This is anyway not critical at all, as the complexity of the circuit increases as $\log_2 P$, being P the period. The result of the comparator corresponds to the new value for the $i-th$ node x_i^* and will be then stored in the Status Register.

After completion of the *Units Updating* phase, the Control Unit starts the *Violation Test* phase: for each i ranging from 0 to $N-1$, the Violation Detection logic checks for constraint satisfaction. Considering the case of the MIS problem, the Violation Detection logic, shown if Figure 3, computes $x_i \bigvee_j a_{ij}x_j$, that results to be 1 in case of violation. If this happens, the Control Unit stops the violation test, performs the *Weights Strengthening* phase, by incrementing the value of W by 1, and starts a new iteration of the main computation cycle. Otherwise, if the violation test phase completes without violation, the Control Unit stops the process: this means that the algorithm is done and the Register Status contains a valid solution.

3.2 FPGA Implementation

The proposed architecture has been developed on a low-cost FPGA chip (Xilinx's *Spartan 3* series); the adopted chip (XC3S-400) contains 400,000 system gates. Figure 4 shows the schematic of an implementation of the architecture. On this chip, the largest architecture that could be hosted is capable to handle a maximum graph size of 256 nodes ($N = 256$), as reported in Table 1. However, being the architecture design almost completely scalable, it is very simple to extend the architecture to larger values of N, in order to solve the problem on larger graphs. Considering for instance the adoption of high-end FPGA chips, like Xilinx's *Virtex* series, the proposed architecture can be easily extended to $N = 1024$ or more [6]. In fact, the main bottleneck on complexity, in terms of number of gates, is the memory bank containing A, whose size increases as N^2. The complexity of the other circuits grows at most linearly with N.

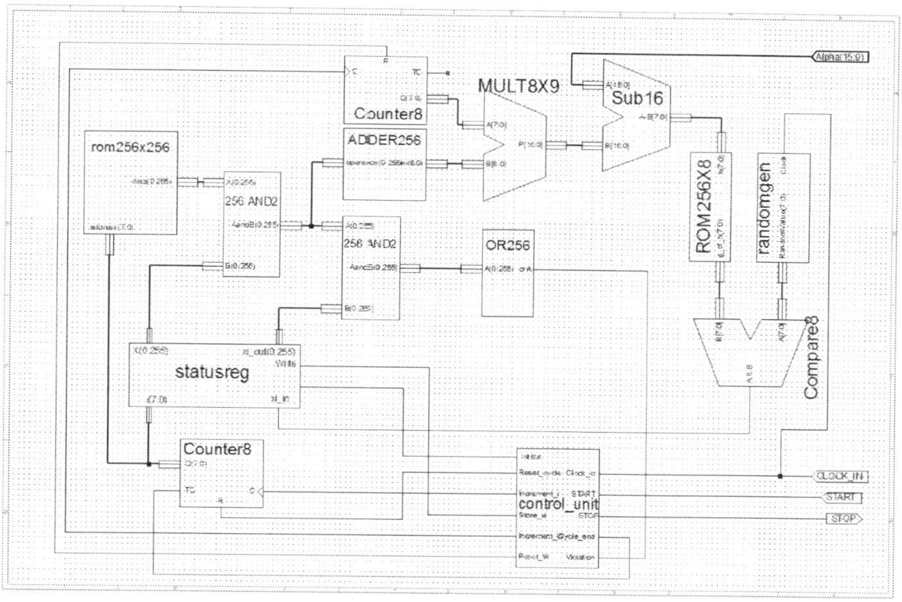

Fig. 4. The schematic of the implemented architecture, for MIS problems, $N = 256$, as implemented on a Xilinx FPGA chip

3.3 Computational Performance

The timing constraints of the implemented circuit, computed by the *place&route* process on the *Spartan 3* chip, allow a maximum clock frequency $f_{MAX} = 21.578MHz$. The circuit has been therefore tested with a conservative clock frequency $f_C = 20MHz$. In order to evaluate the computational speed of the developed architecture, the computation times have been compared to those obtained with a software implementation of the algorithm on a PC (Pentium P4, 2.8 GHz). The values reported in Table 2 are mean values obtained over 100 randomly-generated graphs with 256 nodes and different densities. As the table shows, the proposed architecture featured a speed-up ratio, over the software execution, of about 90 times.

Table 1. Synthesis report corresponding to the implementation of the described architecture (MIS problem, $N = 256$) on a *Xilinx Spartan3-XC400* FPGA chip

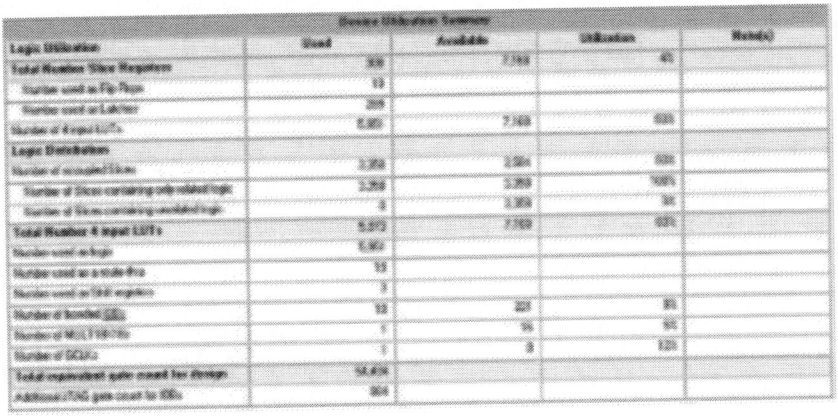

Table 2. Computing times of the algorithm (MIS, $N = 256$), executed by software and by the implemented architecture. The average speed-up ratio is about 90.

N	Graph Density	Iterations	Time (SW)	Time (FPGA@20MHz)
256	0.2	5.09	8.73 ms	0.097 ms
256	0.5	9.71	16.7 ms	0.186 ms
256	0.8	24.72	42.4 ms	0.475 ms

References

1. Grossi, G.: A discrete adaptive stochastic neural model for constrained optimization. In: International Conference on Artificial Neural Networks - ICANN (1), pp. 641–650 (2006)
2. Garey, M.R., Johnson, D.S.: Computers and Intractability. A Guide to the Theory of NP-Completeness. W. H. Freeman & Co., San Francisco, CA (1979)
3. Karp, R.M.: Complexity of Computer Computations. In: Reducibility among Combinatorial Problems, pp. 85–103. Plenum Press, New York (1972)
4. Boros, E., Hammer, P.L.: Pseudo-boolean optimization. DAMATH: Discrete Applied Mathematics and Combinatorial Operations Research and Computer Science 123 (2002)
5. Glauber, R.J.: Time-dependent statistics of the Ising model. Journal of Mathematical Physics 4(2), 294–307 (1963)
6. Patel, H.: Synthesis and implementation strategies to accelerate design performances. Technical report, Xilinx (2005)

Global Robust Stability of Competitive Neural Networks with Continuously Distributed Delays and Different Time Scales

Yonggui Kao* and QingHe Ming

Department of Mathematics, ZaoZhuang University,
Zaozhuang 277160, People's Republic of China
ygkao2006@yahoo.com.cn

Abstract. The dynamics of cortical cognitive maps developed by self-organization must include the aspects of long and short-term memory. The behavior of such a neural network is characterized by an equation of neural activity as a fast phenomenon and an equation of synaptic modification as a slow part of the neural system, besides, this model bases on unsupervised synaptic learning algorithm. In this paper, using theory of the topological degree and strict Liapunov functional methods, we prove existence and uniqueness of the equilibrium of competitive neural networks with continuously distributed delays and different time scales, and present some new criteria for its global robust stability.

1 Introduction

Dynamic neural networks, which contain both feedforward and feedback connections between the neural layers, play an important role in visual processing, pattern recognition, neural computing, and control. Moreover, biological networks possess synapses whose synaptic weights vary in time. Thus, a competitive neural network model with a combined activity and weight dynamics have been paid much attention [1−4]. These neural network models describe the dynamics of both the neural activity levels, the short-term memory (STM), and the dynamics of unsupervised synaptic modifications, the long-term memory (LTM). That is, there are two time scales in these neural networks, one corresponds to the fast changes of the neural network states, another corresponds to the slow changes of the synapses by external stimuli. The neural network can be mathematically expressed as

$$\text{STM}: \quad \epsilon \dot{x}_j(t) = -a_j x_j(t) + \sum_{i=1}^{N} D_{ij} f(x_i(t)) + B_j \sum_{i=1}^{p} m_{ij}(t) y_i \tag{1}$$

$$\begin{aligned} \text{LTM}: \quad & \dot{m}_{ij}(t) = -m_{ij}(t) + y_i f(x_j(t)) \\ & i = 1, 2, \cdots, p, \quad j = 1, 2, \cdots, N \end{aligned} \tag{2}$$

* Corresponding author.

J. Marques de Sá et al. (Eds.): ICANN 2007, Part I, LNCS 4668, pp. 569–578, 2007.

where $x_j(t)$ is the neuron current activity level, $f(x_j(t))$ is the output of neurons, $m_{ij}(t)$ is the synaptic efficiency, y_i is the constant external stimulus, D_{ij} represents the connection weight between the ith neuron and the jth neuron, $a_j > 0$ is a positive function, representing the time constant of the neuron, B_j is the strength of the external stimulus, ϵ is the time scale of STM state[2].

The competitive neural networks with different time scales are extensions of Grossbergs shunting network [5] and Amaris model for primitive neuronal competition [6]. For neural network models without considering the synaptic dynamics, their stability have been extensively analyzed, Cohen and Grossberg [7] found a Lyapunov function for such a neural network, and derived some sufficient conditions ensuring absolute stability. For the competitive neural network with two time scales, the stability was studied in [1-4]. In [1], the theory of singular perturbations was employed to derive a condition for global asymptotic stability of the neural networka (1) and (2). In [2], the theory of flow invariance was used to prove the boundedness of solutions of the neural network, and a sufficient condition was derived based on Lyapunov method. A special case of these neural networks was given in [8]. In [4], a single discrete delay was introduced into a generalized multi-time scale competitive neural network model with nonsmooth functions, and the conditions for global exponential stability of the neural networks were derived.

In the applications and designs of neural networks, neural networks usually has a spatial extent due to the presence of an amount of parallel pathways with a variety of axon sizes and lengths. Thus, the delays in artificial neural networks are usually continuously distributed [9-11, 24-32]. Moreover, there are often some unavoidable uncertainties such as modelling errors, external perturbations and parameter fluctuations, which can cause the network to be unstable. So, it is essential to investigate the globally robust stability of the network with errors and perturbations. To the best of our knowledge, only some limited works are done on this problem [12-20]. In [12], the authors derived a novel result based on a new matrix-norm inequality technique and Lyapunov method, and gave some comments on those in [17]. In [13], the authors viewed the uncertain parameters as perturbations and gave some testable results for robust stability. In [14], several criteria are presented for global robust stability of neural networks without delay. In [15,16], global robust stability of delayed interval Hopfield neural networks is investigated with respect to the bounded and strictly increasing activation functions. Several M-matrix conditions to ensure the robust stability are given for delayed interval Hopfield neural networks. But, in all these models, only neural activation dynamics is considered, the synaptic dynamics has not involved.

So, in this paper, we introduce the continuously distributed delays into the competitive neural networks (1) and (2) with different time scales, where both neural activation and synaptic dynamics are taken into consideration. To our knowledge, such neural networks with continuously distributed delays have never been reported in the literature. We allow the model has non-differentiable and unbounded functions, and use the theory of the topological degree and strict

Liapunov functional methods to prove the existence and uniqueness of the equilibrium, and derive a new sufficient condition ensuring global robust stability of the networks.

Consider a competitive neural network model with continuously distributed delays described by the following differential-difference equations

$$\begin{aligned}
\text{STM}: \quad & \epsilon\dot{x}_i(t) = -a_i x_i(t) + \sum_{j=1}^{N} D_{ij} f_j(x_j(t)) \\
& \quad + \sum_{j=1}^{N} D_{ij}^T f_j\left(\int_{-\infty}^{t} K_{ij}(t-s)x_j(s)ds\right) + B_i \sum_{j=1}^{p} m_{ij}(t) y_j \qquad (3) \\
\text{LTM}: \quad & \dot{m}_{ij}(t) = -m_{ij}(t) + y_j f_i\left(\int_{-\infty}^{t} K_{ij}(t-s)x_i(s)ds\right) \qquad (4) \\
& i = 1, 2, \cdots, N, \quad j = 1, 2, \cdots, p
\end{aligned}$$

where m_{ij} is the synaptic efficiency, y_i is the external stimulus, D_{ij}^T represents the synaptic weight of delayed feedback. For convenience of expression, we interchange the indexes i and j in (1) and (2), symbols appeared in (1) and (2) have the same meaning, but the output of each neuron is allowed to have different relations $f_j(x_j(t))$, and the delay kernels $K_{ij} : [0, \infty) \rightarrow R$ are continuous, integrable and there exit nonnegative constants k_{ij} such that

$$\int_0^{\infty} |K_{ij}(u)| du \leq k_{ij}, \int_0^{\infty} u K_{ij}(u) du < \infty.$$

After setting $S_i(t) = \sum_{j=1}^{N} m_{ij}(t) y_j = \mathbf{y}^T \mathbf{m}_i(t)$, $\mathbf{y} = (y_1, y_2, \cdots, y_p)^T$, $\mathbf{m}_i(t) = (m_{1i}, m_{2i}, \cdots, m_{pi})^T$, and summing up the LTM (4) over j, the network (3) and (4) can be rewritten as the state-space form

$$\begin{aligned}
\text{STM}: \quad & \epsilon\dot{x}_i(t) = -a_i x_i(t) + \sum_{j=1}^{N} D_{ij} f_j(x_j(t)) \\
& \quad + \sum_{j=1}^{N} D_{ij}^T f_j\left(\int_{-\infty}^{t} K_{ij}(t-s)x_j(s)ds\right) + B_i S_i(t) \qquad (5) \\
\text{LTM}: \quad & \dot{S}_{ij}(t) = -S_i(t) + |\mathbf{y}|^2 f_i\left(\int_{-\infty}^{t} K_{ij}(t-s)x_i(s)ds\right) \qquad (6) \\
& i = 1, 2, \cdots, N, \quad j = 1, 2, \cdots, N
\end{aligned}$$

without loss of generality, the input stimulus vector y is assumed to be normalized with unit magnitude $|y|^2 = 1$. , and the fast time scale parameter ϵ is also assumed to be unit, then the network is simplified to

$$\begin{aligned}
\text{STM}: \quad & \dot{x}_i(t) = -a_i x_i(t) + \sum_{j=1}^{N} D_{ij} f_j(x_j(t)) \\
& \quad + \sum_{j=1}^{N} D_{ij}^T f_j\left(\int_{-\infty}^{t} K_{ij}(t-s)x_j(s)ds\right) + B_i S_i(t) \qquad (7) \\
\text{LTM}: \quad & \dot{S}_i(t) = -S_i(t) + f_i\left(\int_{-\infty}^{t} K_{ij}(t-s)x_i(s)ds\right) \qquad (8) \\
& i = 1, 2, \cdots, N, \quad j = 1, 2, \cdots, N
\end{aligned}$$

For convenience, Let

$$x_{i+N}(t) = s_i(t), \quad i = 1, 2, \cdots, N$$

$$W_{ij} = \begin{cases} D_{ij}, & 1 \leq i, j \leq N; \\ B_i, & 1 \leq i \leq N, j = N + i; \\ 0, & else \end{cases}$$

$$W_{ij}^T = \begin{cases} D_{ij}^T, & 1 \leq i, j \leq N; \\ 1, & N < i \leq 2N, i = N + j; \\ 0, & else \end{cases}$$

$$a_i = \begin{cases} a_i, & 1 \leq i \leq N; \\ 1, & N + 1 \leq i \leq 2N, \end{cases}$$

$$f_i(x) = \begin{cases} f_i(x), & 1 \leq i \leq N; \\ x_i, & N + 1 \leq i \leq 2N \end{cases}$$

then we can rewrite the system (7) and (8) as the following models:

$$\dot{x}_i(t) = -a_i x_i(t) + \sum_{j=1}^{2N} W_{ij} f_j(x_j(t)) + \sum_{j=1}^{2N} W_{ij}^T f_j\left(\int_{-\infty}^{t} K_{ij}(t-s) x_j(s) ds\right) \qquad (9)$$
$$i = 1, 2, \cdots, 2N.$$

Throughout this paper,we always suppose that:

$$\begin{cases} A_I = \{A = diag(a_1, a_2, \cdots, a_{2N}) : \underline{A} \leq A \leq \overline{A}, \text{i,e}, \\ \qquad \underline{a}_i \leq a_i \leq \bar{a}_i, \ i = 1, 2, \cdots, 2N, \forall\, A \in A_I, \} \\ W_I = \{W = (W_{ij})_{2N \times 2N} : \underline{W} \leq W \leq \overline{W}, \text{i,e}, \\ \qquad \underline{W}_{ij} \leq W_{ij} \leq \overline{W}_{ij}, \ i = 1, 2, \cdots, 2N, \forall\, W \in W_I, \} \\ W_I^T = \{W^T = (W_{ij}^T)_{2N \times 2N} : \underline{W}^T \leq W^T \leq \overline{W}^T, \text{i,e}, \\ \qquad \underline{W}_{ij}^T \leq W_{ij}^T \leq \overline{W}_{ij}^T, \ i = 1, 2, \cdots, 2N, \forall\, W^T \in W_I^T, \} \end{cases} \qquad (10)$$

In the sequel, we first apply the theory of the topological degree to derive a sufficient condition guaranteeing existence of the equilibrium for the competitive neural networks with distributed delays, then we prove that the condition for existence and uniqueness of equilibrium also ensure the global robust stability of the competitive neural networks. In the remaining sections of this paper, we need the following matrix notations:

$$x = (x_1, x_2, \cdots, x_{2N})^T, \quad [x]^+ = (|x_1|, |x_2|, \cdots .|x_{2N}|)^T, \ B^+ = (|b_{ij}|)_{2N \times 2N},$$
$$B = (b_{ij})_{2N \times 2N} > 0; \quad b_{ij} > 0. \quad B = (b_{ij})_{2N \times 2N} \geq 0; \quad b_{ij} \geq 0.$$

2 Main Results and Proof

Definition1. *system (7) with (8) is called robust stable or global robust stable if its unique equilibrium* $x^* = (x_1^*, x_2^*, \cdots, x_{2N}^*)^T$ *is stable or global stable for each* $A = diag(a_1, a_2, \cdots, a_{2N}) \in A_I$, $W = (w_{ij})_{2N\times 2N} \in W_I$, $W^T = (w_{ij}^T)_{2N\times 2N} \in W_I^T$.

Theorem 1. *Assume that*
(T_1) $|f_j(y_1) - f_j(y_2)| \le \sigma_i|y_1 - y_2|$, *for all* $y_1, y_2 \in R$, $i = 1, 2, \cdots, 2N$;
(T_2) $C = A_0 - W_0^+\sigma - W_0^T\sigma K$ *is an M-matrix, where* $W_0^+ = (|(W_0^+)_{ij}|)_{2N\times 2N}$, $W_0^T = (|(W_0^T)_{ij}|)_{2N\times 2N}, |(W_0^+)_{ij}| = \max\{|\underline{W}_{ij}|, |\overline{W}_{ij}|\}, |(W_0^T)_{ij}| = \max\{|\underline{W}_{ij}^T|, |\overline{W}_{ij}^T|\}, i,j = 1, 2, \cdots, 2N$, $A_0 = diag(\underline{a}_1, \underline{a}_2, \cdots, \underline{a}_{2N}), K = (k_{ij})_{2N\times 2N}$. *Then, the system (7) with (8) has a unique equilibrium which is global robust stable.*

Proof. Part 1 The existence of the equilibrium.
By (T_1), it follows that

$$|f_j(y)| \le \sigma_j|y| + |f_j(0)| \quad j = 1, 2, \cdots, 2N, \ \ \forall y \in R. \tag{11}$$

Let
$$h(x) = Ax - Wf(x) - W^T f(x) = 0 \tag{12}$$

where $A \in A_I$, $W \in W_I$, $W^T \in W_I^T$. It is obvious that the solutions of (12) are the equilibria of system of (7) with (8). Let we define homotopic mapping

$$H(x, \lambda) = \lambda h(x) + (1-\lambda)x \quad \lambda \in J = [0, 1] \tag{13}$$

We have

$$\begin{aligned} |H_i(x,\lambda)| &= |\lambda(a_i x_i - \sum_{j=1}^{2N} W_{ij} f_j(x_j) - \sum_{j=1}^{2N} W_{ij}^T f_j(\textstyle\int_{-\infty}^t K_{ij}(t-s)x_j(s)ds)) \\ &\quad + (1-\lambda)x_i| \\ &\ge [1 + \lambda(\underline{a}_i - 1)]|x_i| - \lambda[\sum_{j=1}^{2N}(|W_{ij}|\sigma_j + |W_{ij}^T|\sigma_j k_{ij})|x_j| \\ &\quad + \sum_{j=1}^{2N}(|W_{ij}| + |W_{ij}^T|)|f_j(0)|] \end{aligned} \tag{14}$$

That is

$$\begin{aligned} H^+ &\ge [E + \lambda(A_0 - E)][x]^+ - \lambda(W_0^+\sigma + W_0^T\sigma K)[x]^+ - \lambda(W_0^+ + W_0^T)[f(0)]^+) \\ &= (1-\lambda)[x]^+ + \lambda(A_0 - W_0^+\sigma - W_0^T\sigma K)[x]^+ - \lambda(W_0^+ + W_0^T)[f(0)]^+ \end{aligned}$$

where $H^+ = (|H_1|, |H_2| \cdots, |H_{2N}|)^T, [x]^+ = (|x_1|, |x_2|, \cdots, |x_{2N}|)^T, [f(0)]^+ = (|f_1(0)|, |f_2(0)|, \cdots, |f_{2N}(x)|)^T$, E is an identity matrix.

Since $C = A_0 - W_0^+\sigma - W_0^T\sigma K$ is an M-matrix [21], we have $(A_0 - W_0^+\sigma - W_0^T\sigma K)^{-1} \geq 0$ (nonnegative matrix) and there exists a $Q = (Q_1, Q_2, \cdots, Q_{2N})^T > 0$, namely $Q_i > 0, i = 1, 2, \cdots, 2N$, such that $(A_0 - W_0^+\sigma - W_0^T\sigma K)Q > 0$. Let

$$U(R_0) = \{x : [x]^+ \leq R_0 = Q + (A_0 - W_0^+\sigma - W_0^T\sigma K)^{-1}(W_0^+ + W_0^T)[f(0)]^+\}, \quad (15)$$

then, $U(R_0)$ is not empty and it follows from (15) that for any $x \in \partial U(R_0)$ (boundary of $U(R_0)$)

$$\begin{aligned} H^+ &\geq (1-\lambda)[x]^+ + \lambda(A_0 - W_0^+\sigma - W_0^T\sigma K)[x]^+ - \lambda(W_0^+ + W_0^T)[f(0)]^+ \\ &= (1-\lambda)[x]^+ + \lambda(A_0 - W_0^+\sigma - W_0^T\sigma K)([x]^+ \\ &\quad - (A_0 - W_0^+\sigma - W_0^T\sigma K)^{-1}(W_0^+ + W_0^T)[f(0)]^+) \\ &= (1-\lambda)[x]^+ + \lambda(A_0 - W_0^+\sigma - W_0^T\sigma K)Q > 0, \qquad \lambda \in [0,1] \end{aligned} \quad (16)$$

That is

$$H(x,\lambda) \neq 0 \ \text{ for } x \in \partial U(R_0), \quad \lambda \in [0,1]$$

So, from homotopy invariance theorem [22], we have

$$d(h, U(R_0), 0) = d(H(x,1), U(R_0), 0) = d(H(x,0), U(R_0), 0) = 1$$

where $d(h, U(R_0), 0)$ denotes topological degree [23]. By topological degree theory, we can conclude that equation (12) has at least one solution in $U(R_0)$. That is, system (7) with (8) has at least an equilibrium.

Part 2 Uniqueness of equilibrium and its global robust stability.

Let $x^* = (x_1^*, x_2^*, \cdots, x_{2N}^*)^T$ be an equilibrium of system (7) with (8) and $x(t) = (x_1(t), x_2(t), \cdots, x_{2N}(t))^T \neq (x_1^*, x_2^*, \cdots, x_{2N}^*)^T$ is any solution of system (7) with (8). Rewrite system (9) as

$$\begin{aligned} &\tfrac{d}{dt}(x_i(t) - x^*) = -a_i(x_i(t) - x_i^*) + \sum_{j=1}^{2N} W_{ij}[f_j(x_j(t)) - f_j(x_j^*)] \\ &+ \sum_{j=1}^{2N} W_{ij}^T[f_j(\textstyle\int_{-\infty}^t K_{ij}(t-s)x_j(s)ds) - f_j(\int_{-\infty}^t K_{ij}(t-s)x_j^*ds)] \\ &\quad i = 1, 2, \cdots, 2N. \end{aligned} \quad (17)$$

Since $C = A_0 - W_0^+\sigma - W_0^T\sigma K$ is an M-matrix, there exists $r_i > 0$, $i = 1, 2, \cdots, 2N$, such that

$$r_i\underline{a}_i - \sum_{j=1}^{2N} r_j(|(W_0^+)_{ji}|\sigma_i + |(W_0^T)_{ji}|\sigma_i k_{ji}) > 0.$$

Define a Lyapunov function

$$V(t) = \sum_{i=1}^{2N} r_i\{|x_i(t) - x_i^*| + \sum_{j=1}^{2N} |W_{ij}^T| \int_0^\infty K_{ij}(s) \int_{t-s}^t |f_j(x_j(s)) - f_j(x_j^*)| dt ds\} \quad (18)$$

its upper right Dini-derivative along the solution of (9) can be calculated as

$$
\begin{aligned}
D^+ V(t)\mid_{(3)} \leq & \sum_{i=1}^{2N} r_i\{-a_i|x_i(t)-x_i^*| + \sum_{j=1}^{2N} |W_{ij}|\cdot|f_j(x_j(t))-f_j(x_j^*)| \\
+ & \sum_{j=1}^{2N} |W_{ij}^T|\cdot|[f_j(\textstyle\int_{-\infty}^t K_{ij}(t-s)x_j(s)ds) - f_j(\int_{-\infty}^t K_{ij}(t-s)x_j^* ds)| \\
+ & \sum_{j=1}^{2N} |W_{ij}^T|\textstyle\int_0^\infty K_{ij}(s)|f_j(x_j(t))-f_j(x_j^*)|ds \\
- & \sum_{j=1}^{2N} |W_{ij}^T|\textstyle\int_0^\infty K_{ij}(s)|f_j(x_j(t-s))-f_j(x_j^*(t-s))|ds\} \\
= & \sum_{i=1}^{2N} r_i\{-a_i|x_i(t)-x_i^*| + \sum_{j=1}^{2N}|W_{ij}||f_j(x_j(t))-f_j(x_j^*)| \\
+ & \sum_{j=1}^{2N}|W_{ij}^T|\textstyle\int_0^\infty K_{ij}(s)|f_j(x_j(t))-f_j(x_j^*)|ds\} \qquad (19) \\
\leq & \sum_{i=1}^{2N} r_i\{-a_i|x_i(t)-x_i^*| + \sum_{j=1}^{2N}(|W_{ij}|\sigma_j + |W_{ij}^T|\sigma_j k_{ij})|(x_j(t))-(x_j^*)|\} \\
= & -\sum_{i=1}^{2N}\{r_i a_i|x_i(t)-x_i^*| - \sum_{j=1}^{2N} r_j(|W_{ji}|\sigma_i + |W_{ji}^T|\sigma_i k_{ji})|x_i(t)-x_i^*|\} \\
= & -\sum_{i=1}^{2N}\{r_i a_i - \sum_{j=1}^{2N} r_j(|W_{ji}|\sigma_i + |W_{ji}^T|\sigma_i k_{ji})\}|x_i(t)-x_i^*| \\
\leq & -\sum_{i=1}^{2N}\{r_i \underline{a}_i - \sum_{j=1}^{2N} r_j(|(W_0^+)_{ji}|\sigma_i + |(W_0^T)_{ji}|\sigma_i k_{ji})\}|x_i(t)-x_i^*| \\
\leq & -m\sum_{i=1}^{2N}|x_i(t)-x_i^*| < 0
\end{aligned}
$$

where $m = \min_{1\leq i\leq 2N}\{r_i\underline{a}_i - \sum_{i=1}^{2N} r_j(|(W_0^+)_{ji}|\sigma_i + |(W_0^T)_{ji}|\sigma_i k_{ji})\} > 0$.

From (19), we know that equilibrium x^* is stable, and we have

$$V(t) + m\int_0^t \sum_{i=1}^{2N} |x_i(s)-x_i^*|ds \leq V(0) \qquad (t\geq 0)$$

which means that

$$\int_0^t \sum_{i=1}^{2N}|x_i(s)-x_i^*|ds < \infty \ \text{ and } \int_0^t |x_i(s)-x_i^*|ds < \infty, \quad t\in[0,\infty]$$

Hence, there exists a constant $M^* > 0$, such that

$$|x_i(t)-x_i^*| \leq M^*, \quad i = 1,2,\cdots,2N. \quad t\in[0,\infty). \tag{20}$$

From (17) and (20), we obtain that

$$-(|\underline{a}_i| + |\overline{a}_i| + \sum_{j=1}^{2N}(|(W_0^+)_{ij}|\sigma_j + |(W_0^T)_{ij}|\sigma_j k_{ij}))M^* \leq \tfrac{d}{dt}|x_i(t)-x_i^*| \leq$$
$$(|\underline{a}_i| + |\overline{a}_i| + \sum_{j=1}^{2N}(|(W_0^+)_{ij}|\sigma_j + |(W_0^T)_{ij}|\sigma_j k_{ij}))M^*$$

That is $\frac{d}{dt}|x_i(t) - x_i^*|$ $(i = 1, 2, \cdots, 2N)$ is upper bounded on $[0, \infty)$, $|x_i(t) - x_i^*|$ $(i = 1, 2, \cdots, 2N)$ is uniformly continuous on $[0, \infty)$. By Barbalat's Lemma[24], we have

$$\lim_{t\to+\infty} \sum_{i=1}^{2N} |x_i(t) - x_i^*| = 0.$$

That is $\lim_{t\to\infty} |x_i(t) - x_i^*| = 0$, $\lim_{t\to\infty} x_i(t) = x_i^*$, $i = 1, 2, \cdots, 2N$. Since all the solutions of system (9) tend to x^* as $t \to \infty$ for any values of the coefficients in (9), i.e, the system described by (7) with (8) has a unique equilibrium which is global robust stable and the theorem is proved.

Corollary 1. *If system* (7) *with* (8) *satisfies* (T_1) *and one of the following*

(T_3) $C = A_0 - W_0^+\sigma - W_0^T\sigma K$ *is a matrix with strictly diagonal dominance of column (or row);*

(T_4) $a_i - |(w_0^+)_{ii}|\sigma_i - |(w_0^T)_{ii}|\sigma_i k_{ii} > \frac{1}{r_i} \sum_{j=1,j\neq i}^{2N} r_j(|(w_0^+)_{ji}|\sigma_i + |(w_0^+)_{ji}|\sigma_i k_{ji})$,

$\underline{a}_i > 0, r_i > 0, \sigma_i > 0, k_{ij} > 0, \gamma_i > 0, i = 1, 2, \ldots, 2N,\ i = 1, 2, \cdots, 2N;$

(T_5) $\max_{1\le i\le 2N} \frac{1}{\underline{a}_i} \sum_{j=1}^{2N} (|(w_0^+)_{ij}|\sigma_i + |(w_0^+)_{ij}|\sigma_i k_{ij}) < 1 \qquad (\underline{a}_i > 0);$

(T_6) $\frac{1}{2\underline{a}_i} \sum_{i=1}^{2N} [|(w_0^+)_{ij}|\sigma_j + |(w_0^+)_{ji}|\sigma_i + |(w_0^T)_{ij}|\sigma_j k_{ij} + |(w_0^T)_{ji}|\sigma_i k_{ji}] < 1, \quad i = 1, 2, \cdots, 2N$; *then, there is a unique equilibrium of system* (7) *with* (8), *which is global robust stable.*

In fact, any one of the condition (T_3), (T_4), (T_5) or (T_6) implies that $C = A_0 - W_0^+\sigma - W_0^T\sigma K$ is an M-matrix.

3 Conclusion

In this paper, we have presented some simple and new sufficient conditions for existence, uniqueness of the equilibrium point, and its global robust stability of the equilibrium point of competitive neural networks with continuously distributed delays and different time scales by means of the topological degree theory and Lyapunov-functional method [21, 22]. The sufficient conditions, which are easily verifiable, have a wider adaptive range. These have an important of leadings significance in the design and application of reaction-diffusion neural networks with delays. Furthermore, This method can easily be extended to competitive neural networks with different time scales and S-type distributed delays.

References

1. Meyer-Base, A., Ohl, F., Scheich, H.: Singular perturbation analysis of competitive neural networks with different time-scales. Neural Computation 8, 1731–1742 (1996)
2. Meyer-Baese, A., Pilyugin, S.S., Chen, Y.: Global Exponential Stability of Competitive Neural Networks With Different Time Scale. IEEE Tran. Neural Networks 14, 716–719 (2003)

3. Meyer-Baese, A., et al.: Local exponential stability of competitive neural networks with different time scales. Engineering Applications of Artificial Intelligence 17, 227–232 (2004)
4. Lu, H., He, Z.: Global exponential stability of delayed competitive neural networks with different time scales. Neural Networks 18, 243–250 (2005)
5. Grossberg, S.: Competition,Decision and Consensus. J.Math.Anal.Applicat 66, 470–493 (1978)
6. Amari, S.: Competitive and Cooperater Aspects in Dynamics of Neural Excition and Self-organization. Competition Cooperation Neural Networks 20, 1–28 (1982)
7. Cohen, A.M., Grossberg, S.: Absolute stability of global pattern formation and parallel memory storage by competitive neural networks. IEEE Transactions on Systems, Man, and Cybernetics 13, 815–826 (1983)
8. Lemmon, M., Kumar, M.: Emulating the dynamics for a class of laterally inhibited neural networks. Neural Networks 2, 193–214 (1989)
9. Zhang, Q., Wei, X.P., Xu, J.: Global exponential stability of Hopfield neural networks with continuously distributed delays. Phys.Lett. A 315(6), 431–436 (2003)
10. Zhang, Q., MA, R.N., XU, J.: Global Exponential Convergence Analysis of Hopfield Neural Networks with Continuously Distributed Delays. Communications in Theoretical Physics 39(3), 381–384 (2003)
11. Zhang, Q., MA, R.N., XU, J.: Global stability of bidirectional associative memory neural networks with continuously distributed delays. Science in China, Ser.F 46(5), 327–334 (2003)
12. Cao, J., Huang, D.S., Qu, Y.: Global robust stability of delayed recurrent neural networks. Chaos Solitons Fractals 23(1), 221–229 (2005)
13. Ye, H., Michel, A.N.: Robust stability of nonlinear time-delay systems with applications to neural networks. IEEE Trans. Circuits Syst. I 43, 532–543 (1996)
14. Hu, S.Q., Wang, J.: Global exponential stability of continuous-time interval neural networks. Phys. Rev. E 65, 36–133 (2002)
15. Liao, X.F., Yu, J.B.: Robust stability for interval Hopfield neural networks with time delay. IEEE Trans. Neural Netw. 9, 1042–1045 (1998)
16. Liao, X.F., Wong, K.W., Wu, Z., Chen, G.: Novel robust stability for interval-delayed hopfield neural networks. IEEE Trans. Circuits Syst. I 48, 1355–1359 (2001)
17. Arik, S.: Global robust stability of delayed neural networks. IEEE Trans. Circuits Syst. I 50(1), 156–160 (2003)
18. Cao, J., Chen, T.: Globally exponentially robust stability and periodicity of delayed neural networks. Chaos Solitons Fractals 22(4), 957–963 (2004)
19. Wang, L., Gao, Y.: Global exponential robust stability of reaction-diffuion internal neural networks with time-varing delays. Physics letters A 350, 342–348 (2006)
20. Wang, L.S., Gao, Y.Y: On global robust stability for interval Hopfield neural networks with time delays. Ann. Differential Equations 19(3), 421–426 (2003)
21. Liao, X.X.: Theory,Methods and Analysis of Stability, Huazhong University of Science and Technology Publishing House, Wuhan (1996)
22. Guo, D.J., Sun, J.X., liu, Z.L.: Functional Methods of Nonlinear Ordinary Differential Equations. Shangdong Science Press, Jinan (1995)
23. Gopalsamy, K.: Stability and Oscillation in Delay Differential Equation of Population Dynamics, Dordrecht Netherlands. Kluwer Academic Publishers, Boston, MA (1992)

24. Kao, Y., Gao, C.: Global Stability of Bidirectional Associative Memory Neural Networks with Variable Coecients and S-Type Distributed Delays. In: King, I., Wang, J., Chan, L., Wang, D. (eds.) ICONIP 2006. LNCS, vol. 4232, pp. 598–607. Springer, Heidelberg (2006)
25. Kao, Y., Gao, C. (eds.): Global Asymptotic Stability of Cellular Neutral Networks With Variable Coefficients and Time-Varying Delays. LNCS, vol. 4491, pp. 923–924. Springer, Heidelberg (2007)
26. Zhang, Q., Wei, X., Xu, J.: Asymptotic stability of transiently chaotic neural networks. Nonlinear Dynamics 42(4), 339–346 (2005)
27. Zhang, Q., Wei, X., Xu, J.: A new global stability result for delayed neural networks. Nonlinear Analysis: Real World Applications 8(3), 1024–1028 (2007)
28. Zhang, Q., Wei, X., Xu, J.: Global exponential stability for nonautonomous cellular neural networks with delays. Phys.Lett. A 351, 153–160 (2005)
29. Zhang, Q., Wei, X., Xu, J.: An improved result for complete stability of delayed cellular neural networks. Automatica 41(2), 333–337 (2005)
30. Zhang, Q., Ma, R., Xu, J.: Delay-independent stability in cellular neural networks with infinite delays. Chinese Journal of Electronics 12(2), 189–192 (2003)
31. Zhang, Q., Wei, X., Xu, J.: Global asymptotic stability of cellular neural networks with infinite delay. Neural Network World 15(6), 579–589 (2005)
32. Zhang, Q., Zhou, C.: Dynamics of Hopfield neural networks with continuously distributed delays. Dynamics of Continuous, Discrete and Impulsive Systems-Series A-Mathematical Analysis, Part 2 Suppl. S 13, 541–544 (2006)

Nonlinear Dynamics Emerging in Large Scale Neural Networks with Ontogenetic and Epigenetic Processes

Javier Iglesias, Olga K. Chibirova, and Alessandro E.P. Villa

Grenoble Institut des Neurosciences-GIN, Centre de Recherche
Inserm U 836-UJF-CEA-CHU
NeuroHeuristic Research Group, University Joseph Fourier, Grenoble, France
{Javier.Iglesias, Olga.Chibirova, Alessandro.Villa}@ujf-grenoble.fr
http://www.neuroheuristic.org/

Abstract. We simulated a large scale spiking neural network characterized by an initial developmental phase featuring cell death driven by an excessive firing rate, followed by the onset of spike-timing-dependent synaptic plasticity (STDP), driven by spatiotemporal patterns of stimulation. The network activity stabilized such that recurrent preferred firing sequences appeared along the STDP phase. The analysis of the statistical properties of these patterns give hints to the hypothesis that a neural network may be characterized by a particular state of an underlying dynamical system that produces recurrent firing patterns.

1 Introduction

The adult pattern of neuronal connectivity in the cerebral cortex is determined by the expression of genetic information, developmental differentiation and by epigenetic processes associated to plasticity and learning. During the early stages of development, excessive branches and synapses are initially formed and distributed somewhat diffusely (Innocenti, 1995). This over-growth phase is generally followed by massive synaptic pruning (Rakic et al., 1986) of only a selected subset of the connections initially established by a neuron. Trigger signals able to induce selective synaptic pruning could be associated to patterns of activity that depend on the timing of action potentials (Catalano and Shatz, 1998). Spike timing dependent synaptic plasticity (STDP) is a change in the synaptic strength based on the ordering of pre- and post-synaptic spikes. It has been proposed as a mechanism to explain the reinforcement of synapses repeatedly activated shortly before the occurrence of a post-synaptic spike (Bell et al., 1997). An equivalent mechanism is keen to explain the weakening of synapses strength whenever the pre-synaptic cell is repeatedly activated shortly after the occurrence of a post-synaptic spike (Karmarkar and Buonomano, 2002).

The study of the relation between synaptic efficacy and synaptic pruning suggests that the weak synapses may be modified and removed through competitive "learning" rules (Chechik et al., 1999). Despite the plasticity of these

J. Marques de Sá et al. (Eds.): ICANN 2007, Part I, LNCS 4668, pp. 579–588, 2007.

phenomena it is rationale to suppose that whenever the same information is presented in the network the same pattern of activity is evoked in a circuit of functionally interconnected neurons, referred to as “cell assembly”. Cell assemblies interconnected in this way would be characterized by recurrent, above chance levels, ordered sequences of precise (in the order of few ms) interspike intervals referred to as spatiotemporal patterns of discharges or preferred firing sequences (Abeles, 1991). Such precise firing patterns have been associated with specific behavioral processes in rats (Villa et al., 1999) and primates (Shmiel et al., 2006). Precise firing patterns have also been associated to deterministic nonlinear dynamics (Villa, 2000) and their detection in noisy time series could be performed by selected algorithms (Tetko and Villa, 1997; Asai et al., 2006).

In the current study we assume that developmental and/or learning processes are likely to potentiate or weaken certain pathways through the network and let emerge cell assemblies characterized by recurrent firing patterns. We investigate whether or not deterministic dynamics can be observed in the activity of a network by the analysis of the preferred firing sequences. The rationale is that a certain network, here identified by the simulation run, may be characterized by a particular state of an underlying dynamical system that produces recurrent firing patterns.

2 Neural Network Model

The complete neural network model is described in details elsewhere (Iglesias, Eriksson, Grize, Tomassini and Villa, 2005). A sketch description of the model with specific model parameters related to the current study follows below.

We assume that at time zero of the simulation the network is characterized by two types of integrate-and-fire units and by its maximum over growth in terms of connectivity. The total amount of units is 10,000 (8,000 excitatory and 2,000 inhibitory) laid down on a 100×100 2D lattice according to a space-filling quasi-random Sobol distribution. Two sets of 400 excitatory units (i.e., 800 units overall) were randomly selected among the 8,000 excitatory units of the network. The units belonging to these sets are the “sensory units” of the network, meaning that in addition to sending and receiving connections from the other units of both types they receive an input from the external stimulus.

The model features cell death mechanisms that may be provoked by: (i) an excessive firing rate and (ii) the loss of all excitatory inputs. An excessive firing rate is assumed to correspond to the biological effect known as glutamate neurotoxicity (Choi, 1988). During an initial phase called “early developmental phase”, at each time step and for each unit a maximum firing rate was arbitrarily determined following a parameter search procedure described in (Iglesias and Villa, 2006). Whenever the rate of discharges of a unit exceeds the maximum allowed rate the cell had a probability to die according to a “genetically” determined probability function. A dead unit is characterized by the absence of any spiking activity.

At the end of the early developmental phase, the synaptic plasticity (Iglesias, Eriksson, Pardo, Tomassini and Villa, 2005) is enabled. It is assumed *a priori* that modifiable synapses are characterized by discrete activation levels (Montgomery and Madison, 2004) that could be interpreted as a combination of two factors: the number of synaptic *boutons* between the pre- and post-synaptic units and the changes in synaptic conductance. In the current study we attributed a fixed activation level (meaning no synaptic modification) $A_{ji}(t) = 1$, to *(inh, exc)* and *(inh, inh)* synapses while activation levels were allowed to take one of $A_{ji}(t) = \{0, 1, 2, 4\}$ for *(exc, exc)* and *(exc, inh)*, $A_{ji}(t) = 0$ meaning that the projection was permanently pruned out. For $A_{ji}(t) = 1$, the post-synaptic potentials were set to $0.84\ mV$ and $-0.8\ mV$ for excitatory and inhibitory units, respectively. The projections from and to "dead" units undergo a decay of their synapses leading eventually to their pruning when $A_{ji}(t) = 0$. Other projections may be pruned due to synaptic depression driven by STDP and also leading to $A_{ji}(t) = 0$. Thus, some units that survived the early phase can also remain without any excitatory input. The loss of all excitatory inputs also provokes the cell death and these units stop firing (even in presence of background activity) immediately after the pruning of the last excitatory afference from within the network.

Two sets of 400 "sensory units" are stimulated by patterned activity organized both in time and space described elsewhere (Iglesias and Villa, 2006). This stimulus is assumed to correspond to content-related activity generated elsewhere in the brain. The overall stimulus duration is set to 100 ms, followed by a period without stimulation that lasted 1900 ms. Thus, the rate of stimulation was 0.5 $stim/s$.

3 Simulations

The values of the set of parameters specified above were kept constant throughout all this study. Because of a high sensitivity to the initial conditions of the pattern of connectivity we repeated the very same analysis with 30 different random generator seeds. Thus, each different seed generated a different connectivity (although keeping the same rules), different stimulus patterns and spontaneous activity patterns. Each simulation run lasted 10^5 discrete time steps (T_{end}), with 1 time step corresponding to 1 ms in the model, that means $100,000\ ms$ as total duration of a simulation run. The states (spiking/not spiking) of all units were updated synchronously. After spiking, the membrane potential was reset to its resting potential, and the unit entered an absolute refractory period lasting 3 and 2 time steps for excitatory and inhibitory units, respectively. Starting at time zero and throughout all the simulation run each unit received a background activity following an independent Poisson process of 5 $spikes/s$ on average.

The early developmental phase, characterized by cell death provoked by excessive firing rate, begins at time $t = 0$ and lasts until $t = T_{edp}$, that was fixed at 700 ms for this study. The spike timing dependent plasticity is enabled at $t = T_{edp} + 1$. At time $t = 1001\ ms$ the first stimulation is applied, lasting 100 ms

until $t = 1100$ ms. Between $t = 1101$ ms and $t = 3000$ ms only the background activity is getting into the network. At time $t = 3001$ ms another stimulation is applied and so forth until the end of the simulation run. Overall this corresponds to 50 presentations of the stimulus along one simulation run.

The events belonging to the univariate time series corresponding to the spike train may have two different origins. Spikes may be simply associated to the Poisson background noise whenever the cell excitability is high and close enough to the threshold of activation. The other spikes can be produced by the convergence of synchronous activity (i.e., temporal summation of excitatory postsynaptic potentials) generated within the network. In order to study the time series of the spikes associated to the network activity those spikes associated to the background process were discarded from the spike trains and the so-called "effective spike trains" were extracted (Hill and Villa, 1997). Such effective spike trains were searched for the occurrence of spatiotemporal firing patterns, also called preferred firing sequences, by means of the "pattern grouping algorithm" (PGA) (Tetko and Villa, 2001). For the present study PGA was set to find patterns formed by three (triplets) or four spikes (quadruplets), with a significance level $p = 0.10$, provided the entire pattern did not last more than 800 ms and was repeated with a jitter accuracy of ± 5 ms.

All occurrences of all patterns detected by PGA within the same simulation run were merged together even if they were found in different units. This operation is aimed at detecting a possible underlying dynamical system that is associated to a specific network, thus encompassing the activity of several units belonging to the functional cell assembly. Such particular time series is referred to as "network patterned spike train", NP-spike train.

4 Results

At time $t = T_{end} = 100,000$ ms the units characterized by more than four active excitatory input projections that did not belong to the sets of "sensory units" were selected and their effective spike trains were analyzed. In one out of the 30 simulations, no spatiotemporal firing patterns could be found. In the other 29 simulations, 147 spatiotemporal firing patterns were detected, corresponding to 5672 spikes distributed into 61 triplets and 86 quadruplets. A single pattern could repeat between 5 (the minimum rate set in the algorithm) and 185 times. The vast majority of the patterns were composed by events belonging to the same neuron. We found only 3 triplets and 2 quadruplets composed by spikes produced by two different units.

Figure 1 illustrate two examples of patterns detected in the simulation run $S008$. The pattern $< 23E5, 23E5, 23E5$; $154 \pm 3.5, 364 \pm 3.0 >$ was composed by spikes produced by one unit labeled here #23E5 (Fig. 1a). This notation means that the pattern starts with a spike of unit $\#23E5$, followed 154 ± 3.5 ms later by a second spike of the same unit, and followed by a third spike 364 ± 3.0 ms after the pattern onset. We observed 15 repetitions of this pattern during the whole simulation run, but their occurrence was limited to the first half of the

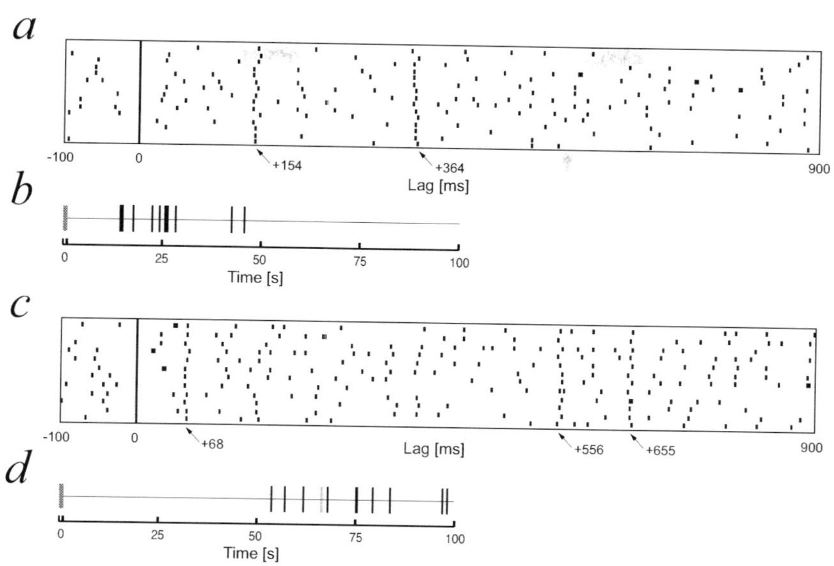

Fig. 1. **(a)**: Spatiotemporal pattern $< 23E5, 23E5, 23E5\ ;\ 154 \pm 3.5, 364 \pm 3.0 >$. Raster plot showing 15 repetitions of the pattern aligned on the pattern start; **(b)**: Pattern occurrence timing plot of pattern (a) : each vertical tick represents the timing of the pattern onset; **(c)**: Spatiotemporal pattern $< B9B, B9B, B9B, B9B\ ;\ 68 \pm 2.0, 556 \pm 1.0, 655 \pm 4.0 >$ Raster plot showing 12 repetitions of the pattern aligned on the pattern start; **(d)**: Pattern occurrence timing plot of pattern (c)

simulation, and in particular in the period $[15 - 28]$ *seconds* (Fig. 1b). No correlation could be found between the timing of the spatiotemporal pattern and the stimulation onset. Figure 1c shows the occurrences of another pattern observed in the same network. This pattern labeled as $< B9B, B9B, B9B, B9B\ ;\ 68 \pm 2.0, 556 \pm 1.0, 655 \pm 4.0 >$ corresponds to a quadruplet that occurred only during the second half of the simulation.

Figure 2 shows the distribution of the onset times of all 147 patterns along the simulation duration. This histogram shows that patterns could start any time after the simulation onset. However, the majority of the patterns tended to appear and repeat above chance levels only during specific periods of the network life, thus suggesting the possibility of shifts in the temporally organized dynamics.

We investigated the distribution of the intra-pattern intervals, corresponding to the interval between the first and second event of the pattern (Figure 3a) and between the second and third event of the pattern (Figure 3b). In the case of quadruplets the intervals of the corresponding sub-patterns (the triplets) were considered for this analysis.

After detecting all patterns the NP-spike trains were constructed for each simulation run. In 19/30 simulation runs we found at least 4 different patterns, each of them repeating many times (Table 1). In these cases it appeared interesting to analyze the patterns that overlapped each other. For each pattern,

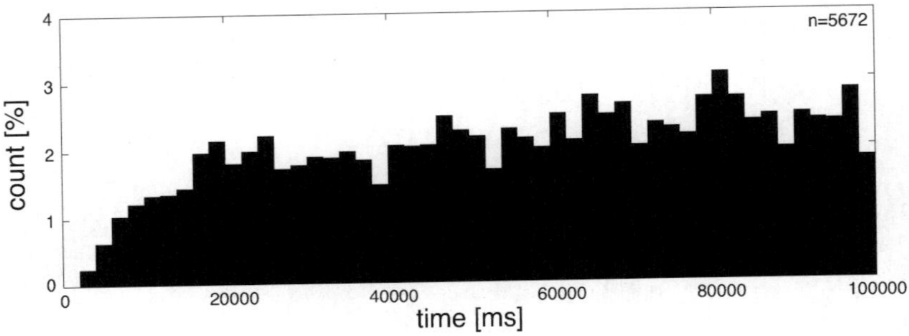

Fig. 2. Relative histogram of the onset time of all 147 patterns. Bin size: 2000 ms.

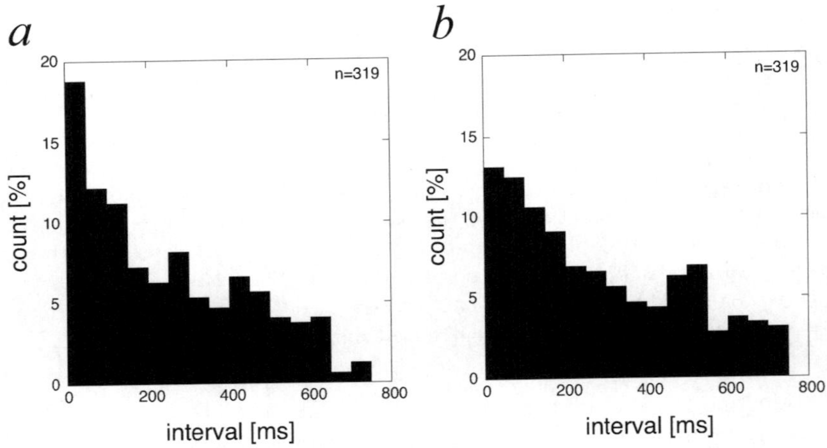

Fig. 3. Normalized distribution of the intervals among the triplets and quadruplets spikes. (*a*): intervals between the first and the second spike; (*b*): intervals between the second and the third spike. Bin size: 50 ms.

an overlapping factor was calculated to describe the probability for a pattern occurrence to overlap another occurrence from the same (self-overlap) or a different (cross-overlap) pattern. The number of overlaps o_i of each pattern i was incremented each time the onset spike of the jth pattern appeared between the first and the last spike of ith pattern. Then, for each kth network (i.e., for each simulation run), the normalized overlapping factor O_k was calculated as follows:

$$O_k = \frac{\sqrt{\sum o_i}}{n_k} \tag{1}$$

where o_i is the number of overlaps for each pattern i and n_k is the total number of occurrences for all the patterns found in the kth network. Table 1 shows that on average the overlapping factor was higher in the networks characterized by

Table 1. 147 patterns were observed in the activity of $N_{TOT} = 30$ distinct networks. Networks are arbitrarily divided into four classes clusterized by the number of different patterns found in each one. *Class 1*: less than 4 patterns; *Class 2*: between 4 an 6 patterns; *Class 3*: between 7 an 9 patterns; *Class 4*: at least 10 patterns. N: number of networks; $\hat{O}$: average overlapping factor for each class; $\hat{R}$: average number of pattern repetitions.

	N	$\widehat{O}$	$\widehat{R}$
Class 1	11	0.04	57.5
Class 2	13	0.06	41.0
Class 3	3	0.08	17.2
Class 4	3	0.07	33.8

7 to 9 different patterns. However, this result should not be considered too strongly because many more simulation runs are needed in order to extract significant statistical information.

The network $S008$ belonging to Class 3 was selected for the plot of the consecutive interspike intervals. Return maps correspond to the plots of the nth interspike-interval against the $(n-1)^{th}$ interspike-interval. Figure 4 shows, for two different time scales, the return map computed on the simulation NP-spike train as scatter plots (fig. 4a,b) and trajectories (fig. 4c,d). Notice the appearance of an ordered structure in the panels with scales $[0-250]$ ms (fig. 4b,d).

5 Discussion

We simulated a large scale spiking neural network, with the time resolution of 1 ms, characterized by a brief initial phase of cell death (Iglesias and Villa, 2006). The addition of this feature greatly improved the stability of the network while maintaining its ability to produce spatiotemporal firing patterns as shown in this study. During this phase the units that exceeded a certain threshold of firing had an increasing probability to die with the passing of time until 700 time units. After the stop of the massive cell death, spike timing dependent plasticity (STDP) and synaptic pruning were made active. Selected sets of units were activated by regular repetitions of a spatiotemporal pattern of stimulation. During the STDP phase, the cell death could occur only if a unit became deafferented, i.e. it lost all its excitatory afferences because of synaptic pruning.

We recorded the spike trains of all excitatory units that were not directly stimulated and that were surviving at the arbitrary end of the simulation set at $t = 100$ *seconds*. In these spike trains we searched for preferred firing sequences that occurred beyond random expectation (Tetko and Villa, 2001). We found 147 patterns in 30 different network configurations determined by different random seeds. It is interesting to notice that the effect of the initialization is important because only 6/30 networks were characterized by at least 7 detectable firing patterns. Moreover, we often observed patterns that occurred only during a limited time of the network life, usually in the order of 30 to 50 *seconds*. Such

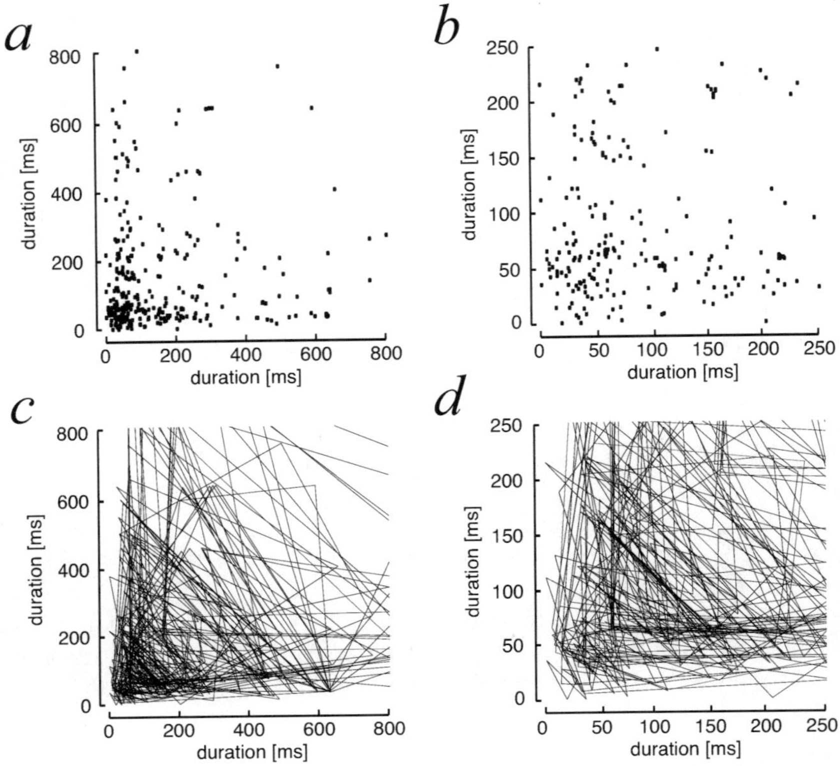

Fig. 4. Fingerprint of the dynamical activity of network $S008$ characterized by an overlapping factor equal to 0.09. **(*a*)**: Return map computed for intervals between **(*a*)** 0 and 800 ms and **(*b*)** focusing around 0 and 250 ms; **(*c,d*)**: Trajectories for the same data and intervals.

an example is illustrated by onset dynamics of the two patterns of Fig. 1. This observation might suggest that the network dynamics giving rise to a pattern occurrence may be suddenly disrupted by the continuous STDP-driven pruning. However, there is an extended overlap between patterns occurrence as shown by the overlapping factor described in this study. The process of patterns appearance looks more like a pattern giving the hand to another pattern but keeping the transition rather smooth.

The results presented here give some hints to the hypothesis that an underlying deterministic dynamical system, detected by the observation of an excess of preferred firing sequences, may appear under the drive of differentiation and learning processes. The fingerprints of the system observed in the return maps are suggestive of the presence of attractors and closed loops (Fig. 4). We believe that the current study opens new perspectives in this investigation but we are aware of the necessity of future studies before any firm conclusion on the relations between preferred firing sequences and deterministic nonlinear dynamical systems.

Acknowledgments. This work was partially funded by the European Community Future and Emerging Technologies Complex Systems program, grant #034632 (PERPLEXUS), and New and Emerging Science and Technology program, grant #043309 (GABA).

References

Abeles, M.: Corticonics: Neural Circuits of the Cerebral Cortex, 1st edn. Cambridge University Press, Cambridge (1991)

Asai, Y., Yokoi, T., Villa, A.E.P.: Detection of a dynamical system attractor from spike train analysis. In: Kollias, S., Stafylopatis, A., Duch, W., Oja, E. (eds.) ICANN 2006. LNCS, vol. 4131, pp. 623–631. Springer, Heidelberg (2006)

Bell, C.C., Han, V.Z., Sugawara, Y., Grant, K.: Synaptic plasticity in a cerebellum-like structure depends on temporal order. Nature 387(6630), 278–281 (1997)

Catalano, S.M., Shatz, C.J.: Activity-dependent cortical target selection by thalamic axons. Science 281(5376), 559–562 (1998)

Chechik, G., Meilijson, I., Ruppin, E.: Neuronal regulation: A mechanism for synaptic pruning during brain maturation. Neural Computation 11, 2061–2080 (1999)

Choi, D.W.: Glutamate neurotoxicity and diseases of the nervous system. Neuron 1(8), 623–634 (1988)

Hill, S., Villa, A.E.: Dynamic transitions in global network activity influenced by the balance of excitation and inhibition. Network: computational neural networks 8, 165–184 (1997)

Iglesias, J., Eriksson, J., Grize, F., Tomassini, M., Villa, A.E.: Dynamics of pruning in simulated large-scale spiking neural networks. BioSystems 79(1), 11–20 (2005)

Iglesias, J., Eriksson, J., Pardo, B., Tomassini, M., Villa, A.E.: Emergence of oriented cell assemblies associated with spike-timing-dependent plasticity. In: Duch, W., Kacprzyk, J., Oja, E., Zadrożny, S. (eds.) ICANN 2005. LNCS, vol. 3696, pp. 127–132. Springer, Heidelberg (2005)

Iglesias, J., Villa, A.: Neuronal cell death and synaptic pruning driven by spike-timing dependent plasticity. In: Kollias, S., Stafylopatis, A., Duch, W., Oja, E. (eds.) ICANN 2006. LNCS, vol. 4131, pp. 953–962. Springer, Heidelberg (2006)

Innocenti, G.M.: Exuberant development of connections, and its possible permissive role in cortical evolution. Trends in Neurosciences 18(9), 397–402 (1995)

Karmarkar, U.R., Buonomano, D.V.: A model of spike-timing dependent plasticity: one or two coincidence detectors? J. Neurophysiol. 88(1), 507–513 (2002)

Montgomery, J.M., Madison, D.V.: Discrete synaptic states define a major mechanism of synapse plasticity. Trends in Neurosciences 27(12), 744–750 (2004)

Rakic, P., Bourgeois, J., Eckenhoff, M.F., Zecevic, N., Goldman-Rakic, P.S.: Concurrent overproduction of synapses in diverse regions of the primate cerebral cortex. Science 232(4747), 232–235 (1986)

Shmiel, T., Drori, R., Shmiel, O., Ben-Shaul, Y.: Temporally precise cortical firing patterns are associated with distinct action segments. Journal of Neurophysiology 96(5), 2645–2652 (2006)

Tetko, I.V., Villa, A.E.: A comparative study of pattern detection algorithm and dynamical system approach using simulated spike trains. In: Gerstner, W., Hasler, M., Germond, A., Nicoud, J.-D. (eds.) ICANN 1997. LNCS, vol. 1327, pp. 37–42. Springer, Heidelberg (1997)

Tetko, I.V., Villa, A.E.: A pattern grouping algorithm for analysis of spatiotemporal patterns in neuronal spike trains. 1. detection of repeated patterns. Journal of Neuroscience Methods 105, 1–14 (2001)

Villa, A.E.: Time and the Brain. vol. 2. Harwood Academic Publishers (2000)

Villa, A.E., Tetko, I., Hyland, B., Najem, A.: Spatiotemporal activity patterns of rat cortical neurons predict responses in a conditioned task. Proc. Natl. Acad. Sci. USA 96, 1106–1111 (1999)

Modeling of Dynamics Using Process State Projection on the Self Organizing Map

Juan J. Fuertes-Martínez, Miguel A. Prada*, Manuel Domínguez-González, Perfecto Reguera-Acevedo[1], Ignacio Díaz-Blanco, and Abel A. Cuadrado-Vega[2]

[1] Instituto de Automática y Fabricación, Universidad de León, Escuela de Ingenierías. Campus Universitario de Vegazana, León, 24071, Spain
{jjfuem,mapram,mdomg,prega}@unileon.es

[2] Departamento de Ingeniería Eléctrica, Electrónica de Computadores y Sistemas, Universidad de Oviedo, Edificio Departamental 2 - Campus de Viesques, Gijón, 33204, Spain
{idiaz,cuadrado}@isa.uniovi.es

Abstract. In this paper, an approach to model the dynamics of multivariable processes based on the motion analysis of the process state trajectory is presented. The trajectory followed by the projection of the process state onto the 2D neural lattice of a Self-Organizing Map (SOM) is used as the starting point of the analysis. In a first approach, a coarse grain cluster-level model is proposed to identify the possible transitions among process operating conditions (clusters). Alternatively, in a finer grain neuron-level approach, a SOM neural network whose inputs are 6-dimensional vectors which encode the trajectory (T-SOM), is defined in a top level, where the KR-SOM, a generalization of the SOM algorithm to the continuous case, is used in the bottom level for continuous trajectory generation in order to avoid the problems caused in trajectory analysis by the discrete nature of SOM. Experimental results on the application of the proposed modeling method to supervise a real industrial plant are included.

Keywords: Self–organizing maps; trajectory analysis; clustering; visualization techniques; process dynamics.

1 Introduction

Visualization and dimension reduction techniques [1,2] are successfully used to process and interpret vast amounts of data with a high dimensionality. The aim of these techniques is to project those data onto a visualizable low-dimensional space where humans can reason using images. It makes possible to take advantage of the human ability to detect patterns, characteristics, trends, etc. A suitable dimension-reduction technique is based on the Self-Organizing Maps (SOM).

* M. A. Prada was supported by a grant from the *Consejería de Educación de la Junta de Castilla y León* and the European Social Fund.

J. Marques de Sá et al. (Eds.): ICANN 2007, Part I, LNCS 4668, pp. 589–598, 2007.

The SOM [3] is a self-organized unsupervised neural network where neurons are organized in a low dimensional (2D or 3D) grid. Each neuron is represented by a $d-$dimensional weight vector $\mathbf{m}_i$, which can be seen as a coordinate vector in the $d-$dimensional input space, and a position $\mathbf{g}_i$ in the low-dimensional grid of the output space. Neurons are connected to the adjacent ones according to a neighborhood relationship that defines the topology or structure of the map. During training, the SOM grid folds onto the input data sets like a flexible net. As a consequence, the $d-$dimensional coordinates of the grid neurons are modified and the zones with a higher density data gather a larger quantity of neurons. The neurons act as codebook vectors that divide the space in a finite collection of Voronoi regions.

The 2-dimensional structure of the neural grid makes the SOM an outstanding visualization tool to extract knowledge about the static nature of the input space data. For instance, the distance maps [4] reveal the intrinsic structure of the data. The average distance is closely related with the neuron density and, therefore, the data density in a certain region of the input space. The component planes [5] are another example that is used to describe process variables and to visually detect correlations between variables. An overview and categorization of visualization methods applied to SOM is presented in [6].

The dynamic features of multivariable data, when they represent the time-variable state of a process, can also be analyzed using SOM [7,8]. The successive projection of the best matching unit (BMU) of the current process state onto the neural grid defines a trajectory on a 2D space. According to the usual projection algorithm, an input vector $\mathbf{x}(t)$ is projected onto a neuron using equation:

$$S(\mathbf{x}(t)) = \mathbf{g}_{c(t)} \tag{1}$$

where

$$c(t) = \arg\min_i \|\mathbf{x}(t) - \mathbf{m}_i\| \tag{2}$$

and $\mathbf{m}_i$ are the neuron weights in the input space.

It is possible to define models associated to the dynamic behavior of the process by means of the interpretation of the trajectory. Information about velocity, acceleration, direction changes and transitions between process conditions is implicit in the trajectory. A suitable learning mechanism must be selected in order to transform this unstructured information in useful knowledge. This has been one of the research efforts of this paper, which has resulted in a method to model the process dynamics. The obtained models have been used to supervise a real industrial installation and to detect fault trends.

2 Cluster-Level Dynamic Model

Some authors have used SOM as a clustering algorithm, using a two-level approach [9,10]. In this approach, the data set is first clustered using the SOM (first abstraction level). The SOM neurons represent near datasets in the input space after the training stage. As the set of neurons is much larger than the

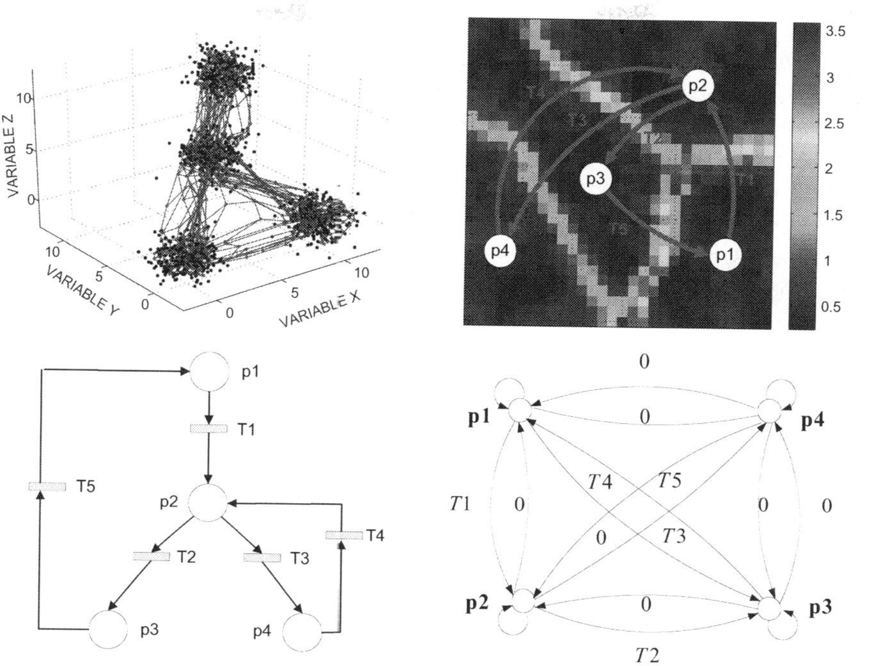

Fig. 1. Cluster-level dynamic model. In the topmost figure, a SOM trained with example data and distance map with transitions among clusters. On the bottom, equivalent Petri net and Markov chain.

expected number of clusters, the SOM is clustered (second abstraction level) using visual inspection or algorithms for agglomerative or partitive clustering. A first approach to the dynamic model can be achieved from the analysis of the process trajectory on the neural lattice, when it represents a cluster structure. These clusters of neurons can be considered operating conditions of the process.

The transitions among clusters experienced by the process trajectory in the 2D visualization space are conceptually related to widespread models such as Petri nets [11] or Markov chains [12]. And thus, these models can be used to store information about the process conditions (clusters) and the transition probabilities among them. These probabilities of transition between every pair of clusters are proportional to the number of times that a trajectory passes from a cluster to another. In Fig. 1, this idea is shown schematically using the mere observation of the distance map as clustering mechanism. Map areas with a high neural density ("valleys") indicate process clusters, whereas high values of the distance matrix ("mountain ranges") define cluster borders.

Although these models are simple, they have a wide applicability for process supervision and anomaly detection purposes. For instance, the simultaneous visualization of the distance map, the transition probabilities among clusters, and the current state of a process (defined by its current BMU) facilitates online supervision, helping to predict future behaviors (those process conditions with a non-null transition probability), and to detect faults and deviations, caused by abrupt changes in the temporal evolution of process variables with regard to the ones used to obtain the model. These faults are detected because they produce accesses of the system to inaccessible conditions with a null probability in the model.

Nevertheless, even though these models provide useful information about cluster-level process transitions, they do not give information about neuron-level transitions. A learning mechanism of the neuron transitions would make possible to extend the supervision capabilities by detecting small faults which are not easily detectable by the cluster-level model. These faults are subtle variations in the velocity and/or acceleration of change in some process variables. These faults may be detectable because they form trajectories inside each cluster which differ from the ones defined by the training data, even when the process has the same transition dynamics among clusters.

3 Neuron-Level Dynamic Model

This section presents the proposed model for neuron-level trajectory learning. The 2D trajectories generated by the SOM from the process data are encoded as sequences of 6−dimensional vectors, and a top level SOM neural network (from now on T-SOM) is used as a learning module. The election of this kind of network is supported by its ability to create new visualization tools.

3.1 Trajectory Encoding

Some authors [13,14,15] have developed trajectory learning modules, where the trajectories are represented by means of discrete state sequences (called *flow vectors*). These vectors represent either the current position (x, y) on the 2D map; or position and velocity (x, y, dx, dy); or position, velocity and any additional information $(x, y, dx, dy, a_1, a_2, ..., a_n)$. The flow vectors constitute the input space of the trajectory learning module. In the supervision stage, if this model identifies unusual behaviors (e.g. trajectory vectors not found in the training stage), the operator is informed.

In the most immediate approach to obtain the movement flow vectors on the neural grid, the coordinates of the vector would be determined by the BMU position whereas velocity and acceleration would be computed from two and three consecutive BMUs, respectively. However, the processing of these vectors presents serious problems. Due to the intrinsic discrete nature of the SOM and its magnification property (areas with a high density of data are amplified), the trajectory created by this projection algorithm (2) is discontinuous and, sometimes, apparently random when the samples are highly concentrated in the input space. Furthermore, although the discontinuous trajectory describes the global behavior of the process, it is not suitable to describe small changes. This is a serious drawback, since the modeling method exposed in this paper has been used for supervision of industrial processes, which are essentially continuous

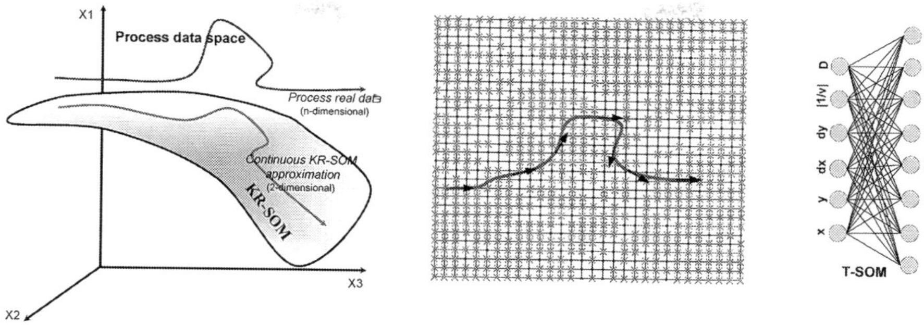

Fig. 2. Sequence of acquisition of the trajectory and T-SOM network structure

and have tendencies or limit cycles that play a crucial role to determine their condition.

The discontinuity problem of the SOM can be minimized by increasing the number of neurons in the map. Nevertheless, this solution would also increase the computational requirements of the system. Another solution to this problem would be the use of a continuous generalization of the SOM projection algorithm, such as the ones presented in [16] or [17]. In these continuous generalizations of the SOM algorithm, the discrete grid of neurons is replaced by a continuous manifold. Thus, new process data are not exclusively mapped to nodes on grid, but to any points of the continuous 2D space spanned by the neural grid.

The Kernel Regression SOM (KR-SOM), defined in [16], has been used in this work to obtain the 2-dimensional continuous trajectory that is used to define the flow vectors which encode the process dynamics. These vectors form the input space of the T-SOM learning module, defined in section 3.2. The KR-SOM is a continuous extension of the SOM projection algorithm using a General Regression Neural Network (GRNN) [18]. The GRNN makes a mapping $S_{\mathbf{x}_i \to \mathbf{y}_i} : X \to Y$ from an input space X to an output space Y, using the following equation:

$$\mathbf{y}(\mathbf{x}) = S_{\mathbf{x}_i \to \mathbf{y}_i}(\mathbf{x}) = \frac{\sum_{i=1}^{n} \exp\left(-\frac{(\mathbf{x}-\mathbf{x}_i)^T(\mathbf{x}-\mathbf{x}_i)}{2\sigma^2}\right)\mathbf{y}_i}{\sum_{i=1}^{n} \exp\left(-\frac{(\mathbf{x}-\mathbf{x}_i)^T(\mathbf{x}-\mathbf{x}_i)}{2\sigma^2}\right)} = \frac{\sum_i \Phi(\mathbf{x}-\mathbf{x}_i)\mathbf{y}_i}{\sum_i \Phi(\mathbf{x}-\mathbf{x}_i)} \tag{3}$$

where σ is the kernel bandwidth. Using this approach, it is possible to obtain a continuous mapping of the input space onto the visualization space of SOM.

$$\mathbf{y}(\mathbf{x}) = S_{\mathbf{m}_i \to \mathbf{g}_i}(\mathbf{x}) = \frac{\sum_i \Phi(\mathbf{x}-\mathbf{m}_i)\mathbf{g}_i}{\sum_i \Phi(\mathbf{x}-\mathbf{m}_i)} \tag{4}$$

where $\mathbf{y}(\mathbf{x})$ is a point of the visualization space $\mathbf{V}$ and $\mathbf{x}$ is a point of the input space $\mathbf{F}$. In a similar way, it is possible to do a mapping from the visualization space to the input space $S_{\mathbf{g}_i \to \mathbf{m}_i} : \mathbf{V} \to \mathbf{F}$.

$$\mathbf{x}(\mathbf{y}) = S_{\mathbf{g}_i \to \mathbf{m}_i}(\mathbf{y}) = \frac{\sum_i \Phi(\mathbf{y}-\mathbf{g}_i)\mathbf{m}_i}{\sum_i \Phi(\mathbf{y}-\mathbf{g}_i)} \tag{5}$$

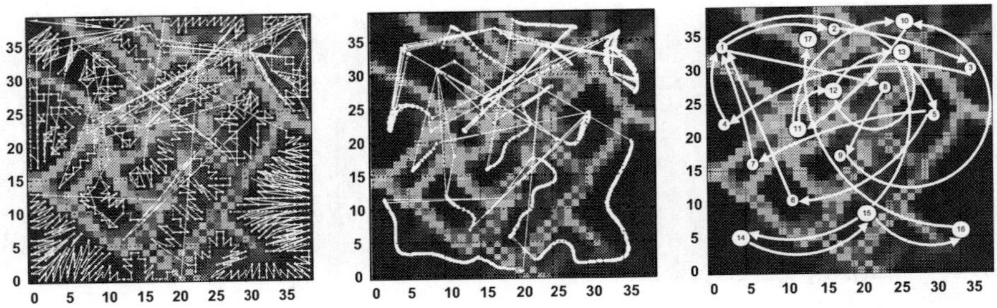

Fig. 3. Distance map of the industrial process with continuous and discrete trajectories, and cluster-level dynamic model

Equations (4) and (5) provide a method to link the input and the visualization space in a continuous way that is the basis to define the continuous trajectories.

3.2 T-SOM Learning Module

The proposed approach for trajectory learning is similar to the ones described in [13] and [14]. Fig. 2 schematically shows the sequence followed to obtain the flow vectors which represent the trajectory, and the network architecture.

The input layer of the T-SOM is composed by six neurons: the four ones used in [13] (both the system position in the 2D space and its velocity unit vector in the KR-SOM algorithm), and two new neurons, one for the inverse of the module of the velocity vector, and another one for the module of the system velocity in the input space. The flow vector that encodes the trajectory is, therefore, $(x, y, dx, dy, |1/v|, D)$. The inverse of the velocity in the fifth component of the vector is used to give more weight to small displacements than to abrupt ones, which are already considered by global models, such as the cluster-level model presented in section 2. The sixth component of the vector provides information about a feature that is not implicit in the projection map: the velocity of the process in the $d-$dimensional input space. Since the SOM magnifies regions with a high density of data, an abrupt displacement in the 2-dimensional output space might correspond to a minimum one in the $d-$dimensional input space and vice versa. For that reason, the sixth component must be considered. It is obtained by subtracting two consecutive positions in the $d-$dimensional input space, assuming a fixed sampling time.

$$D(t) = \|\mathbf{x}(t) - \mathbf{x}(t-1)\| \tag{6}$$

The learning method of the decision module is a SOM neural network that models all the possible instantaneous displacements on the map. The information about the difference between the correct trajectory, learnt by the model, and the current one is very useful for online supervision. The selection of a SOM network for the learning module makes it easier to visually interpret the results, thanks to the visualization tools that can be used:

- Component planes of variables x, y, dx, dy, $|1/v|$ and D.
- Visual representation of the residuals of the flow vectors. The residual, in this case, will represent the deviation between the approximation of the flow vector in the T-SOM model and the real vector. The residual will be non-null when the process follows new non-modeled trajectories.

The simultaneous observation of the component planes, the visual representation of residuals and the current trajectory, contrasted to the one approximated by the model, projected on the distance map, will allow us to deduce if the process is going through a significant change of state and detect erroneous flow vectors in the trajectory, which reveal a potential fault. In section 4, these visualization tools and their associated interpretation are used for an experiment with an industrial plant.

4 Experimental Results

The modeling method proposed in this paper has been tested with some experiments carried out on an industrial plant [19]. This plant manages 30 physical variables (the main ones are level, pressure, flow and temperature) sampled each 20 ms. The SOM was trained using the batch training algorithm with a 30-dimensional input space. The training lenght was 50 epochs. A 40×40 grid with a rectangular topology and a gaussian neighborhood $h_{ij}(t) = e^{-\frac{d(i,j)}{\sigma(t)^2}}$, where $\sigma(t)$ was made to decrease monotonically from 10 to 1, was used. In the rightmost picture of Fig. 3, the cluster structure of the plant and the possible transitions between clusters (cluster-level model) are represented on a distance map. The transitions are represented using arrows. The trajectories followed by the process, when the SOM (leftmost picture) and the KR-SOM (center picture) projection algorithms are used, are also represented in Fig. 4. In the first case, the figure shows the discontinuities caused by the discrete nature of the SOM and the apparently random movement within process conditions, such as condition 1 or 4, due to the magnification property of the SOM algorithm. In the second case, the trajectory is continuous, except for the characteristic discontinuities of the process. These discontinuities are due to the hybrid nature of the industrial process and appear when the process moves to another process condition (or cluster). Some examples of these intrinsic discontinuities are: opening or shutting off electric valves, start or stop of pumps or resistors, etc.

Fig. 4 shows portions of the KR-SOM trajectories followed by the industrial process in the presence of two fault situations, which were induced by varying the flow rate (during the level control stage, cluster 3) and the dynamics of the frequency converter (during the pressure control stage, clusters 12-13) in the abovementioned industrial plant. The trajectories followed by the projection of the process state in the training stage are different from the ones observed when the process experiences these faults. This experiment illustrates the usefulness of this analysis. The graphs that show the temporal development of the variable that causes the fault are also presented on the leftmost pictures of Fig. 4.

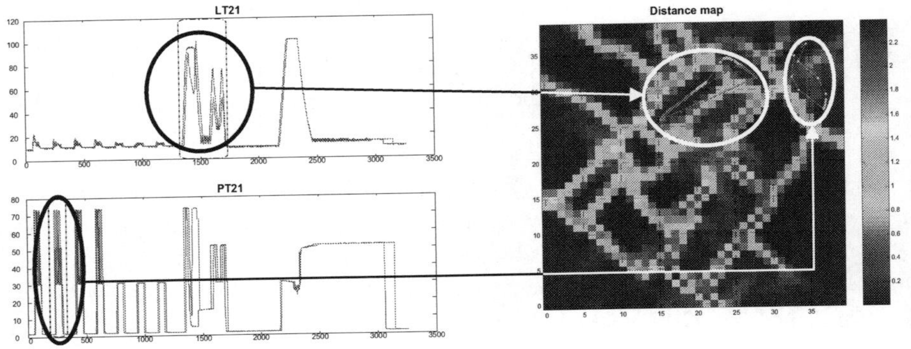

Fig. 4. Detection of two faults by means of the visualization of the trajectory followed by the industrial process. The variable LT21, which causes the first fault, is depicted in the top left chart. The variable PT21, which causes the second fault, is depicted in the bottom left chart.

The flow vectors that constitute the input space of T-SOM were obtained from the trajectory followed by the process, according to the KR-SOM model. The trajectory learning module was trained, using a 20×20 grid with a rectangular topology and a GRNN kernel width $\sigma = 1$.

The information about residuals is also a valuable tool. A graphic representation of residuals can be used to facilitate the visualizations. This representation uses a color code (such as blue tones for negative values, green tones for null values, and red tones for positive ones) in a graph where the vertical axis corresponds with the vector components and the horizontal axis represents time. The colored bands will therefore correspond with the deviation of variables x, y, dx, dy, $|1/v|$ and D through time. This tool is a useful method for semi-automatic

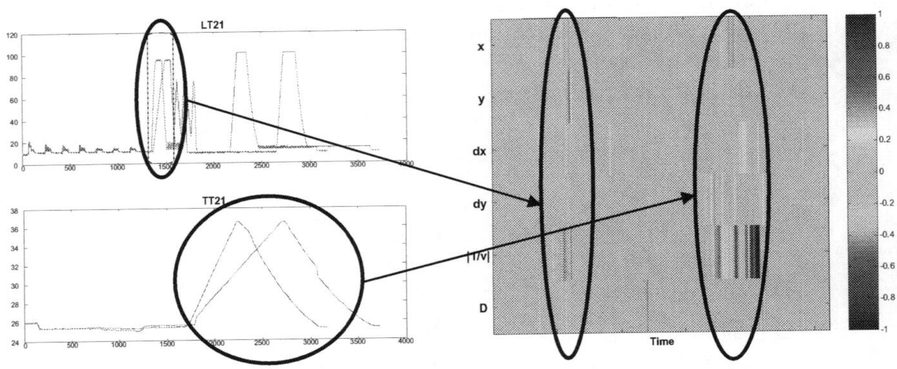

Fig. 5. Detection of two faults by means of residual visualization. The variable LT21, which causes the first fault, is depicted in the top left chart. The variable TT21, which causes the second fault, is depicted in the bottom left chart.

anomaly detection, as it can be used online to represent the current residual for every sample. Fig. 5 shows how the T-SOM model detects a slight deviation in the process behavior caused by a change of dynamics in the temperature control stage. This change is, in turn, the consequence of a loss of power in the heating resistors. It can be inferred, after observing the band corresponding to component $|1/v|$ in the residual representation, that the heating velocity decreases. Similar results are observed when the fault is caused by a change in the setting time of the controlled level variable.

5 Conclusions

In this paper, a method for modeling of the dynamics of multivariable processes, focused on the motion analysis of the state projected on the neural SOM grid, is presented. A first cluster-level approach uses discrete event models, such as Petri nets or Markov chains. Even though these models contain useful information about the possible transitions among clusters, they do not provide detailed information about transitions inside a cluster (transitions among neurons).

This reason led us to define the T-SOM model, used for trajectory learning. These trajectories are the ones followed by the processes on the KR-SOM maps (a continuous generalization of SOM), and they are encoded as 6-dimensional flow vectors. The visualization tools obtained from this model are especially useful for process supervision. Particularly, the visual representation of residuals, calculated as the difference between the modeled flow vector and the real one, is useful to detect changes in the process dynamics.

The proposed method was applied to a real industrial plant, where some changes were induced in the dynamics of some of the most representative variables (flow, level, pressure and temperature), allowing to detect and characterize changes in the process dynamics.

Acknowledgements. This research has been financed by the Spanish Ministerio de Educación y Ciencia under grant DPI2006-13477-C02-02.

References

1. Keim, D.A.: Information visualization and visual data mining. IEEE Trans. Vis. Comput. Graph. 8(1), 1–8 (2002)
2. de Oliveira, M.C.F., Levkowitz, H.: From visual data exploration to visual data mining: A survey. IEEE Transactions on Visualization and Computer Graphics 9(3), 378–394 (2003)
3. Kohonen, T., Oja, E., Simula, O., Visa, A., Kangas, J.: Engineering applications of the self-organizing map. Proceedings of the IEEE 84(10), 1358–1384 (1996)
4. Ultsch, A., Siemon, H.P.: Kohonen's Self Organizing Feature Maps for Exploratory Data Analysis. In: INNC Paris 90, pp. 305–308. Universitat Dortmund (1990)
5. Tryba, V., Metzen, S., Goser, K.: Designing basic integrated circuits by self-organizing feature maps. In: Neuro-Nîmes '89. Int. Workshop on Neural Networks and their Applications, Nanterre, France, ARC; SEE, EC2, pp. 225–235 (1989)

6. Vesanto, J.: SOM-based data visualization methods. Intelligent Data Analysis 3(2), 111–126 (1999)
7. Kasslin, M., Kangas, J., Simula, O.: Process state monitoring using self-organizing maps. In: Aleksander, I., Taylor, J. (eds.) Artificial Neural Networks, 2, vol. II, pp. 1531–1534. North-Holland, Amsterdam, Netherlands (1992)
8. Domínguez, M., Reguera, P., Fuertes, J.J., Díaz, I., Cuadrado, A.A.: Internet-based remote supervision of industrial processes using Self-organizing maps. Engineering Applications of Artificial Intelligence (2007) (articl. in press)
9. Vesanto, J., Alhoniemi, E.: Clustering of the self-organizing map. IEEE Transactions on Neural Networks 11(3), 586–600 (2000)
10. Flexer, A.: On the use of self-organizing maps for clustering and visualization. Intelligent-Data-Analysis 5, 373–384 (2001)
11. Murata, T.: Petri nets: properties, analysis, and applications. Proceedings of the IEEE 77(4), 541–580 (1989)
12. Rabiner, L.: A tutorial on hidden markov models and selected applications in speech recognition. Proceedings of the IEEE 77, 257–286 (1989)
13. Johnson, N., Hogg, D.: Learning the distribution of object trajectories for event recognition. Image Vision Comput. 14(8), 609–615 (1996)
14. Owens, J., Hunter, A.: Application of the self-organising map to trajectory classification. In: Proceedings Third IEEE International Workshop on Visual Surveillance, pp. 77–83. IEEE Computer Society Press, Los Alamitos, CA, USA (2000)
15. Hu, W., Xie, D., Tan, T.: A hierarchical self-organizing approach for learning the patterns of motion trajectories. IEEE-NN 15, 135–144 (2004)
16. Díaz Blanco, I., Díez González, A.B., Cuadrado Vega, A.A.: Complex process visualization through continuous feature maps using radial basis functions. In: Dorffner, G., Bischof, H., Hornik, K. (eds.) ICANN 2001. LNCS, vol. 2130, pp. 443–449. Springer, Heidelberg (2001)
17. Walter, J., Nölker, C., Ritter, H.: The psom algorithm and applications. In: Proc. Symposion Neural Computation 2000, Berlin, pp. 758–764 (2000)
18. Specht, D.F.: A general regression neural network. IEEE Transactions on Neural Networks 2(6), 568–576 (1991)
19. Domínguez, M., Fuertes, J.J., Reguera, P., González, J.J., Ramón, J.M.: Maqueta industrial para docencia e investigación [Industrial scale model for training and research]. Revista Iberoamericana de Automática e Informática Industrial 1(2), 58–63 (2004)

Fixed Points of the Abe Formulation of Stochastic Hopfield Networks*

Marie Kratz[1], Miguel Atencia[2], and Gonzalo Joya[3]

[1] SAMOS-MATISSE. Université Paris I (France)
[2] Departamento de Matemática Aplicada. Universidad de Málaga (Spain)
[3] Departamento de Tecnología Electrónica. Universidad de Málaga (Spain)
Campus de Teatinos, 29071 Málaga, Spain
matencia@ctima.uma.es

Abstract. The stability of stochastic Hopfield neural networks, in the Abe formulation, is studied. The aim is to determine whether the ability of the deterministic system to solve combinatorial optimization problems is preserved after the addition of random noise. In particular, the stochastic stability of the attractor set is analyzed: vertices, which are feasible points of the problem, should be stable, whereas interior points, which are unfeasible, should be unstable. Conditions on the noise intensity are stated, so that these properties are guaranteed. This theoretical investigation establishes the foundations for practical application of stochastic networks to combinatorial optimization.

1 Introduction

Hopfield neural networks [1,2] have been studied for decades, both as a theoretical model and as a practical solver of important problems. In particular, continuous Hopfield networks [2] were successfully applied to the solution of combinatorial optimization problems [3]. Although this success was controversial [4], several enhancements of the original model were proposed, among which the Abe formulation [5] was proved to present favourable properties [6]: under mild assumptions, the only stable fixed points of the system are vertices, whereas interior points, which are unfeasible, cannot be stable. This theoretical result has obvious implications concerning the practical application of the network as an efficient algorithm.

In this paper we consider the Abe formulation, but now the eventual presence of random noise in weights and biases is considered. The aim is to establish results that are parallel to those of the deterministic case. The interest of this research is twofold: on one hand, stochastic noise could model inaccuracies in the components of a hardware implementation; on the other hand, the stochastic model could present a dynamical behaviour that would be more favourable for combinatorial optimization.

* This work has been partially supported by the Spanish Ministerio de Educación y Ciencia, Project No. TIN2005-01359. The comments of the anonymous reviewers are gratefully acknowledged.

J. Marques de Sá et al. (Eds.): ICANN 2007, Part I, LNCS 4668, pp. 599–608, 2007.

A considerable number of recent contributions have dealt with the stochastic Hopfield formulation (e.g. [7,8] and references therein), which is in sharp contrast with the almost unexplored analysis of the Abe formulation. Only in [9], the conditions for stochastic convergence of the Abe formulation were stated, but the determination of the form of the attractor set was not pursued. In this work, the stability of the system is guaranteed, by defining the *same* Lyapunov function as in the deterministic case. From the computational point of view, this is a favourable property, that results in optimization of the target function being preserved under stochastic perturbations.

In Section 2, the main properties of the deterministic model are recalled, emphasizing the consequences for combinatorial optimization applications. The main results of the paper are comprised in Section 3, stating the conditions on the noise intensity, so that vertices belong to the attractor set whereas interior points are unstable. Finally, Section 4 includes the conclusions and some directions for further research.

2 Deterministic Model

2.1 Hopfield Networks and the Abe Formulation

The continuous Hopfield network [2] is a bio-inspired system, which is mathematically formulated as a system of Ordinary Differential Equations (ODEs). A slightly modified formulation was proposed by Abe [5], leading to the following definition:

$$\frac{d\,u_i}{d\,t} = net_i\ ; \qquad s_i(t) = \tanh\left(\frac{u_i(t)}{\beta}\right)\ ; \qquad i = 1, \ldots, n \tag{1}$$

where the constant β regulates the slope of the hyperbolic tangent function and the term net_i is linear—affine, to be more precise—in each of the variables s_i:

$$net_i = \sum_{j=1}^{n} w_{ij}\, s_j - b_i \tag{2}$$

where the weights w_{ij} and the biases b_i are constant parameters. The Abe formulation has been suggested to be better suited to solve combinatorial optimization problems [6], thus it is the object of this contribution.

2.2 Stability of Fixed Points and Combinatorial Optimization

From a dynamical viewpoint, Hopfield networks are systems that can be proved to be stable by defining a suitable Lyapunov function. Assume that the weight matrix is symmetric ($w_{ij} = w_{ji}$) and no self-weights exist ($w_{ii} = 0$), then the following is a Lyapunov function of the system given by Equations (1) and (2):

$$V(s) = -\frac{1}{2}\sum_{i=1}^{n}\sum_{j=1}^{n} w_{ij}\, s_i\, s_j + \sum_{i=1}^{n} b_i\, s_i = -\frac{1}{2} s^\top A\, s + b^\top s \tag{3}$$

The critical condition for V being a Lyapunov function is $\frac{dV}{dt} \leq 0$, which results from the straightforward computation of the chain rule.

One of the first practical applications of Hopfield networks was the solution of combinatorial optimization problems [3]. The procedure for optimization proceeds by formally matching the target function and the Lyapunov function given by Equation (3), which allows to compute the weights and biases. Then, the resulting system is allowed to evolve, while the condition $\frac{dV}{dt} \leq 0$ implies that the network converges towards a—possibly local—minimum of the target function. When the function V is regarded as a target function, the assumption of symmetry ($w_{ij} = w_{ji}$) is not restrictive, since both weights correspond to the coefficient of the same term $s_i\, s_j$.

In combinatorial optimization problems, the feasible points are placed at vertices of the hypercube, i.e. vectors s so that $|s_i| = 1\ \forall i$. Note that, as far as the feasible set is concerned, the assumption that self weights w_{ii} vanish is sensible, since self weights would correspond to constant terms $s_i^2 = 1$ that are negligible in the target function. The crucial advantage of the Abe formulation consists in the fact that vertices are fixed points, so that $\frac{ds_i}{dt} = 0\ \forall i$, whereas additional fixed points may occur at interior points, which satisfy $net_i = 0$; further, with some mild assumptions, vertices are the only stable fixed points of the system [6]. In particular, in non-trivial systems, vertices and interior points are disjoint sets, i.e. $|s_i| = 1$ and $net_i = 0$ cannot occur simultaneously. We will keep this assumption in the rest of the paper.

In short, the aim of this paper is to determine if the characterization of stable fixed points is preserved after the addition of random noise to the system.

3 Stochastic Stability of Fixed Points

Roughly speaking, the aim of this section is proving that the function given by Equation (3) is a Lyapunov function of the Hopfield network, defined in Equation (1), when the presence of stochastic noise is considered. Once the convergence to some set is proved, the form of this attractor set is ascertained.

3.1 Stochastic Model

In order to carry out the stochastic analysis, the presence of random noise must be formalized, by means of the techniques of Stochastic Differential Equations (SDEs) [10,11]. Let $(\Omega, \mathcal{F}, P)$ be a complete probability space, on which an m-dimensional Brownian motion or Wiener process W is defined $W = \{W(t), t \geq 0\}$ and let $\mathcal{F} = (\mathcal{F}_t, t \geq 0)$ denote the corresponding natural filtration, i.e. $\mathcal{F}_t$ is the σ-algebra generated by the processes $W(s)$ with $0 \leq s \leq t$. Then an additive stochastic perturbation, as in [7], can be considered in Equation (1), resulting in the following SDE:

$$d\,u_i = net_i\, dt + (\sigma(u)\, d\,W_t)_i \tag{4}$$

where $\sigma(x) = (\sigma_{ij}(x))_{n \times m}$ is the noise intensity matrix defined on $\mathbb{R}^n$. As in the deterministic case, it is convenient to rewrite Equation (4) in terms of the variables s_i, which remain bounded. This reformulation is performed by means of the stochastic analogue of the chain rule, namely the general Itô formula, resulting in the following expression:

$$\begin{aligned} d\,s_i &= \frac{1}{\beta}\,(1-s_i^2)\,d\,u_i - \frac{1}{\beta^2}\,s_i\,(1-s_i^2)\,\left(\sigma^\top(u)\,\sigma(u)\right)_{ii}\,dt, \\ &= \frac{1}{\beta}\,(1-s_i^2)\left\{\left[net_i - \frac{s_i}{\beta}\sum_{k=1}^{m}\sigma_{ik}^2\,(u)\right]dt + \sum_{k=1}^{m}\sigma_{ik}\,(u)\,dW_k(t)\right\} \end{aligned} \tag{5}$$

where $u(t) = \beta\,\text{argtanh}(s(t))$ and $x^\top$ denotes the vector transpose. The explicit statement of Equation (5) is more compact and results in simpler proofs than the ones used for Hopfield SDEs.

The differential operator L associated to Equation (5) is defined by:

$$\begin{aligned} L =& \frac{1}{\beta}\sum_{i=1}^{n}(1-s_i^2)\left(net_i - \frac{s_i}{\beta}\sum_{k=1}^{m}\sigma_{ik}^2\,(u)\right)\frac{\partial}{\partial s_i} \\ &+ \frac{1}{2\,\beta^2}\sum_{i=1}^{n}\sum_{j=1}^{n}(1-s_i^2)(1-s_j^2)\left(\sum_{k=1}^{m}\sigma_{ik}(u)\,\sigma_{jk}(u)\right)\frac{\partial^2}{\partial s_i s_j} \end{aligned} \tag{6}$$

Hence, for the function V defined in Equation (3), which was the Lyapunov function of the deterministic system, we obtain:

$$\begin{aligned} LV(s) =& -\frac{1}{\beta}\sum_{i=1}^{n}(1-s_i^2)\,net_i^2 \\ &- \frac{1}{2\,\beta^2}\sum_{i\neq j}(1-s_i^2)(1-s_j^2)\,w_{ij}\sum_{k=1}^{m}\sigma_{ik}\,\sigma_{jk} \\ &+ \frac{1}{\beta^2}\sum_{i=1}^{n}s_i(1-s_i^2)\,net_i\sum_{k=1}^{m}\sigma_{ik}^2 \end{aligned} \tag{7}$$

In the deterministic case, i.e. when $\sigma \equiv 0$, as recalled in Section 2 above, it has been proved [12] that, under some assumptions, the state s converges to the vertices: $|s_i| = 1$. Therefore, our aim is to prove, in the stochastic case defined by Equation (5), that the system still exhibits some kind of convergence to some set K, to be determined.

3.2 Stochastic Stability and Existence of Attractor

The conditions for convergence, which were stated in [9], are here reproduced for completeness, in a slightly more organized way. First of all, it must be determined whether we can assure $LV \leq 0$, which is the stochastic equivalent of the Lyapunov condition $\frac{dV}{dt} \leq 0$ of the deterministic case. The required result stems from the following lemma:

Lemma 1. *Let $(\lambda_i)_i$ be the eigenvalues of the symmetric weight matrix W and define the extreme eigenvalues $\lambda_{min} = \min_i \lambda_i$, $\lambda_{max} = \max_i \lambda_i$. Assume $\lambda_{min} \leq 0$. If the following condition holds:*

$$(\sigma^\top \sigma)_{ii} \leq \frac{2\,\beta\, net_i^2}{|\lambda_{min}|\,(1 - s_i^2) + 2\,|s_i\, net_i|} \qquad \forall i,\ \forall s \in \mathbb{R}^n \tag{8}$$

then $LV(s) \leq 0$.

The proof results from a direct algebraic calculation and it is therefore omitted. Note that we will only be concerned by the case $\lambda_{min} \leq 0$ since the matrix $W = (w_{ij})_{ij}$ associated to the weights satisfies by hypothesis trace$(A) = 0$ which implies that $\lambda_{min} < 0 < \lambda_{max}$.

Stochastic convergence, unlike standard Lyapunov stability, requires further technicalities in addition to the condition $LV \leq 0$, in order to guarantee the convergence to the limit set $\{s : LV(s) = 0\}$. Hence, consider the function V defined by Equation (3) and observe that $(s_t)_t$ is an Itô diffusion, whereas the corresponding L is defined by Equation (7). Then, the following relation results [10]:

$$V(s(t)) = V(s_0) + \int_0^t LV(s(r))dr + M_t \tag{9}$$

where M_t is defined by

$$M_t = -\frac{1}{\beta}\int_0^t \sum_{i=1}^n (1 - s_i^2(r))net_i \sum_{k=1}^n \sigma_{ik}\,(u_r)\,dW_k(r) = -\sum_{k=1}^m \int_0^t H_k(s(r))\,dW_k(r) \tag{10}$$

with $H_k(s(r)) = \frac{1}{\beta}\sum_{i=1}^n (1 - s_i^2(r))\, net_i\, \sigma_{ik}\,(u)$. Consequently, the limit set K can be defined as:

$$K = \{s : LV(s) = 0\} \cap \{s : H_k(s) = 0\,, \forall\, 1 \leq k \leq m\} \tag{11}$$

The analogue of Theorem 2.1 in [7], for the Abe formulation, is stated now:

Theorem 1. *Assume $LV(s) \leq 0$, $\forall s \in \mathbb{R}^n$. Then the set K defined in Equation (11) is not empty and for any initial value $s_0 \in [-1, 1]$, the solution $s(t; s_0) = s(t, \omega; s_0)$ of Equation (5) satisfies*

$$\liminf_{t\to\infty} d(s(t; s_0), K) = 0 \quad a.s. \tag{12}$$

with $d(y, K) = \min_{z\in K} |y - z|$. Moreover, if for any $s \in K$, there exists a neighbourhood of s such that every point $r \neq s$ of this neighbourhood satisfies $V(r) \neq V(s)$, then, for any initial value s_0,

$$\lim_{t\to\infty} s(t; s_0) \in K \quad a.s. \tag{13}$$

3.3 Comparison of the Limit Sets of the Deterministic and the Stochastic Systems

The next step in this research is to compare the attractors of the deterministic and the stochastic cases. Note that, strictly speaking, the attractor of the stochastic network, as defined by Equation (11), is a random set, since the trajectories $s(w,t)$ are stochastic processes, so it is defined in a different space as the deterministic attractor. However, for simplicity of notation, we drop the dependance on w and refer to the stochastic attractor as a normal set, whereas the conditions on the noise will be explicitly stated when necessary. Recall from Section 2 above that the fixed points of the deterministic case are either vertices or interior points, which leads to defining the following set K_0:

$$K_0 = \{s : |s| = 1\} \cup \{s : net_i = 0, \forall i \in J \subset \{1, \cdots, n\}, |s_j| = 1, \forall j \in J^c\} \quad (14)$$

where J^c is the complementary set of J. It is also noticeable that, due to the hyperbolicity assumption, s_i can not satisfy at the same time both condition $net_i = 0$ and $|s_i| = 1$. As mentioned above, it has been proved that, under certain conditions, the vertices are the only stable fixed points of the system defined by Equation (1), which is a favourable property as far as combinatorial optimization is concerned.

In the following, we analyze the relationship between the stochastic limit set K, as defined in Equation (11), and the deterministic set K_0, given by Equation (14). In particular, a long term goal is to find a condition on the noise in order to obtain $K = K_0$. Notice that $K_0 \subset \{s : H_k(s) = 0, \forall k\}$ holds.

- Consider first the case of the **vertices** $K_v = \{s : |s_i| = 1, \forall i\}$. From the definition of the deterministic attractor, Equation (14), it is obvious that $K_v \subset K_0$.

  Consider a vertex $s \in K_v$ and assume that the noise intensity σ is bounded in a neighbourhood of s:

$$(\sigma^\top \sigma)_{ii} \leq \frac{\beta \, net_i^2}{\varepsilon + |net_i|} \sim \beta \, |net_i| \quad \forall i, \, \forall s \in \mathbb{R}^n \quad (15)$$

  with $0 \leq \varepsilon < 1$. Such condition implies $LV(s) \leq 0$, through the application of Lemma 1. Now, since $LV(s) = 0$ and $H_k(s) = 0$, $\forall k$, when s is a vertex, we deduce:

$$K_v \subset K \quad (16)$$

  Note that the condition given by Equation (15) is very weak since the parameter β can always be chosen according to the noise in order to satisfy it. Moreover, it could be weakened by taking different parameters β_i for each i.
- Consider next the **interior points** $K_J = K_0 - K_v$. Given a subset of indices $J \subset \{1, 2, \cdots, n\}$, the interior points s are defined by $K_J = \{s : net_i = 0, \forall i \in J$ and $|s_j| = 1, \forall j \in J^c\}$. When J is a proper subset, these interior points are referred to as **mixed points**, since $|s_i| = 1$ for some indices i. Anyway, the analysis for mixed and interior points is identical.

Assume that the noise satisfies the condition of Lemma 1, given by Equation (8), so that $LV(s) \leq 0$ holds. In the neighbourhood of interior points, this condition reduces to:

$$(\sigma^\top \sigma)_{ii} \leq \frac{2\beta}{|\lambda_{min}|}|net_i|^2 \quad \forall i \in J \tag{17}$$

Then, this assumption on the noise results in $(\sigma^\top \sigma)_{ii} = 0, \forall j \in J$. From the definition of $LV(s)$, in Equation (7), upper and lower bounds of $LV(s)$ can be computed:

$$-\frac{\lambda_{max}}{2\beta^2}\sum_{i \in J}(1-s_i^2)(\sigma^\top \sigma)_{ii} \leq LV(s) \leq 0 \leq \frac{\lambda_{min}}{2\beta^2}\sum_{i \in J}(1-s_i^2)(\sigma^\top \sigma)_{ii}, \tag{18}$$

Combining these bounds with the previous assumption on the noise at an interior point, the identity $LV(s) = 0$ results. To summarize, the following holds:

$$\sigma_{ik} = 0, \forall i \in J, \forall k = 1, \cdots, n \Rightarrow K_J \in K \tag{19}$$

Note that this condition on the noise was anyway necessary, for these interior points s to satisfy $ds_t = 0$.

– Finally, combining the two previous items, we have proved:

$$K_0 \subset K \tag{20}$$

In other words, under stated conditions on the noise, fixed points of the system are also stochastic equilibria of the network.

3.4 Stability of the Fixed Points of the Network

Once proved that the property of a point being an equilibrium is preserved under the addition of stochastic noise, it remains to consider whether the same can be said on stability. That is, we study if a stable (respectively unstable) fixed point of the network is stochastically stable (respectively unstable).

Exponential Instability

Consider again the stochastic network given by Equation (4), which we rewrite for convenience as:

$$d\,u_t = f(u_t)\,dt + \sigma_{u_t}\,dW_t, \tag{21}$$

where $f(x) = net_x = W\tanh\left(\frac{x}{\beta}\right) - b$. Let e denote an equilibrium point, i.e. which satisfies $f(e) = 0$ and $\sigma_e = 0$. Firstly, let us give a useful lemma, which is a direct application of known results ([13], Theorem 4.3.5) by choosing a Lyapunov function $V(x) = (x-e)^\top Q(x-e)$:

Lemma 2. *Assume there exists a positive definite diagonal matrix* $Q = diag(q_i;\ 1 \le i \le n)$*. Let* μ *and* $\rho > 0$ *be real numbers, such that,* $\forall x \in \mathbb{R}^n$*,*

$$2(x-e)^\top Q\, f(x) + \mathrm{trace}(\sigma_x Q \sigma_x^\top) \ge \mu (x-e)^\top Q (x-e) \tag{22}$$

and

$$|(x-e)^\top Q\, \sigma_x|^2 \le \rho \left((x-e)^\top Q (x-e) \right)^2 \tag{23}$$

Then, the solution of the system defined by Equation (21) *satisfies*

$$\liminf_{t\to\infty} \frac{\log |x(t;x_0) - e|}{t} \ \ge\ \frac{\mu}{2} - \rho \quad a.s., \quad whenever\ x_0 \neq e. \tag{24}$$

In particular, if $\rho < \mu/2$*, then the stochastic neural network given by Equation* (21) *is a.s. exponentially unstable.*

Now, since we have

$$\begin{aligned}
(x-e)^\top Q\, f(x) &= \sum_{i=1}^{n} (x_i - e_i)\, q_i \left(W \tanh\left(\frac{x}{\beta}\right) - b \right)_i \\
&= \sum_{i=1}^{n}\sum_{j=1}^{n} (x_i - e_i)\, q_i\, w_{ij} \left(\tanh\left(\frac{x_j}{\beta}\right) - \tanh\left(\frac{e_j}{\beta}\right) \right) \\
&\ge - \sum_{i=1}^{n}\sum_{j=1}^{n} |w_{ij}|\, q_i\, (x_i - e_i) \left(\tanh\left(\frac{x_j}{\beta}\right) - \tanh\left(\frac{e_j}{\beta}\right) \right) \\
&\ge - \frac{1}{\beta} \sum_{i=1}^{n}\sum_{j=1}^{n} |w_{ij}|\, q_i\, (x_i - e_i)\, (x_j - e_j) \qquad (25) \\
&= - \frac{1}{2\beta} \sum_{i,j} (x_i - e_i) |w_{ij}| (q_i + q_j)(x_j - e_j) \\
&\ge - \frac{\lambda_{max}(\tilde{W})}{2\beta} (x-e)^\top (x-e), \\
&\ge - \frac{\lambda_{max}(\tilde{A})}{2\beta \min\limits_{1\le i\le n} q_i} V(x)
\end{aligned}$$

where $\tilde{W} = (\tilde{w}_{ij})$, $\tilde{w}_{ij} = |w_{ij}|(q_i + q_j)$ and the second inequality stems from the mean value theorem and from the fact that $0 \le g'(x) \le \frac{1}{\beta}$. Consequently, one can conclude that:

$$LV(u_t) \ \ge\ -\frac{\lambda_{max}(\tilde{W})}{\beta \min\limits_{1\le i\le n} q_i} V(u_t) \ +\ \mathrm{trace}(\sigma_{u_t} Q\, \sigma_{u_t}^\top). \tag{26}$$

Assume also the following condition:

$$\exists \mu \ge 0 \quad \text{s.t.} \quad \mathrm{trace}(\sigma_x Q \sigma_x^\top) \ \ge\ \mu V(x),\ \forall x \in \mathbb{R}^n, \tag{27}$$

which, together with the conditions given by Equations (23) and (26), allows for the application of Lemma 2, leading to the following theorem:

Theorem 2. *Under the above conditions, whenever* $u_0 \neq e$,

$$\liminf_{t\to\infty} \frac{\log |u(t;u_0) - e|}{t} \geq \frac{1}{2}\left(\mu - \frac{\lambda_{max}(\tilde{A})}{\beta \min_{1\leq i\leq n} q_i}\right) - \rho \quad a.s. \qquad (28)$$

Our ongoing task is to obtain explicit conditions on the noise under which this last result applies when considering points of K_J.

Exponential Stability

Similarly as in the previous paragraph, the following lemma can be stated:

Lemma 3. *Assume there exists a positive definite diagonal matrix* $Q = diag(q_i;\ 1 \leq i \leq n)$. *Let* μ *and* $\rho > 0$ *be real numbers, such that,* $\forall x \in \mathbb{R}^n$,

$$2(x-e)^\top Q\, f(x) + \mathrm{trace}(\sigma_x Q \sigma_x^\top) \leq \mu (x-e)^\top Q (x-e) \qquad (29)$$

and

$$|(x-e)^\top Q\, \sigma_x|^2 \geq \rho\left((x-e)^\top Q(x-e)\right)^2. \qquad (30)$$

Then, the solution of the system given by Equation (21) *satisfies*

$$\limsup_{t\to\infty} \frac{\log |x(t;x_0) - e|}{t} \leq \frac{\mu}{2} - \rho \quad a.s., \quad \text{whenever } x_0 \neq e. \qquad (31)$$

In particular, if $\rho > \mu/2$, *then the stochastic neural network given by Equation* (21) *is a.s. exponentially stable.*

Applying this lemma to our network provides, under the conditions of Lemma 3 and the equivalent of Equation (26), the following theorem:

Theorem 3. *Whenever* $u_0 \neq e$,

$$\limsup_{t\to\infty} \frac{\log |u(t;u_0) - e|}{t} \leq \frac{1}{2}\left(\mu - \frac{\lambda_{min}(\tilde{A})}{\beta \min_{1\leq i\leq n} q_i}\right) - \rho \quad a.s. \qquad (32)$$

Our ongoing task is to prove that this last result applies to the points of K_v.

4 Conclusions

In this paper, the stability analysis of continuous Hopfield networks, in the Abe formulation, is replicated when the presence of random noise is considered. The analysis of stochastic terms requires the formalism of Stochastic Differential Equations. With these techniques, the form of the attractor set is ascertained, and the analysis clarifies the relation between the stability of fixed points of the deterministic network and the stochastic stability of the equilibria of the stochastic network. It is proved that the vertices, which are feasible points of combinatorial optimization problems, belong to the attractor set, whereas interior, unfeasible, points, become unstable. The conditions required on the stochastic noise, in order to prove these results, are stated in the paper, and they are not particularly strong.

We are currently developing this theoretical research in order to weaken the requirements of the theorems and provide conditions that are more applicable and more explicit. Also, the form of the attractor set should be more precisely established. However, the main direction for further research is the practical application of the results to a real-world problem. On one hand, in a computational implementation, the stochastic equations should be discretized by means of numerical methods, which require the analysis of the preservation of the dynamical properties under discretization. On the other hand, a hardware implementation could be an efficient optimization algorithm for large size problems, as long as the influence of random noise on analog electric circuitry is considered. Thus, the stochastic formulation will be carefully considered in order to determine whether it is an accurate model of electric circuits. In particular, the presence of multiplicative, rather than additive, noise, is an appealing direction of research, since it could result in a more realistic representation of actual devices.

References

1. Hopfield, J.: Neural Networks and physical systems with emergent collective computational abilities. Proc. Natl. Acad. Sci. USA 79, 2554–2558 (1982)
2. Hopfield, J.: Neurons with graded response have collective computational properties like those of two-state neurons. Proc. Natl. Acad. Sci. USA 81, 3088–3092 (1984)
3. Tank, D., Hopfield, J.: 'Neural' Computation of Decisions in Optimization Problems. Biological Cybernetics 52, 141–152 (1985)
4. Wilson, G., Pawley, G.: On the Stability of the Travelling Salesman Problem Algorithm of Hopfield and Tank. Biological Cybernetics 58(1), 63–70 (1988)
5. Abe, S.: Theories on the Hopfield Neural Networks. In: International Joint Conference on Neural Networks, vol. 1, pp. 557–564 (1989)
6. Joya, G., Atencia, M.A., Sandoval, F.: Hopfield Neural Networks for Optimization: Study of the Different Dynamics. Neurocomputing 43(1-4), 219–237 (2002)
7. Hu, S., Liao, X., Mao, X.: Stochastic Hopfield neural networks. Journal of Physics A: Mathematical and General 36(9), 2235–2249 (2003)
8. Liu, Y., Wang, Z., Liu, X.: On global exponential stability of generalized stochastic neural networks with mixed time-delays. Neurocomputing 70, 314–326 (2006)
9. Kratz, M., Atencia, M.A., Joya, G.: Stochastic analysis of the Abe formulation of Hopfield networks. In: ESANN 2005. Proc. European Symposium on Artificial Neural Networks, pp. 55–60 (2005)
10. Øksendal, B.: Stochastic Differential Equations. Springer, Heidelberg (2003)
11. Kushner, H.J.: Stochastic Stability and Control. Academic Press, London (1967)
12. Atencia, M.A., Joya, G., Sandoval, F.: Dynamical Analysis of Continuous Higher Order Hopfield Networks for Combinatorial Optimization. Neural Computation 17(8), 1802–1819 (2005)
13. Mao, X.: Stochastic Differential Equations and Applications. Horwood (1998)

Visualization of Dynamics Using Local Dynamic Modelling with Self Organizing Maps

Ignacio Díaz-Blanco[1], Abel A. Cuadrado-Vega[1], Alberto B. Diez-González[1], Juan J. Fuertes-Martínez[2], Manuel Domínguez-González[2], and Perfecto Reguera-Acevedo[2]

[1] University of Oviedo, Area de Ingeniería de Sistemas y Automática, Campus de Viesques s/n, 33204, Gijón, Asturias, Spain
[2] Instituto de Automática y Fabricación, Universidad de León, Escuela de Ingenierías. Campus Universitario de Vegazana, León, 24071, Spain

Abstract. In this work, we describe a procedure to visualize nonlinear process dynamics using a self-organizing map based local model dynamical estimator. The proposed method exploits the topology preserving nature of the resulting estimator to extract visualizations (planes) of insightful dynamical features, that allow to explore nonlinear systems whose behavior changes with the operating point. Since the visualizations are obtained from a dynamical model of the process, measures on the goodness of this estimator (such as RMSE or AIC) are also applicable as a measure of the trustfulness of the visualizations. To illustrate the application of the proposed method, an experiment to analyze the dynamics of a nonlinear system on different operating points is included.

Keywords: System identification, self-organizing map, information visualization.

1 Introduction

Information Visualization techniques have experienced a growing interest for the analysis and interpretation of large volumes of multidimensional data. Borrowed from other areas, such as bioinformatics, socioeconomics or medicine, in last years these techniques have been also used for modelling and supervision of industrial processes from data and previous knowledge. One powerful data visualization method is the self-organizing map (SOM), [1] that allows to project data in a smooth manner on a 2D (or 3D) space that can be subject of graphical representation. This link between process data and a 2D space may be used to obtain 2D maps of the process states that allow to represent different features of the process in an insightful manner, helping to exploit available data and knowledge in an efficient way for process understanding [2].

However, while the SOM has been mostly used to model static relationships among the input variables, it can also be used to model nonlinear dynamical processes using local models that describe the dynamical behavior in a small region in which the process behavior can be considered approximately linear. The

J. Marques de Sá et al. (Eds.): ICANN 2007, Part I, LNCS 4668, pp. 609–617, 2007.

idea of assigning a dynamical model to each unit was already described in the early 90's by Kohonen as *operator maps* in [3], and variations of it using different training approaches were later proposed in [4,5,6]. One particularly efficient variant of this approach was proposed in [7] and later, with slight modifications, in [8], with excellent results, for identification and control problems in systems whose dynamics change over the operating regime.

Surprisingly, while the basic SOM has proven to be an excellent tool for exploratory data analysis, very little work has been published on using it to visualize the dynamical behavior of processes in an explicit way. In this paper, we suggest the use of the local linear modelling approach proposed in [8] introducing the notion of *selectors of dynamics* for the visualization of meaningful dynamical features of the local models, exploiting two major characteristics of this approach: the superior estimation accuracy of this method with respect to global approaches, that supports the accuracy of the resulting visualizations, and the topological order of the resulting local models –which occurs under certain mild hypotheses– that ensures a proper visualization.

2 Local Linear Modelling of Dynamics

Let's consider the following parametric Nonlinear AutoRegresive with eXogenous inputs (p-NARX) system that express the present output as a function of past outputs and inputs as well as of a set of parameters.

$$y(k) = f(y(k-1), ..., y(k-n_y), u(k), ..., u(k-n_u), p_1(k), ..., p_p(k)) \quad (1)$$

This model can be expressed in a more compact way as

$$y(k) = f(\varphi(k), \mathbf{p}(k)) \quad (2)$$

where $\varphi(k) = [y(k-1), \cdots, y(k-n), u(k), \cdots, u(k-m)]^T$ is a vector of known data at sample k, $\mathbf{p} = [p_1(k), \cdots, p_p(k)]^T$ is a vector of parameters, and $f(\cdot, \cdot)$ is a given functional relationship that may be linear or nonlinear. Model (2) describes a dynamic relationship between the process inputs and outputs determined by the values of $p_1(k), ..., p_p(k)$.

2.1 Clustering Dynamics

Let's consider a set of available process variables $\mathbf{s}(k) = [s_1(k), \cdots, s_s(k)]^T \in \mathcal{S}$ that are known to discriminate different dynamical process behaviors, that is, such that $\mathbf{p}(i) \neq \mathbf{p}(j) \Longrightarrow \mathbf{s}(i) \neq \mathbf{s}(j)$; under this hypothesis, these variables can be regarded as *selectors of dynamics*, since information regarding the dynamic relationship between the process' inputs and outputs is preserved in the selectors.

In a first stage, a SOM with N units is trained in the selectors space $\mathcal{S}$ to cluster the process dynamics. Each unit i is associated to a s-dimensional prototype vector $\mathbf{m}_i$ in $\mathcal{S}$ and a position vector on a low dimensional regular grid, $\mathbf{g}_i$, in the output space. For each vector $\mathbf{s}(k)$ the best matching unit is computed as

$$c(k) = \arg\min_i \{\mathbf{s}(k) - \mathbf{m}_i(t)\} \tag{3}$$

Then, an adaptation stage is performed according to

$$\mathbf{m}_i(t+1) = \mathbf{m}_i(t) + \alpha(t)h(c(k), i)\left[\mathbf{s}(k) - \mathbf{m}_i(t)\right] \tag{4}$$

where $\alpha(t)$ is the learning rate and $h(.,.)$ is the neighborhood function. After convergence, the result is a set of prototype vectors that divide $\mathcal{S}$ into a finite set of Voronoi regions each of which defines a different dynamic behavior.

2.2 Local Model Estimation

Once a SOM has been trained with the $\{\mathbf{s}(k)\}$, a model i may be estimated for each prototype $\mathbf{m}_i$ using all the pairs $\{y(k), \varphi(k)\}$ such that

$$\|\mathbf{g}(c(k)) - \mathbf{g}(i)\| \leq \sigma_{\text{loc}} \quad \text{where} \quad c(k) = \arg\min_i \{\mathbf{s}(k) - \mathbf{m}_i(t)\} \tag{5}$$

that is, those pairs whose corresponding selectors $\mathbf{s}(k)$ are mapped onto a neighborhood of $\mathbf{m}_i$ of width σ_{loc}. A particular choice can be a linear model

$$y(k) = \mathbf{p}_i^T \varphi(k) + \varepsilon(k), \quad \text{for neuron } i \tag{6}$$

whose parameters $\mathbf{p}_i$ can be obtained using least squares to fit the aforementioned pairs $\{y(k), \varphi(k)\}$. This corresponds to a local ARX model associated to neuron i, that can be rewritten using the difference equation notation

$$y(k) = \sum_{j=1}^{n_y} a_j^i y(k-j) + \sum_{j=0}^{n_u} b_j^i u(k-j) + \varepsilon(k) \tag{7}$$

or as a transfer function

$$G(z, \mathbf{p}_i) = \frac{Y(z)}{U(z)} = \frac{\sum_{j=0}^{n_u} b_j^i z^{-j}}{1 - \sum_{j=1}^{n_y} a_j^i z^{-j}} \tag{8}$$

being $\mathbf{p}_i = [a_1^i, ..., a_{n_y}^i, b_0^i, b_1^i, ..., b_{n_u}^i]$

2.3 Retrieval

Once the model is trained, a local model $f(\varphi(k), \mathbf{p}_i)$ is assigned to each neuron i. The problem of retrieval is stated as to get $y(k)$ given $\varphi(k)$ and the dynamic selectors $\mathbf{s}(k)$. This is accomplished in two steps: 1) obtain the best matching unit, $c(k) = \arg\min_i\{\mathbf{s}(k) - \mathbf{m}_i(t)\}$ and 2) apply the local model $y(k) = f(\varphi(k), \mathbf{p}_{c(k)})$. Note that depending on the problem (one or multiple step ahead prediction) data vector $\varphi(k)$ may contain real or estimated past outputs.

3 Visualization of Dynamics

3.1 Election of the Dynamic Selectors

After the steps outlined in sections 2.1 and 2.2, the result is a set of prototype vectors $\mathbf{m}_i$ in the space of dynamic selectors, along with companion vectors $\mathbf{p}_i$ containing the model parameters, $\{\mathbf{m}_i, \mathbf{p}_i\}$.

A key issue for visualization is topology preservation. Since the prototype vectors $\mathbf{m}_i$ have been directly obtained from the SOM algorithm (3),(4), they may be expected to be properly ordered for visualization as largely seen in the literature [1,9]. But, what about the $\mathbf{p}_i$?. Under the hypothesis of discriminating dynamic selectors stated in sec. 2.1 –i.e., different dynamical behaviors never occur for the same values of the dynamic selectors– and requiring a somewhat smooth relationship between the selectors and the process dynamics, the SOM topology preservation will maintain this smoothness on the maps by transitivity. In other words, smooth transitions between the dynamics of nearby prototypes will emerge. This means that not only component planes obtained from prototype vectors $\mathbf{m}_i$ but also planes of the elements of $\mathbf{p}_i$, or physically insightful transformations of them, will be smooth and susceptible to be visualized.

Despite these apparently restrictive conditions a proper choice of the dynamic selectors can be often done from problem domain knowledge. This is often the case for nonlinear processes whose local dynamics at sample k, (say, $\mathbf{p}(k)$) depend on the current working point (say, $\mathbf{s}(k)$). Also, proper choices can be made with little prior knowledge inspired on the Takens embedding theorem, as suggested in [7] and [8], such as using an embedded vector with delayed versions of one or more process outputs as dynamic selectors

$$\mathbf{s}(k) = [y(k-1), y(k-2), ..., y(k-q)]^T \tag{9}$$

Note however, that the dynamic selectors do not need to be restricted to the embedding space of the outputs as suggested in [8]. Indeed, other process variables –specially those suspected to define working points with different dynamics– may be included in $\mathbf{s}(k)$ to evaluate their influence on the process dynamics by means of the SOM visualizations.

3.2 Visualization of Dynamic Features

The parameter vectors $\mathbf{p}_i$ convey information about the process dynamics. As seen in eqs. (7) and (8), in the linear case the $\mathbf{p}_i$ is equivalent to a difference equation or a transfer function. A trivial approach to visualize the process dynamics is to visualize the component planes of the elements of $\mathbf{p}_i$. Each component plane [1] is obtained by assigning to each node $\mathbf{g}_i$ in the SOM grid a gray or color level proportional to the scalar values of the elements a_j^i and b_j^i of $\mathbf{p}_i$ associated to node i. To get some insight on the dynamics, however, there are other possible representations that can be expressed as parametric transformations

$T(\mathbf{p}_i, \theta_1, \theta_2, ...)$ of the elements of $\mathbf{p}_i$. A useful representation, for instance, is the *power density* of a transfer function at normalized frequency θ

$$T(\mathbf{p}_i, \theta) = \left| \frac{\sum_{j=0}^{n_u} b_j^i e^{-j\theta}}{1 - \sum_{j=1}^{n_y} a_j^i e^{-j\theta}} \right|^2 \tag{10}$$

Of course any other straightforward transformations in the same idea of (10) are suitable such as peak gains, resonant frequencies, bandwidth, etc.

4 Experimental Results

The proposed method was tested on simulated data of a nonlinear system consisting of a pendulum of mass m and length l mounted on a car whose basis is subject to an acceleration u (see fig. 1).

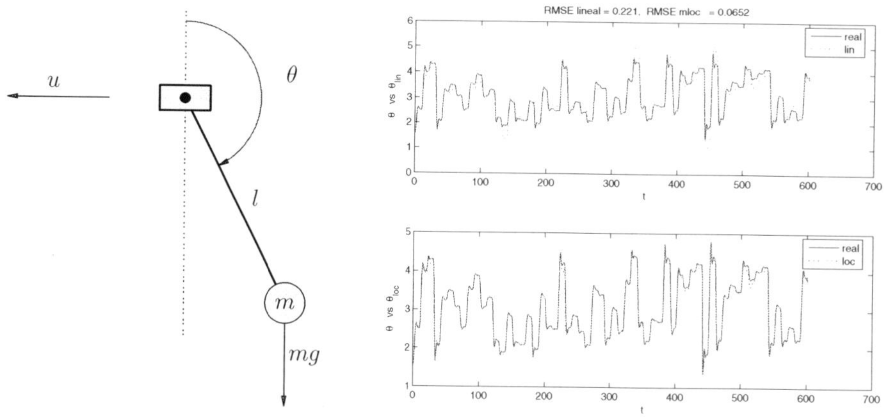

Fig. 1. On the left, the schematics of the pendulum; on the right, comparison of global linear model against local models on test data

$$J\ddot{\theta} + B\dot{\theta} = mgl \sin\theta + mlu \cos\theta \tag{11}$$

where J is the moment of inertia w.r.t. the pivot, B is the friction coefficient and $g = 9.8\text{m/s}^2$ is gravity acceleration. Choosing $\mathbf{x} = [x_1, x_2]^T = [\theta, \dot{\theta}]^T$, and $y = \theta$, it can be expressed in state space form

$$\dot{x}_1 = x_2 \tag{12}$$

$$\dot{x}_2 = \frac{1}{J}(mgl \sin x_1 + mlu \cos x_1 - Bx_2) \tag{13}$$

$$y = x_1 \tag{14}$$

that reveals a nonlinear state space dynamics of the type $\dot{\mathbf{x}} = f(\mathbf{x}, u)$ showing different local behaviors on different state space regions. The linearized model for

Table 1. Comparison of different model orders against validation data

	n_u	n_y	σ_{loc}	RMSE (loc)	RMSE (lin)	AIC (loc)
1	1	1	2	0.163	0.252	−3.59
2	1	1	3	0.143	0.252	−3.84
3	1	2	2	0.168	0.328	−3.52
4	1	2	3	0.145	0.328	−3.81
5	1	3	2	1.18	0.532	0.396
6	1	3	3	0.231	0.532	−2.87
7	2	1	2	0.0557	0.222	−5.72
8	2	1	3	0.06	0.222	−5.58
9	2	2	2	0.0545	0.223	−5.76
→ 10	2	2	3	0.0528	0.223	−5.86
11	2	3	2	18.8	0.493	5.94
12	2	3	3	0.534	0.493	−1.18
13	3	1	2	0.0528	0.235	−5.82
14	3	1	3	0.0645	0.235	−5.46
15	3	2	2	0.108	0.26	−4.38
16	3	2	3	0.451	0.26	−1.57
17	3	3	2	1.18	0.257	0.418
18	3	3	3	100	0.257	100

small movements around an equilibrium point defined by θ_0, in transfer function form is

$$G(s, \theta_0) = \frac{ml \cos \theta_0}{Js^2 + Bs - mgl \cos \theta_0} \tag{15}$$

that, as seen, depends on θ_0. This system is nonlinear and exhibits different local dynamic behaviors depending on the working point defined by θ. An embedded vector with delayed versions of the output, as shown in eq. (9), was selected as dynamic selectors, as suggested in [7]. As in [8], good results were achieved with low values of q, and we finally selected $q = 2$ to yield

$$\mathbf{s}(k) = [\theta(k-1), \theta(k-2)]^T \tag{16}$$

Conceptually, this allows to take into account both the current angle and speed of the pendulum, $[\theta, \dot{\theta}]^T$, since both states can be derived through linear transformations of the elements of $\mathbf{s}(k)$ as $\theta \approx \theta(k-1)$ and $\dot{\theta} \approx \frac{\theta(k-1)-\theta(k-2)}{\Delta t}$

To evaluate the proposed method, the system was simulated under random steps of the car acceleration u of 5 seconds duration each, with a sample time $t_s = 0.5$ sec. and a total simulation time of 5000 sec., resulting in vectors $\theta(k)$ and $u(k)$ of 10000 samples. A 20×20 SOM was trained using the batch algorithm on vectors $\mathbf{s}(k) = [\theta(k-1), \theta(k-2)]^T$ with a PCA normalization and a gaussian neighborhood $h_{ij}(t) = e^{-\frac{d(i,j)}{\sigma(t)^2}}$ where $\sigma(t)$ was made to decrease monotonically from 7 to 0.3.

Once the SOM was trained, local ARX models in the form (7) were trained for all combination of orders $n_y = \{1, 2, 3\}$ and $n_u = \{1, 2, 3\}$ and using local model

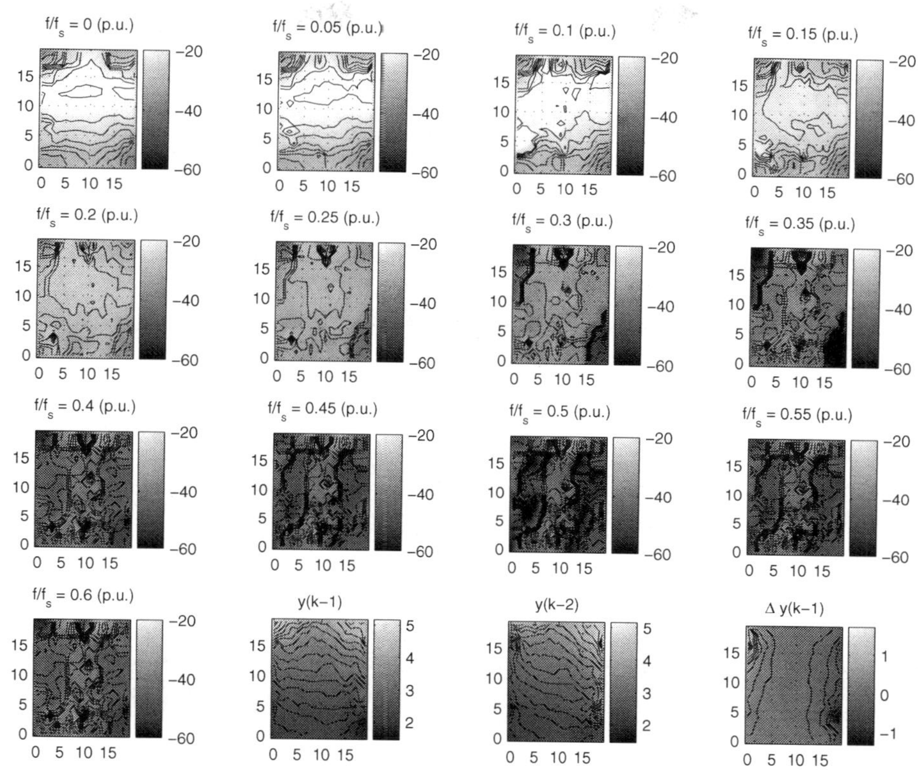

Fig. 2. Frequency response maps. Frequencies are expressed in p.u. with respect to the sample rate f_s; gains at each frequency are expressed in dB.

widths $\sigma_{\text{loc}} = \{2, 3\}$ in eq. (5). For each pair (n_y, n_u) the model was evaluated against validation data, and the Akaike's information criterion (AIC)

$$AIC = \log\left(\frac{1}{N}\sum_{i=1}^{N}[y(i) - \hat{y}(i)]^2\right) + 2d/N \tag{17}$$

was computed for all the local models to select the best model order, where $d = n_y + n_u + 2$ is the total number of model parameters including a bias term, and N is the number of estimation data. From table 1 the best performance is achieved by the 10th model type consisting of ARX(2,2) using a local model width $\sigma_{\text{loc}} = 3$. The prediction ability of this model on test data is compared to the global linear model in fig. 1, showing that it clearly outperforms the global linear model. This suggests that information regarding the system's dynamical behavior is present in the dynamical model based on the local approach. Since the local models are topologically ordered by the SOM, meaningful properties for single local ARX models (such as e.g. spectral densities at given frequencies) can be visualized for all the models by means of SOM maps, providing a comprehensive description of all the dynamic modes of the nonlinear system –see fig. 2.

As a matter of fact, the dc gain can be computed at $s = 0$ in the linearized model (15) for all equilibrium points as $G(0) = -1/g$, regardless the value of the equilibrium point θ_0. However, in the nonlinear system, the angle θ for a constant acceleration u_0 can be shown to be $\theta_0 = \arctan(-u_0/g)$ at steady state, that is, the dc gain is $\frac{\arctan(-u_0/g)}{u_0}$, that approximately coincides with $-1/g$ for small values of u_0, near the working point $\theta_0 = \pi$. This real behavior is shown in the frequency map for $\omega = 0$ (dc gain), where the dc gain of the system is approximately -20 dB ($= 0.1 \approx 1/g$) for values of θ around π, but decreases as the pendulum approaches to the horizontal positions θ around $\pi/2$ or $3\pi/2$.

Finally, other more subtle dynamical relationships can also be observed, such as the relationship between map of angular speed ($\Delta y(k-1) \stackrel{\text{def}}{=} y(k-1)-y(k-2)$) and the harmonics at $0.3f_s$ and $0.35f_s$, for which it seems that the nonlinear model of the pendulum exhibits surprisingly lower gains for large angular speeds regardless the sign of this speed.

5 Conclusions

In this work we have proposed a method to visualize the local dynamical behaviors of nonlinear processes using the SOM local linear approach described in [8] with some assumptions that require some prior knowledge –usually available from problem domain– about the variables that can have an influence on the process dynamics. Since the visualizations are derived from a dynamical model used for estimation, the accuracy of the visual descriptions of dynamics may be measurable, in some sense, using standard performance indices (such as RMSE or AIC) of the estimator on test data. Experimental results with simulated data of a nonlinear system are included to show its application.

While the proposed method has proven to work well under reasonable conditions in other problems, some unsolved questions still remain, such as the requirable conditions for adequate persistence of excitation of the training data, as well as the interpretability of deficient rank or locally unstable models, that may arise even for a good accuracy of the underlying estimator.

Acknowledgement. This research has been financed by the Spanish Ministerio de Educación y Ciencia under grant DPI2006-13477-C02-01.

References

1. Kohonen, T.: The self-organizing map. Proceedings of the IEEE 78(9), 1464–1480 (1990)
2. Alhoniemi, E., Hollmén, J., Simula, O., Vesanto, J.: Process monitoring and modeling using the self-organizing map. Integrated Computer Aided Engineering 6(1), 3–14 (1999)
3. Kohonen, T.: Generalizations of the self-organizing map. In: IJCNN'93-Nagoya. Proceedings of 1993 International Joint Conference on Neural Networks, vol. 1, pp. 457–462 (1993)

4. Vesanto, J.: Using the SOM and local models in time-series prediction. In: WSOM. Proceedings of the Workshop Self-Organizing Maps, June 4-6, pp. 209–214. Helsinki University of Technology, Finland (1997)
5. Fontenla-Romero, O., Alonso-Betanzos, A., Castillo, E., Principe, J., Guijarro-Berdinas, B.: Local Modeling Using Self-Organizing Maps and Single Layer Neural Networks. In: Proceedings of the International Conference on Artificial Neural Networks, pp. 945–950 (2002)
6. Barreto, G., Araujo, A.: Identification and control of dynamical systems using the self-organizing map. Neural Networks, IEEE Transactions 15(5), 1244–1259 (2004)
7. Principe, J.C., Wang, L., Motter, M.A.: Local dynamic modeling with self-organizing maps and applications to nonlinear system identification and control. Proceedings of the IEEE 86(11), 2240–2258 (1998)
8. Cho, J., Principe, J.C., Erdogmus, D., Motter, M.A.: Modeling and inverse controller design for an unmanned aerial vehicle based on the self-organizing map. IEEE Transactions on Neural Networks 17(2), 445–460 (2006)
9. Kohonen, T., Oja, E., Simula, O., Visa, A., Kangas, J.: Engineering applications of the self-organizing map. Proceedings of the IEEE 84(10), 1358–1384 (1996)

Comparison of Echo State Networks with Simple Recurrent Networks and Variable-Length Markov Models on Symbolic Sequences*

Michal Čerňanský[1] and Peter Tiňo[2]

[1] Faculty of Informatics and Information Technologies, STU Bratislava, Slovakia
[2] School of Computer Science, University of Birmingham, United Kingdom
cernansky@fiit.stuba.sk, P.Tino@cs.bham.ac.uk

Abstract. A lot of attention is now being focused on connectionist models known under the name "reservoir computing". The most prominent example of these approaches is a recurrent neural network architecture called an echo state network (ESN). ESNs were successfully applied in more real-valued time series modeling tasks and performed exceptionally well. Also using ESNs for processing symbolic sequences seems to be attractive. In this work we experimentally support the claim that the state space of ESN is organized according to the Markovian architectural bias principles when processing symbolic sequences. We compare performance of ESNs with connectionist models explicitly using Markovian architectural bias property, with variable length Markov models and with recurrent neural networks trained by advanced training algorithms. Moreover we show that the number of reservoir units plays a similar role as the number of contexts in variable length Markov models.

1 Introduction

Echo state network (ESN) [1,2] is a novel recurrent neural network (RNN) architecture based on a rich reservoir of potentially interesting behavior. The reservoir of ESN is the recurrent layer formed of a large number of sparsely interconnected units with non-trainable weights. Under certain conditions RNN state is a function of finite history of inputs presented to the network - the state is the "echo" of the input history. ESN training procedure is a simple adjustment of output weights to fit training data. ESNs were successfully applied in some sequence modeling tasks and performed exceptionally well [3,4]. On the other side part of the community is skeptic about ESNs being used for practical applications [5]. There are many open questions, as noted for example by the author of ESNs [6]. It is still unclear how to prepare the reservoir with respect to the task, what topologies should be used and how to measure the reservoir quality for example.

Many commonly used real-world data with a time structure can be expressed as a sequence of symbols from finite alphabet - symbolic time series. Since their emergence the neural networks were applied to symbolic time series analysis. Especially popular is to use connectionist models for processing of complex language structures. Other works

* This work was supported by the grants APVT-20-030204 and VG-1/4053/07.

J. Marques de Sá et al. (Eds.): ICANN 2007, Part I, LNCS 4668, pp. 618–627, 2007.

study what kind of dynamical behavior has to be acquired by RNNs to solve particular tasks such as processing strings of context-free languages, where counting mechanism is needed [7,8]. Some researchers realized that even in an untrained randomly initialized recurrent network considerable amount of clustering is present. This was first explained in [9] and correspondence to a class of variable length Markov models was shown in [10].

Some attempts were made to process symbolic time series using ESNs with interesting results. ESNs were trained to stochastic symbolic sequences and a short English text in [2] and ESNs were compared with other approaches including Elman's SRN trained by simple BP algorithm in [11]. Promising resulting performance was achieved, superior to the SRN. In both works results of ESNs weren't compared with RNNs trained by advanced algorithms.

2 Methods

2.1 Recurrent Neural Networks

RNNs were successfully applied in many real-life applications where processing time-dependent information was necessary. Unlike feedforward neural networks, units in RNNs are fed by activities from previous time steps through recurrent connections. In this way contextual information can be kept in units' activities, enabling RNNs to process time series.

Elman's simple recurrent network (SRN) proposed in [12] is probably the most widely used RNN architecture. Context layer keeps activities of hidden (recurrent) layer from previous time step. Input layer together with context layer form extended input to the hidden layer. Elman's SRN composed of 5 input, 4 hidden a 3 output units is shown in Fig. 1a.

Common algorithms usually used for RNN training are based on gradient minimization of the output error. Backpropagation through time (BPTT) [13,14] consists of unfolding a recurrent network in time and applying the well-known backpropagation algorithm directly. Another gradient descent approach, where estimates of derivatives needed for evaluating error gradient are calculated in every time step in forward manner,

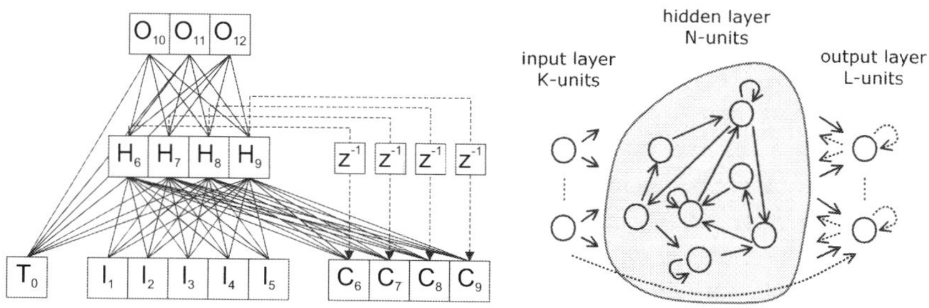

Fig. 1. (a) Elman's SRN and (b) Jaeger's ESN architectures

is the real-time recurrent learning (RTRL) [15,14]. Probably the most successful training algorithms are based on the Kalman filtration (KF) [16]. The standard KF can be applied to a linear system with Gaussian noise. A nonlinear system such as RNNs with sigmoidal units can be handled by extended KF (EKF). In EKF, linearization around current working point is performed and then standard KF is applied. In case of RNNs, algorithms similar to BPTT or RTRL can be used for linearization. Methods based on the Kalman filtration outperform common gradient-based algorithms in terms of in terms of robustness, stability, final performance and convergence, but their computational requirements are usually much higher.

2.2 Echo State Networks

Echo state networks represent a new powerful approach in recurrent neural network research [1,3]. Instead of difficult learning process, ESNs are based on the property of untrained randomly initialized RNN to reflect history of seen inputs - here referred to as "echo state property". ESN can be considered as a SRN with a large and sparsely interconnected recurrent layer - "reservoir" of complex contractive dynamics. Output units are used to extract interesting features from this dynamics, thus only network's output connections are modified during learning process. A significant advantage of this approach is that computationally effective linear regression algorithms can be used for adjusting output weights.

The network includes input, hidden and output "classical" sigmoid units (Fig. 1b). The reservoir of the ESN dynamics is represented by hidden layer with partially connected hidden units. Main and essential condition for successful using of the ESNs is the "echo state" property of their state space. The network state is required to be an "echo" of the input history. If this condition is met, only network output weights adaptation is sufficient to obtain RNN with high performance. However, for large and rich reservoir of dynamics, hundreds of hidden units are needed. When $\mathbf{u}(t)$ is an input vector at time step t, activations of internal units are updated according to

$$\mathbf{x}(t) = f\left(\mathbf{W}^{\mathbf{in}} \cdot \mathbf{u}(t) + \mathbf{W} \cdot \mathbf{x}(t-1) + \mathbf{W}^{\mathbf{back}} \cdot \mathbf{y}(t-1)\right), \tag{1}$$

where f is the internal unit's activation function, $\mathbf{W}$, $\mathbf{W}^{\mathbf{in}}$ and $\mathbf{W}^{\mathbf{back}}$ are hidden-hidden, input-hidden, and output-hidden connections' matrices, respectively. Activations of output units are calculated as

$$\mathbf{y}(t) = f\left(\mathbf{W}^{\mathbf{out}} \cdot [\mathbf{u}(t), \mathbf{x}(t), \mathbf{y}(t-1)]\right), \tag{2}$$

where $\mathbf{W}^{\mathbf{out}}$ is output connections' matrix.

Echo state property means that for each internal unit x_i there exists an echo function e_i such that the current state can be written as $x_i(t) = e_i(u(t), u(t-1), \ldots)$ [1]. The recent input presented to the network has more influence to the network state than an older input, the input influence gradually fades out. So the same input signal history $u(t), u(t-1),$ will drive the network to the same state $x_i(t)$ in time t regardless the network initial state.

2.3 Variable Length Markov Models

As pointed out in [10], the state space of RNNs initialized with small weights is organized in Markovian way prior to any training. To assess, what has been actually learnt during the training process it is always necessary to compare performance of the trained RNNs with Markov models.

Fixed order Markov model is based on the assumption that the probability of symbol occurrence depends only on the finite number of m previous symbols. In the case of the predictions task all possible substrings of length m are maintained by the model. Substrings are prediction contexts of the model and for every prediction context the table of the next symbol probabilities is associated. Hence the memory requirements grow exponentially with the model order m.

To solve some limitations of fixed order Markov models variable length Markov models (VLMMs) were proposed [17,18]. The construction of the VLMM is a more complex task, contexts of various lengths are allowed. The probability of the context is estimated from the training sequence and rare and other unimportant contexts are not included in the model.

2.4 Models Using Architectural Bias Property

Several connectionist models directly using Markovian organization [10] of the RNN's state space were suggested. Activities of recurrent neurons in an recurrent neural network initialized with small weights are grouped in clusters [9]. The structure of clusters reflects the history of inputs presented to the network. This behavior has led to the idea described in [19] where prediction models called neural prediction machine (NPM) and fractal prediction machine (FPM) were suggested. Both use Markovian dynamics of untrained recurrent network. In FPM, activation function of recurrent units is linear and weights are set deterministically in order to create well-defined state space dynamics. In NPM, activation functions are nonlinear and weights are randomly initialized to small values as in regular RNN. Instead of using classical output layer readout mechanism, NPM and FPM use prediction model that is created by extracting clusters from the network state space. Each cluster corresponds to different prediction context with the next symbol probabilities.

More precisely, symbol presented to the network drives the network to some state (activities on hidden units). The state belongs to some cluster and the context corresponding to this cluster is used for the prediction. The context's next symbol probabilities are estimated during training process by relating the number of times that the corresponding cluster is encountered and the given next symbol is observed.

Described prediction model can be created also using activities on recurrent units of the trained RNN. In this article we will refer to this model as NPM built over the trained RNN. RNN training process is computationally demanding and should be justified. More complex dynamics than simple fixed point attractor-based one should be acquired. Hence prediction context of NPM built over the trained RNN usually do not follow Markovian architectural bias principles.

3 Experiments

3.1 Datasets

We present experiments with two symbolic sequences. The first one was created by symbolization of activations of laser in chaotic regime and chaotic nature of the original "real-world" sequence is also present in the symbolic sequence. The second dataset contains words generated by simple context free grammar. The structure and the recursion depths are fully controlled by the designer [10] in this case.

The Laser dataset was obtained by quantizing activity changes of laser in chaotic regime, where relatively predictable subsequences are followed by hardly predictable events. The original real-valued time series was composed of 10000 differences between the successive activations of a real laser. The series was quantized into a symbolic sequence over four symbols corresponding to low and high positive/negative laser activity change. The first 8000 symbols are used as the training set and the remaining 2000 symbols form the test data set [20].

Deep recursion data set is composed of strings of context-free language L_G. Its generating grammar is $G = (\{R\}, \{a, b, A, B\}, P, R)$, where R is the single non-terminal symbol that is also the starting symbol, and a, b, A, B are terminal symbols. The set of production rules P is composed of three simple rules: $R \rightarrow aRb|R \rightarrow ARB|R \rightarrow e$ where e is the empty string. This language is in [7] called palindrome language. The training and testing data sets consist of 1000 randomly generated concatenated strings. No end-of-string symbol was used. Shorter strings were more frequent in the training set than the longer ones. The total length of the training set was 6156 symbols and the length of the testing set was 6190 symbols.

3.2 Performance of ESNs

In this section the predictive performance of ESNs is evaluated on the two datasets. Symbols were encoded using one-hot-encoding, i.e. all input or target activities were set to 0, except the one corresponding to given symbol, which was set to 1. Predictive performance was evaluated by means of a normalized negative log-likelihood (NNL) calculated over the test symbol sequence $S = s_1 s_2 \ldots s_T$ from time step $t = 1$ to T as

$$NNL = -\frac{1}{T} \sum_{t=1}^{T} \log_{|A|} p(t), \tag{3}$$

where the base of the logarithm is the alphabet size, and the $p(t)$ is the probability of predicting symbol s_t in the time step t. For NNL error calculation the activities on output units were first adjusted to chosen minimal activity $o_{\min}$ set to 0.001 in this experiment, then the output probability $p(t)$ for NNL calculation could be evaluated:

$$\hat{o}_i(t) = \begin{cases} o_{\min} & \text{if } o_i(t) < o_{\min} \\ o_i(t) & \text{otherwise} \end{cases}, p(t) = \frac{\hat{o}_i(t)}{\sum_j \hat{o}_j(t)}, \tag{4}$$

where $o_i(t)$ is the activity of the output unit i in time t.

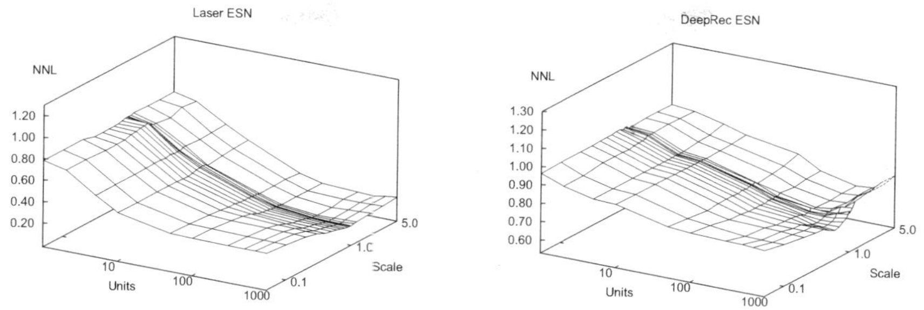

Fig. 2. Performance of ESNs with different unit counts and different values of spectral radius

ESNs with hidden unit count varying from 1 to 1000 were trained using recursive least squares algorithm. Symbols were encoded using one-hot-encoding, i.e. all input or target activities were set to 0, except the one corresponding to given symbol, which was set to 1. Hidden units had sigmoidal activation function and linear activation function was used for output units. Reservoir weight matrix was rescaled to different values of spectral radius from 0.01 to 5. The probability of creating input and threshold connections was set to 1.0 in all experiments and input weights were initialized from interval $(-0.5, 0.5)$. Probability of creating recurrent weights was 1.0 for smaller reservoirs and 0.01 for larger reservoirs. It was found that this parameter has very small influence to the ESN performance (but significantly affects simulation time).

As can be seen from the plots in Fig. 2 results are very similar for wide range of spectral radii. More units in the reservoir results in better prediction. To better asses the importance of reservoir parameterization several intervals for reservoir weights' values were tested starting from $(-0.01, 0.01)$ and ending by $(-1.0, 1.0)$. Also several probabilities of recurrent weight existence were tested from 0.01 to 1.00. Of course no spectral radius rescaling was done in this type of experiments. Various probabilities and intervals for reservoir weights did not influence the resulting performance a lot, hence no figures are shown in the paper. For small weight range and low probability the information stored in the reservoir faded too quickly so the differentiation between points corresponding to long contexts was not possible. This effect was more prominent for the Laser dataset where storing long contexts is necessary to achieve good prediction and hence resulting performance of ESN with weight range of $(-0.1, 0.1)$ and probability 0.01 are worse for higher unit count in the reservoir. Also high probability and wide interval are not appropriate. In this case ESN units are working in the saturated part of its working range very closed to 0.0 and 1.0. Differentiating between states is difficult and hence for example for weight range of $(-1.0, 1.0)$ and probability of 1.0 and higher unit count such as 300 unsatisfactory performance is achieved. For higher values of unit count the performance is worse since the saturation is higher, not because of the overtraining. But for wide range of combinations of these parameters very similar results were obtained. This observation is in accordance with the principles of Markovian architectural bias. Fractal organization of the recurrent neural network state space is scale free and as long as the state space dynamics remains contractive the clusters reflecting the history of the symbols presented to the network are still present.

3.3 Recurrent Neural Networks

In the experiments of this section we show how classical RNNs represented by Elman's SRN perform on the two datasets. Gradient descent approaches such as backpropagation through time or real-time recurrent learning algorithms are widely used by researchers working with symbolic sequences. In some cases even simple backpropagation algorithm is used to RNN adaptation [11,21]. On the other hand, techniques based on the Kalman filtration used for recurrent neural network training on real-valued time series have already shown their potential.

We provide results for standard gradient descent training techniques represented by simple backpropagation and backpropagation through time algorithms and for extended Kalman filter adopted for RNN training with derivatives calculated by BPTT-like algorithm. 10 training epochs (one epoch – one presentation of the training set) for EKF were sufficient for reaching the steady state, no significant NNL improvement has occurred after 10 epochs in any experiment. 100 training epochs for BP and BPTT were done. We improved training by using scheduled learning rate. We used linearly decreasing learning rate in predefined intervals. But no improvements made the training as stable and fast as the EKF training (taking into account the number of epochs). Although it may seem that further training (beyond 100 epochs) may result in better performance, most of BPTT runs started to diverge in higher epochs.

For NNL calculation the value $p(t)$ is obtained by normalizing activities of output units and choosing normalized output activity corresponding to the symbol s_t. NNL performance was evaluated on the test dataset every 1000 training steps.

We present mean and standard deviations of 10 simulations for Elman's SRN with 16 hidden units in Fig. 3. Unsatisfactory simulations with significantly low performance were thrown away. This was usually the case of BP and BPTT algorithms that seems to be much more influenced by initial weight setting and are sensitive to get stuck in local minima or to diverge in later training phase. Generally for all architectures, NNL performances of RNNs trained by EKF are better. It seems to be possible to train RNN by BPTT to have similar performance as the networks trained by EKF, but it usually required much more overhead (i.e. choosing only few from many simulations, more than

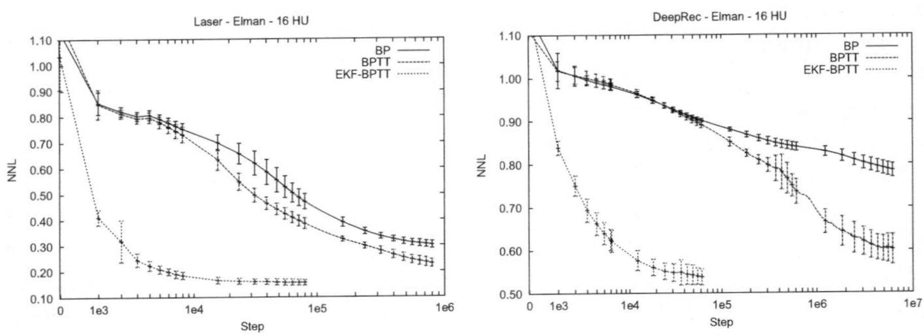

Fig. 3. Performance of Elman's SRN with 16 hidden units trained by BP, BPTT and EKF-BPTT training algorithms

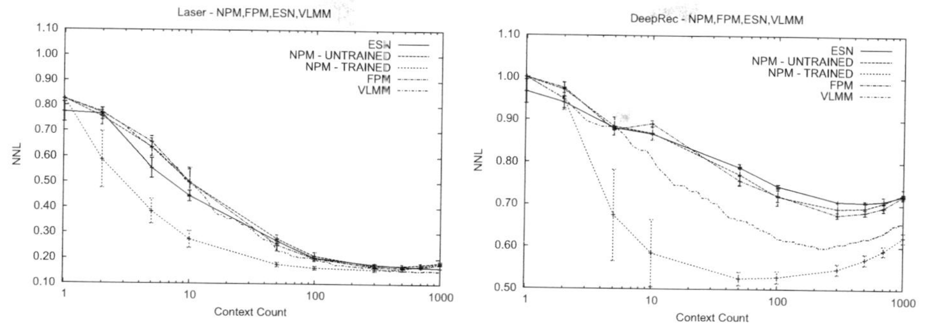

Fig. 4. Performance of ESNs compared to FPMs, NPMs and VLMMs

one thousand of training epochs, extensive experimenting with learning and momentum rates). Also EKF approach to training RNNs on symbolic sequences shows higher robustness and better resulting performance. BP algorithm is too week to give satisfactory results. NNL performances are significantly worse in comparing with algorithms that take into account the recurrent nature of network architectures.

Extended Kalman filter shows much faster convergence in terms of number of epochs and resulting NNLs are better. Standard deviation of results obtained by BPTT algorithm are high revealing BPTT's sensitivity to initial weight setting and to get stuck to local minimum. Although computationally more difficult, extended Kalman filter approach to training recurrent networks on symbolic sequences shows higher robustness and better resulting performance.

3.4 Markov Models and Methods Explicitly Using Architectural Bias Property

To assess what has been actually learnt by the recurrent network it is interesting to compare the network performance with Markov models and models directly using architectural bias of RNNs. Fractal prediction machines were trained for the next symbol prediction task on the two datasets. Also neural prediction machines built over the untrained SRN and also SRN trained by EKF-BPTT with 16 hidden units are tested and results are compared with VLMMs and ESNs. Prediction contexts for all prediction machines (FPMs and NPMs) were identified using K-means clustering with cluster count varying from 1 to 1000.

10 simulation were performed and mean and standard deviation are shown in plots. Neural prediction machines uses dynamics of different networks from previous experiments for each simulation. For fractal prediction machines internal dynamics is deterministic. Initial clusters are set randomly by K-menas clustering hence slightly different results are obtained for each simulation also for FPMs. VLMMs were constructed with the number of context smoothly varying from 1 context (corresponding to the empty string) to 1000 contexts. Results are shown in Fig. 4.

The first observation is that the ESNs have the same performance as other models using architectural bias properties and that the number of hidden units plays very similar role as the number of contexts of FPMs and NPMs built over untrained SRN. For Laser dataset incrementing the number of units resulted in prediction improvement. For

Deep recursion dataset and higher units count (unit counts above 300) ESN model is overtrained exactly as other models. ESN uses linear readout mechanism and the more dimensional state space we have the better hyper-plane can be found with respect to the desired output.

Training can improve the state space organization so better NPM models can be extracted form the recurrent part of the SRN. For Laser dataset the improvement is present for models with small number of context. For higher values of context count the performance of the NPMs created over the trained SRN is the same as for other models. But the carefully performed training process using advanced training algorithm significantly improves the performance of NPMs built over the trained SRN for the Deep recursion dataset.

Significantly better results were achieved by VLMMs on Deep recursion dataset than with ESN or methods based on Markovian architectural bias properties. The reason is in a way how a VLMM tree is constructed. VLMM is built incrementally and the context importance is influenced by Kullback-Leibler divergence between the next symbol distributions of the context and its parent context, context being extended by symbol concatenation. No such mechanism that would take into account the next symbol distribution of the context exists in models based on Markovian architectural bias. Prediction contexts correspond to the clusters that are identified by quantizing the state space. Clustering is based on vectors occurrences (or probabilities) and the distances between vectors. To prove this idea experiments with modified VLMM were performed. Node importance was given by its probability and its length and this type of VLMMs achieved results almost identical to methods based on Markovian architectural bias properties.

4 Conclusion

Extensive simulation using ESNs were made and ESNs were compared with the carefully trained SRNs, with other connectionist models using Markovian architectural bias property and with VLMMs. Multiple parameters for ESN reservoir initialization were tested and the resulting performance wasn't significantly affected. Correspondence between the number of units in ESN reservoir and the context count of FPM, NPM models and Markov models was shown. According to our results ESNs are not able to beat Markov barrier when processing symbolic time series. Carefully trained RNNs or VLMMs can achieve better results on certain datasets. On the other side computational expensive training process may not be justified on other datasets and models such as ESNs can perform just as well as thoroughly trained RNNs.

References

1. Jaeger, H.: The "echo state" approach to analysing and training recurrent neural networks. Technical Report GMD 148, German National Research Center for Information Technology (2001)
2. Jaeger, H.: Short term memory in echo state networks. Technical Report GMD 152, German National Research Center for Information Technology (2001)

3. Jaeger, H.: Adaptive nonlinear system identification with echo state networks. In: Becker, S., Thrun, S., Obermayer, K. (eds.) Advances in Neural Information Processing Systems, vol. 15, pp. 593–600. MIT Press, Cambridge, MA (2003)
4. Jaeger, H., Haas, H.: Harnessing nonlinearity: predicting chaotic systems and saving energy in wireless communication. Science 304(5667), 78–80 (2004)
5. Prokhorov, D.: Echo state networks: Appeal and challenges. In: IJCNN 2005. Proceedings of International Joint Conference on Neural Networks, Montreal, Canada, pp. 1463–1466 (2005)
6. Jaeger, H.: Reservoir riddles: Suggestions for echo state network research. In: IJCNN 2005. Proceedings of International Joint Conference on Neural Networks, Montreal, Canada, pp. 1460–1462 (2005)
7. Rodriguez, P.: Simple recurrent networks learn contex-free and contex-sensitive languages by counting. Neural Computation 13, 2093–2118 (2001)
8. Bodén, M., Wiles, J.: On learning context free and context sensitive languages. IEEE Transactions on Neural Networks 13(2), 491–493 (2002)
9. Kolen, J.: The origin of clusters in recurrent neural network state space. In: Proceedings from the Sixteenth Annual Conference of the Cognitive Science Society, pp. 508–513. Lawrence Erlbaum Associates, Hillsdale, NJ (1994)
10. Tiňo, P., Čerňanský, M., Beňušková, Ľ.: Markovian architectural bias of recurrent neural networks. IEEE Transactions on Neural Networks **15**(1), 6–15 (2004)
11. Frank, S.L.: Learn more by training less: Systematicity in sentence processing by recurrent networks. Connection Science (2006) (in press)
12. Elman, J.L.: Finding structure in time. Cognitive Science 14(2), 179–211 (1990)
13. Werbos, P.: Backpropagation through time; what it does and how to do it. Proceedings of the IEEE 78, 1550–1560 (1990)
14. Williams, R.J., Zipser, D.: Gradient-based learning algorithms for recurrent networks and their computational complexity. In: Chauvin, Y., Rumelhart, D.E. (eds.) Back-propagation: Theory, Architectures and Applications, pp. 433–486. Lawrence Erlbaum Publishers, Hillsdale, N.J (1995)
15. Williams, R.J., Zipser, D.: A learning algorithm for continually running fully recurrent neural networks. Neural Computation 1, 270–280 (1989)
16. Williams, R.J.: Training recurrent networks using the extended Kalman filter. In: IJCNN 1992. Proceedings of International Joint Conference on Neural Networks, Baltimore, vol. 4, pp. 241–246 (1992)
17. Ron, D., Singer, Y., Tishby, N.: The power of amnesia. Machine Learning 25, 117–149 (1996)
18. Machler, M., Bühlmann, P.: Variable length Markov chains: methodology, computing and software. Journal of Computational and Graphical Statistics 13, 435–455 (2004)
19. Tiňo, P., Dorffner, G.: Recurrent neural networks with iterated function systems dynamics. In: International ICSC/IFAC Symposium on Neural Computation (1998)
20. Tiňo, P., Dorffner, G.: Predicting the future of discrete sequences from fractal representations of the past. Machine Learning 45(2). 187–218 (2001)
21. Farkaš, I., Crocker, M.: Recurrent networks and natural language: exploiting self-organization. In: Proceedings of the 28th Cognitive Science Conference, Vancouver, Canada, pp. 1275–1280 (2006)

Data Fusion and Auto-fusion for Quantitative Structure-Activity Relationship (QSAR)

Changjian Huang, Mark J. Embrechts, N. Sukumar, and Curt M. Breneman

RECCR Center, Rensselaer Polytechnic Institute, Troy, NY 12180, USA

Abstract. Data fusion originally referred to the process of combining multi-sensor data from different sources such that the resulting information/model is in some sense better than would be possible when these sources where used individually. In this paper the data fusion concept is extended to molecular drug design. Rather than using data from different sensor sources, different descriptor sets are used to predict activities or responses for a set of molecules. Data fusion techniques are applied in order to improve the predictive (QSAR) model on test data. In this case this type of data fusion is referred to as auto-fusion. An effective auto-fusion functional model and alternative architectures are proposed for a predictive molecular design or QSAR model to model and predict the binding affinity to the human serum albumin.

1 Introduction

Data fusion is an emerging engineering discipline [1] to integrate data from different sources in a predictive model. The basic idea of data fusion is to leverage redundant multi-source data or multi-algorithms to achieve a better performance with the fused system as could be obtained by merging the performance of each system individually, i.e. the whole is better than the sum of the parts.

A commonly referenced data fusion model is the Joint Directors of Laboratories (JDL) model [2] which provides a function-oriented model that was first applied to military projects. Subsequent revisions to the original model refine the fusion terms and function levels [3,4].Although the JDL model has been commonly applied in various different applications, there is no common one-fits-all architecture for these systems because of their diverse nature. Recent data fusion research addressed time series and image based data analysis, but only a few publications concern biometrics or chemometrics. Roussel introduces a data fusion method employing Bayesian inference and shows a significant improvement for a French grape wine classification [5]. Ginn [6] and Hert [7] apply data fusion to combine molecular similarity measures in chemical structure database searches leading to an enhanced performance. With the wide availability of several physically different descriptor sets for the same molecules, and the rapid development of novel predictive QSAR methods, data fusion is a promising methodology that can lead to improved QSAR models. The purpose of this paper is to establish a general data fusion framework for QSAR and to establish procedures and guidelines for data fusion in QSAR modeling.

J. Marques de Sá et al. (Eds.): ICANN 2007, Part I, LNCS 4668, pp. 628–637, 2007.

This paper is organized as follows: section 2 provides an overview of QSAR and data fusion, followed by Sect. 3 describing the performance metrics used in this study. Section 4 proposes an auto-fusion functional model and three alternative architectures for QSAR applications. Section 5 describes a data fusion case study relevant to QSAR for the prediction of the binding affinity to the human serum albumin. Section 6 summarizes findings and conclusions.

2 Background

2.1 QSAR

QSAR refers to a discipline in computational chemistry that addresses the modeling of biological activities or chemical reactivity based on the quantitative description for the chemical structure of molecules. QSAR relies on the basic assumption that molecules with similar physiochemical properties or structures will have similar activities. This is the so called Structure-Activity Relationship (SAR) principle. A key issue is how to formulate quantitative descriptions for molecular structures from which activities or responses can be inferred. The problem is complex because there are several activities such as toxicity, solubility, reaction ability and etc.

A second assumption inherent to QSAR modeling is that molecular properties can be calculated or measured more accurately and cheaply than activities. This is the key motivation for QSAR requiring a mathematical model to infer molecular properties from physical and quantum properties. Once a tentative model is created, all available physicochemical properties such as electronic, steric and hydrophobic properties can be generated by existing programs. Molecular compounds can be described by a vector of properties or descriptors. Associated activities or responses can be modeled or predicted for these compounds based on multivariate statistical and machine learning methods. Popular methods in chemometrics are partial-least squares (PLS), support vector machines (SVM) and kernel-based methods such as kernel partial-least squares (K-PLS).

2.2 Data Fusion

From the late 1970s and throughout the 1980s, automation of data fusion have been intensively studied mainly for defense related applications [8,9,10]. In 1986, the U.S. Joint Directors of Laboratories (JDL) Data Fusion Working Group (DFS) was established to unify the terminology and procedures used in data fusion and resulted in the popular JDL model citeWhite1987-Fusion,White1988-model,White1991-Fusion,Kessler1992-Functional. The JDL model creates a common data fusion framework and categorizes different types of fusion processes based on their functionality.Therefore it is referred to as a functional model rather than an architectural paradigm. The practical implementation of a data fusion system varies depending on the particular nature of the application problem of interest. For that reason there is a need to integrate, tail or further decompose function levels into an overall processing flow that is appropriate for a particular application area.

In 1998, Steinberg [3] proposed extensions to the JDL model [11], including a standardized performance evaluation and other methods related to a system engineering approach and architectural paradigms for data fusion. In 2004, Steinberg and Bowman [4], Llinas and Bowman [12] further refined the definition of the functional data fusion levels to capture the significant differences and to broaden the applicability of a formalized data fusion approach to a wider variety of different problem areas.

3 Performance Metrics

The R^2 and root mean squared error ($RMSE$) are general measures for assessing the multivariate regression models for QSAR. A typical QSAR problem can involve a relative small number of molecules, in which case a leave-one-out cross validation(LOO CV) or Monte-Carlo cross validation (MCCV) [13] can be employed for model selection. The R^2 and $RMSE$ metrics are defined below:

$$R_{cv}^2 = 1 - \frac{PRESS}{SST} = 1 - \frac{\sum\limits_{i\in V} (y_i - \hat{y}_i)^2}{\sum\limits_{i\in V} (y_i - \bar{y}_i)^2}, \quad R_{cv}^2 \in [0, 1] \tag{1}$$

$$RMSE_{cv} = \sqrt{\frac{\sum\limits_{i\in V} (y_i - \hat{y}_i)^2}{n}}, \quad RMSE_{cv} \in [0, \infty] \tag{2}$$

where $\hat{y}_i$ is the predicted value of the ith object in validation set V, and n is the number of objects in V. For external test, R^2 and $RMSE$ are defined similarly, but i belongs to the test set. We will also use the metric of r^2, the squared correlation coefficient between vectors of response y_i and the predicted value $\hat{y}_i$. Unlike r^2, R^2 actually includes additional bias information. Two alternative metrics $q^2 = 1 - r^2$ and $Q^2 = 1 - R^2$ will be used to assess the performance of cross validation data or external test data. For most datasets and good models, the q^2 value is very close to the Q^2 metric. When $q^2 < 0.4$, $Q^2 < 0.4$ and the $RMSE$ is small enough, the model is considered to have predictive power.

4 Auto-fusion Method for QSAR

QSAR models are often built for data sets with limited number of data. Modern descriptor generation algorithms, such as RECON, PEST and MOE, can be employed to extract different descriptor sets for the same set of molecules. These descriptor sets can then be used with appropriate data fusion techniques. The concept where different descriptor sets are used for the same molecules in a data fusion procedure is referred to as an auto-fusion system. The challenge for data fusion is to determine how to combine varying quality data in terms of value for modeling to generate reliable and accurate predictions.

4.1 Functional Model for Data Fusion

Similar to the JDL model, QSAR predictive modeling can be decomposed in several independent levels of functionalities as illustrated in Fig.1:

Level 0: Data alignment - Data standardization process with transformation of $x_i' = \frac{x_i - \bar{x}}{S_x}$, where $\bar{x}$ and S_x are sample mean and sample standard deviation respectively;

Level 1: Feature extraction - Feature extraction algorithm and processing which extracts features from the input data and transfers to other functionalities, such as data alignment and modeling.

Level 2: Modeling - Machine learning methods used for modeling of the available data set, such as SVM, PLS and K-PLS.

Level 3: Model selection and performance estimation (MS&PE) - MS is mainly used for selecting appropriate parameters for learning algorithms in order to maximize the generalization power of the final model. As mentioned in Sect. 3, different cross-validation methods can be used for the MS task. With respect to PE, several different performance metrics are used to assess the performance on external test data or the cross-validation sets.

Level 4: Prediction - the prediction process is mainly used for external test or prediction of unseen data. Training, validation, and test data should be handled consistently with regard to standardization, variable selection and feature extraction. Predicted results need to be restored into the original data space.

4.2 Architectural Paradigm for Data Fusion

Architecture in data fusion addresses the issue of deciding where the fusion must take place in the data flow of the whole system. Analogous with the general JDL model, three alternative auto-fusion architectures are defined for QSAR applications as illustrated in Fig. 2. A detailed description of each follows:

Data level fusion - Fig. 2(a) shows the combination of several sources of raw data generated by different descriptor generation algorithms, such as PEST and MOE, to produce new (raw) data. There is empirical evidence that merely concatenating multidimensional descriptors often only marginally improves the model performance. Data alignment is necessary before implementing predictive modeling for multi-source data.

Feature level fusion - as illustrated in Fig. 2(b), each descriptor generator algorithm provides calculated descriptors from which a feature vector is extracted right after the data alignment. Feature level fusion combines various feature vectors from different descriptors into a new single data set followed by a common model selection, modeling and prediction procedure. The feature extraction can be based on PCA, PLS and/or independent component analysis (ICA).

Hybrid data-decision level fusion - this architecture follows a hybrid scheme, combining data and decision level fusion as illustrated in Fig. 2(c). The procedure from beginning till model selection is identical with the data level fusion. Bagging methods (Breiman 1996b) as a decision level fusion method can now be incorporated in the combined new raw data set.

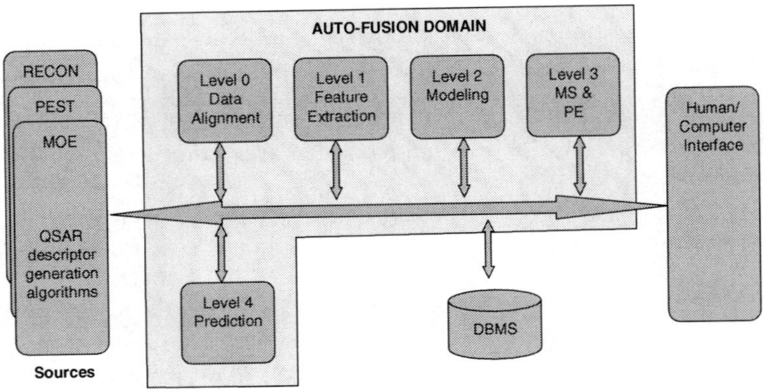

Fig. 1. Functional model of auto-fusion for QSAR

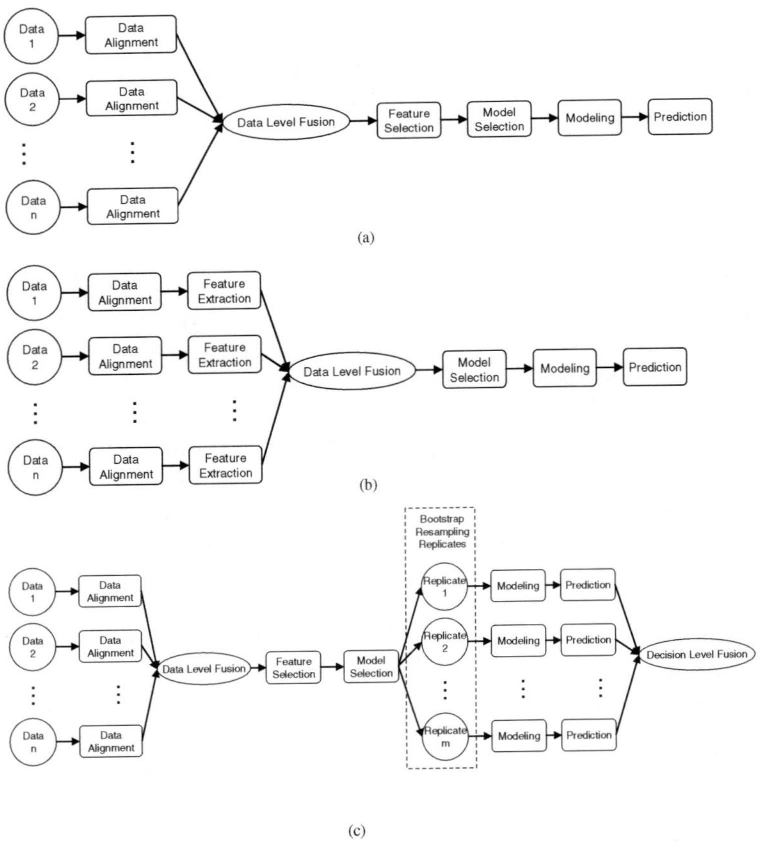

Fig. 2. Three alternative auto-fusion architectures for QSAR

5 Experimental Results

5.1 Problem Description

The binding strength of molecules with the human serum albumin (HSA), a main transporter in the bloodstream, is regarded as the main indicator for the effectiveness for a molecule or drug to reach the target. QSAR models to predict the binding affinities to HSA are part of a computer-aided drug design process [14]. In this study, auto-fusion methods are applied to boost the performance of a QSAR model for the same compounds used in [14], but different descriptor sets were generated as detailed in the next section.

5.2 Albumin Data Set: Description

The original data set includes 95 drugs and druglike compounds. 94 are used in the modeling procedure: captopril is excluded from this study as motivated by [14]. There are two types of descriptors sets that were calculated for the selected molecules: property-encoded surface translator (PEST) descriptors and QuaSAR descriptors [15].

PEST descriptors are an extension of transferable atom equivalent (TAE) and wavelet coefficient descriptors (WCD). TAE descriptors are electron-density derived descriptors, such as kinetic energy density and gradient, local average ionization potential, electrostatic potentials etc. [15]. WCD descriptors are derived from RECON descriptors [16] and are the discrete wavelet transform representation of the atomic RECON histogram properties of the molecule. PEST is a group of hybrid shape/property descriptors which encoded both TAE and shape information. PEST descriptors can be split up into wavelet, histogram, shape and statistics based descriptors. In addition to the PEST descriptors, 2D and i3D MOE (molecular operating environment) descriptors are calculated by the QuaSAR package, the QSAR/QSPR functions in the MOE software.

The generated descriptors sorted by category are: Histogram (100), Wavelet (320), Shape (396), Statistics (80) and MOE (189). The HSA response is the same as in [14]. The data is split into a training set of 84 molecules and an external test set of 10 molecules following [14].

5.3 Classical QSAR Solutions

Classical PLS and K-PLS algorithms are applied to generate a QSAR model for each individual group of descriptors. The experimental results including both cross-validation and external test results as summarized in Tab. 2. Table 1 summarizes the used acronyms. It can be concluded that:

1. The sorted sigma value for each individual group of descriptors are Statistics ($\sigma = 5$) < MOE($\sigma = 12$) = Shape($\sigma = 12$) < Histogram($\sigma = 25$) < Wavelet($\sigma = 100$). The large σ-value used for the wavelet descriptors implies that the Wavelet descriptors have an almost linear relationship with the

Table 1. Acronyms of performance results

Acronym	Description
csThrd	Cousin feature threshold;
olThrd	Times of sigma threshold. 'N' denotes no such preprocessing;
nlvs	Number of latent variables for modeling;
sig	Sigma value of Gaussian kernel;
nrun	For MCCV, nrun represents the number of iterations;
nv	Size of validation set;
Ext Test	External test

Table 2. Performance of classical QSAR modeling (csThrd=85%, olThrd=N; MCCV: nv=10, nrun=100)

Descriptors	q2	Q2	RMSE	Mode	Parameters
Histogram	0.5187	0.5323	0.4065	PLS,LOO CV	nlvs = 4
	0.5331	0.5449	0.4113	PLS,MCCV	nlvs = 4
	0.4332	0.5402	0.476	PLS,Ext Test	olThrd=4, nlvs = 4
	0.523	0.5375	0.4084	KPLS,LOO CV	nlvs = 4, sig = 25
	0.5493	0.5717	0.4237	KPLS,MCCV	nlvs = 4, sig = 25
	0.3288	0.4562	0.4374	KPLS,Ext Test	nlvs = 4, sig = 25
Wavelet	0.5227	0.5233	0.403	PLS,LOO CV CV	nlvs = 2
	0.5558	0.5563	0.4156	PLS,MCCV	nlvs = 2
	0.3735	0.3828	0.4006	PLS,Ext Test	nlvs = 2
	0.5292	0.5296	0.4054	KPLS,LOO CV	nlvs = 2, sig = 100
	0.5237	0.5244	0.4	KPLS,MCCV	nlvs = 2, sig = 100
	0.3748	0.3845	0.4016	KPLS,Ext Test	nlvs = 2, sig = 100
Shape	0.5291	0.5307	0.4058	PLS,LOO CV	nlvs = 2
	0.4815	0.4812	0.3869	PLS,MCCV	nlvs = 3, nv =10, nrun = 100
	0.5448	0.5731	0.4903	PLS,Ext Test	nlvs = 2
	0.532	0.5658	0.4871	PLS,Ext Test	nlvs = 3
	0.5141	0.5147	0.3997	KPLS,LOO CV	nlvs = 2, sig = 12
	0.4573	0.4564	0.3767	KPLS,LOO CV	nlvs = 3, sig = 13
	0.5133	0.5148	0.3963	KPLS,MCCV	nlvs = 2, sig = 11
	0.4519	0.4524	0.3678	KPLS,MCCV	nlvs = 3, sig = 15
	0.4905	0.5229	0.4683	KPLS,Ext Test	nlvs = 2, sig = 12
	0.5414	0.5746	0.4909	KPLS,Ext Test	nlvs = 3, sig = 13
Statistics	0.7312	0.7344	0.4774	PLS,LOO CV	nlvs = 1
	0.7437	0.7456	0.4811	PLS,MCCV	nlvs = 1
	0.544	0.5778	0.4923	PLS,Ext Test	nlvs = 1
	0.7284	0.7324	0.4768	KPLS,LOO CV	nlvs = 1, sig = 7
	0.681	0.6993	0.4619	KPLS,MCCV	nlvs = 2, sig = 5
	0.5977	0.6112	0.5063	KPLS,Ext Test	nlvs = 1, sig = 7
	0.2876	0.3073	0.359	KPLS,Ext Test	nlvs = 2, sig = 5
MOE	0.3055	0.3064	0.3084	PLS,LOO CV	nlvs = 3
	0.3223	0.3236	0.317	PLS,MCCV	nlvs = 3
	0.2242	0.4496	0.4342	PLS,Ext Test	nlvs = 3
	0.2891	0.2893	0.2997	KPLS,LOO CV	nlvs = 3, sig = 12
	0.3294	0.3326	0.3154	KPLS,MCCV	nlvs = 3, sig = 12
	0.166	0.1672	0.2648	KPLS,Ext Test	nlvs = 3, sig = 12

Table 3. Additional acronyms of performance results

Acronym	Description
H	Histogram descriptors;
W	Wavelet descriptors;
B	Shape descriptors;
S	Statistics descriptors;
PEST	The combination of H, W, B and S descriptors;
M	MOE descriptors;
nrun	For MCCV, nrun represents the number of iterations;
	For Bagging, nrun represents the number of resampling bootstrap replicates;
Bagging	Bagging method

Table 4. Performance of auto-fusion for albumin data (csThrd=85%, olThrd=N; MCCV: nv=10, nrun=100; Bagging: nrun=25). B(4) denotes 4 features extracted from Shape descriptors, M(9) denotes 9 features extracted from MOE descriptors and so on.

Fusion Level	q2	Q2	RMSE	Mode	Parameters
Data Level	0.3457	0.3472	0.3283	PLS,LOO CV	PEST+M, nlvs=3
	0.3761	0.3769	0.3421	PLS,MCCV	PEST+M, nlvs=3
	0.1526	0.2022	0.2912	PLS,Ext Test	PEST+M, nlvs=3
	0.3385	0.3412	0.3254	KPLS,LOO CV	PEST+M, nlvs=3, sig=42
	0.3529	0.3586	0.3275	KPLS,MCCV	PEST+M, nlvs=3, sig=42
	0.1887	0.192	0.2838	KPLS,Ext Test	PEST+M, nlvs=3, sig=42
	0.2944	0.2975	0.3039	PLS,LOO CV	B+M, nlvs=3
	0.3153	0.3154	0.3129	PLS,MCCV	B+M, nlvs=3
	0.1817	0.2406	0.3177	PLS,Ext Test	B+M, nlvs=3
	0.2712	0.2736	0.2914	KPLS,LOO CV	B+M, nlvs=3, sig=25
	0.2989	0.3011	0.3001	KPLS,MCCV	B+M, nlvs=3, sig=25
	0.135	0.1537	0.2539	KPLS,Ext Test	B+M, nlvs=3, sig=25
Feature Level	0.3216	0.3226	0.3164	PLS,LOO CV	B(4)+M(9), nlvs=3
	0.3195	0.3239	0.3275	PLS,MCCV	B(4)+M(9), nlvs=3
	0.2059	0.2496	0.3236	PLS,Ext Test	B(4)+M(9), nlvs=4
	0.2919	0.2922	0.3011	KPLS,LOO CV	B(4)+M(9), nlvs=3, sig=12
	0.3195	0.3239	0.3112	KPLS,MCCV	B(4)+M(9), nlvs=3, sig=12
	0.1275	0.1372	0.2399	KPLS,Ext Test	B(4)+M(9), nlvs=3, sig=12
	0.2879	0.2885	0.2992	KPLS,LOO CV	B(4)+M(17), nlvs=3, sig=12
	0.3194	0.3224	0.3105	KPLS,MCCV	B(4)+M(17), nlvs=3, sig=12
	0.1233	0.1267	0.2305	KPLS,Ext Test	B(4)+M(17), nlvs=3, sig=12
	0.3037	0.3046	0.3075	KPLS,LOO CV	B(4)+S(2)+M(17), nlvs=3, sig=12
	0.3147	0.3153	0.3146	KPLS,MCCV	B(4)+S(2)+M(17), nlvs=4, sig=12,
	0.1108	0.1241	0.2281	KPLS,Ext Test	B(4)+S(2)+M(17), nlvs=4, sig=12
	0.45	0.4521	0.3746	KPLS,LOO CV	H(10)+W(8), nlvs=3, sig=30
	0.4444	0.453	0.3681	KPLS,MCCV	H(10)+W(8), nlvs=3, sig=30
	0.236	0.2809	0.3432	KPLS,Ext Test	H(10)+W(8), nlvs=3, sig=30
Hybrid Data-	0.1379	0.1413	0.2435	PLS,Bagging	PEST+M, nlvs=3
Decision Level	0.2642	0.3438	0.3797	KPLS,Bagging	PEST+M, nlvs=3, sig=42
	0.1616	0.1795	0.2743	PLS,Bagging	B+M, nlvs=3
	0.122	0.1517	0.2522	KPLS,Bagging	B+M, nlvs=3, sig=25

response. This can be validated by comparing the performance of a PLS model with a K-PLS model for the Wavelet descriptors. In this case the PLS model actually gives a better performance than the K-PLS model.
2. The sorted best generalization power of each model in terms of Q^2 are: MOE (0.1672) > Statistics (0.2876) > Wavelet (0.3828) > Histogram (0.4562) > Shape (0.5229).
3. When $\sigma < 25$, K-PLS has generally better performance than PLS. This also implies that the corresponding descriptors have a nonlinear relationship with the response.

5.4 Auto-fusion for Albumin

The auto-fusion experiments are conducted for three architectures respectively. Specifically, PCA is used for feature extraction. Both cross validation and external test results are summarized in Tab. 4. Table 3 lists additional acronyms used. It is evident that:

1. When combining Shape and MOE descriptors under data level and hybrid data-decision level fusion architectures, the performance is better than the best performance based on the individual descriptors, i.e. MOE descriptors in this case. However, this is not the case when combining PEST and MOE descriptors. How to combine different data sets with different predictive qualities is crucial for the data fusion system;
2. In most cases, a hybrid data-decision level fusion shows a better performance than that of a data level fusion. Feature level fusion is consistently better than either one of them, and consequently better than any individual group of descriptors;
3. The optimal number of latent variables and the Parzen window σ can change while combining descriptor sets. Combining descriptors with similar performance and similar σ-values generally improves the overall performance. In feature level fusion, the number of latent variables selected for each group of descriptors is related to the number of latent variables for the individual experiment. In the experiment for both Shape and MOE descriptors the same number of latent variables and similar σ-values were used in the individual experiments. The data fusion model leads to an improved prediction performance for the same parameter settings.

6 Conclusion

In this paper, an auto-fusion framework for QSAR applications is proposed. By examining the performance of the HSA data set, we draw the following conclusions: 1) Auto-fusion method consistently yield an improved prediction performance; 2) Selecting specific descriptors sets to be combined is crucial for fusion systems; 3) Among the five generated descriptors mentioned in Sect. 5.2, Statistics, MOE and Shape descriptors show a strong nonlinear relationship with the response, while the Histogram and Wavelet descriptors exhibit a linear relationship.

Acknowledgment

This publication was made possible by Roadmap Grant Number 1 P20 HG003899-01 from the National Human Genome Research Institute (NGHRI) at the National Institutes of Health (NIH). Its contents are solely the responsibility of the authors and do not necessarily represent the official views of NGHRI.

References

1. Hall, D.L., McMullen, S.A.H.: Mathematical Techniques in Multisensor Data Fusion, 2nd edn. Artech House Publishers (2004)
2. Kessler, et. al.: Functional description of the data fusion process. Technical report (1992)
3. Steinberg, A.N., Bowman, C.L., White, F.E.: Revisions to the JDL data fusion model. In: Joint NATO/IRIS Conference Proceedings, Quebec, October (1998), In: Sensor Fusion: Architectures, Algorithms, and Applications, Proceedings of the SPIE, Vol.3719, 1999 (1998)
4. Steinberg, A.N., Bowman, C.L.: Rethinking the JDL data fusion model. In: Proceedings of the 2004 MSS National Symposium Sensor on Sensor and Data Fusion, vol. 1 (2004)
5. Roussel, S., Bellon-Maurel, V., Roger, J.M., Grenier, P.: Fusion of aroma, FT-IR and UV sensor data based on the bayesian inference. application to the discrimination of white grape varieties. Chemometrics and Intelligent Laboratory Systems 65, 209–219 (2003)
6. Ginn, C.M.R., Willett, P., Bradshaw, J.: Combination of molecular similarity measures using data fusion. Perspectives in Drug Discovery and Design 20 (2000)
7. Hert, J., Willett, P., Wilton, D.J.: New methods for ligand-based virtual screening: Use of data fusion and machine learning to enhance the effectiveness of similarity searching. Journal of Chemical Information and Modeling 46, 462–470 (2006)
8. Shafer, G.: A Mathematical Theory of Evidence. Princeton University Press, Princeton (1976)
9. Rogers, W.: An evaluation of statistical matching. Journal of Business and Economic Statistics 2(1), 91–102 (1984)
10. Hall, D.L., Llinas, J.: Handbook of Multisensor Data Fusion. CRC Press, USA (2001)
11. White, F.: A model for data fusion. In: Proc. 1st National Symposium on Sensor Fusion (1988)
12. Llinas, J., Bowman, C., Rogova, G., Steinberg, A., Waltz, E., White, F.: Revisiting the JDL data fusion model II. In: Proceedings 7th International Conference on Information Fusion, Stockholm, Sweden (2004)
13. Xu, Q.S., Liang, Y.Z., Du, Y.P.: Monte Carlo cross-validation for selecting a model and estimating the prediction error in multivariate calibration. Journal of Chemometrics 18(2), 112–120 (2004)
14. Colmenarejo, G., Alvarez-Pedraglio, A., Lavandera, J.L.: Cheminformatic models to predict binding affinities to human serum albumin. Journal of Medicinal Chemistry 44, 4370–4378 (2001)
15. Breneman, C.M., Sundling, C.M., Sukumar, N., Shen, L., Katt, W.P., Embrechts, M.J.: New developments in PEST shape/property hybrid descriptors. Journal of Computer-Aided Molecular Design 17, 231–240 (2003)
16. Rhem, M.O.: RECON: An Algorithm for Molecular Charge Density Reconstruction Using Atomic Charge Density Fragments. PhD thesis, Rensselaer Polytechnic Institute (1996)

Cluster Domains in Binary Minimization Problems

Leonid B. Litinskii

Centre of Optical Neural Technologies SRISS RAS, Moscow
litin@mail.ru

Abstract. Previously it was found that when minimizing a quadratic functional depending on a great number of binary variables, it is reasonable to use aggregated variables, joining together independent binary variables in blocks (domains). Then one succeeds in finding deeper local minima of the functional. In the present publication we investigate an algorithm of the domains formation based on the clustering of the connection matrix.

Keywords: Neural networks, discrete minimization, domain dynamics.

1 Introduction

We discuss the problem of minimization of a quadratic functional depending on N binary variables $s_i = \{\pm 1\}$:

$$E(\mathbf{s}) = -(\mathbf{Js}, \mathbf{s}) = -\sum_{i,j=1}^{N} J_{ij} s_i s_j \xrightarrow[\mathbf{s}]{} \min. \qquad (1)$$

We suppose that the connection matrix $\mathbf{J} = (J_{ij})_1^N$ is a symmetric one with zero diagonal elements: $J_{ij} = J_{ji}$, $J_{ii} = 0$. We make use of physical terminology [1]: the binary variables $s_i = \{\pm 1\}$ will be called *spins,* N-dimensional vectors $\mathbf{s} = (s_1, s_2, ..., s_N)$ will be called *configurations*, and the characteristic $E(\mathbf{s})$, which has to be minimized will be called *the energy* of the state $\mathbf{s}$.

The commonly known procedure of minimization of the functional (1) is as follows: one randomly searches through N spins, and to each spin the sign of the local field $h_i(t)$ acting on this spin is assigned, where

$$h_i(t) = \sum_{j=1}^{N} J_{ij} s_j(t). \qquad (2)$$

In other words, if the current state of the spin $s_i(t)$ coincides with the sign of $h_i(t)$ (if the spin is *satisfied*: $s_i(t)h_i(t) \geq 0$), one does not change the value of the spin: $s_i(t+1) = s_i(t)$; but if $s_i(t)$ and $h_i(t)$ are of opposite signs (if the spin is *unsatisfied*: $s_i(t)h_i(t) < 0$), in the next moment one turns the spin over: $s_i(t+1) = -s_i(t)$.

J. Marques de Sá et al. (Eds.): ICANN 2007, Part I, LNCS 4668, pp. 638–647, 2007.

It is well known that the energy of the state decreases when an unsatisfied spin turns over: $E(\mathbf{s}(t)) > E(\mathbf{s}(t+1))$. Sooner or later the system finds itself in the state, which is an energy minimum (may be this is a local minimum). In this state all the spins are satisfied and evolution of the system ends. In what follows the aforementioned algorithm will be called *the random dynamics* (in the neural networks theory it is called *the asynchronous dynamics* [2]).

Another well known minimization procedure is the synchronous dynamics, when each time all the unsatisfied spins turn over simultaneously [2]. This approach is rarely used in minimization problems. First, this is because it is impossible to guarantee the monotonous decrease of the functional $E(\mathbf{s})$. Second, synchronous dynamics is characterized by the presence of limit cycles of the length 2 (the system sequentially finds itself in one of the two states $\mathbf{s}_1$ and $\mathbf{s}_2$). Due to limit cycles the basin of attraction of local minima decreases.

In the papers [3], [4] a generalization of random dynamics, *the domain dynamics*, was proposed. The essence of the generalization is joining the spins together in larger blocks or, as they were called by the authors, *domains*. During the evolution not single spin turns over, but the whole block/domain. As for the rest the domain approach is the same as the standard random dynamics. It was found that the domain dynamics offers a lot of advantages comparing with other dynamic approaches. The domain dynamics is k^2 times quicker than the random dynamics, where k is the averaged length of a domain (the number of spins joined in one block). Moreover, when using the domain dynamics the energy $E(\mathbf{s})$ decreases monotonically, and no limit cycles appear. Computer simulations [4] showed that the domain dynamics allowed one to obtain deeper local minima of the functional (1) than the standard random dynamics. However, there is an open question: how one has to choose the domains to obtain the deepest minima? In the paper we discuss this question.

Note. The idea of the domain dynamics is so natural that, probably, it was proposed more than once by different authors. In particular, preparing this publication we found out that the domain dynamics is practically equivalent to the block-sequential dynamics presented in [5]. We note that in [5] the block-sequential dynamics was analyzed with regard to the problem of increasing the storage capacity of the Hopfield model. As far as we know, in [3], [4] the domain dynamics was used for minimization of the functional (1) for the first time.

2 The Domain Dynamics

For simplicity of presentation all definitions are given for the case when the first k spins are joined in the 1st domain, the k following spins are joined in the 2nd domain and so on; the last k spins are joined in the last nth domain:

$$\mathbf{s} = (\underbrace{s_1,\ldots,s_k}_{\text{1st domain}}, \underbrace{s_{k+1},\ldots,s_{2k}}_{\text{2nd domain}},\ldots,\underbrace{s_{(n-1)k+1},\ldots,s_N}_{n\text{th domain}}). \tag{3}$$

In Eq. (3) $N=kn$, and $\mathbf{s}$ is an arbitrary configuration.

The total action onto the first domain from all other domains is the superposition of all interactions of the spins, which do not belong to the first domain, with the spins of the first domain. Then *the local domain field* acting onto the spin belonging to the first domain is equal to

$$h_i^{(d)}(t) = \sum_{j=k+1}^{N} J_{ij}s_j(t) = h_i(t) - \sum_{j=1}^{k} J_{ij}s_j(t), \ \forall\, i \le k, \tag{4}$$

where $h_i(t)$ is the local field (2). In other words, the local domain field acting on the ith spin is obtained by means of elimination of the influence of the spins belonging to the same domain from the local field $h_i(t)$ (2). It is clear that the energy of interaction of the first domain with all other domains is equal to

$$E_1(t) = -F_1(t) = -\sum_{i=1}^{k} s_i(t)h_i^{(d)}(t).$$

In the same way we define the local domain field acting onto the spins belonging to the lth domain,

$$h_i^{(d)}(t) = h_i(t) - \sum_{j=(l-1)k+1}^{lk} J_{ij}s_j(t), \ \forall\ 1+(l-1)k \le i \le lk, \ l \in [2,n].$$

The energy of interaction of the lth domain with other domains is:

$$E_l(t) = -F_l(t) = -\sum_{i=(l-1)k+1}^{lk} s_i(t)h_i^{(d)}(t), \quad l = 2,..,n.$$

Then *the domain energy* of the state $\mathbf{s}(t)$ is

$$E^{(d)}(t) = \sum_{l=1}^{n} E_l(t) = -\sum_{l=1}^{n} F_l(t). \tag{5}$$

The energy (1), which we have to minimize, differs from the domain energy by the sum of the energies $E_l^{(in)}$ that characterize the inter-domain interactions only:

$$E(t) = E^{(d)}(t) + \sum_{l=1}^{n} E_l^{(in)}(t) = E^{(d)}(t) - \sum_{l=1}^{n}\ \sum_{i=(l-1)k+1}^{lk}\ \sum_{j=(l-1)k+1}^{lk} J_{ij}s_i(t)s_j(t), \tag{6}$$

After that we can define *the domain dynamics* [3],[4]: one randomly searches through n domains, and if for the lth domain the inequality $F_l(t) \ge 0$ is fulfilled, the domain remains unchanged; but if $F_l(t) < 0$, in the next moment one turns over the lth domain – all the spins of the domain receive the opposite sings simultaneously: $s_i(t+1) = -s_i(t), \ i \in [1+(l-1)k, lk]$.

From Eq. (5) it is easy to obtain that when using the domain dynamics, the domain energy of the state decreases monotonically: if in the moment t the mth domain is turned over, the domain energy goes down by the value $4|F_m(t)|$: $E^{(d)}(t+1) = E^{(d)}(t) - 4|F_m(t)|$. Note, in the same time the energy $E(\mathbf{s}(t))$ (1) goes down just by the same value. This follows from Eq. (6) and the obvious fact that simultaneous overturn of all spins belonging to the mth domain does not change the inter-domain energy $E_m^{(in)}$.

As a result of the aforementioned procedure sooner or later the system finds itself in *the domain local minimum*. In this state for all the domains the inequality $F_l(t) \geq 0$ is fulfilled, and the domain evolution ends. However, the local domain minimum is not necessarily a minimum of the functional (1). That is why in this state one has "to defrost" the domains and use the standard random dynamics. The dynamic system has the possibility to descend deeper into the minimum of the functional (1).

The successive use of the domain dynamics and after that the standard random dynamics is connected with the following argumentation. Minimization of the functional (1) can be interpreted as the sliding of the dynamic system down a hillside that is dug up by shallow local minima. In the case of the random dynamics two successive states of the system differ by an opposite sign of only one binary coordinate, namely those that is turned over during the given step of evolution. It is evident that under this dynamics the system sticks in the first occurring local minimum. On the contrary, under the domain dynamics two successive states of the system differ by opposite signs of some spin coordinates at once. The domain dynamics can be likening to sliding down by more "large-scale steps". It can be expected that such way of motion allows the system to leave aside a lot of shallow local minima, where it can stick in the case of standard random dynamics.

As noted above, for simplicity we gave all the definitions supposing that all the domains are of the same length k. In the general case domains can be of different lengths k_l, and any spins can be joined in a domain. In this connection there are some questions: does the choice of the domains influence the results of minimization? How the domains have to be organized? Have they been invariable for a given matrix **J**, or have they been organized randomly at every step of evolution? Generally, are there any argumentations, which can help? In the next section we formulate our recipe of the domain formation and present arguments proving it. In Section 4 we show the results of computer simulation.

3 Cluster Approach of Domain Formation

For the Hopfield model it is known the situation when domains are naturally appeared due to specific properties of the Hebbian connection matrix **J** [6]. Let us cite the corresponding results, accentuating the points we need. In the end of this Section we formulate the recipe of the domain formation for the general case.

Let us have N M-dimensional vector-columns with binary coordinates $\mathbf{x}_i \in \mathbf{R}^M$, $x_i^{(\mu)} = \pm 1$, $i = 1,...,N$, $\mu = 1,..,M$. Vector-columns $\mathbf{x}_i$ are numerated by subscripts $i \in [1, N]$, and their coordinates by superscripts $\mu \in [1, M]$.

Let us construct (MxN)-matrix **X**, whose columns are M-dimensional vectors $\mathbf{x}_i$:

$$\mathbf{X} = (\mathbf{x}_1 ... \mathbf{x}_i ... \mathbf{x}_N) : \mathbf{R}^N \to \mathbf{R}^M .$$

In the theory of neural networks **X** is called *the pattern matrix*. It is used to construct the Hebbian matrix that is (NxN)-matrix of scalar products of the vectors $\mathbf{x}_i$:

$$J_{ij}=\frac{(1-\delta_{ij})}{M}(\mathbf{x}_i,\mathbf{x}_j)=\frac{(1-\delta_{ij})}{M}\sum_{\mu=1}^{M}x_i^{(\mu)}x_j^{(\mu)},\ 1\le i,j\le N.$$

The matrix elements J_{ij} are cosines of the angles between the corresponding vectors: $|J_{ij}|\le 1$.

We shall examine a special case, when the number of *different* vector-columns in the matrix **X** is not *N*, but a smaller number *n*: $\mathbf{x}_1...\neq..\mathbf{x}_l...\neq..\mathbf{x}_n$, $n<N$. Then each vector-column $\mathbf{x}_i$ will be repeated some times. Let k_l be the number of repetitions of the vector-column $\mathbf{x}_l$: $\sum_1^n k_l=N$. (The subscripts *l*, *m* are used to numerate different vector-columns and related characteristics.) Without loss of generality it can be assumed that the first k_1 vector-columns of the matrix **X** coincide with each other, the next k_2 vector-columns coincide with each other, and so on; the last k_n vector-columns coincide with each other too:

$$\mathbf{X}=\begin{pmatrix} x_1^{(1)} & \cdots & x_1^{(1)} & x_2^{(1)} & \cdots & x_2^{(1)} & & x_n^{(1)} & \cdots & x_n^{(1)} \\ x_1^{(2)} & \cdots & x_1^{(2)} & x_2^{(2)} & \cdots & x_2^{(2)} & \cdots & x_n^{(2)} & \cdots & x_n^{(2)} \\ \cdots & \cdots & \cdots & \cdots & \cdots & \cdots & & \cdots & \cdots & \cdots \\ \underbrace{x_1^{(M)} \quad \cdots \quad x_1^{(M)}}_{k_1} & & & \underbrace{x_2^{(M)} \quad \cdots \quad x_2^{(M)}}_{k_2} & & & & \underbrace{x_n^{(M)} \quad \cdots \quad x_n^{(M)}}_{k_n} & & \end{pmatrix},\ k_l\ge 1. \tag{7}$$

The corresponding Hebbian matrix consists of $n\times n$ blocks. The dimensionality of the (lm)th block is equal to $(k_l\times k_m)$, and all its elements are equal to the same number J_{lm} that is the cosine of the angle between vectors $\mathbf{x}_l$ and $\mathbf{x}_m$ ($l,m=1,...,n$). On the main diagonal there are quadratic $(k_l\times k_l)$-blocks; these blocks consist of ones only (with the exception of the diagonal, where all the elements are equal to zero):

$$\mathbf{J}=\begin{pmatrix} \overbrace{\begin{matrix}0&1&1\\1&\ddots&1\\1&1&0\end{matrix}}^{k_1} & \overbrace{\begin{matrix}J_{12}&\cdots&J_{12}\\\vdots&J_{12}&\vdots\\J_{12}&\cdots&J_{12}\end{matrix}}^{k_2} & \cdots & \overbrace{\begin{matrix}J_{1n}&\cdots&J_{1n}\\\vdots&J_{1n}&\vdots\\J_{1n}&\cdots&J_{1n}\end{matrix}}^{k_n} \\ \begin{matrix}J_{21}&\cdots&J_{21}\\\vdots&J_{21}&\vdots\\J_{21}&\cdots&J_{21}\end{matrix} & \begin{matrix}0&1&1\\1&\ddots&1\\1&1&0\end{matrix} & \cdots & \begin{matrix}J_{2n}&\cdots&J_{2n}\\\vdots&J_{2n}&\vdots\\J_{2n}&\cdots&J_{2n}\end{matrix} \\ \vdots & \vdots & \ddots & \vdots \\ \begin{matrix}J_{n1}&\cdots&J_{n1}\\\vdots&J_{n1}&\vdots\\J_{n1}&\cdots&J_{n1}\end{matrix} & \begin{matrix}J_{n2}&\cdots&J_{n2}\\\vdots&J_{n2}&\vdots\\J_{n2}&\cdots&J_{n2}\end{matrix} & \cdots & \begin{matrix}0&1&1\\1&\ddots&1\\1&1&0\end{matrix} \end{pmatrix},\ |J_{lm}|<1. \tag{8}$$

It was found out [6] that such organization of the connection matrix imposes the form of local minima of the functional (1), namely: local minima certainly have the block-constant form

$$\mathbf{s} = (\underbrace{s_1 \ \ldots \ s_1}_{k_1}, \underbrace{s_2 \ \ldots \ s_2}_{k_2}, \ldots, \underbrace{s_n \ \ldots \ s_n}_{k_n}), \ s_l = \pm 1, \ l = 1, \ldots, n. \tag{9}$$

In other words, the first k_1 coordinates of the local minimum have to be identical; the next k_2 coordinates have to be equal to each other, and so on. In Eq. (9) fragmentation into blocks of constant signs is defined by the block structure of the Hebbian matrix (or, and that is the same, by the structure of the repeating columns in the pattern matrix **X** (7)). The proof that the local minima have the block-constant form (9) is based on the fact that spins from one block have the same connections with all other spins, and under the action of the local fields they behave in the same way [6].

Not all configurations of the form (9) are the local minima of the functional (1), but it is necessary to look for the local minima among these configurations only. In other words, for minimization of the functional (1) it is senseless to examine configurations with different signs inside blocks of constant signs. It makes sense to examine configurations of the form (9) only. All spins from the same block are satisfied or unsatisfied simultaneously. In the last case it is senseless to turn over unsatisfied spins from the block separately, because this leads to a nonsensical configuration (for which inside a block of constant sign there are coordinates with different signs). It turned out, that in this case it is possible to turn over the whole unsatisfied block simultaneously. This leads to a decrease of the energy (1). In other words, here the blocks of the constant signs play the role of *natural* domains.

Let us formulate the general rule of "the correct domains" formation: *A domain consists of spins whose inter-connections are stronger, then their connections with other spins. The values of spins belonging to the same domain have to be equal.*

Due to evident relation of the general rule to the well known problem of the clustering of the symmetric matrix [7], [8], this recipe will be called *the cluster principle* of the domains formation. The validity of the cluster principle is based on the fact that strongly connected spins have to interact with the rest of the spins similarly. Therefore, we have a good chance that under action of an external field the strongly connected spins will behave themselves similarly.

4 The Results of Computer Simulations

1. In first series of the computer experiments we used the Hebbian matrices of the form (8) with extremely strong inter-group connections. The external parameters of the problem were as follows: the dimensionality of the problem was N=1000, the number of patterns was M=60, the number of domains was n=40, sizes of domains k_l were random numbers from the interval [1, 45]: $\sum_{l=1}^{40} k_l = 1000$; the coordinates $x_l^{(\mu)}$ in the matrix (7) took on the values ± 1 equiprobable.

On the main diagonal of the Hebbian matrix (8) there were 40 $(k_l \times k_l)$-blocks from the ones only: $J_{ll}^{(in)} = 1$. Matrix elements outside these blocks were random quantities with the mean values equal to 0 and the dispersions equal to $1/M$: $< J_{lm}^{(out)} >= 0$, $\sigma(J_{lm}^{(out)}) = 1/\sqrt{M}$. The cluster principle of the domains formation provides us with 40 domains of the type (9) generated by 40 groups of strongly connected spins.

Altogether 200 such matrices **J** were generated, and for each matrix the functional (1) was minimized using 3 dynamic approaches:

1) *RANDOM***:** the standard random dynamics was set going from 1000 random start configurations;

2) *Random Domains* (*DM-RND*): the domain dynamics was set going from the same random configurations with n=40 *random* domains of the identical size k=25 (25 spins with random numbers were included in a domain; inside a domain the spins had the values ±1 equiprobable);

3) *Cluster Domains* (*DM-CLS*)**:** the domain dynamics was set going from 1000 random start configurations of block-constant form (9).

When a domain local minimum was achieved, the domains were "defrosted" and under the random dynamics the system went down to a deeper local minimum of the functional (1). Thus, for each matrix we obtained 3000 local minima. Then we found the deepest among them and calculated the frequency of the deepest minimum determination for each of the three dynamics: RANDOM, DM-RND and DM-CLS. The dynamics, for which the frequency of the deepest local minimum determination is the largest, has to be declared the best.

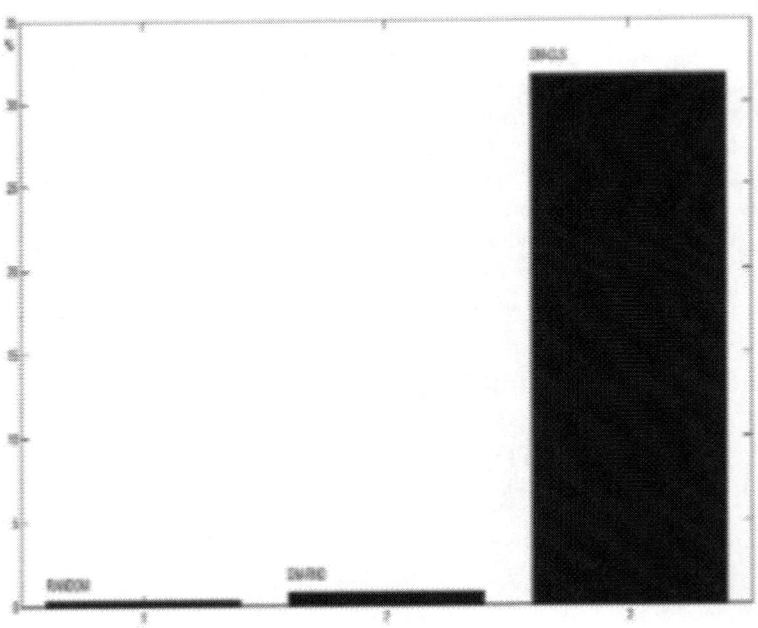

Fig. 1. The average frequency of the deepest local minimum determination for all dynamics

In Fig.1 the results averaged over 200 random matrices are shown for all the three dynamics. Along the abscissa axis the three dynamics are marked off, along the ordinate axis we show the averaged frequency of the deepest minimum determination (in percentages). It is seen that in average the domain dynamics with cluster domains (DM-CLS) leads to the deepest minimum more then *30 times frequently* than the random domain dynamics (DM-RND) or the standard random dynamics (RANDOM). Here the preference of the cluster domain approach is evident.

2. In the described experiments the connections between spins inside "the correct" groups were equal to one, because all k_l vector-columns $\mathbf{x}_l$ in one group were the same. What happens if these groups of vectors are slightly "diluted": If the vector-columns inside "the correct" groups are not identical, but differ slightly? How this affects the result of minimization?

The aforementioned experiments were repeated for "diluted" groups of vector-columns $\mathbf{x}_l$. In this case the first k_1 vector-columns of the matrix (7) were not identical, but they were obtained as a result of multiplicative distortion of the vector $\mathbf{x}_1$: with the probability b the coordinates of the vector $\mathbf{x}_1$ (independently and randomly) were multiplied by -1. Analogously, the next k_2 vector-columns of the matrix $\mathbf{X}$ were obtained by multiplicative distortion of random vector $\mathbf{x}_2$, and so on.

Then the connections between spins inside "the correct" groups were random quantities with the mean value $< J_{ij}^{(in)} >= (1-2b)^2$. Thus, the distortion probability b characterized the level of inhomogeneity of "the correct" spin groups: the larger b, the less the mean value of the inter-group connection, the greater the inhomogeneity of "the correct" group of spins.

Table 1. The mean value $< J_{ij}^{(in)} >$ of the inter-group connection as function of b

b	0.02	0.05	0.1	0.2
$< J_{ij}^{(in)} >$	0.92	0.81	0.64	0.36

The values of b and corresponding mean values of the inter-group connections are given in Table 1. Note, only for b=0.02 and b=0.05 "the correct" groups of spins can be regarded as strongly connected. Indeed, in these cases the mean values of the inter-group connections are $< J_{ij}^{(in)} >\approx 0.9$ and $< J_{ij}^{(in)} >\approx 0.8$ respectively. In other words, the angles between M-dimensional vectors $\mathbf{x}_i$ from "the correct" groups are less than 45°. Such groups of vectors still can be regarded as compact, and the relative spins can be regarded as strongly connected. However, already for b=0.1 we have $< J_{ij}^{(in)} >= 0.64$, and for b=0.2 the mean value of the inter-group connection becomes entirely small: $< J_{ij}^{(in)} >= 0.36$. For such mean values of the matrix elements, the groups we mechanically continue to consider as "correct" groups, in fact are aggregations of slightly connected spin sub-groups. The spins inside these sub-groups can be strongly connected, but in the same time the sub-groups are connected rather slightly. Note, when combining some slightly connected sub-groups into one large group, in fact we organize random domains. One might expect that the more b, the more the results for cluster domain dynamics resemble the results for random domain dynamics.

In Fig.2 for all three types of the dynamics it is shown how the mean frequency of the deepest minimum determination depends on the parameter b. We see that when b increases, the results for the cluster domain dynamics (DM-CLS) progressively less differ from the results for the random domain dynamics (DM-RND), particularly beginning from b=0.1. It can be expected, since when b increases, "the correct" groups of spins resemble the random domains progressively.

The obtained results are the evidence of the productivity of the cluster principle of domains formation. The number and the structure of domains are defined as a result of the connection matrix clustering. A review of clustering methods can be found in [7], [8].

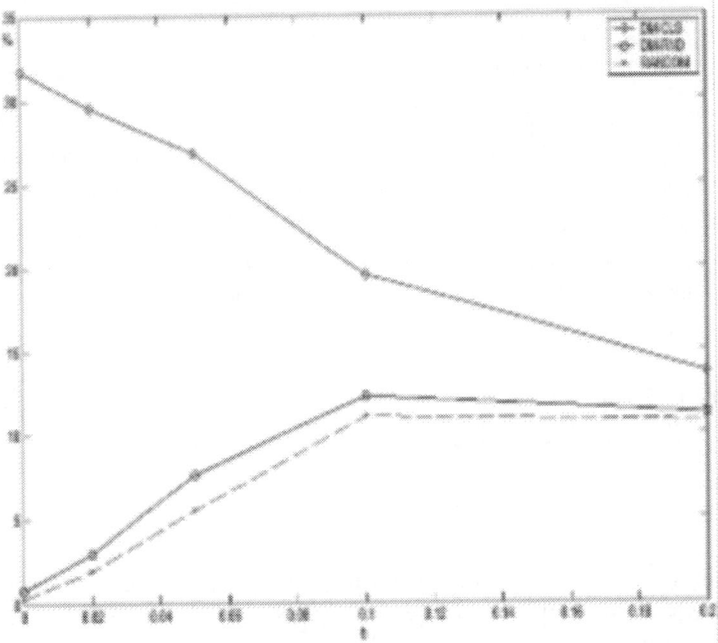

Fig. 2. The mean frequency of the deepest minimum detecting as function of the distortion parameter b for all the three dynamics

Acknowledgments. Author is grateful to Artem Murashkin, who realized computer experiments for this paper. The work was done in the framework of the project «Intellectual computer systems» (the program 2.45) under financial support of Russian Basic Research Foundation (grant 06-01-00109).

References

1. Hartmann, A.K., Rieger, H.: Optimization Algorithms in Physics. Wiley-VCH, Berlin (2001)
2. Hertz, J., Krogh, A., Palmer, R.: Introduction to the Theory of Neural Computation. Addison-Wesley, Reading (1991)

3. Kryzhanovskii, B.V., Magomedov, B.M., Mikaelyan, A.L.: A Domain Model of a Neural Network. Doklady Mathematics 71(2), 310–314 (2005)
4. Kryzhanovsky, B., Magomedov, B.: On the Probability of Finding Local Minima in Optimization Problems. In: Proceedings of IJCNN'2006, Vancouver, pp. 5882–5887 (2006)
5. Herz, A.V.M., Marcus, C.M.: Distributed dynamics in neural networks. Physical Review E47, 2155–2161 (1993)
6. Litinskii, L.B.: Direct calculation of the stable points of a neural network. Theoretical and Mathematical Physics 101(3), 1492–1501 (1994)
7. Xu, R., Wunsh II, D.: Survey of Clustering Algorithms. IEEE Transactions on Neural Networks 16(3), 645–678 (2005)
8. Li, T., Zhu, S., Ogihara, M.: Algorithms for clustering high dimensional and distributed data. Intelligent Data Analysis 7, 305–326 (2003)

MaxSet: An Algorithm for Finding a Good Approximation for the Largest Linearly Separable Set

Leonardo Franco, José Luis Subirats, and José M. Jerez

Departamento de Lenguajes y Ciencias de la Computación,
Universidad de Málaga,
Campus de Teatinos S/N, 29071 Málaga, Spain
{lfranco, jlsubirats, jja} @lcc.uma.es

Abstract. Finding the largest linearly separable set of examples for a given Boolean function is a NP-hard problem, that is relevant to neural network learning algorithms and to several problems that can be formulated as the minimization of a set of inequalities. We propose in this work a new algorithm that is based on finding a unate subset of the input examples, with which then train a perceptron to find an approximation for the largest linearly separable subset. The results from the new algorithm are compared to those obtained by the application of the Pocket learning algorithm directly with the whole set of inputs, and show a clear improvement in the size of the linearly separable subset obtained, using a large set of benchmark functions.

1 Introduction

We propose in this work a new algorithm for finding a linearly separable (LS) subset of examples for a non-linearly separable problem. Related versions of this problem appears in different contexts as machine learning, computational geometry, operation research, etc. [1,2,3,4]. In particular, the problem is very relevant in the field of neural networks given that individual neurons can only compute linearly separable functions. Many constructive algorithms for the design of neural architectures are based on algorithms to find a maximum cardinality set of linearly separable examples [4]. Within the neural networks community, the very well known perceptron algorithm [5], that works well for finding the synaptic weights for a linearly separable problem, has been adapted several times in order to be applied to the case of non-linearly separable problems [6,7]. Among the different modifications, the most popular is the Pocket algorithm introduced by Gallant [6]. The problem of finding the largest LS subset is NP-hard and thus different approximate solutions and heuristics have been proposed to deal with it [3,4,8]. In this paper we introduce an algorithm for finding an approximation to the largest linearly separable subset that is based on finding a set of unate examples. The algorithm then use this subset to train a perceptron using the Pocket algorithm. The new and original contribution of the algorithm

J. Marques de Sá et al. (Eds.): ICANN 2007, Part I, LNCS 4668, pp. 648–656, 2007.

consists in a method for finding an unate approximation of the input function to speed up and improve the process of obtaining a LS set, working as a kind of preprocessing step before applying the Pocket algorithm,. In agreement with this previous idea, it has been shown by Jacob & Mishchenko [9] that obtaining a unate decomposition of a function improves its synthesis procedure in logic design. We have compared the results of the MaxSet algorithm against the most popular of the used algorithms, the Pocket algorithm, trained with the whole sets of examples.

2 Mathematical Preliminaries

A Boolean function of n variables is defined as a mapping $f : \{0,1\}^n \rightarrow \{0,1\}$. A Boolean function is completely specified by the output of the function, $f(x)$, in response to each of the 2^n input examples.

A Boolean function is unate if all its variables are unate. A variable is unate if it is only positive or negative unate but not both. A variable is positive or negative unate according to the change on the output of the function that a change in the variable produces, i.e., if a variable x_i is positive unate then $f(x_0,\ldots,x_i = 0,\ldots,x_{n-1}) \leq f(x_0,\ldots,x_i = 1,\ldots,x_{n-1})$ for all x_j with $j \neq i$ (conversely, for a negative unate variable). The influence of a variable x_i is defined as the number of inputs vectors $x \in \{0,1\}^n$ such that when the ith component is negated (flipped from 0 to 1 or viceversa) the output of the function changes its value. Unate variables have positive or negative influences but not both, case in which are named binate.

It has been demonstrated that all threshold functions are unate but the converse is not true (for example, the function $F(x_0, x_1, x_2, x_3) = x_0\, x_1 + x_2\, x_3$ is unate but it is not threshold).

3 The MaxSet Algorithm

The general idea of the algorithm is to construct an unate approximation of the input function that is as similar as possible to the original function. In order to construct this approximation the algorithm selects a variable at a time, adding to the unate set pair of examples those that produce a positive (or negative) influence. The selection of the variable and the sign of the influence is decided according to the largest number of pairs available at that time with the same influence, in order to maximize the cardinality of the selected set. The optimal solution for obtaining the unate approximation, consists in maximizing the whole set of variables simultaneously, and it is a NP-hard multiple optimization problem. Our method, instead, select variables each at a time, searching for the largest possible subset at each step.

For a clear explanation of the algorithm we will follow a flow diagram shown in Fig. 2 and an example of the application of the algorithm on a non-LS function of 4 variables. The Boolean function selected for the example is shown in the

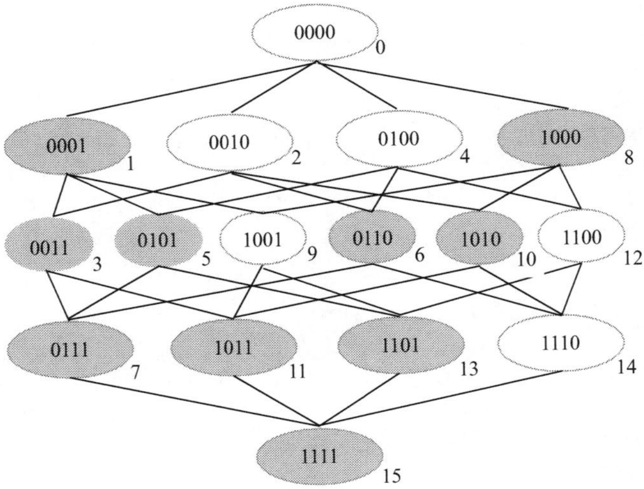

Fig. 1. The partially ordered set (Poset) of the Boolean function used as an example in the text to illustrate the application of the algorithm. The input bits of the examples are shown inside the ovals and the corresponding outputs are indicated by the color of the oval (white corresponds to 0 and grey indicates 1). The decimal value of the inputs is indicated outside the ovals, as it is used in the text. Links between examples connect pair of nearest neighboring examples, whose input bits are at Hamming distance of 1.

partially ordered set (Poset) depicted in Figure 1. The Poset shows the examples ordered in levels according to the weight of the input (number of 1's) where links between nearest neighboring examples are also shown. In the figure (Fig. 1), the output value of each of the examples is indicated by the color of the boxes (white:0, grey:1). It is also shown below each of the examples, the decimal order of the examples used to refer to the examples.

The first step of the algorithm consists in allocating all pairs of nearest neighboring examples (pairs of examples at Hamming distance 1) to one of the 4 variables of the function (the one in which the examples of the pair differ) and to one of the three possible sets : positive, negative and neutral influence sets. The pair is included in a given set according to the influence of the variable in which the pair differs. For example, the pair of examples with decimal indexes 0 and 1 (inputs 0000 and 0001), with output value 0 and 1 respectively will be included in the positive set corresponding to the the last bit variable, x_0. This is because the pair of examples defines the variable x_0 as having positive influence, because a change in the variable from 0 to 1 makes the output of the example to change in an increasing way (positive).

As mentioned, the algorithm starts with the allocation of all possible pairs of neighboring examples to a table in which every pair is assigned to a variable and to a category out of the three possible ones: positive, negative and neutral, indicated by $(+, -, N)$. The allocation of the pairs of neighboring examples for the case of the function that we are using to illustrate the method is shown in Table

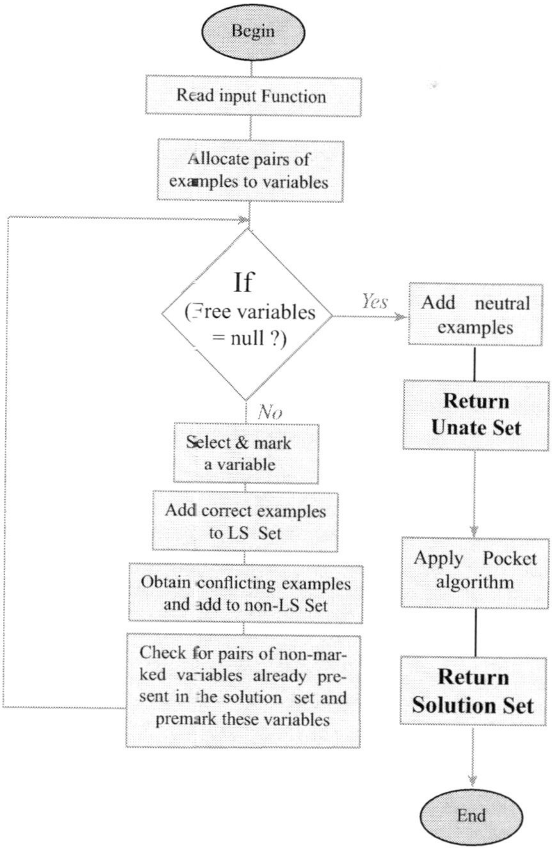

Fig. 2. Flow diagram of the MaxSet algorithm introduced in this work to find an approximation to the largest LS subset of a non-LS input function

1. The function is defined by the truth vector (0,1,0,1,0,1,1,1,1,0,1,1,0,1,0,1), that contains the output values of the function in response to the 16 inputs arranged in decimal order. The table has 4 columns corresponding to the four variables of the function and three rows for the three types of influences.

After all the pairs have been allocated into the table, the algorithm continues by selecting a set of examples belonging to a positive or negative category with the largest cardinality. The neutral sets are only considered at the end of the algorithm after all variables have been defined as positive or negative. For the function that we are considering, the first selected set is the examples with positive influence corresponding to the variable x_0, as this is the largest set containing 10 examples. The selected examples are then included in the LS set. For the function of the example, this step means that examples 0,1,2,3,4,5,12,13,14 and 15 are included in the LS set. After a set of examples have been selected, the algorithm normally checks if there are possible conflicting examples, but this checking step is only applied after a second variable has been selected and the process is described below. Thus, the algorithm continues by selecting a second variable in the same way as it was done for the first one, selecting the largest subset corresponding to a negative or positive non-selected variable. For the example under analysis, the new selected subset is the positive examples corresponding to variable x_1 that will add to the LS set a total of 3 **new** examples (the examples 6,9 and 11, as example 4 is already included). After the second subset (and variable) have been selected, a checking procedure is applied in order to avoid possible future conflicts. The checking procedure analyzes the non-selected pairs of example with opposite influence of the selected sets that have an example included in the LS set. For the Boolean function of the example, the variables x_0 and x_1 have been already selected with positive influence, implying that we should look for the negative sets of these two variables for probable conflicting examples. The way to look for conflicts is to search for pairs included in these non selected sets of variables x_0 and x_1 in which one of the examples of the pair belongs to the LS set. For example, the pair $8-9$ with negative influence on variable x_0 contains the example number 9 that has been already included in the LS set. The checking procedure marks example 8 as a forbidden example and includes it in the Non-LS set (avoiding its future selection), because an eventual selection of example 8 will imply that the variable x_0 would have both positive and negative influence, and this is not possible for a unate or a LS set of examples. The algorithm also checks for pairs of examples already included in the LS set, that belong to non-marked variables. In the example, the pairs 3-6 and 9-13 belonging to the variable x_2 with positive influence are already included in the solution set, and also the pairs 1-9 and 6-14 corresponding to the variable x_3 with negative influence. On the contrary, the example 8 that has been marked as a conflictive example, makes the algorithm to eliminate the pairs 8-12 and 0-8 in the negative x_2 variable and in the positive x_3 variable respectively. The main loop of the algorithm is applied again, but this time can only consider the pre-marked variables, and thus the pre-marked positive part of variable x_2 is included to the LS set. No new examples are added to the LS set as the examples of the pairs with positive influence for variable x_2 are already part of it. Any of the variables x_2 or x_3 could have been selected as they contain the same number of possible examples, but the algorithm is set to select the first one, in this case the x_2. The checking procedure now marks example 10 as forbidden example and thus the pair 2-10 is deleted from the positive part of variable x_3. Lastly, the algorithm marks the only left variable, x_3 as negative as it has been already pre-marked, and the main loop of the procedure finishes. The unate set contains the examples $0, 1, 2, 3, 4, 5, 6, 9, 12, 13, 14, 15$, while the forbidden examples 8 and 10 belong to the non-LS set. Examples 7 and 11 are not yet selected, meaning that we can include them in the unate set, as they only involve neutral influence pairs, and thus the final solution set includes the 14 examples $0, 1, 2, 3, 4, 5, 6, 7, 9, 11, 12, 13, 14, 15$.

The procedure then finishes by applying the pocket algorithm with the selected unate set that as a result will give a linearly separable set. If the unate test set was, indeed, linearly separable the pocket algorithm will then indicate so but if the unate set happened to be unate but not linearly separable the pocket algorithm will select only some examples of this candidate set. The procedure ends by giving as output the LS separable set found by the pocket algorithm. If a LS function is needed instead of a LS subset, all that is needed is just to test the examples not already included in the LS set to find their outputs. In the case of the function of the example, the unate set found is also linearly separable and thus the pocket algorithm is able to learn it all.

Table 1. An example of the application of the MaxSet algorithm to a non-LS Boolean function of 4 variables. All the 32 possible pairs of neighboring examples are included in the table. (See text for details).

	x_0	x_1	x_2	x_3
+	0 – 1 2 – 3 4 – 5 12 – 13 14 – 15	4 – 6 9 – 11	3 – 6 9 – 13	0 – 8 2 – 10
–	8 – 9		8 – 12 10 – 14	1 – 9 6 – 14
N	6 – 7 10 – 11	0 – 2 1 – 3 5 – 7 8 – 10 12 – 14 13 – 15	0 – 4 1 – 5 4 – 7 11 – 15	3 – 11 4 – 12 5 – 13 7 – 15

4 Testing the Algorithm on a Set of Benchmark Functions

The algorithm described above was implemented in $C\#$ and tested on different sets of Boolean function. The efficiency of the algorithm to find large LS sets and the computational time required for the implementation was compared to the results obtained using the pocket algorithm applied to the original set of examples. An initial test was carried out with all the 65536 Boolean functions of 4 variables with the aim of checking that the implementations of the algorithms used in the test was working properly, and also to see how the new algorithms perform with LS functions. The test confirmed that both algorithms found the correct solution (i.e., the whole set of input examples was LS) for the 1881 LS functions of 4 variables that exist in this case. The average size of the LS subset across all the functions of 4 variables was 12.72 (79.5 % of the 16 examples) for the new algorithm, while 10.79 (67.4 %) examples resulted from the application of the Pocket algorithm (16 examples constitute the whole set of inputs for functions of 4 variables). The Pocket algorithm was implemented for this case setting a maximum of 200 iterations each time a new set of weights was found.

Table 2. Results of the implementation of the MaxSet and Pocket algorithms on a set of 54 non-LS Boolean functions used in logic synthesis (See text for details)

Function	Inputs	Fraction LS MaxSet	CPU Time (Seconds)	Fraction LS Pocket	CPU Time (Seconds)
cm82af	5	0.75	0	0.50	1
cm82ag	5	0.75	0	0.50	1
parity5	5	0.50	0	0.50	1
z4ml25	7	0.75	0	0.50	1
z4ml26	7	0.75	0	0.52	1
z4ml27	7	0.75	0	0.50	1
f51m44	8	0.80	0	0.50	1
f51m45	8	0.80	0	0.54	1
f51m46	8	0.80	0	0.53	1
f51m47	8	0.78	0	0.47	1
f51m48	8	0.81	0	0.53	1
f51m49	8	0.88	0	0.56	1
f51m50	8	0.75	0	0.50	1
9symml52	9	0.82	0	0.74	1
alu2k	10	0.53	0	0.54	1
alu2l	10	0.51	0	0.52	1
alu2m	10	0.75	0	0.50	1
alu2p	10	0.75	0	0.72	1
x2l	10	0.88	0	0.79	1
x2p	10	0.94	4	0.80	4
x2q	10	0.83	1	0.82	1
cm85am	11	0.65	3	0.96	3
cm152al	11	0.55	1	0.58	1
cm151am	12	0.79	5	0.79	5
cm151an	12	0.75	4	0.77	4
alu4o	14	0.50	59	0.46	60
alu4p	14	0.56	70	0.51	71
alu4r	14	0.53	64	0.50	65
alu4s	14	0.75	42	0.51	43
alu4u	14	0.78	90	0.73	91
alu4v	14	0.38	85	0.94	87
cm162ao	14	0.92	56	0.84	57
cm162ap	14	0.92	69	0.85	70
cm162aq	14	0.93	63	0.86	64
cm162ar	14	0.93	87	0.87	89
cup	14	0.81	41	0.88	43
cuq	14	0.81	41	0.87	43
cuv	14	0.94	380	0.93	382
cux	14	0.83	118	0.96	120
cm163aq	16	0.92	757	0.83	773
cm163ar	16	0.93	774	0.79	793
cm163as	16	0.93	793	0.86	811
cm163at	16	0.94	931	0.86	948
parityq	16	0.50	983	0.50	1000
pm1a0	16	0.94	741	0.88	758
pm1c0	16	0.97	788	0.94	806
tcona0	17	0.75	5645	0.76	5712
tconb0	17	0.79	7441	0.75	7513
tconc0	17	0.76	5294	0.75	5368
tcond0	17	0.75	6212	0.75	6287
tcone0	17	0.72	5256	0.75	5331
tconf0	17	0.75	7476	0.75	7548
tcong0	17	0.76	5877	0.75	5952
tconh0	17	0.75	5952	0.75	6015
Average		0.77	1041	0.70	1054

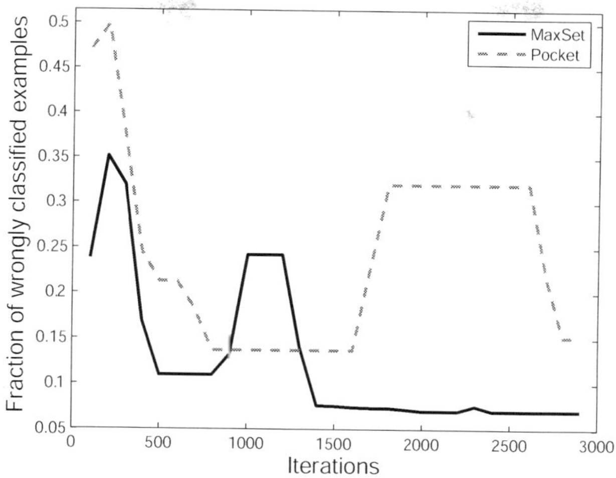

Fig. 3. The fraction of incorrect classified examples vs. the number of iterations for the function cm163aq using the MaxSet and Pocket algorithms

A further test was carried out using a set of 54 benchmarks functions used in logic synthesis. The functions considered where all non-linearly separable, and have a number of input variables ranging from 5 to 17. In Table 2, we report the results for the 54 functions indicating in the first column the name of the function, followed by the number of input variables. The rest of the columns shows the results for the size of the linearly separable set found, indicated as a fraction of the whole set of examples, and the CPU time in seconds used for the calculations for the case of the new algorithm and for the application of the Pocket algorithm directly with the whole set of examples. In 40 out of the 54 cases considered, the MaxSet algorithm found the largest LS set, while in 11 cases the best results were obtained with the original dataset using the Pocket algorithm (In the remaining 3 functions the results were the same.) On average the application of the MaxSet algorithm leads to LS sets that cover on average 76.7% of the total set of examples, while the result from the application of the Pocket algorithm was 70%, this means a 10.00% improvement on the size of the LS subset found (equivalent to a 23% reduction on the average fraction of errors). The previous comparisons were done by selecting similar computational times for both algorithms (1041 and 1054 seconds on average per function).

We also shown in Fig. 3 an example of the learning process for both algorithms (MaxSet and Pocket) for the case of the 16 input variables Boolean function cm163aq. It can be shown from the graph, that the MaxSet algorithm performs better both in terms of size of the obtained set and in terms of the number of iterations needed. Fig. 3 shows the result of the fraction of wrongly classified examples as a function of the number of iterations, where the values have been averaged across 100 repetitions of the learning process to eliminate random fluctuations. Moreover, it can be seen that the dynamic of the learning process for the case of the MaxSet has a similar shape to the one obtained from the Pocket algorithm (trained with the whole sets of examples), but it is a re-scaled version of it, where both the fraction of errors and the number of iterations are reduced approximately by half.

5 Discussion

We introduced in this work a new algorithm for the problem of finding a large subset of LS examples from a non-LS set. The new procedure works by finding first a unate set of examples (the new contribution of this work), to then train a perceptron through the Pocket algorithm with this unate set. The results over a large set of benchmark functions show that for similar computational times the MaxSet algorithm outperforms in most cases the standard Pocket algorithm and lead to an average 11.94% improvement on the size of the LS subset found (equivalent to a 24% reduction in the fraction of errors made). An important aspect of the introduced MaxSet algorithm, regarding the obtention of an unate subset, is that this part of the method can in principle be combined with any other alternative method in order to obtain the final linearly separable solution set.

The extension of the method to partially defined Boolean functions is not straightforward but it is being developed under similar lines and it will be reported elsewhere.

References

1. Johnson, D.S., Preparata, F.P.: The densest hemisphere problem. Theoretical Computer Science 6, 93–107 (1978)
2. Greer, R.: Trees and hills: Methodology for maximizing functions of systems of linear relations. Annals Discrete Mathematics 22 (1984)
3. Amaldi, E., Pfetsch, M.E., Trotter, L.E.: On the maximum feasible subsystem problem, IISs and IIS-hypergraphs. Mathematical Programming 95, 533–554 (2003)
4. Elizondo, D.: Searching for linearly separable subsets using the class of linear separability method. In: Proceedings of the IEEE International Joint Conference on Neural Networks, pp. 955–960. IEEE Computer Society Press, Los Alamitos (2004)
5. Rosenblatt, F.: Principles of Neurodynamics. Spartan Books, New York (1962)
6. Gallant, S.I.: Perceptron-based learning algorithms. IEEE Trans. on Neural Networks 1, 179–192 (1990)
7. Frean, M.: Thermal perceptron learning rule. Neural Computation 4, 946–957 (1992)
8. Marchand, M., Golea, M.: An Approximation Algorithm to Find the Largest Linearly Separable Subset of Training Examples. In: Proceedings of the 1993 Annual Meeting of the International Neural Network Society, vol. 3, pp. 556–559. Erlbaum Associates, Hillsdale, NJ (1993)
9. Jacob, J., Mischenko, A.: Unate Decomposition of Boolean Functions. In: Proceedings of the International Workshop on Logic and Synthesis, California (2001)

Generalized Softmax Networks for Non-linear Component Extraction

Jörg Lücke and Maneesh Sahani

Gatsby Computational Neuroscience Unit, UCL, London WC1N 3AR, UK

Abstract. We develop a probabilistic interpretation of non-linear component extraction in neural networks that activate their hidden units according to a softmax-like mechanism. On the basis of a generative model that combines hidden causes using the max-function, we show how the extraction of input components in such networks can be interpreted as maximum likelihood parameter optimization. A simple and neurally plausible Hebbian Δ-rule is derived. For approximately-optimal learning, the activity of the hidden neural units is described by a generalized softmax function and the classical softmax is recovered for very sparse input. We use the bars benchmark test to numerically verify our analytical results and to show competitiveness of the derived learning algorithms.

1 Introduction

Neural network models that can be applied to unsupervised data classification, i.e. clustering, have been well understood in terms of generative models. For mixture models [1], the learning rule was shown to involve the softmax (see, e.g., [2] for an overview) as activation function for the hidden units. Other than clustering, recent neural network models have been shown to be competitive in extracting input components. These networks, e.g. [3,4,5], also use softmax-like mechanisms but a probabilistic understanding in terms of multiple causes models has not yet been developed. In this paper we will show how and under which conditions learning in feed-forward networks with Hebbian plasticity can be understood as maximization of the data likelihood under a non-linear generative model.

2 The Neural Network Model

Consider the network of Fig. 1 which consists of D input units with values $y_1, \ldots, y_D$ and H hidden units with values $g_1, \ldots, g_H$. An input $\boldsymbol{y}$ to the system is represented by activities of input units. Via connections between the input and hidden layer that are parameterized by (W_{id}) the input unit activities determine the activities of the hidden units, $g_i = g_i(\boldsymbol{y}, W)$. The parameters (W_{id}) will be called weights.

Given a set of input patterns we want to adjust the weights W such that the activities of the hidden units appropriately represent the input patterns. Learning will thus depend on the distribution of input patterns. A standard approach in neural network modeling is to use a Δ-rule with divisive normalization. In the case of Hebbian learning the Δ-rule reads:

J. Marques de Sá et al. (Eds.): ICANN 2007, Part I, LNCS 4668, pp. 657–667, 2007.

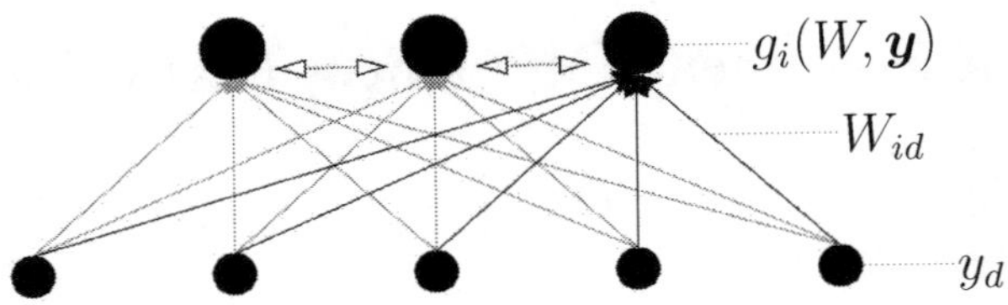

Fig. 1. Architecture of a two layer neural network. Input is represented by values y_1 to y_D of D input units (small black circles). Depending on the input and on the parameters W the activities of the hidden units g_1 to g_H (big black circles) are computed. The dotted horizontal arrows symbolize lateral information exchange that is in general required to compute the functions g_1 to g_H. After the g_i are computed the parameters (W_{id}) are modified using a Δ-rule.

$$\Delta W_{id} = \epsilon\, g_i(\boldsymbol{y}, W)\, y_d \quad \text{and} \quad W_{id}^{\text{new}} = C \frac{W'_{id} + \Delta W_{id}}{\sum_{d'} (W'_{id'} + \Delta W_{id'})}, \tag{1}$$

where W'_{id} denote the old values. Divisive normalization (right-hand-side equation) prevents the weights from growing infinitely and represents the most commonly used constraint for learning in artificial neural networks. The learning rule (1) can directly be associated with Hebbian learning between neural units g_1 to g_H and input units y_1 to y_D. The dotted horizontal arrows in Fig. 1 symbolize lateral information transfer that is generally required to compute the activities $g_i(\boldsymbol{y}, W)$. Neural networks that fit into the framework of Fig. 1 and use a learning rule like (1) or similar are frequently found in the literature and their application domains typically lie in the fields of clustering and component extraction [2,3,4,5,6]. Especially since generative models are used to study unsupervised learning, the class of neural network models that perform clustering could be embedded into a probabilistic framework. Learning in these networks can be regarded as optimization of the data likelihood under a generative model. This relationship in turn determines the function $g_i(\boldsymbol{y}, W)$ in (1) which has to be used for optimal learning. For clustering tasks the related generative models are mixture models and the function $g_i(\boldsymbol{y}, W)$ is the softmax operation, see e.g. [1,2]. A limit case of the softmax is a winner-take-all mechanisms which is likewise frequently applied to clustering tasks [2]. For networks such as [3,4] that extract input components, the situation is different. Linear systems such as independent (ICA) or principal component analysis (PCA) do not seem to represent the right correlates. ICA and PCA have been shown to fail in non-linear tasks, see [7], in which neural network models consistent with the above framework, e.g. [3,4,5], succeed in extracting the elementary input components. The learning dynamics used in these networks are manifold and comparison with other component extraction systems is usually only possible on the basis of benchmark tests. A probabilistic understanding similar to that for mixture models has not yet been developed. Furthermore, some network models perform clustering if data is generated by single causes, and component extraction if input consists of combinations of components, e.g. [3,4]. From the generative point of view this seems contradictory.

For a probabilistic analysis of networks (1), a generative model will be defined whose parameter update rules for likelihood maximization can be related to the delta rule in

(1). Before we define a generative model in the next section let us express the on-line update rule (1) in terms of a batch mode update for a set of N input patterns $\boldsymbol{y}^{(1)}, \ldots, \boldsymbol{y}^{(N)}$. By repeatedly applying (1) we get for slowly changing W:

$$W_{id} \approx \frac{W'_{id} + \epsilon \sum_n g_i(W, \boldsymbol{y}^{(n)})\, y_d^{(n)}}{\frac{1}{C}\sum_{d'}\left(W'_{id'} + \epsilon \sum_n g_i(W, \boldsymbol{y}^{(n)})\, y_{d'}^{(n)}\right)} \approx C \frac{\sum_n g_i(W, \boldsymbol{y}^{(n)})\, y_d^{(n)}}{\sum_{d'}\sum_n g_i(W, \boldsymbol{y}^{(n)})\, y_{d'}^{(n)}} . \quad (2)$$

The smaller the change of W the better the Δ-rule (1) is approximating (2).

3 A Non-linear Generative Model

Consider a generative model with D input units $y_1, \ldots, y_D$ and H hidden units $s_1, \ldots, s_h$. The model is parameterized by $\Theta = (\pi, W)$. The scalar π parameterizes the distributions of the hidden units which are assumed to be independent and Bernoulli distributed:

$$p(\boldsymbol{s} \,|\, \pi) = \prod_h p(s_h \,|\, \pi), \text{ where } p(s_h \,|\, \pi) = \pi^{s_h} (1-\pi)^{1-s_h} . \quad (3)$$

For given $\boldsymbol{s}$ we take the activities of the input units to be independent. The parameters W determine the influence of a particular $\boldsymbol{s}$ on the distributions of the input units. The values of y_d we take to be independent given $\boldsymbol{s}$ and distributed according to a Poisson distributions:

$$p(\boldsymbol{y} \,|\, \boldsymbol{s}, W) = \prod_d p(y_d \,|\, \overline{W}_d(\boldsymbol{s}, W)), \text{ where } p(y_d \,|\, w) = e^{-w} \frac{w^{y_d}}{y_d!} . \quad (4)$$

The Poisson distribution is a natural choice for non-negative and noisy input. E.g., if we take the causes to be objects with gray values W_{id}, all input patterns are non-negative. Furthermore, in the context of neural networks and Hebbian learning (1), a representation of non-negative values is straightforward. For the function $\overline{W}_d$ in (4) the most common choice is a linear superposition of the weights multiplied by the activities of the hidden units, $\overline{W}_d(\boldsymbol{s}, W) = \sum_h s_h W_{hd}$. Algebraically, this kind of superposition is advantageous for the computation of derivatives of $p(\boldsymbol{y} \,|\, \boldsymbol{s}, W)$ as required for optimization methods. However, for many applications linear superposition is difficult to motivate. E.g., for visual input it does not seem natural to assume linearity because occlusion is not modeled appropriately. For instance, if the causes are objects with the same gray value, linear superposition results in an error in the region where the objects overlap. Alternatively, the maximum function would represent the right model in this case and seems to be a better approximation for occlusion-like scenarios in general. Neural network models that fit into the framework (1) have been shown to be competitive in occlusion-like scenarios [3,4,5]. For our purposes the max-operation therefore seems to be preferable to linear superposition and we use as function $\overline{W}_d$ in (4):

$$\overline{W}_d(\boldsymbol{s}, W) := \max_h\{s_h W_{hd}\} . \quad (5)$$

Using the equality $\overline{W}_d(\boldsymbol{s}, W) = \lim_{\rho\to\infty} \overline{W}^{\rho}_d(\boldsymbol{s}, W) = \lim_{\rho\to\infty} (\sum_h (s_h W_{hd})^{\rho})^{\frac{1}{\rho}}$ we define:

$$\mathcal{A}_{id}(\boldsymbol{s}, W) := \lim_{\rho\to\infty} \left(\frac{\partial}{\partial W_{id}} \overline{W}^{\rho}_d(\boldsymbol{s}, W) \right) = \lim_{\rho\to\infty} \frac{s_i\,(W_{id})^{\rho}}{\sum_h s_h (W_{hd})^{\rho}}, \tag{6}$$

$$\text{which implies: } \mathcal{A}_{id}(\boldsymbol{s}, W)\, f(\overline{W}_d(\boldsymbol{s}, W)) = \mathcal{A}_{id}(\boldsymbol{s}, W)\, f(W_{id})\,. \tag{7}$$

f can be any well-behaved function. Equation (7) holds because both sides are zero whenever $\overline{W}_d(\boldsymbol{s}, W) \neq W_{id}$.

4 Maximum Likelihood

Given a set of data $Y = \{\boldsymbol{y}^{(n)}\}_{n=1,\ldots,N}$ we want to find the set of parameters W that maximizes the likelihood of the data under the above generative model. Finding the optimal π is straightforward but for the purposes of this paper we will concentrate on the difficult part of optimizing the parameters W. Following the EM procedure, this amounts to maximizing the lower bound $Q(W, W')$:

$$Q(W, W') = \sum_{n=1}^{N} \sum_{\boldsymbol{s}} p(\boldsymbol{s}\,|\,\boldsymbol{y}^{(n)}, \Theta')\, \log\left(p(\boldsymbol{y}^{(n)}\,|\,\boldsymbol{s}, W)\, p(\boldsymbol{s}\,|\,W)\right), \tag{8}$$

where Θ' are the old parameters and where $\sum_{\boldsymbol{s}}$ is the sum over all binary vectors of length H. Using Bayes' rule the posterior probabilities are given by:

$$p(\boldsymbol{s}\,|\,\boldsymbol{y}^{(n)}, \Theta') = \frac{p(\boldsymbol{s}\,|\,\pi)\, p(\boldsymbol{y}^{(n)}\,|\,\boldsymbol{s}, W')}{\sum_{\tilde{\boldsymbol{s}}} p(\tilde{\boldsymbol{s}}\,|\,\pi)\, p(\boldsymbol{y}^{(n)}\,|\,\tilde{\boldsymbol{s}}, W')} \quad \text{(E-step)}, \tag{9}$$

where $p(\boldsymbol{s}\,|\,\pi)$ and $p(\boldsymbol{y}^{(n)}\,|\,\boldsymbol{s}, W')$ are given by (3) and (4), respectively. To account for divisive normalization we use constraint optimization of $Q(W)$:

$$\frac{\partial}{\partial W_{id'}} Q(W, W') + \sum_{h=1}^{H} \lambda_h \frac{\partial}{\partial W_{id'}} G_h(W) \stackrel{!}{=} 0\,, \tag{10}$$

where $G_h(W) := \sum_{d=1}^{D} W_{hd} - C = 0$ is a constraint that keeps the sum over the weights for each hidden unit h constant. If we insert (8) with (4) into (10), we obtain by using (6) and (7):

$$\sum_n \sum_{\boldsymbol{s}} p(\boldsymbol{s}\,|\,\boldsymbol{y}^{(n)}, \Theta')\, \mathcal{A}_{id'}(\boldsymbol{s}, W)\, \frac{y^{(n)}_{d'} - W_{id'}}{W_{id'}} + \lambda_i = 0\,. \tag{11}$$

Equations (11) can be solved for the Lagrange multipliers by summing over d' and by using $\sum_d W_{hd} = C$. Applying some straightforward algebra, we obtain a set of non-linear equations for the parameter update:

$$W_{id} = \frac{\sum_n \langle \mathcal{A}_{id}(\boldsymbol{s}, W)\rangle\, y^{(n)}_d}{\frac{1}{C} \sum_{n,d'} \langle \mathcal{A}_{id}(\boldsymbol{s}, W)\rangle\, y^{(n)}_{d'} - \sum_n \left((\sum_{d'} \langle \mathcal{A}_{id'}(\boldsymbol{s}, W)\rangle\, \frac{W_{id'}}{C}) - \langle \mathcal{A}_{id}(\boldsymbol{s}, W)\rangle \right)}, \tag{12}$$

$$\text{where } \langle \mathcal{A}_{id}(\boldsymbol{s}, W)\rangle = \sum_{\boldsymbol{s}} p(\boldsymbol{s}\,|\,\boldsymbol{y}^{(n)}, \Theta')\, \mathcal{A}_{id}(\boldsymbol{s}, W). \tag{13}$$

Equation (12) can be simplified if we assume that the hidden causes are relatively homogeneously covering the input space. More precisely, if we parameterize the actually generating causes by W^{gen}, let us assume that all $W^{\text{gen}}_{id} > 0$ of cause i can be covered by the same number of $W^{\text{gen}}_{cd} \geq W^{\text{gen}}_{id}$:

$$W^{\text{gen}}_{id} > 0 \Rightarrow \sum_{c \neq i} \mathcal{H}(W^{\text{gen}}_{cd} - W^{\text{gen}}_{id}) \approx b_i \,, \tag{14}$$

where $\mathcal{H}$ is the Heaviside function. b_i is the number of causes that can cover cause i. Fig. 2C,D illustrate this condition. For both examples condition (14) is fulfilled. E.g. for Fig. 2D $b_i = 0$ for all horizontal causes and $b_i = 1$ for the vertically oriented cause. If the model parameters W satisfy condition (14), which can be expected at least close to the maximum likelihood solution, we obtain with (12) and after rearranging terms:

$$W_{id} = C \, \frac{\sum_n \langle \mathcal{A}_{id}(\boldsymbol{s}, W) \rangle \, y_d^{(n)}}{\sum_{d'} \sum_n \langle \mathcal{A}_{id'}(\boldsymbol{s}, W) \rangle \, y_{d'}^{(n)}} \quad \text{(M-step)}. \tag{15}$$

Equation (15) can be used as a fixed point equation. Inserting the old values of W on the right-hand-side to obtain new values of W, approaches the right solution in numerical simulations.

5 E-Step Approximation

The M-step (15) together with expectations (13) represent an optimization algorithm for the generative model (3) to (5). However, for practical applications we are still confronted with the problem of an exponential cost for computing the sufficient statistics (13). For our model let us assume that most inputs have been generated by just few hidden causes. To approximate for sparse input we first group terms in (13) according to the number of active hidden units:

$$\langle \mathcal{A}_{id}(\boldsymbol{s}, W) \rangle = \sum_{\boldsymbol{s}} p(\boldsymbol{s} \,|\, \boldsymbol{y}^{(n)}, \Theta') \, \mathcal{A}_{id}(\boldsymbol{s}, W) \tag{16}$$

$$= \sum_{h} p(\boldsymbol{s}_h \,|\, \boldsymbol{y}^{(n)}, \Theta') \, \mathcal{A}_{id}(\boldsymbol{s}_h, W) + \sum_{\substack{a,b \\ a<b}} p(\boldsymbol{s}_{ab} \,|\, \boldsymbol{y}^{(n)}, \Theta') \, \mathcal{A}_{id}(\boldsymbol{s}_{ab}, W) + \sum_{\substack{a,b,c \\ a<b<c}} \ldots \,, \tag{17}$$

where $\boldsymbol{s}_h := (0, \ldots, 0, 1, 0, \ldots, 0)$ 1 at h's position,
$\boldsymbol{s}_{ab} := (0, \ldots, 0, 1, 0, \ldots, 0, 1, 0, \ldots, 0)$ 1 at a's and b's position $a \neq b$,

and analogously for $\boldsymbol{s}_{abc}$ etc. Note that $\mathcal{A}_{id}(\boldsymbol{0}, W) = 0$ because of (6). Each of the posterior probabilities $p(\boldsymbol{s} \,|\, \boldsymbol{y}^{(n)}, \Theta')$ above implicitly contains a similar sum over $\boldsymbol{s}$ for normalization whose terms may be grouped in the same way:

$$p(\boldsymbol{s} \,|\, \boldsymbol{y}^{(n)}, \Theta') = \frac{p(\boldsymbol{s}, \boldsymbol{y}^{(n)} \,|\, \Theta')}{\sum_{a} p(\boldsymbol{s}_a, \boldsymbol{y}^{(n)} \,|\, \Theta') + \sum_{\substack{a,b \\ a<b}} p(\boldsymbol{s}_{ab}, \boldsymbol{y}^{(n)} \,|\, \Theta') + \sum_{\substack{a,b,c \\ a<b<c}} \ldots} \,, \tag{18}$$

where we have used that fact that the update rule (15) remains unchanged if we remove all inputs equal to zero, i.e., all inputs with $\sum_d y_d^{(n)} = 0$. Combining (16) and (18) yields:

$$\langle \mathcal{A}_{id}(\boldsymbol{s}, W)\rangle = \frac{\sum_a p(\boldsymbol{s}_a, \boldsymbol{y}^{(n)} \,|\, \Theta')\, \mathcal{A}_{id}(\boldsymbol{s}_a, W) + \sum_{\substack{a,b\\ a<b}} p(\boldsymbol{s}_{ab}, \boldsymbol{y}^{(n)} \,|\, \Theta')\, \mathcal{A}_{id}(\boldsymbol{s}_{ab}, W) + \dots}{\sum_a p(\boldsymbol{s}_a, \boldsymbol{y}^{(n)} \,|\, \Theta') + \sum_{\substack{a,b\\ a<b}} p(\boldsymbol{s}_{ab}, \boldsymbol{y}^{(n)} \,|\, \Theta') + \dots}. \tag{19}$$

If we assume that the significant posterior probability mass will concentrate on vectors, $\boldsymbol{s}$, with only a limited number of non-zero entries, the expanded sums in both numerator and denominator of (19) may be truncated without significant loss. We will use an approximation of $\langle \mathcal{A}_{id}(\boldsymbol{s}, W)\rangle$ that corresponds to a truncation of (19) after terms of order two in the denominator and first order terms in the numerator. As numerical simulations show, this is the approximation of lowest possible order which still results in a system that can be used for component extraction. After inserting (3) to (5) into (19) and by using $\mathcal{A}_{id}(\boldsymbol{s}_h, W) = \delta_{ih}$ the approximation reads:

$$\langle \mathcal{A}_{id}(\boldsymbol{s}, W)\rangle \approx \frac{\exp(I_i^{(n)})}{\sum_h \exp(I_h^{(n)}) + \frac{\overline{\pi}}{2} \sum_{\substack{a,b\\ a\neq b}} \exp(I_{ab}^{(n)})}, \quad \overline{\pi} := \left(\frac{\pi}{1-\pi}\right)^{\beta}, \tag{20}$$

$$I_i^{(n)} = \beta \sum_d \left(\log(W_{id})\, y_d^{(n)} - W_{id}\right),\; I_{ab}^{(n)} = \beta \sum_d \left(\log(\tilde{W}_d^{ab})\, y_d^{(n)} - \tilde{W}_d^{ab}\right) \tag{21}$$

where $\tilde{W}_d^{ab} = \max(W_{ad}, W_{bd})$.The parameter β has been introduced to implement an annealing procedure [8] in later simulations. The M-step (15) together with the approximate E-step (20) is a learning algorithm that we will refer to as *approximate EM*.

If we now compare (15) with (2), we find that a neural network (1) can be used for parameter optimization by choosing $g_i(W, \boldsymbol{y}^{(n)}) = \langle \mathcal{A}_{id}(\boldsymbol{s}, W)\rangle$. Note that in general g_i would have to depend on d. However, if we use approximation (20), $\langle \mathcal{A}_{id}(\boldsymbol{s}, W)\rangle$ only depends on i. For the neural network this implies that the change of the weight W_{id} only depends on the activities of the pre- and post-synaptic units:

$$\Delta W_{id} = \epsilon\, g_i\, y_d \quad \text{with} \quad g_i = \frac{\exp(I_i)}{\sum_h \exp(I_h) + \frac{\overline{\pi}}{2} \sum_{\substack{a,b\\ a\neq b}} \exp(I_{ab})} \tag{22}$$

and divisive normalization ($\sum_d W_{id} = C$). Note that the function g_i in (22) is similar to the classical softmax [2] but has an additional term in the denominator. For very sparse input, i.e. $\overline{\pi} \ll 1$, the classical softmax is recovered.

The intuition for the learning rule is as follows: The system tries to explain a given input pattern using its current state of the model parameters W. If one hidden unit explains the input better than any combination of two units, this unit is modified. If the input is better explained by a combination of two units, learning is penalized. Eqns. (22) describe an easy and computationally efficient method to maximize the data likelihood under the generative model (3) to (5).

6 Simulations

It remains to be shown that the approximations we have introduced result in parameter updates that indeed approximately maximize the data likelihood. That the likelihood is always increased is only guaranteed if we use the exact M-step (12) and the exact sufficient statistics (13). Although our approximations are derived from the exact M- and E-steps there is no guarantee that they always result in an increase of the likelihood. Furthermore, even in the case of exact EM, learning can converge to local optima. To

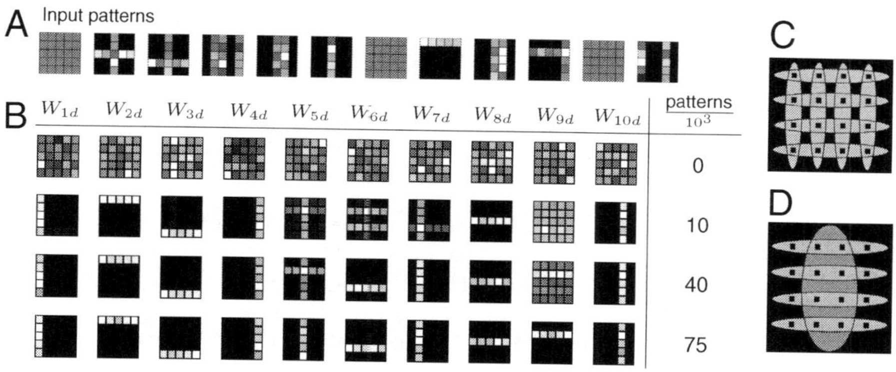

Fig. 2. Bars test with 5x5 pixels, 10 bars, and $\pi = \frac{2}{10}$. **A** 12 input patterns from the set of $N = 500$ input patterns used. **B** Update of the parameters W using the neural network NN_{slow}. **C** Sketch of the hidden causes of a bars test with 8 bars. The black squares symbolize the pixels. The bars test satisfies condition (14). **D** Exemplarily another distribution of hidden causes that satisfies the condition.

verify that the derived approximate EM and neural networks with Δ-rule (22) are indeed able to extract hidden causes, we use the bars benchmark test for component extraction [3,4,5,6]. In the bars test a bar appears with fixed probability at one of, e.g., 10 positions. To test the algorithms we use generative parameters W^{gen} in the form of $H = 10$ horizontal and vertical bars. Values W_{id}^{gen} that correspond to a bar are set to $W_{id}^{\text{gen}} = 10$, values that correspond to the background are zero, $W_{id}^{\text{gen}} = 0$. We generate $N = 500$ input patterns according to the generative model (3) to (5) with $\pi = \frac{2}{10}$. A subset of the resulting input patterns is displayed in Fig. 2A. The Poisson distribution (4) results in noisy bars. In Fig. 2B the parameters W are displayed in the case of using a neural network (22) with $\epsilon = 0.1$, which we will refer to as NN_{slow}. As can be seen, the system's parameters W converge from randomly initialized values to values similar to those of the generating parameters W^{gen}. In Fig. 2B learning time is measured in terms of the number of patterns that are randomly drawn from a set of $N = 500$. For the same bars test settings, the development of the data likelihood is plotted in Fig. 3 for ten trials with NN_{slow}(back lines). Note that for the likelihood plots we have used the same set of $N = 500$ input patterns for all trials. In all other simulations we use a newly generated set of 500 patterns for each trial. In Fig. 3 it can be observed that the network increases the likelihood up to a value that closely corresponds to the likelihood of the generating weights W^{gen}. The likelihoods of ten trials using approximate EM, (15) and (20), are

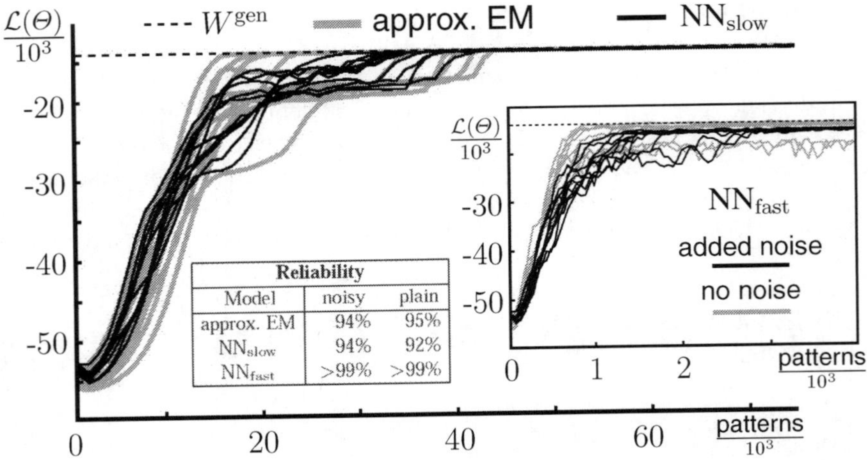

Reliability		
Model	noisy	plain
approx. EM	94%	95%
NN_{slow}	94%	92%
NN_{fast}	>99%	>99%

Fig. 3. Development of log-likelihood values during learning for the same bars test as in Fig. 2. Time courses for approximate EM (broad grey lines) and NN_{slow} (black lines) approach the likelhood value of the actually generating weights W^{gen} (dashed line). Note that one EM iteration corresponds to $N = 500$ patterns. The small plot shows log-likelihood values for NN_{fast} (black). Likelihoods for NN_{fast} without Gaussian noise on the weights is plotted for comparison (gray). The table shows the probability of each system to extract all bars in a given trial, i.e., its reliability. Values are given for bars with Poisson noise and without noise.

plotted for comparison (broad gray lines). Note that EM and NN_{slow} essentially behave in the same way as could be expected on the basis of our analytical results. NN_{slow} updates W after every pattern presentation whereas EM uses all $N = 500$ patterns for each update. In Fig. 3, $N = 500$ patterns therefore correspond to one EM iteration. For NN_{slow} the online update introduces stochasticity. This effects learning only marginally, however. Compared to approx. EM, the likelihoods are slightly noisier and convergence times seem to vary less.

For learning we use an annealing proceedure, i.e., we use $\beta < 1$ in (21) and increase β to one in the course of learning. For EM and NN_{slow} we use the same cooling schedule, which remains unchanged throughout this paper. We increase β in (21) by decreasing a 'temperature' T using $\beta = \frac{1}{T}$. We use $T_1 = 18$ for the first 50 iterations and linearly cool to $T = 1$ in the subsequent 100 iterations[1]. For later comparison we introduce another neural network version based on (22). Instead of a cooling schedule we use a fixed temperature $T = 16$ and add Gaussian noise for each weight at each update: $\Delta W_{id} = \epsilon\, g_i\, y_d + \sigma\, \eta$. Using $\epsilon = 1.0$ and $\sigma = 0.02$ we obtain a neural network that requires relatively few pattern presentations for learning ($< 1,000$). We therefore refer to the network as NN_{fast}. Ten time-courses of likelihood values are shown for NN_{fast} in Fig. 3 (small plot, black lines). Because of the additional noise the final likelihood values are somewhat lower than those of the generating weights. If a network with the same

[1] For NN_{slow} this translates to: 25×10^3 pattern presentations with $T_1 = 18$ and linear cooling to $T = 1$ within the next 50×10^3 patterns.

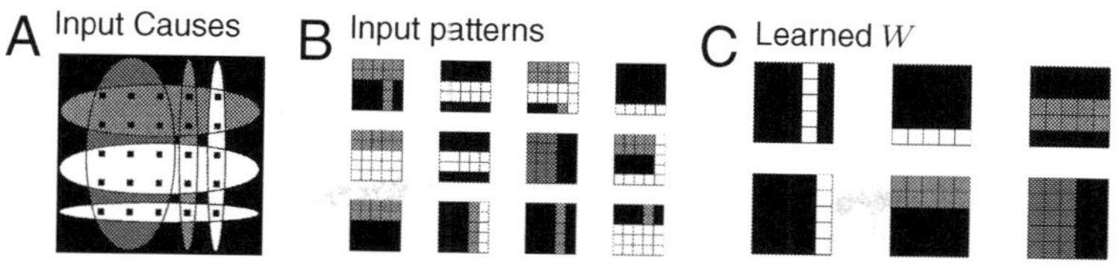

Fig. 4. Bars test with different bar intesities and widths. **A** Scematic visualization of the six bars. **B** 12 examples of the $N = 500$ input patterns used. **C** Learned parameters W using $\mathrm{NN_{slow}}$ (from black for $W_{id} = 0$ to white for $W_{id} = 10$). Note that $\sum_d W_{id} = 50$ for all weights.

parameters as $\mathrm{NN_{fast}}$ but without added noise ($\sigma = 0$) is used, likelihood values are higher (small plot, gray lines) but the system converges to local optima relatively often.

For comparison with other systems in the literature we will use the same bars test as above but instead of bars with Poisson noise we use non-noisy bars. In the bars test, a very prominent criterion for comparison is the so called *reliablity* of the system [3,4,5,6,9,10]. It is defined as the probability to find all bars in a given trial. If a system tends to converge to local optima, its reliability is low. The table in Fig. 3 summarizes the reliability values for approx. EM, $\mathrm{NN_{slow}}$, and $\mathrm{NN_{fast}}$ for bars with and without Poisson noise. Each reliability value was computed on the basis of 100 trials and each trail starts with a different randomly initialized W. For both bars test versions the same algorithms and the same cooling schedule were used (the same fixed T in the case of $\mathrm{NN_{fast}}$). Using the criterion of reliability, $\mathrm{NN_{fast}}$ shows the best performance. The reason is the additional Gaussian noise and noise introduced by updating online. The combination of both sources of noise drives the system out of shallow optima. Furthermore, $\mathrm{NN_{fast}}$ is the fastest system in terms of required pattern presentations. It needs less than $1,000$ patterns to find all bars in the majority of 100 simulations[2]. Note, however, that in some trials learning time is much longer. In terms of likelihood values, approx. EM and $\mathrm{NN_{slow}}$ are to be preferred. These system also have the advantage of a well defined stopping criterion.

Regarding values of reliability, the systems' performance is very robust if their parameter π in (20) or (22) is chosen very different from the actually generating parameter $\pi^{\mathrm{gen}} = \frac{2}{10}$ (note that it was not neccessary to distinguished between π and π^{gen}, so far). In the beginning of learning the small exponent $\beta \approx 0.05$ in (20) results in $\overline{\pi}$ values close to one for a large range of different π. Close to the end of the cooling schedule, for $\beta \approx 1$, different π values do likewise only marginally effect learning because the second term in the denominator of (20) is either very small or very large compared to the first. In contrast, the likelihoods are relatively sensitive to different values of π. Using an EM approach similar to the one for W, the value for π can, however, be infered.

Finally, the reliabilities of the algorithms are found to be robust w.r.t. violations of assumptions (14) and $\sum_d W_{id}^{\mathrm{gen}} = C$ for the generating weights. An example of a bars test that violates both assumptions is shown in Fig. 4. $\mathrm{NN_{slow}}$ finds all bars with 96% reliability in this case.

[2] The system is taken to have found all bars if each hidden unit represents a different bar. This is usually already the case for intermediate likelihood values.

7 Discussion

Generative model and neural network approaches have both been suggested for the extraction of non-linear components. In this paper we have analytically shown that Hebbian learning in feed-forward neural networks can correspond to parameter updates in a generative model using approximate EM. The distinguishing feature of this generative model is an explicit non-linearity in the form of a max operation.The appropriate activation rule for the networks turns out to be the generalized softmax rule (20). The generalization drops back to the classical softmax for very sparse input.

The analytical results were verified in numerical simulations using the bars benchmark test. We found that the derived neural networks increase the likelihood under the generative model and avoid local optima in the great majority of trials. The derived learning algorithm NN_{fast} performs best if learning time and system reliability are used for comparison. Using a high learning rate and additional noise the network model NN_{fast} avoids all local optima and needs few pattern presentations for learning. The same is reported, e.g., for the networks [3,4] which both fit into the framework considered here (Eqn. 1 and Fig. 1). Furthermore, the generalized softmax (20) offers an explanation of why networks such as [3] and [4] can also be successfully applied to clustering tasks. In contrast to many neural networks, algorithms that are based on generative models are by definition well interpretable probabilistically. They have, however, been criticized [3,4,7] for frequently failing to extract the true causes. E.g., for the same bars test as used here, the models [9] and [10] find all bars in just 27% and 69% (even though bar overlap is excluded) of trials, respectively. Instead of treating them as separate or competing approaches, we have in this paper shown that generative and neural network approaches can come together in the form of competitive learning algorithms that are neurally plausible, reliable, and probabilistically fully interpretable.

Acknowledgment. This work was funded by the Gatsby Charitable Foundation.

References

1. McLachlan, G., Peel, D.: Finite Mixture Models. Wiley, Chichester (2000)
2. Yuille, A.L., Geiger, D.: Winner-take-all networks. In: Arbib, M.A. (ed.) The handbook of brain theory and neural networks, pp. 1228–1231. MIT Press, Cambridge (2003)
3. Spratling, M.W., Johnson, M.H.: Preintegration lateral inhibition enhances unsupervised learning. Neural Computation 14, 2157–2179 (2002)
4. Lücke, J., von der Malsburg, C.: Rapid processing and unsupervised learning in a model of the cortical macrocolumn. Neural Computation 16, 501–533 (2004)
5. Spratling, M.W.: Learning image components for object recognition. Journal of Machine Learning Research 7, 793–815 (2006)
6. O'Reilly, R.C.: Generalization in interactive networks: The benefits of inhibitory competition and Hebbian learning. Neural Computation 13, 1199–1241 (2001)

7. Hochreiter, S., Schmidhuber, J.: Feature extraction through LOCOCODE. Neural Computation 11, 679–714 (1999)
8. Ueda, N., Nakano, R.: Deterministic annealing EM algorithm. Neural Networks 11(2), 271–282 (1998)
9. Saund, E.: A multiple cause mixture model for unsupervised learning. Neural Computation 7, 51–71 (1995)
10. Dayan, P., Zemel, R.S.: Competition and multiple cause models. Neural Computation 7, 565–579 (1995)

Stochastic Weights Reinforcement Learning for Exploratory Data Analysis

Ying Wu[1], Colin Fyfe[1], and Pei Ling Lai[2]

[1] Applied Computational Intelligence Research Unit,
The University of Paisley, Scotland
ying.wu,colin.fyfe@paisley.ac.uk

[2] Southern Taiwan University of Technology, Tainan, Taiwan
pei_ling_lai@hotmail.com

Abstract. We review a new form of immediate reward reinforcement learning in which the individual unit is deterministic but has stochastic synapses. 4 learning rules have been developed from this perspective and we investigate the use of these learning rules to perform linear projection techniques such as principal component analysis, exploratory projection pursuit and canonical correlation analysis. The method is very general and simply requires a reward function which is specific to the function we require the unit to perform. We also discuss how the method can be used to learn kernel mappings and conclude by illustrating its use on a topology preserving mapping.

1 Introduction

Reinforcement learning describes a group of techniques for parameter adaptation based on a methodology which envisages an agent making an exploratory investigation of its environment with a view to identifying the optimal strategy for actions within the environment where optimal is defined in terms of the reward which an agent can gain from taking actions in the environment. It is often thought of as lying between the extremes of supervised learning in which the best action to take is known and the parameters are adjusted to make this more likely in future and unsupervised learning in which the learning algorithm must self-organize the parameters in order to perform the specific exploratory data analysis task at hand.

However there has always been a stream of research which has described methods of performing supervised learning tasks using reinforcement learning methods [13] and more recently, [10]. A somewhat less prominent stream has been using reinforcement learning methods for unsupervised learning tasks [9]. The REINFORCE method of [13] has recently been used for unsupervised projection techniques [4], for kernel projection techniques [3] and for topology preserving mappings [7]. In this paper, we investigate the reinforcement learning method of [10] for unsupervised projection techniques.

J. Marques de Sá et al. (Eds.): ICANN 2007, Part I, LNCS 4668, pp. 668–676, 2007.

2 Immediate Reward Reinforcement Learning

[14,13] investigated a particular form of reinforcement learning in which reward for an action is immediate which is somewhat different from mainstream reinforcement learning [12,5]. Williams [13] considered a stochastic learning unit in which the probability of any specific output was a parameterised function of its input, $\mathbf{x}$. For the ith unit, this gives

$$P(y_i = \zeta|\mathbf{w}_i, \mathbf{x}) = f(\mathbf{w}_i, \mathbf{x}) \tag{1}$$

where, for example,

$$f(\mathbf{w}_i, \mathbf{x}) = \frac{1}{1 + \exp(- \parallel \mathbf{w}_i - \mathbf{x} \parallel^2)} \tag{2}$$

Williams [13] considers the learning rule

$$\Delta w_{ij} = \alpha_{ij}(r_{i,\zeta} - b_{ij}) \frac{\partial \ln P(y_i = \zeta|\mathbf{w}_i, \mathbf{x})}{\partial w_{ij}} \tag{3}$$

where α_{ij} is the learning rate, $r_{i,\zeta}$ is the reward for the unit outputting ζ and b_{ij} is a reinforcement baseline which in the following we will take as the reinforcement comparison, $b_{ij} = \overline{r} = \frac{1}{K}\sum r_{i,\zeta}$ where K is the number of times this unit has output ζ. ([13], Theorem 1) shows that the above learning rule causes weight changes which maximises the expected reward.

[13] gave the example of a Bernoulli unit in which $P(y_i = 1) = p_i$ and so $P(y_i = 0) = 1 - p_i$. Therefore

$$\frac{\partial \ln P(y_i)}{\partial p_i} = \begin{cases} -\frac{1}{1-p_i} & \text{if } y_i = 0 \\ \frac{1}{p_i} & \text{if } y_i = 1 \end{cases} = \frac{y_i - p_i}{p_i(1 - p_i)} \tag{4}$$

[9] applies the Bernoulli model to (unsupervised) clustering with

$$p_i = 2(1 - f(\mathbf{w}_i, \mathbf{x})) = 2(1 - \frac{1}{1 + \exp(- \parallel \mathbf{w}_i - \mathbf{x} \parallel^2)}) \tag{5}$$

The environment identifies the p_{i^*} which is maximum over all output units and y_{i^*} is then drawn from this distribution. Rewards are given such that

$$r_i = \begin{cases} 1 \text{ if } i = i^* \text{ and } y_i = 1 \\ -1 \text{ if } i = i^* \text{ and } y_i = 0 \\ 0 \qquad \text{if } i \neq i^* \end{cases} \tag{6}$$

This is used in the update rule

$$\Delta w_{ij} = \alpha r_i(y_i - p_i)(x_j - w_{ij}) \tag{7}$$
$$= \alpha|y_i - p_i|(x_j - w_{ij}) \text{ for } i = i^* \tag{8}$$

which is shown to perform clustering of the data set. Such methods have recently been used for unsupervised projection techniques [4], for kernel projection techniques [3] and for topology preserving mappings [7].

3 Stochastic Weights

An alternative view of supervised networks with reinforcement learning is given in [10]: the networks are composed of deterministic nodes but stochastic synapses. There are two variants of this method: in the first, [10] uses a set of weights, w_{ij}, each of which is drawn from a Gaussian distribution $N(\mu_{ij}, \sigma_{ij}^2)$ with mean μ_{ij} and variance, σ_{ij}^2. Then the output of this deterministic neuron is also a Gaussian random variable with mean $\mu_{y_i} = \sum_j \mu_{ij} x_j$ and variance $\sigma_{y_i}^2 = \sum_j \sigma_{ij}^2 x_j^2$. With this model, [10] shows that the learning rule

$$\Delta\mu_{ij} = \eta r \frac{(y_i - \mu_{y_i}) x_j}{\sigma_{y_i}^2} \tag{9}$$

where η is the learning rate, follows the gradient of r, the reward given to the neuron. i.e. we are performing gradient ascent on the reward function by changing the parameters using (9). Often a simplified form is used by incorporating $\sigma_{y_i}^2$ into the learning rate:

$$\Delta\mu_{ij} = \eta r (y_i - \mu_{y_i}) x_j \tag{10}$$

[10] also investigate an additional term which they say enables the rule to escape from local optima:

$$\Delta\mu_{ij} = \eta [r(y_i - \mu_{y_i}) + \lambda(1 - r)(-y_i - \mu_{y_i})] x_j \tag{11}$$

[10] denotes (10) and (11) by rules A1 and A2 respectively.

In this first view, the neuron's output is a function of two independent sets of variables, the inputs and the weights. In the second view, the input is deemed to be fixed and only the synapses are thought of as stochastic. Then $p(w_{ij}) = Z \exp(-\frac{(w_{ij} - \mu_{ij})^2}{2\sigma_{ij}^2})$ which leads to a second set of learning rules:

$$\Delta\mu_{ij} = \eta r (w_{ij} - \mu_{ij}) \tag{12}$$

which again may include the local optima avoidance term to give

$$\Delta\mu_{ij} = \eta [r(w_{ij} - \mu_{ij}) + \lambda(1 - r)(w_{ij} - \mu_{ij})] \tag{13}$$

[10] denotes (12) and (13) by rules B1 and B2 respectively. These rules are tested on a number of standard problems in [10]. In this paper, we use these rules for exploratory data analysis.

4 Exploratory Data Analysis

The method is quite general: each agent is deemed to be taking actions in the environment (which consists of the data to be explored) in order to maximise his reward in this environment. Each agent has a set of parameters which determine the Gaussian distribution from which the actions will be drawn. We illustrate the method with Principal Component Analysis (PCA) and then discuss the reward functions for other linear projections. We then discuss Kernel methods and illustrate with Kernel Canonical Correlation Analysis. We finally illustrate a topology preserving mapping.

4.1 Principal Component Analysis

Principal Component Analysis (PCA) finds the linear filter onto which projections of a data set have greatest variance. Thus the agent is rewarded in proportion to its variance. Let the agent currently select actions from $N(\mu_t, \sigma_t^2)$ and let $\mathbf{w}_t$ be the action which the agent has selected at time instant t. For centered data, $\mathbf{x} \in X$, we can simply take $r = (\mathbf{w}_t^T \mathbf{x})^2$ for principal component analysis or the more complex but equally effective $r = \frac{1}{1+\exp(-\gamma(\mathbf{w}_t^T \mathbf{x})^2)}$. In Figure 1, we illustrate convergence with the first of these on an artificial data set in which each element of $\mathbf{x}$, $x_i, i = 1, ..., 5$, is drawn from $N(0, i^2)$ i.e. the first principal component filter is (0,0,0,0,1). Note that in (10), $y = \mathbf{w}_t^T \mathbf{x}$ while $\mu_{y_i} = \mu_t^T \mathbf{x}$

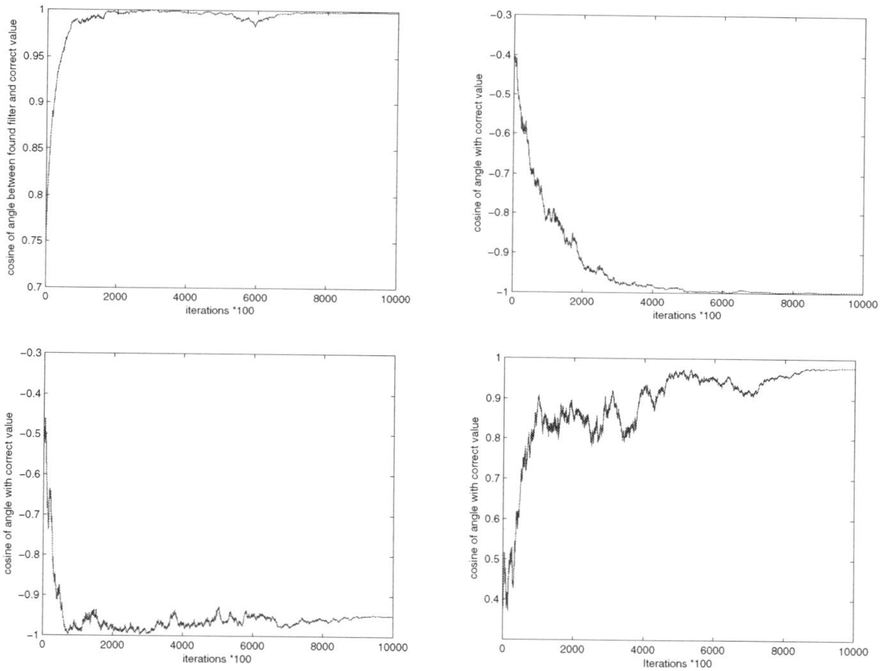

Fig. 1. Convergence with PCA reward function $r = (\mathbf{w}_t^T \mathbf{x})^2$. Top left: convergence with rule A1 (10). Top Right:convergence with rule A2 (11). Bottom left: convergence with rule B1 (12). Bottom Right: convergence with rule B2 (13).

The method clearly finds the first principal component in all cases, however the A rules seem to be more efficient than the B rules.

However we are often interested in finding more than the first principal component of a data set. This may be easily done by this method by incorporating a Gram-Schmidt orthogonalisation into the reward function rather than the learning rule. Thus we have

$$r_k = (\mathbf{w}_k^T \mathbf{x})^2 - \gamma \sum_{j=1}^{k-1} (\mathbf{w}_k^T \mu_j)^2 \tag{14}$$

where we have used the subscript k to denote the kth filter. Results from a similar data set as before and using the rule A2 are shown in Table 1. Again we see all principal components have been found with great accuracy. The reward function for this table was

$$r_k = \frac{1}{1 + \exp(-(\mathbf{w}_k^T \mathbf{x})^2 - \gamma \sum_{j=1}^{k-1} ()\mathbf{w}_k^T \mu_j)^2)} \tag{15}$$

Table 1. Each column corresponds to one of the principal component filters found by the method. The principal components have been found in order of magnitude of the variance of their projections.

-0.0152	0.0212	0.0747	0.0502	**0.9947**
0.0012	-0.0370	-0.0192	**0.9952**	0.0209
0.0222	0.0393	**0.9932**	0.0174	-0.0399
-0.0305	**0.9974**	-0.0341	-0.0073	-0.0397
-0.9992	0.0418	-0.0803	-0.0819	0.0839

Note also that we could have used minimization of the mean squared error as the basis of our reward function for PCA.

4.2 Other Linear Projections

The method is quite general. We have used the method to perform

- Minor Component Analysis: finding the filter with least variance has been used in regression and in spectral clustering. The reward function is then $r = \exp(-(\mathbf{w}_t^T \mathbf{x})^2)$. Exemplar convergence for the first minor component is shown in Figure 2.
- Exploratory Projection Pursuit (EPP): this method searches for the filter which produces the most "interesting" filter where "interesting" is defined in terms of the statistics of the data. If we are seeking outliers (which will be identified by positive kurtosis) we may use $r = (\mathbf{w}_t^T \mathbf{x})^4$. If we are seeking a projection which will reveal clusters, this is often shown by negative kurtosis.
- Independent Component Analysis (ICA): this method relies on the fact that mixtures of signals are more Gaussian than the independent signals and so again a measure like kurtosis can be used as the reward function.
- Canonical Correlation Analysis (CCA): this method searches for the linear filter within two data sets simultaneously which maximises the correlations between the respective projections. Thus we may use $r = (\mathbf{w}_1^T \mathbf{x}_1 - \mu_1^T \mathbf{x}_1)(\mathbf{w}_2^T \mathbf{x}_2 - \mu_2^T \mathbf{x}_2)$ where we have used the subscripts 1 and 2 to identify

the two data streams. Similar to PCA, we can extend the reward function to ensure that the output variance is 1 and to ensure orthogonality in the vectors. This leads to reward functions such as

$$r_k = \frac{1}{1+\exp(- \parallel \mathbf{w}_{1,k}\mathbf{x}_1 - \mathbf{w}_{2,k}\mathbf{x}_2) \parallel^2 + \gamma_1(\parallel \mathbf{w}_{1,k}\mathbf{x}_1 \parallel^2 -1) + \gamma_2 \sum_{j=1}^{k-1}(\mathbf{w}_{k,1}\mathbf{w}_{j,1})^2)} \tag{16}$$

where the second term in the denominator is ensuring the variance is 1 while the last term ensures that the new canonical vectors are orthogonal to the old. We illustrate this learning rule on an artificial data set similar to that used in [8]: we create two random Gaussian vectors, one 4 dimensional, the other 3 dimensional where all samples are drawn iid from N(0,1). We add an additional sample from N(0,1) to the first elements of each vector to create a correlation and then normalise so that there are no variance effects. Similarly we add a second sample from N(0,0.5) to the second element of each data set so that there is most correlation between the first element of each data set and the remaining correlation is found in the second elements of the data sets. Results using (16) are shown in Table 2.

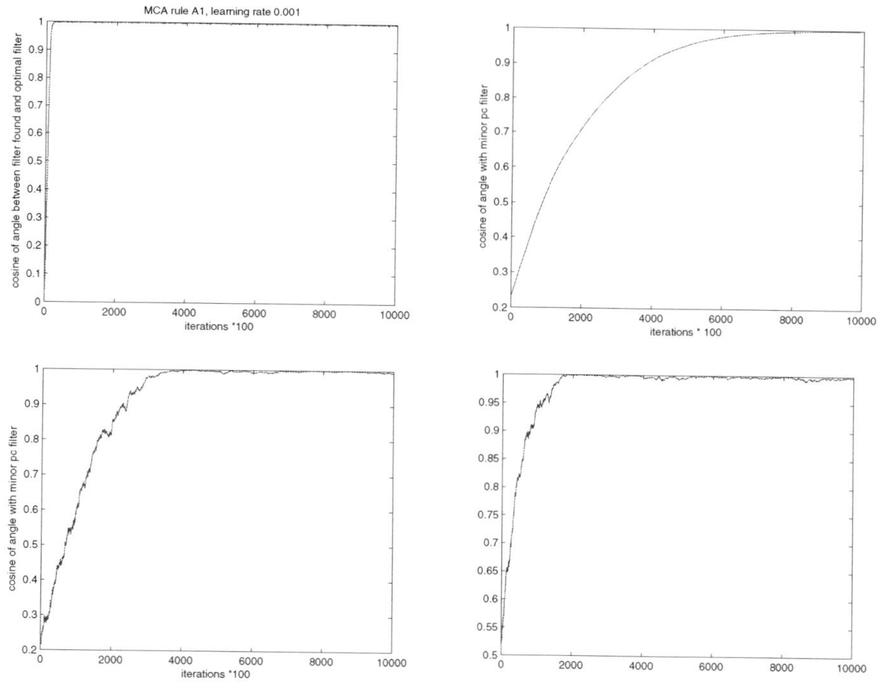

Fig. 2. Convergence of the minor component algorithm. Top left: convergence with rule A1. Top Right:convergence with rule A2. Bottom left: convergence with rule B1. Bottom Right: convergence with rule B2.

Table 2. The first two canonical correlation vectors for the artificial data

	μ_1	μ_2
1st CC	**0.9979**	**0.9998**
	0.0087	0.0142
	-0.0592	-0.0136
	0.0235	
2nd CC	-0.0029	-0.0221
	0.9989	**0.9990**
	-0.0402	0.0399
	0.0109	

4.3 Kernel Mappings

Kernel methods are a recent addition to the data analysts tool package. The data is first mapped to a feature space using some (nonlinear) mapping $\phi()$. In this feature space a linear operation is performed which because of the nonlinear mapping is equivalent to some nonlinear operation on the original data. The most interesting part of kernel methods is that you do not need to ever work with or even know the mapping $\phi()$; the whole operation can be performed with the scalar product matrix $K = \phi(\mathbf{x})^T\phi(\mathbf{x})$. Examples include Kernel Principal Component Analysis (KPCA) [11] and Kernel Canonical Correlation Analysis [8].

All of the above methods can be performed in Kernel spaces and so the mean vector is of length N as are the samples from this vector. To perform Kernel Principal Component Analysis (KPCA) a batch algorithm uses $r = \mathbf{w}^T K\mathbf{w}$, where K is the kernel matrix.

To perform Kernel Canonical Correlation Analysis, we have two kernel matrices each of size $N \times N$. If we decide to use an online reward function (taking each data sample as it comes), we may use $r = \mathbf{w}_1^T K_1(current)^T K_2(current)\mathbf{w}_2$, where we have used $K_i(current)$ to denote the column of the appropriate kernel matrix corresponding to the current sample.

4.4 Topology Preserving Manifolds

Topology preserving maps [6] can also be thought of as clustering techniques but clustering techniques which attempt to capture some structure in a data set by creating a visible relationship between nearby clusters. Most research has been into methods which modify an adaptive clustering rule in a way that preserves neighbouring relationships. However an alternative [1,2] is to create a latent space of points $\mathbf{t}_1, ..., \mathbf{t}_K$ which *a priori* have some structure such as lying equidistantly placed on a line or at the corners of a grid. These are then mapped non-linearly to the data space; the nonlinearity is essential since the data clusters need not be constrained to lie on a line. Thus we [1,2] map the latent points through a set of basis functions, typically squared exponentials centered in latent space, and then map the output of the basis functions through a set of weights to points, $\mathbf{m}_1, ..., \mathbf{m}_K$, in data space. Let there be M basis functions, $\phi_j(), j = 1, ..., M$,

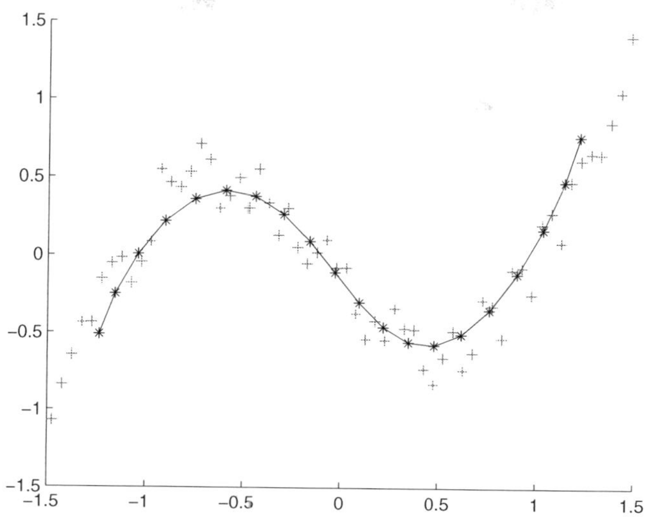

Fig. 3. The data are shown by red '+'s; the latent points' projections are shown by black '*'s

with centres, μ_j in latent space; therefore, using $\mathbf{w}_j$ as the weight from the jth basis function to data space,

$$\mathbf{m}_k = \sum_{j=1}^{M} \mathbf{w}_j \phi_j(\mathbf{t}_k) = \sum_{j=1}^{M} \mathbf{w}_j \exp(-\beta \parallel \mu_j - \mathbf{t}_k \parallel^2), \forall k \in \{1, ..., K\} \tag{17}$$

Since $\frac{\partial \mathbf{m}_k}{\partial w_{ij}} = \phi_j(\mathbf{t}_k)$, with the Gaussian model A1 above

$$\Delta\mu_{i^*} = \eta r(\mathbf{w} - \mu_{y_{i^*}})\phi(i^*) \tag{18}$$

where i^* identifies the agent with closest centre, $\mathbf{m}_{i^*}$ to the current data point, $\mathbf{x}$. $\mathbf{w}$ is a sample from $\mu_{y_{i^*}}$ and the reward is $r = \exp(- \parallel \mathbf{x} - \mathbf{w} \parallel)$.

We illustrate (Figure 3) a one dimensional latent space mapped to a two dimensional data set, $\mathbf{x} = (\mathbf{x}_1, \mathbf{x}_2)$, in which $\mathbf{x}_1$ evenly divides the interval $[-\pi/2, \pi/2]$ and $\mathbf{x}_2 = \mathbf{x}_1 + 1.25 \sin(\mathbf{x}_1) + \mu$ where μ is noise from a uniform distribtion in [0,0.5]; we see that the one dimensional nature of the manifold has been identified and the latent points' projections into data space have maintained their ordering. This simulation used rule A2 and we can see one of the effects of the local minimum avoidance term which is a drawing in of the map to the centre of the data. However there are local optima with these maps (the Kohonen SOM often exhibits twists across the data) and so in practice it is bestto use this term but with a decay of the λ parameter during the course of the simulation till rule A1 is reached.

5 Conclusion

We have extended our previous work on using immediate reward reinforcement learning for exploratory data analysis with a new set of learning rules which have been shown to optimise a reward function. The new rules are based on a view of the reinforcement learning unit as a deterministic unit but with stochastic parameters. We have illustrated our new set of rules performing principal component analysis, minor component analysis, exploratory projection pursuit, canonical correlation analysis, independent component analysis and kernel versions of these linear exploratory data methods. We have further illustrated how to create a topology-preserving mapping with these methods.

It is an open question as to whether these new rules perform better or worse than the existing immediate reward reinforcement learning rules (or indeed standard temporal difference methods). This will be the subject of future investigations. However the long term promise is of a set of automata which investigate a data set using trial and error reinforcement methods just as a human analyst would when presented with a new data set. The integration of these methods into a single reinforcement learning system is also a topic we are currently studying.

References

1. Bishop, C.M., Svensen, M., Williams, C.K.I.: Gtm: The generative topographic mapping. Neural Computation (1997)
2. Fyfe, C.: Two topographic maps for data visualization. Data Mining and Knowledge Discovery (2007) ISSN 1384-5810
3. Fyfe, C., Lai, P.L.: Immediate reward reinforcement learning for projective kernel methods. In: ESANN2007. 14th European Symposium on Artificial Neural Networks (2007)
4. Fyfe, C., Lai, P.L.: Reinforcement learning reward functions for unsupervised learning. In: ISNN 2007. 4th International Symposium on Neural Networks (2007)
5. Kaelbling, L.P., Littman, M.L., Moore, A.W.: Reinforcement learning: A survey. Journal of Artificial Intelligence Research 4, 237–285 (1996)
6. Kohonen, T.: Self-Organising Maps. Springer, Heidelberg (1995)
7. Lai, P.L., Fyfe, C.: Reinforcement learning for topographic mappings. In: 20th International Joint Conference on Artificial Neural Networks (2007)
8. Lai, P.L., Fyfe, C.: Kernel and nonlinear canonical correlation analysis. International Journal of Neural Systems 10(5), 365–377 (2001)
9. Likas, A.: A reinforcement learning approach to on-line clustering. Neural Computation (2000)
10. Ma, X., Likharev, K.K.: Global reinforcement learning in neural networks with stochastic synapses. IEEE Transactions on Neural Networks 18(2), 573–577 (2007)
11. Scholkopf, B., Smola, A., Muller, K.-R.: Nonlinear component analysis as a kernel eigenvalue problem. Neural Computation 10, 1299–1319 (1998)
12. Sutton, R.S., Barto, A.G.: Reinforcement Learning: an Introduction. MIT Press, Cambridge (1998)
13. Williams, R.: Simple statistical gradient-following algorithms for connectionist reinforcement learning. Machine Learning 8, 229–256 (1992)
14. Williams, R.J., Pong, J.: Function optimization using connectionist reinforcement learning networks. Connection Science 3, 241–268 (1991)

Post Nonlinear Independent Subspace Analysis

Zoltán Szabó, Barnabás Póczos, Gábor Szirtes, and András Lőrincz

Department of Information Systems, Eötvös Loránd University,
Pázmány P. sétány 1/C, Budapest H-1117, Hungary
{szzoli,pbarn}@cs.elte.hu,szirtes@inf.elte.hu,andras.lorincz@elte.hu

Abstract. In this paper a generalization of Post Nonlinear Independent Component Analysis (PNL-ICA) to Post Nonlinear Independent Subspace Analysis (PNL-ISA) is presented. In this framework sources to be identified can be multidimensional as well. For this generalization we prove a separability theorem: the ambiguities of this problem are essentially the same as for the linear Independent Subspace Analysis (ISA). By applying this result we derive an algorithm using the mirror structure of the mixing system. Numerical simulations are presented to illustrate the efficiency of the algorithm.

1 Introduction

Independent Component Analysis (ICA) has become one of the most popular research line in signal processing. It aims to solve the following (so called 'cocktail party') problem: let us assume there are D 1-dimensional, independent hidden sources and we can only observe their linear mixture through D microphones. The goal is to recover the original sources from the mixed signals. In spite of its simplicity this model has been successfully applied to, for example, feature extraction problems, denoising or biomedical signal processing. For a recent review on ICA see e.g. [1,2]. The strong assumption on the linearity can be relaxed by assuming component-wise distortion resulting in a post nonlinear extension (PNL-ICA) of the ICA [3]. This direction has recently gained much attention, for a review see [4]. Another generalization of the original ICA problem is called Multidimensional Independent Component Analysis (MICA) [5], or Independent Subspace Analysis (ISA) (in this paper we use the latter abbreviation). In the ISA framework the assumption on independence is relaxed in the sense that it only requires that groups of the sources be independent so the hidden components can now be multidimensional. Several methods have already been proposed to solve this problem [5,6,7,8,9,10,11,12,13,14,15,16], and it has also been used in EEG analysis [6].

In this paper we show that both generalizations can be treated at the same time (Post Nonlinear Independent Subspace Analysis, PNL-ISA). PNL-ISA problem arises, for example, if the world is observed through a layer of artificial neural networks: neurons mix and sum input signals and then pass the sums through non-linearities.

We prove that multidimensional sources can still be recovered even if the observations are component-wise nonlinear distortions of the linear mixtures. To

J. Marques de Sá et al. (Eds.): ICANN 2007, Part I, LNCS 4668, pp. 677–686, 2007.

do so, first we introduce the PNL-ISA problem in Section 2. The ambiguities of the problem are analyzed in Section 3, where a theorem about the separability of the components is proven. Applying this theorem we propose an efficient algorithm in Section 4 that solves the PNL-ISA task. In Section 5 the efficiency of the algorithm is illustrated on several problems. Our results are recapped in Section 6.

2 The PNL-ISA Model

Let us define the PNL-ISA task. Let us assume the observations are post non-linear mixtures of multidimensional independent sources:

$$\mathbf{x}(t) = \mathbf{f}[\mathbf{A}\mathbf{s}(t)], \tag{1}$$

where $\mathbf{s}(t)$ is the concatenation of the components $\mathbf{s}_m(t) \in \mathbb{R}^d$ that is $\mathbf{s}(t) = [\mathbf{s}_1(t); \ldots; \mathbf{s}_M(t)] \in \mathbb{R}^D$, $(D = dM)$. (For simplicity let the dimension of each component be the same d.) Our assumptions are the following:

1. Source $\mathbf{s}$ is *d-independent*, that is $I(\mathbf{s}_1, \ldots, \mathbf{s}_M) = 0$, where I denotes the mutual information, and $\mathbf{s}(t)$ is i.i.d in t.
2. matrix $\mathbf{A} \in Gl(D)$ (that is it is of size $D \times D$ and it is invertible) and it is 'mixing'. By 'mixing' we mean the following: decomposing matrix $\mathbf{A}$ into blocks of size $d \times d$ ($\mathbf{A} = [\mathbf{A}_{ij}]_{i,j=1,\ldots,M}, \mathbf{A}_{ij} \in \mathbb{R}^{d\times d}$), then for any index $i \in \{1, \ldots, M\}$ there exist a pair of indices $j \neq k \in \{1, \ldots, M\}$ for which matrices $\mathbf{A}_{ij}$ and $\mathbf{A}_{ik}$ are invertible.
3. function $\mathbf{f} : \mathbb{R}^D \to \mathbb{R}^D$ is a component-wise transformation that is $\mathbf{f}(\mathbf{z}) = [f_1(z_1); \ldots; f_D(z_D)]$ and $\mathbf{f}$ is invertible.

The PNL-ISA problem is to estimate the hidden source components $\mathbf{s}_m$ knowing only the observations $\mathbf{x}(t)$ $(t = 1, \ldots, T)$. For $d = 1$ we get back the PNL-ICA problem, while for choosing $\mathbf{f}$ as identity the ISA problem is recovered.

3 Ambiguities of PNL-ISA

To solve the PNL-ISA problem it is important to see to what extent we can expect to regain the true sources. The ambiguities of the ICA, PNL-ICA and the ISA problems are well known: in ICA, hidden sources can be recovered up to a scalar multiplier and permutation [17]. In addition to these ambiguities there is an additive scalar term in the PNL-ICA problem [18], while in the ISA problem the $\mathbf{s}_m$ components can be recovered up to the permutation ambiguity between the components and invertible transformations within each subspace [19].

Because of the PNL assumption the hidden sources can be estimated using the mirror structure of the mixing system, that is $\hat{\mathbf{s}} = \mathbf{W}\mathbf{g}(\mathbf{x})$ ($\mathbf{W} \in Gl(D)$, $\mathbf{g} : \mathbb{R}^D \to \mathbb{R}^D$, where $\mathbf{g}$ is a component-wise transformation). It has to be shown, however, that the d-independence of the resulting $\hat{\mathbf{s}}$ unequivocally means that the true $\mathbf{s}$ has been found. The following separability theorem shows that indeed this is the case. This statement can be considered as an extension of the results in [20] for the case $d \geq 1$. The proof of the theorem will be based on Lemmas 3.4 and 3.5 of [21]. Due to the limited space these lemmas will only be cited. Let $C^2(V,\mathbb{R})$ and $C^\omega(V,\mathbb{R})$ denote the V (open) $\subseteq \mathbb{R}^n \to \mathbb{R}$ 2-times continuously differentiable and analytic functions, respectively.

Theorem 1 (PNL-ISA Ambiguities with Locally-Constant Nonzero C^2 Densities). *Let (i)* $\mathbf{A}, \mathbf{W} \in Gl(D)$, *be mixing matrices; (ii) let* $\mathbf{s}$ *be as above, with at most one* $\mathbf{s}_m$ *gaussian component, with existing covariance matrix, with somewhere locally constant density function* $p_S \in C^2(\mathbb{R}^D,\mathbb{R})$, *(iii)* $\mathbf{h}: \mathbb{R}^D \to \mathbb{R}^D$ *is a component-wise bijection with coordinate functions in* $C^\omega(\mathbb{R},\mathbb{R})$. *In this case, if* $\mathbf{e} := [\mathbf{e}_1;\ldots;\mathbf{e}_M] = \mathbf{Wh}(\mathbf{As})$ *is d-independent* ($\mathbf{e}_m \in \mathbb{R}^d$) *with somewhere locally constant density function, then (i)* $\mathbf{h}(\mathbf{x}) = \mathbf{Lx} + \mathbf{p}$, *where* $\mathbf{L} \in Gl(D)$ *is a diagonal matrix and* $\mathbf{p} \in \mathbb{R}^D$, *and (ii) components* $\mathbf{e}_m$ $(m = 1,\ldots,M)$ *recover the hidden sources up to permutation and invertible transformations within each subspace (and maybe up to a constant translation).*

Proof. To prove the first statement it suffices to show that if $\mathbf{e} := \mathbf{Wh}(\mathbf{As})$ is d-independent then derivatives h'_{mi} are constant for $\forall(i,m) \in \{1,\ldots,d\} \times \{1,\ldots,M\}$ where $\mathbf{h}_m$ is part of $\mathbf{h}$ that belongs to subspace m and h_{mi} is the ith coordinate function of $\mathbf{h}_m : \mathbb{R}^d \to \mathbb{R}^d$. It directly implies the second part of the statement as if $\mathbf{Bs} + \mathbf{Wp}$ (where $\mathbf{B}$ denotes the matrix product $\mathbf{WLA}$) is d-independent then $\mathbf{Bs}$ is also d-independent. In turn, because of the separability properties of the linear ISA, $\mathbf{B}$ can recover the hidden components up to permutation and invertible transformations within the subspaces (and maybe up to a constant translation within the subspaces).

To prove the first part, let p_E and p_S denote the density functions of $\mathbf{e}$ and $\mathbf{s}$, respectively. Based on the transformation rule of the density functions, p_E can be given as

$$p_E[\mathbf{Wh}(\mathbf{As})] = |\det(\mathbf{W})|^{-1}\left(\prod_{m=1}^{M}|\det(\mathbf{h}'_m[(\mathbf{As})_m])|^{-1}\right)|\det(\mathbf{A})|^{-1}p_S(\mathbf{s}), \quad (2)$$

where $(\mathbf{As})_m \in \mathbb{R}^d$ is part of $\mathbf{As}$ belonging to subspace m. In addition, $\mathbf{e}$ is d-independent implying that $p_E(\cdot)$ is *d-separated*, that is it can be rewritten as

$$p_E = \otimes_{m=1}^{M} p_m, \quad (3)$$

where functions p_m are of $\mathbb{R}^d \to \mathbb{R}$. Let us choose point $\mathbf{s}^0$, for which $p_S(\mathbf{s}^0) > 0$. Then, there exists an open neighborhood $U \subseteq \mathbb{R}^D$ of this point, where $p_S|_U > 0$, and $p_S|_U \in C^2(U,\mathbb{R})$. Let us define $r(\mathbf{s}) := \ln\left[|\det(\mathbf{W})|^{-1}|\det(\mathbf{A})|^{-1}p_S(\mathbf{s})\right]$ on set U. It can be shown using Eqs. (2)-(3) that the following relation holds

$$r(\mathbf{s}) = \ln\left[\left(\prod_{m=1}^{M}|\det(\mathbf{h}'_m[(\mathbf{As})_m])|\right)\left(\prod_{m=1}^{M}p_m([\mathbf{Wh}(\mathbf{As})]_m)\right)\right] \quad (4)$$

$$= \sum_{m=1}^{M}\ln[|\det(\mathbf{h}'_m[(\mathbf{As})_m])|] + \xi_m([\mathbf{Wh}(\mathbf{As})]_m), \quad (\mathbf{s} \in U) \quad (5)$$

where $\xi_m := \ln(p_m)$, locally at $\mathbf{s}^0_m$. p_S is d-separated, thus function $r \in C^2(U, \mathbb{R})$ is *linearly d-separated*. In other words, it can be written as a direct sum, so $\partial_i \partial_j r \equiv 0$, where $\lfloor \frac{i}{d} \rfloor \neq \lfloor \frac{j}{d} \rfloor$ (i,j correspond to indices in different subspaces). However, since p_m and therefore $\xi_m = \ln(p_m)$ is locally constant, this relation holds (without loss of generality on set U) when ignoring terms ξ_m, since their derivatives are 0. Now, following the reasoning of Lemma 3.5 [21] for the d-separated function $\left[\otimes_{m=1}^M \det(\mathbf{h}'_m)\right](\mathbf{As})$, where $\mathbf{A}$ is mixing ($|\cdot|$s were dropped since functions $\mathbf{h}_m$ were assumed to be invertible), we can see that each function $g_m(\mathbf{v}) = \det[\mathbf{h}'_m(\mathbf{v})] (\neq 0)$ satisfies a differential equation

$$g_m H_{g_m} - \nabla g_m (\nabla g_m)^* \equiv \mathbf{C}_m g_m^2 \tag{6}$$

on set $U_m := \mathbf{A}_m(U)$, where $\mathbf{A}_m := [\mathbf{A}_{m1}, \ldots, \mathbf{A}_{mM}]$, $\mathbf{C}_m \in \mathbb{R}^{d \times d}$, ∇ stands for the gradient, H is the Hessian and $*$ denotes transposition. For this reason - similarly to Lemma 3.4 [21] – functions g_m can be given in the form $g_m(\mathbf{v}) = e^{\mathbf{v}^* \mathbf{D}_m \mathbf{v} + \mathbf{b}_m^* \mathbf{v} + c_m}$ $(\mathbf{v} \in U_m)$ with suitable $\mathbf{D}_m \in \mathbb{R}^{d \times d}, \mathbf{b}_m \in \mathbb{R}^d, c_m \in \mathbb{R}$. Furthermore, as functions $\mathbf{h}_m$ are assumed to act on each coordinate separately as h_{mi}, g_m can be written as $g_m = \otimes_{i=1}^d h'_{mi}$, we arrive at

$$h'_{mi}(t) = \pm e^{d_{mi} t^2 + b_{mi} t + c_{mi}} \quad (d_{mi}, b_{mi}, c_{mi} \in \mathbb{R}) \tag{7}$$

locally, exploiting that $h'_{mi} \equiv 0$ is not allowed. Based on our assumption on h_{mi} it holds on all $\mathbb{R} (\ni t)$.

Let us introduce the following notation: $\mathbf{y} = [\mathbf{y}_1; \ldots; \mathbf{y}_M] = \mathbf{As}$, where $\mathbf{y} \in \mathbf{A}(U)$. Following the above reasoning for the inverse system $\mathbf{s} = \mathbf{A}^{-1} \mathbf{h}^{-1} (\mathbf{W}^{-1} \mathbf{y})$, h_{mi}^{-1} can be given similar to (7), with other constants. However, if both $h_{mi}{}'$ and $(h_{mi}^{-1})'$ are of exponential type then it follows that $d_{mi} = 0$ and $b_{mi} = 0$ $[\forall (m, i) \in \{1, \ldots, M\} \times \{1, \ldots, d\}]$. Therefore, h_{mi}s are affine, that is $h_{mi}(z) = l_{mi}(z) + p_{mi}$ $(l_{mi}, p_{mi} \in \mathbb{R})$, what we wanted to demonstrate. □

4 Algorithm (Sketch)

The theorem proven above implies that by using an appropriate transformation $\mathbf{g}$ acting on each coordinate separately, the d-independence of the estimation $\hat{\mathbf{s}} = \mathbf{Wg}(\mathbf{x})$ solves the PNL-ISA task. The estimation can be done in two steps:

1. Estimate $\mathbf{g}$: according to the d-dependent central limit theorem [22], term $\mathbf{As}$ can be considered as an approximately gaussian variable, so $\mathbf{g}$ can be approximated as a 'gaussianization' transformation (see [23,24] for $d = 1$).
2. Estimate $\mathbf{W}$: apply a linear ISA method on the result of the 'gaussianization' transformation.

5 Illustrations

Now we illustrate the efficiency of the algorithm presented in Section 4. Test databases are described in Section 5.1. To evaluate the solutions we use a performance measure given in Section 5.2. The numerical results are summarized in Section 5.3.

5.1 Databases

To test our PNL-ISA method, we have created 4 datasets ($\mathbf{s}$) shown in Fig. 1.

In dataset *3D-geom* (see Fig. 1(a)) the components $\mathbf{s}_m$ were random variables uniformly distributed over 3-dimensional geometrical shapes ($d = 3$). There were 6 hidden components ($M = 6$), so the total dimension of the hidden source $\mathbf{s}$ was $D = 18$. In dataset *celebrities* the components $\mathbf{s}_m$ were generated using cartoons of celebrities.[1] The cartoons were interpreted as density functions and the 2-dimensional coordinate pairs ($d = 2$) were sampled with frequency proportional to the pixel intensity at the given coordinates. We used 10 hidden sources ($M = 10$), resulting in $D = 20$ (see Fig. 1(c)). The dataset $A\omega$ is scalable, the number of components (M) can be set within a quite broad interval. Here, components $\mathbf{s}_m$ are random variables of uniform distribution over the letters of the English and the Greek alphabets ($M \leq 26+24 = 50$, $d = 2$), see Fig. 1(b). While in the datasets defined so far variable $\mathbf{s}$ was i.i.d in accord with the PNL-ISA model, in the dataset *IFS* this constraint has been relaxed.[2] Here, components $\mathbf{s}_m$ are realizations of IFS based 2-dimensional ($d = 2$) self-similar structures. For all m we have chosen the following triple: $(\{\mathbf{h}_k\}_{k=1,\ldots,K}, \mathbf{p} = (p_1, \ldots, p_K), \mathbf{v}_1\}$, where (i) $\mathbf{h}_k : \mathbb{R}^2 \to \mathbb{R}^2$ are affine transformations in the form $\mathbf{h}_k(\mathbf{z}) = \mathbf{C}_k\mathbf{z} + \mathbf{d}_k$ ($\mathbf{C}_k \in \mathbb{R}^{2\times 2}, \mathbf{d}_k \in \mathbb{R}^2$), (ii) $\mathbf{p}$ is a distribution over the indices $\{1, \ldots, K\}$ ($\sum_{k=1}^K p_k = 1, p_k \geq 0$), and (iii) for the initial value we chose $\mathbf{v}_1 := (\frac{1}{2}, \frac{1}{2})$. We generated T samples in the following way: (i) $\mathbf{v}_1$ is given ($t = 1$), (ii) an index $k(t) \in \{1, \ldots, K\}$ was drawn according to the distribution $\mathbf{p}$ and the next sample is generated as $\mathbf{v}_{t+1} := \mathbf{h}_{k(t)}(\mathbf{v}_t)$. The resulting series $\{\mathbf{v}_1, \ldots, \mathbf{v}_T\}$ was taken as a hidden source component $\mathbf{s}_m$ and this way we generated 9 components ($M = 9$, $D = 18$) to make the *IFS* dataset (see Fig. 1(d)).

5.2 Performance Measure

Let $E[\cdot]$ denote the expectation. In the sense of Theorem 1, in ideal case $\mathbf{s} \in \mathbb{R}^D \mapsto \hat{\mathbf{s}} = \mathbf{Wg}[\mathbf{f}(\mathbf{As})] \in \mathbb{R}^D$ is an affine transformation residing within the subspaces. For this reason, the linear transformation $\mathbf{G}$ that optimally approximates the relation $\mathbf{s} - E[\mathbf{s}] \mapsto \hat{\mathbf{s}} - E[\hat{\mathbf{s}}]$, resides also within the subspaces and so it is a *block-permutation matrix*. This block-permutation structure can be measured by the normalized version of the Amari-error [25] adapted to the ISA task [21]. Let us decompose matrix $\mathbf{G} \in \mathbb{R}^{D\times D}$ into blocks of size $d \times d$: $\mathbf{G} = [\mathbf{G}_{ij}]_{i,j=1,\ldots,M}$. Let g_{ij} denote the sum of the absolute values of matrix $\mathbf{G}_{ij} \in \mathbb{R}^{d\times d}$. Now, the following term

$$r(\mathbf{G}) := \frac{1}{2M(M-1)} \left[\sum_{i=1}^{M} \left(\frac{\sum_{j=1}^{M} g_{ij}}{\max_j g_{ij}} - 1 \right) + \sum_{j=1}^{M} \left(\frac{\sum_{i=1}^{M} g_{ij}}{\max_i g_{ij}} - 1 \right) \right] \quad (8)$$

denotes the Amari-index that takes values in [0,1]: for an ideal block-permutation matrix $\mathbf{G}$ it takes 0; for the worst case it takes 1.

[1] http://www.smileyworld.com

[2] IFS stands for Iterated Function System.

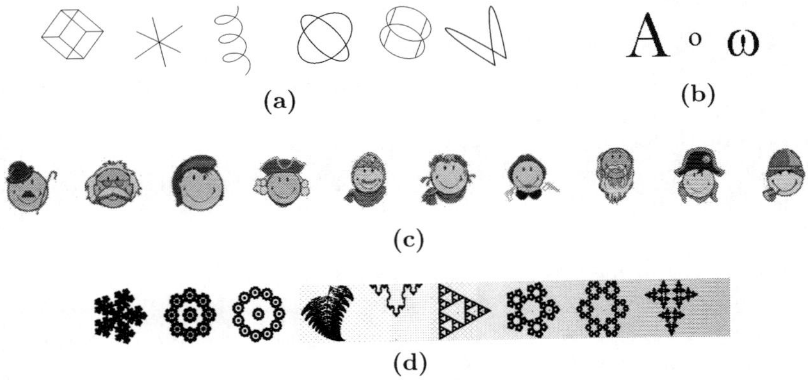

Fig. 1. Illustration of the test datasets. (a): *3D-geom* set. The hidden components $\mathbf{s}_m$ are random variables of uniform distribution over 3-dimensional geometric shapes: $d = 3$, $M = 6$, $D = 18$. (b): *A*ω set. Here, components $\mathbf{s}_m$ are 2-dimensional variables of uniform distribution over the shape of the letters of the English and Greek alphabets: $d = 2$, $M \leq 50$. (c): *celebrities* dataset. The components are 2-dimensional variables with sampling frequency proportional to the pixel intensity of the corresponding coordinates of 'celebrities' cartoons: $d = 2$, $M = 10$, $D = 20$. (d): *IFS* dataset. Here components $\mathbf{s}_m$ are not i.i.d. variables anymore, instead they are self-similar structures generated from iterated function systems: $d = 2$, $M = 9$, $D = 18$.

5.3 Simulations

In this section the results of our numerical simulations on the above defined datasets are presented. We focused on the following questions:

1. The error of the source estimation as a function of (i) the sample size, (ii) the dimension (D) of the task.
2. Estimation of D when it is not known beforehand.

The gaussianization (see Section 4) was based on the ranks of samples [23], in the solution of the ISA task we followed the method described in [14]. The goodness of the estimation was measured by the Amari-index introduced in Section 5.2. For a given sample size T the goodness of 50 random runs $(\mathbf{A}, \mathbf{s}, \mathbf{f})$ were averaged. $\mathbf{A}$ was a random orthogonal matrix. The nonlinear functions f_i have been generated as

$$f_i(z) = c_i[a_i z + \tanh(b_i z)] + d_i, \tag{9}$$

that is they are mixtures of random, scaled and translated *id* and tanh functions. Here $a_i \in [0, 0.5]$, $b_i \in [0, 5]$, $d_i \in [0, 2]$ are random variables of uniform distribution, c_i take ± 1 values with probability $\frac{1}{2}, \frac{1}{2}$.

First we studied the Amari-index as a function of the sample size. For *3D-geom*, *celebrities* and *IFS* the dimension and the number of the components d, M and the dimension of the task D were fixed while for *A*ω M has been chosen as $2, 3, 4, 10, 20, 50$. As for small M the gaussian assumption on $\mathbf{As}$ is less likely, we expect

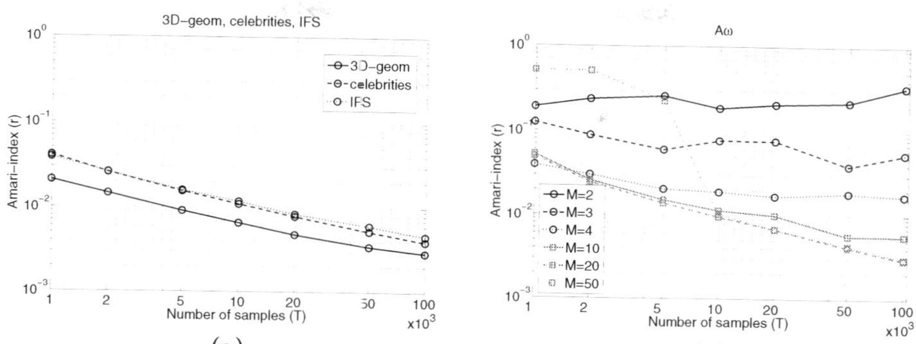

(a) (b)

Fig. 2. Average Amari-index as a function of the sample size, on loglog scale. (a): datasets *3D-geom*, *celebrities* and *IFS*. (b): $A\omega$ with different number of components (M). For $T = 100,000$, the exact errors are shown in Table 1 and Table 2.

to see deterioration of the goodness of the estimation in this range. In all cases the sample size T was chosen between $1,000$ and $100,000$. The Amari-index as a function of the sample size is shown in Fig. 2. For $T = 100,000$, the exact errors are shown in Table 1 and Table 2. As an example, the estimation results for *IFS* and *3D-geom* datasets are illustrated in Fig. 3.

Table 1. Amari-index for *3D-geom*, *celebrities* and *IFS* datasets: average ± std. Sample size: $T = 100,000$. The error as a function of sample size T is plotted in Fig. 2(a).

3D-geom	celebrities	IFS
0.29%(±0.05)	0.40%(±0.03)	0.46%(±0.06)

Table 2. Amari-index for dataset $A\omega$, as a function of the number of components M: values shown are average ± std. Sample size: $T = 100,000$. The error as a function of sample size T is plotted in Fig. 2(b).

$M = 2$	$M = 3$	$M = 4$	$M = 10$	$M = 20$	$M = 50$
33.20%(±39.42)	5.37%(±8.82)	1.71%(±0.52)	0.56%(±0.50)	0.30%(±0.03)	0.30%(±0.01)

As Fig. 2(a) shows, the dependency of the Amari-index on the sample size T follows a power-law for datasets *3D-geom*, *celebrities* and *IFS*, $r(T) \propto T^{-c}$ ($c > 0$). This can be seen as a linear decrease on the loglog scale. The slope of the lines are about the same for the different datasets. Table 1 summarizes the average estimation errors for sample size $100,000$.

In Fig. 2(b) similar power law relation can be found for dataset $A\omega$ as well, with increasing D. For $M \geq 3$, (that is when $D \geq 6$) the PNL-ISA solution is already efficient. It can also be seen that for the case of $M = 50$ number of components (that is the dimension of the task is $D = 100$) we need at least 10,000

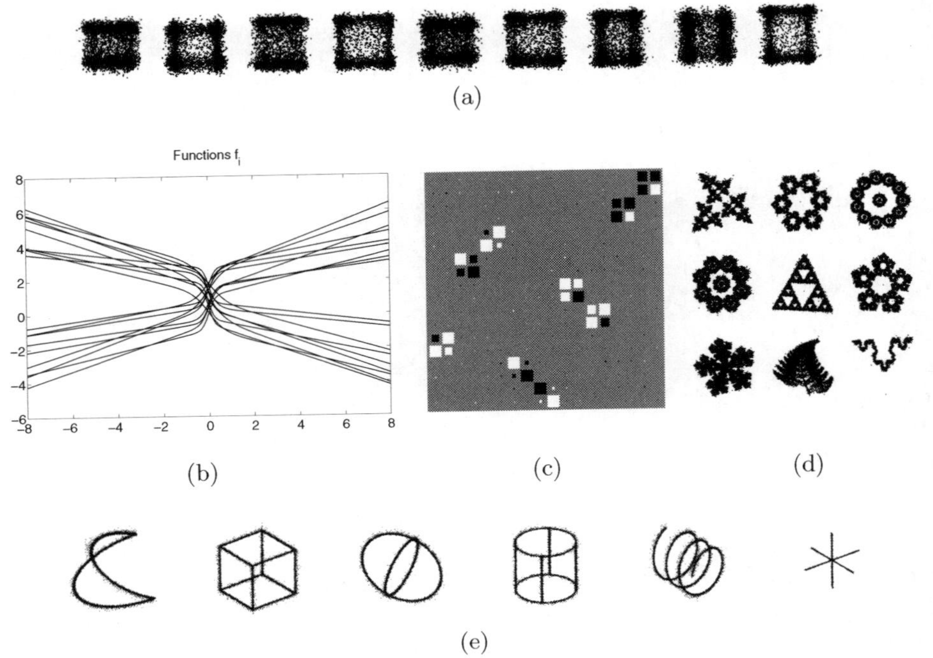

Fig. 3. Illustration of the PNL-ISA estimation on the dataset *IFS* and *3D-geom*, in (a)-(d) and (e), respectively. Sample size: $T = 100,000$. (a): the observed mixed $\mathbf{x}$ signal. (b): the nonlinear f_i functions. (c) the Hinton-diagram of $\mathbf{G}$, ideally it is a block-permutation matrix with blocks of size 2×2. (d): the estimated hidden components ($\hat{\mathbf{s}}_m$). (e): the same as (d), but for the *3D-geom* dataset.

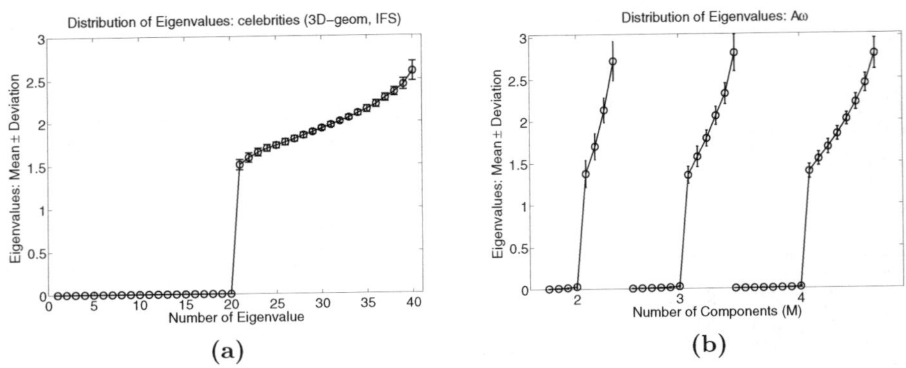

Fig. 4. Estimation of the dimension of the hidden source $\mathbf{s}$. The ordered eigenvalues of the covariance matrix of the transformed signal $\mathbf{g}(\mathbf{x})$ are plotted. Results are averaged over 50 runs. (a): dataset *celebrities*; results for *3D-geom* and *IFS* are similar. (b): eigenvalues for the dataset Aω, at different number of components M.

samples to get a more reliable estimation, while for $3 \leq M < 50$ $2,000 - 5,000$ samples are sufficient.

Next we studied to what extent we can guess the overall dimension D of the hidden source $\mathbf{s}$ when it is not given beforehand. The dimension of the observation $\mathbf{x}$ was set to $D_x = 2D$ (D of course is not available for the algorithm). The mixing matrix $\mathbf{A} \in \mathbb{R}^{D_x \times D}$ was generated by first creating a random orthogonal matrix of size $D_x \times D_x$ then choosing its first D columns. Gaussianization has been done on the observations and then we studied the eigenvalues of the covariance matrix of the resulting transformed signal $\mathbf{g}(\mathbf{x})$: ideally there are D positive and $D_x - D$ (almost) 0 values. We show the ordered eigenvalues on dataset *celebrities* averaged over 50 runs in Fig. 4. It can be seen that exactly half of the eigenvalues are near 0, then there is a big leap. (For datasets *3D-geom IFS* we have got similar results, data is not shown.) Figure 4(b) shows the results corresponding to dataset $A\omega$ for different number of components M: only the $M \leq 4$ cases are illustrated, but in the whole range of $2 \leq M \leq 50$ there is a sharp transition similar to the results gained for the other 3 datasets.

6 Conclusions

In this paper we introduced the PNL-ISA problem as a common extension of the Post Nonlinear Independent Component Analysis (PNL-ICA) and the Independent Subspace Analysis (ISA). We have shown the ambiguities of the PNL-ISA task are essentially the same as in the linear ISA task (up to a constant translation in each subspace). We derived an algorithm based on our separability results. We also demonstrated the efficiency of the algorithm on different datasets. Our simulations revealed that the error of the estimation of the hidden sources decreases in a power law fashion as the sample size increases. This tendency is even more characteristic when the overall dimension of the task (D) increases. Interestingly, our algorithm can recover the sources in cases when the assumptions of the PNL-ISA problem are violated: even non i.i.d self-similar hidden components can be recovered. In addition, we demonstrated that the dimension of the hidden source can also be estimated.

References

1. Cichocki, A., Amari, S.: Adaptive blind signal and image processing. John Wiley & Sons, West Sussex, England (2002)
2. Hyvärinen, A., Karhunen, J., Oja, E.: Independent Component Analysis. John Wiley & Sons, West Sussex, England (2001)
3. Taleb, A., Jutten, C.: Source separation in post-nonlinear mixtures. IEEE Transactions on Signal Processing 10(47), 2807–2820 (1999)
4. Jutten, C., Karhunen, J.: Advances in blind source separation (BSS) and independent component analysis (ICA) for nonlinear systems. International Journal of Neural Systems 14(5), 267–292 (2004)
5. Cardoso, J.: Multidimensional independent component analysis. In: Proc. of ICASSP 1998, vol. 4, pp. 1941–1944 (1998)

6. Akaho, S., Kiuchi, Y., Umeyama, S.: MICA: Multimodal independent component analysis. In: Proc. of IJCNN 1999, vol. 2, pp. 927–932 (1999)
7. Hyvärinen, A., Hoyer, P.O.: Emergence of phase and shift invariant features by decomposition of natural images into independent feature subspaces. Neural Computation 12, 1705–1720 (2000)
8. Hyvärinen, A., Köster, U.: FastISA: A fast fixed-point algorithm for independent subspace analysis. In: Proc. of ESANN, pp. 371–376 (2006)
9. Vollgraf, R., Obermayer, K.: Multi-dimensional ICA to separate correlated sources. In: Proc. of NIPS 2001, vol. 14, pp. 993–1000 (2001)
10. Bach, F.R., Jordan, M.I.: Beyond independent components: Trees and clusters. Journal of Machine Learning Research 4, 1205–1233 (2003)
11. Póczos, B., Lőrincz, A.: Independent subspace analysis using k-nearest neighborhood distances. In: Duch, W., Kacprzyk, J., Oja, E., Zadrożny, S. (eds.) ICANN 2005. LNCS, vol. 3697, pp. 163–168. Springer, Heidelberg (2005)
12. Póczos, B., Lőrincz, A.: Independent subspace analysis using geodesic spanning trees. In: Proc. of ICML 2005, vol. 119, pp. 673–680 (2005)
13. Theis, F.J.: Blind signal separation into groups of dependent signals using joint block diagonalization. In: Proc. of ISCAS 2005, vol. 6, pp. 5878–5881 (2005)
14. Szabó, Z., Lőrincz, A.: Real and complex independent subspace analysis by generalized variance. In: Proc. of ICARN, pp. 85–88 (2006)
15. Nolte, G., Meinecke, F.C., Ziehe, A., Müller, K.-R.: Identifying interactions in mixed and noisy complex systems. Physical Review E 73(051913) (2006)
16. Theis, F.J.: Towards a general independent subspace analysis. In: Proc. of NIPS, vol. 19 (2006)
17. Comon, P.: Independent component analysis, a new concept? Signal Processing 36, 287–314 (1994)
18. Achard, S., Jutten, C.: Identifiability of post nonlinear mixtures. IEEE Signal Processing Letters 12(5), 423–426 (2005)
19. Theis, F.J.: Uniqueness of complex and multidimensional independent component analysis. Signal Processing 84(5), 951–956 (2004)
20. Theis, F.J.: A new concept for separability problems in source separation. Neural Computation 16, 1827–1850 (2004)
21. Theis, F.J.: Multidimensional independent component analysis using characteristic functions. In: Proc. of EUSIPCO (2005)
22. Petrov, V.: Central limit theorem for m-dependent variables. In: Proc. of the All-Union Conf. on Probability Theory and Mathematical Statistics, pp. 38–44 (1958)
23. Ziehe, A., Kawanabe, M., Harmeling, S., Müller, K.-R.: Blind separation of post-nonlinear mixtures using linearizing transformations and temporal decorrelation. Journal of Machine Learning Research 4(7-8), 1319–1338 (2004)
24. Solé-Casals, J., Jutten, C., Pham, D.: Fast approximation of nonlinearities for improving inversion algorithms of PNL mixtures and wiener systems. Signal Processing 85, 1780–1786 (2005)
25. Amari, S., Cichocki, A., Yang, H.H.: A new learning algorithm for blind signal separation. Advances in Neural Information Processing Systems 8, 757–763 (1996)

Algebraic Geometric Study of Exchange Monte Carlo Method

Kenji Nagata[1] and Sumio Watanabe[2]

[1] Department of Computational Intelligence and Systems Science, Tokyo Institute of Technology, MailBox R2-5,4259, Nagatsuta, Midori-ku, Yokohama, 226-8503 Japan
kenji.nagata@cs.pi.titech.ac.jp
[2] P&I Lab.,Tokyo Institute of Technology
swatanab@pi.titech.ac.jp

Abstract. In hierarchical learning machines such as neural networks, Bayesian learning provides better generalization performance than maximum likelihood estimation. However, its accurate approximation using Markov chain Monte Carlo (MCMC) method requires huge computational cost. The exchange Monte Carlo (EMC) method was proposed as an improved algorithm of MCMC method. Although its effectiveness has been shown not only in Bayesian learning but also in many fields, the mathematical foundation of EMC method has not yet been established. In this paper, we clarify the asymptotic behavior of symmetrized Kullback divergence and average exchange ratio, which are used as criteria for designing the EMC method.

1 Introduction

A lot of learning machines with hierarchical structures such as neural networks, hidden Markov models are widely used for pattern recognition, gene analysis and many other applications. For these hierarchical learning machines, Bayesian learning is proven to provide better generalization performance than maximum likelihood estimation [12][13].

In Bayesian learning, we need to compute the expectation over the Bayesian posterior distribution, which cannot be performed exactly. Therefore, Bayesian learning requires some approximation methods. One of the well-known approximation methods is Markov chain Monte Carlo (MCMC) method. The MCMC method is well-known algorithm to generate a sample sequence which converges to a target distribution. However, it requires huge computational cost to generate the sample sequence, in particular, in the Bayesian posterior distribution for a hierarchical learning machine.

Recently, various improvements of MCMC methods have been developed based on the idea of extended ensemble, which are surveyed in [6]. Multicanonical method [2] and simulated tempering [8] belong to this category called the extended ensemble method. This idea gives us a general strategy to overcome the problem of huge computational cost. The exchange Monte Carlo (EMC) method is well known as one of the extended ensemble methods [4]. This method is to

J. Marques de Sá et al. (Eds.): ICANN 2007, Part I, LNCS 4668, pp. 687–696, 2007.

generate a sample sequence from the joint distribution, which consists of many distributions with different temperatures. Its algorithm is based on two steps of MCMC simulations. One is the conventional update of MCMC simulation for each distribution. The other is the exchange process between two neighboring sequences with a certain probability. The EMC method has been successfully applied to not only Bayesian neural network learning in the literature [7] but also an optimization problem [5][10] and a protein-folding problem [11].

When we design the EMC method, the setting of temperature is very important to make this algorithm efficient [3]. The values of temperatures have close relation to the exchange ratio and its average, which is the acceptance ratio of exchange process. The symmetrized Kullback divergence between two neighboring distributions is used as a criterion for setting temperatures because this Kullback divergence has relation to the average exchange ratio [6]. However, the mathematical relation between the symmetrized Kullback divergence and the average exchange ratio has not been clarified. Moreover, there are two types of exchange ratio, Metropolis type and heat bath type. It often becomes a problem which type should be used.

In this paper, we mathematically clarify the symmetrized Kullback divergence and the average exchange ratio for Metropolis type and for heat bath type. This analytic result gives us the mathematical relation between the symmetrized Kullback divergence and the average exchange ratio, the optimal setting of temperatures, and the properties of the average exchange ratio for Metropolis type and for heat bath type.

This paper consists of five chapters. In Chapter 2, we explain the framework of EMC method and the design of EMC method. In Chapter 3, the main result of analysis for the EMC method is described. Discussion and Conclusion are followed in Chapter 4 and 5.

2 Background

2.1 Exchange Monte Carlo Method

In this section, we introduce the well-known EMC method.

Suppose that $w \in R^d$ and our aim is to generate a sample sequence from the following target probability distribution with a energy function $H(w)$ and a probability distribution $\varphi(w)$,

$$p(w) = \frac{1}{Z(n)} \exp(-nH(w))\varphi(w),$$

where $Z(n)$ is the normalization constant. In Bayesian learning, in order to calculate the expectation over the Bayesian posterior distribution, a sample sequence from the posterior distribution is needed. Then, the number n, the function $H(w)$ and the probability distribution $\varphi(w)$ respectively correspond to the number of training data, the log likelihood and the prior distribution. The EMC method treats a compound system which consists of K non-interacting sample sequences from the system concerned. The k-th sample sequence $\{w_k\}$ converges to the following probability distribution

$$p(w|t_k) = \frac{1}{Z(nt_k)} \exp(-nt_k H(w))\varphi(w) \quad (1 \le k \le K),$$

where $t_1 < t_2 < \cdots < t_K$. Given a set of the temperatures $\{t\} = \{t_1, \cdots, t_K\}$, the joint distribution of $\{w\} = \{w_1, w_2, \cdots, w_K\}$ is expressed by a simple product formula,

$$p(\{w\}) = \prod_{k=1}^{K} p(w_k|t_k). \tag{1}$$

The EMC method is based on two types of updating in constructing a Markov chain. One is the conventional updates based on the Metropolis algorithm for each target distribution $p(w_k|t_k)$. The other is the position exchange between two sequences, that is, $\{w_k, w_{k+1}\} \to \{w_{k+1}, w_k\}$. The transition probability u is determined by the detailed balance condition for the joint distribution (1). We have two types of position exchange. One is the Metropolis type as follows,

$$\begin{aligned} u_1 &= \min(1, r) \\ r &= \frac{p(w_{k+1}|t_k)p(w_k|t_{k+1})}{p(w_k|t_k)p(w_{k+1}|t_{k+1})} = \exp(n(t_{k+1} - t_k)(H(w_{k+1}) - H(w_k))). \end{aligned} \tag{2}$$

The other is the heat bath type,

$$\begin{aligned} u_2 &= \frac{p(w_{k-1}|t_k)p(w_k|t_{k+1})}{p(w_k|t_k)p(w_{k+1}|t_{k+1}) + p(w_{k+1}|t_k)p(w_k|t_{k+1})} \\ &= \frac{1}{2}\left\{1 + \tanh\left(\frac{1}{2}n\left(t_{k+1} - t_k\right)\left(H(w_{k+1}) - H(w_k)\right)\right)\right\}. \end{aligned} \tag{3}$$

Hereafter, we call u_1 and u_2 exchange ratios. Under these updates, the joint distribution of Eq.(1) is invariant because these updates satisfy the detailed balance condition for the distribution of Eq.(1) [4].

Consequently, the following two steps are carried out alternately:

1. Each sequence is generated simultaneously and independently for a few iteration by a conventional MCMC method.
2. Two positions are exchanged with the exchange ratio u_1 or u_2.

The advantage of EMC method is to accelerate the convergence of sample sequence comparing the conventional MCMC method. The conventional MCMC method requires huge computational cost to generate a sample sequence from the target distribution because this algorithm is based on local updating. The EMC method can realize the efficient sampling by preparing a simple distribution such as a normal distribution, which is easy for sample sequence to converge. In practical, we set the temperature of target distribution as $t_K = 1$, and $\varphi(w)$ is sought to be a simple distribution that is easy to sample from and $t_1 = 0$.

2.2 Exchange Ratios for EMC Method

When we design the EMC method, the setting of temperatures is very important to make the EMC method efficient. As we can see in Eq.(2) and Eq.(3), temperature has close relation to the exchange ratio. Therefore, temperatures are very important parameters in adjusting the exchange ratio and its average.

For the efficient EMC method, each sample needs to wander over the whole temperature region. Moreover, the time for a sample to move from end to end (from t_1 to t_K) is good to be short. Therefore, it is not efficient for the interval of neighboring temperatures to be large, which leads to the low average exchange ratio. On the contrary, in order to make the average exchange ratio high, the interval of neighboring temperatures has to be very small, that is, the total number K of temperature needs to be large. Therefore, this setting is not also efficient because it needs huge cost to generate a sample from each distribution. Consequently, the set of temperatures needs to optimize so that the average exchange ratios for any neighboring temperatures become not low and not too high.

As a criterion for the setting of temperature, the following symmetrized Kullback divergence $I(t_k, t_{k+1})$ is used [6],

$$I(t_k, t_{k+1}) = \int p(w_k|t_k) \log \frac{p(w_k|t_k)}{p(w_k|t_{k+1})} dw_k + \int p(w_{k+1}|t_{k+1}) \log \frac{p(w_{k+1}|t_{k+1})}{p(w_{k+1}|t_k)} dw_{k+1}.$$

This function has the following property, $E[\log r] = -I(t_k, t_{k+1})$, where $E[\log r]$ means the average of $\log r$ over the joint distribution $p(w_k|t_k) \times p(w_{k+1}|t_{k+1})$. Moreover, when the free energy $F(nt)$ is defined by

$$F(nt) = -\log \int \exp(-ntH(w))\varphi(w)dw,$$

the following equation is satisfied in small interval of temperatures, $I(t_k, t_{k+1}) = \left.\frac{\partial^2 F}{\partial t^2}\right|_{t=t_k} (t_{k+1} - t_k)^2$. By using these properties, the implementation approach of temperatures based on the value of $\sqrt{\partial^2 F / \partial t^2}$ has been proposed.

As above mentioned, the average exchange ratio, the symmetrized Kullback divergence and the free energy are used as criteria for the setting of temperatures. However, the theoretical properties of these functions are not clarified. Hence, in practical, the simulations of EMC method have to be carried out in order to obtain the value of each function. However, since it is typically difficult to check the convergence of EMC simulation, the accuracy of experimental value of each function is not clear.

In this paper, we show the analytical results for the symmetrized Kullback divergence and for the average exchange ratio in the low temperature limit, that is, $n \to \infty$. This result gives us the criteria for the setting of temperature and for checking the convergence of EMC simulations.

3 Main Result

In this section, we show the main result of this paper. We assume that $t > 0$ and $t + \Delta t > 0$, and consider the EMC method between the two distributions, $p_1(w) = p(w|t)$ and $p_2(w) = p(w|t + \Delta t)$, where the real number Δt is not necessarily small. For these distributions, the symmetrized Kullback divergence I is rewritten as follows,

$$I = \int p_1(w_1) \log \frac{p_1(w_1)}{p_2(w_1)} dw_1 + \int p_2(w_2) \log \frac{p_2(w_2)}{p_1(w_2)} dw_2.$$

As we can see in Eq.(2) and Eq.(3), the exchange ratios u_1 and u_2 are functions of w_1 and w_2. Hence, we define the average exchange ratios J_1 and J_2 as the expectations of exchange ratios u_1 and u_2 over the joint distribution $p_1(w_1) \times p_2(w_2)$ as follows,

$$J_1 = \int\int u_1 p_1(w_1) p_2(w_2) dw_1 dw_2$$
$$J_2 = \int\int u_2 p_1(w_1) p_2(w_2) dw_1 dw_2.$$

In a lot of learning machines such as neural networks, normal mixtures and hidden Markov models, the Bayesian posterior distribution does not converge to the normal distribution as $n \to \infty$ because the Hessian of log likelihood $H(w)$ is not positive definite. We can assume $H(w) \geq 0$ and $H(w_0) = 0 (\exists w_0)$ without loss of generality. The zeta function of $H(w)$ and $\varphi(w)$ is defined by $\zeta(z) = \int H(w)^z \varphi(w) dw$, where z is a one complex variable. Then $\zeta(z)$ is a holomorphic function in the region of $Re(z) > 0$, and can be analytically continued to the meromorphic function on the entire complex plane, whose poles are all real, negative, and rational numbers [1]. We also define the rational number $-\lambda$ as the largest pole of zeta function $\zeta(z)$ and the natural number m as its order. If the Hessian matrix $\left(\frac{\partial^2 H(w)}{\partial w_i \partial w_j}\right)$ is positive definite for arbitrary w, it holds that $\lambda = \frac{d}{2}$, $m = 1$. Otherwise, λ and m can be calculated by using the resolution of singularities in algebraic geometry [1]. In fact, there are some studies to calculate the value λ and m for a certain energy function $H(w)$ and a probability distribution $\varphi(w)$ [13].

Then, the following lemma holds [12].

Lemma 1. *The state density function $V(s)$ $(s > 0)$ has the following asymptotic expansion for $s \to 0$,*

$$V(s) = \int \delta(s - H(w)) \varphi(w) dw \cong c s^{\lambda - 1} (-\log s)^{m-1},$$

where c is a constant.

In our previous work, we analyzed the symmetrized Kullback divergence and the average exchange ratio for Metropolis type [9]. However, the condition $\Delta t \ll t$ is assumed in this analysis. In this paper, we analyze these functions without this assumption.

Firstly, we show the theorem about the symmetrized Kullback divergence and the average exchange ratio for Metropolis type.

Theorem 1. *The symmetrized Kullback divergence I and the average exchange ratio J_1 for Metropolis type respectively converge to the following values as $n \to \infty$,*

$$I \to \lambda \frac{(\Delta t)^2}{t(t+\Delta t)}$$

$$J_1 \to \begin{cases} \left(1+\frac{\Delta t}{t}\right)^{\lambda} \frac{2\Gamma(2\lambda)}{\Gamma(\lambda)^2} A(\lambda, \frac{\Delta t}{t}) & (\text{if } \Delta t \geq 0) \\ \left(1-\frac{\Delta t}{t+\Delta t}\right)^{\lambda} \frac{2\Gamma(2\lambda)}{\Gamma(\lambda)^2} A(\lambda, -\frac{\Delta t}{t+\Delta t}) & (\text{if } \Delta t < 0), \end{cases}$$

where $A(\lambda, \frac{\Delta t}{t})$ is defined by

$$A\left(\lambda, \frac{\Delta t}{t}\right) = \int_0^1 \frac{s^{\lambda-1}}{(1+\frac{\Delta t}{t}+s)^{2\lambda}} ds.$$

(outline of the proof) This theorem is simply proven by using the almost same analysis of [9] without applying the Taylor expansion about $\frac{\Delta t}{t} = 0$. (Q.E.D)

From Theorem 1, we can make the symmetrized Kullback divergence constant over the various temperatures by the temperature setting that the value $\frac{\Delta t}{t}$ or $\frac{\Delta t}{t+\Delta t}$ is constant, that is, the set $\{t_k\}$ of temperature is set as geometric progression. Moreover, under this setting, the average exchange ratio for Metropolis type becomes constant over the various temperatures.

Secondly, we reveal the average exchange ratio J_2 for heat bath type. We show its proof in Appendix.

Theorem 2. *The average exchange ratio J_2 for heat bath type converges to the following value as $n \to \infty$,*

$$J_2 \to \begin{cases} \frac{1}{2}\left(1+\left(1+\frac{\Delta t}{t}\right)^{\lambda} \frac{B(\lambda, \frac{\Delta t}{t})}{\Gamma(\lambda)^2}\right) & (\text{if } \Delta t \geq 0) \\ \frac{1}{2}\left(1+\left(1-\frac{\Delta t}{t+\Delta t}\right)^{\lambda} \frac{B(\lambda, -\frac{\Delta t}{t+\Delta t})}{\Gamma(\lambda)^2}\right) & (\text{if } \Delta t < 0), \end{cases}$$

where the function $B\left(\lambda, \frac{\Delta t}{t}\right)$ is defined by,

$$B\left(\lambda, \frac{\Delta t}{t}\right) = \int_0^{\infty} ds_1 \int_0^{\infty} ds_2 \tanh\left(\frac{1}{2}\frac{\Delta t}{t}(s_2 - s_1)\right) e^{-s_1} e^{-(1+\frac{\Delta t}{t})s_2} s_1^{\lambda-1} s_2^{\lambda-1}.$$

Theorem 2 claims that the average exchange ratio for heat bath type is also constant over the various temperatures by setting the value $\frac{\Delta t}{t}$ constant. Consequently, if we set the value $\frac{\Delta t}{t}$ constant, the symmetrized Kullback divergence and the average exchange ratio for both type are clarified to be constant over the various temperatures.

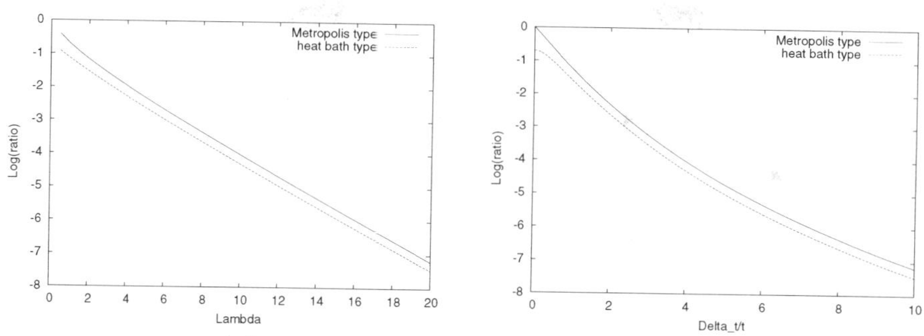

(a) (b)

Fig. 1. The theoretical value of average exchange ratio (a) for the value λ with $\frac{\Delta t}{t} = 2.0$, and (b) for the value $\frac{\Delta t}{t}$ with $\lambda = 5.0$. The solid line is for Metropolis type and the dashed line is for heat bath type.

4 Discussion

In this paper, we analyzed the symmetrized Kullback divergence and the average exchange ratio for Metropolis type and for heat bath type in the low temperature limit. As above mentioned, the EMC simulations have to be carried out in order to obtain the value of these functions. However, the accuracy of these experimental values are not clear because of the difficulty of checking the convergence of EMC simulations. Our result gives us the accurate value of these functions without the EMC simulations, which leads to the criteria for the setting of temperature. Moreover, since the accuracy of the experimental values of these functions can be evaluated by using our result, we can check the convergence of EMC simulation, which is very important for the EMC method.

Let us discuss three points in association to this paper.

Firstly, we discuss the relation between the shape of target distribution and setting of temperature. Figure 1(a) shows the relation between the value λ and the theoretical value of average exchange ratio for each type with $\frac{\Delta t}{t} = 2.0$. According to this figure, the EMC method works efficiently for the target distribution with small value λ, which requires high average exchange ratio. On the other hand, by comparing two distributions whose ground states are one point and analytic set in the sample space, the latter distribution is well known to have smaller value λ than the former distribution [13]. Consequently, the EMC method works efficiently for the target distribution with the energy function whose ground state is an analytic set. As an example of such distributions, the Bayesian posterior distribution in hierarchical learning machines such as neural networks and normal mixtures is well known, and this theorem shows the availability of EMC method for the Bayesian learning in hierarchical learning machines.

Secondly, we discuss the properties of average exchange ratio for both type. When we design the EMC method, which type should be used as the exchange ratio, Metropolis type or heat bath type, often becomes a problem. By comparing each exchange ratio, the computational cost is almost equal. Therefore, we should select the type to make the average exchange ratio high for the same setting of temperature. Figure 1(b) shows the relation between the value $\frac{\Delta t}{t}$ and the theoretical value of average exchange ratio for each type with $\lambda = 5.0$. According to Figure 1(a) and 1(b), the average exchange ratio for Metropolis type is clarified to be higher than that for heat bath type in the low temperature limit. This claims the effectiveness of Metropolis type for the EMC method.

Finally, we discuss the implementation of EMC method. This theorem gives us the implementation approach of optimal temperatures in order to make the average exchange ratio constant over the various temperatures. However, the optimum value of average exchange ratio is not clarified, which leads to the implementation of optimal number K of temperature. Moreover, since the EMC method includes the algorithm of conventional MCMC method, the implementation of conventional MCMC method should be considered in the future.

5 Conclusion

In this paper, we analytically calculated the symmetrized Kullback divergence and the average of exchange ratio, and clarified the relation between the symmetrized Kullback divergence and the exchange ratio. as a result, the following properties are clarified,

1. When the symmetrized Kullback divergence for arbitrary temperature t is constant, the average exchange ratio is also constant over the various temperatures.
2. Then, the set of temperature $\{t_k\}$ is set as geometric progression.

As the future works, verifying the theoretical result in this study by some experiments, constructing the implementation approach of EMC method, and applying these results to the practical problems such as the Bayesian learning should be addressed.

Acknowledgement

This work was supported by the Ministry of Education, Science, Sports and Culture, Grant-in-Aid for JSPS Research Fellows 18-5809 and for Scientific Research 18079007.

References

1. Atiyah, M.F.: Resolution of singularities and division of distributions. Communications of Pure and Applied Mathematics 13, 145–150 (1970)
2. Berg, B.A., Neuhaus, T.: Multicanonical algorithms for first order phase transitions. Physics Letter B 267(2), 249–253 (1991)
3. Hukushima, K.: Domain Wall Free Energy of Spin Glass Models: Numerical Method and Boundary Conditions. Physical Review E 60, 3606–3613 (1999)

4. Hukushima, K., Nemoto, K.: Exchange Monte Carlo Method and Application to Spin Glass Simulations. Journal of Physical Society of Japan 65(6), 1604–1608 (1996)
5. Hukushima, K.: Extended ensemble Monte Carlo approach to hardly relaxing problems. Computer Physics Communications 147, 77–82 (2002)
6. Iba, Y.: Extended Ensemble Monte Carlo. International Journal of Modern Physics C 12, 623–656 (2001)
7. Liang, F.: An effective Bayesian neural network classifier with a comparison study to support vector machine. Neural Computation 15, 1959–1989 (2003)
8. Marinari, E., Parisi, G.: Simulated tempering: a new Monte Carlo scheme. Europhysics Letters 19(6), 451–455 (1992)
9. Nagata, K., Watanabe, S.: Analysis of Exchange Ratio for Exchange Monte Carlo Method. In: FOCI'07. Proc. of The First IEEE Symposium on Foundation of Computational Intelligence, pp. 434–439. IEEE Computer Society Press, Los Alamitos (2007)
10. Pinn, K., Wieczerkowski, C.: Number of Magic Squares from Parallel Tempering Monte Carlo. International Journal of modern Physics 9(4), 541–546 (1998)
11. Sugita, Y., Okamoto, Y.: Replica-exchange molecular dynamics method for protein folding. Chemical Physics Letters 314(1-2), 141–151 (1999)
12. Watanabe, S.: Algebraic Analysis for Nonidentifiable Learning Machines. Neural Computation 13, 899–933 (2001)
13. Yamazaki, K., Watanabe, S.: Singularities in mixture models and upper bounds of stochastic complexity. Neural networks 16(7), 1029–1038 (2003)

Appendix: Proof of Theorem 2

In Appendix, we show the proof of Theorem 2 about the average exchange ratio for heat bath type.

(outline of the proof) The average exchange ratio J_2 is expressed by the definition of exchange ratio u_2 as follows,

$$J_2 = \frac{1}{2}\Bigg(1 + \int dw_1 \int dw_2 \tanh\left(\frac{n\Delta t}{2}(H(w_2) - H(w_1))\right) \times \frac{e^{-ntH(w_1)}\varphi(w_1)}{Z(nt)} \frac{e^{-n(t+\Delta t)H(w_2)}\varphi(w_2)}{Z(n(t+\Delta t))}\Bigg).$$

Then, the normalization constant $Z(nt)$ is given from Lemma 1,

$$Z(nt) = \frac{c(\log nt)^{m-1}}{(nt)^{\lambda-1}}\Gamma(\lambda).$$

Hence, by defining

$$J_2^* = \int dw_1 \int dw_2 \tanh\left(\frac{n\Delta t}{2}(H(w_2) - H(w_1))\right) \times e^{-ntH(w_1)}\varphi(w_1)e^{-n(t+\Delta t)H(w_2)}\varphi(w_2),$$

we obtain

$$J_2 = \frac{1}{2}\left(1 + \frac{J_2^*}{Z(nt)Z(n(t+\Delta t))}\right).$$

The function J_2^* is expressed by using the Dirac delta function as follows,

$$\begin{aligned} J_2^* &= \int_0^\infty ds_1 \int_0^\infty ds_2 \tanh\left(\frac{n\Delta t}{2}(s_2 - s_1)\right) e^{-nts_1} e^{-n(t+\Delta t)s_2} \\ &\quad \times \int dw_1 \delta(s_1 - H(w_1))\varphi(w_1) \int dw_2 \delta(s_2 - H(w_2))\varphi(w_2) \\ &\cong \int_0^\infty ds_1 \int_0^\infty ds_2 \tanh\left(\frac{n\Delta t}{2}(s_2 - s_1)\right) e^{-nts_1} e^{-n(t+\Delta t)s_2} \\ &\quad \times c s_1^{\lambda-1}(-\log s_1)^{m-1} c s_2^{\lambda-1}(-\log s_2)^{m-1}. \end{aligned}$$

In the case that $\Delta t \geq 0$, the function J_2^* is given by putting $s_1' = nts_1$ and $s_2' = nts_2$ as follows,

$$J_2^* \cong \frac{c^2(\log nt)^{2(m-1)}}{(nt)^{2\lambda}} B\left(\lambda, \frac{\Delta t}{t}\right).$$

In the case that $\Delta t < 0$, the function J_2^* is given by putting $s_1' = n(t+\Delta t)s_1$ and $s_2' = n(t+\Delta t)s_2$ as follows,

$$J_2^* \cong \frac{c^2(\log n(t+\Delta t))^{2(m-1)}}{(n(t+\Delta t))^{2\lambda}} B\left(\lambda, -\frac{\Delta t}{t+\Delta t}\right).$$

Consequently, the average exchange ratio J_2 is given by

$$\begin{aligned} J_2 &= \frac{1}{2}\left(1 + \frac{J_2^*}{Z(nt)Z(n(t+\Delta t)}\right) \\ &\to \begin{cases} \frac{1}{2}\left(1 + \left(1 + \frac{\Delta t}{t}\right)^\lambda \frac{B(\lambda, \frac{\Delta t}{t})}{\Gamma(\lambda)^2}\right) & (\text{if } \Delta t \geq 0) \\ \frac{1}{2}\left(1 + \left(1 - \frac{\Delta t}{t+\Delta t}\right)^\lambda \frac{B(\lambda, -\frac{\Delta t}{t+\Delta t})}{\Gamma(\lambda)^2}\right) & (\text{if } \Delta t < 0), \end{cases} \end{aligned}$$

which completes the theorem (Q.E.D).

Solving Deep Memory POMDPs with Recurrent Policy Gradients

Daan Wierstra[1], Alexander Foerster[1], Jan Peters[2], and Jürgen Schmidhuber[1]

[1] IDSIA, Galleria 2, 6928 Manno-Lugano, Switzerland
[2] University of Southern California, Los Angeles, CA, USA

Abstract. This paper presents Recurrent Policy Gradients, a model-free reinforcement learning (RL) method creating limited-memory stochastic policies for partially observable Markov decision problems (POMDPs) that require long-term memories of past observations. The approach involves approximating a policy gradient for a Recurrent Neural Network (RNN) by backpropagating return-weighted characteristic eligibilities through time. Using a "Long Short-Term Memory" architecture, we are able to outperform other RL methods on two important benchmark tasks. Furthermore, we show promising results on a complex car driving simulation task.

1 Introduction

Policy gradient (PG) methods have been among the few successful algorithms for solving real world Reinforcement Learning (RL) tasks (e.g. see [1,2,3,4,5]) as they can deal with continuous states and actions. After being introduced by Williams [6] and Gullapalli [7], they have recently been extended to deal more efficiently [8] with complex high-dimensional tasks [9]. Although sensitive to local minima in policy representation space, an attractive property of this class of algorithms is that they are guaranteed to converge to at least a locally optimal policy, even using function approximators such as neural networks [10]. Furthermore, unlike most regular RL approaches, they provide a natural framework for learning policies with *continuous*, high-dimensional actions. Provided the choice of policy representation is powerful enough, PGs can tackle arbitrary RL problems. Unfortunately, most PG approaches have only been used to train policy representations of, typically, a few dozen parameters at most. Surprisingly, the obvious combination with standard backpropagation techniques has not been extensively investigated (a notable exception being the SRV algorithm [7,11]). In this paper, we address this shortcoming, and show how PGs can be naturally combined with backpropagation, and BackPropagation Through Time (BPTT) [12] in particular, to form a powerful RL algorithm capable of training complex neural networks with large numbers of parameters.

RL tasks in realistic environments typically need to deal with incomplete and noisy state information resulting from partial observability such as encountered in POMDPs. Furthermore, they often need to deal with non-Markovian

J. Marques de Sá et al. (Eds.): ICANN 2007, Part I, LNCS 4668, pp. 697–706, 2007.

problems where there are dependencies onto significantly earlier states. Both POMDPs and non-Markovian problems largely defy traditional value function based approaches and require complex state estimators with internal memory based on accurate knowledge of the system. In reinforcement learning, we can rarely assume a good memory-based state estimator as the underlying system is usually unknown.

A naive alternative to using memory is learning *reactive stochastic policies* [13] which simply map observations to probabilities of actions. The underlying assumption is that state-information does not play a crucial role during most parts of the problem and that using random actions can prevent the actor from getting stuck in an endless loop for ambiguous observations (e.g., the scenario of a blind robot rolling forever against the wall would only happen for a stationary, deterministic reactive policy). In general, this strategy is clearly suboptimal, and instead algorithms that use some form of memory still seem essential.

However, if we cannot assume a perfect model and a precisely estimated state, an optimal policy needs to be both stochastic and memory-based for realistic environments. As the memory is always an abstraction or imperfect, i.e., limited, we will focus on learning what we define as *limited-memory stochastic policies*, that is, policies that map limited memory states to probability distributions on actions. Work on policy gradient methods with memory has been scarce so far, largely limited to finite state controllers [14,15]. In this paper, we extend this approach to more sophisticated policy representations capable of representing memory using an RNN architecture called Long Short-Term Memory (LSTM), see [16]. We develop a new reinforcement learning algorithm that can effectively learn memory-based policies for deep memory POMDPs. We present *Recurrent* Policy Gradients (RPG), an algorithm which backpropagates the estimated return-weighted *eligibilities* backwards in time through recurrent connections in the RNN. As a result, policy updates can become a function of any event in the history. We show that the presented method outperforms other RL methods on two important RL benchmark tasks with different properties: continuous control in a non-Markovian double pole balancing environment, and discrete control on the deep memory T-maze [17] task. Moreover, we show promising results in a complex car driving simulation.

The structure of the paper is as follows. The next section describes the Recurrent Policy Gradient algorithm. The subsequent sections describe our experimental results using RPGs with memory and conclude with a discussion.

2 Recurrent Policy Gradients

In this section we describe our basic algorithm for using memory in a policy gradient setting. First, we briefly summarize reinforcement learning terminology, then we briefly review the policy gradient framework. Next, we describe the particular type of recurrent neural network architecture used in this paper, Long Short-Term Memory. Subsequently, we show how the methods can be combined to form Recurrent Policy Gradients.

2.1 Reinforcement Learning – Generalized Problem Statement

First let us introduce the RL framework used in this paper and the corresponding notation. The environment produces a state g_t at every time step. Transitions from state to state are governed by a probability function $p(g_{t+1}|a_{1:t}, g_{1:t})$ unknown to the agent but dependent upon all previous actions $a_{1:t}$ executed by the agent and all previous states $g_{1:t}$ of the system. Note that most reinforcement learning papers need to assume Markovian environments – we will later see that we do not need to for policy gradient methods with an internal memory. Let r_t be the reward assigned to the agent at time t, and let o_t be the corresponding observation produced by the environment. We assume that both quantities are governed by fixed distributions $p(o|g)$ and $p(r|g)$, solely dependent on state g.

In the more general reinforcement setting, we require that the agent has a memory of the generated experience consisting of finite episodes. Such episodes are generated by the agent's operations on the (stochastic) environment, executing action a_t at every time step t, after observing observation o_t and special 'observation' r_t (the reward) which both depend solely on g_t. We define the *observed history*[1] h_t as the string or vector of observations and actions up to moment t since the beginning of the episode: $h_t = \langle o_0, a_0, o_1, a_1, \ldots, o_{t-1}, a_{t-1}, o_t \rangle$. The complete history H includes the unobserved states and is given by $H_T = \langle h_T, g_{0:T} \rangle$. At any time t, the actor optimizes $R_t = (1-\gamma)^{-1} \sum_{k=t}^{\infty} r_k \gamma^{t-k-1}$ which is the *return* at at time t where $0 < \gamma < 1$ denotes a discount factor.

The expectation of this return R_t at time $t = 0$ is also the measure of quality of our policy and, thus, the objective of reinforcement learning is to determine a policy which is optimal with respect to the expected future discounted rewards or expected return $J = E[R_0] = (1-\gamma)^{-1} \lim_{T\to\infty} \mathbf{E}\left[\sum_{t=0}^{T-1} \gamma^t r_t\right]$. For the average reward case where $\gamma \to 1$ or, equivalently, $(1-\gamma)^{-1} \to \infty$, this expression remains true analytically but needs to be replaced by $J = \lim_{T\to\infty} \mathbf{E}[\sum_{t=0}^{T-1} r_t/T]$ in order to be numerically feasible.

An optimal or near-optimal policy in a non-Markovian or partially observable Markovian environment requires that the action a_t is taken depending on the *entire* preceding history. However, in most cases, we will not *need* to store the whole string of events but only sufficient statistics $T(h_t)$ of the events which we call the limited memory of the agents past. Thus, a stochastic policy π can be defined as $\pi(a|h_t) = p(a|T(h_t); \theta)$, implemented as an RNN with weights θ and stochastically interpretable output neurons. This produces a probability distribution over actions, from which actions a_t are drawn $a_t \sim \pi(a|h_t)$.

2.2 The Policy Gradient Framework

The type of RL algorithm we employ in this paper falls in the class of policy gradient algorithms, which, unlike many other (notably TD) methods, update the agent's policy-defining parameters θ directly by estimating a gradient in the direction of higher (average or discounted) reward.

[1] Note that such histories are also called path or trajectory in the literature.

Now, let $R(H)$ be some measure of the total reward accrued during a history (e.g., $R(H)$ could be the average of the rewards for the average reward case or the discounted sum for the discounted case), and let $p(H|\theta)$ be the probability of a history given policy-defining weights θ, then the quantity the algorithm should be optimizing is $J = \int_H p(H|\theta)R(H)dH$. This, in essence, indicates the expected reward over all possible histories, weighted by their probabilities under π. Using gradient ascent to update parameters θ of policy π, we can write

$$\nabla_\theta J = \int \nabla_\theta p(H) R(H) dH = \int \frac{p(H)}{p(H)} \nabla_\theta p(H) R(H) dH = \int p(H) \nabla_\theta \log p(H) R(H) dH$$

using the "likelihood-ratio trick" and the fact that $\nabla_\theta R(H) = 0$ for a single, fixed H. Taking the sample average as Monte Carlo (MC) approximation of this expectation by taking N trial histories we get

$$\nabla_\theta J = \mathbf{E}_H \left[\nabla_\theta \log p(H) R(H) \right] \approx \frac{1}{N} \sum_{n=1}^{N} \nabla_\theta \log p(H^n) R(H^n).$$

which is a fast approximation of the policy gradient for the current policy with the convergence speed of $O(N^{-1/2})$ to the true gradient independent of the number of parameters of the policy (i.e., number of elements of the gradient).

Probabilities of histories $p(H)$ are dependent on an unknown initial state distribution, on unknown observation probabilities per state, and on unknown state transition function $p(g_{t+1}|a_{1:t}, g_{1:t})$. But at least the agent knows its own action probabilities, so the log derivative for agent parameters θ in $\nabla_\theta \log p(h)$ can be acquired by first realizing that the probability of a particular history is the product of all actions and observations given subhistories:

$$p(H_T) = p(\langle o_0, g_0 \rangle) \prod_{t=1}^{T} p(\langle o_t, g_t \rangle | h_{t-1}, a_{t-1}, g_{0:t}) \pi(a_{t-1}|h_{t-1})$$

Taking the log-derivative results into transforming this large product into a sum $\log p(H_T) = (\text{const}) + \sum_{t=0}^{T} \log \pi(a_t|h_t)$: where most parts are not affected by θ, i.e., are constant. Thus, when taking the derivative of this term, we obtain

$$\nabla_\theta \log p(H_T) = \sum_{t=0}^{T} \nabla_\theta \log \pi(a_t|h_t).$$

Substituting this term into our MC approximation results in a gradient estimator which only requires observed variables. However, if we make use of the fact that future actions do not depend on past rewards, we can show that these terms can be omitted from the gradient estimate (see [5] for details). Thus, an unbiased gradient estimator is given by

$$\nabla_\theta J \approx \frac{1}{N} \sum_{n=1}^{N} \sum_{t=0}^{T} \nabla_\theta \log \pi(a_t|h_t^n) R_t^n$$

which yields the desired gradient estimator which only has observable variables.

Nevertheless, an important problem with this Monte Carlo approach is the often high variance in the gradient estimate. For example, if $R(h) = 1$ for all h, the variance can be given by $\sigma_T^2 = E[\sum_{t=0}^{T}(\nabla_\theta \log \pi(a_t|h_t^n))^2]$ which grows linearly with T. One way to tackle such problems and reduce this variance is to include a *baseline* b (first introduced by Williams [6]) into the gradient estimate $\nabla_\theta J \approx \frac{1}{N}\sum_{n=1}^{N} \nabla_\theta \log p(h_t^n)(R_t^n - b)$ where b is typically the expected average reward or some kind of value function. Due to the likelihood-ratio trick $\int p(H)\nabla_\theta \log p(H)R(H)dH = \nabla_\theta \int p(H)bdH = \nabla_\theta 1 = 0$, we can guarantee that $E[\sum_{n=1}^{N} \nabla_\theta \log p(H_t^n)b] = 0$ and, thus, the baseline can only reduce the variance but not bias the gradient in any way [6].

Another way to reduce the variance of the gradient is to reduce the planning horizon of the agent, i.e., the episode length can be truncated or we can use a discount factor $\hat{\gamma}$ which is lower than the true γ as in the GPOMDP algorithm [18] where $0 < \hat{\gamma} < \gamma$. Unfortunately, here can be a big trade-off as $\hat{\gamma} < \gamma$ biases the gradient, but when $\hat{\gamma} \to \gamma$, this bias vanishes while the gradient estimate variance can increase significantly.

2.3 LSTM Recurrent Function Approximators

RNNs have attracted some attention in the past decade because of their simplicity and potential power. However, though powerful in theory, they turn out to be quite limited in practice due to their inability to capture long-term time dependencies – they suffer from the problem of *vanishing gradient* [19], the fact that the gradient signal vanishes as the error signal is propagated back through time. Because of this, events more than 10 time steps apart can typically not be related.

One method purposely designed to avoid this problem is Long Short-Term Memory (LSTM [16]), which constitutes a special RNN architecture capable of capturing long term time dependencies. The defining feature of this architecture is that it consists of a number of *memory cells*, which can be used to store activations arbitrarily long. Access to the memory cell is *gated* by units that learn to open or close depending on the context.

LSTM networks have been shown to outperform other RNNs on numerous time series requiring the use of deep memory [20]. Therefore, they seem well-suited for usage in PG algorithms for complex, deep memory requiring tasks. Whereas RNNs are usually used to *predict*, we use them to *control* an agent directly, to represent a controller's policy receiving observations and producing action probabilities at every time step.

2.4 Introducing the Recurrency in Recurrent Policy Gradients

Typically, PG algorithms learn to map observations to action probabilities, i.e. they learn stochastic reactive policies. As noted before, this is clearly suboptimal for all but the simplest partial observability problems. We would like to equip our algorithm with adaptable memory, using LSTM to map histories or *memory states* to action probabilities. Unlike earlier methods, our method makes full use of the backpropagation technique while doing this: whereas most if not all published and experimentally tested PG methods (as far as the authors are aware) estimate parameters θ individually, we use *eligibility-backpropagation through time* (as opposed to standard error-backpropagation or BPTT [12]) to update all parameters conjunctively, yielding solutions that better generalize over complex histories. Using this method, we can map *histories* to actions instead of *observations* to actions.

In order to estimate the gradient for a history-based approach, we map histories h_t to action probabilities by using LSTM's internal state representation. Backpropagating return-weighted eligibilities [6] affects the policy such that it makes histories that were better than other histories (in terms of reward) more likely by reinforcing the probabilities of taking similar actions for similar histories.

Recurrent Policy Gradients are architecturally equal to supervised RNNs, however, the output neurons are interpreted as a probability distribution. It takes, at every time step during the forward pass of BPTT, as input observation o_t and reward r_t. Together with the recurrent connections, these produce outputs $\pi(h_t)$, representing the probability distribution on actions.

Only the output part of the neural network is interpreted stochastically. This allows us, during the backward pass, to only estimate the eligibilities of the output units at every time step. The gradient on the other parameters θ can be derived efficiently via eligibility backpropagation through time, treating output eligibilities like we would treat normal errors ('deltas') in an RNN trained with gradient descent. Also, by having only stochastic output units, we do not have to compute complicated gradients on stochastic internal (belief) states such as is done in [14,15] – eligibility backpropagation through time disambiguates relevant hidden state automatically, when possible.

3 Experiments

We carried out experiments on three fundamentally different problem domains. The first task, double pole balancing with incomplete state information, is a *continuous* control task that has been a benchmark in the RL community for many years. The second task, the T-maze, is a difficult discrete control task that requires remembering its initial observation until the end of the episode. The third task involves a complex car racing simulator, called Torcs.

All experiments were carried out with 10-cell LSTMs. The baseline estimator used was simply a moving average of the return received at any time step.

3.1 Continuous Control: (PO)MDP (Double) Pole Balancing

This task involves trying to balance a pole hinged on a cart that moves on a finite track (see Figure 1). The single control consists of the force F applied to the cart (in Newtons), and observations usually include the cart's position x, the pole's angle θ and velocities $\dot{x}$ and $\dot{\theta}$. It provides a perfect testbed for algorithms

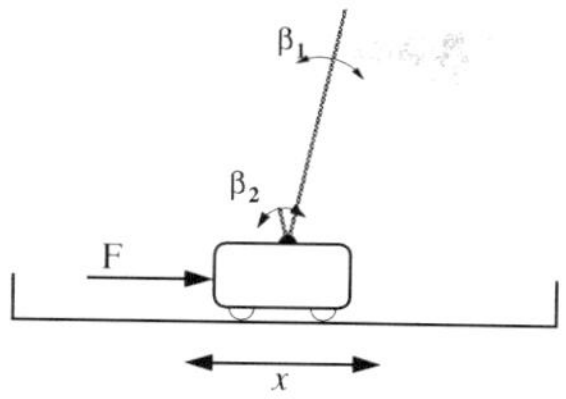

	Markov	non-Markov
1 pole	863 ± 213	1893 ± 527
2 poles	4981 ± 1386	5649 ± 1548

Fig. 1. The non-Markov double pole balancing task. The results show the mean and standard deviation of the number of evaluations until the success criterion.

focussing on learning fine control in continuous state and action spaces. However, recent successes in the RL field have made the standard pole balancing setup too easy and therefore obsolete. To make the task more challenging, we (1) remove velocity information $\dot{x}$ and $\dot{\theta}$ such that the problem becomes non-Markov, and (2) add a second pole to the same cart, of length 1/10th of the original one. This yields non-Markovian double pole balancing [21], a truly challenging task that has not been solved by any other single agent RL method but RPGs.

We applied RPGs to the pole balancing task, using a Gaussian output structure, consisting of a mean μ (which was interpreted linearly) and a standard deviation σ (which was scaled with the logistic function in order to prevent variances from being negative) where eligibilities were calculated according to [6]. We use a learning rate $\alpha\sigma^2$ (as suggested by Williams [6]) and $\alpha = 0.001$, $momentum = 0.9$ and $\gamma = 0.99$. Initial parameters θ were initialized randomly between -0.01 and 0.01. Reward was always 0.0, except for the last time step when one of the poles falls over, where it is -1.0.

A run was considered a *success* when the pole(s) did not fall over for 10,000 time steps. Figure 1 shows results averaged over 20 runs. RPGs clearly outperform earlier PG methods, even by orders of magnitude (compare [15]'s finite state controller).

3.2 Discrete Control: The Long Term Dependency T-Maze

The second experiment was carried out on the T-maze [17] (see Figure 2). Designed to test an RL algorithm's ability to correlate events far apart in history, it involves having to *learn* to remember the observation from the first time step until the episode ends. At the first time step, it starts at position **S** and perceives the **X** either north or south – meaning that the goal state **G** is in the north or south part of the T-junction, respectively. Additionally, the agent perceives its

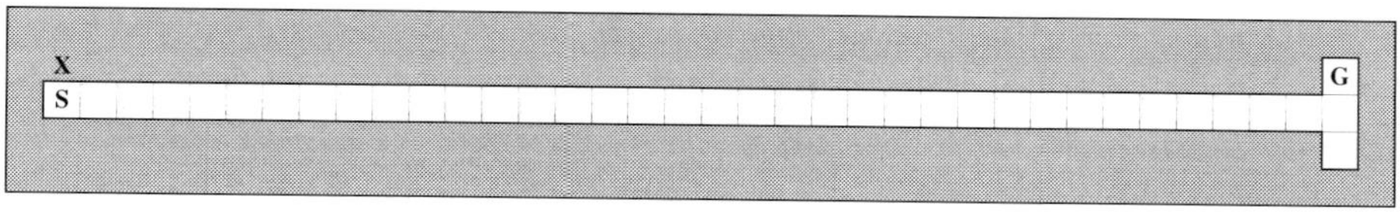

Fig. 2. The T-maze task. In this example, N, the length of the alley, is 35

immediate surroundings. The agent has four possible actions: North, East, South and West. These discrete actions are represented in the network as a softmax layer. When the agent makes the correct decision at the T-junction, i.e. go south if the **X** was south and north otherwise, it receives a reward of 4.0, otherwise a reward of -0.1. In both cases, this ends the episode. Note that the corridor length N can be increased to make the problem more difficult, since the agent has to learn to remember the initial 'road sign' for $N+1$ time steps. In Figure 2 we see an example T-maze with corridor length 35.

Corridor length N was systematically varied from 10 to 100, and for each length 10 runs were performed. Training was performed in batches of 20 normalizing the gradient to length 0.3. Discount factor $\gamma = 0.98$ was used. In Figure 3 the results are displayed, in addition to other algorithms' results taken from [17]. We can see that RPGs clearly outperform the value-based methods. This might be due to the difference in complexity of a simple policy versus an unnecessarily complex value function.

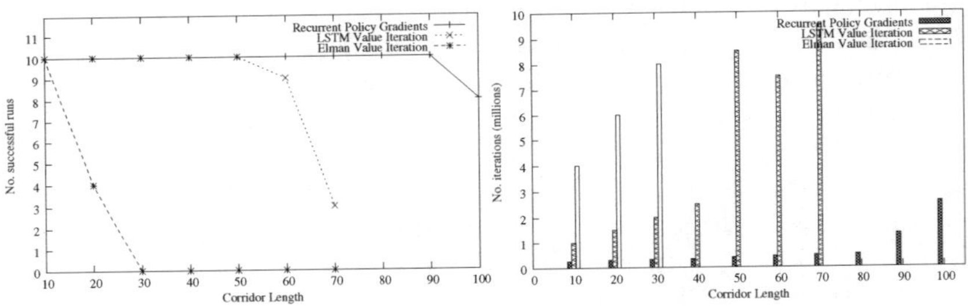

Fig. 3. T-maze results. The left chart shows the number of successful runs for $N = 10, \ldots, 100$. Recurrent Policy Gradients' performance starts to degrade at length $N = 100$. The right plot shows the number of average iterations required to solve the task, averaged over the successful runs. RPGs clearly outperform other RL methods on this task, to the best of the authors' knowledge. (The results for the Value Iteration based algorithms are taken from [17]).

4 The TORCS Car Racing Simulator

In order to test our algorithm's performance on a more complicated environment, we carried out experiments on the TORCS [22] car racing simulator. Observations are speed, steering angle, position and look-ahead-distance. The agent has to learn to drive and stay on the road while maintaining high speed. Its reward consists of the speed, but it gets punished for getting off the track. Maximum speed starts off with 10 km/h, and is gradually increased.

Our current experiments, on 5 trials, show the agent can consistently learn to steer and stay on the road after just under 2 minutes of real-time behavior. The agent can learn to drive safely – not getting off track – up to 70 km/h, after which its behavior destabilizes. The fastest lap time is just under 3 minutes, which is still twice as slow as a trained human player or our preprogrammed agent.

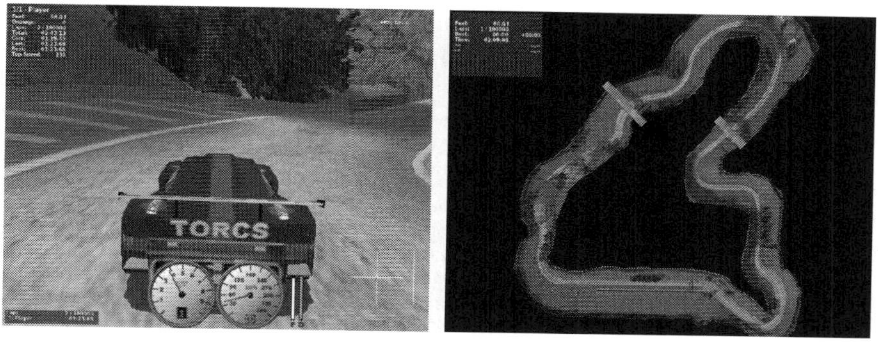

Fig. 4. The TORCS racing car simulator

5 Discussion

We introduced a new policy gradient method equipped with memory capable of memorizing events from arbitrarily far in the past. The approach involves computing and backpropagating a policy gradient through time with LSTM representing memory, thus updating a policy which maps event histories to action probabilities. The approach outperformed other RL methods on two important benchmarks. We think Recurrent Policy Gradients constitute one of the most efficient RL algorithms to date on difficult non-Markovian tasks.

Acknowledgments

This research was funded by SNF grant 200021-111968/1 and by the 6th FP of the EU (project number IST-511931).

References

1. Benbrahim, H., Franklin, J.: Biped dynamic walking using reinforcement learning. Robotics and Autonomous Systems Journal (1997)
2. Moody, J., Saffell, M.: Learning to Trade via Direct Reinforcement. IEEE Transactions on Neural Networks 12(4), 875–889 (2001)
3. Prokhorov, D.: Toward effective combination of off-line and on-line training in adp framework. In: ADPRL. Proceedings of the IEEE International Symposium on Approximate Dynamic Programming and Reinforcement Learning, IEEE Computer Society Press, Los Alamitos (2007)
4. Baxter, J., Bartlett, P., Weaver, L.: Experiments with infinite-horizon, policy- gradient estimation. Journal of Artificial Intelligence Research 15, 351–381 (2001)
5. Peters, J., Schaal, S.: Policy gradient methods for robotics. In: IROS. Proceedings of the IEEE/RSJ International Conference on Intelligent Robots and Systems, Beijing, China, pp. 2219–2225 (2006)
6. Williams, R.J.: Simple statistical gradient-following algorithms for connectionist reinforcement learning. Machine Learning 8, 229–256 (1992)

7. Gullapalli, V.: A stochastic reinforcement learning algorithm for learning real-valued functions. Neural Networks 3(6), 671–692 (1990)
8. Schraudolph, N., Yu, J., Aberdeen, D.: Fast online policy gradient learning with smd gain vector adaptation. In: Weiss, Y., Schölkopf, B., Platt, J. (eds.) Advances in Neural Information Processing Systems, vol. 18, MIT Press, Cambridge, MA (2006)
9. Peters, J., Vijayakumar, S., Schaal, S.: Natural actor-critic. In: Gama, J., Camacho, R., Brazdil, P.B., Jorge, A.M., Torgo, L. (eds.) ECML 2005. LNCS (LNAI), vol. 3720, pp. 280–291. Springer, Heidelberg (2005)
10. Sutton, R., McAllester, D., Singh, S., Mansour, Y.: Policy gradient methods for reinforcement learning with function approximation (2001)
11. Gullapalli, V.: Reinforcement learning and its application to control (1992)
12. Werbos, P.: Back propagation through time: What it does and how to do it. Proceedings of the IEEE 78, 1550–1560 (1990)
13. Singh, S.P., Jaakkola, T., Jordan, M.I.: Learning without state-estimation in partially observable markovian decision processes. In: International Conference on Machine Learning, pp. 284–292 (1994)
14. Aberdeen, D.: Policy-Gradient Algorithms for Partially Observable Markov Decision Processes. PhD thesis, Australian National University (2003)
15. Meuleau, N., Peshkin, L., Kim, K.-E., Kaelbling, L.P.: Learning finite-state controllers for partially observable environments. In: UAI '99. Proc. Fifteenth Conference on Uncertainty in Artificial Intelligence, pp. 427–436. Morgan Kaufmann, San Francisco (1999)
16. Hochreiter, S., Schmidhuber, J.: Long short-term memory. Neural Computation 9(8), 1735–1780 (1997)
17. Bakker, B.: Reinforcement learning with long short-term memory. In: Advances in Neural Information Processing Syst., vol. 14 (2002)
18. Baxter, J., Bartlett, P.: Direct gradient-based reinforcement learning (1999)
19. Hochreiter, S., Bengio, Y., Frasconi, P., Schmidhuber, J.: Gradient flow in recurrent nets: the difficulty of learning long-term dependencies. In: Kremer, S.C., Kolen, J.F. (eds.) A Field Guide to Dynamical Recurrent Neural Networks, IEEE Press, NJ, New York (2001)
20. Schmidhuber, J.: RNN overview (2004), `http://www.idsia.ch/~juergen/rnn.html`
21. Wieland, A.: Evolving neural network controllers for unstable systems. In: Proceedings of the International Joint Conference on Neural Networks, Seattle, WA, pp. 667–673. IEEE Service Center, Piscataway, NJ (1991)
22. Torcs: Torcs, the open racing car simulator (2007),`http://torcs.sourceforge.net/`

Soft Clustering for Nonparametric Probability Density Function Estimation

Ezequiel López-Rubio, Juan Miguel Ortiz-de-Lazcano-Lobato, Domingo López-Rodríguez, and María del Carmen Vargas-González

School of Computer Engineering
University of Málaga
Campus de Teatinos, s/n. 29071
Málaga Spain
Phone: (+34) 95 213 71 55
Fax: (+34) 95 213 13 97
{ezeqlr,jmortiz}@lcc.uma.es, dlopez@ctima.uma.es

Abstract. We present a nonparametric probability density estimation model. The classical Parzen window approach builds a spherical Gaussian density around every input sample. Our method has a first stage where hard neighbourhoods are determined for every sample. Then soft clusters are considered to merge the information coming from several hard neighbourhoods. Our proposal estimates the local principal directions to yield a specific Gaussian mixture component for each soft cluster. This leads to outperform other proposals where local parameter selection is not allowed and/or there are no smoothing strategies, like the manifold Parzen windows.

Keywords: Probability density estimation, nonparametric modeling, soft clustering, Parzen windows.

1 Introduction

The estimation of the unknown probability density function (PDF) of a continuous distribution from a set of data points forming a representative sample drawn from the underlying density is a problem of fundamental importance to all aspects of machine learning and pattern recognition (see [1], [2] and [3]).

Parametric approaches make assumptions about the unknown distribution. They consider, a priori, a particular functional form for the PDF and reduce the problem to the estimation of the required functional parameters. On the other hand, nonparametric methods make less rigid assumptions. Popular nonparametric methods include the histogram, kernel estimation, nearest neighbor methods and restricted maximum likelihood methods, as can be found in [4], [5], [6] and [7].

The kernel density estimator, also commonly referred as the Parzen window estimator, [9], places a Gaussian kernel on each data point of the training set. Then, the PDF is approximated by summing all the kernels, which are multiplied by a normalizing factor. Thus, this model can be viewed as a finite mixture model (see [8]) where the number of mixture components will equal the number of points in the data

J. Marques de Sá et al. (Eds.): ICANN 2007, Part I, LNCS 4668, pp. 707–716, 2007.

sample. Parzen windows estimates are usually built using a 'spherical Gaussian' with a single scalar variance parameter σ^2, which spreads the density mass equally along all input space directions and gives too much probability to irrelevant regions of space and too little along the principal directions of variance of the distribution. This drawback is partially solved in Manifold Parzen Windows algorithm [10], where a different covariance matrix is calculated for each component. The covariance matrix is estimated by considering a hard neighbourhood of each input sample. We propose in Section 2 to build soft clusters to share the information among neighbourhoods. This leads to filter the input noise by smoothing the estimated parameters.

We present in section 3 a method that automatizes the selection of the right dimensionality of the manifold. The asymptotical convergence of the proposed method is formally proven in Section 4. We show some experimental results in section 5, where the selection achieved by our method produces more precise estimations than the Manifold Parzen Windows and other approaches. Finally, Section 6 is devoted to conclusions.

2 The Smooth Parzen Windows Method

Let $\mathbf{x}$ be an D-dimensional random variable and $p()$ an arbitrary probability density function over $\mathbf{x}$ which is unknown and we want to estimate. The training set of the algorithm is formed by N samples of the random variable. For each training sample $\mathbf{x}_i$ we build a hard Q-neighbourhood H_i with the Q nearest neighbours of $\mathbf{x}_i$, including itself. Hence H_i is interpreted as a random event which happens iff the input belongs to that neighbourhood. The knowledge about the local structure of the distribution around $\mathbf{x}_i$ is obtained when we calculate the mean vector $\boldsymbol{\mu}$ and the correlation matrix $\mathbf{R}$:

$$\boldsymbol{\mu}(H_i) = E[\mathbf{x} \mid H_i] = \frac{1}{Q} \sum_{\mathbf{x}_j \in H_i} \mathbf{x}_j \tag{1}$$

$$\mathbf{R}(H_i) = E[\mathbf{x}\mathbf{x}^T \mid H_i] = \frac{1}{Q} \sum_{\mathbf{x}_j \in H_i} \mathbf{x}_j \mathbf{x}_j^T \tag{2}$$

Now we present a smoothing procedure to merge the information from different hard neighbourhoods. We define a soft cluster i by a random event S_i which verifies when the input belongs to cluster i. Each hard neighbourhood H_j contributes to S_i with a weight w_{ij}:

$$w_{ij} = P[H_j \mid S_i] \tag{3}$$

So, we have

$$\forall i \in \{1,2,\ldots,M\}, \quad \sum_{j=1}^{N} w_{ij} = 1 \tag{4}$$

where the number of soft clusters M may be different from the number of hard neighbourhoods N. We can infer the structure of the soft cluster by merging the information from the hard neighbourhoods:

$$\boldsymbol{\mu}(S_i) = E[\mathbf{x} \mid S_i] = \sum_j P[H_j \mid S_i] E[\mathbf{x} \mid H_j] = \sum_{j=1}^{N} w_{ij} \boldsymbol{\mu}(H_j) \tag{5}$$

$$\mathbf{R}(S_i) = E[\mathbf{x}\mathbf{x}^T \mid S_i] = \sum_j P[H_j \mid S_i] E[\mathbf{x}\mathbf{x}^T \mid H_j] = \sum_{j=1}^{N} w_{ij} \mathbf{R}(H_j) \tag{6}$$

In order to define a Gaussian distribution we need the estimation of the covariance matrix $\mathbf{C}$ for each soft cluster:

$$\mathbf{C}(S_i) = E[(\mathbf{x} - \boldsymbol{\mu}(S_i))(\mathbf{x} - \boldsymbol{\mu}(S_i))^T \mid S_i] = \mathbf{R}(S_i) - \boldsymbol{\mu}(S_i)\boldsymbol{\mu}(S_i)^T \tag{7}$$

Finally, we need a method to determine the merging weights w_{ij}. We propose two approaches:

a) If $M=N$, we can perform the smoothing by replacing the 'hard' model at the data sample $\mathbf{x}_i$ by a weighted average of its neighbours ranked by their distance to $\mathbf{x}_i$. Here the model at $\mathbf{x}_i$ has the maximum weight, and their neighbours $\mathbf{x}_j$ have a weight which is a decreasing function of the distance from $\mathbf{x}_i$ to $\mathbf{x}_j$:

$$\omega_{ij} = \exp\left(-\frac{\left\| \mathbf{x}_i - \mathbf{x}_j \right\|^2}{\psi^2} \right) \tag{8}$$

$$w_{ij} = \frac{\omega_{ij}}{\sum_{k=1}^{N} \omega_{ik}} \tag{9}$$

where ψ is a parameter to control the width of the smoothing. Please note that $\omega_{ii}=1$.

b) We may use the fuzzy c-means algorithm [11] to perform a soft clustering. This algorithm partitions the set of training data into M clusters so it minimizes the distance within the cluster. The objective function is:

$$J = \sum_{i=1}^{M} \sum_{j=1}^{N} m_{ij}^{\phi} d_{ij}^{2} \tag{10}$$

where ϕ is the fuzzy exponent which determines the degree of fuzzyness, and d_{ij} is the distance between training sample $\mathbf{x}_j$ and the centroid of cluster i.

The degrees of membership of training sample j to soft cluster i are obtained as m_{ij}, which can be regarded as the probability of training sample j belonging to cluster i. In this approach the weights w_{ij} of the local models that we merge to yield the model of cluster i are computed as follows:

$$w_{ij} = \frac{m_{ij}}{\sum_{k=1}^{N} m_{ik}} \tag{11}$$

3 Dynamic Parameter Selection

Once we have the estimations of the mean vectors $\mu(S_i)$ and covariance matrices $\mathbf{C}(S_i)$ for each soft cluster S_i, it is needed to obtain a Gaussian distribution from them which is both accurate and small sized.

First we incorporate the capacity of estimating the intrinsic dimensionality. A second level of automated adaptation to the data will be added by means of a parameter γ. This parameter will enable the method to select the right noise level.

3.1 The Explained Variance Method

The explained variance method considers a variable number, K_i, of eigenvalues and their corresponding eigenvectors to be kept which is computed independently for each cluster S_i. Through the training process the method ensures that a minimum amount of variance is conserved in order to satisfy the level of accuracy, $\alpha \in [0,1]$, chosen by the user.

The lost variance when no directions are kept, V_0, can be defined as:

$$V_0 = \sum_{p=1}^{D} \lambda_i^p \tag{12}$$

with λ_i^p the p-th eigenvalue of the covariance matrix $\mathbf{C}(S_i)$, which are supposed to be sorted in decreasing order, and D the dimension of the training samples.

In any other situation the discarded variance, V_Z, depends on the number, Z, of principal directions conserved:

$$V_Z = \sum_{p=Z+1}^{D} \lambda_i^p \tag{13}$$

Let $V_0 - V_Z$ be the amount of error eliminated when we conserve information about the Z principal directions. Then

$$K_i = \min\left\{ Z \in \{0,1,...,D\} \mid V_0 - V_Z \geq \alpha \; V_0 \right\} \tag{14}$$

Substitution of (**12**) and (**13**) into (**14**) yields

$$K_i = \min\left\{ Z \in \{0,1,...,D\} \mid \sum_{p=1}^{D} \lambda_i^p - \sum_{p=Z+1}^{D} \lambda_i^p \geq \alpha \sum_{p=1}^{D} \lambda_i^p \right\} \tag{15}$$

By expressing the sum of variances as the trace of $\mathbf{C}(S_i)$ we can simplify (**15**):

$$K_i = \min\left\{ Z \in \{0,1,\ldots,D\} \mid \sum_{p=1}^{Z} \lambda_i^p \geq \alpha \; trace(\mathbf{C}(S_i)) \right\} \tag{16}$$

3.2 The Qualitative Parameter γ

We propose to select the noise variance parameter σ^2 as follows:

$$\sigma^2(S_i) = \gamma \cdot \lambda_i^{K_i} \tag{17}$$

where $\gamma \in [0,1]$ and $\lambda_i^{K_i}$ is the last of the preserved eigenvalues of $\mathbf{C}(S_i)$, i.e. the smallest of the first K_i largest eigenvalues. The estimated noise variance $\sigma^2(S_i)$ is added to the first K_i largest eigenvalues to yield the estimated variances of the K_i principal directions, while the D–K_i trailing directions have estimated variances of $\sigma^2(S_i)$.

3.3 Smooth Parzen Windows with Qualitative Parameters

The proposed algorithm is designed to estimate an unknown density distribution *p()* which the *N* samples of the training dataset are generated from. The generated estimator will be formed by a mixture of *M* Gaussians, one for each soft cluster:

$$\hat{p}(\mathbf{x}) = \frac{1}{M}\sum_{i=1}^{M} N_i(\mathbf{x}) \tag{18}$$

$$N_i(\mathbf{x}) = \exp\left(-\frac{1}{2}(a_i + b_i)\right) \tag{19}$$

$$a_i = D\log(2\pi) + (D - K_i)\log\left(\lambda_i^p + \sigma^2(S_i)\right) + \sum_{p=1}^{K_i}\log\left(\lambda_i^p + \sigma^2(S_i)\right) \tag{20}$$

$$b_i = \frac{1}{\sigma^2(S_i)}\left\|\mathbf{x} - \boldsymbol{\mu}(S_i)\right\|^2 + \sum_{p=1}^{K_i}\left(\frac{1}{\lambda_i^p} - \frac{1}{\sigma^2(S_i)}\right)\left\|\left(\mathbf{u}_i^p\right)^T\left(\mathbf{x} - \boldsymbol{\mu}(S_i)\right)\right\|^2 \tag{21}$$

where $\mathbf{u}_i^p$ is the eigenvector corresponding to the *p*-th largest eigenvalue of $\mathbf{C}(S_i)$.

3.4 Summary

The training algorithm can be summarized as follows:

1. For each training sample, compute the mean vector $\boldsymbol{\mu}(H_i)$ and correlation matrix $\mathbf{C}(H_i)$ of its hard neighbourhood H_i with equations (**1**) and (**2**).
2. Estimate the merging weights w_{ij} either by the distance method (**9**) or the fuzzy c-means algorithm (**11**).

3. Compute the mean vectors $\mu(S_i)$ and covariance matrices $\mathbf{C}(S_i)$ of each soft cluster S_i following (**5**) and (**7**).
4. Extract the eigenvalues and eigenvectors from $\mathbf{C}(S_i)$ and estimate the dimensionality of the underlying manifold K_i, by means of (**16**).
5. Use (**17**) to calculate $\sigma^2(S_i)$, the noise variance for the discarded directions.
6. Store each local model, i. e., the first K_i eigenvectors and eigenvalues, the local noise level $\sigma^2(S_i)$ and the mean vector $\mu(S_i)$.

4 Convergence Proof

In this section we prove that our estimator $\hat{p}(\ )$ converges to the true density function $p()$ in the limit $N\rightarrow\infty$ and $M\rightarrow\infty$.

Lemma 1. Every local Gaussian $N_i(\mathbf{x})$ tends to the D-dimensional Dirac delta function $\delta(\mathbf{x}-\mu(S_i))$ as $N\rightarrow\infty$ and $M\rightarrow\infty$.

Proof. In the limit $N\rightarrow\infty$ and $M\rightarrow\infty$ the clusters S_i reduce their volume to zero. This means that $\sigma^2(S_i)\rightarrow 0$ and $\lambda_i^p\rightarrow 0$ for all i and p. Hence the Gaussians $N_i(\mathbf{x})$ are confined to a shrinking volume centered at $\mu(S_i)$, because the variances in each direction are $\lambda_i^p+\sigma^2(S_i)$ or $\sigma^2(S_i)$, but they continue to integrate to 1. So, we have that $N_i(\mathbf{x})\rightarrow\delta(\mathbf{x}-\mu(S_i))$.

Theorem 1. The expected value of the proposed estimation tends to the true probability density function as $N\rightarrow\infty$ and $M\rightarrow\infty$.

Proof. The expectation is w.r.t. the underlying distribution of the training samples, which is the true probability density function $p()$:

$$E[\hat{p}(\mathbf{x})]=\frac{1}{M}\sum_{i=1}^{M}E[N_i(\mathbf{x})] \tag{22}$$

Since $N_i(\mathbf{x})$ are independent and identically distributed random variables we get

$$E[\hat{p}(\mathbf{x})]=E[N_i(\mathbf{x})]=\int p(\mathbf{y})N_{\mathbf{y}}(\mathbf{x})d\mathbf{y} \tag{23}$$

where $N_{\mathbf{y}}()$ is a Gaussian centered in $\mathbf{y}$. Then, by Lemma 1, if $N\rightarrow\infty$ and $M\rightarrow\infty$ then $N_{\mathbf{y}}()$ shrinks to a Dirac delta:

$$E[\hat{p}(\mathbf{x})]\rightarrow\int p(\mathbf{y})\delta(\mathbf{x}-\mathbf{y})d\mathbf{y} \tag{24}$$

So, the expectation of the estimation converges to a convolution of the true density with the Dirac delta function. Then,

$$E[\hat{p}(\mathbf{x})]\rightarrow p(\mathbf{x}) \tag{25}$$

Theorem 2. The variance of the proposed estimation tends to zero as $N\rightarrow\infty$ and $M\rightarrow\infty$.

Proof. The variance is w.r.t. the underlying distribution of the training samples, which is the true probability density function $p()$:

$$\mathrm{var}[\hat{p}(\mathbf{x})] = \mathrm{var}\left[\frac{1}{M}\sum_{i=1}^{M} N_i(\mathbf{x})\right] \tag{26}$$

Since $N_i(\mathbf{x})$ are independent and identically distributed random variables we get

$$\mathrm{var}[\hat{p}(\mathbf{x})] = \frac{1}{M}\mathrm{var}[N_i(\mathbf{x})] \tag{27}$$

By the properties of variance and (**23**) we obtain

$$\mathrm{var}[\hat{p}(\mathbf{x})] = \frac{1}{M}\left(E\left[(N_i(\mathbf{x}))^2\right] - E[N_i(\mathbf{x})]^2\right) = \frac{1}{M}\left(E\left[(N_i(\mathbf{x}))^2\right] - E[\hat{p}(\mathbf{x})]^2\right) \tag{28}$$

By definition of expectation

$$\mathrm{var}[\hat{p}(\mathbf{x})] = \frac{1}{M}\left(\int p(\mathbf{y})(N_{\mathbf{y}}(\mathbf{x}))^2 d\mathbf{y} - E[\hat{p}(\mathbf{x})]^2\right) \tag{29}$$

where again $N_{\mathbf{y}}()$ is a Gaussian centered in $\mathbf{y}$. We can bound the integral of the above equation with the help of (**23**), and so we get

$$\mathrm{var}[\hat{p}(\mathbf{x})] \leq \frac{\sup(N(\cdot))E[\hat{p}(\mathbf{x})]}{M} \to 0 \text{ as } N\to\infty \text{ and } M\to\infty \tag{30}$$

5 Experimental Results

This section shows some experiments we have designed in order to study the quality of the density estimation achieved by our method. We call it SmoothDist when the distance weighting is used, and SmoothFuzzy when we use fuzzy c-means. Vincent and Bengio's method is referred as MParzen, the original Parzen windows method (with isotropic Gaussian kernels) is called OParzen, and finally the Mixtures of Probabilistic PCA model of Tipping and Bishop [12] is called MPPCA. For this purpose the performance measure we have chosen is the average negative log likelihood

$$ANLL = -\frac{1}{T}\sum_{i=1}^{T}\log\hat{p}(\mathbf{x}_i) \tag{31}$$

where $\hat{p}()$ is the estimator, and the test dataset is formed by T samples $\mathbf{x}_i$.

5.1 Experiment on 2D Artificial Data

A training set of 100 points, a validation set of 100 points and a test set of 10000 points were generated from the following distribution of two dimensional (x,y) points:

$$x = 0.04t\sin(t) + \varepsilon_x, \quad y = 0.04t\cos(t) + \varepsilon_y \tag{32}$$

where $t \quad U(3,15), \varepsilon_x \quad N(0,0.01), \varepsilon_y \quad N(0,0.01), U(a,b)$ is uniform in the interval (a,b) and $N(\mu,\sigma)$ is a normal density.

We have optimized separately all the parameters of the five competing models with disjoint training and validation sets. The performance of the optimized models has been computed by 10-fold cross-validation, and the results are shown in Table 1, with the best result marked in bold. It can be seen that our models outperform the other three in density distribution estimation.

Table 1. Quantitative results on the espiral dataset (standard deviations in parentheses)

Method	Optimized parameters used	ANLL on test set
SmoothDist	M=100, α=0.9, Q=4, γ=0.03, ψ=0.001	–1.5936 (0.2557)
SmoothFuzzy	M=100, α=0.1, Q=4, γ=0.05	**–1.6073 (0.3293)**
OParzen	M=100, σ^2=0.0001	1.0817 (1.3357)
MParzen	M=100, K=2, Q=4, σ^2=1.6E–5	–0.9505 (0.3301)
MPPCA	M=4, K=1	0.2473 (0.0818)

Fig. 1. Density estimation for the 2D artificial dataset. From left to right and from top to bottom: SmoothDist, SmoothFuzzy, OParzen, MParzen and MPPCA.

Figure 1 shows density distribution plots corresponding to the five models. Darker areas represent zones with high density mass and lighter ones indicate the estimator has detected a low density area.

We can see in the plots that our models have less density holes (light areas) and less 'bumpiness'. This means that our model represents more accurately the true distribution, which has no holes and is completely smooth. We can see that the quantitative ANLL results agree with the plots, because the lowest values of ANLL match the best-looking plots. So, our model outperforms clearly the other three considered approaches.

5.2 Density Estimation Experiment

A density estimation experiment has been designed, where we have chosen three multidimensional datasets from the UCI Repository of Machine Learning Databases [13]. As in the previous experiment, we have optimized all the parameters of the five competing models with disjoint training and validation sets. The parameters for the density estimator of each dataset have been optimized separately. Table 2 shows the results of the 10-fold cross-validation, with the winning models in bold. Our two proposals show a superior performance.

Table 2. ANLL on test set (standard deviations in parentheses)

Dataset	SmoothDist	SmoothFuzzy	OParzen	MParzen	MPPCA
auto-mpg	**25.4 (0.8)**	26.3 (2.4)	29.3 (3.8)	30.1 (3.7)	33.1 (1.7)
cloud	24.9 (0.7)	**24.8 (0.6)**	36.9 (3.5)	37.7 (3.5)	31.6 (1.0)
housing	25.7 (1.2)	**24.9 (1.7)**	31.2 (2.2)	32.1 (2.2)	43.4 (4.9)

5.3 Classification Experiment

We have selected three classification benchmarks from the UCI Repository of Machine Learning Databases [13] to perform a classification experiment. We have considered Bayesian classifiers which are built by estimating the probability density function of each class separately. Then, a test pattern is assigned to the class which yields the largest probability density. We have optimized the model parameters separately, as in the previous experiment. In this case we have optimized the models independently for each class of each database. The results of the 10-fold cross-validation are shown in Table 3, and the winning model for each database is shown in bold. We can see that our two approaches outperform the other three.

Table 3. Successful classification percentages on test set (standard deviations in parentheses)

Database	SmoothDist	SmoothFuzzy	OParzen	MParzen	MPPCA
glass	**69.2 (11.5)**	66.7 (8.5)	63.0 (14.6)	64.0 (10.4)	46.4 (11.2)
liver	67.4 (5.9)	**68.8 (7.2)**	60.6 (7.9)	62.6 (7.8)	59.2 (10.9)
pima	**69.9 (5.0)**	61.6 (5.4)	66.0 (3.9)	62.0 (4.9)	62.9 (5.6)

6 Conclusions

We have presented a probability density estimation model. It is based in the Parzen window approach. Our proposal builds local models for a hard neighbourhood of each training sample. Then soft clusters are obtained by merging these local models. A local Gaussian density is developed by selecting independently for each soft cluster the best number of retained dimensions and the best estimation of noise variance. This allows our method to represent input distributions more faithfully than the Manifold Parzen window model, which is an improvement of the original Parzen window method. Computational results show the superior performance of our method.

Acknowledgements

This work was partially supported by the Ministry of Education and Science of Spain under Projects TIN2005-02984 and TIN2006-07362.

References

1. Bishop, C.: Neural Networks for Pattern Recognition. Oxford University Press, Oxford (1995)
2. Silverman, B.: Density Estimation for Statistics and Data Analysis. Chapman and Hall, New York (1986)
3. Vapnik, V.N.: Statistical Learning Theory. John Wiley & Sons, New York (1998)
4. Izenman, A.J.: Recent developments in nonparametric density estimation. Journal of the American Statistical Association 86(413), 205–224 (1991)
5. Lejeune, M., Sarda, P.: Smooth estimators of distribution and density functions. Computational Statistics & Data Analysis 14, 457–471 (1992)
6. Hjort, N.L., Jones, M.C.: Locally Parametric Nonparametric Density Estimation. Annals of Statistics 24(4), 1619–1647 (1996)
7. Hastie, T., Loader, C.: Local regression: Automatic kernel carpentry. Statistical Science 8, 120–143 (1993)
8. McLachlan, G., Peel, D.: Finite Mixture Models. Wiley, Chichester (2000)
9. Parzen, E.: On the Estimation of a Probability Density Function and Mode. Annals of Mathematical Statistics 33, 1065–1076 (1962)
10. Vincent, P., Bengio, Y.: Manifold Parzen Windows. Advances in Neural Information Processing Systems 15, 825–832 (2003)
11. Bezdek, J.C.: Numerical taxonomy with fuzzysets. J. Math. Biol. 1, 57–71 (1974)
12. Tipping, M.E., Bishop, C.M.: Mixtures of Probabilistic Principal Components Analyzers. Neural Computation 11, 443–482 (1999)
13. Newman, D.J., Hettich, S., Blake, C.L., Merz, C.J.: UCI Repository of machine learning databases. University of California, Department of Information and Computer Science, Irvine, CA (1998), http://www.ics.uci.edu/~mlearn/MLRepository.html

Vector Field Approximation by Model Inclusive Learning of Neural Networks

Yasuaki Kuroe[1] and Hajimu Kawakami[2]

[1] Department of Information Science, Kyoto Institute of Technology
Matsugasaki, Sakyo-ku, Kyoto 606-8585, Japan
kuroe@kit.ac.jp

[2] Department of Electronics and Informatics, Ryukoku University
1-5, Yokotani, Ohe-cho, Seta, Ohtsu 520-21, Japan
kawakami@rins.ryukoku.ac.jp

Abstract. The problem of vector field approximation arises in the wide range of fields such as motion control, computer vision and so on. This paper proposes a method for reconstructing an entire continuous vector field from a sparse set of sample data by training neural networks. In order to make approximation results possess inherent properties of vector fields and to attain reasonable approximation accuracy with computational efficiency, we include a priori knowledge on inherent properties of vector fields into the learning problem of neural networks, which we call model inclusive learning. An efficient learning algorithm of neural networks is derived. It is shown through numerical experiments that the proposed method makes it possible to reconstruct vector fields accurately and efficiently.

1 Introduction

The problem of approximating a vector field arises in numerous problems in control, instrumentation and signal processing areas. Typical examples are estimations of optical flow in computer vision, force fields in biological motor control, contact surfaces of robot manipulators and velocity fields in quantitative flow visualization. They are all approximation problems of reconstructing an entire continuous vector field from a sparse set of data (measured vectors data). In approximation problems, it becomes a problem how to attain reasonable accuracy with reasonable computation efficiency. Furthermore it is essentially important to make the approximation results possess inherent properties of a target object. One promising method is to introduce a priori knowledge on an object into the formulation of approximation problems. Along this line F. A. Mussa-Ivaldi proposed a method for approximating a vector field [2]. He defines basis fields by using derivatives of the Green functions and proposes a method to reconstruct a vector field by superposition of the basis fields.

In this paper we propose a method for reconstructing an entire continuous vector field from a sparse set of sample data by neural networks. In order to make approximation results possess inherent properties of vector fields and attain reasonable approximation accuracy with computational efficiency, we include a priori knowledge on inherent properties of vector fields into the learning problem of neural networks, which we call model inclusive learning. We formulate the learning problem in such a way

J. Marques de Sá et al. (Eds.): ICANN 2007, Part I, LNCS 4668, pp. 717–726, 2007.

that the knowledge on vector fields, "any vector field is composed of the sum of two vector fields: irrotational vector field and solenoidal vector field" is embedded into the approximation problem. An efficient learning algorithm for the formulation is derived in a systematic way by introducing the adjoint models of neural networks. In order to check the performance of the proposed method numerical experiments have been done. It is shown that the proposed method makes it possible to reconstruct vector fields more accurately and efficiently.

2 Problem Formulation of Vector Field Approximation

For the purpose of simplicity, we consider a two dimensional vector field and discuss a method for reconstructing its entire field from a set of sparse sample data. Let

$$\boldsymbol{F}(\boldsymbol{x}) := \begin{bmatrix} F_x(\boldsymbol{x}) \\ F_y(\boldsymbol{x}) \end{bmatrix}$$

be a vector field considered in this paper. Here $\boldsymbol{x} = [x, y]^T$ denotes a position on the two dimensional plane. The problem is to reconstruct the vector field $\boldsymbol{F}(\boldsymbol{x}) \in \mathbb{R}^2$ from a given set of sample data $\{\boldsymbol{F}(\boldsymbol{x}_p)\}$ where $\boldsymbol{x}_p$ denotes a sample point.

A usual and conventional method to solve this problem by neural networks is as follows. Considering that $\boldsymbol{F}(\boldsymbol{x})$ is a nonlinear mapping with two inputs and two outputs, we prepare a two-input and two-output neural network. The network is trained such that, for a given set of sample data $\{\boldsymbol{F}(\boldsymbol{x}_p)\}$, $\boldsymbol{x}_p$ is fed to the input of the network and its output comes close to the data $\boldsymbol{F}(\boldsymbol{x}_p)$. Figure 1 shows a block diagram of this learning method. The backpropagation [1] is usually utilized in order to adjust the weights of the neural network so as to minimize the square error:

$$\sum_p \{(O_1 - F_x(\boldsymbol{x_p}))^2 + (O_2 - F_y(\boldsymbol{x_p}))^2\}$$

where O_1 and O_2 are the outputs of the neural network.

Note that $\boldsymbol{F}(\boldsymbol{x})$ is not merely a nonlinear function but a vector field. If a priori knowledge on inherent properties of vector fields can be embedded into the learning problem, the approximation result can possess the inherent properties, and moreover approximation accuracy and learning efficiency can be improved. In this paper we propose a new learning formulation of neural networks for reconstructing vector fields which we call model inclusive learning. We formulate the learning problem in such a way that a model of the knowledge on the inherent property of vector fields is included in the learning loop of neural networks.

It is known that any vector field $\boldsymbol{F}(\boldsymbol{x}) \in \mathbb{R}^2$ is composed of the sum of two vector fields as follows.

$$\boldsymbol{F}(\boldsymbol{x}) = \boldsymbol{C}(\boldsymbol{x}) + \boldsymbol{S}(\boldsymbol{x}), \quad \boldsymbol{x} = [x, y]^T \tag{1}$$

where $\boldsymbol{C}(\boldsymbol{x})$ and $\boldsymbol{S}(\boldsymbol{x})$ are an irrotational vector field and a solenoidal vector field, respectively, and satisfy the following relations.

$$\text{curl}(\boldsymbol{C}) = \nabla \times \boldsymbol{C}(\boldsymbol{x}) = 0, \qquad \text{div}(\boldsymbol{S}) = \nabla \cdot \boldsymbol{S}(\boldsymbol{x}) = 0 \tag{2}$$

Here we introduce scalar functions $U_1(\boldsymbol{x})$ and $U_2(\boldsymbol{x})$, and express the vector fields $\boldsymbol{C}(\boldsymbol{x})$ and $\boldsymbol{S}(\boldsymbol{x})$ as follows.

$$\boldsymbol{C}(\boldsymbol{x}) = \alpha \nabla U_1(\boldsymbol{x}), \;\; \boldsymbol{S}(\boldsymbol{x}) = \begin{pmatrix} 0 & \beta \\ -\beta & 0 \end{pmatrix} \nabla U_2(\boldsymbol{x}) \tag{3}$$

where α and β are scalars. Since the above equation (3) satisfies eq.(2), the vector field $\boldsymbol{F}(\boldsymbol{x})$ can be expressed by using the scalar functions $U_1(\boldsymbol{x})$ and $U_1(\boldsymbol{x})$ as follows.

$$\boldsymbol{F}(\boldsymbol{x}) = \boldsymbol{C}(\boldsymbol{x}) + \boldsymbol{S}(\boldsymbol{x}) = \begin{bmatrix} \alpha \dfrac{\partial U_1(\boldsymbol{x})}{\partial x} + \beta \dfrac{\partial U_2(\boldsymbol{x})}{\partial y} \\ \alpha \dfrac{\partial U_1(\boldsymbol{x})}{\partial y} - \beta \dfrac{\partial U_1(\boldsymbol{x})}{\partial x} \end{bmatrix} =: \begin{bmatrix} F_x(\boldsymbol{x}) \\ F_y(\boldsymbol{x}) \end{bmatrix} \tag{4}$$

Considering the relation (4), we now give a new learning formulation for reconstructing vector fields by neural networks. We formulate the learning problem in such a way that a neural network reconstructs the vector field $\boldsymbol{F}(\boldsymbol{x})$ with the relation (4) being satisfied. For this purpose we train the neural network such that the vector field $\boldsymbol{F}(\boldsymbol{x})$ itself is not realized on the neural network as is in the conventional method (Fig. 1), but the scalar functions $U_1(\boldsymbol{x})$ and $U_2(\boldsymbol{x})$ satisfying the relation (4) are realized on the network as its input-output relation. Figure 2 shows a block diagram of the proposed model-inclusive learning method.

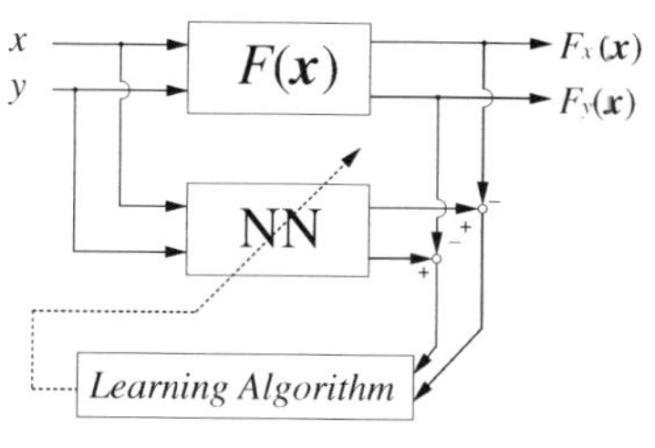

Fig. 1. Conventional learning method of vector fields

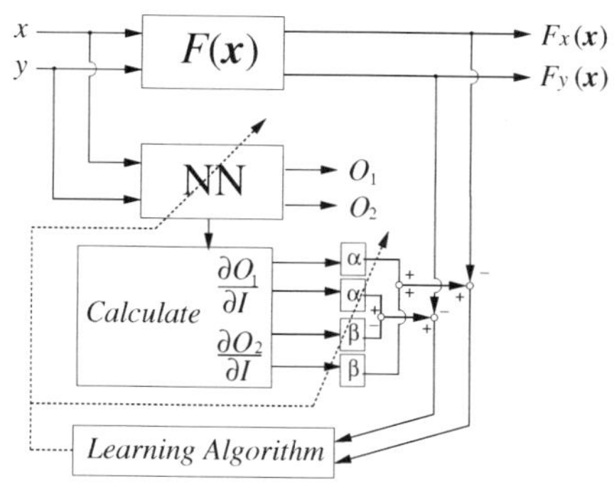

Fig. 2. Proposed model-inclusive learning method of vector fields

We prepare a neural network with two inputs denoted by $\boldsymbol{I} = [I_1, I_2]^T$ and two outputs denoted by $\boldsymbol{O} = [O_1, O_2]^T$. For a set of sample data $\{\boldsymbol{F}(\boldsymbol{x}_p)\} = \{[F_x(\boldsymbol{x}_p), lF_y(\boldsymbol{x}_p)]^T\}$, the network is given the position data $\boldsymbol{x}_p = [x_p, y_p]^T$ as input data ($\boldsymbol{I} = \boldsymbol{x}_p$) and is trained such that the outputs O_1 and O_2 come close to $U_1(\boldsymbol{x}_p)$ and $U_2(\boldsymbol{x}_p)$ satisfying the relation (4), respectively. This can be realized in the following manner. We input the position data $\boldsymbol{x}_p = [x_p, y_p]^T$ to the neural network and derive the derivatives of the output of the neural network with respect to the input:

$$\left.\frac{\partial O_k}{\partial \boldsymbol{I}}\right|_{\boldsymbol{I}=\boldsymbol{x}_p} = \left[\left.\frac{\partial O_k}{\partial I_1}\right|_{I_1=x_p}, \left.\frac{\partial O_k}{\partial I_2}\right|_{I_2=y_p} \right]^T, \quad (k = 1, 2).$$

According to eq.(4), their linear combination

$$\left(\alpha \left. \frac{\partial O_1}{\partial I_1} \right|_{I_1 = x_p} + \beta \left. \frac{\partial O_2}{\partial I_2} \right|_{I_2 = y_p}, \alpha \left. \frac{\partial O_1}{\partial I_2} \right|_{I_2 = y_p} - \beta \left. \frac{\partial O_2}{\partial I_1} \right|_{I_1 = x_p} \right)^T$$

is calculated and the result is compared with the given vector data $F(\boldsymbol{x}_p) = [F_x(\boldsymbol{x}_p), F_y(\boldsymbol{x}_p)]^T$. The neural network is trained so as to minimize the error between them. Define the performance index by

$$J = \frac{1}{2} \sum_p \left\{ \left\{ \left(\alpha \left. \frac{\partial O_1}{\partial I_1} \right|_{I_1 = x_p} + \beta \left. \frac{\partial O_2}{\partial I_2} \right|_{I_2 = y_p} \right) - F_x(\boldsymbol{x}_p) \right\}^2 + \left\{ \left(\alpha \left. \frac{\partial O_1}{\partial I_2} \right|_{I_2 = y_p} - \beta \left. \frac{\partial O_2}{\partial I_1} \right|_{I_1 = x_p} \right) - F_y(\boldsymbol{x}_p) \right\}^2 \right\} \quad (5)$$

where $\sum_p$ stands for the summation with respect to the set of data. By minimizing the performance of index thus introduced, the scalar functions $U_1(\boldsymbol{x})$ and $U_2(\boldsymbol{x})$ satisfying eq.(4) are realized on the neural network. As the result an approximation of the vector field $\boldsymbol{F}(\boldsymbol{x})$ is obtained. Thus the problem is reduced to finding the parameters of the neural network which minimize J. Note that, since the parameters α and β in eq.(4) are not known a prior, we choose not only weights of connections w_{ij} but also α and β as the learning parameters. In order to search the values of the learning parameters which minimize J, we use gradient based methods, in which several useful algorithms are available: the steepest decent algorithm, the conjugate gradient algorithm, the quasi-Newton algorithm and so on. For instance, the steepest decent algorithm is given by the following iteration.

$$\boldsymbol{w}^{(l+1)} = \boldsymbol{w}^{(l)} - \eta \frac{\partial J}{\partial \boldsymbol{w}^{(l)}} \quad (6)$$

where $\boldsymbol{w} = [\omega_{ij}, \alpha, \beta]^T$, l is the iteration number and $\eta > 0$ is a learning rate. In gradient based algorithms, it becomes a key how to compute the gradients of performance index, $\partial J/\partial w_{ij}$, $\partial J/\partial \alpha$, and $\partial J/\partial \beta$ efficiently and accurately. In order to derive an algorithm for computing them in a systematic manner, the adjoint models of the neural network is introduced.

3 Models of Neural Network and Its Adjoint

We consider a general class of neural networks, arbitrarily connected neural networks, for the purpose of generality. In order to derive an algorithm in a systematic manner, the adjoint neural network [3],[4] is introduced. First we give the mathematical models of the neural network and its adjoint.

We consider the neural network denoted by N as shown in Fig.3. The network comprises N_u neuron units, which are arbitrarily connected, two external inputs $I_\ell(\ell = 1, 2)$

and two external outputs $O_k(k = 1, 2)$. The mathematical model of the neural network N is given by

$$u_i = \sum_{j=1}^{N_u} w_{ij} y_j + \sum_{\ell=1}^{2} \delta_{\ell i}^{I} I_\ell, \quad y_i = f(u_i), \quad O_k = \sum_{j=1}^{N_u} \delta_{kj}^{O} y_j,$$
$$i = 1, 2, \cdots, N_u, \qquad k = 1, 2 \tag{7}$$

where u_i and y_i are the total input value and the output of the i-th neuron, respectively, and w_{ij} is the weight from the j-th neuron to the i-th neuron, $f(\cdot)$ is a nonlinear output function such as sigmoidal functions, and $\delta_{\ell i}^{I}$ and δ_{kj}^{O} take values 1 or 0. If the ℓ-th external input is connected to the i-th neuron, $\delta_{\ell i}^{I} = 1$, otherwise $\delta_{\ell i}^{I} = 0$. If the output of the j-th neuron is connected to the k-th external output, $\delta_{kj}^{O} = 1$, otherwise $\delta_{kj}^{O} = 0$.

The adjoint neural network denoted by $\hat{N}$ is derived in the following manner as shown in Fig.4. The directions of all signal arrows in the neural network N are inverted and the external inputs $\boldsymbol{I}$ are replaced by the external outputs $\hat{\boldsymbol{I}}$, the external output $\boldsymbol{O}$ is replaced by the external inputs $\hat{\boldsymbol{O}}$, the branching points are replaced by the summing points vice versa, and all the elements are replaced by the elements with the same characteristics except that nonlinear elements are replaced by their linearized elements. The mathematical model of the adjoint neural network is given by

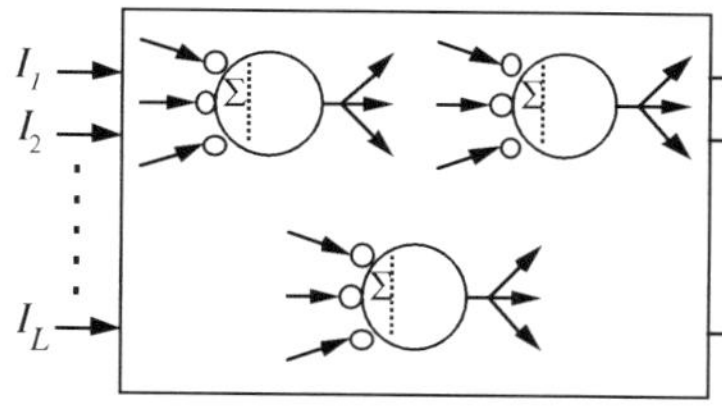

Fig. 3. Neural Network N $(L = 2, K = 2)$

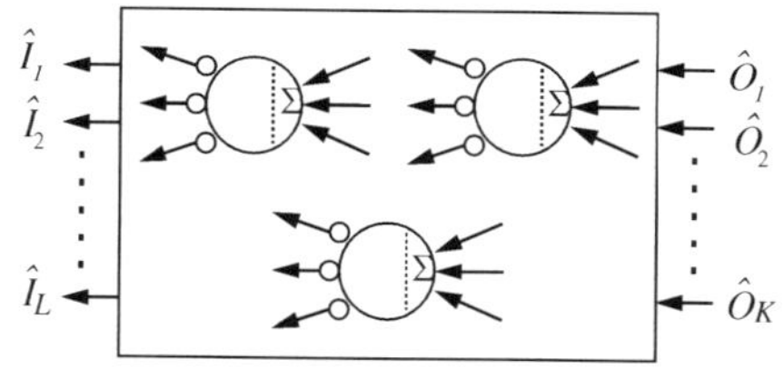

Fig. 4. Adjoint Neural Network $\hat{N}$ $(L = 2, K = 2)$

$$\hat{y}_i = \sum_{j=1}^{N_u} w_{ji} \hat{u}_j + \sum_{k=1}^{2} \delta_{ki}^{O} \hat{O}_k, \quad \hat{u}_i = \frac{\partial f}{\partial u_i} \hat{y}_i, \quad \hat{I}_\ell = \sum_{j=1}^{N_u} \delta_{\ell j}^{I} \hat{u}_j,$$
$$i = 1, 2, \cdots, N_u, \qquad \ell = 1, 2 \tag{8}$$

where the variables with ^ denote the variables of the adjoint model corresponding to those of the neural network N.

4 Learning Method

It can be proved that the relation like the Tellegen's theorem in electric circuits holds between the variables in the neural network and its adjoint. We call it the extended Tellegen's theorem [3]. The learning algorithm can be derived by using this theorem. Its derivation in detail is omitted because of the limitation of the space, only the result is shown here.

4.1 Derivation of $\partial \boldsymbol{O}/\partial \boldsymbol{I}$

In our problem formulation of learning vector fields as shown in Fig.4, the first thing to do is to evaluate the derivatives of the output of the neural network with respect to the input $\partial \boldsymbol{O}/\partial \boldsymbol{I}$. For this purpose, we input the position data to the external inputs of the neural network, $\boldsymbol{I} = \boldsymbol{x}_p$, and construct the corresponding adjoint neural network given by eq.(8). In order to obtain $\partial O_1/\partial \boldsymbol{I}$ we specify the external inputs $\hat{\boldsymbol{O}}$ of the adjoint neural network $\hat{N}$ as:

$$[\hat{O}_1, \hat{O}_2]^T = [1, 0]^T. \tag{9}$$

We denote the adjoint network $\hat{N}$ with the external input given by eq.(9) by $\hat{N}^1$. It can be shown by using the extended Tellegen's theorem that $\partial O_1/\partial I_\ell$ ($\ell = 1, 2$) are obtained from the external output of $\hat{N}^1$ as:

$$\left[\left.\frac{\partial O_1}{\partial I_1}\right|_{I_1=x_p}, \left.\frac{\partial O_1}{\partial I_2}\right|_{I_2=y_p} \right]^T = \left[\hat{I}_1^1, \hat{I}_2^1\right]^T. \tag{10}$$

In order to obtain $\partial O_2/\partial \boldsymbol{I}$ we specify the external inputs $\hat{\boldsymbol{O}}$ of the adjoint neural network $\hat{N}$ as:

$$[\hat{O}_1, \hat{O}_2]^T = [0, 1]^T. \tag{11}$$

We denote the adjoint network $\hat{N}$ with the external input given by eq.(11) by $\hat{N}^2$. $\partial O_2/\partial I_\ell$ ($\ell = 1, 2$) are obtained from the external output of $\hat{N}^2$ as follows:

$$\left[\left.\frac{\partial O_2}{\partial I_1}\right|_{I_1=x_p}, \left.\frac{\partial O_2}{\partial I_2}\right|_{I_2=y_p} \right]^T = \left[\hat{I}_1^2, \hat{I}_2^2\right]^T. \tag{12}$$

4.2 Derivation of the Gradients $\partial J/\partial w_{ij}$, $\partial J/\partial \alpha$ and $\partial J/\partial \beta$

We now show a method for computing the gradients $\partial J/\partial w_{ij}$, $\partial J/\partial \alpha$ and $\partial J/\partial \beta$. For this purpose the additional three types of networks, denoted by N'^k, N'' and $\hat{N}''$ are introduced. The network N'^k ($k = 1, 2$) is defined as the adjoint of the network $\hat{N}^k$ ($k = 1, 2$) and constructed by the same procedure as the network $\hat{N}$ is constructed from N. The mathematical model of N'^k is given by

$$u_i'^k = \sum_{j=1}^{N_u} w_{ij} y_j'^k + \sum_{\ell=1}^{2} \delta_{\ell i}^I I_\ell'^k, \quad y_i'^k = \frac{\partial f}{\partial u_i} y_i'^k, \quad O_m'^k = \sum_{j=1}^{N_u} \delta_{mj}^O y_j'^k,$$
$$i = 1, 2, \cdots, N_u, \qquad m = 1, 2, \tag{13}$$

Note that the external output $\hat{\boldsymbol{I}}^k$ of $\hat{N}^k$ is replaced by the external input $\boldsymbol{I}'^k$ of N'^k. We give the external inputs $\boldsymbol{I}'^k$ of the network N'^k as follows. By using the derivatives of the output of the neural network with respect to the input, $\partial O_k/\partial \boldsymbol{I}$ ($k = 1, 2$), which are obtained in the previous subsection, their linear combination

$$\left(\alpha \left.\frac{\partial O_1}{\partial I_1}\right|_{I_1=x_p} + \beta \left.\frac{\partial O_2}{\partial I_2}\right|_{I_2=y_p}, \alpha \left.\frac{\partial O_1}{\partial I_2}\right|_{I_2=y_p} - \beta \left.\frac{\partial O_2}{\partial I_1}\right|_{I_1=x_p} \right)^T$$
$$= (\alpha \hat{I}_1^1 + \beta \hat{I}_2^2, \alpha \hat{I}_2^1 - \beta \hat{I}_1^2)^T$$

is calculated according to eq.(4). The results are corresponding to an approximation of the given data $\boldsymbol{F}(\boldsymbol{x}_p) = [F_x(\boldsymbol{x}_p), F_y(\boldsymbol{x}_p)]^T$ of the vector field. Then we calculate the errors between them and give the external inputs of the networks N'^1 and N'^2:

$$I_1'^1 = \alpha\left\{\left(\alpha\hat{I}_1^1 + \beta\hat{I}_2^2\right) - F_x(\boldsymbol{x}_p)\right\}, \quad I_2'^1 = \alpha\left\{\left(\alpha\hat{I}_2^1 - \beta\hat{I}_1^2\right) - F_y(\boldsymbol{x}_p)\right\} \quad (14)$$

$$I_1'^2 = -\beta\left\{\left(\alpha\hat{I}_2^1 - \beta\hat{I}_1^2\right) - F_y(\boldsymbol{x}_p)\right\}, \quad I_2'^2 = \beta\left\{\left(\alpha\hat{I}_1^1 + \beta\hat{I}_2^2\right) - F_x(\boldsymbol{x}_p)\right\}. \quad (15)$$

The network N'' is defined to be just the same network as the original neural network N except that the number of the external outputs O_j'' in N'' is N_u and given by

$$O_i'' = u_i, \qquad i = 1, 2, \cdots, N_u. \quad (16)$$

The network $\hat{N}''$ is defined as the adjoint of the network N''. The mathematical model of $\hat{N}''$ is given by

$$\hat{y}_i'' = \sum_{j=1}^{N_u} w_{ji}\hat{u}_j'', \quad \hat{u}_i'' = \frac{\partial f}{\partial u_i}\hat{y}_i'' + \hat{O}_i'', \quad \hat{I}_\ell'' = \sum_{j=1}^{N_u} \delta_{\ell j}^I \hat{u}_j'',$$
$$i = 1, 2, \cdots, N_u, \qquad \ell = 1, 2. \quad (17)$$

Note that the external output O'' of N'' is replaced by the external input $\hat{O}''$ of $\hat{N}''$. We give the inputs of the network $\hat{N}''$ by using the signals in the networks N'^k and $\hat{N}^k$ as follows.

$$\hat{O}_i''^k = \frac{\partial^2 f}{\partial u_i^2}\hat{y}_i^k u_i'^k, \qquad i = 1, 2, \cdots, N_u \quad (18)$$

The network $\hat{N}''$ with the external input eq.(18) is denoted by $\hat{N}''^k$. The extended Tellegen's theorem brings that $\partial J/\partial w_{ij}$ can be evaluated by using the variables of the networks N, $\hat{N}^k$, N'^k and $\hat{N}''^k$ as follows.

$$\frac{\partial J}{\partial w_{ij}} = \sum_p \sum_{k=1}^{2} (\hat{u}_i^k y_j'^k + y_j \hat{u}_i''^k) \quad (19)$$

Furthermore the gradients of the performance index J with respect to the parameters α and β are obtained by using the external outputs of the adjoint network $\hat{N}^k$ as follows.

$$\frac{\partial J}{\partial \alpha} = \sum_p \left\{\{(\alpha\hat{I}_1^1 + \beta\hat{I}_2^2) - F_x(\boldsymbol{x}_p)\}\hat{I}_1^1 + \{(\alpha\hat{I}_2^1 - \beta\hat{I}_1^2) - F_y(\boldsymbol{x}_p)\}\hat{I}_2^1\right\} \quad (20)$$

$$\frac{\partial J}{\partial \beta} = \sum_p \left\{\{(\alpha\hat{I}_1^1 + \beta\hat{I}_2^2) - F_x(\boldsymbol{x}_p)\}\hat{I}_2^2 - \{(\alpha\hat{I}_2^1 - \beta\hat{I}_1^2) - F_y(\boldsymbol{x}_p)\}\hat{I}_1^2\right\} \quad (21)$$

By using the gradients of the performance index J, a gradient-based learning iteration can be performed. Note that the approximation of the target vector field $\boldsymbol{F}(\boldsymbol{x}) = [F_x(\boldsymbol{x}), F_y(\boldsymbol{x})]^T$ is obtained by $[\alpha\hat{I}_1^1 + \beta\hat{I}_2^2, \alpha\hat{I}_2^1 - \beta\hat{I}_1^2]^T$ after the learning iteration converges.

5 Numerical Experiments

In this section we show some results of the numerical experiments in order to demonstrate the performance of the proposed method. In the following examples we use a three-layer feedforward neural network with 5 hidden units and the Davidon-Fletcher-Powell method [5] to search values of the parameters which minimize J.

5.1 Example 1

First we consider the vector field shown in Fig.5 as a target vector field. We reconstruct the vector field from the set of data consisting of 10 sample points shown in Fig.6. By

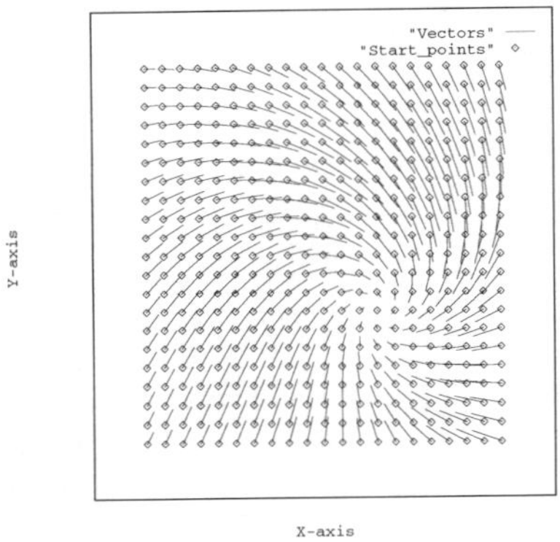

Fig. 5. Target vector field

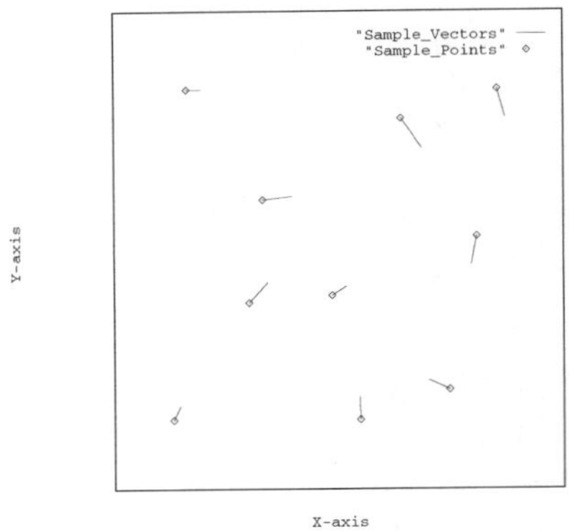

Fig. 6. Sampled vector data

changing the initial values of the learning parameters randomly, learning experiments were carried out over 20 times by using the proposed method. For the purpose of comparison the experiments by using the conventional learning method shown in Fig.1 are also carried out. Figure 7 shows an example of the convergence behavior of the proposed learning method. In the figure the variation of the performance index J versus the number of the learning iterations is plotted. Figures 8 and 9 show the reconstructed vector fields by the proposed method and the conventional method, respectively. By comparing Figs.8 and 9 with Fig.5, it is seen that the result obtained by the proposed method achieves more approximation accuracy with reflecting the smoothness of the target vector field than that by the conventional method. In particular, in the left part of the reconstructed vector field obtained by the conventional method, approximation accuracy deteriorates.

Table 1 summarizes the approximation accuracy of both the learning methods. The approximation accuracy is estimated by calculating the mean square errors of vector components and the vector directions. It is observed that the proposed learning method can reconstruct the vector field more accurately, in particular, in vector directions.

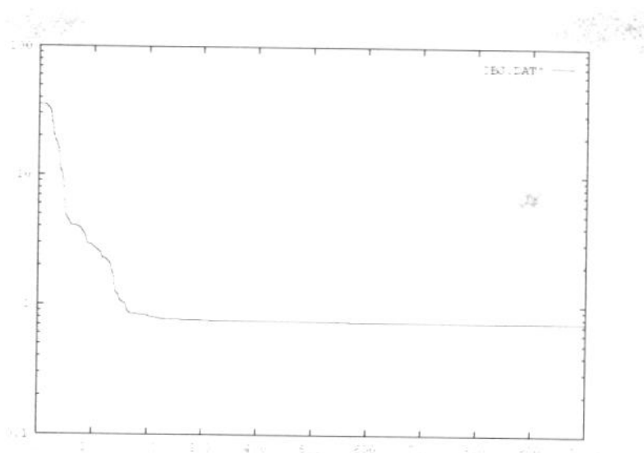

Fig. 7. Values of the performance index during the learning iteration

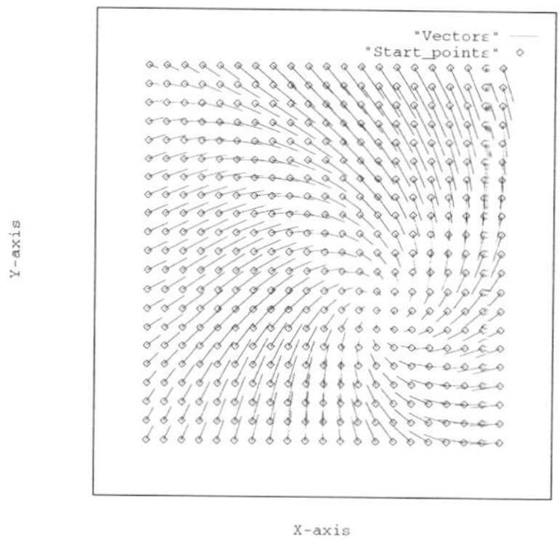

Fig. 8. Reconstructed vector field by the proposed method

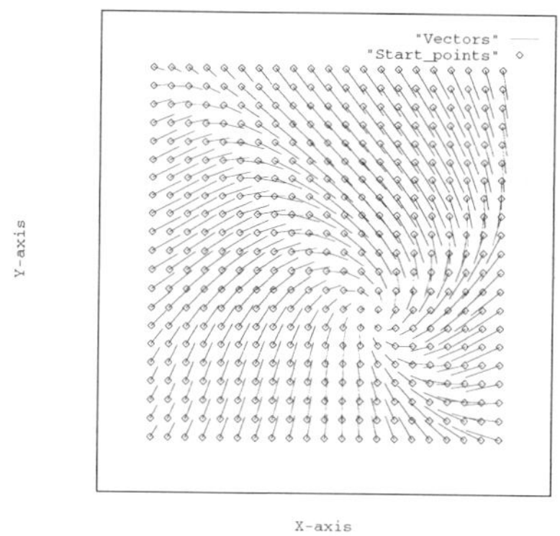

Fig. 9. Reconstructed vector field by the conventional method

5.2 Example 2

The proposed learning method can be applied to a shape recovering problem of a three dimensional object with smooth surface. The problem is to reconstruct the whole surface shape from a given set of data of the surface gradient vectors. This problem can be solved by using the proposed learning method with letting the parameters $\beta = 0$ and the obtained scalar function $U_1(\boldsymbol{x})$ corresponds to the recovered surface. Figure 10 shows example data of the surface gradient vectors of an object for the experiment. From this

Table 1. Summary of approximation accuracy

	Proposed method		Conventional method	
	Vector Component	Vector Direction $[rad]$	Vector Component	Vector Direction $[rad]$
Mean square error (average)	5.5851×10^{-1}	1.6662×10^{-1}	2.8377	2.9134×10^{-1}
Mean square error (best)	3.0789×10^{-1}	9.7027×10^{-2}	3.6610×10^{-1}	1.2233×10^{-1}

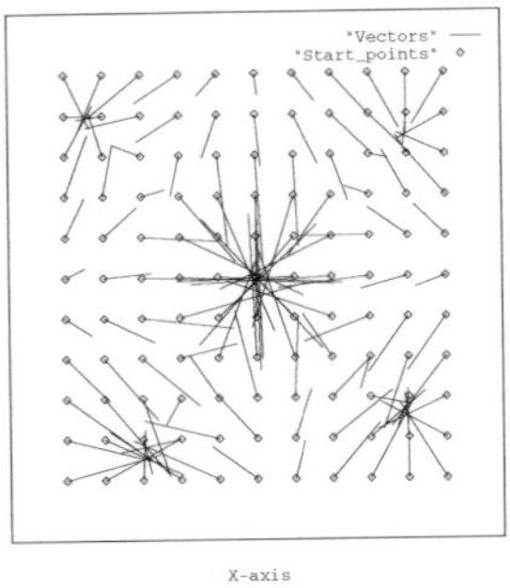

Fig. 10. Example data of the surface gradient vectors

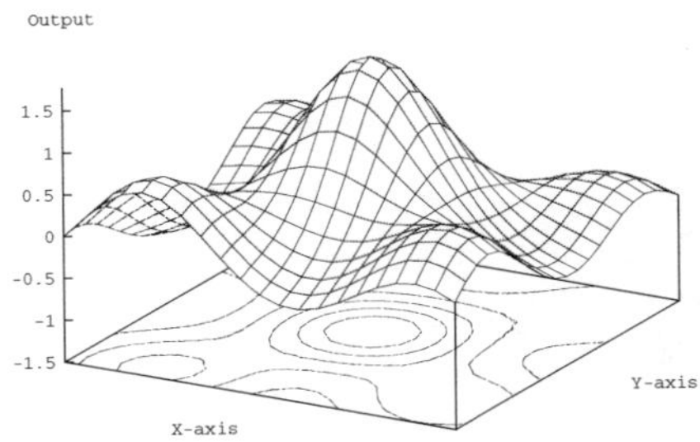

Fig. 11. Recovered Surface

data we recover the surface shape by using the proposed learning method. The obtained result is shown in Fig.11 in which the solid line is the target surface and the dashed line is the recovered surface and their contour lines are also shown. It is observed that the recovered surface agrees with its target well.

6 Conclusion

In this paper a method of approximating a vector field from a sparse set of sample data by training neural networks was discussed. We proposed a model inclusive learning of neural networks for approximating vector fields, in which the model of the inherent property of vector fields, "any vector field is composed of the sum of two vector fields, irrotational vector field and solenoidal vector field" is included into the problem formulation. The proposed method makes it possible to easily extract the component of irrotational vector field and that of solenoidal vector field from the resultant approximation of a target vector field.

We derived an efficient learning algorithm for the proposed formulation in a systematic way by introducing the adjoint model of neural networks. It should be stressed that the proposed learning algorithm is general and is easily implemented because all the steps of the algorithm is represented in terms of solving some networks with the same topology.

References

1. Rumelharlt, D.E., Hinton, G.E., Williams, R.J.: Learning Internal Representation by Error Backpropagation. In: Parallel Distributed Processing, vol. 1, MIT Press, Cambridge (1986)
2. Mussa-Ivaldi, F.A.: From basis functions to basis fields: vector field approximation from sparse data. Biological Cybernetics 67, 479–489 (1992)
3. Saehan, S., Iwagami, T., Kuroe, Y.: Learning Algorithm in Neural Networks Based on Adjoint Neuron Model Technical. Report. of IEICE, NC 90-30, pp. 1–9 (1989)
4. Kuroe, Y., Nakai, Y., Mori, T.: 'A Learning Method of Nonlinear Mappings by Neural Networks with Considering Their Derivatives. In: Proc. IJCNN, Nagoya,Japan, pp. 528–531 (1993)
5. Luenberger, D.G.: Introduction to Linear and Nonlinear Programming, pp. 194–197. Addison-Wesley, Reading (1973)

Spectral Measures for Kernel Matrices Comparison

Javier González and Alberto Muñoz

Universidad Carlos III de Madrid, c/ Madrid 126, 28903 Getafe, Spain
{javier.gonzalez,alberto.munoz}@uc3m.es

Abstract. With the emergence of data fusion techniques (kernel combinations, ensemble methods and boosting algorithms), the task of comparing distance/similarity/kernel matrices is becoming increasingly relevant. However, the choice of an appropriate metric for matrices involved in pattern recognition problems is far from trivial.

In this work we propose a general spectral framework to build metrics for matrix spaces. Within the general framework of matrix pencils, we propose a new metric for symmetric and semi-positive definite matrices, called Pencil Distance (PD). The generality of our approach is demonstrated by showing that the Kernel Alignment (KA) measure is a particular case of our spectral approach.

We illustrate the performance of the proposed measures using some classification problems.

Keywords: Kernel Matrix, Kernel Alignment, Procrustes, Simultaneous Diagonalization.

1 Introduction

In this work we propose a general spectral framework to build metrics to compare distance/similarity/kernel matrices. Increasingly interest has focused in the last years in the development of pattern recognition techniques that make use of several sources of information [9,8]. In this context, the comparison of matrices involved in pattern recognition is a key issue.

However, the choice of an appropriate metric for matrices is conditioned by the particular problem under study. For instance, consider the problem of kernel combination in the context of Support Vector Machine (SVM) classification [11]. In [9,10] the authors propose several SVM kernels following the expression $K = \frac{1}{M}\sum_{i=1}^{M} K_i + f(K_1, K_2, \ldots, K_M, y)$, where y is a vector containing the labels of the data points, and K_m are the kernel matrices to be combined ($K_m(i,j) = K_m(x_i, x_j)$, where x_i, $x_j \in X$, the sample). The function f depends on the differences of information among the K_i kernels. Thus, a metric to quantify such kernel matrices differences is needed. A concrete example arises when a set of web pages has to be classified [7]. In a case like this, the co-citation matrix and the terms by document matrix convey complementary information and should be combined. We need to quantify the common information between the two

J. Marques de Sá et al. (Eds.): ICANN 2007, Part I, LNCS 4668, pp. 727–736, 2007.

matrices in order to build an appropriate metric for this case that eliminates the existing redundancy.

The use of standard metrics inherited from Euclidean geometry may not be appropiate for many pattern recognition (PR) problems. For instance, consider a isometry in $\mathbb{R}^n$, given by a matrix A. Let K a kernel matrix evaluated on a finite data set. K transforms under A as $K' = AKA^{-1}$. In this case $d(K, K') = \|K - K'\| \neq 0$, but since we have simply performed a change of basis $d(K, K')$ should equal 0.

The use of eigenvalues tends to avoid these problems because the spectrum of a matrix is invariant under many common transformations in PR. An example is the **Spectral Norm** [4]. Given two kernels K_1 and K_2, then $d(K_1, K_2)$ is given by

$$d(K_1, K_2) = \sqrt{\lambda^1_{max}(K_1, K_2)} \tag{1}$$

where $\lambda^1_{max}(K_1, K_2)$ represents the largest eigenvalue of $K_1 - K_2$. In addition, even the well known Frobenius (Euclidean) norm can be bounded in terms of the eigenvalues of the involved matrices [12].

The usual similarity [1] measure for kernel matrices is the Kernel Alignment (KA) [2]. It can be interpreted as a measure of linear relationship between the corresponding kernel matrices. Given two kernel matrices K_1 and K_2 evaluated on the sample X, the empirical alignment is defined as

$$A(K_1, K_2) = \frac{\langle K_1, K_2 \rangle_F}{\sqrt{\langle K_1, K_1 \rangle_F \langle K_2, K_2 \rangle_F}}. \tag{2}$$

where $\langle K_1, K_2 \rangle_F = \sum_{i,j} K_1(i,j) K_2(i,j)$ represents the Frobenius product. This measure has several interesting properties and it has been used to optimize linear kernels combinations [7].

In this paper we study a new methodology for kernel comparison through the use of simultaneous diagonalization. This tool allows us to represent two or more kernels in the same base and go for new metric definitions.

This paper is organized as follows. In Section 2 a new kernel distance based on the simultaneous diagonalization of matrices is proposed. In Section 3 we show how the Kernel Alignment can be expressed in terms of the spectrum of the involved kernels. Section 4 illustrates how to use the spectral approach to detect redundant information between kernels. Finally, Section 5 concludes with some experimental results.

2 Pencil Distance

In this section we work with generalized eigenvalues. Given two matrices A and B, the generalized eigenvalues are the roots of the polynomial $det(A - \lambda B) = 0$ (while the eigenvalues of A are the roots of $det(A - \lambda I) = 0$).

[1] In the rest of the paper we will deal both with distance and dissimilarity measures. It is immediate to transform a distance into a similarity and conversely.

Next we afford the task of diagonalizing two kernels in the same base of vectors. The resulting eigenvalues will be used to define a new kernel distance measure.

2.1 Kernel Matrix Pencils

Definition 1. *Given two matrices A and B, the matrix-valued function $L(\lambda) = A - \lambda B$ is called* ***matrix pencil****. The Pencil is represented through the pair (A, B).*

Definition 2. *A pencil (A, B) is called* ***definite pencil*** *if the matrices A and B are symmetric and positive definite.*

Any kernel matrix K can be expressed by $K = \sum_i d_i v_i v_i^T$ where v_i are its eigenvectors and d_i are the corresponding eigenvalues. The equivalent process in pencil theory is simultaneous diagonalization. Since any matrix kernel is both symmetric and definite positive, (K_1, K_2) is a definite pencil. The following theorem ensures the existence of bases in which both kernels diagonalize.

Theorem 1. *[13] Let be (K_1, K_2) a definite pencil. then there is a nonsingular matrix V such that the matrices A and B can be simultaneously diagonalized.*

$$
\begin{aligned}
V^T K_1 V &= \Lambda \\
V^T K_2 V &= \Sigma
\end{aligned}
\tag{3}
$$

being $\Lambda = diag(\lambda_1, ..., \lambda_n)$, $\Sigma = diag(\sigma_1, ..., \sigma_n)$ and where the generalized eigenvalues λ_i/σ_i are real and finite.

Thus the diagonalized versions of K_1 and K_2 are, respectively, $K_1' = \Lambda$ and $K_2' = \Sigma$. Notice that the base of vectors given by the columns of V is not necessarily orthonormal. This base is not unique but the values λ_i/σ_i are invariant under the choice of the diagonalization basis. In many real applications it is interesting to perform simultaneous diagonalization for $\Sigma = I$ [3]. Several algorithms have been proposed for this case [6]. To solve the equation $K_1 v = \lambda K_2 v$ for λ we will calculate the eigenvalues of $K_2^{-1} K_1 v = \lambda v$. As we show in this section, only pencil eigenvalues are needed to build kernel metrics.

2.2 A Pencil Distance (PD) Measure for Matrices

The new similarity kernel measure is based on the generalized eigenvalues of the pencil (K_1, K_2). Since K_1 and K_2 are interchangeable, the measure should be invariant to the order. In our case $\Sigma = Id$, and we have $\lambda_i/\sigma_i = \lambda_i$. It is clear that, if the order of the matrices in the pencil changes, the corresponding new eigenvalues become $1/\lambda_i$. For this reason, any measure based on the numbers $\lambda_1, ..., \lambda_r$ should be invariant under reciprocication of the eigenvalues. Then, the values to consider are $\lambda_i^* = \frac{1+\lambda_i}{\sqrt{1+\lambda_i^2}}$, where we use the notation $\lambda_i^* = (1/\lambda_i)^*$.

Once K_1 and K_2 are expressed in the same base, we can define a a kernel distance in terms of the generalized eigenvalues:

$$PD(K_1, K_2) = \sum_{i=1}^{r} \left(\lambda_i^* - 2/\sqrt{2}\right)^2, \tag{4}$$

where r is the number of different from zero generalized eigenvalues. This **Pencil Distance** is equivalent to $\|\Lambda - I\|_F$ once the correction under reciprocication has been applied to the diagonal matrices Λ and I.

3 Kernel Alignment Is a Spectral Similarity

Consider the spectral decomposition of K_1 and K_2:

$$\begin{aligned} K_1 &= UD_1U^T = UD_1^{1/2}D_1^{1/2}U^T = U^*(U^*)^T \\ K_2 &= VD_2V^T = VD_2^{1/2}D_2^{1/2}V^T = V^*(V^*)^T. \end{aligned} \tag{5}$$

where $U^* = UD_1^{1/2}$ and $V^* = VD_2^{1/2}$. Canonical Correlations [5] can be applied in order to estimate the degree of similarity of U^* and V^*. This procedure calculates the angles between the spaces respectively generated by the columns of U^* and V^* by searching for maximal linear correlations over combinations of the variables. Unfortunately, if both kernels are full rank, the spanned spaces are the same and differences cannot be found. Nevertheless, the technique can be generalized calculating the sum of the cross correlations among the variables of the two basis. This can be done with the squared Frobenius norm of the matrix $(U^*)^TV^*$. Normalizing and rewriting in terms of the original decompositions we can define the following spectral similarity:

$$S_1(K_1, K_2) = \frac{\|D_1^{1/2}U^TVD_2^{1/2}\|_F^2}{\|D_1\|_F\|D_2\|_F} \tag{6}$$

Proposition 1. *The kernel alignment is equivalent to the S_1 measure.*

Proof. Let $K_1 = UD_1U^T$ and $K_2 = VD_2V^T$. Then:

$$\begin{aligned} S_1(K_1, K_2) &= \frac{\|D_1^{1/2}U^TVD_2^{1/2}\|_F^2}{\|D_1\|_F\|D_2\|_F} = \frac{\sum_{ij}\lambda_i\mu_j\left\langle u_i, v_j\right\rangle_F^2}{\sqrt{\sum_i\lambda_i^2}\sqrt{\sum_j\mu_j^2}} \\ &= \frac{\sum_{ij}\lambda_i\mu_j\left\langle u_i, v_j\right\rangle_F^2}{\sqrt{\sum_i\lambda_i\lambda_j\left\langle u_i, u_j\right\rangle}\sqrt{\sum_j\mu_i\mu_j\left\langle v_i, v_j\right\rangle}} \\ &= \frac{\left\langle\sum_i\lambda_iu_iu_i^T, \sum_j\mu_jv_jv_j^T\right\rangle}{\sqrt{\sum_i\lambda_i\lambda_j\left\langle u_iu_i^T, u_ju_j^T\right\rangle}\sqrt{\sum_j\mu_i\mu_j\left\langle v_iv_i^T, v_jv_j^T\right\rangle}} \\ &= \frac{\left\langle K_1, K_2\right\rangle_F}{\sqrt{\left\langle K_1, K_1\right\rangle_F\left\langle K_2, K_2\right\rangle_F}} = A(K_1, K_2) \end{aligned} \tag{7}$$

4 Analysis of Kernel Redundances Based on Simultaneous Diagonalization

In Data Fusion is essential to work with procedures that eliminates redundant information when two or more sources of information have to be combined. Simultaneous Diagonalization and Joint Diagonalization (in the case of more than two kernels) provide a thecretical framework for this purpose.

Working with Matrix Pencils involves the computation of a base on vectors in which both kernels diagonalize and more information about the problem can be obtained. The use of kernel matrices eigenvalues has proven to be useful for the definitions of kernel distances. Eigenvalues can also be used to obtain information about the data set of variables. For instance, in Principal Component Analysis (PCA) a diagonalization of the covariance matrix is performed in order to obtain more representative components of the data. In this case, the eigenvalues can be interpreted as a the weights of the new variables.

A detailed analysis of the kernels redundancy can be done. The diagonalization of K_1 and K_2, produces two matrices $\Lambda = V^T K_1 V$ and $\Sigma = V^T K_2 V$. In this analysis, $V = \{v_1, ..., v_n\}$ is the new base where both kernels diagonalize. The components of matrices $\Lambda = diag(\lambda_1, ..., \lambda_n)$ and $\Sigma = diag(\sigma_1, ..., \sigma_n)$ can be interpreted, generalizing the previous PCA example:

- $\lambda_i = 0$ and $\sigma_i = 0$: the vector v_i is irrelevant to both kernels structures. That is, the i-th variable is in the null space of both K_1 and K_2.
- $\lambda_i = 0$ and $\sigma_i \neq 0$: in this case v_i is in the null space of K_1 but it is a relevant component for K_2. Similar considerations apply for the case $\lambda_i \neq 0$ and $\sigma_i = 0$.
- $\lambda_i \neq$ and $\sigma_i \neq 0$: here v_i is a variable of the new base relevant for both kernels. The generalized eigenvalue λ_i/σ_i measures relative weight that the metric induced by K_1 gives tc the i-th variable in relation to K_2 when using the common base to express the data points.

5 Experiments

In this section we study the validity of the spectral approach in two ways. First, we show that the Pencil Distance performs as good as the kernel alignment in an illustrative example where the behaviour of the kernels in terms of the data structure is known. In a second example, an analysis of redundant information in kernels is performed applying simultaneous diagonalization.

5.1 Body Mass Index (BMI) Example

The Body Mass Index (BMI) is a corporal index based on the weight and height of persons. It is a fast and inexpensive method for the assesment of overweight given by $weight/height^2$ (using kilograms and meters). The BMI induces the following taxonomy in human beings:

– Below 20: Underweight.
– 20-25: Normal.
– Above-30: Overweight.

In this experiment we consider a sample of 150 data with three apparent clusters. The BMI averages for each cluster are, respectively, 18, 22.5 and 28.5 (see Figure 1).

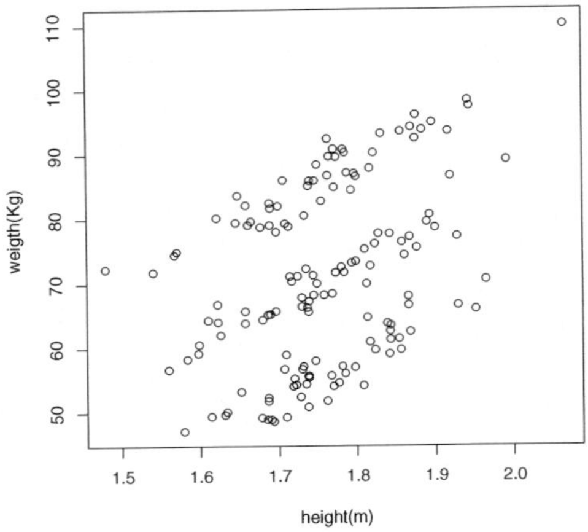

Fig. 1. Body Mass Index example data

For the present case, six representations of the data are used. They are summarized in Table 1. The goal of this experiment is to compare the mesures PD and $KA = S_1$ in a case in which six linear kernels $K_1, ..., K_6$ were calculated according the representations of Table 1 (Figure 1 corresponds to K_2). The example is favorable for the use of KA, because only linear transformations are involved and KA, being a correlation measure, is invariant under linear transformations.

Table 1. Six different representations of the data

Representation	Variables	units
1	(weigth, heigth)	kilos(normalized), meters(normalized)
2	(weigth, heigth)	kilos, meters
3	(weigth, heigth)	grams, meters
4	(weigth, heigth)	grams, centimeters
5	(weigth, heigth)	Mahalanobis transformation
6	$MBI = kg/m^2$	None

In this example, the reference kernel is assumed to be the one calculated with the BMI because it perfectly evidences the cluster structure of the data. In other words, some representations on Table 1 may be affected by the choice of the units, but the BMI is independent of the unit scaling. In order to estimate the similarity between the kernels $K_1, ..., K_5$ with K_6, a k-means algorithm ($k = 3$) was applied to the six data representations. After clustering, the number of points that were missclassified with respect to taxonomy induced by K_6 is summarized in Table 2.

Table 2. Missclassified points for the six kernel representations with respect to the three true clusters calculated using the BMI

Kernel	**1**	**2**	**3**	**4**	**5**	**6**
Errors	60	29	98	29	0	0

Based on Table 2, the ranking of kernels (regarding their similarity to K_6), should be (begining with the most similar) $K_5 \rightarrow (K_4 = K_2) \rightarrow K_1 \rightarrow K_3$ in decreasing order. Given that the measures involved in the kernels calculations are not bounded, it is convenient to perform some previous normalization. Thus we produce two standarized versions for each kernel K_i, $i = 1, ..., 6$ given by:

$$K_i^A(\mathbf{x}, \mathbf{y}) = \frac{K_i(\mathbf{x}, \mathbf{y}) - min(K_i(\mathbf{x}, \mathbf{y}))}{max(K_i(\mathbf{x}, \mathbf{y})) - min(K_i(\mathbf{x}, \mathbf{y}))} \quad (8)$$

$$K_i^B(\mathbf{x}, \mathbf{y}) = \frac{K_i(\mathbf{x}, \mathbf{y})}{\sqrt{K_i(\mathbf{x}, \mathbf{x})}\sqrt{K_i(\mathbf{y}, \mathbf{y})}} \quad (9)$$

Results of the experiment are shown in Table 3. It is clear that, kernel alignment and PD are equivalent. Since KA is a similarity and PD a distance, the corresponding values of the table should be interpreted in a diferent way: Kernels close to K_6 should show large values for the alignment and small values for the PD.

Table 3. Results for the three measures over the battery of 6 kernels with respect to K_6. Two normalization were applied to the kernels.

Norm.	**Measure**	K_1	K_2	K_3	K_4	K_5	K_6
A	**KA**	0.961	0.973	0.940	0.973	0.993	1.000
	PD	0.049	0.011	0.159	0.011	0.000	0.000
B	**KA**	0.395	0.259	0.007	0.256	0.741	1.000
	PD	0.030	0.038	0.119	0.038	0.001	0.000

In order to show graphically the concordance of the results for alignment and PD, Figure 2 shows a scatterplot of the values for each kernel. The relationship is non linear but it is always decreasing, just as one should expect for this example. Thus, we can concluded that DP performs as well as KA, the best available measure for this particular example.

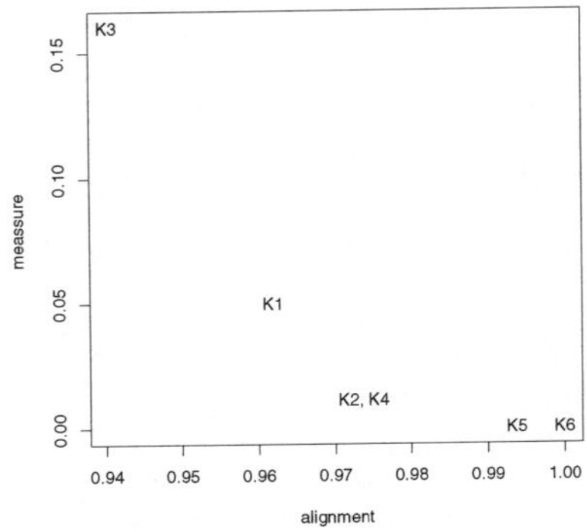

Fig. 2. Scatterplot of the alignment and the new measure for the six kernels. This representation corresponds to first method of normalization 1.

5.2 Example 2

In order to validate the effectiveness of the simultaneous diagonalization detecting possible redundancies among kernels, the following experiment was developed. Consider a data base with 5 observations and three orthogonal variables x_1, x_2 and x_3. Let K_1, K_2 and K_3 be three linear kernels calculated using the variable sets $\{x_1\}$, $\{x_1, x_2\}$ and $\{x_2, x_3\}$ respectively. Notice that in this example K_1 and K_3 are based on independent variable sets while K_2 and K_3 share the variable x_2.

The goal of this experiment is to show the utility of the spectral approach (via simultaneous diagonalization) to detect the redundances when using the kernel battery $\{K_1, K_2, K_3\}$. First, we perform simultaneous diagonalization with K_1 and K_2. This produces a common base of eigenvectors $V = \{v_1, v_2, v_3, v_4, v_5\}$ and two diagonal matrices $D_1 = \Sigma = diag(11.58, 0, 0, 0, 0)$ and $D_2 = \Lambda = diag(0, 0, 0, 3.33, 1.19)$, where $span\{v_1, v_4, v_5\} = span\{x_1, x_2, x_3\}$. Let K be the kernel defined by $K = V^T(D_1 + D_2)V = K_1 + K_2$. Then K is the linear kernel calculated using the whole variable set $\{x_1, x_2, x_3\}$, and is the (direct sum) of K_1 and K_2, as one can expect because there is no redundancy.

Now consider the diagonalization with kernels K_2 and K_3 where the variable x_2 is present in both kernels. The new diagonal matrices are $D_1' = diag(11.58, 0, 0.33, 0, 0)$ and $D_2 = diag(0, 0, 0.33, 0, 1.19)$. In this new diagonalization neither the base V of eigenvectors nor D_2 change. Now $K_1 + K_2$ does not match the original K because $K_1 + K_2$ gives variable v_3 twice its weight (the sum doubles the eigenvalue). A general expresion for recovering K from the information in this example is $K^* = V^T D^* V$ where D^* is a diagonal matrix such that $d_{i,i} = max(\lambda_i, \sigma_i)$. This expression makes $K = K^*$ and includes the previous example as a particular case, allowing for a new framework for data fusion, that will be studied in the next future.

6 Conclusions and Future Work

In this paper we propose an spectral framework for the definition of metrics for matrix spaces. We propose a new Pencil Distance (PD) based on the simultaneous diagonalization of kernel matrices. The new measure is easy to calculate and is proven to be consistent with the Kernel Alignment in a significant example. In addition, simultaneous diagonalization is used for the task of detecting redundant information in data fusion. Given two kernels, redundancies can be detected through the use eigenvalues and the composition of better kernels is possible. Future work will include the use of the ideas of this paper to define an algebra of kernels, where operations such as sum and difference of kernels can be defined in order to define new kernel combination techniques in the context of classification problems.

Acknowledgments

This work was partially supported by Spanish Goverment grants 2006-03563-001, 2004-02934-001/002 and Madrid Goverment grant 2007-04084-001.

References

1. Bach, F.R., Jordan, M.I.: Kernel Principal Components Analysis. Journal of Machine Learning Research 3, 1–48 (2002)
2. Cristianini, N., Shawe-Taylor, J.: On the Kernel Target Alignment. Journal of Machine Learning Research 5, 27–72 (2002)
3. Epifanio, I., Gutierrez, J., Malo, J.: Linear transform for simultaneous diagonalization of covariance and perceptual metric matrix in image coding. Pattern Recognition 36, 799–1811 (2003)
4. Golub, G., Loan, C.V.: Matrix Computations. University Press, Baltimore (1997)
5. Hotelling, H.: Relation between two Sets of Variables. Biometrika 28, 32177 (1936)
6. Hua, Y.: On SVD estimating Generalized Eigenvalues of Singular Matrix Pencils in Noise. IEEE Transactions on Signal Processing 39(4), 892–900 (1991)
7. Joachims, T., Cristianini, N., Shawe-Taylor, J.: Composite Kernels for Hipertext Categorisasion. In: Proceedings of the International Conference of Machine Learning, pp. 250–257 (2002)
8. Lanckriet, G.R.G., Bartlett, P., Cristianini, N., Ghaoui, L., E y Jordan, M.I.: Learning the Kernel Matrix with Semidefinite Programming. Journal of Machine Learning Research 5, 27–72 (2002)
9. de Diego, I.M., Muñoz, A., Moguerza, J. M.: On the Combination of Kernels for Support Vector Classifiers. Working paper, 05-45-08, Statistics and Econometrics Series, University Carlos III (2005)
10. de Diego, I.M., Moguerza, J.M., Muñoz, A.: Combining Kernel Information for Support Vector Classification. In: Roli, F., Kittler, J., Windeatt, T. (eds.) MCS 2004. LNCS, vol. 3077, pp. 102–111. Springer, Heidelberg (2004)
11. Moguerza, J., Muñoz, A.: Support Vector Machines with Applications. Statistical Science 21(3), 322–336 (2006)

12. Omladic, M., Semrl, P.: On the distance between Normal Matrices. Proceedings of the American Mathematical Society 110, 591–596 (1990)
13. Parlett, N.: Beresford. The Symmetric Eigenvalue Problem. Classics in Applied Mathematics. SIAM (1997)
14. Schölkopf, B., Smola, A., Muller, K.R.: Kernel Principal Component Analysis. In: Advanced in Kernel Methods-Support Vector Learning, pp. 327–352. MIT Press, Cambridge (1999)
15. Schölkopf, B., Smola, A.J., Müller, K.R.: Nonlinear Component Analysis as a Kernel Eigenvalue Problem. Neural Computation 10, 1299–1319 (1998)

A Novel and Efficient Method for Testing Non Linear Separability

David Elizondo[1], Juan Miguel Ortiz-de-Lazcano-Lobato[2], and Ralph Birkenhead[1]

[1] School of Computing
De Montfort University
The Gateway
Leicester, LE1 9BH
United Kingdom
{elizondo,rab}@dmu.ac.uk
[2] School of Computing
University of Málaga
Málaga, Spain
jmortiz@lcc.uma.es

Abstract. The notion of linear separability is widely used in machine learning research. Learning algorithms that use this concept to learn include neural networks (Single Layer Perceptron and Recursive Deterministic Perceptron), and kernel machines (Support Vector Machines). Several algorithms for testing linear separability exist. Some of these methods are computationally intense. Also, several of them will converge if the classes are linearly separable, but will fail to converge otherwise. A fast and efficient test for non linear separability is proposed which can be used to pretest classification data sets for non linear separability thus avoiding expensive computations. This test is based on the convex hull separability method but does not require the computation of the convex hull.

1 Introduction

Classification type problems can be tackled using neural networks. Linear models, where they can be used, are robust against noise and most likely will not overfit. Non linear models (such as backpropagation) work well where the data is not linearly separable.

Two sub-sets X and Y of $\mathbb{R}^d$ are said to be linearly separable (LS) if there exists a hyperplane P of $\mathbb{R}^d$ such that the elements of X and those of Y lie on opposite sides of it. Figure (1) shows an example of both a LS (a) and a NLS (b) set of points. Squares and circles denote the two classes. A study on several methods for testing linear separability and their complexities is given in [1].

Linear separability is an important topic in the domain of machine learning. In real applications, the data is often linearly separable. For such problems, using backpropagation is an overkill, with thousands of iterations needed to get to the point where linear separation can bring us fast. Furthermore, multilayer linear neural networks, such as the Recursive Deterministic Perceptron [2,3] can always linearly separate, in a deterministic way, two or more classes (even if the two classes are not linearly separable). The idea behind the construction of a RDP is to augment the affine dimension of the input

J. Marques de Sá et al. (Eds.): ICANN 2007, Part I, LNCS 4668, pp. 737–746, 2007.

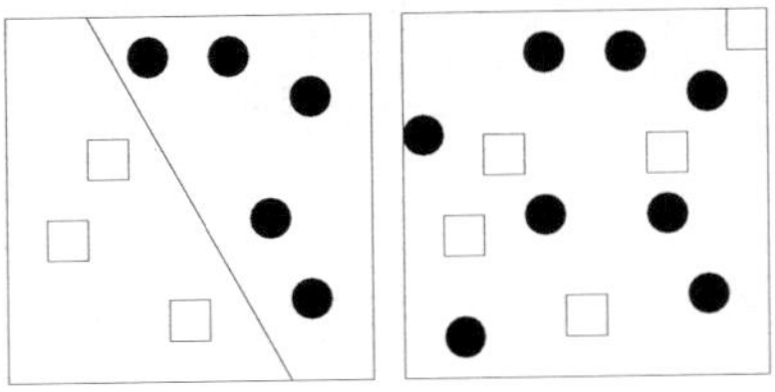

Fig. 1. LS (a) and a non-LS (b) set of points

vector by adding to these vectors the outputs of a sequence of intermediate neurons as new components. Each intermediate neuron corresponds to a single layer perceptron and it is characterised by a hyperplane which linearly separates a LS subset, taken from the non-LS (NLS) original set, and the remaining points in the input vector.

This paper presents a novel and efficient test for non linear separability which can be used to pretest classification data sets. By performing this test, more expensive computations, carried out by the methods for testing linear separability, can be avoided. Also, some linear separability methods work well when the given classification problem is linearly separable, but can run into very large calculations for non linearly separable problems. This test is based on the convex hull separability method but does not require the calculation of the convex hull. The paper is divided into five sections. In the second section some standard notations and definitions together with some general properties related to them are given. Section three describes the fast method for testing non linear separability. Results obtained on classification data using the proposed fast method and the convex hull classification methods sets are presented and compared in section four. Finally some conclusions are given in section five.

2 Preliminaries

The following standard notions are used ([3]):

Let $E, F \subset I\!R^d$, $\boldsymbol{u}, \boldsymbol{v}, \boldsymbol{w} \in I\!R^d$ and $r, t \in I\!R$,

- $Card(E)$ stands for the cardinality or number of elements of a set E. For example, the cardinality of the set $E = \{(1,2), (-1,2.5), (2.3,-2), (3.2,1.2), (3.4,0.1)\}$ is equal to five.
- $E \setminus F$ is the set of elements which belong to E and do not belong to F. If $E = \{(1,2), (-1,2.5), (2.3,-2), (3.2,1.2), (3.4,0.1)\}$ and $F = \{(1,2), (-1,2.5), (2.2,-1.3)\}$, then $E \setminus F$ is equal to $\{(2.3,-2), (3.2,1.2), (3.4,0.1)\}$.
- $E \oplus F$ is the set of elements of the form $\boldsymbol{e} + \boldsymbol{f}$ with $\boldsymbol{e} \in E$ and $\boldsymbol{f} \in F$. If $E = \{(1,2), (-1,2.5)\}$ and $F = \{(2.2,3), (3.1,1)\}$, then $E \oplus F$ is equal to $\{(3.2,5), (4.1,3), (1.2,5.5), (2.1,3.5)\}$.
- $E \ominus F$ stands for $E \oplus -(F)$. If $E = \{(1,2), (-1,2.5)\}$ and $F = \{(2.2,3), (3.1,1)\}$, then $E \ominus F$ corresponds to $\{(-1.2,-1), (-2.1,1), (-3.2,-0.5), (-4.1,1.5)\}$.

The fact that two sub-sets X and Y of $I\!R^d$ are linearly separable is denoted by $X \mid\mid Y$ or $X \mid\mid Y\ (P)$ or $X \mid\mid_P Y$. Thus if $X \mid\mid Y\ (\mathcal{P}(\boldsymbol{w}, t))$, then ($\forall \boldsymbol{x} \in X,\ \boldsymbol{w}\boldsymbol{x}^T + t > 0$ and $\forall \boldsymbol{y} \in Y,\ \boldsymbol{w}\boldsymbol{y}^T + t < 0$) or ($\forall \boldsymbol{x} \in X,\ \boldsymbol{w}\boldsymbol{x}^T + t < 0$ and $\forall \boldsymbol{y} \in Y,\ \boldsymbol{w}\boldsymbol{y}^T + t > 0$).

2.1 Definitions and Properties

For the theoretical representations, column vectors are used to represent points. However, for reasons of saving space, where concrete examples are give, row vectors are used to represent points. We also introduce the notions of convex hull and linear separability.

Definition 1. *A subset D of $I\!R^d$ is said to be convex if, for any two points $\mathbf{p}_1$ and $\mathbf{p}_2$ in D, the segment $[\mathbf{p}_1, \mathbf{p}_2]$ is entirely contained in D.*

Definition 2. *Let S be a subset of $I\!R^d$, the convex hull of S, denoted by $CH(S)$, is the smallest convex subset of $I\!R^d$ containing S.*

Property 1. [4] Let S be a subset of $I\!R^d$
• $CH(S) = \{t_1\boldsymbol{x}_1 + \ldots + t_k\boldsymbol{x}_k \mid \boldsymbol{x}_1, \ldots, \boldsymbol{x}_k \in S,\ t_1, \ldots, t_k \in [0,1]\ \ and\ \ t_1 + \ldots + t_k = 1\}$.
• If S is finite, then there exists $\boldsymbol{a}_1, \ldots, \boldsymbol{a}_k \in I\!R^d$ and $b_1, ..., b_k \in I\!R$ such that $CH(S) = \{\boldsymbol{x} \in I\!R^d \mid \boldsymbol{a}_i^T\boldsymbol{x} \geq b_i\ \ for\ \ 1 \leq i \leq k\}$. Thus, $CH(S)$ is the intersection of k half spaces.

2.2 The Convex Hull Method for Testing Linear Separability

Lemma 1. *Let $X, Y \subset I\!R^d, X \parallel Y\ (P) \Leftrightarrow CH(X) \parallel CH(Y)\ (P)$.*

Lemma 2. *Let $X, Y \subset I\!R^d$, then $-CH(X) = CH(-X)$ and $CH(X) \oplus CH(Y) = CH(X \oplus Y)$. Thus, $CH(X \ominus Y) = CH(X) \ominus CH(Y)$.*

Lemma 3. *Let X, Y be two finite subsets of $I\!R^d$, $CH(X) \parallel CH(Y) \Leftrightarrow CH(X) \cap CH(Y) = \emptyset$.*

Theorem 1. *Let X, Y be two finite subsets of $I\!R^d$, then $X \parallel Y$ iff $CH(X) \cap CH(Y) = \emptyset$.*

Property 1. *Let S be a finite subset of $I\!R^d$, then there exists $\boldsymbol{x} \in S$ such that $S - \{\boldsymbol{x}\} \parallel \{\boldsymbol{x}\}$.*

This property states that, there exists a point $\boldsymbol{x} \in S$ which is LS from the rest of the points in S.

Remark
The lemmas 1, 2, and 3 and the property 1 are intuitively evident and their proofs are not very important for the global comprehension of this work. A more general result than the one shown in theorem 1 can be found in [5]

Property 2. *Let X, Y be a finite subset of $I\!R^d$, and assume that $CH(X \ominus Y) = \{\boldsymbol{x} \in I\!R^n \mid \boldsymbol{v}_i^T\boldsymbol{x} \geq b_i\}$ then,*

- *if for all i $b_i \leq 0$, then $X \not\parallel Y$,*
- *if there exists i such that $b_i > 0$ then, $X \parallel Y\ (\mathcal{P}(\boldsymbol{v_i}, t))$, where $t = -\frac{\beta+\alpha}{2}$, $\alpha = Max(\{\boldsymbol{v}_i^T\boldsymbol{y} \mid \boldsymbol{y} \in Y\})$, and $\beta = Min(\{\boldsymbol{v}_i^T\boldsymbol{x} \mid \boldsymbol{x} \in X\}$.*

Proof

- if for all i, $b_i \leq 0$ then, $\mathbf{0} \in CH(X \ominus Y)$, thus, $CH(X) \cap CH(Y) \neq \emptyset$, and thus, by the theorem 1, it can be concluded that $X \not\parallel Y$,
- if there exists i such that $b_i > 0$ then, for every $\boldsymbol{x} \in X$ and $\boldsymbol{y} \in Y$ then $\boldsymbol{v}_i^T(\boldsymbol{x} - \boldsymbol{y}) \geq b_i > 0$, that is $\boldsymbol{v}_i^T\boldsymbol{x} > \boldsymbol{v}_i^T\boldsymbol{y}$; let $\alpha = Max(\{\boldsymbol{v}_i^T\boldsymbol{y} \mid \boldsymbol{y} \in Y\})$, and $\beta = Min(\{\boldsymbol{v}_i^T\boldsymbol{x} \mid \boldsymbol{x} \in X\}$, so, $\alpha < \beta$
  thus, $\forall \boldsymbol{y} \in Y, \boldsymbol{v}_i^T\boldsymbol{y} - \frac{\alpha+\beta}{2} \leq \alpha - \frac{\alpha+\beta}{2} = \frac{\alpha-\beta}{2} < 0$,
  $\forall \boldsymbol{x} \in X, \boldsymbol{v}_i^T\boldsymbol{x} - \frac{\alpha+\beta}{2} \geq \beta - \frac{\alpha+\beta}{2} = \frac{\beta-\alpha}{2} > 0$,
  then, $X \parallel Y$ $(\mathcal{P}(\boldsymbol{v}_i, -\frac{\alpha+\beta}{2}))$.

The algorithm 1 ([3,6]) is based in this property.

Given $X, Y \subset I\!R^d$, the convex hull separability algorithm, presented in table 1, computes $\boldsymbol{w} \in I\!R^d$ and $t \in I\!R$ *(if they exist)* such that $X \parallel Y$ $(\mathcal{P}(\boldsymbol{w}, t))$. Where w and t represent the weight vector and threshold value which make the hyperplane that linearly separates X and Y. To start with, the convex hull $CH(X \ominus Y)$ can be defined as the set of points which verify the constraints $\{\boldsymbol{x} \in I\!R^n \mid \boldsymbol{v}_i^T\boldsymbol{x} \geq b_i$ **for** $1 \leq i \leq k\}$ where k represents the number of vertices in $CH(X \ominus Y)$. Two cases are then possible:

1. for all i $b_i \leq 0$ which means that the two classes are NLS,
2. there exists i such that $b_i > 0$, which means that the two classes are LS: with this value of b_i the values of α and β are then computed. They correspond respectively to: the maximum value of the vector product of $\boldsymbol{v}_i^T\boldsymbol{y}$ with $\boldsymbol{y}$ belonging to the finite set of points of the second class Y, and the minimum value of the vector product of $\boldsymbol{v}_i^T\boldsymbol{x}$ with $\boldsymbol{x}$ belonging to the finite set of points of the first class X. These two values are then used to compute the threshold value t which is defined by $-\frac{\beta+\alpha}{2}$. The set of weights $\boldsymbol{w}$ that represents the perceptron which linearly separates the two classes X and Y, corresponds to $\boldsymbol{v}_i$.

The 2 class 2 dimension classification problem shown in figure 2-a will be used to illustrate the Convex Hull separability algorithm.

Table 1. The Convex Hull separability algorithm

CHSA(X, Y, $\boldsymbol{w}$, t)
– data: two data set vectors, X and Y representing two classes
– result: a weight vector, $\boldsymbol{w}$ and a threshold value t which separates the two classes if $X \parallel Y$)
Begin
Compute $\boldsymbol{v}_1, \ldots, \boldsymbol{v}_k \in I\!R^d$, $b_1, \ldots, b_k \in I\!R$ such that $CH(X \ominus Y) = \{\boldsymbol{z} \in I\!R^n \mid \boldsymbol{v}_i^T\boldsymbol{z} \geq b_i$ for $1 \leq i \leq k\}$.
If $(\forall i \leq k,\ b_i \leq 0)$ **Then** not($X \parallel Y$)
Else
  Begin
    chose i such that $b_i > 0$;
    $\alpha := Max(\{\boldsymbol{v}_i^T\boldsymbol{y} \mid \boldsymbol{y} \in Y\})$;
    $\beta := Min(\{\boldsymbol{v}_i^T\boldsymbol{x} \mid \boldsymbol{x} \in X\}$;
    $\boldsymbol{w} := \boldsymbol{v}_i$;
    $t := -\frac{\beta+\alpha}{2}$;
    (i.e. $X \parallel Y$ $(\mathcal{P}(\boldsymbol{w}, t))$ *)*
  End;
End

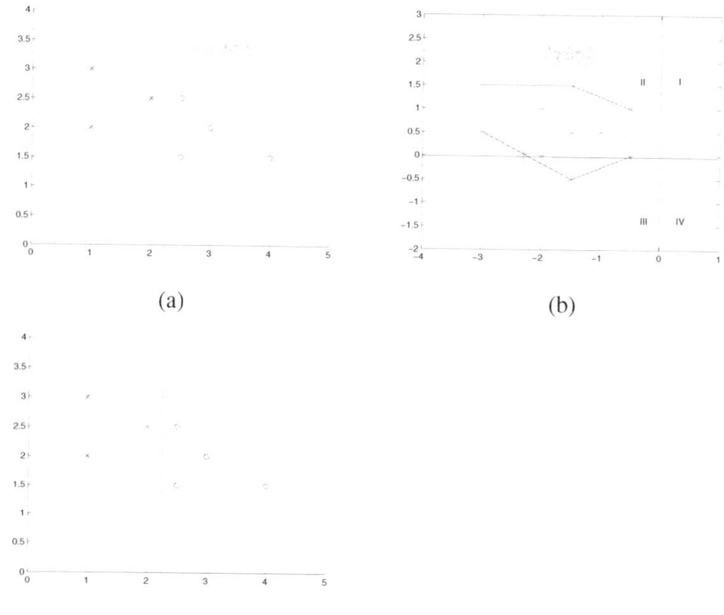

Fig. 2. (a) XY PLOT of the 2-D classification problemLS, (b) Convex Hull of $(X \ominus Y)$ of the 2-D classification problem (I, II, III, IV represent the 1st, 2nd, 3rd and 4th quadrants respectively), (c) Hyperplane that linearly separates the 2-D classification problem (dotted line)

Let $X = \{(1,2),(2,2.5),(1,3)\}$ and $Y = \{(2.5,1.5),(2.5,2.5),(3,2),(4,1.5)\}$ represent the input patterns for the two classes which define these classification problem. The next task is to find out if $X \parallel Y$.

To start with:

$$(X \ominus Y) = \{(-1.5,0.5),(-1.5,-0.5),(-2,0),(-3,0.5),\\ (-0.5,1),(-0.5,0),(-1,0.5),(-2,1),\\ (-1.5,1.5),(-1.5,0.5),(-2,1),(-3,1.5)\}$$

has to be calculated.

The convex hull of $(X \ominus Y)$ is then calculated:

$$CH(X \ominus Y) = \{(-0.5,0),(-0.5,1),(-1.5,1.5),\\ (-3,1.5),(-3,0.5),(-1.5,-0.5)\}$$

Thus, k, the number of vertices in the CH, is equal to 6. Next, all the vectors $\boldsymbol{v}_k$ which are perpendicular to each of the facets that form the CH together with the corresponding values of b_k are computed (see figure 2-b).

$$\begin{aligned} \boldsymbol{v}_0 &= (-1,\ 0)\ b_0 = 0.5, & \boldsymbol{v}_1 &= (-1,-1)\ b_1 = -0.5 \\ \boldsymbol{v}_2 &= (\ 0,-1)\ b_2 = -1.5, & \boldsymbol{v}_3 &= (\ 1,\ 0)\ b_3 = -3.0 \\ \boldsymbol{v}_4 &= (\ 2,\ 3)\ b_4 = -4.5, & \boldsymbol{v}_5 &= (-1,\ 1)\ b_5 = 0.5 \end{aligned}$$

Any of the $\boldsymbol{v}_k, b_k$ couples having a value $b_k > 0$ ($k = 0$ and 5 in this example) can now be selected. $k = 0$ was selected. Thus, the values of α and β can now be calculated.

$$\alpha := Max(\{\boldsymbol{v}_i^T \boldsymbol{y} \mid \boldsymbol{y} \in Y\}) := -2.5$$
$$\beta := Min(\{\boldsymbol{v}_i^T \boldsymbol{x} \mid \boldsymbol{x} \in X\} := -2$$

The values of $\boldsymbol{w}$ and t can now be computed:

$$\boldsymbol{w} := \boldsymbol{v}_i \quad := (-1, 0)$$
$$t := -\frac{\beta+\alpha}{2} := 2.25$$

Figure 2-c shows that this hyperplane linearly separates the two classes X and Y.

There are other methods based on computational geometry that can be found in the literature for testing linear separability. An example of this is the Open Hemisphere method [7].

3 Novel Efficient Method for Testing Non Linear Separability

This section describes a fast method for testing classification data sets for non linear separability. The method is based on the convex hull separability algorithm described in the previous section. Two sets X, Y are not linearly separable if the origin is inside the convex hull $CH(X \ominus Y)$. A way to find out if the origin is inside the convex hull is to calculate the convex hull itself. This can be an expensive calculation specially for large data sets in high dimensions. Instead of performing this expensive calculation a fast way of finding if an n dimensional classification problem is non linearly separable is to check if there exists at least one point of $CH(X \ominus Y)$ in each of the 2^n quadrants in n dimensions of the input space. If this condition is true, it can safely be concluded that the two classes are non linearly separable. Otherwise, no conclusion can be made, and further analysis is required.

The following steps summarise an algorithm for implementing the method for testing non linear separability:

1.- Obtain the difference set *Dif* by computing $X \ominus Y$.
2.- Convert all elements of *Dif* to binary. The value zero is used as threshold.
3.- Calculate a set of unique binary values by means of removing the repeated binarized vectors.
4.- If the number of samples in this unique set equals the number of quadrants in the input space then the method finishes and states the datasets are non linearly separable. Otherwise, the method finishes and states nothing about the non separability.

4 Experimental Setup

Ninety linearly separable and a ninety non linearly separable data sets, containing samples for two classes, were generated (figure 3(a,b)). The input dimension of the data sets was varied from two all the way to ten dimensions with increments of one. Ten data sets were generated for each input dimension. Each data set contained 200 samples for dimensions less than eight. For dimensions of eight and greater, 2^n samples were generated, where n represents the input space dimension. The method is not restricted to the number of samples but larger samples allow for a more thorough comparison. The data sets were tested for non-linear/linear separability using the convex hull, and the perceptron neural network algorithms. They were also tested for non linear separability using the quadrant method. The choice of convex hull algorithm came from the fact that the quadrant method is based on this method. The more standard perceptron neural network algorithm ([8]) was used as a reference. Two different implementations of the quadrant method were used, but, because of lack of space, results are only given for the method which was more efficient and consistent. The three methods were also tested using ninety non linearly separable data sets based on two concentric and independent hyper spheres (figure 3(c)). The results obtained with these data sets were very similar to the ones given in table 2 and therefore were not included. The meshed spiral 2D classification problem [9] was also used for comparison purposes (figure 3(d)).

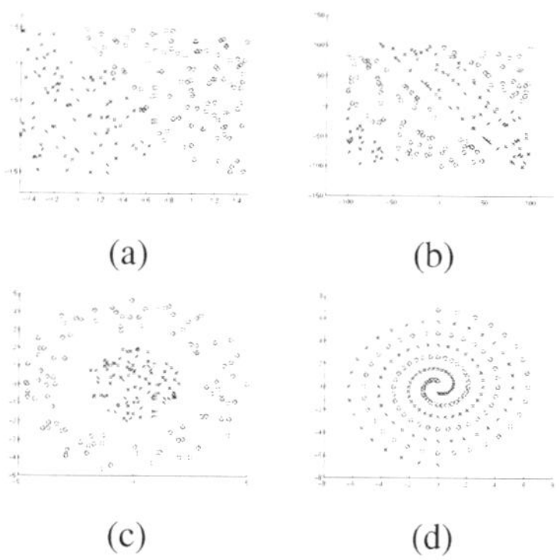

Fig. 3. LS (a) and non-LS (b,c,d) data sets

5 Results and Conclusions

A comparison, based on their time of convergence, of the three methods for testing for non linear separability (convex hull, perceptron and quadrant) is now given. Tables 2 and 3 present the results in terms of the average convergence time for ten data sets generated for each dimension (2 to 10 dimensions) like figures 3(a) and 3(b).

It can be clearly seen that the quadrant method gives a dramatic time improvement both for the linearly separable and the non linearly separable data sets. For the non linearly separable data sets, the quadrant method performs 70 and up to over 3500 times faster than the convex hull method and 30 and up to over 2000 times faster than the perceptron method. A similar tendency, with a smaller gain in time, can be observed with the linear data sets were the quadrant method is 1.5 and up to 30 times faster than the convex hull method, and 0.5 to up to 230 times faster than the perceptron method. This can be explained by the fact that both the convex hull and the perceptron algorithms will converge if the two classes are linearly separable. Both these algorithms depend on a number of iterations, specified by the user, being reached before concluding that two classes are not linearly separable. This leaves open the possibility of the two classes being linearly separable, but neither of the two algorithms finding a solution because the number of maximum iterations has been reached.If the number of iterations is not restricted, both algorithms will end in an infinite loop for non linearly separable data sets.

Table 2. Results obtained with the three non linear separability testing methods and ten,non linearly separable data sets, for two to ten dimensional data sets in terms of the convergence time (seconds) using 3(a)

	Quadrant				Convex Hull				Perceptron			
Dim	Max	Min	Mean	Std	Max	Min	Mean	Std	Max	Min	Mean	Std
2	0.07	0.01	0.02	0.02	105.72	0	31.66	50.96	27.15	26.52	26.83	0.19
3	0.02	0.01	0.01	0	106.31	0.01	21.26	44.8	27.15	26.52	26.83	0.19
4	0.03	0.02	0.02	0	106.96	0.01	52.88	55.74	27.65	27.03	27.4	0.19
5	0.03	0.03	0.03	0	123.95	0.01	96.78	34.53	27.66	26.73	27.36	0.29
6	0.05	0.03	0.03	0.01	107.61	0.01	74.07	51.12	27.9	27.36	27.61	0.18
7	0.04	0.04	0.04	0	108.69	0.01	95.51	33.6	28.13	27.28	27.77	0.27
8	0.08	0.07	0.08	0	116.05	0.01	91.01	47.98	32.85	32.07	32.34	0.28
9	0.6	0.47	0.52	0.05	151.33	0.02	135.12	47.48	53.26	51.55	52.62	0.54
10	3.12	2.94	3.01	0.06	228.88	221.24	224.84	2.99	95.24	94.34	94.73	0.32

To further illustrate the advantages of the quadrant method over other non linear separability testing methods, the two spiral problem was used [9]. As can be seen in figures 3(d), this is a highly non linearly separable data set. Table 4 gives the convergence times obtained with the different methods on this data set. It should be observed that although the convex hull converged within 0.0272 seconds, it provided an invalid answer suggesting that the problem was linearly separable and providing, as expected, a non separable plane. The perceptron neural network stopped after 22 seconds without finding a plane and, therefore, concluding that the two classes are not linearly separable. The SVM method [10] was used as a check on linear separability for 3(c) and 3(d).

The quadrant method is deterministic and has proven to give consistent dramatic convergence time improvements over other methods. The current implementation of the method has some limitations with respect to the input dimension and the number of samples in the data set. This is because this implementation needs to have enough samples to be able to cover all quadrants in the input space. Other ways of finding out if the origin is inside the convex hull of $X \ominus Y$ (X and Y representing the samples in the two classes of a classification problem) could be studied to overcome this limitation. An extension of this work could include the calculation of the hyperplane when the two classes are linearly separable. Another area of further study could include the use of this method to help cut down the time required to build cascade linear neural networks such as the RDP. This way, the number of times that the algorithms for testing linear separability need to be called, could be diminished. Another extension of this work will

Table 3. Results obtained with the three non linear separability testing methods and ten, linearly separable data sets, for two to ten dimensional data sets in terms of the convergence time (seconds) using 3(b)

	Quadrant				Convex Hull				Perceptron			
Dim	Max	Min	Mean	Std	Max	Min	Mean	Std	Max	Min	Mean	Std
2	0.01	0.01	0.01	0	0.8	0.04	0.24	0.25	23.15	0	2.33	7.32
3	0.01	0.01	0.01	0	1.09	0.09	0.41	0.29	23.5	0	3.08	7.47
4	0.02	0.02	0.02	0	0.8	0.22	0.53	0.18	0.67	0	0.12	0.21
5	0.03	0.03	0.03	0	0.87	0.17	0.49	0.23	4.41	0.01	0.51	1.38
6	0.03	0.03	0.03	0	1.14	0.38	0.76	0.27	26.43	0.01	3.27	8.27
7	0.04	0.04	0.04	0	1.13	0.47	0.71	0.23	0.46	0.01	0.08	0.14
8	0.08	0.07	0.07	0	2.21	0.69	1.11	0.42	5.2	0.01	0.55	1.64
9	0.6	0.46	0.51	0.04	3.12	1.4	2.42	0.57	0.71	0.05	0.28	0.24
10	3.6	2.97	3.07	0.19	6.85	4.11	5.22	0.89	92.31	0.43	10.81	28.71

Table 4. Results obtained with the two spiral 2D classification problem in terms of the convergence time using the quadrant method and three linearly separability testing methods

	Quadrant	Convex Hull	Perceptron	SVM Spiral
Time	0.0807	0.0272	21.9629	0.687

be to compare the performance of the quadrant method with real data sets and to include some results obtained with other methods for testing linear separability such as support vector machines or the simplex algorithms ([1]).

References

1. Elizondo, D.: A survey of methods for testing linear separability. Transactions on Neural Networks (IEEE) 17(2), 330–344 (2006)
2. Tajine, M., Elizondo, D.: Enhancing the perceptron neural network by using functional composition. Computer Science Department, Université Louis Pasteur, Strasbourg, France, Tech. Rep. 96-07 (1996)
3. Elizondo, D.A.: The recursive determinist perceptron (rdp) and topology reduction strategies for neural networks. Ph.D. dissertation, Université Louis Pasteur, Strasbourg, France (January 1997)
4. Preparata, F.P., Shamos, M.: Computational Geometry. An Introduction. Springer, New York (1985)
5. Stoer, J., Witzgall, C.: Convexity and Optimization in finite dimensions I. Springer, Berlin (1970)
6. Tajine, M., Elizondo, D.: New methods for testing linear separability. Neurocomputing 47, 161–188 (2002)
7. Johnson, D.S., Preparata, F.P.: The densest hemisphere problem. Theoretical Computer Science 6, 93–107 (1978)

8. Rosenblatt, F.: Principles of Neurodynamics. Spartan, Washington D.C (1962)
9. Lang, K.J., Witbrock, M.J.: Learning to tell two spirals apart. In: D., T., G., H., T., S. (eds.) Procedings of the 1988 Connectionist Models Summer School, Morgan Kaufmann, San Francisco (1988)
10. Cortes, C., Vapnik, V.: Support-vector network. Machine Learning 20, 273–297 (1995)

A One-Step Unscented Particle Filter for Nonlinear Dynamical Systems

Nikolay Y. Nikolaev[1] and Evgueni Smirnov[2]

[1] Goldsmiths College, University of London, London SE14 6NW, United Kingdom
n.nikolaev@gold.ac.uk
[2] MICC-IKAT, Maastricht University, Maastricht 6200 MD, The Netherlands
smirnov@micc.unimaas.nl

Abstract. This paper proposes a one-step unscented particle filter for accurate nonlinear estimation. Its design involves the elaboration of a reliable one-step unscented filter that draws state samples deterministically for doing both the time and measurement updates, without linearization of the observation model. Empirical investigations show that the one-step unscented particle filter compares favourably to relevant filters on nonlinear dynamic systems modelling.

1 Introduction

Particle filters (PF) [3][4][5][9] are efficient computational tools for sequential Bayesian estimation. Applied to the task of dynamic system modelling, they approximate the posterior distribution of the unknown state parameter by averaging over its realization as a set of particles. Once the posterior is obtained, there can be generated forecasts from unseen inputs. PF are suitable for practical tasks because they can work well with nonlinear models and in relaxed circumstances, like non-Gaussianity and non-stationarity.

The generic PFs are however sensitive to the magnitude of outliers, which disturbs their performance on real-world data. One strategy for improving them is through more accurate particle sampling as suggested by Sigma Point Particle Filters (SPPF) [12]. During the time update step they calculate the predictive density of each particle by deterministic sampling using derivative free Unscented Kalman filters (UKF) [13] or Central Difference Filters (CDF) [10]. In the next measurement update step they compute unfortunately the posterior with the standard equations relying on linearization of the observation model which causes two serious problems: 1) with the increase of model nonlinearities the accuracy decreases; and 2) the linearization may lead to uncorrelated state and observation, and then the state is not updated [14]. Therefore, SPPF do not have the potential to infer a large class of system models.

The recent One-step Unscented Kalman Filter (OUKF) [2][14] provided the idea to alleviate such performance failure by doing the measurement update in only one step, without linearizing the observation model through derivatives.

J. Marques de Sá et al. (Eds.): ICANN 2007, Part I, LNCS 4668, pp. 747–756, 2007.

A critique to the OUKF is that it uses Gaussian quadrature as a technique to determine the sampling points, which limits its usefulness because it requires too many points to achieve accurate approximation.

This paper proposes an efficient One-step Unscented Filter (OUF) that requires a smaller number of particles to achieve second-order accuracy of the state mean and variance, compared to the OUKF [2][14]. A distinguishing characteristics of the OUF is that during both the time and measurement updates it samples the state deterministically using the unscented transform [7]. The OUF is more reliable than the UKF [13] because it performs sampling to compute the measurement update. Another contribution is the design of a One-step Unscented Particle Filter (OUPF) for approximating posteriors with complex shapes, which often arise in practical tasks. Empirical investigations show that the one-step unscented particle filter compares favorably to relevant filters on modeling nonstationary series dynamics and stochastic volatility modeling.

This paper is organized as follows. Section two presents the approach to unscented filtering and elaborates the general one-step unscented filter. The next section three develops the OUPF, and section four offers the empirical study with related filters. Finally a brief discussion and a conclusion are provided.

2 The One-Step Unscented Filter

2.1 Sequential Bayesian Estimation

The problem of sequential Bayesian estimation is to infer the state of a system that describes the latent dynamics of a provided series of discrete noisy observations $D = \{y_t\}_{t=1}^{T}$. Assuming univariate dependencies, a dynamical system can be formally defined by the couple of equations:

$$\begin{aligned} x_t &= g(x_{t-1}, u_t) \quad \text{/state model/} \\ y_t &= f(x_t, v_t) \quad\;\; \text{/observation model/} \end{aligned} \tag{1}$$

where x_t is the state, y_t is the observation at time t, g is the state transition function driven by state noise u_t, and f is the measurement function driven by observational noise v_t. Both functions f and g can be nonlinear.

Solution to this problem is the posterior state distribution $p(x_t|\mathbf{y}_{0:t})$, and, more precisely, its two characteristics the mean and the variance. The posterior distribution of the system state x_t, as a sufficient statistics of the whole state history $\mathbf{x}_{0:t}$ up to the particular time instant t, can be obtained by a probabilistic filter which applies recursively the Bayes' theorem as follows:

$$p(x_t|\mathbf{y}_{0:t}) = \frac{p(y_t|x_t)p(x_t|\mathbf{y}_{0:t-1})}{p(y_t|\mathbf{y}_{0:t-1})} \tag{2}$$

where $p(x_t|\mathbf{y}_{0:t})$ is the state posterior (filtering distribution), $p(y_t|x_t)$ is the likelihood of the data, $p(x_t|\mathbf{y}_{0:t-1})$ is the state prior (predictive distribution), and $p(y_t|\mathbf{y}_{0:t-1})$ is the evidence. The state prior is defined by the integral:

$p(x_t|\mathbf{y}_{0:t-1}) = \int p(x_t|x_{t-1}, \mathbf{y}_{0:t-1})p(x_{t-1}|\mathbf{y}_{0:t-1})dx_{t-1}$, where $p(x_{t-1}|\mathbf{y}_{0:t-1})$ is the previous filtering density, and $p(x_t|x_{t-1}, \mathbf{y}_{0:t-1})$ is the transition density.

Adopting a nonlinear observation model leads to nonstandard distributions (without a predominant mode) which are analytically intractable in most cases. This difficulty can be alleviated using sampling to approximate the integrals of the distributions of interest, instead of evaluating them by means of linearization with the derivatives of the system equations. There are two main approaches to sampling for this purpose: stochastic and deterministic [2]. Here we focus attention on deterministic sampling which is supposed to be more accurate. A recent popular approach to deterministic sampling for integral evaluation is given by the unscented transform [7].

2.2 Unscented Kalman Filtering

The unscented transform [7] suggests to pick state samples, called sigma-points, from carefully selected location points. The spread of the location points is determined in such a way so as to obtain a density estimate with the same statistical properties (mean and variance) as the true, unknown state distribution. The usefulness of drawing sigma-points is in that they enable to achieve higher accuracy of approximation of the state mean and variance, which is up to the second order for any kind of model nonlinearity.

The UKF [13] implements algorithmically the numerically stable scaled version of the unscented transform [8], which attains high fitting accuracy by choosing symmetric sigma-points at locations set as follows:

$$\begin{aligned} \mathcal{X}_{0,t-1} &= \widehat{x}_{t-1} \\ \mathcal{X}_{i,t-1} &= \widehat{x}_{t-1} + \left(\sqrt{(L+\lambda)\,P_{t-1}}\right),\ i = 1, ..., L \\ \mathcal{X}_{i,t-1} &= \widehat{x}_{t-1} - \left(\sqrt{(L+\lambda)\,P_{t-1}}\right),\ i = L+1, ..., 2L \end{aligned} \tag{3}$$

where L is the dimension of the state variable, $\widehat{x}_{t-1}$ is the previous mean state estimate, and P_{t-1} is the covariance of the state distribution.

The probability distribution of the state random variable is modelled by weighted averaging over these sigma-points, after propagating them through the state transition function. Each sigma-point is associated with a corresponding weight $\mathcal{S}_{i,t} = \{W_i, \mathcal{X}_{i,t-1}\}$, $i = 0, ..., 2L$, and such that $\sum_{i=0}^{2L} W_i = 1$. The weights are computed by specific formulae:

$$\begin{aligned} W_0^m &= \lambda/(\lambda+1) \\ W_0^c &= \lambda/(\lambda+1) + (1 - \alpha^2 + \beta) \\ W_i^m &= W_i^c = 1/(2(L+\lambda)),\ i = 1, ..., 2L \end{aligned} \tag{4}$$

where α, β, and λ are scaling parameters. The parameters α and β are chosen to control the spread of the distribution. The parameter λ is calculated by the formula $\lambda = \alpha^2(L+\kappa) - L$, using another parameter κ selected to scale the spread with respect to the mean of the distribution.

The time update step performed by the UKF passes the sigma-points $\mathcal{X}_{i,t-1}$, generated around the mean from the previous time instant $\widehat{x}_{t-1}$, through the state transition function $\mathcal{X}_{i,t|t-1} = f(\mathcal{X}_{i,t-1}, v_t)$, and, next, estimates the predicted state distribution $\widehat{p}(x_t|\mathbf{y}_{0:t-1}) = \mathcal{N}(\widehat{x}_{t|t-1}, P_{t|t-1})$ as follows:

$$\widehat{x}_{t|t-1} = \sum_{i=0}^{2L} W_i^m \mathcal{X}_{i,t|t-1} \tag{5}$$

$$P_{t|t-1} = \sum_{i=0}^{2L} W_i^c \left(\mathcal{X}_{i,t|t-1} - \widehat{x}_{t|t-1}\right)\left(\mathcal{X}_{i,t|t-1} - \widehat{x}_{t|t-1}\right)^T \tag{6}$$

where $\widehat{x}_{t|t-1}$ is the predicted state mean, and $P_{t|t-1}$ is the predicted variance.

2.3 One-Step Measurement Update

The idea for doing a measurement update in one step [2][14] is elaborated here by developing yet another OUF. This OUF simulates the integral of the state posterior by sampling instead of evaluating it using linearization. Thus, the state mean and variance are obtained in one step, not in the typical two steps of linearized approximation of the joint posterior through the derivatives of the observation model, and, next, linear filtering with the Kalman equations. The novelty is in doing deterministic sampling from predicted distribution $\widehat{p}(x_t|\mathbf{y}_{0:t-1})$ using the unscented transform. We apply the sigma-point approach again to the predicted state mean $\widehat{x}_{t|t-1}$ to approximate directly the posterior $\widehat{p}(x_t|\mathbf{y}_{0:t})$.

In order to capture the effect from the given target at the particular moment on the state distribution, this distribution can be simulated by sigma-points $\mathcal{S}_{i,t} = \{W_i, \mathcal{X}_{i,t|t}\}$, $i = 0, ..., 2L$, around the predicted state mean $\widehat{x}_{t|t-1}$:

$$\begin{aligned} \mathcal{X}_{0,t|t} &= \widehat{x}_{t|t-1} \\ \mathcal{X}_{i,t|t} &= \widehat{x}_{t|t-1} + \left(\sqrt{(L+\lambda)\, P_{t|t-1}}\right), \ i = 1, ..., L \\ \mathcal{X}_{i,t|t} &= \widehat{x}_{t|t-1} - \left(\sqrt{(L+\lambda)\, P_{t|t-1}}\right), \ i = L+1, ..., 2L \end{aligned} \tag{7}$$

where $P_{t|t-1}$ is the variance obtained in the time step.

The approximation of the posterior absorbs the effect of the given target y_t through the likelihood. The likelihood integral may be envisioned analytically tractable [14], and it can be estimated with the outputs $\mathcal{Y}_t$ produced by passing the sigma-points through the measurement function:

$$\begin{aligned} \mathcal{Y}_t &= f(\mathcal{X}_{i,t|t}, v_t) \\ \widehat{p}(y_t|\mathcal{X}_{i,t|t}) &= \int \delta\left(\mathcal{Y}_t - y_t\right) p(v_t) dv_t \end{aligned} \tag{8}$$

where δ is the Dirac delta function. There are standard density functions for computing $\widehat{p}(y_t|\mathcal{X}_{i,t|t})$, like the normal probability density for example.

The statistical mean and variance of the posterior probability distribution of the state $\widehat{p}(x_t|\mathbf{y}_{0:t}) = \mathcal{N}(\widehat{x}_{t|t}, P_{t|t})$ are obtained with the following expressions:

$$\widehat{x}_{t|t} = \sum_{i=0}^{2L} W_i^m \mathcal{X}_{i,t|t} \widehat{p}(y_t|\mathcal{X}_{i,t|t})/Z_t \tag{9}$$

$$P_{t|t} = \left[\sum_{i=0}^{2L} W_i^c \mathcal{X}_{i,t|t}^2 \widehat{p}(y_t|\mathcal{X}_{i,t|t})/Z_t \right] - \widehat{x}_{t|t}^2 \tag{10}$$

where the normalizing constant is $Z_t = \sum_{i=0}^{2L} W_i^m \widehat{p}(y_t|\mathcal{X}_{i,t|t})$.

The OUF is reliable from a theoretical point of view as it can deal with uncorrelated states and observations, therefore it can handle a larger class of models than the UKF [13]. The OUF is efficient because relying on the unscented transform it requires less points than Gaussian quadrature.

3 Unscented Particle Filtering

Based on the above developments, we propose a One-step Unscented Particle Filter (OUPF) for non-parametric estimation of the state. The state is realized by a set of particles, associated with weights corresponding to their probability mass, so that the sum of the weighted particles approximates the posterior. OUPF is an extension to the (SPPF) [12] in that it uses the general OUF for importance sampling, which makes it more robust.

The particle filtering involves two stages: 1) importance sampling with updating of the particles and their weights when a data point arrives; and 2) propagating the particles via resampling. The importance sampling uses the OUF to generate improved states $\widehat{x}_{t|t}^{(i)}$ that serve as means for the proposal distribution π, and, next, it draws states $x_t^{(i)}$ from this proposal $\pi(x_t^{(i)}|\widehat{x}_{t|t}^{(i)}) = \mathcal{N}(x_t^{(i)}; \widehat{x}_{t|t}^{(i)}, P_{t|t})$. After that, the particle weights are updated from $w_{t-1}^{(i)}$ to $w_t^{(i)}$ with the following modification equation:

$$w_t^{(i)} \approx w_{t-1}^{(i)} \frac{p(y_t|x_t^{(i)}) p(x_t^{(i)}|x_{t-1}^{(i)})}{\pi(x_t^{(i)}|\widehat{x}_{t|t}^{(i)})} \tag{11}$$

where the prior is $p(x_t^{(i)}|x_{t-1}^{(i)}) = \mathcal{N}(x_t^{(i)}; x_{t-1}^{(i)}, P_{t-1}^{(i)})$, the model likelihood is $p(y_t|x_t^{(i)}) \approx \exp(-0.5\sigma^{-2} \sum_{t=1}^{T} (y_t - y_t^{(i)})^2)$, σ is the observational noise variance, and T is the series size. Thus, a particular weight reflects the density in the state space in the vicinity of its corresponding particle.

The empirical estimate of the state posterior is obtained as follows:

$$\widetilde{p}(x_t|\mathbf{y}_{0:t}) \simeq \sum_{i=1}^{N} w_t^{(i)} \delta(x_t - x_t^{(i)}) \tag{12}$$

where the weights $w_t^{(i)}$ are taken normalized by $w_t^{(i)} = w_t^{(i)} / \sum_{i=1}^{N} w_t^{(i)}$.

The resampling stage is necessary to promote only the most promising particles [5], because otherwise the population often becomes occupied by many particles with negligible weights. Resampling is performed by selecting particles with high weights and discarding some of those with insignificant weights, using different techniques. Algorithms using such techniques are known as PF with Sampling Importance Resampling (SIR).

Overall, the OUPF tends to capture better the statistics of the true state posterior than SPPF as it has been found experimentally during preliminary research. The OUPF is especially useful when the shape of the state posterior density is complex and hard to model. Such situations arise in various practical tasks, for example in control and financial engineering applications.

4 Empirical Investigations

Research has been conducted with two dynamical systems: a nonlinear observation model using non-Gaussian state noise, and a nonlinear stochastic volatility model whose state model is of unknown variance. Although producing quite irregular series the first system can be estimated with sigma-point filters, so we run the SPPF [12], the PF (SIR) [5][9], the PF_{MCMC} [1], and OUPF. PF_{MCMC} is an improved generic particle filter which makes additional Markov Chain Monte Carlo (MCMC) steps on each particle by sampling from the transition prior, and accepting the move by Metropolis-Hastings steps [1].

The considered stochastic volatility model, however, can not be inferred by standard filters using linearized Kalman equations for the measurement update. That is why, we run the PF [5][9], the PF_{MCMC} [1], OUPF and another APF [11] filter. The APF is a benchmark Auxiliary Particle Filter [11], especially for financial tasks, that boosts particles with higher predictive likelihood to reduce the variance of the state in order to attain a better posterior. All studied particle filters used systematic resampling [9], populations of 200 samples.

Modeling Nonstationary Series Dynamics. A nonstationary time series was generated using the following non-linear observation model [12]:

$$
\begin{aligned}
x_t &= 1 + \phi_1 x_t + \sin(\omega\pi(t-1)) + v_t \\
y_t &= \begin{cases} \phi_2 x_t^2 + n_t & k < 30 \\ \phi_3 x_t - 2 + n_t & k \geq 30 \end{cases}
\end{aligned}
\tag{13}
$$

where we used non-Gaussian (Gamma) state noise $v_t \in \Gamma(3,1)$, Gaussian observation noise $n_t \in \mathcal{N}(0, 1.0e-5)$, and scalar parameters: $\omega = 0.04$, $\phi_1 = 0.5$, $\phi_2 = 0.2$, $\phi_3 = 0.5$. A sequence of $t = 60$ values was generated. The SPPF was implemented using the UKF filter to produce results comparable to previous research [12]. The UKF was tuned with $\alpha = 1$, $\beta = 0$, and $\kappa = 2$. There were performed 100 independent runs with each filter, every run using re-generated series and randomly re-initialized particles.

Table 1. Accuracies of modeling the unknown state in the non-stationary observation model, obtained after averaging over 100 independent runs with each filter

Algorithm	RMSE	Var
PF	0.4844	0.049
PF_{MCMC}	0.4685	0.045
SPPF	0.0961	0.006
OUPF	0.0922	0.005

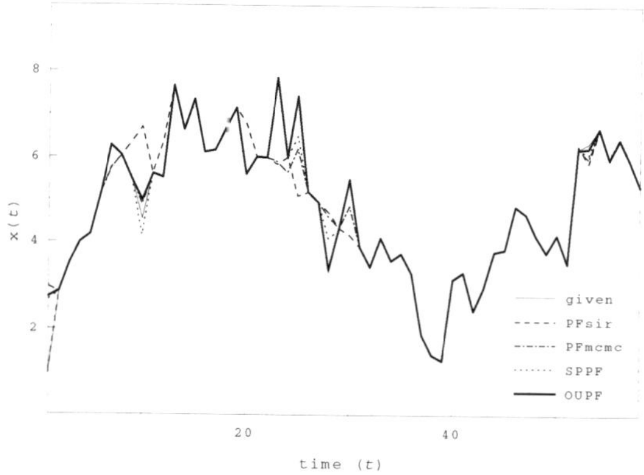

Fig. 1. A segment from the estimated state by the studied particle filters

Table 1 provides the average rooted mean squared errors (RMSE) of the state estimate and its standard deviation. Figure 1 illustrates the quality of learning achieved by the studied filters on this task. The evolution of the RMSE state errors, averaged over independent 100 runs, is shown in the next Figure 2 to provide empirical evidence for their convergence characteristics.

Stochastic Volatility Modeling. A stochastic volatility model describes how stock returns depend on the volatility (variance of the returns). We consider a nonlinear model of the log returns $y_t \equiv \log(S_t) - \log(S_{t-1})$, from stock prices S_t and S_{t-1}, whose noise is driven by the log of the volatility $x_t \equiv \log(v)$ at time t as follows [6][14]:

$$\begin{aligned} x_t &= \gamma + \phi(x_{t-1} - \gamma) + \varepsilon_t \\ y_t &= \eta_t \exp(x_t/2) + \mu \end{aligned} \tag{14}$$

where the state noise ε_t is univariate Gaussian $\varepsilon_t = \mathcal{N}(0, q)$, and the observation noise is proportional to Gaussian $\eta_t = \mathcal{N}(0, 1)$. This is a dynamic model of the returns in which the volatility is referred to as state.

A time series of log returns was made by generating 1000 successively random values for the volatility x_t and passing them through the autoregressive function

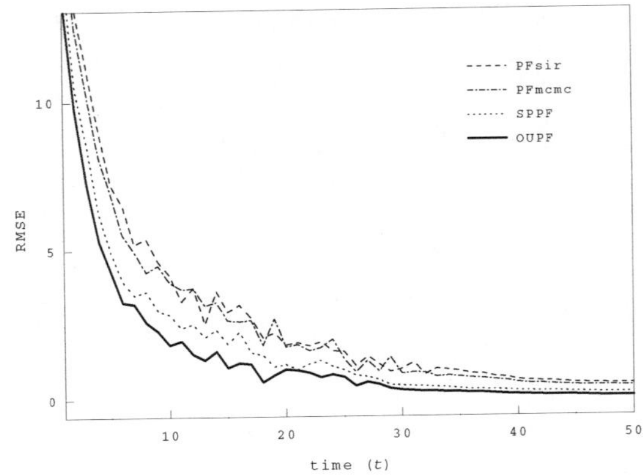

Fig. 2. Convergence of the studied particle filters, measured as the RMSE state error from the beginning to the corresponding point in the series

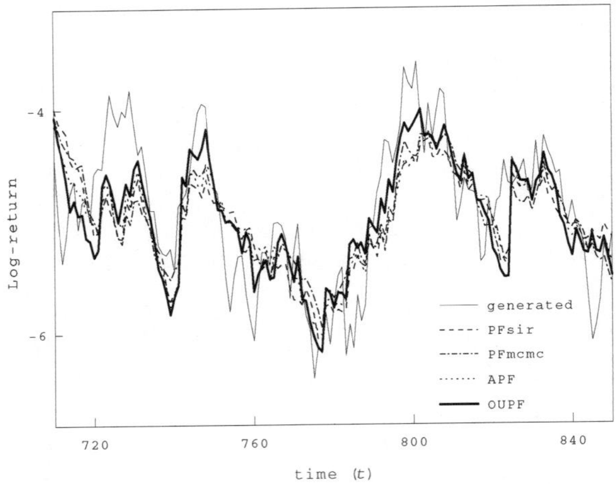

Fig. 3. Sequentially estimated volatility (state) by the studied particle filters

using parameters $\gamma = log(0.01)$ and $\phi = 0.95$. This volatility was further contaminated by state noise ε_t of variance $q = sqrt(0.1)$. The log returns y_t were produced assuming a mean value $\mu = 0.0015$.

The above equations (14) define a state-space model which may be expected to be learnable using standard Kalman filters, however there is an intrinsic problem that prevents this. The problem is that the returns and the volatility become uncorrelated when treated by linearization with standard filters, and such filters fail as the Kalman gain becomes zero. Even the derivative-free filters UKF [13] and CDF [10] can not infer the volatility using this general log volatility model.

The synthesized series was divided into three subseries of increased size: 250, 500 and 1000 training points (log returns). Table 2 offers the results, calculated by averaging over 100 runs conducted with each filter on each subseries.

Table 2. Accuracies of modeling the unknown state in the stochastic volatility model, obtained after averaging over 100 runs with each filter

Algorithm	1-250		1-500		1-1000	
	RMSE	Var	RMSE	Var	RMSE	Var
PF	0.4921	0.0145	0.5215	0.0135	0.5341	0.0123
PF_{MCMC}	0.4857	0.0140	0.5203	0.0132	0.5374	0.0121
APF	0.4721	0.0096	0.5188	0.0082	0.5218	0.0075
OUPF	0.4702	0.0054	0.5016	0.0051	0.5134	0.0048

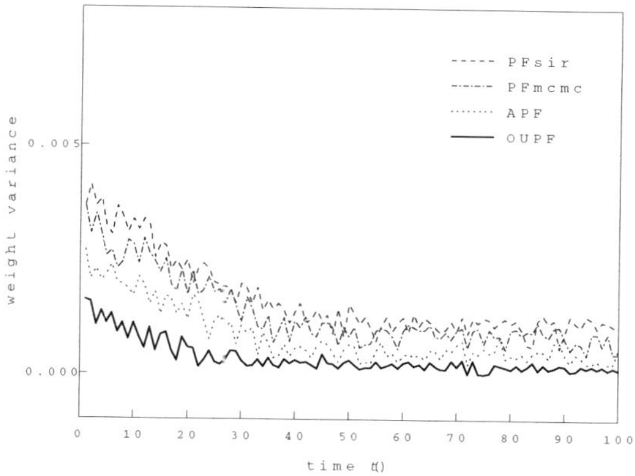

Fig. 4. Standard deviation (variance) of the particle weights, recorded during particular runs with each of the studied filters

The volatility curves learned by the filters are given in Figure 3, along with the curve of the true (artificially generated) volatility. It can be seen that the OUPF curve is closest to the true volatility than all other curves.

The variance of the weights is an important characteristic that reflects the ability of the filter to achieve high precision despite the disturbing effects of the adopted sampling scheme. Figure 4 shows that OUPF operates with lower weights variance than the APF, PF and PF_{MCMC} filters, which explains why it has achieved better results.

5 Discussion and Conclusion

This paper presented an unscented filter that can be incorporated to make efficient particle filters for sequential estimation of distributions with complex shapes. The OUF has the capacity to estimate a larger class of models than the previous SPPF relying on the popular sigma-point filters UKF [13] and CDF [10]. Although achieving good results on the addressed problems, it should be noted that particle filters and OUPF are sensitive to the initial conditions.

It has been found experimentally that the designed OUPF gives a more accurate alternative to the benchmark APF on stochastic volatility modelling. Future research aims to apply the OUPF on other financial engineering tasks that assume state-space descriptions.

References

1. Andrieu, C., de Freiras, J.F.G., Doucet, A.: Sequantial MCMC for Bayesian Model Selection. In: Proc. IEEE Higher Order Statistics Workshop, pp. 130–134. IEEE Computer Society Press, Los Alamitos (1999)
2. Bolviken, E., Storvik, G.: Deterministic and Stochastic Particle Filters. In: Doucet, A., et al. Sequantial Monte Carlo Methods in Practice, pp. 97–116 (2001)
3. de Freitas, J.F.G., Niranjan, M., Gee, A.H., Doucet, A.: Sequantial Monte Carlo Methods to Train Neural Networks. Neural Computation 12, 955–993 (2000)
4. Doucet, A., de Freitas, N., Gordon, N.: Sequantial Monte Carlo Methods in Practice. Springer, New York (2001)
5. Gordon, N.J., Salmond, D.G., Smith, A.F.M.: A Novel Approach to Nonlinear/non-Gaussian Bayesian State Estimation. Proceedings IEE-F Radar, Sonar and Navigation 140, 107–113 (1993)
6. Hull, J., White, A.: The Pricing of Options on Assets with Stochastic Volatilities. The Journal of Finance 42, 281–300 (1987)
7. Julier, S.J., Uhlmann, J.K.: A New Extension of the Kalman Filter to Nonlinear Systems. In: Proc. SPIE Int. Soc. Opt. Eng., Orlando, FL, vol. 3068, pp. 182–193 (1997)
8. Julier, S.J.: The Scaled Unscented Transformation. In: Proc. American Control Conf., vol. 6, pp. 4555–4559 (2002)
9. Kitagawa, G.: Monte Carlo Filter and Smoother for Non-Gaussian Nonlinear State Space Models. J. Computational and Graphical Stat. 5, 1–25 (1996)
10. Norgaard, M., Poulsen, N.K., Ravn, O.: New Developments in State Estimation for Nonlinear Systems. Automatica 36, 1627–1638 (2000)
11. Pitt, M.K., Shephard, N.: Filtering via Simulation: Auxiliary Particle Filters. J. American Stat. Assoc. 94, 590–599 (1999)
12. van der Merwe, R., Doucet, A., de Freitas, N., Wan, E.: The Unscented Particle Filter. In: Leen, T.K., Dietterich, T.G., Tresp, V. (eds.) Advances in Neural Inf. Processing Systems (NIPS13), pp. 584–590 (2001)
13. van der Merwe, R.: Sigma-point Kalman Filters and Probabilistic Inference in Dynamic State-Space Models. PhD Thesis, OGI School of Science and Engineering, Oregon Health and Science University (2004)
14. Zoeter, O., Ypma, A., Heskes, T.: Improved Unscented Kalman Smoothing for Stock Volatility Estimation. In: Barros, A., Principe, J., Larsen, J., Adali, T., Douglas, S. (eds.) Machine Learning for Signal Processing. Proc. of the 14th IEEE Signal Processing Society Workshop, pp. 143–152. IEEE Press, NJ (2004)

Spike-Timing-Dependent Synaptic Plasticity to Learn Spatiotemporal Patterns in Recurrent Neural Networks

Masahiko Yoshioka[1], Silvia Scarpetta[1,2,3], and Maria Marinaro[1,2,3]

[1] Department of Physics, "E.R. Caianiello", University of Salerno, 84081 Baronissi SA, Italy
[2] INFN, 84100 Salerno SA, Italy
[3] IIASS, Vietri sul Mare, 84019 Vietri sul Mare SA, Italy

Abstract. Assuming asymmetric time window of the spike-timing-dependent synaptic plasticity (STDP), we study spatiotemporal learning in recurrent neural networks. We first show numerical simulations of spiking neural networks in which spatiotemporal Poisson patterns (i.e., random spatiotemporal patterns generated by independent Poisson process) are successfully memorized by the STDP-based learning rule. Then, we discuss the underlying mechanism of the STDP-based learning, mentioning our recent analysis on associative memory analog neural networks for periodic spatiotemporal patterns. Order parameter dynamics in the analog neural networks explains time scale change in retrieval process and the shape of the STDP time window optimal to encode a large number of spatiotemporal patterns. The analysis further elucidates phase transition due to destabilization of retrieval state. These findings on analog neural networks are found to be consistent with the previous results on spiking neural networks. These STDP-based spatiotemporal associative memory possibly gives some insights into the recent experimental results in which spatiotemporal patterns are found to be retrieved at the various time scale.

1 Introduction

In the variety of sensory systems, spatiotemporal activities of neurons reflect their sensory inputs[1,2]. It has also been revealed that in the central nervous system, short-term memory is represented by spatiotemporal patterns of neurons[3]. There are the growing evidences that support a pivotal role of spatiotemporal neural activities in the biological information processing. However, the learning mechanism of spatiotemporal patterns in neural networks is still not well understood. There has been a long history of studies on associative memory neural networks[4,5,6,7,8,9,10,11,12,13,14,15,16,17,18,19,20,21], and it is well known that in discrete-time spin neural networks with synchronous update rule $x_i(t+1) = \mathrm{sgn}(\sum_j J_{ij} x_j(t))$ $(i = 1, \ldots, N)$ learning of binary spatiotemporal patterns $\xi_i(t) = \pm 1$ $(t = 1, 2, \ldots)$ is easily attained by the application of the

J. Marques de Sá et al. (Eds.): ICANN 2007, Part I, LNCS 4668, pp. 757–766, 2007.

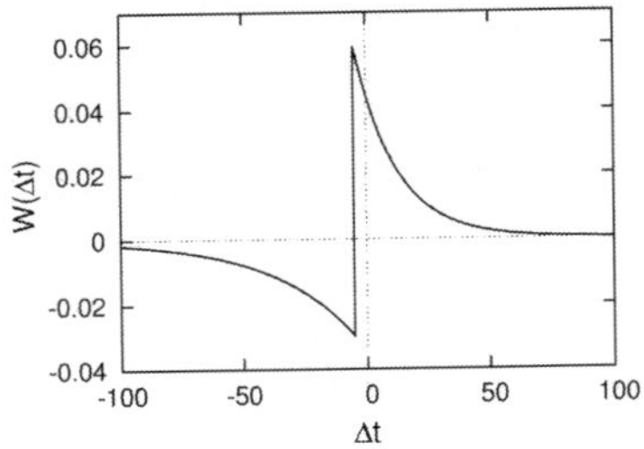

Fig. 1. The schematic shape of the time window of spike-timing-dependent synaptic plasticity (STDP) given by Eq. (2) with parameters $\tau_- = 33.7$, $\tau_+ = 16.8$, $\tau_0 = -5$, and $r = 1$. The experimental studies have revealed that weights of some types of synapses are modified according to $\Delta J \propto W(\Delta t)$, where ΔJ represents change of synaptic weight and $\Delta t = t_{\mathrm{post}} - t_{\mathrm{pre}}$ denotes spike time difference between presynaptic and postsynaptic neurons.

asymmetrically modified Hebb learning rule $J_{ij} = (1/N)\sum_t \xi_i(t+1)\xi_j(t)$. Nevertheless, the modified Hebb learning rule does not work well in most of realistic continuous-time neural networks, including asynchronous spin neural networks, analog neural networks, and spiking neural networks. Several attempts have been made to enhance the applicability of the modified Hebb learning rule to spatiotemporal learning[8,9]. However, as far as we know, a satisfactory level of spatiotemporal learning has not yet been obtained from the improvement of the modified Hebb learning rule.

In the present study, we give a simple solution for spatiotemporal learning, introducing a different approach from the modified Hebb learning rule. We assume the spike-timing-dependent synaptic plasticity (STDP) in spatiotemporal learning rule and realize associative memory neural networks for spatiotemporal patterns. The STDP is the experimental finding on a synaptic plasticity observed in many parts of the real nervous system[22,23,24,25]. In these experiments, synaptic weights of some types of neurons are found to be modified according to spike time delay between presynaptic and postsynaptic neurons; when a presynaptic neuron fires before a postsynaptic neuron, the synaptic weight is strengthened, while if the neurons fire in the opposite order, the synaptic weight is weakened. This relation of synaptic plasticity to spike time deference is expressed by the asymmetric time window $W(\Delta t)$ described in Fig. 1. In our study, we incorporate this biologically plausible asymmetric time window into a learning rule so that neural networks act as spatiotemporal associative memory.

The outcome of the application of the STDP to spatiotemporal learning is remarkable. In the present paper, we first show numerical simulations in which random spatiotemporal spike sequences are successfully memorized in spiking neural networks. These random sequences are generated by independent Poisson process. This means that for successful retrieval, some neurons must fire multiple times during retrieval process, while others must keep a silent state. Our numerical simulation reveals that the STDP-based learning rule easily realizes such complicated control of spiking neural networks. As shown in our previous analyses, retrieval state of the STDP-based associative memory can be analyzed exactly both in spiking neural networks and in analog neural networks when periodic spatiotemporal patterns are assumed [17,18,21]. In the present study, we summarize these analytical results for periodic patterns and discuss the underlying mechanism of the STDP-based spatiotemporal learning.

The present paper is organized as follows. In Sec. 2, we introduce the STDP-based learning of Poisson patterns. Then, in Sec. 3, we summarize our previous analysis on the STDP-based learning of periodic patterns in analog neural networks. We compare these analytical results on analog neural networks with the previous analysis on spiking neurons in Sec. 4. Finally, in Sec. 5, we give a brief summary and discuss the implication of the present analytical studies for some experimental results.

2 The Spike-Timing-Dependent Synaptic Plasticity (STDP) and Spatiotemporal Learning of Poisson Patterns

The experimental studies on the STDP has revealed that in the real nervous system, weights of some types of synapses are modified according to

$$\Delta J \propto W(\Delta t), \tag{1}$$

where ΔJ represents degree of modification of synaptic weight and $\Delta t = t_{\text{post}} - t_{\text{pre}}$ represents the spike time difference between presynaptic and postsynaptic neurons. Typically, the time window $W(\Delta t)$ takes the asymmetric shape with the negative part and the positive part as described in Fig. 1. In the present study, we assume that the STDP time window is given by

$$W(\Delta t) = \begin{cases} -(r/\tau_-)\, e^{(\Delta t - \tau_0)/\tau_-} & \Delta t \le \tau_0 \\ (1/\tau_+)\, e^{-(\Delta t - \tau_0)/\tau_+} & \tau_0 < \Delta t \end{cases}, \tag{2}$$

where parameters τ_- and τ_+ represent decay time of the STDP time window and parameter τ_0 denotes time shift. Parameter r is introduced to control the size of the negative part of the STDP time window.

Let us discuss how the STDP contributes to learning of spatiotemporal Poisson patterns in networks of spiking neurons. We assume that N spiking neurons defined by the Hodgkin-Huxley equations are interconnected by chemical synapses J_{ij}. Then, we consider learning of three random spatiotemporal spike patterns ($\mu = 1, 2, 3$) generated by homogeneous Poisson process as shown in Fig. 2(a). We carry out encoding of the three patterns by the STDP-based learning rule

$$J_{ij} \propto \sum_{\mu=1}^{P} \sum_{k_1} \sum_{k_2} W\left(t_i^\mu(k_1) - t_j^\mu(k_2)\right), \tag{3}$$

where $t_i^\mu(k)$ represents the k-th spike timing of neuron i in spatiotemporal pattern μ. Fig. 2(b) represents result of numerical simulation, in which pulsed external input is given at time $t = 0$ to force network retrieve the initial part of the

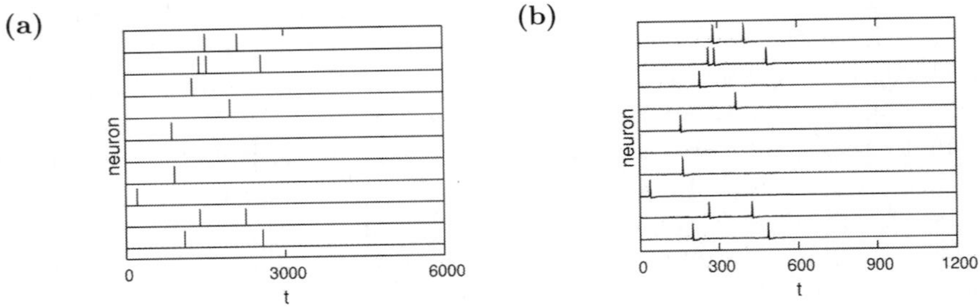

Fig. 2. The STDP-based learning of spatiotemporal Poisson patterns in Hodgkin-Huxley type of spiking neural networks. (a) The first spatiotemporal spike pattern to be memorized ($\mu = 1$) is displayed (Only first ten neurons in all 10000 neurons are described.). In the interval $0 \leq t \leq 3000$, random spike timings are generated by homogeneous Poisson process with a constant rate $\lambda = 1/3000$. Three random Poisson patterns ($\mu = 1, 2, 3$) are generated in the same manner and encoded by the STDP-based learning rule (3) in a network of 10000 spiking neurons. (b) Behavior of membrane potentials (i.e., internal potentials) of spiking neurons in a numerical simulation are plotted as a function of time. At $t = 0$, we give pulsed external input to force the network to retrieve the initial part of the spatiotemporal pattern that is shown in (a) . For $t > 0$, the network retrieve the memorized pattern autonomously (Note that we give no external input for $t > 0$).

first pattern ($\mu = 1$). Then, after $t = 0$, neurons in Fig. 2(b) emit autonomous spikes in the exact accordance with Fig. 2(a). This means that the partial external input at $t = 0$ evokes the retrieval of all the part of the first pattern ($\mu = 1$). The other patterns ($\mu = 2, 3$) can also be retrieved if we apply the corresponding initial inputs. In this manner, the STDP-based learning rule enables spiking neural networks to act as spatiotemporal associative memory.

3 Analysis of Retrieval Process of Periodic Spatiotemporal Patterns in Analog Neural Networks

As shown in the previous studies on spiking neurons [17,18], the underlying mechanism of spatiotemporal learning of spike timings can be explained by the analytical evaluation of time-dependent synaptic electric currents arising in spiking neurons. However, the evaluated synaptic electric currents, which depend on various parameters such as decay time constants of chemical synapses and the shape of the STDP time window, are not necessarily easy to understand. For a simpler explanation of essential mechanism of spatiotemporal learning, we here summarize our recent study on analog neural networks that memorize periodic spatiotemporal patterns[21]. The assumption of periodic patterns in analog neural networks allow us to derive order parameter dynamics, which would ease understanding of some key features of the STDP-based associative memory for spatiotemporal patterns.

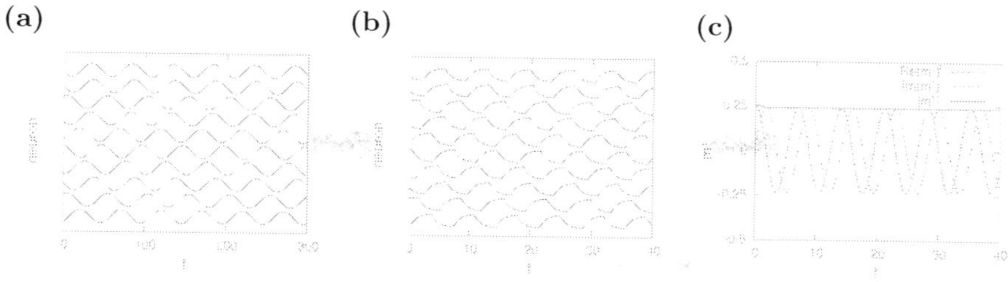

Fig. 3. A numerical simulation of the STDP-based learning of periodic spatiotemporal patterns in an analog neural network. (a) Three periodic spatiotemporal patterns η_i^μ ($\mu = 1, 2, 3$) are generated randomly by Eq. (6) with frequency $\omega = 0.02 \times 2\pi$. Figure (a) describes the behavior of the first ten neurons in the first pattern($\mu = 1$). All the three random patterns are encoded in an analog neural network according to the learning rule (8). (b) Setting the initial condition $x_i(0)$ close to $\eta_i^1(0)$, we carry out a numerical simulation. Figure (b) describes the behavior of the first ten neurons in this numerical simulation. (c) The time evolution of overlap m^1 defined by Eq. (10) is plotted as a function of time. We assume a network of 10000 analog neurons that are defined by Eq. (4) with the Heaviside function $F(u) = H(u)$.

Let us consider analog neural networks defined by

$$dx_i/dt = -x_i + F(h_i), \quad i = 1, \dots, N, \tag{4}$$

where x_i represents the activity of neuron i and transfer function $F(h)$ denotes the input-output relationship of neurons. Local fields h_i, which are analogous to synaptic electric currents in spiking neurons, are defined by

$$h_i = \sum_j J_{ij} x_j. \tag{5}$$

In these analog neural networks, we consider encoding periodic spatiotemporal patterns

$$\eta_i^\mu = (1/2)\left(1 + \cos(\omega t - \phi_i^\mu)\right), \quad i = 1, \dots, N, \ \mu = 1, \dots, P, \tag{6}$$

where ω represents the frequency of periodic spatiotemporal patterns, and phase shift ϕ_i^μ is chosen randomly from the uniform distribution within the interval $[0, 2\pi)$. An example of a randomly generated pattern is illustrated in Fig 3(a). We carry out learning of the periodic patterns according to the STDP-based learning

$$J_{ij} \propto \sum_{\mu=1}^P \int_{-\infty}^{\infty} \int_{-\infty}^{\infty} \eta_i^\mu(t_1) W(t_1 - t_2) \eta_j^\mu(t_2) dt_1 dt_2. \tag{7}$$

Then, we arrive a simple learning rule

$$J_{ij} = (1/N) \sum_{\mu=1}^P \mathrm{Re}\left(a \xi_i^\mu \xi_j^{\mu *}\right) + b/N, \tag{8}$$

where ξ_i^μ represents $\exp\left(i\phi_i^\mu\right)$, and a complex number a and a real number b are defined by

$$a = \int_{-\infty}^{\infty} W(t) e^{-i\omega t} dt \quad \text{and} \quad b = 2P \int_{-\infty}^{\infty} W(t) dt. \tag{9}$$

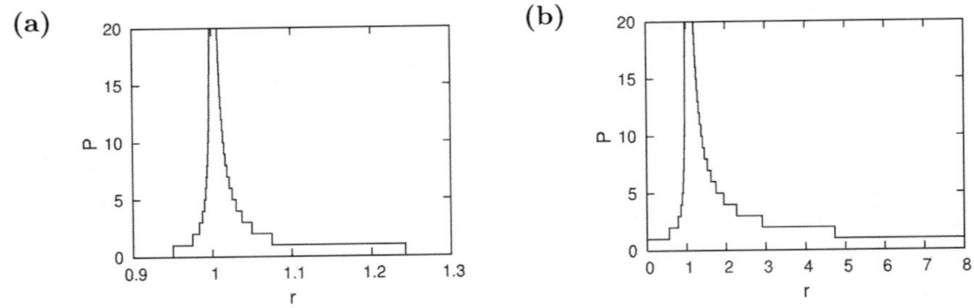

Fig. 4. (a) The storage capacity (the upper limit of pattern number for successful retrieval) in the associative memory analog neural networks for periodic spatiotemporal patterns is plotted as a function of r. Parameter r controls the size of the negative part in the STDP time window (see Eq. (2)). (b) The same as (a) for associative memory spiking neurons for periodic spatiotemporal patterns.

Figure 3(b) explains a result of a numerical simulation of an analog neural network based on the learning rule (8). The first pattern described in Fig. 3(a) is successfully retrieved in the numerical simulation in Fig. 3(b).

To quantify degree of retrieval of patterns, it is useful to define overlaps

$$m^\mu = (1/N) \sum_i \xi_i^\mu x_i. \tag{10}$$

During retrieval of the first pattern, overlap m^1 shows periodic oscillation with the constant amplitude as shown in Fig. 3(c), while $|m^2|$ and $|m^3|$ take the considerably small value, reflecting non-retrieval of the second and the third patterns. In such a case, the local field (5) is rewritten in the form

$$h_i = \mathrm{Re}\left(a\xi_i^1 m^{1*}\right) + bX, \tag{11}$$

where $X = (1/N)\sum_i x_i$. The oscillation of m^1 causes oscillation of local field h_i with phase shift $-\arg(a\xi_i^1)$, which brings about oscillation of x_i with the high correlation with ξ_i^1. Then, this oscillation of x_i sustains oscillation of m^1. When we assume a finite number of patterns $P = \mathcal{O}(1)$ and an infinite number of neurons $N \to \infty$, this dynamics of overlaps is expressed as

$$dm^\mu/dt = -m^\mu + \langle \xi^\mu F\left(\textstyle\sum_\nu \mathrm{Re}\left(a\xi^\nu m^{\nu*}\right) + bX\right)\rangle, \quad \mu = 1, \ldots, P, \tag{12}$$

$$dX/dt = -X + \langle F\left(\textstyle\sum_\nu \mathrm{Re}\left(a\xi^\nu m^{\nu*}\right) + bX\right)\rangle. \tag{13}$$

where $\langle \ldots \rangle$ represents the average over $\xi^\mu = \exp(i\phi^\mu)$. The retrieval process of the analog neural networks thus can be investigated in the analytical manner.

3.1 The Optimal Shape of the STDP Time Window to Encode a Large Number of Spatiotemporal Patterns

Stationary solutions in the dynamics (12) and (13) clarify the various important features of the spatiotemporal analog neural networks. For example, substitution

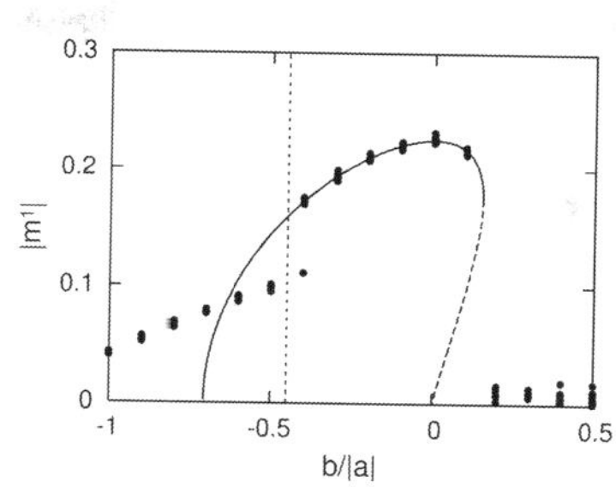

Fig. 5. The stability analysis of stationary solution in the associative memory analog neural networks. The solid line represents stationary solution with negative $|\lambda_1|$ and $|\lambda_2|$, and the dotted line represents stationary solution with positive $|\lambda_1|$ and/or positive $|\lambda_2|$. $|\kappa|$ takes a negative value only beyond the critical point indicated by the dotted vertical line. Dots represent results of numerical simulations conducted with $P = 2$. Eigenvalue κ well explains the phase transition observed in the numerical simulations.

of the stationary solution of the form $m^1 = |m^1| \exp(i\tilde{\omega}t)$, $m^\mu = 0$, $(2 \leq \mu)$, and $d|m^1|/dt = dX/dt = 0$ yields $\tilde{\omega} = -\tan \arg(a)$, implying that retrieval of patterns occurs at the different frequency than oscillation frequency of encoded pattern ω. Note that such time scale change in retrieval occurs also in learning of Poisson patterns shown in Fig. 2 (Compare the time scales of Fig. 2(a) and Fig. 2(b)). It also turns out that the parameter b yields the strong influence on the pattern-number dependence of the stationary solution since parameter b defined in Eqs. (9) is proportional to P as well as the time average of the time window $\overline{W} = \int W(t)dt$. In Fig. 4(a), we plot the analytically derived storage capacity P^c as a function of r, where parameter r controls the negative part of the STDP time window as defined in Eq. (2). In general, the retrieval state disappears with finite P since increase of bX term in the local field (11) interferes the existence of meaningful stationary solution. At $r = 1$, however, we see the divergence of the storage capacity P^c since the average $\overline{W} = 1 - r$ vanishes at $r = 1$. The storage capacity of the analog neural networks is hence maximized when the area of the negative part of the STDP time window has the same size as the positive part so that the time average $\overline{W}$ takes the zero value.

3.2 Stability of Retrieval State for Single-Pattern Learning ($P = 1$) and for Multiple-Pattern Learning ($2 \leq P$)

Not all of stationary solutions in the dynamics (12) and (13) are stable. To discuss the meaningful solution for retrieval state, we need to analyze stability of stationary solutions. It is interesting that the present analog neural networks show the different stability condition between the cases of single-pattern learning ($P = 1$) and of multiple-pattern learning ($2 \leq P$).

When a single pattern is encoded ($P = 1$), linear stability of the stationary solution is evaluated from the dynamics $\frac{d}{dt}\left(\delta X\ \delta|m^1|\right)^{\mathrm{T}} = \mathbf{A}\left(\delta X\ \delta|m^1|\right)^{\mathrm{T}}$, where δX and $\delta|m^1|$ represent deviation from the stationary solution. When two

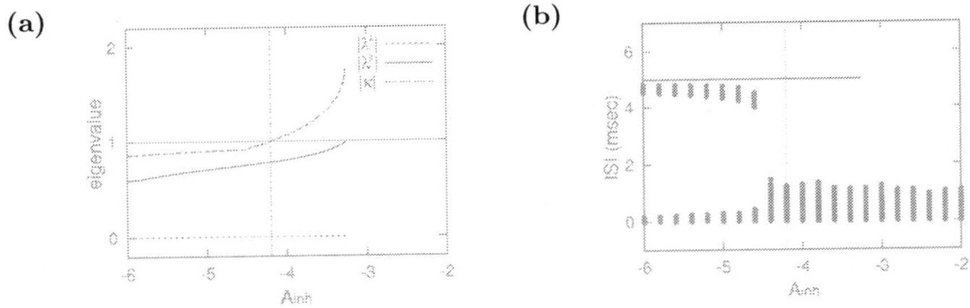

Fig. 6. Analysis and numerical simulations of associative memory spiking neural networks for periodic spatiotemporal patterns. In this spiking neural network, periodic external inhibition is applied to all of spiking neurons so as to discretize spike timings. (a) Absolute values of eigenvalues $\lambda^{\perp}$, $\lambda^{\parallel}$, and κ are plotted as a function of intensity of inhibition A_{inh}. The dotted vertical line represents the critical point at which κ takes the zero value. (b) ISIs of all neurons in a numerical simulation are plotted as a function of A_{inh} (For the ISIs of all neurons, see Refs. [17,18].) The thick line represents analytically derived ISIs, and the vertical line represents the critical point due to κ.

eigenvalues λ_1 and λ_2 of matrix $\mathbf{A}$ have negative real parts, stationary solution is stable.

When we encode a multiple number of patterns ($2 \leq P$), the linear stability analysis yields $P+1$ eigenvalues $\lambda_1, \lambda_2, \ldots, \lambda_{P+1}$. It can be shown that these eigenvalues include $(P-1)$-fold eigenvalue κ. Thus, the stability of stationary solution in multiple-pattern learning is determined by real parts of eigenvalues λ_1, λ_2, and $(P-1)$-fold eigenvalue κ.

Under a certain condition, the $(P-1)$-fold eigenvalue κ causes phase transition as shown in Fig. 5. The solid line in Fig. 5 represents stationary solution with negative $|\lambda_1|$ and $|\lambda_2|$, while the solution with the dotted line has positive $|\lambda_1|$ and/or positive $|\lambda_2|$. Therefore, the solution with the solid line is stable for single-pattern learning ($P = 1$). For multiple-pattern learning ($2 \leq P$), however, the solution with the solid line is stable only beyond the critical point indicated by the dotted vertical line since $|\kappa|$ takes the positive value below the vertical line. Dots in Fig. 5 represents results of numerical simulations in which two periodic patterns are encoded. The critical point determined by the eigenvalue κ well explains the sudden change of the amplitude of overlap in the numerical simulations.

4 Analysis of Retrieval Process of Periodic Spatiotemporal Patterns in Spiking Neural Networks

Analysis of spiking neural networks requires the different treatment of mean field interactions than the overlap dynamics in analog neural networks[17,18]. Nevertheless, some of the features of the analog neural networks have also been found in associative memory spiking neural networks for periodic spatiotemporal patterns. Here, we briefly summarize these similarities between analog and spiking neural networks in learning of periodic spatiotemporal patterns.

It has been shown that term proportional to $\overline{W}$ and P appears also in synaptic electric currents in the spiking neurons. As a result, the storage capacity of the spiking neural networks shown in Fig. 4(b) diverges at $r = 1$ in the same manner as Fig. 4(a) . This highlights the general importance of $\overline{W} = 0$ in encoding a large number of spatiotemporal patterns.

Figure 6(a) explains the result of the stability analysis of associative memory of spiking neural network for periodic patterns in which periodic external inihibition is applied to all of spiking neurons so as to discretize spike timings. When a single pattern is encoded, stability of stationary solution is characterized by two types of eigenvalues; eigenvalue $\lambda^{\perp}$ characterizes stability of synchronization in discretized spike timings and eigenvalue $\lambda^{\parallel}$ characterizes stability in dynamics among clusters of synchronized neurons. In addition, with a multiple number of patterns, degenerate eigenvalue κ appears as in the case of analog neural networks. Note that $|\kappa|$ exceeds one at the point indicated by the dotted vertical line in Fig. 6(a). This critical point due to κ causes the phase transition in numerical simulations in Fig. 6(b), following the same manner as Fig. 5.

5 Discussion

We have studied the STDP-based spatiotemporal learning in recurrent neural networks. First, we have introduced the general ideas of the STDP-based spatiotemporal learning, demonstrating the numerical simulation of spiking neural networks that memorize spatiotemporal Poisson patterns (Fig. 2). Then, we discuss the some key features of the STDP-based spatiotemporal learning, following our recent analysis on learning of periodic patterns in analog neural networks. Stationary solution of the order parameter dynamics (12) and (13) in the analog neural networks elucidates that spatiotemporal patterns encoded by the STDP-based learning rule are retrieved at the different time scale than the original one. It also has turned out that the storage capacity is maximized when the negative part of the STDP time window has the same size as the positive part (Fig. 4). The further analysis has shown the influence of the eigenvalues charactering stability of stationary solution on retrieval performance (Fig. 5). Then, we have briefly compared these analytical results on analog neural networks with spiking neural networks (Figs. 4 and 6).

It is remarkable that analog neural networks and spiking neural networks show the similar retrieval properties in the learning of periodic patterns. It has also been found that the STDP-based learning permits analog neural network to memorize spatiotemporal Poisson patterns of bursting activities [21]. Wave forms of analog neuron activities during retrieval of spatiotemporal patterns, however, differ from those of memorized patterns as shown in Fig. 3, implying the difficulty of the STDP-based learning in encoding amplitude of slowly changing neural activities. The dynamics of overlaps (12) and (13) is useful to understand retrieval of periodic patterns in analog neural networks, but seems to be difficult to be applied to the case of Poisson patterns. The different approach discussed in the studies on spiking neural networks [17,18] possibly gives the satisfactory approximate of the case of Poisson patterns, especially when spikes during retrieval are sufficiently sparse. Eigenvalue κ in Fig. 6(a) is not discussed in Ref. [18] since some restrictions on perturbations are assumed in that stability analysis. We have derived eigenvalue κ from the more general stability analysis without the restrictions.

The retrieval frequency of the analog neural network $\tilde{\omega} = -\tan\arg(a)$ can take a negative value, that is to say, retrieval of spatiotemporal patterns in analog neural networks can occur in the reversed order with respect to time. Interestingly, such odd retrieval of reversed spatiotemporal patterns has recently been observed in the Hippocampus of the awaking rat[26]. The normal order replay at the faster time scale has also been found to occur in sleeping rats[3]. The time scale change discussed in our present study possibly explains various time scales in memory replay in the real nervous system.

References

1. MacLeod, K., Laurent, G.: Science 274, 976 (1996)
2. Lorenzo, P.M.D., Hallock, R.M., Kennedy, D.P.: Behavior Neuroscience 117, 1423 (2003)
3. Nádasdy, Z., Hirase, H., Czurkó, A., Csicsvari, J., Buzsáki, G.: J. Neurosci. 19, 9497 (1999)
4. Hopfield, J.J.: Proc. Natl. Acad. Sci. USA 79, 2554 (1982)
5. Amit, D.J., Gutfreund, H., Sompolinsky, H.: Phys. Rev. Lett. 55, 1530 (1985)
6. Amari, S.: IEEE Transactions on Computers C-21, 1197 (1972)
7. Matsumoto, N., Okada, M.: Neural Comp. 14, 2883 (2002)
8. Sompolinsky, H., Kanter, I.: Phys. Rev. A 57, 2861 (1986)
9. Coolen, A.C.C., Ruijgrok, T.W.: Phys. Rev. A 38, 4253 (1988)
10. Cook, J.: J. Phys. A 22, 2057 (1989)
11. Arenas, A., Vicente, C.J.P.: Europhys. Lett. 26, 79 (1994)
12. Kühn, R., Bös, S., van Hemmen, J.L.: Phys. Rev. A 43, 2084 (1991)
13. Amari, S., Maginu, K.: Neural Netw. 1, 63 (1988)
14. Nishimori, H., Nakamura, T., Shiino, M.: Phys. Rev. A 41, 3346 (1990)
15. Shiino, M., Fukai, T.: J. Phys. A 25, L375 (1992)
16. Okada, M.: Neural Netw. 8, 833 (1995)
17. Yoshioka, M.: Phys. Rev. E 65, 011903 (2001)
18. Yoshioka, M.: Phys. Rev. E 66, 061913 (2002)
19. Scarpetta, S., Zhaoping, L., Hertz, J.: Neural Comp. 14, 2371 (2002)
20. Scarpetta, S., Marinaro, M.: Hippocampus 15, 979 (2005)
21. Yoshioka, M., Scarpetta, S., Marinaro, M.: Phys. Rev. E 75 (2007) 051917
22. Markram, H., Lübke, J., Frotscher, M., Sakmann, B.: Science 275, 213 (1997)
23. Bi, G.Q., Poo, M.M.: J. Neurosci. 18, 10464 (1998)
24. Zhang, L.I., Tao, H.W., Holt, C.E., Harris, W.A., Poo, M.M.: Nature 395, 37 (1998)
25. Bi, G.Q., Poo, M.M.: Annu. Rev. Neurosci. 24, 139 (2001)
26. Foster, D.J., Wilson, M.A.: Nature 440, 680 (2006)

A Distributed Message Passing Algorithm for Sensor Localization

Max Welling and Joseph J. Lim

Bren School of Information and Computer Science
UC Irvine, CA 92697-3425, USA
{welling, josephjl}@ics.uci.edu

Abstract. We propose a fully distributed message passing algorithm based on expectation propagation for the purpose of sensor localization. Sensors perform noisy measurements of their mutual distances and their relative angles. These measurements form the basis for an iterative, local (i.e. distributed) algorithm to compute the sensor's locations including uncertainties for these estimates. This approach offers a distributed, computationally efficient and flexible framework for information fusion in sensor networks.

1 Introduction

Sensor localization and information fusion have become an active area of research in recent years. It has become clear that distributed approaches will become increasingly important to handle the computational challenges in large scale networks. Many such approaches have been proposed in the past (e.g. [5,3,4] and many more), but very few can handle uncertainty in a distributed computational setting. The most well-known approach (not presented as a distributed algorithm) is based on the Fisher information [3], but this method can be shown to grossly under-estimate uncertainty for high noise levels. The state-of-the-art in this field is presumably "non-parametric belief propagation" that propagates sample sets over the edges the graph and can as such handle non-gaussian distributions [1]. The price being paid for this flexibility is that the algorithm is computationally intensive, a factor that is not unimportant in applications where energy is on a tight budget.

In this paper we propose a method which trades computation with accuracy. The proposed method maintains and communicates Gaussian estimates (which can be represented by 5 real numbers) instead of sample sets. The price being paid is obviously that it cannot represent multi-modal distributions. The underlying algorithm is expectation propagation (EP), a form of belief propagation that has recently been introduced in the context of Bayesian modeling [2]. We have adapted that framework to the problem of sensor localization and resolved some numerical issues in the process.

2 A MRF Model for Sensor Localization

Our general approach is based on constructing a Markov random field model (MRF) for the joint distribution over all sensor locations $\{\mathbf{x}_i,\ i = 1..N\}$ with $\mathbf{x} \in \mathbb{R}^2$. The general

J. Marques de Sá et al. (Eds.): ICANN 2007, Part I, LNCS 4668, pp. 767–775, 2007.

expression for a MRF with pairwise interactions is given by,

$$P(\mathbf{x}_1, .., \mathbf{x}_N) \propto \prod_{ij \in \mathcal{E}} \psi_{ij}(\mathbf{x}_i, \mathbf{x}_j) \prod_{i \in \mathcal{V}} \psi_i(\mathbf{x}_i) \tag{1}$$

where $\psi_{ij}(\mathbf{x}_i, \mathbf{x}_j)$ denotes an interaction potential between sensors i and j and $\psi_i(\mathbf{x}_i)$ a node potential which could for instance encode an absolute measurement of sensor position.

The problem that we need to solve consist of two parts: 1) finding suitable expressions for the potential functions and 2) solving for the marginal distribution $p(\mathbf{x}_i),\ \forall i$. We'll first deal with the first question.

Interactions between sensors consist of two types in this paper: noisy distance measurements between sensors and noisy angle measurements between sensors. The noisy distance measurement is modeled using a gamma distribution for the *squared* distances. The reason we use squared distances is computational. More precisely we use,

$$\psi_{ij}^{\text{dist}}(\mathbf{x}_i, \mathbf{x}_j) \propto r_{ij}^{2(\alpha_{ij}-1)} \exp[-\beta_{ij} r_{ij}^2] \tag{2}$$

where $r_{ij} = ||\mathbf{x}_i - \mathbf{x}_j||$ and α_{ij} and β_{ij} are free parameters. The interaction between angles is based on the Von-Mises distribution,

$$\psi_{ij}^{\text{angle}}(\mathbf{x}_i, \mathbf{x}_j) \propto \exp[\kappa \cos(\theta_{ij} - \rho_{ij})] \tag{3}$$

where $\theta_{ij} = \arctan_2(\mathbf{x}_i - \mathbf{x}_j)$ denotes the angle[1] between the sensors, measured relative to absolute north (this requires the sensors to carry a compass) . The constant κ controls the variance of this distribution while ρ_{ij} denotes its mean.

In the following we will assume that the variance of these distributions is determined offline, i.e. the distribution over measurement errors is assumed known. On the other hand, a single measurement will determine the value for the mode of the distribution. Given a particular observation $r_{ij} = d_{ij}, \theta_{ij} = \phi_{ij}$ on the edge ij and given a variance v on squared distance measurements and ν on relative angle measurements we derive the following values for the parameters,

$$\beta_{ij} = (d_{ij}^2 + \sqrt{d_{ij}^4 + 4v})/(2v) \tag{4}$$

$$\alpha_{ij} = 1 + \beta_{ij} d_{ij}^2 \tag{5}$$

$$\rho_{ij} = \phi_{ij} \tag{6}$$

$$\kappa = F(\nu) \tag{7}$$

where $F(\nu)$ is obtained by inverting the relation $\nu = 1 - \frac{I_1(\kappa)^2}{I_0(\kappa)^2}$ with I_j a Bessel function of order j. As mentioned, these parameter settings will place the mode of the distributions at the observed values and sets the variance equal to the given values.

In this paper we will not use any node potentials. This implies that we don't have access to any absolute location information and that the localization can only be successful up to a global translation of the coordinate frame. Note that we do have access

[1] We use the "four quadrant inverse tangent" $\arctan_2(\cdot)$ so that $\mathbb{R}$ is mapped to $[-\pi, \pi]$.

to absolute north and that the absolute orientation can indeed be recovered. In this paper we will assume that the translational degrees of freedom will be determined by a separate procedure. To align results with ground truth we will perform a global transformation which minimizes root mean squared error (RMSE). By adding node potentials we could in fact have avoided this separate phase.

The final MRF is now defined as the product of all the interaction potentials, i.e.

$$P(\mathbf{x}_1, .., \mathbf{x}_N) \propto \prod_{ij \in \mathcal{E}} \psi_{ij}^{\text{dist}}(\mathbf{x}_i, \mathbf{x}_j) \psi_{ij}^{\text{angle}}(\mathbf{x}_i, \mathbf{x}_j) \tag{8}$$

We will be interested in computing the marginal distribution,

$$P(\mathbf{x}_i) = \int d\mathbf{x}_1..d\mathbf{x}_{i-1} d\mathbf{x}_{i+1}..d\mathbf{x}_N P(\mathbf{x}_1, .., \mathbf{x}_N) \tag{9}$$

This problem is intractable, but we will employ an effective approximation technique to find Gaussian projections of these marginals.

3 Localization by Expectation Propagation

The computation of the marginal distributions $P(\mathbf{x}_i)$ is intractable because $\mathbf{x}$ is continuous and the interaction potentials are nonlinear. Expectation propagation offers an approximate solution. It sends messages over the edges of the communication graph (corresponding to pairs of sensors that can exchange information). Each message is loosely interpreted as an estimate by sensor i of the distribution over locations for sensor j as approximated by a normal distribution. This estimate is based on similar estimates from sensor i'th neighbors but excluding sensor j in that estimate (this is done to avoid evidence coming from j to directly flow back to j). For more details we refer to [2].

Applied to our problem we maintain the following quantities: 1) Gaussian estimates of the marginal distributions over sensor locations $b_i(\mathbf{x}_i)$ and 2) Gaussian messages $M_{ij}(\mathbf{x}_j)$. Messages and marginals relate in a simple way,

$$b_i(\mathbf{x}_i) \propto \prod_{j \in \mathcal{N}_i} M_{ji}(\mathbf{x}_i) \tag{10}$$

The algorithm iteratively updates messages and marginals as we show in the algorithm box below. Note that the computation involves a "projection operator" $\mathcal{P}$, which converts a general distribution in the plane to the closest normal distribution in KL-divergence by matching the first and second moments. The updates also involve a difficult integration over $\mathbf{x}_j$ which should be done numerically in general. The constant $\gamma \in [0, 1)$ is a damping factor and should be increased if the system has difficulty converging. We note that we have used $\gamma = 0.7$ in all our experiments which was sufficient for the algorithm to consistently converge.

In applying the general form of the EP algorithm to sensor localization one runs into some numerical issues. Firstly, the integral has to be performed in a four dimensional space which may be computationally taxing. Fortunately, one can analytically

Algorithm 1. EP Algorithm

1. Initialize:

1A. $M_{j\to i}(\mathbf{x}_i)$ as random Gaussian distributions

1B. $b_i(\mathbf{x}_i) \propto \prod_{j\in\mathcal{N}_i} M_{j\to i}(\mathbf{x}_i)$

2. Repeat until convergence: update messages $(j \to i)$.

2A. $P_{ij}(\mathbf{x}_i) = \mathcal{P}\int d\mathbf{x}_j\ \psi_{ij}(\mathbf{x}_i,\mathbf{x}_j)\frac{b_i(\mathbf{x}_i)b_j(\mathbf{x}_j)}{M_{i\to j}(\mathbf{x}_j)M_{j\to i}(\mathbf{x}_i)}$

2B. $b_{i,\text{new}}(\mathbf{x}_i) = P_{ij}(\mathbf{x}_i)^{1-\gamma} b_i(\mathbf{x}_i)^{\gamma}$

2C. $M_{j\to i,\text{new}}(\mathbf{x}_i) = \left(\frac{P_{ij}(\mathbf{x}_i)}{b_i(\mathbf{x}_i)}\right)^{1-\gamma} M_{j\to i}(\mathbf{x}_i)$

integrate out two dimensions by observing that all interaction potentials only depend on the relative coordinates $\mathbf{x}_- = \mathbf{x}_i - \mathbf{x}_j$ and not on the coordinates $\mathbf{x}_+ = \frac{1}{2}(\mathbf{x}_i + \mathbf{x}_j)$. The algorithm detailed below first transforms to these new coordinates, performs the integral over $\mathbf{x}_+$ analytically while the two- dimensional integral over $\mathbf{x}_-$ is performed numerically (e.g. by importance sampling). Finally, the the results are converted back to the space $\mathbf{x}_i, \mathbf{x}_j$ and the messages are updated.

Numerical instabilities can also occur because we need to perform an integral over a product of local densities and then renormalize. If two of these terms peak at different locations and decay fast, their product is very small (e.g. smaller than machine precision) everywhere. The result is that we cannot reliably compute the normalization constant. We have circumvented this problem by merging the exponential term of the gamma distribution in eqn.2 with the normal distributions for the messages and marginals (see step 2A. of the "EP Algorithm" box). We then sample from the resulting Gaussian distribution and evaluate the remaining terms at the sampled values in order to approximate the integral. The final algorithm is provided in the algorithm box below.

4 Experimental Results

In the following experiments we will study the influence of various parameters on localization accuracy. Accuracy is quantified using the root mean squared error (RMSE),

$$RMSE = \sqrt{\frac{2}{N(N-1)}\sum_{i<j} ||\mathbf{y}_i - \mathbf{y}_j||^2} \tag{11}$$

Before computing the error we first align the estimates by computing the optimal rotation and translation as follows: first translate both the true sensors locations and the predicted sensor locations to the origin: $\mathbf{x}_i \to \mathbf{x}_i - \mu$. Call $\mathbf{x}$ the $2 \times N$ matrix of true centered locations and $\mathbf{y}$ the predicted locations. Next decompose $\mathbf{x}\mathbf{y}^T$ using an SVD: $USV^T = \mathbf{x}\mathbf{y}^T$ and transform $\mathbf{y} \to \mathbf{y}' = \mathbf{y}UV^T$.

The error in the uncertainty ellipse was assessed using the following measure,

$$E_C = \frac{1}{N}\sum_i \frac{|\text{Trace}(C_i^{\text{est}}) - \text{Trace}(C_i^{\text{true}})|}{\text{Trace}(C_i^{\text{true}})} \tag{12}$$

Algorithm 2. SLEEP Algorithm

1. Input: $\gamma,\ \ \alpha_{ij},\beta_{ij},\kappa,\rho_{ij}\ \forall ij,$ Output: $\Sigma_{i,\text{new}},\mu_{i,\text{new}}\ \forall i$

2. Initialize: $V_{i\to j},\nu_{i\to j}\ \ \forall(i\to j)$

2A. $\Sigma_i^{-1} = \sum_{j\in\mathcal{N}(i)} V_{j\to i}^{-1}\ \ \forall i$

2B. $\mu_i = \Sigma_i(\sum_{j\in\mathcal{N}(i)} V_{j\to i}^{-1}\nu_{j\to i})\ \ \forall i$

3. Repeat until convergence:

choose edge ij according to schedule and update.

3A. Remove messages $i\to j$ and $j\to i$:

3Aa. $\Sigma_{i\backslash j}^{-1} = \Sigma_i^{-1} - V_{ji}$

3Ab. $\Sigma_{j\backslash i}^{-1} = \Sigma_j^{-1} - V_{ij}$

3Ac. $\mu_{i\backslash j} = \Sigma_{i\backslash j}(\Sigma_i^{-1}\mu_i - V_{ji}^{-1}\nu_{ji})$

3Ad. $\mu_{j\backslash i} = \Sigma_{j\backslash i}(\Sigma_j^{-1}\mu_j - V_{ij}^{-1}\nu_{ij})$

3B. Transform to relative coordinates:

3Ba. $\Sigma_+ = \Sigma_{i\backslash j} + \Sigma_{j\backslash i}$

3Bb. $\Sigma_- = \Sigma_{i\backslash j} - \Sigma_{j\backslash i}$

3Bc. $\mu_+ = \frac{1}{2}(\mu_{i\backslash j} + \mu_{j\backslash i})$

3Bd. $\mu_- = \mu_{i\backslash j} - \mu_{j\backslash i}$

3C. Absorb Gamma distribution into Gaussian:

3Ca. $C = (I + 2\beta\Sigma_+)^{-1}$

3Cb. $K = C\Sigma_+$

3Cc. $\mathbf{s} = C\mu_-$

3D. Compute the following moments:

3Da. $Z = \int d\mathbf{x}_-\ \exp[\kappa\cos(\arctan_2(\mathbf{x}_-) - \rho_{ij})]\ ||\mathbf{x}_-||^{2\alpha-2}\mathcal{N}_{\mathbf{x}_-}[\mathbf{s},K]$

3Db. $E[\mathbf{x}_-] = \int d\mathbf{x}_-\ \exp[\kappa\cos(\arctan_2(\mathbf{x}_-) - \rho_{ij})]\ ||\mathbf{x}_-||^{2\alpha-2}\ \mathbf{x}_-\mathcal{N}_{\mathbf{x}_-}[\mathbf{s},K]/Z$

3Dc. $E[\mathbf{x}_-\mathbf{x}_-^T] = \int d\mathbf{x}_-\ \exp[\kappa\cos(\arctan_2(\mathbf{x}_-) - \rho_{ij})]\ ||\mathbf{x}_-||^{2\alpha-2}\ \mathbf{x}_-\mathbf{x}_-^T\mathcal{N}_{\mathbf{x}_-}[\mathbf{s},K]/Z$

3E. Compute moments for $\mathbf{x}_+$ variables:

3Ea. $\Lambda = \frac{1}{2}\Sigma_-\Sigma_+^{-1},\quad \delta = \mu_+ - \mathbf{\Lambda}\mu_-$

3Eb. $E[\mathbf{x}_+] = \delta + \mathbf{\Lambda}E[\mathbf{x}_-]$

3Ec. $E[\mathbf{x}_+\mathbf{x}_+^T] = \frac{1}{4}(\Sigma_+ - \Sigma_-\Sigma_+^{-1}\Sigma_-) + \delta\delta^{\mathbf{T}} + \delta E[\mathbf{x}_-]^{\mathbf{T}}\mathbf{\Lambda}^{\mathbf{T}} + \mathbf{\Lambda}E[\mathbf{x}_-]\delta^{\mathbf{T}} + \mathbf{\Lambda}E[\mathbf{x}_-\mathbf{x}_-^{\mathbf{T}}]\mathbf{\Lambda}^{\mathbf{T}}$

3Ed. $E[\mathbf{x}_+\mathbf{x}_-^T] = \delta E[\mathbf{x}_-]^{\mathbf{T}} + \mathbf{\Lambda}E[\mathbf{x}_-\mathbf{x}_-^{\mathbf{T}}]$

3Ee. $E[\mathbf{x}_-\mathbf{x}_+^T] = E[\mathbf{x}_+\mathbf{x}_-^T]^T$

3F. Transform back to sensor coordinates:

3Fa. $E[\mathbf{x}_i] = E[\mathbf{x}_+] + \frac{1}{2}E[\mathbf{x}_-]$

3Fb. $E[\mathbf{x}_j] = E[\mathbf{x}_+] - \frac{1}{2}E[\mathbf{x}_-]$

3Fc. $E[\mathbf{x}_i\mathbf{x}_i^T] = E[\mathbf{x}_+\mathbf{x}_+^T] + \frac{1}{4}E[\mathbf{x}_-\mathbf{x}_-^T] + \frac{1}{2}(E(\mathbf{x}_+\mathbf{x}_-^T] + E(\mathbf{x}_-\mathbf{x}_+^T])$

3Fd. $E[\mathbf{x}_j\mathbf{x}_j^T] = E[\mathbf{x}_+\mathbf{x}_+^T] + \frac{1}{4}E[\mathbf{x}_-\mathbf{x}_-^T] - \frac{1}{2}(E(\mathbf{x}_+\mathbf{x}_-^T] + E(\mathbf{x}_-\mathbf{x}_+^T])$

3G. Compute candidate sensor locations and covariances:

3Ga. $\mu_{i,\text{cand}} = E[\mathbf{x}_i]$

3Gb. $\mu_{j,\text{cand}} = E[\mathbf{x}_j]$

3Gc. $\Sigma_{i,\text{cand}} = E[\mathbf{x}_i\mathbf{x}_i^T] - \mu_{i,\text{cand}}\mu_{i,\text{cand}}^T$

3Gd. $\Sigma_{j,\text{cand}} = E[\mathbf{x}_j\mathbf{x}_j^T] - \mu_{j,\text{cand}}\mu_{j,\text{cand}}^T$

3H. Compute new sensor locations and covariances:

3Ha. $\Sigma_{i,\text{new}}^{-1} = (1-\gamma)\Sigma_{i,\text{cand}}^{-1} + \gamma\Sigma_i^{-1}$

3Hb. $\Sigma_{j,\text{new}}^{-1} = (1-\gamma)\Sigma_{j,\text{cand}}^{-1} + \gamma\Sigma_j^{-1}$

3Hc. $\mu_{i,\text{new}} = \Sigma_{i,\text{new}}((1-\gamma)\Sigma_{i,\text{cand}}^{-1}\mu_{i,\text{cand}} + \gamma\Sigma_i^{-1}\mu_i)$

3Hd. $\mu_{j,\text{new}} = \Sigma_{j,\text{new}}((1-\gamma)\Sigma_{j,\text{cand}}^{-1}\mu_{j,\text{cand}} + \gamma\Sigma_j^{-1}\mu_j)$

3I. Compute new messages:

3Ia. $V_{ji,\text{new}}^{-1} = V_{ji}^{-1} + (1-\gamma)[\Sigma_{i,\text{new}}^{-1} - \Sigma_i^{-1}]$

3Ib. $V_{ij,\text{new}}^{-1} = V_{ij}^{-1} + (1-\gamma)[\Sigma_{j,\text{new}}^{-1} - \Sigma_j^{-1}]$

3Ic. $\nu_{ji,\text{new}} = V_{ji,\text{new}}(V_{ji}^{-1}\nu_{ji} + (1-\gamma)[\Sigma_{i,\text{new}}^{-1}\mu_{i,\text{new}} - \Sigma_i^{-1}\mu_i])$

3Id. $\nu_{ij,\text{new}} = V_{ij,\text{new}}(V_{ij}^{-1}\nu_{ij} + (1-\gamma)[\Sigma_{j,\text{new}}^{-1}\mu_{j,\text{new}} - \Sigma_j^{-1}\mu_j])$

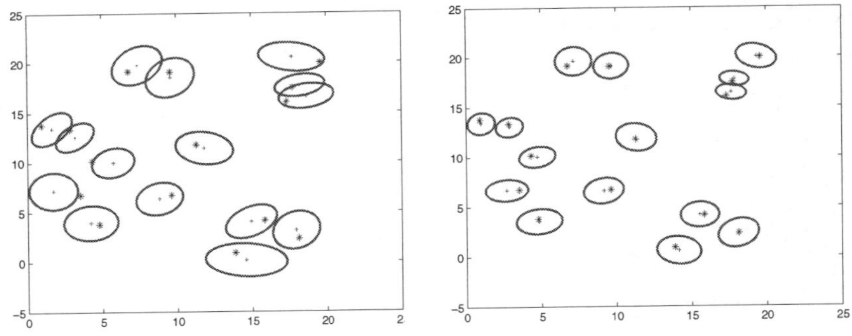

Fig. 1. Left: Localization results with $v = 0.2$ and infinite communicating range but ignoring angle information. Right: same as left but now including angle information with $\kappa = 3$.

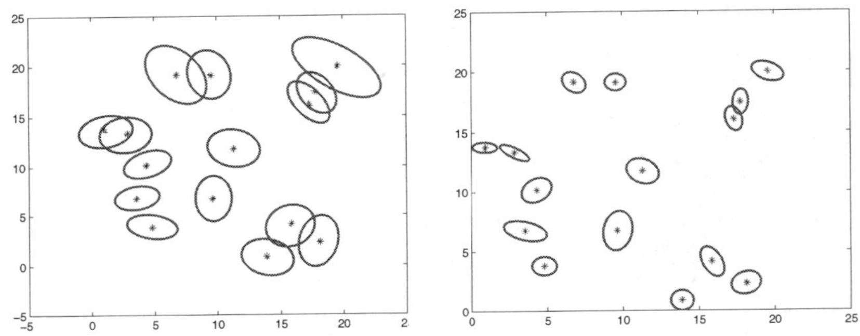

Fig. 2. Left: Ground truth Localization with $v = 0.2$ and infinite communicating range but ignoring angle information. Right: same as left but now including angle information with $\kappa = 3$.

where C_i is the covariance of sensor i. Note that we divide by Trace(C_i^{true}) to divide out the scale of the variance so that smaller variances do not automatically imply smaller error (i.e. we try to quantify the error in the size and shape of the ellipse).

In our experiments we first placed sensors randomly on a square. Then, we perturbed the true distances into noisy distances by sampling them from a log-normal distribution centered at the true distance and with variance v. We used a log-normal distribution because it was shown in [4] to provide a good fit to actual distance measurements. A similar procedure was followed to obtain noisy measurements for the angles which we were sampled from a Von-Mises distribution with a certain value for κ.

We estimated ground truth for the sensor location uncertainty by drawing many samples according to the procedure above and for each of them minimizing the log-probability of the sensor locations using a standard optimization package[2]. Note that this procedure can not be used in a non-simulated environment because we typically only have access to a single sample of a distance measurement. For each marker in a plot we ran the algorithm 30 times and computed average performance measures and their standard deviations shown as error-bars.

[2] http://www.kyb.tuebingen.mpg.de/bs/people/carl/code/minimize.old/minimize.m

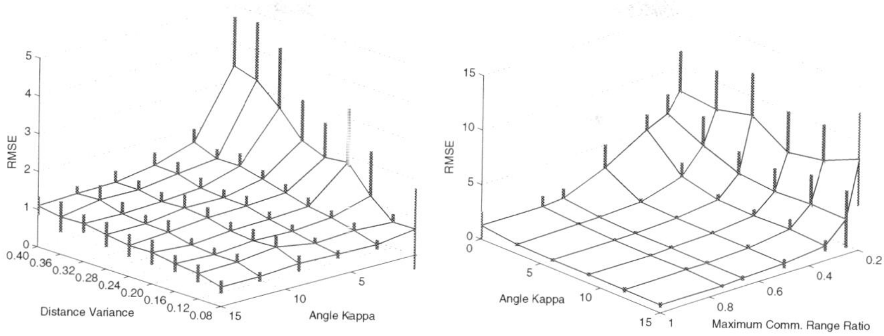

Fig. 3. Left: RMSE as a function of variance in squared distance measurements (v) and κ. All sensors communicate. κ is inversely related to the variance in the angle measurement, i.e. larger κ implies smaller variance. Right: RMSE as a function of communication range and κ ($v = 0.2$). A communication range of 1 is equal to the diagonal spanning the box in which the sensors are placed. In both plot there are $N = 15$ sensors.

In the following examples we vary over the settings of a number variables. The standard settings are given by: $v = 0.2$, $N = 15$ and infinite communication range.

In figure 1 we show two runs of our algorithm, one without angle information (left) and one with angle information (right). It can clearly be observed that the angle information is helpful in disambiguating sensors which are nearby. Figure 2 shows the corresponding ground truth uncertainty regions.

In figure 3 (left) we show RMSE as a function of the variance in distance measurements and as a function of κ which is inversely related to variance in angle measurements. One can observe that even small values of κ around $\kappa = 2$ (corresponding to a std. of approximately $\pi/4$ in angle) can be very helpful in improving the accuracy of the location estimates.

In figure 3 (right) we plot RMSE as a function of the communication range (the fraction of the maximal (diagonal) distance in our 20 by 20 window) and as a function of κ. Again, angle information can help building good maps and is especially helpful when the communication range is small.

Figure 4 (left) shows RMSE for varying κ and N (the number of sensors). As expected, more sensors are helpful and bigger κ is helpful. For values of $\kappa > 3$ we see diminishing returns.

Finally, we studied the error in the uncertainty itself. Figures 4 (right) and 5 (left and right) show the dependence of E_C w.r.t. κ, v, N and communication range. While decreasing the variance in the distance measurements improves the estimates for uncertainty, the trend for the variance in angle measurements in reversed: less variance makes the estimation harder. We suspect that decreasing variance will improve E_C if the location estimates themselves are bad because in that case estimating uncertainty is essentially impossible. However, if the sensor map is in the right ballpark then there may be another effect which offsets this. Decreasing the variance implies that the sensors are more tightly coupled and that loopy inference by EP deteriorates as evidence over-counting becomes a problem. This assessment is supported by the fact that more

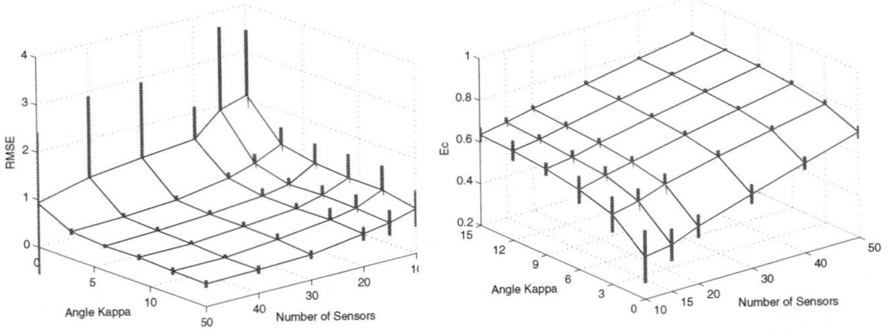

Fig. 4. Left: RMSE as a function of the number of sensors N and κ ($v = 0.2$ and $N = 15$). Right: E_C as a function of the number of sensors N and κ ($v = 0.2$ and $N = 15$).

sensors (which all communicate) also has a detrimental effect on E_C presumably because it creates more tight cycles in the communication graph.

We have also compared our algorithm to a current state-of-the-art algorithm, namely the "anchor-free localization" (AFL) algorithm discussed in [5]. Since this algorithm was only defined for distance measurements we ignored angle information as well. This is a distributed algorithm based on quadratic penalties for deviations between the noisy observed distances and distances measured in the sensor map. However, unlike SLEEP, AFL has a preprocessing phase where it unfolds the map approximately, avoiding local minima. We have enhanced AFL with a computation of the inverse Fisher information in order to compute uncertainty in localization [3]. In all our experiments we have found that our algorithm works significantly better than AFL both in terms of RMSE as well as E_C. These results will be published elsewhere in full detail and are left out due to space constraints.

5 Discussion

We presented a flexible distributed computational paradigm for sensor localization. Various kinds of information are easily integrated by adding potential functions and adapting EP to handle inference. In this paper we have presented an explicit algorithm that incorporates noisy distance and angle information.

The experiments have shown that the algorithm is able to competitively estimate both location and uncertainty in location. We have also shown that approximate inference in graphs with many tight loops begins to break down in terms of estimating uncertainty (but not in terms of the location estimates themselves). This suggests interesting avenues for future research. For instance, one could try to develop distributed implementations for generalized EP algorithms.

There are a few other issues that need further scrutiny. We are currently using importance sampling to do the required numerical integration but this may not be optimal in terms of computational efficiency. We plan to apply quadrature integration to improve upon this. Also, we plan to extend the current model with more sources of information. For instance one may want to simultaneously infer location and, say, the concentration of a chemical in the atmosphere. Extensions to dynamic models where robots move around in their environment are also under investigation.

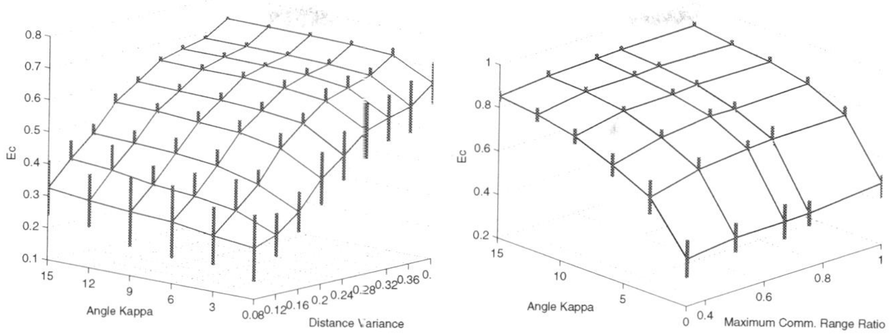

Fig. 5. E_C as a function of variance in squared distance measurements (v) and κ. All sensors communicate. Right: E_C as a function of communication range and κ ($v = 0.2$). In both plot there are $N = 15$ sensors.

Acknowledgements

We thanks Tom Minka for important suggestions and Alex Ihler for interesting discussions.

References

1. Ihler, A.T., Fisher III, J.W., Moses, R.L., Willsky, A.S.: Nonparametric belief propagation for self-calibration in sensor networks. In: Information Processing in Sensor Networks (2004)
2. Minka, T.: Expectation propagation for approximate Bayesian inference. In: UAI, pp. 362–369 (2001)
3. Moses, R.L., Krishnamurthy, D., Patterson, R.: A self-localization method for wireless sensor networks. EURASIP Journal on Applied Signal Processing 4, 348–358 (2003)
4. Peng, R., Sichitiu, M.L.: Robust, probabilistic, constraint-based localization for wireless sensor networks. pp. 541–550 (2005)
5. Priyantha, N.B., Balakrishnan, H., Demaine, E., Teller, S.: Anchor-free distributed localization in sensor networks.Technical Report TR-892, MIT LCS (April 2003)

An Analytical Model of Divisive Normalization in Disparity-Tuned Complex Cells

W. Stürzl[1], H.A. Mallot[2], and A. Knoll[1]

[1] Robotics and Embedded Systems, Technical University of Munich, Germany
[2] Cognitive Neuroscience, University of Tübingen, Germany
stuerzl@in.tum.de

Abstract. Based on the energy model for disparity-tuned neurons, we calculate probability density functions of complex cell activity for random-dot stimuli. We investigate the effects of normalization and give analytical expressions for the disparity tuning curve and its variance. We show that while normalized and non-normalized complex cells have similar tuning curves, the variance is significantly lower for normalized complex cells, which makes disparity estimation more reliable. The results of the analytical calculations are compared to computer simulations.

1 Introduction to the Binocular Energy Model

An overview of the binocular energy model [1] and an extension consisting of an additional normalization operation is shown in Fig. 1. The model is motivated by neurophysiological recordings from disparity-tuned neurons in the visual cortex of mammals. It is similar to the spatio-temporal energy model for motion perception described in [2].

In the first stage, left and right images, $I_l(x,y)$ and $I_r(x,y)$ respectively, are convolved with pairs of orthogonal filters,

$$S_{al} = \iint f_{al}(\xi,\eta)\, I_l(\xi,\eta)\, d\xi d\eta \ , \quad S_{ar} = \iint f_{ar}(\xi,\eta)\, I_r(\xi,\eta)\, d\xi d\eta \ , \tag{1}$$

$$S_{bl} = \iint f_{bl}(\xi,\eta)\, I_l(\xi,\eta)\, d\xi d\eta \ , \quad S_{br} = \iint f_{br}(\xi,\eta)\, I_r(\xi,\eta)\, d\xi d\eta \ . \tag{2}$$

The linear receptive fields defined by the functions $f_{al}, f_{bl}, f_{al}, f_{bl}$ can be modelled as Gabor functions, see e.g. [3],

$$G(\boldsymbol{x}-\boldsymbol{x}_0|\boldsymbol{k},\hat{\boldsymbol{\Sigma}},\phi) = e^{-\frac{1}{2}(\boldsymbol{x}-\boldsymbol{x}_0)^\top \hat{\boldsymbol{\Sigma}}^{-1}(\boldsymbol{x}-\boldsymbol{x}_0)} \cos(\boldsymbol{k}^\top(\boldsymbol{x}-\boldsymbol{x}_0)-\phi) \ . \tag{3}$$

$\boldsymbol{x}_0$ denotes the center position of a simple cell's receptive field in the image or retina, $\boldsymbol{k}$ determines the spatial frequency and the orientation. The monocular receptive fields have a phase difference of 90°, i.e. $\phi_{bl} = \phi_{al} + \frac{1}{2}\pi$, $\phi_{br} = \phi_{ar} + \frac{1}{2}\pi$.

Binocular simple cells combine the corresponding left and right filter responses (see Fig. 1),

$$S_a = S_{al} + S_{ar} \ , \quad S_b = S_{bl} + S_{br} \ , \tag{4}$$

J. Marques de Sá et al. (Eds.): ICANN 2007, Part I, LNCS 4668, pp. 776–787, 2007.

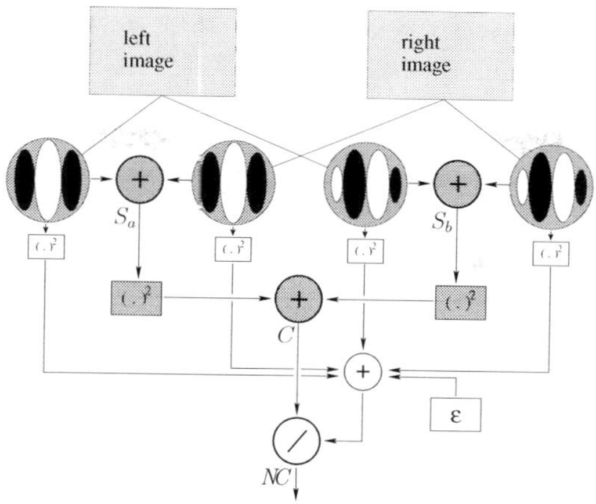

Fig. 1. Overview of the binocular energy model. The "classical" model, Eqs. (1)–(5), is depicted as gray panels. The white panels show the additional operations of the normalization step, Eq. (6). See text for details.

and output the squared sum to the complex cell. Mathematically equivalent but biologically more plausible, one can use four units instead, that have sign-inverted receptive fields and a half-squaring non-linearity [4].

Finally, binocular complex cells integrate responses of (at least) two simple cells,

$$C = S_a^2 + S_b^2 \ . \tag{5}$$

Binocular Normalization

The "classical" energy model has been extended by adding monocular and binocular divisive normalization operations [5]. In this paper, we will examine the effect of binocular normalization as depicted in Fig. 1. We define a normalized complex cell by

$$NC := \frac{C}{(S_{al}^2 + S_{ar}^2 + S_{bl}^2 + S_{br}^2) + \varepsilon} = \frac{S_a^2 + S_b^2}{(S_{al}^2 + S_{ar}^2 + S_{bl}^2 + S_{br}^2) + \varepsilon} \ . \tag{6}$$

$\varepsilon \geq 0$ is a small normalization constant. For high local image contrast (estimated by $S_{al}^2 + S_{ar}^2 + S_{bl}^2 + S_{br}^2$) the effect of ε is negligible. Eq. (6) is a simple variant of the binocular normalization model proposed in [5] which enables us to show analytically, that the variance of normalized complex cell activity is reduced, and disparity detection is thus enhanced.

Similar models for "divisive normalization" or "gain control" have been used to describe several properties of visual cortex neurons, like response saturation and cross-orientation inhibition [6,7,8].

2 Probability Density Functions for Complex Cells

In this section, we will calculate probability density functions, tuning curves and variances of disparity-tuned complex cells of the "classical" energy model (which we will call "non-normalized complex cells"), and of normalized complex cells defined in Eq. (6). We will assume random dot stereo images as stimuli.

2.1 Receptive Field Functions of the Position-Shift Type

In the following, we will consider vertically oriented receptive field functions, i.e. we set $\boldsymbol{k} = k\,\boldsymbol{e}_x$ and $\hat{\boldsymbol{\Sigma}} = \mathrm{diag}(\sigma_x^2, \sigma_y^2)$ in Eq. (3),

$$G(\boldsymbol{x} - \boldsymbol{x}_0 | k, \sigma_x^2, \sigma_y^2, \phi) = e^{-\frac{(x-x_0)^2}{2\sigma_x^2} - \frac{(y-y_0)^2}{2\sigma_y^2}} \cos(k(x - x_0) - \phi) \ . \quad (7)$$

In this paper, we will use receptive fields of the position-shift type [5,9]: Receptive fields in the right image are shifted by D relative to the left receptive fields. To simplify notation, we set $\phi_{al} = \phi_{ar} = 0$ and use the fact that the receptive field function (7) is separable. Eqs. (1),(2) are then transformed into one-dimensional integrals (using $\sigma := \sigma_x$),

$$S_{al} = \int e^{-\frac{\xi^2}{2\sigma^2}} \cos(k\xi) I_l(\xi)\, d\xi \ , \quad S_{ar} = \int e^{-\frac{(\xi-D)^2}{2\sigma^2}} \cos(k(\xi - D)) I_r(\xi)\, d\xi \ , \quad (8)$$

$$S_{bl} = \int e^{-\frac{\xi^2}{2\sigma^2}} \sin(k\xi) I_l(\xi)\, d\xi \ , \quad S_{br} = \int e^{-\frac{(\xi-D)^2}{2\sigma^2}} \sin(k(\xi - D)) I_r(\xi)\, d\xi \ , \quad (9)$$

where we have defined the one-dimensional images

$$I_{l/r}(\xi) := \int e^{-\frac{(y-y_0)^2}{2\sigma_y^2}} I_{l/r}(\xi + x_0, y) dy \ . \quad (10)$$

The receptive fields of the phase-shift type can be obtained by replacing $\exp(-\frac{(x-x_0-D)^2}{2\sigma^2})$ with $\exp(-\frac{(x-x_0)^2}{2\sigma^2})$ in Eqs. (8) and (9), i.e. the corresponding receptive fields are located in the same positions, but have phase shifts.

2.2 Random Dot Stereo Stimuli

In the following, we do not restrict our derivations to a single stereo stimulus, but rather assume that images are generated by a random process. In the simplest case, each pixel is an identically and independently distributed random variable with mean I_0 and standard deviation σ_I^2. In the continuous limit we will use

$$\langle \bar{I}(\xi)\bar{I}(\xi') \rangle = I_0^2 + \sigma_I^2 \delta(\xi - \xi') \ . \quad (11)$$

We will also assume that the output of the linear filters is zero for an input with constant pixel values (or intensity) over the receptive fields. While this is true for the odd receptive field function, $\int e^{-\frac{\xi^2}{2\sigma^2}} \sin(k\xi) d\xi = 0$, it does not hold exactly for the even function, $\int e^{-\frac{\xi^2}{2\sigma^2}} \cos(k\xi) d\xi = \sqrt{2\pi}\sigma \exp(-\frac{1}{2}k^2\sigma^2)$. This can be accounted for by replacing $\cos(k\xi)$ with $(\cos(k\xi) - \exp(-\frac{1}{2}k^2\sigma^2))$ in Eq. (8). However, in order to simplify the analytical calculations, we do not use this, but ignore constant offsets (which is equal to setting $I_0 = 0$).

We will analyze the case where left and right receptive fields "look" at image parts that are shifted versions of each other, i.e. over the whole receptive field there is a constant disparity d. We do not consider occlusions and variation in depth. However, we take (sensor) noise into account described by identically and independently distributed random variables. Thus, left and right input is modelled as

$$I_l(x) = \bar{I}(x) + \nu_l(x) \ , \qquad I_r(x) = \bar{I}(x-d) + \nu_r(x) \ . \tag{12}$$

Throughout this paper, we will assume $\langle \nu_u(x) \rangle = 0$ and $\langle \nu_u(x)\nu_v(x') \rangle = \sigma_n^2 \times \delta(x-x')\delta_{u,v}$, $u, v \in \{l, r\}$.

Substituting (12) into (8) and (9), and defining $\tilde{d} := d - D$, we obtain

$$S_{al} = \int e^{-\frac{\xi^2}{2\sigma^2}} \cos(k\xi)\bar{I}(\xi)\, d\xi + \int e^{-\frac{\xi^2}{2\sigma^2}} \cos(k\xi)\nu_l(\xi)\, d\xi \ , \tag{13}$$

$$S_{ar} = \int e^{-\frac{(\xi+\tilde{d})^2}{2\sigma^2}} \cos(k(\xi+\tilde{d})\bar{I}(\xi) d\xi + \int e^{-\frac{(\xi-D)^2}{2\sigma^2}} \cos(k(\xi-D))\nu_r(\xi)\, d\xi \ , \tag{14}$$

$$S_{bl} = \int e^{-\frac{\xi^2}{2\sigma^2}} \sin(k\xi)\bar{I}(\xi)\, d\xi + \int e^{-\frac{\xi^2}{2\sigma^2}} \sin(k\xi)\nu_l(\xi)\, d\xi \ , \tag{15}$$

$$S_{br} = \int e^{-\frac{(\xi+\tilde{d})^2}{2\sigma^2}} \sin(k(\xi+\tilde{d}))\bar{I}(\xi)\, d\xi + \int e^{-\frac{(\xi-D)^2}{2\sigma^2}} \sin(k(\xi-D))\nu_r(\xi)\, d\xi \ . \tag{16}$$

2.3 Probability Density Function for the Linear Stage

Since the elements of the vector $\boldsymbol{S} := (S_{al}, S_{ar}, S_{bl}, S_{br})^\top$ are sums of independently and identically distributed random variables, the joint probability density function can be approximated by a Gaussian according to the central limit theorem, i.e.

$$P_{\boldsymbol{S}}(\boldsymbol{S}|\tilde{d}) \approx \frac{1}{\sqrt{(2\pi)^4 \det \hat{\boldsymbol{\Sigma}}(\tilde{d})}} \exp(-\tfrac{1}{2}(\boldsymbol{S} - \langle \boldsymbol{S} \rangle)^\top \hat{\boldsymbol{\Sigma}}(\tilde{d})^{-1}(\boldsymbol{S} - \langle \boldsymbol{S} \rangle)) \ . \tag{17}$$

As discussed in Sect. 2.2, we assume $\langle \boldsymbol{S} \rangle = \boldsymbol{0}$. The covariance matrix is given by

$$\hat{\boldsymbol{\Sigma}}(\tilde{d}) = \langle (\boldsymbol{S} - \langle \boldsymbol{S} \rangle)(\boldsymbol{S} - \langle \boldsymbol{S} \rangle)^\top \rangle = \langle \boldsymbol{S}\boldsymbol{S}^\top \rangle = \begin{pmatrix} a & c & 0 & e \\ c & a & -e & 0 \\ 0 & -e & b & d \\ e & 0 & d & b \end{pmatrix} , \tag{18}$$

$$a := \langle S_{al}^2 \rangle = \langle S_{ar}^2 \rangle = \iint \langle \bar{I}(x)\bar{I}(x') \rangle f_a(x) f_a(x') dx dx' + \sigma_n^2 \int f_a(x)^2 dx \ , \tag{19}$$

$$b := \langle S_{bl}^2 \rangle = \langle S_{br}^2 \rangle = \iint \langle \bar{I}(x)\bar{I}(x') \rangle f_b(x) f_b(x') dx dx' + \sigma_n^2 \int f_b(x)^2 dx \ , \tag{20}$$

$$c := \langle S_{al} S_{ar} \rangle = \iint \langle \bar{I}(x)\bar{I}(x') \rangle f_a(x) f_a(x' + \tilde{d}) dx dx' \ , \tag{21}$$

$$d := \langle S_{bl} S_{br} \rangle = \iint \langle \bar{I}(x)\bar{I}(x') \rangle f_b(x) f_b(x' + \tilde{d}) dx dx' \ , \tag{22}$$

$$e := \langle S_{al} S_{br} \rangle = -\langle S_{ar} S_{bl} \rangle = \iint \langle \bar{I}(x)\bar{I}(x') \rangle f_a(x) f_b(x' + \tilde{d}) dx dx' \ , \tag{23}$$

$$\langle S_{al} S_{bl} \rangle = \langle S_{ar} S_{br} \rangle = 0 \ . \tag{24}$$

By means of the orthogonal 4×4 matrix

$$\hat{\boldsymbol{O}} := \begin{pmatrix} \hat{\boldsymbol{O}}_2 & \mathbf{0} \\ \mathbf{0} & \hat{\boldsymbol{O}}_2 \end{pmatrix} = \hat{\boldsymbol{O}}^{-1} = \hat{\boldsymbol{O}}^\top \ , \quad \hat{\boldsymbol{O}}_2 := \frac{1}{\sqrt{2}} \begin{pmatrix} 1 & -1 \\ -1 & 1 \end{pmatrix} \ , \tag{25}$$

we can calculate the probability density function of $\boldsymbol{u} := (u_a, v_a, u_b, v_b)^\top = \hat{\boldsymbol{O}}\mathbf{S} = \frac{1}{\sqrt{2}}(S_{al} + S_{ar},\, S_{al} - S_{ar},\, S_{bl} + S_{br},\, S_{bl} - S_{br})^\top$,

$$P_{\boldsymbol{u}}(\boldsymbol{u}|\,\tilde{d}) = \frac{1}{\sqrt{(2\pi)^4 \det \hat{\boldsymbol{\Lambda}}(\tilde{d})}} \exp(-\tfrac{1}{2}\boldsymbol{u}^\top \hat{\boldsymbol{\Lambda}}(\tilde{d})^{-1}\boldsymbol{u}) \ , \tag{26}$$

$$\text{with } \hat{\boldsymbol{\Lambda}}(\tilde{d}) = \hat{\boldsymbol{O}}\hat{\boldsymbol{\Sigma}}(\tilde{d})\,\hat{\boldsymbol{O}}^\top = \begin{pmatrix} a+c & 0 & 0 & -e \\ 0 & a-c & e & 0 \\ 0 & e & b+d & 0 \\ -e & 0 & 0 & b-d \end{pmatrix} \ . \tag{27}$$

2.4 Non-normalized Complex Cells

According to the energy model, a disparity-tuned complex cell combines the output of two simple cells with orthogonal receptive field functions, i.e.

$$C = S_a^2 + S_b^2 = 2(u_a^2 + u_b^2) \ . \tag{28}$$

Using

$$P_{u_a u_b}(u_a, u_b|\,\tilde{d}) = \iint P_{\boldsymbol{u}}(\boldsymbol{u}|\,\tilde{d}) dv_a dv_b = \frac{\exp\left(-\frac{u_a^2}{2(a+c)} - \frac{u_b^2}{2(b+d)}\right)}{\sqrt{4\pi^2(a+c)(b+d)}} \ , \tag{29}$$

the corresponding probability density function is (for $a + c > b + d$)

$$P_C(C|\tilde{d}) = \int_{-\pi}^{\pi} P_{u_a u_b}(\sqrt{C/2}\cos\phi, \sqrt{C/2}\cos\phi|\,\tilde{d})|\det\left(\frac{\partial(u_a, u_b)}{\partial(C,\phi)}\right)|d\phi \tag{30}$$

$$= \frac{\exp\left(-\frac{C(a+c+b+d)}{8(a+c)(b+d)}\right)\mathcal{I}_0\left(\frac{C((a+c)-(b+d))}{8(a+c)(b+d)}\right)}{4\sqrt{(a+c)(b+d)}} \ . \tag{31}$$

$\mathcal{I}_0$ is a modified Bessel function of the first kind.

The mean of complex cell activity ("tuning curve") and its variance are given by

$$\langle C(\tilde{d})\rangle = \int_0^\infty C\, P_C(C|\tilde{d})\, dC = 2(a+b+c+d) \ , \tag{32}$$

$$\text{Var}[C(\tilde{d})] = \langle C^2\rangle - \langle C\rangle^2 = 8(a+c)^2 + 8(b+d)^2 \ . \tag{33}$$

Eqs. (19)–(23) can be further evaluated for random dot-stimuli,

$$a = \tfrac{1}{2}\sqrt{\pi}\sigma(\sigma_I^2 + \sigma_n^2)\left(1 + \exp(-k^2\sigma^2)\right) \ , \tag{34}$$

$$b = \tfrac{1}{2}\sqrt{\pi}\sigma(\sigma_I^2 + \sigma_n^2)\left(1 - \exp(-k^2\sigma^2)\right) \ , \tag{35}$$

$$c = \tfrac{1}{2}\sqrt{\pi}\sigma\sigma_I^2 \exp\left(-\frac{\tilde{d}^2}{4\sigma^2}\right)\left(\cos(k\tilde{d}) + \exp(-k^2\sigma^2)\right) \ , \tag{36}$$

$$d = \tfrac{1}{2}\sqrt{\pi}\sigma\sigma_I^2 \exp\left(-\frac{\tilde{d}^2}{4\sigma^2}\right)\left(\cos(k\tilde{d}) - \exp(-k^2\sigma^2)\right) \ , \tag{37}$$

$$e = \tfrac{1}{2}\sqrt{\pi}\sigma\sigma_I^2 \exp\left(-\frac{\tilde{d}^2}{4\sigma^2}\right)\sin(k\tilde{d}) \ . \tag{38}$$

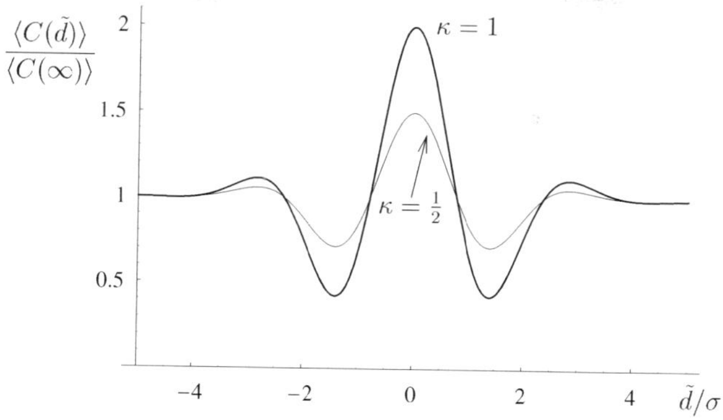

Fig. 2. Tuning curves $\langle C(\tilde{d})\rangle$ of non-normalized complex cells with $k\sigma = 2$ for random-dot stimuli, see Eq. (39). Depth of modulation is $\kappa = 1$ and $\kappa = \frac{1}{2}$ (thin curve). As can be seen from Eq. (40), these curves also show the corresponding standard deviations $\sqrt{\text{Var}[C(\tilde{d})]} \approx \langle C(\tilde{d})\rangle$.

If we assume that $\exp(-k^2\sigma^2) \ll 1$, we can use the approximation $b+d \approx a+c$. A typical value for the visual cortex is $ks \approx 2$, see [10], and thus $\exp(-k^2\sigma^2) \approx \exp(-4) \approx 0.018$.

For the tuning curve of the non-normalized complex cell, we finally have

$$\langle C(\tilde{d})\rangle \approx 4(a+c) = \langle C(\infty)\rangle \left(\kappa \exp\left(-\frac{\tilde{d}^2}{4\sigma^2}\right)\cos(k\tilde{d}) + 1\right) , \qquad (39)$$

where we have defined $\langle C(\infty)\rangle := 2\sqrt{\pi}\sigma(\sigma_I^2 + \sigma_n^2)$ (asymptotic amplitude of complex cell activity for $\tilde{d} \to \infty$) and $\kappa := \frac{\sigma_I^2}{\sigma_I^2+\sigma_n^2}$ ("depth of modulation"). For $\kappa = 1$, Eq. (39) matches the results in [11,12,13]. In addition, we find that the variance of the non-normalized complex cell, Eq. (33), is proportional to the square of the expected response,

$$\text{Var}[C(\tilde{d})] \approx 16(a+c)^2 = \langle C(\tilde{d})\rangle^2 . \qquad (40)$$

The probability density function can be approximated for $a+c \approx b+d$ by

$$P_C(C|\tilde{d}) \approx \frac{1}{4(a+c)} \exp\left(-\frac{C}{4(a+c)}\right) = \frac{1}{\langle C(\tilde{d})\rangle} \exp\left(-\frac{C}{\langle C(\tilde{d})\rangle}\right) . \qquad (41)$$

Fig. 2 shows tuning curves for $\kappa = 1$ (no noise, $\sigma_n^2 = 0$) and $\kappa = \frac{1}{2}$ (high noise, $\sigma_n^2 = \sigma_I^2$). The corresponding probability density functions, calculated from Eq. (31), are shown in Fig. 3. The maxima at $C = 0$, $\tilde{d}/\sigma \approx 1.5$ correspond to the minima of the tuning curves in Fig. 2.

Except for very high C-values, it is difficult to recognize $\tilde{d} \approx 0$ from the response of the complex cell. This is due to the fact that the complex cell activity also depends on stimulus contrast. As is shown in the next section, this can be significantly reduced by means of divisive contrast normalization.

2.5 Normalized Complex Cells

In this section we analyze the properties of contrast normalized complex cells. According to Eq. (6) and (4) the normalized complex cell is computed as

$$NC = \frac{(S_{al}+S_{ar})^2 + (S_{bl}+S_{br})^2}{(S_{al}^2+S_{ar}^2+S_{bl}^2+S_{br}^2)+\varepsilon} = \frac{2(u_a^2+u_b^2)}{(u_a^2+v_a^2+u_b^2+v_b^2)+\varepsilon} \quad . \tag{42}$$

$NC \in [0,2]$ since $v_a^2 + v_b^2 \geq 0$.

Using the approximation $b \pm d \approx a \pm c$ that is valid for $\exp(-k^2\sigma^2) \ll 1$, we have

$$\hat{\boldsymbol{A}}(\tilde{d})^{-1} \approx \frac{1}{a^2-c^2-e^2} \begin{pmatrix} a-c & 0 & 0 & e \\ 0 & a+c & -e & 0 \\ 0 & -e & a-c & 0 \\ e & 0 & 0 & a+c \end{pmatrix} \quad . \tag{43}$$

Substituting $(u_a, u_b) = \sqrt{w}(\cos\phi, \sin\phi)$ and $(v_a, v_b) = \sqrt{z}(\cos\psi, \sin\psi)$ the normalized complex cell is given by

$$NC = \frac{2w}{w+z+\varepsilon} \ , \tag{44}$$

and the corresponding probability function can be calculated according to

$$P_{NC}(NC|\,\tilde{d}) = \int_0^\infty\int_0^\infty P_{wz}(w,z)\,\delta\Big(NC - \frac{2w}{w+z+\varepsilon}\Big)dwdz \ , \tag{45}$$

$$\text{with} \quad P_{wz}(w,z|\,\tilde{d}) = \int_{-\pi}^{\pi}\int_{-\pi}^{\pi} \frac{1}{4} P_{\boldsymbol{u}}(w,\phi,z,\psi)d\psi d\phi \tag{46}$$

$$\approx \frac{1}{8\pi A}\int_{-\pi}^{\pi} e^{-\frac{(a-c)w+(a+c)z-2e\sqrt{wz}\sin\phi}{2A}} d\phi \ , \tag{47}$$

with $A := a^2-c^2-e^2$. The delta-function in Eq. (45) ensures that the integration is restricted to the domain where $NC - \frac{2w}{w+z+\varepsilon} = 0$, Eq. (44).

Using $\delta(f(x)) = \sum_i |f'(x_i)|^{-1}\delta(x-x_i)$, where $\{x_i\}$ are the roots of f, we find

$$\delta\Big(NC - \frac{2w}{w+z+\varepsilon}\Big) = \frac{2(z+\varepsilon)}{(2-NC)^2}\,\delta\Big(w - \frac{NC}{2-NC}(z+\varepsilon)\Big) \ , \tag{48}$$

and the integration over w in Eq. (45) yields, with $n := \frac{NC}{2-NC}$,

$$P_{NC}(NC|\,\tilde{d}) \approx \frac{2}{(2-NC)^2}\int_0^\infty (z+\varepsilon)P_{wz}\Big(n(z+\varepsilon),z\Big)dz$$

$$= \frac{e^{-\frac{(a-c)n\varepsilon}{2A}}}{4\pi(2-NC)^2 A}\int_{-\pi}^{\pi}\int_0^\infty (z+\varepsilon)e^{\frac{2e\sqrt{n(z+\varepsilon)z}\sin\phi-((a-c)n+a+c)z}{2A}} dzd\phi \ . \tag{49}$$

We could not find a way to integrate (49) analytically for $\varepsilon > 0$. However, with the approximation $\sqrt{n(z+\varepsilon)z} \approx \sqrt{n}z$, integration over z and ϕ yields

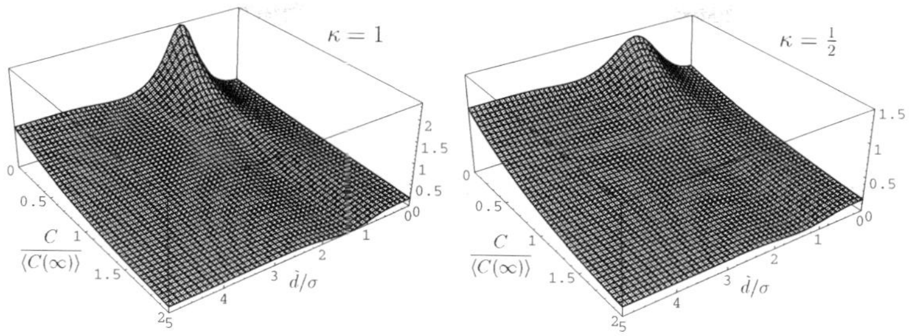

Fig. 3. $P_C(C|\,\tilde{d}\,)$ for $k\sigma = 2$. Because of symmetry, only the range $\tilde{d}/\sigma \geq 0$ is shown.

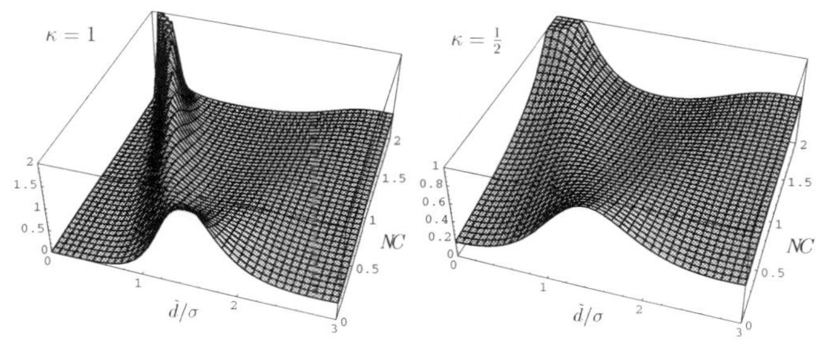

Fig. 4. Probability density functions of normalized complex cells, Eq. (53) for different noise levels, $\kappa = 1 \iff \sigma_n^2 = 0$ and $\kappa = \frac{1}{2} \iff \sigma_n^2 = \sigma_I^2$

$$P_{NC}(NC|\,\tilde{d}\,) \approx \frac{e^{-\frac{(a-c)NC\,\varepsilon}{2(2-NC)A}}}{2g[NC]^{\frac{1}{2}}}\Big(\frac{(a+(1-NC)c)A}{g[NC]} + \frac{\varepsilon}{2-NC}\Big) \ , \tag{50}$$

$$\text{with} \quad g[NC] := (a+(1-NC)c)^2 - e^2 NC(2-NC) \ . \tag{51}$$

Eq. (50) is exact for $\varepsilon = 0$, and with $\kappa := \frac{\sigma_I^2}{\sigma_I^2+\sigma_n^2}$ we find

$$P_{NC}(NC|\,\tilde{d};\, \varepsilon = 0) = \frac{(a^2-c^2-e^2)(a+(1-NC)c)}{2\big((a+(1-NC)c)^2 - e^2 NC(2-NC)\big)^{\frac{3}{2}}} \tag{52}$$

$$= \frac{\big(1-\kappa^2 e^{-\frac{\tilde{d}^2}{2\sigma^2}}\big)\big(1+\kappa(1-NC)e^{-\frac{\tilde{d}^2}{4\sigma^2}}\cos(k\tilde{d})\big)}{2\Big(\big(1+\kappa(1-NC)e^{-\frac{\tilde{d}^2}{4\sigma^2}}\cos(k\tilde{d})\big)^2 - \kappa^2 NC(2-NC)e^{-\frac{\tilde{d}^2}{2\sigma^2}}\sin^2(k\tilde{d})\Big)^{\frac{3}{2}}} \ . \tag{53}$$

Fig. 4 shows the probability density function described by Eq. (53), again for $k\sigma = 2$. Compared to the non-normalized complex cell (see Fig. 3), there is – at least for low noise levels ($\sigma_n^2 \ll \sigma_I^2 \iff \kappa \approx 1$) – a much stronger link between small disparities and high activity ($NC \approx 2$). Thus, if one observes high activity then there is $\tilde{d} \approx 0$ with high probability.

Eq. (53) has simple expressions for the following two cases,

$$\tilde{d} \gg \sigma : \ P_{NC}(NC|\,\tilde{d} \gg \sigma;\, \varepsilon = 0) \approx \frac{1}{2} \ , \tag{54}$$

$$\tilde{d}=0: \; P_{NC}(NC|\,\tilde{d}=0;\varepsilon=0)=\frac{1-\kappa^2}{2[1+\kappa(1-NC)]^2} \; . \tag{55}$$

In the limit of zero noise, i.e. $\sigma_n^2 \to 0$, we find

$$\lim_{\kappa\to 1} P_{NC}(NC|\,\tilde{d}=0;\varepsilon=0)=\delta(2-NC) \; , \tag{56}$$

since $\int_0^2 P_{NC}(NC|\,\tilde{d}=0;\varepsilon=0)\,dNC=1$, and $\lim_{\kappa\to 1} P_{NC}(NC|\,\tilde{d}=0;\varepsilon=0)=0$ for $0 \le NC < 2$.

Tuning Curve and Variance. The tuning curve is given by

$$\langle NC(\tilde{d})\rangle = \int_0^2 NC\, P_{NC}(NC|\,\tilde{d};\varepsilon=0)\,dNC = 1+\cos(k\tilde{d})\kappa e^{-\frac{\tilde{d}^2}{4\sigma^2}} h[\kappa e^{-\frac{\tilde{d}^2}{4\sigma^2}}] \; , \tag{57}$$

$$\text{with} \quad h[u] := \frac{1}{u^2} - \left(\frac{1}{u^3}-\frac{1}{u}\right)\operatorname{artanh}(u) \; . \tag{58}$$

Using $\operatorname{artanh}(u)=\sum_{k=0}^{\infty}\frac{1}{2k+1}u^{2k+1}$, one can show that (for $|u|\le 1$)

$$h[u] = 1-\sum_{k=1}^{\infty}\frac{1}{2k+1}\left(u^{2(k-1)}-u^{2k}\right) = \sum_{k=0}^{\infty}\left(\frac{1}{2k+1}-\frac{1}{2k+3}\right)u^{2k} \; . \tag{59}$$

Since $h[0]=\frac{2}{3}\le h[u]\le h[1]=1$, we obtain

$$\tfrac{2}{3}|\cos(k\tilde{d})|\kappa e^{-\frac{\tilde{d}^2}{4\sigma^2}} \le |\langle NC(\tilde{d})\rangle - 1| \le |\cos(k\tilde{d})|\kappa e^{-\frac{\tilde{d}^2}{4\sigma^2}} \; , \tag{60}$$

and for $|\delta| \ll 2\sigma$ the tuning curve can be approximated by

$$\langle NC(\tilde{d})\rangle \approx 1+\cos(k\tilde{d})\kappa e^{-\frac{\tilde{d}^2}{4\sigma^2}} \; . \tag{61}$$

Since $\langle C(\tilde{d})\rangle/\langle C(\infty)\rangle = \cos(k\tilde{d})\kappa e^{-\frac{\tilde{d}^2}{4\sigma^2}}$, we find that normalized and non-normalized complex cell have similar tuning curves – as can be seen in Fig. 5. The approximation $\langle NC(\tilde{d})\rangle \approx 1+\frac{2}{3}\kappa\cos(k\tilde{d})e^{-\frac{\tilde{d}^2}{4\sigma^2}}+\frac{1}{3}\kappa^3\cos(k\tilde{d})e^{-\frac{3\tilde{d}^2}{4\sigma^2}}$ has been obtained using the first three terms in Eq. (59), i.e. $h[u]\approx\frac{2}{3}+\frac{1}{3}u^2$.

Using

$$\langle NC(\tilde{d})^2\rangle := \int_0^2 NC^2\, P_{NC}(NC|\,\tilde{d};\varepsilon=0)\,dNC \; , \tag{62}$$

the variance in the response of the non-normalized complex cell is

$$\operatorname{Var}[NC(\tilde{d})] = \langle NC(\tilde{d})^2\rangle - \langle NC(\tilde{d})\rangle^2 = (1-\kappa^2 e^{-\frac{\tilde{d}^2}{2\sigma^2}})H[\kappa^2 e^{-\frac{\tilde{d}^2}{4\sigma^2}}, k\tilde{d})] \; , \tag{63}$$

$$\text{with } H[u,v] := (\cos^2 v - \sin^2 v)\frac{1}{u^2} + \sin^2 v\frac{1}{u^3}\operatorname{artanh}(u) \; . \tag{64}$$

One can show that $\frac{1}{3}(1-\kappa^2 e^{-\frac{\tilde{d}^2}{2\sigma^2}}) < \operatorname{Var}[NC(\tilde{d})] < \frac{1}{3}$.

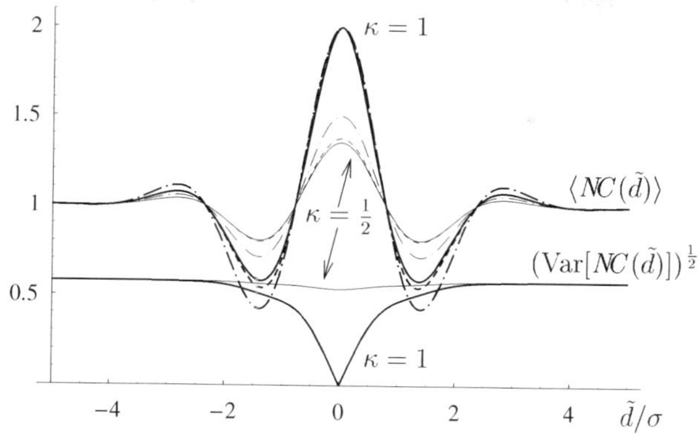

Fig. 5. Tuning curve $\langle NC(\tilde{d})\rangle$ of the normalized complex cell (continuous curve), Eq. (57), and the approximations $1+\kappa\cos(k\tilde{d})e^{-\frac{\tilde{d}^2}{4\sigma^2}} = \langle C(\tilde{d})\rangle/\langle C(\infty)\rangle$ (dash-dotted), and $1+\frac{2}{3}\kappa\cos(k\tilde{d})e^{-\frac{\tilde{d}^2}{4\sigma^2}} + \frac{1}{3}\kappa^3\cos(\kappa\tilde{d})e^{-\frac{3\tilde{d}^2}{4\sigma^2}}$ (dashed), for $k\sigma = 2$. The noise levels are $\sigma_n^2 = 0$ ($\kappa = 1$, bold curves) and $\sigma_n^2 = \sigma_I^2$ ($\kappa = \frac{1}{2}$, thin curves). The dash-dotted curves are identical to the curves in Fig 2. Also shown is the standard deviation $(\mathrm{Var}[NC(\tilde{d})])^{\frac{1}{2}}$.

By means of Taylor expansion of artanh(u), we have found a good approximation, $\mathrm{Var}[NC(\tilde{d})] \approx (1-\kappa^2 e^{-\frac{\tilde{d}^2}{2\sigma^2}})\Big(\frac{1}{3}+\kappa^2 e^{-\frac{\tilde{d}^2}{2\sigma^2}}[\frac{1}{5}\sin^2(k\tilde{d}) + \frac{7}{45}\cos^2(k\tilde{d})]\Big)$.

Fig. 5 also shows the standard deviation $(\mathrm{Var}[NC(\tilde{d})])^{1/2}$ for zero noise ($\kappa = 1$) and strong noise ($\sigma_n^2 = \sigma_I^2$, $\kappa = \frac{1}{2}$). The low variance for small $\tilde{d}$ (and low noise) enhances the detectability of stimulus disparities D that are close to the neuron's preferred disparity d (compare with Fig. 2 and $\mathrm{Var}[C(\tilde{d})] \approx \langle C(\tilde{d})\rangle^2$).

3 Comparing Analytical Results with Simulations

As a control of the analytically derived results we performed computer simulations with discrete images and applied circular Gabor filters of bandwidth two, more specifically we set $\sigma_x = \sigma_y = 5/2$ pixel and $k = 2/3\,\text{pixel}^{-1}$. The (separable) 2D Gabor filters are approximated by two pairs of row and column filters. The size of these filters is 13×1 and 1×13 (corresponding to a 13×13 filter mask in 2D), and their elements are integer values. As "stimuli" we used random dot images consisting of black and white pixels. Disparities were simulated by shifting the right image with respect to the left image. For each disparity, mean and standard deviation of complex cells were computed. As can be seen in Fig. 6, the analytically derived equations fit the data quite well.

4 Discussion, Limitations of the Model

We have shown analytically that simple contrast normalization can greatly enhance disparity detection for binocular neurons of the energy type. While the

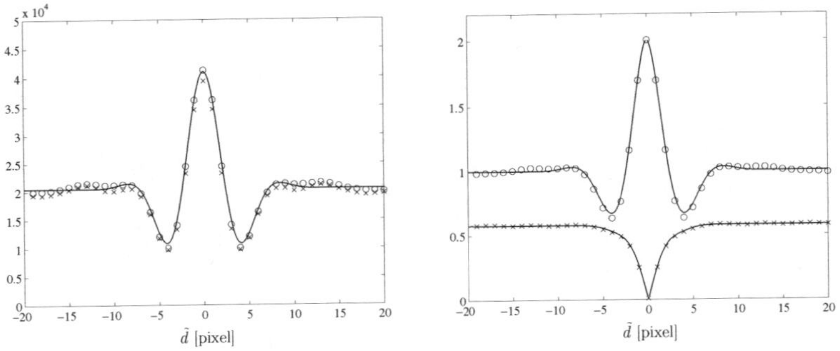

Fig. 6. Simulated disparity tuning curves (circles) and standard deviation (crosses) for non-normalized complex cell (left) and normalized complex cell (right). The continuous curves show the analytical results, Eqs. (39),(40) and (57),(63) respectively.

"classical" model predicts that the standard deviation is proportional to the mean firing rate for random-dot stimuli, divisive normalization significantly reduces the standard deviation, in particular close to the preferred disparity. The simple expression for contrast normalization given by Eq. (6) has been chosen in order to make analytical calculations more convenient. For the same reason, i.e. to keep the model simple, monocular normalization which is known to play a major role in gain control [14], was not included. In addition, although we have taken sensor noise (with constant variance) into account, our model does not include intrinsic response variability of the disparity tuned neuron. This has to be considered when comparing the calculated variances to measured data. Response variances to dynamic random dot stimuli have been reported to be approximately proportional to the mean firing rate [15,16].

References

1. Ohzawa, I., DeAngelis, G., Freeman, R.: Stereoscopic depth discrimination in the visual cortex: Neurons ideally suited as disparity detectors. Science 249, 1037–1041 (1990)
2. Adelson, E., Bergen, J.: Spatiotemporal energy models for the perception of motion. J. Opt. Soc. Am. A 2, 284–299 (1985)
3. Anzai, A., Ohzawa, I., Freeman, R.: Neural mechanisms for processing binocular information. I. Simple cells. J. Neurophysiol. 82, 891–908 (1999)
4. Anzai, A., Ohzawa, I., Freeman, R.: Neural mechanisms for processing binocular information. II. Complex cells. J. Neurophysiol. 82, 909–924 (1999)
5. Fleet, D., Heeger, D., Wagner, H.: Modeling binocular neurons in primary visual cortex. In: Jenkin, M., Harris, L. (eds.) Computational and Biological Mechanisms of Visual Coding, pp. 103–130. Cambridge University Press, Cambridge (1996)
6. Heeger, D.: Normalization of cell responses in cat striate cortex. Vis. Neurosci. 9, 181–197 (1992)
7. Carandini, M., Heeger, D., Movshon, J.A.: Linearity and normalization in simple cells of the macaque primary visual cortex. The Journal of Neuroscience 17, 8621–8644 (1997)

8. Wainwright, M., Schwartz, O., Simoncelli, E.: Natural image statistics and divisive normalization: Modeling nonlinearity and adaptation in cortical neurons. In: Rao, R., Olshausen, B., Lewicki, M. (eds.) Probabilistic Models of the Brain: Perception and Neural Function, MIT Press, Cambridge (2002)
9. Anzai, A., Ohzawa, I., Freeman, R.: Neural mechanisms for encoding binocular disparity: Receptive field position versus phase. J. Neurophysiol. 82, 874–890 (1999)
10. DeValois, R., DeValois, K.: Spatial Vision. Oxford University Press, Oxford (1988)
11. Zhu, Y., Qian, N.: Binocular receptive field models, disparity tuning, and characteristic disparity. Neural Computation 8, 1611–1641 (1996)
12. Tsai, J., Victor, J.: Reading a population code: a multi-scale neural model for representing binocular disparity. Vision Research 43, 445–466 (2003)
13. Tanabe, S., Doi, T., Umeda, K., Fujita, I.: Disparity-tuning characteristics of neuronal responses to dynamic random-dot stereograms in macaque visual area V4. J. Neurophysiol. 94, 2683–2699 (2005)
14. Truchard, A., Ohzawa, I., Freeman, R.: Contrast gain control in the visual cortex: Monocular versus binocular mechanisms. The Journal of Neuroscience 20, 3017–3032 (2000)
15. Read, J., Cumming, B.: Testing quantitative models of binocular disparity selectivity in primary visual cortex. J. Neurophysiol. 90, 2795–2817 (2003)
16. Prince, S., Pointon, A., Cumming, B., Parker, A.: Quantitative analysis of the responses of v1 neurons to horizontal disparity in dynamic random-dot stereograms. J. Neurophysiol. 87, 191–208 (2002)

Automatic Design of Modular Neural Networks Using Genetic Programming

Naser NourAshrafoddin, Ali R. Vahdat, and M.M. Ebadzadeh

Amirkabir University of Technology (Tehran Polytechnic)

Abstract. Traditional trial-and-error approach to design neural networks is time consuming and does not guarantee yielding the best neural network feasible for a specific application. Therefore automatic approaches have gained more importance and popularity. In addition, traditional (non-modular) neural networks can not solve complex problems since these problems introduce wide range of overlap which, in turn, causes a wide range of deviations from efficient learning in different regions of the input space, whereas a modular neural network attempts to reduce the effect of these problems via a divide and conquer approach. In this paper we are going to introduce a different approach to autonomous design of modular neural networks. Here we use genetic programming for automatic modular neural networks design; their architectures, transfer functions and connection weights. Our approach offers important advantages over existing methods for automated neural network design. First it prefers smaller modules to bigger modules, second it allows neurons even in the same layer to use different transfer functions, and third it is not necessary to convert each individual into a neural network to obtain the fitness value during the evolution process. Several tests were performed with problems based on some of the most popular test databases. Results show that using genetic programming for automatic design of neural networks is an efficient method and is comparable with the already existing techniques.

Keywords: Modular neural networks, evolutionary computing, genetic programming, automatic design.

1 Introduction

The human brain is the most elaborate information processing system known. Researchers believe that it derives most of its processing power from the huge numbers of neurons and connections. It is also believed that the human brain contains about 10^{11} neurons, each of which is connected to an average of 10^4 other neurons. This amounts to a total of 10^{15} connections. This sophisticated biologic system teaches us some important principles regarding design of NNs:

- Despite its massive connections it is relatively small, and highly organized [1].
- Subsequent layers of neurons, arranged in a hierarchical fashion, form increasingly complex representations.

J. Marques de Sá et al. (Eds.): ICANN 2007, Part I, LNCS 4668, pp. 788–798, 2007.

- Structuring of the brain into distinct streams or pathways allows for the independent processing of different information types and modalities [1].
- Modules containing little more than a hundred cells, also known as *minicolumns*, are the basic functional modular units of the cerebral cortex [2].

We propose to capture these global and local structural regularities by modeling the brain in a framework of modular neural networks (MNNs) [3]. Rather than starting from a fully connected network and then letting the desired functions emerge with prolonged learning, we advocate an approach that proceeds from a structured modular network architecture. Utilizing this method of designing, NNs achieved will also be scalable, i.e. expansion modules can be added or redundant modules can be removed if necessary. By choosing the right modular structure, networks with a wide range of desirable characteristics can be developed, as will be outlined in section 3.

Non-modular NNs tend to introduce high internal interference because of the strong coupling among their hidden-layer weights, i.e. all neurons try to affect overall output of the NN. On the other hand, complex tasks tend to introduce a wide range of overlap which, in turn, causes a wide range of deviations from efficient learning in the different regions of the input space. A modular NN attempts to reduce the effect of these problems via a divide and conquer approach. It, generally, decomposes the large size/high complexity task into several sub-tasks; each one is handled by a simple, fast and efficient module. Then, sub-solutions are integrated via a multi-module decision-making strategy. Hence, modular NNs, generally, proved to be more efficient than the non-modular alternatives [4].

All problems regarding manual design of NNs have led the researchers to invent techniques for automating the design of NNs. Some automatic approaches have been proposed for this task. A recent group of these approaches are based on evolutionary algorithms (EAs) [5], [6]. Some methods of EAs used for automatic design of NNs are genetic algorithm (GA) and evolutionary programming (EP). In GAs, both the structure and parameters of a network are represented as a (usually) fixed-length string, and a population of such strings (or genotypes) is evolved, mainly by recombination. However, one problem with this approach is the size of the required genotype bit string. On the other hand, EP evolves by mutation only, operating directly on the network components. It has been suggested that this approach is much more suited to network design than GAs [5].

Another problem faced by evolutionary training of NNs is the *permutation* problem, also known as the *competing convention* problem. It is caused by the many-to-one mapping from the representation (genotype) to the actual NN (phenotype), since two NNs that order their hidden nodes differently in their chromosomes will still be functionally equivalent. For example, NNs shown in Fig. 1 are functionally equivalent, but they have different chromosomes. In general, any permutation of the hidden nodes will produce functionally equivalent NNs with different chromosome representations. The permutation problem makes crossover operator very inefficient and ineffective for producing good offspring. That is why EP works better than GA in evolution of NNs.

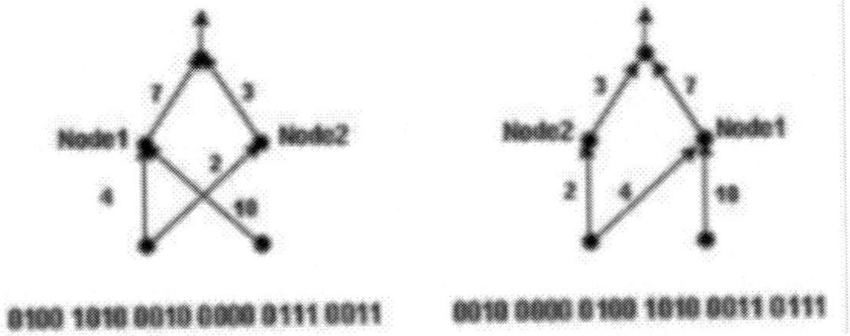

Fig. 1. Permutation problem. Two equivalent NNs and their corresponding non-equivalent binary representations.

The problems mentioned above are both due to the representations and codings used, and not to the EAs themselves; thus, with suitable coding schemes, it is possible to develop non-deceptive EA-based NN design algorithms. A very well-known and efficient approach which suppresses these problems is genetic programming (GP).

GP [7] is an automatic domain-independent method for solving problems. Starting with many randomly created solutions, it applies the Darwinian principle of natural selection, recombination (crossover), mutation, gene duplication, gene deletion, gene insertion and certain mechanisms of developmental biology. It thus breeds an improved population over many generations. GP starts from a high-level statement of a problem's requirements and attempts to produce a solution that solves the problem.

The remainder of the paper is organized as follows: in section 2 we will review almost all the previous attempts to automatic design of NNs using various implementations of GP. In section 3 we will describe our approach in details and also some important implementation points are stated. In section 4 some experiments and their results are explained and in section 5 we conclude our work.

2 Related Works

After presenting the recent evolutionary approach, *Genetic Programming,* by Koza [7], [8], exploiting genetic programming approach to automatic design of NNs is, perhaps, first introduced in [9]. Koza and Rice show how to find both weights and architecture for a NN using a population of LISP symbolic expressions (S-expressions) of varying size and shape. They use the function set $F = \{P, W, +, -, *, \%\}$ and terminal set $T = \{R, D0, D1, \ldots, Dn\}$ to produce trees and its corresponding S-expressions. Function *P* denotes a neuron with linear threshold transfer function, and function *W* is the weighting function used to give a weight to a signal going into a *P* function. +, –, and * are ordinary arithmetic operators and % is protected division operator which returns zero in case of a division by zero. *R* is the random floating point constant and *D0, D1,* …, and *Dn* are network input signals. A sample S-expression which performs odd-parity (exclusive-or) perfectly, in their work, is shown below:

(P (W (+ 1.1 0.741) (P (W 1.66 D0) (W –1.387 D1)))
(W (+ 1.2 1.584) (P (W 1.19 D1) (W –0.989 D0)))

This S-expression can be graphically presented as the tree below in Fig.2(a). This S-expression can be further converted to the form that one would typically see in the NN literature, such as Fig. 2 (b):

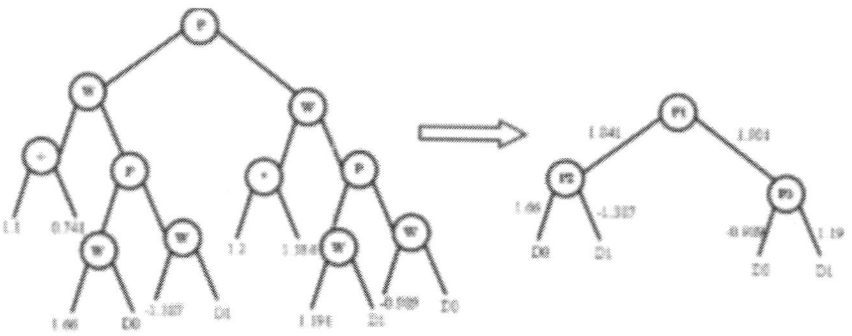

Fig. 2. (a) The tree of a simple NN corresponding to above S-expression. (b) and its corresponding NN [9].

Gruau [10] has developed a compact cellular growth (constructive) algorithm based on symbolic S-expressions called *cellular encoding* that are evolved by genetic programming. A scheme similar to cellular encoding, *edge encoding* by Luke [11] grows network graphs using edges instead of nodes. While the cellular encoding approach evaluates each grammar tree node in a breadth first search manner, thus ensuring parallel execution, edge encoding is designed to work using depth first search. The cellular encoding tends to create graphs with a large number of connections which must then be pruned using respective operators. Edge encoding, using its depth first search approach, favors graphs with fewer connections. Ritchie et al. [12] implemented a GP-based algorithm for optimizing NN architecture (GPNN) and compared its ability to model and detect gene-gene interactions with a traditional back-propagation NN (BPNN). The rules used for their implementation are consistent with those described by Koza and Rice [9]. Using simulated data, they demonstrated that their GPNN was able to model nonlinear interactions as well as a traditional BPNN and also achieved better predictive ability.

3 Proposed Method

Here we develop a GP-optimized NN design approach in an attempt to improve upon the trial-and-error process of choosing an optimal architecture for a pure feed-forward modular NN. This method utilizes more important inputs from a larger pool of variables, optimizes the weights, and the connectivity of the network including the number of hidden layers and the number of nodes in the hidden layer. Thus, the algorithm attempts to generate appropriate modular NN for a given dataset.

3.1 Function and Terminal Sets

Here we use different transfer functions for different neurons. These transfer functions include: linear (***purelin***), hard limit (***hardlim***), symmetric hard limit (***hardlims***), log-sigmoid (***logsig***), and tan-sigmoid (***tansig***). Utilizing these transfer functions means that there is no necessity to use similar transfer functions for different neurons in a NN even in the same layer.

Arithmetic functions used are ***plus***, ***minus***, and ***times*** functions, which are simple arithmetic operators. For division operation we use a protected function (***pdivide***) which returns the dividend when divisor is zero. ***ppower*** returns the same result as power function unless the result is a complex value. If this is the case, it returns zero. ***psqrt*** returns simple square root of its argument. If its argument is less than zero, the function returns zero. ***Exp*** returns exponential value of its two arguments and ***plog*** returns logarithm of absolute value of its argument, and if its argument is zero the function returns zero. Furthermore ***min*** and ***max*** operators are also added which return their minimum and maximum arguments respectively. So the function set used in this work is, ***F = {hardlim, hardlims, purelin, logsig, tansig, plus, minus, times, pdivide, ppower, psqrt, exp, plog, min, max}***.

We use a random floating point constant generator in addition to *n* input signals (according to number of dataset features) to form our terminal set. Therefore the terminal set is ***T = {R, D0, …, Dn}*** where ***Di*** is the ***i***th feature of the dataset and ***n*** is the number of dataset features. There is no need for declaring the number of input signals to the algorithm beforehand, instead the algorithm recognizes the number of

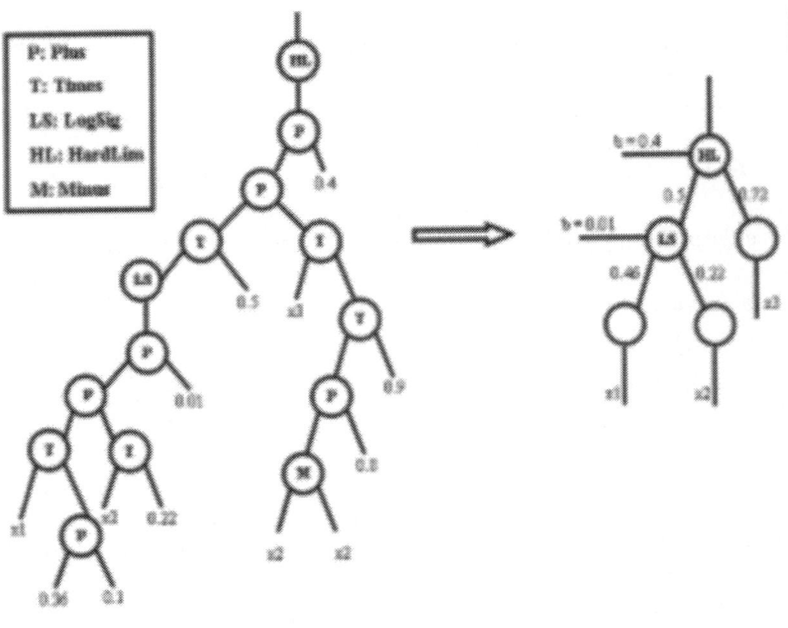

Fig. 3. A sample tree individual and its equivalent neural network

input signals (dataset features) automatically and adds them to the terminal set. Once a terminal ***R*** is used in a tree, a random constant number is replaced for ***R*** and this random number will not change until the end of the algorithm. A sample individual and its corresponding NN are shown below in Fig. 3.

To further simplify the evaluation of each tree as a NN, in our implementation each NN is a MATLAB executable string. A codification of the tree in Fig. 3 is:

HardLim (Plus (Plus (Times (LogSig (Plus (Plus (Times (x1, 0.46), Times(x2, 0.22)), 0.01)), 0.5),Times(x3, 0.72)), 0.4))

Since each tree should represent a NN, all combinations of functions and terminals can not be used and there are some rules about creation and manipulation of trees in this work, e.g. the first node of the tree (root node) must be a transfer function.

To avoid the *bloat* problem (a phenomenon consisting of an excessive tree growth without the corresponding improvement in fitness) we should continuously supervise depth and size of trees - depth denotes number of levels of a tree and size denotes number of nodes in a tree. To do this we have set some restrictions on depth and size of the trees. The maximum allowed depth of each tree is limited to 15 levels.

To generate the initial population we use *ramped half-and-half* method. Standard tree crossover and mutation are also used as diversity operators. For tree crossover, random nodes are chosen from both parent trees, and the respective branches are swapped creating two new offspring. For tree mutation, a random node is chosen from the parent tree and is substituted by a new random tree created with the terminals and functions available. This new random tree is created using the *grow* method. For parent selection we use *lexicography* selection method which is very similar to tournament method. The main difference is that, if two individuals are equally fit, the shortest one (the tree with fewer nodes) is chosen as the best. For survivor selection we use *total elitism* method. The best individuals from both parents and children are chosen to fill the new population.

3.2 Fitness (Evaluation) Function

A fast and easy-calculating criterion for evaluating performance of a NN in a classification problem is to use its classification error. The evolutionary process demands the assignation of a fitness value to each genotype. This value is the result of evaluation of the NN with the pattern set representing the problem. Since each individual corresponds to a NN, so at first glance it seems that every individual should be converted into a NN (with its weights already set, thus it does not need to be trained) to evaluate its fitness value. But the interesting point in our work is that there is no need to convert each tree to a NN. Due to our string based implementations each individual is represented as a MATLAB executable string which can be readily evaluated. Therefore upon reproducing a tree and applying the inputs to it, the fitness value of the corresponding NN is available. The fitness function is sum of the absolute values between expected output ($\boldsymbol{y}_i$) and real output ($\boldsymbol{o}_i$) of each individual NN (N is the number of input samples).

$$Fitness = \sum_{i=1}^{N} \left| y_i - o_i \right| \tag{1}$$

Calculating fitness may be a time consuming task, and during the evolutionary process the same tree is certainly to be evaluated more than once. To avoid this, some evaluations are kept in memory for future use, in case their results are needed again.

4 Experimental Results

To show the effectiveness of our proposed method we have conducted several experiments on some of the most popular knowledge-extraction datasets of different complexity from *UCI Machine Learning Repository*. Briefly, the goal here is to classify a number of samples in a number of different classes i.e. prediction about an attribute of the database, given some other attributes.

First dataset used here is *iris* dataset. It contains 150 samples in 3 classes and each sample has 4 features. Second dataset is *wine* dataset which contains 178 samples in 3 classes, each with 13 features. And the third dataset is *glass* dataset with 214 samples in 6 classes, each with 10 features. We used 70% of samples for fitness evaluation (training error) of the individuals and the remained 30% for performing tests (generalization error). We show our setup for two experiments and results obtained in these experiments.

In the first experiment, performed by utilizing the proposed GP approach, we try to produce a single tree (NN) for each dataset, which is able to classify all samples of its respective dataset. Therefore the expected output of this NN is class number of the sample to be classified.

Optimal parameters of the algorithm which have been obtained after some runs are as follows:

- Population size: 150 individuals
- Initial population by: Ramped half-and-half
- Termination condition (number of generations): 50 generations
- Maximum depth of each tree: 15 levels
- Maximum number of nodes in each tree: No constraint
- Parent selection method: Lexicography with 10% of population as competitors
- Survivor selection method: Total elitism

Two other important parameters are the probabilities for crossover and mutation operators. In our implementation these two parameters are variable and can be updated by the algorithm whenever needed [13]. If an individual generated by crossover operator yields a good fitness value (e.g. it will be the best individual so far), the probability of crossover will increase. The same rule applies for mutation operator. Minimum and maximum limits set for these two probabilities are 0.1 and 0.9 respectively. Table 1 shows the characteristics of the best NN and average errors found in this experiment for classification of all samples in each dataset for 10 runs.

Table 1. Characteristics of the best NNs found in first experiment for classification of all samples in each dataset

Dataset name	# of classes	# of training samples	# of test samples	Average training error (Fitness)	Average generalization error	Depth (# of levels)	Size (# of nodes)
Iris	3	105	45	0.1278	0.1150	15	59
Wine	3	125	53	0.4689	0.5600	15	61
Glass	6	152	62	0.7129	0.9360	14	68

Second experiment is conducted so that a NN (module) is evolved, using GP, for each class of the dataset. Inputs to these modules are all samples of the dataset, and the output is either 0 or 1, deciding whether the sample belongs to the corresponding class of the module or not. In the next step, another NN is evolved, as in Fig. 4, which receives the outputs of the previous modules and outputs the final class number of the sample. Briefly, the features of each sample are fed into the modules and the number of its class is produced as the output of the last NN. For all three datasets the same procedure is accomplished with the same set of GP parameters as in first experiment.

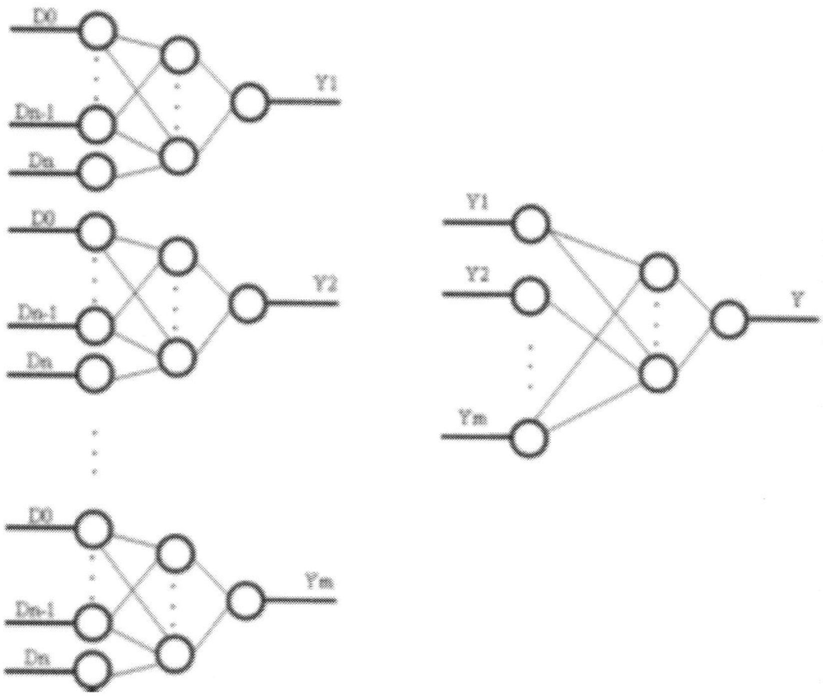

Fig. 4. The general structure of modular NNs generated with our method

Tables 2, 3 and 4 show the characteristics of the best single NNs for each class and also the final modular neural network found in this experiment to classify all samples of each dataset. These results are the average of 10 runs of the algorithm.

As can be seen in Table 1, when a single GP-optimized NN is used for classification of *iris* dataset, training and generalization errors are not very promising. Average training and generalization errors for this dataset are 0.1278 and 0.1150 respectively. These NNs decide the class number of each sample.

Table 2. Characteristics of the best NN modules and final modular NN found in second experiment for classification of all samples of *iris* dataset

Class	# of training samples	# of test samples	Average training error (Fitness)	Average test error	Depth (# of levels)	Size (# of nodes)
1	35	15	0	---	10	29
2	35	15	0.1232	---	15	135
3	35	15	0.1357	---	13	75
all	105	45	0.0228	0.0542	15	114

Table 3. Characteristics of the best NN modules and final modular NN found in second experiment for classification of all samples of *wine* dataset

Class	# of training samples	# of test samples	Average training error (Fitness)	Average test error	Depth (# of levels)	Size (# of nodes)
1	41	18	0.0731	---	7	21
2	50	21	0.1044	---	15	76
3	34	14	0.0028	---	14	79
all	125	53	0.0365	0.0891	11	32

Table 4. Characteristics of the best NN modules and final modular NN found in second experiment for classification of all samples of *glass* dataset:

Class	# of training samples	# of test samples	Average training error (Fitness)	Average test error	Depth (# of levels)	Size (# of nodes)
1	49	21	0.0693	---	10	28
2	12	5	0.4333	---	4	8
3	54	22	0.0546	---	12	37
4	9	4	0.5555	---	8	17
5	7	2	0.4285	---	4	9
6	21	8	0.0952	---	6	20
all	152	62	0.0125	0.0539	13	70

On the other hand, using the second method (modular neural network), the average training and generalization errors for *iris* dataset as shown in last row of Table 2 are reduced to 0.0228 and 0.0542, respectively. These results show that the second approach is far more effective than the first method. For *wine* dataset the same methods are applied. The average training and generalization error of the first method are 0.4689 and 0.5600 respectively. Using the second method these errors are reduced to 0.0365 and 0.0891 respectively. The average training and generalization errors for *glass* dataset with the first method are 0.7129 and 0.9360 while with the second approach these errors are reduced to 0.0125 and 0.0539.

5 Conclusion

By comparison of tables 1 through 4, it can be concluded that using a single NN for classification of all samples of a dataset - although it is evolved by an evolutionary algorithm - is not a good idea and this NN doesn't yield a good generalization error. On the other hand utilizing the structure of a modular neural network with some modules according to the number of classes in each dataset and then feeding their outputs into a new GP-optimized NN to decide for the class number of each sample yields much better results. The reason is that since a module is evolved for each class, due to the relative similarity of the samples in each class, learning the input space is easier. In fact in this method each module classifies a single class from other classes and a final NN decides the sample's class number.

In summary the most important features of this work are:

1. No knowledge of the architecture or connection weights of the NNs is known a-priori.
2. Smaller NNs are preferred to larger NNs with the same performance.
3. Neurons, even in the same layer, are not forced to use the same transfer function and a variety of transfer functions are available for each neuron.
4. Since we have used a string-based representation for our implementation, there is no need to convert chromosomes of the population to real NNs in order to calculate their fitness value.

References

[1] Murre, J.M.J.: Transputers and neural networks: An analysis of implementation constraints and performance. IEEE Transactions on Neural Networks 4, 284–292 (1993)
[2] Eccles, J.C.: Neuroscience 6, 1839–1855 (1981)
[3] Murre, J.: Learning and Categorization in Modular Neural Networks, Harvester–Wheatcheaf (1992)
[4] Auda, G.: Cooperative modular neural network classifiers, PhD thesis, University of Waterloo, Systems Design Engineering Department, Canada (1996)
[5] Yao, X.: Evolving Artificial Neural Networks. Proceedings of the IEEE 87(9) (1999)
[6] Back, T., Fogel, D.B., Michalewicz, Z.: Evolutionary Computation 1: Basic Algorithms and Operators, and Evolutionary Computation 2: Advanced Algorithms and Operators. IOP Publishing Ltd (2000)

[7] Koza, J.R.: Genetic Programming: On the Programming of Computers by Means of Natural Selection. MIT Press, Cambridge, MA (1992)

[8] Koza, J.R.: Genetic Programming: A Paradigm for Genetically Breeding Populations of Computer Programs to Solve Problems. Stanford University Computer Science Department Technical Report STAN-CS-90-1314 (June 1990)

[9] Koza, J.R., Rice, J.P.: Genetic Generation of Both Weights and Architecture for a Neural Network. In: Proceedings of the International Joint Conference on Neural Networks, vol. II, pp. 397–404 (1991)

[10] Gruau, F.: Automatic definition of modular neural networks. Adaptive Behaviour 3(2), 151–183 (1995)

[11] Luke, S., Spector, L.: Evolving graphs and networks with edge encoding: Preliminary report. In: Koza, J.R. (ed.) Late Breaking Papers at the Genetic Programming Conference Stanford University, July 28-31, pp. 117–124 (1996)

[12] Ritchie, M.D., White, B.C., Parker, J.S., Hahn, L.W., Moore, J.H.: Optimization of neural network architecture using genetic programming improves detection and modeling of gene-gene interactions in studies of human diseases. BMC Bioinformatics 2003 4(28) (2003)

[13] Davis, L.: Adapting operator probabilities in genetic algorithms. In: Schaffer, J.D. (ed.) Proceedings of the Third International Conference on Genetic Algorithms, pp. 61–69. Morgan Kaufmann, San Mateo, CA (1989)

Blind Matrix Decomposition Via Genetic Optimization of Sparseness and Nonnegativity Constraints

Kurt Stadlthanner[1], Fabian J. Theis[1], Elmar W. Lang[1], Ana Maria Tomé[2], and Carlos G. Puntonet[3]

[1] Institute of Biophysics, University of Regensburg, 93040 Regensburg, Germany
elmar.lang@biologie.uni-regensburg.de
[2] DETI / IEETA, Universidade de Aveiro, 3810-Aveiro, Portugal
[3] DATC, Universidad de Granada, E-18071 Granada, Spain

Abstract. Nonnegative Matrix Factorization (NMF) has proven to be a useful tool for the analysis of nonnegative multivariate data. However, it is known not to lead to unique results when applied to nonnegative Blind Source Separation (BSS) problems. In this paper we present first results of an extension to the NMF algorithm which solves the BSS problem when the underlying sources are sufficiently sparse. As the proposed target function is discontinuous and possesses many local minima, we use a genetic algorithm for its minimization. Application to a microarray data set will be considered also.

1 Introduction

Environmental stimuli cause the induction or repression of genes in living cells via a corresponding up- or down-regulation of the amount of messenger RNA (mRNA). Different experimental conditions may thus result in different characteristic expression patterns. Recently, high throughput methods like microarrays have become available and allow to measure whole genom wide gene expression profiles [2]. Intelligent data analysis tools are needed to unveil the information hidden in those microarray data sets [21].

Matrix factorization techniques seem promising to go beyond simple clustering and decompose such data sets into component profiles which might be indicative of underlying biological processes [15], [23], [3].

These unsupervised analysis methods decompose any given data matrix into a product of two matrices with desired properties. The latter are imposed as constraints on the matrix decomposition procedure. These techniques can equivalently be regarded as expanding the given data vectors into a basis of desired property. Famous among such projection methods are the following:

1. **PCA**: Principal Component Analysis [8] projects data into a space spanned by mutually orthogonal principal components (PCs). With microarray data the latter are called *eigenarrays*.

J. Marques de Sá et al. (Eds.): ICANN 2007, Part I, LNCS 4668, pp. 799–808, 2007.

2. **ICA**: Independent Component Analysis [11], [5] decomposes the data into statistically independent components (ICs). With microarray data, ICA extracts *expression modes*, the ICs which define corresponding groups of induced and repressed genes.
3. **NMF**: Non-negative Matrix Factorization [13] decomposes a given data matrix into a product of two strictly non-negative matrices. Applied to microarrays, NMF extracts so called *metagenes*. Non-negativity constraints seem natural as the raw fluorescence intensities measured with microarrays can have non-negative values only. Thus this technique alleviates some of the problems which arise with PCA and ICA which both yield negative entries in their component expression profiles. The latter have no obvious interpretation. In most applications so far this fact is gently ignored and the negative entries are turned positive by simply considering only the absolute values of the entries to the component profiles.

2 Matrix Factorization and Blind Source Separation

In the field of modern data analysis mathematical transforms of the observed data are often used to unveil hidden causes. Especially in situations where different observations of the same process are available matrix factorization techniques have proven useful [14]. Thereby, the $M \times T$ observation matrix $\mathbf{X}$ is decomposed into a $M \times N$ matrix $\mathbf{A}$ and a $N \times T$ matrix $\mathbf{S}$

$$\mathbf{X} = \mathbf{AS} \tag{1}$$

Here, it is assumed that M observations, consisting of T samples, constitute the rows of $\mathbf{X}$ and that $M << T$ and $M \leq N$. Obviously, the decomposition is highly non-unique and can only be solved uniquely if additional conditions constraining the row vectors of $\mathbf{H}$ or the column vectors of $\mathbf{W}$ are given.

One application of matrix factorization is linear blind source separation (BSS), where the observed microarray signals $\mathbf{X}$ are considered a weighted superposition of N underlying *eigenmodes*. If the latter form the rows of the $N \times T$ matrix $\mathbf{S}$, then each element a_{ij} of the mixing matrix $\mathbf{A}$ represents the weight with which the j-th source signal contributes to the i-th observed signal.

3 Sparse Nonnegative Blind Source Separation

3.1 Non-negative Matrix Factorization

A natural constraint to many real world problems is reflected in nonnegative matrix factorization (NMF) where the source matrix $\mathbf{S}$, the mixing matrix $\mathbf{A}$ as well as the observation matrix $\mathbf{X}$ are assumed to be strictly nonnegative. But NMF does not yield a unique decomposition, hence additional constraints are needed.

Considering gene expression profiles, if underlying *eigenmodes* should be indicative of ongoing biological processes in cells, then it may be expected that only a small number of genes are highly up- or down-regulated within a single process, i.e. many expression levels are close-to-average. Hence sparse coding seems to provide an additional natural constraint to NMF. Such sparse coding concepts have since long been discussed in the vision community [20], [22], [12] [16] and have already been exploited successfully in NMF based image analysis methods [9] as well as in other BSS algorithms [17].

An extension to standard NMF is presented in this paper. This sparse NMF (sNMF)algorithm seeks a nonnegative matrix factorization of the data matrix, for which the matrix $\mathbf{S}$ has the largest number of close-to-zero components in every *eigenmode*. This approach uniquely solves the matrix decomposition problem up to the usual scaling and permutation indeterminacies.

The basic idea of this sparse NMF (sNMF) algorithm is to estimate two nonnegative matrices $\hat{\mathbf{A}}$ and $\hat{\mathbf{S}}$ such that

1. $\hat{\mathbf{A}}$ and $\hat{\mathbf{S}}$ are both nonnegative
2. the rows of the matrix $\hat{\mathbf{S}}$ are as sparsely encoded as possible, i.e. contains as many close-to-zero components as possible
3. the reconstruction error $F(\mathbf{A}, \mathbf{S}) = ||\mathbf{X} - \hat{\mathbf{A}}\hat{\mathbf{S}}||^2$ of the estimated observations is as small as possible.

To solve this problem algorithmically, we minimize the following cost function $E(\hat{\mathbf{A}}, \hat{\mathbf{S}})$

$$E(\hat{\mathbf{A}}, \hat{\mathbf{S}}) = \frac{1}{2M}||n(\mathbf{X}) - n(\hat{\mathbf{A}}\hat{\mathbf{S}})||^2 - \frac{\lambda}{N}\sum_{n=1}^{N}\sigma_\tau(\hat{\mathbf{s}}_n), \tag{2}$$

where the function $n(..)$ normalizes the row vectors of $\mathbf{X}$ and $\hat{\mathbf{A}}\hat{\mathbf{S}}$ according to $x_{ij} = (x_{ij})(\sum_{k=1}^{T} x_{ik}^2)^{-1/2}$. Further σ denotes an appropriate sparseness measure, λ represents a Lagrange weighting factor, and $\hat{\mathbf{s}}_n$ denotes the n-th row of the matrix $\hat{\mathbf{S}}$.

3.2 Sparseness Measure

For the sparse nonnegative BSS problem at hand, we define the sparseness σ_τ of a row vector $\mathbf{s}$ as the ratio of the number of its close-to-zero elements and the total number of components

$$0 \leq \sigma_\tau(\mathbf{s}_n) = \frac{\text{number of elements of } \mathbf{s}_n \leq \tau \cdot \mathbf{s}_n^{max}}{\text{number of elements of } \mathbf{s}_n} \leq 1, \tag{3}$$

where $\mathbf{s}_n^{max}$ is the maximum value of $\mathbf{s}_n$, $\tau \in [0, 1]$ and we set all vector components s_n $le\tau$ to zero elements. The Lagrange parameter λ is used to balance the factorization of the data matrix and the sparseness requirement. As both terms in the cost function are normalized, results should be robust against varying sizes of $\mathbf{X}$ and $\mathbf{S}$ as soon as an appropriate λ has been obtained.

With NMF, a different sparseness measure is used in the literature which defines the spareness of a vector by the ratio of its L_1 norm to its L_2 norm [9].

Such a sparseness measure is, however, inappropriate for microarrays as the following illustrative example demonstrates.

Two nonnegative random sources $\mathbf{s}_1^{org}$ and $\mathbf{s}_2^{org}$ consisting of 1000 data points were generated with 90 % of the components of the first source signal and 80 % of the components of the second source signal being zeros. These two sources were normalized and then used to constitute the rows of the source matrix $\mathbf{S}^{org}$. The matrix of observations $\mathbf{X}$ was obtained by mixing the sources with the following mixing matrix $\mathbf{A}^{org} = \begin{bmatrix} 5 & 1 \\ 6 & 1 \end{bmatrix}$. Note, that as the first source signal is dominating in both of the mixtures the following alternative factorization of the observation matrix $\mathbf{X}$ is feasible. First, the original mixing matrix $\mathbf{A}^{org}$ can be replaced by the pseudo mixing matrix

$$\mathbf{A}^{pseu} = \begin{bmatrix} 0 & 1 \\ 1 & 1 \end{bmatrix}. \quad (4)$$

Correspondingly, the original source matrix $\mathbf{S}^{org}$ has then to be replaced by the matrix $\mathbf{S}^{pseu}$, the row vectors of which constitute the following pseudo source signals.

$$\mathbf{s}_1^{pseu} = \mathbf{s}_1^{org}, \quad (5)$$

$$\mathbf{s}_2^{pseu} = 5\mathbf{s}_1^{org} + \mathbf{s}_2^{org}. \quad (6)$$

Table 1. The L_1/L_2-norm sparseness σ and σ_τ as defined in Eq. (3) ($\tau = 0$) of the original and the pseudo sources

	$\sigma(\mathbf{s})$	$\sigma_\tau(\mathbf{s})$
$\mathbf{s}_1^{org}$	0.76	0.90
$\mathbf{s}_2^{org}$	**0.62**	**0.80**
$\mathbf{s}_1^{pseu}$	0.76	0.90
$\mathbf{s}_2^{pseu}$	**0.69**	**0.72**

Obviously, these matrices also factorize $\mathbf{X}$, i.e. $\mathbf{X} = \mathbf{A}^{pseu}\mathbf{S}^{pseu}$ still holds, but the number of zero elements in the second pseudo source signal $\mathbf{s}_2^{pseu}$ is about 8% lower than that of the original source signal $\mathbf{s}_2^{org}$ (cf. Tab. 1). Hence, a smaller value of the cost function is obtained with the matrices $\mathbf{A}^{org}$ and $\mathbf{S}^{org}$ than with the matrices $\mathbf{A}^{pseu}$ and $\mathbf{S}^{pseu}$ if the sparseness measure $\sigma_{\tau=0}$ is used in Eq. (3). Accordingly, the matrices $\mathbf{A}^{org}$ and $\mathbf{S}^{org}$ would be recovered correctly.

In contrast, the L_1/L_2 - norm sparseness measure σ assigns a higher sparseness value to the second pseudo source signal $\mathbf{s}_2^{pseu}$ than to the original source signal $\mathbf{s}_2^{org}$ (cf. Tab. 1). This would lead to a smaller value of the cost function with the matrices $\mathbf{A}^{pseu}$ and $\mathbf{S}^{pseu}$ than with the matrices $\mathbf{A}^{org}$ and $\mathbf{S}^{org}$ if sparsity is measured by σ in Eq. (3). Accordingly, a sparse BSS algorithm based on the latter sparseness measure would fail to recover the original source signal matrix $\mathbf{S}^{org}$ and mixing matrix $\mathbf{A}^{org}$, respectively.

4 Genetic Algorithm Based Optimization

4.1 Fitness Function

As mentioned above, the cost or fitness function defined in Eq. 2 is discontinuous hence cannot by optimized by techniques based on gradient descent. Furthermore, it possesses many local minima which suggests to use a Genetic Algorithm (GA) [19], [4] for its minimization.

For the minimization of the fitness function in Eq. 2 the M^2 elements of the solution matrix $\hat{\mathbf{A}}$ have to be determined. Taking advantage of the scaling indeterminacy inherent in the matrix decomposition model (1) we may assume that the columns of the original mixing matrix $\mathbf{A}$ are normalized such that $a_{ii} = 1 \ \forall \ i$. Hence, only the $M^2 - M$ off diagonal elements of the matrix $\hat{\mathbf{A}}$ have to be determined by the GA. Accordingly, each of the N_{ind} individuals of the GA comprises $M^2 - M$ nonnegative genes. During the optimization procedure, however, we allow the genes to turn negative as we have observed in our experiments that otherwise the GA often fails to find the global minimum of the fitness function.

In every generation of the GA, the fitness of each individual for the optimization task has to be computed in order to determine the number of offsprings it will be allowed to produce. In order to compute the fitness function values, for every individual a matrix $\hat{\mathbf{A}}_{-}$ is generated with its off elements consisting of the genes as stored in the individual and with unit diagonal elements. As in the next step of the algorithm the matrix $\hat{\mathbf{A}}_{-}$ has to be inverted, care must be taken that it does not become singular. Therefore, we replace matrices $\hat{\mathbf{A}}_{-}$ with a conditional number with respect to inversion which is higher than a user defined threshold τ_{sing} by a random matrix (also with unit values on its diagonal) with a conditional number lower than τ_{sing}. Finally, the genes of the corresponding individual are adjusted accordingly.

Next, the matrices $\hat{\mathbf{A}}$ and $\hat{\mathbf{S}}$ are needed in order to evaluate the target function (2). For this purpose, the inverse $\hat{\mathbf{W}}_{-} = \hat{\mathbf{A}}_{-}^{-1}$ is computed and the matrices $\hat{\mathbf{S}}$ and $\hat{\mathbf{A}}$ are then obtained by setting the negative elements of the matrices $\hat{\mathbf{A}}_{-}$ and $\hat{\mathbf{S}}_{-} = \hat{\mathbf{W}}_{-}\mathbf{X}$, respectively, to zero.

After assigning a fitness function value to each individual, they are arranged in ascending order of their fitness function values $F(p^{(i)})$, $i = 1, \ldots, N_{ind}$ which are determined by

$$F(p^{(i)}) = 2 - \mu + 2(\mu - 1)\frac{p^{(i)} - 1}{N_{ind} - 1}, \tag{7}$$

where $p^{(i)}$ is the position of individual i in the ordered population. The scalar parameter μ is chosen between 1.1 and 2.0 and denotes the selective pressure towards the fittest individuals.

4.2 Genetic Operators

Stochastic Universal Sampling (SUS) [19] is used to determine the absolute number of offsprings an individual may produce. The latter are created in a two step procedure. In the first crossover step, two individuals are chosen at random according to the SUS criterion. They are used to create a new individual by uniform crossover. In the second mutation step, these offsprings are mutated by altering a certain fraction r_{mut} of the genes at random in the range of $[0, m_{max}]$. Finally the parent individuals are replaced by their offsprings applying an elitist reinsertion scheme. Hence, only the $(1 - r_{elit})N_{ind}$ less fittest parent individuals are replaced by their fittest offsprings. In order to prevent the algorithm from converging prematurely we make use of the concept of multiple populations. Thereby, a number N_{pop} of populations, each consisting of N_{ind} individuals, are evolving independently in parallel and are only allowed to exchange their fittest individuals after every T_{ex}-th generation. The calculations have been performed using the functions provided by the *Genetic Algorithm Toolbox* [?] except from the mutation operator which has been implemented separately.

4.3 Parallelization

A general problem of the proposed sNMF-GA algorithm is its tendency to converge prematurely to suboptimal solutions. This problem can be overcome to a large extent, however, if instead of one large population several smaller subpopulations are used which only exchange individuals from time to time. Another reason to parallelize the GA in sNMF-GA is that the evaluation of the fitness function may become very time consuming because of the evaluation of the sparseness in every step. Hence, the implementation of the sNMF-GA algorithm is designed such that it can be run on several computers in parallel. As the implementation of the sNMF-GA algorithm is fully written in the C programming language, the routines of the MPICH2 [10] implementation of the Message Passing Interface standard [1] for the communication between computers can be used advantageously. The run time of the GA can be reduced by a factor close to $1/N_{nod}$ if a cluster consisting of N_{nod} nodes is available.

4.4 Algorithm Repetitions

Despite the use of the mutation operator and multiple populations, the algorithm still could not recover the source and mixing matrix after its first run in most cases. In order to keep the computational load of the algorithm bearable, this problem could not be overcome by simply increasing the number N_{ind} of individuals and N_{pop} of populations to arbitrarily large values. But satisfying results could still be obtained by applying the algorithm repeatedly. As usually, the algorithm is provided with the observation matrix $\mathbf{X}$ in its first run which is then decomposed into first estimates of the source matrix $\hat{\mathbf{S}}^{(1)}$ and the mixing matrix $\hat{\mathbf{A}}^{(1)}$, i.e. $\mathbf{X} = \hat{\mathbf{A}}^{(1)}\hat{\mathbf{S}}^{(1)}$. In order to make use of the suboptimal results already achieved, during the next run the matrix $\hat{\mathbf{S}}^{(1)}$ is provided to the algorithm instead of the matrix $\mathbf{X}$. The matrix $\hat{\mathbf{S}}^{(1)}$ is then factorized into the matrices $\hat{\mathbf{A}}^{(2)}$ and $\hat{\mathbf{S}}^{(2)}$, which means that the matrix $\mathbf{X}$ can now be factorized as $\mathbf{X} = \hat{\mathbf{A}}^{(1)}\hat{\mathbf{A}}^{(2)}\hat{\mathbf{S}}^{(2)}$. This procedure is repeated K times until the newly determined mixing matrix $\mathbf{A}^{(K)}$ differs only marginally from the identity matrix.

With this procedure the final estimates of the mixing matrix $\hat{\mathbf{A}}$ and of the source matrix $\hat{\mathbf{S}}$ are determined as

$$\hat{\mathbf{A}} = \prod_{j=1}^{K} \hat{\mathbf{A}}^{(j)} \quad \text{and} \quad \hat{\mathbf{S}} = \hat{\mathbf{S}}^{(K)} \tag{8}$$

respectively, as the matrix $\mathbf{X}$ can be factorized as

$$\mathbf{X} = \prod_{j=1}^{K} \hat{\mathbf{A}}^{(j)} \hat{\mathbf{S}}^{(K)} \tag{9}$$

4.5 Application of sNMF to Microarray Data

This subsection deals with the sNMF analysis of microarray data. Results will be compared with those achieved by fastICA. The data sets to be analyzed were recorded during an investigation of *Pseudo-Xanthoma Elasticum* (PXE), an inherited connective tissue disorder characterized by progressive calcification and fragmentation of elastic fibers in the skin, the retina, and the cardiovascular system.

During the investigations $M = 8$ microarray experiments have been carried out. In the first and second experiment the PXE fibroblasts were incubated in Bovine Serum Albumin (BSA). The incubation time was three hours in the former and 24 hours in the latter experiment. In the third experiment the PXE fibroblasts were incubated for three hours in an environment with a high concentration of the Transcription Growth Factor beta (TGF-β) and in the fourth experiment the cells were incubated for 24 hours in a medium enriched with Interleukin 1β. The same experiments were then repeated with a control group of normal fibroblasts. Gene expression levels were recorded on an Affymetrix HG-U133 plus 2.0 microarray chip. This *in-situ* chip is capable of detecting 54675 genes in parallel and makes use of the probe pair strategy. Hence, each measured expression value was accompanied by a detection call. If in all experiments the detection call of a particular gene amounted to "absent", the gene was removed from all data sets. After this procedure only 10530 genes remained in each data set. These data sets were used to constitute the 8×10530 observation matrix $\mathbf{X}$ which was then decomposed into the matrices $\hat{\mathbf{A}}$ and $\hat{\mathbf{S}}$ by the proposed sNMF algorithm. For the genetic algorithm the number of sub-populations was $N_{sub} = 56$, the total number of iterations was $T_{total} = 2500$ and the number of algorithm repetitions was $N_{rep} = 8$. The sparseness measure σ_τ was set to 0.30. As the overall number of individuals was large ($N_{ind} = 2800$), the algorithm was run in parallel on a cluster of 28 computers.

After 8 iterations the algorithm converged to a stable estimate of the matrix of *eigengenes*. The corresponding *eigenmodes* turned out to be rather sparse as can be seen in Tab. 2. Hence, the sNMF algorithm deals with the expression levels of a small number of genes only to reconstruct the data matrix $\mathbf{X}$. It is assumed that these genes are the most important representatives of cellular processes related with the experiments conducted. This assumption reflects a widely used praxis in genetics which says that the most highly expressed genes are the most typical for a specific cellular process [15].

After the sNMF analysis, each row of the matrix **S** should ideally consist of an expression pattern indicative of a group of genes most important to the cellular processes related to the disease under investigation. However, many more translational processes are occurring simultaneously in a biological cell expressing genes not related to the disease investigated. Hence, each estimated source is expected to also contain signatures from various other cellular processes. Despite this highly overcomplete setting, the algorithm succeeded in grouping the majority of genes which are related with calcium ion binding (344 in total) and hence with the disease signatures of PXE into the sixth estimated source (see Tab. **??**). Furthermore, the calcium ion binding related genes in the sixth source seem to be specific for only one biological process as maximally 14 % of them could be found in any of the remaining component signals. To assign genes to molecular functions, the publicly available Gene Ontology database [6, 7] was queried.

Table 2. Sparseness $\sigma_{\tau=0}$ of the estimated sources and number of genes related with calcium ion binding (#(cib)) for each of the 8 *eigenmodes* estimated with sNMF

Source	1	2	3	4	5	6	7	8
$\sigma_{\tau=0}$	0.996	0.971	0.983	0.995	0.984	0.954	0.989	0.985
#(cib)	0	13	7	0	4	**35**	9	6

In [18] the same PXE data set was analyzed by means of the fastICA algorithm. In this analysis the following procedure was carried out: first the data matrix **X** was fed into fastICA with the gaussian nonlinearity being used. The independent components extracted display both positive and negative expression levels. The latter, of course, miss any obvious biological meaning - a fact often gently neglected in the literature. Furthermore, there are no close-to-zero entries in the component signals. Thus all genes seemingly participate in all translational processes observed. To get rid of these negative expression levels which lack any biological interpretation, in [18] a further postprocessing step was undertaken. In this step two clusters were formed for each source. For this purpose the maximum positive (s_{max}) and the minimum negative expression level (s_{min}) encountered in the component signals was determined. The first cluster then consisted of all those genes which had positive expression levels higher than $0.26s_{max}$ while the second group consisted of genes with expression levels lower than $-0.26s_{min}$. Finally, the molecular function of the genes in each cluster was determined by enquiring the GO database. In this case, at most 22 genes related with calcium ion binding could be grouped into one cluster as can be seen in Tab. 3. Furthermore, other clusters (e.g. clusters 1, 2 and 13) also contained significant amounts of genes related with calcium ion binding. sNMF thus achieved a better grouping of calcium ion binding genes than fastICA. These results are

Table 3. Number of genes related with calcium ion binding (#(cib)) for each of the 16 clusters formed during the analysis of the PXE data by fastICA

cluster	1	2	3	4	5	6	7	8	9	**10**	11	12	13	14	15	16
#(cib)	9	8	3	2	5	1	0	4	1	**22**	2	4	12	3	0	1

rather preliminary, though. Further cooperation with biologists is necessary to investigate the gene regulatory networks and biochemical pathways in which the calcium ion binding genes grouped into the sixth component signal by the sNMF algorithm are involved.

Acknowledgements

Financial support by Siemens AG, Corporate Technology, Munich, the DFG (GRK 638: Nonlinearity and Nonequilibrium in Condensed Matter), the DAAD-GRICES Acçôes Integradas Lusc - Alemãs and the DAAD Acciones Integradas Hispano - Alemanas is gratefully acknowledged.

References

1. The Message Passing Interface (MPI) standard, http://www.mpi-forum.org
2. Baldi, P., Hatfield, W.: DNA Microarrays and Gene Epression. Cambridge University Press, Cambridge (2002)
3. Chiapetta, P., Roubaud, M.C., Torrésani, B.: Blind source separation and the analysis of microarray data. J. Comp. Biology 11, 1090–1109 (2004)
4. Chipperfield, A., Fleming, P., Pohlheim, H., Fonseca, C.: Genetic Algorithm Toolbox. University of Sheffield
5. Cichocki, A., Amari, S.-I.: Adaptive Blind Signal and Image Processing. John Wiley & Sons, England (2002)
6. The Gene Onkology Consortium Gene ontologie: Tool for the unification of biology. Nature Genetics 25, 25–29 (2000)
7. The Gene Onkology Consortium. Creating the gene ontology resource: design and implementation. Genome Research 11, 1425–1433 (2001)
8. Diamantaras, K.I., Kung, S.Y.: Principal Component Neural Networks, Theory and Applications. Wiley, Chichester (1996)
9. Hoyer, P.O.: Non-negative matrix factorization with sparseness constraints. J. Machine Learning Research 5, 1457–1469 (2004)
10. http://www.mpi-forum.org/docs/mpi-20-html/mpi2report.html MPI-2: Extensions to the Message-Passing Interface, www.mpi-forum.org
11. Hyvärinen, A., Karhunen, J., Oja, E.: Independent Componenet Analysis. John Wiley & Sons, England (2001)
12. Hyvärinen, A., Oja, E., Hoyer, P., Hurri, J.: Image feature extraction by sparse coding and independent component analysis. In: Proc. ICPR 1998 (1998)
13. Lee, D.D., Seung, H.S.: Learning the parts of objects by non-negative matrix factorization. Nature 401, 788–791 (1999)
14. Lee, D.D., Seung, H.S.: Algorithms for non-negative matrix factorization. In: Advances in Neural Information Processing 13 (NIPS'2000), Masachusetts, MIT Press, Washington (2001)

15. Lee, S.-I., Batzoglou, S.: Application of independent component analysis to microarrays. Genome Biology 4, R76.1–R76.21 (2003)
16. Lewicki, M.S., Sejnowski, T.J.: Learning overcomplete representations. Neural Computation 12, 337–365 (2000)
17. Li, Y., Cichocki, A., Amari, S.: Analysis of sparse representation and blind source separation. Neural Computation 16, 1193–1234 (2004)
18. Lutter, D., Stadlthanner, K., Theis, F.J., Lang, E.W., Tomé, A.M., Becker, B., Vogt, T.: Analyzing gene expression profiles with ica. In: Ruggiero, C. (ed.) Proc. BIOMED 2006, Canada, Acta Press (2006)
19. Michalewicz, Z.: Genetic Algorithms + Data Structures = Evolution Programs. Springer, Berlin (1999)
20. Ohlshausen, B.A., Field, D.J.: Natural image statistics and efficient coding. Network: Computation in Neural Systems 7, 333–339 (1996)
21. Quackenbush, J.: Computational analysis of microarray data. Nature 2, 418–427 (2001)
22. Ruderman, D.: The statistics of natural images. Network: Computations in Neural Systems 5, 517–548 (1994)
23. Saidi, S.A, Holland, C.M., Kreil, D.P., MacKay, D.J.C., Charnock-Jones, D.S., Print, C.G., Smith, S.K.: Independent component analysis for gene arrays. Oncogene 23, 6677–6683 (2004)

Meta Learning Intrusion Detection in Real Time Network

Rongfang Bie[1,*], Xin Jin[1], Chuanliang Chen[1], Chuan Xu[1], and Ronghuai Huang[2]

[1] College of Information Science and Technology,
Beijing Normal University, Beijing 100875, P.R. China
rfbie@bnu.edu.cn, xinjin4@yahoo.com,
byronccl@126.com, abram_win@hotmail.com,
[2]School of Education Technology,
Beijing Normal University, Beijing 100875, P.R. China
huangrh@bnu.edu.cn

Abstract. With the rapid increase in connectivity and accessibility of computer systems over the internet which has resulted in frequent opportunities for intrusions and attacks, intrusion detection on the network has become a crucial issue for computer system security. Methods based on hand-coded rule sets are laborous to build and not very reliable. This problem has led to an increasing interest in intrusion detection techniques based upon machine learning or data mining. However, traditional data mining based intrusion detection systems use single classifier in their detection engines. In this paper, we propose a meta learning based method for intrusion detection by MultiBoosting multi classifiers. MultiBoosting can form decision committees by combining AdaBoost with wagging. It is able to harness both AdaBoost's high bias and variance reduction with wagging's superior variance reduction. Experiments results show that MultiBoosting can improve the detection performance of state-of-art machine learning based intrusion detection techniques. Furthermore, we present a Symmetrical Uncertainty (SU) based method for reducing network connection features to make MultiBoosting more efficient in real-time network environment, in the meanwhile, keep the detection performance unundermined and in some cases, even further improved.

1 Introduction

Computer security has become a critical issue with the rapid increase in connectivity and accessibility of computer systems over the internet which has resulted in frequent opportunities for intrusions and attacks. Intrusion Detection Systems (IDS) is very important for computer security [24]. IDS can identify attacks while they are occurring on a network, in an attempt to defend the network. IDS were introduced by Anderson [20] and formalized later by Denning [21]. Network Intrusion Detection Systems (NIDS) use audit data such as network packet data, especially network packet headers and payloads. Most NIDS generally collects network packets through using Network Interface Card (NIC) and takes passive analysis. NIDS detects attacks

* Corresponding author.

J. Marques de Sá et al. (Eds.): ICANN 2007, Part I, LNCS 4668, pp. 809–816, 2007.

that host-based detection systems miss, e.g., many IP-based Denial-Of-Service (DoS) attacks and fragmented packet (Teardrop) attacks [28][23]. In this paper, we will concentrate our attention on NIDS.

Most NIDS are based on hand-crafted signatures that are developed by manual coding of expert knowledge [13]. The major problem with this approach is that these IDSs rely heavily on human analysts to differentiate intrusive from non-intrusive network traffic. The large and growing amount of data confronts the analysts with an overwhelming task, making the automation of aspects of this task necessary [9]. Recently, there has been an increased interest in data mining based approaches to building detection models for IDSs. The methods include Naïve Bayes [6], K-means Nearest Neighbor [29], Support Vector Machine (SVM) [29][28][25] (Least Squares SVM [13], Robust SVM [15]), Decision Tree (ID3 [12], C4.5 [14]), Radial Basis Function (RBF) networks [4], Improved Iterative Scaling (IIS) [27], etc.

In this paper we describe a meta mining based method for intrusion detection by MultiBoosting traditional classifiers based on network connection features. We also present a Symmetrical Uncertainty (SU) based method for selecting relevant connection features to improve the detection performance.

The rest of this paper is structured as follows. Section 2 describes SU based feature ranking method for selecting relevant connection features. Section 3 presents the meta learning method. Section 4 presents the experimental results. Conclusions are covered in Section 5.

2 Symmetrical Uncertainty of Network Features

Network intrusion detection relies on building network state models from a given set of labeled connection feature vectors. However, using large multidimensional feature vectors is costly in terms of training and operational complexity, memory demands, etc. Therefore, one would prefer to select relevant network features in the feature vector [7].

In this paper we describe a Symmetrical Uncertainty (SU) based method for selecting relevant features for network intrusion detection. SU has been described in books on information theory and in numerical recipes, by Press et al.

We define each network feature as variable X and the attack class as variable C. We first calculate the entropy of X. Entropy, a commonly used measure in the information theory, characterizes the purity of an arbitrary collection of samples. The entropy of X will be obtained by

$$H(X) = -\sum_{x \in X} p(x) \log_2(p(x)) \tag{1}$$

where $p(x)$ is the marginal probability density function for the variable X. The entropy of X after observing C is:

$$H(X \mid C) = -\sum_{c} p(c) \sum_{x} p(x \mid c) \log_2(p(x \mid c)) \tag{2}$$

where $p(x|c)$ is the conditional probability of x given c.

Given the entropy as a criterion of impurity in a training set *S*, Information Gain (IG) can reflect additional information about *X* provided by *C* that represents the amount by which the entropy of *X* decreases.

$$IG = H(C) - H(X \mid C) = H(C) + H(X) - H(C, X) \tag{3}$$

where, *H*(*C*, *X*) is the joint entropy of *C* and *X*, it is calculated from the joint probabilities of all combinations of values of *C* and *X*.

A well-known weakness of the IG criterion is that it is biased in favor of attributes with more values even when they are not more informative. The Symmetrical Uncertainty (SU) criterion compensates for the inherent bias of IG by dividing it by the sum of the entropies of *C* and *X* [8]:

$$SU = 2.0 \times IG / (H(C) + H(X)) \tag{4}$$

Due to the correction factor 2.0, SU coefficient takes values which are normalized to the range [0, 1]. A value of SU = 0 indicates that the attribute *X* and the class *C* have no association; the value 1 for the symmetrical uncertainty coefficient indicates that *X* can completely predicts *C*.

3 Meta Learning

3.1 MultiBoosting

MultiBoosting [5] can be considered as wagging (which is in turn a variant of bagging) committees formed by AdaBoost. A decision is made as to the size of sub-committees, and how many sub-committees should be formed for a single run.

In the absence of an a-priori reason for selecting any specific values for those factors, MultiBoost takes a single committee size *T* and sets the number of sub-committees and the size of those sub-committees to square of *T*. Note that both these values must be whole numbers, it is necessary to round off the result.

Deriving the values is achieved by setting a target final sub-committee member index, where each member of the final committee is given an index by starting from one. Due to too great or too low error, that allows the premature termination of boosting one sub-committee to lead to an increase in the size of the next sub-committee.

An additional sub-committee is added with a target of completing the full complement of committee members if the last sub-committee is prematurely terminated. When this additional sub-committee also fails to reach this target, this process is repeated, adding further sub-committees until the target total committee size is achieved. More information about MultiBoosting algorithm is presented in [5].

In our experiments, MultiBoosting is combined with the following base learners.

3.2 Base Learners

Radial Basis Function (RBF) Network: The Radial Basis Function (RBF) Network has shown a great promise in this sort of problems because of its faster learning capacity [16]. The input to RBF Network is a vector $\boldsymbol{x}$ of extracted features from the network connection. We use the *k*-means clustering [17] algorithm to provide the basis functions. Then RBF Network learns a logistic regression on top of that. For classification it uses the given number of clusters per class [3].

Support Vector Machine (SVM): Being developed in 1995 by Vapnik and co-workers, the Support Vector Machines (SVM) is a relatively new machine learning formulation [19]. SVM have been successfully applied to a wide range of pattern recognition problems [18]. SVM can learn linear rules described by a weighted vector and a threshold. Geometrically, we can find two parts representing the two classes with a hyper-plane with normal and distance from the origin [28]. In the case that linear separation is impossible, the technique of kernel will be used to automatically inject the training samples into a higher-dimensional space, and to learn a separator in that space. In our experiments, we use the Linear kernel function.

Naive Bayes: Naive Bayes [1] based intrusion detection problem can be described as follows. For a given sample we search for a class c_i that maximizes the posterior probability $P(c_i|x;\theta^{'})$, by applying Bayes rule. Then x can be classified by computing.

$$c_l = \arg\max_{c_i \in C} P(c_i \mid \theta^{'}) P(x \mid c_i;\theta^{'}) \tag{5}$$

Decision Tree (DT): The most popular DT learning algorithm is Quinlan's C4.5 [2]. For network intrusion detection, DT will construct a tree where the nodes correspond to connection features/attribute test, the links (to attribute values and the leaves) to the attack types. To classify an incoming network connection represented as a feature vector we start at the root of the tree and follow the path corresponding to the vector's values until a leaf node is reached and the classification is obtained.

4 Experimental Results

The following experiments are based on ten runs of 10-fold cross-validation on the KDDCUP99 Network Intrusion Detection Benchmark Dataset, available on [11, 22, 26]. We investigated all the 41 network session features [10].Performance is expressed in terms of the standard measure in the intrusion detection literature, *Detection Rate*.

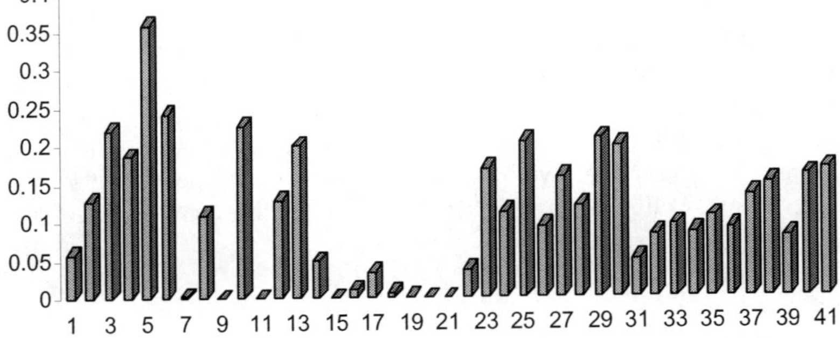

Fig. 1. SU scores of the 41 network features. X-axis represents the feature ID. Y-axis represents the SU score.

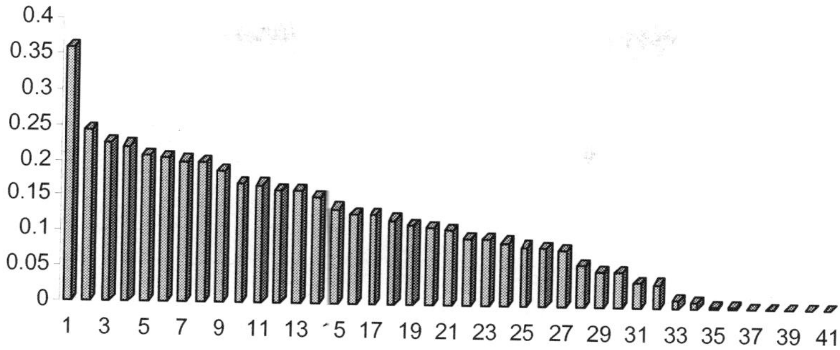

Fig. 2. Sorted SU scores of the 41 network features. X-axis represents the sorted ranking index according to the SU score of the features. For example, '1' represents the rank 1 feature, whose ID is 5, which can be seen from Fig 2. Y-axis represents the SU score.

The *detection rate* of an intrusion class is defined by the ratio of the number of connections which are correctly predicted as the class and the number of all connections which are of the class (some of these connections may be incorrectly predicted to be other class). *Detection rate* ranges [0%, 100%], the higher the better. We use *detection rate* of a class to evaluate the ability of the learning methods to find all the samples which belong to the attack type.

Fig. 1 shows the SU ranking scores computed between each feature and the class. Fig. 2 shows the sorted results. For feature selection we select the top 29, 33, 36 features respectively in different experiments. We choose the top 29 features because they have a SU value of greater than 0.05. In other experiments, we try 33 for 0.01, and 36 for 0 (there are several features having an SU value of 0, which means that they are utterly irrelevant features).

Fig. 3 shows the *detection rate* of MultiBoosting and four baseline classifiers (SVM, RBF Network, C4.5 and Naïve Bayes) with the 41 original features and the SU selected features.

The results show that by using MultiBoosting, all the four base classifiers' performance is greatly improved. With the original features, MultiBoosting improve the detection rate of SVM by 1.16%, RBF Network by 2.21%, C4.5 by 1.69% and Naïve Bayes by 2.23%.

The results also show that by using SU based feature selection, we can use only 29 features instead of the original 41features to achieve still high performance. Actually, in many cases, the performance is even improved. For example, using the top 33 features, the detection rate of MultiBoosting with RBF Network is improved from95.96% to 98.25%, compared to using the full original 41 features. The reason is that SU based method can filter out irrelevant features which may impair the detection ability of the classifiers. This will have two advantages: 1. further improve the detection rate by selecting relevant features and rejecting irrelevant features. 2. make the MultiBoosting based method more suitable for real time network by simplifying the models with reduced features.

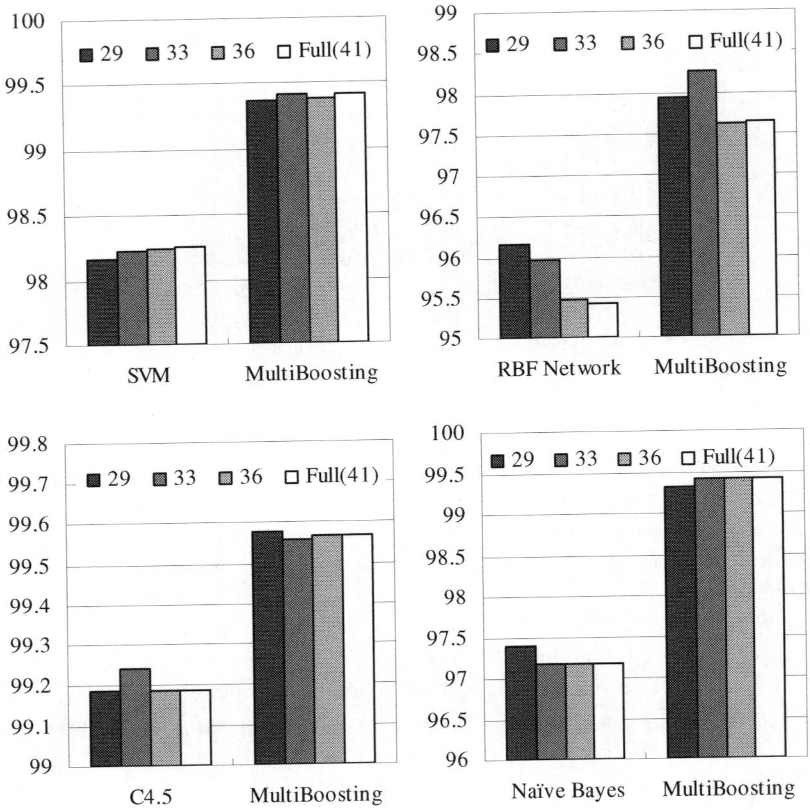

Fig. 3. Detection rate of MultiBoosting and four baseline classifiers (SVM, RBF Network, C4.5 and Naïve Bayes) with the full 41 original features and the SU selected 29, 33 and 36 features. Y-axis represent the detection rate, we omit the '%' symbol.

5 Conclusions

The contribution of this paper is two folds.

Firstly, we propose a meta learning based method for network intrusion detection by MultiBoosting [5] multi classifiers. Comparison results based on the standard DARPA benchmark data sets, using the *detection rate* performance measure, indicate that MultiBoosting can greatly improve the detection performance of traditional machine learning intrusion detection methods.

Secondly, we propose to use Symmetrical Uncertainty (SU) for reducing network connection features. Experimental results show two advantages of this method: the one is that SU can further improve the detection rate by selecting relevant features and rejecting irrelevant features. The other is that SU can make the MultiBoosting based method more efficient in real-time network environment by simplifying the models with reduced features.

Acknowledgments

This work was supported by the National Science Foundation of China under the Grant No. 60273015 and No. 10001006. This research was also supported by the Foundation of Chinese National 985 Project "Educational Information and Technology Platform" in Beijing Normal University.

References

1. Schneider, K.-M.: Comparison of Event Models for Naive Bayes Anti-Spam E-Mail Filtering. In: Proceedings of the 10th Conference of the European Chapter of the Association for Computational Linguistics, Budapest, Hungary, pp. 307–314 (April 2003)
2. Quinlan, R.: C4.5: Programs for Machine Learning. Morgan Kaufmann Publishers, San Mateo, CA (1993)
3. Witten, I., Frank, E.: Data Mining –Practical Machine Learning Tools and Techniques with Java Implementation. Morgan Kaufmann, San Francisco (2000)
4. Zhang, Z., Shen, H.: Online Training of SVMs for Real-time Intrusion Detection. In: AINA'04. 18th International Conference on Advanced Information Networking and Applications, p. 568 (2004)
5. Geoffrey, I.: Webb: MultiBoosting: A Technique for Combining Boosting and Wagging. Machine Learning 40(2), 159–196 (2000)
6. BenAmor, N., Benferhat, S., ElOuedi, Z.: Naive Bayes vs Decision Trees in Intrusion Detection Systems. In: Handschuh, H., Hasan, M.A. (eds.) SAC 2004. LNCS, vol. 3357, Springer, Heidelberg (2004)
7. Ganchev, T., Zervas, P., Fakotakis, N., Kokkinakis, G.: Benchmarking Feature Selection Techniques on the Speaker Verification Task. In: 5th International Symposium on Communication Systems, Network and Digital Signal Processing (July 19-21, 2006)
8. Hall, M.A., Smith, L.A.: Practical feature subset selection for machine learning. In: Proceedings of the 21st Australian Computer Science Conference, pp. 181–191 (1998)
9. Stolfo, S., Fan, W., Lee, W., Prodromidis, A., Chan, P.: Cost-based Modeling for Fraud and Intrusion Detection: Results from the JAM Project. In: DISCEX '00. Proceedings of the 2000 DARPA Information Survivability Conference and Exposition (2000)
10. KDDCUP99 Dataset Task Description (Accessed 2006), http://kdd.ics.uci.edu/databases/kddcup99/task.html
11. KDDCUP99 Network Intrusion Detection Benchmark Dataset (Accessed 2006), http://kdd.ics.uci.edu/databases/kddcup99/kddcup99.html
12. Sinclair, S.M.C., Pierce, L.: An Application of Machine Learning to Network Intrusion Detection. In: Proceedings of the 15th Annual Computer Security Applications Conference, Phoenix, AZ, USA, pp. 371–377 (1999)
13. Kim, B.-J., Kim II, K.: Two-Tier Based Intrusion Detection System. In: Wang, L., Jin, Y. (eds.) FSKD 2005. LNCS (LNAI), vol. 3614, pp. 27–29. Springer, Heidelberg (2005)
14. Sabhnani, M., Serpen, G.: Application of Machine Learning Algorithms to KDD Intrusion Detection Dataset within Misuse Detection Context. In: MLMTA03. Proceedings of the International Conference on Machine Learning, Models, Technologies and Applications, Las Vegas, NV, pp. 209–215 (June 2003)
15. Hu, W., Liao, Y., Vemuri, V R.: Robust Support Vector Machines for Anomaly Detection in Computer Security. In: Proceddings of Conference on Machine Learining and Application (2003)

16. Lee, C.-C., Chung, P.-C., Tsai, J.-R., Chang, C.-I: Robust Radial Basis Function Neural Networks. IEEE Transactions on Systems, Man, and Cybernetics-Part B: Cybernetics 29(6) (1999)
17. Wang, H., et al.: Clustering by Pattern Similarity in Large Data sets. In: SIGMOD, pp. 394–405 (2002)
18. Guo, G., Li, S.Z., Chan, K.: Face Recognition by Support Vector Machines. In: Fourth IEEE International Conference on Automatic Face and Gesture Recognition, pp. 196–201. IEEE Computer Society Press, Los Alamitos (2000)
19. Vapnik, V.N.: Statistical learning theory. Adaptive and learning systems for signal processing, communications, and control, Wiley, New York (1998)
20. Anderson, J.P.: Computer security threat monitoring and surveillance. Technical Report, James P Anderson Co. Fort Washington, PA (April 1980)
21. Denning, D.E.: An intrusion-detection model. IEEE Transactions on Software Engineering SE-13(2), 222–232 (1987)
22. Nguyen, B.V.: An Application of Support Vector Machines to Anomaly Detection. Research in Computer Science - Support Vector Machine, report, Fall (2002)
23. Vigna, G., Kemmerer, R.: Netstat: a network based intrusion detection system. Journal of Computer Security 7(1) (1999)
24. Symantec.com: Symantec internet security threat report highlights rise in threats to confidential information. (Accessed 2006), Available at http://www.symantec.com/press/2005/n050321.html
25. Sung, A., Mukkamala, S.: Identifying important features for intrusion detection using support vector machines and neural networks. In: Symposium on Applications and the Internet, pp. 209–216 (2003)
26. Kendall, K.: A Database of Computer Attacks for the Evaluation of Intrusion Detection Systems. Master's Thesis, Massachusetts Institute of Technology (1998)
27. Jin, X., Huang, R., Bie, R.: Detecting Network Attacks via Improved Iterative Scaling. In: INDIN07. Proceedings of the 5th IEEE International Conference on Industrial Informatics, Vienna, Austria, July 23-26 (2007)
28. Kim, D.S., Park, J.S.: Network-Based Intrusion Detection with Support Vector Machines. In: Kahng, H.-K. (ed.) ICOIN 2003. LNCS, vol. 2662, pp. 747–756. Springer, Heidelberg (2003)
29. Kaplantzis, S., Mani, N.: A Study on Classification Techniques for Network Intrusion Detection. In: NCS06. Proceedings of the IASTED International Conference on Networks and Communication Systems (2006)

Active Learning to Support the Generation of Meta-examples

Ricardo Prudêncio[1] and Teresa Ludermir[2]

[1] Departament of Information Science, Federal University of Pernambuco, Av. dos Reitores, s/n - CEP 50670-901 - Recife (PE) - Brazil
prudencio.ricardo@gmail.com
[2] Center of Informatics, Federal University of Pernambuco, Pobox 7851 - CEP 50732-970 - Recife (PE) - Brazil
tbl@cin.ufpe.br

Abstract. Meta-Learning has been used to select algorithms based on the features of the problems being tackled. Each training example in this context, i.e. each meta-example, stores the features of a given problem and the performance information obtained by the candidate algorithms in the problem. The construction of a set of meta-examples may be costly, since the algorithms performance is usually defined through an empirical evaluation on the problem at hand. In this context, we proposed the use of Active Learning to select only the relevant problems for meta-example generation. Hence, the need for empirical evaluations of the candidate algorithms is reduced. Experiments were performed using the classification uncertainty of the k-NN algorithm as the criteria for active selection of problems. A significant gain in performance was yielded by using the Active Learning method.

1 Introduction

In several domains of application, there are different algorithms that can be applied to a single problem. In Machine Learning, for instance, different learning algorithms have been proposed to solve learning problems. An important question that arises in such domains is the appropriate selection of algorithms for each problem at hand [1]. The most traditional strategies to algorithm selection involve, in general, expert knowledge, or costly procedures of empirical evaluation [2]. Meta-Learning [3] arises in this context as an alternative solution, capable of automatically acquire knowledge to be used in algorithm selection.

The knowledge in Meta-Learning is acquired from a set of *meta-examples* that store the experience obtained in the application of a number of candidate algorithms in problems investigated in the past. More specifically, each meta-example is related to a given problem and stores: (1) the features that describe the problem; and (2) information about the performance obtained by the candidate algorithms when applied to the problem (e.g. the best algorithm, error rates, execution times,...). A meta-learner is a learning model that receives as input a set of such meta-examples, and generates knowledge relating features of the problems and the performance of the candidate algorithms.

J. Marques de Sá et al. (Eds.): ICANN 2007, Part I, LNCS 4668, pp. 817–826, 2007.

A limitation of Meta-Learning is related to the process of generating meta-examples. In order to generate a meta-example from a given problem, it is necessary to perform an empirical evaluation (e.g. cross-validation) to collect the performance information of the candidate algorithms. Although the proposal of Meta-Learning is to perform this empirical evaluation only in a limited number of problems, the cost of generating a whole set of meta-examples may be high, depending, for instance, on the number and complexity of the candidate algorithms, the used methodology of empirical evaluation and the amount of data available in the problems.

Considering the above context, we present here an original proposal in which Active Learning [4] is used to select problems for meta-example generation. Active Learning is a paradigm of Machine Learning, used in domains in which it is hard or costly to acquire training examples [5]. The main motivation of Active Learning is to reduce the number of training examples, at same time maintaining the performance of the learning algorithms. In our proposal, it corresponds to reduce the set of meta-examples by selecting only the more relevant problems, consequently, reducing the number of empirical evaluations performed on the candidate algorithms.

In order to evaluate the viability of our proposal, we implemented a prototype in which the k-NN (k-Nearest Neighbors) algorithm is used as meta-learner, and a method based on uncertainty of classification [5] is used to select problems for meta-example generation. The prototype was evaluated in a case study which corresponds to the task of selecting two specific algorithms for time series forecasting problems: the Time Delay Neural Network (TDNN) [6] and the Simple Exponential Smoothing (SES) [7]. Experiments performed on a set of 99 problems revealed a gain in the meta-learner performance by using the implemented Active Learning method, compared to a random procedure for selecting problems.

The remaining of this paper is organized as follows. Section 2 brings a brief presentation of Meta-Learning, followed by section 3 which describes, in more details, the proposed solution and the implemented prototype. Section 4 presents the performed experiments and obtained results. Finally, section 5 concludes the paper by presenting some future work.

2 Meta-learning

In different knowledge areas, it is observed the existence of several algorithms that compete to be applied to specific classes of problems. In such situations, both empirical and theoretical results have shown that there is no algorithm uniformly superior, independently on the problem being tackled [8,9,10].

Meta-Learning is a framework that defines techniques to assist algorithm selection for learning problems (usually classification and regression problems) [3]. In a strict formulation of Meta-Learning, each training example (or *meta-example*) is related to an individual learning problem investigated in the past and stores: (1) a set of features (called *meta-attributes*) that describes the problem; and (2) a class attribute which indicates the best algorithm for the problem, among a set of candidate algorithms. The *meta-learner* in the strict formulation is simply a classifier algorithm which predicts the best algorithm for a given problem based on its descriptive meta-attributes [8].

The meta-attributes usually are statistical and information theory measures of the problem's dataset, such as number of training examples and attributes, correlation between attributes, class entropy, presence of outliers in the dataset, among others [11]. The class label stored in a meta-example is usually defined by empirically evaluating each candidate algorithm on the problem. This can be performed, for instance, via a cross-validation experiment using the available problem's dataset. The accuracy of each classifier is estimated by cross-validation, and the class label is assigned to a problem according to the algorithm with highest estimated accuracy.

Although the strict Meta-Learning has been investigated by different authors (see for instance [1,8,12,13,14,15]), other Meta-Learning techniques have been proposed to provide more informative solutions to algorithm selection. In [16], the authors proposed a meta-learner not only to predict the best algorithm but also to predict the applicability of each candidate algorithm to the new problems being tackled. In [9], the NOEMON system combined different strict meta-learners in order to provide rankings of the candidate algorithms. In [10], the authors applied instance-based learning to provide rankings of algorithms, taking into account the predicted accuracy and execution time of the algorithms. In [17], the authors used a regression model as meta-learner in order to predict the numerical value of the accuracy for each candidate algorithm.

The concepts and techniques of Meta-Learning were originally evaluated to select algorithms for classification and regression problems. However, in recent years, Meta-Learning has been extrapolated to other domains of application [12,14,18]. In [12,14], for instance, the authors proposed the use of Meta-Learning to select algorithms for time series forecasting. In [18], the authors applied Meta-Learning to support the construction of planning systems. Considering these applications, Meta-Learning can be viewed as a more general framework to algorithm selection, and it is expected to be useful in problems related to a large range of domains.

3 Active Learning for Meta-example Generation

An important step in the generation of meta-examples is to estimate the performance of the candidate algorithms on a set of problems. The evaluation of performance is accomplished by applying a pre-defined methodology of experiments (e.g. hold-out, cross-validation,...) to the available datasets. The information resulting from the performance evaluation is used to define the target attribute in the Meta-Learning task (e.g. the class corresponding to the best algorithm).

The generation of meta-examples may be a costly process depending on a number of aspects, such as the methodology of experiments adopted to performance evaluation, the number of problems available to generate meta-examples, and the number and complexity of the candidate algorithms.

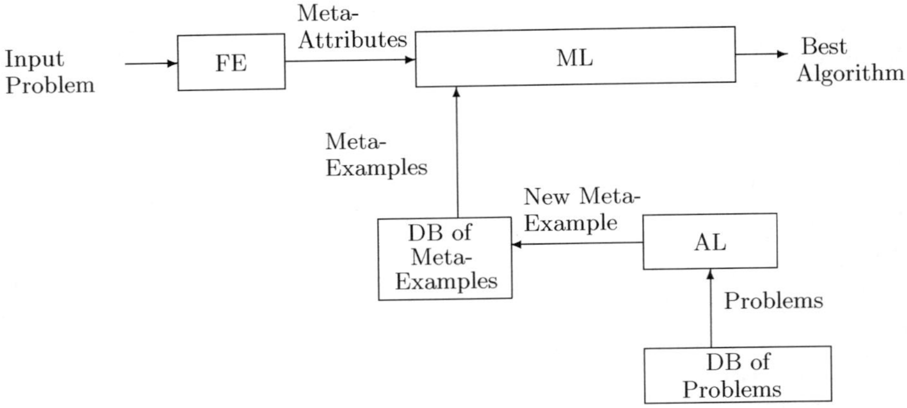

Fig. 1. System Architecture

In the above context, we proposed here the use of Active Learning to support the generation of training examples for Meta-Learning. Active Leaning is a paradigm of Machine Learning in which the learning algorithm has some control over the inputs on which it trains [4]. The main objective of this paradigm is to reduce the number of training examples, at same time maintaining the performance of the learning algorithm. Active Learning is ideal for learning domains in which the acquisition of labeled examples is a costly process, such as image recognition [5], text classification [19] and information filtering [20].

The main motivation of the use of Active Learning in our context is to select only the more relevant problems for Meta-Learning, and hence, to reduce the number of empirical evaluations on the candidate algorithms. Figure 1 presents the general architecture of system following our proposal. The system has three different phases: meta-example generation, training and use, described as follows.

In the meta-example generation, the Active Learning (AL) module selects from a base of problems, those ones considered the most relevant for the Meta-Learning task. The selection of problems is performed based on a pre-defined criteria implemented in the module. The candidate algorithms are then empirically evaluated on the selected problems, in order to collect the performance information related to the algorithms. Each generated meta-example (composed by meta-attributes and performance information) is then stored in an appropriate database.

In the training phase, the Meta-Learner (ML) acquires knowledge from the database of meta-examples generated by the AL module. This knowledge associates meta-attributes to the performance of the candidate algorithms. The acquired knowledge may be refined as more meta-examples are provided by the AL module.

In the use phase, given a new input problem, the Feature Extractor (FE) module extracts the values of the meta-attributes. According to these values, the ML module predicts the best candidate algorithm. For that, it uses the knowledge previously acquired as a result of the training phase.

In order to evaluate the proposal, we implemented a prototype to select between two candidate algorithms for time series forecasting problems: the Time-Delay Neural Network (TDNN) [6] and the Simple Exponential Smoothing model (SES) [7]. Both algorithms were used for short-term forecasting of stationary time series (with no trend or seasonality). According to [14], both algorithms are indicated to forecast stationary time series, however the quality of the forecasts provided by the algorithms may be very different depending on the time series at hand.

In the current prototype, the k-Nearest Neighbors (k-NN) algorithm was used in the ML module, and an Active Learning method based on classification uncertainty of the k-NN [5] is used in the AL module. In the next sections, we provide more details of the implemented prototype.

3.1 Feature Extractor

In our proposal, the FE module implements p meta-attributes $X_1, \ldots, X_p$ which correspond to features that describe the input problems. Hence, each problem e is described by a vector $\mathbf{x} = (x^1, \ldots, x^p)$ in which $x^j = X_j(e), (j = 1, \ldots, p)$.

In the implemented prototype, each input problem consists of a time series to be forecasted, and $p = 10$ features were used as meta-attributes:

1. Length of the time series (X_1): number of observations of the series.
2. Mean of the absolute values of the 5 first autocorrelations (X_2): high values of this feature suggests that the value of the series at a time point is very dependent of the values in recent past points.
3. Test of significant autocorrelations (X_3): presence of at least one significant autocorrelation taking into account the first 5 ones.
4. Significance of the first, second and third autocorrelation (X_4, X_5 and X_6): indicates significant dependences in more recent past points.
5. Coefficient of variation (X_7): measures the degree of instability in the series.
6. Absolute value of the skewness and kurtosis coefficient (X_8 and X_9): measure the degree of non-normality in the series.
7. Test of Turning Points for randomness (X_{10}): The presence of a very large or a very low number of turning points in a series suggests that the series is not generated by a purely random process.

All implemented features are directly computed from the avaliable series data, which has the advantage of avoiding subjective analysis, such as visual inspection of plots.

3.2 Meta-learner

The Meta-Learning task in the current prototype corresponds to the strict formulation described in Section 2, i.e. the meta-learner is a conventional classifier that uses the meta-attributes to predict the best candidate algorithm. In the implemented prototype, we used the k-NN algorithm as the strict meta-learner. In

[10], the authors presented some advantages of using instance-based algorithms, such as the k-NN, as meta-learners. Instance-based algorithms are extensible: once a new meta-example becomes available, it can be easily integrated without the need to initiate re-learning. According to [10], this is relevant for algorithm selection since the user typically starts with a small set of meta-data that increases steadily with time.

In this section, we describe more formally the meta-learner used in the prototype. Let $E = \{e_1, \ldots, e_n\}$ be the set of n problems used to generate a set of n meta-examples $ME = \{me_1, \ldots, me_n\}$. Each meta-example is related to a single problem and stores the values of p features $X_1, \ldots, X_p$ for the problem and the value of a class attribute C, which indicates the best algorithm for the problem, among L candidates. In our prototype, we have $p = 10$ meta-attributes describing problems and $L = 2$ candidate algorithms (TDNN and SES).

Let $D = \{c_1, \ldots, c_L\}$ be the domain of the class attribute C where each class label $c_l \in D$ represents a candidate algorithm. In this way, each meta-example $me_i \in ME$ is represented as the pair $(\mathbf{x}_i, C(e_i))$ storing: (1) the description $\mathbf{x}_i$ of the problem e_i, where $\mathbf{x}_i = (x_i^1, \ldots, x_i^p)$ and $x_i^j = X_j(e_i)$; and (2) the class label associated to e_i, i.e. $C(e_i) = c_l$, where $c_l \in D$.

Given a new input problem described by the vector $\mathbf{x} = (x^1, \ldots, x^p)$, the k-NN meta-learner retrieves k meta-examples from ME, according to the distance between meta-attributes. The distance function (*dist*) implemented in the prototype was the unweighted L_1-Norm, defined as:

$$dist(\mathbf{x}, \mathbf{x}_i) = \sum_{j=1}^{p} \frac{|x^j - x_i^j|}{max_i(x_i^j) - min_i(x_i^j)} \quad (1)$$

The prediction of the class label for the new problem (i.e. the prediction of the best algorithm) is performed according to the number of occurrences (votes) of each $c_l \in D$ in the class labels associated to the retrieved meta-examples.

3.3 Active Learning

As seen, the ML module predicts the best algorithms by using a set of meta-examples generated from *labeled* problems, i.e. the problems in which the best candidate algorithm is known. The AL module, described in this section, receives a set of *unlabeled* problems, i.e. the problems in which the candidate algorithms were not yet evaluated and, hence, the best algorithm for each problem is not known. Therefore, the main objective of the AL module is to incrementally select unlabeled problems to be used for generating new meta-examples.

In the prototype, the AL module implements a method for selecting unlabeled problems which is based on the criteria of *classification uncertainty* of the k-NN algorithm [5]. In this criteria, initially, the k-NN algorithm classifies each unlabeled example by using the available labeled examples. A degree of uncertainty of the provided classification is assigned for each unlabeled example. Finally, the unlabeled example with the highest classification uncertainty is then selected.

The classification uncertainty of the k-NN algorithm is defined in [5] as the ratio of: (1) the distance between the unlabeled example and its nearest labeled neighbor; and (2) the sum of the distances between the unlabeled example and its nearest labeled neighbors of different classes. A high value of uncertainty indicates that the unlabeled example has nearest neighbors with similar distances but conflicting labeling. Hence, once the unlabeled example is labeled, it is expected that the uncertainty of classification in its neighborhood should be reduced.

In our context, let E be the set of labeled problems, and let $\widetilde{E}$ be the set of unlabeled problems. Let E_l be the subset of labeled problems associated to the class label c_l, i.e. $E_l = \{e_i \in E | C(e_i) = c_l\}$. Given the set E, the classification uncertainty of k-NN for each $\widetilde{e} \in \widetilde{E}$ is defined as:

$$\mathcal{S}(\widetilde{e}|E) = \frac{\min_{e_i \in E} dist(\widetilde{\mathbf{x}}, \mathbf{x}_i)}{\sum_{l=1}^{L} \min_{e_i \in E_l} dist(\widetilde{\mathbf{x}}, \mathbf{x}_i)} \tag{2}$$

In the above equation, $\widetilde{\mathbf{x}}$ is the description of problem $\widetilde{e}$. The AL module then selects, for generating a new meta-example, the problem $\widetilde{e}^* \in \widetilde{E}$ with highest uncertainty:

$$\widetilde{e}^* = argmax_{\widetilde{e} \in \widetilde{E}} \mathcal{S}(\widetilde{e}|E) \tag{3}$$

Finally, the selected problem is labeled (i.e. the class value $C(\widetilde{e}^*)$ is defined), through the empirical evaluation of the candidate algorithms using the avaliable data of the problem.

In our prototype, the labeling of a time series is performed through the empirical evaluation of TDNN and SES in forecasting the series. For this, a hold-out experiment was performed, as described in [13]. Given a time series, its data was divided into two parts: the fit period and the test period. The test period consists on the last 30 points of the time series and the fit period consists on the remaining data. The fit data was used to calibrate the parameters of both models TDNN and SES. Both calibrated models were used to generate one-step-ahead forecasts for the test data. Finally, the class attribute was assigned as the model which obtained the lowest mean absolute forecasting error on the test data.

4 Experiments and Results

In the performed experiments, we used 99 time series collected from the Time Series Data Library (TSDL)[1]. This repository contains time series data from several domains, most of them used as benchmark problems in the forecasting field. Both algorithms (TDNN and SES) were empirically evaluated for forecasting each series (as seen in section 3.3), and hence, 99 meta-examples were generated for the experiments with Meta-Learning.

The prototype was evaluated for different configurations of the k-NN meta-learner (with k = 1, 3, 5, 7, 9 and 11 nearest neighbors). For each configuration,

[1] TSDL - http://www-personal.buseco.monash.edu.au/~hyndman/TSDL

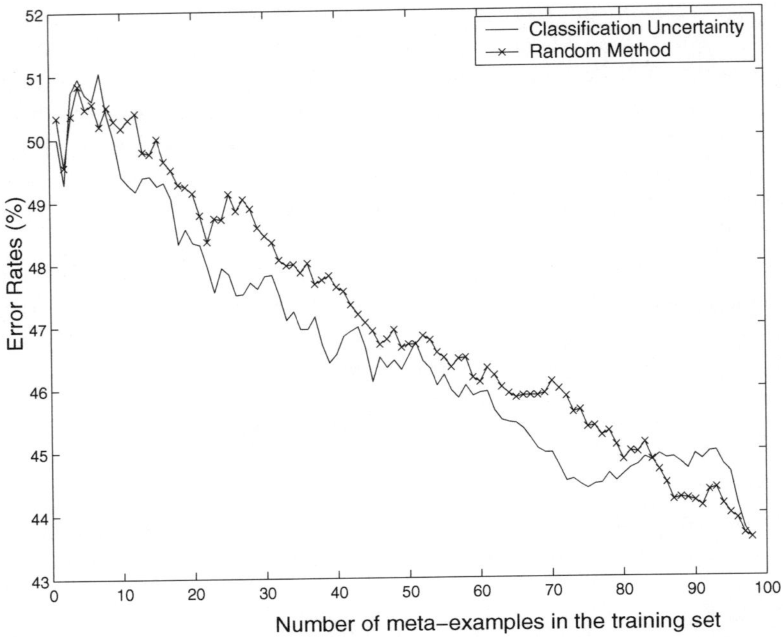

Fig. 2. Average curves of error rates obtained by k-NN meta-learner for both the classification uncertainty and the random active learning methods

a leave-one-out experiment was performed to evaluate the performance of the meta-learner, also varying the number of meta-examples provided by the Active Learning module. This experiment is described just below.

At each step of leave-one-out, one problem is left out for testing the ML module, and the remaining 98 problems are considered as candidates to generate meta-examples. The AL module progressively includes one meta-example in the training set of the ML module, up to the total number of 98 training meta-examples. At each included meta-example, the ML module is judged on the test problem left out, receiving either 1 or 0 for failure or success. Hence, a curve with 98 binary judgments is produced for each test problem. Finally, the curve of error rates obtained by ML can be computed by averaging the curves of judgments over the 99 steps of the leave-one-out experiment.

As a basis of comparison, the same above experiment was applied to each configuration of k-NN, but using in the AL module a random method for selecting unlabeled problems. According to [5], despite its simplicity, the random method has the advantage of performing a uniform exploration of the example space.

Figure 2 presents the curve of error rates obtained by the k-NN meta-learner averaged across the different configurations of the parameter k. The figure presents the average curve obtained when both methods were used: the classification uncertainty (described in section 3.3) and the random method. As it is expected, for both methods, the error rate obtained by the ML module decreased as the number of meta-examples in the training set increased. However, the error rates obtained by deploying the classification uncertainty method were, in general, lower than the error rates obtained by deploying the random method. In fact, from 8 to 84 meta-examples included in the training set, the classification uncertainty method steadily achieved better performance compared to the random method.

5 Conclusion

In this paper, we presented an original work that proposes the use of Active Learning to enhance the generation of examples for Meta-Learning. In order to verify the viability of our proposal, we implemented a prototype in which the classification uncertainty criteria is used to select meta-examples for a k-NN meta-learner. The prototype was evaluated in a task of selecting two candidate algorithms for time series forecasting, and the experiments results were promising.

We highlight here that, despite the originality of the proposal, it still has some limitations that will be dealt with in future work. Although the classification uncertainty method obtained good results in the investigated Meta-Learning task, other Active Learning methods for k-NN have presented very competitive results in tasks related to other contexts [5].

Another question to investigate is the use of Active Learning not only for strict k-NN meta-learners, but also for other Meta-Learning techniques (as those cited in section 2). The choice of the strict k-NN meta-learner in our prototype was due to the fact that different authors in literature have focused their efforts on this kind of meta-learner or on similar techniques. However, other Meta-Learning techniques have been achieved good results in algorithm selection and hence, they also have to be investigated in the context of the current proposal.

References

1. Kalousis, A., Gama, J., Hilario, M.: On data and algorithms - understanding inductive performance. Machine Learning 54(3), 275–312 (2004)
2. Kalousis, A., Hilario, M.: Representational issues in meta-learning. In: Proceed. of the 20th International Conference on Machine Learning, pp. 313–320 (2003)
3. Giraud-Carrier, C., Vilalta, R., Brazdil, P.: Introduction to the special issue on meta-learning. Machine Learning 54(3), 187–193 (2004)
4. Cohn, D., Atlas, L., Ladner, R.: Improving Generalization with Active Learning. Machine Learning 15, 201–221 (1994)
5. Lindenbaum, M., Markovitch, S., Rusakov, D.: Selective sampling for nearest neighbor classifiers. Machine Learning 54, 125–152 (2004)
6. Lang, K., Hinton, G.: Time-delay neural network architecture for speech recognition. In CMU Technical Report CS-88-152, Carnegie-Mellon University, Pittsburgh, PA (1988)
7. Brown, R.G.: Smoothing, Forecasting and Prediction. Prentice-Hall, Englewood Cliffs (1963)
8. Aha, D.: Generalizing from case studies: a case study. In: Proceedings of the 9th International Workshop on Machine Learning, pp. 1–10 (1992)

9. Kalousis, A., Theoharis, T.: Noemon: Design, implementation and performance results of an intelligent assistant for classifier selection. Intelligent Data Analysis 3(5), 319–337 (1999)
10. Brazdil, P., Soares, C., da Costa, J.: Ranking learning algorithms: Using IBL and meta-learning on accuracy and time results. Machine Learning 50(3), 251–277 (2003)
11. Engels, R., Theusinger, C.: Using a data metric for preprocessing advice for data mining applications. In: ECAI-98. Proceedings of the 13th European Conference on Artificial Intelligence, pp. 430–434 (1998)
12. Prudêncio, R., Ludermir, T.: Selection of models for time series prediction via meta-learning. In: Proceedings of the 2nd International Conference on Hybrid Systems, pp. 74–83 (2002)
13. Prudêncio, R., Ludermir, T., de Carvalho, F.: A modal symbolic classifier to select time series models. Pattern Recognition Letters 25(8), 911–921 (2004)
14. Prudêncio, R., Ludermir, T.: Meta-learning approaches to selecting time series models. Neurocomputing 61, 121–137 (2004)
15. Leite, R., Brazdil, P.: Predicting relative performance of classifiers from samples. In: Proceedings of the 22nd International Conference on Machine Learning (2005)
16. Michie, D., D. J. S., Taylor, C.C., (eds.): Machine Learning, Neural and Statistical Classification. Ellis Horwood, New York (1994)
17. Bensusan, H., Alexandros, K.: Estimating the predictive accuracy of a classifier. In: Proceedings of the 12th European Conference on Machine Learning, pp. 25–36 (2001)
18. Tsoumakas, G., Vrakas, D., Bassiliades, N., Vlahavas, I.: Lazy adaptive multicriteria planning. In: Proceed. of the 16th European Conference on Artificial Intelligence, pp. 693–697 (2004)
19. Tong, S., Koller, D.: Support vector machine active learning with applications to text classification. Journal of Machine Learning Research 2, 45–66 (2002)
20. Sampaio, I., Ramalho, G., Corruble, V., Prudêncio, R.: Acquiring the preferences of new users in recommender systems - The role of item controversy. In: Proceedings of the ECAI 2006 Workshop on Recommender Systems, pp. 107–110 (2006)

Co-learning and the Development of Communication*

Viktor Gyenes and András Lőrincz

Eötvös Loránd University, Pázmány P. sétány 1/C, Budapest, Hungary, H-1117
gyenesvi@inf.elte.hu, andras.lorincz@elte.hu

Abstract. We investigate the properties of coupled co-learning systems during the emergence of communication. Co-learning systems are more complex than individual learning systems because of being dependent on the learning process of each other, thus risking divergence. We developed a neural network approach and implemented a concept that we call reconstruction principle, which we found adequate for overcoming the instability problem. Experimental simulations were performed to test the emergence of both compositional and holistic communication. The results show that compositional communication is favorable when learning performance is considered, however it is more error-prone to differences in the conceptual representations of the individual systems. We show that our architecture enables the adjustment of the differences in the individual representations in case of compositional communication.

1 Introduction

The emergence and evolution of communication has gained significant research attention in the past decade. Multi-agent simulations are popular to model the coordinated development of natural languages. The inherent property of the development of a coordinated communication system that *multiple agents* participate in it poses extra difficulties to the algorithms aiming to model it.

When communication evolves, it should be the result of a *negotiation process* between many parties. During this process, certain new items are invented by individuals and accepted and learned by the others. Who invents things and who accepts them should neither be predefined nor one-sided. All agents take part in both of these tasks, that is, they teach and learn simultaneously. To let the whole process converge to a useful communication system, agents have to adapt to each other, not only to the task to be learned; their learning depends on that of the others. The complexity of the problem is that learning concerns hidden variables different for each agent while learning is inherently coupled.

Most work done in the field of language emergence is motivated by modelling natural language evolution. Here, we take a broader view: we consider the optimization of information transfer among the agents as a process of negotiation

* Research has been supported by the New Ties project (EC FET grant No. 003752). Any opinions, findings and conclusions or recommendations expressed in this material are those of the authors and do not necessarily reflect the views of the EC.

J. Marques de Sá et al. (Eds.): ICANN 2007, Part I, LNCS 4668, pp. 827–837, 2007.

about a 'language'. This approach is more general and may be relevant for the encoding of information in different kinds of distributed sensory and computational systems. One of the motivations of our work is to encapsulate the difficulties of parallel learning for agents that have different conceptual representations.

Here, we address this problem. We model the agents with so called reconstruction networks. We provide a neural implementation of what we call *reconstruction principle*, and argue that it is efficient for making co-learning stable.

2 Related Work

When modelling natural language evolution, it is a natural idea to involve knowledge transfer from generation to generation, like in the Iterated Learning Model of Kirby and colleagues [1,2]: the new generation of language users learns the language from the previous generation and then the old generation is replaced by the new one. An interesting conclusion of the model is that the compositional nature of language might be the result of the learning bottleneck imposed when language has to pass from one generation to the other. Vogt [3] also builds on the Iterated Learning Model and combines it with language games [4] to model the emergence of compositional languages when agents aim to communicate about their observations. An interesting aspect of this work is that it deals with the conceptual representations of the agents upon which they build their language, which is strongly related to the symbol grounding problem [5], and also the compositional nature of language. Smith [6] also considers the development of individual, distinct meaning structures and examines its effect on the evolved language. All of these models apply learning from generation to generation, and thus teachers are fixed. This way these models avoid the problem of co-learning.

Cangelosi [7] uses artificial neural networks trained by a genetic algorithm to develop a language in an agent system that aims to differentiate between edible and poisonous food items and emphasizes that the evolution of language requires the parallel evolution of the ability of language understanding and production. He also considers the parallel development of input categorization and language. Hutchins and Hazlehurst aim to invent a shared lexicon [8] utilizing feedforward connectionist networks that model language learning agents.

The work of Oliphant and Batali [9] is very close to ours regarding the reconstruction principle. They model the development of a stable coordinated communication system using a method that they call the 'obverter' procedure in which agents observe each other and try to maximize their chances to communicate successfully, instead of simply imitating the others. They provide mathematical considerations about the convergence of their method. The underlying idea is very similar to generative or reconstruction networks [10], which – in the field of vision – would claim that vision is inverse graphics [11].

Our architecture can be seen as a neural network implementation of the 'obverter' learner that also generalizes it for *combinatorial/compositional* internal representations and communication. Up to our best knowledge, no neural network approach has incorporated this idea, only 'imitator' approaches exist.

Central to our methodology is the idea of Cangelosi that production and understanding must be maintained in parallel. Our framework enables the learners to have distinct conceptual representations. We investigate the properties of both compositional and non-compositional (holistic) communication systems. We treat the problem of co-learning, and restrict our methods to local Hebbian learning for the individual systems.

3 Methods

This section details our network architecture and the learning methods applied. The general context of the learning is a signalling game, in which the networks observe inputs and the learning task is to co-develop a language (agree on a set of signals) to communicate the observations. This helps investigating communication related issues without being effected by other environmental factors, and gives us freedom to vary related parameters and test various settings.

3.1 Network Architecture

We model the agents by three-layer neural networks, the architecture is depicted in Fig. 1: the input layer of the network receives the observation ($x \in R^m$), which is processed and an internal representation ($h \in \{0,1\}^n$) is formed. In our model, this transformation (G) represents the extraction of features, resulting in a *combinatorial* internal representation, whose components indicate the presence/absence of the features. The $x \to h$ transformation is modelled as follows. Each input x (belonging to a finite set for now) is assigned a vector $f_G(x) \in [0,1]^n$ of real values as if indicating the degree (or probability) of the presence of various features. The internal representation $h \in \{0,1\}^n$ is generated from $f_G(x)$ by rounding to 0 or 1. Adjustment of concept formation is modelled by letting the values in $f_G(x)$ be tuned. A particular property of this characterization is that inputs are categorized into multiple categories: each feature represents being the member of a category (see, Fig. 1).

Two other transformations govern the communication related behavior of the network. The network can generate an 'utterance' u from its internal representation by means of transformation Q. We let $u \in \{0,1\}^{2^n}$, so that the utterance may contain combinatorially many signals. Furthermore, the network also has another transformation, W, that we call 'parsing' or 'understanding an utterance', since it yields some internal representation based on an utterance.

In our studies, we used two methods for generation and parsing. The first method performs the linear transformations Q and W that are followed by a nonlinearity σ rounding the values to 0 or 1: $u = \sigma(Qh)$, $h = \sigma(Wu)$. The other method implements the so called *reconstruction principle*. In this case, the network generates an utterance u_g for a given internal representation h such that when it is parsed (by the network itself) the resulting internal representation is closest to the original vector h. That is, the network tries to reconstruct the internal representation from its own intended output, and chooses an utterance that reconstructs the internal representation the best. The same principle is

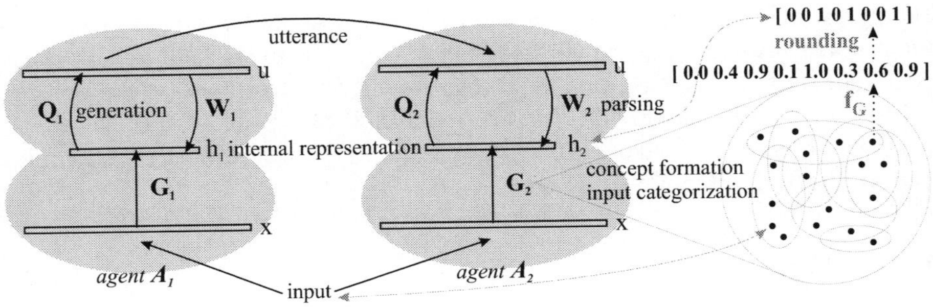

Fig. 1. Network architecture. Left: two agents, A_1 and A_2, structure of the input set and the transformation to internal representation. Right: each input belongs the certain (feature) sets with certain probabilities. For each input, these probabilities are rounded to make the internal representation.

used for parsing, except the roles are changed: given an utterance u, the network chooses an internal state h, that when transformed back to an utterance, yields an utterance closest to u. Below, this idea is formalized for the generation of utterance u_g from internal representation h, and for creating an approximate $\tilde{h}$ via parsing utterance u, respectively:

$$u_g = argmin_u \parallel h - \sigma(Wu) \parallel_{L_2}^2 \tag{1}$$

$$\tilde{h} = argmin_h \parallel u_g - \sigma(Qh) \parallel_{L_2}^2 \tag{2}$$

3.2 Optimization Method

The minimization tasks (1) and (2) are combinatorial optimization problems, since we restricted the vectors h and u to have only $0-1$ components. To solve these problem, we use the cross-entropy (CE) method [12], which is a generic approach to combinatorial optimization. The CE method maintains a parameterized probability distribution, from which it iteratively randomly generates solution samples, evaluates them according to the cost function, and continuously updates the probability distribution, until convergence.

The CE algorithm nicely fits into the reconstruction network frame. We used the multi-dimensional Bernoulli distribution as the probability density function for generating random samples of n-dimensional $0-1$ valued vectors. Initial guesses of the probability distribution are also provided by the linear transformations of the networks. The cost function is the reconstruction error. The original batch version of the CE method generates a population of random samples, and chooses the best p $(=5)$ percent, which is used to update the density function. We modified the method to make it online: the reconstruction error is considered as a Gaussian variable with given mean and standard deviation. We updated the distribution if the error fell into the best p percent of the most recent samples. The online CE algorithm of Table 1 finds $\tilde{h}$ that minimizes $\parallel u - \sigma(Qh) \parallel_{L_2}^2$.

Table 1. Pseudo-code for the cross-entropy reconstruction algorithm

$m = 0,\ d = 0$	// mean and standard deviation of errors
$\alpha, \beta \in [0, 1]\ (= 0.1)$	// update rates
$q = 1.648$	// 95% percentile of normal distribution
$p = \frac{Wu}{max(Wu)}$	// initial probability distribution
$min = \infty$	// initial minimum value
until convergence or a fixed iteration count	
generate a random sample h from p	
$e = \parallel u - \sigma(Qh) \parallel^2_{L_2}$	// calculate reconstruction error
$m \leftarrow (1-\beta)m + \beta e$	// update mean error
$d \leftarrow (1-\beta)d + \beta(m-e)^2$	// update standard deviation of errors
if ($e < m - qd$)	// if error falls to the best 5%
$p \leftarrow (1-\alpha)p + \alpha h$	// update probability distribution
end if	
if ($e < min$)	
$min \leftarrow e,\ \tilde{h} \leftarrow h$	// update current minimum and best solution
end if	
end	

We note, that if the matrices Q and W are well tuned, then the initial guess for the probability density function becomes sharp, and the algorithm converges very quickly. In this case, the algorithm essentially behaves as a simple feedforward linear transformation. However, we found that the whole combinatorial reconstruction algorithm is needed for proper training. We note too that the L_2 norm is equivalent to the L_1 norm for our case, because the vectors are $0-1$ valued. The known property of the CE method that it easily finds *combinatorial* solutions is exploited during the learning of compositional communication.

3.3 Network Training

The training of the matrices Q and W is Hebbian and has certain 'quasi-supervised flavor': networks are presented with observations, from which they generate internal representations. One of the networks, say agent A_1 generates utterance u_1, which is then sent to the another agent. That is, the output is not supplied externally but generated by one of the networks. Then each network has an internal representation-utterance pair and can use it to update its transformation matrices. The update is Hebbian, it uses the negative gradient of the squared reconstruction error. For agent A_i (i=1,2) we have:

$$\Delta Q_i = \varepsilon\ (u_1 - Q_i h_i) h_i^T = \varepsilon\ e_i^u h_i^T\ , \tag{3}$$

$$\Delta W_i = \varepsilon\ (h_i - W_i u_1) u_1^T = \varepsilon\ e_i^h u_1^T\ , \tag{4}$$

where $\varepsilon \in [0, 1]$ is some update factor, e_i^h and e_i^u denote the errors at the internal representation and utterance level, respectively. Note that the vector u_1 is the same for each agent, but vector h_i, the matrices Q_i and W_i may be different.[1]

Feature extraction can be tuned at the listener (agent A_2 in the present example), because internal representation h_2 is available and its estimation $\tilde{h}_2$ can be computed from the utterance. Let us suppose that the input was x, then:

$$f_{G_2}(x) \leftarrow \theta\left(f_{G_2}(x) + \varepsilon(\tilde{h}_2 - h_2)\right) , \tag{5}$$

where $f_{G_2}(x)$ is the vector containing the values that generates the internal representation h_2 for input x. Function θ clamps the values to [0,1]. We used this model for adjusting feature extraction in our illustrations. The key is that the error term $\tilde{h} - h$ is available in (5) in our frame.

4 Computer Simulations

We start with the simple case when networks develop the *same internal representation* ($G_1 = G_2 = I$, the identity transformation; $h_1 = h_2$), merely to investigate language emergence independently from differences in internal representations. Next, we investigate the effect of *different internal representations*.

4.1 Test Scenarios

The following experimental scenarios were studied:

- We studied non-combinatorial 'languages', where the utterances were forced to have only one nonzero element, and also combinatorial 'languages', where utterances were let to have arbitrary combinations of nonzero entries.
- We studied generation and parsing methods using simple linear transformations Q and W followed by a rounding nonlinearity, i.e., without the reconstruction algorithm. In the non-combinatorial case, following the linear transformation the maximum valued component was set to 1, others to 0.
- We systematically varied the size of the internal representation, and the number of networks to see how the learning scales with these factors.

In each episode of learning, two random selected networks participated in communication. An input was selected randomly, and one of the networks generated an utterance to it, the other parsed it, and then both of them updated their transformations. To decide whether a consistent language had emerged, we defined a performance matrix: the relative frequency of the usage of each signal for each input was calculated from communications about the given input. We say that a consistent language developed, if all networks produced consistently the same signals for the same input. In the non-compositional case, each state was required to be denoted by a different signal. In the compositional case, we

[1] The matrices Q_i and W_i are initialized to have random values between 0 and 1.

call a language consistent, if an utterance denoting an internal representation with certain features are composed of utterances referring to those features, and all the networks use the same combinations. We also recorded how often the parsing agent could reconstruct the same internal representation from the utterance it received as it generated from its observation ($\tilde{h} = h$) towards the end of a series of communication episode; this is the communication success rate (it happens to be 1 for a consistent language).

4.2 Results

First we tested how the combinatorial and non-combinatorial methods behave as a function of the size of the internal state and the number of agents. We evaluated the percentage of the runs when the method converged to a consistent language, and the average number of learning episodes that agents needed to reach that. We found that if the reconstruction principle was applied, learning reached a consistent state and 100% communication success in all of the cases, both for combinatorial and for holistic languages. The number of episodes needed to reach an agreement is shown in Fig. 2. It can be seen that the combinatorial method needs reasonably smaller number of learning episodes as the state size and the agent count increases. We observed that when combinatorial solution was allowed then *compositional* language developed in all of the cases.

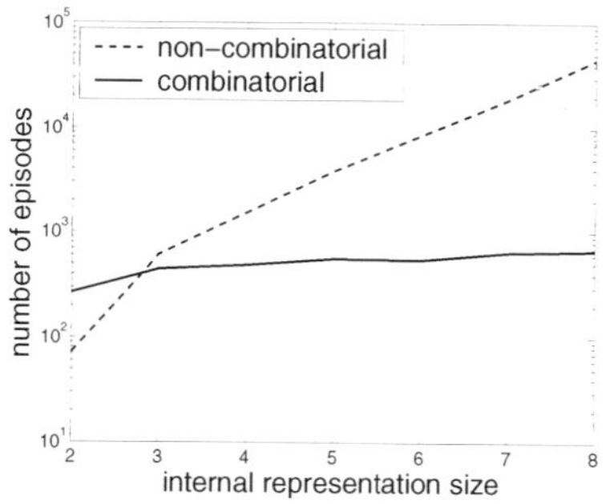

(a) Varying state size

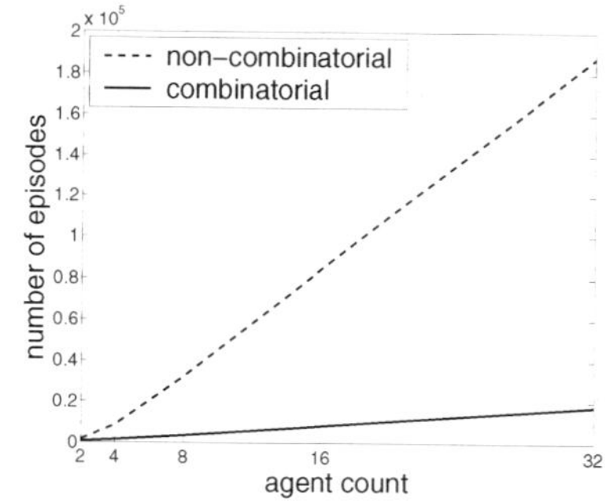

(b) Varying agent count

Fig. 2. The effect of varying internal representation size and agent number. (a) the y axis is on a logarithmic scale: slope is slow for compositional languages, but it is exponential for holistic ones as a function of the size of the internal representation (b) the y axis is linear, size of the internal representation is 4. Average of 100 runs.

We also investigated how the learning changes when the reconstruction principle was not applied. Surprisingly, in the non-combinatorial case, this method was never sufficient to develop a consistent language. For the combinatorial case, Fig. 3 shows the results: without the reconstruction principle both the ratio of consistent languages and the communication success drops drastically with the size of internal representation, and also with the agent count (not shown here).

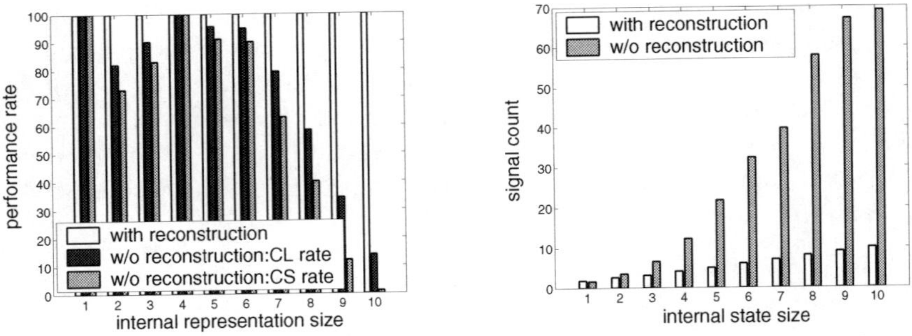

(a) Learning performance (b) Number of signals used

Fig. 3. Using and not using the reconstruction principle. CS: communication success, CL: consistent language. (a) sharp drop for larger state sizes and (b) strong increase in the number of developed signals without reconstruction. Result of 100 runs.

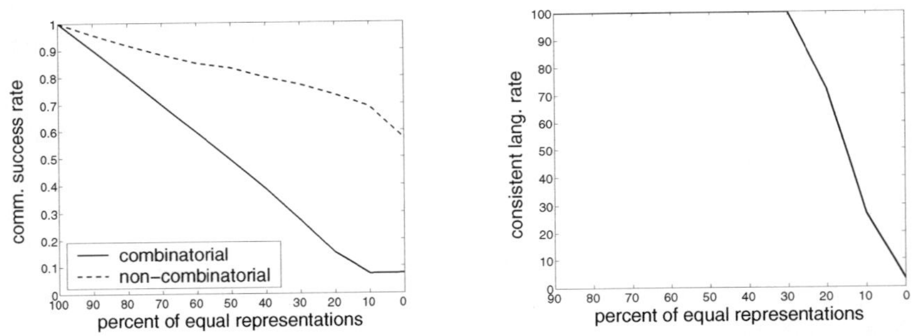

(a) Without adjusting concept formation (b) Adjusting concept formation

Fig. 4. The effect of differences in the internal representations. Size of representation: 6. (a) drop of performance is more serious in the case of compositional languages. (b) learning becomes successful for compositional communication with probability 1 if initial differences are not too large. Result of 10 runs.

Furthermore, the reconstruction principle also has an intriguing effect on the number of signals used by the agents. Theoretically, an n-component state can be communicated using the combination of n signals. This lower bound was reached with reconstruction, but was significantly exceeded without it.

To see how learning behaves when agents have different internal representations, we have explicitly generated feature sets (i.e., G transformations) for a finite number of inputs. Differences between agents' G transformations were systematically introduced. In this case, totally consistent language can not develop, agents can not agree because the categorization of the observations differs. We let learning run for a sufficiently large number of episodes, and during the last 1000 episodes, we evaluated what fraction of the communication episodes were successful (the parsing network was able to reconstruct the same internal state from the utterance as it developed from its input). As expected, communication success rate drops as the discrepancies between internal representations increase. The drop is faster for the combinatorial case (Fig. 4a).

However, when feature values were adapted according to (5) then after some communication tuning episodes, the language development converged for *compositional* languages. Consistent communication developed in a broad domain where the initial differences were not too large (Fig. 4b).

5 Discussion

We observed that at the beginning of the learning, the networks have synonymous signals for denoting components of the state. First '*they learn*' to understand each others' signals, and later '*they refine*' their dictionaries to single common signals for any given component. When the reconstruction principle is not in effect, this negotiation is not successful and the number of signals often increases. However, when reconstruction is utilized, negotiation is accomplished by adaptation to signals used by the other parties.

The obverter learning procedure [9] applies the same idea as our reconstruction principle. In [9] they prove that the best strategy for agents is to produce utterances that maximize the chance of other agents understanding it. They argue that agents do not have access to what others would understand, so it seems a good idea to produce utterances that the agent itself would understand well. This idea is exploited in our reconstruction network. Our algorithm goes beyond the ideas described in [9], because we can handle compositions, and we can work with individuals having distinct conceptual representations. The necessity of an obvious feature of our model, that production and understanding are dependent on each other and evolve simultaneously, has been emphasized by Cangelosi [7].

In [13] Kirby argues that compositional languages emerge due to the learning bottleneck effect of linguistic knowledge transfer from generation to generation. He claims that compositional languages are favored because they are easier to pass to the next generation since fewer observations are enough to learn them because of their compressed nature. Our simulations indicate that there is another reason why compositional languages are favored, namely that they are easier to agree upon. Nonetheless, the reason is the same as that of Kirby; their compressed nature enables faster negotiation, since only the signals referring to components (instead of their combinations) need to be agreed on. Actually, we have observed, that relatively few categorization samples are enough to agree on a consistent language (about 20 in case of an 8-component representation).

Smith [6] and Vogt [3] both use discrimination games, by which agents develop a categorial representation of observations. When experimenting with the effect of different representations, Smith comes to a conclusion that the overall success of communication seems to be directly related to the amount of shared meaning structure in the agents. This conjecture is strengthened by our results, and is also present in the conclusions of Cangelosi. However, [7] only deals with holistic communication in a scenario where there is an evolutionary pressure for agents to develop *similar* internal representations. We had shown that holistic communication is more resistant to representational differences. The underlying reason is probably the compactness and generalizing capability of compositional communication: the misunderstanding of utterances generalizes across similar internal representations. To let successful communication emerge even in the case of different internal representations, we adjusted the representations themselves. The adjustments are based on the differences between the representations induced by the utterance and the one generated from the agent's own observation. In case of a compositional language, the utterance is a projection of how the other agent categorizes the observation, and this information can be used to alter an agents own categorization. This might turn out to be an important feature of compositional languages, since holistic languages lack this information.

It must be noted, that we did not aim to model the development of compositional *syntax*, only that of a compositional *lexicon*. Syntactic structures (e.g. the order of signals) could compress information even further.

The use of the powerful cross-entropy stochastic optimization for the individual examples allowed us to restrict learning to a local Hebbian rule derived from the reconstruction principle, which is an important feature of our approach. Furthermore, no other method reported 100% success even for many agents.

To conclude, a neural network approach [10] was adapted in a novel way to effectively implement ideas about the development of a combinatorial communication system. The reconstruction principle [11] has a central role in our approach. It makes the negotiation process of the parties convergent in the case of identical meaning structures. Experiments show that communication is sensitive to the differences in the conceptual representation, and compositional communication has the advantage that it carries a potential to adjust it.

References

1. Smith, K., Kirby, S., Brighton, H.: Iterated learning: a framework for the emergence of language. Artif. Life 9, 371–386 (2003)
2. Kirby, S., Hurford, J.: The emergence of linguistic structure: An overview of the iterated learning model. In: Simulating the Evolution of Language, pp. 121–148. Springer, London (2002)
3. Vogt, P.: The emergence of compositional structures in perceptually grounded language games. Artif. Intell. 167, 206–242 (2005)
4. Steels, L.: Grounding symbols through evolutionary language games. In: Simulating the Evolution of Language, pp. 211–226. Springer, Heidelberg (2002)
5. Harnad, S.: The symbol grounding problem. Physica D 42, 335–346 (1990)
6. Smith, A.D.M.: Establishing communication systems without explicit meaning transmission. In: Kelemen, J., Sosík, P. (eds.) ECAL 2001. LNCS (LNAI), vol. 2159, pp. 381–390. Springer, Heidelberg (2001)
7. Cangelosi, A., Parisi, D.: The emergence of a language in an evolving population of neural networks. Conn. Sci. 10, 83–97 (1998)
8. Hutchins, E., Hazlehurst, B.: How to invent a lexicon: the development of shared symbols in interaction. In: Artificial Societies, pp. 157–189. UCL Press, London (1995)

9. Oliphant, M., Batali, J.: Learning and the emergence of coordinated communication. Newslett. Center Res. Lang. 11, 1–46 (1997)
10. Ballard, D.H., Hinton, G.E., Sejnowski, T.J.: Parallel visual computation. Nature 306, 21–26 (1983)
11. Horn, B.K.P.: Understanding image intensities. Artif. Intell. 8, 201–231 (1977)
12. Rubinstein, R.Y., Kroese, D.P.: The Cross-Entropy Method. In: Information Science and Statistics, Springer, New York (2004)
13. Kirby, S.: Learning, bottlenecks and infinity: a working model of the evolution of syntactic communication. In: Proceedings of the AISB'99 Symposium on Imitation in Animals and Artifacts, pp. 121–129 (1999)

Models of Orthogonal Type Complex-Valued Dynamic Associative Memories and Their Performance Comparison

Yasuaki Kuroe and Yuriko Taniguchi

Department of Information Science, Kyoto Institute of Technology
Matsugasaki, Sakyo-ku, Kyoto 606-8585, Japan
kuroe@kit.ac.jp

Abstract. Associative memories are one of the popular applications of neural networks and several studies on their extension to the complex domain have been done. One of the important factors to characterize behavior of a complex-valued neural network is its activation function which is a nonlinear complex function. In complex-valued neural networks, there are several possibilities in choosing an activation function because of a wide variety of complex functions. This paper proposes three models of orthogonal type dynamic associative memories using complex-valued neural networks with three different activation functions. We investigate their behavior as associative memories theoretically. Comparisons are also made among these three models in terms of dynamics and storage capabilities.

1 Introduction

In recent years, there have been increasing research interests of artificial neural networks and many efforts have been made on applications of neural networks to various fields. As applications of the neural networks spread more widely, developing neural network models which can directly deal with complex numbers is desired in various fields. Several models of complex-valued neural networks have been proposed and their abilities of information processing have been investigated.

One of the most useful and most investigated areas of applications of neural networks addresses implementations of associative memories. Among them associative memories of self-correlation type are easy to implement and have been extensively studied. Some models of complex-valued associative memories of self-correlation type have been proposed [1,2,3,4,5,6].

One of the important factors to characterize behavior of a complex-valued neural network is its activation function which is a nonlinear complex function. In the real-valued neural networks, the activation is usually chosen to be a smooth and bounded function such as a sigmoidal function. In the complex region, however, there are several possibilities in choosing an activation function because of a variety of complex functions. In [7] the properties that a suitable activation should possess are discussed for complex-valued backpropagation of complex-valued feedforward neural networks. [8] discusses the properties of activation functions from the standpoint of existence of an energy function for complex-valued recurrent neural networks.

J. Marques de Sá et al. (Eds.): ICANN 2007, Part I, LNCS 4668, pp. 838–847, 2007.

The purpose of this paper is to present models of orthogonal type dynamic associative memories using complex-valued neural networks and to investigate their qualitative behaviors theoretically. We propose three models of orthogonal type complex-valued associative memories with three different activate functions. These three models are orthogonal type counterparts of self-correlation type complex-valued associative memories proposed in [2,9].

We treat the models of complex-valued associative memories as nonlinear dynamical systems and study their qualitative behavior theoretically. In particular, we investigate the structures and asymptotic behavior of solution orbits near each memory pattern. Comparisons are also made among these three models in terms of asymptotic behavior of solution orbits near each memory pattern and storage capabilities.

In the following, the imaginary unit is denoted by i ($i^2 = -1$). The n-dimensional complex (real) space is denoted by $\mathbb{C}^n (\mathbb{R}^n)$ and the set of $n \times m$ complex (real) matrices is denoted by $\mathbb{C}^{n\times m}(\mathbb{R}^{n\times m})$. For $A \in \mathbb{C}^{n\times m}$ ($\boldsymbol{a} \in \mathbb{C}^n$), its real and imaginary parts are denoted by A^R ($\boldsymbol{a}^R$) and A^I ($\boldsymbol{a}^I$), respectively.

2 Complex-Valued Associative Memories of Orthogonal Type

2.1 Models

Let m be the number of memory patterns to be stored and each memory pattern be an N dimensional complex vector, denoted by $\boldsymbol{s}^{(\gamma)} \in \mathbb{C}^N$, $\gamma = 1, 2, \cdots, m$. Suppose that $\{\boldsymbol{s}^{(1)}, \boldsymbol{s}^{(2)}, \cdots, \boldsymbol{s}^{(m)}\}$ are linearly independent vectors.

Consider a complex-valued neural network described by difference equations of the form:

$$x_j[t+1] = f(\sum_{k=1}^{N} w_{jk} x_k[t]), \qquad j = 1, 2, \cdots, N \tag{1}$$

where $x_j[t] \in \mathbb{C}$ is the output of the jth neuron at time t, $w_{jk} \in \mathbb{C}$ is the connection weight from the kth neuron to the jth neuron and $f(\cdot)$ is the activation function which is a nonlinear complex function ($f : \mathbb{C} \to \mathbb{C}$).

Let us determine the weight matrix $\boldsymbol{W} = \{w_{jk}\} \in \mathbb{C}^{N\times N}$ and the activation function $f(\cdot)$ so that the neural network (1) can store the set of memory vectors $\{\boldsymbol{s}^{(1)}, \boldsymbol{s}^{(2)}, \cdots, \boldsymbol{s}^{(m)}\}$ and act as a dynamic associative memory.

The weight matrix $\boldsymbol{W}$ is determined after the model of conventional real-valued associative memories of orthogonal type as follows.

$$\boldsymbol{W} := \boldsymbol{S}\boldsymbol{S}^{+} \in \mathbb{C}^{N\times N} \tag{2}$$

where

$$\boldsymbol{S} := (\boldsymbol{s}^{(1)} \ldots \boldsymbol{s}^{(m)}) \in \mathbb{C}^{N\times m}$$
$$\boldsymbol{S}^{+} := (\boldsymbol{S}^{*}\boldsymbol{S})^{-1}\boldsymbol{S}^{*} \in \mathbb{C}^{m\times N}$$

Note that $\boldsymbol{S}^{+}$ is the pseudo-inverse matrix of the matrix $\boldsymbol{S}$.

One of the important factors to characterize behavior of a complex-valued neural network is its activation function $f(\cdot)$. In the real-valued neural networks, the activation function is usually chosen to be a smooth and bounded function such as sigmoidal functions. In the complex region, however, there are several possibilities in choosing an activation function because of a variety of complex functions. As a candidate of the activation function $f(\cdot)$, we will choose a complex function which satisfies the following conditions:

(i) $f(\cdot)$ is a smooth and bounded function by analogy with the sigmoidal function of real-valued neural networks, and
(ii) each memory vector $\boldsymbol{s}^{(\gamma)}$ becomes an equilibrium point of the network (1).

From the definition of the weight matrix $\boldsymbol{W}$ given by (2), the following condition holds:

$$\boldsymbol{W}\boldsymbol{s}^{(\gamma)} = \boldsymbol{s}^{(\gamma)}, \quad \gamma = 1, 2, \cdots, m. \tag{3}$$

Therefore the condition (ii) is accomplished if the following condition hold.

$$f(s_j^{(\gamma)}) = s_j^{(\gamma)}, \ j = 1, 2, \cdots, N, \ \gamma = 1, 2, \cdots, m. \tag{4}$$

which implies that each element of each memory vector is a fixed point the activation function. In regard to the condition (i), we recall the Liouville's theorem, which says that 'if $f(u)$ is analytic at all $u \in \mathbb{C}$ and bounded, then $f(u)$ is a constant function'. Since a suitable $f(u)$ should be bounded, it follows from the theorem that if we choose an analytic function for $f(u)$, it is constant, which is clearly not suitable [7]. In place of analytic function we choose functions which have the continuous partial derivatives $\partial f^R/\partial u^R$, $\partial f^R/\partial u^I$, $\partial f^I/\partial u^R$ and $\partial f^I/\partial u^I$ where $f(u) = f^R(u^R, u^I) + i f^I(u^R, u^I)$.

In complex-valued neural networks the following three classes of complex nonlinear functions $f(\cdot)$ are considered as a candidate of activation functions. Let us express u in the polar representation as $u = re^{i\theta}$.

- Type A: $f(u) = f^R(u^R) + i f^I(u^I)$ (5)
- Type B: $f(u) = \psi(r) \exp\{i\phi(\theta)\}$ (6)
- Type C: $f(u) = \psi(r) \exp\{i\theta\}$ (7)

In Type A, $f^R(\cdot)$ and $f^I(\cdot)$ are nonlinear real functions, $f^R : R \to R$ and $f^I : R \to R$. In Type B and C, $\psi(\cdot)$ and $\phi(\cdot)$ are nonlinear real functions, $\psi : R_{0+} \to R_{0+}$, $\phi : R \to R$, where $R_{+_0} = \{x \mid x \geq 0 \ x \in R\}$. Note that in Type A the real and imaginary parts of an input go through nonlinear functions separately, and in Type B the magnitude and the phase of an input go through nonlinear functions separately, and in Type C only the magnitude of an input go through the nonlinear function with the phase being unchanged. Most of the activation functions yet proposed for the models of complex-valued neural networks belong to either of Type A or C.

Noting the condition (4), we choose the following three complex functions as the activation function of (1).

$$f(u) := \frac{\eta u^R}{\eta - 1 + \sqrt{2}|u^R|} + i\frac{\eta u^I}{\eta - 1 + \sqrt{2}|u^I|} \qquad \text{for Type A,} \tag{8}$$

$$f(u) := \frac{\eta|u|}{\eta - 1 + |u|} \exp\left[i\{\arg u - \frac{1}{2^n}\sin(2^n \arg u)\}\right], -\pi \leq \arg u < \pi \qquad \text{for Type B,} \tag{9}$$

$$f(u) := \frac{\eta u}{\eta - 1 + |u|} = \frac{\eta r}{\eta - 1 + r}\exp\{i\theta\} \qquad \text{for Type C} \tag{10}$$

where η is a real constant number satisfying $\eta - 1 > 0$ and n is a natural number. The functions (8) and (10) have been often used as activation functions in complex-valued neural networks. The function (9) is obtained by modifying the discrete complex-valued activation function based on the complex-signam function used in [6] so as to satisfy the condition (i), that is, it becomes a continuous and smooth function. The functions (8), (9) and (10) are all not analytic, but have the continuous partial derivatives $\partial f^R/\partial u^R$, $\partial f^R/\partial u^I$, $\partial f^I/\partial u^R$ and $\partial f^I/\partial u^I$ and are bounded. Note also that, they all satisfies (4), which makes all the memory vectors $\boldsymbol{s}^{(\gamma)}$, $\gamma = 1, 2, \cdots, m$ be equilibrium points of the complex-valued neural network (1) with the weight matrix (2).

We call the complex-valued associative memory described by (1), (2) and (8) Model A, by (1), (2) and (9) Model B, and by (1), (2) and (10) Model C, respectively.

2.2 Memory Patterns

For Models A, B and C of complex-valued associative memories, each memory vector $\boldsymbol{s}^{(\gamma)}$ to be stored is demanded to become an equilibrium point of the network (1) and (2), which is accomplished if the condition (4) holds. This requirement makes each element of the memory vectors $\boldsymbol{s}^{(\gamma)}, \gamma = 1, 2, \cdots, m$ be allowed only to take restricted values. In Model A, each element s_j^γ of the memory vectors $\boldsymbol{s}^{(\gamma)}$ is allowed only to take the values

$$s_j^\gamma \in \{e^{i\pi/4}, e^{i3\pi/4}, e^{i5\pi/4}, e^{i7\pi/4}\}, \quad j = 1, 2, \cdots, N \tag{11}$$

and, in Model B, each element s_j^γ of the memory vectors $\boldsymbol{s}^{(\gamma)}$ is allowed only to take the values

$$s_j^\gamma \in \{e^{i0\pi/2^n}, e^{i\pi/2^n}, \ldots, e^{i(2^n-1)\pi/2^n}\}, \quad j = 1, 2, \cdots, N. \tag{12}$$

In Model C each element s_j^γ of the memory vectors $\boldsymbol{s}^{(\gamma)}$ is allowed to take all the values satisfying

$$|s_j^{(\gamma)}| = 1. \tag{13}$$

Figure 1 shows the values which each element s_j^γ of the memory vectors is allow to take in Models A, B and C, respectively. Note that, in Model B, the number of the allowed values can be increased arbitrarily by increasing n.

By choosing such values for each element, each memory vector $\boldsymbol{s}^{(\gamma)}$ becomes an equilibrium point of the network (1) and (2). For Models A and B, it is easy to check

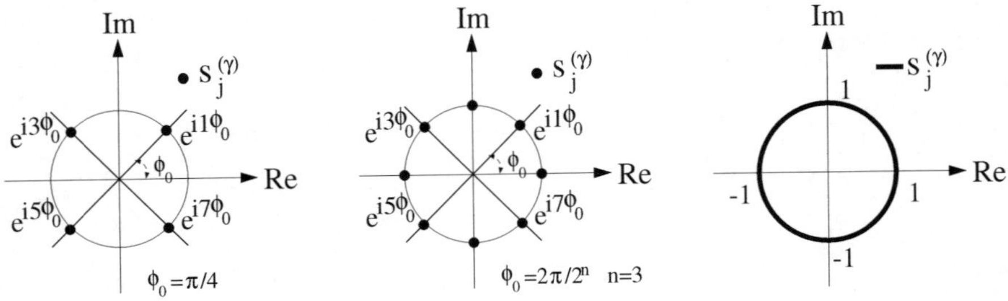

(a) model A (b) Model B ($n = 3$) (c) Model C

Fig. 1. The values which each element s_j^γ of the memory vectors is allowed to take

that each memory vector is an isolated equilibrium point of the network. On the other hand, in Model C, each memory vector $\boldsymbol{s}^{(\gamma)}$ is not an isolated equilibrium point of the network as is discussed in [2], because the function (10) satisfies

$$e^{i\alpha} s_j^{(\gamma)} = f(\sum_{k=1}^{N} w_{jk} e^{i\alpha} s_k^{(\gamma)}), \qquad j = 1, 2, \cdots, N \tag{14}$$

for any real number α. This implies that $\boldsymbol{s}^{(\gamma)}$ is a point in the set of equilibrium points defined by

$$\Phi^{(\gamma)} = \{e^{i\alpha} \boldsymbol{s}^{(\gamma)} : \forall \alpha \in \mathbb{R}\} \subset \mathbb{C}^N \tag{15}$$

which is a closed curve (hyper circle) in the complex N dimensional state space $\mathbb{C}^N$. From this fact we identify all the points in the equilibrium set $\Phi^{(\gamma)}$ and regard $\Phi^{(\gamma)}$ as a memory pattern for Model C [2].

3 Qualitative Analysis of Behavior Near Memory Patterns

In this section we study qualitative behavior of Models A, B and C of the complex-valued associative memories. Especially, we investigate asymptotic behavior of solution orbits near each memory pattern $\boldsymbol{s}^{(\gamma)}$.

3.1 Linearized Models Near Memory Patterns

The qualitative behavior of a nonlinear dynamical system near an equilibrium point can be studied via linearization with respect to that point. We will derive the linearized model at each memory vector $\boldsymbol{s}^{(\gamma)}$. Note that the activation functions $f(u)$ in (8), (9) and (10) are not differentiable with respect to u for all $u \in \mathbb{C}$, but their real and imaginary parts, f^R and f^I, are continuously partially differentiable with respect to u^R and u^I. It is, therefore, possible to derive the linearized model by separating each model into its real and imaginary parts.

We linearize Models A and B at each memory vector $\boldsymbol{s}^{(\gamma)}$ and Model C at a point $\boldsymbol{q}^{(\gamma)} = e^{i\alpha}\boldsymbol{s}^{(\gamma)}$ in each equilibrium set $\Phi^{(\gamma)}$ where α is a real number. Let $\Delta x_i[t] := x_i[t] - s_i^{(\gamma)}$ and define $\boldsymbol{y}[t] \in \mathbb{R}^{2N}$ by

$$\boldsymbol{y}[t] = (\Delta x_1^R[t], \Delta x_2^R[t], \cdots, \Delta x_N^R[t], \Delta x_1^I[t], \Delta x_2^I[t], \cdots, \Delta x_N^I[t])^T \quad (16)$$

where $(\cdot)^T$ denotes the transpose of $(\cdot)$. The linearized model for Models A and B is obtained as

$$\boldsymbol{y}[t+1] = J(\boldsymbol{s}^{(\gamma)})\boldsymbol{y}[t], \quad \gamma = 1, 2, \cdots, m \quad (17)$$

where $J(\boldsymbol{s}^{(\gamma)}) = F(\boldsymbol{s}^{(\gamma)})Y \in \mathbb{R}^{2N\times 2N}$. The matrices $F(\boldsymbol{s}^{(\gamma)}) \in \mathbb{R}^{2N\times 2N}$ and $Y \in \mathbb{R}^{2N\times 2N}$ are given as follows.

$$F(\boldsymbol{s}^{(\gamma)}) = \begin{bmatrix} F_{RR} & F_{RI} \\ F_{IR} & F_{II} \end{bmatrix} \qquad Y = \begin{bmatrix} W^R & -W^I \\ W^I & W^R \end{bmatrix}$$

where

$$F_{RR} = \mathrm{diag}\left(\left.\frac{\partial f^R}{\partial u^R}\right|_{u=s_1^{(\gamma)}}, \left.\frac{\partial f^R}{\partial u^R}\right|_{u=s_2^{(\gamma)}}, \cdots, \left.\frac{\partial f^R}{\partial u^R}\right|_{u=s_N^{(\gamma)}} \right)$$

$$F_{RI} = \mathrm{diag}\left(\left.\frac{\partial f^R}{\partial u^I}\right|_{u=s_1^{(\gamma)}}, \left.\frac{\partial f^R}{\partial u^I}\right|_{u=s_2^{(\gamma)}}, \cdots, \left.\frac{\partial f^R}{\partial u^I}\right|_{u=s_N^{(\gamma)}} \right)$$

$$F_{IR} = \mathrm{diag}\left(\left.\frac{\partial f^I}{\partial u^R}\right|_{u=s_1^{(\gamma)}}, \left.\frac{\partial f^I}{\partial u^R}\right|_{u=s_2^{(\gamma)}}, \cdots, \left.\frac{\partial f^I}{\partial u^R}\right|_{u=s_N^{(\gamma)}} \right)$$

$$F_{II} = \mathrm{diag}\left(\left.\frac{\partial f^I}{\partial u^I}\right|_{u=s_1^{(\gamma)}}, \left.\frac{\partial f^I}{\partial u^I}\right|_{u=s_2^{(\gamma)}}, \cdots, \left.\frac{\partial f^I}{\partial u^I}\right|_{u=s_N^{(\gamma)}} \right)$$

and the elements of $F(\boldsymbol{s}^{(\gamma)})$ are given by

$$\left.\frac{\partial f^R}{\partial u^R}\right|_{u=s_j^\gamma} = \left.\frac{\partial f^I}{\partial u^I}\right|_{u=s_j^\gamma} = 1 - \frac{1}{\eta}, \qquad \left.\frac{\partial f^R}{\partial u^I}\right|_{u=s_j^\gamma} = \left.\frac{\partial f^I}{\partial u^R}\right|_{u=s_j^\gamma} = 0$$

for Model A and

$$\left.\frac{\partial f^R}{\partial u^R}\right|_{u=s_j^\gamma} = \left(1 - \frac{1}{\eta}\right)(s_j^{\gamma R})^2, \qquad \left.\frac{\partial f^I}{\partial u^I}\right|_{u=s_j^\gamma} = \left(1 - \frac{1}{\eta}\right)(s_j^{\gamma I})^2$$

$$\left.\frac{\partial f^R}{\partial u^I}\right|_{u=s_j^\gamma} = \left.\frac{\partial f^I}{\partial u^R}\right|_{u=s_j^\gamma} = \left(1 - \frac{1}{\eta}\right) s_j^{\gamma R} s_j^{\gamma I}$$

for Model B. The linearized model for Model C is obtained as the same form as (17) with $J(\boldsymbol{s}^{(\gamma)}) = F(\boldsymbol{s}^{(\gamma)})Y$ being replaced by $J(\boldsymbol{q}^{(\gamma)}) = F(\boldsymbol{q}^{(\gamma)})Y$, where the elements of $F(\boldsymbol{q}^{(\gamma)})$ are given by

$$\left.\frac{\partial f^R}{\partial u^R}\right|_{u=q_j^\gamma} = 1 + \frac{-1 + (q_j^{\gamma I})^2}{\eta}, \qquad \left.\frac{\partial f^I}{\partial u^I}\right|_{u=q_j^\gamma} = 1 + \frac{-1 + (q_j^{\gamma R})^2}{\eta}$$

$$\left.\frac{\partial f^R}{\partial u^I}\right|_{u=q_j^\gamma} = \left.\frac{\partial f^I}{\partial u^R}\right|_{u=q_j^\gamma} = \frac{-q_j^{\gamma R} q_j^{\gamma I}}{\eta}$$

3.2 Structure of Solution Orbits Near Memory Patterns

We now analyze the qualitative behavior and structure of the solution orbits near each memory pattern. This can be done by investigating the eigenvalues and eigenvectors of the coefficient matrix $J(\boldsymbol{s}^{(\gamma)})$ (or $J(\boldsymbol{q}^{(\gamma)})$) of the linearized model (17). Let $\lambda_i(J(\boldsymbol{s}^{(\gamma)}))$ $(\lambda_i(J(\boldsymbol{q}^{(\gamma)})))$ be the ith eigenvalue of $J(\boldsymbol{s}^{(\gamma)})$ $(J(\boldsymbol{q}^{(\gamma)}))$.

The following theorems are obtained for Models A.

Theorem A.1. *All the eigenvalues of the coefficient matrix $\boldsymbol{J}(\boldsymbol{s}^{\gamma})$ of Model A are real and satisfy the following condition:*

$$|\lambda_j(\boldsymbol{J}(\boldsymbol{s}^{\gamma}))| \leq \frac{\eta - 1}{\eta} < 1, \quad j = 1, 2, \cdots, 2N. \tag{18}$$

Theorem A.2. *The matrix $\boldsymbol{J}(\boldsymbol{s}^{\gamma})$ of Model A has $2m$ eigenvalues $(\eta-1)/\eta$ and $2(N-m)$ eigenvalues 0. Two of the corresponding eigenvectors to the eigenvalue $(\eta-1)/\eta$ are*

$$\boldsymbol{r}^{(\gamma)} := \left((\boldsymbol{s}^{(\gamma)R})^T, (\boldsymbol{s}^{(\gamma)I})^T\right)^T \tag{19}$$

$$\boldsymbol{p}^{(\gamma)} := \left((\{i\boldsymbol{s}^{(\gamma)}\}^R)^T, (\{i\boldsymbol{s}^{(\gamma)}\}^I)^T\right)^T. \tag{20}$$

The following theorems are obtained for Model B.

Theorem B.1. *All the eigenvalues of the coefficient matrix $\boldsymbol{J}(\boldsymbol{s}^{\gamma})$ of Model B satisfy the following condition:*

$$|\lambda_j(\boldsymbol{J}(\boldsymbol{s}^{\gamma}))| \leq \frac{\eta - 1}{\eta} < 1, \quad j = 1, 2, \cdots, 2N \tag{21}$$

Theorem B.2. *The matrix $\boldsymbol{J}(\boldsymbol{s}^{\gamma})$ of Model B has at least one eigenvalues $(\eta-1)/\eta$ and at least one eigenvalue 0. The corresponding eigenvector to the eigenvalue $(\eta-1)/\eta$ is $\boldsymbol{r}^{(\gamma)}$ defined by (19) and the corresponding eigenvector to the eigenvalue 0 is $\boldsymbol{p}^{(\gamma)}$ defined by (20).*

For Model C the following theorems are obtained.

Theorem C.1. *All the eigenvalues of the coefficient matrix $\boldsymbol{J}(\boldsymbol{q}^{\gamma})$ of Model C are real and satisfy the following condition.*

$$|\lambda_j(\boldsymbol{J}(\boldsymbol{q}^{\gamma}))| \leq 1, \quad j = 1, 2, \ldots, 2N \tag{22}$$

Theorem C.2. *The matrix $J(\boldsymbol{q}^{(\gamma)})$ has at least one eigenvalue 1 and at least one eigenvalue $(\eta-1)/\eta$. It also has $2(N-m)$ eigenvalues 0. The corresponding eigenvector to the eigenvalue 1 is*

$$\boldsymbol{p}^{(\gamma)} = ((\{i\boldsymbol{q}^{(\gamma)}\}^R)^T, (\{i\boldsymbol{q}^{(\gamma)}\}^I)^T)^T \tag{23}$$

and that to the eigenvalue $(\eta-1)/\eta$ is

$$\boldsymbol{r}^{(\gamma)} = ((\boldsymbol{q}^{(\gamma)R})^T, (\boldsymbol{q}^{(\gamma)I})^T)^T. \tag{24}$$

Note that Theorems A.1 to C.2 hold for all the memory vectors $\boldsymbol{s}^{(\gamma)}$ $(\boldsymbol{q}^{(\gamma)})$, $\gamma = 1, 2, \cdots, m$. The proofs of these theorems are omitted due to the space limitation.

3.3 Discussions

It is known that qualitative behavior of solutions near an equilibrium point of a nonlinear dynamical system is determined by its linearized model at the point if it is hyperbolic (in a discrete time system, the coefficient matrix of the linearized model has no eigenvalues of unit modulus) [10]. From the theorems obtained in the previous subsection, all the memory vectors $\boldsymbol{s}^{(\gamma)}, \gamma = 1, 2, \cdots, m$ are hyperbolic in Models A and B, whereas all the memory vectors $\boldsymbol{q}^{(\gamma)}, \gamma = 1, 2, \cdots, m$ are not hyperbolic in Model C. For each model, it is required that all embedded memory vectors are at least asymptotically stable equilibrium points of the network for correct recalling as associative memories.

From Theorems A.1 and B.1, each memory vector $\boldsymbol{s}^{(\gamma)}$ of Model A and B is an asymptotically stable equilibrium point because the magnitude of all the eigenvalues of the coefficient matrix $\boldsymbol{J}(\boldsymbol{s}^{(\gamma)})$ is less than one. Therefore every solution starting in the neighborhood of the vector $\boldsymbol{s}^{(\gamma)}$ tends to $\boldsymbol{s}^{(\gamma)}$ as $t \to \infty$.

For Model C, further study is needed to investigate the stability of the memory patterns $\Phi^{(\gamma)}, \gamma = 1, 2, \cdots, m$ because all the memory vectors $\boldsymbol{q}^{(\gamma)}, \gamma = 1, 2, \cdots, m$ are not hyperbolic. The structure of solution orbits near memory patterns $\Phi^{(\gamma)}$ can be investigated based on the center manifold theory [10] for Model C. It can be shown that, if $J(\boldsymbol{q}^{(\gamma)})$ has only one eigenvalue 1, each memory pattern $\Phi^{(\gamma)}$ itself is the center manifold of each point $\boldsymbol{q}^{(\gamma)} \in \Phi^{(\gamma)}$ and is asymptotically stable, that is, every solution starting in the neighborhood of $\Phi^{(\gamma)}$ tend to $\Phi^{(\gamma)}$ as $t \to \infty$.

Therefore it is concluded that in Model A and B all embedded memory vectors $\boldsymbol{s}^{(\gamma)}, \ \gamma = 1, 2, \cdots, m$ are asymptotically stable and they all can be recalled. On the other hand, in Model C, all the embedded memory vectors are not necessarily asymptotically stable and it is necessary to evaluate eigenvalues of the coefficient matrices $\boldsymbol{J}(\boldsymbol{q}^{\gamma})$ to check their stability.

Let us compare the obtained results with those obtained in [9] for the self-correlation type complex-valued associative memories. The difference between the models of associative memories in this paper and those in [9] is the choice of memory patterns to be stored $\boldsymbol{s}^{(\gamma)} \in \mathbb{C}^N, \ \gamma = 1, 2, \cdots, m$ and the determination of weight matrix $\boldsymbol{W} = \{w_{jk}\} \in \mathbb{C}^{N \times N}$. For the self-correlation type complex-valued associative memories in [9], memory patterns $\boldsymbol{s}^{(\gamma)} \in \mathbb{C}^N, \ \gamma = 1, 2, \cdots, m$ are chosen so as to satisfies the following orthogonal relations.

$$\boldsymbol{s}^{(\gamma)*}\boldsymbol{s}^{(l)} = \begin{cases} N, \gamma = l \\ 0, \gamma \neq l \end{cases} \tag{25}$$

$$|s_j^{(\gamma)}| = 1, \quad j = 1, 2, \cdots, N \tag{26}$$

for all $\gamma, l = 1, 2, \cdots, m$ where $\boldsymbol{s}^*$ is the conjugate transpose of $\boldsymbol{s}$ and $s_j^{(\gamma)}$ is the jth element of $\boldsymbol{s}^{(\gamma)}$, and the weight matrix $\boldsymbol{W}$ is determined by the sum of the autocorrelation matrix of each memory vector $\boldsymbol{s}^{(\gamma)}$:

$$\boldsymbol{W} = \frac{1}{N} \sum_{\gamma=1}^{m} \boldsymbol{s}^{(\gamma)} \boldsymbol{s}^{(\gamma)*} \tag{27}$$

Theorems A.1, A.2, B.1, B.2, C.1 and C.2 are equivalent to the corresponding theorems obtained in [9] for the self-correlation type complex-valued associative memories. Therefore the qualitative behavior and structure of solution orbits near each memory pattern of the self-correlation type and those of orthogonal type complex-valued associative memories are much the same.

Note that, it is required that for the complex-valued associative memory of self-correlation type that memory patterns $\{\boldsymbol{s}^{(1)}, \boldsymbol{s}^{(2)}, \cdots, \boldsymbol{s}^{(m)}\}$ are orthogonal each other, whereas the orthogonality of the memory patterns is not required for the complex-valued associative memory of orthogonal type. The requirement of the orthogonality of the memory patterns makes differences among Models A, B and C on how many vectors can be chosen such orthogonal vectors in N dimensional space $\mathbb{C}^N$ [9]. For Model C of the complex-valued associative memory of self-correlation type it can be shown that, for any N there always exist N vectors satisfying the conditions (25) and (26) in the N dimensional space $\mathbb{C}^N$. On the other hand, there not always exit N vectors in the N dimensional space $\mathbb{C}^N$ for Model A and Model B of the complex-valued associative memory of self-correlation type. The problems of how many vectors we can choose such orthogonal vectors for Models A and B are equivalent to the existence problems of the complex Hadamard matrices and the generalized Hadamard matrices, respectively, which are partially solved but not completely solved.

4 Conclusions

In this paper we presented models of orthogonal type dynamic associative memories using complex-valued neural networks and studied their qualitative behavior theoretically. One of the important factors to characterize behavior of a complex-valued neural network is its activation function. We presented three kinds of models with three different activation functions and investigate the structures and asymptotic behavior of solution orbits near each memory pattern. We compared three models in terms of asymptotic behavior of solution orbits near each memory pattern and storage capabilities. We also compared the results for complex-valued associative memories of orthogonal type obtained in this paper with those for the complex-valued associative memories of self-correlation type obtained in [9].

References

1. Hirose, A.: Dynamics of fully complex-valued neural networks. Electronics Letters 28(16), 1492–1494 (1992)
2. Kuroe, Y., Hashimoto, N., Mori, T.: Qualitative analysis of a self-correlation type complex-valued associative memories. Nonlinear Analysis 47, 5795–5806
3. Kuroe, Y., Hashimoto, N., Mori, T.: Qualitative Analysis of Continuous Complex-Valued Associative Memories. In: Dorffner, G., Bischof, H., Hornik, K. (eds.) ICANN 2001. LNCS, vol. 2130, pp. 843–850. Springer, Heidelberg (2001)
4. Nemoto, I., Kono, T.: Complex-valued neural network. The Trans. of IEICE J74-D-II(9), 1282–1288 (1991) (in Japanese)
5. Noest, A.J.: Phaser neural network. In: Anderson, D.Z. (ed.) Neural Informaion Processing Systems, pp. 584–591. AIP, New York (1988)

6. Jankowski, S., Lozowski, A., Zurada, J.M.: Complex-valued multistate neural associative memory. IEEE Trans. Neural Networks 7(6), 1491–1496 (1996)
7. Georgiou, G.M., Koutsougeras, C.: Complex Domain Backpropagation. IEEE Transactions on Circuits and Systems-II 39(5), 330–334 (1992)
8. Kuroe, Y., Yoshida, M., Mori, T.: On Activation Functions for Complex-Valued Neural Networks - Existence of Energy Functions -; Artificial Neural Networks and Neural Information Processing - ICANN/ICONIP 2003, Okyay Kaynak et. In: Kaynak, O., Alpaydın, E., Oja, E., Xu, L. (eds.) ICANN 2003 and ICONIP 2003. LNCS, vol. 2714, pp. 985–992. Springer, Heidelberg (2003)
9. Kuroe, Y., Taniguchi, Y.: Models of Self-Correlation Type Complex-Valued Associative Memories and Their Dynamics. In: Duch, W., Kacprzyk, J., Oja, E., Zadrożny, S. (eds.) ICANN 2005. LNCS, vol. 3696, pp. 185–192. Springer, Heidelberg (2005)
10. Sastry, S.: Nonlinear Systems Analysis, Stability, and Control. Springer, Heidelberg (1999)

Dynamics of Discrete-Time Quaternionic Hopfield Neural Networks

Teijiro Isokawa[1], Haruhiko Nishimura[2], Naotake Kamiura[1], and Nobuyuki Matsui[1]

[1] Division of Computer Engineering, Graduate School of Engineering, University of Hyogo, Japan
{isokawa, kamiura, matsui}@eng.u-hyogo.ac.jp
[2] Graduate School of Applied Informatics, University of Hyogo, Japan
haru@ai.u-hyogo.ac.jp

Abstract. We analyze a discrete-time quaternionic Hopfield neural network with continuous state variables updated asynchronously. The state of a neuron takes quaternionic value which is four-dimensional hypercomplex number. Two types of the activation function for updating neuron states are introduced and examined. The stable states of the networks are demonstrated through an example of small network.

1 Introduction

Quaternion is a four-dimensional hypercomplex number system discovered by W.R.Hamilton [1]. This is extensively used in several fields, such as modern mathematics, physics, computer graphics, and so on [2]. One of the benefits by the use of quaternions is that it can treat and operate three or four dimensional vector as one entity, so the effective information processing can be achieved by the operations for quaternionic variables.

In this respect, there have been growing the studies concerning the introduction of quaternions into neural networks. The computational ability for a single quaternionic neuron was exhibited in [3]. There have been several multilayer perceptron models with their learning algorithm (back-propagation method) in quaternion domain, and several applications of them have also been proposed [4,5,6].

However, only a few investigations concerning recurrent neural networks in quaternion domain are performed. The existence conditions of energy function in the Hopfield-type quaternionic network was explored in the case of continuous value of neuron states with continuous time [7]. The case of bipolar value of the neuron states with discrete time was also studied and its stability was shown explicitly [8].

In this paper, we further explore the properties of the Hopfield neural networks where the neuron states take continuous values and are updated in discrete time. Two types of the activation function for updating the neuron states are introduced. For both activation functions, it is proved that the energy decreases

J. Marques de Sá et al. (Eds.): ICANN 2007, Part I, LNCS 4668, pp. 848–857, 2007.

monotonically with respect to the change of the neuron state. The attractors of neuron states are demonstrated in the network with three neurons.

2 Quaternion Algebra

Quaternions form a class of hypercomplex numbers that consist of a real number and three kinds of imaginary number, $\boldsymbol{i}, \boldsymbol{j}, \boldsymbol{k}$. Formally, a quaternion number is defined as a vector $\boldsymbol{x}$ in a 4-dimensional vector space,

$$\boldsymbol{x} = x^{(e)} + x^{(i)}\boldsymbol{i} + x^{(j)}\boldsymbol{j} + x^{(k)}\boldsymbol{k} \tag{1}$$

where $x^{(e)}$, $x^{(i)}, x^{(j)}$, and $x^{(k)}$ are real numbers. $\boldsymbol{H}$, the division ring of quaternions, thus constitutes the four-dimensional vector space over the real numbers with the bases $1, \boldsymbol{i}, \boldsymbol{j}, \boldsymbol{k}$. It is also written using 4-tuple or 2-tuple notations as

$$\boldsymbol{x} = (x^{(e)}, x^{(i)}, x^{(j)}, x^{(k)}) = (x^{(e)}, \vec{x}) \tag{2}$$

where $\vec{x} = \{x^{(i)}, x^{(j)}, x^{(k)}\}$. In this representation $x^{(e)}$ is the scalar part of $\boldsymbol{x}$ and $\vec{x}$ forms the vector part. The quaternion conjugate is defined as

$$\boldsymbol{x}^* = (x^{(e)}, -\vec{x}) = x^{(e)} - x^{(i)}\boldsymbol{i} - x^{(j)}\boldsymbol{j} - x^{(k)}\boldsymbol{k}. \tag{3}$$

Quaternion bases satisfy the following identities, known as the Hamilton rules:

$$\boldsymbol{i}^2 = \boldsymbol{j}^2 = \boldsymbol{k}^2 = \boldsymbol{ijk} = -1,\ \boldsymbol{ij} = -\boldsymbol{ji} = \boldsymbol{k},\ \boldsymbol{jk} = -\boldsymbol{kj} = \boldsymbol{i},\ \boldsymbol{ki} = -\boldsymbol{ik} = \boldsymbol{j}. \tag{4}$$

From these rules it follows immediately that multiplication of quaternions is not commutative.

The operations between quaternions, $\boldsymbol{p} = (p^{(e)}, \vec{p}) = (p^{(e)}, p^{(i)}, p^{(j)}, p^{(k)})$ and $\boldsymbol{q} = (q^{(e)}, \vec{q}) = (q^{(e)}, q^{(i)}, q^{(j)}, q^{(k)})$, are defined as follows. The addition and subtraction of quaternions are defined, in the same manner as those of complex-valued numbers or vectors, by

$$\boldsymbol{p} \pm \boldsymbol{q} = (p^{(e)} \pm q^{(e)},\ \vec{p} \pm \vec{q}) = (p^{(e)} \pm q^{(e)}, p^{(i)} \pm q^{(i)}, p^{(j)} \pm q^{(j)}, p^{(k)} \pm q^{(k)}). \tag{5}$$

With regard to the multiplication, the product between $\boldsymbol{p}$ and $\boldsymbol{q}$, notation $\boldsymbol{p} \otimes \boldsymbol{q}$, is determined by Eq. (4) as

$$\boldsymbol{p} \otimes \boldsymbol{q} = (p^{(e)}q^{(e)} - \vec{p} \cdot \vec{q},\ p^{(e)}\vec{q} + q^{(e)}\vec{p} + \vec{p} \times \vec{q}) \tag{6}$$

where $\vec{p} \cdot \vec{q}$ and $\vec{p} \times \vec{q}$ denote the dot and cross products respectively between three dimensional vectors $\vec{p}$ and $\vec{q}$. The conjugate of product holds the relation of

$$(\boldsymbol{p} \otimes \boldsymbol{q})^* = \boldsymbol{q}^* \otimes \boldsymbol{p}^*. \tag{7}$$

The quaternion norm of $\boldsymbol{x}$, notation $|\boldsymbol{x}|$, is defined by

$$|\boldsymbol{x}| = \sqrt{\boldsymbol{x} \otimes \boldsymbol{x}^*} = \sqrt{{x^{(e)}}^2 + {x^{(i)}}^2 + {x^{(j)}}^2 + {x^{(k)}}^2}. \tag{8}$$

3 Quaternionic Hopfield Neural Network

3.1 Quaternionic Neuron Model

We introduce the formalism of the neural network in which all the variables are represented by quaternion numbers. We start with the quaternion-valued neuron model. The action potential of the neuron p at time t and output state of the neuron p at time $(t+1)$ are defined as follows:

$$\boldsymbol{s}_p(t) = \sum_q \boldsymbol{w}_{pq} \otimes \boldsymbol{x}_q(t) - \boldsymbol{\theta}_p, \tag{9}$$

$$\boldsymbol{x}_p(t+1) = \boldsymbol{f}(\boldsymbol{s}_p(t)), \tag{10}$$

where $\boldsymbol{x}, \boldsymbol{w}_{pq}, \boldsymbol{\theta} \in \boldsymbol{H}$ are the input to the neuron, the connection weight from neuron q to neuron p, and the threshold, respectively. The activation function $\boldsymbol{f}$ determines the output of the neuron.

Two types of quaternionic functions for $\boldsymbol{f}$ are introduced in this paper. One is designed so that each quaternionic component is updated independently and defined by

$$\boldsymbol{f}_1(\boldsymbol{s}) = f_1^{(e)}(s^{(e)}) + f_1^{(i)}(s^{(i)})\boldsymbol{i} + f_1^{(j)}(s^{(j)})\boldsymbol{j} + f_1^{(k)}(s^{(k)})\boldsymbol{k}, \tag{11}$$

where each real-valued function $f_1^{(\alpha)}$ $(\alpha = \{e, i, j, k\})$ is set to

$$f_1^{(e)}(s) = f_1^{(i)}(s) = f_1^{(j)}(s) = f_1^{(k)}(s) = \tanh(s/\epsilon), \tag{12}$$

where $\epsilon > 0$ is a steepness parameter. This is a straightforward extension of the conventional updating scheme in real-valued neurons and often used in the quaternionic multilayer perceptron [4,5,6].

The other is an extension to the quaternion domain of the activation function proposed in [9] in the complex domain and defined as

$$\boldsymbol{f}_2(\boldsymbol{s}) = \frac{a\boldsymbol{s}}{1+|\boldsymbol{s}|}, \tag{13}$$

where a is a real-valued constant.

3.2 Energy Function

The neurons are connected to each other in the network, as in real-valued Hopfield neural network. We introduce an energy function of the network that consists of N neurons as

$$\boldsymbol{E}(t) = -\frac{1}{2}\sum_{p=1}^{N}\sum_{q=1}^{N} \boldsymbol{x}_p^*(t) \otimes \boldsymbol{w}_{pq} \otimes \boldsymbol{x}_q(t) + \frac{1}{2}\sum_{p=1}^{N}\left(\boldsymbol{\theta}_p^* \otimes \boldsymbol{x}_p(t) + \boldsymbol{x}_p \otimes \boldsymbol{\theta}_p^*(t)\right) + \sum_{p=1}^{N} G(\boldsymbol{x}_p(t)), \tag{14}$$

where $G(\boldsymbol{x}(t))$ is a scalar function that satisfies

$$\frac{\partial G(\boldsymbol{x})}{\partial x^{(\alpha)}} = g^{(\alpha)}(x^{(e)}, x^{(i)}, x^{(j)}, x^{(k)}) \quad (\alpha = \{e, i, j, k\}). \tag{15}$$

Here, $\boldsymbol{g}(\boldsymbol{x})$ is the inverse function of $\boldsymbol{f}(\boldsymbol{x})$ and represented as

$$\begin{aligned} \boldsymbol{g}(\boldsymbol{x}) &= \boldsymbol{f}^{-1}(\boldsymbol{x}) \\ &= g^{(e)}(x^{(e)}, x^{(i)}, x^{(j)}, x^{(k)}) + g^{(i)}(x^{(e)}, x^{(i)}, x^{(j)}, x^{(k)})\boldsymbol{i} \\ &\quad + g^{(j)}(x^{(e)}, x^{(i)}, x^{(j)}, x^{(k)})\boldsymbol{j} + g^{(k)}(x^{(e)}, x^{(i)}, x^{(j)}, x^{(k)})\boldsymbol{k}, \end{aligned} \tag{16}$$

so that the following relation is hold from Eq. (10):

$$\boldsymbol{s}_p(t) = \boldsymbol{f}^{-1}(\boldsymbol{x}_p(t+1)) = \boldsymbol{g}(\boldsymbol{x}_p(t+1)). \tag{17}$$

$\boldsymbol{E}$ should be a real-valued function ($\boldsymbol{E} = \boldsymbol{E}^*$) and this is always satisfied by the condition for the connection weight, $\boldsymbol{w}_{pq} = \boldsymbol{w}_{qp}^*$. This network allows self-connections and their values take real, $\boldsymbol{w}_{pp} = \boldsymbol{w}_{pp}^* = (w_{pp}^{(e)}, \boldsymbol{0})$.

As in the case of the real-valued Hopfield network, the energy E of this network also never increases with time when the state of a neuron changes. Let $E_r(t)$ be the contribution of a given neuron r to the energy E at time t, then it is given as

$$\begin{aligned} E_r(t) = &-\frac{1}{2}\left\{\left(\sum_{p=1}^{N} \boldsymbol{x}_p^*(t) \otimes \boldsymbol{w}_{pr}\right) \otimes \boldsymbol{x}_r(t) + \boldsymbol{x}_r^*(t) \otimes \left(\sum_{q=1}^{N} \boldsymbol{w}_{rq} \otimes \boldsymbol{x}_q(t)\right)\right. \\ &\left. - \boldsymbol{x}_r^*(t) \otimes \boldsymbol{w}_{rr} \otimes \boldsymbol{x}_r(t)\right\} + \frac{1}{2}\left(\boldsymbol{\theta}_r^* \otimes \boldsymbol{x}_r(t) + \boldsymbol{x}_r^* \otimes \boldsymbol{\theta}_r(t)\right) + G\left(\boldsymbol{x}_r(t)\right) \\ = &-\frac{1}{2}\left\{\left(\sum_{p=1}^{N} \boldsymbol{w}_{rp} \otimes \boldsymbol{x}_p(t) - \boldsymbol{\theta}_r\right)^* \otimes \boldsymbol{x}_r(t)\right. \\ &\left. + \boldsymbol{x}_r^*(t) \otimes \left(\sum_{q=1}^{N} \boldsymbol{w}_{rq} \otimes \boldsymbol{x}_q(t) - \boldsymbol{\theta}_r\right) - \boldsymbol{x}_r^*(t) \otimes \boldsymbol{w}_{rr} \otimes \boldsymbol{x}_r(t)\right\} \\ &+ G\left(\boldsymbol{x}_r(t)\right) \end{aligned}$$

Considering the relation in Eq. (9) we obtain

$$\begin{aligned} E_r(t) &= -\frac{1}{2}\{\boldsymbol{s}_r^*(t) \otimes \boldsymbol{x}_r(t) + \boldsymbol{x}_r^*(t) \otimes \boldsymbol{s}_r(t) - \boldsymbol{x}_r^*(t) \otimes \boldsymbol{w}_{rr} \otimes \boldsymbol{x}_r(t)\} + G\left(\boldsymbol{x}_r(t)\right) \\ &= -Re\left\{\boldsymbol{x}_r^*(t) \otimes \boldsymbol{s}_r(t)\right\} + \frac{1}{2} w_{rr}^{(e)} \boldsymbol{x}_r^*(t) \otimes \boldsymbol{x}_r(t) + G\left(\boldsymbol{x}_r(t)\right). \end{aligned} \tag{18}$$

Suppose that only the state of the neuron r is asynchronously updated at time $(t+1)$ according to Eqs. (9) and (10). Its contribution to the energy E becomes

$$E_r(t+1) = -Re\left\{\boldsymbol{x}_r^*(t+1) \otimes \boldsymbol{s}_r(t+1)\right\} + \frac{1}{2} w_{rr}^{(e)} \boldsymbol{x}_r^*(t+1) \otimes \boldsymbol{x}_r(t+1)$$

$$+G\left(\boldsymbol{x}_r(t+1)\right)$$
$$= -Re\left\{\boldsymbol{x}_r^*(t+1)\otimes(\boldsymbol{s}_r(t)+\boldsymbol{w}_{rr}\otimes\Delta\boldsymbol{x}_r)\right\}$$
$$+\frac{1}{2}w_{rr}^{(e)}\boldsymbol{x}_r^*(t+1)\otimes\boldsymbol{x}_r(t+1)+G\left(\boldsymbol{x}_r(t+1)\right),$$

where $\Delta\boldsymbol{x}_r = \boldsymbol{x}_r(t+1) - \boldsymbol{x}_r(t)$. The difference of the energy between time $t+1$ and t, ΔE, becomes

$$\begin{aligned}\Delta E &= E_r(t+1) - E_r(t)\\ &= -Re\left(\Delta\boldsymbol{x}_r^*\otimes\boldsymbol{s}_r(t)\right) - w_{rr}^{(e)}\cdot Re\left\{\boldsymbol{x}_r^*(t+1)\otimes\Delta\boldsymbol{x}_r\right\}\\ &\quad+\frac{1}{2}w_{rr}^{(e)}\boldsymbol{x}_r^*(t+1)\otimes\boldsymbol{x}_r(t+1)-\frac{1}{2}w_{rr}^{(e)}\boldsymbol{x}_r^*(t)\otimes\boldsymbol{x}_r(t)\\ &\quad+G(\boldsymbol{x}_r(t+1))-G(\boldsymbol{x}_r(t))\\ &= -Re\left(\Delta\boldsymbol{x}_r^*\otimes\boldsymbol{s}_r(t)\right)-\frac{1}{2}w_{rr}^{(e)}\Delta\boldsymbol{x}_r^*\otimes\Delta\boldsymbol{x}_r+G(\boldsymbol{x}_r(t+1))-G(\boldsymbol{x}_r(t))\\ &= -Re\left(\Delta\boldsymbol{x}_r^*\otimes\boldsymbol{s}_r(t)\right)-\frac{1}{2}w_{rr}^{(e)}|\Delta\boldsymbol{x}_r|^2+G(\boldsymbol{x}_r(t+1))-G(\boldsymbol{x}_r(t)). \qquad (19)\end{aligned}$$

$G(\boldsymbol{x}_r(t))$ can be expanded around $\boldsymbol{x}_r(t+1)$ by the Taylor's theorem as

$$\begin{aligned}G(\boldsymbol{x}_r(t)) &= G(\boldsymbol{x}_r(t+1)) - \sum_\alpha \Delta\boldsymbol{x}_r^{(\alpha)}\frac{\partial G}{\partial x_r^{(\alpha)}}\Bigg|_{x^{(\alpha)}=x_r^{(\alpha)}(t+1)}\\ &\quad+\sum_{\alpha,\beta}\Delta\boldsymbol{x}_r^{(\alpha)}\Delta\boldsymbol{x}_r^{(\beta)}\frac{\partial^2 G}{\partial x_r^{(\alpha)}\partial x_r^{(\beta)}}\Bigg|_{x^{(\alpha)}=\phi^{(\alpha)},x^{(\beta)}=\phi^{(\beta)}},\end{aligned}$$

where $\phi^{(\alpha)}$ is between $x_r^{(\alpha)}(t)$ and $x_r^{(\alpha)}(t+1)$. Hence, the difference of the energy can be written as

$$\begin{aligned}\Delta E &= -Re\left(\Delta\boldsymbol{x}_r^*\otimes\boldsymbol{s}_r(t)\right)-\frac{1}{2}w_{rr}^{(e)}|\Delta\boldsymbol{x}_r|^2\\ &\quad+\sum_\alpha \Delta\boldsymbol{x}_r^{(\alpha)}\frac{\partial G}{\partial x_r^{(\alpha)}}\Bigg|_{x^{(\alpha)}=x_r^{(\alpha)}(t+1)}\\ &\quad-\sum_{\alpha,\beta}\Delta\boldsymbol{x}^{(\alpha)}\Delta\boldsymbol{x}_r^{(\beta)}\frac{\partial^2 G}{\partial x_r^{(\alpha)}\partial x_r^{(\beta)}}\Bigg|_{x^{(\alpha)}=\phi^{(\alpha)},x^{(\beta)}=\phi^{(\beta)}}.\end{aligned}$$

Considering Eqs.(15) and (17), ΔE is reduced to

$$\Delta E = -\frac{1}{2}w_{rr}^{(e)}|\Delta\boldsymbol{x}_r|^2-\sum_{\alpha,\beta}\Delta\boldsymbol{x}^{(\alpha)}\Delta\boldsymbol{x}_r^{(\beta)}\frac{\partial g^{(\beta)}}{\partial x^{(\alpha)}}\Bigg|_{x^{(\alpha)}=\phi^{(\alpha)},x^{(\beta)}=\phi^{(\beta)}}. \qquad (20)$$

We first consider the case that $\boldsymbol{f}_1$(Eq. (11)) is applied for the activation function of neurons. The inverse function of $\boldsymbol{f}_1$, $\boldsymbol{g}_1$, is calculated as

$$\begin{aligned}\boldsymbol{g}_1(\boldsymbol{x}) &= g_1^{(e)}(x^{(e)})+g_1^{(i)}(x^{(i)})\boldsymbol{i}+g_1^{(j)}(x^{(j)})\boldsymbol{j}+g_1^{(k)}(x^{(k)})\boldsymbol{k}\\ g_1^{(\alpha)}(x) &= \tanh^{-1}x = \frac{\epsilon}{2}\ln\frac{1+x}{1-x}.\end{aligned}$$

In this case, Eq. (20) becomes

$$\begin{aligned}\Delta E &= -\frac{1}{2}w_{rr}^{(e)}|\Delta \boldsymbol{x}_r|^2 - \sum_{\alpha}(\Delta x^{(\alpha)})^2\frac{\partial g^{(\alpha)}}{\partial g^{(\beta)}}\\ &= -\frac{1}{2}w_{rr}^{(e)}|\Delta \boldsymbol{x}_r|^2 - \sum_{\alpha}(\Delta \boldsymbol{x}_r^{(\alpha)})^2\cdot\frac{\epsilon}{(1+\phi^{(\alpha)})(1-\phi^{(\alpha)})},\end{aligned}$$

because $\partial g^{\beta}/\partial x^{(\alpha)} = 0$ for $\alpha \neq \beta$ and $\partial g^{(\alpha)}/\partial x^{(\alpha)} = \epsilon/(1+x^{(\alpha)})(1-x^{(\alpha)})$. Furthermore,

$$\min\frac{\epsilon}{(1+\phi^{(\alpha)})(1-\phi^{(\alpha)})} = \epsilon \quad (\text{at } \phi^{(\alpha)} = 0),$$

hence the following inequality holds:

$$\begin{aligned}\Delta E &< -\frac{1}{2}w_{rr}^{(e)}\left|\Delta \boldsymbol{x}_r\right|^2 - \epsilon\sum_{\alpha}(\Delta x_r^{(\alpha)})^2\\ &= -\left(\frac{1}{2}w_{rr}^{(e)} + \epsilon\right)\left|\Delta x_r\right|^2.\end{aligned} \tag{21}$$

The energy with the activation function $\boldsymbol{f}_1$ never increases under the condition of $w_{rr}^{(e)} > -2\epsilon$.

We further consider the case of $\boldsymbol{f}_2$ used for the activation function. In this case, the inverse of $\boldsymbol{f}_2$ is obtained as

$$\begin{aligned}\boldsymbol{g}_2(\boldsymbol{x}) &= \frac{\boldsymbol{x}}{a-|\boldsymbol{x}|}\\ &= g_2^{(e)}(\boldsymbol{x}) + g_2^{(i)}(\boldsymbol{x})\boldsymbol{i} + g_2^{(j)}(\boldsymbol{x})\boldsymbol{j} + g_2^{(k)}(\boldsymbol{x})\boldsymbol{k},\\ g_2^{(\alpha)}(\boldsymbol{x}) &= \frac{x^{(\alpha)}}{a-|\boldsymbol{x}|}.\end{aligned}$$

The differential of $g_2^{(\beta)}(\boldsymbol{x})$ ($\beta = \{e,i,j,k\}$) with respect to $x^{(\alpha)}$ can be calculated as

$$\frac{\partial g_2^{(\beta)}}{\partial x^{(\alpha)}} = \begin{cases}\frac{x^{(\alpha)}x^{(\beta)}}{(a-|\boldsymbol{x}|)^2|\boldsymbol{x}|} & (\alpha \neq \beta)\\ \frac{a|\boldsymbol{x}|-\sum_{\gamma\neq\beta}x^{(\gamma)2}}{(a-|\boldsymbol{x}|)^2|\boldsymbol{x}|} & (\alpha = \beta)\end{cases}.$$

In this case, the difference of the energy (Eq. (20)) becomes

$$\Delta E = -\frac{1}{2}w_{rr}^{(e)}|\Delta \boldsymbol{x}_r|^2 - \sum_{\alpha}(\Delta x^{(\alpha)})^2\frac{a}{(a-|\phi|)^2} \tag{22}$$

where $-a < \phi^{(\alpha)} < a$. Under the condition of $a > 0$, this energy never increases when the state of the neuron changes.

4 An Example: 3 Neurons Network

We introduce a quaternion equivalent of Hebbian rule for embedding patterns to the network. The Hebbian rule is defined as

$$\boldsymbol{w}_{pq} = \frac{1}{4N}\sum_{\mu=1}^{n_p} \boldsymbol{\xi}_{\mu,p} \otimes \boldsymbol{\xi}^*_{\mu,q} \tag{23}$$

where $\boldsymbol{\xi}_{\mu,p} = (\xi^{(e)}_{\mu,p}, \xi^{(i)}_{\mu,p}, \xi^{(j)}_{\mu,p}, \xi^{(k)}_{\mu,p})$ ($\xi^{\{(e),(i),(j),(k)\}} \in \{1, -1\}$) represents the pattern vector for neuron p in the μ-th pattern and n_p is the number of patterns to be stored. The norm of $\boldsymbol{w}_{pq}$ is normalized by the number of neurons and $|\boldsymbol{\xi}|^2 = 4$. This form satisfies the conditions $\boldsymbol{w}_{pq} = \boldsymbol{w}^*_{qp}$ and $\boldsymbol{w}_{pp} \geq 0$:

$$\boldsymbol{w}^*_{qp} = \frac{1}{4N}\sum_{\mu=1}^{n_p} \left(\boldsymbol{\xi}_{\mu,q} \otimes \boldsymbol{\xi}^*_{\mu,p}\right)^* = \frac{1}{4N}\sum_{\mu=1}^{n_p} \boldsymbol{\xi}_{\mu,p} \otimes \boldsymbol{\xi}^*_{\mu,q} = \boldsymbol{w}_{pq},$$

and

$$\boldsymbol{w}_{pp} = \frac{1}{4N}\sum_{\mu=1}^{n_p} \boldsymbol{\xi}_{\mu,p} \otimes \boldsymbol{\xi}^*_{\mu,p} = \frac{1}{4N}\sum_{\mu=1}^{n_p} |\boldsymbol{\xi}_{\mu,p}|^2 = \frac{n_p}{N} > 0.$$

We consider the network that consists of three neurons and one pattern ($n_p = 1$) is to be stored in. The pattern to be stored, $\boldsymbol{\xi}_1 = \{\boldsymbol{\xi}_{1,1}, \boldsymbol{\xi}_{1,2}, \boldsymbol{\xi}_{1,3}\}$, is as follows:

$$\begin{aligned}
\boldsymbol{\xi}_{1,1} &= \xi^{(e)}_{1,1} + \xi^{(i)}_{1,1}\boldsymbol{i} + \xi^{(j)}_{1,1}\boldsymbol{j} + \xi^{(k)}_{1,1}\boldsymbol{k} \\
&= (\ \ 1, \ \ 1, -1, \ \ 1), \\
\boldsymbol{\xi}_{1,2} &= (-1, \ \ 1, -1, -1), \\
\boldsymbol{\xi}_{1,3} &= (\ \ 1, \ \ 1, \ \ 1, -1).
\end{aligned} \tag{24}$$

The connection weight matrix $\boldsymbol{W} = \{\boldsymbol{w}_{pq}\}$ obtained from Eq. (23) is

$$\boldsymbol{W} = \frac{1}{3}\begin{pmatrix} 1 & -\boldsymbol{i} & -\boldsymbol{j} \\ \boldsymbol{i} & 1 & -\boldsymbol{k} \\ \boldsymbol{j} & \boldsymbol{k} & 1 \end{pmatrix}. \tag{25}$$

4.1 In the Case of f_1

We start with the case that $\boldsymbol{f}_1$ is applied for the activation function. From the definition of $\boldsymbol{f}_1$, this function corresponds to the sign function when $\epsilon \to 0$. Then, we examined the stable states of the network.

Table 1 shows a list of fixed points in the network with the weight matrix of Eq. (25). The pattern with $\mu = 1$, i.e. $\boldsymbol{\xi}_1 = \{\boldsymbol{\xi}_{1,1}, \boldsymbol{\xi}_{1,2}, \boldsymbol{\xi}_{1,3}\}$, in this table corresponds to the stored pattern itself (Eq. (24)). The patterns $\boldsymbol{\xi}_\mu$ with $\mu = 9, \cdots, 16$ are respectively the same as quaternionic component-wise inverted ones

Table 1. The degenerate fixed points in the network with the same $\boldsymbol{W}$ of Eq. (25)

μ	$\boldsymbol{\xi}_{\mu,1}$	$\boldsymbol{\xi}_{\mu,2}$	$\boldsymbol{\xi}_{\mu,3}$
1	$(1, 1, -1, 1)$	$(-1, 1, -1, -1)$	$(1, 1, 1, -1)$
2	$(-1, 1, 1, -1)$	$(-1, -1, 1, 1)$	$(-1, -1, -1, -1)$
3	$(-1, 1, -1, -1)$	$(-1, -1, 1, -1)$	$(1, -1, -1, -1)$
4	$(-1, 1, 1, 1)$	$(-1, -1, -1, 1)$	$(-1, 1, -1, -1)$
5	$(-1, 1, -1, 1)$	$(-1, -1, -1, -1)$	$(1, 1, -1, -1)$
6	$(1, 1, 1, -1)$	$(-1, 1, 1, 1)$	$(-1, -1, 1, -1)$
7	$(1, 1, -1, -1)$	$(-1, 1, 1, -1)$	$(1, -1, 1, -1)$
8	$(1, 1, 1, 1)$	$(-1, 1, -1, 1)$	$(-1, 1, 1, -1)$
9	$(-1, -1, 1, -1)$	$(1, -1, 1, 1)$	$(-1, -1, -1, 1)$
10	$(-1, -1, -1, -1)$	$(1, -1, 1, -1)$	$(1, -1, -1, 1)$
11	$(-1, -1, 1, 1)$	$(1, -1, -1, 1)$	$(-1, 1, -1, 1)$
12	$(-1, -1, -1, 1)$	$(1, -1, -1, -1)$	$(1, 1, -1, 1)$
13	$(1, -1, 1, -1)$	$(1, 1, 1, 1)$	$(-1, -1, 1, 1)$
14	$(1, -1, -1, -1)$	$(1, 1, 1, -1)$	$(1, -1, 1, 1)$
15	$(1, -1, 1, 1)$	$(1, 1, -1, 1)$	$(-1, 1, 1, 1)$
16	$(1, -1, -1, 1)$	$(1, 1, -1, -1)$	$(1, 1, 1, 1)$

with $\mu = 1, \cdots, 8$. Energy with respect to these fixed points is all -6. The same weight matrix as Eq. (25) can also be obtained from any pattern in this table, so they seem to form a quaternionic degenerated state.

These degenerated states can be explained by introducing a quaternionic parameter $\boldsymbol{a}_\mu$ of which norm is 1, i.e., $|\boldsymbol{a}_\mu| = 1$. Let $\boldsymbol{\xi}_\mu = \{\boldsymbol{\xi}_{\mu,1}, \cdots, \boldsymbol{\xi}_{\mu,N}\}$ be the μ-th stored pattern and $\boldsymbol{w}_{pq}$ be a connection weight between the p-th neuron and the q-th neuron calculated from the pattern $\boldsymbol{\xi}_\mu$. From the pattern $\boldsymbol{\xi}'_\mu = \{\boldsymbol{\xi}_{\mu,1} \otimes \boldsymbol{a}_\mu, \cdots, \boldsymbol{\xi}_{\mu,N} \otimes \boldsymbol{a}_\mu\}$, the connection weight $\boldsymbol{w}'_{pq}$ can be calculated by Hebbian rule (Eq. (23)) as follows:

$$\boldsymbol{w}'_{pq} = \frac{1}{4N} \sum_{\mu=1}^{n_p} \boldsymbol{\xi}'_{\mu,p} \otimes \boldsymbol{\xi}'^{*}_{\mu,q} = \frac{1}{4N} \sum_{\mu=1}^{n_p} \boldsymbol{\xi}_{\mu,p} \otimes \boldsymbol{a}_\mu \otimes \boldsymbol{a}^*_\mu \otimes \boldsymbol{\xi}^*_{\mu,q} = \boldsymbol{w}_{pq}.$$

Note that $\boldsymbol{a}_\mu \otimes \boldsymbol{a}^*_\mu = |\boldsymbol{a}_\mu|^2 = 1$. This relation shows that the patterns $\boldsymbol{\xi}_\mu \otimes \boldsymbol{a}_\mu$ are also stored in the network when we embed the pattern $\boldsymbol{\xi}_\mu$ into the network. The parameter $\boldsymbol{a}_\mu$ can be any quaternion in the condition of $|\boldsymbol{a}_\mu| = 1$, but the quaternionic component of a stored pattern takes either $+1$ or -1 in this case, hence the number of components in a set $\{\boldsymbol{a}_\mu\}$ is at most 16. This number reflects the number of the degenerated patterns in the network when one stored pattern is embedded. Henceforth we call a set of these patterns '*Multiplet*'. In this network $\{\boldsymbol{\xi}_1, \cdots, \boldsymbol{\xi}_{16}\}$ forms a multiplet. A component in a multiplet, $\boldsymbol{\xi}_\mu$, can be transformed into another component in the same multiplet, $\boldsymbol{\xi}_\nu$ ($\nu \neq \mu$), by multiplying a certain $\boldsymbol{a}_\mu$ ($\boldsymbol{a}_\mu$-transformation). For example, $\boldsymbol{\xi}_5$ can be obtained from $\boldsymbol{\xi}_1$ by $\boldsymbol{a}_\mu$-transformation with $\boldsymbol{a}_\mu = \frac{1}{2}(1, 1, -1, 1)$.

Even if the parameter ϵ takes larger value, the state of the network converges to one of 16 stable points. Each quaternionic element of resultant network states takes bipolar values, but their intensities are lower than ones in the case of $\epsilon \to 0$. Hence, a set of the component in the multiplet can be represented by using real-valued parameter $c > 0$ as $\{c\boldsymbol{\xi}_1, c\boldsymbol{\xi}_2, \cdots, c\boldsymbol{\xi}_{16}\}$. This is due to the steepness of the activation function for updating the state of neurons.

4.2 In the Case of $\boldsymbol{f}_2$

We also examine the stable points of the network in the case that $\boldsymbol{f}_2$ is applied for the activation function. We set $a = 3$ in Eq. (13), so that the stored pattern is the stable point in the network, i.e., $\boldsymbol{x}_p = \boldsymbol{f}_2(\boldsymbol{s}_p) = \boldsymbol{f}_2(\boldsymbol{\xi}_1) = \boldsymbol{\xi}_1$.

This network has the degenerated states same as $\boldsymbol{\xi}_n$ shown in Table 1. These can be explained by the same reason in the case of $\boldsymbol{f}_1$. This network also converges to the states other than $\boldsymbol{\xi}_n$. We show two stable points as examples:

$$\begin{Bmatrix} \boldsymbol{x}_1 \\ \boldsymbol{x}_2 \\ \boldsymbol{x}_3 \end{Bmatrix} = \begin{Bmatrix} (-0.58, 0.58, 0.58, -1.73) \\ (-0.58, -0.58, 1.73, 0.58) \\ (-0.58, -1.73, -0.58, -0.58) \end{Bmatrix}, \begin{Bmatrix} (-1.13, 1.13, -0.38, -1.13) \\ (-1.13, -1.13, 1.13, -0.38) \\ (0.38, -1.13, -1.13, -1.13) \end{Bmatrix}.$$

These patterns are also in the Multiplet because the stored pattern ($\boldsymbol{\xi}_1$) can be obtained by $\boldsymbol{a}_\mu$-transformation from them, and hence more stable patterns constitute the components of the Multiplet than those in the case of $\boldsymbol{f}_1$.

5 Conclusion

In this paper, we constructed quaternionic hopfield networks where the state of the neurons takes continuous value and discrete-time is used for update of the neuron states. Two types of the activation function are introduced, and it has been proven that the energies of the networks for both activation functions always decrease with respect to the state changes of neurons. We examined the properties of the networks through the networks with 3 neurons and one pattern embedded. We have shown that our proposed network has the degenerated states as the stable points called Multiplet and that the type of the activation function affects the members in the Multiplet.

The properties of the network should be explored for larger scale of the networks, i.e., the number of neurons in the network and the number of embedded patterns become larger. This remains for our future work.

Acknowledgements

This study was financially supported by Japan Society for the Promotion of Science (Grant-in-Aid for Young Scientists (B) 19700221).

References

1. Hankins, T.L.: Sir William Rowan Hamilton. Johns Hopkins University Press, Baltimore, London (1980)

2. Mukundan, R.: Quaternions: From Classical Mechanics to Computer Graphics, and Beyond. In: Proceedings of the 7th Asian Technology Conference in Mathematics, pp. 97–105 (2002)
3. Nitta, T.: A Solution to the 4-bit Parity Problem with a Single Quaternary Neuron. Neural Information Processing - Letters and Reviews 5(2), 33–39 (2004)
4. Nitta, T.: An Extension of the Back-propagation Algorithm to Quaternions. In: ICONIP'96. Proceedings of International Conference on Neural Information Processing, vol. 1, pp. 247–250 (1996)
5. Arena, P., Fortuna, L., Muscato, G., Xibilia, M.: Multilayer Perceptrons to Approximate Quaternion Valued Functions. Neural Networks 10(2), 335–342 (1997)
6. Matsui, N., Isokawa, T., Kusamichi, H., Peper, F., Nishimura, H.: Quaternion Neural Network with Geometrical Operators. Journal of Intelligent & Fuzzy Systems 15(3–4), 149–164 (2004)
7. Yoshida, M., Kuroe, Y., Mori, T.: Models of hopfield-type quaternion neural networks and their energy functions. International Journal of Neural Systems 15(1–2), 129–135 (2005)
8. Isokawa, T., Nishimura, H., Kamiura, N., Matsui, N.: Fundamental Properties of Quaternionic Hopfield Neural Network. In: Proceedings of 2006 International Joint Conference on Neural Networks, pp. 610–615 (2006)
9. Georgiou, G.M., Koutsougeras, C.: Complex domain backpropagation. IEEE Transactions on Circuits and Systems II 39(5), 330–334 (1992)

Neural Learning Algorithms Based on Mappings: The Case of the Unitary Group of Matrices

Simone Fiori

Dipartimento di Elettronica, Intelligenza Artificiale e Telecomunicazioni (DEIT), Università Politecnica delle Marche, Via Brecce Bianche, I-60131 Ancona (Italy)
fiori@deit.univpm.it

Abstract. Neural learning algorithms based on optimization on manifolds differ by the way the single learning steps are effected on the neural system's parameter space. In this paper, we present a class counting four neural learning algorithms based on the differential geometric concept of mappings from the tangent space of a manifold to the manifold itself. A learning stepsize adaptation theory is proposed as well under the hypothesis of additiveness of the learning criterion. The numerical performances of the discussed algorithms are illustrated on a learning task and are compared to a reference algorithm known from literature.

1 Introduction

A recent discovery in neural learning theory is that neural learning dynamics may be formulated in terms of differential equations on manifolds [1,4,5,7]. A differential manifold is seen as a continuous curved parameter space, which inherently takes into account the natural constraints that learning the solution of a given problem may be subjected to.

A way to numerically integrate differential equations on manifolds for neural learning is by piece-wise geodesic arcs. Geodesics are defined up to metrization of the base-manifold. However, it has been suggested that other maps may be taken advantage of, while it is not clear whether such maps configure geodesic arcs (upon any appropriate metric). This observation suggests that there exists a more general concept that encompasses a wider class of maps than the class of geodesics.

In the present manuscript, we consider the unitary group $U(p)$ as parameter space and suggest four different kinds of mappings from the tangent space of the group to the group itself. It is worth mentioning that the set $U(p)$ is a classical Lie group, it thus enjoys the properties of an algebraic group as well as the properties of a differential manifold.

Under the hypothesis that network learning criterion is additive, namely, it may be broken down into the sum of terms each of which depends only on one of neural network's output signals, we are able to develop a learning stepsize adaptation rule, as well.

The numerical performances of the discussed algorithms are illustrated on a complex independent component learning task and are compared to those exhibited by the complex-FastICA algorithm known from literature.

J. Marques de Sá et al. (Eds.): ICANN 2007, Part I, LNCS 4668, pp. 858–863, 2007.

2 Mappings-Based Learning Algorithms

The derivation of learning algorithms in the present paper is based on notions from differential geometry [6]. In particular, on the notion of mapping $B : TM \to M$ from the tangent bundle TM of a manifold M to the manifold itself. Mappings come into effect in neural learning whenever an adapting procedure is formulated in terms of optimization of a network's performance criterion under constraints, where the manifold structure accounts for the constraints.

2.1 Mappings-Based Numerical Solution of Gradient-Based Learning Equations

Let us consider an optimization problem that consists in finding a local extremum of a non-linear regular function $f : M \to \mathcal{R}$. The solution $\mathbf{W} = \mathbf{W}(t)$ of the steepest gradient-ascent differential equation:

$$\frac{d}{dt}\mathbf{W}(t) = \nabla_{\mathbf{W}(t)} f \ , \ \text{with } \mathbf{W}(0) = \mathbf{W}_0 \in M \ , \tag{1}$$

will asymptotically tend to one of the local extremum of the function $f(\cdot)$ over manifold M, depending on the boundary value $\mathbf{W}_0$. In the above expression, symbol $\nabla_{\mathbf{W}} f \in T_{\mathbf{W}} M$ denotes the Riemannian gradient of the criterion $f(\cdot)$. If the unitary group is endowed with the canonical bi-invariant metric, then the Riemannian gradient writes:

$$\nabla_{\mathbf{W}} f = \frac{\partial f}{\partial \mathbf{W}} - \mathbf{W} \left(\frac{\partial f}{\partial \mathbf{W}} \right)^H \mathbf{W} \ , \tag{2}$$

with $\frac{\partial f}{\partial \mathbf{W}} \stackrel{\text{def}}{=} \frac{\partial f}{\partial \Re \mathbf{W}} + j \frac{\partial f}{\partial \Im \mathbf{W}}$. Here, symbols $\Re(\cdot)$ and $\Im(\cdot)$ denote real and imaginary part, respectively, symbol j denotes imaginary unit and superscript ^{H} denotes Hermitian (conjugate) transpose. The optimization-type differential-equation (1) may not be solved in closed form and needs to be turned into a discrete-time learning algorithm via a suitable time-discretization scheme. This operation may be effected by the help of mappings as:

$$\mathbf{W}_{n+1} = B_{\mathbf{W}_n}(h_n \nabla_{\mathbf{W}_n} f) \ , \ n \geq 0 \ , \tag{3}$$

where sequence $h_n \in \mathcal{R}$ denotes a suitable learning step-size schedule.

In the present paper, we take as a special example the case of learning by optimization over the Lie group of unitary matrices, namely $U(p) \stackrel{\text{def}}{=} \{\mathbf{W} \in \mathbb{C}^{p \times p} | \mathbf{W}^H \mathbf{W} = \mathbf{I}_p\}$. The Lie algebra associated to the unitary group is denoted as $\mathbf{u}(p) \stackrel{\text{def}}{=} \{\mathbf{S} \in \mathbb{C}^{p \times p} | \mathbf{S} + \mathbf{S}^H = \mathbf{0}_p\}$. It holds $T_{\mathbf{W}} U(p) = \mathbf{W} \cdot \mathbf{u}(p)$ for every $\mathbf{W} \in U(p)$.

For the unitary group of matrices, we consider the following mappings:

Mapping I: The exponential mapping:

$$\exp_{\mathbf{W}}(\mathbf{WS}) = \mathbf{W}\exp(\mathbf{S}) \ . \tag{4}$$

In the above expressions, symbol $\exp(\cdot)$ denotes matrix exponentiation.

Mapping II: The Cayley-transform-based mapping:

$$\mathrm{cay}_{\mathbf{W}}(\mathbf{WS}) \stackrel{\text{def}}{=} \mathbf{W}\,(2\mathbf{I}_p + \mathbf{S})\,(2\mathbf{I}_p - \mathbf{S})^{-1} \ . \tag{5}$$

Mapping III: The polar-decomposition-based mapping:

$$\mathrm{pol}_{\mathbf{W}}(\mathbf{WS}) \stackrel{\text{def}}{=} \mathbf{W}(\mathbf{I}_p + \mathbf{S})\sqrt{(\mathbf{I}_p - \mathbf{S}^2)^{-1}} \ . \tag{6}$$

Mapping IV: The QR-decomposition-based mapping:

$$\mathrm{qrm}_{\mathbf{W}}(\mathbf{WS}) \stackrel{\text{def}}{=} \mathrm{q}(\mathbf{W} + \mathbf{WS}) \ , \tag{7}$$

where $\mathrm{q}(\cdot)$ denotes the 'Q' factor of the QR decomposition of the matrix-argument.

2.2 Learning Stepsize Adaptation for Additive Learning Criteria

We consider the case of training a linear feed-forward neural network described by $\mathbf{y} = \mathbf{W}^H\mathbf{x}$, where $\mathbf{y} \in \mathbb{C}^p$ denotes network response, $\mathbf{x} \in \mathbb{C}^p$ denotes network input and $\mathbf{W} \in \mathbb{C}^{p\times p}$ denotes the network connection pattern.

We make the technical hypothesis that the learning criterion f has the additive structure:

$$f_n \stackrel{\text{def}}{=} f(\mathbf{W}_n) = \sum_{i=1}^{p} \mathbb{E}[G(|y_{i,n}|^2)] = \mathbf{1}_p^T \mathbb{E}[G(|\mathbf{y}_n|^2)] \ , \tag{8}$$

with $G : \mathcal{R}^+ \to \mathcal{R}$ being a differentiable sample-wise learning performance index. Modulus operator $|\cdot|$ as well as function $G(\cdot)$ are supposed to act component-wise. Also, operator $\mathbb{E}[\cdot]$ denotes statistical expectation with respect to the statistics of signal $\mathbf{x}$.

From equation (3), it is readily seen that:

$$\mathbf{y}_{n+1} = \mathbf{W}_{n+1}^H \mathbf{x} = B(h_n \mathbf{S}_n)^H \mathbf{y}_n \ , \tag{9}$$

where symbol $B(\cdot)$ denotes mapping from the identity of the group $U(p)$. The following approximations hold:

$$B(\mathbf{S}) \approx \mathbf{I}_p + \mathbf{S} + \frac{1}{2}\mathbf{S}^2 \ , \ \text{for small-norm } \mathbf{S} \in \mathbf{u}(p) \ , \tag{10}$$

$$G(|\mathbf{y}_{n+1}|^2) \approx G(|\mathbf{y}_n|^2) + 2\Re[(G'(|\mathbf{y}_n|^2) \circ \mathbf{y}_n)^H (\mathbf{y}_{n+1} - \mathbf{y}_n)] \ , \tag{11}$$

where symbol $\circ$ denotes component-wise vector multiplication. Thanks to the above approximations, it is easily seen that:

$$f_{n+1} - f_n \approx -2h_n \Re(\mathbb{E}[(G'(|\mathbf{y}_n|^2) \circ \mathbf{y}_n)^H \mathbf{S}_n \mathbf{y}_n]) + h_n^2 \Re(\mathbb{E}[(G'(|\mathbf{y}_n|^2) \circ \mathbf{y}_n)^H \mathbf{S}_n^2 \mathbf{y}_n]) \ . \tag{12}$$

At optimality it should hold $f_{n+1} - f_n \approx 0$, thus the optimal learning stepsize may be taken as:

$$h_n = \frac{2\Re(\mathbb{E}[(G'(|\mathbf{y}_n|^2) \circ \mathbf{y}_n)^H \mathbf{S}_n \mathbf{y}_n])}{\Re(\mathbb{E}[(G'(|\mathbf{y}_n|^2) \circ \mathbf{y}_n)^H \mathbf{S}_n^2 \mathbf{y}_n])} \ . \tag{13}$$

2.3 Plain Expressions

We deem it appropriate to summarize here the presented algorithm and to provide plain expressions for the quantities of interest.

When the learning criterion function is of the type (8), the Euclidean gradient has expression:

$$\frac{\partial f}{\partial \mathbf{W}} = 2\mathbb{E}[\mathbf{x}(G'(|\mathbf{y}_n|^2) \circ \mathbf{y}_n)^H] , \tag{14}$$

therefore, the quantity $\mathbf{S}_n$ to be inserted in the learning algorithm:

$$\mathbf{W}_{n+1} = B_{\mathbf{W}_n}(h_n \mathbf{W}_n \mathbf{S}_n) , \tag{15}$$

in the present case, is given by:

$$\mathbf{S}_n = \mathbf{W}_n^H(\nabla_{\mathbf{W}_n} f) = 2\mathbb{E}[\mathbf{y}_n(G'(|\mathbf{y}_n|^2) \circ \mathbf{y}_n)^H] - 2\mathbb{E}[(G'(|\mathbf{y}_n|^2) \circ \mathbf{y}_n)\mathbf{y}_n^H] . \tag{16}$$

It is worth noting that, if we set $\mathbf{T}_n \stackrel{\text{def}}{=} \mathbb{E}[\mathbf{y}_n(G'(|\mathbf{y}_n|^2) \circ \mathbf{y}_n)^H]$, then we have the following simpler expressions:

$$\mathbf{S}_n = 2(\mathbf{T}_n - \mathbf{T}_n^H) , \; h_n = \frac{2\Re\,\mathrm{tr}(\mathbf{T}_n\mathbf{S}_n)}{\Re\,\mathrm{tr}(\mathbf{T}_n\mathbf{S}_n^2)} , \tag{17}$$

where symbol $\mathrm{tr}(\cdot)$ denotes the trace of the matrix-argument.

3 Numerical Experiment

As a numerical experiment, we suppose the network input stream to have structure $\mathbf{x} = \mathbf{M}\mathbf{s}$, where $\mathbf{s} \in \mathbb{C}^p$ is formed by statistically independent proper complex-valued signals and $\mathbf{M} \in \mathbb{C}^{p \times p}$ is full-rank. The aim of the neural network here is to separate out the independent signals from their mixtures [3].

It is known that every noiseless full-rank mixture may be reduced to an unitary mixture by whitening the observations, therefore we may restrict our analysis to unitary mixtures without loss of generality. Namely, we shall assume $\mathbf{M} \in U(p)$. In this instance, the separating network may be described by a connection pattern $\mathbf{W} \in U(p)$. A network learning criterion is [2] like (8) with $G(|y|^2) = \log(0.1 + |y|^2)$, $y \in \mathbb{C}$.

The simplest choice for the initial network connection pattern is $\mathbf{W}_0 = \mathbf{I}_p$. The presented experiments were carried out over mixtures of telecommunication-type complex-valued signals (namely, three 4-QAM, three 16-QAM, three 8-PSK and a Gaussian noise). The mixing matrix $\mathbf{M}$ was randomly generated, each part of any entries distribution being uniform. As component extraction performance index, the average inter-channel interference (ICI) was defined as:

$$\mathrm{ICI} \stackrel{\text{def}}{=} \frac{1}{p}\frac{\sum_{ij} P_{ij}^2 - \sum_i \max_k\{P_{ik}^2\}}{\sum_i \max_k\{P_{ik}^2\}} , \tag{18}$$

where the separation product $\mathbf{P}$ is defined by $\mathbf{P}\mathbf{s} \stackrel{\text{def}}{=} \mathbf{y}$. As computational complexity indices, we measured the total flops per learning iteration and the total time required by the algorithms to run on a 1.86 GHz – 512 MB platform. As an index we used to evaluate the numerical behavior of the learning algorithms, an unitarity measure was defined in order to check for the adherence of the network connection matrix to the unitary group, namely $\delta(\mathbf{W}) \stackrel{\text{def}}{=} \|\mathbf{W}^T\mathbf{W} - \mathbf{I}_p\|_{\text{F}}$, where $\|\cdot\|_{\text{F}}$ denotes matrix Frobenius norm.

For further comparison purpose, the numerical behavior of the algorithms based on Mapping I, Mapping II, Mapping III and Mapping IV was compared to the numerical behavior of the complex-FastICA algorithm [2] endowed with the same non-linearity and symmetric orthogonalization.

The computational burden of the five considered algorithms, in terms of flops per iteration and run-time count, is illustrated in Figure 1, which also offers a comparative view of the separation performances of the five algorithms. As it can be readily noted, the mapping-based algorithms behaved satisfactorily: In particular, they nearly behave in the same way, as only slight differences may be noted in performances. The algorithm based on Mapping III shows better trade-off among separation performances, numerical precision and computational burden than the other considered algorithms.

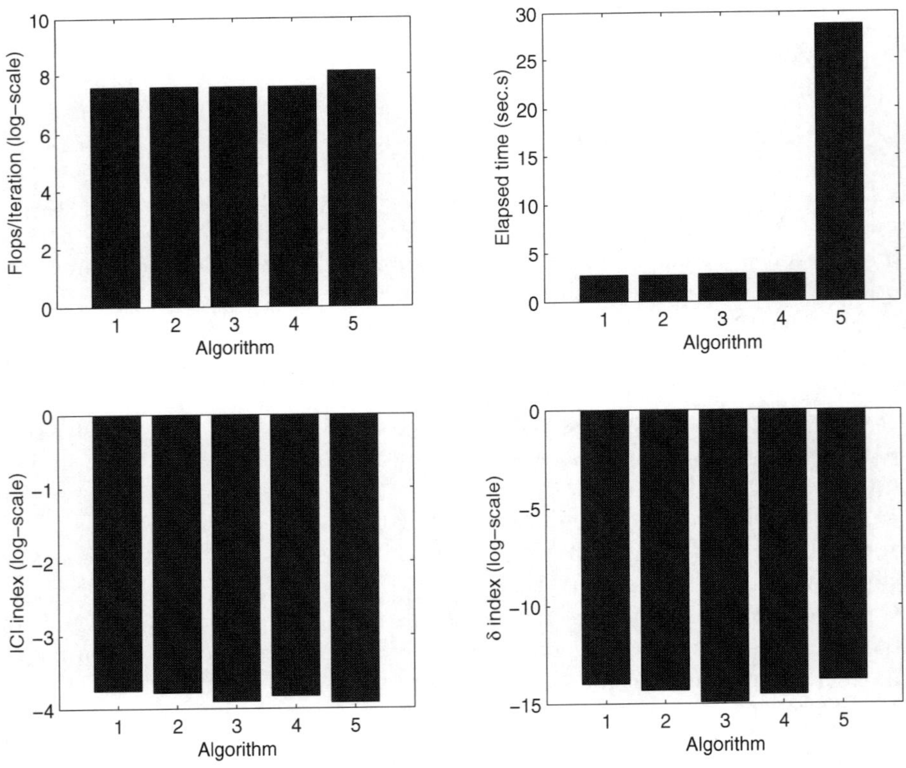

Fig. 1. Values of flops per iteration and time-count as well as of the ICI and δ indices values (both in $\log_{10}$-scale) after learning completion. (Algorithms: 1 – Mapping I, 2 – Mapping II, 3 – Mapping III, 4 – Mapping IV and 5 – complex-FastICA).

4 Conclusion

The aim of the present paper was to illustrate the theory of learning by mappings in the special case of function optimization on the manifold of unitary matrices. Results of a numerical experiment showed that mapping-based algorithms behave satisfactorily and may exhibit better trade-off between learning performances and computational burden over second-order (namely, Newton-like) algorithm.

References

1. Amari, S.-I., Fiori, S.: Editorial: Special issue on Geometrical Methods in Neural Networks and Learning. Neurocomputing 67, 1–7 (2005)
2. Bingham, E., Hyvärinen, A.: A fast fixed-point algorithm for independent component analysis of complex valued signals. International Journal of Neural Systems 10(1), 1–8 (2000)
3. Cichocki, A., Amari, S.-I.: Adaptive Blind Signal and Image Processing. J. Wiley & Sons, England (2002)
4. Fiori, S.: A theory for learning based on rigid bodies dynamics. IEEE Trans. on Neural Networks 13(3), 521–531 (2002)
5. Fiori, S.: Quasi-geodesic neural learning algorithms over the orthogonal group: A tutorial. Journal of Machine Learning Research 6, 743–781 (2005)
6. Gallot, S., Hulin, D., Lafontaine, J.: Riemannian Geometry, 3rd edn. Springer, Heidelberg (2004)
7. Tanaka, T., Fiori, S.: Simultaneous tracking of the best basis in reduced-rank Wiener filter. In: IEEE-ICASSP. International Conference on Acoustics, Speech and Signal Processing, vol. III, pp. 548–551 (May 2006)

Optimal Learning Rates for Clifford Neurons

Sven Buchholz[1], Kanta Tachibana[2], and Eckhard M.S. Hitzer[3]

[1] Institute of Computer Science, University of Kiel, Germany
[2] Graduate School of Engineering, Nagoya University, Japan
[3] Department of Applied Physics, University of Fukui, Japan

Abstract. Neural computation in Clifford algebras, which include familiar complex numbers and quaternions as special cases, has recently become an active research field. As always, neurons are the atoms of computation. The paper provides a general notion for the Hessian matrix of Clifford neurons of an arbitrary algebra. This new result on the dynamics of Clifford neurons then allows the computation of optimal learning rates. A thorough discussion of error surfaces together with simulation results for different neurons is also provided. The presented contents should give rise to very efficient second–order training methods for Clifford Multi-layer perceptrons in the future.

1 Introduction

Neural computation in Clifford algebras, which include familiar complex numbers and quaternions, has recently become an active research field [6,5,11]. Particularly, Clifford-valued Multi-layer perceptrons (CMLPs) have been studied. To our knowledge, no second–order methods to train CMLPs are available yet. This paper aims to provide therefore the needed facts on Hessians and optimal learning rates by studying the dynamics on the level of Clifford neurons. After an introduction to Clifford algebra, we review Clifford neurons as introduced in [4]. Section 4 provides as a general and explicit notion for Hessian matrix of these neurons, which enables easy and accurate computation of optimal learning rates. This is demonstrated by simulation in section 5, after which the paper concludes with a summary.

2 Basics of Clifford Algebra

An algebra is a real linear space that is equipped with a bilinear product. Clifford algebras are of dimension $2^n (n \in \mathbb{N}_0)$ and are generated by so-called quadratic spaces, from which they inherit a metric structure. Every quadratic space has an orthonormal basis. Here we are interested in spaces $\mathbb{R}^{p+q}$ $(p, q \in \mathbb{N}_0)$ with canonical basis $\mathcal{O} := (e_1, ..., e_{p+q})$ that are endowed with a quadratic form Q, such that for all $i, j \in \{1, \ldots, p+q\}$

$$Q(e_i) = \begin{cases} 1, & i \leq p \\ -1, & p+1 \leq i \leq p+q, \end{cases} \tag{1}$$

$$Q(e_i + e_j) - Q(e_i) - Q(e_j) = 0\,, \tag{2}$$

J. Marques de Sá et al. (Eds.): ICANN 2007, Part I, LNCS 4668, pp. 864–873, 2007.

which, in turn, renders $\mathcal{O}$ to be orthonormal. These quadratic spaces, abbreviated as $\mathbb{R}^{p,q}$ hereafter, give now rise to the following definition.

Definition 1 (Clifford Algebra [12]). *The Clifford algebra $\mathcal{C}_{p,q}$ of $\mathbb{R}^{p,q}$ is the 2^{p+q} dimensional associative algebra with unity $1_{\mathbb{R}}$ given by*

(i) *$\mathcal{C}_{p,q}$ is generated by its distinct subspaces $\mathbb{R}$ and $\mathbb{R}^{p,q}$*
(ii) *$xx = Q(x)$ for any $x \in \mathbb{R}^{p,q}$.*

Note the use of juxtaposition for the so-called geometric product of $\mathcal{C}_{p,q}$. Since nothing new is generated in case of $p = q = 0$, $\mathcal{C}_{0,0}$ is isomorphic to the real numbers $\mathbb{R}$. The Clifford algebras $\mathcal{C}_{0,1}$ and $\mathcal{C}_{0,2}$ are isomorphic to complex numbers and quaternions, respectively.

The set of all geometric products of elements of $\mathcal{O}$

$$\mathcal{E} := \{e_{j_1} e_{j_2} \cdots e_{j_r},\ 1 \leq j_1 < \ldots < j_r \leq p+q\} \tag{3}$$

constitutes a basis of $\mathcal{C}_{p,q}$, which follows directly from condition (ii) in Definition 1. This basis can be ordered by applying the canonical order of the power set $\mathcal{P}$ of $\{1, \ldots, p+q\}$, i.e. by using the index set

$$\mathcal{I} := \{\{i_1, \ldots, i_s\} \in \mathcal{P} \mid 1 \leq i_1 < \ldots < i_s \leq p+q\} \tag{4}$$

and then define $e_I := e_{i_1} \ldots e_{i_s}$ for all $I \in \mathcal{I}$. Set $e_\emptyset := 1_{\mathbb{R}}$. A general element x of $\mathcal{C}_{p,q}$

$$x = \sum_I e_I\, x_I := \sum_{I \in \mathcal{I}} e_I\, x_I, \quad (x_I \in \mathbb{R}) \tag{5}$$

is termed multivector. The name stems from the fact that for $k \in \{0, \cdots, p+q\}$

$$\mathcal{C}^k_{p,q} := \{e_I \mid I \in \mathcal{I}, |I| = k\} \tag{6}$$

spans a linear subspace of $\mathcal{C}_{p,q}$ ($\mathcal{C}^0_{p,q} = \mathbb{R}$, $\mathcal{C}^1_{p,q} = \mathbb{R}^{p,q}$ and $\sum_k \mathcal{C}^k_{p,q} = \mathcal{C}_{p,q}$). Projection onto subspaces is defined via the grade operator

$$\langle\cdot\rangle_k : \mathcal{C}_{p,q} \to \mathcal{C}^k_{p,q}, x \mapsto \sum_{\substack{I \in \mathcal{I},\\ |I|=k}} x_I e_I\,, \tag{7}$$

with convention $\langle\cdot\rangle := \langle\cdot\rangle_0$ for the so-called scalar part. Projection onto the I-th component (w.r.t. $\mathcal{E}$) of a multivector, on the other hand, is written as $[\cdot]_I$.

Involutions on $\mathcal{C}_{p,q}$ are linear mappings that when applied twice yield the identity again. Examples are inversion $inv(x) := \sum_{k=0}^{p+q}(-1)^k \langle x\rangle_k$ and reversion $rev(x) := \sum_{k=0}^{p+q}(-1)^{\frac{k(k-1)}{2}}\langle x\rangle_k$. The principal involution defined by $\widetilde{e_I} = rev(e_I)\, e_0^{[I_q]}$ is of main importance for us, whereby $[I_q]$ stands for the number of negative signature vector factors in e_I. Note that $\widetilde{xy} = \widetilde{y}\widetilde{x}$. Any non-null vector of $\mathcal{C}_{p,q}$ has an inverse (w.r.t. the geometric product) and is an element of the Clifford group.

Definition 2 (Clifford group [10]). *The Clifford group associated with the Clifford algebra* $\mathcal{C}_{p,q}$ *is defined by*

$$\Gamma_{p,q} = \{s \in \mathcal{C}_{p,q} \mid \forall x \in \mathbb{R}^{p,q}, s\,x\,inv(s)^{-1} \in \mathbb{R}^{p,q}\}. \tag{8}$$

The action of $\Gamma_{p,q}$ on $\mathbb{R}^{p,q}$ is an orthogonal automorphism [12] that extends to multivectors due to the bilinearity of the geometric product. Since $\mathcal{C}_{0,1}$ is commutative, the fact that complex multiplication can be viewed geometrically as dilation–rotation [14] follows as special case from (8).

3 Clifford Neurons and the Linear Associator

Formally, a Clifford neuron can be easily introduced as follows. Consider the most common real neuron (of perceptron–type with identity as activation function) that computes a weighted sum of its inputs x_i according to

$$y = \sum_i w_i x_i + \theta\,. \tag{9}$$

Using instead multivector weights $w_i \in \mathcal{C}_{p,q}$, a multivector threshold $\theta \in \mathcal{C}_{p,q}$ and replacing real multiplication by the geometric product readily turns (9) into a Clifford neuron (for $\mathcal{C}_{0,0}$, of course, one gets the real neuron back again). Although complex-valued and quaternionic–valued Multi-layer perceptrons (MLPs) are known for a long time (see [9] and [2], respectively), not much research has been carried out on the neuron level itself in the beginning. Only recently, the following two types of Clifford neurons have been studied in greater detail in the literature.

Definition 3 (Basic Clifford Neuron (BCN) [4]). *A Basic Clifford Neuron,* $\mathrm{BCN}_{p,q}$*, computes the following function from* $\mathcal{C}_{p,q}$ *to* $\mathcal{C}_{p,q}$

$$y = wx + \theta. \tag{10}$$

A BCN has only one multivector input and therefore also only multivector weight. Contrary to a real neuron (9) it makes sense to study the one input only case, since this will be in general a multidimensional entity. The general update rule using gradient descent for a BCN with simplified, for notation purposes, error function

$$E = \|error\|^2 := \|d - y\|^2\,, \tag{11}$$

where d stands for desired output, reads

$$\Delta w = -\frac{\partial E}{\partial w} = 2 error\,\widetilde{x}\,. \tag{12}$$

Equation (12) generalizes the various specific update rules given in [4] by using the principal involution of Section 2. The second type of Clifford neuron proposed in [4] mimics the action of the Clifford group.

Definition 4 (Spinor Clifford Neuron (SCN) [4]). *For $p, q \in \mathbb{N}_0,\ p+q>1$ a Spinor Clifford Neuron,* $\mathrm{SCN}_{p,q}$, *computes the following function from $\mathcal{C}_{p,q}$ to $\mathcal{C}_{p,q}$*

$$y = w\,x\,\phi(w) + \theta\,, \tag{13}$$

where ϕ denotes a particular involution (e.g. inversion, reversion,...).

Like a BCN, a SCN has also only one independent weight though a two–sided multiplication takes place. For a SCN with (adapted) error function (11) gradient descent therefore results in a two-stage update procedure. First compute the update of the "right weight" in (13) as

$$w_R := \widetilde{(w\,x)}\,error \tag{14}$$

which then yields $\Delta w = \phi(w_R)$. Again, this is a powerful generalization of the few specific rules known from the literature.

BCNs are the atoms of Clifford MLPs while SCNs are the atoms of Spinor MLPs. Clifford MLPs have already found numerous applications (see e.g. [6,1] and the references therein). Recent applications of Spinor MLPs can be found in [11,5]. All these works, however, rely on a simple modified Back–propagation algorithm. To our knowledge, there are no known general second–order based training algorithms for Clifford MLPs and Spinor MLPs yet. Given the increasing interest in such architectures a high demand for second order methods can be stated. This paper aims to establish the foundations of such forthcoming algorithms by providing general and explicit expressions for describing the dynamics of Clifford neurons in terms of their Hessian matrices.

Hessians (of the error functions) of Clifford neurons are not a complete *terra incognita* though. Both BCNs and SCNs can be viewed, to some extent, as particular Linear Associators (LAs) (see e.g. [3,13] for an introduction to this most simple feed–forward neural network). In the case of BCNs this readily follows from the well–known theorem that every associative algebra of finite dimension is isomorphic to some matrix algebra. For SCNs, however, this additionally, relies on the fact (mentioned in Section 2) that the action of the Clifford group is always an orthogonal automorphism.

Dynamics of the Linear Associator are well understood and [8] is perhaps the first complete survey on the topic. Note that for a LA gradient descent, with an appropriate learning rate η, converges to the global minimum for any convex error function like the sum-of-squares error (SSE). The Hessian $\mathrm{H_{LA}}$ of a LA with error function of SSE-type (of which (11) is a simplified version) is always a symmetric matrix. For batch–learning (see e.g. the textbook [13]) the optimal learning rate η_{opt} is known to be $\frac{1}{\lambda_{max}}$ where λ_{max} is the largest eigenvalue of $\mathrm{H_{LA}}$. Also, $\mathrm{H_{LA}}$ is known to be the auto–correlation matrix of the inputs.

In principle it would be possible to compute Hessians of BCNs from the above facts. There are however several drawbacks that limit this procedure to only low–dimensional algebras and render it impossible for all other cases. First of all, note that in order to proceed in this way one has to know and apply the appropriate matrix isomorphism for every single algebra. There is no general solution and the isomorphisms can get rather complicated. Also, by applying an isomorphism numerical problems may occur for the eigenvalue computation. Even if one succeeds, no explicit structural insights are easily obtained from this procedure since everything is expressed in terms of a particular input. For SCNs the situation will be even more complicated, and, by missing the big picture again, no conclusions of truly general value are obtained.

The next section therefore now offers a new general and explicit solution for computing Hessian matrices of Clifford neurons without the above drawbacks. This is done by invoking the principal involution.

4 Hessian Matrices of Clifford Neurons

Hessian matrices w.r.t. the parameters of a neural architecture allow to compute optimal learning rates. Decoupling of parameters and using an individual optimal learning rate for every single one yields fastest possible convergence [7]. For the two types of Clifford neurons introduced in the previous section it is very natural to decouple the multivector weight w from the multivector threshold θ. The threshold θ of a BCN or SCN is a generic parameter in the sense, that it does not depend on the signature (p, q) of the underlying Clifford algebra. In fact, it can be handled in exactly the same way as the threshold vector of a LA of dimension $p + q$. In what follows we therefore only study Hessians w.r.t. the multivector weight w. To simplify notations, inputs to the neurons are written as x without indexing over the training set.

Theorem 1 (Hessian of a BCN). *The elements of the Hessian matrix of a* BCN *are given by*

$$\partial_{IJ}E = \partial_{JI}E = \frac{\partial E}{\partial w_J \partial w_I} = 2\langle \widetilde{e_I} e_J x \widetilde{x} \rangle \tag{15}$$

where $\widetilde{\cdot}$ *denotes principal involution.*

Proof. Rewriting the error function as

$$\begin{aligned} E = \|d - (wx + \theta)\|^2 &= \sum_I (d_I - ([wx]_I + \theta_I))^2 \underbrace{e_I \widetilde{e_I}}_{1} \\ &= \langle (d - wx - \theta)(d - wx - \theta)\widetilde{\ } \rangle \\ &= \left\langle wx\widetilde{(wx)} \right\rangle - \langle wx(d-\theta)\widetilde{\ } \rangle - \left\langle (d-\theta)\widetilde{(wx)} \right\rangle \\ &= \langle wx\widetilde{x}\widetilde{w} \rangle - 2\left\langle (d-\theta)\widetilde{x}\widetilde{w} \right\rangle \end{aligned} \tag{16}$$

yields

$$\begin{aligned} \frac{\partial E}{\partial w_I} &= \underbrace{\langle e_I x\widetilde{x}\widetilde{w} \rangle}_{\langle (x\widetilde{x}\widetilde{w})\widetilde{\ }\widetilde{e_I} \rangle} + \langle wx\widetilde{x}\widetilde{e_I} \rangle - 2\left\langle (d-\theta)\widetilde{x}\widetilde{e_I} \right\rangle \\ &= 2\left\langle wx\widetilde{x}\widetilde{e_I} \right\rangle - 2\left\langle (d-\theta)\widetilde{x}\widetilde{e_I} \right\rangle \\ &= -2\left\langle (d - wx - \theta)\widetilde{x}\widetilde{e_I} \right\rangle \end{aligned}$$

and finally

$$\frac{\partial E}{\partial w_J \partial w_I} = 2\left\langle e_J x \widetilde{x} \widetilde{e_I} \right\rangle$$
$$= 2\langle \widetilde{e_I} e_J x \widetilde{x} \rangle \,. \tag{17}$$

For every I, J, (17) "picks out" the scalar component of the product resulting from multiplying $\widetilde{e_I} e_J$ with the multivector $x\widetilde{x}$. Consequently, one gets easily the following result for the neurons $\mathrm{BCN}_{0,q}$.

Corollary 1. *The Hessian of* $\mathrm{BCN}_{0,q}$ *is a scalar matrix* aE*, where* E *denotes the identity matrix.*

Of course, every diagonal entry of the Hessian of a $\mathrm{BCN}_{p,q}$ equals the scalar a above. The next corollary gives a complete overview for BCNs of dimension four.

Corollary 2. *Let* $\{x^s = c_0^s e_0 + c_1^s e_1 + c_2^s e_2 + c_{12}^s e_{12}\}_{s=1\ldots S}$ *denote the input set to a* $\mathrm{BCN}_{p,q}$, $p+q=2$. *Set* $a := \sum_{s=1}^{S}(c_0^s c_0^s + c_1^s c_1^s + c_2^s c_2^s + c_{12}^s c_{12}^s)$, $b := 2\sum_{s=1}^{S}(c_0^s c_1^s + c_2^s c_{12}^s)$, $c := 2\sum_{s=1}^{S}(c_0^s c_2^s - c_1^s c_{12}^s)$, $d := 2\sum_{s=1}^{S}(c_0^s c_{12}^s - c_1^s c_2^s)$. *Then the Hessian w.r.t. the SSE-function of the* $\mathrm{BCN}_{0,2}$, *the* $\mathrm{BCN}_{1,1}$, *and the* $\mathrm{BCN}_{2,0}$ *is given by*

$$\mathbf{H}_{\mathrm{BCN}_{0,2}} = \begin{pmatrix} a & 0 & 0 & 0 \\ 0 & a & 0 & 0 \\ 0 & 0 & a & 0 \\ 0 & 0 & 0 & a \end{pmatrix}, \tag{18}$$

$$\mathbf{H}_{\mathrm{BCN}_{1,1}} = \begin{pmatrix} a & b & 0 & d \\ b & a & d & 0 \\ 0 & d & a & -b \\ d & 0 & -b & a \end{pmatrix}, \text{ and} \tag{19}$$

$$\mathbf{H}_{\mathrm{BCN}_{2,0}} = \begin{pmatrix} a & b & c & 0 \\ b & a & 0 & c \\ c & 0 & a & -b \\ 0 & c & -b & a \end{pmatrix}, \tag{20}$$

respectively.

Unfortunately, the Hessian of a SCN does have a more comlicated form and does not depend alone on the received input.

Theorem 2 (Hessian of a SCN). *The elements of the Hessian matrix of a* SCN *are given by*

$$\frac{\partial E}{\partial w_J \partial w_I} = 2\left\langle (e_J x \overline{w} + w x \overline{e_J})\,(e_I x \overline{w} + w x \overline{e_I})^{\widetilde{}} \right\rangle$$
$$-2\left\langle (d - w x \overline{w})\,(e_I x \overline{e_J} + e_J x \overline{e_I})^{\widetilde{}} \right\rangle \tag{21}$$

where $\overline{(\cdot)}$ *stands for the involution of the* SCN.

Proof. From

$$\begin{aligned} E = \|d - wx\overline{w}\|^2 &= \sum_I \left(d_I - (wx\overline{w})_I\right)^2 e_I\widetilde{e_I} \\ &= \langle (d - wx\overline{w})(d - wx\overline{w})^{\sim} \rangle \\ &= \left\langle d\widetilde{d} + wx\overline{w}(wx\overline{w})^{\sim} \right\rangle - 2\left\langle d(wx\overline{w})^{\sim} \right\rangle \end{aligned} \tag{22}$$

follows

$$\begin{aligned} \frac{\partial E}{\partial w_I} &= \langle (e_I x\overline{w} + wx\overline{e_I})(wx\overline{w})^{\sim} \rangle \\ &\quad + \langle (wx\overline{w})(e_I x\overline{w} + wx\overline{e_I})^{\sim} \rangle \\ &\quad - 2\langle d(e_I x\overline{w} + wx\overline{e_I})^{\sim} \rangle \\ &= -2\langle (d - wx\overline{w})(e_I x\overline{w} + wx\overline{e_I})^{\sim} \rangle \end{aligned} \tag{23}$$

and by computing $\partial_{w_J} \frac{\partial E}{\partial w_I}$ follows the assumption.

Hence, if w is of small absolute value the approximation $\mathbf{H}_{\mathrm{SCN_{p,q}}} \approx 2 \cdot \mathbf{H}_{\mathrm{BCN_{p,q}}}$ holds. Due to space limitations, this concludes or study of Hessians of Clifford neurons.

5 Simulations on Optimal Learning Rates

From Theorem 1 direct structural insights on the dynamics of BCNs can be obtained. According to Corollary 1, the error function of a $\mathrm{BCN}_{0,q}$ is always isotropic, its Hessian is a scalar matrix aE, and therefore $\eta = \frac{1}{a}$ is its optimal learning rate. This section now reports simulation results for all four–dimensional BCNs (theoretically covered by Corollary 2) in order to give some illustrations and to get further insights on the dynamics of these neurons. For the quaternionic $\mathrm{BCN}_{0,2}$ 100 points have been generated from a uniform distribution of $[0,1]^4$. Each element of the that way obtained input training set was left multiplied in $\mathcal{C}_{0,2}$ by $1e_0 + 2e_1 + 3e_2 + 4e_{12}$ in order to get the output training set. Using the optimal learning rate $\eta = \eta_{opt}$ (computed as $1/\lambda_{max}$ from (19) and

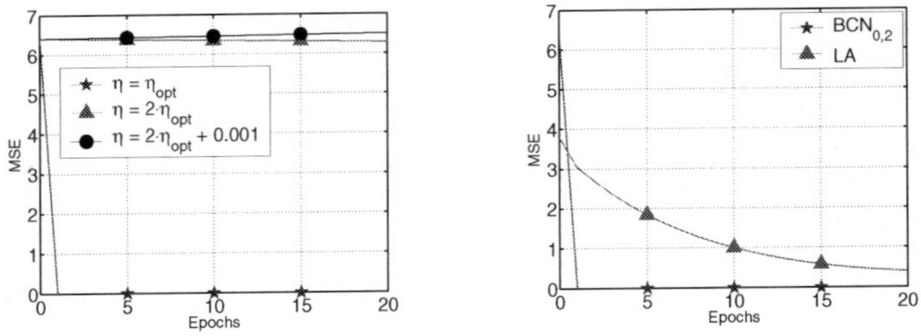

Fig. 1. Left panel shows learning curves of the $\mathrm{BCN}_{0,2}$ for different learning rates η. Right panel shows comparison of the optimal learning curves for the $\mathrm{BCN}_{2,0}$ and the LA. Both architectures are trained with their respective optimal learning rate.

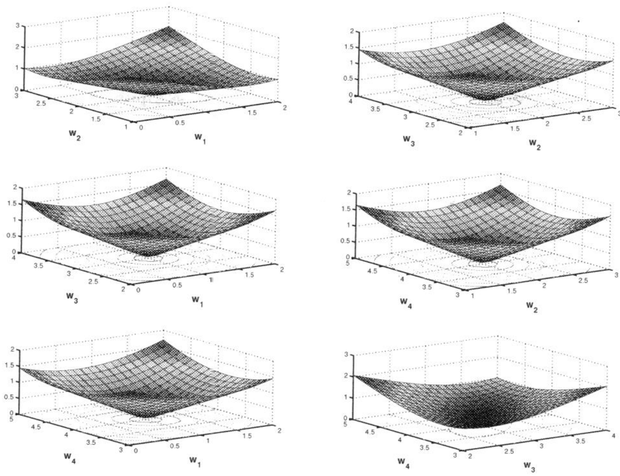

Fig. 2. Plot of error function for the $BCN_{1,1}$. Projections on the six 2D planes that coincide with the coordinate axes of weight space.

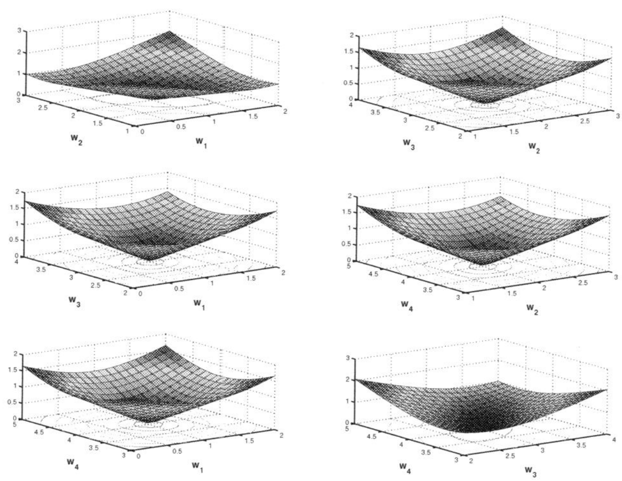

Fig. 3. Plot of error function for the $BCN_{2,0}$. Projections on the six 2D planes that coincide with the coordinate axes of weight space.

(20), respectively) batch-learning converged indeed in one epoch to the global solution. This can be seen from the left panel of figure 1, which also shows obtained learnig curves for $\eta = 2\eta_{opt}$ (slowest convergence) and $\eta = 2\eta_{opt} + 0.001$ (beginning of divergence). Right panel of figure 1 shows optimal learning curve of the BCN $_{0,2}$ versus optimal learning curve of the LA for illustration.

Similar simulations have been carried out for the remaining two BCNs of dimension four. In both cases input set of the $BCN_{0,2}$ has been used again. For

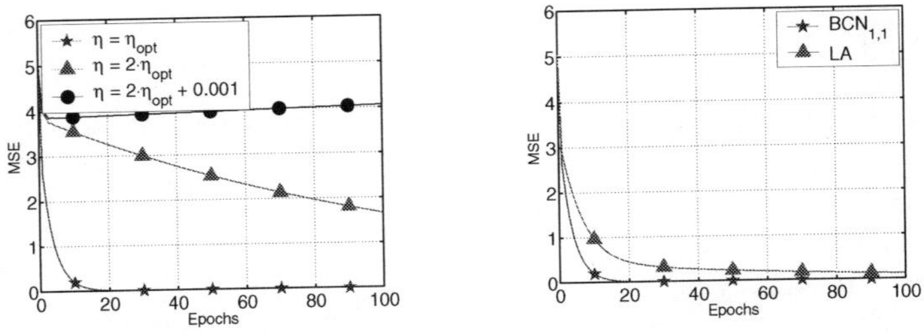

Fig. 4. Left panel shows learning curves of the $\mathrm{BCN}_{1,1}$ for different learning rates η. Right panel shows comparison of the optimal learning curves for the $\mathrm{BCN}_{2,0}$ and the LA. Both architectures are trained with their respective optimal learning rate.

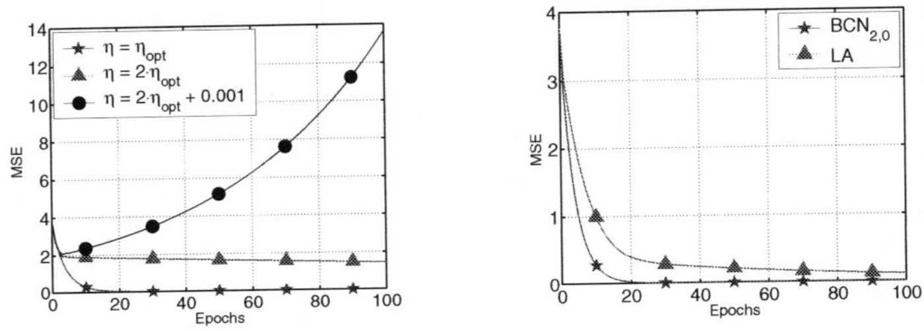

Fig. 5. Left panel shows learning curves of the $\mathrm{BCN}_{2,0}$ for different learning rates η. Right panel shows comparison of the optimal learning curves for the $\mathrm{BCN}_{2,0}$ and the LA. Both architectures are trained with their respective optimal learning rate.

obtaining the output set, it was left multilpied in $\mathcal{C}_{1,1}$ and $\mathcal{C}_{0,2}$ by $1e_0+2e_1+3e_2+4e_{12}$, respectively. Error surfaces obtained for these data are plotted in figure 2 and figure 3, respectively. Both surfaces do have a partial isotropy since isotropic projections do exist. Moreover, the following can be proven from (19),(20) for the projection planes $E_{w_I,w_J}(\mathrm{BCN}_{p,q})$: $E_{w_1,w_4}(\mathrm{BCN}_{1,1}) = E_{w_2,w_3}(\mathrm{BCN}_{1,1}) = E_{w_1,w_3}(\mathrm{BCN}_{2,0}) = E_{w_2,w_4}(\mathrm{BCN}_{2,0})$ and $E_{w_1,w_3}(\mathrm{BCN}_{1,1}) = E_{w_2,w_4}(\mathrm{BCN}_{1,1}) = E_{w_1,w_4}(\mathrm{BCN}_{2,0}) = E_{w_2,w_3}(\mathrm{BCN}_{2,0})$. Note how this nicely resembles the fact that $\mathcal{C}_{1,1}$ and $\mathcal{C}_{0,2}$ are isomorphic algebras. For both neurons a further decoupling of weights might be of interest due to the existence of isotropic planes. Finally, plots of learnig curves are provided by figure 4 and figure 5. Inclusion of an LA in all simulations was done to show that and how Clifford neurons do differ significantly from an LA.

6 Summary

The dynamics of Clifford neurons have been studied in this paper. General and explicit expressions of the Hessian of a BCN and of a SCN have been derived. From the general BCN result structural insights on the dynamics of specific BNCs can be easily obtained. The Hessian of a $\mathrm{BCN}_{0,q}$ has been identified to be a scalar matrix aE, which readily gives $1/a$ as optimal learning rate for such a neuron. Simulations have been carried out to show the stability and accuracy of computation of optimal learning rates from the derived Hessians. We hope that the presented material will give rise to fast second–order training methods for Clifford and Spinor MLPs in the future.

References

1. Arena, P., Fortuna, L., Muscato, G., Xibilia, M.G.: Neural Networks in Multidimensional Domains. LNCIS, vol. 234. Springer, Heidelberg (1998)
2. Arena, P., Fortuna, L., Occhipinti, L., Xibilia, M.G.: Neural networks for quaternion valued function approximation. In: IEEE Int. Symp. on Circuits and Systems, London, vol. 6, pp. 307–310. IEEE Computer Society Press, Los Alamitos (1994)
3. Baldi, P.F., Hornik, K.: Learning in linear neural networks: A survey. IEEE Transactions on Neural Networks 6(4), 837–858 (1995)
4. Buchholz, S.: A theory of neural computation with Clifford algebra. Technical Report Number 0504, Universität Kiel, Institut für Informatik (2005)
5. Buchholz, S., Le Bihan, N.: Optimal separation of polarized signals by quaternionic neural networks. In: EUSIPCO 2006. 14th European Signal Processing Conference, Florence, Italy (September 4-8, 2006)
6. Hirose, A.: Complex–Valued Neural Networks. Studies in Computational Intelligence, vol. 32. Springer, Heidelberg (2006)
7. LeCun, Y., Bottou, L., Orr, G.B., Müller, K.–R.: Efficient BackProp. In: Orr, G.B., Müller, K.-R. (eds.) Neural Networks: Tricks of the Trade. LNCS, vol. 1524, Springer, Heidelberg (1998)
8. LeCun, Y., Kanter, I., Solla, S.A.: Eigenvalues of Covariance Matrices: Application to Neural–Network Learning. Physical Review Letters 66(18), 2396–2399 (1991)
9. Leung, H., Haykin, S.: The Complex Backpropagation Algorithm. IEEE Transactions on Signal Processing 3(9), 2101–2104 (1991)
10. Lounesto, P.: Clifford Algebras and Spinors. Cambridge University Press, Cambridge (1997)
11. Matsui, N., Isokawa, T., Kusamichi, H., Peper, F., Nishimura, H.: Quaternion neural network with geometrical operators. Journal of Intelligent & Fuzzy Systems 15(3–4), 149–164 (2004)
12. Porteous, I.R.: Clifford Algebras and the Classical Groups. Cambridge University Press, Cambridge (1995)
13. Rojas, R.: Neural Networks. Springer, Heidelberg (1996)
14. Yaglom, I.M.: Complex Numbers in Geometry. Academic Press, New York (1968)

Solving Selected Classification Problems in Bioinformatics Using Multilayer Neural Network Based on Multi-Valued Neurons (MLMVN)

Igor Aizenberg[1] and Jacek M. Zurada[2]

[1] Texas A&M University-Texarkana; [2] University of Louisville
igor.aizenberg@tamut.edu (IA),
jacek.zurada@louisville.edu (jmz)

Abstract. A multilayer neural network based on multi-valued neurons (MLMVN) is a new powerful tool for solving classification, recognition and prediction problems. This network has a number of specific properties and advantages that follow from the nature of a multi-valued neuron (complex-valued weights and inputs/outputs lying on the unit circle). Its backpropagation learning algorithm is derivative-free. The learning process converges very quickly, and the learning rate for all neurons is self-adaptive. The functionality of the MLMVN is higher than the one of the traditional feedforward neural networks and a variety of kernel-based networks. Its higher flexibility and faster adaptation to the mapping implemented make it possible to solve complex classification problems using a simpler network. In this paper, we show that the MLMVN can be successfully used for solving two selected classification problems in bioinformatics.

1 Introduction

A multilayer neural network based on multi-valued neurons (MLMVN) has been introduced in [1] and then developed further in [2]. This network consists of multi-valued neurons (MVN). That is a neuron with complex-valued weights and an activation function, defined as a function of the argument of a weighted sum. It maps a whole complex plane onto the unit circle. This activation function was proposed in [3]. The discrete-valued (k-valued) multi-valued neuron was introduced in [4]. It implements such mappings between its n inputs and a single output that can be represented by the k-valued threshold functions. Thus, it is based on the principles of multiple-valued threshold logic over the field of complex numbers formulated in [5] and then developed in [6]. The discrete-valued MVN, its properties and learning are observed in detail in [6]. A continuous-valued MVN has been introduced in [1],[2].

Both discrete-valued and continuous-valued MVNs have a number of wonderful properties. The first important property is that MVN is a complex-valued neuron. Its weights are complex numbers and its inputs and outputs are lying on the unit circle. In the discrete-valued (k-valued) case, they are exactly k^{th} roots of unity. In the continuous-valued case, they are arbitrary numbers with a unitary magnitude that is

J. Marques de Sá et al. (Eds.): ICANN 2007, Part I, LNCS 4668, pp. 874–883, 2007.

they are lying on the unit circle. The MVN's activation function is a function of the argument of the weighted sum. It maps the whole complex plane into the unit circle. This makes the MVN's functionality higher than the one of the traditional neurons (a neuron with a sigmoid activation function, first of all). It is important that MVN learning is reduced to the movement along the unit circle. There is the efficient learning algorithm for MVN, which is based on a simple linear error correction rule. It is derivative-free.

Different applications of MVN have been considered during recent years. MVN is used as a basic neuron in the cellular neural networks [6], as the basic neuron of the associative memories [6],[7]-[10], as the basic neuron in a number of neural classifiers [10]-[12]. In [1],[2], it was suggested to use MVN as a basic neuron of a feedforward neural network, which was called MLMVN. The MLMVN's backpropagation learning algorithm generalizes the one for the single MVN and it is derivative-free. The MLMVN outperforms a classical multilayer feedforward network and different kernel-based networks in terms of learning speed, network complexity, and classification/prediction rate. It has been tested for a number of benchmarks problems (parity *n*, the two spirals, the sonar, and the Mackey-Glass time series prediction [1],[2]). In [13],[14], the MLMVN has been successfully used for solving a problem of blur and blur parameters identification. It is important to mention that since the MLMVN (as well as the single MVN) implements those input/output mappings that are described by multiple-valued (up to infinite-valued) functions, it is an efficient mean for solving the multiple-class classification problems. The results reported in [1],[2] and [13],[14] show that the MLMVN is really an efficient mean for solving classification problems. It is more flexible and adapts faster. This makes its training more efficient in comparison with other solutions.

In this paper, we apply MLMVN to solving several classification problems in bioinformatics. There are multiple-class classification of microarray gene expression data and breast cancer diagnostics. After presenting the basic properties of MLMVN and its backpropagation learning algorithm, we will consider two four-class test cases of microarray gene expression data classification (referred to as "Lung" and "Novartis" [15]) and the test case of breast cancer diagnostics with a famous "Wisconsin breast cancer data base"[1].

2 Multi-Valued Neuron (MVN) and MVN-Based Multilayer Neural Network (MLMVN)

2.1 Multi-Valued Neuron

A discrete-valued multi-valued neuron (MVN) has been introduced in [4] as a neural element based on the principles of multiple-valued threshold logic over the field of complex numbers. It has been comprehensively studied in [6], where its theory, basic

---

[1] This breast cancer databases was obtained from the University of Wisconsin Hospitals, Madison from Dr. William H. Wolberg. There are publicly available from a number of sources, for example, http://www.ics.uci.edu/~mlearn/MLSummary.html

properties, and learning are observed. A discrete-valued MVN performs a mapping between n inputs and a single output. This mapping is described by a multiple-valued (k-valued) function of n variables $f(x_1, ..., x_n)$ with $n+1$ complex-valued weights as parameters:

$$f(x_1, ..., x_n) = P(w_0 + w_1 x_1 + ... + w_n x_n), \tag{1}$$

where $X = (x_1, ..., x_n)$ is a vector of inputs (a pattern vector) and $W = (w_0, w_1, ..., w_n)$ is a weighting vector. The function and variables are the kth roots of unity: $\varepsilon^j = \exp(i2\pi j/K)$, $j = 0, ..., k-1$, where i is an imaginary unity. P is the activation function of the neuron:

$$P(z) = \exp(i2\pi j/k), \text{ if } 2\pi j/k \le \arg z < 2\pi (j+1)/k, \tag{2}$$

where j=0,..., k-1 are the values of k-valued logic, $z = w_0 + w_1 x_1 + ... + w_n x_n$ is a weighted sum, arg z is an argument of the complex number z. Fig. 1 illustrates the definition of the function (2). It divides a complex plane onto k equal sectors and maps the whole complex plane into a subset of points belonging to the unit circle. This is a set of kth roots of unity.

The MVN learning is reduced to the movement along the unit circle. This movement is derivative-free, because any direction along the circle always leads to the target. The shortest way of this movement is completely determined by an error that is a difference between the desired and actual outputs. Let ε^q be a desired output of the neuron (see Fig. 2) and $\varepsilon^s = P(z)$ be an actual output of the neuron. The most efficient MVN learning algorithm is based on the error correction learning rule [6]:

$$W_{r+1} = W_r + \frac{C_r}{(n+1)} (\varepsilon^q - \varepsilon^s) \bar{X}, \tag{3}$$

where X is an input vector, n is a number of neuron's inputs, $\bar{X}$ is a vector with the components complex conjugated to the components of vector X, r is the number of iteration, W_r is a current weighting vector, W_{r+1} is a weighting vector after correction, C_r is a learning rate. Recently it was proposed in [2] to modify (3) as follows:

$$W_{r+1} = W_r + \frac{C_r}{(n+1)|z_r|} (\varepsilon^q - \varepsilon^s) \bar{X}, \tag{4}$$

where z_r is a current value of the weighted sum. A factor $1/|z_r|$ in (4) is a self-adaptive part of the learning rate. When it is used, C_r in (4) should be equal to 1 that makes the learning rate completely self-adaptive. Learning rule (4) makes it possible squeezing a space for the possible values of the weighted sum. Thus, this space can be reduced to the respectively narrow ring, which includes the unit circle inside. This approach can be useful in order to exclude a situation when small changes either in the weights or the inputs lead to a significant change of z.

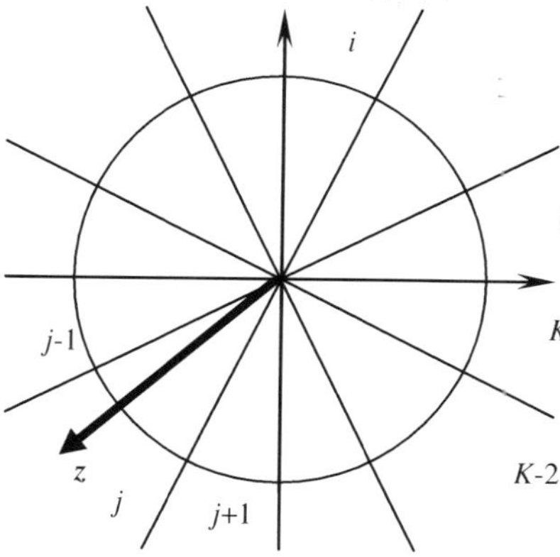

Fig. 1. Geometrical interpretation of the MVN activation function

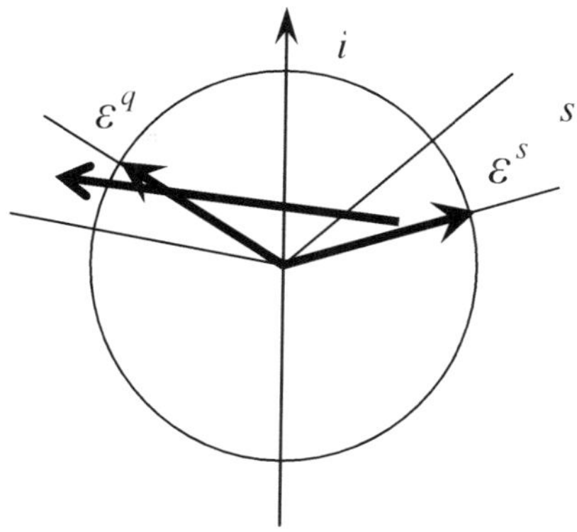

Fig. 2. Geometrical interpretation of the MVN learning rule

The convergence of the learning process based on the rule (3) is proven in [6]. The rule (3) ensures such a correction of the weights that the weighted sum is moving from the sector s to the sector q (see Fig. 2). The direction of this movement is completely determined by the error $\delta = \varepsilon^q - \varepsilon^s$. The correction of the weights according to (3) changes the weighted sum exactly on the value δ. This ensures that a desired output can be reached right after a single step of the weights adjustment for the given input.

The activation function (2) is discrete. It has been recently proposed in [1],[2], to modify the function (2) in order to generalize it for the continuous case in the following way. If $k \to \infty$ in (2) then the angle value of the sector (see Fig. 1) tends to zero. Hence, the function (2) is transformed in this case as follows:

$$P(z) = \exp(i\ (\arg z)) = e^{iArg\ z} = \frac{z}{|z|}, \tag{5}$$

where Arg z is a main value of the argument of the complex number z and $|z|$ is its absolute value.

The function (5) maps the complex plane into a whole unit circle, while the function (2) maps a complex plane just into a discrete subset of the points belonging to the unit circle. Thus, the activation function (5) determines a continuous-valued MVN. The learning rule (4) is modified for the continuous-valued case in the following way [1],[2]:

$$W_{r+1} = W_r + \frac{C_r}{(n+1)|z_r|}\left(T - \frac{z}{|z_r|}\right)\bar{X}, \tag{6}$$

where T is a desired output of the neuron. It should be mentioned that the use of a self-adaptive part of the learning rate (factor $1/|z_r|$) is optional. It is reasonable to use it for the mappings described by the functions with many high jumps. However, it should not be used to learn the smoothed mappings. It should also be omitted from (6) for the learning of the output layer neurons in a feedforward network (see below) because the exact errors for these neurons are known, while for the hidden neurons the exact errors are unknown. Thus, in this case (6) is transformed as follows

$$W_{r+1} = W_r + \frac{C_r}{(n+1)}\left(T - \frac{z}{|z_r|}\right)\overline{X}\,. \tag{7}$$

It is possible to consider two different modifications of the MVN. There are continuous inputs → discrete output and discrete inputs → continuous output, respectively. The first of these modifications can be used for solving those classification problems, where the features of the classified objects are continuous-valued, while there are a fixed finite number of classes. Evidently, in this case, the learning rule (4) can be used for training this type of the neuron. Respectively, the learning rule (6) can be used for training the MVN modification with the continuous-valued output and discrete-valued inputs. In this work, we use the discrete-valued MVN and the "continuous inputs → discrete output" modification.

2.2 Multilayer Neural Network Based on Multi-Valued Neurons (MLMVN)

A multilayer feedforward neural network based on multi-valued neurons (MLMVN) has been proposed in [1],[2]. It refers to the basic principles of the network with a feedforward dataflow through nodes proposed in [16]. The most important is that there is a full connection between the consecutive layers (the outputs of neurons from the preceding layer are connected with the corresponding inputs of neurons from the following layer). The network contains one input layer, m-1 hidden layers and one output layer. Let us use here the following notations. Let T_{km} be a desired output of the kth neuron from the mth (output) layer; Y_{km} be an actual output of the kth neuron from the mth (output) layer. Then the global error of the network taken from the kth neuron of the mth (output) layer is calculated as follows:

$$\delta^*_{km} = T_{km} - Y_{km}\,. \tag{8}$$

Following the backpropagation learning algorithm for the MLMVN proposed in [1],[2], the errors of all the neurons from the network are determined by the global errors of the network (8). It is essential that the global error of the network consists not only of the output neurons errors, but of the local errors of the output neurons and hidden neurons. It means that in order to obtain the local errors for all neurons, the global error must be shared among these neurons.

Let w_i^{kj} be the weight corresponding to the ith input of the kj^{th} neuron (k^{th} neuron of the jth level), Y_{ij} be the actual output of the ith neuron from the jth layer (j=1,…,m), and N_j be the number of the neurons in the jth layer (it means that the neurons from the j+1st layer have exactly N_j inputs). Let $x_1,\ldots,x_n$ be the network inputs.

Hence, the local errors are represented in the following way. The errors of the mth (output) layer neurons are:

$$\delta_{km} = \frac{1}{s_m}\delta^*_{km}, \tag{9}$$

where km is a kth neuron of the mth layer; $s_m = N_{m-1}+1$, i.e. the number of all neurons on the previous layer (layer m-1 which the error is backpropagated to) incremented by 1. The errors of the hidden layers neurons are computed as follows:

$$\delta_{kj} = \frac{1}{s_j}\sum_{i=1}^{N_{j+1}} \delta_{ij+1}(w_k^{ij+1})^{-1}, \tag{10}$$

where kj specifies the kth neuron of the jth layer (j=1,…,m-1); $s_j = N_{j-1}+1,\ j=2,...,m,\ s_1=1$ is the number of all neurons on the layer j-1 incremented by 1. Thus, the equations (9),(10) determine the error backpropagation for the MLMVN. It is worth to stress on its principal distinction from the classical error backpropagation: the MLMVN backpropagation does not depend on the derivative of the activation function. Moreover, both discrete-valued and continuous valued activation functions (2) and (5) that we use here are not differentiable at all as the functions of a complex variable.

A factor $1/s_j$ in (9),(10) ensures sharing of the particular neuron error among all the neurons on which this error depends. It should be mentioned that for the 1st hidden layer the parameter $s_1 = 1$ because there is no previous hidden layer, and there are no neurons the error may be shared with.

It is important that both error backpropagation and the learning process for the MLMVN are organized in the same manner independently of a particular modification of MVN (discrete-valued, continuous-valued, continuous inputs → discrete output and discrete inputs → continuous output).

After the error has been backpropagated, the weights of all neurons of the network can be adjusted using either the learning rule (3) or (7) (depending on the type of MVN and mapping, which we learn) for the output layer and, respectively, either (4) or (6) for the hidden layers. Hence, the following correction rules are used for the weights [1],[2]:

$$\begin{aligned} \tilde{w}_i^{kj} &= w_i^{km} + \frac{C_{km}}{(N_m+1)}\delta_{km}\bar{\tilde{Y}}_{im-1},\ i=1,...,n, \\ \tilde{w}_0^{km} &= w_0^{km} + \frac{C_{km}}{(N_m+1)}\delta_{km}, \end{aligned} \tag{11}$$

for the neurons from the mth (output) layer (k^{th} neuron of the mth layer),

$$\begin{aligned} \tilde{w}_i^{kj} &= w_i^{kj} + \frac{C_{kj}}{(N_j+1)|z_{kj}|}\delta_{kj}\bar{\tilde{Y}}_{ij-1},\ i=1,...,n, \\ \tilde{w}_0^{kj} &= w_0^{kj} + \frac{C_{kj}}{(N_j+1)|z_{kj}|}\delta_{kj}, \end{aligned} \tag{12}$$

for the neurons from the 2nd till m-1st hidden layers (k^{th} neuron of the jth layer (j=2, …, m-1), and

$$\tilde{w}_i^{k1} = w_i^{k1} + \frac{C_{k1}}{(n+1)|z_{k1}|}\delta_{k1}\bar{x}_i,\ i=1,...,n,$$
$$\tilde{w}_0^{k1} = w_0^{k1} + \frac{C_{k1}}{(n+1)|z_{k1}|}\delta_{k1}, \tag{13}$$

for the neurons of the 1st hidden layer, where n is the number of network inputs, N_j is the number of neurons in the jth layer.

It is important that it should be taken $C_{kj} = 1$ in (11)-(13). Thus, the learning rate in the MLMVN training is completely self-adaptive for the hidden neurons and it is not needed for the output neurons.

In general, the MLMVN training process should continue until the comdition of the minimal mean square error will be satisfied:

$$E = \frac{1}{N}\sum_{s=1}^{N}\sum_{k}(\delta_{kms}^{*})^2(W) = \frac{1}{N}\sum_{s=1}^{N}E_s \leq \lambda, \tag{14}$$

where λ determines the precision of learning, E_s is the square error for the sth pattern $X_s = (x_1,...,x_n)$. In particular, in the case when $\lambda = 0$ the equation (14) is transformed to $\forall k, \forall s\ \delta_{kms}^{*} = 0$, which means zero training error.

3 Solving Some Classification Problems in Bioinformatics

We are considering two classification problems. The first problem is classification of microarray gene expression data. For this problem, we use two well-known microarray gene expression data sets "Novartis" and "Lung" described in detail in [15]. Both publicly available datasets consist of multiple classes. The "Lung" data set includes 197 samples with 419 features (genes) that represent the four known classes. The "Novartis" data set includes 103 samples with 697 features that also represent the four known classes. Though feature selection is left outside the scope of this study, it should be noted that any screening or selection of features prior to classification can have significant effect on the result. Since using MLMVN we have to put the inputs on the unit circle, the gene expression data was not used in classification as such. We used a simple linear transform to convert the initial continuous-valued data to the complex numbers lying on the unit circle. If the initial range of the features is $[a,b]$ and $y \in [a,b]$ is the initial feature value then $x = e^{i\varphi}$, where $\varphi = \left(\frac{(y-a)}{(b-a)}6.27\right) \in [0,6.27]$, is the transformed value on the unit circle and $\varphi \in [0,6.27]$ is its argument.

To test the MLMVN as a classifier for both problems, we used the network with one hidden layer and one output layer containing the same number of neurons as the number of classes. Thus, each output neuron works according to the rule "the winner takes it all". Each of them recognizes samples from "its" class while rejecting other samples. Best results for both test data sets are produced by the network with 6 neurons in a single hidden layer (any increase of the hidden neurons amount does not improve the results; on the other hand, the results are a bit worse for a smaller size of hidden layer). Thus, taking into account that we have exactly 4 classes in both classification problems, the network $n \rightarrow 6 \rightarrow 4$ (where n is the number of inputs) has been used. The hidden neurons were continuous-valued (continuous inputs → continuous output), while the output neurons had continuous inputs and a discrete output (k=2 in (2)). If more than one output neuron classifies some sample as "its", we consider it to be a "winner", whose output's argument is closer to $\pi/2$. However, if all neurons reject a sample, this sample can be considered as "unclassified" (false negative classification). During the learning process we directed the weighted sum to the angles $\pi/2$ in the upper semi-plane and $3\pi/2$ in the bottom semi-plane. The domains $\pi/2 \pm \pi/8$ and $3\pi/2 \pm \pi/8$ were considered as acceptable, respectively.

We used a K-folding cross validation with K=5 to separate the data on the training and testing sets. Thus, K=5 training and testing sets have been created. For the "Novartis" data set, 80-84 samples of 103 were used for training and the rest 19-23 ones for testing. For the "Lung" data set, 117-118 samples of 197 were used for training and the rest 39-40 ones for testing.

The training process requires for its convergence just about 1 minute on a PC with Pentium IV 3.8 GHz CPU using a software simulator developed in the Borland Delphi 5 environment and about 500-1000 training epochs starting from the random weighted vectors with the real and imaginary parts belonging to [0, 1].

For the "Novartis" data set there is 98.95% classification rate, and for the "Lung" data set there is 95.94% classification rate. For comparison, the best classification results for the "Novartis" data set by using the k nearest neighbors (kNN) classifier was 97.69% with k = 1. For the "Lung" data set, the best classification accuracy by using the kNN classifier was 92.55% with k = 5. Exactly the same data transformation and subsampling partitions were used for all classifiers.

The next experiment was done using the "Wisconsin breast cancer database". One of the best known classification results for this data set first described in [17] was in turn reported in [18]. There is also detailed description of the data set in [18]. It contains 699 samples drawn from the classes "Benign" (458 samples) and "Malignant" (241 samples). Each entry is described by a 9-dimensional vector of integer-valued features (each element of which has value between 1 and 10). The nine features were clump thickness (x_1), uniformity of cell size (x_2), uniformity of cell shape (x_3), marginal adhesion (x_4), single epithelial cell size (x_5), bare nuclei (x_6), bland chromatin (x_7), normal nucleoli (x_8) and mitoses (x_9). The missing values for the feature x_6 for 16 samples have been replaced with the average equals to 3. The data set was divided into training set containing 350 randomly selected samples and test set containing the rest of 349 input samples like it was done in [18].

Since both inputs and outputs for this experiment are discrete, we used the "discrete input (k=10 in (2) → continuous output" hidden neurons and the "continuous inputs → discrete output (k=2 in (2)" output neuron. This makes it possible to ensure faster convergence of the training algorithm because the network becomes more flexible and adaptive. The best results were obtained using the network 9→8→1 (9 inputs, 8 hidden neurons and a single output neuron). Any increase of the hidden layer size does not improve the results, while the results are a bit worse for reduced hidden layer sizes. The training process converges quickly (just about 2.5-3 minutes and 4000-5000 epochs are required for the mentioned above software simulator). The classification rate for five independent runs (every time the learning process was started from the random weights) is 95.99%. This result is a bit better than 95% reported in [18]. However, a classification method, which was used in [18], consists of two stages. After the first stage, where a standard backpropagation neural network 9→9→1 is trained, fuzzy rules are derived from the results of the first stage and only the fuzzy rules are used for classification. Thus, although the results reported in [18] and those obtained here are comparable, the training process for the MLMVN is single-stage, and hence simpler.

4 Conclusions

In this paper, we documented the use of the multi-valued neural network based on multi-valued neurons for solving two popular bioinformatics classification problems. The MLMVN produces good results both for microarray gene expression data classification and for the breast cancer diagnostics.

References

1. Aizenberg, I., Moraga, C., Paliy, D.: A Feedforward Neural Network based on Multi-Valued Neurons. In: Reusch, B. (ed.) Computational Intelligence, Theory and Applications. Advances in Soft Computing, vol. XIV, pp. 599–612. Springer, Heidelberg (2005)
2. Aizenberg, I., Moraga, C.: Multilayer Feedforward Neural Network Based on Multi-Valued Neurons (MLMVN) and a Backpropagation Learning Algorithm. Soft Computing 11, 169–183 (2007)
3. Aizenberg, N.N., Ivaskiv Yu., L., Pospelov, D.A: About one generalization of the threshold function. Doklady Akademii Nauk SSSR (The Reports of the Academy of Sciences of the USSR) 196(6), 1287–1290 (1971)
4. Aizenberg, N.N., Aizenberg, I.N.: CNN Based on Multi-Valued Neuron as a Model of Associative Memory for Gray-Scale Images. In: Proceedings of the Second IEEE Int. Workshop on Cellular Neural Networks and their Applications, pp. 36–41. Technical University Munich, Germany (1992)

5. Aizenberg, N.N., Ivaskiv Yu, L.: Multiple-Valued Threshold Logic. Naukova Dumka Publisher House, Kiev (1977) (in Russian)
6. Aizenberg, I., Aizenberg, N., Vandewalle, J.: Multi-valued and universal binary neurons: theory, learning, applications. Kluwer Academic Publishers, Dordrecht (2000)
7. Jankowski, S., Lozowski, A., Zurada, J.M.: Complex-Valued Multistate Neural Associative Memory. IEEE Trans. on Neural Networks 7, 1491–1496 (1996)
8. Aoki, H., Kosugi, Y.: An Image Storage System Using Complex-Valued Associative Memory. In: Proc. of the 15th International Conference on Pattern Recognition, vol. 2, pp. 626–629. IEEE Computer Society Press, Los Alamitos (2000)
9. Muezzinoglu, M.K., Guzelis, C., Zurada, J.M.: A New Design Method for the Complex-Valued Multistate Hopfield Associative Memory. IEEE Trans. on Neural Networks 14(4), 891–899 (2003)
10. Aoki, H., Watanabe, E., Nagata, A., Kosugi, Y.: Rotation-Invariant Image Association for Endoscopic Positional Identification Using Complex-Valued Associative Memories. In: Mira, J.M., Prieto, A.G. (eds.) IWANN 2001. LNCS, vol. 2085, pp. 369–374. Springer, Heidelberg (2001)
11. Aizenberg, I., Myasnikova, E., Samsonova, M., Reinitz, J.: Temporal Classification of Drosophila Segmentation Gene Expression Patterns by the Multi-Valued Neural Recognition Method. Journal of Mathematical Biosciences 176(1), 145–159 (2002)
12. Aizenberg, I., Bregin, T., Butakoff, C., Karnaukhov, V., Merzlyakov, N., Milukova, O.: Type of Blur and Blur Parameters Identification Using Neural Network and Its Application to Image Restoration. In: Dorronscro, J.R. (ed.) ICANN 2002. LNCS, vol. 2415, pp. 1231–1236. Springer, Heidelberg (2002)
13. Aizenberg, I., Paliy, D., Astola, J.: Multilayer Neural Network based on Multi-Valued Neurons and the Blur Identification Problem, 2006 IEEE World Congress on Computational Intelligence. In: Proceedings of the 2006 IEEE Joint Conference on Neural Networks, Vancouver, Canada, July 16-21, pp. 1200–1207. IEEE Computer Society Press, Los Alamitos (2006)
14. Aizenberg, I., Paliy, D., Moraga, C., Astola, J.: Blur Identification Using Neural Network for Image Restoration. In: Reusch, B. (ed.) book Computational Intelligence, Theory and Application. Proceedings of the 9th International Conference on Computational Intelligenc, pp. 441–455. Springer, Heidelberg (2006)
15. Monti, S., Tamayo, P., Mesirov, J., Golub, T.: Consensus Clustering: A resampling-based method for class discovery and visualization of gene expression microarray data. Machine Learning 52, 91–118 (2003)
16. Rumelhart, D.E., McClelland, J.L.: Parallel Distributed Processing: Explorations in the Microstructure of Cognition. MIT Press, Cambridge (1986)
17. Mangasarian, O.L., Setiono, R., Wolberg, W.H.: Pattem recognition via linear programming: Theory and application to medical diagnosis. In: Coleman, T.F., Li, Y. (eds.) Large-scale numerical optimization, pp. 22–30. SIAM Publications, Philadelphia (1990)
18. Wettayapi, W., Lursinsap, C., Chu, C.H.: Rrule extraction from neural networks using fuzzy sets. In: ICONIP'O2. Proceedings of the 9th International Conference on Neural Information Processing, vol. 5, pp. 2852–2856 (2002)

Error Reduction in Holographic Movies Using a Hybrid Learning Method in Coherent Neural Networks

Chor Shen Tay, Ken Tanizawa, and Akira Hirose

Department of Electronic Engineering, The University of Tokyo
7-3-1 Hongo, Bunkyo-ku, Tokyo 113-8656, Japan

Abstract. Computer Generated Holograms (CGHs) are commonly used in optical tweezers which are employed in various research fields. Frame interpolation using coherent neural networks (CNNs) based on correlation learning can be used to generate holographic movies efficiently. However, the error that appears in the interpolated CGH images need to be reduced even further so that the method with frame interpolation can be accepted for use generally. In this paper, we propose a new hybrid CNN learning method that is able to generate the movies almost just as efficiently and yet reduces even more error that is present in the generated holographic images as compared to the method based solely on correlation learning.

1 Introduction

Holograms are recordings of information that are used to produce 3-dimensional (3D) holographic images for optical tweezers and other purposes [1]-[4]. Computer generated holograms (CGHs) are holograms which are not created tangibly but are generated using a computer that can be used to produce wavefronts of light with any prescribed amplitude and phase distribution. A specific type of CGH called the kinoform - a phase-only diffractive CGH - which is widely used in many application fields, is considered in this paper.

A method of generating holographic movies efficiently with frame interpolation using coherent neural network (CNN) that deals with the complex-amplitude information with generalization ability in the carrier-frequency domain was proposed by Hirose et al. (previous learning method) [5][6]. Nevertheless, generating kinoforms that are relatively error-free is also crucial in order for the method with frame interpolation to be accepted for common use.

In this paper, we propose and actualize a new hybrid learning method (HLM) using CNN that is capable of reducing the amount of error present in the generated kinoforms. It is also able to generate the movies almost just as efficiently as compared to the previous learning method.

J. Marques de Sá et al. (Eds.): ICANN 2007, Part I, LNCS 4668, pp. 884–893, 2007.

2 Coherent Neural Networks

2.1 Fundamentals of Learning and Information Processing

Figure 1 is a self-homodyne neural circuit, which is the most basic unit of the neural network used in the whole course of the present research [7][8]. There is only a single input $x_m = |x_m| \exp(i\alpha_m)$ into, single output $y_n = |y_n| \exp(i\beta_n)$ from, and multiple connections with different synaptic weights $w_{nm,h} = |w_{nm,h}| \exp(i\theta_{nm,h})$ into the neuron where $i \equiv \sqrt{-1}$. The subscripts m, n are the indices for the input, output vectors respectively while h is the index for the connection path.

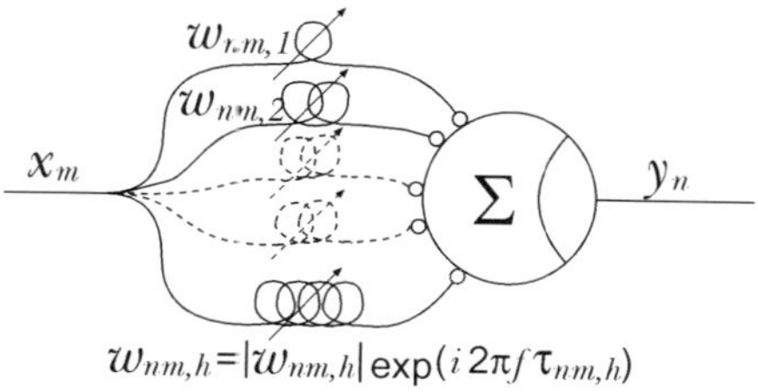

Fig. 1. Self-homodyne neural circuit

We employ an amplitude-phase type activation function defined as

$$y_n = g(u_n) = \tanh(B|u_n|) \exp(i \arg(u_n)) \tag{1}$$

where B is the amplitude gain of the input signals and u_n is the internal state of the neuron which is given as

$$u_n = \sum_m \sum_h (|w_{nm,h}| \exp(i2\pi f \tau_{nm,h}) x_m) \tag{2}$$

Here, f is the carrier frequency of light and $\tau_{nm,h}$ is the time delay of $w_{nm,h}$. $|w_{nm,h}|$ and $2\pi f \tau_{nm,h} (= \theta_{nm,h})$ represent the amplitude and phase of $w_{nm,h}$ respectively.

As the CGH involved with in this paper is the kinoform, we consider only the phase portion of y_n in (1). Also, as neural networks possess generalization ability, the HLM using CNN can change the networks' learning and processing behavior by changing its f. This frequency-domain generalization allows us to generate and interpolate CGH as the network generates the output based on the frequency that the user presents to it, even if f does not correspond to a teacher data set of data [9].

Two different learning rules were used in this paper. The first rule is correlation learning expressed as

$$\mu \frac{d|w_{nm,h}|}{dt} = -|w_{nm,h}| + |\hat{y}_n||x_m| \cos(\hat{\beta}_n - \alpha_m - \theta_{nm,h}) \tag{3}$$

$$\mu \frac{d\tau_{nm,h}}{dt} = \frac{1}{2\pi f} \frac{|\hat{y}_n||x_m|}{|w_{nm,h}|} \sin(\hat{\beta}_n - \alpha_m - 2\pi f \tau_{nm,h}) \tag{4}$$

The second rule is learning by method of steepest descent (SDM) stated as

$$\mu \frac{d|w_{nm,h}|}{dt} = -\left(\gamma |x_m| \cos \theta^{\mathrm{rot}}_{nm,h} - |y_n||\hat{y}_n| \sin(\beta_n - \hat{\beta}_n) \frac{|x_m|}{|u_n|} \sin \theta^{\mathrm{rot}}_{nm,h} \right) \tag{5}$$

$$\mu \frac{d\tau_{nm,h}}{dt} = -\frac{1}{2\pi f} \left(\gamma |x_m| \sin \theta^{\mathrm{rot}}_{nm,h} + |y_n||\hat{y}_n| \sin(\beta_n - \hat{\beta}_n) \frac{|x_m|}{|u_n|} \cos \theta^{\mathrm{rot}}_{nm,h} \right) \tag{6}$$

Here, μ is the time constant of learning, $\gamma = B(1 - |y_n|^2)(|y_n| - |\hat{y}_n| \cos(\beta_n - \hat{\beta}_n))$, $\theta^{\mathrm{rot}}_{nm,h} = \beta_n - \alpha_m - 2\pi f \tau_{nm,h}$, and the $\hat{}$ notation means that the term belongs to the teacher data presented to the neural network for learning.

2.2 Hybrid Learning Method

The proposed HLM comprises two different learning methods carried out in series. A primary reason for using an HLM in generating holographic movies with frame interpolation lies in the fact that each of the component methods within the HLM possesses different areas of strengths and weaknesses that do not coincide with each other: one component method removes the low frequency distortion in the kinoforms more effectively while the other method removes the high frequency noise in the kinoforms more effectively. In addition, we found that the error generated using the HLM is much less than the error generated by each individual component method. The detail is presented below.

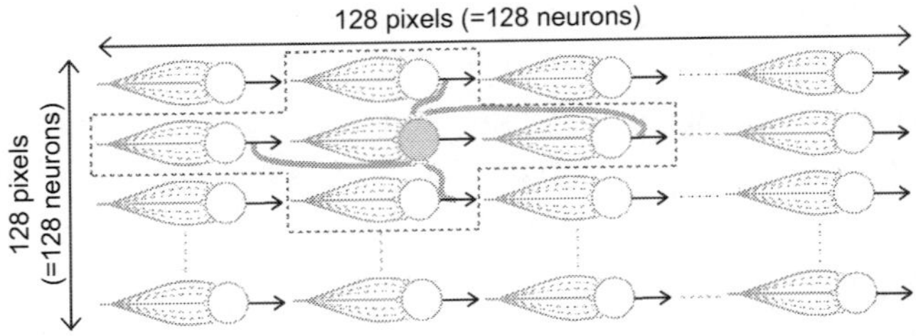

Fig. 2. Structure of neural network with lateral connections to be used in Method 1

The first part of the HLM executes correlation learning using the neural network structure shown in Fig. 2 (method 1). As can be seen for the gray neuron in Fig. 2 as an example, the network in method 1 learns using the standard coherent synaptic connections as well as lateral connections that are output from the adjacent neurons. The weights of the lateral connections are frequency independent in this present experiment and expressed by a simple complex-valued

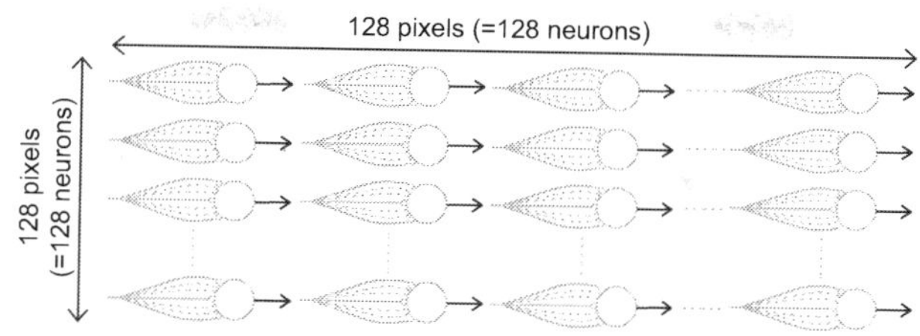

Fig. 3. Structure of previous neural network, which is used also in Method 2

weight $w \in$C, i.e., $w = |w| \exp(i\theta)$, with a frequency-insensitive angle θ instead of $2\pi f \tau$. Actually, we prefer a frequency-sensitive connection, just like the connections shown in Fig. 1, because the total neural behavior should be frequency sensitive [6]. However, we cannot implement it with our present personal computer having 1GB memory area. For the implementation, we need five times large area as well as a longer computation time.

The second part of the HLM carries out learning by SDM using the neural network structure shown in Fig. 3 (method 2), which was used also in previous report [6]. The neural network in method 2 learns using the standard coherent synaptic connections only - the reason why we have used different neural networks for methods 1 and 2 will be explained in Section 3.3.

3 Computer Simulations

3.1 Types of Error Identified

The sources of error in the generated kinoforms are found by calculating the difference or error d_j between the ideal hologram data s_j and the generated hologram data obtained using the respective learning methods l_j, i.e.,

$$d_j = s_j - l_j \tag{7}$$

Diagrams (a) and (d) in Fig. 4 show typical examples of the errors that were identified at the generated teacher and interpolated kinoforms after learning with the previous method. The errors are found using (7) and displayed in grayscale (white: π, black: $-\pi$). Two types of errors can be identified from the diagrams (a) and (d), which are

1. Local, high frequency noise characterized by the rough patterns.
2. Global, low frequency distortion characterized by the warped patterns.

The sources of error identified allowed us to consider possible countermeasures to reduce them from the generated kinoforms.

3.2 Learning Results Obtained Using Method 1

Diagrams (b) and (e) in Fig. 4 show examples of the error present in the generated teacher and generated interpolated kinoforms respectively when method 1

Fig. 4. Comparison of error in generated holograms between different learning methods

is used independently in generating the kinoforms. We find in the both diagrams that the distortion is reduced when we compare them to their counterparts that used the previous learning method shown in diagrams (a) and (d).

However, we can also see that the noise present in diagrams (b) and (e) of Fig. 4 increases (area of rough patterns becomes larger). Figure 5(a) shows the evolution of the mean error values E_v, for learning in method 1, at the teacher frequencies defined as

$$E_v = \frac{1}{\text{SIZE}} \sum_n \left(\hat{\beta} - \arg(\hat{u}_n) \right)^2 \tag{8}$$

where $\hat{\beta}$ is the teacher presented to the neural network, $\hat{u}_n$ is temporary output, and SIZE is the size of the hologram (128 × 128 pixels). The values of E_v obtained at the end of learning for method 1 are much higher than those obtained using the previous learning method. As such, it is possible to draw the conclusion that method 1 is effective against distortion, but weak against noise.

The reason why method 1 is effective against distortion is that the structure of the neural network with the lateral connections in Fig. 2 feedback information from the surrounding neurons into the central neuron while learning. As a result, the learning process is dynamic and incorporating the output from the surrounding adjacent neurons. This is the primary reason why method 1 is effective in reducing the distortion in the generated kinoforms.

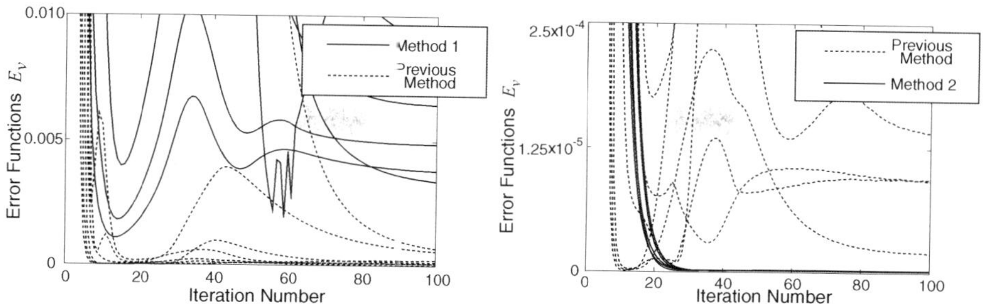

(a) Results obtained using method 1 (b) Results obtained using method 2

Fig. 5. E_v of teacher data vs. learning iteration number

3.3 Learning Results Obtained Using Method 2

Diagrams (c) and (f) in Fig. 4 show examples of the error obtained in the generated kinoforms with method 2. Comparing diagram (c) with diagram (a) in Fig. 4, it can be observed that the error present in the (c) is greatly reduced. Note that a totally error-free diagram will correspond to a pattern that is totally and evenly gray throughout. It is also observable by comparing diagrams (f) and (d) in the same figure that although the noise present in the generated interpolated kinoform of method 2 is significantly less than that of the previous learning method, the distortion becomes even more apparent.

Figure 5(b) shows the mean error E_v at teacher frequencies in method 2 against the learning iteration number carried out. It is apparent from Fig. 5(b) that the E_v using method 2 are less than those obtained using the previous learning method. It is therefore possible to infer from examining Figs. 4(c), 4(f) and 5(b) that although method 2 is effective against reducing the noise present in the holograms, it is also extremely weak against distortion at the same time.

The reason why method 2 does not make use of the neural network structure shown by Fig. 2 is that the lateral connections shown in the network are frequency independent, unlike the standard coherent synaptic connections that are frequency dependent. Because the network behavior should be carrier frequency dependent, the lateral connection also should be frequency dependent. The lateral connection is useful to remove global distortion in the first rough learning because the coherent neuron can possess redundancy in learning in general [10]. Because the structure of Fig. 2 is inconsistent with generalized learning in the frequency-domain and also because the problem of the distortion has already been handled by method 1, method 2 utilizes the neural network structure without lateral connections shown in Fig. 3.

Method 2 is effective against noise as learning by SDM approaches the solution straightforward when we have no local minima. This is different from correlation learning. This difference in learning by SDM in method 2 makes it effective against the noise present in the generated kinoforms.

3.4 Learning Results Obtained Using the HLM

Figure 6 shows the comparison made between the final iteration error E_v of respective learning methods versus carrier frequency at learning frequency points. It can be seen that the E_v obtained using the HLM is the smallest compared to all of the other methods that are discussed in this paper. The error E_v obtained using the HLM is approximately $10^3 \sim 10^5$ times smaller than that obtained for the previous learning method. We can deduce that E_v obtained at the end of the learning iterations correspond and is directly proportional to the noise in the generated kinoforms by comparing Fig. 5(a) with 5(b).

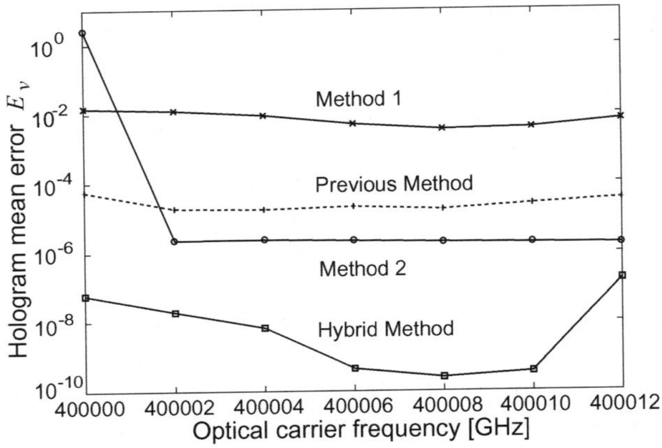

Fig. 6. Comparison of final iteration E_v in different methods

Figure 7 compares examples of the generated teacher diagrams learnt between the different learning methods against that of the ideal diagram. The images on the top-half of the diagrams show the kinoforms (hologram domain), while those at the bottom-half of the diagrams represent the reconstructed images on the image screen (movie frame domain). In this movie, a bright pixel moves from upper left to upper right. Figure 7 shows that the generated diagram (i.e. both the kinoform and frame) using the HLM is very similar to that of the ideal diagram. The fact indicates that almost all of the error has been removed.

Likewise, Fig. 8 shows the comparison made for the generated interpolated diagrams between respective learning methods against that of the ideal diagram in the same movie. It is again observe that there is less noise and distortion in the diagram generated with the HLM than that with the previous learning method.

Figure 9 shows the comparison of the degree of blurring (unsharpness) of the reconstructed image frames generated by the HLM against those by the previous learning method. We define the degree of blurring as

$$E_b = \frac{1}{\mathrm{SIZE}} \sum_{k=1}^{\mathrm{SIZE}} \left(L - \hat{L} \right)^2 \tag{9}$$

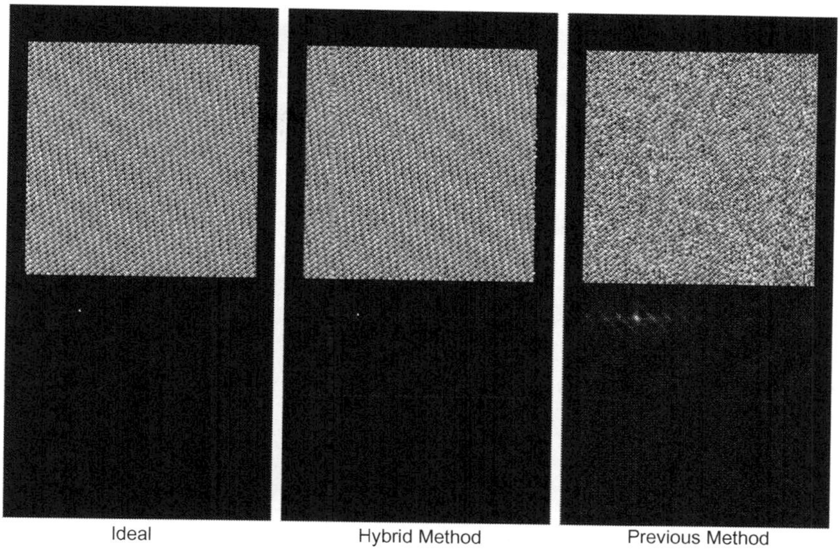

Fig. 7. Comparison of learning-point diagrams (top: holograms, bottom: movie frames)

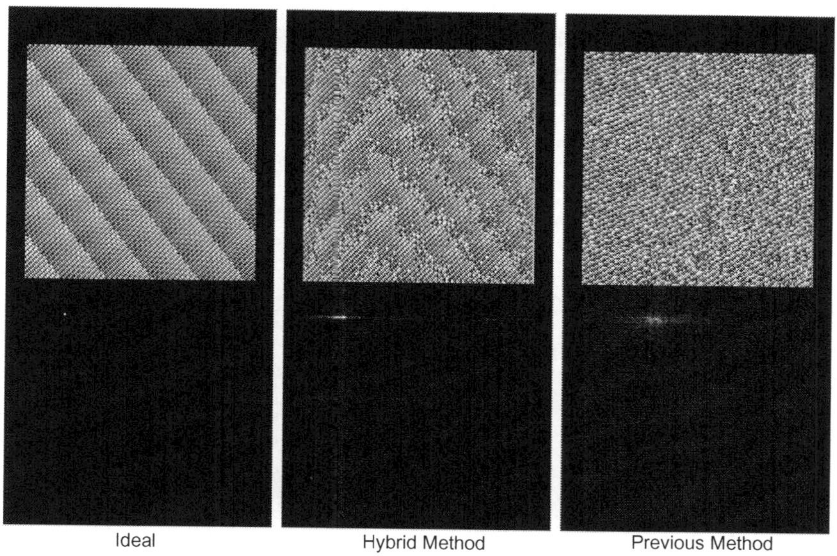

Fig. 8. Comparison of interpolated diagrams (top: holograms, bottom: movie frames)

where L is the luminosity of the individual reconstructed image pixels on the image screen. The degree of blu:ring E_b shows the error in the reconstructed-image domain, whereas the mean erro E_v shows the error level in the holographic domain. The lower the value of E_b, the better we consider the reconstructed images to be.

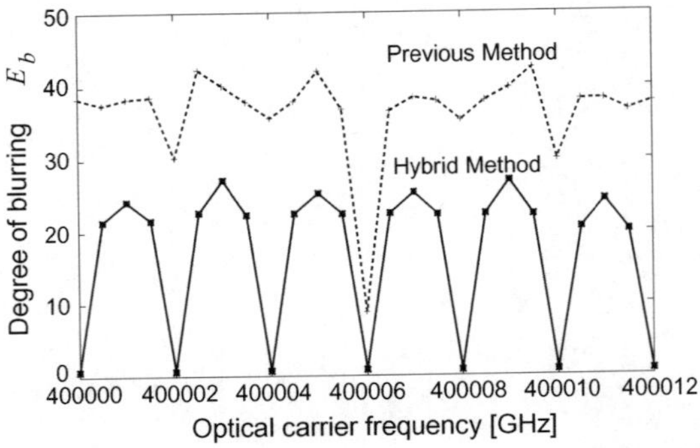

Fig. 9. Comparison of sharpness of reconstructed images

The 25 E_b values in Fig. 9 correspond to generated image frames at 25 frequency points (7 frames at learning frequencies and 18 interpolating frames) when a single pixel moves from upper left to upper right inthe image screen. The ticks on the x-axis (every four points) represent the frequencies used for learning. The rest of the points correspond to the generated interpolated images.

The curves in Fig. 9 clearly shows that the degree of blurring E_b is lower for the image reconstructed with the HLM. Note particularly that the points showing the E_b which correspond to the learning frequency points of the HLM are practically zero. This means numerically that the images are very sharp and identical to the ideal images.

3.5 Calculation Cost to Generate Movies

The amount of time required for generation of the above movie (7 teachers) using the previous learning method is approximately 18.5 minutes on average, while that using the HLM is approximately 21.6 minutes. This corresponds to a 17% increase in calculation time. However, the quality is much better than the previous method and is near to that of kinoforms generated with complete set of object data and FFT.

The calculation cost to generate object data is incomparably higher than other costs because, in most cases, we have to generate them by hand three dimensionally. The proposal provides us to generate a high-quality kinoform movies with a small number of object data.

4 Summary

We have proposed a new hybrid CNN learning method that is able to generate 3D movies almost just as efficiently and yet reduces even more error. We have compared the results of the newly proposed hybrid learning method with that of the previous learning method in this paper. We have confirmed that the hybrid learning method is capable of generating holographic movies almost just as efficiently and with less error as compared to the previous learning method.

References

1. Grier, D.G.: A revolution in optical manipulation. Nature 424, 810–816 (2003)
2. Liesener, J., Reicherter, M., Haist, T., Tiziani, H.J.: Multi-functional optical tweezers using computer-generated holograms. Optics Communications 185(1), 77–82 (2000)
3. Reicherter, M., Haist, T., Wagemann, E.U., Tiziani, H.J.: Optical particle trapping with computer-generated holograms written on a liquid-crystal display. Optics Letters 24(9), 608–610 (1999)
4. Schonbrun, E., et al.: 3D interferometric optical tweezers using a single spatial light modulator. Optics Express 13(10), 3777–3786 (2005)
5. Hirose, A., Higo, T., Tanizawa, K.: Efficient generation of holographic movies with frame interpolation using a coherent neural network. IEICE Electronics Express 3(19), 417–423 (2006)
6. Hirose, A., Higo, T., Tanizawa, K.: Holographic Three-Dimensional Movie Generation with Frame Interpolation Using Coherent Neural Networks. In: WCCI/IJCNN 2006, Vancouver (2006)
7. Kawata, S., Hirose, A.: Coherent optical neural network that learns desirable phase values in the frequency domain by use of multiple optical-path differences. Optics Letters 28(24) (2003)
8. Hirose, A. (ed.): Complex-Valued Neural Networks: Theories and Applications. World Scientific Publishing Co. Pte. Ltd, Singapore (2003)
9. Hirose, A.: Complex-Valued Neural Networks. Springer, Heidelberg (2006)
10. Hirose, A., Eckmiller, R.: Coherent optical neural networks that have optical-frequency-controlled behavior and generalization ability in the frequency domain. Applied Optics 35(5) (1996)

Sparse and Transformation-Invariant Hierarchical NMF

Sven Rebhan[1], Julian Eggert[1], Horst-Michael Groß[2], and Edgar Körner[1]

[1] HONDA Research Institute Europe GmbH
Carl-Legien-Strasse 30, 63073 Offenbach/Main, Germany
[2] Ilmenau Technical University, Dept. of Neuroinformatics and Cognitive Robotics
P.O.B. 100565, 98684 Ilmenau, Germany

Abstract. The hierarchical non-negative matrix factorization (HNMF) is a multilayer generative network for decomposing strictly positive data into strictly positive activations and base vectors in a hierarchical manner. However, the standard hierarchical NMF is not suited for overcomplete representations and does not code efficiently for transformations in the input data. Therefore we extend the standard HNMF by sparsity conditions and transformation-invariance in a natural, straightforward way. The idea is to factorize the input data into several hierarchical layers of activations, base vectors and transformations under sparsity constraints, leading to a less redundant and sparse encoding of the input data.

1 Introduction

The NMF has been introduced by Lee and Seung [1,2] as an efficient factorization method for decomposing multivariant data under the constraint of non-negativity. This results in a parts-based representation, because it allows only additive combination of components. While the standard NMF makes no further assumption on the input data, even so it is often used on inputs containing particular transformation properties, e.g. input images presented at different positions, scales and rotations. The resulting base vectors of the standard NMF then encode each transformation implicitly, which leads to a large amount of redundancy in the base vectors.

Ahn et al. developed a hierarchical multilayered variant of the NMF [3] by stacking multiple layers of NMF networks. This way a hierarchical representation of the input can be learned, where higher layers of the hierarchy code for more complex features composed of less complex features from lower layers. One can interprete this as a more and more abstract representation of the input with increasing hierarchy levels. Despite the interesting property of increasing abstraction in the layers of the network, the main problem of the NMF, the implicit coding of transformations in the base vectors, remains.

By assuming transformation properties in the input data we introduce, based on the work of Eggert et al. [4,5], a hierarchical, sparse and transformation-invariant version of the NMF. It has been shown that the proposed separation of the input data into activations, base vectors and transformations leads to

J. Marques de Sá et al. (Eds.): ICANN 2007, Part I, LNCS 4668, pp. 894–903, 2007.

a sparser and less redundant representation of the input than in the standard NMF. Extending the approach of Eggert et al. in a hierarchical way, we combine the advantage of sparsity and reduced redundancy in the representation with the advantage of growing abstraction in the hierarchical network of Ahn et al.

In Sect. 2 we extend the energy term formulations of [4,5] and derive the update rules for the activations and base vectors. Afterwards we discuss two possible update schemes in Sect. 2.3 and finally present simulation results using the new algorithm in Sect. 3. A short discussion finalizes the paper.

2 Hierarchical Extension to the Sparse, Transformation-Invariant NMF Framework

2.1 Sparse and Transformation-Invariant NMF

First we look at the sparse and transformation-invariant NMF, which serves as the base for our hierarchical extension. The energy term of the sparse and transformation-invariant NMF is defined as the Euclidian distance between the i-th input $\mathbf{V}_i$ and its reconstruction $\mathbf{R}_i$ plus the sparsity term $\lambda_H \cdot g_H(H)$

$$F(H,\hat{W}) = \frac{1}{2}\sum_i ||\mathbf{V}_i - \mathbf{R}_i||^2 + \lambda_H \cdot g_H(H) \tag{1}$$

where $g_H(H)$ is a sparsity function and λ_H is used to control the sparsity in the activations H_i. The reconstruction $\mathbf{R}_i$ itself is gained by linearly overlapping the normalized base vectors $\hat{\mathbf{W}}_j$ transformed by the operators T^m, weighted by the activation $H_i^{j,m}$

$$\mathbf{R}_i = \sum_j \sum_m H_i^{j,m} T^m \hat{\mathbf{W}}_j \; . \tag{2}$$

To avoid the scaling problem described in [4] the base vectors have to be normalized. Now one can calculate the derivation of the energy function with respect to $H_i^{j,m}$ and $\mathbf{W}_j$ and update them according to the standard NMF update rules:

1. Calculate the reconstruction $\mathbf{R}_i$ according to (2).
2. Update the activations according to [1]

$$H_i^{j,m} \leftarrow H_i^{j,m} \odot \frac{\left(T^m \hat{\mathbf{W}}_j\right)^T \mathbf{V}_i}{\left(T^m \hat{\mathbf{W}}_j\right)^T \mathbf{R}_i + \lambda_H \cdot g'_H\left(H_i^{j,m}\right)} \; . \tag{3}$$

3. Calculate the reconstruction $\mathbf{R}_i$ using the new activations according to (2).
4. Update the non-transformed base vectors according to

$$\mathbf{W}_j \leftarrow \mathbf{W}_j \odot \frac{\sum\limits_m \left[(T^m)^T V \left(\mathbf{H}^{j,m}\right)^T + \left[\left(\hat{\mathbf{W}}_j\right)^T (T^m)^T R \left(\mathbf{H}^{j,m}\right)^T \right] \nabla_{\mathbf{W}_j}\left(\hat{\mathbf{W}}_j\right) \right]}{\sum\limits_m \left[(T^m)^T R \left(\mathbf{H}^{j,m}\right)^T + \left[\left(\hat{\mathbf{W}}_j\right)^T (T^m)^T V \left(\mathbf{H}^{j,m}\right)^T \right] \nabla_{\mathbf{W}_j}\left(\hat{\mathbf{W}}_j\right) \right]} \; . \tag{4}$$

[1] Where $\odot$ denotes componentwise multiplication as $\mathbf{C} = \mathbf{A} \odot \mathbf{B} := C_i = A_i \cdot B_i, \forall i$.

5. Return to 1 until convergence.

The terms for updating the activations and base vectors can be found in [4], in addition the transformation matrix T^m from [5] is already included to make the sparse NMF transformation-invariant. Based on the update rules above we formulate the hierarchical energy equation and the corresponding update rules.

2.2 Sparse and Transformation-Invariant Hierarchical NMF

The sparse and transformation-invariant HNMF can be seen, similar as suggested in [3], as a network composed of multiple sparse and transformation-invariant NMF layers. This leads to the following Euclidian energy formulation[2]:

$$F(H^{(L)}, W^{(1)}, \ldots, W^{(L)}) = \frac{1}{2} \sum_i \left|\left| \mathbf{V}_i - R_i^{(1)} \right|\right|^2 , \tag{5}$$

where L denotes the topmost layer of the network. The reconstruction $R^{(l)}$ of the layer l in the hierarchy serves as the activation of the layer (l-1) and is calculated according to the transformation-invariant NMF

$$\mathbf{R}_i^{(l),m_{l-1}} := \mathbf{H}_i^{(l-1),m_{l-1}} = \sum_{m_l} \mathbf{H}_i^{(l),m_l} T^{(l),m_l} W^{(l),m_{l-1}} . \tag{6}$$

This recursive definition reveals that the reconstruction of the input data depends only on the highest layer activations $H^{(L)}$ and all base vectors. Having a closer look at (6) we see that the transformation information m_{l-1} is propagated down the hierarchy.

As [5] shows, sparsity is absolutely necessary in a transformation-invariant NMF network to avoid trivial solutions for the base vectors. Therefore we have to include at least sparsity in the activations. In contrast to the activations of all other layers, which are defined through down-propagation, the activations $H^{(L)}$ are independent. The extended energy formulation reads as

$$F = \frac{1}{2} \sum_i \left|\left| \mathbf{V}_i - R_i^{(1)} \right|\right|^2 + \lambda_H \cdot g_H \left(H^{(L)} \right) . \tag{7}$$

This step additionally requires the normalization of all base vectors $\hat{W}^{(1)}$, $\ldots$, $\hat{W}^{(L)}$. We choose the normalization function for the base vectors as

$$\hat{\mathbf{W}}_{j_l}^{(l),m_{l-1}} = \frac{\mathbf{W}_{j_l}^{(l),m_{l-1}}}{\sum_a \sum_b W_{j_l}^{(l),a,b}} , \tag{8}$$

which normalizes the length of each base vector to one. This leads to

$$\mathbf{R}_i^{(l),m_{l-1}} := \mathbf{H}_i^{(l-1),m_{l-1}} = \sum_{m_l} \mathbf{H}_i^{(l),m_l} T^{(l),m_l} \hat{W}^{(l),m_{l-1}} . \tag{9}$$

[2] Denoted as F from now on for convenience.

To be able to control the arrangement in the base vectors it is useful to include sparsity in the base vectors as well. Therefore we add another sparsity term to the energy function, which is now composed of three elements

$$F = \underbrace{\frac{1}{2} \sum_i \left|\left| \mathbf{V}_i - R_i^{(1)} \right|\right|^2}_{\text{Reconstruction error}} + \underbrace{\lambda_H \cdot g_H \left(H^{(L)} \right)}_{\text{Activation sparsity}} + \underbrace{\sum_l \lambda_W^{(l)} \cdot g_W \left(\hat{W}^{(l)} \right)}_{\text{Base vector sparsity}} . \quad (10)$$

The sparsity terms for the activations and base vectors are chosen as

$$g_H \left(H^{(L)} \right) = \sum_i \sum_{j_L} \sum_{m_L} H_i^{(L),j_L,m_L} \quad (11)$$

$$g_W \left(\hat{W}^{(l)} \right) = \sum_{j_l} \sum_{j_{l-1}} \sum_{m_{l-1}} \hat{W}_{j_l}^{(l),j_{l-1},m_{l-1}} . \quad (12)$$

Starting from the functions above we calculate the gradients with respect to the activations $H^{(L)}$ and base vectors $W^{(l)}$.

In order to formulate the multiplicative update rule we split the remaining two gradient terms into the positive part ∇^+ and the negative part ∇^-.[3] For the highest layer activations we get the following update rule

$$\mathbf{H}_i^{(L),m_L} \leftarrow \mathbf{H}_i^{(L),m_L} \odot \frac{\sum_{m_{L-1}} \left(T^{(L),m_L} \hat{W}^{(L),m_{L-1}} \right)^T \mathbf{V}_i^{(L),m_{L-1}}}{\sum_{m_{L-1}} \left(T^{(L),m_L} \hat{W}^{(L),m_{L-1}} \right)^T \mathbf{R}_i^{(L),m_{L-1}} + \lambda_H} \quad (13)$$

with the substitutions

$$\mathbf{V}_i^{(l+1),m_l} = \sum_{m_{l-1}} \left(T^{(l),m_l} \hat{W}^{(l),m_{l-1}} \right)^T \mathbf{V}_i^{(l),m_{l-1}} \quad (14)$$

$$\mathbf{R}_i^{(l+1),m_l} = \sum_{m_{l-1}} \left(T^{(l),m_l} \hat{W}^{(l),m_{l-1}} \right)^T \mathbf{R}_i^{(l),m_{l-1}} . \quad (15)$$

The reconstruction error is propagated from the bottom to the top layer of the hierarchy and is then used to adjust the activations of the highest layer. The sparsity term itself is an additional constraint which is independent of the reconstruction quality of the network. As a consequence the sparse and transformation-invariant HNMF network has to find a tradeoff between reconstruction quality and sparsity, controlled by the sparsity parameter λ_H.

Performing the same steps for the gradient with respect to $W^{(l)}$, we get

$$\mathbf{W}_{j_l}^{(l),m_{l-1}} \leftarrow \mathbf{W}_{j_l}^{(l),m_{l-1}} \odot \frac{\nabla^-_{W^{(l)}} F}{\nabla^+_{W^{(l)}} F} . \quad (16)$$

[3] This is possible due to the non-negative character of all elements in the equation.

The gradient of the sparsity term $g_H\left(H^{(L)}\right)$ for the highest layer activations is zero, because $H^{(L)}$ is independent of the base vectors. For the two parts of the gradient of (16) we get

$$\nabla^-_{W^{(l)}} F = \mathbf{W_V}^{(l),m_{l-1}}_{j_l} + \sum_k \left[\left(\hat{\mathbf{W}}^{(l),k}_{j_l} \right)^T \left[\mathbf{W_R}^{(l),k}_{j_l} + \lambda^{(l)}_W \right] \right] \hat{\mathbf{W}}^{(l),m_{l-1}}_{j_l} \tag{17}$$

$$\nabla^+_{W^{(l)}} F = \mathbf{W_R}^{(l),m_{l-1}}_{j_l} + \sum_k \left[\left(\hat{\mathbf{W}}^{(l),k}_{j_l} \right)^T \mathbf{W_V}^{(l),k}_{j_l} \right] \hat{\mathbf{W}}^{(l),m_{l-1}}_{j_l} + \lambda^{(l)}_W \tag{18}$$

with the substitutions

$$\mathbf{W_R}^{(l),m_{l-1}}_{j_l} = \sum_{m_l} \left[\left(T^{(l),m_l} \right)^T R^{(l),m_{l-1}} \left(\mathbf{H}^{(l),j_l,m_l} \right)^T \right] \tag{19}$$

$$\mathbf{W_V}^{(l),m_{l-1}}_{j_l} = \sum_{m_l} \left[\left(T^{(l),m_l} \right)^T V^{(l),m_{l-1}} \left(\mathbf{H}^{(l),j_l,m_l} \right)^T \right] . \tag{20}$$

Similar to the substitutions for the activations we see an upwards propagation of the reconstruction error. The sparsity constraint on the base vectors, controlled by the parameter $\lambda^{(l)}_W$, can also be seen in the update function (17) and (18). The normalization of the base vectors which is required by the sparsity in the activations leads to an additional term in the update rule for $W^{(l)}$.

In the next section we discuss two possible update schemes, starting from the update rules (13) to (20).

2.3 Possible Update Schemes

With the extension of the NMF to a hierarchical network new possibilities to update the whole network come along.

1. Update the network by iterating layer by layer
2. Update the whole network by propagating through all layers

In the following we discuss the pro and contra of the two methods.

Update Network Layer by Layer
In this scheme each layer is learned separately, beginning from the lowest layer, as shown in Fig. 1a. The goal is to reconstruct the activations of the layer below as a linear combination of base vectors. Afterwards the layer above is adapted and so on. This means that the activations of each layer are adapted sequentially and are therefore mutually independent. In this sense you get a stack of separate transformation-invariant, sparse NMF networks. A big advantage of this independent relaxation is the fact that all parameters of one layer are independent from the parameters of the other layers, which makes the parameter setting much easier. Another advantage is the extensibility of the framework. After the convergence of the network you can add another layer on top of the

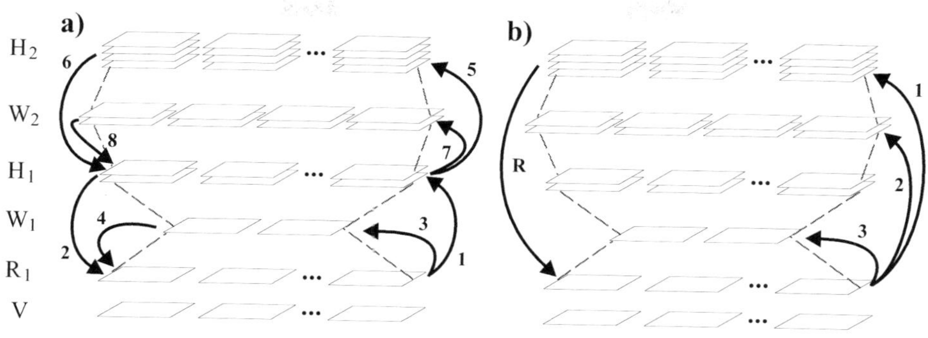

Fig. 1. The graphics above show two possible update sequences for the HNMF.
a) In this scheme each layer is iterated independently. First 1-4 is iterated until convergence, then 5-8 is iterated, trying to reconstruct H_1.
b) In this scheme the whole network is iterated in a combined manner by first calculating 1 followed by the reconstruction R, then 2, the reconstruction R, 3 and finally the reconstruction R until convergence.

existing layers and learn the new one. All other layers can be left untouched. The big disadvantage is that each layer only minimizes the local energy function on the activations of the lower layer. This leads to a smaller number of minima in the energy function, but has the disadvantage that in most cases the global minimum of the whole network is not found (see [3]).

Update Network as a Whole
Contrary to the independent relaxation, in this scheme we learn the base vectors and the highest layer activations simultaneously as shown in Fig. 1b. The sequence of updating is the following:

1. Calculate the reconstruction of the lowest layer by propagating down the highest layer activations through the base vectors by applying (9) iteratively.
2. Propagate the reconstruction error between $R^{(1)}$ and the input $V^{(1)}$ to the highest layer using (14) and (15).
3. Adapt the activations of the highest layer as described in (13).
4. Execute step 1 using the updated activations $H^{(L)}$.
5. Adapt the base vectors of the lowest layer $W^{(1)}$ using (16).
6. Execute step 1 using the updated base vectors.
7. Propagate the reconstruction error between $R^{(1)}$ and the input $V^{(1)}$ to the next higher layer using (14) and (15).
8. Adapt the base vectors of the next higher layer $W^{(l+1)}$ using (16).
9. Repeat step 6 to 8 until the base vectors of all layers are updated.
10. Repeat beginning with step 1 until convergence.

The advantage of this combined relaxation is the minimization of the overall energy function, which leads to a better reconstruction and a sparser representation. One drawback is the introduction of relations between the sparsity parameters by combined relaxation, which makes the selection of the parameters more difficult. Because of the better reconstruction results, we choose the combined update scheme for the experiments we present in the next section.

3 Results

For the following experiments we set up a two layer, sparse and **translation-invariant** hierarchical network. We use only translation for $T^{(l)}$ because this transformation can be coded very efficiently using correlations. The learning of the base vectors is performed with the combined relaxation scheme discussed in Sect. 2.3. The used dataset (see examples in Fig. 2) consists of 162 bar images of 4x4 pixel size. Each of the images is a superposition of up to four horizontal and vertical bars. The horizontal bar can be applied at four different horizontal positions; the vertical bar at four different vertical positions. A complete overlap of two bars is not allowed.

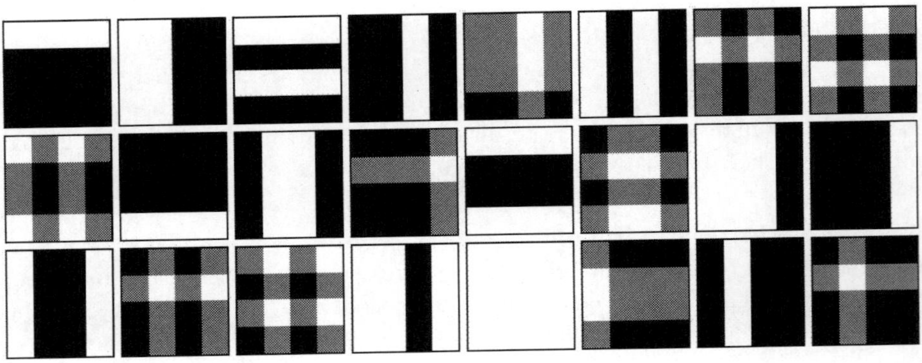

Fig. 2. These are 24 examples for the input dataset, which consists of 162 images. Each image has a size of 4x4 pixel and is a superposition of horizontal and vertical bars.

The task for the network is to find a set of base vectors that encodes the input under a given sparsity constraint. In Fig. 3 the two lower layer base vectors of the translation-invariant, sparse HNMF network are shown. One can see that the network finds the two original bars (one horizontal, one vertical). Based on these vectors, the 64 upper layer base vectors compose more complex structures in order to satisfy the sparsity constraint.

Figure 4 shows the base vectors of the upper layer projected to the input space. The vectors themselves consist of very sparse, sporadic peaks. By increasing the sparsity constraint in the activations, the base vectors get more and more complex, whereas the sparsity in the base vectors leads to a reallocation of the information between the different layers. As a result, the sparsity settings for the network are essential to force a meaningful distribution of the information within the hierarchy.

Fig. 3. Here the two base vectors of the lower layer, which are nearly perfect reconstructions of the original vectors, are shown

Fig. 4. This shows the 16 resulting base vectors of the upper layer projected into the input space. As you can see the set also contains the two vectors of the layer below.

If we have a look at the energy function depicted in Fig. 5 we see three phases in the relaxation process. In the first phase, which comprises the steps 0 to 380, the overall energy decreases very slowly. This is mainly a result of the reconstruction error optimization done by the network. As one can see clearly the minimization of the reconstruction error is taking place in the lower base vectors, because in this phase the higher layer vectors are not changed significantly. In the second phase, including the steps 380 to 600, a minimization of the sparsity penalization is taking place, where large changes in the higher layer base vectors can be seen. We can observe a reorganization in the HNMF network, where the information stored in the upper layer base vectors is transferred down to lower layer vectors. Along with the transfer of information, the base vectors in the upper layer get sparser. This has a major effect on the energy function. The following third phase, starting at step 600, is characterized by using the reorganized structure of the network to optimize both the reconstruction error and the sparsity. This is achieved by modifying the lower layer base vectors and the activations, leaving the upper layer base vectors mostly unchanged.

Through the whole relaxation process the energy function is steadily decreasing with a jump at the beginning of the reorganization phase. This leads to the conclusion that during the process the focus of what should be minimized is shifted between reconstruction and sparsity according to the chosen parameter set. When choosing extreme settings for the sparsity parameters the focus switches to sparsity maximization whereas the minimization of the reconstruction error does not play a role anymore and vice versa.

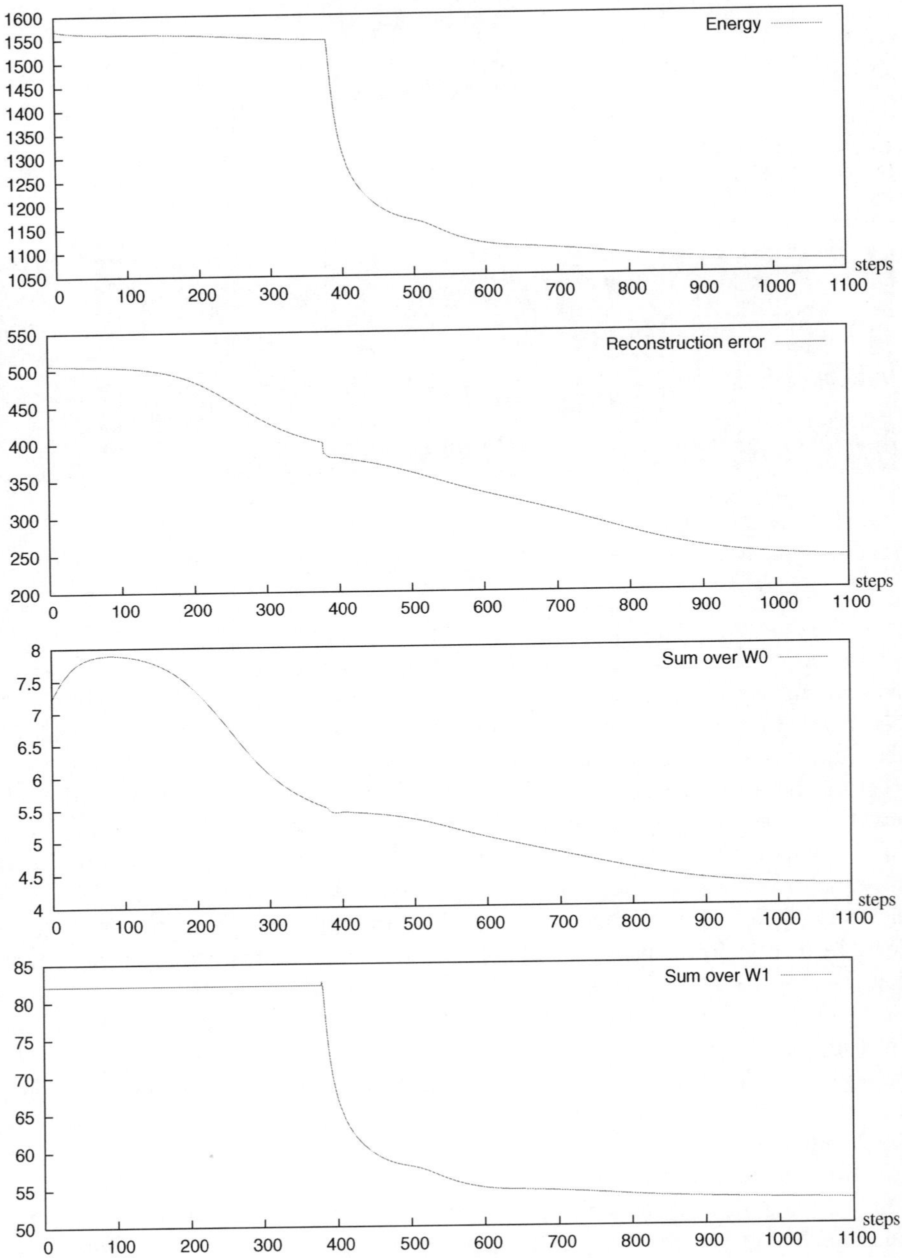

Fig. 5. These plots show the three phases of a two-layer HNMF network relaxation process for (from top to bottom) the energy function, the reconstruction error and the sparsity measure in the base vectors. The phases are: minimization of the reconstruction error, reorganization within the network to meet the sparsity constraint and the combined minimization of the reconstruction error and the sparsity penalization.

4 Conclusion

In this paper we propose the extension of the sparse Overlapping NMF introduced in [4] and [5] to a hierarchical, sparse and transformation-invariant network. This extension was done in a straightforward manner by defining the activations of each layer as a reconstruction of the layer above (see (6)). While other hierarchical approaches as in [3] store input transformations implicitly in the base vectors, our approach encodes the transformations explicitly. This explicit encoding leads to a reduction of redundancy in the base vectors, making the representation sparser and more efficient.

In Sect. 2.3 we discussed two different update schemes and concluded that only a combined relaxation of the whole network leads to a minimization of the overall energy function and is therefore preferable. Using the combined relaxation scheme on bar stimuli, we achieved the results depicted in Sect. 3, which show that the transformation-invariant and sparse HNMF is able to decompose the stimuli into the original parts. In this process the basic parts of the data set are stored in the lowest layer base vectors, whereas the higher layer base vectors compose a more complex and more abstract representation of the input by combining lower layer vectors. The resulting decomposition is a sparse representation of the input, having also very good reconstruction properties. By adapting the sparsity parameters, the network solution can be steered towards a perfect reconstruction or towards a sparse representation, where extreme settings will lead to insensible or trivial solutions.

As a final interesting point we want to mention that the proposed algorithm includes all previous approaches as special cases. To emulate [3] we just set all transformation matrices to unity matrices (no transformation), for [4,5] we take a single layer network and choose the sparsity parameters accordingly. So the transformation-invariant and sparse hierarchical NMF can be seen as a unification of the three mentioned NMF approaches.

References

1. Lee, D.D., Seung, H.S.: Learning the parts of objects with nonnegative matrix factorization. Nature 401, 788–791 (1999)
2. Lee, D.D., Seung, H.S.: Algorithms for Non-negative Matrix Factorization. Advances in Neural Information Processing Systems 13, 556–562 (2001)
3. Ahn, J.-H., Choi, S., Oh, J.-H.: A multiplicative up-propagation algorithm. In: Proceedings of the 21th International Conference, pp. 17–24 (2004)
4. Eggert, J., Körner, E.: Sparse coding and NMF. In: IJCNN 2004. Proceedings of the International Joint Conference on Neural Networks, pp. 2529–2533 (2004)
5. Eggert, J., Wersing, H., Körner, E.: Transformation-invariant representation and NMF. In: IJCNN 2004. Proceedings of the International Joint Conference on Neural Networks, pp. 2535–2539 (2004)

Zero-Lag Long Range Synchronization of Neurons Is Enhanced by Dynamical Relaying

Raul Vicente[1,2], Gordon Pipa[1,2], Ingo Fischer[3], and Claudio R. Mirasso[4]

[1] Max-Planck Institute for Brain Research, Deutschordenstrasse 46, 60528 Frankfurt am Main, Germany
[2] Frankfurt Institute for Advanced Studies, Max-von-Laue Strasse 1, 60438 Frankfurt am Main, Germany
[3] Dept. of Applied Physics and Photonics (TONA) Vrije Universiteit Brussel, Pleinlaan 2, B-1050 Brussel, Belgium
[4] Dept. de Fisica Universitat de les Illes Balears, Crta de Valldemossa km 7.5, E-07071 Palma de Mallorca, Spain

Abstract. How can two distant neural assemblies synchronize their firings at zero-lag even in the presence of non-negligible delays in the transfer of information between them? Here we propose a simple network module that naturally accounts for zero-lag neural synchronization for a wide range of temporal delays. In particular, we demonstrate that isochronous (without lag) millisecond precise synchronization between two distant neurons or neural populations can be achieved by relaying their dynamics via a third mediating single neuron or population.

1 Introduction

Neural synchronization stands today as one of the most promising mechanisms to counterbalance the huge anatomical and functional specialization of the different brain areas [1,2,3]. In particular, it proposes the formation of transiently synchronized neural assemblies during few hundreds of milliseconds as the underlying process to bind several local neural dynamics. Consequently, neural synchrony can provide a dynamic and reconfigurable mechanism for large-scale integration of distributed brain activity and serve as an efficient code to complement the rate modulation and overcome some of its limitations. The synchrony hypothesis has been supported by experimental findings demonstrating that millisecond precise synchrony of neuronal oscillations across well separated cortical areas plays an essential role in visual coherent perception and other high-cognitive tasks [3,4].

However, and albeit more evidence is being accumulated in favor of its functional role as a binding mechanism of distributed neural responses, the physical and anatomical substrate for such a dynamic and precise synchrony, especially zero-lag even in the presence of non-negligible delays, remains unclear [4]. Several mechanisms have been proposed to explain the appearance of synchronization between neural populations. Inhibitory connections and gap junctions have been proposed to increase the stability of the synchronous state [5]. In general, these and other approaches require either the precise tuning of properties such as the

J. Marques de Sá et al. (Eds.): ICANN 2007, Part I, LNCS 4668, pp. 904–913, 2007.

synaptic rise time or rely on rather complex architectures [6] while exhibiting several limitations in the range of synchronization attainable.

In addition, the problem of the communication delays between the neural units involved in the interaction, although fundamental, has been hardly faced [7]. Conduction and synaptic delays between the brain areas which are observed to synchronize can amount to several tens of milliseconds. How under such latency times the reciprocal interactions between two brain regions can lead the associated neural populations come into sync without almost any lag?

Here we propose a simple network motif that is able to naturally lead to the zero-lag synchronization between two arbitrarily distant neural populations. The basic idea is that when two spiking neurons interact not directly but through a third mediating neuronal unit, the redistribution of the dynamics performed by this central unit leads in a robust and self-consistent manner toward the zero-lag synchronization of the outer neurons [8]. This simple network module is expected to exist within the complex functional architecture of the brain and especially within the reciprocal thalamocortical interactions. It is significant that recent studies have demonstrated the constant latency between the thalamus (the main relay unit of sensory information in the brain) and almost any area in the mammalian neocortex [9]. Remarkably, this occurs irrespective of the very different distances that separate the thalamic nuclei and the cortex regions involved. This means that an action potential generated in a thalamic cell will take the same time to reach a cortical neuron independently of the thalamocortical afferent used to propagate the spike. To our purposes, this implies that a thalamocortical circuit is an ideal representation of this network module. With the proposed network motif the synchronous state can be achieved after the exchange of a few spikes and consequently within a few tens of milliseconds. This time scale is perfectly compatible with the required processing times of information found experimentally [4].

2 Methods

In order to test the synchronization properties of such neural circuits we simulated the dynamics of Hodgkin-Huxley (HH) neurons that interact with each other via reciprocal synaptic connections with an intermediate third neuron of the same type.

2.1 Mathematical Model

The dynamics of the membrane potential of each neuron is modeled by the classical Hodgkin-Huxley equations [10] with the addition of appropriate synaptic currents to mimic the chemical coupling between neurons.

The temporal evolution of the voltage across the membrane is given by

$$C\frac{dV}{dt} = -g_{Na}m^3h(V-E_{Na}) - g_K n^4(V-E_k) - g_L(V-E_L) + I_{ext} + I_{syn} \; , \quad (1)$$

where $C = 1\ \mu\text{F/cm}^2$ is the membrane capacitance, the constants $g_{Na} = 120$ mS/cm^2, $g_K = 36$ mS/cm^2, and $g_L = 0.3$ mS/cm^2 are the maximal conductances of the sodium, potassium, and leakage channels, and $E_{Na} = 50$ mV, $E_K = -77$ mV, and $E_L = -54.5$ mV stand for the corresponding reversal potentials. According to Hodgkin and Huxley formulation the voltage-gated ion channels are described by the following set of differential equations

$$\frac{dm}{dt} = \alpha_m(V)(1-m) - \beta_m(V)m\ , \tag{2}$$

$$\frac{dh}{dt} = \alpha_h(V)(1-h) - \beta_h(V)h\ , \tag{3}$$

$$\frac{dn}{dt} = \alpha_n(V)(1-n) - \beta_n(V)n\ , \tag{4}$$

where the gating variables $m(t)$, $h(t)$, and $n(t)$ represent the activation and inactivation of the sodium channels and the activation of the potassium channels, respectively. The experimentally fitted voltage-dependent transition rates are

$$\alpha_m(V) = \frac{0.1(V+40)}{1-\exp\left(-(V+40)/10\right)}\ , \tag{5}$$

$$\beta_m(V) = 4\exp\left(-(V+65)/18\right)\ , \tag{6}$$

$$\alpha_h(V) = 0.07\exp\left(-(V+65)/20\right)\ , \tag{7}$$

$$\beta_h(V) = \left[1+\exp\left(-(V+35)/10\right)\right]^{-1}\ , \tag{8}$$

$$\alpha_n(V) = \frac{(V+55)/10}{1-\exp\left(-0.1(V+55)\right)}\ , \tag{9}$$

$$\beta_n(V) = 0.125\exp\left(-(V+65)/80\right)\ . \tag{10}$$

The synaptic transmission between neurons is modeled by a postsynaptic conductance change with the form of an alpha-function

$$\alpha(t) = \frac{1}{\tau_d - \tau_r}\left(\exp\left(-t/\tau_d\right) - \exp\left(-t/\tau_r\right)\right)\ , \tag{11}$$

where the parameters τ_d and τ_r stand for the decay and rise time of the function and determine the duration of the response. Consequently, the the synaptic current takes the form

$$I_{syn}(t) = -g_{max}\sum_{\tau_l}\sum_{spikes}\alpha\left(t - t_{spike} - \tau_l\right)\left(V(t) - E_{syn}\right)\ , \tag{12}$$

where g_{max} describes the maximal synaptic conductance and the internal sum is extended over the train of presynaptic spikes occurring at t_{spike}. The delays arising from the finite conduction velocity of axons are taken into account through the latency time τ_l in the alpha-function. Thus, the external sum covers the different latencies that arise from the existence of multiple synaptic connections between two different neural populations. Excitatory and inhibitory transmissions are differentiated by setting the synaptic reversal potential to be $E_{syn} = 0$ mV or $E_{syn} = -80$ mV, respectively.

Finally, the external current I_{ext} stimulation is adjusted to a constant value of 10 μA/cm^2. Under such conditions a single Hodgkin-Huxley type neuron enters into a regime of periodic firing with a natural period of $T_{nat} = 14.66$ ms. When inspecting the role of noise, white Gaussian stochastic fluctuations were added to the otherwise constant level of the external current.

2.2 Geometry

We consider a neural circuit composed by two Hodgkin-Huxley neurons interacting with a relay neuron of the same class. This mediating neuron is receiving input from the two outer neurons and projecting output toward them. Latency times are included in the interaction to capture the non-negligible conduction delays occurring in the transmission of spikes. This neural circuit can be thought as a simple model of two neuronal populations (α and β) interacting through a set of neurons whose dynamics is summarized by a single mediating population c with a given transfer function (see Fig. 1).

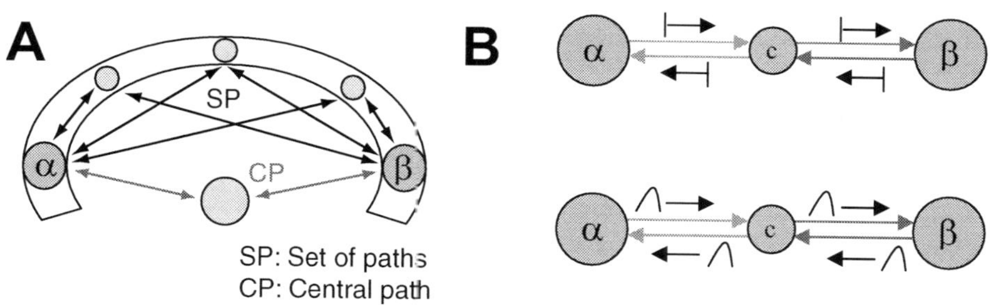

Fig. 1. a) Sketch of a neural circuit with recurrent connections mapped to the interaction between three neural populations (α, β, and a central population c.) b) Interaction between populations α and β through the mediating element c with single (top) or distributed (bottom) synaptic delays.

Individual temporal delays of the arrival of presynaptic potentials (i.e., latency times) were modelled by a gamma distribution to mimic the multiple connections between the neural populations involved in the synchronization process.

The probability density function of the distribution of delays is then

$$f(\tau_l) = \tau_l^{k-1} \frac{\exp(-\tau_l/\theta)}{\theta^k \Gamma(k)} \qquad (13)$$

where k and θ are shape and scale parameters of the gamma distribution. The mean time delay is given by $\hat{\tau}_l = k\theta$.

2.3 Numerical Integration

The set of equations (1-12) is numerically integrated using the Heun method with a time step of 0.02 ms.

Starting from random initial conditions each neuron is first simulated without any synaptic coupling for 200 ms after which frequency adaptation has occurred and each neuron settles into a periodic firing regime with a well defined frequency. The relation between the phases of the oscillatory activities of the neurons at the end of this heating time was entirely determined by the initial conditions. Following this period and once the synaptic transmission has been activated, a simulation time of 3 seconds is recorded. This allowed us to trace the change in the relative timing of the spikes induced by the synaptic coupling in this neural circuit.

2.4 Data Analysis

The strength of the synchronization and the phase-difference between each individual pair of neurons (m, n) were derived by the computation of the order parameter defined as

$$\rho(t) = \frac{1}{2} | \exp(i\phi_m(t)) + \exp(i\phi_n(t))| \, , \tag{14}$$

which takes the value of 1 when two oscillators are moving in-phase and 0 in an anti-phase regime. In order to compute this quantifier it is only necessary to estimate the phases of the individual neural oscillators. An advantage of this method is that one can easily reconstruct the phase of a neuronal oscillation from the train of spikes without the need of recording the full membrane potential time series [11]. The idea behind is that the time interval between two well defined events (such as action potentials) define a complete cycle and the phase increase during this time amounts to 2π. Then, linear interpolation is used to assign a value to the phase during the spike events.

Cross-correlation analysis of the membrane potentials was also computed in order to check the results obtained from the order parameter estimation.

3 Results

Prior to the illustration of the results obtained from the direct simulation of equations (1-12), we first characterize in Fig. 2 the response of a single Hodgkin-Huxley neuron to the arrival of a presynaptic potential and the subsequent change in the postsynaptic conductivity. In particular, we computed how much the period of the oscillation of a single HH neuron is changed as a function of the phase at which this single perturbation is received. This phase response curve (PRC) contains useful information about the synchronization properties of the oscillators involved.

Figure 2 shows the trace of the voltage potential of an unperturbed HH neuron and the phase response curves to alpha-functions with different synaptic rise and

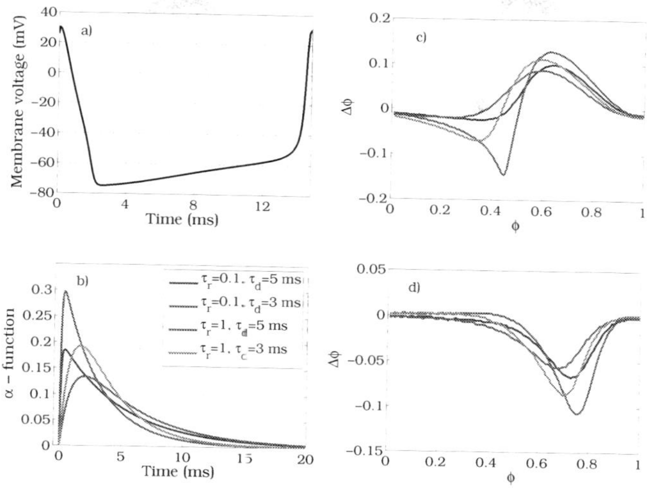

Fig. 2. a) Membrane voltage of an unperturbed HH neuron during a cycle of oscillation. b) Alpha-functions with different rise and decay times indicated in the legend. c) PRCs for excitatory synapses (E_{syn}=0 mV). d) PRCs for inhibitory synapses (E_{syn}=-80 mV). The PRC is defined as $\Delta\phi = 1 - T_{per}/T_{nat}$, where T_{per} stands for the length of the cycle containing the perturbation. ϕ represents the phase at which the perturbation is received. The colors in panels c) and d) code for alpha-functions of the same color shown in b). $g_{max} = 0.2$.

decay times. The Hodgkin and Huxley neuron is known to produce a response such that a perturbation can advance or retard the next spike as it is seen in panel c) where an excitatory coupling was simulated. However, it is observed in panel d) that in the investigated regime, an inhibitory synapse can only retard the firing of the next action potential. It is also noticed that due to the finite duration of the alpha-function the value of the PRCs for excitatory synapses differs from zero in the neighborhood of a spike ($\phi = 0\,, 1$).

The different synchronization characteristics of inhibitory and excitatory synapses in two-coupled neuron models and the role of the synaptic rise and decay times can be attributed to the different features of their respective PRCs [5,12]. Nevertheless, here we propose a network module that naturally provides a robust mechanism for synchronizing two neural populations with zero-phase lag. This effect is based on the dynamical relaying provided by a third population, and opposite to directly-coupled neuron models it turns out to be largely independent of the characteristics of the synaptic transmission. Figure 3 displays the time traces of the membrane potential for the three neurons coupled as in the module of Fig.1 with excitatory synapses.

It is observed that once the synaptic coupling is activated the network module consistently self-organizes toward the state in which the outer neurons synchronize their spikes. The zero-lag synchronization arises as a consequence of the relay and redistribution of excitatory postsynaptic potentials (EPSP) performed by the central neuron. The EPSPs induced in such a neural circuit are responsible

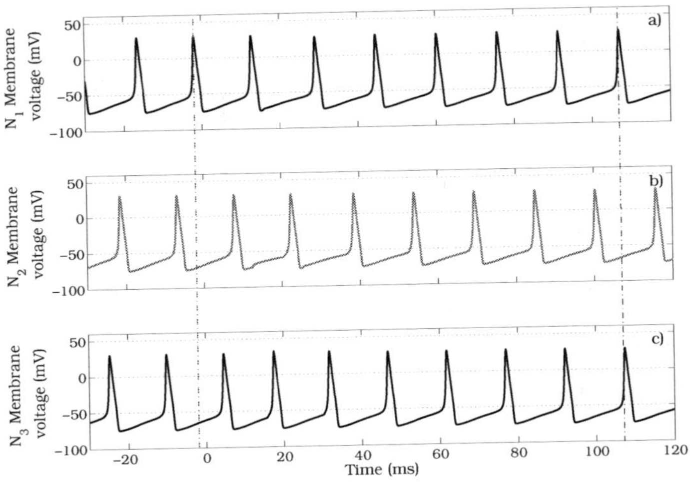

Fig. 3. a), b), and c) show the evolution of membrane potential of the three neurons interacting through the network sketched in Fig.1. Neuron 2 is the relay neuron. The synaptic coupling is activated at $t = 0$ ms with a $g_{max} = 0.5$. The gamma distribution of the latency times is chosen to be at the limit at which it tends to a delta distribution centered in 8 ms. The alpha-function used for the synaptic coupling had a rise time of $\tau_r = 0.1$ ms and a decay time of 3 ms.

to slightly modify the firing timing of the postsynaptic cells in a manner that the outer neurons tend to fire at unison after a few spikes irrespective of the initial conditions. We have checked that this phenomenon also occurs for inhibitory synapses and different synaptic temporal scales.

To inspect if the unavoidable noise sources around a neuron are able to disrupt the synchronization we simulated the dynamics of the HH neurons with independent additive white noise in the external current input of each cell. Figure 4 shows that even in the presence of moderate noise intensities the synchronization process takes place.

At this point, a crucial question is whether this synchronization transition is particular to single latency synaptic pathways or it is maintained for broad distributions of conduction delays. To answer such issue we have represented in Figure 5 the order parameter of the neuronal oscillations of the outer cells for several delay distributions. In particular, we scanned the shape factor and the mean value of the distribution of delays. The results indicate the existence of a broad region of delays (between 2 ms and 9 ms) where for almost any shape factor the outer neurons come into sync. Only distributions with an unrealistic small shape factor (nearly decaying exponentials distributions) are unable to produce synchrony regardless the average delay of the synaptic connections. The drop in the synchronization quality found around $\hat{\tau}_l \sim 10$ ms is associated to an irregular firing state of the neurons where no locking behavior is found.

Thus, zero-phase synchronization can take place in such network module even in the presence of broad distributions in the delays communicating two

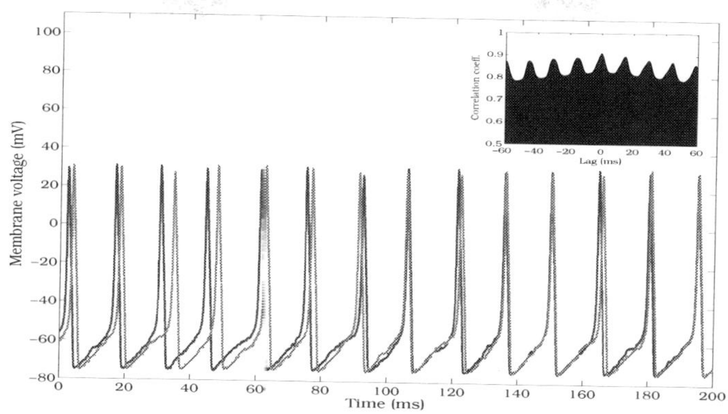

Fig. 4. The large panel shows the temporal evolution of the outer neurons in the presence of noise once synaptic coupling is activated at $t = 0$ ms. The noise level was adjusted to $\sigma = 1\ \mu\text{A ms}^{1/2}/\text{cm}^2$. It is clear that even the spikes become more irregular a good synchronization level is obtained at zero-lag. The inset displays the cross-correlation function of the voltage temporal series for a 3 second simulation.

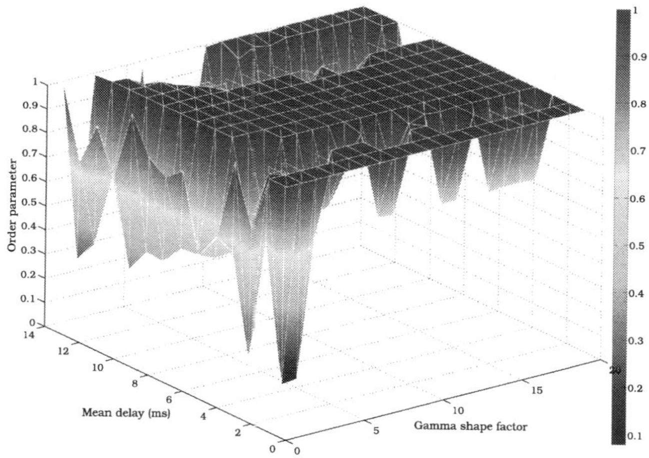

Fig. 5. Order parameter of the phases of the outer neurons as a function of the shape factor and mean delay of the gamma distribution of delays. $g_{max} = 0.2$, $E_{syn} = 0$ mV. The alpha-function used for the synaptic coupling had a rise time of $\tau_r = 0.1$ ms and a decay time of 3 ms.

neural populations. Similar behaviors are found for inhibitory synapses and alpha-functions with different durations demonstrating that the synchronization process in such a network is largely independent of the particular characteristics of the PRCs of the neuron.

Up to now, only symmetrical distributions of delays have been considered in the pathways from the relay neuron to each one of the outer units. However, when different distributions are simulated for each of the pathways we have found that the phase lag between the outer neurons usually deviates from zero with the cell with the shortest mean distance to the relay unit leading the dynamics. Nevertheless, it is quite remarkable that broad distributions (large shape factors) allow for almost zero-lag synchronization even in the presence of differences of several milliseconds in the average delay of both pathways.

4 Discussion

We have introduced a simple and extremely robust network motif that is able to account for the zero-phase synchronization of distant neural elements in a natural way. This robust synchronization arises as a consequence of the relay and redistribution of the dynamics performed by a mediating neuron. As a consequence and in opposition to previous works, neither inhibitory, gap junctions, nor complex networks need to be invoked to provide a stable mechanism of zero-phase correlated activity of neural populations even in the presence of large conduction delays.

As a future direction we are working to provide a more physiologically detailed model of the network motif here presented in order to compare with experimental data. In particular, for the relay or thalamic neuron we are including a description of additional ionic channels (transient Ca^{2+} and K^{+} as well as a persistent Na^{+} conductance). Neural population modeling for each of the three involved pools of neurons is also under current research.

Acknowledgments

Research supported by the GABA project (European Commission, FP6-NEST contract 043309).

References

1. Von der Malsburg, C., Schneider, W.: A neural cocktail-party processor. Biological Cybernetics 54, 29–40 (1986)
2. Rieke, F., Warland, D., De Ruyter van Steveninck, R.: Spikes: Exploring the Neural Code. MIT Press, Cambridge (1997)
3. Varela, F.J., Lachaux, J.P., Rodriguez, E., Martinerie, J.: The brainweb: phase synchronization and large-scale integration. Nature Reviews Neuroscience 2, 229–239 (2001)
4. Singer, W.: Neuronal Synchrony: A Versatile Code for the Definition of Relations? Neuron 24, 49–65 (1999)
5. Van Vreeswijk, C., Abbott, L.F., Ermentrout, B.: When inhibition not excitation synchronizes neural firing. Journal of Computational Neuroscience 1, 313–321 (1994)
6. Lago-Fernandez, L.F., Huerta, R., Corbacho, F., Siguenza, J.A.: Fast response and temporal coherent oscillations in small-world networks. Physical Review Letters 84, 2758–2761 (2000)

7. Freeman, W.: Characteristics of the synchronization of brain activity imposed by finite conduction velocity of axons. International Journal of Bifurcation and Chaos 10, 2307–2322 (2000)
8. Fischer, I., Vicente, R., Buldu, J.M., Peil, M., Mirasso, C.R., Torrent, M.C., Garcia-Ojalvo, J.: Zero-lag synchronization via dynamical relaying. Physical Review Letters 97, 123902(1)–123902(4) (2006)
9. Salami, M., Itami, C., Tsumoto, T., Kimura, F.: Change of conduction velocity by regional myelination yields to constant latency irrespective of distance between thalamus and cortex. PNAS 100, 6174–6179 (2003)
10. Hodgkin, A.L., Huxley, A.F.: A quantitative description of the membrane current and its application to conduction and excitation in nerve. Journal of Physiology 117, 500–544 (1952)
11. Pikovsky, A., Rosenblum, M., Kurths, J.: Synchronization: A universal concept in nonlinear sciences. Cambridge University Press, Cambridge (2001)
12. Goel, P., Ermentrout, B.: Synchrony, stability and firing pattern is pulse-coupled oscillators. Physica D 163, 191–216 (2002)

Polynomial Cellular Neural Networks for Implementing the Game of Life

Giovanni Egidio Pazienza[1], Eduardo Gomez-Ramirez[2], and Xavier Vilasís-Cardona[3]

[1] GRSI, Enginyeria i Arquitectura La Salle, Universitat Ramon Llull, Quatre Camins 2, 08022 Barcelona (Spain)
gpazienza@salle.url.edu

[2] LIDETEA, Posgrado e Investigación, Universidad La Salle, Benjamín Franklin 47, Col. Condesa, 06140 Mexico City (Mexico)
egr@ci.ulsa.mx

[3] LIFAELS, Enginyeria i Arquitectura La Salle, Universitat Ramon Llull, Quatre Camins 2, 08022 Barcelona (Spain)
xvilasis@salle.url.edu

Abstract. One-layer space-invariant Cellular Neural Networks (CNNs) are widely appreciated for their simplicity and versatility; however, such structures are not able to solve non-linearly separable problems. In this paper we show that a polynomial CNN - that has with a direct VLSI implementation - is capable of dealing with the 'Game of Life', a Cellular Automaton with the same computational complexity as a Turing machine. Furthermore, we describe a simple design algorithm that allows to convert the rules of a Cellular Automaton into the weights of a polynomial CNN.

1 Introduction

Cellular Neural Networks (CNNs) [1] combine the features of Neural Networks and Cellular Automata, since neurons exhibit only local connections. CNNs with space-invariant weights have found application in a number of fields - especially in image processing [2] - because of their versatility and fast computation.

Thanks to its topological simplicity, the one-layer space-invariant CNN allows a VLSI implementation [3], but its computational power is restricted by the fact that it cannot solve linearly non-separable problems. In order to extend the capability of representation of CNNs, many authors have proposed modifications to the standard model like space-variant CNNs [4], multilayer CNNs [5], trapezoidal activation functions [6], piecewise-linear discriminant functions [7], and even a supercomputer whose computation core is a CNN [8]. All these methods make the CNN capable of universal computation [9], but they complicate the correspondent hardware implementation.

In our studies we take into consideration a so-called polynomial CNN that is a standard one-layer space-invariant CNN with the addition of a polynomial term. One of the main advantages of polynomial CNNs is that they have a direct VLSI realization [10].

J. Marques de Sá et al. (Eds.): ICANN 2007, Part I, LNCS 4668, pp. 914–923, 2007.

In this paper we prove that the simplest model of polynomial CNN can implement the 'Game of Life' [11], a well-known problem with the same computational power as a universal Turing machine. The core of our work is the introduction of a design algorithm for polynomial CNNs based exclusively on the solution of a system of inequalities. On the one hand, such algorithm allows to find the weights of the network straightforwardly with respect to other learning algorithms for CNNs (e.g. [12]); on the other hand, we set a direct link between the rules of a Cellular Automata and the learning process of a polynomial CNN.

This paper is structured as follows: first, we introduce the theoretical model of the polynomial CNNs; then, we explain the rules 'Game of Life' showing how they can be summarized in a Cartesian system; next, we describe in detail the design algorithm to find the weights of the network; finally, we draw conclusions.

2 Polynomial Cellular Neural Networks

2.1 A Brief Introduction to Cellular Neural Networks

Cellular Neural Networks (CNNs) are a particular kind of neural networks in which neurons are usually arranged in a two-dimensional square grid, thus each cell is locally connected only with its 8 neighbor cells. In our studies we employ a synchronous version of CNNs called Discrete-Time CNNs (DTCNNs) [13], whose dynamic is described as follows

$$x_{ij}(n+1) = \sum_{C(k,l)\in Nr(i,j)} A(i,j;k,l)y_{kl}(n) + \sum_{C(k,l)\in Nr(i,j)} B(i,j;k,l)u_{kl}(n) + i, \tag{1}$$

where u_{ij}, x_{ij} and y_{ij} are the input, the state and the output of the cell in position (i,j), respectively; $Nr(i,j)$ is the set of indexes corresponding to the cell (i,j) and its neighbors; finally, the matrices A and B contain the weights of the network, and the value i is a bias.

In DTCNNs only binary output is allowed, then a suitable activation function is

$$y_{ij}(n) = f(x_{ij}(n)) = \begin{cases} 1, & \text{if } x_{ij}(n) \geq 0 \\ -1, & \text{if } x_{ij}(n) < 0. \end{cases} \tag{2}$$

The most interesting case is the space-invariant CNN, in which weights depend on the difference between cell indexes rather than on their absolute values. In this case the behavior of the network is completely specified by a unique set of weights $\{A, B, i\}$ called *cloning template*. Usually, A and B are 3-by-3 matrices. In the case of a space-invariant network, the Eq. (1) can be written in a compact way by using the Einstein notation

$$x^c(n) = a^c_d y^d(n) + b^c_d u^d(n) + i. \tag{3}$$

According to this notation, when an index appear twice in a single term, it implies that we are summing over all its possible values.

2.2 The Polynomial Model and Its Advantages

It is possible to prove that one-layer space-invariant Cellular Neural Networks cannot solve non-linearly separable problems. However, this drawback can be overcome by a polynomial CNN, whose general discrete-time state equation is

$$x^d(n) = a_e^d y^e(n) + b_e^d u^e(n) + i + g(u^e, y^e), \tag{4}$$

where a polynomial term $g(u^e, y^e)$ is added to the standard DTCNN model. The polynomial CNN was first introduced in [14], and the stability of the network is analysed in [15]. Its applications range from elementary non-linearly separable tasks, like the XOR operation [16], to real-life cases, like the epilepsy seizures prediction [17]. In the simplest case the term $g(u^e, y^e)$ is a second degree polynomial with the following form

$$\begin{aligned} g(u^e, y^e) &= \sum_{i=0}^{2} ((p_i)_e^d (u^e)^i \cdot (q_i)_e^d (y^e)^{2-i}) \qquad (5)\\ &= (p_0)_e^d (1^e) \cdot (q_0)_e^d (y^e)^2 + (p_1)_e^d (u^e) \cdot (q_1)_e^d (y^e) + (p_2)_e^d (u^e)^2 \cdot (q_2)_e^d (1^e), \end{aligned}$$

where $\{(p_0)_i\} = P0$ etc. and $(P0, P1, P2, Q0, Q1, Q2)$ are 3-by-3 matrices. Note that throughout this paper capital letters indicate matrices and lower case letters indicate scalar values.

3 The Game of Life

3.1 The Rules

One of the best known example of Cellular Automaton (CA) is the 'Game of Life' (GoL). It is a no-player game usually played on an infinite two-dimensional grid of square cells, and its evolution is determined only by the initial state. Each cell interacts with its 8 neighbor cells, and at any fixed time it can be either alive (black cell) or dead (white cell). In each time step, the next state of each cell is defined by the following rules

- *Birth*: a cell that is dead at time t becomes alive at time t + 1 only if exactly 3 of its eight neighbors were alive at time t;
- *Survival*: a cell that was living at time t will remain alive at t + 1 if and only if it had exactly 2 or 3 alive neighbors at time t.

The importance of the GoL comes from the fact that this structure has the same complexity as a universal Turing machine [18], therefore any conceivable algorithm can be computed by a GoL-based computer.

3.2 A Visual Representation for the Rules of Semitotalistic CA

In the GoL the next state of the central cell depends only on its present state and the sum of its eight nearest neighbors. This kind of CA are called *semitotalistic*

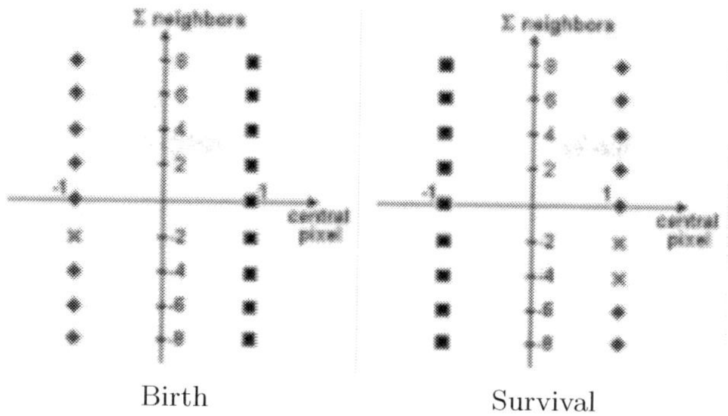

Fig. 1. Birth and Survival rules for the GoL. The red cross corresponds to a 1 (black) for the next state, a blue diamond to a -1 (white), and the black square is a '*don't care*'.

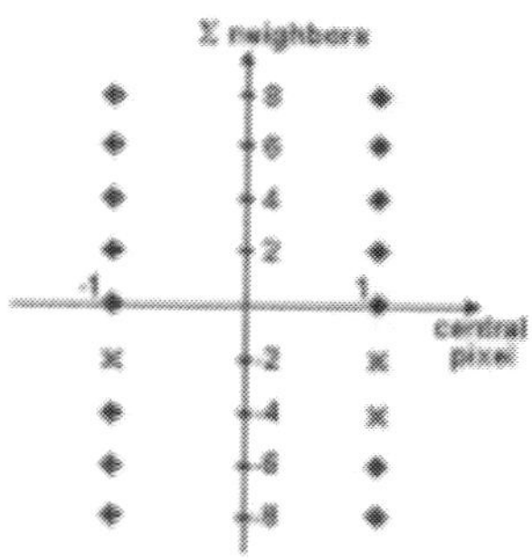

Fig. 2. The Game of Life: a red cross corresponds to a 1 (black) for the next state, and a blue diamond is a -1 (white)

and their rules can be conveniently represented in a Cartesian system, as depicted in Fig. 1. Following the convention that a black pixel is +1 and a white pixel is -1, the x axis corresponds to the central pixel of the 3×3 Cellular Automaton, whereas the y axis is equal to the sum of the 8 outer neighbours. The rules of the CA are represented by associating to each point of the grid - corresponding to a pair (central pixel, sum of neighbours) - a symbol indicating the following state of the cell: a red cross represents a 1 (black) for the next state, a blue diamond represents a -1 (white), and a black square is a '*don't care*'.

As mentioned in the previous section, the GoL can be expressed as the combination of two rules (*Birth* and *Survival*); therefore, the scheme for the GoL is the logic AND of two correspondent schemes (see Fig. 2). Thanks to this representation, it is evident that the GoL is not a linearly separable task.

3.3 Alternative Set of Rules for the GoL

Looking at Fig. 1, it is possible to state that also the *Birth* and the *Survival* rules taken individually are not linearly separable. However, the GoL can be formulated as the logic combination of two linearly separable rules as follows

- *Rule 1*: a cell will be alive in the next state if at least 3 of the 9 cells in its 3×3 neighborhood are alive, otherwise it will be dead;
- *Rule 2*: a cell will be alive in the next state if at most 3 of its 8 neighbors are alive, otherwise it will be dead.

Their schemes are represented in Fig. 3, and hereafter these rules will be used as reference for our work.

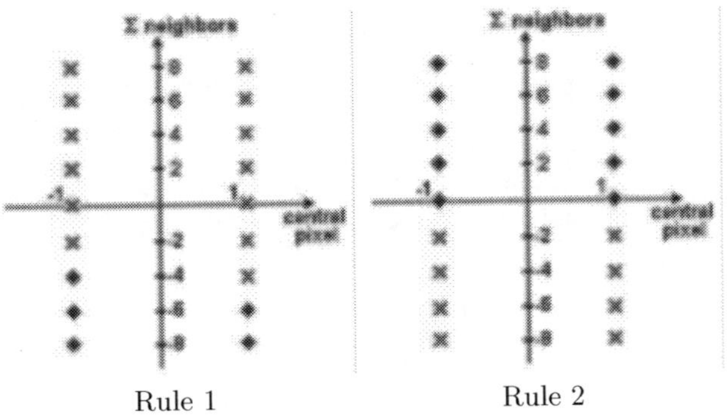

Fig. 3. *Rule 1* and *Rule 2* rules. The red cross corresponds to a 1 (black) for the next state, and a blue diamond is a -1 (white).

4 Design Algorithm for the Polynomial CNN

4.1 Remarks on the Polynomial CNN

The polynomial CNN model taken into consideration comprises 73 free parameters: 9 for the eight matrices (A,B,P0,P1,P2,Q0,Q1,Q2) plus the bias i. This huge design space would be impossible to be handled, unless we make some preliminary analysis of the problem in order to reduce the number of variables.

Firstly, since the GoL is a semitotalistic CA, all the matrices of the CNN model must have central symmetry: we indicate the central element with the subindex c, and the peripheral element with the subindex p. Secondly, it is noteworthy to mention that the GoL is a function of the input pattern only, therefore all the matrices multiplying (y^d) and $(y^d)^2$ - that is A, Q1 and Q2 - must have only the central element. Finally, we focus on the Eq. (5). As the input is binary, the last term of the equation is constant, and thus it can be added to the bias. Also the value $(p_0)^d_e(1^e)$ belonging to the first term is constant, then its central element can be absorbed into the matrix Q0. As for the second term, we can say that the central element of Q2 can be included in the matrix A.

To sum up, the polynomial CNN model for the GoL can be written as

$$x^d(n) = a^d_e y^e(n) + b^d_e u^e(n) + i + (q_0)^d_e (y^e)^2 + (p_1)^d_e (u^e) \cdot (y^d) \tag{6}$$

where

$$A = \begin{pmatrix} 0 & 0 & 0 \\ 0 & a_c & 0 \\ 0 & 0 & 0 \end{pmatrix}; B = \begin{pmatrix} b_p & b_p & b_p \\ b_p & b_c & b_p \\ b_p & b_p & b_p \end{pmatrix}; Q0 = \begin{pmatrix} 0 & 0 & 0 \\ 0 & q_{0c} & 0 \\ 0 & 0 & 0 \end{pmatrix}; P1 = \begin{pmatrix} p_{1p} & p_{1p} & p_{1p} \\ p_{1p} & p_{1c} & p_{1p} \\ p_{1p} & p_{1p} & p_{1p} \end{pmatrix}.$$

In conclusion, it is necessary to focus only on 7 parameters: $a_c, b_c, b_p, q_{0c}, p_{1c}, p_{1p}, i$.

4.2 Dynamic Behavior

In this section we show how it is possible to design a cloning template for the polynomial CNN to perform a step of the GoL on a binary image. All the possible combinations for the GoL are represented in Table 1, in which it is emphasized that the *Output* is the logic AND between the two *Rules*.

Table 1. List of the 18 possible combinations for the GoL

Row	Central pixel	# black neigh.	$\sum_{neigh.}$	*Rule* 1	*Rule* 2	Output
1.	White	0	-8	-	+	- (White)
2.	White	1	-6	-	+	- (White)
3.	White	2	-4	-	+	- (White)
4.	White	3	-2	+	+	+ (Black)
5.	White	4	0	+	-	- (White)
6.	White	5	2	+	-	- (White)
7.	White	6	4	+	-	- (White)
8.	White	7	6	+	-	- (White)
9.	White	8	8	+	-	- (White)
10.	Black	0	-8	-	+	- (White)
11.	Black	1	-6	-	+	- (White)
12.	Black	2	-4	+	+	+ (Black)
13.	Black	3	-2	+	+	+ (Black)
14.	Black	4	0	+	-	- (White)
15.	Black	5	2	+	-	- (White)
16.	Black	6	4	+	-	- (White)
17.	Black	7	6	+	-	- (White)
18.	Black	8	8	+	-	- (White)

Since our goal is to link the rules of the CA with the weights of a polynomial CNN, we write the Eq. (6) as a function of the central pixel u_c and sum of its neighbors $\sum u_p$, so to keep a representation as similar as possible to a Cellular Automaton

$$x^d(n) = u_c(b_c + p_{1c}y_c) + (b_p + p_{1p}y_c)\sum u_p + a_c y_c + q_c y_c^2 + i \qquad (7)$$

The first step of our approach consists in supplying the initial image as input $U = \{(u)_i\}$ of the network, whereas the initial state $Y = \{(y)_i\}$ of the network is composed by a matrix of 0's. Thanks to this mechanism, the first iteration of the CNN can be used to carry out one rule, whereas for the following iterations also the other rule will be taken into consideration.

Therefore, for the first iteration $y_c = 0$ and the state equation is

$$x_d^d(1) = u_c b_c + b_p \sum u_p + i \tag{8}$$

This means that, during the first iteration, we can perform any linearly separable task by setting properly b_c, b_p and i. For instance, in the following we assume that the *Rule* 1 is performed first; therefore, a possible choice for the variable is b_c=1, b_p=1 and $i = 3$, as justified by the Fig. 3. Obviously, when the *Rule* 2 is performed first, the values for the parameters change.

Now, for the sake of simplicity, we look for solutions where $p_{1c} = 0$. To sum up, the Eq. (7) can be simplified as follows

$$x^d(n+1) = u_c + (1 + p_{1p} y_c) \sum u_p + a_c y_c + q_c y_c^2 + 3, \tag{9}$$

and there are only three free parameters (p_{1p}, a_c, q_c).

In the following, we cluster the rows of Table 1 according to the value of *Rule 1* - that is, the result of the first iteration for the polynomial CNN - and *Output*, corresponding to the parameter y_c. Each case gives rise to a number of inequalities that can be collected in a system whose variables are a_c, p_{1p}, and q_c; if the system is compatible, the problem has at least one solution.

Case 1: $\textit{Previous state}(\textit{Rule}\,1) = + \Rightarrow y_c = 1;$

$\textit{Output} = + \ (\textit{black}) \Rightarrow x^d(n+1) > 0;$

$\textit{Rows} : 4, 12, 13 \Rightarrow (u_c, \sum u_p) = \{(-1,-2), (1,-4), (1,-2)\}$

Case 2: $\textit{Previous state}(\textit{Rule}\,1) = - \Rightarrow y_c = -1;$

$\textit{Output} = - \ (\textit{white}) \Rightarrow x^d(n+1) < 0;$

$\textit{Rows} : 1, 2, 3, 10, 11 \Rightarrow$

$\Rightarrow (u_c, \sum u_p) = \{(-1,-8); (-1,-6); (-1,-4); (1,-8); (1,-6)\}$

Case 3: $\textit{Previous state}(\textit{Rule}\,1) = + \Rightarrow y_c = +1;$

$\textit{Output} = - \ (\textit{white}) \Rightarrow x^d(n+1) < 0;$

$\textit{Rows} : 5, 6, 7, 8, 9, 14, 15, 16, 17, 18 \Rightarrow$

$\Rightarrow (u_c, \sum u_p) = \{(-1,0); (-1,2); (-1,4); (-1,6); (-1,8); (1,0); (1,2); (1,4); (1,6); (1,8)\}$

Moreover, for the Case 3 we need to assure that the *Output* -1 (white) is stable after the third iteration. Therefore, it is necessary to take into consideration also the case

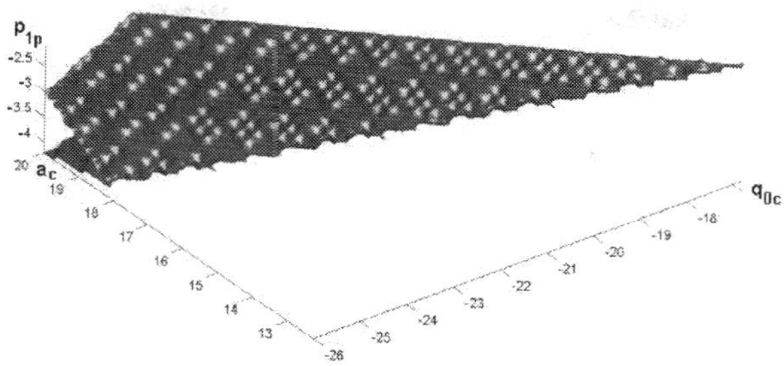

Fig. 4. Space of solutions when $p_{1c} = 0$ and $a_c > 0$

Case 4: *Previous state* $= - \Rightarrow y_c = -1;$

$$\begin{aligned}
&\textit{Output} = -\ (\textit{white}) \Rightarrow x^d(n+1) < 0;\\
&\textit{Rows} : 5, 6, 7, 8, 9, 14, 15, 16, 17, 18 \Rightarrow\\
&\Rightarrow (u_c, \sum u_p) = \{(-1,0); (-1,2); (-1,4); (-1,6); (-1,8); (1,0); (1,2);\\
&\qquad (1,4); (1,6); (1,8)\}
\end{aligned}$$

By solving this system of equations we can get the solution space that is depicted in Fig. 4. Note that it is not a polyhedron but an infinite space.

For example, when the *Rule* 1 is executed during the first iteration and the *Rule* 2 next, a possible solution to the system of inequalities is

$$a_c = 12.9,\ b_c = 1,\ b_p = 1,\ i = 3,\ q_{0c} = -17,\ p_{1c} = 0,\ p_{1p} = -2.1$$

On the other hand, when the *Rule* 2 is executed during the first iteration and the *Rule* 1 next, a possible solution becomes

$$a_c = 40.9,\ b_c = 0,\ b_p = -1,\ i = -2,\ q_{0c} = -26,\ p_{1c} = 1,\ p_{1p} = 4$$

Moreover, it is also possible to demonstrate that there exist only three types of solutions: in the first type all the parameters $(a_c, b_c, b_p, q_{0c}, p_{1c}, p_{1p}, i)$ are different from zero, whereas in the second and in the third types either p_{1c} or b_c, respectively, are 0. The solutions just described belong to the last two classes, then are the simplest possible (in the sense of solutions with more null parameters).

5 Conclusions

Standard one-layer space-invariant Cellular Neural Networks cannot solve linearly separable problems, and this fact represents a limitation for their use. In this paper we show that a one-layer space-invariant polynomial CNN can perform a single step of the 'Game of Life', a Cellular Automaton with the same computational complexity of a universal Turing machine. This result has been achieved thanks to the introduction of a design algorithm for the weights of the network. Such algorithm is very simple with respect to other CNN learning methods, being based on the resolution of a system of linear inequalities. Moreover, it allows to establish a direct link between the rules of a semitotalistic Cellular Automata and the learning procedure of a polynomial CNNs.

In this paper we have focused on the 'Game of Life' because it is a sort of benchmark for testing the capability of representation of polynomial CNNs. However, our long-term objective is to investigate the advantages of polynomial CNNs over standard CNNs for complex image processing tasks. Furthermore, we will soon compare the performances of the two systems by emulating them on a Field Programmable Gate Array (FPGA). In the near future, we are also going to prove theoretically, through the design procedure introduced in this paper, the tight connection between semitotalistic CA and polynomial CNNs.

References

1. Chua, L., Yang, L.: Cellular neural networks: theory. IEEE Trans. Circuits Syst. 35, 1257–1272 (1988)
2. Chua, L., Roska, T.: The cellular neural network CNN and the CNN universal machine: concept, architecture and operation modes. In: Roska, T., Rodríguez-Vázquez, Á. (eds.) Towards the visual microprocessor, Wiley, Chichester, UK (2001)
3. Yang, L., Chua, L., Krieg, K.: VLSI implementation of cellular neural networks. In: ISCAS'90. Proc. IEEE International Symposium on Circuits and Systems, vol. 3, pp. 2425–2427. IEEE Computer Society Press, Los Alamitos (1990)
4. Balsi, M.: Generalized CNN: potentials of a CNN with non-uniform weights. In: CNNA'92. Proc. second IEEE International Workshop on Cellular Neural Networks and their Applications, Munich, Germany, pp. 129–139. IEEE Computer Society Press, Los Alamitos (1992)
5. Yang, Z., Nishio, Y., Ushida, A.: Templates and algorithms for two-layer cellular neural networks neural networks. In: Proc. of IJCNN'02, vol. 2, pp. 1946–1951 (2002)
6. Bilgili, E., Goknar, I., Ucan, O.: Cellular neural networks with trapezoidal activation function. International journal of Circuit Theory and Applications 33, 393–417 (2005)
7. Dogaru, R., Chua, L.: Universal CNN cells. International Journal of Bifurcation and Chaos 9(1), 1–48 (1999)
8. Roska, T., Chua, L.O.: The CNN universal machine: an analogic array computer. IEEE Trans. Circuits Syst. II 40, 163–173 (1993)
9. Chua, L.O., Roska, T., Venetianer, P.: The CNN is universal as the Turing machine. IEEE Trans. Circuits Syst. I 40(4), 289–291 (1993)
10. Laiho, M., Paasio, A., Kananen, A., Halonen, K.A.I.: A mixed-mode polynomial cellular array processor hardware realization. IEEE Trans. Circuits Syst. I 51(2), 286–297 (2004)
11. Berlekamp, E., Conway, J.H., Guy, R.K.: Winning ways for your mathematical plays. Academic Press, New York (1982)
12. Magnussen, H., Nossek, J.: Global learning algorithms for discrete-time cellular neural networks. In: CNNA'94. Proc. third IEEE International Workshop on Cellular Neural Networks and their Applications, Rome, Italy, pp. 165–170. IEEE Computer Society Press, Los Alamitos (1994)

13. Harrer, H., Nossek, J.: Discrete-time cellular neural networks. International Journal of Circuit Theory and Applications 20, 453–467 (1992)
14. Schonmeyer, R., Feiden, D., Tetzlaff, R.: Multi-template training for image processing with cellular neural networks. In: CNNA'02. Proc. of 2002 7th IEEE International Workshop on Cellular Neural Networks and their Applications, Frankfurt, Germany, IEEE Computer Society Press, Los Alamitos (2002)
15. Corinto, F.: Cellular Nonlinear Networks: Analysis, Design and Applications. PhD thesis, Politecnico di Torino, Turin (2005)
16. Gómez-Ramírez, E., Pazienza, G.E., Vilasís-Cardona, X.: Polynomial discrete time cellular neural networks to solve the XOR problem. In: CNNA'06. Proc. 10th International Workshop on Cellular Neural Networks and their Applications, Istanbul, Turkey (2006)
17. Niederhofer, C., Tetzlaff, R.: Recent results on the prediction of EEG signals in epilepsy by discrete-time cellular neural networks DTCNN. In: ISCAS'05. Proc. IEEE International Symposium on Circuits and Systems, pp. 5218–5221. IEEE Computer Society Press, Los Alamitos (2005)
18. Rendell, P.: A Turing machine in Conway's game life (2006), Available: www.cs.ualberta.ca/~{}bulitko/f02/papers/tmwords.pdf

Deterministic Nonlinear Spike Train Filtered by Spiking Neuron Model

Yoshiyuki Asai[1,2], Takashi Yokoi[1], and Alessandro E.P. Villa[2,3,4]

[1] National Institute of Advanced Industrial Science and Technology (AIST), Tsukuba, Japan
{yoshiyuki.asai, takashi-yokoi}@aist.go.jp
[2] NeuroHeuristic Research Group, Information Science Institute
University of Lausanne, Switzerland
{oyasai, avilla}@neuroheuristic.org
[3] Université Joseph Fourier, NeuroHeuristic Research Group, Institut des Neurosciences, Grenoble, F-38043, France
INSERM, U836, Grenoble, F-38043, France
alessandro.villa@ujf-grenoble.fr
http://www.neuroheuristic.org/

Abstract. Deterministic nonlinear dynamics has been observed in experimental electrophysiological recordings performed in several areas of the brain. However, little is known about the ability to transmit a complex temporally organized activity through different types of spiking neurons. This study investigates the response of a spiking neuron model representing three archetypical types (regular spiking, thalamo-cortical and resonator) to input spike trains composed of deterministic (chaotic) and stochastic processes with weak background activity. The comparison of the input and output spike trains allows to assess the transmission of information contained in the deterministic nonlinear dynamics. The pattern grouping algorithm (PGA) was applied to the output of the neuron to detect the dynamical attractor embedded in the original input spike train. The results show that the model of the thalamo-cortical neuron can be a better candidate than regular spiking and resonator type neurons in transmitting temporal information in a spatially organized neural network.

1 Introduction

A cell assembly–a neuronal network composed of functionally related units–may be considered as a highly complex nonlinear dynamical system able to exhibit deterministic chaotic behavior, as suggested by experimental observations (Mpitsos, 1989; Celletti and Villa, 1996; Villa et al., 1998). The time series corresponding to the sequence of the spikes of a neuron is referred to as "spike train" and can be searched for detecting embedded temporal structures. Previous studies (Tetko and Villa, 1997; Asai et al., 2006) showed that deterministic nonlinear dynamics in noisy time series could be detected by applying algorithms aimed at finding preferred firing sequences with millisecond order time precision that

J. Marques de Sá et al. (Eds.): ICANN 2007, Part I, LNCS 4668, pp. 924–933, 2007.

repeated more frequently than expected by chance from simultaneously recorded multivariate time series of neural activities. Then, a neuron belonging to a cell assembly is expected to receive input spike trains having a temporal structure that reflects the underlying nonlinear dynamical system. Neuronal responses to stochastic time series (Bulsara et al., 1991; Takahata et al., 2002) and periodic inputs (Nomura et al., 1994; Pakdaman, 2001) have been intensively studied from the view point of the neural coding. However there are less works shedding light on neuronal response to an input spike train whose inter-spike-intervals have temporal structure related to deterministic process.

The purpose of the present work is to investigate how different types of archetypical neurons (i.e., thalamo-cortical, regular spiking, resonator) react to an input spike train generated by deterministic dynamical system in presence of "noisy" processes (Poissonian and Gaussian). The comparison between the output spike train and the input chaotic dynamics is carried out by applying the pattern grouping algorithm (Tetko and Villa, 2001; Abeles and Gat, 2001) to both time series and evaluating the degree of preservation of the original dynamics.

2 Spiking Neuron Model

The simple spiking neuron (Izhikevich, 2003) considered in this study can reproduce several known types of neuron dynamics. The model is described as follows;

$$\frac{dv}{dt} = 0.04v^2 + 5v + 140 - u + I_{bg} + I_{ext} \qquad (1)$$
$$\frac{du}{dt} = a(bv - u) \ ,$$

with the auxiliary after-spike resetting, $v \leftarrow c$ and $u \leftarrow u + d$ when $v \geq +30\ mV$. The variable v represents the membrane potential (in millivolt units) of the neuron, and u is a membrane recovery variable giving negative feedback to v. The time unit is millisecond. When the membrane potential reaches $+30\ mV$, the state variables are reset with amounts given by parameters c and d, respectively. Note that $v = +30\ mV$ is not a threshold for generating spikes, but it corresponds to the peak of action potentials. Parameters a and b control the time scale of the recovery variable and its sensitivity to the subthreshold fluctuation of the membrane potential.

Three neuron types were considered in this study according to the literature (Izhikevich, 2003; Izhikevich, 2004): (*i*) a neo-cortical neuron of regular spiking (RS) type with parameters $a = 0.02$, $b = 0.2$, $c = -65$, $d = 8$; (*ii*) a thalamo-cortical (TC) neuron with parameters $a = 0.02$, $b = 0.25$, $c = -65$, $d = 2$; (*iii*) a neuron type characterized by its sustained subthreshold oscillation of the membrane potential called resonator (RZ) with parameters $a = 0.1$, $b = 0.25$, $c = -65$, $d = 2$.

I_{bg} is a background activity simulated by square pulses lasting 3 ms distributed according to a Poissonian distribution with an average firing rate of

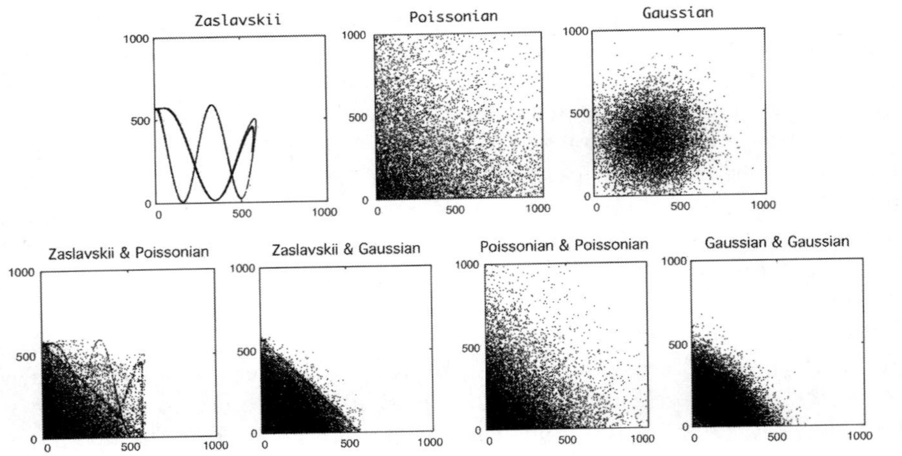

Fig. 1. Return maps of input spike trains. The axes are scaled in *ms* time units. Upper row: single inputs, $N = 10,000$ points, 3 *spikes/s*. Lower row: combined inputs, $N = 20,000$ points, 6 *spikes/s*.

5 *spikes/s*. The amplitude of background pulses was set to 5.5, 1.0 and 0.5 for RS, TC and RZ neurons, respectively. The duration and amplitude of the pulses were selected in order to allow a high membrane excitability with a minimum of spurious spikes generated by the background noise alone. Because of the nonlinearity of the model, few spikes were nevertheless generated by the background noise, corresponding to rates of 0.09, 0.09 and 0.07 *spikes/s* for RS, TC and RZ neurons, respectively.

I_{ext} is an external input spike train to the neuron corresponding to square-like pulse current lasting 1 *ms*. The amplitude of the external pulses is tuned for each neuron type such that a single input pulse can evoke an output spike at resting membrane condition. The amplitude of the external input was set to 17, 5 and 2.5 for RS, TC and RZ neurons, respectively.

3 Input Spike Train

We considered three types of single input spike trains corresponding to a deterministic process generated by the Zaslavskii map and to Poissonian and Gaussian stochastic processes. Ten thousand points ($N = 10,000$) were generated in each series. The Zaslavskii map (Zaslavskii, 1978) is described by the following equation:

$$\begin{cases} x_{n+1} = x_n + v(1 + \mu y_n) + \varepsilon v \mu \cos x_n & (mod.\ 2\pi) \\ y_{n+1} = e^{-\gamma}(y_n + \varepsilon \cos x_n)\ , \end{cases} \tag{2}$$

where $x,\ y \in \mathbf{R}$, the parameters are real numbers with $\mu = \frac{1-e^{-\gamma}}{\gamma}$, $v = \frac{4}{3} \cdot 100$ and initial conditions set to $x_0 = 0.3$ and $y_0 = 0.3$. With this parameter

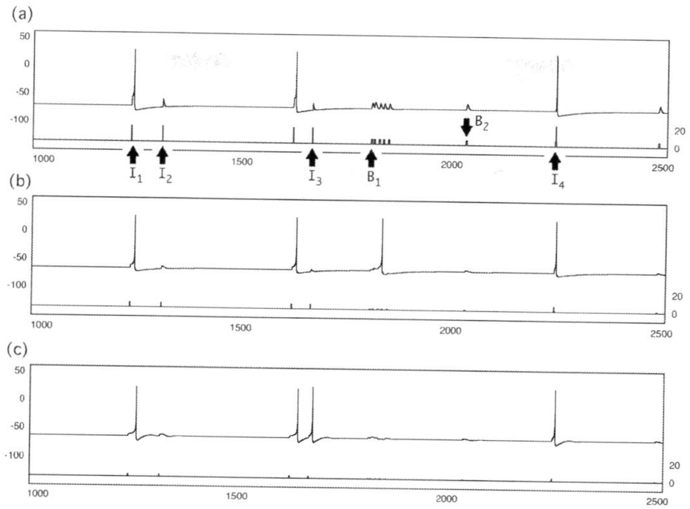

Fig. 2. (a) Regular spiking (RS), (b) thalamo-cortical (TC), (c) Resonator (RZ) neuronal types. The time-series of the membrane potential and of the input spike train are illustrated in the upper and lower traces in each panel. I_1 marks a single input strong enough to evoke a spike in all neuronal types at rest. I_2 marks an input spike arriving shortly after the previous input, such that no spike is generated due to neuron's refractoriness. I_3 marks an input spike arriving also shortly after the previous input but due to peculiar internal RZ neuronal dynamics the temporal summation at neuronal membrane is enough to evoke an output spike. B_1 indicate burst of background activities that can be integrated in the TC neuron such to evoke an output spike. B_2 indicates an isolated single pulse of background activity unable to evoke an output whereas I_4 indicates an input spike superimposed on the background activity.

set, the system exhibits a chaotic behavior. Time series $\{x_n\}$ are generated by iterative calculation. A new time series $\{w_n\}$ is derived by taking the difference between two adjacent values of $\{x_n\}$, and adding a constant C to make all values positive, *i.e.* , $w_n = x_{n+1} - x_n + C$, where $C = \min\{(x_{n+1} - x_n)\} + 0.1$. The time series $\{w_n\}$ were assumed to correspond to the sequence of the inter-spike-intervals. The dynamics was rescaled in milliseconds time units with an average rate of 3 *events/s* (*i.e.*, 3 *spikes/s*) in order to let the *Zaslavskii spike train* be comparable to neurophysiological experimental data. Inter-spike-intervals generated according to the Exponential distribution with an average rate of 3 *events/s* were referred to *Poissonian spike train*. Inter-spike-intervals generated according to a Gaussian distribution with standard deviation set to 150 *ms* and an average rate of 3 *events/s* were referred to *Gaussian spike train*. Return maps of the three single input spike trains, corresponding to the plots of the n-th inter-spike-interval against the $(n-1)$-th inter-spike-interval, are displayed in the upper row of Fig. 1. Notice the typical structure of the attractor of Zaslavskii map in the leftmost panel.

Four types of complex input spike trains were obtained by merging two simple spike trains with the following combinations: Zaslavskii and Poisson spike

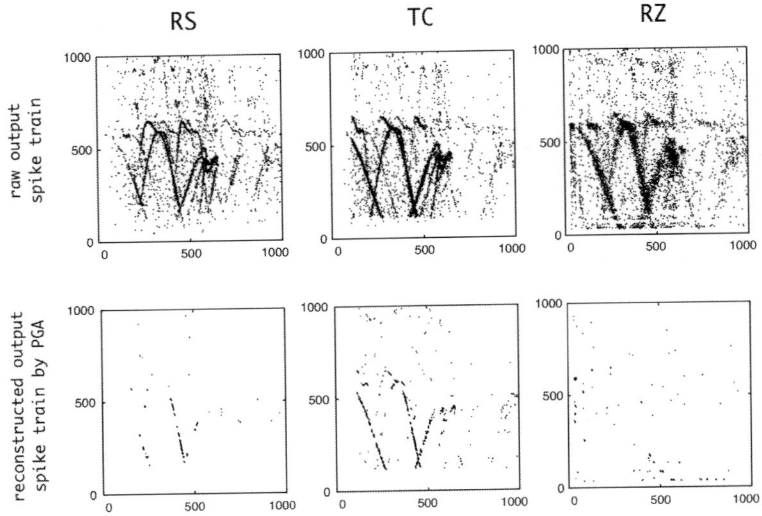

Fig. 3. Return maps of raw output spike trains and reconstructed output spike trains by PGA for an input Zaslavskii spike train with background activity

train, Zaslavskii and Gaussian, Poisson and Poisson, Gaussian and Gaussian. The return maps of the lower row of Fig. 1 show that the Gaussian spike train could completely destroy the Zaslavskii attractor, whereas the attractor could still be recognized in the combination with the Poissonian spike train.

In addition to three simple and four complex inputs, we considered also the inputs in combination with the background activity described above, that means fourteen cases overall (3 simple, 4 complex, 3 simple with background, 4 complex with background).

4 Filtering and Reconstruction of Spike Train

Spatiotemporal firing patterns, also called preferred firing sequences, are defined as a sequences of spikes (formed by three or more) with high temporal precision of the order of milliseconds. The Pattern Grouping Algorithm (PGA) (Tetko and Villa, 2001; Abeles and Gat, 2001) is aimed at finding all sequences of events that repeat at least 5 times within the time series and to estimate the significance of this repetition above chance levels. In the current study the three important parameters controlling the PGA performance were set as follows: the maximal duration of the temporal pattern was 900 ms, the time precision of the events within the temporal pattern (called *jitter*) was 3 ms, the level of significance of each pattern was $p = 0.01$. All events belonging to all recurrent firing sequences detected by PGA are pooled together keeping their original timing and form a subset of the original spike train (Asai et al., 2006).

Table 1. Number of points in the raw and PGA reconstructed ("rec") spike train for the Zaslavskii, Poissonian and Gaussian input spike train with (BG) and without the background activity (no BG). The input spike trains included 10,000 points with a rate of 3 *spikes/s*.

Input spike train		*Regular Spiking*		*Thalamo-Cortical*		*Resonator*	
		raw	rec	raw	rec	raw	rec
Zaslavskii	no BG	7392	2753	8234	4424	8345	3988
	BG	6597	864	7834	2448	7115	359
Poissonian	no BG	6197	0	7488	93	7567	0
	BG	5643	0	7173	0	6724	21
Gaussian	no BG	8354	107	9266	129	9036	174
	BG	7343	0	8716	66	7517	0

5 Differential Effect of Neuronal Type Dynamics

Figure 2 briefly illustrates some examples of different responses to input and background activity as a function of the selected neuronal type. At first we investigated the response of the neuron model to the simple inputs with background activity. In presence of an input Zaslavskii spike train with background activity one can recognise in the return map of the output train of all neuronal types (Fig. 3 upper panels) the sets of points which reflect the original Zaslavskii attractor. Notice also margins without points near the ordinate and abscissa axis in each panel due to the refractoriness of the neuronal dynamics.

In order to investigate how the deterministic component of the original input spike train is transmitted by the neuron we searched recurrent firing sequences in the output spike trains (Fig. 3 lower panels). The method demonstrated its efficacy in removing almost all random noisy points seen in the return maps of the raw neuron outputs and the attractor was clearly visible for the TC neuron and for RS to a lesser extent. In the case of the resonator (RZ) PGA with the current control parameters detected only 5 % of the spikes (Fig. 3 bottom-right) that were contained in the raw output (Fig. 3 upper-right). In fact the RZ neuron is somewhat blurring the original attractor (compare Fig. 3 upper raw) and introduces additional jitter associated to the internal oscillatory dynamics. We applied PGA with jitter equal to 5 ms instead of 3 ms as initially set for all analyses and we could find 22 % spikes in the RZ reconstructed output (figure not shown).

The count of spikes involved in the output spike trains and in the corresponding reconstructed spike train are summarised in Table 1. For all types of neuron, the background activity tended to reduce the number of spikes in the output train. Notice also that spurious patterns were detected by PGA when the neuron receives stochastic inputs, such as Poissonian and Gaussian spike trains, but the amount of these points is very little compared to the other cases.

The response of the neuron to the complex inputs in presence of the background activity is shown in Fig. 4. In the combined Zaslavskii and Poissonian

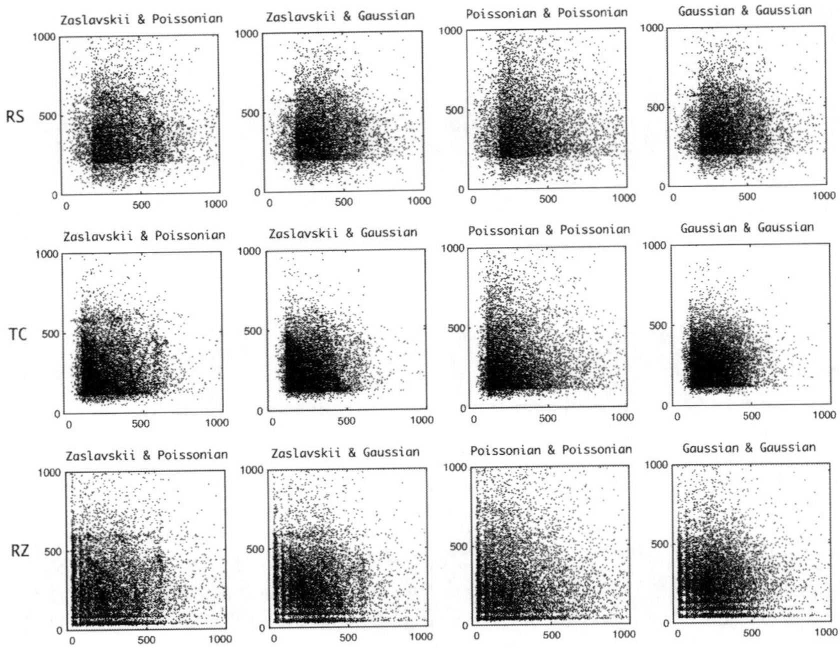

Fig. 4. Return maps of raw output spike trains for combined inputs (as labeled) with background activity. RS: regular spiking neuron; TC: thalamo-cortical neuron; RZ: resonator.

input spike train (see Fig. 1 lower row) the dynamical structure was still present and it appears that all types of neurons are able to preserve some deterministic temporal structure in the inputs (Fig. 4 leftmost column). Interestingly, although the original Zaslavskii attractor was completely destroyed in Zaslavskii and Gaussian spike train (Fig. 1), it appeared that the output spike train preserved a kind of dim shade of the original Zaslavskii attractor in the return maps of TC and RZ neurons and of RS to a lesser extent (Fig. 4 second column). The control analyses with a combination of Gaussian and Poissonian inputs did not show any particular feature other than the known effects due to the internal dynamics, such as the "stripes" of the RZ neuron associated to its subthreshold oscillatory dynamics.

Likewise with a single input we investigated how the deterministic component of the combined original input spike train that is transmitted through the neuron by means of PGA reconstruction (Fig. 5). The method demonstrated its efficacy in removing almost all random noisy points seen in the return maps of the raw neuron outputs and the attractor was clearly visible for the TC neuron and for RS to a lesser extent. In the case of TC neuron, PGA was able to detect several points relevant to the original Zaslavskii attractor (Fig. 5 two leftmost columns).

Table 2 shows that the numbers of points in the reconstructed spike trains of TC neuron is about the half of what was observed for the single input train (Table 1). Conversely, for RS and RZ neurons, few relevant points only could

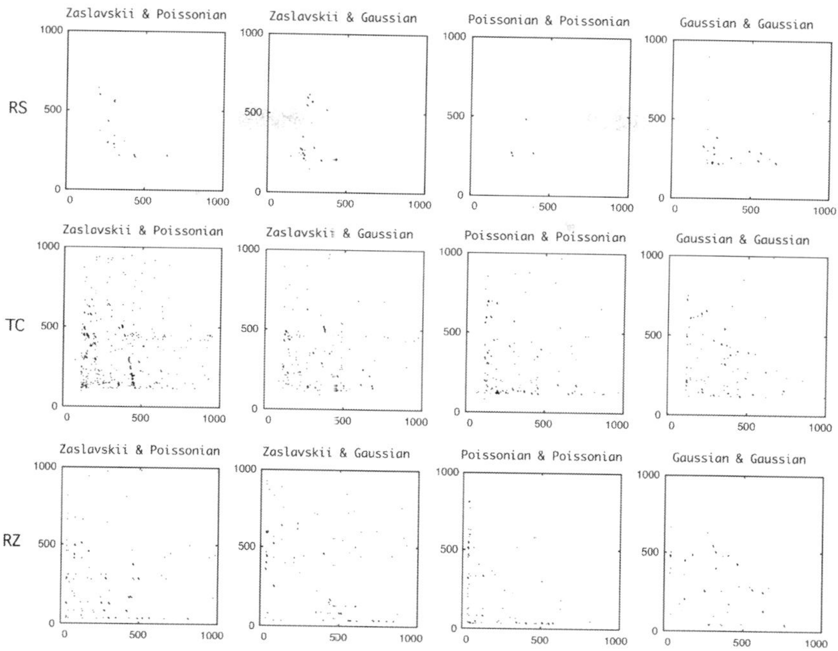

Fig. 5. Return maps of reconstructed output spike trains by PGA for combined inputs (as labeled) with background activity. RS: regular spiking neuron; TC: thalamo-cortical neuron; RZ: resonator.

Table 2. Number of points in the raw and PGA reconstructed ("rec") spike train for the Zaslavskii, Poissonian and Gaussian input spike train with (BG) and without the background activity (no BG). The overall input spike trains included 20,000 points with an overall rate of 6 *spikes/s*.

		Regular Spiking		*Thalamo-Cortical*		*Resonator*	
		raw	rec	raw	rec	raw	rec
Zaslavskii	no BG	9220	1188	12224	1962	12902	1373
& Poisson	BG	8544	194	11725	1632	11987	746
Zaslavskii	no BG	9735	977	12951	1532	13207	1608
& Gaussian	BG	8972	350	12344	978	12111	621
Poissonian	no BG	8836	324	11852	1100	12704	763
& Poissonian	BG	8203	80	11390	777	11863	553
Gaussian	no BG	10002	944	13407	1581	13446	1059
& Gaussian	BG	9178	279	12792	622	12421	393

be detected and their number was close to what is detected in the control tests containing pure stochastic inputs, as indicated also in Table 2. Table 2 shows also that the effect of background is less dramatic than in the single input and that a strong stochastic input is able to produce a significant amount of preferred firing sequences that are only related to the internal dynamics and not the deterministic input.

6 Discussion

We investigated the response of the spiking neuron model to input spike trains characterized by deterministic temporal structure or stochastic properties. Three types of cortical neuron *i.e.* regular spiking (RS), thalamo-cortical (TC) and resonator (RZ) neurons were modeled following (Izhikevich, 2003; Izhikevich, 2004). The effect of the difference in dynamics to transmit the deterministic temporal structure in the input spike train to the output was also examined.

Regular spiking neuron (RS) had a rather long refractory period and hence could not generate neural discharges quickly when the neuron received rapid sequence of input spikes. On the opposite, RZ is characterized by subthreshold oscillatory membrane potential and could respond quickly, but its feature clearly appeared in the stripe patterns of the return map of the output spike train. Latency to generate neural discharge of RZ after receiving an input spike can widely vary according to the neuron's state. Hence weak spikes in the background activity were effective to vary the timing of neural discharges corresponding to input spikes, leading to dim shape of the attractor seen in Fig. 4. Because of this nonlinearity, PGA could detected less spikes in the output of RZ neuron even when simple and complex input spike trains included Zaslavskii spike train with the background activity. TC neuron appeared to be the model that was better able to preserve the input deterministic dynamics. In this case PGA could detect spike sets which corresponded to parts of the original Zaslavskii attractor even with a combined input that included a strong stochastic noise (Fig. 5).

The fact that thalamo-cortical neurons appear as a good candidate for the transmission of temporal information may raise several hypotheses on the role of the thalamo-cortical loop with respect to feed-forward networks formed by cortical regular spiking neurons (Villa, 2002). The current work is only at an incipient state but it clearly address the need to investigate in detail the ability of transmitting detailed temporal information as a function of neuronal internal dynamics. Furthermore, the effect of background activity and the nature of additional stochastic inputs might act as additional controlling parameters on the transmission of temporal information through chains of different types of neurons.

Acknowledgements

This study was partially funded by the bi-national JSPS/INSERM grant SYRNAN and Japan-France Research Cooperative Program.

References

Abeles, M., Gat, I.: Detecting precise firing sequences in experimental data. Journal of Neuroscience Methods 107, 141–154 (2001)

Asai, Y., Yokoi, T., Villa, A.E.P.: Detection of a dynamical system attractor from spike train analysis. In: Kollias, S., Stafylopatis, A., Duch, W., Oja, E. (eds.) ICANN 2006. LNCS, vol. 4131, pp. 623–631. Springer, Heidelberg (2006)
Bulsara, A., Jacobs, E.W., Zhou, T., Moss, F., Kiss, L.: Stochastic resonance in a single neuron model: theory and analog simulation. J. Theor. Biol. 152, 531–555 (1991)
Celletti, A., Villa, A.E.P.: Low dimensional chaotic attractors in the rat brain. Biological Cybernetics 74, 387–394 (1996)
Izhikevich, E.M.: Simple model of spiking neurons. IEEE Transactions on Neural Networks 14, 1569–1572 (2003)
Izhikevich, E.M.: Which model to use for cortical spiking neurons? IEEE Transactions on Neural Networks 15, 1063–1070 (2004)
Mpitsos, G.J.: Chaos in brain function and the problem of nonstationarity: a commentary. In: Basar, E., Bullock, T.H. (eds.) Dynamics of sensory and cognitive processing by the brain, pp. 521–535. Springer, London (1989)
Nomura, T., Sato, S., Segundo, J.P., Stiver, M.D.: Global bifurcation structure of a bonhoeffer-van der pol oscillator driven by periodic pulse trains, Biol. Cybern. vol. 72, pp. 55–67 (1994)
Pakdaman, K.: Periodically forced leaky integrate-and-fire model. Physical Review E 63, 41907 (1907)
Takahata, T., Tanabe, S., Pakdaman, K.: White-noise stimulation of the hodgkin-huxley model. Biol. Cybern. 86, 403–417 (2002)
Tetko, I.V., Villa, A.E.: A comparative study of pattern detection algorithm and dynamical system approach using simulated spike trains. In: Gerstner, W., Hasler, M., Germond, A., Nicoud, J.-D. (eds.) ICANN 1997. LNCS, vol. 1327, pp. 37–42. Springer, Heidelberg (1997)
Tetko, I.V., Villa, A.E.P.: A pattern grouping algorithm for analysis of spatiotemporal patterns in neuronal spike trains. 1. detection of repeated patterns. J. Neurosci. Meth. 105, 1–14 (2001)
Villa, A.E.P.: Cortical modulation of auditory processing in the thalamus. In: Lomber, S.G., Galuske, R.A.W. (eds.) Virtual lesions: Examining Cortical Function with reversible Deactivation, ch. 4, pp. 83–119. Oxford University Press, Oxford (2002)
Villa, A.E.P., Tetko, I.V., Celletti, A., Riehle, A.: Chaotic dynamics in the primate motor cortex depend on motor preparation in a reaction-time task. Current Psychology of Cognition 17, 763–780 (1998)
Zaslavskii, G.M.: The simplest case of a strange attractor. Phys. Let. 69A, 145–147 (1978)

The Role of Internal Oscillators for the One-Shot Learning of Complex Temporal Sequences

Matthieu Lagarde, Pierre Andry, and Philippe Gaussier

ETIS, Neurocybernetic Team, UMR CNRS 8051
2, avenue Adolphe-Chauvin, University of Cergy-Pontoise, France
{lagarde,andry,gaussier}@ensea.fr

Abstract. We present an artificial neural network used to learn online complex temporal sequences of gestures to a robot. The system is based on a simple temporal sequences learning architecture, neurobiological inspired model using some of the properties of the cerebellum and the hippocampus, plus a diversity generator composed of CTRNN oscillators. The use of oscillators allows to remove the ambiguity of complex sequences. The associations with oscillators allow to build an internal state to disambiguate the observable state. To understand the effect of this learning mechanism, we compare the performance of (i) our model with (ii) simple sequence learning model and with (iii) the simple sequence learning model plus a competitive mechanism between inputs and oscillators. Finally, we present an experiment showing a AIBO robot, which learns and reproduces a sequence of gestures.

1 Introduction

Our long term goal is to build an autonomous robot able to learn sensorimotor tasks. Such a system should be (i) able to acquire new "behaviors" : gestures, objects manipulation as sequences combining multimodal elements of different levels. To do this, an autonomous system must (ii) take advantage of information of the associations between vision and motor capabilities. This paper focuses essentially on the first point : learning, predicting and reproduction of complex sensorimotor sequences.

In this scope, solutions based on neural networks are an interesting solution. Neural networks are able to learn sequences using associative mechanisms. Moreover, these networks offer a level of coding (neuron) that takes into account information about the lower sensorimotor system; such systems avoid the use of symbols or information that could separate the sequence learning component from the building of associations between sensation and action. Networks are adapted to online learning favoring easier interactions with humans and other robots. Among these models, chaotic neural networks are based on recurrent network (RN). In [1], a fully connected RN learns a sequence thanks to a single layer of neurons. The dynamics generated by the network help to learn a short sequence. After a few iterations, the learned sequence vanishes progressively. In [2], a random RN (RRN) learns a sequence thanks to a combination of two

J. Marques de Sá et al. (Eds.): ICANN 2007, Part I, LNCS 4668, pp. 934–943, 2007.

layers of neurons. The first layer generates an internal dynamic by means of a RRN. The second layer generates a resonance phenomenon. The network learns short sequences of 7 or 8 states. But this model is highly sensitive to noises or the stimulus variations and does not learn long periods sequences. A similar model is the Echo States Network (ESN) based on RRN for short term memory [3] (STM). Under certain conditions (detailed in [4]), the activation of each neuron in the hidden layer, is a function of the input history presented to the network; this is the echo function. Once again, the idea is to use a "reservoir" of dynamics from which the desired output is learned in conjunction with the effect of the input activity.

In the context of robotics, many models concern gesture learning. By means of nonlinear dynamical systems, [5] develops control policies to approximate the recorded movements and to learn them with a fitting of mixture model using a recursive least square regression technique. In [6], the trajectories of gestures are acquired by the construction of motor skills with a probalistic representation of the movement. Trajectories can be learnt through via points [7] with parallel vector-integration-to-endpoint models [8]. In our work, we wish to be able to re-use and detect subsequences and possibly, combine them. Thus, we need to learn some of the important components of the sequence and not only to approximate the trajectory of the gesture.

In this paper, we present a biologically inspired model of neural network for temporal complex sequences learning. A first approach described in [9] proposes a neural network for the online learning of the timing between events for simple sequences (with non ambiguous states like "A B C"). We propose a model for complex sequences (with ambiguous states like A and B in "A B A C B"). In order to remove the ambiguous states or transitions, we use batteries of oscillators as a reservoir of diversity allowing to separate the inputs appearing repeatedly in the sequence. In section 3, we show results from simulations comparing the performances of 3 different systems involved in the learning and reproduction of the same set of complex sequences : (i) the system described in [9], (ii) this system plus a simple competitive mechanism between the oscillators and the input (showing the effect of adding internal dynamics in order to separate ambiguous states) and (iii) a system optimizing the use of the oscillators by using an associative learning rule in order to recruit new internal states when needed (repetition of the same input state). Section 4 details the application of our model on a real robot for the learning of a complex gesture. Finally, we conclude and point out some open problems.

2 A Model for Timing and Sequence Learning

The architecture (Fig. 1) is based on a neurobiological model [10] inspired from some of the properties of the cerebellum and the hippocampus. This model uses associative learning rules between past inputs memorized as a STM and present inputs in order to learn the timing of simple sequences. "Simple" refers here to the sequences in which the same state appears only once. The main

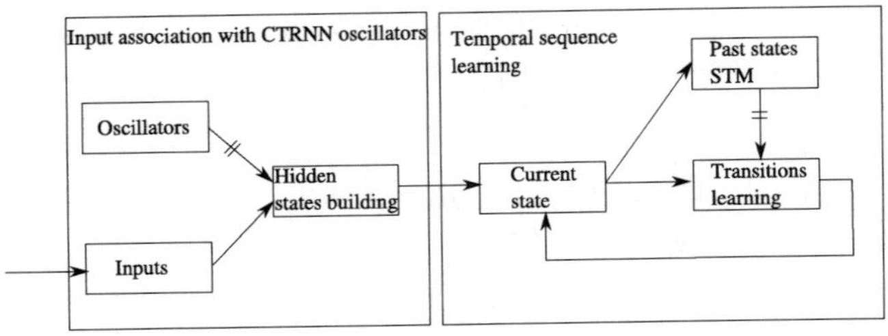

Fig. 1. Complex sequences learning model. Barred links are modifiable connexions. The others are associated to unmodifiable connexions. The left part is detailed in figure 3.A and 3.B. The right part is detailed in figure 4.

advantage of this model is that the associative mechanism also learns the timing of the sequence, which allows accurate predictions of the transitions that compose the sequence. In order to learn complex sequences in which the same state is repeated several times, in our model we have added a mechanism that generates internal dynamics and that can be associated with the repeated inputs of the sequence. The association between the repeated inputs and different activities of the oscillator allows to code hidden states with different and un-ambiguous patterns of activities. As a result, our architecture manages to learn/predict and reproduce complex temporal sequences.

2.1 Generating Internal Diversity

Oscillators are very much used in robotic applications like locomotion using central pattern generator (CPG) [11]. An oscillator is a continuous time recurrent neural network (CTRNN) composed of two neurons (Fig. 2.A). The study on CTRNNs can be found in [12]. This kind of oscillators is known for stability, and resistance to the noises. CTRNN are easy to implement too. A CTRNN coupling two neurons produces an oscillator (Fig. 2.B) :

$$\tau_e.\frac{dx}{dt} = -x + S((w_{ii} * x) - (w_{ji} * y) + w_{e_{const}}) \quad (1)$$

$$\tau_i.\frac{dy}{dt} = -y + S((w_{jj} * y) + (w_{ij} * x) + w_{i_{const}}) \quad (2)$$

with τ_e a time constant for the excitatory neuron and τ_i for the inhibitory neuron. x and y are the activity of the excitatory and the inhibitory neuron respectively. w_{ii} is the weight of the recurrent link of the excitatory neuron, w_{jj} the weight of the recurrent link of the inhibitory neuron. w_{ij} is the weight of the link from the excitatory neuron to inhibitory neuron. w_{ji} is the weight of the link from the inhibitory neuron to excitatory neuron. $w_{e_{const}}$ and $w_{i_{const}}$ are the weights of the links from the constant inputs. And S is the transfer function of each neuron. In our model, we use three oscillators with $\tau_e = \tau_i$.

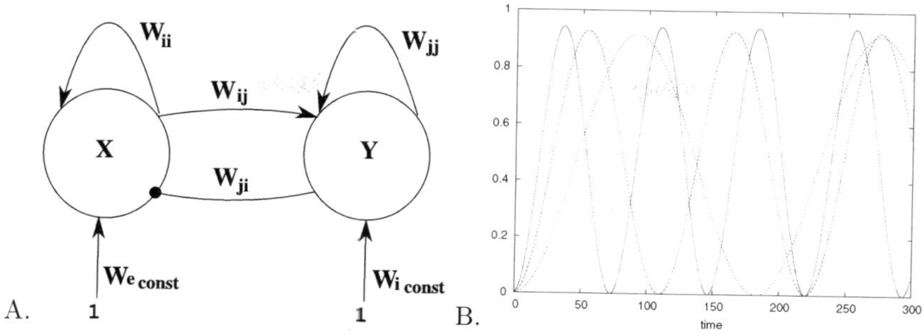

Fig. 2. A. Oscillator model. Left neuron is excitatory and right neuron is inhibitory. Excitatory links are : $W_{ii} = 1$, $W_{jj} = 1$, $W_{ij} = 1$. Inhibitory links is : $W_{ji} = -1$, Constant input value is equal to 1 with constant links $W_{e_{const}} = 0.45$ and $W_{i_{const}} = 0$. Initial activities of neurons are $X(0) = 0$, $Y(0) = 0$.
B. Display of the instantaneous mean frequency activity of 3 oscillators systems with $\tau_1 = 20$ (plain line), $\tau_2 = 30$ (long dashed line), $\tau_3 = 50$ (short dashed line).

2.2 Learning of Internal States

In order to use repeatedly the same input in a given sequence, different configurations of oscillators can be associated with the same input. To understand the generation of diversity and its implication in our learning algorithm, we have tested two mechanisms : a simple competition coupling input states with oscillators (Fig. 3.A) and an associative mechanism based on a learning rule (Fig. 3.B) that recruits neurons according to the activities of the oscillators and the repeated inputs.

Competitive Mechanism. The competition is computed as follow : each neuron ij of the Competition group acts as an neuron performing the logical operator AND between the neurons of the Inputs group and of the Oscillators group :

$$Pot_{ij} = (w_{input_i} * x_{input_i} + w_{osci_j} * x_{osci_j}) - threshold_{ij} \quad (3)$$

with $w_{input_i} = 1$, $w_{osci_j} = 1$, $threshold_{ij} = 1.2$, x_{input_i} the activity of the input at index i and x_{osci_j} the activity of the oscillator at index j.

In a second step, a competition between all neurons ij of the Competition group is applied :

$$Winner_{ij} = \begin{cases} 1 \; if \; ij = Argmax_{ij}(Pot_{ij}) \\ 0 \; otherwise \end{cases} \quad (4)$$

The winner neuron becomes the input of the temporal sequence learning network (subsection 2.3). In this way, a "reservoir" of oscillator neurons can be used as a way to associate the same input with different internal patterns. Intuitively, the simple competition (no learning is required here) allows to directly select different "internal" states corresponding to the same input repeated many times in the sequence. For example in Fig. 3.A, each input (A,B,C,D) can appears up to 3 times (corresponding to the number of oscillators) in the same sequence. Moreover, such a mechanism does not disturb the prediction nor the reproduction of the sequence. Obviously, if the competition between oscillators is an avenue worth exploring, it is still possible to have ambiguity. An input can be associated with same winner oscillator two or more times. Consequently, there is still potential ambiguities on the "internal" states of our model, and some sequences could not be reproduced correctly. A precise measure of this problem corresponds to the probability that the same state can be associated with the same oscillator several times and therefore the "internal" state partially depends of the shape, phase and number of oscillators. Typically, the problem happens when a given state comes back with the same frequency as the selected oscillator. The curves C2 and C3 on figure 5 show the performances of the competitive mechanism. To solve this problem, an associative mechanism allowing to recruit neurons coding "internal" states has been added.

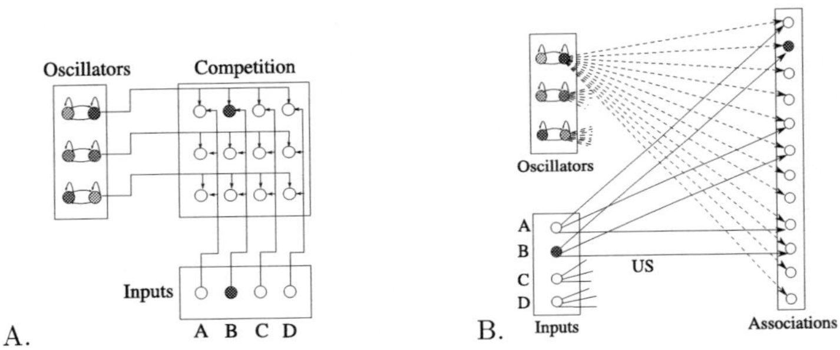

Fig. 3. A. Model of the neural network coupling an input state with an oscillator. All links are fixed connections. B. Model of the neural network used to associate an input state with a configuration of oscillators. Only few links are represented for the legibility. Dashed links are modifiable connections. Solid links are fixed connections.

Associative Mechanism. The learning process of an association between an input state and a configuration of oscillators is :

$$US = w_i * x_i \tag{5}$$

with w_i the weight of the link from input state i, and x_i activity of the input state i. If $US > threshold$, we compute the potential and the activity of the neuron as follow :

$$Pot_j = \sum_{j=0}^{M_{osci}} |(w_j - e_j)| \qquad Act_j = \frac{1}{1 + Pot_j} \tag{6}$$

with M_{osci} the number of oscillators, w_j the weight of the link from oscillator j, and e_j the activity of the oscillator j. The neuron that has the minimum activity is recruited : $Win = Argmin_j(Act_j)$. Initial weigths of connexions have high values. The oscillators configuration is learnt according to the error of distance

$\Delta w_j = \varepsilon(e_j - w_j)$ with ε a learning rate, w_j weight of link from Oscillator j, and e_j activity of oscillator j. The Associations group becomes the new input of the temporal sequence learning network (subsection 2.3). As showed on the figure 3.B, an input allows recruiting 3 different neurons coding "internal" states. They correspond to the connectivity of the unconditional links chosen between the Inputs group and the Association group. The associative mechanism ensures to recruit a new "internal" state for each input (A, B, C or D) from the sequence. The connectivity of links between the Input group and the Association group, has been chosen to have a number of hidden states equal for each input. This allows the comparison between the different models in our simulations. But it could be possible to change the connectivity of the links to allow the recruitement of more hidden states for each repeated input in the sequence. We have tested this mechanism in our architecture in simulation and robotic application.

2.3 Temporal Sequences Learning

This part of model is based on a schematic representation of hippocampus [10] (Fig. 4). DG represents past state (STM), and develops a temporal activity spectrum. CA3 links allow pattern completion and recognition between incoming state from EC and previous state maintained in DG. We suppose the DG activity can be modelled as follow :

$$Act_{j,l}^{DG}(t) = \frac{1}{m_j} \cdot \exp - \frac{(t - m_j)^2}{2 \cdot \sigma_j} \tag{7}$$

with $Act_{j,l}^{DG}$ the activity of the cell at index l on the line j, t the time, m_j a time constant and σ_j the standard deviation. Neurons on one line share their activity in the time and represent a temporal trace of EC. Learning of an association is on the weights of links between CA3 and DG. The normalization of the activity coming from DG neurons is performed due to the normalization of the DG-CA3 weights.

$$W_{CA3(i,j)}^{DG(j,l)} = \begin{cases} \frac{Act_{j,l}^{DG}}{\Sigma_{j,l}(Act_{j,l}^{DG})^2} & if\ Act_j^{DG} \neq 0 \\ unchanged & otherwise \end{cases} \tag{8}$$

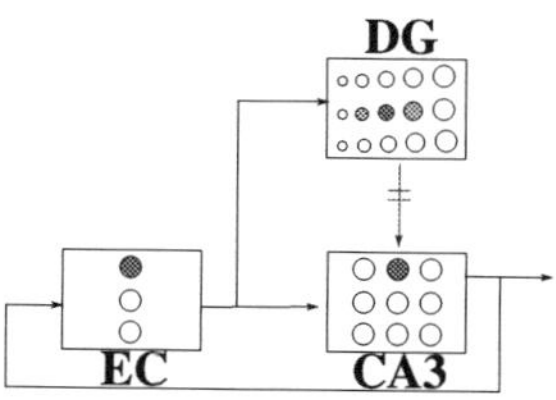

Fig. 4. Representation of hippocampus. Entorhinal Cortex (EC) receives inputs and transmits them to Dentate Gyrus (DG) and CA3 pyramidal cells. Between the DG group and the CA3 group there are fully connected with modifiable connections. Between the EC group and CA3 group, and the EC group and DG group, there are fixed one to neighborhood connections.

Interestingly, this model has the property to work when a same input comes several times continously. Thanks to the derivative group EC, a repeated input is stored during the total time of its presence. Consequently, two successive states are not ambiguous for the system ("A A" = "A").

3 Simulation Results

A temporal sequence of states is rarely replayed two times exactly with the same rhythm. Time can vary between two states especially when demonstrating a sequence to a robot. In our simulations we apply a time variation between states and observe the consequences on three architectures. The first architecture is the model of simple sequences learning presented in subsection 2.3. The second is the same model plus the competitive mechanism presented in subsection 2.2. The third architecture is the same as the first one plus the associative mechanism (Fig. 1) seen in subsection 2.2. References sequences are generated to be successfully reproduced by the second architecture with a timing variation of 0%. All architectures are trained with the same sequences and the same maximum of timing variation (0%, 5% or 10%), but with a time variation randomly chosen between 0 and the maximum variation of the time. In our experiments, to bootstrap a sequence, we provide the first state. Consequently, this state will not be ambiguous in the sequences. For example, a complex sequences can be "D B C B A C A B" : "D" is the starting state and it will not be repeated after. Fig. 5 shows the performances of each architecture. We can see that the first architecture (subsection 2.3) has very good performances with sequences of 3 and 4 states, because those sequences have no repeated states (simple sequences). With sequences having more than 4 states, the performances fall drastically , because there is at least one state

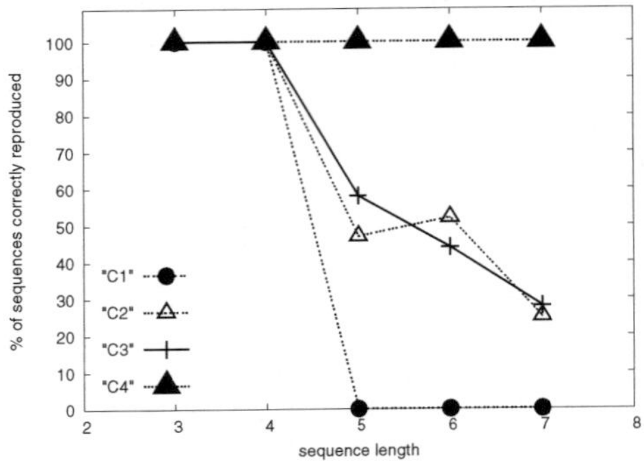

Fig. 5. C1 : first architecture : simple sequences learning. The results are the same with time variation of 0%, 5% and 10%. C2 : second architecture : complex sequences learning with a competitive mechanism and a time variation of 5%. C3 : second architecture : complex sequences learning with a competitive mechanism and a time variation of 10%. C4 : current architecture : complex sequences learning with an associative mechanism. The results are the same timing variation of 0%, 5% and 10%.

repeated in the sequences. We can see the time variation has no effect on the performances. The architecture can not reproduce them, because CA3 group learns two transitions and, thus it predicts two states for each repeated input. The second architecture using competitive mechanism, has better performances, but, as we have seen previously in subsection 2.2, ambiguous internal states can appear and reduce this gain of sequences correctly reproduced. Consequently, like the first architecture, the CA3 group learns two "internal" states and, thus, it predicts two states from one input repeated. We can see, the performances change according to the timing variation between states in the sequences : a same input from a given sequence can be associated with two different oscillators and, consequently a different "internal" state wins. Thanks to the recruitment mechanism, the third architecture, has the best performances : 100% with all tested sequences. There are not ambiguous states or "internal" states. The time variation has no effect on the performances of the model.

4 Robotic Application

The robot used in our experiments is an Aibo ERS7 (Sony). In our application, we use only the front left leg, in a passive movement mode to learn a sequence of gestures. The sequence to be learned and reproduced is showed Fig. 6.A. In this application, we test the third architecture previously described.

During learning, we manipulate the front left leg of the robot passively (Fig. 6.B). During the execution of the movement, the neural network learns online, and in one shot the succession of the joints orientation thanks to the motors feedback information of its leg (proprioceptive signal). Hence, the inputs of our model are the orientations/angles of the leg. The recorded motors information while learning are shown in Fig. 7.X-*learning* (horizontal movements) and Fig. 7.Y-*learning* (vertical movements).

To initiate the reproduction of the sequence by the robot, we give the first state of the sequence ("down"). As Aibo can not be manipulated when motors are activated, we send the command directly to the robot. Next, Aibo plays the sequence autonomously (Fig. 7, top). With the starting state, our model predicts the next orientation and send the corresponding command to the robot.

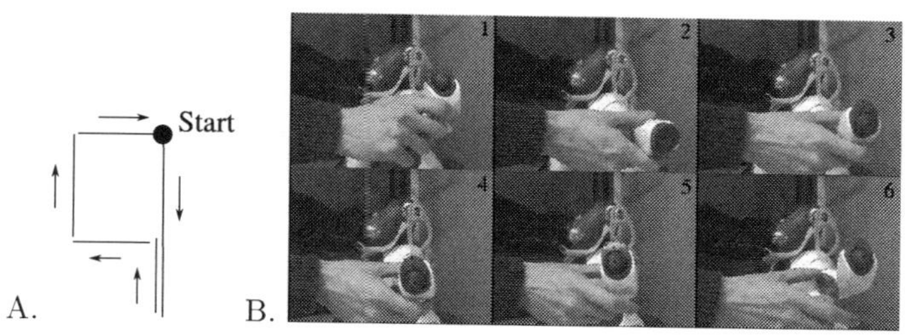

Fig. 6. A. Representation of desired sequence. It begins from the start point. B. We manipulate Aibo passively. It learns the succession of orientations of the movement from these front left leg motor information.

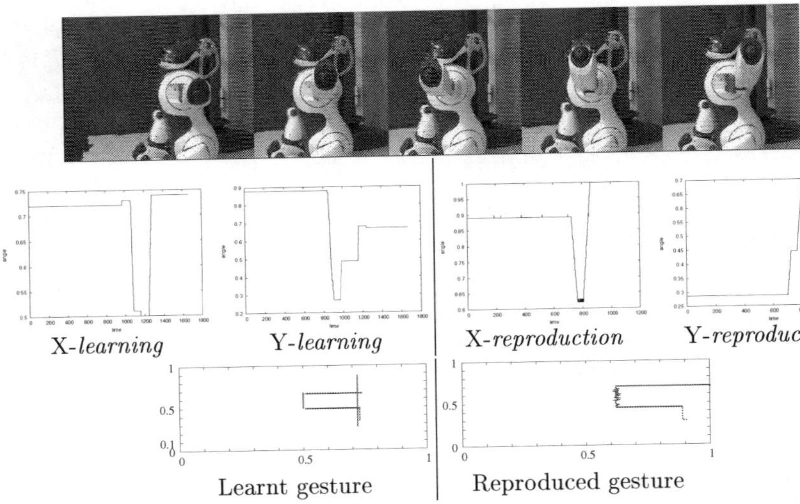

Fig. 7. Top : Aibo reproduces the learnt sequence. Middle : X-*learning* and Y-*learning* are respectively the horizontal and the vertical motors information while robot learns the sequence. X-*reproduction* and Y-*reproduction* are respectively the horizontal and the vertical motors information during the reproduction of the sequence. On the figure Y-*reproduction*, the first movement is not reproduced (not predicted), but given by the user in order to trigger the recall it is our bootstrap state to start the sequence. X-axis are the time and Y-axis are the angles of the motors.

5 Conclusions and Discussions

We have proposed a model of neural network for the learning of complex temporal sequences. This model introduces an associative mechanism taking advantage of a diversity generator composed of oscillators. These oscillators are based on coupled CTRNN. This model is efficient in the frame of autonomous robotics and succeed in learning in one shot the timing of sequences of gestures.

During the robotics application, we have noticed that the robot reproduces the sequence with a different amplitude of the movement. This effect comes from the speed of the displacement of the leg of Aibo. In our application, the speed of the reproduction is a predefined constant different from the user dynamic during learning. The rhythm of the sequence is respected thanks to atemporal group of neurons. A possible improvement would be to add a model like CPG [5] for each movements ("up", "down", "left" and "right") composing sequences with variable speeds. In our model, the number of neurons coding the associations between the inputs and the oscillators, represents the size of the "short term memory". In our simulations and application, the sequences learnt do not saturate the "memory". It would be interesting to analyze the behavior of the neural network with longer sequences, and test the limitations of the system when the neural limit has been reached by the recruitment mechanism. In the present system, it would mean that the already recruited neurons could be erased in order to encode new states.

In further works, this sequence learning model will complete a model for imitation based on low level sensorimotors capabilities and vision [13]. In this way, the robot will learn sensorimotors capabilities based on its vision and learn a demonstrated gesture from human or robot by imitation and reproduce it.

Acknowledgements

This work was supported by the French Region Ile de France, the network of excellence HUMAINE and the FEELIX GROWING european project. (FP6 IST-045169)

References

1. Molter, C., Salihoglu, U., Bersini, H.: Learning cycles brings chaos in continuous hopfield networks. In: IJCNN. Proceedings of the International Joint Conference on Neural Networks conference (2005)
2. Daucé, E., Quoy, M., Doyon, B.: Resonant spatio-temporal learning in large random neural networks. Biological Cybernetics 87, 185–198 (2002)
3. Jaeger, H.: Short term memory in echo state networks. Technical Report GMD Report 152, German National Research Center for Information Technology (2001)
4. Jaeger, H.: The "echo state" approach to analysing and training recurrent neural networks. Technical Report GMD Report 148, German National Research Center for Information Technology (2001)
5. Ijspeert, A.J., Nakanishi, J., Shibata, T., Schaal, S.: Nonlinear dynamical systems for imitation with humanoid robots. In: Humanoids2001. Proceedings of the IEEE/RAS International Conference on Humanoids Robots, pp. 219–226 (2001)
6. Calinon, S., Billard, A.: Learning of Gestures by Imitation in a Humanoid Robot. In: dautenhahn, K., nehaniv, c.l. (eds.), Cambridge University Press, Cambridge (2006) (in press)
7. Hersch, M., Billard, A.: A biologically-inspired model of reaching movements. In: Proceedings of the 2006 IEEE/RAS-EMBS International Conference on Biomedical Robotics and Biomechatronics, Pisa (2006)
8. Bullock, D., Grossberg, S.: Neural dynamics of planned arm movements: Emergent invariants and speed-accuracy properties during trajectory formation. Psychological Review 95, 49–90 (1988)
9. Ans, B., Coiton, Y., Gilhodes, J.C., Velay, J.L.: A neural network model for temporal sequence learning and motor programming. Neural Networks 7(9), 1461–1476 (1994)
10. Gaussier, P., Moga, S., Banquet, J.P., Quoy, M.: From perception-action loops to imitation processes. Applied Artificial Intelligence (AAI) 1(7), 701–727 (1998)
11. Ijspeert, A.J.: A neuromechanical investigation of salamander locomotion. In: AMAM 2000. Proceedings of the International Symposium on Adaptive Motion of Animals and Machines (2000)
12. Yamauchi, B., Beer, R.D.: Sequential behaviour and learning in evolved dynamical neural networks. Adapt. Behav. 2(3), 219–246 (1994)
13. Andry, P., Gaussier, P., Nadel, J., Hirsbrunner, B.: Learning invariant sensorimotor behaviors: A developmental approach to imitation mechanisms. Adaptive behavior 12(2), 117–138 (2004)

Clustering Limit Cycle Oscillators by Spectral Analysis of the Synchronisation Matrix with an Additional Phase Sensitive Rotation

Jan-Hendrik Schleimer and Ricardo Vigário

Adaptive Informatics Research Centre
Helsinki University of Technology
P.O. Box 5400, FIN-02015 Espoo, Finland
schleime@cis.hut.fi, rvigario@cis.hut.fi

Abstract. Synchrony is a phenomenon present in many complex systems of coupled oscillators. It is often important to cluster those systems into subpopulations of oscillators, and characterise the interactions therein. This article derives the clustering information, based on an eigenvalue decomposition of the complex synchronisation matrix. A phase sensitive post-rotation is proposed, to separate classes of oscillators with similar frequencies, but with no physical interaction.

1 Introduction

Synchronisation can arise in complex systems comprised of natural or artificial non-linear oscillators. This is the case, *e.g.*, in chemical oscillators [1] or technical devices, like Josephson's junctions [2]. It can be seen as a by-product of physical interactions between oscillators. In particular, the study of neuronal communication can benefit from a better characterisation of synchronisation phenomena between elements of the nervous system.

Neural coding schemes, relying on the relative timing of responses between neurons, emphasise the relevance of synchronisation phenomena in nervous systems [3]. Essentially, synchronisation is either caused by a common input or by interactions between neurons. In this paper, the focus is on the later "internal synchronisation", since its role is less well understood. In many neuron models, stable limit cycle attractors exist. Based on the results of perturbation theory, the interaction between these self-sustained oscillators can be reduced, to take place on the phase ϕ_j associated with each neuron. See [4] for an explanation of how to perform the reductions of interacting dynamical systems to a phase description.

Here, we address the problem of clustering a population of oscillators into segregated subpopulations, exhibiting high internal interactions. Approaches to solve this problem have often assumed different frequencies for the various subpopulations, usually neglecting phase information. These assumptions pose a restriction to the analysis of the dynamic world of natural systems, where

J. Marques de Sá et al. (Eds.): ICANN 2007, Part I, LNCS 4668, pp. 944–953, 2007.

communication can be unrelated to the natural frequency of the constituent oscillators.

Such problem is of particular relevance in systems with adaptive coupling strengths. They typically show distributed values for their internal frequencies, which are independent from the interactions they temporarily form. Existing algorithmic solutions, only focusing on frequencies, and neglecting phase information, would therefore have difficulties disambiguating the population.

Here, we present a solution that makes explicit use of phase information, extracted from known models of physical interactions. The approach relies on a post-rotation of the eigenvectors of the synchronisation matrix. The derivation of the method is illustrated with simulated examples.

2 Background

A generally accepted model, describing the dynamics of a weakly interacting population of m oscillators, expressed as the evolution of their associated phase variable $\phi_j(t)$, is

$$\frac{d\phi_j}{dt} = \omega_j + \sum_{k=1}^{m} \Gamma_{jk}(\phi_j, \phi_k). \tag{1}$$

This reduction of a d-dimensional oscillator to a one-dimensional phase equation can be employed whenever the differential equation describing the oscillators display attracting limit cycles. The role of the phase variable can be intuitively grasped as the position of the oscillator on the limit cycle, disregarding its exact position in the d-dimensional space. The natural frequency of each oscillator is denoted as ω_j, and Γ is a coupling function, resulting from the phase reduction of the interaction terms. In particular, if $\Gamma_{jk}(\phi_j, \phi_k) = \Gamma(\phi_k - \phi_j)$ is odd, the coupling is attractive and can lead to synchronisation. For a simple global coupling $\Gamma_{jk}(\phi_j, \phi_k) = \frac{\kappa}{m}\sin(\phi_k - \phi_j)$ with large enough κ, a meanfield (also called order-parameter) $\varrho e^{\Phi} = \frac{1}{m}\sum_k e^{\mathrm{i}\varphi_k}$ can be introduced so that its magnitude ϱ describes the synchronisation state of the system [1]. This special choice is called the Kuramoto model. With the Hilbert transform, the phase information can be obtained for electrophysiological measurements, from which one can calculate ϱ, and investigate synchronisation. The method is not restricted to single neurons, but can also be applied to recordings that summarise whole ensembles, like local field potentials and even EEG and MEG. Nevertheless, the interpretation of the results, in terms of interacting oscillators gets more difficult.

In case of non-global coupling, the population can split into synchronous subpopulations which is not reflected in the parameter ϱ. With the subpopulations having no phase relation between each other, ϱ can be very small, which suggests no synchronisation at all.

Generalisations of the order-parameter can measure the tendency of the population to segregate into clusters, in theoretical situations [5]. However, we proceed with an alternative way to study phase synchronisation, which is adequate for clustering oscillatory data. Consider the bivariate statistical quantities

$$q_{jk} = \left\langle e^{\mathrm{i}(\phi_j - \phi_k)} \right\rangle , \tag{2}$$

which we gather into a Hermitian matrix $\boldsymbol{Q} \in \mathbb{C}^{m\times m}$. One immediate difference to the order-parameter is that, while $\varrho > 0$ indicates entrainment in the Kuramoto model, only $q_{jk} = 1$ truly implies a constant phase lag, *i.e.*, perfect entrainment between the two oscillators.

The expectation in Eq. (2) can be approximated with a temporal average under certain ergodicity assumptions and for a system in an equilibrium state (stationarity holds). So henceforth, we set $\langle x \rangle = \frac{1}{T}\int_0^T x(t)dt$ for sufficiently long observation period T.

If we postulate that synchronisation arises from physical interactions between the populations, can we infere which oscillators interact, based solely on $\boldsymbol{Q}$?

3 Spectral Analysis of the Synchronisation Matrix

It has been pointed out recently that an eigenvalue decomposition of the matrix formed by the absolute values $|q_{jk}|$ can help identifying subpopulations of synchronous oscillators [6]. This approach can be understood as spectral clustering, when $|q_{jk}|$ are perceived as the elements of the connectivity matrix.

It is also possible to motivate an eigenvalue decomposition of $\boldsymbol{Q}$ directly, by properties of wave superposition and cancellation, illustrated in the following examples.

Example 1. Examine the maximum of the expected square amplitude $|y|^2 = yy^*$ of two superimposed waves $y = \alpha e^{\mathrm{i}\omega t} + \beta e^{\mathrm{i}(\omega t + \vartheta)}$, which have the same frequency, but are phase shifted by ϑ. Impose the constraint $|\alpha|^2 + |\beta|^2 = 1$ for $\alpha, \beta \in \mathbb{C}$. From the Lagrangian formulation $L = \langle |y|^2 \rangle - \lambda(|\alpha|^2 + |\beta|^2 - 1)$, and the conditions $\frac{\partial L}{\partial u^*} = 0|_{u=\alpha,\beta,\lambda}$, three equations follow[1]

$$(\lambda - 1)\alpha = \beta e^{\mathrm{i}\vartheta}, \qquad (\lambda - 1)\beta = \alpha e^{-\mathrm{i}\vartheta}, \qquad |\alpha|^2 + |\beta|^2 = 1.$$

One can then conclude that $|\alpha| = |\beta| = \frac{1}{\sqrt{2}}$ and $\beta = \alpha e^{-\mathrm{i}\vartheta}$, $\forall \alpha = \frac{1}{\sqrt{2}} e^{\mathrm{i}\psi}$, for any arbitrary ψ. That is, as intuition suggests, the maximum is obtained if the weights compensate for the relative phase shift between the oscillators, prior to the superimposition.

Example 2. Superimpose three oscillators with frequencies ω_1, $\omega_2 = \omega_1$ and $\omega_3 = \omega_1 + \varpi$, for $\varpi > 0$, subject to $|\alpha|^2 + |\beta|^2 + |\gamma|^2 = 1$. Examining the amplitudes of α, β and γ at the extrema of L (as in *Ex. 1*), with $y = \alpha e^{\mathrm{i}\omega_1 t} + \beta e^{\mathrm{i}\omega_2 t} + \gamma e^{\mathrm{i}\omega_3 t}$, leads to the relations

$$|\alpha| = |\beta|, \qquad |\gamma| \leqslant \frac{2(|\alpha| + |\beta|)}{T\varpi},$$

where T is the observation period. This means that, for non-zero $|\alpha|$ and $|\beta|$, the magnitude of γ will be inversely related to the product $T\varpi$. Specifically, $|\gamma|$

[1] We use the complex gradient operator as in [7].

decreases with the increase of the frequency difference. For small differences, a compensating large T is required.

These examples show that, assuming one has a large enough observation time, a maximisation scheme as above select frequency-entrained oscillators, and provide information about the relative phase offset between them.

We can take *Ex. 1 & 2* as rationale, and search for a complex superposition $y(t) = \boldsymbol{v}^{\mathsf{H}}\boldsymbol{z}(t)$ of all amplitude-normalised oscillators $z_k = e^{\mathrm{i}\phi_k}$, obtained from the data through Hilbert transform, such that the average square magnitude $E = \langle yy^* \rangle \geqslant 0$ is maximal. If $v_k \in \mathbb{C}$, its phase can compensate for any phase shift of the oscillators. With the magnitude $|v_k|$, oscillators of similar frequency can be selected. An unconstrained projection vector $\boldsymbol{v}$ would allow to maximise E *ad infinitum*, so the additional restriction $\|\boldsymbol{v}\|^2 = 1$ needs to be imposed. For a population of oscillators that form entrained subsets, a maximum can be expected, if the largest such population is selected, and their phase shifts compensated by v_k.

One can rewrite $E = \boldsymbol{v}^{\mathsf{H}} \left\langle \boldsymbol{z}\boldsymbol{z}^{\mathsf{H}} \right\rangle \boldsymbol{v} = \boldsymbol{v}^{\mathsf{H}}\boldsymbol{Q}\boldsymbol{v}$. The Lagrangian, including the constraint is

$$L = \boldsymbol{v}^{\mathsf{H}}\boldsymbol{Q}\boldsymbol{v} - \lambda(\boldsymbol{v}^{\mathsf{H}}\boldsymbol{v} - 1). \tag{3}$$

All solutions to the necessary condition $\nabla L = 0$ must fulfil the eigenvalue problem

$$\boldsymbol{Q}\boldsymbol{v} = \lambda \boldsymbol{v}. \tag{4}$$

The largest eigenvalue is the maximum of L. The matrix $\boldsymbol{Q}$ represents a Hermitian operator that defines a quadratic hermitian form, $0 \leqslant (\boldsymbol{v}^{\mathsf{H}}\boldsymbol{Q}\boldsymbol{v}) \in \mathbb{R},\ \forall \boldsymbol{v} \in \mathbb{R}$, *i.e.*, it is positive semidefinite and has, according to the spectral theorem, an orthonormal basis given by the eigenvalue decomposition

$$\boldsymbol{Q} = \boldsymbol{V}\boldsymbol{\Lambda}\boldsymbol{V}^{\mathsf{H}} = \sum_{k=1}^{m} \lambda_k \boldsymbol{v}_k \boldsymbol{v}_k^{\mathsf{H}}, \tag{5}$$

with m orthogonal complex eigenvectors $\boldsymbol{V} = [\boldsymbol{v}_1, \cdots, \boldsymbol{v}_m]$ and positive real eigenvalues $\lambda_k = \boldsymbol{v}_k^{\mathsf{H}}\boldsymbol{Q}\boldsymbol{v}_k$. Without loss of generality, let them be ordered decreasingly $\lambda_1 \geqslant \cdots \geqslant \lambda_m$. Since, for a matrix such as $\boldsymbol{Q}$, one has $\sum_k \lambda_k = \mathrm{tr}(\boldsymbol{Q}) = m$, the eigenvalues measure the effective size of a subpopulation. The eigenvector basis is unique only if all eigenvalues are distinct, $\forall j \neq k : \lambda_j \neq \lambda_k$. Multiplicity of eigenvalues can arise if two subpopulations of equal size exist.

Note that smaller non-zero eigenvalues $\lambda_k \neq 0$ can correspond to local maxima, if the Hessian $\nabla\nabla L = \boldsymbol{Q} - \lambda_k \boldsymbol{I}$ is negative semidefinite; or saddle points, if the Hessian is indefinite. Those eigenvalues can often be associated with smaller synchronous subpopulations.

In [6], the condition $\lambda \geqslant 1$, known as Kaiser-Guttman rule in factor analysis or spectral clustering, was used to identify significant synchronisation clusters.

3.1 Phase Oscillator View on the Synchronisation Matrix

Let us first clarify the relation between values of λ_1, and the dynamical system that can produce one synchronous meanfield (Eq. (1) with $\Gamma_{jk} = \frac{\kappa}{m}\sin(\phi_k - \phi_j)$).

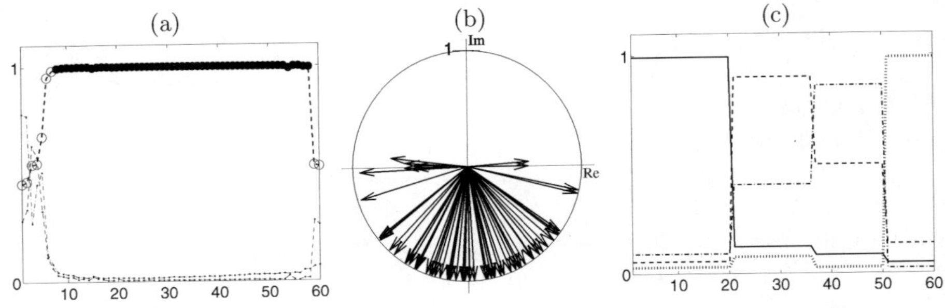

Fig. 1. (a) Abscissa: number of oscillator (ordered by frequency, •: entrained / ○: free); Ordinate: Amplitudes of scaled coefficients of the eigenvectors, $\sqrt{\lambda_j}\boldsymbol{v}_j$, from *Ex. 3*. The maximum achievable for the coefficents is one. (b) Vectors in the complex plane, showing the phase angles of the coefficients of the largest eigenvector. (c) Four clusters with different internal frequencies as in *Ex. 4*. Coefficients of different eigenvectors are connected with different linestyles for visibility.

When the largest eigenvalue dominates the expansion in Eq. (5), the synchronisation matrix can be approximated by dropping relatively small terms, resulting in $\boldsymbol{Q} \approx \lambda_1 \boldsymbol{v}_1 \boldsymbol{v}_1^{\mathsf{H}}$. In [8], it is noted that if the model allows the introduction of a non-zero magnitude meanfield, *cf.* Sec. 1, then the coefficients in $\boldsymbol{Q}$, for two synchronised oscillators j and k factorise

$$q_{jk} = \left\langle e^{\mathrm{i}(\phi_j - \phi_k)} \right\rangle = \underbrace{\left\langle e^{\mathrm{i}(\phi_j - \Phi)} \right\rangle}_{Q_j^{(1)}} \left\langle e^{-\mathrm{i}(\phi_k - \Phi)} \right\rangle . \tag{6}$$

In that case a theoretical solution ($T \to \infty$) to $E = \lambda_1$ is given by the choice

$$v_{k1} = \begin{cases} \frac{1}{\sqrt{\lambda_1}} \left\langle e^{\mathrm{i}(\phi_k - \Phi + \vartheta)} \right\rangle & \text{if } \dot{\phi}_k = \dot{\Phi}, \\ 0 & \text{else }, \end{cases} \qquad \text{for any constant } \vartheta. \tag{7}$$

This indeterminacy of the phase offset is also known in quantum systems. Combining Eq. (7) with Eq. (6), the factorisation reads

$$q_{jk} = Q_j^{(1)} (Q_k^{(1)})^* = (\sqrt{\lambda_1} v_{j1})(\sqrt{\lambda_1} v_{k1}^*).$$

Taking the squared absolute value, confirms the relation $|Q_j^{(1)}|^2 = \lambda_1 |v_{j1}|^2$, found in [6].

The following example illustrates that $\sqrt{\lambda_1}|v_{j1}|$ gives information about the involvement of oscillator j in a cluster of synchronous neurons. It further shows that the phase angle of v_{j1} holds the relative phase offset.

Example 3. Consider a population of $m = 60$ phase oscillators, simulated with Kuramoto's equation (Sec. 2). For this choice it can be analytically shown that, if $|\omega_j - \Omega| \leqslant \kappa\varrho$, they are entrained. Otherwise they are "free" [1]. The natural frequencies, ω_j, have to be drawn from a symmetric distribution, say, the normal distribution $\mathcal{N}(\Omega, 1)$, with $\Omega = 10$. The global coupling strength is set to $\kappa = 1.9$. The time evolution is simulated simply with Euler's forward method and random initialisations. Only samples taken from the period, in which the system has reached equilibrium, are used to estimate the matrix $\boldsymbol{Q}$. The eigenvalues for the decomposition of $\boldsymbol{Q}$ are $\lambda_1 = 54.8, \lambda_2 = 1.8, \lambda_3 = 1.2, \lambda_4 = 1.1, \ldots$, revealing one large synchronisation cluster. From Fig. 1a one sees that the oscillators j, for which the coefficients $\sqrt{\lambda_1}|v_{j1}|$ are close to one, correspond to those that are theoretically predicted to be entrained (Fig. 1a, •). Yet, the numerical solution does not yield a zero amplitude for all free oscillators, like in Eq. (7), (Fig. 1a, ∘). A reason for that is the finite sample size and the fact that the relative phase velocity of almost entrained oscillators goes through a bottleneck. More speciffically, for "just not entrained oscillators", *e.g.*, with a too high frequency $\omega - \Omega = \kappa\varrho + \varepsilon$, and small $\varepsilon > 0$, a rather long observation period is required. This is due to the fact that the velocity in the rotational frame of the meanfield $d(\phi - \Phi)/dt$ reduces dramatically near $\phi = \Phi \pm \frac{\pi}{2}$, if $\varepsilon \rightarrow 0$. This is the case because

$$\frac{d(\phi_k - \Phi)}{dt} = \kappa\varrho\underbrace{(1 - \sin(\phi_k - \Phi))}_{\text{zero at } \pm\frac{\pi}{2}} + \varepsilon,$$

which is easily derived from the formulas in [1, Cpt. 5].

Also note in Fig 1b that information about the relative phase of the oscillators is conserved in the phase of the coefficients of the eigenvectors, as we have theoretically motivated.

Example 4. Let us now take a population of $m = 60$ oscillators comprising sub-populations with four distinctly distributed frequencies. Assume a local coupling function $\Gamma_{jk}(\phi_j, \phi_k) = K_{jk}\sin(\phi_k - \phi_j)$. The coupling strength matrix K_{jk} is chosen as a block-diagonal matrix, with $c = 1, \ldots, 4$ blocks, such that

$$K_{jk} = \begin{cases} \kappa/|N(c)| & \text{if } j, k \in N(c), \\ 0 & \text{else,} \end{cases} \qquad \text{for } \bigcup_{c=1}^{4} N(c) = \{1, \ldots, m\}.$$

$N(c)$ denotes the index set of the cth block. In this way, an oscillators is connected to only one of the four subpopulations. The overall strength is chosen $\kappa = 3$, so that all oscillators within the cluster get entrained. With similar reasoning as in *Ex 3* we reach the result shown in Fig. 1c. 4 clusters with different frequencies. As a decision rule, an oscillator is associated to the cluster for which it has the largest involvement [6]

$$c = \underset{k}{\operatorname{argmax}}\{\sqrt{\lambda_k}|v_{jk}|\}. \tag{8}$$

4 Rotating the Eigenvectors with a Phase Constrain

A problem with maximising Eq. (3) is that oscillators with very similar frequencies will agregate in the same cluster, regardless of whether or not they are really

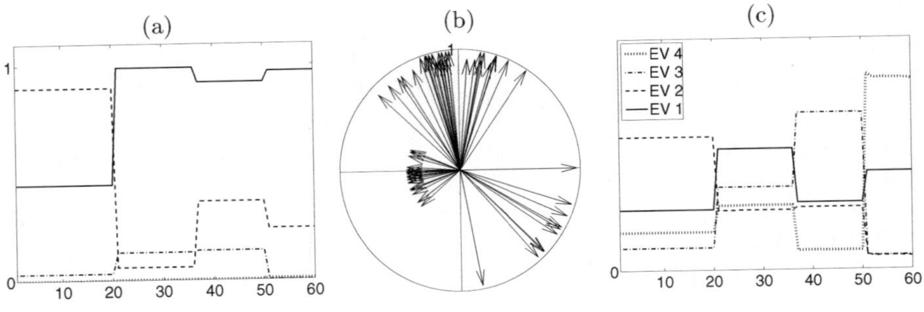

Fig. 2. (a) Ordinate: Amplitude of the scaled eigenvectors $\sqrt{\lambda}_j \boldsymbol{v}_j$ of the four largest eigenvectors of $\boldsymbol{Q}$, from *Ex 5.* (b) Phase angles of eigenvector coefficients of the largest population. (c) Rotated eigenvectors $\boldsymbol{U}$, that allow correct clustering.

entrained. This means that, if there are several non-interacting subpopulations with similar meanfield frequencies, they will appear as grouped together. Such situations can happen by plain coincidence; because two unconnected oscillators have similar natural frequency; or for the reason illustrated in the following example.

Example 5. Considering a similar set of oscillators as in *Ex. 4*, now with the individual frequencies drawn from a global distribution $\omega_j \sim \mathcal{N}(10, 1), \forall j = 1, \ldots, m$, across all subpopulations. In neurobiological terms, this means that the neuron's system parameters, which determine its natural frequency, do not depend on the synaptic connections it has formed. If, like in *Ex. 3*, we estimate the eigenvalues of $\boldsymbol{Q}$, we get $\lambda_1 = 40.9, \lambda_2 = 18.5, \lambda_3 = 0.5, \ldots$ According to the Kaiser-Guttman criterion, only two clusters are present. Such result is clearly illustrated in Fig. 2a. Yet, the data was generated with four distinct subpopulations.

If the coupling is strong enough within an interacting subpopulation c, a meanfield can be introduced

$$\varrho^{(c)} e^{\mathrm{i}\Phi^{(c)}} = \frac{1}{m} \sum_{k \in N(c)} e^{\mathrm{i}\phi_k}. \tag{9}$$

Then Eq. (1) can be rewritten as

$$\dot{\phi}_j = \omega_j + \kappa \varrho^{(c)} \sin(\Phi^{(c)} - \phi_j), \qquad \forall j \in N(c). \tag{10}$$

For any subpopulation at equilibrium, the entrainment condition reads

$$\phi_j - \Phi^{(c)} = \sin^{-1}\left(\frac{\omega_j}{\kappa \varrho^{(c)}}\right). \tag{11}$$

This means that oscillators with $|\phi_j - \Phi^{(c)}| > \frac{\pi}{2}$ are not entrained to the cluster c. In Fig. 2b, however, there are oscillators belonging to cluster 1 which do not fulfil this constraint. Instead we find them dispersed all around the unit circle.

In general, phases within one cluster are forced towards each other, through the coupling term. On the other hand, detuning caused by differences in natural frequencies opposes this trend, *cf.* Eq. (10). This is not accounted for in the spectral analysis, presented in Sec. 3, which lead to $\boldsymbol{v}_k$.

A solution to this problem can be sought in the following manner. Assume there are d clusters. If we denote a new set of vectors $\boldsymbol{U} = [\boldsymbol{u}_1, \ldots, \boldsymbol{u}_d]$, defining the clusters in much the same way as $\boldsymbol{v}_c$, one can formulate the criterion

$$\left| \sum_{j=1}^{m} u_{jk} \right|, \qquad \forall k = 1, \ldots, d.$$

This criterion measures the compactness of the phase angles in one cluster, much like the meanfield does Eq. (9).

We now search for a linear transformation between $\tilde{\boldsymbol{V}} = [\boldsymbol{v}_1, \cdots, \boldsymbol{v}_d]$ and $\boldsymbol{U}$, which takes the restrictions presented in *Ex. 5* into account. Keep in mind that $\tilde{\boldsymbol{V}}$ is composed of only the eigenvalues defining the d clusters.

Such transformation can be written as $\boldsymbol{U} = \tilde{\boldsymbol{V}}\boldsymbol{W}$. The magnitude of the new loadings, $|u_{jc}| = |\sum_k v_{jk} w_{kc}|$, indicates the jth oscillator's degree of membership to cluster c. Since we do not want to alter the phase of the old loadings, v_{jk}, $\boldsymbol{W} \in \mathbb{R}^{d \times d}$. Further note that $\boldsymbol{W}$ should be restricted to be orthonormal, since the total amount of oscillators should not be altered

$$\operatorname{tr}(\boldsymbol{U}^{\mathsf{H}}\boldsymbol{C}\boldsymbol{U}) = \sum_{k=1}^{d} \lambda_k = m, \qquad \text{iff } \boldsymbol{W}^{\mathsf{T}}\boldsymbol{W} = \boldsymbol{I}.$$

To obtain a rotation with the desired property, one can maximise the following cost function

$$J = \sum_{k=1}^{d} \left| \sum_{j=1}^{m} u_{jk} \right|, \qquad \text{w.r.t. } \boldsymbol{W}. \tag{12}$$

The gradient of J w.r.t. an entry of the rotation matrix w_{jk} is given as

$$\frac{\partial J}{\partial w_{ij}} = \Re\left(\bar{u}_j \sum_k v_{ki} \right) + \Im\left(\bar{u}_j \sum_k v_{ki} \right), \qquad \text{where } \bar{u}_j = \sum_k u_{kj}. \tag{13}$$

4.1 Flow on the Stiefel Manifold

An elegant and efficieant way to incorporate the orthonormalisty constrain required above is derived from differential geometric concepts. The restriction $\boldsymbol{W}^{\mathsf{T}}\boldsymbol{W} = \boldsymbol{I}$ to gether with $J(\boldsymbol{W}) \neq J(\boldsymbol{W}\boldsymbol{R})$, for any rotation matrix, $\boldsymbol{R}$, defines the Stiefel manifold $\mathcal{S}$ of matrices. One can project the gradient $\boldsymbol{G} = \partial J / \partial \boldsymbol{W}$ (from Eq. (13)) onto the tangent space of the Stiefel manifold by [9]

$$\boldsymbol{\Pi} = \boldsymbol{W} \operatorname{skew}(\boldsymbol{W}^{\mathsf{T}}\boldsymbol{G}) + (\boldsymbol{I} - \boldsymbol{W}\boldsymbol{W}^{\mathsf{T}})\boldsymbol{G},$$

in order for a gradient based optimisation of J w.r.t. to $\boldsymbol{W}$ to stay on $\mathcal{S}$. Alternatively, one can also modify the gradient to account for the geometry of the Stiefel manifold directly. This leads to [9]

$$\boldsymbol{G}_{\mathcal{S}} = \boldsymbol{G} - \boldsymbol{W}(\boldsymbol{G})^{\mathsf{T}}\boldsymbol{W}.$$

Let us revisit *Ex. 5*, and perform a rotation defined by the stationary point of

$$\frac{d\boldsymbol{W}}{dt} = \boldsymbol{G} - \boldsymbol{W}(\boldsymbol{G})^{\mathsf{T}}\boldsymbol{W}.$$

The result $\boldsymbol{U} = \tilde{\boldsymbol{V}}\boldsymbol{W}$ is depicted in Fig. 2c. For the jth oscillator the decision rule $c = \text{argmax}_k\{u_{jk}\}$, instead of Eq. (8), then leads to the correct clustering.

5 Concluding Remarks

In this paper we address the problem of clustering phase synchronous oscillators. Particular focus is given to subpopulations with similar frequencies, but no physical cross-interactions.

The approach is based on a phase sensitive rotation of the eigenvectors obtained from a decomposition of the complex synchronisation matrix. The rotation is determined by optimising a criterion motivated by a model of interacting phase oscillators.

In earlier work [6], the magnitude $\lambda_k v_{jk}^2$ was utilised, based on a real valued synchronisation matrix. This effectively neglects all phase information available in the complex version synchronisation matrix. However, that phase information can be utilised to disambiguate non-interacting oscillators and solve our particular clustering problem. Therefore, it is proposed to use $\sqrt{\lambda_k} v_{jk}$ instead. This comes with no additional computational costs, but with the clear benefit of retaining the phase information. Such information can be used to define the rotation that finally allows the clustering. Post-rotations of loading matrices are common in factor analysis, *cf.*, the varimax rotation [10].

References

1. Kuramoto, Y.: Chemical Oscillations, Waves and Turbulences. Springer, Heidelberg (1984)
2. Josephson, B.D.: The discovery of tunnelling supercurrents. Rev. Mod. Phys. 46(2), 251–254 (1974)
3. Singer, W.: Neuronal synchrony: A versatile code for the definition of relations? Neuron 24, 49–65 (1999)
4. Brown, E., Moehlis, J., Holmes, P.: On the phase reduction and response dynamics of neuroal osicllator populations. Neural Computation 16, 673–715 (2004)
5. Golomb, D., Hanssel, D.: The number of synaptic inputs and the synchrony of large, sparse neuronal networks. Neural Computation 12(5), 1095–1139 (2000)
6. Allefeld, C., Müller, M., Kurths, J.: Eigenvalue decomposition as a generalized synchronization cluster analysis. Int J. Bifurcation & Chaos (NDES '05) (2007)

7. Brandwood, D.H.: A complex gradient operator and its application in adaptive array theory. IEE Proceedings, Part F - Communications, Radar and Signal Processing 130, 11–16 (1983)
8. Allefeld, C., Kurths, J.: An approach to multivariate phase synchronization analysis and its application to event-related potentials. Int. J. Bif. & Chaos 14(2), 417–426 (2004)
9. Edelman, A., Arias, T.A., Smith, S.T.: The geometry of algorithms with orthogonality constraints. SIAM Journal on Matrix Analysis and Applications 20(2), 303–353 (1998)
10. Richman, M.B.: Rotation of principal components. Journal of Climatology 6, 293–335 (1985)

Control and Synchronization of Chaotic Neurons Under Threshold Activated Coupling

Manish Dev Shrimali[1,2], Guoguang He[1,2], Sudeshna Sinha[3], and Kazuyuki Aihara[1,2]

[1] Aihara Complexity Modelling Project, ERATO, JST, Tokyo 151 0064, Japan
[2] Institute of Industrial Science, The University of Tokyo, Tokyo 153 8505, Japan
[3] The Institute of Mathematical Sciences, CIT Campus, Taramani, Chennai 600 113, India

Abstract. We have studied the spatiotemporal behaviour of threshold coupled chaotic neurons. We observe that the chaos is controlled by threshold activated coupling, and the system yields synchronized temporally periodic states under the threshold response. Varying the frequency of thresholding provides different higher order periodic behaviors, and can serve as a simple mechanism for stabilising a large range of regular temporal patterns in chaotic systems. Further, we have obtained a transition from spatiotemporal chaos to fixed spatiotemporal profiles, by lengthening the relaxation time scale.

Keywords: Chaotic dynamics, Control and Synchronization.

1 Introduction

In the last two decades, many control techniques have been proposed including the OGY [1] for controlling chaos in chaotic dynamical systems. In this paper, we focus on the network of chaotic neuron model that has been proposed in studies of the squid giant axon and the Hodgkin–Huxley equation [2]. We have applied the threshold activated coupling to chaotic neurons to control chaos [3,4,5]. This mechanism works in marked contrast to the OGY method. In the OGY method the chaotic trajectories in the vicinity of unstable fixed points are controlled onto these points. In threshold control, on the other hand, the system does not have to be close to any particular fixed point before implementing the control. Here the trajectory merely has to exceed the prescribed threshold. So the control transience is typically very short. The chaotic neurons are controlled with the threshold activated coupling and spatially synchronized temporal patterns with different periods are obtained. We have investigated the effect of the variation of the relaxation time on the spatiotemporal characteristics of the threshold coupled chaotic neurons. We have also obtained a wide range of stable cyclic behavior of threshold coupled chaotic neurons by simply varying the frequency of control.

The threshold-activated coupling of chaotic systems are relevant for certain mechanical systems like chains of nonlinear springs, as also for some biological

J. Marques de Sá et al. (Eds.): ICANN 2007, Part I, LNCS 4668, pp. 954–962, 2007.

systems, such as synaptic transmissions among neurons [6]. It is also relevant to population migrations as it is reasonable to model the population of an area (state at a site) as a nonlinear map and when this population exceeds a certain critical amount the "excess" population moves to a neighboring area (site). The threshold mechanism is also reminiscent of the Bak-Tang-Wiesenfeld cellular automata algorithm [7], or the "sandpile" model, which gives rise to self organized criticality (SOC). The model system studied here is however significantly different, the most important difference being that the threshold mechanism now occurs on a *nonlinearly evolving substrate*, i.e. there is an *intrinsic deterministic dynamics* at each site. So the local chaos here is like an *internal* driving or perturbation, as opposed to *external* perturbation/driving in the sandpile model, which is effected by dropping "sand" from outside.

The paper is organized as follows. In Sec.II, a model of chaotic neuron is described. The threshold activated coupling is introduced as a control mechanism for chaotic neurons and the effect of relaxation time scale on the spatiotemporal properties of threshold coupled chaotic neurons is studied. In Sec.III, we have studied the threshold activated coupling at the varying time interval to obtain different temporal patterns of spatially synchronized neurons. The conclusion are presented in the last section.

2 Chaotic Neuron Model

We study the following network of chaotic neuron model proposed by Aihara [2].

$$\begin{aligned} y_{n+1}(i) &= ky_n(i) - \alpha f(y_n(i)) + a, \\ x_{n+1}(i) &= f(y_{n+1}(i)), \end{aligned} \quad (1)$$

where, the sigmod function, $f(x) = 1/[1 + exp(-x/\varepsilon)]$, is the output function of the neuron; and ε is the steepness parameter of the sigmoid function. The internal state of the ith neuron is $y_t(i)$ at time t, $x_t(i)$ is the output of the neuron at time t, k is the decay parameter of the refractoriness, and α is the refractory scaling parameter.

In Fig. 1(a), we have shown the output x of each chaotic neuron for the network size $N = 100$ as a function of parameter α. Other parameters of the map are $k = 0.5$, $a = 1.0$, and $\varepsilon = 0.04$. There are regions in parameter space around $\alpha \sim 1.1$ and ~ 1.4, where the system is chaotic.

2.1 Threshold Mechanism

Now, on this nonlinear network a threshold activated coupling is incorporated [3,4,5]. The coupling is triggered when a site in the network exceeds the critical value y^* i.e. when a certain site $y_n(i) > y^*$. The super critical site then relaxes (or "topples") by transporting the excess $\delta = (y_n(i) - y^*)$ equally to its two neighbors:

$$y_n(i) \rightarrow y^*$$

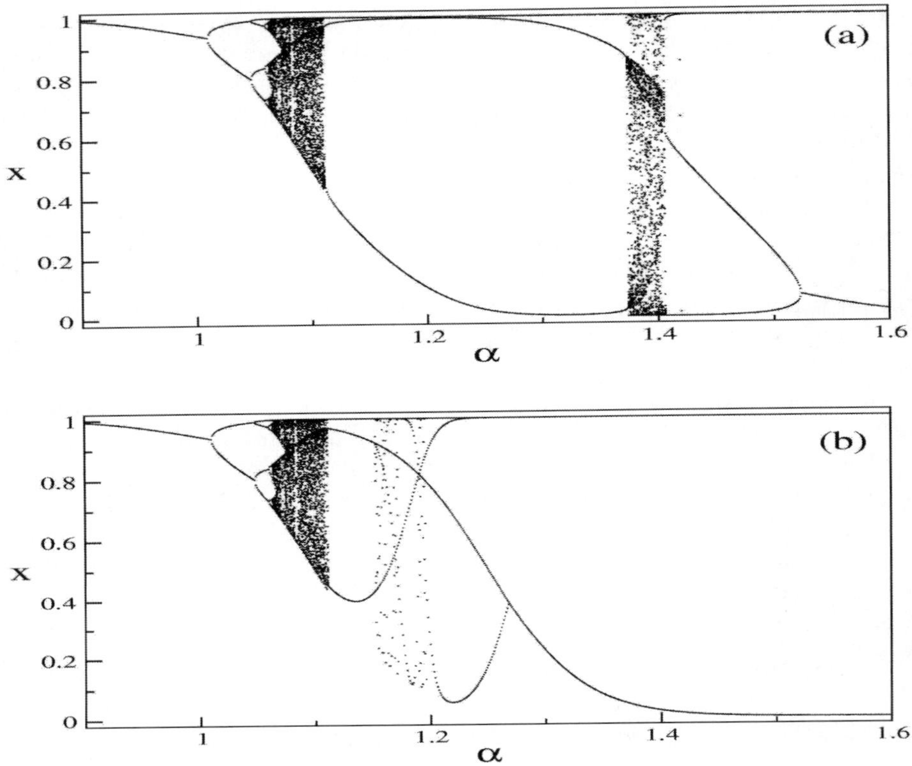

Fig. 1. The output of each neuron x as a function of bifurcation parameter α for $N = 100$ uncoupled neurons (a) and threshold $y^* = 0.5$ coupled neurons (b)

$$
\begin{aligned}
y_n(i-1) &\rightarrow y_n(i-1) + \delta/2 \\
y_n(i+1) &\rightarrow y_n(i+1) + \delta/2
\end{aligned}
\qquad (2)
$$

The process above occurs in parallel, i.e. all supercritical sites at any instant relax simultaneously, according to Eqs. 2, and this constitutes one relaxation time step. After r such relaxation steps, the system undergoes the next chaotic update. In some sense then, time n associated with the chaotic dynamics is measured in units of r. The relaxation of a site may initiate an avalanche of relaxation activity, as neighboring sites can exceed the threshold limit after receiving the excess from a supercritical site, thus starting a domino effect. This induces a bi-directional transport to the two boundaries of the array. These boundaries are open so that the "excess" may be transported out of the system.

The spatiotemporal behavior of the threshold coupled chaotic systems under different threshold levels has been investigated both numerically and analytically, specifically, for the case of networks of chaotic logistic maps [3,4,5]. There exist many *phases* in threshold space ($0 < x^* < 1$), i.e. for $x^* \leq 3/4$ the dynamics goes to a fixed point. When $0.75 < x^* < 1.0$, the dynamics is attracted to *cycles*

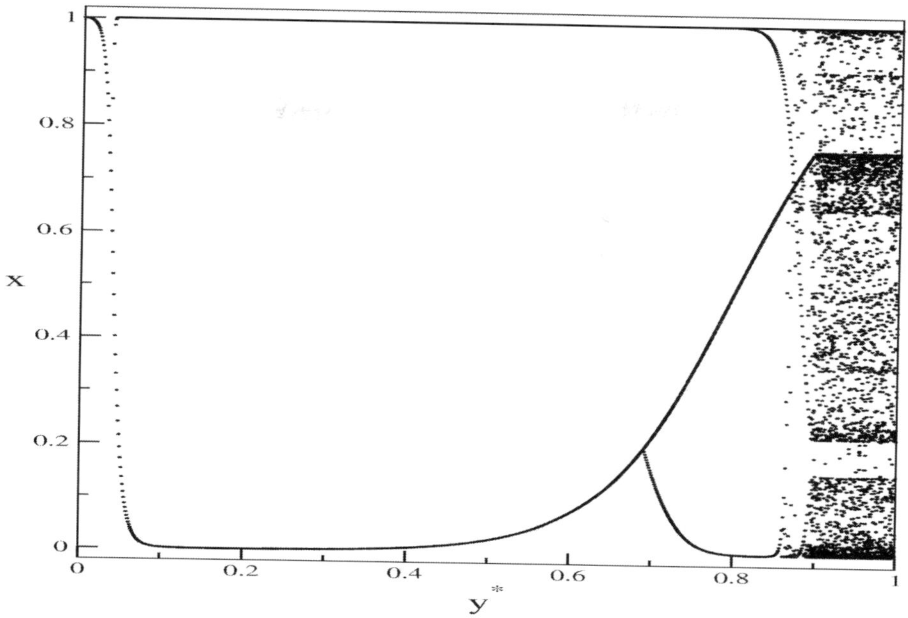

Fig. 2. The output of threshold coupled neurons x as a function of threshold y^* for the fixed value of $\alpha = 1.4$

whose periodicities depend on the value of x^*. By tuning the threshold level x^* one thus obtain spatiotemporal cycles of different orders.

In Fig. 1(b), we have shown the output x of each chaotic neuron as a function of parameter α with a threshold $y^* = 0.5$. With the threshold activated coupling, the chaos in the chaotic neurons is controlled around $\alpha \sim 1.4$. In Fig. 2, we have shown the output x of each chaotic neuron as a function of threshold y^* for the fixed value of $\alpha = 1.4$. There are two period–doubling bifurcations near $y^* \sim 0$ and ~ 0.7.

For $\alpha = 1.4$, we get controlled period-2 output dynamics of the network with the threshold values $y^* < 0.7$. The network of chaotic neurons is synchronized with period two and all sites are taking values close to 0 and 1 alternatively for $y^* < 0.7$ and $\alpha = 1.4$, where the system is chaotic in the absence of threshold activated coupling. In Fig. 3, the space-time density plot of $x_n(i)$ is shown for fixed values of $\alpha = 1.4$, $y^* = 0.5$ and $N = 100$ after the transient dynamics.

2.2 Relaxation Timescale

Note however, that the dynamical outcome crucially depends the relaxation time r, i.e. on the timescales for autonomously updating each site and propagating the threshold-activated coupling between sites. For sufficiently large value of relaxation time, i.e. in the limit $r \to \infty$, the system is fully relaxed (subcritical) before the subsequent dynamical update. So the time scales of the two

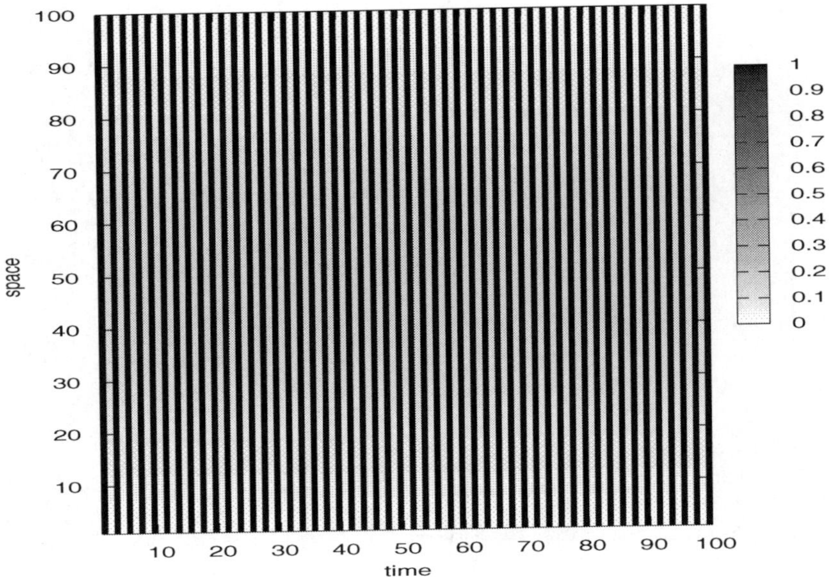

Fig. 3. Space-time density plots of an array of threshold-coupled chaotic neurons, with threshold value $y^* = 0.5$, size $N = 100$, and relaxation time $r = 1000$. The x-axis denotes the time and the y-axis denotes the site index.

processes, the intrinsic chaotic dynamics of each site and the threshold-activated relaxation, are separable. Here the relaxation mechanism is much faster than the chaotic evolution, enabling the system to relax completely before the next chaotic iteration. This scenario is similar to the SOC model, where the driving force (perturbation) is very dilute, e.g. in the sandpile model the avalanche of activity, initiated by an external "sand grain" dropped on the pile, ceases before the next "sand grain" perturbs the pile.

At the other end of the spectrum is the limit of very small r where the local dynamics and the coupling take place simultaneously. It is evident that lowering r essentially allows us to move from a picture of separated relaxation processes to one where the relaxation processes can overlap, and disturbances do not die out before the subsequent chaotic update. It was observed in [8] that for short relaxation times the system is driven to spatiotemporal chaos. This is due to the fast driving rate of the system which does not provide enough time to spatially distribute the perturbations and allow the excess to propagate to the open boundaries. However large r gives the system enough time to relax, and allows the excess to be transported out of the system through the open ends. So for large r the system displays very regular behavior for a large range of threshold values.

In Fig. 4, the bifurcation diagram of the output state of each neuron is shown as a function of threshold y^* for different values of relaxation time r. Fig. 5 shows a example of dynamical transition of threshold coupled chaotic neurons from a fixed temporal behavior to spatiotemporal chaos as r becomes smaller. There

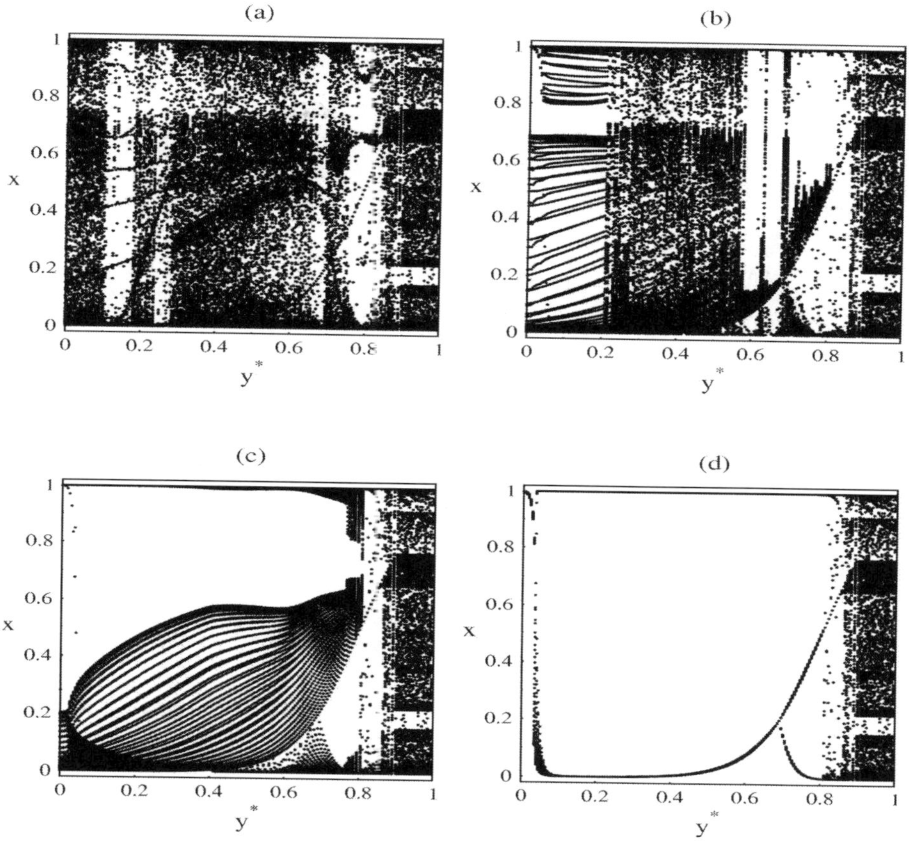

Fig. 4. Bifurcation diagram of the output state of each neuron with respect to threshold y^* for an array of threshold–coupled chaotic neurons. Size N=50 and relaxation times are (a) $r = 1$, (b) $r = 50$, (c) $r = 100$, and (d) $r = 1000$.

is a transition from the spatiotemporal fixed point to spatiotemporal chaos as we decrease the relaxation time. These transitions result from the competition between the rate of intrinsic driving arising from the local chaotic dynamics and the time required to propagate the threshold–activated coupling.

3 Thresholding at Varying Interval

Now, we implement the threshold mechanism at varying intervals n_c, with $1 < n_c \leq 20$, i.e. the thresholding frequency ranges from once every two iterates of the chaotic neuron maps to once every 20 iterates. We find that for all n_c in this range the chaotic neuron maps gets controlled onto an exact and stable orbit of periodicity $p \geq n_c$, where p is an integer multiple of n_c [9].

In Fig. 6 the temporal behavior of threshold coupled chaotic neurons is shown for thresholding at different intervals n_c. It is clear evident that thresholding at

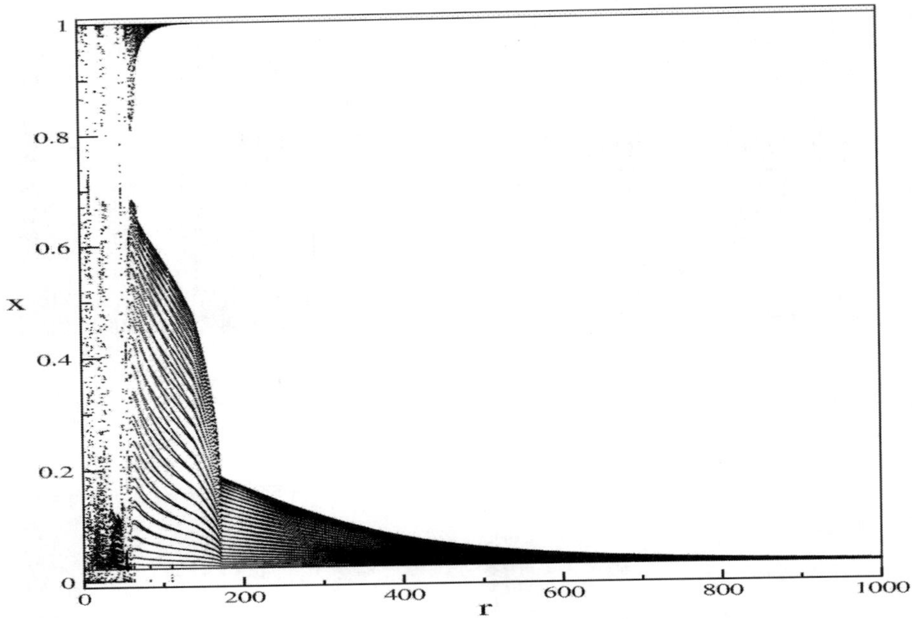

Fig. 5. Bifurcation diagram of the output state x of each neuron with respect to the relaxation time r for an array of threshold–coupled chaotic neurons with threshold $y^* = 0.5$ and system size $N = 50$

different frequencies yields different temporal periods. For instance one obtain temporal periods $15, 5, 7$, and 18 with $n_c = 3, 5, 7$ and 9 respectively. In fact it can serve as a control parameter for selecting a different cyclic behaviors from the chaotic neuron dynamics. This has particular utility in obtaining higher order periods, which are difficult to obtain otherwise.

4 Conclusion

We have studied the spatiotemporal behaviour of threshold coupled chaotic neurons. We observe that the chaos is controlled by threshold activated coupling, and the system yields synchronized temporally periodic states under the threshold response. Varying the frequency of thresholding provides different higher order periodic behaviors, and can serve as a simple mechanism for stabilising a large range of regular temporal patterns in chaotic systems. Further, we have obtained a transition from spatiotemporal chaos to fixed spatiotemporal profiles, by lengthening the relaxation time scale.

Threshold-coupled chaotic systems have the capacity to directly and flexibly implement fundamental logic and arithmetic operations [10]. Such extended dynamical systems offer much scope of parallelism, allowing rapid solutions of

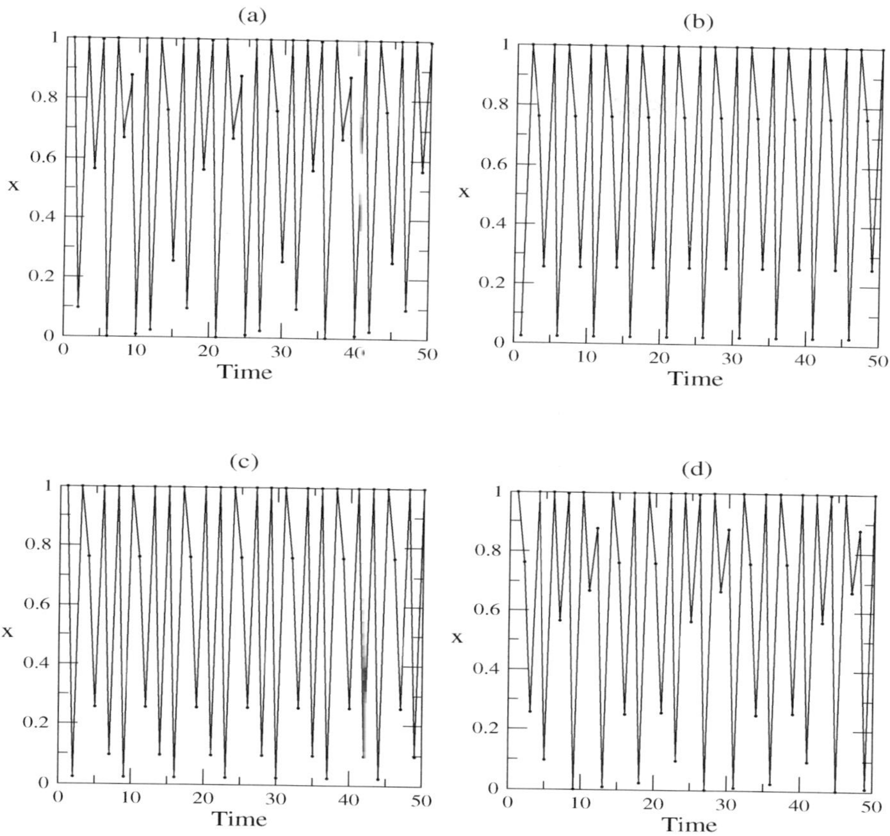

Fig. 6. The temporal behavior of an array of spatially synchronized threshold-coupled logistic maps, with threshold value $y^* = 0.5$, size $N = 100$, and relaxation time $r = 10000$ for (a) $n_c = 3$, (b) $n_c = 5$, (c) $n_c = 7$, and (d) $n_c = 9$. The spatially synchronized neurons have temporal periods $15, 5, 7$, and 18 for $n_c = 3, 5, 7$ and 9 respectively.

certain problems utilizing the collective responses and collective properties of the system [11,12]. So the varied temporal and spatial responses of the array of model neurons studied here, can also be potentially harnessed to accomplish different computational tasks, and the system may be used for information processing [13].

Note that in terms of practical implementation, the threshold mechanism has been implemented in electrical circuits [14]. Parallel distributed processing with spatio–temporal chaos, on the basis of a model of chaotic neural networks, has also been proposed [13], and a mixed analog/digital chaotic neuro–computer prototype system has been constructed for quadratic assignment problems (QAPs) [15]. So our model system of threshold-coupled neurons, which combines threshold mechanisms and neuronal units, should be readily realizable with electrical circuits.

References

1. Ott, E., Grebogi, C., Yorke, J.A.: Controlling chaos. Phys. Rev. Letts. 64, 1196–1199 (1990)
2. Aihara, K., Takebe, T., Toyoda, M.: Chaotic neural networks. Phys. Lett. A 144, 333–340 (1990)
3. Sinha, S., Biswas, D.: Adaptive dynamics on a chaotic lattice. Phys. Rev. Letts. 71, 2010–2013 (1993)
4. Sinha, S.: Unidirectional adaptive dynamics. Phys. Rev. E 49, 4832–4842 (1994)
5. Sinha, S.: Chaos and Regularity in Adaptive Lattice Dynamics. Int. Journ. Mod. Phys. B 9, 875–931 (1995)
6. Aihara, K., Matsumoto, G.: Chaotic oscillations and bifurcations in squid giant axons. In: Holden, A.R. (ed.) Chaos, pp. 257–269. Manchester University Press, Manchester (1986)
7. Bak, P., Tang, C., Weisenfeld, K.: Self-organized criticality: An explanation of the 1/f noise. Phys. Rev. Letts. 59, 381–384 (1987)
8. Mondal, A., Sinha, S.: Spatiotemporal Consequences of Relaxation Timescales in Threshold Coupled Systems. Phys. Rev. E 73, 026215 (2006)
9. Sinha, S.: Using thresholding at varying intervals to obtain different temporal patterns. Phys. Rev. E 63, 036212 (2001)
10. Sinha, S., Ditto, W.L.: Dynamics based computation. Phys. Rev. Letts. 81, 2156–2159 (1998)
11. Sinha, S., Munakata, T., Ditto, W.L.: Parallel computing with extended dynamical systems. Phys. Rev. E 65, 036214 (2002)
12. Munakata, T., Sinha, S., Ditto, W.L.: Chaos Computing: Implementation of Fundamental Logical and Arithmetic Operations and Memory by Chaotic Elements. IEEE Trans. on Circuits and Systems 49, 1629–1633 (2002)
13. Aihara, K.: Chaos engineering and its application to parallel distributed processing with chaotic neural networks. Proceedings of the IEEE 90, 919–930 (2002)
14. Murali, K., Sinha, S.: Experimental realization of chaos control by thresholding. Phys. Rev. E 68, 016210 (2003)
15. Horio, Y., Ikeguchi, T., Aihara, K.: A mixed analog/digital chaotic neuro-computer system for quadratic assignment problems. Neural Networks 18, 505–513 (2005)

Neuronal Multistability Induced by Delay

Cristina Masoller, M.C. Torrent, and Jordi García-Ojalvo

Departament de Física i Enginyeria Nuclear, Universitat Politècnica de Catalunya,
Colom 11, E-08222 Terrassa, Barcelona, Spain

Abstract. Feedback circuits are important for understanding the emergence of patterns of neural activity. In this contribution we study how a delayed circuit representing a recurrent synaptic connection interferes with neuronal nonlinear dynamics. The neuron is modeled using a Hodgkin-Huxley type model in which the firing pattern depends on subthreshold oscillations, and the feedback is included as a time delayed linear term in the membrane voltage equation. In the regime of subthreshold oscillations the feedback amplifies the oscillation amplitude, inducing threshold crossings and firing activity that is self regularized by the delay. We also study a small neuron ensemble globally coupled through the delayed mean field. We find that the firing pattern is controlled by the delay. Depending on the delay, either all neurons fire spikes, or they all exhibit subthreshold activity, or the ensemble divides into clusters, with some neurons displaying subthreshold activity while others fire spikes.

1 Introduction

Time-delayed feedback mechanisms are relevant in many biological systems. Excitable gene regulatory circuits [1], human balance [2,3], and eye movements [4,5] are just a few examples. Many feedback loops have been proposed to explain patterns of neural activity. A well known recurrent circuit is in the hippocampal CA3 region, that is known to be involved in associative memory recall [6,7]. A neuron might experience recurrent excitatory or inhibitory feedback though an auto-synapse, and/or though a circuit of synaptic connections involving other neurons. Since recurrent connections require the propagation of action potentials along the synaptic path, finite signal transmission velocity, and processing time in synapses lead to a broad spectrum of conduction and synaptic delay times, ranging from few to several hundreds of milliseconds. Since neural spike frequencies can exceed 10 Hz, these delays times can be much longer than the characteristic inter-spike interval.

Within the framework of neuron rate-equation models, a recurrent feedback circuit has been studied by adding to the membrane potential equation a term proportional to the potential at an earlier time, $\eta V(t-\tau)$ [8]. Here η is the synaptic strength and τ is the delay time, that is the sum of conduction and synaptic delays. While this is a very simplified approach, it has been successful for understanding characteristic delayed feedback-induced phenomena, such as multi-stability [9,10].

J. Marques de Sá et al. (Eds.): ICANN 2007, Part I, LNCS 4668, pp. 963–972, 2007.

Early experiments on the response of a single neuron in a recurrent excitatory loop were performed by Diez-Martinez and Segundo [11], who studied a pacemaker neuron in the crayfish stretch receptor organ. By having each spike trigger electronically a brief stretch after a certain delay, they showed that with increasing delay the discharge patterns transformed from periodic spikes to trains of spikes separated by silent intervals. Pakdaman et al. [12] interpreted this behavior as due to neuronal adaptation mechanisms, that decreased sensitivity along successive firings. By studying models of various levels of complexity (an integrate and fire, a leaky integrator and a rate-equation model including membrane conductances), with and without adaptation to repeated stimuli, Pakdaman and coworkers found that models including adaptation predicted a dynamics that was similar to that observed experimentally in the crayfish receptor, exhibiting tonic firings for short delays, and multiplets or bursts for longer delays. The influence of noise on a single neuron with a delayed recurrent synaptic connection was analyzed by Vibert et al. [13]. The noise-induced inter-spike interval irregularity was found to decrease when the delay increased above the natural firing period: for short delays noise irregularizes the firing period, while for long delays, the neuron fires with a mean period equal to the delay, as observed without noise.

Recently, interest on delayed feedback circuitry has focused on its influence on neural "spontaneous" or self synchronous activity. Spontaneous synchrony is known to be part of consciousness and perception processes in the brain [14], but also of diseases such as epilepsy. Rosenblum and Pikovsky [15] proposed the use linear delayed feedback to control synchrony in ensembles of globally coupled neurons. It was shown that by variations of the delay time of the coupling, neural synchrony can be either enhanced or suppressed [16,17]. Popovych et al. [18] extended this method to the case of nonlinear delayed feedback, that is, linear delayed feedback nonlinearly combined with the instantaneous signal, showing that nonlinear feedback cannot reinforce synchronization, which can be relevant for applications requiring robust de-synchronization.

With the aim of providing further insight into the influence of a recurrent connection in a single neuron, and global delayed coupling in a neuron ensemble, we study a small neuron ensemble composed by a few neurons that are globally coupled through their delayed mean field, and compare their activity with that of a single neuron that has a recurrent delayed synaptic connection. We show that, as for the single neuron, the firing pattern of the ensemble is controlled not only by the coupling strength but also by the delay time. Depending on the delay, either all neurons fire spikes, or they all exhibit subthreshold activity, or the ensemble divides into clusters, with some neurons displaying subthreshold activity while others fire spikes.

This paper is organized as follows. Section II presents the single neuron model and the coupling scheme of the neuron ensemble. Section III presents results of the numerical simulations, where the control parameter is the delay time and the dynamics of a single neuron with a recurrent connection is compared with that of a globally coupled neuron ensemble. Section IV presents a summary and the conclusions.

2 Model

To simulate the firing activity of a single neuron we use a model proposed by Braun et al. [19] that was developed on the basis of experimental data from shark electroreceptors [20] and mammalian cold receptors [21]. The model is a flexible neuronal pattern generator that produces different types of firing patterns, that are of relevance also in cortical neurons [22]. The temporal sequence of spikes indicates that the activity of these neurons depends on *subthreshold oscillations* of the membrane potential. In electro-receptors and in the upper temperature range of cold receptors there is an irregular sequence of spikes which, however, shows a multi-modal inter-spike interval distribution that suggest the existence of subthreshold oscillations which operate below but near the spike-triggering threshold. In this situation it essentially depends on noise whether a spike is triggered or not but the subthreshold oscillation period is still reflected in the basic rhythm of the discharge. External stimuli can alter the frequency and/or the amplitude of the oscillations, thus inducing pronounced changes of the neuron firing pattern. In contrast, in the low temperature range of cold receptors, irregular single spikes can be recorded, whose histogram of inter-spike intervals do not have a distinct modal structure but seems to reflect pacemaker activity under random fluctuations.

The model proposed by Braun et al. has been studied by several authors. As a function of the temperature different *deterministic* firing patterns have been identified [23], including coexistence of spikes and subthreshold oscillations (spikes with skippings), tonic spiking and bursting patterns. This rich dynamic behavior is due to the interplay of two sets of de- and re- polarizing ionic conductances that are responsible for spike generation and slow-wave potentials [24]. The influence of noise was studied by Feudel et al. [25], who showed that the model predictions are in good agreement with data of electro-physiological experiments with the caudal photoreceptor of the crayfish. Neiman et al. [26] demonstrated experimentally that electroreceptor cells in the paddlefish contain an intrinsic oscillator that can be synchronized with an external signal, and simulations based on Braun et al. model with a periodic external stimulus yield good agreement with the observations. Noise induced synchronization [27] and anticipated synchronization [28] have also been demonstrated. The effect of delayed recurrent feedback was analyzed by Sainz-Trapaga et al. [29], who showed that the feedback can modify the amplitude of the subthreshold oscillations in a way such that they operate slightly above threshold, therefore leading to feedback-induced spikes.

The rate equation for the potential voltage across the membrane, V, is [19]:

$$C_M \dot{V} = -I_{Na} - I_K - i_{sd} - I_{sr} - I_l + \eta V(t-\tau), \qquad (1)$$

where C_M is the capacitance, I_{Na} and I_K are fast sodium and potassium currents, I_{sd} and I_{sr} are additional slow currents. These four currents depend on the temperature T as described in [19]. I_l is a passive leak current. Further details and definitions of the other quantities can be found in [19]. The last term in the r.h.s. of Eq. (1) accounts for the recurrent synaptic connection. $V(t-\tau)$ is the membrane potential at the earlier time, $t-\tau$, η is the synaptic strength and τ the delay time. Because we are interested in feedback-induced patterns, we do not include any noise source in the equations, and consider parameters such that the neuron, without feedback, displays only subthreshold oscillations.

We compare the dynamics of a neuron with feedback, with that of N neurons globally coupled through their delayed mean field. The rate-equation for the membrane potential of the ith neuron of the ensemble, V_i, is:

$$C_M \dot{V}_i = -I_{Na,i} - I_{K,i} - I_{sd,i} - I_{sr,i} - I_{l,i} + \eta V_T(t-\tau), \tag{2}$$

where $V_T(t-\tau) = (1/N)\sum_{i=1}^{N} V_i(t-\tau)$ is the delayed mean field and the other variables have the same meaning as in Eq.(1). This coupling scheme resembles that studied by Rosenblum and Pikovsky [15], but in [15] the neuron ensemble ($N = 2000$ Hindmarsh-Rose neurons in the regime of chaotic bursting) was coupled by two terms, one proportional to the instantaneous mean field, and the other, proportional to the delayed mean field. The authors found that, for certain parameters, the delayed mean field destroyed the synchrony among the neurons induced by the instantaneous coupling, without affecting the oscillations of individual neurons. With the aim of providing further insight into the effect of delayed coupling, here we consider a small neuron ensemble that is coupled only through its delayed mean field.

3 Results

The parameters used in the simulations are such that the neurons, in the absence of feedback or coupling, display subthreshold oscillations of period $T_0 \sim 130$ ms (the temperature parameter is set to 35^0C, and other parameters are as in [19]). To integrate Eq. (1) [Eq.(2)] it is necessary to specify the initial value of the potential V [V_i with $i = 1, N$] on the time interval $[-\tau, 0]$. It is known that the neuron exhibits multistability as different initial conditions lead, after a transient time, to different stable firing patterns [9,10]. Here the initial conditions are such that the neuron is oscillating in its natural cycle when the feedback begins to act (i.e., the feedback starts when the neuron is at a random phase of the cycle). For the neuron ensemble, the initial conditions are such that the neurons oscillate independently one of another (i.e., they are at random, different phases of the cycle) when the coupling starts.

We begin by considering a single neuron with self-feedback. Due to the excitable nature of the dynamics it can be expected that even weak feedback strengths can be a strong perturbation to the neuron subthreshold oscillations. The feedback can amplify the amplitude of the oscillations, inducing threshold-crossings and giving rise to firing activity that can be self-regularized by the delay time. This is indeed observed in Fig. 1, that depicts the amplitude, $A = \max(V) - \min(V)$, and the frequency of the neuronal oscillations vs. the delay for fixed synaptic strength. When η is positive, Fig. 1(a), the oscillation amplitude is diminished, with respect to the natural oscillation amplitude, for

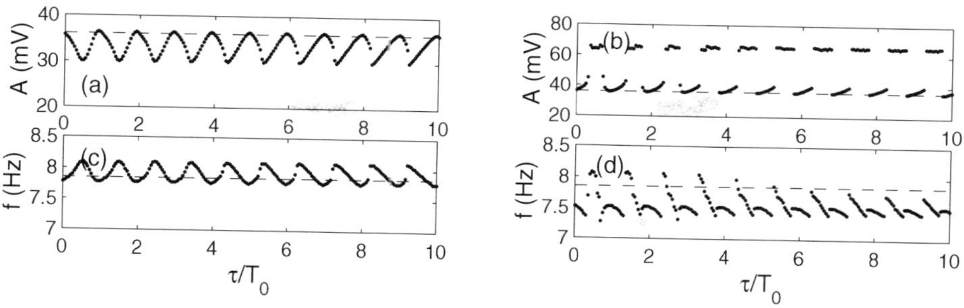

Fig. 1. (a), (b) Amplitude of the neuron oscillations vs. the delay time, normalized to the subthreshold oscillation period, T_0, for feedback strength $\eta = 0.001$ (a) and $\eta = -0.001$ (b). The dashed line indicates the amplitude of the natural subthreshold oscillations (in the absence of feedback). (c), (d) Frequency of the oscillations vs. the normalized delay time for $\eta = 0.001$ (c) and $\eta = -0.001$ (d). The dashed line indicates the frequency of the natural oscillations.

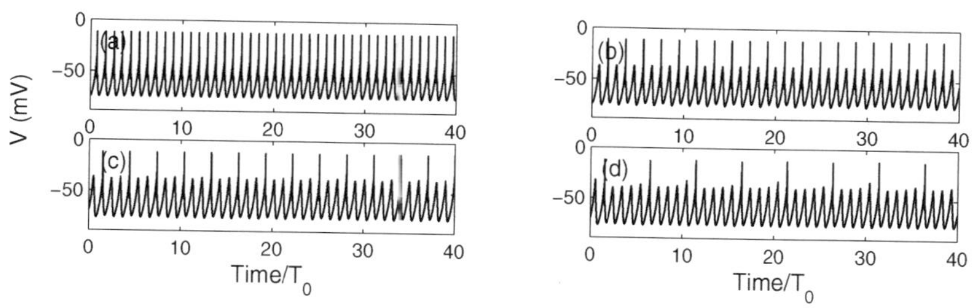

Fig. 2. Oscillatory waveforms for increasing values of the delay time and negative feedback strength, $\eta = -0.001$. $\tau/T_0 = 2.4$ (a), 2.5 (b), 2.65 (c) and 2.7 (d).

all delay values, but there is a non-monotonic relationship of the amplitude with the delay: the oscillation amplitude is maximum (minimum) for $\tau \sim nT_0$ [$\tau \sim (n+1/2)T_0$] with n integer. When η is negative, Fig. 1(b), the oscillation amplitude is enhanced with respect to the natural amplitude, and the neuron fires spikes; however, the feedback is not strong enough to induce firings for all delay values; there are feedback-induced spikes only in "windows" of the delay centered at $\tau \sim (n+1/2)T_0$ with n integer. The frequency of the neuronal oscillations [Figs. 1(c), 1(d)] is also modified by the feedback: for delays longer than a few oscillation periods, the frequency decreases with τ in a piece-wise linear way, increasing abruptly at certain delay values.

In the windows of delay values where feedback-induced spikes occur, the firing pattern is governed by the value of the delay: at the beginning of the window the firings are frequent [tonic spikes are displayed in Fig. 2(a)], they become increasingly sporadic as the delay increases [spikes with skippings are displayed in Figs. 2(b)-2(d)], until they disappear at the end of the window. The process repeats itself in the next window, that is separated by an interval $\sim T_0$.

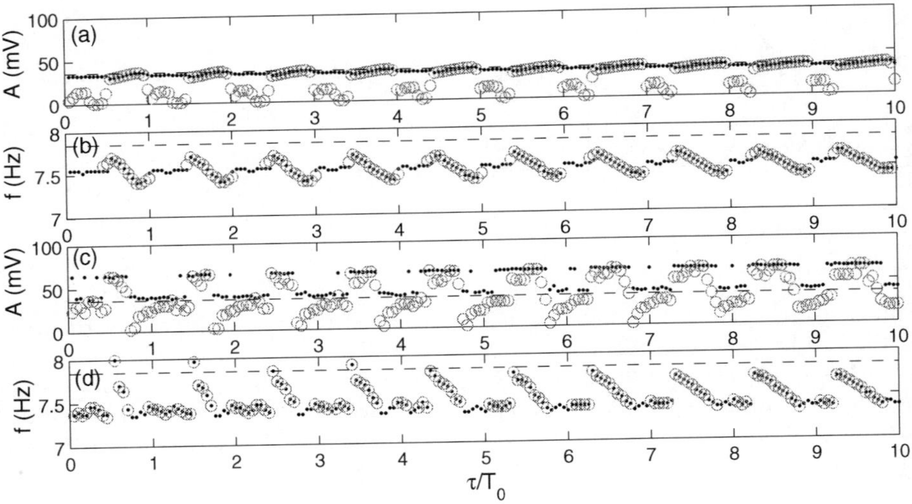

Fig. 3. Amplitude (a), (c) and frequency (b), (d) of the mean field oscillations (gray circles) vs. τ/T_0, for $\eta = 0.001$ (a), (b); and $\eta = -0.001$ (c),(d). The black dots display the amplitude and the frequency of one neuron of the ensemble. The dashed lines indicate the amplitude and frequency of the natural oscillations.

Next, we consider the ensemble of N neurons under the influence of delayed global coupling. we present results for $N = 5$ neurons, but similar patterns are observed for other values of N.

Figure 3 displays the amplitude and the frequency of the oscillations of the mean field (black dots) vs. the delay time for fixed coupling strength. For comparison, the amplitude and the frequency of the oscillations of one of the neurons of the array are also displayed (gray circles). For both, positive and negative coupling strength, it can be observed that the mean field oscillation amplitude [Figs. 3(a), 3(c)] exhibits periodic features at delay times separated by T_0, similar to those observed in Fig. 1 for a single neuron with a recurrent connection. In windows of τ separated by T_0 the amplitude of the oscillations of the mean field decreases to close to zero, revealing that the neurons organize their activity such that they oscillate out of phase, and even in perfect antiphase, keeping the mean field nearly constant. The frequency of both, the mean field and one neuron of the array has a piece-wise linear dependence with the delay [Figs. 3(b), 3(d)]. In the regions of out of phase behavior the frequency of the individual neurons is nearly constant, equal to the natural frequency $f_0 = T_0^{-1}$, while the frequency of the mean field is nf_0 with $n \geq 2$ integer (not shown because of the scale).

For positive coupling strength all the neurons display subthreshold oscillations. The oscillations can be either in-phase, out-of-phase, or in perfect antiphase depending on the delay τ. A few examples are displayed in Fig. 4: in Fig. 4(a) the neuron oscillate inphase; in Fig. 4(b) two neurons display in-phase oscillations (i.e., they form a cluster); in Fig. 4(c) the ensemble splits into two

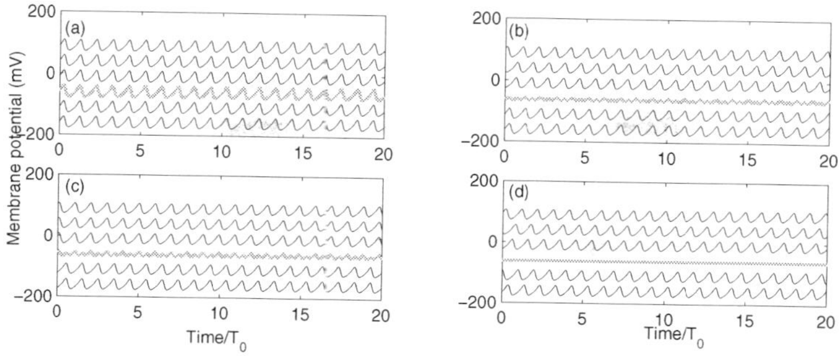

Fig. 4. Oscillatory waveforms for $\eta = 0.001$ and $\tau/T_0 = 6.9$ (a), 7.0 (b), 7.1 (c), and 7.3 (d). The black lines display the oscillation of the individual neurons (displaced vertically for clarity) and the grey lines, the collective mean field.

clusters, and the oscillations of the clusters are in antiphase, in Fig. 4(d) the five neurons display perfect antiphase behavior that leaves the mean field constant.

For negative coupling strength the ensemble displays more complex behavior: depending on τ either all the neurons fire spikes, or they all display subthreshold oscillations, or some neurons display subthreshold oscillations while others fire spikes. The neuronal oscillations can be either in-phase or out-of-phase depending on the delay. A few examples are displayed in Fig. 5: in Fig. 5(a) the neurons fire synchronized spikes; in Fig. 5(b) four neurons fire out of phase spikes, while the other displays subthreshold oscillations; in Fig. 5(c) the five neurons display perfect antiphase subthreshold behavior that leaves the mean field constant; in Fig. 5(d) the ensemble splits into two clusters, one fires synchronized spikes while the other displays subthreshold oscillations; in Fig. 5(e) the ensemble splits into two clusters that alternate their firing pattern; in Fig. 5(f) the neurons synchronize their firings again, but the pattern is different from that of Fig. 5(a).

The firing pattern varies not only with τ, but also with the initial conditions of the neurons, that is, with the positions of the neurons in the subthreshold oscillation cycle when the coupling begins to act. For most values of the delay, different initial positions lead to different firing patterns, and there is multistability of solutions with the coexistence of in-phase and out-of-phase behaviors. As an example, Figs. 6(a) and 6(b) display, for the same parameters as Figs. 5(a) and 5(b), different firing patterns, that occur with different (random) initial conditions. However, for specific values of the delay, the basins of attraction of inphase firings and out of phase subthreshold oscillations are very wide, and almost all initial conditions lead to these states. Fig. 6(c) and 6(d) display the mean field oscillation amplitude and the oscillation amplitude of one neuron of the array, respectively, for 8 random initial conditions. For τ/T_0 =5.5, 6.5 and 7.5, it is observed that all initial conditions lead to large amplitude oscillations of both, the mean field and one neuron of the array, indicating inphase firings, while for τ/T_0 =5.8, 6.8 and 7.8, almost all initial conditions lead to small amplitude oscillations, of both, the mean field and one neuron of the array, reveling out of phase subthreshold oscillations.

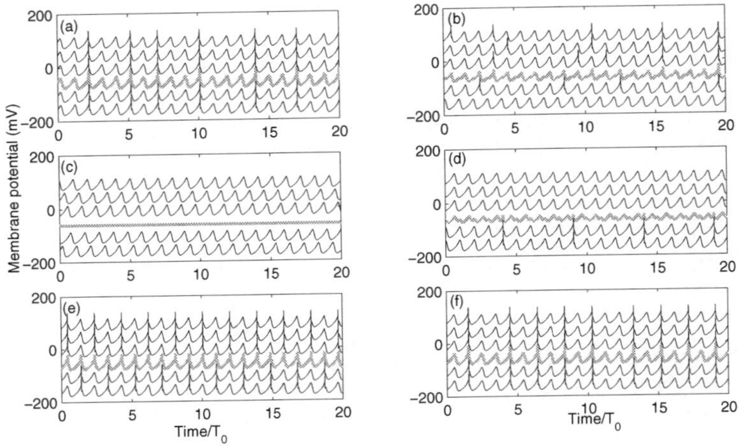

Fig. 5. Oscillatory waveforms for $\eta = -0.001$ and $\tau/T_0 = 6.6$ (a), 6.7 (b), 6.8 (c), 7.0 (d), 7.3 (e), and 7.5 (f)

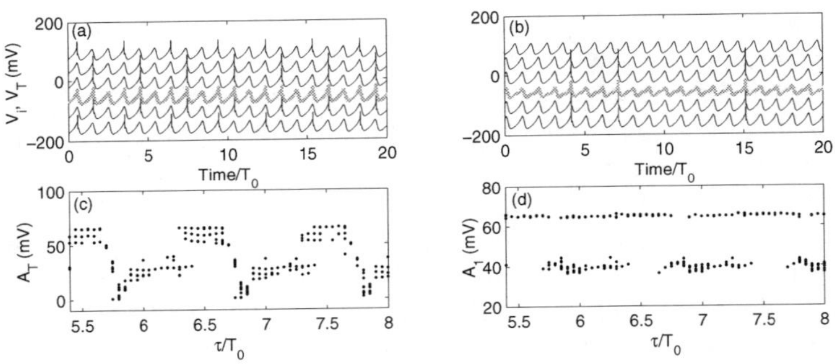

Fig. 6. (a), (b) Oscillatory waveforms for $\eta = -0.001$, $\tau/T_0 = 6.6$ (a), 6.7 (b), and initial conditions different from those of Figs. 5(a), 5(b). (c), (d) Amplitude of the oscillations of the mean field (c) and of the oscillations of one neuron of the ensemble (d) for 8 random initial conditions.

4 Summary and Conclusions

We studied the dynamics of a neuron under the influence of a delayed feedback circuit representing a recurrent synaptic connection. The neuron was modeled using a Hodgkin-Huxley type model with parameters corresponding to the subthreshold oscillation regime, and the feedback was included as a time delayed linear term in the membrane voltage equation. We found that weak positive feedback strengths reduce the amplitude of the subthreshold oscillations, while weak negative feedback strengths amplify the oscillation amplitude, inducing threshold-crossings and firing activity for certain values of the delay time, which are related to integer multiples of the subthreshold oscillation period, T_0. We also studied the firing pattern of a small ensemble of neurons globally coupled through the delayed mean field, and found a rich variety of different behaviors, with the firing pattern controlled by the delay time of the mutual coupling. For certain intervals of the delay, related to T_0, the neurons synchronize their oscillations, either subthreshold oscillations (for positive coupling strength), or firing activity (for negative coupling strength). Outside the synchronization regions, depending on the value of the delay time, either the neurons exhibit out of phase oscillations, or the ensemble divides into clusters, with the clusters exhibiting anti-phased oscillations. As the delay is increased for a fixed coupling strength, the different regimes repeat themselves in a periodic sequence (synchronization, out of phase, cluster behavior, and synchronization) with a periodicity approximately equal to T_0. These results can be of relevance for a deeper understanding of the role played by time delays in weakly coupled neuronal ensembles.

Acknowledgments. This research was supported by "Ramon y Cajal" program (Spain) and the GABA project (European Commission, FP6-NEST 043309).

References

1. Suel, G.M., Garcia-Ojalvo, J., Liberman, L.M., Elowitz, M.B.: An excitable gene regulatory circuit induces transient cellular differentiation. Nature 440, 545 (2006)
2. Stepan, G., Kollar, L.: Math. Comp. Modelling 31, 199 (2000)
3. Cabrera, J.L., Milton, J.G.: On-off intermittency in a human balancing task. Phys. Rev. Lett. 89, 158702 (2002)
4. Moschovakis, A.K., Scudder, C.A., Highstein, S.M.: The microscopic anatomy and physiology of the mammalian saccadic system. Prog. Neurobiol. 50, 133 (1996)
5. Mergenthaler, K., Engbert, R.: Modeling the control of fixational eye movements with neurophysiological delays. Phys. Rev. Lett. 98, 138104 (2007)
6. Debanne, D., Gahwiler, B.H., Thompson, S.M.: Long-term synaptic plasticity between pairs of individual CA3 pyramidal cells in rat hippocampal slice cultures. J. of Physiol. (London) 507, 237–247 (1998)
7. Nakazawa, K., Quirk, M.C, Chitwood, R.A., Watanabe, M., Yeckel, M.F., Sun, L.D., Kato, A., Carr, C.A., Johnston, D., Wilson, M.A., Tonegawa, S.: Requirement for hippocampal CA3 NMDA receptors in associative memory recall. Science 297, 211–218 (2002)
8. Plant, R.E.: A Fitzhugh differential-difference equation modeling recurrent neural feedback. SIAM J. Appl. Math. 40, 150–162 (1981)
9. Foss, J., Longtin, A., Mensour, B., Milton, J.: Multiestability and delayed recurrent loops. Phys. Rev. Lett. 76, 708–711 (1996)
10. Foss, J., Milton, J.: Multistability in recurrent neural loops arising from delay. J. Neurophysiol. 84, 975–985 (2000)
11. Martinez, O.D., Segundo, J.P.: Behavior of a single neuron in a recurrent excitatory loop. Biological Cybernetics 47, 33–41 (1983)
12. Pakdaman, K., Vibert, J.F., Bousssard, E., Azmy, N.: Single neuron with recurrent excitation: Effect of the transmission delay. Neural Networks 9, 797–818 (1996)
13. Vibert, J.F., Alvarez, F., Pham, J.: Effects of transmission delays and noise in recurrent excitatory neural networks. BioSystems 48, 255–262 (1998)
14. Rodriguez, E., George, N., Lachaux, J.P., Martinerie, J., Renault, B., Varela, F.J.: Perception's shadow: long-distance synchronization of human brain activity. Nature 397, 430–433 (1999)

15. Rosenblum, M.G., Pikovsky, A.S.: Phys. Rev. Lett. 92, 114102 (2004)
16. Rosenblum, M., Pikovsky, A.: Delayed feedback control of collective synchrony: an approach to suppression of pathological brain rhythms. Phys. Rev. E 70, 041904 (2004)
17. Rosenblum, M., Tukhlina, N., Pikovsky, A., Cimponeriu, L.: Delayed feedback suppression of collective rhythmic activity in a neuronal ensemble. Int. J. Bif. Chaos 16, 1989 (2006)
18. Popovych, O.V., Hauptmann, C., Tass, P.A.: Effective desynchronization by nonlinear delayed feedback. Phys. Rev. Lett. 94, 164102 (2005)
19. Braun, H.A., Huber, M.T., Dewald, M., Schafer, K., Voigt, K.: Int. J. Bif. Chaos Appl. Sci. Eng. 8, 881–889 (1998)
20. Braun, H.A., Wissing, H., Schafer, K., Hirsch, M.C.: Oscillation and noise determine signal transduction in shark multimodal sensory cells. Nature 367, 270–273 (1994)
21. Braun, H.A., Bade, H., Hensel, H.: Static and dynamic discharge patterns of bursting cold fibers related to hypothetical receptor mechanisms. Pflügers Arch. 386, 1–9 (1980)
22. Braun, H.A., Voigt, K., Huber, M.T.: Oscillations, resonances and noise: basis of flexible neuronal pattern generation. Biosystems 71, 39–50 (2003)
23. Braun, W., Eckhard, B., Braun, H.A., Huber, M.: Phase-space structure of a thermoreceptor. Phys. Rev. E 62, 6352–6360 (2000)
24. Braun, H.A., Huber, M.T., Anthes, N., Voigt, K., Neiman, A., Pei, X., Moss, F.: Interactions between slow and fast conductances in the Huber/Braun model of cold-receptor discharges. Neurocomputing 32, 51–59 (2000)
25. Feudel, U., Neiman, A., Pei, X., Wojtenek, W., Braun, H., Huber, M., Moss, F.: Homoclinic bifurcation in a Hodgkin-Huxley model of thermally sensitive neurons. Chaos 10, 231–239 (2000)
26. Neiman, A., Pei, X., Russell, D., Wojtenek, W., Wilkens, L., Moss, F., Braun, H.A., Huber, M.T., Voigt, K.: Synchronization of the noisy electrosensitive cells in the paddlefish. Phys. Rev. Lett. 82, 660–663 (1999)
27. Zhou, C.S., Kurths, J.: Noise-induced synchronization and coherence resonance of a Hodgkin-Huxley model of thermally sensitive neurons. Chaos 13, 401–409 (2003)
28. Ciszak, M., Calvo, O., Masoller, C., Mirasso, C.R., Toral, R.: Anticipating the response of excitable systems driven by random forcing. Phys. Rev. Lett. 90, 204102 (2003)
29. Sainz-Trapaga, M., Masoller, C., Braun, H.A., Huber, M.T.: Influence of time-delayed feedback in the firing pattern of thermally sensitive neurons. Phys. Rev. E 70, 031904 (2004)

Author Index

Lecture Notes in Computer Science

Sublibrary 1: Theoretical Computer Science and General Issues

For information about Vols. 1–4420
please contact your bookseller or Springer

Vol. 4743: P. Thulasiraman, X. He, T.L. Xu, M.K. Denko, R.K. Thulasiram, L.T. Yang (Eds.), Frontiers of High Performance Computing and Networking ISPA 2007 Workshops. XXIX, 536 pages. 2007.

Vol. 4742: I. Stojmenovic, R.K. Thulasiram, L.T. Yang, W. Jia, M. Guo, R.F. de Mello (Eds.), Parallel and Distributed Processing and Applications. XX, 995 pages. 2007.

Vol. 4736: S. Winter, M. Duckham, L. Kulik, B. Kuipers (Eds.), Spatial Information Theory. XV, 455 pages. 2007.

Vol. 4708: L. Kučera, A. Kučera (Eds.), Mathematical Foundations of Computer Science 2007. XVIII, 764 pages. 2007.

Vol. 4707: O. Gervasi, M.L. Gavrilova (Eds.), Computational Science and Its Applications – ICCSA 2007, Part III. XXIV, 1205 pages. 2007.

Vol. 4706: O. Gervasi, M.L. Gavrilova (Eds.), Computational Science and Its Applications – ICCSA 2007, Part II. XXIII, 1129 pages. 2007.

Vol. 4705: O. Gervasi, M.L. Gavrilova (Eds.), Computational Science and Its Applications – ICCSA 2007, Part I. XLIV, 1169 pages. 2007.

Vol. 4703: L. Caires, V.T. Vasconcelos (Eds.), CONCUR 2007 – Concurrency Theory. XIII, 507 pages. 2007.

Vol. 4697: L. Choi, Y. Paek, S. Cho (Eds.), Advances in Computer Systems Architecture. XIII, 400 pages. 2007.

Vol. 4688: K. Li, M. Fei, G.W. Irwin, S. Ma (Eds.), Bio-Inspired Computational Intelligence and Applications. XIX, 805 pages. 2007.

Vol. 4684: L. Kang, Y. Liu, S. Zeng (Eds.), Evolvable Systems: From Biology to Hardware. XIV, 446 pages. 2007.

Vol. 4683: L. Kang, Y. Liu, S. Zeng (Eds.), Intelligence Computation and Applications. XVII, 663 pages. 2007.

Vol. 4681: D.-S. Huang, L. Heutte, M. Loog (Eds.), Advanced Intelligent Computing Theories and Applications. XXVI, 1379 pages. 2007.

Vol. 4671: V. Malyshkin (Ed.), Parallel Computing Technologies. XIV, 635 pages. 2007.

Vol. 4668: J.M. de Sá, L.A. Alexandre, W. Duch, D.P. Mandic (Eds.), Artificial Neural Networks – ICANN 2007, Part I. XXXI, 978 pages. 2007.

Vol. 4666: M.E. Davies, C.J. James, S.A. Abdallah, M.D Plumbley (Eds.), Independent Component Analysis and Blind Signal Separation. XIX, 847 pages. 2007.

Vol. 4664: J. Durand-Lose, M. Margenstern (Eds.), Machines, Computations, and Universality. X, 325 pages. 2007.

Vol. 4649: V. Diekert, M.V. Volkov, A. Voronkov (Eds.), Computer Science – Theory and Applications. XIII, 420 pages. 2007.

Vol. 4647: R. Martin, M. Sabin, J. Winkler (Eds.), Mathematics of Surfaces XII. IX, 509 pages. 2007.

Vol. 4644: N. Azemard, L. Svensson (Eds.), Integrated Circuit and System Design. XIV, 583 pages. 2007.

Vol. 4641: A.-M. Kermarrec, L. Bougé, T. Priol (Eds.), Euro-Par 2007 Parallel Processing. XXVII, 974 pages. 2007.

Vol. 4639: E. Csuhaj-Varjú, Z. Ésik (Eds.), Fundamentals of Computation Theory. XIV, 508 pages. 2007.

Vol. 4638: T. Stützle, M. Birattari, H.H. Hoos (Eds.), Engineering Stochastic Local Search Algorithms. X, 223 pages. 2007.

Vol. 4628: L.N. de Castro, F.J. Von Zuben, H. Knidel (Eds.), Artificial Immune Systems. XII, 438 pages. 2007.

Vol. 4627: M. Charikar, K. Jansen, O. Reingold, J.D.P. Rolim (Eds.), Approximation, Randomization, and Combinatorial Optimization. XII, 626 pages. 2007.

Vol. 4624: T. Mossakowski, U. Montanari, M. Haveraaen (Eds.), Algebra and Coalgebra in Computer Science. XI, 463 pages. 2007.

Vol. 4619: F. Dehne, J.-R. Sack, N. Zeh (Eds.), Algorithms and Data Structures. XVI, 662 pages. 2007.

Vol. 4618: S.G. Akl, C.S. Calude, M.J. Dinneen, G. Rozenberg, H.T. Wareham (Eds.), Unconventional Computation. X, 243 pages. 2007.

Vol. 4616: A. Dress, Y. Xu, B. Zhu (Eds.), Combinatorial Optimization and Applications. XI, 390 pages. 2007.

Vol. 4613: F.P. Preparata, Q. Fang (Eds.), Frontiers in Algorithmics. XI, 348 pages. 2007.

Vol. 4600: H. Comon-Lundh, C. Kirchner, H. Kirchner (Eds.), Rewriting, Computation and Proof. XVI, 273 pages. 2007.

Vol. 4599: S. Vassiliadis, M. Berekovic, T.D. Hämäläinen (Eds.), Embedded Computer Systems: Architectures, Modeling, and Simulation. XVIII, 466 pages. 2007.

Vol. 4598: G. Lin (Ed.), Computing and Combinatorics. XII, 570 pages. 2007.

Vol. 4596: L. Arge, C. Cachin, T. Jurdziński, A. Tarlecki (Eds.), Automata, Languages and Programming. XVII, 953 pages. 2007.

Vol. 4595: D. Bošnački, S. Edelkamp (Eds.), Model Checking Software. X, 285 pages. 2007.

Vol. 4590: W. Damm, H. Hermanns (Eds.), Computer Aided Verification. XV, 562 pages. 2007.

Vol. 4588: T. Harju, J. Karhumäki, A. Lepistö (Eds.), Developments in Language Theory. XI, 423 pages. 2007.

Vol. 4583: S.R. Della Rocca (Ed.), Typed Lambda Calculi and Applications. X, 397 pages. 2007.

Vol. 4580: B. Ma, K. Zhang (Eds.), Combinatorial Pattern Matching. XII, 366 pages. 2007.

Vol. 4576: D. Leivant, R. de Queiroz (Eds.), Logic, Language, Information and Computation. X, 363 pages. 2007.

Vol. 4547: C. Carlet, B. Sunar (Eds.), Arithmetic of Finite Fields. XI, 355 pages. 2007.

Vol. 4546: J. Kleijn, A. Yakovlev (Eds.), Petri Nets and Other Models of Concurrency – ICATPN 2007. XI, 515 pages. 2007.

Vol. 4545: H. Anai, K. Horimoto, T. Kutsia (Eds.), Algebraic Biology. XIII, 379 pages. 2007.

Vol. 4533: F. Baader (Ed.), Term Rewriting and Applications. XII, 419 pages. 2007.

Vol. 4528: J. Mira, J.R. Álvarez (Eds.), Nature Inspired Problem-Solving Methods in Knowledge Engineering, Part II. XXII, 650 pages. 2007.

Vol. 4527: J. Mira, J.R. Álvarez (Eds.), Bio-inspired Modeling of Cognitive Tasks, Part I. XXII, 630 pages. 2007.

Vol. 4525: C. Demetrescu (Ed.), Experimental Algorithms. XIII, 448 pages. 2007.

Vol. 4514: S.N. Artemov, A. Nerode (Eds.), Logical Foundations of Computer Science. XI, 513 pages. 2007.

Vol. 4513: M. Fischetti, D.P. Williamson (Eds.), Integer Programming and Combinatorial Optimization. IX, 500 pages. 2007.

Vol. 4510: P. Van Hentenryck, L.A. Wolsey (Eds.), Integration of AI and OR Techniques in Constraint Programming for Combinatorial Optimization Problems. X, 391 pages. 2007.

Vol. 4507: F. Sandoval, A. Prieto, J. Cabestany, M. Graña (Eds.), Computational and Ambient Intelligence. XXVI, 1167 pages. 2007.

Vol. 4501: J. Marques-Silva, K.A. Sakallah (Eds.), Theory and Applications of Satisfiability Testing – SAT 2007. XI, 384 pages. 2007.

Vol. 4497: S.B. Cooper, B. Löwe, A. Sorbi (Eds.), Computation and Logic in the Real World. XVIII, 826 pages. 2007.

Vol. 4494: H. Jin, O.F. Rana, Y. Pan, V.K. Prasanna (Eds.), Algorithms and Architectures for Parallel Processing. XIV, 508 pages. 2007.

Vol. 4493: D. Liu, S. Fei, Z. Hou, H. Zhang, C. Sun (Eds.), Advances in Neural Networks – ISNN 2007, Part III. XXVI, 1215 pages. 2007.

Vol. 4492: D. Liu, S. Fei, Z. Hou, H. Zhang, C. Sun (Eds.), Advances in Neural Networks – ISNN 2007, Part II. XXVII, 1321 pages. 2007.

Vol. 4491: D. Liu, S. Fei, Z.-G. Hou, H. Zhang, C. Sun (Eds.), Advances in Neural Networks – ISNN 2007, Part I. LIV, 1365 pages. 2007.

Vol. 4490: Y. Shi, G.D. van Albada, J. Dongarra, P.M.A. Sloot (Eds.), Computational Science – ICCS 2007, Part IV. XXXVII, 1211 pages. 2007.

Vol. 4489: Y. Shi, G.D. van Albada, J. Dongarra, P.M.A. Sloot (Eds.), Computational Science – ICCS 2007, Part III. XXXVII, 1257 pages. 2007.

Vol. 4488: Y. Shi, G.D. van Albada, J. Dongarra, P.M.A. Sloot (Eds.), Computational Science – ICCS 2007, Part II. XXXV, 1251 pages. 2007.

Vol. 4487: Y. Shi, G.D. van Albada, J. Dongarra, P.M.A. Sloot (Eds.), Computational Science – ICCS 2007, Part I. LXXXI, 1275 pages. 2007.

Vol. 4484: J.-Y. Cai, S.B. Cooper, H. Zhu (Eds.), Theory and Applications of Models of Computation. XIII, 772 pages. 2007.

Vol. 4475: P. Crescenzi, G. Prencipe, G. Pucci (Eds.), Fun with Algorithms. X, 273 pages. 2007.

Vol. 4474: G. Prencipe, S. Zaks (Eds.), Structural Information and Communication Complexity. XI, 342 pages. 2007.

Vol. 4459: C. Cérin, K.-C. Li (Eds.), Advances in Grid and Pervasive Computing. XVI, 759 pages. 2007.

Vol. 4449: Z. Horváth, V. Zsók, A. Butterfield (Eds.), Implementation and Application of Functional Languages. X, 271 pages. 2007.

Vol. 4448: M. Giacobini (Ed.), Applications of Evolutionary Computing. XXIII, 755 pages. 2007.

Vol. 4447: E. Marchiori, J.H. Moore, J.C. Rajapakse (Eds.), Evolutionary Computation, Machine Learning and Data Mining in Bioinformatics. XI, 302 pages. 2007.

Vol. 4446: C. Cotta, J. van Hemert (Eds.), Evolutionary Computation in Combinatorial Optimization. XII, 241 pages. 2007.

Vol. 4445: M. Ebner, M. O'Neill, A. Ekárt, L. Vanneschi, A.I. Esparcia-Alcázar (Eds.), Genetic Programming. XI, 382 pages. 2007.

Vol. 4436: C.R. Stephens, M. Toussaint, D. Whitley, P.F. Stadler (Eds.), Foundations of Genetic Algorithms. IX, 213 pages. 2007.

Vol. 4433: E. Şahin, W.M. Spears, A.F.T. Winfield (Eds.), Swarm Robotics. XII, 221 pages. 2007.

Vol. 4432: B. Beliczynski, A. Dzielinski, M. Iwanowski, B. Ribeiro (Eds.), Adaptive and Natural Computing Algorithms, Part II. XXVI, 761 pages. 2007.

Vol. 4431: B. Beliczynski, A. Dzielinski, M. Iwanowski, B. Ribeiro (Eds.), Adaptive and Natural Computing Algorithms, Part I. XXV, 851 pages. 2007.

Vol. 4424: O. Grumberg, M. Huth (Eds.), Tools and Algorithms for the Construction and Analysis of Systems. XX, 738 pages. 2007.

Vol. 4423: H. Seidl (Ed.), Foundations of Software Science and Computational Structures. XVI, 379 pages. 2007.

Vol. 4422: M.B. Dwyer, A. Lopes (Eds.), Fundamental Approaches to Software Engineering. XV, 440 pages. 2007.

Vol. 4421: R. De Nicola (Ed.), Programming Languages and Systems. XVII, 538 pages. 2007.

SPRINGER NATURE

GPSR Compliance

The European Union's (EU) General Product Safety Regulation (GPSR) is a set of rules that requires consumer products to be safe and our obligations to ensure this.

If you have any concerns about our products, you can contact us on ProductSafety@springernature.com

In case Publisher is established outside the EU, the EU authorized representative is:

Springer Nature Customer Service Center GmbH
Europaplatz 3
69115 Heidelberg, Germany

FSC
www.fsc.org
MIX
Papier | Fördert gute Waldnutzung
FSC® C083411

Kolibri 360 GmbH
Ferdinand-Jühlke-Straße 7
99095 Erfurt, Deutschland
produktsicherheit@kolibri360.de